Epigenetics: Development and Disease

SUBCELLULAR BIOCHEMISTRY

SERIES EDITOR
J. ROBIN HARRIS, University of Mainz, Mainz, Germany

ASSISTANT EDITORS
B.B. BISWAS, University of Calcutta, Calcutta, India
P. QUINN, King's College London, London, UK

Recent Volumes in this Series

Volume 48	**The Cronin Family of Proteins**	
	Edited by Christoph S. Clemen and Ludwig Eichinger	
Volume 49	**Lipids in Health and Disease**	
	Edited by Peter J. Quinn and Xiaoyuan Wang	
Volume 50	**Genome Stability and Human Diseases**	
	Edited by Heinz-Peter Nasheuer	
Volume 51	**Cholesterol Binding and Cholesterol Transport Proteins**	
	Edited by Robin J. Harris	
Volume 52	**A Handbook of Transcription Factors**	
	Edited by Tim Hughes	
Volume 53	**Endotoxins: Stricture, Function and Recognition**	
	Edited by Xiaoyuan Wang and Peter J. Quinn	
Volume 54	**Conjugation and Deconjugation of Ubiquitin Family Modifiers**	
	Edited by Marcus Groettrup	
Volume 55	**Purinergic Regulation of Respiratory Diseases**	
	Edited by Maryse Picher and Richard C. Boucher	
Volume 56	**Water Soluble Vitamins**	
	Edited by Olaf Stanger	
Volume 57	**Aging Research in Yeast**	
	Edited by Michael Breitenbach, Michal S. Jazwinski and Peter Laun	
Volume 58	**Phosphoinositides I: Enzymes of Synthesis and Degradation**	
	Edited by Tamas Balla, Matthias Wymann and John D. York	
Volume 59	**Phosphoinositides II: The Diverse Biological Functions**	
	Edited by Tamas Balla, Matthias Wymann and John D. York	
Volume 60	**Adherens Junctions: From Molecular Mechanisms to Tissue Development and Disease**	
	Edited by Tony Harris	

For further volumes:
http://www.springer.com/series/6515

Tapas K. Kundu
Editor

Epigenetics: Development and Disease

 Springer

Editor
Tapas K. Kundu
Molecular Biology and Genetics Unit
Jawaharlal Nehru Centre
 for Advanced Scientific Research
Bangalore, India

ISSN 0306-0225
ISBN 978-94-007-4524-7   ISBN 978-94-007-4525-4 (eBook)
DOI 10.1007/978-94-007-4525-4
Springer Dordrecht Heidelberg New York London

Library of Congress Control Number: 2012946199

© Springer Science+Business Media Dordrecht 2013
This work is subject to copyright. All rights are reserved by the Publisher, whether the whole or part of the material is concerned, specifically the rights of translation, reprinting, reuse of illustrations, recitation, broadcasting, reproduction on microfilms or in any other physical way, and transmission or information storage and retrieval, electronic adaptation, computer software, or by similar or dissimilar methodology now known or hereafter developed. Exempted from this legal reservation are brief excerpts in connection with reviews or scholarly analysis or material supplied specifically for the purpose of being entered and executed on a computer system, for exclusive use by the purchaser of the work. Duplication of this publication or parts thereof is permitted only under the provisions of the Copyright Law of the Publisher's location, in its current version, and permission for use must always be obtained from Springer. Permissions for use may be obtained through RightsLink at the Copyright Clearance Center. Violations are liable to prosecution under the respective Copyright Law.
The use of general descriptive names, registered names, trademarks, service marks, etc. in this publication does not imply, even in the absence of a specific statement, that such names are exempt from the relevant protective laws and regulations and therefore free for general use.
While the advice and information in this book are believed to be true and accurate at the date of publication, neither the authors nor the editors nor the publisher can accept any legal responsibility for any errors or omissions that may be made. The publisher makes no warranty, express or implied, with respect to the material contained herein.

Printed on acid-free paper

Springer is part of Springer Science+Business Media (www.springer.com)

INTERNATIONAL ADVISORY EDITORIAL BOARD

R. Bittman, Queens College, City University of New York, New York, USA
D. Dasgupt, Saha Institute of Nuclear Physics, Calcutta, India
A. Holzenburg, Texas A&M University, Texas, USA
S. Rottem, The Hebrew University, Jerusalem, Israel
M. Wyss, DSM Nutritional Products Ltd., Basel, Switzerland

To

A teacher, a mentor and a source of constant inspiration, **Prof. G. Padmanaban**, *distinguished biotechnologist and former Director of Indian Institute of Science, Bangalore-12, India.*

Foreword

"Our interest in and understanding of the concept of epigenetics has increased dramatically in the past decade, with the general perception of epigenetics having evolved from one of a phenomenon considered to originate from anomalous and disparate patterns of inheritance to one that is linked to a variety of normal and disease-related physiological processes through specific molecular mechanisms. Notably in this regard, the field of epigenetics has come a long way from an early predominant emphasis on DNA methylation to the current inclusion of chromosomal histone modifications and, more recently, non-coding RNAs in epigenetic regulatory events.

Epigenetic phenomena are intimately related to chromatin structure and organization and thereby influence gene expression. The importance of epigenetics for development and cell differentiation is increasingly clear and underlying mechanisms are being unraveled. Similarly, epigenetic changes are now being linked to early events in the pathogenesis of diseases such as cancer, diabetes, and many others. These revelations have sparked efforts to develop new generation therapeutics against components of the epigenetic machinery for the treatment of complex multifactorial diseases.

This collection entitled *Epigenetics: Development and Disease* very effectively covers the above-mentioned aspects of epigenetics research, along with considerations of the evolution of the epigenetic machinery and the role of epigenetics in transcriptional regulation, in five separate parts. The various chapters in these parts have been written by experts who themselves have contributed significantly to their respective fields. Although there are other books with similar titles, this book provides a comprehensive update on the role of epigenetics in development and disease, efforts to develop therapeutics for some of these diseases and the role of epigenetics in transcriptional regulation. Consideration of the latter topic is especially important in view of the probable key role of transcription factors in the initial induction or establishment of many epigenetic changes or states – as dramatically evidenced by the ability of small subsets of ectopic transcription factors to reprogram somatic cells to pluripotent states through epigenetic changes.

Last but not least, the editor, Tapas K. Kundu, himself is an active scientist in the field and deserves a great deal of appreciation for his excellent job in conceiving and bringing to fruition this book. Students, as well as established investigators, will find the book to be a stimulating overview of the field."

Laboratory of Biochemistry
and Molecular Biology
The Rockefeller University
1230 York Avenue, New York, NY 10065

Robert G. Roeder, Ph.D.
Arnold and Mabel Beckman Professor

Preface

The field of 'Epigenetics' has moved on from the Waddington concept proposed in the year 1942; the definition has undergone a constant expansion and the scope of this subject has broadened over the years. The actual resurrection of the field can be marked by the discovery of the first histone acetyltransferase, GCN5, in the year 1995 by David Allis' group. Although the activity of histone modifications (acetylation and methylation) and its role in the transcriptional activation was discovered by Vincent Allfrey in 1964 in a very elegant manner, its significance could only be appreciated by the scientific community after the identification of the GCN5 acetyltransferase and the subsequent expansion of the histone modifying enzymes family. Initially epigenetics and DNA methylation mediated gene regulation was thought to be synonymous. However, this concept has now been replaced with the understanding that these modifications along with DNA methylation form the basis of epigenetic phenomenon. However, all the histone modifications need not be involved at the same time in this event. Furthermore, it has also been realised that several non-histone proteins which can harbor the similar modifications such as acetylation and methylation also form an integral component of the epigenetic network.

I was fortunate to be associated with the growth of the field since 1996 during my days in the Roeder (Robert G Roeder) Laboratory in the Rockefeller University, where a majority of my work was towards understanding the mechanism of transcriptional regulation by histone acetyltransferase complexes and their recruitment in the activator dependent transcription from chromatin. Coincidentally, at the same time I was also a part of the discovery of the first p300 and PCAF acetyltransferase activity specific inhibitors (a collaboration with Philip Cole's group). It is during this time that I got the opportunity to interact with Vincent Allfrey, David Allis and Jerry Workman. Interaction with Vincent Allfrey was really memorable. Vincent's approach towards the discovery of histone modifications was really a bold step in the late 1960s when the use of radioactive material was difficult even at the Rockefeller University. However, Vincent was confident and optimistic about the histone modification field and its link to epigenetics. When I met him for the last time, it was coincident with Elizabeth Pennisi's article in *Science*, highlighting the discovery of acetylation; the last line of which read "Vincent Allfrey should be pleased".

I found that indeed Vincent was really happy. Vincent passed away soon in the year 2002.

Back home in India, I continued in the field, focussing on the regulation of chromatin dynamics by non-histone chromatin proteins, histone chaperones and also small molecule modulators of histone modifying enzymes. At this juncture, in 2007, I got an opportunity to edit a volume of Subcellular Biochemistry entitled 'Chromatin and Disease' (Vol. No. 41). While editing this volume I started realizing that chromatin function is tightly linked to epigenetic phenomenon and that the next volume must be on 'Epigenetics'. I thank the editorial board members, especially, Robin Harris and Dipak Dasgupta, who were so forthcoming and encouraging that I took the responsibility to edit the present volume entitled 'Epigenetics: Development and Disease'.

Epigenetics is not the monopoly of eukaryotes. During the course of evolution, as the genomic organization became more complex and evolved into systematically arranged chromatin structure, epigenetic machineries also started appearing as early as in Archaea. It is interesting to learn that soon after protozoans, all the four core histones along with different variants, ATP dependent remodeling systems and histone modifying enzymes are involved in genome function and thereby in the process of differentiation. In higher eukaryotes, epigenetically regulated dynamic chromatin function is the fundamental basis of differentiation and development. One of the basic cellular processes through which the epigenetic machineries operate is transcriptional regulation. Besides RNA Polymerase II driven transcription, RNA Polymerase I and RNA Polymerase III mediated transcription also requires histone modifications and promoter methylation. The non-coding RNA transcripts transcribed by RNA Polymerase III themselves function as one of the components of epigenetic machineries.

Cellular homeostasis is often disturbed in pathophysiological conditions. Thus in different diseases, inflammatory to infectious, epigenetic marks are altered during the disease progression. The consequent alteration of gene expression network is also remarkable. However, it is not yet established whether altered epigenetic marks are a cause or result of the diseased microenvironment. Nevertheless, the altered epigenetic marks are emerging as targets of new generation therapeutics, some of which are already in the advanced stages of drug development. Considering these facts, the present volume has been organized into five different parts: (i) Epigenetics and Evolution, (ii) Developmental Epigenetics, (iii) Epigenetics and transcription regulation, (iv) Epigenetics and Disease and (v) Understanding of Epigenetics: A Chemical Biology Approach and Epigenetic Therapy.

Experts from all over the globe (14 countries) have contributed excellent articles covering the thoughts expressed above. They have modified their article based on the reviewers' and my comments as and when they were requested, in spite of their heavily loaded schedule. I express my heartfelt thanks to all the contributors for their great effort. Several of my present laboratory colleagues and a few who have left the laboratory and gone abroad to pursue their further research career have contributed immensely to make this volume a reality, among whom I must acknowledge B Ruthrotha Selvi (presently at MRC HGU, Edinburgh, UK) and my present

Preface

lab colleagues Sujata Kumari and D. Karthigeyan. I also acknowledge Parijat Senapati and Snehajyoti Chatterjee without whose constant effort for more than a year, in the process of sending out invitation letters, time to time communications, organising the articles and giving several scientific inputs, publishing this volume would have been impossible. I, my research team as well as all the contributors greatly acknowledge all the reviewers who have worked so hard from behind the scenes for their valuable comments to improve each article.

This book is dedicated to Prof. G. Padmanaban, who is not only associated in all the scientific ventures I am involved in but has also played an active role in maturing the idea of the whole book through several discussions. I am at a loss of words to express my gratitude towards him. Last but not the least, I thank the past and present staff members of Springer, Max Haring, and Marlies Vlot, who worked hard with us to bring this volume for all of you. All the contributors and myself hope that this volume will be useful to students who are learning chromatin biology and epigenetics, the teachers and researchers of the field, and also scientists from pharmaceutical industries.

Bangalore-64   Tapas K. Kundu

Acknowledgment

I thank the reviewers who spent their precious time and greatly contributed through their valuable comments on each chapter.

Anne Laurence Boutillier
Université de Strasbourg, Strasbourg, France

B. Ruthrotha Selvi
Medical Research Council Human Genetics Unit, University of Edinburgh, Edinburgh, UK

David Engelke
University of Michigan, Michigan, USA

Dipak Dasgupta
Saha Institute of Nuclear Physics, Kolkata, India

Jayasha Shandilya
University at Buffalo (SUNY), New York, USA

Malancha Ta
Manipal Hospital, Bangalore, India

Parag Sadhale
Indian Institute of Science, Bangalore, India

Philippe Bouvet
Université de Lyon, Lyon, France

Uday Kumar Ranga
Jawaharlal Nehru Centre for Advanced Scientific Research, Bangalore, India

Saman Habib
Central Drug Research Institute, Lucknow, India

Sanjeev Galande
Indian Institute of Science Education and Research (IISER), Pune, India

Sanjeev Khosla
Centre for Cellular and Molecular Biology, Hyderabad, India

Sunil Manna
Centre for DNA Fingerprinting and Diagnostics, Hyderabad, India

Sunil Mukherjee
International Centre for Genetic Engineering and Biotechnology, New Delhi, India

Venkatesh Swaminathan
Stowers Institute for Medical Research, Kansas City, USA

Contents

Part I Epigenetics and Evolution

1 **Chromatin Organization, Epigenetics and Differentiation: An Evolutionary Perspective**.. 3
Sujata Kumari, Amrutha Swaminathan, Snehajyoti Chatterjee, Parijat Senapati, Ramachandran Boopathi, and Tapas K. Kundu

2 **Secondary Structures of the Core Histone N-terminal Tails: Their Role in Regulating Chromatin Structure** 37
Louis L. du Preez and Hugh-G. Patterton

3 **Megabase Replication Domains Along the Human Genome: Relation to Chromatin Structure and Genome Organisation** 57
Benjamin Audit, Lamia Zaghloul, Antoine Baker, Alain Arneodo, Chun-Long Chen, Yves d'Aubenton-Carafa, and Claude Thermes

4 **Role of DNA Methyltransferases in Epigenetic Regulation in Bacteria** 81
Ritesh Kumar and Desirazu N. Rao

Part II Developmental Epigenetics

5 **Metabolic Aspects of Epigenome: Coupling of S-Adenosylmethionine Synthesis and Gene Regulation on Chromatin by SAMIT Module** 105
Kazuhiko Igarashi and Yasutake Katoh

6 **Epigenetic Regulation of Male Germ Cell Differentiation** 119
Oliver Meikar, MatteoDa Ros, and Noora Kotaja

7 **Epigenetic Regulation of Skeletal Muscle Development and Differentiation** 139
Narendra Bharathy, Belinda Mei Tze Ling, and Reshma Taneja

8   Small Changes, Big Effects: Chromatin Goes Aging........................... 151
    Asmitha Lazarus, Kushal Kr. Banerjee,
    and Ullas Kolthur-Seetharam

9   Homeotic Gene Regulation: A Paradigm for Epigenetic
    Mechanisms Underlying Organismal Development 177
    Navneet K. Matharu, Vasanthi Dasari, and Rakesh K. Mishra

Part III Epigenetics and Transcription Regulation

10 Basic Mechanisms in RNA Polymerase I Transcription
    of the Ribosomal RNA Genes.. 211
    Sarah J. Goodfellow and Joost C.B.M. Zomerdijk

11 The RNA Polymerase II Transcriptional Machinery
    and Its Epigenetic Context 237
    Maria J. Barrero and Sohail Malik

12 RNA Polymerase III Transcription – Regulated
    by Chromatin Structure and Regulator of Nuclear
    Chromatin Organization 261
    Chiara Pascali and Martin Teichmann

13 The Role of DNA Methylation and Histone Modifications
    in Transcriptional Regulation in Humans 289
    Jaime L. Miller and Patrick A. Grant

14 Histone Variants and Transcription Regulation 319
    Cindy Law and Peter Cheung

15 Noncoding RNAs in Chromatin Organization
    and Transcription Regulation: An Epigenetic View 343
    Karthigeyan Dhanasekaran, Sujata Kumari,
    and Chandrasekhar Kanduri

16 Chromatin Structure and Organization: The Relation
    with Gene Expression During Development and Disease................... 373
    Benoît Moindrot, Philippe Bouvet, and Fabien Mongelard

Part IV Epigenetics and Disease

17 Cancer: An Epigenetic Landscape 399
    Karthigeyan Dhanasekaran, Mohammed Arif, and Tapas K. Kundu

18 Epigenetic Regulation of Cancer Stem Cell Gene Expression............ 419
    Sharmila A. Bapat

Contents

**19 Role of Epigenetic Mechanisms in the Vascular
Complications of Diabetes**.. 435
Marpadga A. Reddy and Rama Natarajan

**20 Epigenetic Changes in Inflammatory
and Autoimmune Diseases**.. 455
Helene Myrtue Nielsen and Jörg Tost

**21 Epigenetic Regulation of HIV-1 Persistence
and Evolving Strategies for Virus Eradication** 479
Neeru Dhamija, Pratima Rawat, and Debashis Mitra

22 Epigenetics in Parkinson's and Alzheimer's Diseases 507
Sueli Marques and Tiago Fleming Outeiro

23 Cellular Redox, Epigenetics and Diseases 527
Shyamal K. Goswami

**Part V Understanding of Epigenetics: A Chemical Biology
Approach and Epigenetic Therapy**

**24 Stem Cell Plasticity in Development and Cancer:
Epigenetic Origin of Cancer Stem Cells** 545
Mansi Shah and Cinzia Allegrucci

25 Histone Acetylation as a Therapeutic Target 567
B. Ruthrotha Selvi, Snehajyoti Chatterjee, Rahul Modak,
M. Eswaramoorthy, and Tapas K. Kundu

26 DNA Methylation and Cancer 597
Gopinathan Gokul and Sanjeev Khosla

27 Role of Epigenetics in Inflammation-Associated Diseases 627
Muthu K. Shanmugam and Gautam Sethi

**28 *Plasmodium falciparum*: Epigenetic Control
of *var* Gene Regulation and Disease** 659
Abhijit S. Deshmukh, Sandeep Srivastava, and Suman Kumar Dhar

Index 683

Abbreviations

5-Aza-CdR	5-aza-2'-deoxycytidine
5hmc	5-hydroxymethylcytosine
5mC	5-methylcytosine
7SKsnRNP	7SK small nuclear ribonuclear protein
AAP	ambiant air pollution
ABC	ATP-binding cassette
AML	acute myeloid leukaemia
APC	antigen-presenting cell
APOBEC3G	Apolipoprotein B mRNA-editing enzyme-catalytic polypeptide-like protein 3G
APP	Amyloid beta (Aβ) precursor protein
AR	Androgen receptor
ARE	Antioxidant response element
ART	Anti-retroviral therapy
ASC	adult stem cells
ATF2	Activating transcription factor 2
ATM	Ataxia telangiectasia mutated
ATR	Ataxia telangiectasia related gene
ATRX	Alpha thalassemia/mental retardation syndrome X-linked
BBB-	Blood brain barrier
BRCA1	Breast cancer 1, early onset
BRG1	Brahma-related gene 1
CAF-1	Chromatin assembly factor-1
CAP	catabolite activator protein
CARM1	Coactivator-associated arginine methyltransferase 1
CAST	CD3 epsilon-associated signal transducer
CBF-1	C-promoter Binding Factor-1
CCL19	C-C motif chemokine 19
CCL21	C-C motif chemokine 21
CcrM	cell cycle regulated methyltransferase
CD	chron's disease

Cdk9	Cyclin-Dependent Kinase 9
CDKN2	Cyclin dependent kinase 2
CENP-A	Centromere protein A
CF	core factor
CHD7	Chromodomain helicase DNA-binding protein 7
ChIP	Chromatin immunoprecipitation
CNS	conserved non-coding sequence
CNS	central nervous system
CpG	cytosine-phosphate-guanine
CREB	c-AMP response element binding protein
CRFs	Chromatin reassembly factors
CSB	Cockayne syndrome B protein
CSC-	Cancer stem cell
CSP-	Carbon nanospheres
CTCF	CCCTC-binding factor
CTD	Carboxy-Terminal Domain
CTIP2	Chicken ovalbumin upstream promoter transcription factor interacting proteins 2
CTK7A-	Sodium 4-(3,5-bis(4-hydroxy-3-methoxystyryl)-1H-pyrazol-1-yl)benzoate
CTL	Cytotoxic T-Lymphocytes
CTPB-	N-(4-Chloro-3-trifluoromethyl-phenyl)-2-ethoxy-6-pentadecyl-benzamide
DAC	5-deoxy-azacytidine
Dam	DNA adenine methyltransferase
DAPK1	Death associated protein kinase 1
Dcm	DNA cytosine methyltransferase
DEP	diesel exhaust particles
DNMT	DNA methyltransferase
DNMT1/3A/3B/3L	DNA methyltransferase 1/3A/3B/3L
DR3	death receptor 3
DSB-	Double strand break
EGCG	(−)-Epigallocatechin gallate
EMT	epithelial-to-mesenchymal transitions
eNoSC	energy-dependent nucleolar silencing complex
EPC	epithelial progenitor cells
ER	Estrogen receptor
ESC	embryonic stem cells
ESCC-	Esophageal squamous cell carcinoma
ETS	external transcribed spacer
FACT	facilitates chromatin transcription
FGF	fibroblast growth factor
GADD45α	growth arrest and DNA damage inducible protein 45 alpha
GFP	Green Fluorescent Protein
GMCSF	Granulocyte macrophage colony stimulating factor

Abbreviations

H2AK119u ubiquitinated	ubiquitinated histone H2A at lysine 119 histone H2A at lysine 119
H3K27me3	trimethylated lysine 27 on histone H3
H3K4me1/2/3	mono-di-tri-methylated histone H3 at lysine 4
H3K64me3	trimethylated histone 3 at lysine 64
H3K9ac	acetylated histone H3 at lysine 9
H3K9me1/2/3	mono-di-tri-methylated histone H3 at lysine 9
H4K20me3	trimethylated histone 4 at lysine 20
HAT	histone acetyltransferases
HATi	histone acetyltransferases inhibitor
HCC-	Hepatocellular carcinoma
HDAC	Histone deacetylase
HDACi	histone deacetylase inhibitor
HDMs	histone demethylases
HEXIM1	HMBA Inducible protein 1
HEXIM2	HMBA Inducible protein 2
HIV-	Human immune deficiency virus
HMBA	Hexa Methylene Bis Acetamide
HMG	high mobility group
HMTs	histone methyltransferases
HP1α	Heterochromatin protein-1α
HPC	high CpG promoter
HS	hypersensitive sites
HSC	hematopoietic stem cells
IBD	inflammatory bowel disease
ICM	inner cell mass
IG	isogarcinol
IGS	intergeneic spacer
IKK	IκB kinase
IL	interleukin
INO80	Inositol-requiring protein 80
iPSC	induced pluripotent stem cells
ISWI	Imitation SWI
ITS	internal transcribed spacer
JMJD3	jumonji-domain-containing protein histone deacetylase 3
KAT-	Lysine (K) acetyltransferase
KDAC-	Lysine deacetylase
LARP7	La related protein
LAT	Latency associated transcripts
lncRNAs	long non coding RNAs
LPC	low CpG promoter
LSF	Late SV40 Factor
MBD	methyl-binding domain
MBD2	Methyl-CpG-binding domain protein 2
MBD3	methyl-CpG binding domain protein 3

MBP	myelin basic protein
MDMs	Monocytes derived macrophages
MEPCE	Methyl phosphate capping enzyme
MHC	major histocompatibility complex
miRNAs	microRNAs
MLL	Histone methyl transferase
MOZ-	Monocytic leukaemia zinc-finger protein
mRNAs	messenger RNAs
MS	multiple sclerosis
MZ	monozygotic
ncRNAs	non coding RNAs
NER	nucleotide-excision repair
NF-κB	Nuclear factor-κB
NFAT	nuclear factor of activated T cell
NK	natural killer
NOD	non obese diabetic
NoRC	nucleolar remodelling complex
NORs	nucleolar organiser regions
NPM-	Nucleophosmin
NSC	neural stem cells and progenitors
ORC-	Origin recognition complex
p300/CBP	E1A binding protein p300/CREB-binding protein
Paf1c	Polymerase-associated factor 1 complex
PAF53	RNA Polymerase I associated factor 53
PARP-1	poly-ADP-ribose-polymerase 1
PBMC	peripheral blood mononuclear cell
PCAF	p300/CBP-Associated Factor
PcG	polycomb group of proteins
PEPCK-	Phosphoenolpyruvate carboxykinase
PGC	primordial germ cells
PGC-1α	peroxisome proliferator activated receptor gamma coactivator 1
PIC	pre-initiation complex
PMA	Phorbol 12-Myristate 13-Acetate
Pol	RNA polymerase
PRC2	Polycomb repressive complex 2
PRMT	Protein arginine methyl transferases
PRMT6	Protein arginine methyl transferase 6
pRNA	promoter RNA
pTEFb	Positive Transcription Elongation Factor b
PTM-	Post translational modification
PTRF	Pol I and transcript release factor
r	ribosomal
RA	rheumatic arthritis
RbBp5	retinoblastoma-binding protein 5
RDR2	RNA-dependent RNA polymerase 2

RFX1	regulatory factor X 1
RISC	RNA induced silencing complex
R–M	restriction – modification
RNS	Reactive nitrogen species
RORγt	RAR-related orphan receptor γ
ROS	Reactive oxygen species
RSS	RNA silencing suppressor
SAHA-	Suberoylanilide hydroxamic acid
SAM	S-adenosyl methionine
SCF-β-TrCP	Skp1-Cul1-F- box ligase containing the F-box protein β-transducin repeat-containing protein (βTrCP)
SCID	Severe combined immunodeficiency
shRNA	Short-hairpin RNA
siRNA	Small interfering RNA
SIRT	sirtuin
SL1	selectivity factor-1
SLE	systemic lupus erythematosus
SNF2	sucrose non-fermentable 2 chromatin remodeller
SWI/SNF	SWItch/Sucrose nonfermentable
SWR	Swi2/Snf2-related ATPase
T1D	type 1 diabetes
TAF1A	TAF_I48
TAF1B	TAF_I63
TAF1C	TAF_I110
TAF1D	TAF_I41
TAFII250	TATA binding protein associated factor 250
TAFs	TBP-associated factors
TAP	Transporter associated with antigen presentation
TBP	TATA-box binding protein
TCR	T cell receptor
TET	ten-eleven-translocation
TGF-β	transforming growth factor beta
Th	T helper cell
TIF1A/B	transcription initiation factor 1A/B
TIMP-3	Tissue inhibitor of metalloproteinase 3
TNF-α	Tumor necrosis factor alpha
TRBP	TAR RNA binding protein
TRD	target recognition domain
Treg	regulatory T cell
TrxG	tritorax group of proteins
TSA	trichostatin A
TSS	Transcription start site
TTF-I	transcription termination factor
UBF	upstream binding factor
UC	ulcerative colitis

UCE	upstream control element
UPE	upstream promoter element
vmiRNA	Viral miRNA
VPA	Valproic acid
vsr	very short repair
WRN	Werner's syndrome helicase
WSTF	Williams syndrome transcription factor
WT	Wild-type
YY1	Ying Yang Protein 1
ZBG	zinc-binding group

Part I
Epigenetics and Evolution

Chapter 1
Chromatin Organization, Epigenetics and Differentiation: An Evolutionary Perspective

Sujata Kumari, Amrutha Swaminathan, Snehajyoti Chatterjee, Parijat Senapati, Ramachandran Boopathi, and Tapas K. Kundu

Abstract Genome packaging is a universal phenomenon from prokaryotes to higher mammals. Genomic constituents and forces have however, travelled a long evolutionary route. Both DNA and protein elements constitute the genome and also aid in its dynamicity. With the evolution of organisms, these have experienced several structural and functional changes. These evolutionary changes were made to meet the challenging scenario of evolving organisms. This review discusses in detail the evolutionary perspective and functionality gain in the phenomena of genome organization and epigenetics.

1.1 Introduction

Epigenetics is a phenomenon which operates beyond the information present in the DNA sequence. However, information underlying the DNA sequence also plays an important role in the organization of epigenetic elements. Therefore, DNA sequence elements such as repetitive elements especially CpG islands are the functional components of chromatin organization as well as epigenetic machinery. Different histones and nonhistone proteins are dynamic packaging components of the versatile genome. Covalent modifications of DNA and these proteins as well as ATP remodelling factors contribute to the storage, maintenance and propagation of epigenetic information. As a consequence of diverse evolutionary pathways, these DNA and

---

S. Kumari • A. Swaminathan • S. Chatterjee • P. Senapati
• R. Boopathi • T.K. Kundu (✉)
Transcription and Disease Laboratory, Molecular Biology and Genetics Unit (MBGU),
Jawaharlal Nehru Centre for Advanced Scientific Research (JNCASR), Jakkur Post,
Bangalore 560064, India
e-mail: tapas@jncasr.ac.in

protein elements have evolved gradually towards complexity to facilitate the highly specific signal response and genome function. To begin with, we shall discuss about DNA elements.

1.2 DNA Elements: The Thread of Life

The number of genes, amount of DNA and collinearity of evolutionary hierarchy remains an unsolved puzzle. In prokaryotes, DNA sequence information is maximally used, whereas in higher organisms there is a large amount of DNA which is apparently functionless. However, this concept is rapidly changing in light of recent discoveries of small nuclear RNAs. Although it is beyond the scope of this review to discuss evolution of DNA elements from prokaryotes to higher eukaryotes, however, to discuss epigenetic phenomena, it is important to understand evolution of CpG islands and methylation.

Prokaryotes are devoid of CpG islands. Even in lower eukaryotes there are no defined CpG islands. In cold blooded animals primitive type of CpG islands are observed. The higher vertebrates (warm blooded animals) have well defined CpG islands. CpG islands get methylated and demethylated. Methylation of CpG islands is the founding phenomenon of epigenetic operations. In the heavily methylated eukaryotic genome, CpG islands in promoter regions have been kept free of methylation which is an interesting mechanism of gene expression regulation.

CpG Islands (CGIs) are not just the presence of CpG dinucleotides. Characteristically, CGIs are around 1,000 bp stretches of DNA that possess more than 60% G+C base composition and are devoid of DNA methylation. In mammals 1% of the genome is CGIs. Interestingly, half of the mammalian CGIs are not associated with the annotated promoters and are thus termed "orphan" CGIs. Highly evolved CGIs are found in warm blooded vertebrates. The other eukaryotes, such as *Drosophila, C. elegans* and yeast donot have typical CpG Islands. However, since in these organisms, DNA methylation system is almost absent, CpG sequences are unmethylated throughout the genome. In cold blooded animals, 'CGI-like' sequences are observed. Although initial reports suggested that mouse genome has far fewer CGIs as compared to the human genome, recent data reveals that the number is almost similar (for human 25,495 and mice 23,201) (Deaton and Bird 2011).

CGIs play important roles in the genome organization. Several experimental data have shown that CGIs are relatively nucleosome deficient or possess a structure of transcriptionally active chromatin. For example, histones H3 and H4 are found to be hyperacetylated and histone H1 is almost depleted at CGIs. Furthermore, genome-wide studies have shown that H3K4me3 is a predominant mark of CGIs associated with promoter regions. This category of CGIs is generally unmethylated, but another type of CpG Island which are non-gene CGIs and contain AluI repeat elements, are generally methylated (Ohlsson and Kanduri 2002). These islands are transcriptionally silent and possess highly compact chromatin structure.

The existence of a new class of unique CGIs that are methylated on both alleles has also been reported. The unmethylated or differentially methylated CGIs are possibly not only important for the active transcription start site but may also be the nucleation site for other physiological phenomenon such as spermiogenesis (Kundu and Rao 1996). However, methylation of CGIs plays an important role in the compact chromatin organization and genomic instability.

CGIs are important elements for transcription regulation. Mammalian transcription factor binding sites are generally GC rich. Sixty percent human protein coding genes' transcription start sites are located in CGIs. All the mammalian housekeeping genes' promoters contain CpG islands at their 5′ end. Several tissue-specific genes also have CpG islands. CGIs at promoter sites are unmethylated. In a heavily methylated mammalian genome, how these CGIs are kept methylation free is not yet well understood. Interestingly, CGI containing promoters often lack TATA boxes. As in every case, there are exceptions to this generalization and the examples are α globin, MyoD1 and erythropoietin which are CGI containing promoters and possess TATA boxes (Juven-Gershon et al. 2008). CGI chromatin at the promoter sites is predominantly acetylated and H3K4me3 modified (Fig. 1.1a). The protein Cfp1, a component of Set1 complex, specifically binds to unmethylated CGIs, and thus trimethylates the lysine 4 of histone H3 (Voo et al. 2000; Lee and Skalnik 2005). Recent evidence suggest that the promoters of miRNAs also posses CGIs with similar epigenetic state based on the signals. Orphan CGIs, may also possess active chromatin characteristics during development, suggesting their yet undefined role in the development (Illingworth et al. 2010).

As CGIs are important for chromatin dynamics and transcription regulation, they have profound effect on the maintenance of pluripotency and stem cell differentiation. The divalent mark of H3K4me3 (activation) and H3K27me3 (repression) at CGIs is the hall mark of undifferentiated ES cells (Fig. 1.1c). During the process of differentiation, some genes become completely active and possess only active marks while some others are completely silenced acquiring H3K27me3 and DNA methylation at CGIs (Azuara et al. 2006). Mechanistically, polycomb group protein complex 2 (PRC2) mediates the H3K27me3 methylation (Fig. 1.1b). Hypothetically, the PcG protein Ezh2 (catalytic component of PRC2), recruits the DNA methyltransferase to silence the CGI area of the genome (Viré et al. 2006).

The aberrant methylation of CGIs is predominant in several cancers. Several tumor suppressor genes harbor cancer specific methylation at CGIs. However, cancer specific methylation is not the monopoly of tumor suppressor genes. It is yet to be established whether CGI methylation is the cause or result of malignancy. Along with CGI methylation, threefold over-representation of H3K27me3 at CGIs was observed in malignant colorectal cancers (Illingworth et al. 2010). Cancer specific methylation equally affect annotated promoters as well as orphan CGIs. Role of PRC2 complex is being implicated in cancer specific hypermethylation (Illingworth et al. 2010). An interesting question or possibility is emerging: Does the DNA methylation mediated "pseudo-pluripotent state" favour the indefinite proliferation ability of a cancer cell? The role of CGI methylation in cancer stem cells is yet to be elucidated.

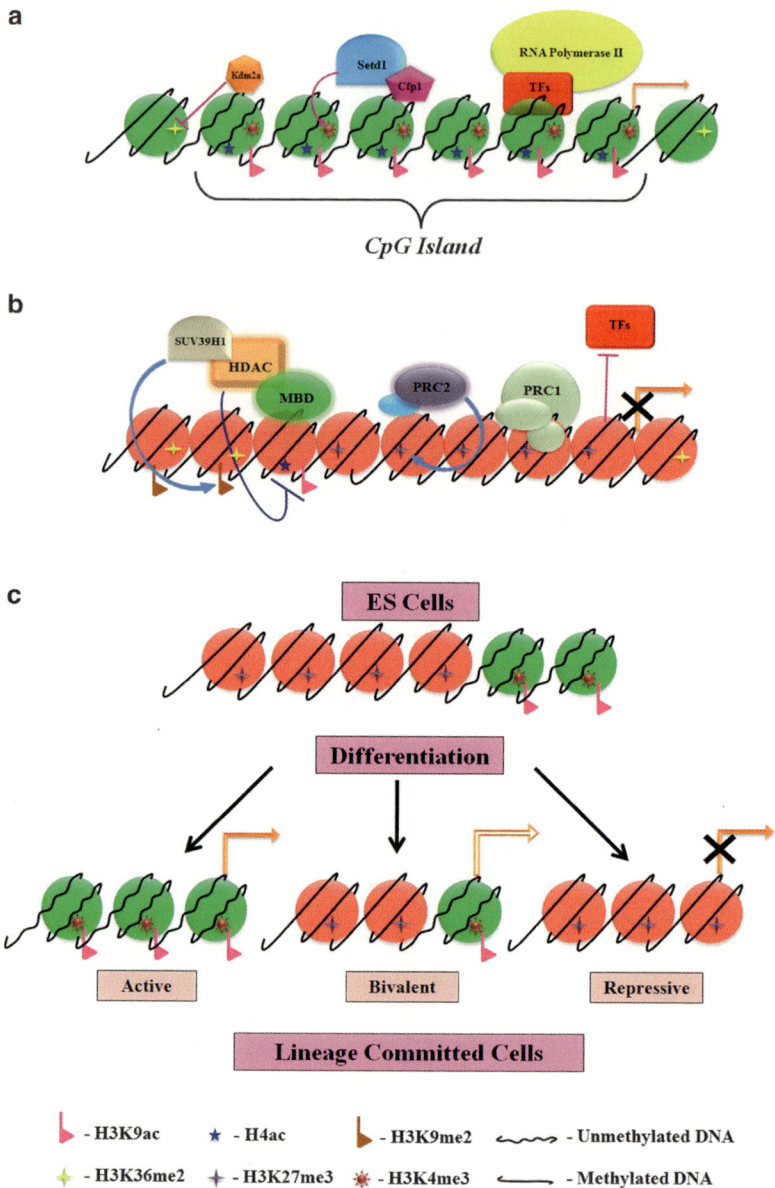

Fig. 1.1 CpG islands (CGIs) and transcription regulation. (**a**) Unmethylated CGIs in the promoter indicates transcriptionally active promoter. Cfp1 protein of Set complex binds to unmethylated DNA and trimethylated lysine4 of histone H3. At promoters bearing H3K4me3 and other active acetylation marks, RNA polymerase machinery is recruited which follows transcriptional activation. (**b**) Transcriptionally repressed promoter is characterized by methylated CGIs. Polycomb group protein complex 2 (PRC2) and other repressive complex containing histone deacetylases add repressive H3K27me3 marks and erase active acetylation marks respectively. (**c**) Undifferentiated ES cells possess divalent marks of H3K4me3 (*active*) and H3K27me3 (*repressive*) at CGIs. During lineage commitment, these marks are redistributed in such a way that required genes acquire active marks while others are silenced

1.3 Protein Elements (Histones and Histone Like Proteins)

There are different DNA binding proteins (predominantly sequence non specific binders) which compact the genomic DNA, not only to accommodate a huge amount of DNA in a tiny space but also to regulate gene expression.

1.3.1 Bacterial/Archeal Chromatin and Architectural Proteins

These proteins are basic in nature and belong to diverse classes. Histones are such proteins which form the basis of genome organization in eukaryotes. Histones have undergone evolution as histone fold motifs are also found in prokaryotic architectural proteins. It is now apparent that the relative amount of DNA bound proteins, presence/absence of histones and/or histone folds thereby affecting nucleosome stability have become principle criteria to resolve dichotomy between bacteria and higher life forms rather than counting for cellular architecture differences among them. Low protein:DNA ratio (<0.5) containing chromatin like structure is found in eubacteria, dinoflagellates, bacteriophage heads and mitochondria. The eubacterial chromatin is prone to aggregation and tends to form 'compactosome' (labile nucleosomes) (Kellenberger and Arnold-Schulz-Gahrnen 1992). This chromatin is very fragile and thus difficult to subject it to biochemical analysis. These DNA binding proteins lack histone folds. On the contrary, eukaryotic and archeal chromatin have higher protein:DNA ratio (>0.5) and the DNA is bound with regularly distributed protein partners and thus possesses more regular structure. Chromatin of *Halobacterium salinarium*, a halophilic archaebacterium was found to be composed of two types of DNA: protein free DNA and DNA assembled as nucleosomes forming fibers of 17–20 nm depicting the classical 'beads on a string' form (Shioda et al. 1989; Takayanagi et al. 1992). Chromatin compaction in non-nucleosome containing organisms like bacteria occurs via macromolecular crowding due to their higher concentrations in the cytoplasm (Zimmerman and Murphy 1996 and references therein; Bloomfield 1996). On the other hand, some organisms presumably utilize both nucleosomal packaging and crowding mechanisms at different time points in their life cycle to organize the genome (Drlica and Bendich 2000; Hildebrandt and Cozzarelli 1995).

1.3.1.1 Nucleoid Associated Proteins (NAPs)

E. coli nucleoid is irregularly shaped and poorly resembles the nucleus of higher eukaryotes. The nucleoid harbours genetic material organised as condensed circular chromosome having supercoiled domains. Nucleoid isolated from exponentially growing *E. coli* cells using a modified protocol showed fivefold more protein weight relative to DNA (Murphy and Zimmerman 1997). Apart from their well known DNA architectural roles, NAPs are also believed to act as buffers of DNA superhelical structures (reviewed in Rimsky and Travers 2011). Different NAPs use different means to apply structural constraints on the DNA. Accordingly, these can be classified as DNA bridging and DNA bending proteins. In the following sections we shall discuss a few of these proteins in brief.

(a) **H-NS (Histone Like Nucleoid Structuring Protein)** is a non-specific DNA binding protein but shows preference to curved or bent DNA (Dame et al. 2001; Jordi et al. 1997). H-NS is a DNA bridging protein composed of two domains separated by a flexible linker. The C- terminal domain binds to DNA whereas the N-terminal domain is involved in dimerization. Dimeric H-NS bridges adjacent DNA duplexes (Dame et al. 2000, 2005a). The binding between DNA duplexes could be by the DNA binding domain of each monomer partner extending in opposite directions or the dimers can bind to the same strand of DNA (Badaut et al. 2002; Dorman 2004). Studies show that H-NS has approximately 1% genome coverage during exponential growth phase. *In vivo* evidence for architectural role of H-NS comes from over expression studies of H-NS which results in a compacted genome (Spurio et al. 1992).

(b) **SMC (Structural Maintenance of Chromosomes) complexes** are V shaped, and have large molecular weight (150–200 kDa) homodimers (Losada and Hirano 2005; Nasmyth and Haering 2005; Melby et al. 1998). The monomeric structure consists of two anti parallel coiled coils which join the extreme terminals of the protein. Two dimers can join through the apex region of the coils and form a V shaped dimer. This association is through a flexible hinge that provides the dimer an angular orientation. The globular region formed by association of the N and C terminus is called as the 'head' and contains the ATP binding domain. Dimers associate to form multimers usually yielding ring like structures or rossette structures. Stability of these structures depends on ATP binding and/or presence of protein co-factors (Hirano and Hirano 2004; Graumann 2001; Hopfner et al. 2000). The hinge region of SMC contains three consecutive lysine residues which probably help in DNA binding (Strunnikov 2006; Hirano and Hirano 2006). They associate with multiple DNA duplexes to yield these structures in the presence of different cofactors like ScpA, ScpB, MukE, MukF etc. These proteins indirectly contribute to chromatin organisation by affecting DNA topology (Lindow et al. 2002).

(c) **Lrp (Leucine-responsive regulatory protein)** is a small Leucine responsive protein with an N-terminal DNA binding domain (by virtue of three alpha helices) (Leonard et al. 2001) and leucine responsive C-terminus which are connected by a hinge region. Functional unit of Lrp is an octamer formed by four dimers resulting in a disc-like structure (Calvo and Matthews 1994; Brinkman et al. 2003). Lrp recognises consensus DNA binding sequence but can also bind with high affinity to sub-optimal binding sites (Cui et al. 1995). Modes of interaction of Lrps with DNA involves bridging mechanisms and wrapping of DNA around Lrp disc like structure remotely resembling the DNA-histone octamer interaction in the nucleosomes of higher eukaryotes (Beloin et al. 2003; Jafri et al. 1999; Thaw et al. 2006).

(d) **IHF (Integration host factor)** is an abundant non specific DNA bending protein. Its association with DNA can lead reduction in the length of DNA by 30% (Ali et al. 2001). It is composed of two subunits, a and b, which share a similarity of upto 25% in their secondary structure. IHF is a heterodimer and its crystal structure reveals two flexible arms protruding out of the main body (Rice et al. 1996). A beta ribbon extends as an arm from the main body on both sides and inserts into the minor groove of DNA (Rice et al. 1996; Swinger and Rice 2004). The beta ribbon has a conserved proline residue which intercalates and causes hydrophobic interactions between DNA bases resulting in narrowing of the minor groove. IHF can cause sharp bends of upto 180° in DNA (Dame et al. 2005b; Dhavan et al. 2002; Rice et al. 1996; Lorenz et al. 1999). The negatively charged bent DNA is stabilised by the positively charged amino acid residues in the IHF main body (Swinger and Rice 2004).

(e) **HU (Histone-like protein from *Eschrichia coli* strain U93)** is a non specific DNA bending protein related to IHF. It is a dimeric protein and exists as two isoforms HUα and HUβ encoded by two homologous genes hupA and hupB in most bacteria. However, in enteric bacteria heterodimeric HU$\alpha\beta$ is formed. Structurally, HU is similar to IHF in having a globular domain with two protruding beta ribbon arms (Swinger et al. 2003). The mode of interaction of HU with DNA is exactly the same except for the fact that they induce two kinks in different planes resulting in underwinding of the DNA double helices (Swinger et al. 2003). HU induced bends are flexible unlike IHF and can stabilize a range of different bend angles (Paull et al. 1993; Swinger and Rice 2004; Swinger et al. 2003). Owing to its random non specific DNA bending, overall DNA length reduction has been estimated upto 50% (Van Norrt et al. 2004). Another mode of HU- DNA association has been exemplified by *in vitro* studies where higher concentration of HU leads to its binding to DNA in a spiral manner forming stiff superhelical filaments (Van Noort et al. 2004). *In vivo* authenticity of this model is being addressed (Dame and Goosen 2002; Van Noort et al. 2004).

(f) **Fis (Factor for inversion stimulation)** is yet another DNA binding and DNA bending protein which is abundant in the early exponential phase. It functions as a dimer. Each monomer consists of four helices, two of which form the hydrophobic core body and the other two are involved in DNA bending (Johnson et al. 2005; Pan et al. 1996). It recognises a 15 bp consensus sequence but can also bind nonspecifically. Two arms of the Fis dimer bends DNA and fit themselves into the major groove. Fis binding sites are present at the upstream of many RNA operons where it acts as a transcriptional activator. Another mode of Fis DNA interaction is described as its binding to the DNA crossing point. At higher concentration it coats the DNA and results in DNA compaction (Schneider et al. 2001). Interaction between DNA bound to Fis dimers results in formation of DNA loops (Skoko et al. 2005). Thus Fis, via its DNA bending (non specific binding) and DNA looping properties contributes to nucleoid compaction.

(g) **DPS (DNA-binding proteins from starved cells)** As the name suggests, its role matches with its requirement during starvation. Its association with DNA protects it from UV light, redox insult and thermal shock (Minsky et al. 2002). DPS belongs to ferritin superfamily. It forms large oligomers like dodecamers with negatively charged surfaces. Interestingly, there are no DNA binding domains in the protein and due to negative charge of its oligomeric state is very unlikely to interact with DNA. Nevertheless, DPS binds to DNA non specifically.

A unique mode of DPS-DNA complex has been proposed. DPS-DNA complexes have been shown to form hexagonally packed 2-D arrays (Frenkiel-Krispin et al. 2004) with pores. These pores form a small positively charged patch by virtue of three lysine residues from three neighbouring N-termini of different dodecamers. This positive patch allows DNA interaction (Grant et al. 1998). 3-D arrays showing significant compaction have also been proposed via dodecamer-dodecamer interaction. Unlike eukaryotic histones which are exclusively involved in chromatin organization, none of the NAPs seem to be solely responsible for Nucleoid organization. Mutant studies show that these proteins can compensate for each other (Kano and Imamoto 1990; Paull and Johnson 1995) while some antagonize the other (Dame and Goosen 2002); For example DNA bending proteins (IHF, HU and Fis) locally antagonize effects of H-NS or Lrp.

Interestingly, NAPs dynamically appear and disappear based on the state of cellular life. This phenomena is also found in protozoans where highly differentiated cannonical histones and variants are involved. In each growth phase like early exponential, exponential and stationary phase, at least one of the DNA bending proteins predominates probably to antagonize the activity of other DNA binding proteins.

1.3.1.2 Archeal Histones

Most of the archeal chromosomal proteins are unique and do not share homology with bacterial proteins. Nevertheless, there are few exceptions. HTa protein isolated from *Thermoplasma acidophilum* is similar to HU family proteins (Oberto and Rouvie`re-Yaniv 1996; DeLange et al. 1981). HTa is a DNA binding protein that stabilizes DNA under thermal stress. *Sulfolobus* species possess different small, acid soluble, DNA binding polypeptides. These bind to minor groove of the DNA causing a bend and gets coated around the bent DNA (Robinson et al. 1998; Agback et al. 1998). DNA bending protein MC1 was identified from *Methanosarcina barkeri* which introduces kinks in the DNA without altering its contour length (Toulmé et al. 1995).

Archeal histones were first discovered in *Methanothermus fervidus* belonging to *Methanobacterium* clade (Sandman et al. 1990). Now it is known that almost all lineages of euryarchaeota possess histones (Sandman et al. 1997). Archeal histone-DNA assembly showed almost classical 'beads on a string' structure (Pereira et al. 1997; Sandman et al. 1990). Most of the archeal histones contain a single histone fold (represented by three alpha helices, $\alpha 1$-$\alpha 3$ and two intermittent loops, L1 and L2) (Fig. 1.2a). Archeal histones lack the N-terminal tail and C-terminal extensions or tails. There are two distinct histone like proteins in *M. fervidus* named as HmfA and HmfB. Archeal histones are grouped in two groups A and B based on their N-terminal residue. In archeal histones A, N-formyl methionine is replaced by alanine or glycine at position 1 whereas archeal histones B retain formyl methionine at position 1. Amino acid composition of HmfA and B differs which results in different tetramers

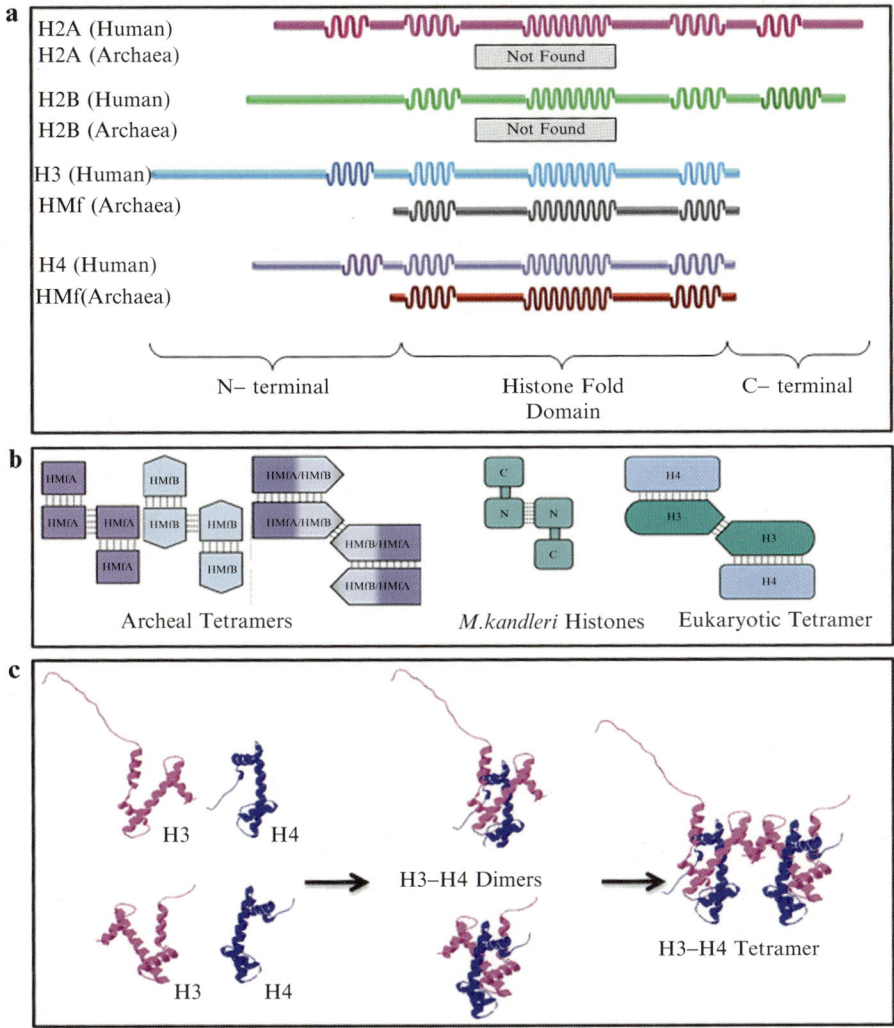

Fig. 1.2 Archeal and eukaryotic histones. (**a**) Line diagram of eukaryotic core histones depicting their histone fold domains and N terminal and C terminal extensions. (**b**) A proposed evolutionary route for transition from archeal to eukaryotic tetramers. *M. kandleri* (an archaebacterium) histones (doublet) are proposed to be intermediate in this transition. (**c**) Assembly of eukaryotic histone tetramer. Eukaryotic tetramer resembles that of archeal tetramers

with different DNA topology. HMfB has been shown to cause greater degree of DNA compaction than HMfA. Consistent with this, abundance of these two proteins differ in different life cycle stages (Sandman et al. 1994). Number of histone genes in euryarchaeota lineage goes upto 176 with the exception of *Thermoplasma* having no histone genes. Instead it has HU like proteins. Archeal histones HmfA and HmfB can form both homodimers and heterodimers. A tetramer of these proteins has been shown to protect 60 bp of DNA indicating single round wrapping of DNA (Pereira et al. 1997).

Collectively, organization of DNA elements with the help of compacting proteins is a common phenomenon for all organisms. In archaebacteria the first step towards the ordered organization of genome with histones could be observed.

1.3.2 Eukaryotic Chromatin

Larger genome and more sophisticated circuit in eukaryotes necessitate ordered and more efficient genome organization. In eukaryotes, linear chromosomes are formed by several levels of DNA compaction. But here also, fundamental unit is nucleosome formed by DNA wrapped around histone proteins. There are four eukaryotic nucleosome core histones namely H2A, H2B, H3 and H4. They are highly conserved and are encoded by different histone genes found in clusters. Eukaryotic histones have a common globular domain and N and C terminal extended tails which are not essential for nucleosome assembly (Fig. 1.2a). These tails interact with regulatory proteins. Eukaryotic histones form exclusive H2A-H2B and H3-H4 heterodimers and not other homodimers. H3 and H2B proteins additionally have five adjacent amino acids that are not found in H4 and archeal histones. Thus H3 and H2B have larger alpha helix $\alpha 1$ and loop L1 than H4 and H2A resulting in asymmetrical H3-H4 and H2A-H2B dimer (Luger et al. 1997).

Crystal structure of nucleosome core particle revealed 146 bp DNA wrapped around the histone octamer core. Histone octamer consists of four histone dimers (H2A-H2B as dimers and H3-H4 as a tetramer) around which DNA is wound 1.65 times. Histone tails extend out beyond the nucleosome. Linker histones do not share the histone fold of canonical core histones. These histones bind near the dyad axis, entry-exit point of DNA in nucleosome thus organizing additional 20 bp resulting in a complex called chromatosome (Hayes et al. 1994). However, exact location of linker histone H1 is yet to be elucidated. During nucleosome assembly, two H3-H4 dimers associate to form a tetramer (Fig. 1.2c). This tetramer recognises nucleosome positioning signal and binds to DNA. The H2A-H2B dimers flank the tetramer from both side followed by DNA wrapping around it (Wolffe 1992; Luger et al. 1997). Interface of two H3 monomers in central H3-H4 tetramer is designated as the dyad axis of symmetry.

1.3.3 Evolution of Eukaryotic Chromatin

Archeal and eukaryal histones share homology both in amino acid sequence and their 3-D structure (Starich et al. 1996; Zhu et al. 1998; Luger and Richmond 1998).

Archeal histones do not contain N and C terminal extensions beyond the histone fold and thus are shorter than eukaryal histones. Eukaryotic histone tails are not required for nucleosome assembly and can be proteolytically chopped without affecting nucleosome structure (Ausio et al. 1989). Tails of H3 and H4 make contacts with adjacent nucleosomes and thus result in tighter packaging. Dimerization properties of histone proteins seem to be conserved to some extent. Hydrophobic interactions among α-helices of the monomers stabilize the dimer structure. Residues of the hydrophobic surface are identical in HMfA and HMfB and thus are capable of forming both homodimers and heterodimers. On the contrary, hydrophobic amino acid residues present in H2A-H2B and H3-H4 corresponding monomers is complementary and not identical explaining their restricted heterodimer formation. Nevertheless, these residues are present at the same location as that of archeal histones. From an evolutionary perspective, transition might have occurred due to mutation in the hydrophobic patch which restrained the formation of homodimer and allowed only heterodimerization.

Discovery of 'doublet histones' in *Methanopyrus kandleri* was the major advancement in understanding evolutionary link between archeal and eukaryal nucleosomes (Slesarev et al. 1998). *Methanopyrus kandleri* histone has duplicated histone folds encoded by single gene. So, unlike other archeal histones, only one type of dimer and tetramer is possible here (Fig. 1.2b). *M. kandleri* histone dimerizes via its C-terminal helices of each N-terminal histone fold in the doublet. The N-terminal histone fold of the dimer structurally correspond to H3 whereas C-terminal to that of H4 (Fahrner et al. 2001). Several residues in doublet histones show deviation from singlet histones (Reviewed in Malik and Henikoff 2003). The N-terminal domains of doublet histones show a conserved change of lysine to methionine as compared to singlet histones. The change remains persistent in eukaryotic H3, emphasizing common constraint in these two classes of dimers.

Phylogenetic studies regarding order of origin of eukaryotic octameric histones is confusing. Nevertheless, tetramer formation is conserved and is observed both with archeal and eukaryal histones. Archeal histones can form tetramers but not the higher forms. Archeal tetramers resemble H3-H4 tetramer of the eukaryotic nucleosomes (Pereira and Reeve 1998) indicating that H3 and H4 have evolved before H2A and H2B. Specifically, comparison between H3 and HMfB sequence reveals identical hydrophobic core in tetramers formed by these proteins (Luger and Richmond 1998). Also, self dimerization of H3 in octamer resembles that of *M. kandleri* doublet histones whereas H2A self dimerization seems unusual. These observations lead to two possibilities of eukaryotic histone appearance: (1) duplication of already divergent pair of histone genes followed by sequence divergence and (2) simultaneous duplication and divergence of one gene. Presence of additional unique five amino acids in α helix of H3 and H2B argues for the former possibility. Appearance of H2A-H2B dimers have led to the octameric form of the nucleosome. The degree of packaging of DNA has increased as octamer is capable of accommodating nearly two turns of DNA when compared to single turn in tetramers. This transition seems to provide selection benefit to the eukaryotes where high degree and rapid condensation/decondensation of chromatin is required.

Centromeric CenH3 containing nucleosomes better known as hemisomes, donot contain conventional octameric histone core as studied in budding yeast and *Drosophila*. In budding yeast, Cse4-H4 tetramer has been reported at the centromere where a nonhistone protein Scm3 is shown to replace histone H2A-H2B dimers (Mizuguchi et al. 2007). In *Drosophila melanogaster* interphase cells, centromeric nucleosomes are composed of one copy each of CenH3, H4, H2A and H2B. These histones form heterotypic tetramers and associate with approximately 120 bp DNA (Dalal et al. 2007). Like archeal nucleosomes, these hemisomes might represent an evolutionary link to ancient form of nucleosomes.

Protozoa and lower eukaryotes like yeast share conserved nucleosome core particle structure with minor differences. Crystal structure reveals slightly destabilized nucleosome core particle in *Saccharomyces cerevisiae* (White et al. 2001). Yeast nucleosomes have shorter repeat length than higher organisms and are more closely spaced (Horz and Zachau 1980). *Saccharomyces cerevisiae* genome contains two genes for each of the histone proteins unlike the histone gene cluster found in higher eukaryotes. Yeast histones are divergent and have sequence differences throughout the protein body including flexible tails (Baxevanis and Landsman 1998). No linker histone H1 gene could be detected in yeast or *Plasmodium*. However, linker histone homolog protein Hho1p has been reported in *Saccharomyces cerevisiae*. The *Hho1p* is not essential as its null mutant is not lethal. Possibly, it does not play any significant role in yeast chromatin organization (Patterton et al. 1998).

Like higher eukaryotes, *Plasmodium falciparum* chromatin is composed of nucleosomal building blocks. However, higher order chromatin organization is lacking probably due to absence of linker histones. *Plasmodium* genome encodes all four canonical histone genes and additional histone variants genes as well (discussed later). Like bacterial chromosomal proteins, *Plasmodium* histones also show life cycle specific expression pattern. Histones are highly expressed in the late trophozoite and schizont stages when DNA synthesis is maximum (Miao et al. 2006).

Histone fold does not seem to be restricted to histone proteins. In eukaryotes, multiple transcription factors like TAFII, PCAF, CBF etc. possess histone folds (Kokubo et al. 1994; Birck et al. 1998; Ogryzko et al. 1998; Ouzounis and Kyrpides 1996). Relative position of DNA binding elements in these histone fold containing proteins differ from that of eukaryotic histones indicating different functions of these proteins. Apart from histones, there are chromatin interacting nonhistone proteins which confer fluidity to the genome. These nonhistone chromatin proteins are not being discussed here due to space constraints.

Chromatin is a nonstatic entity committed to perform multifaceted activities. Conserved structure of histones and their interactions with each other and DNA restrict their functional diversity. Spatial and temporal operations require additional levels of regulation. The chromatin remodelling machineries and covalent modifications of histones are the major players to perform this job. Recently, histone variants have also been shown to modulate various chromatin related phenomena like transcription, silencing, DNA repair, heterochromatinization, meiosis etc. Histone variants harbour minor changes in their primary sequences as compared to canonical histones but result in alteration in chromatin structure (Dryhurst et al. 2004;

Table 1.1 List of histones and histone variants in different organisms

Organisms	Core histones	Core histone variants
Archaea	HMfA, HMfB	–
Yeast	H2A, H2B, H3, H4	H3.3, Cse4, Htz1, H2B1, H2B2
Plasmodium	H2A, H2B, H3, H4	H2Az, H2Bv, H3.3, CenH3
Drosophila	H2A, H2B, H3, H4	H3.3, Cid, H2Av
Mammals	H2A, H2B, H3, H4	H3.1, H3.2, H3.3, CENPA, H2AZ, H2AX, macroH2A, H2A.Bbd, TH2B, H2BFW, SubH2BV, H2BL1, H2BL2

Ramaswamy et al. 2005). Histone variants are encoded by single copy genes located outside the canonical histone genes cluster. Unlike histones, variants genes contain introns and are transcribed as polyadenylated mRNAs. Variants have acquired differential expression pattern contrary to ubiquitous histone expression. Histone variants seem to have evolved from their counterparts with different propensities and through multiple independent events (Thatcher and Gorovsky 1994). Among the four canonical histones, H4 is invariant and most of the H2B variants are involved in spermatogenesis process (Zalensky et al. 2002; Churikov et al. 2004; Aul and Oko 2002; Govin et al. 2007). H2A has a rich family of variants in several different organisms and H3 has a few but functionally important variants. Heterogenity in histone variant propensity is obvious. Basically, only one partner of the allowed heterodimer (H2A-H2B and H3-H4) is varied at a time. It seems only one of the histone partner is selected for variation that will cause least perturbation in nucleosome structure (Reviewed in Malik and Henikoff 2003; Pusarla and Bhargava 2005). Table 1.1 shows the list of histone variants in different organisms.

In eukaryotes, some DNA sequences show high preference for nucleosome assembly which are termed as nucleosome positioning sequences (NPS). It is believed that H3-H4 tetramer recognises NPS and initiates nucleosome assembly (Dong and van Holde 1991; Spangenberg et al. 1998). In archaea too, NPS have been reported. Six or multiple repeats of CTG have been shown to be recognised by archeal histones (Sandman and Reeve 1999).

1.3.4 Evolution of Nucleosome Positioning

It has long been debated whether DNA sequence determines nucleosome deposition or not. After thirty years of research in this field, it has now become clear that DNA sequence does influence nucleosome positioning. DNA wrapped around nucleosomes is far more sharply bent than unstressed DNA which means significant free energy would be needed for the stability of the nucleosomal organization. Certain DNA sequences could reduce the amount of free energy needed by having an inherent bendedness or flexibility. Indeed it has been shown that certain dinucleotides present at the right positions might help in bending. Work by Segal et al. showed that certain dinucleotides AA/TT/TA exhibited a ~10 bp periodicity; the dinucleotide GC showed the same periodicity, however, GC was out of phase with AA/TT/TA. These nucleosome signatures were obtained by comparing genomic sequences from several organisms, (eg. chicken, yeast, mouse) as well as random *in vitro* synthesized DNA which showed maximum affinity for the histone octamer (Segal et al. 2006).

The nucleosome positioning pattern has been more refined now and has been derived by three independent approaches: analysis of nucleosome DNA sequences, deformational properties of the dinucleotide stacks of nucleosome DNA, and Shannon N-gram extension for genomic sequences (Trifonov 2011b). The pattern derived from all these three approaches were found to be the same when about 160,000 nucleosome DNA sequences from *C. elegans* were analyzed (Gabdank et al. 2009). The nucleosome positioning sequences followed the pattern CGRAAATTTYCG (where R represents purines A or G and Y represents pyrimidines C or T) or YRRRRRYYYYYR in binary form. The CGRAAATTTYCG pattern indicates the positions where certain dinucleotide combinations should be present in the DNA to make the bending energetically less expensive and thus thermodynamically more favourable when wrapped around a histone octamer. The CG dinucleotides reside at the minor grooves which are oriented inwards contacting the surface of the histone octamer whereas the central AT dinucleotides that are five bases away (half-period of DNA), are positioned at the dyad axis of the structural DNA repeat in the minor grooves oriented outwards. Theoretically expressed, the nucleosome positioning pattern reflects the deformational properties of DNA and rules of base stacking interactions which explain why weak bases (W) should be at the minor grooves facing outwards and the strong bases (S) should be at the minor grooves contacting the octamer. The nucleosome positioning pattern finally comes down to ….SSSSWWWWWWSSSSWWWWWW…. sequence for an AT rich eukaryotic genome (S stands for strong dinucleotides CC, GG, GC and CG and W stands for weak dinucleotides AA, TT, AT and TA). Further, the pattern has perfect complementary symmetry due to which the complete identity of the two strands of the DNA duplex can be derived (Trifonov 2011a). This pattern was also found to be completely consistent with the nucleosome DNA patterns of other eukaryotes as established earlier and hence it can be considered to be a consensus nucleosome DNA sequence across eukaryotic species (Rapoport et al. 2011). The actual nucleosomal DNA sequences are not exact matches of the repeating pattern. They are usually rather weak resemblances to the pattern, as an exact match will lead to highly positioned nucleosomes throughout the genome and the information coding capacity of the sequences would be significantly lower. However, the nucleosome DNA sequences form different organisms do show specific dinucleotide sequences at positions matching the nucleosome positioning pattern (Trifonov 2011a).

Recent work on the analysis of DNA sequences in prokaryotes and archaea also showed the periodical patterns of (AAAAATTTTT)n and (GAAAATTTTC)n respectively which are the same as the eukaryotic nucleosome positioning pattern YRRRRRYYYYYR. This study found that although prokaryotes and some species of archaea (Crenarchaea) lack nucleosomes, their sequences do possess the nucleosome positioning code (Rapoport and Trifonov 2011). Archaea that possess primitive nucleosomes (Euryarchaea) also have been shown to possess the same nucleosome positioning sequence as eukaryotes (Bailey et al. 2000). Some prokaryotic histone like proteins such as HU, H-NS, Fis etc. have also been shown to recognise specific DNA sequences. H-NS binds more preferentially to AAATT sequence (Bouffartigues et al. 2007) whereas Fis binds to AAAWTTT sequence (Hengen et al. 1997) which is a part of the nucleosome positioning motif. It has now been proposed that although prokaryotes do not have DNA folded into nucleosomes, they do have nucleosome positioning sequences which appear to have evolved before the prokaryote-eukaryote separation (Rapoport and Trifonov 2011).

Nucleosomal arrangement of genome not only serves the packaging purpose but it encompasses functional significance as well. Nucleosomes inherently possess repressive properties, which the genome cannot withstand throughout the cell cycle stages. Genome has to follow ordered opening and closing for proper functioning. Several factors like remodelers, DNA and histone modifying enzymes and histone chaperones effect fluidity and dynamicity of chromatin. There is a very precise and complicated interplay among these effectors to fine tune the life processes which is not yet fully understood.

1.3.5 Evolution of Lysine Acetyltransferases

Acetylation, the most widely studied posttranslational modification of histones and nonhistone proteins is catalyzed by different classes of lysine (K)acetyltransferases (KATs). The type A KATs are nuclear whereas type B KATs are cytoplasmic. Lysine acetylation is a highly conserved phenomenon observed from archaea, protozoan to humans. The Haloarchaeal species *H. Volcanii* genome contains GCN5 family acetyltransferase (Pat1, Pat2 and Elp3) (Altman-Price and Mevarech 2009). The Alba family of proteins in *S. solfataricus* also gets acetylated by *Salmonella* protein acetyltransferase (PAT) at lysine 16 (Bell and Grogan 2002; Eichler et al. 2005). While some KATs like Elp3 are present in both archaea and eukaryotes suggesting the occurrence of these enzymes early in evolution (Reeve 2003), some enzymes are restricted to the eukaryotes, like the MYST family of KATs and p300/CBP family of KATs. p55, a type A nuclear KAT is present in the protozoan *Tetrahymena*. This protein is surprisingly a homolog of GCN5 and contains a highly conserved bromodomain (Brownell et al. 1996). Apart from archaea and protozoans, two classes of KATs are found in malarial parasites. The *P. falciparum* genome codes for a GCN5 (PfGCN5) and pfMYST (shares homology with yeast Esa1) which is essential for malarial infectivity (Cui and Miao 2010) (Table 1.2).

In eukaryotes, KATs are present in functionally distinct protein complexes. The yeast GCN5 is found in two high molecular mass complexes, SAGA and Ada. The SAGA complex is known to activate transcription and acetylates nucleosomal histone H3 and H2B (Eberharter et al. 1999; Grant et al. 1997). Esa1 in yeast is essential for growth (Smith et al. 1998) whereas Rtt109 is essential for acetylation

Table 1.2 Evolution of different Lysine acetyltransferases from archaea to humans

Organism	PCAF/GCN5	CBP/p300	MYST
Archaebacteria	H. volcanii contains three protein acetyltransferases (Pat1, Pat2 and Elp3) belonging to the GCN5 family of acetyltransferases	–	–
Protozoa	p55 (Homolog of GCN5)	–	–
Plasmodium	PfGCN5	–	PfMYST
Yeast	yGCN5	Rtt109 (Homolog of p300)	Sas3 and Esa1
Drosophila	GCN5 containing KAT complexes	dCBP (No p300 is identified)	dMOF
Human	PCAF/GCN5	p300/CBP	Tip60 and MOZ, HBO1 and MORF

of histone H3K56. *Drosophila* lacks p300 homolog but CBP is present and deletion of dCBP results in cell cycle arrest and stalled DNA replication (Smolik and Jones 2007). GCN5 in *Drosophila* is present in two different complexes STAGA and ATAC and is involved in acetylation of histone H3 and H4 (Kusch et al. 2003; Martinez et al. 1998). Chameau is a MYST family KAT present in *Drosophila* required for maintenance of Hox gene silencing (Grienenberger et al. 2002). In mammals, p300/CBP has diverse function including acetyltransferase activity, E3 ubiquitin ligase and transcriptional coactivation. PCAF/GCN5 mediated acetylation of histones and non histone proteins are essential for transcriptional activation and also muscle differentiation. Tip60 (homolog of Esa1), a MYST family KAT (in mammals) mediated acetylation is essential for DNA damage response. Remarkably, among all the KATs, the Esa1/Tip60 seems to be an essential gene. Evolutionarily, this KAT is conserved from protozoans to human. However, the p300/CBP family of KAT seems to be more mammalian specific. Deletion of either of them is not lethal but knockout of both is life threatening (Yao et al. 1998; Kasper et al. 2006).

1.3.6 Evolution of ATP Dependent Chromatin Remodelers

Apart from chromatin modification systems, the ATP dependent chromatin remodelling machineries also significantly contribute to chromatin dynamics by mobilising the nucleosomes on DNA based on epigenetic signals. Chromatin remodelers are classified into four families: SWI/SNF (switching defective/sucrose nonfermenting), INO80 (inositol requiring 80), ISWI (imitation switch) and CHD (chromodomain, helicase, DNA binding) family (Clapier and Cairns 2009). Among these complexes the INO80 family of chromatin remodelers, seems to be highly conserved and functionally significant. Due to space constraint, we will discuss only about the INO80 family of chromatin remodelers in this section. INO80 is a class of chromatin remodelers comprising of INO80 and SWR1 complexes. These complexes remodel chromatin by interacting with the nucleosome and perform nucleosome sliding along the DNA or exchange histones within the nucleosome (Bao and Shen 2011). The INO80 complex also recruits various regulatory factors to the nucleosome. Homologs and orthologs of this family are present in yeast, *Drosophila*, plants and mammals. The ATPase subunit of INO80 possesses the maximum homology among different organisms. INO80 and SWR1 complexes are composed of 15 and 14 subunits respectively with approximately 1.2–1.5 MDa mass of each. The various members of this family of proteins are Ino80 (INOsitol requiring), Rvb1, Rvb2, Arp4 (Actin-related protein 4), Arp5, Arp8, actin, Nhp10 (Non-histone protein 10), Ies1 (Ino Eighty Subunit 1), Taf14 (TATA-binding protein-associated factor) Ies2, Ies3, Ies4, Ies5 and Ies6. One distinguished characteristic of INO80 subfamily is the presence of RuvB-like helicases in the complexes. RuvB-like helicases are especially present in bacteria as DNA repair factor. Yeast INO80 ATPase domain bears significant resemblances with yeast Snf2/Swi2 family of DNA-dependent ATPases. One additional feature in INO80 ATPase domain is the presence of a spacer inside the conserved ATPase domain (Ebbert et al. 1999). TELY motif present in the N-terminal domain of INO80 is conserved in yeast, *Drosophila* and human and is known to be the interacting residue for actin, Arp4, Arp8 and Taf14/Anc1 (Shen et al. 2000, 2003). The SWR1 complex is composed of 14 members and shares homology with INO80 subunits (Fig. 1.3a). The members of SWR1 complex are Swr1, Swc2/Vp372, Swc3, Swc4/Eaf2/God1, Swc5/Aor1, Swc6/Vps71, Swc7, Yaf9 (yeast homolog of the human leukemogenic protein AF9), Bdf1 (bromodomain Factor), Act1/actin, Arp4, Arp6, Rvb1 and Rvb2 (Mizuguchi et al. 2004; Wu et al. 2005; Bao and Shen 2007). Actin, Arp4, Rvb1 and Rvb2 are the common shared subunits between INO80 and SWR1 complexes. Rvb1 and Rvb2 proteins are class of AAA+ ATPases (ATPases associated with a variety of cellular activities) and are related closely to the Holliday junction-resolving RUVB proteins present in bacteria (Shen et al. 2000). Rvb1 and Rvb2 are crucial components of the INO80 and SWR1 remodeling complexes, Tip60 acetyltransferase complex (Ikura et al. 2000) and c-MYC complex. The Ies2 and Ies6 proteins of INO80 complex are also highly conserved from yeast to mammals (Fig. 1.3b) including humans and have distinct roles in transcription regulation (Cai et al. 2007; Shen et al. 2003). Recent report suggests that Iec1 is another novel component of INO80 complex and bears sequence similarity to the *Drosophila* and human counterparts. Iec1 is essential for transcriptional regulation and with other components of INO80 complex (arp8, ies6, and ies2) plays crucial roles in DNA damage repair (Hogan et al. 2010).

The *Drosophila* ortholog of INO80 is Pho-dINO80 and that of NuA4 complex is Tip60. *Drosophila* SWR1 complex comprises of BAP55 and Arp. These proteins are also present in BAP and PBAP complexes. The SWR1 complexes functions in the exchange of H2A variants and is conserved throughout evolution. *Drosophila* Tip60 complex helps in the exchange of H2A for H2Av, which is a variant of H2A

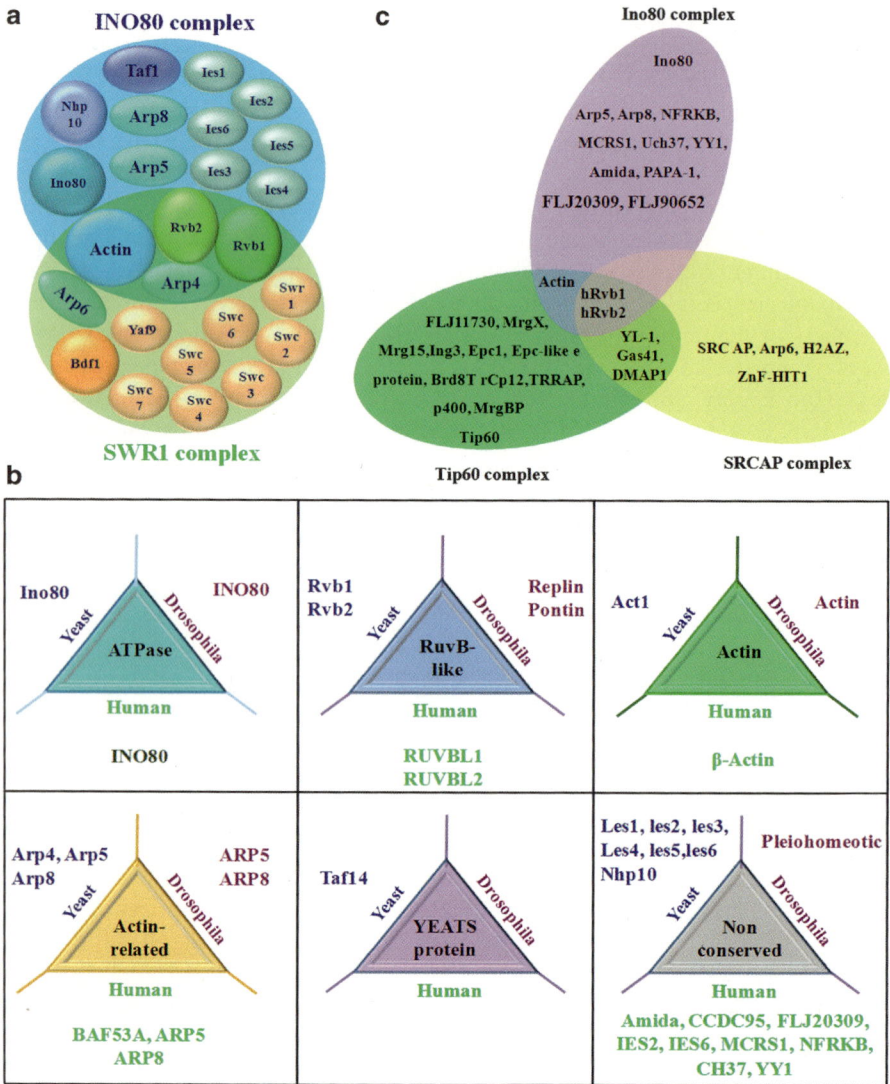

Fig. 1.3 (**a**) Components of yeast INO80 (inositol requiring 80) and SWR1 complexes. The structural-functional homologs are common among the two complexes. (**b**) Homologous factors of INO80 complex among yeast (*Saccharomyces cerevisiae*), *Drosophila* and humans. Each *triangle* represents different subunit types of INO80 complex. (**c**) Protein subunits of Ino80, Tip60 and SRCAP complexes in humans

having resemblances with both H2AZ and H2AX (Kusch et al. 2004). SRCAP complex, the mammalian homologue of SWR1 also catalyzes the exchange of H2A to H2AZ (Ruhl et al. 2006; Wong et al. 2007). Based on a number of shared subunits, *Drosophila* Tip60 (dTip60) is similar to yeast SWR1, but also contains additional seven subunits including a histone acetyltransferase subunit (Tip60). dTip60 is presumed to be an evolutionary product of combined SWR1 and NuA4 complexes in yeast (Hargreaves and Crabtree 2011). In humans the related Tip60 complex is Tip60/ TRRAP complex which possesses three distinct enzymatic activities: histone H4/H2A acetyltransferase, ATP-dependent H2AZ-H2B histone dimer exchange, and DNA helicase. Tip60/TRRAP complex is considered to be the merge of NuA4 and SWR1 complexes through evolution by fusion of Swr1 and Eaf1 (Auger et al. 2008). The major components of human Tip60/TRRAP complex are Swr1-related ATPase, p400, and Tip60. Arp4 and/or actin are similar throughout evolution and are common in yeast, *Drosophila* and humans (Fig. 1.3c). The Swr1 homologue in *Drosophila* is Domino and SRCAP and p400 in humans (Cai et al. 2005). The human homologues of yeast Rvb1 and Rvb2 are RUVBL1 and RUVBL2 while they are named Reptin and Pontin in *Drosophila*. Tip60 in all the organisms are recruited to DNA double strand breaks presumably by an Arp4 dependent mechanism. Subsequently in yeast, NuA4 acetylates histone H4 at break points (Downs et al. 2004), in *Drosophila* Tip60 acetylates phosphorylated H2Av at DNA break points and catalyses exchange with H2A (Kusch et al. 2004). In humans, Tip60 acetylates γH2AX and upon DNA damage facilitates its removal (Ikura et al. 2007). In addition to the INO80 ATPase and other core subunits, human INO80 complex contains Gli-Kruppel zinc finger transcription factor Yin-Yang 1 (YY1), the deubiquitylating enzyme Uch37 and nuclear factor related to κB (NFRKB). *Drosophila melanogaster* polycomb group protein Pleiohomeotic (Pho) is the ortholog of human YY1 and also contains Uch37 and NFRKB (Conaway and Conaway 2009). Significant role of INO80 and SWR1 has been identified in regulation of telomere structure (Yu et al. 2007). Thus, INO80 family of chromatin remodelers function in various cellular processes like DNA damage repair, telomere stability and cell cycle checkpoint. One of the major features of this class of complexes is the recognition of histone variants and recruitment to various structures like DSB sites, holliday junctions and telomeres etc. This function of the complexes is facilitated by the presence of the unique RuvB like proteins and split ATPase domains in various organisms.

Apart from chromatin modifications and remodeling, histone chaperones also significantly contribute in the chromatin dynamics and thereby the operation of epigenetic phenomenon.

1.3.7 Evolution of Histone Chaperones

Histone chaperones are histone interacting proteins that are involved in all types of histone metabolism such as histone storage, transport and assembly and disassembly of histones on chromatin. Due to their involvement in nucleosome assembly and disassembly, histone chaperones have been found to play important roles in essential processes such as replication, transcription and repair. Several histone chaperones have evolved from yeast to mammals and exhibit conservation of sequence and function. Due to space constraint, we will only be discussing about the Nucleoplasmin family.

Nucleoplasmin was the first histone chaperone to be identified. It was identified as a factor in *Xenopus* egg extracts that binds to histones and assembles them onto the DNA in a cell free nucleosome assembly system (Laskey et al. 1978). Since then, many other homologs of Nucleoplasmin have been identified and they comprise the Nucleoplasmin/Nucleophosmin (NPM) family of histone chaperones. Nucleoplasmin homologs are present throughout the animal kingdom but are not found in lower unicellular organisms such as yeast (Frehlick et al. 2007). Based on the protein sequences they are divided into four groups, NPM1, NPM2, NPM3 and invertebrate NPM like protein (Frehlick et al. 2007).

The Nucleoplasmin family members are characterized by the presence of a conserved N-terminal domain (also called the core domain) and a highly divergent C-terminal domain. The N-terminal domain is responsible for oligomerization of the subunits as well as for histone chaperone activity in some members. It contains an acidic stretch A1 which is important for sperm decondensation activity of the members (Salvany et al. 2004). However, human NPM1 differs from the other family members in the fact that histone chaperone activity resides in its C-terminal domain (Swaminathan et al. 2005) and the N-terminal domain has a rudimentary A1 consisting of two amino acids which might not be functional. Moreover, human, mouse and rat NPM2 also donot possess the A1 acidic stretch (Frehlick et al. 2007). The C-terminal domain contains one or more acidic tracts (A2 and/or A3) which are responsible for histone binding (Swaminathan et al. 2005). Nucleoplasmin family members possess a classic bipartite nuclear localization signal $KRX_{10}KKK$ where X is any amino acid (Dingwall et al. 1987). Apart from these domains, some members like NPM1 have additional Nuclear Export Signal (NES), Nucleolar Localization Signal (NoLS) and a nucleic acid binding domain that binds to RNA and is required for rRNA cleavage (Hingorani et al. 2000).

The earliest occurrence of the Nucleoplasmin family has been found in the invertebrates and they are known as Nucleoplasmin like proteins (NLPs). The best characterized among them is the *Drosophila* NLP (dNLP) (Ito et al. 1996). dNLP has been shown to bind to core histones and exhibits nucleosome assembly activity but additionally requires ATP unlike other members and the help of atleast one other assembly activity. Moreover, dNLP was not able to promote sperm decondensation unlike other members of the family (Ito et al. 1996). NPM1 (Nucleophosmin) or NO38 in *Xenopus* is a nucleolar protein. The human homolog Nucleophosmin was first identified as a nucleolar phosphoprotein (Kang et al. 1974; Prestayko et al. 1974) and since then has been extensively studied. The histone chaperone activity of NPM1 was demonstrated later (Okuwaki et al. 2001) where it was shown to bind both H2A-H2B dimer as well as the H3-H4 tetramer without any preference (Swaminathan et al. 2005). Unlike its human homolog, the *Xenopus* NPM1 (NO38) shows a preference towards H3-H4 tetramer (Namboodiri et al. 2004). NPM1 also binds to linker histone H1 and possesses the linker histone chaperone activity (Gadad et al. 2011). NPM1 has been implicated in many other cellular processes which are reviewed in (Grisendi et al. 2006; Okuwaki 2008). NPM2 or Nucleoplasmin of *Xenopus* is the founder member of the family and has been extensively studied (Laskey et al. 1978). NPM2 is found only in eggs and oocytes (Mills et al. 1980) and binds to H2A-H2B dimer preferentially (Dutta et al. 2001). NPM2 is involved in storage of H2A-H2B dimer in the oocytes and sperm chromatin decondensation following fertilization. It displaces the sperm basic proteins and deposits H2A-H2B dimer on to the paternal chromatin (Frehlick et al. 2007). NPM3 was identified later (MacArthur and Shackleford 1997) and was shown to regulate the ribosome biogenesis (Huang et al. 2005) and histone chaperone activity of NPM1 (Gadad et al. 2010). NPM3 binds to all the core histones, yet, in *in vitro* supercoiling assays, it did not show any nucleosome assembly (Gadad et al. 2010). However, it may be involved in the regulation of sperm chromatin decondensation (McLay and Clarke 2003).

The crystal structures of the N-terminal domain of *Xenopus* Nucleoplasmin, NO38, human NPM1, NPM2 and dNLP have been solved and all of them were shown to form a pentameric structure. However, the functional form of all the members is a decamer formed by head-to-head association of two pentamers (Dutta et al. 2001; Namboodiri et al. 2003, 2004; Lee et al. 2007; Platonova et al. 2011). Biochemical and modeling studies suggest that each decamer can bind to five histone octamers (Dutta et al. 2001). Further, in case of human NPM2, it was shown that it forms decamers when it binds to both H3-H4 tetramer and H2A-H2B dimer simultaneously. However, when it binds to H2A-H2B dimer alone, it remains as a pentamer (Platonova et al. 2011).

NPM family has been shown to undergo functional evolution and diversification (Eirín-López et al. 2006). Phylogenetic analyses on the NPM family have revealed that the NPM family members of a single type (such as NPM1, NPM2, NPM3 or invertebrate NLP) cluster together by type rather than by species in the phylogenetic tree. The NPM1 and NPM-like lineage showed to have a monophyletic origin whereas NPM2 and NPM3 showed a polyphyletic origin having evolved independently in mammals as compared to the other vertebrates. The NPM-like lineage from invertebrates was found to be phylogenetically closer to NPM1. NPM sequences were found to have diverged extensively through silent substitutions indicating the presence of selection acting on specific residues especially glutamate and aspartate in the acidic tracts (Eirín-López et al. 2006). The acidic residues in the acidic tracts are important for their interactions with histones and the chaperone activity and hence, are maintained by selection. Even histones are known to evolve similarly by a selection process by silent substitutions operating at the nucleotide level in order to conserve the basic residues (Frehlick et al. 2007).

1.3.8 Epigenetics and Differentiation

As discussed in earlier sections, genome organising and modifying machinery have gone through systematic development through the course of evolution. This sequence of events is well-documented in the phenomenon of differentiation, which is associated with global changes in genome organisation and is characterised by fluctuations in gene expression. The role of chromatin modifying machinery in genome organisation, in modulating the balance between "stemness" and differentiation in multiple lineages has been studied extensively. This section will briefly discuss the conserved molecular role of the biochemical phenomena of histone modifications, and DNA methylation during differentiation, taking the neural system as an example.

1.3.8.1 Epigenetics and Neural Differentiation

Till few decades ago, neurons were not known to regenerate, but the scenario changed with the discovery of neural stem cells (NSCs), which are primarily generated in the subgranular zone (SGZ) of the hippocampal dentate gyrus, and the subventricular zone (SVZ). The NSCs can differentiate into three principal types of cells- Neurons, Astrocytes and Oligodendrocytes. The role of epigenetic modulators in the process of neural differentiation has been well-elucidated over the past two decades. The switch from neurogenesis to astrogliogenesis during mid to late gestation period is primarily modulated by the JAK-STAT pathway controlled by DNMT1 (Fan et al. 2001, 2005; Namihira et al. 2009). The methylation of STAT3-binding element (CpG) in the GFAP promoter occurs during neurogenesis (Bonni et al. 1997) and astrogliogenesis is accompanied by demethylation of these elements in the GFAP promoter, along with other astrocyte-specific gene promoters (Shimozaki et al. 2005).

The histone deacetylases 1 and 2 (HDAC 1 and 2) modulate neural differentiation (Montgomery et al. 2009), and are associated with both induction as well as regulation of differentiation process. While it was observed that treatment of NSCs with VPA, an HDAC inhibitor, induced neuronal differentiation (Hsieh et al. 2004), HDAC activity is required for oligodendrocyte differentiation, and during myelination (Marin-Husstege et al. 2002; Shen et al. 2005). However, paradoxically HDACs are known to be primarily associated with gene repression, thus the mechanism of HDAC action in the differentiation process is yet to be understood. The involvement of specific miRNA which are regulated by HDACs cannot be ruled out. Recently, the role of miRNA in the neural differentiation has indeed been implicated. In the process of neural specification, miR-124, a brain specific miRNA (Lagos-Quintana et al. 2002), is upregulated during differentiation of progenitor cells to mature neurons (Deo et al. 2006). This miRNA is required for silencing of non-neural genes in neural cells (Lim et al. 2005; Conaco et al. 2006). The miR-124 along with miR-9, regulates the STAT3 pathway mediated glial-specific gene expression, by inhibiting STAT3 phosphorylation, and hence, astrocyte differentiation (Krichevsky et al. 2006), indicating multiple roles of the miRNA in regulation of gene expression and lineage specification.

1.3.8.2 Epigenetic Control: Linking Invertebrate and Mammalian Differentiation

Chromatin in stem cells is distinctly different from differentiated cells. Though the gross structure and organ systems vary to a large extent between species, the epigenetic factors regulating development seem to show significant conservation with diversification. The comparative analysis of the chromatin signatures of mammalian embryonic stem cells shows a high degree of conservation. The characteristic open, pliable chromatin was observed in early embryos, with DNA methylation-dependent transcription regulation in the later stages during organogenesis and differentiation (Bogdanovic et al. 2011). Histone modifications are seen to be involved both in transcriptional regulation and chromatin organization during development, across species. As seen with human embryonic stem cell (hESC) differentiation, there is a characteristic shift from the abundance of transcriptionally active to repressive marks during differentiation in *Xenopus* (Schneider et al. 2011). The overall role of histone acetylation and HDAC activity in influencing *Xenopus* development has been studied (Shechter et al. 2009; Almouzni et al. 1994). Specifically, in zebrafish, HDAC1 plays the role of a transcriptional activator in neurogenesis and CNS development (Harrison et al. 2011). H3 acetylation in *Drosophila*, shows a stage-specific increase during gastrulation, associated with high levels of transcription (Harisanova and Ralchev 1986). Similarly, histone methylation patterns (both lysine and arginine methylation) in the three organisms have been shown to be associated with differentiation (Peng et al. 2009; Fujii et al. 2011; Tao et al. 2011; Cakouros et al. 2008; Villar-Garea and Imhof 2008). However, histone modification patterns differ among organisms, as shown in the recent study comparing murine ES cells and *Xenopus* pluripotent cells (Schneider et al. 2011). Although, both murine and *Xenopus* pluripotent cells share high H3K4me2/me3 levels, the H3K27me3 mark is more than 100-fold enriched in murine ES cells as compared to *Xenopus* blastulae. Also, combinatorial tri and higher methylated states of H3 K27 and H3 K36 were significantly more enriched in murine than in *Xenopus* pluripotent cells. The functional significance of these organism-specific epigenetic modifications is yet to be elucidated.

As in mammals, DNMTs have been observed to regulate gene expression in a stage-specific manner in other vertebrates as well. For instance, DNMT1 depletion in *Xenopus* is seen to lead to premature differentiation and apoptosis associated with development (Stancheva and Meehan 2000; Stancheva et al. 2001) and DNA hypomethylation is observed during differentiation (Talwar et al. 1984). Similarly, DNMTs, in zebrafish have been shown to be essential in regulating various developmental pathways, including neurogenesis (Rai et al. 2006), pancreatic beta cell differentiation (Anderson et al. 2009), liver, retina and lens development (Rai et al. 2007; Tittle et al. 2011). Significantly, studies during *Xenopus* development have led to the hypothesis that DNMT1 have other functions than solely DNA methylation itself (Hashimoto et al. 2003).

The information summarized above suggests the conserved nature of chromatin modifying machinery, and its role in differentiation. Notably, HDACs, DNA methylation and H3 methylation play essential roles during development and regulate lineage-specific differentiation. The presence of the global open chromatin structure, followed by large scale chromatin organization and remodeling, associated with changes in DNA methylation and histone modification can hence be considered a phenomenon conserved over the classes, ranging from invertebrates to vertebrates.

1.3.9 Perspective

With the increasing complexity of genome structure and function the organizational and modification or remodelling machineries of chromatin has also achieved inconspicuous diversity during the course of evolution. Nevertheless, overall pathway of their development could be traced significantly. In the light of availability of whole genome databases of DNA sequence and epigenetic modifications, our understanding of this pathway will soon be clearer. Collectively, these informations should be highly useful not only to perceive the evolutionary process but also to understand disease biology.

Acknowledgements We thank Department of Biotechnology, Government of India (for Programme Support Grant on Chromatin and Disease, Grant No. Grant/DBT/CSH/GIA/1957/2011-12) and JNCASR for financial assistance. TKK is a recipient of the Sir J.C. Bose Fellowship (Department of Science and Technology, Government of India). SK, PS and AS are research fellows of Council of Scientific and Industrial Research (CSIR), Government of India.

References

Agback P, Baumann H, Knapp S, Ladenstein R, Härd T (1998) Architecture of nonspecific protein-DNA interactions in the Sso7d-DNA complex. Nat Struct Biol 5:579–584
Ali BM, Amit R, Braslavsky I, Oppenheim AB, Gileadi O, Stavans J (2001) Compaction of single DNA molecules induced by binding of integration host factor (IHF). Proc Natl Acad Sci U S A 98:10658–10663
Almouzni G, Khochbin S, Dimitrov S, Wolffe AP (1994) Histone acetylation influences both gene expression and development of *Xenopus laevis*. Dev Biol 165:654–669
Altman-Price N, Mevarech M (2009) Genetic evidence for the importance of protein acetylation and protein deacetylation in the halophilic archaeon *Haloferax volcanii*. J Bacteriol 191:1610–1617
Anderson RM, Bosch JA, Goll MG, Hesselson D, Dong PD, Shin D, Chi NC, Shin CH, Schlegel A, Halpern M, Stainier DY (2009) Loss of Dnmt1 catalytic activity reveals multiple roles for DNA methylation during pancreas development and regeneration. Dev Biol 334:213–223
Auger A, Galarneau L, Altaf M, Nourani A, Doyon Y, Utley RT, Cronier D, Allard S, Côté J (2008) Eaf1 is the platform for NuA4 molecular assembly that evolutionarily links chromatin acetylation to ATP-dependent exchange of histone H2A variants. Mol Cell Biol 28:2257–2270

Aul RB, Oko RJ (2002) The major subacrosomal occupant of bull spermatozoa is a novel histone H2B variant associated with the forming acrosome during spermiogenesis. Dev Biol 242:376–387

Ausio J, Dong F, vanHolde KE (1989) Use of selectively trypsinized nucleosome core particles to analyze the role of the histone tails in the stabilization of the nucleosome. J Mol Biol 206:451–463

Azuara V, Perry P, Sauer S, Spivakov M, Jørgensen HF, John RM, Gouti M, Casanova M, Warnes G, Merkenschlager M, Fisher AG (2006) Chromatin signatures of pluripotent cell lines. Nat Cell Biol 8:532–538

Badaut C, Williams R, Arluison V, BouVartigues E, Robert B, Buc H, Rimsky S (2002) The degree of oligomerization of the H-NS nucleoid structuring protein is related to specific binding to DNA. J Biol Chem 277:41657–41666

Bailey KA, Pereira SL, Widom J, Reeve JN (2000) Archaeal histone selection of nucleosome positioning sequences and the procaryotic origin of histone-dependent genome evolution. J Mol Biol 303:25–34

Bao Y, Shen X (2007) INO80 subfamily of chromatin remodeling complexes. Mutat Res 618:18–29

Bao Y, Shen X (2011) SnapShot: chromatin remodeling: INO80 and SWR1. Cell 144:158–158. e152

Baxevanis AD, Landsman D (1998) Histone sequence database: New histone fold family members. Nucleic Acids Res 26:372–375

Bell GD, Grogan DW (2002) Loss of genetic accuracy in mutants of the thermoacidophile *Sulfolobus acidocaldarius*. Archaea 1:45–52

Beloin C, Jeusset J, Revet B, Mirambeau G, Le Hegarat F, Le Cam E (2003) Contribution of DNA conformation and topology in righthanded DNA wrapping by the *Bacillus subtilis* LrpC protein. J Biol Chem 278:5333–5342

Birck C, Poch O, Romier C, Ruff M, Mengus G, Lavigne AC, Davidson I, Moras D (1998) Human TAFII28 and TAFII18 interact through a histone fold encoded by atypical evolutionary conserved motifs also found in the SPT3 family. Cell 94:239–249

Bloomfield VA (1996) DNA condensation. Curr Opin Struct Biol 6:334–341

Bogdanovic O, Long SW, van Heeringen SJ, Brinkman AB, Gómez-Skarmeta JL, Stunnenberg HG, Jones PL, Veenstra GJ (2011) Temporal uncoupling of the DNA methylome and transcriptional repression during embryogenesis. Genome Res 21:1313–1327

Bonni A, Sun Y, Nadal-Vicens M, Bhatt A, Frank DA, Rozovsky I, Stahl N, Yancopoulos GD, Greenberg ME (1997) Regulation of gliogenesis in the central nervous system by the JAK-STAT signaling pathway. Science 278:477–483

Bouffartigues E, Buckle M, Badaut C, Travers A, Rimsky S (2007) H-NS cooperative binding to high-affinity sites in a regulatory element results in transcriptional silencing. Nat Struct Mol Biol 14:441–448

Brinkman AB, Ettema TJ, de Vos WM, van der Oost J (2003) The Lrp family of transcriptional regulators. Mol Microbiol 48:287–294

Brownell JE, Zhou J, Ranalli T, Kobayashi R, Edmondson DG, Roth SY, Allis CD (1996) Tetrahymena histone acetyltransferase A: a homolog to yeast Gcn5p linking histone acetylation to gene activation. Cell 84:843–851

Cai Y, Jin J, Florens L, Swanson SK, Kusch T, Li B, Workman JL, Washburn MP, Conaway RC, Conaway JW (2005) The mammalian YL1 protein is a shared subunit of the TRRAP/TIP60 histone acetyltransferase and SRCAP complexes. J Biol Chem 280:13665–13670

Cai Y, Jin J, Yao T, Gottschalk AJ, Swanson SK, Wu S, Shi Y, Washburn MP, Florens L, Conaway RC, Conaway JW (2007) YY1 functions with INO80 to activate transcription. Nat Struct Mol Biol 14:872–874

Cakouros D, Mills K, Denton D, Paterson A, Daish T, Kumar S (2008) dLKR/SDH regulates hormone-mediated histone arginine methylation and transcription of cell death genes. J Cell Biol 182:481–495

Calvo JM, Matthews RG (1994) The leucine-responsive regulatory protein, a global regulator of metabolism in Escherichia coli. Microbiol Rev 58:466–490

Churikov D, Siino J, Svetlova M, Zhang K, Gineitis A, Morton Bradbury E, Zalensky AO (2004) Novel human testis-specific histone H2B encoded by the interrupted gene on the X chromosome. Genomics 84:745–756

Clapier CR, Cairns BR (2009) The biology of chromatin remodeling complexes. Annu Rev Biochem 78:273–304

Conaco C, Otto S, Han JJ, Mandel G (2006) Reciprocal actions of REST and a microRNA promote neuronal identity. Proc Natl Acad Sci U S A 103:2422–2427

Conaway RC, Conaway JW (2009) The INO80 chromatin remodeling complex in transcription, replication and repair. Trends Biochem Sci 34:71–77

Cui L, Miao J (2010) Chromatin-mediated epigenetic regulation in the malaria parasite *Plasmodium falciparum*. Eukaryot Cell 9:1138–1149

Cui Y, Wang Q, Stormo GD, Calvo JM (1995) A consensus sequence for binding of Lrp to DNA. J Bacteriol 177:4872–4880

Dalal Y, Wang H, Lindsay S, Henikoff S (2007) Tetrameric structure of centromeric nucleosomes in interphase Drosophila cells. PLoS Biol 5:e218

Dame RT, Goosen N (2002) HU: promoting or counteracting DNA compaction? FEBS Lett 529:151–156

Dame RT, Wyman C, Goosen N (2000) H-NS mediated compaction of DNA visualised by atomic force microscopy. Nucleic Acids Res 28:3504–3510

Dame RT, Wyman C, Goosen N (2001) Structural basis for preferential binding of H-NS to curved DNA. Biochimie 83:231–234

Dame RT, Luijsterburg MS, Krin E, Bertin PN, Wagner R, Wuite GJ (2005a) DNA bridging: a property shared among H-NS-like proteins. J Bacteriol 187:1845–1848

Dame RT, van Mameren J, Luijsterburg MS, Mysiak ME, Janicijevic A, Pazdzior G, van derVliet PC, Wyman C, Wuite GJ (2005b) Analysis of scanning force microscopy images of protein induced DNA bending using simulations. Nucleic Acids Res 33:e68

Deaton AM, Bird A (2011) CpG islands and the regulation of transcription. Genes Dev 25:1010–1022

DeLange RJ, Williams LC, Searcy DG (1981) A histone-like protein (HTa) from *Thermoplasma acidophilum*. J Biol Chem 256:905–911

Deo M, Yu JY, Chung KH, Tippens M, Turner DL (2006) Detection of mammalian microRNA expression by in situ hybridization with RNA oligonucleotides. Dev Dyn 235:2538–2548

Dhavan GM, Crothers DM, Chance MR, Brenowitz M (2002) Concerted binding and bending of DNA by *Escherichia coli* integration host factor. J Mol Biol 315:1027–1037

Dingwall C, Dilworth SM, Black SJ, Kearsey SE, Cox LS, Laskey RA (1987) Nucleoplasmin cDNA sequence reveals polyglutamic acid tracts and a cluster of sequences homologous to putative nuclear localization signals. EMBO J 6:69–74

Dong F, van Holde KE (1991) Nucleosome positioning is determined by the (H3-H4)2 tetramer. Proc Natl Acad Sci U S A 88:10596–10600

Dorman CJ (2004) H-NS: a universal regulator for a dynamic genome. Nat Rev Microbiol 2:391–400

Downs JA, Allard S, Jobin-Robitaille O, Javaheri A, Auger A, Bouchard N, Kron SJ, Jackson SP, Côté J (2004) Binding of chromatin-modifying activities to phosphorylated histone H2A at DNA damage sites. Mol Cell 16:979–990

Drlica K, Bendich AJ (2000) Chromosome, bacterial. In: Lederberg J (ed) Encylcopedia of microbiology. Academic, San Diego

Dryhurst D, Thambirajah AA, Ausio J (2004) New twists on H2A.Z: a histone variant with a controversial structural and functional past. Biochem Cell Biol 82:490–497

Dutta S, Akey IV, Dingwall C, Hartman KL, Laue T, Nolte RT, Head JF, Akey CW (2001) The crystal structure of nucleoplasmin-core: implications for histone binding and nucleosome assembly. Mol Cell 8:841–853

Ebbert R, Birkmann A, Schüller HJ (1999) The product of the SNF2/SWI2 paralogue INO80 of *Saccharomyces cerevisiae* required for efficient expression of various yeast structural genes is part of a high-molecular-weight protein complex. Mol Microbiol 32:741–751

Eberharter A, Sterner DE, Schieltz D, Hassan A, Yates JR, Berger SL, Workman JL (1999) The ADA complex is a distinct histone acetyltransferase complex in *Saccharomyces cerevisiae*. Mol Cell Biol 19:6621–6631

Eichler JF, Cramer JC, Kirk KL, Bann JG (2005) Biosynthetic incorporation of fluorohistidine into proteins in *E. coli*: a new probe of macromolecular structure. Chembiochem 6:2170–2173

Eirín-López JM, Frehlick LJ, Ausió J (2006) Long-term evolution and functional diversification in the members of the nucleophosmin/nucleoplasmin family of nuclear chaperones. Genetics 173:1835–1850

Fahrner RL, Cascio D, Lake JA, Slesarev A (2001) An ancestral nuclear protein assembly: crystal structure of the *Methanopyrus kandleri* histone. Protein Sci 10:2002–2007

Fan G, Beard C, Chen RZ, Csankovszki G, Sun Y, Siniaia M, Biniszkiewicz D, Bates B, Lee PP, Kuhn R, Trumpp A, Poon C, Wilson CB, Jaenisch R (2001) DNA hypomethylation perturbs the function and survival of CNS neurons in postnatal animals. J Neurosci 21:788–797

Fan G, Martinowich K, Chin MH, He F, Fouse SD, Hutnick L, Hattori D, Ge W, Shen Y, Wu H, ten Hoeve J, Shuai K, Sun YE (2005) DNA methylation controls the timing of astrogliogenesis through regulation of JAK-STAT signaling. Development 132:3345–3356

Frehlick LJ, Eirín-López JM, Ausió J (2007) New insights into the nucleophosmin/nucleoplasmin family of nuclear chaperones. Bioessays 29:49–59

Frenkiel-Krispin D, Ben-Avraham I, Englander J, Shimoni E, Wolf SG, Minsky A (2004) Nucleoid restructuring in stationary-state bacteria. Mol Microbiol 51:395–405

Fujii T, Tsunesumi S, Yamaguchi K, Watanabe S, Furukawa Y (2011) Smyd3 is required for the development of cardiac and skeletal muscle in zebrafish. PLoS One 6(8):e23491

Gabdank I, Barash D, Trifonov EN (2009) Nucleosome DNA bendability matrix (*C. elegans*). J Biomol Struct Dyn 26:403–411

Gadad S, Shandilya J, Kishore A, Kundu T (2010) NPM3, A member of the nucleophosmin/nucleoplasmin family, enhances activator-dependent transcription. Biochemistry 49:1355–1357

Gadad SS, Senapati P, Syed SH, Rajan RE, Shandilya J, Swaminathan V, Chatterjee S, Colombo E, Dimitrov S, Pelicci PG, Ranga U, Kundu TK (2011) The multifunctional protein nucleophosmin (NPM1) is a human linker histone H1 chaperone. Biochemistry 50:2780–2789

Govin J, Escoffier E, Rousseaux S, Kuhn L, Ferro M, Thévenon J, Catena R, Davidson I, Garin J, Khochbin S, Caron C (2007) Pericentric heterochromatin reprogramming by new histone variants during mouse spermiogenesis. J Cell Biol 176:283–294

Grant PA, Duggan L, Côté J, Roberts SM, Brownell JE, Candau R, Ohba R, Owen-Hughes T, Allis CD, Winston F, Berger SL, Workman JL (1997) Yeast Gcn5 functions in two multisubunit complexes to acetylate nucleosomal histones: characterization of an Ada complex and the SAGA (Spt/Ada) complex. Genes Dev 11:1640–1650

Grant RA, Filman DJ, Finkel SE, Kolter R, Hogle JM (1998) The crystal structure of Dps, a ferritin homolog that binds and protects DNA. Nat Struct Biol 5:294–303

Graumann PL (2001) SMC proteins in bacteria: condensation motors for chromosome segregation? Biochimie 83:53–59

Grienenberger A, Miotto B, Sagnier T, Cavalli G, Schramke V, Geli V, Mariol MC, Berenger H, Graba Y, Pradel J (2002) The MYST domain acetyltransferase Chameau functions in epigenetic mechanisms of transcriptional repression. Curr Biol 12:762–766

Grisendi S, Mecucci C, Falini B, Pandolfi PP (2006) Nucleophosmin and cancer. Nat Rev Cancer 6:493–505

Hargreaves DC, Crabtree GR (2011) ATP-dependent chromatin remodeling: genetics, genomics and mechanisms. Cell Res 21:396–420

Harisanova NT, Ralchev KH (1986) Histones and histone acetylation during the embryonic development of *Drosophila hydei*. Cell Differ 19:115–124

Harrison MR, Georgiou AS, Spaink HP, Cunliffe VT (2011) The epigenetic regulator Histone Deacetylase 1 promotes transcription of a core neurogenic programme in zebrafish embryos. BMC Genomics 12:24

Hashimoto H, Suetake I, Tajima S (2003) Monoclonal antibody against dnmt1 arrests the cell division of xenopus early-stage embryos. Exp Cell Res 286:252–262

Hayes JJ, Pruss D, Wolffe AP (1994) Contacts of the globular domain of histone H5 and core histones with DNA in a "chromatosome". Proc Natl Acad Sci U S A 91:7817–7821

Hengen PN, Bartram SL, Stewart LE, Schneider TD (1997) Information analysis of Fis binding sites. Nucleic Acids Res 25:4994–5002

Hildebrandt ER, Cozzarelli NR (1995) Comparison of recombination in vitro and in *E. coli* cells: measure of the effective concentration of DNA in vivo. Cell 81:331–340

Hingorani K, Szebeni A, Olson MO (2000) Mapping the functional domains of nucleolar protein B23. J Biol Chem 275:24451–24457

Hirano M, Hirano T (2004) Positive and negative regulation of SMC–DNA interactions by ATP and accessory proteins. EMBO J 23:2664–2673

Hirano M, Hirano T (2006) Opening closed arms: long-distance activation of SMC ATPase by hinge-DNA interactions. Mol Cell 21:175–186

Hogan CJ, Aligianni S, Durand-Dubief M, Persson J, Will WR, Webster J, Wheeler L, Mathews CK, Elderkin S, Oxley D, Ekwall K, Varga-Weisz PD (2010) Fission yeast Iec1-ino80-mediated nucleosome eviction regulates nucleotide and phosphate metabolism. Mol Cell Biol 30:657–674

Hopfner KP, Karcher A, Shin DS, Craig L, Arthur LM, Carney JP, Tainer JA (2000) Structural biology of Rad50 ATPase: ATPdriven conformational control in DNA double-strand break repair and the ABC-ATPase superfamily. Cell 101:789–800

Horz W, Zachau HG (1980) Deoxyribonuclease II as a probe for chromatin structure. I. Location of cleavage sites. J Mol Biol 144:305–327

Hsieh J, Nakashima K, Kuwabara T, Mejia E, Gage FH (2004) Histone deacetylase inhibition-mediated neuronal differentiation of multipotent adult neural progenitor cells. Proc Natl Acad Sci U S A 101:16659–16664

Huang N, Negi S, Szebeni A, Olson MO (2005) Protein NPM3 interacts with the multifunctional nucleolar protein B23/nucleophosmin and inhibits ribosome biogenesis. J Biol Chem 280:5496–5502

Ikura T, Ogryzko VV, Grigoriev M, Groisman R, Wang J, Horikoshi M, Scully R, Qin J, Nakatani Y (2000) Involvement of the TIP60 histone acetylase complex in DNA repair and apoptosis. Cell 102:463–473

Ikura T, Tashiro S, Kakino A, Shima H, Jacob N, Amunugama R, Yoder K, Izumi S, Kuraoka I, Tanaka K, Kimura H, Ikura M, Nishikubo S, Ito T, Muto A, Miyagawa K, Takeda S, Fishel R, Igarashi K, Kamiya K (2007) DNA damage-dependent acetylation and ubiquitination of H2AX enhances chromatin dynamics. Mol Cell Biol 27:7028–7040

Illingworth RS, Gruenewald-Schneider U, Webb S, Kerr AR, James KD, Turner DJ, Smith C, Harrison DJ, Andrews R, Bird AP (2010) Orphan CpG islands identify numerous conserved promoters in the mammalian genome. PLoS Genet 6:e1001134

Ito T, Bulger M, Kobayashi R, Kadonaga JT (1996) Drosophila NAP-1 is a core histone chaperone that functions in ATP-facilitated assembly of regularly spaced nucleosomal arrays. Mol Cell Biol 16:3112–3124

Jafri S, Evoy S, Cho K, Craighead HG, Winans SC (1999) An Lrp type transcriptional regulator from Agrobacterium tumefaciens condenses more than 100 nucleotides of DNA into globular nucleoprotein complexes. J Mol Biol 288:811–824

Johnson RC, Johnson LM, Schmidt JW, Gardner JF (2005) Major nucleoid proteins in the structure and function of the *Escherichia coli* chromosome. In: Patrick Higgins N (ed) The bacterial chromosome. ACM Press, Washington

Jordi BJ, Fielder AE, Burns CM, Hinton JC, Dover N, Ussery DW, Higgins CF (1997) DNA binding is not sufficient for H-NS-mediated repression of proU expression. J Biol Chem 272:12083–12090

Juven-Gershon T, Hsu JY, Theisen JW, Kadonaga JT (2008) The RNA polymerase II core promoter – the gateway to transcription. Curr Opin Cell Biol 20:253–259

Kang YJ, Olson MO, Busch H (1974) Phosphorylation of acid-soluble proteins in isolated nucleoli of Novikoff hepatoma ascites cells. Effects of divalent cations. J Biol Chem 249:5580–5585

Kano Y, Imamoto F (1990) Requirement of integration host factor (IHF) for growth of *Escherichia coli* deficient in HU protein. Gene 89:133–137

Kasper LH, Fukuyama T, Biesen MA, Boussouar F, Tong C, de Pauw A, Murray PJ, van Deursen JM, Brindle PK (2006) Conditional knockout mice reveal distinct functions for the global transcriptional coactivators CBP and p300 in T-cell development. Mol Cell Biol 26:789–809

Kellenberger E, Arnold-Schulz-Gahrnen B (1992) Chromatins of low protein content: special features of their compaction and condensation. FEMS Microbiol Lett 100:361–370

Kokubo T, Gong DW, Wootton JC, Horikoshi M, Roeder RG, Nakatani Y (1994) Molecular cloning of Drosophila TFIID subunits. Nature 367:484–487

Krichevsky AM, Sonntag KC, Isacson O, Kosik KS (2006) Specific microRNAs modulate embryonic stem cell-derived neurogenesis. Stem Cells 24:857–864

Kundu TK, Rao MR (1996) Zinc dependent recognition of a human CpG island sequence by the mammalian spermatidal protein TP2. Biochemistry 35:15626–15632

Kusch T, Guelman S, Abmayr SM, Workman JL (2003) Two Drosophila Ada2 homologues function in different multiprotein complexes. Mol Cell Biol 23:3305–3319

Kusch T, Florens L, Macdonald WH, Swanson SK, Glaser RL, Yates JR, Abmayr SM, Washburn MP, Workman JL (2004) Acetylation by Tip60 is required for selective histone variant exchange at DNA lesions. Science 306:2084–2087

Lagos-Quintana M, Rauhut R, Yalcin A, Meyer J, Lendeckel W, Tuschl T (2002) Identification of tissue-specific microRNAs from mouse. Curr Biol 12:735–739

Laskey RA, Honda BM, Mills AD, Finch JT (1978) Nucleosomes are assembled by an acidic protein which binds histones and transfers them to DNA. Nature 275:416–420

Lee JH, Skalnik DG (2005) CpG-binding protein (CXXC finger protein 1) is a component of the mammalian Set1 histone H3- Lys4 methyltransferase complex, the analogue of the yeast Set1/COMPASS complex. J Biol Chem 280:41725–41731

Lee HH, Kim HS, Kang JY, Lee BI, Ha JY, Yoon HJ, Lim SO, Jung G, Suh SW (2007) Crystal structure of human nucleophosmin-core reveals plasticity of the pentamer-pentamer interface. Proteins 69:672–678

Leonard PM, Smits SH, Sedelnikova SE, Brinkman AB, de Vos WM, van der Oost J, Rice DW, RaVerty JB (2001) Crystal structure of the Lrp-like transcriptional regulator from the archaeon *Pyrococcus furiosus*. EMBO J 20:990–997

Lim LP, Lau NC, Garrett-Engele P, Grimson A, Schelter JM, Castle J, Bartel DP, Linsley PS, Johnson JM (2005) Microarray analysis shows that some microRNAs downregulate large numbers of target mRNAs. Nature 433:769–773

Lindow JC, Britton RA, Grossman AD (2002) Structural maintenance of chromosomes protein of *Bacillus subtilis* aVects supercoiling in vivo. J Bacteriol 184:5317–5322

Lorenz M, Hillisch A, Goodman SD, Diekmann S (1999) Global structure similarities of intact and nicked DNA complexed with IHF measured in solution by Xuorescence resonance energy transfer. Nucleic Acids Res 27:4619–4625

Losada A, Hirano T (2005) Dynamic molecular linkers of the genome: the first decade of SMC proteins. Genes Dev 19:1269–1287

Luger K, Richmond TJ (1998) DNA binding within the nucleosome core. Curr Opin Struct Biol 8:33–40

Luger K, Mäder AW, Richmond RK, Sargent DF, Richmond TJ (1997) Crystal structure of the nucleosome core particle at 2.8 Å resolution. Nature 389:251–260

MacArthur CA, Shackleford GM (1997) Npm3: a novel, widely expressed gene encoding a protein related to the molecular chaperones nucleoplasmin and nucleophosmin. Genomics 42:137–140

Malik HS, Henikoff S (2003) Phylogenomics of the nucleosome. Nat Struct Biol 10:882–891

Marin-Husstege M, Muggironi M, Liu A, Casaccia-Bonnefil P (2002) Histone deacetylase activity is necessary for oligodendrocyte lineage progression. J Neurosci 22:10333–10345

Martinez E, Kundu TK, Fu J, Roeder RG (1998) A human SPT3-TAFII31-GCN5-L acetylase complex distinct from transcription factor IID. J Biol Chem 273:23781–23785

McLay DW, Clarke HJ (2003) Remodelling the paternal chromatin at fertilization in mammals. Reproduction 125:625–633

Melby TE, Ciampaglio CN, Briscoe G, Erickson HP (1998) The symmetrical structure of structural maintenance of chromosomes (SMC) and MukB proteins: long, antiparallel coiled coils, folded at a Xexible hinge. J Cell Biol 142:1595–1604

Miao J, Fan Q, Cui L, Li J, Li J, Cui L (2006) The malaria parasite *Plasmodium falciparum* histones: organization, expression, and acetylation. Gene 369:53–65

Mills AD, Laskey RA, Black P, De Robertis EM (1980) An acidic protein which assembles nucleosomes in vitro is the most abundant protein in Xenopus oocyte nuclei. J Mol Biol 139:561–568

Minsky A, Shimoni E, Frenkiel-Krispin D (2002) Stress, order and survival. Nat Rev Mol Cell Biol 3:50–60

Mizuguchi G, Shen X, Landry J, Wu WH, Sen S, Wu C (2004) ATP-driven exchange of histone H2AZ variant catalyzed by SWR1 chromatin remodeling complex. Science 303:343–348

Mizuguchi G, Xiao H, Wisniewski J, Smith MM, Wu C (2007) Nonhistone Scm3 and histones CenH3-H4 assemble the core of centromere-specific nucleosomes. Cell 129:1153–1164

Montgomery RL, Hsieh J, Barbosa AC, Richardson JA, Olson EN (2009) Histone deacetylases 1 and 2 control the progression of neural precursors to neurons during brain development. Proc Natl Acad Sci U S A 106:7876–7881

Murphy LD, Zimmerman SB (1997) Isolation and characterization of spermidine nucleoids from *Escherichia coli*. J Struct Biol 119:321–335

Namboodiri VM, Dutta S, Akey IV, Head JF, Akey CW (2003) The crystal structure of Drosophila NLP-core provides insight into pentamer formation and histone binding. Structure 11:175–186

Namboodiri VM, Akey IV, Schmidt-Zachmann MS, Head JF, Akey CW (2004) The structure and function of Xenopus NO38-core, a histone chaperone in the nucleolus. Structure 12:2149–2160

Namihira M, Kohyama J, Semi K, Sanosaka T, Deneen B, Taga T, Nakashima K (2009) Committed neuronal precursors confer astrocytic potential on residual neural precursor cells. Dev Cell 16:245–255

Nasmyth K, Haering CH (2005) The structure and function of SMC and kleisin complexes. Annu Rev Biochem 74:595–648

Oberto J, Rouvie`re-Yaniv J (1996) Serratia marcescens contains a heterodimeric HU protein like *Escherichia coli* and *Salmonella typhimurium*. J Bacteriol 178:293–297

Ogryzko VV, Kotani T, Zhang X, Schiltz RL, Howard T, Yang XJ, Howard BH, Qin J, Nakatani Y (1998) Histone-like TAFs within the PCAF histone acetylase complex. Cell 94:35–44

Ohlsson R, Kanduri C (2002) New twists on the epigenetics of CpG islands. Genome Res 12:525–526

Okuwaki M (2008) The structure and functions of NPM1/Nucleophsmin/B23, a multifunctional nucleolar acidic protein. J Biochem 143:441–448

Okuwaki M, Matsumoto K, Tsujimoto M, Nagata K (2001) Function of nucleophosmin/B23, a nucleolar acidic protein, as a histone chaperone. FEBS Lett 506:272–276

Ouzounis CA, Kyrpides NC (1996) Parallel origins of the nucleosome core and eukaryotic transcription from Archaea. J Mol Evol 42:234–239

Pan CQ, Finkel SE, Cramton SE, Feng JA, Sigman DS, Johnson RC (1996) Variable structures of Fis–DNA complexes determined by Xanking DNA–protein contacts. J Mol Biol 264:675–695

Patterton HG, Landel CC, Landsman D, Peterson CL, Simpson RT (1998) The biochemical and phenotypic characterization of hho1p, the putative linker histone H1 of *Saccharomyces cerevisiae*. J Biol Chem 273:7268–7276

Paull TT, Johnson RC (1995) DNA looping by Saccharomyces cerevisiae high mobility group proteins NHP6A/B. Consequences for nucleoprotein complex assembly and chromatin condensation. J Biol Chem 270:8744–8754

Paull TT, Haykinson MJ, Johnson RC (1993) The nonspecific DNA-binding and -bending proteins HMG1 and HMG2 promote the assembly of complex nucleoprotein structures. Genes Dev 7:1521–1534

Peng JC, Valouev A, Swigut T, Zhang J, Zhao Y, Sidow A, Wysocka J (2009) Jarid2/Jumonji coordinates control of PRC2 enzymatic activity and target gene occupancy in pluripotent cells. Cell 139:1290–1302

Pereira SL, Reeve JN (1998) Histones and nucleosomes in Archaea and Eukarya: a comparative analysis. Extremophiles 2:141–148

Pereira SL, Grayling RA, Lurz R, Reeve JN (1997) Archaeal nucleosomes. Proc Natl Acad Sci U S A 94:12633–12637

Platonova O, Akey IV, Head JF, Akey CW (2011) Crystal structure and function of human nucleoplasmin (Npm2): a histone chaperone in oocytes and embryos. Biochemistry 50:8078–8089

Prestayko AW, Olson MO, Busch H (1974) Phosphorylation of proteins of ribosomes and nucleolar preribosomal particles in vivo in Novikoff hepatoma ascites cells. FEBS Lett 44:131–135

Pusarla RH, Bhargava P (2005) Histones in functional diversification Core histone variants. FEBS J 272:5149–5168

Rai K, Nadauld LD, Chidester S, Manos EJ, James SR, Karpf AR, Cairns BR, Jones DA (2006) Zebra fish Dnmt1 and Suv39h1 regulate organ-specific terminal differentiation during development. Mol Cell Biol 26:7077–7085

Rai K, Chidester S, Zavala CV, Manos EJ, James SR, Karpf AR, Jones DA, Cairns BR (2007) Dnmt2 functions in the cytoplasm to promote liver, brain, and retina development in zebrafish. Genes Dev 21:261–266

Ramaswamy A, Bahar I, Ioshikhes I (2005) Structural dynamics of nucleosome core particle: comparison with nucleosomes containing histone variants. Proteins 58:683–696

Rapoport AE, Trifonov EN (2011) "Anticipated" nucleosome positioning pattern in prokaryotes. Gene 488:41–45

Rapoport AE, Frenkel ZM, Trifonov EN (2011) Nucleosome positioning pattern derived from oligonucleotide compositions of genomic sequences. J Biomol Struct Dyn 28:567–574

Reeve JN (2003) Archaeal chromatin and transcription. Mol Microbiol 48:587–598

Rice PA, Yang S, Mizuuchi K, Nash HA (1996) Crystal structure of an IHF–DNA complex: a protein-induced DNA U-turn. Cell 87:1295–1306

Rimsky S, Travers A (2011) Pervasive regulation of nucleoid structure and function by nucleoid-associated proteins. Curr Opin Microbiol 14:136–141

Robinson H, Gao YG, McCrary BS, Edmondson SP, Shriver JW, Wang AH (1998) The hyperthermophile chromosomal protein Sac7d sharply kinks DNA. Nature 392:202–205

Ruhl DD, Jin J, Cai Y, Swanson S, Florens L, Washburn MP, Conaway RC, Conaway JW, Chrivia JC (2006) Purification of a human SRCAP complex that remodels chromatin by incorporating the histone variant H2A.Z into nucleosomes. Biochemistry 45:5671–5677

Salvany L, Chiva M, Arnan C, Ausió J, Subirana JA, Saperas N (2004) Mutation of the small acidic tract A1 drastically reduces nucleoplasmin activity. FEBS Lett 576:353–357

Sandman K, Reeve JN (1999) Archaeal nucleosome positioning by CTG repeats. J Bacteriol 181:1035–1038

Sandman K, Krzycki JA, Dobrinski B, Lurz R, Reeve JN (1990) HMf, a DNA-binding protein isolated from the hyperthermophilic archaeon *Methanothermus fervidus*, is most closely related to histones. Proc Natl Acad Sci U S A 87:5788–5791

Sandman K, Grayling RA, Dobrinski B, Lurz R, Reeve JN (1994) Growth-phase-dependent synthesis of histones in the archaeon *Methanothermus ferviidus*. Proc Natl Acad Sci U S A 91:12624–12628

Sandman K, Pereira SL, Reeve JN (1997) Diversity of prokaryotic chromosomal proteins and the origin of the nucleosome. Cell Mol Life Sci 54:1350–1364

Schneider R, Lurz R, Luder G, Tolksdorf C, Travers A, Muskhelishvili G (2001) An architectural role of the *Escherichia coli* chromatin protein FIS in organising DNA. Nucleic Acids Res 29:5107–5114

Schneider TD, Arteaga-Salas JM, Mentele E, David R, Nicetto D, Imhof A, Rupp RA (2011) Stage-specific histone modification profiles reveal global transitions in the Xenopus embryonic epigenome. PLoS One 6:e22548

Segal E, Fondufe-Mittendorf Y, Chen L, Thåström A, Field Y, Moore IK, Wang JP, Widom J (2006) A genomic code for nucleosome positioning. Nature 442:772–778

Shechter D, Nicklay JJ, Chitta RK, Shabanowitz J, Hunt DF, Allis CD (2009) Analysis of histones in *Xenopus laevis*. I. A distinct index of enriched variants and modifications exists in each cell type and is remodeled during developmental transitions. J Biol Chem 284:1064–1074

Shen X, Mizuguchi G, Hamiche A, Wu C (2000) A chromatin remodelling complex involved in transcription and DNA processing. Nature 406:541–544

Shen X, Ranallo R, Choi E, Wu C (2003) Involvement of actin-related proteins in ATP-dependent chromatin remodeling. Mol Cell 12:147–155

Shen S, Li J, Casaccia-Bonnefil P (2005) Histone modifications affect timing of oligodendrocyte progenitor differentiation in the developing rat brain. J Cell Biol 169:577–589

Shimozaki K, Namihira M, Nakashima K, Taga T (2005) Stage- and site-specific DNA demethylation during neural cell development from embryonic stem cells. J Neurochem 93:432–439

Shioda M, Sugimori K, Shiroga T (1989) Nucleosome-like structures associated with chromosomes of the archaebacterium *Halobacterium salinarium*. J Bacteriol 171:4514–4517

Skoko D, Yan J, Johnson RC, Marko JF (2005) Low-force DNA condensation and discontinuous high-force decondensation reveal a loop-stabilizing function of the protein Fis. Phys Rev Lett 95:208101

Slesarev AI, Belova GI, Kozyavkin SA, Lake JA (1998) Evidence for an early prokaryotic origin of histones H2A and H4 prior to the emergence of eukaryotes. Nucleic Acids Res 26:427–430

Smith ER, Eisen A, Gu W, Sattah M, Pannuti A, Zhou J, Cook RG, Lucchesi JC, Allis CD (1998) ESA1 is a histone acetyltransferase that is essential for growth in yeast. Proc Natl Acad Sci U S A 95:3561–3565

Smolik S, Jones K (2007) Drosophila dCBP is involved in establishing the DNA replication checkpoint. Mol Cell Biol 27:135–146

Spangenberg C, Eisfeld K, Stünkel W, Luger K, Flaus A, Richmond TJ, Truss M, Beato M (1998) The mouse mammary tumor virus promoter positioned on a tetramer of histones H3 and H4 binds nuclear factor 1 and OTF1. J Mol Biol 278:725–739

Spurio R, Durrenberger M, Falconi M, La Teana A, Pon CL, Gualerzi CO (1992) Lethal overproduction of the Escherichia coli nucleoid protein H-NS: ultramicroscopic and molecular autopsy. Mol Gen Genet 231:201–211

Stancheva I, Meehan RR (2000) Transient depletion of xDnmt1 leads to premature gene activation in Xenopus embryos. Genes Dev 14:313–327

Stancheva I, Hensey C, Meehan RR (2001) Loss of the maintenance methyltransferase, xDnmt1, induces apoptosis in Xenopus embryos. EMBO J 20:1963–1973

Starich MR, Sandman K, Reeve JN, Summers MF (1996) NMR structure of HMfB from the hyperthermophile, Methanothermus fer6idus, confirms that this archaeal protein is a histone. J Mol Biol 255:187–203

Strunnikov AV (2006) SMC complexes in bacterial chromosome condensation and segregation. Plasmid 55:135–144

Swaminathan V, Kishore A, Febitha K, Kundu T (2005) Human histone chaperone nucleophosmin enhances acetylation-dependent chromatin transcription. Mol Cell Biol 25:7534–7545

Swinger KK, Rice PA (2004) IHF and HU: Xexible architects of bent DNA. Curr Opin Struct Biol 14:28–35

Swinger KK, Lemberg KM, Zhang Y, Rice PA (2003) Flexible DNA bending in HU–DNA cocrystal structures. EMBO J 22:3749–3760

Takayanagi S, Morimura S, Kusaoke H, Yokoyama Y, Kano K, Shioda M (1992) Chromosomal structure of the halophilic archaebacterium *Halobacterium salinarium*. J Bacteriol 174:7207–7216

Talwar S, Pocklington MJ, Maclean N (1984) The methylation pattern of tRNA genes in *Xenopus laevis*. Nucleic Acids Res 12:2509–2517

Tao Y, Neppl RL, Huang ZP, Chen J, Tang RH, Cao R, Zhang Y, Jin SW, Wang DZ (2011) The histone methyltransferase Set7/9 promotes myoblast differentiation and myofibril assembly. J Cell Biol 194:551–565

Thatcher H, Gorovsky MA (1994) Phylogenetic analysis of the core histones H2A, H2B, H3 and H4. Nucleic Acids Res 22:174–179

Thaw P, Sedelnikova SE, Muranova T, Wiese S, Ayora S, Alonso JC, Brinkman AB, Akerboom J, van der Oost J, RaVerty JB (2006) Structural insight into gene transcriptional regulation and eVector binding by the Lrp/AsnC family. Nucleic Acids Res 34:1439–1449

Tittle RK, Sze R, Ng A, Nuckels RJ, Swartz ME, Anderson RM, Bosch J, Stainier DY, Eberhart JK, Gross JM (2011) Uhrf1 and Dnmt1 are required for development and maintenance of the zebrafish lens. Dev Biol 350:50–63

Toulmé F, LeCam E, Teyssier C, Delain E, Sautière P, Maurizot JC, Culard F (1995) Conformational changes of DNA minicircles upon the binding of the archaebacterial histone-like protein MC1. J Biol Chem 270:6286–6291

Trifonov EN (2011a) Cracking the chromatin code: precise rule of nucleosome positioning. Phys Life Rev 8:39–50

Trifonov EN (2011b) Thirty years of multiple sequence codes. Genomics Proteomics Bioinformatics 9:1–6

van Noort J, Verbrugge S, Goosen N, Dekker C, Dame RT (2004) Dual architectural roles of HU: formation of Xexible hinges and rigid Wlaments. Proc Natl Acad Sci U S A 101:6969–6974

Villar-Garea A, Imhof A (2008) Fine mapping of posttranslational modifications of the linker histone H1 from *Drosophila melanogaster*. PLoS One 3:e1553

Viré E, Brenner C, Deplus R, Blanchon L, Fraga M, Didelot C, Morey L, Van Eynde A, Bernard D, Vanderwinden JM, Bollen M, Esteller M, Di Croce L, de Launoit Y, Fuks F (2006) The polycomb group protein EZH2 directly controls DNA methylation. Nature 439:871–874

Voo KS, Carlone DL, Jacobsen BM, Flodin A, Skalnik DG (2000) Cloning of a mammalian transcriptional activator that binds unmethylated CpG motifs and shares a CXXC domain with DNA methyltransferase, human trithorax, and methyl-CpG binding domain protein 1. Mol Cell Biol 20:2108–2121

White CL, Suto RK, Luger K (2001) Structure of the yeast nucleosome core particle reveals fundamental changes in internucleosome interactions. EMBO J 20:5207–5218

Wolffe A (1992) Chromatin: structure and function. Academic, San Diego

Wong MM, Cox LK, Chrivia JC (2007) The chromatin remodeling protein, SRCAP, is critical for deposition of the histone variant H2A.Z at promoters. J Biol Chem 282:26132–26139

Wu WH, Alami S, Luk E, Wu CH, Sen S, Mizuguchi G, Wei D, Wu C (2005) Swc2 is a widely conserved H2AZ-binding module essential for ATP-dependent histone exchange. Nat Struct Mol Biol 12:1064–1071

Yao TP, Oh SP, Fuchs M, Zhou ND, Ch'ng LE, Newsome D, Bronson RT, Li E, Livingston DM, Eckner R (1998) Gene dosage-dependent embryonic development and proliferation defects in mice lacking the transcriptional integrator p300. Cell 93:361–372

Yu EY, Steinberg-Neifach O, Dandjinou AT, Kang F, Morrison AJ, Shen X, Lue NF (2007) Regulation of telomere structure and functions by subunits of the INO80 chromatin remodeling complex. Mol Cell Biol 27:5639–5649

Zalensky AO, Siino JS, Gineitis AA, Zalenskaya IA, Tomilin NV, Yau P, Bradbury EM (2002) Human testis/sperm-specific histone H2B (hTSH2B). Molecular cloning and characterization. J Biol Chem 277:43474–43480

Zhu W, Sandman K, Lee GE, Reeve JN, Summers MF (1998) NMR structure and comparison of the archaeal histone HFoB from the mesophile *Methanobacterium formicicum* with HMfB from the hyperthermophile *Methanothermus ferviidus*. Biochemistry 37:10573–10580

Zimmerman SB, Murphy LD (1996) Macromolecular crowding and the mandatory condensation of DNA in bacteria. FEBS Lett 390:245–248

Chapter 2
Secondary Structures of the Core Histone N-terminal Tails: Their Role in Regulating Chromatin Structure

Louis L. du Preez and Hugh-G. Patterton

Abstract The core histone N-terminal tails dissociate from their binding positions in nucleosomes at moderate salt concentrations, and appear unstructured in the crystal. This suggested that the tails contributed minimally to chromatin structure. However, *in vitro* studies have shown that the tails were involved in a range of intra- and inter-nucleosomal as well as inter-fibre contacts. The H4 tail, which is essential for chromatin compaction, was shown to contact an adjacent nucleosome in the crystal. Acetylation of H4K16 was shown to abolish the ability of a nucleosome array to fold into a 30 nm fibre. The application of secondary structure prediction software has suggested the presence of extended structured regions in the histone tails. Molecular Dynamics studies have further shown that sections of the H3 and H4 tails assumed α-helical and β-strand content that was enhanced by the presence of DNA, and that post-translational modifications of the tails had a major impact on these structures. Circular dichroism and NMR showed that the H3 and H4 tails exhibited significant α-helical content, that was increased by acetylation of the tail. There is thus strong evidence, both from biophysical and from computational approaches, that the core histones tails, particularly that of H3 and H4, are structured, and that these structures are influenced by post-translational modifications. This chapter reviews studies on the position, binding sites and secondary structures of the core histone tails, and discusses the possible role of the histone tail structures in the regulation of chromatin organization, and its impact on human disease.

L.L. du Preez • H.-G. Patterton (✉)
Advanced Biomolecular Research Cluster, University of the Free State,
PO Box 339, Bloemfontein 9300, South Africa
e-mail: patterh@ufs.ac.za

2.1 Introduction

2.1.1 The Need for DNA Packaging

The total length of the DNA in a single diploid human cell is approximately 2 m, a length that must fit into a cell nucleus that is roughly 10 μm in diameter. To accomplish this, the DNA is packaged into arrays of nucleosomes. Each nucleosome is formed by spooling about 168 bp of DNA in two negative superhelical turns onto a histone octamer, which is composed of two copies of each of the core histones H2A, H2B, H3 and H4. A fifth histone, linker histone H1, binds to the outside of the structure, close to the point of DNA entry and exit. The H1 causes partial charge neutralisation of the linker DNA, which connects adjacent nucleosomes in the array (Van Holde 1989).

This array of nucleosomes is further condensed into a 30 nm fibre, which is composed of a helical arrangement of nucleosomes. No definitive structural detail is currently available on the 30 nm fibre. The 30 nm fibre undergoes additional levels of folding to form higher order structures, culminating in the condensed structures observed electron microscopically in the metaphase chromosome (Woodcock and Dimitrov 2001).

Although the packaging of DNA into chromatin solves the problem of fitting an extended, poly-anionic, linear polymer into the confined space of a eukaryotic nucleus, a significant problem is introduced in that the DNA molecule also becomes masked from most of the proteins and enzymes that must interact with it as part of its genetic function. Thus, to allow access to the DNA molecule, eukaryotic cells have evolved intricate mechanisms whereby the chromatin is locally and reversibly decondensed. Mechanisms involved in this local decondensation include the structural perturbation of chromatin structures by ATP-dependent chromatin remodelling enzymes, the deposition of different histone isotypes, and the reversible chemical modification of the "histone tails".

2.1.2 Histone Tails: More Than Just Fashionable

The histone tails are seemingly unstructured extensions of the core histones beyond the central histone fold domains (Arents et al. 1991; Luger et al. 1997), and contribute approximately 38% of the core histone mass to the histone octamer (Fig. 2.1). The chemical modification of the histone was first observed by Phillips (1961, 1963) and by Murray (1964). The Mirsky group subsequently showed that acetylation of histones facilitated synthesis of RNA in cell-free extracts (Allfrey et al. 1964). These initial observations defined the beginning of a field that has become known as *Epigenetics* and its high-throughput application, *Epigenomics*. A substantial scientific literature has since developed describing an extensive range of modifications (Fig. 2.1) and the function of these modifications [see Kouzarides (2007) and Kundu (2007) for reviews].

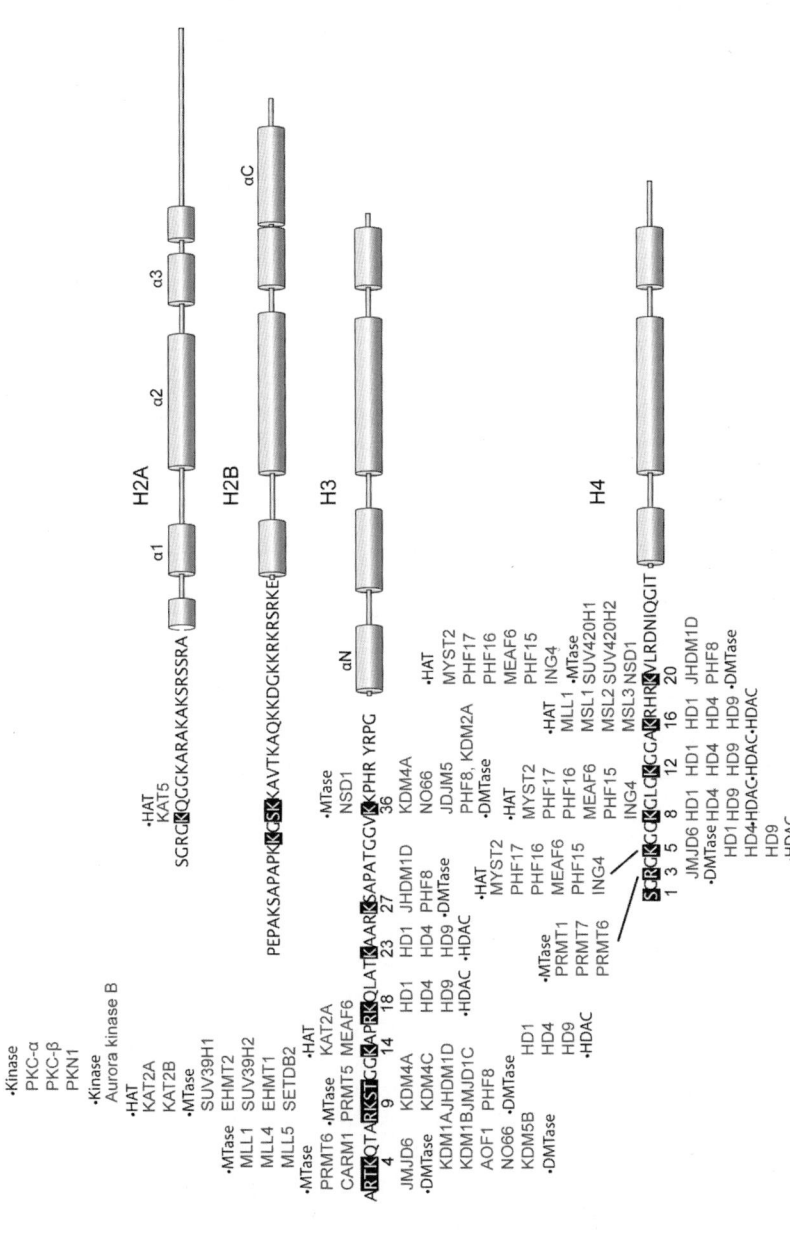

Fig. 2.1 The sites of epigenetic modification in the core histone tails. The core histone tail sequences are shown, as well as the central histone folds with the additional α-N and α-C helices of H3 and H2B. The individual residues that are sites of epigenetic modification are indicated. Human enzymes in the UniProt database (Apweiler et al. 2004) that were annotated with gene ontology terms indicating specificity towards each of these different residues, are identified (http://www.uniprot.org; accessed January 2011)

Early indications were that modified residues served as molecular beacons for the recruitment of specific proteins to such flagged areas of the genome (Strahl and Allis 2000). For instance, it was shown that the heterochromatin-associated protein HP1 was recruited to regions marked for transcriptional silencing by tri-methylated K9 of histone H3 (Bannister et al. 2001). Regulatory effects of one modification on another modification in the same tail or in the tail of a different histone were also discovered, termed *cis*-tail and *trans*-tail pathways, respectively. For instance, phosphorylation of S10 of H3 was shown to inhibit demethylation of the mono- and di-methylated K4, thus maintaining an "active" epigenetic signal (Forneris et al. 2005). Ubiquitination of K123 of histone H2B required a sequence motif in the H2A tail, which is in close proximity to H2B K123 in the nucleosome, and may be involved in the recruitment of the ubiquitination machinery (Zheng et al. 2010). Ubiquitination of H2B K123, in turn, was required for recruitment of the methyltransferase complexes for the subsequent methylation of H3 K4 by Set1 (Dover et al. 2002) and of H3 K79 by Dot1 (Briggs et al. 2002), marks associated with transcriptional activation. This system of molecular flags and interdependencies is known as the histone code, proposing that histone modification represent a template for direct "read-out" by other proteins who then perform specific chromatin-associated functions (Strahl and Allis 2000).

2.1.3 Histone Tails: Beyond the Histone Code

Although there are many instances where this histone code model is an accurate description of biochemical functions *in vivo*, instances were also observed where histone tail modifications represented more than simple molecular beacons. The most striking observation involved K16 of histone H4. *In vitro* data showed that deacetylation of K16 was required for full compaction of chromatin into a condensed fibre in the presence of a linker histone (Robinson et al. 2008). In the absence of H1, acetylation of H4 K16 was also shown to inhibit formation of a condensed structure in a reconstituted nucleosome array, although the relationship of this array to the canonical 30 nm fibre was not determined (Shogren-Knaak et al. 2006; Allahverdi et al. 2011). This represented an example where a histone tail modification had a significant effect on chromatin structure, but was not involved in the recruitment of any protein to accomplish the structural effect in an *in vitro* system composed of purified and defined components. A genome-wide gene expression analysis also demonstrated a redundant, cumulative effect for mutations of K5, K8 and K12 of histone H4 to arginine, designed to mimic the unacetylated state of lysine. The H4 K16R mutation, on the other hand, had a transcriptional effect that was independent of the state of K5, K8 and/or K12, suggesting that acetylation played a fundamentally different functional role in these two groups of residues (Dion et al. 2005). Also, unlike K5 and K12 of H4, which showed a strong correlation between acetylation state and gene expression, there was little correspondence between the acetylation state of H4 K16 in nucleosomes adjacent to the transcription start site, and the average transcriptional activity of genes (Liu et al. 2005).

Biochemical studies have shown that some chemical modifications of amino acid residues in peptides caused significant changes in the secondary structure of the peptides (Wang et al. 2001). However, very little attention has been given to the possible effect of epigenetic modifications on the secondary structures of the histone tails, and the impact this may have on the association of the tails in chromatin. The fact that the histone tails appeared unstructured in X-ray crystallographic studies, most likely due to the dissociation of the tails from binding sites under conditions of moderate salt (Walker 1984; Luger et al. 1997), may have contributed to an impression that they were structurally unimportant. However, the finding that deacetylation of H4 K16 was required for full compaction of the chromatin fibre (Shogren-Knaak et al. 2006; Robinson et al. 2008; Allahverdi et al. 2011), renewed interest in a direct structural role of the histone tails in chromatin. In this Chapter we review the literature on the effect of amino acid residue modifications on the secondary structures of peptides including histone tails, citing biophysical, biochemical and *in silico* computational studies. We finally discuss how this may impact on chromatin structure and the epigenetic basis of human disease.

2.2 Chromatin Structure

2.2.1 *A Model for the 30 nm Fibre*

A series of images recorded of chromatin in the presence of a linker histone at increasing salt concentrations showed the systematic compaction of the chromatin through successively more condensed structures, reaching a fibre of approximately 30 nm diameter as a compaction limit (Thoma et al. 1979). This most condensed state of packaging of the nucleosomes relative to each other was termed the "30 nm fibre", which may undergo additional levels of folding into higher-order structures and helices (Woodcock and Dimitrov 2001).

Despite significant effort spanning many decades, there is still no agreement on the exact structural arrangement of nucleosomes in the 30 nm fibre. One model proposed a continuation of the folding of a linear array of nucleosomes into a contact helix or solenoid, where each neighbour in the solenoid was also adjacent in the linear array (Finch and Klug 1976). An alternative model suggested that the fibre was assembled in a manner that placed neighbouring nucleosomes consecutively on opposite sides of the fibre axis, to form a two-start helix, with the linker DNA running through the fibre centre (Worcel et al. 1981; Woodcock et al. 1984; Williams et al. 1986). This latter model has received strong experimental support from cross-linking (Dorigo et al. 2004) and X-ray crystallographic studies (Schalch et al. 2005). Many variations of these two central proposals exist, mainly based on the connectivity between nucleosomes in the fibre (Daban and Bermudez 1998; Robinson et al. 2006).

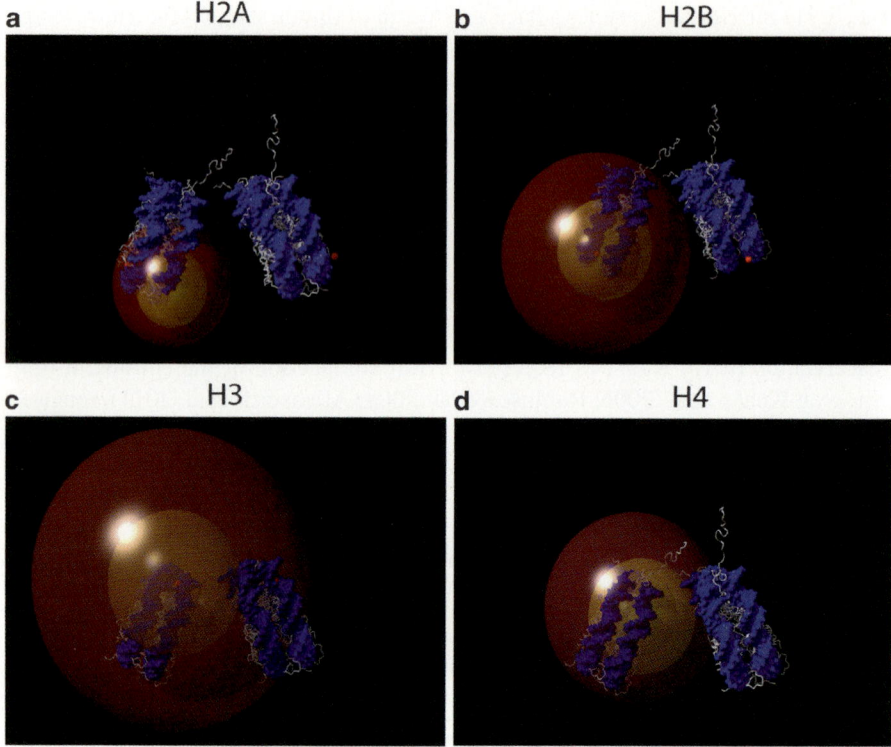

Fig. 2.2 Reach of the N-terminal core histone tails in chromatin. The reach of each of the N-terminal tails of the core histones (**a**) H2A, (**b**) H2B, (**c**) H3 and (**d**) H4 is shown. The volume that can be swept out by each tail is represented by a *sphere centred* on the defined start of each tail (Luger et al. 1997), with the tail maximally extended (3.3 Å per residue) or with the full length of the tail in an α-helical conformation (1.5 Å per residue), represented by the outer (*red*) and inner (*yellow*) sphere in each panel, respectively. An idealised 30 nm fibre, independent of any connectivity model, is shown with the two nucleosomes rotated by 60° on the fiber axis, and with an internucleosomal rise of 20 Å. Note that the radii of the spheres assume free and unhindered rotation of the tails, which is not always physically possible. The H3 tail, for instance, would have to bend back over the nucleosomal DNA to approach the anterior side of the nucleosome on the "outside" of the fibre, a geometric path that would significantly decrease its reach in that direction

2.2.2 Position of the H2A and H2B Tails

Irrespective of differences in connectivity, all models place the site of DNA entry and exit of the nucleosome, pointing inwards towards the fibre axis (Woodcock and Dimitrov 2001). This places the base of the tails at specific spatial positions within the fibre, and imposes a constraint on the possible sites of interaction of the histone tails, both within and between nucleosomes of the same fibre, as well as to different fibres. Figure 2.2 shows the maximal reach of the N-terminal tails of the core histones with the tail either fully extended, or with the full length in an α-helical conformation in a hypothetical, idealised fibre. This provides the maximal and minimal reach of the tails, respectively. It is clear that both the N-terminal tails of histones H2A and H2B have limited or no access to an adjacent nucleosome in the idealised fibre structure, but may be involved in fibre-fibre contacts, and should be accessible to *trans*-acting proteins, even in the condensed 30 nm fibre. It is therefore interesting that the sterically accessible H2B K123 is ubiquitinated as a prelude to methylation of the K4 in the histone H3 tail.

2.2.3 Position of the H3 Tail

The histone H3 and H4 N-terminal tails appear to be able to contact distal positions within the same nucleosome as well as neighbouring nucleosomes (see Fig. 2.2). The histone H3 tail is the most extensive, and exits the nucleosome between the two DNA superhelical gyres close to the pseudo-dyad axis (Luger et al. 1997). If the H3 tail continued on its exit trajectory, it would point towards the 30 nm fibre axis, and may approach nucleosomes on the other side of the fibre (see Fig. 2.2).

There exists substantial evidence that the lysine-rich tail of linker histone H1 is associated with the inter-nucleosomal linker DNA in the fibre centre (reviewed in Caterino and Hayes 2011). Since fibre compaction required histone H1 (Thoma et al. 1979; Robinson et al. 2008) as well as the N-terminal tail of histone H4 (Dorigo et al. 2003; Robinson et al. 2008), but not the H3 tail, it appears unlikely that the H3 tail contributed to any significant partial charge neutralisation of the linker DNA in the fibre centre, or acted as a nucleosome-nucleosome stabilisation scaffold, such as the H4 tail. Thus, the possibility of the strong binding of the extended histone H3 tail to the DNA in the fibre centre appears remote. In fact, many studies suggested that the H3 tail was readily accessible in chromatin, including in an H1-containing, condensed fibre. In native H1-containing chromatin, the H3 tail remained the most susceptible to trypsin cleavage (Harborne and Allan 1983). It was also shown that recombinant PCAF, which preferentially acetylated K14 of H3 (Schiltz et al. 1999), could still acetylate the H3 tail in condensed chromatin lacking H1 (Herrera et al. 2000). Furthermore, HP1 was specifically bound to the H3 tail tri-methylated at K9 in condensed heterochromatin (Bannister et al. 2001). In a silenced *MATa*-specific gene in *Saccharomyces cerevisiae*, Tup1, which bound at a density of two Tup1 molecules per nucleosome (Ducker and Simpson 2000), was associated with the H3 tail in the repressive chromatin structure (Edmondson et al. 1996). Also, a substituted cysteine residue, close to the tip of the H3 tail, could be cross-linked from one oligonucleosome array to another array (Kan et al. 2007).

All these studies are consistent with a histone H3 tail that is exposed for binding by proteins. Thus, the H3 tail may either continue on its exit trajectory and appear on the side of the central, crossed-linker stack, between the two nucleosome helices in the two-start helix model. Alternatively, it could follow a curved path over the nucleosomal DNA gyre, protruding into the space between two neighbouring nucleosomes in the 30 nm fiber. In either of these two possibilities, the H3 tail could be bound by sequence specific proteins, or the tail could bind to the originating or to an adjacent nucleosome.

2.2.4 Position of the H4 Tail

The location of the H4 N-terminal tail on the lateral surface of the nucleosome places it in a position where it can easily be extended to contact the lateral surface of the adjacent nucleosome in the chromatin fibre (see Fig. 2.2). Such a contact was, in fact, observed in the crystal structure of *Xenopus* histones reconstituted onto human α-satellite DNA repeats (Luger et al. 1997). Clear contacts were observed to an acidic patch on the nucleosome surface, constituted by H2A E56, E61, E64, D90, E91 and E92 as well as H2B E110. The importance of this observed contact was shown by the absolute requirement for an intact H4 tail by a reconstituted fibre to condense fully in the absence of histone H1 (Dorigo et al. 2003). None of the other core histone tails were required for full compaction (Moore and Ausio 1997; Dorigo et al. 2003). Also, nucleosome arrays reconstituted with the human histone variant H2A-Bbd (Chadwick and Willard 2001), which lacks three glutamic acid residues that forms part of the acidic patch of H2A, did not condense to the same degree as nucleosome arrays reconstituted with H2A (Zhou et al. 2007).

Interestingly, the contact of the H4 tail to the lateral surface of a nucleosome does not appear to require a single docking surface, such as the acidic patch. This was shown by a peptide comprised of residues 1–23 of the Kaposi's sarcoma-associated herpes virus latency-associated nuclear antigen (LANA). When LANA was bound to the acidic patch of a nucleosome (Barbera et al. 2006), this association did not abolish the histone H4-dependent compaction of a nucleosome array (Chodaparambil et al. 2007), suggesting that the H4 tail could still bind to the adjacent nucleosome in the presence of bound LANA. This degeneracy in H4 binding was also demonstrated by the non-saturable nature of association of an H4 peptide with the nucleosome surface, suggesting that many binding sites existed for the H4 tail on the lateral nucleosome surface (Chodaparambil et al. 2007). The binding of the LANA peptide, in contrast, was found to be saturable (Chodaparambil et al. 2007). The H4 tail interaction was, nevertheless, sensitive to chemical modification. It was shown that 30% acetylation of the histone H4 N-terminal tail resulted in the inability of a 61-mer nucleosome array, containing linker histone H1, to fully condense *in vitro* (Robinson et al. 2008). Taken together, these studies provide a very strong argument that the N-terminal tail of histone H4 was bound to an adjacent nucleosome in the chromatin fibre, and that this interaction, which could be disrupted by acetylation of H4 K16, was essential to fully condense the chromatin into a 30 nm fibre structure.

2.3 Histone Tail Associations

2.3.1 N-H2A, H2A-C and N-H2B

Numerous studies have made use of chemical cross-linking to identify DNA and protein sites that can be contacted by the histone tails in the nucleosome and in a condensed fibre by using reagents that are either freely diffusible (Sperling and Sperling 1978; Jackson 1999) or immobilised (Lee and Hayes 1997). The Hayes group have developed a technique where a photo-activatable azidophenacylbromide (ACP) is linked to a uniquely engineered cysteine residue (Lee and Hayes 1997). The conjugated ACP group forms a reactive nitrene upon UV irradiation, cross-linking to spatially proximal DNA or protein molecules, and allowing the mapping of the contact positions of the region containing the substituted cysteine (Lee and Hayes 1997). Using this technique, it was shown that the conjugated A12C of H2A cross-linked to approximately symmetrical positions 4 helical turns removed from the pseudo-dyad in reconstituted and purified nucleosome cores (Lee and Hayes 1997). This is expected from the close proximity of H2A A12 to the DNA in the crystal structure (Davey et al. 2002). In a reconstituted di-nucleosome, cross-linking of residue 12 of the H2A tail was almost exclusively within the same nucleosome (Zheng and Hayes 2003). The more distal portion of the H2A tail, mapped with a G2C substitution, was found to cross-link to two sites approximately 5 bp to either side of the A12C cross-linking position, in agreement with a less constrained motion of the tail further removed from the relatively immobile tail base (Lee and Hayes 1997). This larger freedom of movement of the tail tip was also consistent with the cross-linking of almost 20% of H2A residue 2 to the neighbouring nucleosome in a di-nucleosome template (Zheng and Hayes 2003). Using a zero-length cross-linker, Bradbury and colleagues showed that the H2A C-terminal tail could be cross-linked to the DNA at the pseudo-dyad axis (Usachenko et al. 1994), in agreement with the exit location of this tail from the nucleosome (Luger et al. 1997). Residue 2 of H2B was shown to participate in inter-nucleosomal contacts (Zheng and Hayes 2003).

2.3.2 N-H3

The N-terminal tails of histone H3 and H4 made predominantly intra-nucleosomal contacts in a reconstituted di-nucleosome (Zheng and Hayes 2003). In a 13-mer nucleosome array, it was also found that the H3 tail was exclusively cross-linked intra-nucleosomally at 0 mM Mg^{2+}, but at higher concentrations of Mg^{2+}, where the 13-mer array became more condensed, an increase in inter-nucleosomal cross-links were observed (Zheng et al. 2005). A large proportion of the H3 tail, spanning residues 6–24, could be inter-nucleosomally cross-linked, but not the region of residue 35, close to the base of the tail, which is in agreement with the probable reach of these regions in the H3 tail (Zheng et al. 2005). Looking at the ability of the H3 tail to make contacts between different nucleosome array molecules, it was found that the entire region spanning residue 6–35 could be efficiently cross-linked within the same reconstituted 12-mer oligonucleosome. Long-range inter-array cross-linking was only detected at higher Mg^{2+} concentrations, ionic conditions suggested to promote self-association of the individual arrays (Kan et al. 2007). This inter-array cross-linking efficiency was increased by the presence of H1, the binding of which may have limited unproductive associations of the H3 tail (Kan et al. 2007). As expected, distal parts of the H3 tail could be cross-linked more efficiently to neighbouring nucleosome arrays compared to regions close to the tail base (Kan et al. 2007). Acetylation of the H3 tail, studied in K->Q substitution mutants, required at least 4 modified residues to display a reduced inter-array cross-linking efficiency, an effect that disappeared at elevated Mg^{2+} concentrations (Kan et al. 2007). The intra-array cross-linking did not appear sensitive to the K->Q substitutions (Kan et al. 2007).

2.3.3 N-H4

The Mirzabekov group showed that H18 of H4 could be cross-linked to the DNA approximately 15 bp from the pseudo-dyad in a nucleosome core particle (Ebralidse et al. 1988). More recently, using reconstituted nucleosome arrays, it was shown that at 0 mM Mg^{2+} the H4 tail cross-linked exclusively within the originating array (Kan et al. 2009). An increase in inter-array cross-links was observed at elevated Mg^{2+} concentrations (Kan et al. 2009). Although an H2A-H4 cross-link, expected from binding of the H4 tail to the H2A-H2B acidic patch, was demonstrated by the simultaneous appearance of fluorescently labelled H2A and tritiated H4 in a higher mobility electrophoretic band, this band was also present in cross-linked mononucleosomes, suggesting that this interaction also occurred intra-nucleosomally (Kan et al. 2009). This cross-link was severely diminished by the presence of the LANA peptide, previously shown to bind in the H2A-H2B acidic pocket (Barbera et al. 2006; Chodaparambil et al. 2007). Interestingly, although tetra-acetylation of H4 reduced fibre self-association, no acetylation dependent difference in inter-fibre cross-linking efficiency was detectable (Kan et al. 2009). In the presence of H1, however, inter-fibre cross-linking was enhanced, and a clear decrease was detected with tetra-acetylated H4 tail (Kan et al. 2009).

2.4 Histone Tail Structure

Many different techniques have been used to study the structure of the histone N-terminal tails. These include the biophysical methods of circular dichroism (CD), nuclear magnetic resonance (NMR) and other forms of spectrometry, and computational methods including secondary structure prediction and molecular dynamics (MD).

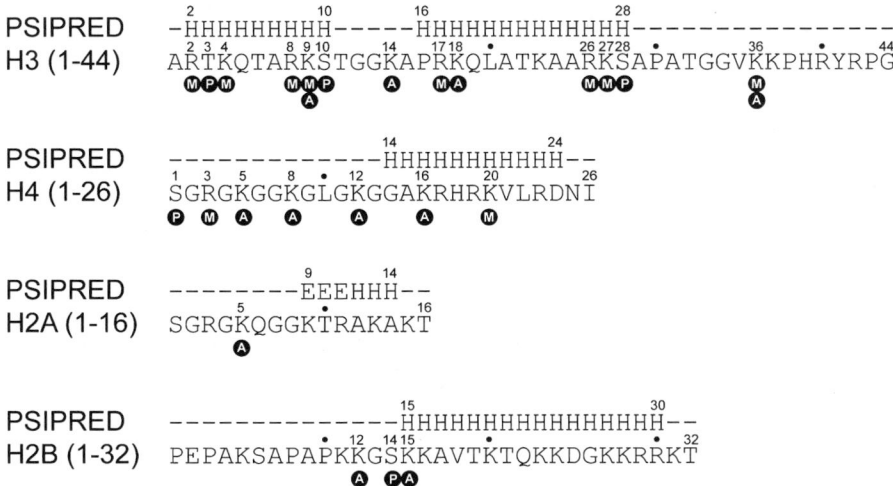

Fig. 2.3 Predicted secondary structures of the core histone tails. The secondary structures predicted by PSIPRED (Jones 1999) are shown above the sequence of each of the four core histone tails with α-helix, β-strand, and random coil regions represented by the symbols "H", "E" and "-", respectively. Sites of epigenetic modification as well as the types of modification are indicated

2.4.1 Secondary Structure Prediction

Secondary structure predictions are often used to obtain insight into the secondary structures of proteins of unknown structure based solely on sequence, and have predictive accuracies in excess of 75% (McGuffin et al. 2000) that are continually being improved by algorithmic advances. The secondary structure predictions for the unmodified, major human core histones using PSIPRED (Jones 1999) are shown in Fig. 2.3.

Two α-helical segments are predicted for H3 spanning 9 residues from R2 to S10, and 13 residues from P16 to S28, respectively. Interestingly, known post-translational modifications (PTMs) appear to be clustered at the predicted α-helix termini, and in both bases serine, which can be phosphorylated (Wei et al. 1998; Goto et al. 1999), are present at the C-terminal end of the predicted α-helices.

In the case of H4 a single 11-residue α-helical segment is predicted spanning from G14 to D24. This segment contains K16, known to be required in a de-acetylated state to allow condensation of the 30 nm fibre *in vitro* (Robinson et al. 2008).

A 3-residue β-strand segment from K9 to R11 followed by a 3-residue α-helical segment spanning A12 to A14 is predicted for the H2A tail, and a single 16 residue α-helical segment is predicted, stretching from K15 to R30, in the case of H2B.

It is therefore clear, based on the propensity of amino acid residues to assume defined secondary structures, that the N-terminal tails of the core histones are likely to be highly structured.

2.4.2 Molecular Dynamics

MD is a molecular mechanics technique that involves the modelling of molecular systems using potential energy functions, and has been widely applied to biomolecular systems over the last 30 years, prominently so in the study of protein folding pathways (Adcock and McCammon 2006).

The application of MD in elucidating the structure of N–terminal histone tails has been limited, and has only been applied to the H3 and H4 tails at the time of writing. Most early work was based on coarse-grained models (Arya and Schlick 2006; Korolev et al. 2006) that were used to study chromatin folding, and did therefore not provide any structural detail on the histone tails. Recently there has been an increase in all-atom MD studies of the tails, and with the development of force field parameters for most of the predominant PTMs (Grauffel et al. 2010), more studies are likely to follow.

LaPenna and co-workers simulated a 25-residue H3 tail peptide in the presence and absence of 10 bp of DNA (LaPenna et al. 2006). The peptide exhibited a wide range of structures with a high α– and 3_{10}-helical content in the presence of DNA. In agreement with the secondary structure prediction (see Fig. 2.3), most of the residues, except for residues 10–15, were found in a helical structure. No β-strand content was observed. The presence of DNA increased the average helical content in the peptide, and resulted in compact, rod-like structures, despite only 4–5 bp of DNA directly interacting with the peptide (LaPenna et al. 2006).

Liu and Duan incorporated PTMs into their MD study of the H3 tail (Liu and Duan 2008), using an 18-residue H3 variant identical to the major H3, except for 2 N-terminal glycine residues. Five PTM states were studied in the H3 peptide: unmodified, K4me2, K9me2, K4me2-K9me2, and K4Ac-K9Ac-K14Ac. The peptides preferred α-helical regions with a similar structure: a shared α-helix between K9 and T11 with the rest in an extended conformation. The singly di-methylated peptides did not differ significantly from the unmodified peptide. The doubly di-methylated peptide, however, showed a decrease in α-helical and an increase in β-strand content, although the biological relevance of a simultaneous K4 and K9 methylation is questionable. The acetylated peptide showed a decrease in helical content compared to the unmodified peptide, and exhibited a β-hairpin as the most populated structure (Liu and Duan 2008). It is thus evident that "cross-talk" between different modification groups may have a structural basis, where combinations of modifications may stabilize specific secondary structural distributions in the tail that could influence binding of the tails in chromatin.

Lins and Röthlisberger conducted MD studies on tetra- and un-acetylated 23-residue N-terminal H4 peptides (Lins and Röthlisberger 2006). The starting conformation for the two peptides was a canonical α-helix, which was found to be more stable in the tetra–acetylated peptide than in the un-acetylated peptide. A small β-hairpin was formed that spanned residues 4–12 in the tetra-acetylated peptide, which remained stable for approximately 2 ns of a 20 ns simulation (Lins and Röthlisberger 2006). Taken together with results from the previous studies, the histone tails seem able to stably accommodate secondary structures other than only α-helices. This opens the possibility that modifications to residues may be a way of changing the transition of the tails to different secondary structures on the fly, impacting on tail binding and, consequently, chromatin structure, and could thus provide a mechanism for genetic control.

In the most recent MD study, Yang and Arya investigated the effect of K16 acetylation in a 25-residue H4 tail peptide (Yang and Arya 2011). An α-helical region was formed and stabilized between residue 15 and 20 in the unmodified peptide. An α-helix was formed in the same region in the K16Ac peptide, but, in contrast to another study (Lins and Röthlisberger 2006), the helix exhibited a significantly reduced stability (Yang and Arya 2011). It is, however, important to note that the authors of the MD studies used a wide range of different simulation protocols and techniques, which makes the comparison of results between studies difficult.

Nevertheless, MD studies suggested that both H3 and H4 tail peptides preferred helix-rich structures. PTMs changed the stability of these structures, and β-strands were also observed in some cases. These studies therefore underscore a possibly critical role in PTMs tipping the balance between different secondary structures in the histone tails, which may have a major impact on the function of these tails.

2.4.3 Biophysical Methods

Parello and co-workers compared CD spectra obtained from a native and two selectively proteolyzed nucleosome core particles (NCP) to investigate the secondary structure of the N-terminal tails (Banères et al. 1997). Clostripain was used to produce a "half-proteolyzed" NCP that lacked the H3 and H4 tails, and a "fully proteolyzed" NCP, that lacked all four core histone tails. The authors established that approximately 60% of the residues in the H3 and H4 tails were in an α-helical conformation, and contributed about 35% to the α-helical content in the whole NCP. It was confirmed that these contributions corresponded to the tails in the bound state in the nucleosome. The individual contributions of the H3 and H4 tails to α-helical content could, however, not be resolved. The H2A and H2B tails were found to be in a random coil conformation. A subsequent NMR study also showed that 31 residues of the H2B tail were unstructured (Nunes et al. 2009).

Ausio and co-workers investigated the contribution of the histone tails to the secondary structure of the octamer, and the effect that acetylation of the tails had on this contribution (Wang et al. 2000). The contribution of the tails to the overall α-helical content of the octamer was calculated at 17% by comparing the α-helical content of trypsin digested octamer with an undigested octamer. This value was about half of that reported by Parello and colleagues (Banères et al. 1997), and was attributed to the use of different experimental conditions. Consequently, it was shown that the overall α-helical content of the nucleosome increased by about 3% as a result of acetylation. This translated to an increase of about 17% in the α-helicity of the tails. An H4 tail peptide corresponding to residue 1–23 was isolated as mono-, di-, tri- and tetra-acetylated isomers, and analysed by CD in an aqueous solution and in trifluoroethanol (TFE), a known stabilizer of α-helices. The unmodified peptide showed an α-helical content of 17% in TFE, which increased to about 24% in the tetra-acetylated peptide in the same solvent. In the aqueous solution, the isolated peptides exhibited CD spectra consistent with a random coil conformation, suggesting that the chemical environment of the histone tails played a major role in their structural conformations.

In a combined NMR and CD study Lee and co-workers also showed that a 27-residue synthetic H4 peptide had no defined structure in aqueous solution at physiological pH (Bang et al. 2001). However, a pH dependent structural transition was observed at an acidic pH for the native peptide. None of the peptides displayed any regular secondary structures. The acetylated form of this peptide seemed insensitive to pH change, and exhibited two regions of turn-like structures at L10-G13 and R19-L22.

2.5 Histone Tails and Human Health

A link between chromatin and human disease is long established. In recent times thousands of studies have been published reporting the role of epigenetics in human disease. This role is varied and fundamental. Epigenetics was shown to be involved in development, trans-generational inheritance, memory formation, psychiatric disorders, autism spectrum disorders, carcinogenesis, cardiovascular diseases and a slew of heritable diseases including Fragile X syndrome, Friedreich's ataxia, Machado-Joseph disease, spinocerebellar ataxia, Huntington's disease and myotonic dystrophy, to provide but a significantly truncated representative list. Epigenetics have also been implicated in longevity in model eukaryotic organisms (Dang et al. 2009). Many excellent reviews have recently appeared on epigenetics and human health (Watanabe and Maekawa 2010; Luco et al. 2011; Kurdistani 2011). Because of the extensive role of epigenetics in human disease, modulators of epigenetic modifications suitable for therapy have become pharmacologically highly prized (Kundu 2007). A multitude of modifiers, including deactylase and demethylate inhibitors, are currently in various phases of clinical trials, and many show extremely promising results.

Many of the epigenetic therapeutic agents direct a change in gene expression level of numerous genes, where misexpression is associated with a diseased state. The precise mechanism whereby the epigenetic modification alters gene expression level is often not fully understood. Some modifiers are now known to induce structural transitions in the core histone tails. For instance, the binding of Ni^{2+} to the sequence 15-AKRHRK-20 in the tail of H4 showed a drastic structural shift in the conformation of the peptide (Zoroddu et al. 2000). The binding of Ni^{2+} to a 22-residue H4 tail peptide had the same effect as acetylation on the α-helical content of the peptide (Zoroddu et al. 2009). This is an interesting observation since Ni is a known carcinogen which seems to act on the epigenetic level. This suggests that the epigenetic link between some human diseases and chromatin may not simply be the chemical modifications of the core histone tails that subsequently act as binding surfaces for transcription-related enzymes, but may also occur due to changes in the stable secondary structures of the histone tails which may impact not only on transcription, but also other genetic processes of the DNA molecule.

2.6 Conclusions

There is significant evidence that the core histone tails are partially structured (Banères et al. 1997; Wang et al. 2000), and that they are involved in intra- and internucleosomal as well as in inter-fibre contacts (Zheng et al. 2005; Kan et al. 2007, 2009). It seems likely that the H4 tail binds to the lateral surface of an adjacent nucleosome in chromatin (Luger et al. 1997), and may act as a molecular tether, stabilising the architecture of the 30 nm fibre (Dorigo et al. 2003; Robinson et al. 2008). It is further known that acetylation of H4 K16 abolished formation of the 30 nm fibre (Robinson et al. 2008). Although this may simply involve a reduced electrostatic attraction between the acidic surface and the acetylated lysine residue, it is also possible that acetylation may disrupt secondary structures required for docking to the acidic patch or to sites in its vicinity. Alternatively, acetylation may stabilise an extended α-helix, diminishing the reach of the H4 tail, and limiting contact to the adjacent nucleosome.

Although no H3 mediated inter-nucleosome contacts were seen in X-ray crystallographic studies, this tail was, nevertheless, shown to bind intra-nucleosomally as well as between fibres (Zheng et al. 2005; Kan et al. 2007). The predicted presence of two α-helices, demarcated by clusters of sites targeted for epigenetic modification, appears intriguing. Although, clearly, the recognition and binding of specific protein domains such as chromo and bromo domains to methylated and acetylated lysine residues are well established, and recruit proteins that serve crucial biochemical functions, the cross-linking data suggests that the H3 as well as the H2A and H2B tails are also involved in binding to DNA and/or protein surfaces in chromatin (Lee and Hayes 1997; Zheng and Hayes 2003; Kan et al. 2007). The binding of chromatin-associated proteins and enzymes to the histone tails may therefore only reflect a part of the functionality of the tails, which may also make a direct structural contribution to chromatin organization. One may therefore speculate that specific PTMs, stabilizing a specific distribution of secondary structures, are required for binding of the tail in chromatin. Removal of these PTMs may destabilise the structure, disrupt binding, and allow subsequent association of other regulatory proteins with the released tail. Conversely, specific PTMs may favour defined structures that allow an exact binding in chromatin, which may then provide a combined molecular surface that is recognised and bound by other regulatory factors. It is thus evident from the studies cited above that our understanding of the biochemical role of the core histone tails is incomplete, and that the tails may be multi-functional molecular entities that impact on chromatin structure and genetic function in a way that is only partially appreciated. This opens the exciting possibility of a different angle on the role of epigenetics in human disease, and the development of therapies that target histone tail structures and patterns of association as opposed to only the enzymes that are recruited by a fraction of the epigenetic marks.

References

Adcock SA, McCammon JA (2006) Molecular dynamics: survey of methods for simulating the activity of proteins. Chem Rev 106:1589–1615

Allahverdi A, Yang R, Korolev N, Fan Y, Davey CA, Liu CF, Nordenskiöld L (2011) The effects of histone H4 tail acetylations on cation-induced chromatin folding and self-association. Nucleic Acids Res 39:1680–1691

Allfrey VG, Faulkner RR, Mirsky AE (1964) Acetylation and methylation of histones and their possible role in the regulation of RNA synthesis. Proc Natl Acad Sci U S A 51:786–794

Apweiler R, Bairoch A, Wu CH, Barker WC, Boeckmann B, Ferro S, Gasteiger E, Huang H, Lopez R, Magrane M, Martin MJ, Natale DA, O'Donovan C, Redaschi N, Yeh LS (2004) UniProt: the Universal Protein knowledgebase. Nucleic Acids Res 32:D115–D119

Arents G, Burlingame RW, Wang BC, Love WE, Moudrianakis EN (1991) The nucleosomal core histone octamer at 3.1 Å resolution: a tripartite protein assembly and a left-handed superhelix. Proc Natl Acad Sci U S A 88:10148–10152

Arya G, Schlick T (2006) Role of histone tails in chromatin folding revealed by a mesoscopic oligonucleosome model. Proc Natl Acad Sci U S A 103:16236–16241

Banères JL, Martin A, Parello J (1997) The N tails of histones H3 and H4 adopt a highly structured conformation in the nucleosome. J Mol Biol 273:503–508

Bang E, Lee CH, Yoon JB, Lee DW, Lee W (2001) Solution structures of the N-terminal domain of histone H4. J Pept Res 58:389–398

Bannister AJ, Zegerman P, Partridge JF, Miska EA, Thomas JO, Allshire RC, Kouzarides T (2001) Selective recognition of methylated lysine 9 on histone H3 by the HP1 chromo domain. Nature 410:120–124

Barbera AJ, Chodaparambil JV, Kelley-Clarke B, Joukov V, Walter JC, Luger K, Kaye KM (2006) The nucleosomal surface as a docking station for Kaposi's sarcoma herpesvirus LANA. Science 311:856–861

Briggs SD, Xiao T, Sun ZW, Caldwell JA, Shabanowitz J, Hunt DF, Allis CD, Strahl BD (2002) Gene silencing: trans-histone regulatory pathway in chromatin. Nature 418:498

Caterino TL, Hayes JJ (2011) Structure of the H1 C-terminal domain and function in chromatin condensation. Biochem Cell Biol 89:35–44

Chadwick BP, Willard HF (2001) A novel chromatin protein, distantly related to histone H2A, is largely excluded from the inactive X chromosome. J Cell Biol 152:375–384

Chodaparambil JV, Barbera AJ, Lu X, Kaye KM, Hansen JC, Luger K (2007) A charged and contoured surface on the nucleosome regulates chromatin compaction. Nat Struct Mol Biol 14:1105–1107

Daban JR, Bermudez A (1998) Interdigitated solenoid model for compact chromatin fibers. Biochemistry 37:4299–4304

Dang W, Steffen KK, Perry R, Dorsey JA, Johnson FB, Shilatifard A, Kaeberlein M, Kennedy BK, Berger SL (2009) Histone H4 lysine 16 acetylation regulates cellular lifespan. Nature 459:802–807

Davey CA, Sargent DF, Luger K, Maeder AW, Richmond TJ (2002) Solvent mediated interactions in the structure of the nucleosome core particle at 1.9 Å resolution. J Mol Biol 319:1097–1113

Dion MF, Altschuler SJ, Wu LF, Rando OJ (2005) Genomic characterization reveals a simple histone H4 acetylation code. Proc Natl Acad Sci U S A 102:5501–5506

Dorigo B, Schalch T, Bystricky K, Richmond TJ (2003) Chromatin fiber folding: requirement for the histone H4 N-terminal tail. J Mol Biol 327:85–96

Dorigo B, Schalch T, Kulangara A, Duda S, Schroeder RR, Richmond TJ (2004) Nucleosome arrays reveal the two-start organization of the chromatin fiber. Science 306:1571–1573

Dover J, Schneider J, Tawiah-Boateng MA, Wood A, Dean K, Johnston M, Shilatifard A (2002) Methylation of histone H3 by COMPASS requires ubiquitination of histone H2B by Rad6. J Biol Chem 277:28368–28371

Ducker CE, Simpson RT (2000) The organized chromatin domain of the repressed yeast a cell-specific gene *STE6* contains two molecules of the corepressor Tup1p per nucleosome. EMBO J 19:400–409

Ebralidse KK, Grachev SA, Mirzabekov AD (1988) A highly basic histone H4 domain bound to the sharply bent region of nucleosomal DNA. Nature 331:365–367

Edmondson DG, Smith MM, Roth SY (1996) Repression domain of the yeast global repressor Tup1 interacts directly with histones H3 and H4. Genes Dev 10:1247–1259

Finch JT, Klug A (1976) Solenoidal model for superstructure in chromatin. Proc Natl Acad Sci U S A 73:1897–1901

Forneris F, Binda C, Vanoni MA, Battaglioli E, Mattevi A (2005) Human histone demethylase LSD1 reads the histone code. J Biol Chem 280:41360–41365

Goto H, Tomono Y, Ajiro K, Kosako H, Fujita M, Sakurai M, Okawa K, Iwamatsu A, Okigaki T, Takahashi T, Inagaki M (1999) Identification of a novel phosphorylation site on histone H3 coupled with mitotic chromosome condensation. J Biol Chem 274:25543–25549

Grauffel C, Stote RH, Dejaegere A (2010) Force field parameters for the simulation of modified histone tails. J Comput Chem 31:2434–2451

Harborne N, Allan J (1983) Modulation of the relative trypsin sensitivities of the core histone 'tails'. FEBS Lett 155:88–92

Herrera JE, Schiltz RL, Bustin M (2000) The accessibility of histone H3 tails in chromatin modulates their acetylation by P300/CBP-associated factor. J Biol Chem 275:12994–12999

Jackson V (1999) Formaldehyde cross-linking for studying nucleosomal dynamics. Methods 17:125–139

Jones DT (1999) Protein secondary structure prediction based on position-specific scoring matrices. J Mol Biol 292:195–202

Kan PY, Lu X, Hansen JC, Hayes JJ (2007) The H3 tail domain participates in multiple interactions during folding and self-association of nucleosome arrays. Mol Cell Biol 27:2084–2091

Kan PY, Caterino TL, Hayes JJ (2009) The H4 tail domain participates in intra- and internucleosome interactions with protein and DNA during folding and oligomerization of nucleosome arrays. Mol Cell Biol 29:538–546

Korolev N, Lyubartsev AP, Nordenskiöld L (2006) Computer modeling demonstrates that electrostatic attraction of nucleosomal DNA is mediated by histone tails. Biophys J 90:4305–4316

Kouzarides T (2007) Chromatin modifications and their function. Cell 128:693–705

Kundu TT (2007) Small molecular modulators in epigenetics: implications in gene expression and therapeutics. In: Kundu TT, Dasgupta D (eds) Chromatin and disease. Springer, New York

Kurdistani SK (2011) Histone modifications in cancer biology and prognosis. Prog Drug Res 67:91–106

LaPenna G, Furlan S, Perico A (2006) Modeling H3 histone N-terminal tail and linker DNA interactions. Biopolymers 83:135–147

Lee KM, Hayes JJ (1997) The N-terminal tail of histone H2A binds to two distinct sites within the nucleosome core. Proc Natl Acad Sci U S A 94:8959–8964

Lins RD, Röthlisberger U (2006) Influence of long-range electrostatic treatments on the folding of the N-terminal H4 histone tail peptide. J Chem Theory Comput 2:246–250

Liu H, Duan Y (2008) Effects of post-translational modifications on the structure and dynamics of histone H3 N-terminal peptide. Biophys J 94:4579–4585

Liu CL, Kaplan T, Kim M, Buratowski S, Schreiber SL, Friedman N, Rando OJ (2005) Single-nucleosome mapping of histone modifications in *S. cerevisiae*. PLoS Biol 3:e328

Luco RF, Allo M, Schor IE, Kornblihtt AR, Misteli T (2011) Epigenetics in alternative pre-mRNA splicing. Cell 144(1):16–26

Luger K, Mader AW, Richmond RK, Sargent DF, Richmond TJ (1997) Crystal structure of the nucleosome core particle at 2.8 Å resolution. Nature 389:251–260

McGuffin LJ, Bryson K, Jones DT (2000) The PSIPRED protein structure prediction server. Bioinformatics 16:404–405

Moore SC, Ausio J (1997) Major role of the histones H3-H4 in the folding of the chromatin fiber. Biochem Biophys Res Commun 230:136–139

Murray K (1964) The occurrence of epsilon-N-methyl lysine in histones. Biochemistry 3:10–15

Nunes AM, Zavitsanos K, Del CR, Malandrinos G, Hadjiliadis N (2009) Interaction of histone H2B (fragment 63–93) with Ni(ii). An NMR study. Dalton Trans 11:1904–1913

Phillips DMP (1961) Acetyl groups as N-terminal substituents in calf-thymus histones. Biochem J 80:40P

Phillips DMP (1963) The presence of acetyl groups in histones. Biochem J 87:258–263

Robinson PJ, Fairall L, Huynh VA, Rhodes D (2006) EM measurements define the dimensions of the "30-nm" chromatin fiber: evidence for a compact, interdigitated structure. Proc Natl Acad Sci U S A 103:6506–6511

Robinson PJ, An W, Routh A, Martino F, Chapman L, Roeder RG, Rhodes D (2008) 30 nm chromatin fibre decompaction requires both H4-K16 acetylation and linker histone eviction. J Mol Biol 381:816–825

Schalch T, Duda S, Sargent DF, Richmond TJ (2005) X-ray structure of a tetranucleosome and its implications for the chromatin fibre. Nature 436:138–141

Schiltz RL, Mizzen CA, Vassilev A, Cook RG, Allis CD, Nakatani Y (1999) Overlapping but distinct patterns of histone acetylation by the human coactivators p300 and PCAF within nucleosomal substrates. J Biol Chem 274:1189–1192

Shogren-Knaak M, Ishii H, Sun JM, Pazin MJ, Davie JR, Peterson CL (2006) Histone H4-K16 acetylation controls chromatin structure and protein interactions. Science 311:844–847

Sperling J, Sperling R (1978) Photochemical cross-linking of histones to DNA nucleosomes. Nucleic Acids Res 5:2755–2773

Strahl BD, Allis CD (2000) The language of covalent histone modifications. Nature 403:41–45

Thoma F, Koller T, Klug A (1979) Involvement of histone H1 in the organization of the nucleosome and of the salt-dependent superstructures of chromatin. J Cell Biol 83:403–427

Usachenko SI, Bavykin SG, Gavin IM, Bradbury EM (1994) Rearrangement of the histone H2A C-terminal domain in the nucleosome. Proc Natl Acad Sci U S A 91:6845–6849

Van Holde KD (1989) Chromatin. Academic, New York

Walker IO (1984) Differential dissociation of histone tails from core chromatin. Biochemistry 23:5622–5628

Wang X, Moore SC, Laszckzak M, Ausio J (2000) Acetylation increases the alpha-helical content of the histone tails of the nucleosome. J Biol Chem 275:35013–35020

Wang X, He C, Moore SC, Ausio J (2001) Effects of histone acetylation on the solubility and folding of the chromatin fiber. J Biol Chem 276:12764–12768

Watanabe Y, Maekawa M (2010) Methylation of DNA in cancer. Adv Clin Chem 52:145–167

Wei Y, Mizzen CA, Cook RG, Gorovsky MA, Allis CD (1998) Phosphorylation of histone H3 at serine 10 is correlated with chromosome condensation during mitosis and meiosis in *Tetrahymena*. Proc Natl Acad Sci U S A 95:7480–7484

Williams SP, Athey BD, Muglia LJ, Schappe RS, Gough AH, Langmore JP (1986) Chromatin fibers are left-handed double helices with diameter and mass per unit length that depend on linker length. Biophys J 49:233–248

Woodcock CL, Dimitrov S (2001) Higher-order structure of chromatin and chromosomes. Curr Opin Genet Dev 11:130–135

Woodcock CL, Frado LL, Rattner JB (1984) The higher-order structure of chromatin: evidence for a helical ribbon arrangement. J Cell Biol 99:42–52

Worcel A, Strogatz S, Riley D (1981) Structure of chromatin and the linking number of DNA. Proc Natl Acad Sci U S A 78:1461–1465

Yang D, Arya G (2011) Structure and binding of the H4 histone tail and the effects of lysine 16 acetylation. Phys Chem Chem Phys 13:2911–2921

Zheng C, Hayes JJ (2003) Intra- and inter-nucleosomal protein-DNA interactions of the core histone tail domains in a model system. J Biol Chem 278:24217–24224

Zheng C, Lu X, Hansen JC, Hayes JJ (2005) Salt-dependent intra- and internucleosomal interactions of the H3 tail domain in a model oligonucleosomal array. J Biol Chem 280:33552–33557

Zheng S, Wyrick JJ, Reese JC (2010) Novel trans-tail regulation of H2B ubiquitylation and H3K4 methylation by the N terminus of histone H2A. Mol Cell Biol 30:3635–3645

Zhou J, Fan JY, Rangasamy D, Tremethick DJ (2007) The nucleosome surface regulates chromatin compaction and couples it with transcriptional repression. Nat Struct Mol Biol 14:1070–1076

Zoroddu MA, Kowalik-Jankowska T, Kozlowski H, Molinari H, Salnikow K, Broday L, Costa M (2000) Interaction of Ni(II) and Cu(II) with a metal binding sequence of histone H4: AKRHRK, a model of the H4 tail. Biochim Biophys Acta 1475:163–168

Zoroddu MA, Medici S, Peana M (2009) Copper and nickel binding in multi-histidinic peptide fragments. J Inorg Biochem 103:1214–1220

Chapter 3
Megabase Replication Domains Along the Human Genome: Relation to Chromatin Structure and Genome Organisation

Benjamin Audit, Lamia Zaghloul, Antoine Baker, Alain Arneodo, Chun-Long Chen, Yves d'Aubenton-Carafa, and Claude Thermes

Abstract In higher eukaryotes, the absence of specific sequence motifs, marking the origins of replication has been a serious hindrance to the understanding of (i) the mechanisms that regulate the spatio-temporal replication program, and (ii) the links between origins activation, chromatin structure and transcription. In this chapter, we review the partitioning of the human genome into megabased-size replication domains delineated as N-shaped motifs in the strand compositional asymmetry profiles. They collectively span 28.3% of the genome and are bordered by more than 1,000 putative replication origins. We recapitulate the comparison of this partition of the human genome with high-resolution experimental data that confirms that replication domain borders are likely to be preferential replication initiation zones in the germline. In addition, we highlight the specific distribution of experimental and numerical chromatin marks along replication domains. Domain borders correspond to particular open chromatin regions, possibly encoded in the DNA sequence, and around which replication and transcription are highly coordinated. These regions also present a high evolutionary breakpoint density, suggesting that susceptibility to breakage might be linked to local open chromatin fiber state. Altogether, this chapter presents a compartmentalization of the human genome into replication domains that are landmarks of the human genome organization and are likely to play a key role in genome dynamics during evolution and in pathological situations.

B. Audit (✉) • A. Arneodo
Université de Lyon, F-69000 Lyon, France

Laboratoire de Physique, CNRS, Ecole Normale Supérieure de Lyon,
F-69007 Lyon, France
e-mail: benjamin.audit@ens-lyon.fr

L. Zaghloul • A. Baker
Université de Lyon, F-69000 Lyon, France

Laboratoire Joliot-Curie, CNRS, Ecole Normale Supérieure de Lyon,
F-69007 Lyon, France

C. Chen • Y. d'Aubenton-Carafa • C. Thermes
Centre de Génétique Moléculaire, CNRS, F-91198, Gif-sur-Yvette, France

3.1 Introduction

DNA replication is an essential genomic function responsible for the accurate transmission of genetic information through successive cell generations. According to the so-called "replicon" paradigm derived from prokaryotes (Jacob et al. 1963), this process starts with the binding of some "initiator" protein to a specific "replicator" DNA sequence called *origin of replication*. The recruitment of additional factors initiate the bi-directional progression of two divergent replication forks along the chromosome. In eukaryotic cells, this event is initiated at a number of replication origins and propagates until two converging forks collide at a *terminus of replication* (Bell and Dutta 2002). In general, metazoan replication origins are rather poorly defined and initiation may occur at multiple sites distributed over a thousand of base pairs (Gilbert 2001). Moreover, random initiation of replication was repeatedly observed in *Drosophila* and *Xenopus* early embryo cells, suggesting that any DNA sequence can function as a replicator (Hyrien and Méchali 1993; Coverley and Laskey 1994; Sasaki et al. 1999). Thus, although it is clear that some sites consistently act as replication origins in most eukaryotic cells, the mechanisms that select these sites and the sequences that determine their location remain elusive in many cell types (Bogan et al. 2000; Gilbert 2004). The need to fulfill specific requirements that result from cell diversification may have led multicellular eukaryotes to develop various epigenetic controls over the replication origin selection rather than to conserve specific replication sequence (Méchali 2001; Demeret et al. 2001; McNairn and Gilbert 2003).

In that context, we report on what we learned about the human replication program starting from an *in silico* analysis of strand composition asymmetry (skew) profiles (Brodie of Brodie et al. 2005; Touchon et al. 2005). We describe a wavelet-based method to systematically detect replication domains labeled N-domains as their skew profile displays a N-like shape attributed to mutational asymmetries intrinsic to the replication process (Arneodo et al. 2007, 2011; Audit et al. 2007; Huvet et al. 2007; Baker et al. 2010; Chen et al. 2011). We present a comparative analysis of the detected N-domains with experimental replication data, in particular with preferential replication initiation zones identified by a multi-scale peak detection analysis of high-resolution replication timing profile (Chen et al. 2010; Hansen et al. 2010). This study corroborates N-border as putative replication origins of the germline. Finally, we examine the distribution of genes and of numerical and experimental chromatin marks within N-domains. Altogether these results make a compelling case that N-domain borders are "islands" of open chromatin defining units of coordinated replication and gene transcription.

3.2 Uncovering Replication Skew N-domain Along the Human Genome

The existence of replication associated strand asymmetries has been mainly established in bacterial genomes (Lobry 1995, 1996; Mrázek and Karlin 1998; Frank and Lobry 1999; Rocha et al. 1999; Tillier and Collins 2000). $S_{GC} = (G-C)/(G+C)$

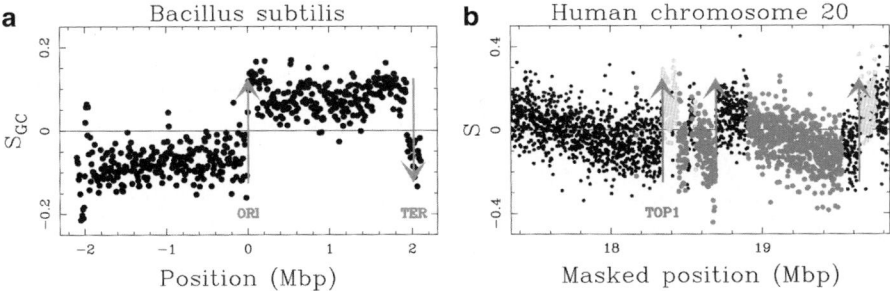

Fig. 3.1 (**a**) Skew profile S_{GC} along the genome of *Bacillus subtillis* calculated in 8192 bp windows. The *upward* (resp. downward) *arrow* marks the unique origin of replication (ORI) (resp. terminus (TER)) where a transition from negative to positive (resp. positive to negative) skew values is observed (Lobry 1996). (**b**) Skew profiles $S = S_{TA} + S_{GC}$ along a fragment of chromosome 20 including the TOP1 origin calculated in 1 kbp windows of the repeat-masked sequence (Touchon et al. 2005; Arneodo et al. 2007). *Upward arrows* mark upward jumps in the skew profile that have been identified as putative replication origins (Brodie of Brodie et al. 2005; Touchon et al. 2005). *Black*, intergenic regions; *light gray*, sense (+) genes; *dark grey*, antisense (−) genes

and $S_{TA} = (T - A)/(T + A)$ skews abruptly switch sign (over few kbp) from negative to positive values at the replication origin and in the opposite direction from positive to negative values at the replication terminus. This step-like profile is characteristic of the replicon model (Jacob et al. 1963) (Fig. 3.1a). In eukaryotes, the existence of compositional biases is unclear. Most attempts to detect the replication origins from strand compositional asymmetry have been inconclusive (Mrázek and Karlin 1998; Bulmer 1991; Francino and Ochman 2000). Even though, strand asymmetries associated with replication have been observed in the subtelomeric regions of *Saccharomyces cerevisiae* chromosomes (Gierlik et al. 2000).

As shown in Fig. 3.1b for TOP1 replication origin (Brodie of Brodie et al. 2005; Touchon et al. 2005; Arneodo et al. 2007, 2011), most of the experimentally known replication origins in the human genome correspond to rather sharp (over several kbp) transitions from negative to positive skew values (S_{TA}, S_{GC} as well as $S = S_{TA} + S_{GC}$) that clearly emerge from the noisy background. Moreover, sharp upward jumps of amplitude $\Delta S > 15\%$, similar to the ones observed for the known replication origins, also exist at many other locations along the human chromosomes (Figs. 3.1b and 3.2a). But the most striking feature is the fact that in between two neighboring major upward jumps, not only does the noisy S profile not present any comparable downward sharp transition as observed for the bacterial replicon (Fig. 3.1a), rather it displays a remarkable decreasing linear behavior. At chromosome scale, we thus get jagged S profiles that have the aspect of a succession of Ns (Brodie of Brodie et al. 2005; Touchon et al. 2005; Arneodo et al. 2007). In genic regions, we observe the superimposition of this replication N-shaped profile and of the step-like profiles resulting from transcription associated strand asymmetry (Touchon et al. 2003, 2004; Nicolay et al. 2007; Baker et al. 2010), that appear as upward ((+) genes) and downward ((−) genes) blocks standing out from the replication pattern (see Fig. 3.1b, rhs of TOP1). The observation that some of the segments between two successive skew upward jumps are entirely intergenic (Fig. 3.1b, lhs of

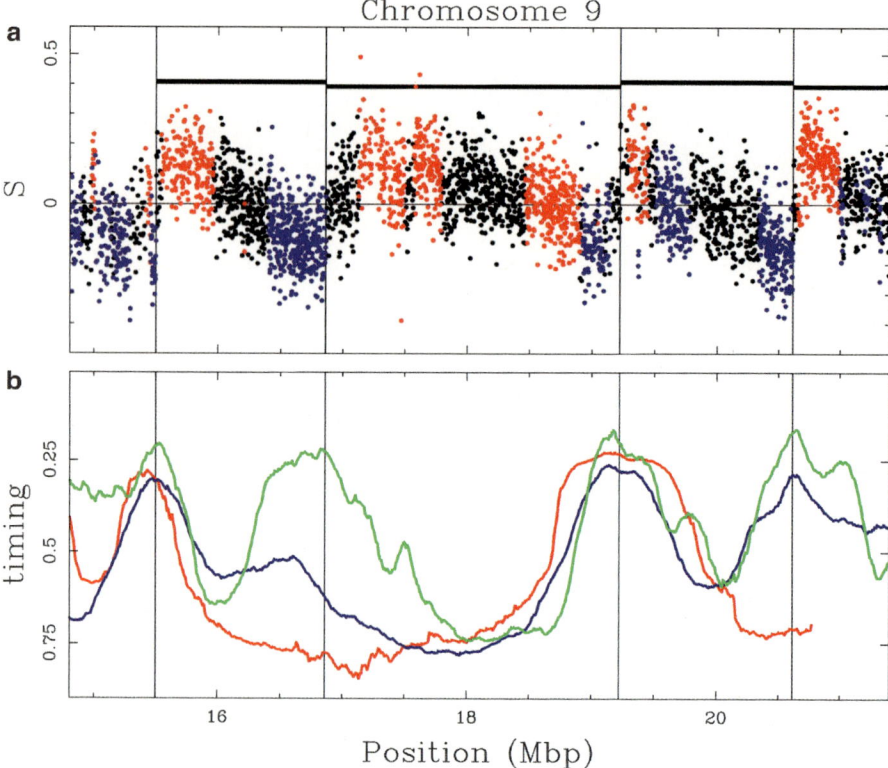

Fig. 3.2 Compositional skew and replication timing profiles along a fragment of Human chromosome 9. (**a**) Skew profiles $S = S_{TA} + S_{GC}$ calculated in 1 kbp windows of the masked-sequence (Touchon et al. 2005; Arneodo et al. 2007, 2011). *Horizontal black lines* mark the replication N-domains delineated using a multi-scale methodology (Sect. 3.6.1) (Audit et al. 2007; Huvet et al. 2007; Baker et al. 2010). *Vertical lines* mark the corresponding putative replication initiation zones. *Black*, intergenic regions; *red*, sense (+) genes; *blue*, antisense (−) genes. (**b**) Mean replication timing determined in BG02 ESC (*green*), K562 (*red*) and GM06990 (*blue*) cell lines (Hansen et al. 2010)

TOP1), clearly illustrates that the N-shape profiles result solely from replication and not from some particular distribution of transcription units (Brodie of Brodie et al. 2005; Touchon et al. 2005; Arneodo et al. 2007, 2011; Huvet et al. 2007). Importantly, it is observed that strand asymmetry profiles are conserved between a number of human regions and their homologous loci in the mouse and dog genomes (Touchon et al. 2005). This conservation between regions that have strongly diverged during evolution further supports the association between N-shaped skew profiles and replication in mammalian germline cells (Brodie of Brodie et al. 2005; Touchon et al. 2005; Arneodo et al. 2007, 2011). On shorter evolutionary time scale, recent computation of nucleotide substitution rates in the human lineage since divergence with chimpanzee shows that the differences between complementary rates (mutational asymmetry between the complementary strands) change sign when crossing replication initiation zones identified as abrupt upward jumps in the skew profile (Chen et al. 2011). This analysis demonstrates that the mutational patterns observed in these regions fully explain the jumps in the S skew profile and provides strong evidence in favor of a model based on replication errors and strand-biased repair to explain the origins of N-domains.

We developed a multi-scale procedure to detect replication N-domains along the skew S profiles (Sect. 3.6.1). As illustrated in Fig. 3.2a for a fragment of human chromosome 9 that contains adjacent replication domains, this wavelet-based methodology provides a very efficient way of delineating the N-shaped components induced by replication along the skew S profile. Applying this procedure to the 22 human autosomes from human May 2004 (hg17) assembly, we delineated 678 replication domains of mean length $L = 1.2 \pm 0.6$ Mbp, containing 18% of the genes (Audit et al. 2007; Huvet et al. 2007; Baker et al. 2010). There are 1,060 different N-domains borders since in 296 cases a border in shared by two consecutive domains. Here, we used the LiftOver coordinate conversion tool from the UCSC website to map N-domains to Human March 2006 (hg18) coordinates and we kept only the N-domains that had exactly the same size before and after conversion. This resulted in the assignment on hg18 of 663 unambiguous N-domains delimited by 1,040 borders and spanning 28.3% of the sequenced genome length. To our knowledge, these loci correspond to the largest set of human replication initiation zones available to date (Baker et al. 2010).

3.3 DNA Replication Data Corroborate Human Replication Skew N-domain Predictions

3.3.1 N-domain Borders Correspond to Experimental Replication Origins Mapped on ENCODE Pilot Regions

Previous analyses of nucleotide strand compositional asymmetries have shown that, out of the 9 experimentally identified replication origins, 7 (78%) presented an upward jump in the asymmetry profile analog to those bordering N-domains (Brodie of Brodie et al. 2005; Touchon et al. 2005). Recently, the localization of replication origins has been experimentally investigated along 1% of the human genome (ENCODE pilot regions) by hybridization to Affymetrix ENCODE tiling arrays of purified small nascent DNA strands and of restriction fragments containing small replication bubble (The ENCODE Project Consortium 2007; Cadoret et al. 2008). Out the 7 N-domain borders that reside within an ENCODE region, 4 match an experimental replication origin ($P = 4\,10^{-3}$): 3 to the bubble trapping dataset (Bubble-HL, $P=0.017$) and 1 to a nascent strand purification dataset (NS-HL-2, $P=0.09$) (Table 3.1). Actually, as previously noted (Cadoret et al. 2008), a second N-domain border is located within 1 kbp of a NS-HL-2 origin (Table 3.1). Hence, there is direct experimental evidence that 5/7 (71%) N-domain borders

Table 3.1 Correspondence between N-domain borders and experimental replication origins datasets along ENCODE pilot regions (The ENCODE Project Consortium 2007; Cadoret et al. 2008)

Method	Number	Coverage	Match with N-domain borders (P-value)
Bubble-HL	234	8.6%	3 (**0.017**)
NS-GM	758	1.0%	0
NS-HL-1	434	0.6%	0
NS-HL-2	282	1.4%	1 (0.093)
All		11%	4 (**0.004**)
NS+1kbp-HL-2		3.2%	2 (**0.019**)
+Bublle-HL		11.2%	5 (**0.0003**)

Main characteristics of replication origin prediction along ENCODE pilot regions based on purified restriction fragments containing replication bubble (Bubble) (Mesner et al. 2006) or purified small nascent strands (NS) (Gerbi and Bielinsky 1997). First column indicates the experimental method (Bubble or NS) and the cell type (HL: HeLa cells and GM: GM06990 cell lines). They are two independent NS-HL datasets labelled 1 and 2. "All" corresponds to the four initial datasets considered together. NS+1kbp-HL-2 corresponds to the NS-HL-2 dataset when extending replication origins by 1 kbp on both sides. +Bublle-HL corresponds to merging the NS+1kbp-HL-2 and Bubble-HL datasets. We provide the number of replication origins, their total coverage of ENCODE pilot regions, the number of N-domain borders out of 7 within ENCODE regions that match with one of the experimental replication origins and the corresponding P-value using a binomial test (Audit et al. 2009)

correspond to active replication origins at a few kbp resolution (Audit et al. 2009). These results are all the more significant considering the rather low overlap between the experimental datasets. For example, only 69 (25%) of the NS-HL-2 origins overlap with a Bubble-HL origins, only 4 (1.4%) with the second nascent strand dataset in HeLa cells (NS-HL-1) and only 12 (4.3%) even when extending NS-HL-1 and NS-HL-2 origins by 1 kbp on both sides.

3.3.2 N-domain Borders Colocate with Peaks in Replication TimingProfiles

The recent advance in the experimental characterization of replication programme provides genome-wide timing data in several human cell types (Woodfine et al. 2005; Desprat et al. 2009; Chen et al. 2010; Hansen et al. 2010; Ryba et al. 2010; Yaffe et al. 2010). We examined the average timing profiles computed from Repli-Seq data (Chen et al. 2010; Hansen et al. 2010) for 5 cell lines including one embryonic stem cell line (BG02), a fibroblast (BJ), a lymphoblastoid (GM06990), an erythroid (K562) and an HeLa cell line. They all present numerous peaks pointing towards early replication time (Figs. 3.2b and 3.6a). The regions at the tip of the peaks are on average replicated earlier than their surrounding regions and thus harbor replication initiation zones highly active in the corresponding cell type. In the 6 Mbp regions analyzed in Fig. 3.2, we observed a strong correspondence between the germline putative replication origins at N-domain borders and the initiation zones

3 Megabase Replication Domains Along the Human Genome...

Table 3.2 Correspondence between N-domain borders and peaks delineated along genome-wide timing profiles (Chen et al. 2010; Hansen et al. 2010). For each cell line, we indicate the number of timing peaks, the cumulative genome coverage of the regions ± 100 kbp around the peaks, the number of N-domain borders located within these regions and the corresponding proportion of so-confirmed N-domain borders. "All" corresponds to the five timing peak datasets considered together. In the six cases the number of N-domain borders with a distance to a timing peak smaller than 100 kbp is highly significant (P-value < 10^{-15} using a binomial test)

Cell line	Number of timing peaks (Genome cov.)	Number of co-localization with N-domain borders (N-domain border prop.)
BG02	1,690 (12.5%)	333 (32%)
BJ	981 (7.3%)	185 (18%)
GM06990	706 (5.2%)	167 (16%)
Hela	1,556 (11.5%)	292 (28%)
K562	795 (5.8%)	176 (17%)
All	5,728 (28.0%)	590 (57%)

pointed to by the timing peaks. We also noticed a strong conservation of timing peak location between cell lines in this region. In order to perform a systematic comparison between these loci containing strong replication origins and N-domain borders, we developed a multi-scale methodology for the detection of peaks along timing profiles (Sect. 3.6.2). For the 5 cell lines considered, we delineated between 706 (for GM06990) and 1,690 (for BG02) timing peaks (Table 3.2). When comparing the observed distribution of the distances of N-domain borders to the closest timing peak (considering the 5 cell lines) to the expected distribution for uniformly distributed borders, we observed a significant excess of short distances (175 kbp) (Fig. 3.3a). This result provides a first quantitative evidence for the co-localisation of N-domain borders with the active replication initiation zones at timing peaks in different cell lines.

The spatial resolution of the timing profiles being ~100 kbp, we carried on the analysis considering that N-domain borders and timing peaks coincide if they are within 100 kbp from each other. In BG02, the regions ±100 kbp of the timing peaks cover 12.5% of the genome but contain 32% of the N-domain borders. Consistently, we observed a significant over-representation of N-domain borders in the regions ± 100 kbp of timing peaks for all 5 cell lines (Table 3.2). Altogether, 57% of N-domain borders are associated (at 100 kbp resolution) with an active replication initiation regions in at least one of the considered cell lines. Moreover, N-domain borders and BG02 timing peaks present a very similar degree of association with the timing peaks of the 4 other cell lines: 55% (575/1,040) of N-domain borders and 54% (908/1,690) of BG02 timing peaks are not associated within 100 kbp with a peak in one of the other 4 cell lines and only 2.9% (30/1,040) and 3.7% (62/1,690) respectively are also found in the 4 other cell lines. This observation is a strong indication that N-domain borders mark replication initiation zones active in the germline.

Finally, we observed that the probability of BG02 timing peaks to correspond to a N-domain border increases with their conservation level in the 4 other cell lines,

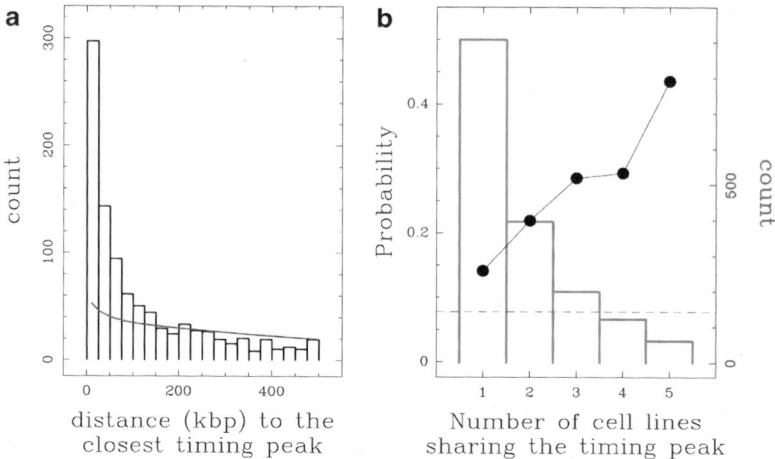

Fig. 3.3 (a) N-domain border distances to the closest replication timing peak from one of the 5 cell lines considered; (*black*) the histogram corresponds to observe counts in 25 kbp bins; (*grey*) the *solid line* marks the expected counts for uniformly distributed N-domain borders. (b) BG02 timing peaks are classified according to the number of cell lines where a corresponding timing peak is present within 100 kbp (from 1 for BG02 specific peaks to 5 when a peak is observed within 100 kbp in the 4 other cell lines); (*black bullet*) for each BG02 peak category, we report the proportion of peaks corresponding to a N-domain within 100 kbp; the *dashed line* marks the expected proportion (7.7%) for uniformly distributed N-domain borders; (*grey*) the histogram corresponds to the number of BG02 peaks in each category

from 14% for BG02 specific peaks to 44% for peaks also observed in the 4 other cell lines (Fig. 3.3b). Similar results are generally obtained when considering the conservation of timing peaks (Chen et al. 2011). This illustrates that replication initiation zones have very diverse conservation status relative to cell differentiation, from cell specific initiation zones to apparently generally active zones (Hiratani et al. 2008, 2010).

3.4 Open Chromatin Encoded in the DNA Sequence is the Signature of Germline Initiation Zones

Our knowledge of the mechanisms that control the spatio-temporal program of replication remains sparse and understanding how origins are distributed along the genome and how their activation is controlled and coordinated constitutes one of the main challenges of molecular biology (Schwaiger and Schubeler 2006). For years, the small number (~30) of well established origins in the human genome and more generally in mammalian genomes, has been an obstacle to fully appreciate the genome-wide organization of replication in relation to gene expression and local chromatin structure – much remains to be understood about the impact of the DNA sequence on origin activity in human cells in parallel to epigenetic controls (Bogan et al. 2000; Méchali 2001; Gerbi and Bielinsky 2002; McNairn and

Gilbert 2003; Lemaitre et al. 2005; Courbet et al. 2008; Hamlin et al. 2008). In that context, the 663 N-domains provide an unprecedented opportunity to analyze the human replication program in relation to chromatin state and gene organization (Huvet et al. 2007; Audit et al. 2009; Arneodo et al. 2011). Indeed as it has been argued in the previous section, the comparison of recent high-resolution replication data with N-domains provided clear experimental evidence for their relationship to the replication program (Audit et al. 2007; Huvet et al. 2007; Chen et al. 2011). N-domains borders are associated to regions that replicate earlier in the S phase than their surrounding regions and are thus likely associated with early replicating origins, while N-domain central regions are late replicating (Fig. 3.2). Hence, most N-domains correspond to units of replication where timing decreases when going from borders to center (Audit et al. 2007; Huvet et al. 2007; Baker et al. 2012).

3.4.1 Active Replication Initiation Regions at N-domain Borders are Hypersensitive to DNase I Digestion

High-throughput sequencing and whole-genome tiled strategies have been developed to identify DNase I hypersensitive sites (HS) as markers of open chromatin across the genome (Sabo et al. 2006; Boyle et al. 2008). When mapping DNase I HS determined in GM06990 inside the 663 replication N-domains, we observed that the mean site coverage is maximum at the N-domain extremities and decreases significantly from the extremities to the center that is rather insensitive to DNase I cleavage ($\sim 1/3$ of the genome average; Fig. 3.4a, b). This decrease extends over ~ 150 kbp whatever the size of replication N-domains suggesting that N-domain extremities are at the center of an open chromatin region with a ~ 300 kbp mean characteristic size (Audit et al. 2009). We observed that the over representation of GM06990 DNase I HS is much stronger at the 167 N-domain borders that co-localize (± 100 kbp) with a GM06990 timing peak (3.5 the genome average; Fig. 3.4a) compared to the 873 other N-domain borders (1.5 the genome average; Fig. 3.4b). These results illustrate the dynamic coupling between the activity of a replication initiation zones and the chromatin status of the surrounding region (Baker et al. 2012).

3.4.2 DNA Sequence Codes for the Accumulation of Nucleosome Free Regions Around N-domain Borders

Previous analysis revealed that promoter regions for protein-coding genes are extremely hypersensitive to DNase I digestion (Boyle et al. 2008). These regions were shown to be nucleosome depleted (Boyle et al. 2008; Ozsolak et al. 2007; Heintzman et al. 2007; Barski et al. 2007; Schones et al. 2008), very much like the NFRs observed at yeast promoters (Lee et al. 2007; Yuan et al. 2005). Recent numerical studies revealed that, to a large extent, these NFRs are coded in the DNA

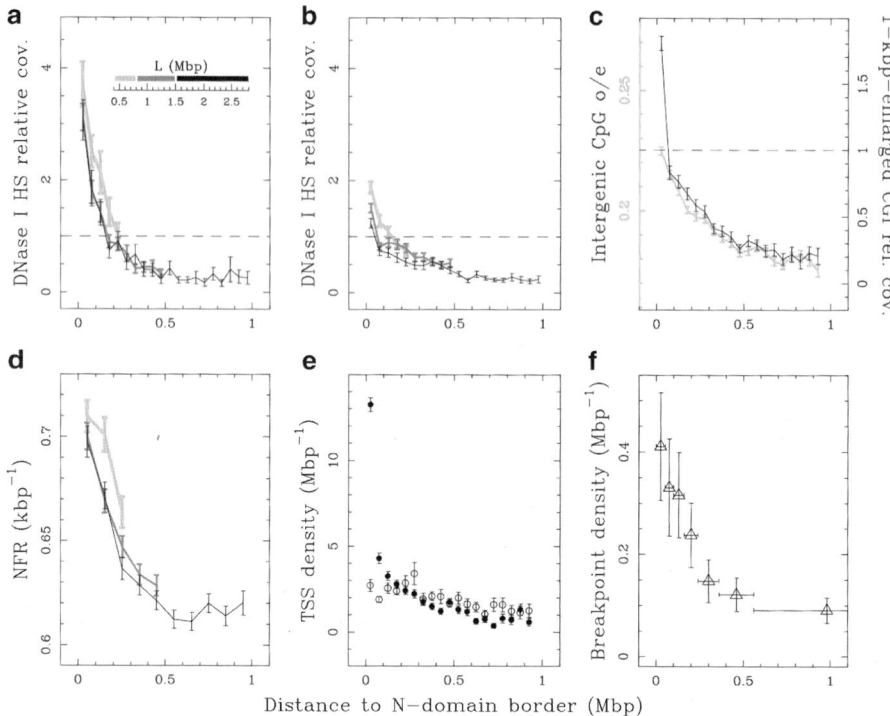

Fig. 3.4 Genome organization and chromatin marks distribution within N-domains. (**a, b**) DNase I hypersensitive sites (HS) coverage relative to the genomic average. DNase I sensitivity data were obtained in GM06990 cell line; N-domain borders were classified accordingly: N-domain borders corresponding to a GM06990 timing peak in (**a**) and other N-domain borders in (**b**) (Table 3.2); the three shades of *grey* corresponds to three N-domain size categories: $L<0.8$ Mbp (*light grey*), $0.8<L<1.5$ Mbp (*dark grey*) and $L>1.5$ Mbp (*black*); for each dataset, the average profile determined in 50 kbp windows is plotted as a function of the distance to the closest N-domain border; the *horizontal dashed line* represent the genomic average of DNase I HS coverage. (**c**) 1-kbp-enlarged CGI coverage normalized by the genomic average (*black*) and CpG ratio computed for intergenic regions (*grey*) determined in 50 kbp windows; the average profiles are plotted as a function of the distance to the closest N-domain border; the *horizontal dashed line* represent the genomic average of 1-kbp-enlarged CGI coverage. (**d**) In silico NFR density determined in 50 kbp windows with GC content $<41\%$; the average profiles are plotted as a function of the distance to the closest N-domain border for the same N-domain size categories as in (**a, b**). (**e**) Average transcription start site (TSS) density determined in 50 kbp windows as a function of the distance to the closest N-domain border for CpG rich (*dots*) and CpG poor (*circles*) TSS. (**f**) Intergenic breakpoint density as a function of the distance to the closest N-domain border; horizontal bars represent the distance range used for each data point such that they all correspond to a~equal number of intergenic breakpoints. In (**a–f**) vertical bars represent standard errors of the mean

sequence via high energy barriers that impair nucleosome formation (Vaillant et al. 2007; Miele et al. 2008; Mavrich et al. 2008; Chevereau et al. 2009; Vaillant et al. 2010). Furthermore, these excluding genomic energy barriers were shown to play a fundamental role in the collective nucleosomal organization observed over rather large distances along the chromatin fiber (Vaillant et al. 2007, 2010). Here we used the same physical modeling of nucleosome formation energy based on sequence-dependent bending properties as previously introduced for modeling nucleosome occupancy profiles in the yeast genome (Vaillant et al. 2007; Miele et al. 2008). Since the GC content of *S. cerevisiae* is rather homogeneous around 39% as compared to the heterogeneous isochore structure of the human genome (Bernardi 2001), we restricted our modeling of nucleosome positioning to the light isochores L1 and L2 (GC <41%). Combining the nucleosome occupancy probability profile and the original energy profile, we identified nucleosome NFRs as the genomic energy barriers that are high enough to induce a nucleosome depleted region in the nucleosome occupancy profile. We checked that the average of an experimental genome-wide nucleosome occupancy profile (Schones et al. 2008) presented a clear nucleosomal depletion at the *in silico* predicted NFR positions (Audit et al. 2009). This indicates that the regions depleted in nucleosome in vivo are likely to be encoded, at least to some extent, in the DNA sequence. The distribution of NFRs along the 663 N-domains shows a mean density profile that is maximum at N-domain extremities (~0.7 NFR/kbp) and that decreases from extremities to center where some NFR depletion is observed (~0.62 NFR/kbp) (Fig. 3.4d). This decay over a characteristic length scale ~150 kbp is strikingly similar to that displayed by DNase I HS coverage (Fig. 3.4a, b).

Altogether, these results show that the NFR density profile displays the same characteristic increase around N-domain borders, as the experimental DNase I HS coverage profile (Audit et al. 2009). If this correlation was expected, the fact that we recovered it using a sequence-based modeling of nucleosome occupancy suggests that putative replication origins that border the N-domains are situated within regions of accessible open chromatin state that are likely to be encoded in the DNA sequence via excluding energy barriers that inhibit nucleosome formation and participate to the collective ordering of the nucleosome array (Vaillant et al. 2007, 2010; Arneodo et al. 2011).

3.4.3 DNA Hypomethylation Is Associated with N-domain Borders

Cytosine DNA methylation is a mediator of gene silencing in repressed heterochromatic regions, while in potentially active open chromatin regions, DNA is essentially unmethylated (Bird and Wolffe 1999). DNA methylation is continuously distributed in mammalian genomes with the notable exceptions of CpG islands (CGIs), short unmethylated regions rich in CpGs, and of certain promoters and transcription start sites (TSS) (Suzuki and Bird 2008). Since there was no genome wide map of DNA methylation available, we investigated the distribution of DNA methylation using instead indirect estimators calculated directly from the genomic sequence. Methyl-cytosines being hypermutable, prone to deamination to thymines, we considered the CpG observed/expected (CpG o/e) ratio as an estimator of DNA methylation (Bird 2002). Using data from the Human Epigenome Project (Eckhardt et al. 2006), we confirmed that hypomethylation in sperm corresponded to high values of the CpG o/e outside CGIs (data not shown). We also observed that CGIs' majoritary hypomethylated state spreaded out 1 kbp around the annotated CGIs, so that the sequence coverage by CGIs enlarged 1 kbp at both extremities provided a complementary marker for hypomethylated regions (Audit et al. 2009). When averaging over the 663 N-domains, the overall 1-kbp-enlarged CGI coverage (Fig. 3.4c) presents a maximum at origins positions, as the signature of hypomethylation, and decreases over a characteristic distance ~150 kbp, similar to the one found for DNase I HS coverage and NFR density profiles (Fig. 3.4a, b, d), from the extremities to the center of N-domains where CGI coverage is ~5 times less than the genome average. This observation is consistent with the hypothesis (Antequera and Bird 1999) that CGIs are protected from methylation due to there co-localisation with replication origins. The complementary analysis using the CpG o/e ratio as hypomethylation marker provided exactly the same diagnosis (data no shown).

N-domain borders correspond to a high concentration of genes (see below) (Huvet et al. 2007), TSS density profiles presenting, as expected, a strong similarity with 1-kbp-enlarged CGI coverage (Fig. 3.4c, e). Since the other open chromatin markers analyzed here have also been associated, at least to some extent, with genes (e.g. in CD4[+] cells, 16% of all DNase I HS are in the first exon or at the TSS of a gene, and 42% are found inside a gene (Boyle et al. 2008)), we reproduced the analysis of their distribution along the N-domains after masking the genes extended by 2 kbp at both extremities and the CGIs. The fact that the CpG o/e profiles (Fig. 3.4c), as well as the mean DNase I HS coverage and NFR density (data not shown) still present the decaying behavior over ~150 kbp, demonstrates that the excess observed around the putative replication origins does not simply reflects the rather packed gene organization at the N-domain borders but more likely an hypomethylated open chromatin state where CpG o/e is correlated with DNase I HS coverage and NFR density (Audit et al. 2009).

3.4.4 Coordination of Replication and Transcription at N-domain Borders

In higher-eukaryotes, extensive connections have been established between replication timing, genome organization and gene transcriptional state; early replication tends to co-localize with active transcription and, in mammals, to gene-dense GC rich isochores (Goldman et al. 1984; Schübeler et al. 2002; MacAlpine et al. 2004; Woodfine et al. 2005; Farkash-Amar et al. 2008; Hiratani et al. 2008; Karnani et al. 2007) and to transcription initiation early in development (Sequeira-Mendes et al. 2009). Correspondingly, N-domain borders are at the heart of a remarkable gene organization (Huvet et al. 2007): in a close neighborhood, genes are abundant and broadly expressed and their transcription is mainly directed away from the borders (data not shown). All these features weaken progressively with the distance to domain borders. This prefential orientation was interpreted as a gene coorientation with the movement of replication fork originating from N-domain borders (Huvet et al. 2007).

CGIs were linked to the presence of a promoter active in the germline in a single-gene analysis in mouse (Macleod et al. 1998) and genes with a CGI associated promoter where shown to be expressed in early embryo using a dataset of ~400 human genes with a mouse ortholog (Ponger et al. 2001). Hence, we used CpG enrichment at the promoter as an indicator of germline expression. We observed that the distribution of CpG-rich promoters presents a strong enrichment at N-domains borders, whereas the density of CpG-poor genes does not present such a dependence with the distance to N-domain border. The preference for N-domain border proximity is thus specific to the set of genes more likely to be expressed in the germline, indicating that regions around N-domain borders are likely to be transcriptionally active in the germline. Thus the open chromatin regions around N-domain borders are permissive to transcription whereas N-domain central regions appear transcriptionally quiescent and further illustrates the likely coordination of replication and transcription at N-domain borders in the germline.

3.4.5 Open Chromatin Around N-domain Borders are Potentially Fragile Regions Involved in Chromosome Instability

Since chromatin accessibility and openness are possible factors responsible for fragility and instability, N-domain borders could also play a key role in genome dynamics during evolution. and genome instability in pathologic situations like cancer. We analyzed the distribution of evolutionary breakpoint regions for the mammalian lineage (Lemaitre et al. 2008) along N-domains. We observed that breakpoints appear more frequently near N-domain borders than in their central regions (Fig. 3.4f), suggesting that the distribution of large-scale rearrangements in mammals reflects a mutational bias towards regions of high transcriptional activity and replication initiation (Lemaitre et al. 2009). Furthermore, the fact that chromosome anomalies involved in the tumoral process like at the RUNX1T1 oncogene locus coincide with replication N-domain extremities (data not shown) raises the possibility that the replication origins detected *in silico* are potential candidate loci susceptible to breakage in some cancer cell types.

3.5 Conclusion

To summarize, we have found that replication associated mutational asymmetry is responsible for Mbp sized N-shaped domains along the nucleotide compositional asymmetry profiles of human chromosomes. We systematically detected these

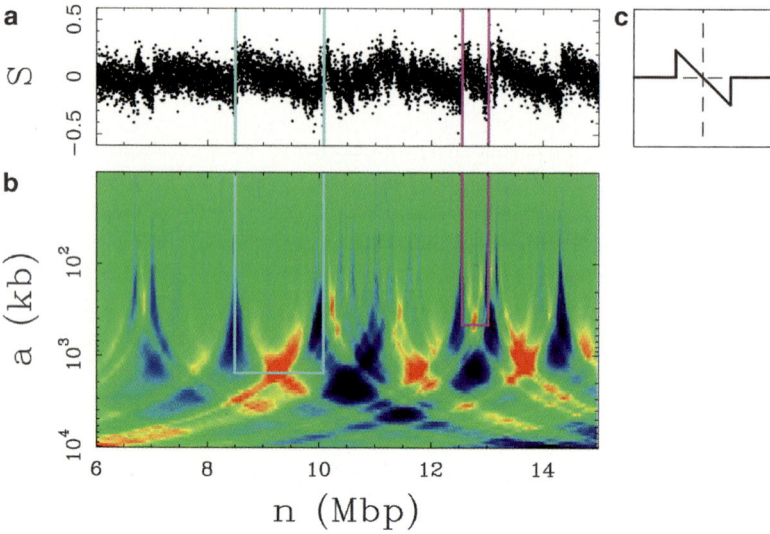

Fig. 3.5 Multi-scale detection of replication skew N-domains. (**a**) Skew profile S of a 9 Mbp repeat-masked fragment of human chromosome 21. (**b**) WT of S using the N-shaped wavelet illustrated in (**c**); the WT is color-coded from *dark-blue* (min; negative values) to *red* (max; positive values) through *green* (null values). *Light-blue* and purple lines illustrate the detection of two replication domains of significantly different sizes. Note that in (b), *blue* cone-shape areas signing upward jumps point at small scale (*top*) towards the putative replication origins and that the vertical positions of the WT maxima (*red areas*) corresponding to the two indicated replication domains match the distance between the putative replication origins (1.6 Mbp and 470 kbp respectively)

replication N-domains using an adapted wavelet-based methodology (Fig. 3.5). The 663 N-domains cover ~1/3 of the human genome and are bordered by 1,040 putative replication origins of the germline. These origins significantly overlap with experimental replication origins determined over ENCODE pilot regions (1% of the human genome) (Table 3.1). We developed a multi-scale analysis of high-resolution replication timing profiles allowing us to characterize in 5 cell lines preferential replication initiation zones as peaks pointing towards early time (Fig. 3.6). The analysis of the co-localization between N-domain borders and these timing peaks provides further genome-wide experimental evidence that N-domain borders are certainly preferential replication initiation zones mostly active in the early S phase, whereas N-domain central regions replicated more likely in late S phase (Table 3.2 and Fig. 3.2). These experimental verifications of *in silico* replication origin predictions are even more convincing when considering that, on top of the limitations due to experimental resolution, putative origin predictions concern only the replication origins that are well positioned and active in germline cells which does not guaranty that they are also active in somatic cells. Reciprocally, there is no guaranty that the replication origins that are active in this particular cell line are also active in the germline. In that respect, the results reported in this work provide interesting estimate of the proportion of preferential replication initiation

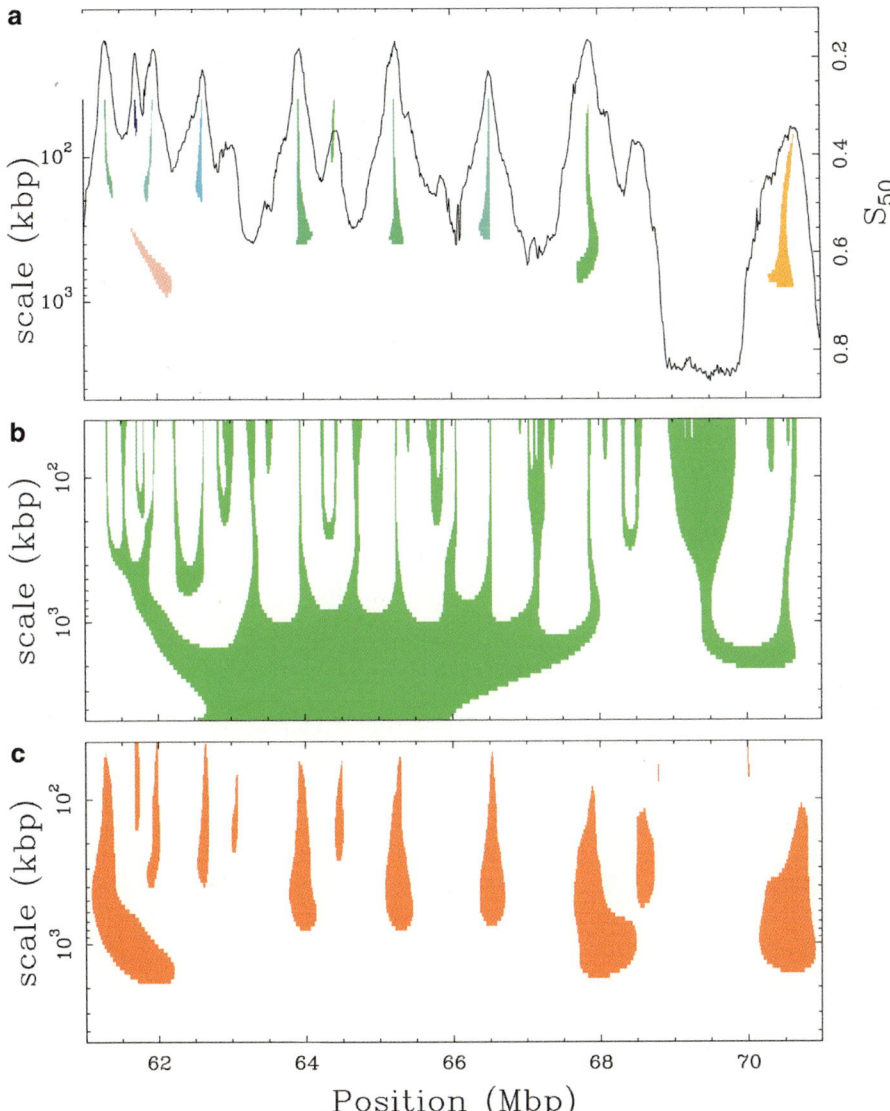

Fig. 3.6 (a) Timing profile normalised between 0 (start of S phase) and 1 (end of S phase). The color patches are space-scale representations of the detected peaks (see main text). (b) Regions of the space-scale half plane where the timing profile is flat according to: $\left| a^{-1/2} T_{g^{(1)}}[f](x,a) \right| < c_1$ with $c_1 = 0.01$. (c) Regions of the space-scale half plane where the timing profile present a significant negative curvature according to $a^{-1/2} T_{g^{(2)}}[f](x,a) < -c_2$ with $c_2 = 0.03$

zones that are active across several cell lineage (Fig. 3.3b). Our findings show that these replication initiation zones are located within a ~300 kbp region extremely sensitive to DNase I cleavage, presenting hypomethylation marks, suggesting that these regions present an open chromatin structure (Fig. 3.4a, b). This accessible chromatin organization is to some extent encoded in the DNA sequence via an enrichment of nucleosome excluding energy barriers (NFR) (Fig. 3.4d). The additional observation that the high gene density around N-domain borders is specific to CpG-rich TSS genes further suggests that this local chromatin structure is associated with transcriptional activity (Fig. 3.4e). The fact that experimental open chromatin signature obtained in a cell line appears specific to N-domain borders associated to timing peaks of the same cell line illustrates the correlation between replication activity and chromatin state. It raises the question of what are the determinants of this coordinated changes across cell differentiation.

In metazoans, recognition of replication origins by the origin recognition complex (ORC) does not involve simple consensus DNA sequence. Initiation sites do not share commmon genetic entities but seem to be favored by various factors that can differ from one origin to another and be required or dispensable under different conditions (Gilbert 2004). Specification of initiation sites can be favored by negatively supercoiled DNA (Remus et al. 2004) (possibly resulting from the removal or displacement of nucleosomes), by interacting proteins that chaperone ORC to specific chromatin sites (Schepers et al. 2001), by the transcriptional activity (Danis et al. 2004) or by open chromatin to which ORC might bind in a non-specific way (Vashee et al. 2003). A recent study performed on 283 replication origins identified in the ENCODE pilot regions showed that, besides a strong association with CGIs, only 29% overlap a DNase I hypersensitive site and that half of these origins do not present open chromatin epigenetic marks and are not associated with active transcription (Cadoret et al. 2008). The particular open chromatin state associated with N-domain borders suggest that these putative early replication origins present properties that are only shared by a subset of origins. These properties likely contribute to the specification of this peculiar subset of origins. The typical inter-origin distance in the human somatic cells has been estimated to be of the order of 50–100 kbp (Conti et al. 2007; Cadoret et al. 2008), a value significantly smaller than the typical size (1 Mbp) of N-domains. We propose that replication would initiate in early S phase at these privileged open chromatin locations and that the replication timing gradients observed from N-domain borders (Audit et al. 2007; Huvet et al. 2007; Baker et al. 2012) would correspond to the diverging replication forks progression triggering secondary origins in a "domino cascade" manner (Hyrien and Goldar 2010; Arneodo et al. 2011). As structural defects (bursts of "openness") in the chromatin fiber, these replication initiation zones might also be central to the tertiary structure of eukaryotic chromatin into rosette-like structures (St-Jean et al. 2008). The present data suggest that they are likely to be associated with structuring chromatin elements playing an essential role in the spatio-temporal replication program.

3.6 Multi-scale Analysis of Compositional Asymmetry and Replication Timing Profiles

3.6.1 Delineating N-shaped Replication Domains Using Wavelets

The continuous wavelet transform (WT) is a space-scale analysis which consists in expanding signals in terms of wavelets that are constructed from a single function, the analyzing wavelet ψ, by means of dilations and translations (Muzy et al. 1994; Arneodo et al. 1995, 2002; Mallat 1998):

$$T_\psi[S](b,a) = \frac{1}{a}\int S(x)\psi\left(\frac{b-x}{a}\right)dx. \qquad (3.1)$$

where b and a (>0) are the space and scale parameters respectively. The wavelet coefficient $T_\psi[S](b,a)$ quantifies to which extent, around position b over a distance a, S has a similar shape as the analyzing wavelet ψ. Thus, using an adapted N-shaped analyzing wavelet constituted by a linearly decreasing segment between two upward jumps ($\psi(x) = -(x-1/2)$ for x [0,1] and 0 elsewhere, Fig. 3.5c), we deploy the WT as an objective segmentation strategy of the human genome into candidate replication domains where the skew S displays a characteristic N-shaped pattern (Figs. 3.1b and 3.2a) (Audit et al. 2007; Huvet et al. 2007; Baker et al. 2010). The space-scale location of significant maxima values in the 2d WT decomposition (red areas in Fig. 3.5b) indicates the middle position (spatial location) of candidate replication domains whose size is given by the scale location. In order to avoid false positives, we then check that there does exist a well-defined upward jump at each domain extremity. These jumps appear in Fig. 3.5b as blue cone-shape areas pointing at small scale to the upward jumps positions where are located the putative replication origins.

But, the overall observed skew S also contains some contribution induced by transcription that generates step-like blocks corresponding to (+) and (−) genes (Touchon et al. 2003, 2004; Huvet et al. 2007; Nicolay et al. 2007). Hence, when superimposing the replication N-shaped and transcription step-like skew profiles, we get the following theoretical skew profile in a replication domain (Audit et al. 2007; Baker et al. 2010):

$$S(x') = S_R(x') + S_T(x') = -2\delta \times (x' - 1/2) + \sum_{\text{gene}} c_g \chi_g(x'), \qquad (3.2)$$

where position x' within the domain has been rescaled between 0 and 1, $\delta > 0$ is the replication bias, χ_g is the characteristic function for the gth gene (1 when x' points within the gene and 0 elsewhere) and c_g is its transcriptional bias calculated on the Watson strand (likely to be positive for (+) genes and negative for (−) genes). The objective is thus to detect human replication domains by delineating, in the noisy S profile obtained at 1 kbp resolution (Fig. 3.2a), all chromosomal loci where S is well fitted by the theoretical skew profile Eq. 3.2. In order to enforce strong compatibility with the mammalian replicon model (Sect. 3.2), we only retain the domains the most likely to be bordered by putative replication origins, namely those that are delimited by upward jumps corresponding to a transition from a negative S value $< -3\%$ to a positive S value $> +3\%$. Also, for each domain so-identified, we use a least-square fitting procedure to estimate the replication bias δ, and each of the gene transcription bias c_g. The resulting χ^2 value is then used to select the candidate domains where the noisy S profile is well described by Eq. 3.2.

3.6.2 Multiscale Detection of Peaks in Replication Timing Profiles

The simple intuitive idea allowing for effective detection of peaks in a noisy profile f is to delineate positions x along the signal that are a local extrema $(f'(x) \sim 0)$ and present a strong (negative) curvature $(f''(x) \ll 0)$ as expected at the tip of a peak symmetrical about a vertical axis. In order to avoid the confusion between "true" peaks and those induced by the presence of a noisy background, the rates of signal variation have to be estimated over a sufficiently large number of data points. This can be achieved using the continuous wavelet transform (WT) which provides a powerful framework for the estimation of signal variations over different length scales (Mallat 1998; Arneodo et al. 2002).

When using as the analyzing wavelet the derivatives of the Gaussian function, namely $g^{(n)}(x) = d^n g^{(0)}(x)/dx^n$, with $g^{(0)}(x) = \frac{1}{\sqrt{2\pi}} e^{-x^2/2}$, then the WT of a profile f takes the following expression:

$$T_{g^{(n)}}[f](x,a) = a^n \frac{d^n}{dx^n}(g_a^{(0)} * f)(x), \qquad (3.3)$$

where x and a (>0) are the space and scale parameters respectively. Equation 3.3 shows that the WT computed with $g^{(n)}$ is proportional to the nth derivative of the profile f smoothed by a dilated version $g_a^{(0)}(x) = \frac{1}{a} g^{(0)}(x/a)$ of the Gaussian function. This property is at the heart of various applications of the WT microscope as a very efficient multi-scale singularity tracking technique (Mallat 1998; Arneodo et al. 2002). When the profile of f is the graph of a Brownian motion i.e. the increments of f are independent, identically distributed Gaussian variables, then the WT at scale a is Gaussian with a standard deviation proportional to $a^{1/2}$. Hence, the amplitude of the fluctuations of $a^{-1/2}T_g^{(n)}[f](x,a) \sim \mathcal{N}(0,\sigma_o)$ are independent of the scale of analysis.

The basic principle of the detection of peaks in the replication timing profiles with the WT is illustrated in Fig. 3.6. In a first step, we determine (i) the regions of the space-scale half plane candidates to be a local extrema applying the following thresholding of the WT of f using $g^{(1)}$: $|a^{-1/2}T_g^{(1)}[f](x,a)| < c_1$ (Fig. 3.6b) and (ii) the regions of strong concavity applying the following thresholding of the WT of f using $g^{(2)}$: $a^{-1/2}T_g^{(2)}[f](x,a) < -c_2$ (Fig. 3.6c). In this way, both thresholds on the first and second derivatives are uniform with respect to the fluctuations for a Brownian profile. In a second step, we determine the connected regions of the space-scale half plane where both requirements are fulfilled (color regions in Fig. 3.6a). Finally, connected regions that have a scale extension (ratio between the region largest and smallest scales) smaller than 1.74 are disregarded in order to guaranty the existence of a well defined peak robust with respect to the scale of observation.

3.7 Material

3.7.1 Sequence and Annotation Data

Sequence and annotation data were retrieved from the Genome Browsers of the University of California Santa Cruz (UCSC) (Karolchik et al. 2003). Analyses were performed using the human genome assembly of March 2006 (NCBI36 or hg18). As human gene coordinates, we used the UCSC Known Genes table. When several genes presenting the same orientation overlapped, they were merged into one gene whose coordinates corresponded to the union of all the overlapping gene coordinates, resulting in 23,818 distinct genes. We used CpG islands (CGIs) annotation provided in UCSC table "cpgIslandExt".

3.7.2 CpG Observed/Expected Ratio

CpG observed/expected ratio (CpG o/e) was computed as $\frac{n_{CpG}}{L-l} \times \frac{L^2}{n_C n_G}$, where n_C, n_G and n_{CpG} are the number of C, G and dinucleotides CG counted along the sequence, L is the number of non-masked nucleotides of the sequence and l the number of masked nucleotide gaps plus one, i.e., $L-l$ is the number of dinucleotide sites.

3.7.3 Determining Mean Replication Timing Profiles

We determined the mean replication timing profiles along the complete human genome using Repli-Seq data (Hansen et al. 2010; Chen et al. 2010). For ESC cell line (BG02), a lymphoblastoid cell lines (GM06990), a fibroblast cell line (BJ), and erythroid K562 cell line, Repli-Seq tags for 6 FACS fractions were obtained directly from the authors (Hansen et al. 2010). For the HeLa cell line we computed the mean replication timing (MRT) instead of computing the S50 (median replication timing) as in (Chen et al. 2010).

3.7.4 DNase I Hypersensitive Site Data

We used the DNase I sensitivity measured genome-wide (Sabo et al. 2006). Data corresponding to Release 3 (Jan 2010) of the ENCODE UW DNaseI HS track, were downloaded from the UCSC FTP site: ftp://hgdownload.cse.ucsc.edu/goldenPath/hg18/encodeDCC/wgEncodeUwDnaseSeq/.

We plotted the coverage by DNase I hypersentive sites identified as signal peaks at a false discovery rate threshold of 0.5% within hypersensitive zones delineated using the HotSpot algorithm ("wgEncodeUwDnaseSeqPeaks" tables). When several replicates were available, data were merged.

3.7.5 Genome-Wide Nucleosome Positioning Data

We used the genome-wide map of nucleosome positioning in resting human CD4⁺ T cells obtained from direct sequencing of nucleosome ends using the Solexa high-throughput sequencing technique (Schones et al. 2008). Nucleosome score profiles for human genome assembly hg18 were downloaded from http://dir.nhlbi.nih.gov/papers/lmi/epigenomes/hgtcellnucleosomes.html.

3.7.6 CpG-Rich and CpG-Poor Promoters

We noticed that several promoters that were not overlapping a CGI yet had an enrichment in CpG content at the promoter and displayed the same characteristics in terms of Pol II binding and H3K4me3 enrichment as genes with a CGI associated promoter. Hence following previous work (Saxonov et al. 2006), we considered the CpG o/e at the promoter calculated in a 1 kbp window centered on the TSS rather than its association to a tabulated CGI, whose definition is based on *ad-hoc* criteria. Thanks to a clear bimodal distribution of CpG o/e at the promoters, we could objectively separate genes based on CpG-rich (CpG o/e > 0.48) and CpG-poor (CpG o/e < 0.48) promoters. According to this criteria 65% of the genes are CpG-rich.

Acknowledgements We thank G. Chevereau, G. Guilbaud, O. Hyrien, H. Julienne, O. Rappailles and C. Vaillant for helpful discussions. This work was supported by ACI IMPBio2004, the PAI Tournesol and the Agence Nationale de la Recherche under project HUGOREP (ANR PCV 2005) and REFOPOL (ANR BLANC SVSE6).

References

Antequera F, Bird A (1999) CpG islands as genomic footprints of promoters that are associated with replication origins. Curr Biol 9:R661–R667

Arneodo A, Audit B, Decoster N, Muzy JF, Vaillant C (2002) Wavelet based multifractal formalism: application to DNA sequences, satellite images of the cloud structure and stock market data.

In: Bunde A, Kropp J, Schellnhuber HJ (eds) The science of disasters: climate disruptions, heart attacks, and market crashes. Springer, Berlin, pp 26–102
Arneodo A, Bacry E, Muzy JF (1995) The thermodynamics of fractals revisited with wavelets. Physica A 213:232–275
Arneodo A, d'Aubenton-Carafa Y, Audit B, Brodie of Brodie EB, Nicolay S, St-Jean P, Thermes C, Touchon M, Vaillant C (2007) DNA in chromatin: from genome-wide sequence analysis to the modeling of replication in mammals. Adv Chem Phys 135:203–252
Arneodo A, Vaillant C, Audit B, Argoul F, d'Aubenton Carafa Y, Thermes C (2011) Multi-scale coding of genomic information: from DNA sequence to genome structure and function. Phys Rep 498:45–188
Audit B, Nicolay S, Huvet M, Touchon M, d'Aubenton-Carafa Y, Thermes C, Arneodo A (2007) DNA replication timing data corroborate in silico human replication origin predictions. Phys Rev Lett 99:248102
Audit B, Zaghloul L, Vaillant C, Chevereau G, d'Aubenton-Carafa Y, Thermes C, Arneodo A (2009) Open chromatin encoded in DNA sequence is the signature of "master" replication origins in human cells. Nucleic Acids Res 37:6064–6075
Baker A, Audit B, Chen CL, Moindrot B, Leleu A, Guilbaud G, Rappailles A, Vaillant C, Goldar A, Mongelard F et al (2012) Replication fork polarity gradients revealed by megabase-sized U-shape replication timing domains in human cell lines. PLOS Comput Biol 8:e1002443
Baker A, Nicolay S, Zaghloul L, d'Aubenton-Carafa Y, Thermes C, Audit B, Arneodo A (2010) Wavelet-based method to disentangle transcription- and replication-associated strand asymmetries in mammalian genomes. Appl Comput Harmon Anal 28:150–170
Barski A, Cuddapah S, Cui K, Roh TY, Schones DE, Wang Z, Wei G, Chepelev I, Zhao K (2007) High-resolution profiling of histone methylations in the human genome. Cell 129:823–837
Bell SP, Dutta A (2002) DNA replication in eukaryotic cells. Annu Rev Biochem 71:333–374
Bernardi G (2001) Misunderstandings about isochores. Part 1. Gene 276:3–13
Bird A (2002) DNA methylation patterns and epigenetic memory. Genes Dev 16:6–21
Bird AP, Wolffe AP (1999) Methylation-induced repression–belts, braces, and chromatin. Cell 99:451–454
Bogan JA, Natale DA, Depamphilis ML (2000) Initiation of eukaryotic DNA replication: conservative or liberal? J Cell Physiol 184:139–150
Boyle AP, Davis S, Shulha HP, Meltzer P, Margulies EH, Weng Z, Furey TS, Crawford GE (2008) High-resolution mapping and characterization of open chromatin across the genome. Cell 132:311–322
Brodie of Brodie EB, Nicolay S, Touchon M, Audit B, d'Aubenton-Carafa Y, Thermes C, Arneodo A (2005) From DNA sequence analysis to modeling replication in the human genome. Phys Rev Lett 94:248103
Bulmer M (1991) Strand symmetry of mutation rates in the beta-globin region. J Mol Evol 33:305–310
Cadoret JC, Meisch F, Hassan-Zadeh V, Luyten I, Guillet C, Duret L, Quesneville H, Prioleau MN (2008) Genome-wide studies highlight indirect links between human replication origins and gene regulation. Proc Natl Acad Sci USA 105:15837–15842
Chen CL, Duquenne L, Audit B, Guilbaud G, Rappailles A, Baker A, Huvet M, d'Aubenton Carafa Y, Hyrien O, Arneodo A et al (2011) Replication-associated mutational asymmetry in the human genome. Mol Biol Evol 28:2327–2337
Chen CL, Rappailles A, Duquenne L, Huvet M, Guilbaud G, Farinelli L, Audit B, d'Aubenton Carafa Y, Arneodo A, Hyrien O et al (2010) Impact of replication timing on non-CpG and CpG substitution rates in mammalian genomes. Genome Res 4:447–457
Chevereau G, Palmeira L, Thermes C, Arneodo A, Vaillant C (2009) Thermodynamics of intragenic nucleosome ordering. Phys Rev Lett 103:188103
Conti C, Sacca B, Herrick J, Lalou C, Pommier Y, Bensimon A (2007) Replication fork velocities at adjacent replication origins are coordinately modified during DNA replication in human cells. Mol Biol Cell 18:3059–3067
Courbet S, Gay S, Arnoult N, Wronka G, Anglana M, Brison O, Debatisse M (2008) Replication fork movement sets chromatin loop size and origin choice in mammalian cells. Nature 455:557–560

Coverley D, Laskey RA (1994) Regulation of eukaryotic DNA replication. Annu Rev Biochem 63:745–776

Danis E, Brodolin K, Menut S, Maiorano D, Girard-Reydet C, Méchali M (2004) Specification of a DNA replication origin by a transcription complex. Nat Cell Biol 6:721–730

Demeret C, Vassetzky Y, Méchali M (2001) Chromatin remodelling and DNA replication: from nucleosomes to loop domains. Oncogene 20:3086–3093

Desprat R, Thierry-Mieg D, Lailler N, Lajugie J, Schildkraut C, Thierry-Mieg J, Bouhassira EE (2009) Predictable dynamic program of timing of DNA replication in human cells. Genome Res 19:2288–2299

Eckhardt F, Lewin J, Cortese R, Rakyan VK, Attwood J, Burger M, Burton J, Cox TV, Davies R, Down TA, et al (2006) DNA methylation profiling of human chromosomes 6, 20 and 22. Nat Genet 38:1378–1385

Farkash-Amar S, Lipson D, Polten A, Goren A, Helmstetter C, Yakhini Z, Simon I (2008) Global organization of replication time zones of the mouse genome. Genome Res 18:1562–1570

Francino MP, Ochman H (2000) Strand symmetry around the beta-globin origin of replication in primates. Mol Biol Evol 17:416–422

Frank AC, Lobry JR (1999) Asymmetric substitution patterns: a review of possible underlying mutational or selective mechanisms. Gene 238:65–77

Gerbi SA, Bielinsky AK (1997) Replication initiation point mapping. Methods 13:271–280

Gerbi SA, Bielinsky AK (2002) DNA replication and chromatin. Curr Opin Genet Dev 12:243–248

Gierlik A, Kowalczuk M, Mackiewicz P, Dudek MR, Cebrat S (2000) Is there replication-associated mutational pressure in the Saccharomyces cerevisiae genome? J Theor Biol 202:305–314

Gilbert DM (2001) Making sense of eukaryotic DNA replication origins. Science 294:96–100

Gilbert DM (2004) In search of the holy replicator. Nat Rev Mol Cell Biol 5:848–855

Goldman MA, Holmquist GP, Gray MC, Caston LA, Nag A (1984) Replication timing of genes and middle repetitive sequences. Science 224:686–692

Hamlin JL, Mesner LD, Lar O, Torres R, Chodaparambil SV, Wang L (2008) A revisionist replicon model for higher eukaryotic genomes. J Cell Biochem 105:321–329

Hansen RS, Thomas S, Sandstrom R, Canfield TK, Thurman RE, Weaver M, Dorschner MO, Gartler SM, Stamatoyannopoulos JA (2010) Sequencing newly replicated DNA reveals widespread plasticity in human replication timing. Proc Natl Acad Sci USA 107:139–144

Heintzman ND, Stuart RK, Hon G, Fu Y, Ching CW, Hawkins RD, Barrera LO, Calcar SV, Qu C, Ching KA et al (2007) Distinct and predictive chromatin signatures of transcriptional promoters and enhancers in the human genome. Nat Genet 39:311–318

Hiratani I, Ryba T, Itoh M, Rathjen J, Kulik M, Papp B, Fussner E, Bazett-Jones DP, Plath K, Dalton S et al (2010) Genome-wide dynamics of replication timing revealed by in vitro models of mouse embryogenesis. Genome Res 20:155–169

Hiratani I, Ryba T, Itoh M, Yokochi T, Schwaiger M, Chang CW, Lyou Y, Townes TM, Schubeler D, Gilbert DM (2008) Global reorganization of replication domains during embryonic stem cell differentiation. PLoS Biol 6:e245

Huvet M, Nicolay S, Touchon M, Audit B, d'Aubenton-Carafa Y, Arneodo A, Thermes C (2007) Human gene organization driven by the coordination of replication and transcription. Genome Res 17:1278–1285

Hyrien O, Goldar A (2010) Mathematical modelling of eukaryotic DNA replication. Chromosome Res 18:147–161

Hyrien O, Méchali M (1993) Chromosomal replication initiates and terminates at random sequences but at regular intervals in the ribosomal DNA of Xenopus early embryos. EMBO J 12:4511–4520

Jacob F, Brenner S, Cuzin F (1963) On the regulation of DNA replication in bacteria. Cold Spring Harb Symp Quant Biol 28:329–342

Karnani N, Taylor C, Malhotra A, Dutta A (2007) Pan-S replication patterns and chromosomal domains defined by genome-tiling arrays of ENCODE genomic areas. Genome Res 17:865–876

Karolchik D, Baertsch R, Diekhans M, Furey TS, Hinrichs A, Lu YT, Roskin KM, Schwartz M, Sugnet CW, Thomas DJ et al (2003) The UCSC genome browser database. Nucleic Acids Res 31:51–54

Lee W, Tillo D, Bray N, Morse RH, Davis RW, Hughes TR, Nislow C (2007) A high-resolution atlas of nucleosome occupancy in yeast. Nat Genet 39:1235–1244

Lemaitre JM, Danis E, Pasero P, Vassetzky Y, Mechali M (2005) Mitotic remodeling of the replicon and chromosome structure. Cell 123:787–801

Lemaitre C, Tannier E, Gautier C, Sagot MF (2008) Precise detection of rearrangement breakpoints in mammalian chromosomes. BMC Bioinformatics 9:286

Lemaitre C, Zaghloul L, Sagot MF, Gautier C, Arneodo A, Tannier E, Audit B (2009) Analysis of fine-scale mammalian evolutionary breakpoints provides new insight into their relation to genome organisation. BMC Genomics 10:335

Lobry JR (1995) Properties of a general model of DNA evolution under no-strand-bias conditions. J Mol Evol 40:326–330

Lobry JR (1996) Asymmetric substitution patterns in the two DNA strands of bacteria. Mol Biol Evol 13:660–665

MacAlpine DM, Rodriguez HK, Bell SP (2004) Coordination of replication and transcription along a Drosophila chromosome. Genes Dev 18:3094–3105

Macleod D, Ali RR, Bird A (1998) An alternative promoter in the mouse major histocompatibility complex class II I-Abeta gene: implications for the origin of CpG islands. Mol Cell Biol 18:4433–4443

Mallat S (1998) A wavelet tour of signal processing. Academic, New York

Mavrich TN, Ioshikhes IP, Venters BJ, Jiang C, Tomsho LP, Qi J, Schuster SC, Albert I, Pugh BF (2008) A barrier nucleosome model for statistical positioning of nucleosomes throughout the yeast genome. Genome Res 18:1073–1083

McNairn AJ, Gilbert DM (2003) Epigenomic replication: linking epigenetics to DNA replication. Bioessays 25:647–656

Méchali M (2001) DNA replication origins: from sequence specificity to epigenetics. Nat Rev Genet 2:640–645

Mesner LD, Crawford EL, Hamlin JL (2006) Isolating apparently pure libraries of replication origins from complex genomes. Mol Cell 21:719–726

Miele V, Vaillant C, d'Aubenton-Carafa Y, Thermes C, Grange T (2008) DNA physical properties determine nucleosome occupancy from yeast to fly. Nucleic Acids Res 36:3746–3756

Mrázek J, Karlin S (1998) Strand compositional asymmetry in bacterial and large viral genomes. Proc Natl Acad Sci USA 95:3720–3725

Muzy JF, Bacry E, Arneodo A (1994) The multifractal formalism revisited with wavelets. Int J Bifurc Chaos 4:245–302

Nicolay S, Brodie of Brodie EB, Touchon M, Audit B, d'Aubenton-Carafa Y, Thermes C, Arneodo A (2007) Bifractality of human DNA strand-asymmetry profiles results from transcription. Phys Rev E 75:032902

Ozsolak F, Song JS, Liu XS, Fisher DE (2007) High-throughput mapping of the chromatin structure of human promoters. Nat Biotechnol 25:244–248

Ponger L, Duret L, Mouchiroud D (2001) Determinants of CpG islands: expression in early embryo and isochore structure. Genome Res 11:1854–1860

Remus D, Beall EL, Botchan MR (2004) DNA topology, not DNA sequence, is a critical determinant for Drosophila ORC-DNA binding. EMBO J 23:897–907

Rocha EP, Danchin A, Viari A (1999) Universal replication biases in bacteria. Mol Microbiol 32:11–16

Ryba T, Hiratani I, Lu J, Itoh M, Kulik M, Zhang J, Schulz TC, Robins AJ, Dalton S, Gilbert DM (2010) Evolutionarily conserved replication timing profiles predict long-range chromatin interactions and distinguish closely related cell types. Genome Res 20:761–770

Sabo PJ, Kuehn MS, Thurman R, Johnson BE, Johnson EM, Cao H, Yu M, Rosenzweig E, Goldy J, Haydock A et al (2006) Genome-scale mapping of DNase I sensitivity in vivo using tiling DNA microarrays. Nat Methods 3:511–518

Sasaki T, Sawado T, Yamaguchi M, Shinomiya T (1999) Specification of regions of DNA replication initiation during embryogenesis in the 65-kilobase DNApolalpha-dE2F locus of Drosophila melanogaster. Mol Cell Biol 19:547–555

Saxonov S, Berg P, Brutlag DL (2006) A genome-wide analysis of CpG dinucleotides in the human genome distinguishes two distinct classes of promoters. Proc Natl Acad Sci USA 103:1412–1417

Schepers A, Ritzi M, Bousset K, Kremmer E, Yates JL, Harwood J, Diffley JF, Hammerschmidt W (2001) Human origin recognition complex binds to the region of the latent origin of DNA replication of epstein-barr virus. EMBO J 20:4588–4602

Schones DE, Cui K, Cuddapah S, Roh TY, Barski A, Wang Z, Wei G, Zhao K (2008) Dynamic regulation of nucleosome positioning in the human genome. Cell 132:887–898

Schübeler D, Scalzo D, Kooperberg C, van Steensel B, Delrow J, Groudine M (2002) Genome-wide DNA replication profile for Drosophila melanogaster: a link between transcription and replication timing. Nat Genet 32:438–442

Schwaiger M, Schubeler D (2006) A question of timing: emerging links between transcription and replication. Curr Opin Genet Dev 16:177–183

Sequeira-Mendes J, Diaz-Uriarte R, Apedaile A, Huntley D, Brockdorff N, Gomez M (2009) Transcription initiation activity sets replication origin efficiency in mammalian cells. PLoS Genet 5:e1000446

St-Jean P, Vaillant C, Audit B, Arneodo A (2008) Spontaneous emergence of sequence-dependent rosettelike folding of chromatin fiber. Phys Rev E 77:061923

Suzuki MM, Bird A (2008) DNA methylation landscapes: provocative insights from epigenomics. Nat Rev Genet 9:465–476

The ENCODE Project Consortium (2007) Identification and analysis of functional elements in 1% of the human genome by the ENCODE pilot project. Nature 447:799–816

Tillier ER, Collins RA 2000 The contributions of replication orientation, gene direction, and signal sequences to base-composition asymmetries in bacterial genomes. J Mol Evol 50:249–257

Touchon M, Arneodo A, d'Aubenton-Carafa Y, Thermes C (2004) Transcription-coupled and splicing-coupled strand asymmetries in eukaryotic genomes. Nucleic Acids Res 32:4969–4978

Touchon M, Nicolay S, Arneodo A, d'Aubenton-Carafa Y, Thermes C (2003) Transcription-coupled TA and GC strand asymmetries in the human genome. FEBS Lett 555:579–582

Touchon M, Nicolay S, Audit B, Brodie of Brodie EB, d'Aubenton-Carafa Y, Arneodo A, Thermes C (2005) Replication-associated strand asymmetries in mammalian genomes: toward detection of replication origins. Proc Natl Acad Sci USA 102:9836–9841

Vaillant C, Audit B, Arneodo A (2007) Experiments confirm the influence of genome long-range correlations on nucleosome positioning. Phys Rev Lett 99:218103

Vaillant C, Palmeira L, Chevereau G, Audit B, d'Aubenton-Carafa Y, Thermes C, Arneodo A (2010) A novel strategy of transcription regulation by intra-genic nucleosome ordering. Genome Res 20:59–67

Vashee S, Cvetic C, Lu W, Simancek P, Kelly TJ, Walter JC (2003) Sequence-independent DNA binding and replication initiation by the human origin recognition complex. Genes Dev 17:1894–1908

Woodfine K, Beare DM, Ichimura K, Debernardi S, Mungall AJ, Fiegler H, Collins VP, Carter NP, Dunham I (2005) Replication timing of human chromosome 6. Cell Cycle 4:172–176

Yaffe E, Farkash-Amar S, Polten A, Yakhini Z, Tanay A, Simon I (2010) Comparative analysis of DNA replication timing reveals conserved large-scale chromosomal architecture. PLoS Genet 6:e1001011

Yuan GC, Liu YJ, Dion MF, Slack MD, Wu LF, Altschuler SJ, Rando OJ (2005) Genome-scale identification of nucleosome positions in S. cerevisiae. Science 309:626–630

Chapter 4
Role of DNA Methyltransferases in Epigenetic Regulation in Bacteria

Ritesh Kumar and Desirazu N. Rao

Abstract In prokaryotes, alteration in gene expression was observed with the modification of DNA, especially DNA methylation. Such changes are inherited from generation to generation with no alterations in the DNA sequence and represent the epigenetic signal in prokaryotes. DNA methyltransferases are enzymes involved in DNA modification and thus in epigenetic regulation of gene expression. DNA methylation not only affects the thermodynamic stability of DNA, but also changes its curvature. Methylation of specific residues on DNA can affect the protein-DNA interactions. DNA methylation in prokaryotes regulates a number of physiological processes in the bacterial cell including transcription, DNA mismatch repair and replication initiation. Significantly, many reports have suggested a role of DNA methylation in regulating the expression of a number of genes in virulence and pathogenesis thus, making DNA methlytransferases novel targets for the designing of therapeutics. Here, we summarize the current knowledge about the influence of DNA methylation on gene regulation in different bacteria, and on bacterial virulence.

4.1 Introduction

DNA methylation is known to play a critical role in epigenetic gene regulation in prokaryotes and eukaryotes (Wion and Casadesus 2006; Marinus and Casadesus 2009). While it is N^6 adenine methylation which brings about the epigenetic control in bacteria, it is methylation of cytosine at C^5 position in eukaryotes (Marinus and Casadesus 2009; Bestor 2000). The microenvironment in which bacteria grows is highly dynamic. Epigenetic regulation of gene expression helps the bacteria to cope up with the changing environment. Bacteria respond to changes in the environment,

R. Kumar • D.N. Rao (✉)
Department of Biochemistry, Indian Institute of Science, Bangalore 560012, India
e-mail: rithesh@biochem.iisc.ernet.in; dnrao@biochem.iisc.ernet.in

such as nutrient availability, temperature, pH, osmolarity by regulating gene expression (Wion and Casadesus 2006). A regulated expression of genes according to stimuli is critical for successful survival. For pathogenic bacteria it is very essential to cope up with host immune response and to colonize different microenvironments and regulated expression of some critical genes plays a vital role in enhancing the adaptability.

It has been known for a long time that DNA in different organisms has methylated bases N^6- methyladenine, C^5 methylcytosine and N^4 methylcytosine (Jeltsch 2002). All the three modified bases could be found in bacteria but eukaryotes are known to have only C^5 methylcytosine (Jeltsch 2002). Methylation of the bases does not affect the Watson/Crick pairing properties of adenine and cytosine. However, addition of the methyl group can be easily detected by proteins interacting with the DNA. Thus, methylation adds extra information on DNA without changing the sequence (Wion and Casadesus 2006).

It has been shown that C^5 cytosine methylation in eukaryotes plays significant role in epigenetic regulation (Nan et al. 1998; Jorgensen and Bird 2002; Klose and Bird 2006). Methylation of CG sites by eukaryotic DNA methyltransferases is involved in gene regulation. A number of genes in eukaryotes are known to have CG sites in their promoter region and the methylation status of CG sites affects the gene expression (Lewis et al. 1992; Klose and Bird 2006). Methylation of CG sites not only abrogates the binding of number of transcription factors, but methylated CG sites also recruit 5-methylcytosine binding proteins, which represses transcription (Nan et al. 1998). In prokaryotes it is the N^6 methyl adenine which is known to affect the DNA protein interaction and thus, affecting the gene expression (Casadesu's and Low 2006). DNA adenine methyltransferase (Dam) is the most studied methyltransferase involved in epigenetic regulation in bacteria. Dam recognizes 5' GATC 3' and methylates adenine on both strands (Fälker et al. 2007). It is the hemimethylated form of DNA formed just after the DNA replication of a fully methylated DNA, which acts as a signal by different DNA interacting proteins. For example, in case of replication it is SeqA protein which interacts with GATC sites in the hemimethylated DNA clustered in the origin of replication (*oriC*) and checks the replication initiation (Braun et al. 1985; Yamaki et al. 1988). Methyl directed mismatch repair protein MutH interacts with GATC sites in hemimethylated DNA and cuts the unmethylated strand to ensure that methylated strand is used as a parental strand for repair-associated DNA synthesis (Marinus 1996). Cell cycle regulated methyltransferase (CcrM) is another well studied DNA methyltransferase involved in gene regulation (Reisenauer et al. 1999; Reisenauer and Shapiro 2002). Sequencing of *Helicobacter pylori* strains have unveiled that *H. pylori* genome is rich in DNA methyltransferases associated with restriction enzymes. They also contain a number of solitary methyltransferases different from Dam and CcrM (McClain et al. 2009; Vitkute et al. 2001; Tomb et al. 1997). Recent findings have shown that methyltransferases associated with R-M systems can affect the gene expression possibly by promoter methylation (Srikhanta et al. 2005). A number of methyltransferases associated with differential gene expression are components of Type III R-M systems and few of them exhibit phase variation thus, adding an extra dimension in the gene regulation in bacteria (Srikhanta et al. 2005, 2010).

4.2 Restriction-Modification System in Bacteria

DNA methylation was discovered in the context of restriction- modification system (R-M). R-M systems were discovered as the consequence of observations made in the early 1950s on host-controlled variation of bacterial viruses (Luria and Human 1952; Bertani and Weigle 1953). Prokaryotic restriction-modification (R-M) systems were first recognized in *Escherichia coli* nearly 50 years ago (Arber and Dussoix 1962) and are now known to be ubiquitous among bacterial species. In general, R-M systems consist of two components having distinct enzymatic activities: first, a restriction endonuclease that cleaves DNA at a specific recognition sequence, and second, a DNA methyltransferase that methylates DNA at the same site and thus prevents cleavage by the cognate restriction enzyme (Raleigh and Brooks 1998). Restriction endonucleases occur ubiquitously among bacteria, archaea (Bickle and Kruger 1993; Raleigh and Brooks 1998; Roberts et al. 2003, 2007; Sistla and Rao 2004; Pingoud et al. 2005), in viruses of certain unicellular algae (Van Etten 2003), and they are usually accompanied by a modification enzyme of identical specificity. Together, the two activities form a R-M system- the prokaryotic equivalent of an immune system. Their main function is to protect their host genome against foreign DNA. The host DNA is resistant to cleavage, as these sites are modified by cognate methyltransferase. It was proposed that R-M system has functions other than host protection, including maintenance of species identity among bacteria (Jeltsch 2003) and generation of genetic variation (Arber 2000, 2002). Restriction endonucleases of *Chlorella* viruses may have a nutritive function by helping in degradation of host DNA or preventing infection of a cell by another virus (Van Etten 2003). Other functions have also been suggested, such as involvement in recombination and transposition (Carlson and Kosturko 1998; Heitman 1993; McKane and Milkman 1995). Many R-M systems can also be considered as selfish DNA elements (Naito et al. 1995; Kobayashi 2004). In general, bacteria and archaea harbour multiple types of R-M systems. For example, in *H. pylori* more than 20 putative R-M systems, comprising greater than 4% of the total genome, have been identified in two completely sequenced *H. pylori* strains (Lin et al. 2001). These enzymes represent the largest family of functionally related enzymes. Based on the number and organization of subunits, regulation of their expression, cofactor requirements, catalytic mechanism, and sequence specificity, restriction enzymes have been classified into four different types. The broader classification includes Types I, II, III, and IV (Wilson and Murray 1991; Roberts et al. 2003, 2007; Sistla and Rao 2004; Pingoud et al. 2005).

4.2.1 DNA Methyltransferases

Post-replicative base methylation is the most common DNA modification in bacteria. C^5-methylcytosine and N^6-methyladenine are found in the genomes of many

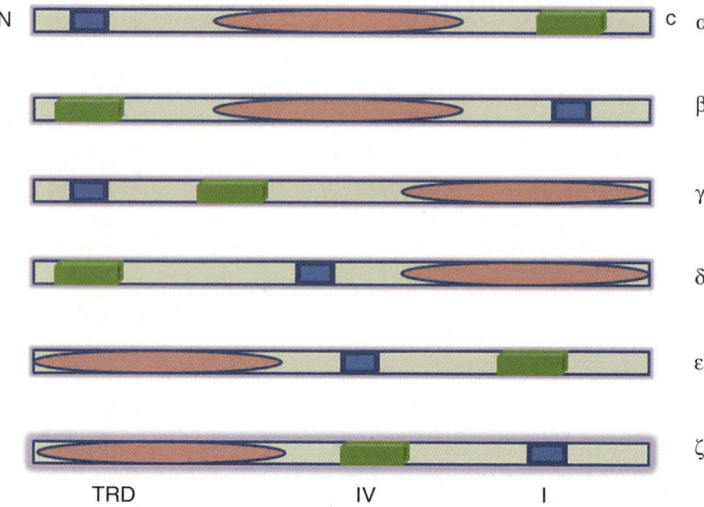

Fig. 4.1 Schematic diagram representing the arrangement of conserved motifs in the primary structure of DNA MTases. The linear arrangements of three consecutive motifs, AdoMet-binding domain, motif I; the catalytic domain, motif IV; the TRD, target recognition domain are shown

fungi, bacteria and protists, whereas N^4-methyl-cytosine is found only in bacteria (Cheng 1995). N^6-methyl-adenine is also present in archaeal DNA (Barbeyron et al. 1984; Cheng 1995). Two classes of DNA methyltransferases perform base modifications in bacterial genomes: those associated with R–M systems (Bickle and Kruger 1993), and solitary methyltransferases that do not have a restriction-enzyme counterpart. Based on the chemistry of the methylation reaction catalyzed, DNA methyltransferases are classified as endocyclic methyltransferases, which transfer the methyl group from AdoMet to C^5 position of cytosine (^{m5}C), and exocyclic amino methyltransferases, which transfer the methyl group to the exocyclic amino group of adenine (^{m6}A) or cytosine (^{m4}C), respectively (Bheemanaik et al. 2006). DNA methyltransferases of all three types contain conserved regions, which are responsible for catalysis and AdoMet binding, and variable regions known as target recognition domains (TRD), which determine the substrate specificity of a particular enzyme. Ten conserved amino acid motifs (I–X) were found in C^5 methyltransferases, the motif order being always constant (Posfai et al. 1989; Kumar et al. 1994). DNA methyltransferases are subdivided further into six groups (namely α, β, γ, ζ, δ and ε) (Fig. 4.1), according to the linear arrangements of three conserved motifs, the AdoMet-binding domain (FXGXG), the TRD (target recognition domain) and the catalytic domain (D/N/S)PP(Y/F) motifs (Malone et al. 1995; Bheemanaik et al. 2006).

4.2.2 Orphan DNA Methyltransferases

Most of the DNA methyltransferases known are the component of R-M systems. However, some of the methyltransferases lack cognate restriction enzymes (Casadesu's and Low 2006). They are termed as orphan or solitary methyltransferases. Solitary methyltransferases include the N^6-adenine methyltransferases Dam (**DNA a**denine **m**ethyltransferase) and CcrM (**c**ell **c**ycle- **r**egulated **m**ethyltransferase), and the Dcm (**DNA c**ytosine **m**ethyltransferase) (Marinus 1996; Reisenauer et al. 1999; Low et al. 2001; Lobner-Olesen et al. 2005; Wion and Casadesus 2006; Marinus and Casadesus 2009).

E. coli DNA adenine methyltransferase (EcoDam) recognizes the sequence 5'-GATC-3' and methylates at the N^6 position of the adenine (Geier and Modrich 1979). EcoDam showed preference towards the *in vivo* substrate hemimethylated DNA (Herman and Modrich 1982). EcoDam was able to methylate denatured DNA substrates and single stranded synthetic oligonucleotides with lesser rates. Methylation of the GATC sequence was shown to be modulated by the nature of the three base pairs flanking both sides of the site (Bergerat et al. 1989). However, it was demonstrated using short, synthetic duplexes, that EcoDam does not prefer hemimethylated DNA substrates to unmethylated substrates (Bergerat et al. 1989).

It was observed that a functional monomeric EcoDam methylates only one strand of the DNA in each binding event (Urig et al. 2002). They showed that EcoDam scans 3000 GATC sites per binding event randomly and methylates GATC sites on DNA processively. On the contrary, Mashhoon et al. (2004) demonstrated that *E. coli* Dam methylates adjacent GATC sites in a distributive manner and thus controls the transcription of the adjacent genes.

T4 DNA adenine methyltransferase (T4Dam) from bacteriophage T4, recognizes the sequence 5'-GATC-3' and methylates the adenine residue at the N^6 position (Schlagman and Hattman 1983; Hattman and Malygin 2004). It is capable of methylating unmethylated, hemimethylated GATC sequences and also methylates noncanonical sites but less efficiently (Schlagman et al. 1988; Kossykh et al. 1995). Cell-cycle-regulated methyltransferase (CcrM) is another N^6 adenine methyltransferase without a cognate restriction enzyme. CcrM has been identified in number of bacterial species and it recognizes the sequence 5' GANTC 3' and methylates adenine (Zweiger et al. 1994). Unlike Dam which belongs to γ -group, CcrM belongs to β-group of DNA methyltransferases, and shares homology with HinfI methylase of *Haemophilus influenzae* (Reisenauer et al. 1999).

Dcm is another orphan methyltransferase which methylates cytosine at C^5 position. The product of the *dcm* gene is the only DNA C^5-cytosine methyltransferase of *E. coil* K-12 which catalyses transfer of a methyl group to the C^5 position of the inner cytosine residue of the cognate sequence CCA/TGG. Till date there are no reports on the role of cytosine methylation in epigenetic regulation. However it has been shown that DNA cytosine- methyltransferase (Dcm) is associated with very short repair (*vsr*) endonuclease (Sohail et al. 1990).

4.3 DNA Methylation Dependent Regulatory Systems

Methylation of the specific DNA sequences by solitary methyltransferases is involved in bacterial physiology, pathogenesis, and host–pathogen interactions. It was proposed that N^6-methyl-adenine can act as an epigenetic signal for DNA–protein interactions in bacteria, much similar way as C^5-methylcytosine in eukaryotes (Wion and Casadesus 2006). Dam methylation provides signals for DNA protein interactions (Polaczek et al. 1998; Messer and Noyer-Weidner 1988). Dam is found to be involved in many of the vital cellular processes: Chromosome replication and nucleoid segregation, Dam-directed mismatch repair, regulation of transposition, phase variation, bacteriophage infection and bacterial conjugation (Fig. 4.2) (Marinus 1996; Reisenauer et al. 1999; Low et al. 2001; Lobner-Olesen et al. 2005).

Transposition of insertion sequences play an important role in the biology of bacteria. It has been shown that the methylation pattern near or in the transposon sequences control the rate of transposition (Dodson and Berg 1989; Reznikoff 1993). In IS*10* promoter, a GATC site overlaps with −10 module and a second GATC site is present at the end of the transposon. Methylation of GATC site in the promoter hinders the binding of RNA polymerase and methylation of GATC at the end of transposon inhibits tansposase activity at these ends (Roberts et al. 1985). Similar to IS*10*, transposition of IS*50* and Tn*5* is also controlled by DNA methylation (McCommas and Syvanen 1988; Dodson and Berg 1989).

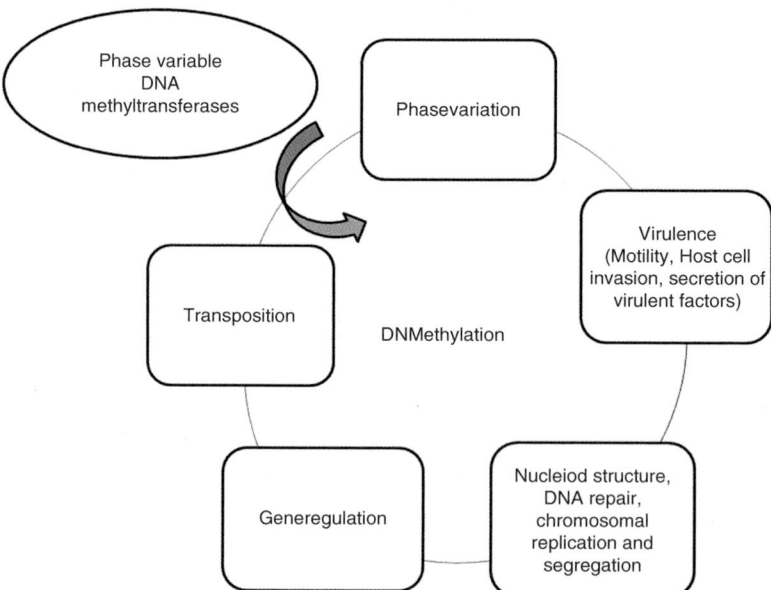

Fig. 4.2 Overview of the roles of N^6-methyl adenine in bacteria

A number of bacteriophages like T-even, PI, and Mu have *dam*, which methylate GATC sites in their genome (Blaisdell et al. 1996). It has been shown that methylation by Dam regulates the expression of *cre* gene of phage P1 and the *mom* gene of phage Mu (Sternberg et al. 1986; Hattman and Sun 1997). Dam methylation is also significant in packaging of P1 DNA into capsids. It has been shown that a P1 *dam⁻* mutant produces only 5% of the normal titre on infection of a *dam⁻* host (Yarmolinski and Sternberg 1988).

Dam methylation is important for the conjugal transfer of the virulence plasmid of *Salmonella enterica*. Dam methylation suppresses the transfer (*tra*) operon of the *Salmonella* virulence plasmid (pSLT) and F sex factor (Torreblanca et al. 1999). The effect of Dam methylation on *tra* operon is indirect. Dam methylation controls the expression of regulatory genes *traJ* and *finP*, which in turn affects the operon (Camacho and Casadesus 2002).

It was shown that Dam plays major role in virulence gene expression of *S. enterica* (Heithoff et al. 1999, 2001; Garcia-Del Portillo et al. 1999), *H. influenzae* (Watson et al. 2004), and *Yersinia pseudotuberculosis* (Taylor et al. 2005). Dam-overproducing strains of *Y. pseudotuberculosis* and *Yersinia enterocolitica* show increased secretion of Yops (Yersinia outer proteins), a group of virulence proteins that are involved in inhibition of phagocytosis and pro-inflammatory cytokine release (Julio et al. 2001, 2002; Fa¨lker et al. 2005). Overproduction of *E. coli* Dam methylase has also been shown to attenuate virulence in *Pasteurella multocida* (Chen et al. 2003). It was therefore proposed that Dam methylation might possibly regulate virulence gene expression in many other pathogens. It has been reported that mutation of the solitary N^6 adenine methyltransferase Cj1461 affects several phenotypes related to virulence in *Campylobacter jejuni*, suggesting that epigenetic regulation may play a role in *C. jejuni* pathogenesis (Kim et al. 2008). CcrM, a solitary methyltransferase is essential and participates in cell-cycle regulation in *Caulobacter crescentus*, *Brucella*, *Agrobacterium* and *Sinorhizobium*, (Reisenauer and Shapiro 2002; Zweiger et al. 1994; Marczynski and Shapiro 2002; Robertson et al. 2000). However, no biological function has yet been demonstrated for Dcm (Wion and Casadesus 2006). *E. coli* very short patch (*vsr*) gene is in a transcriptional unit with *dcm*. Intriguingly, deletion of *dam* in *E. coli* reduces VSP repair (Bell and Cupples 2001).

4.3.1 DNA Methylation: Critical for Replication and Repair

Initiation of DNA replication is highly regulated process in bacteria and DNA methylation plays a significant role in regulating the initiation process. In a number of bacteria it has been shown that Dam methylation is important for regulation of a number of processes, like replication and repair (Casadesu's and Low 2006). DnaA protein initiates the replication by binding to *oriC*. OriC and promoter of *dnaA* is rich in GATC sequence and the methylation status of these GATC sites controls the initiation of replication (Bogan and Helmstetter 1997; Campbell and Kleckner 1990). A first round of replication results in hemimethylated DNA and hemimethylated DNA acts as a binding site for SeqA protein. Sequestration by SeqA prevents the reinitiation of replication more than once per cell cycle (Lu et al. 1994). For an efficient initiation, *oriC* and *dnaA* promoter should be fully methylated. DNA methylation is therefore, not essential for DNA replication but regulates the timing of replication initiation. Hemimethylated state of *oriC* acts as a signal for nucleoid segregation. The hemimethylated origin of replication of both daughter chromosomes bind to proteins involved in segregation (Ogden et al. 1988). During DNA replication, DNA polymerase can incorporate wrong bases and this results in mismatch bases in DNA. DNA mismatch repair system corrects the base mismatches that can occur during the synthesis of new unmethylated strand (Au et al. 1992). DNA methylation by Dam acts as a marker for the repair machinery to distinguish between methylated parental strand and unmethylated newly synthesized strand. Any alteration in expression of Dam in bacterial cell (increase or decrease in Dam concentration in the cell) results in a mutator phenotype (Løbner-Olesen et al. 2003; Calmann and Marinus 2003; Oshima et al. 2002; Julio et al. 2002). Deletion of *dam* in bacterial cell results in the loss of strand discrimination by mismatch repair proteins, resulting in the use of parental strand as a template for repair with a 50% probability (Modrich and Lahue 1996; Chen et al. 2003). On the other hand, overproduction of Dam leads to premature methylation of newly synthesised strand, therefore preventing MutH action on a mismatch (Marinus 1996). It must be noted that Dam expression is tightly regulated in cell at the transcriptional level, involves multiple promoters, and is growth rate regulated (Marinus 1996).

4.3.2 Regulation of Gene Expression

DNA methylation can affect gene expression in two ways. First, methylation of cognate recognition sequences in the promoter regions can increase, decrease, or have no change in transcription initiation by affecting the interaction between RNA polymerase and promoters. A second mechanism involves the competition between Dam and regulatory proteins (Cap, Lrp or OxyR) for overlapping sites in or near promoters/ regulatory regions on DNA (Low et al. 2001). A number of studies with the knockout strains of *dam* have shown that the *dam* deletion affects the transcript levels of a number of genes involved in different pathways such as motility, virulence, lipid and aminoacid metabolism and cofactor biosynthesis (Table 4.1) (Heithoff et al. 2001; Oshima et al. 2002; Watson et al. 2004).

The CcrM methyltransferase from *Caulobacter crescentus* plays an essential role in cell cycle. *C. crescentus* has two different cell forms: replicating stalked form and non-replicating swarmer form (Marczynski and Shapiro 2002). Cell division yields one motile swarmer cell and one non-motile stalked cell. CcrM is only produced only at the late stages of replication in the stalked cell (Reisenauer et al. 1999). Activation of Caulobacter replication origin (*Cori*) requires complete methylation, thus synthesis of CcrM at the later stages of DNA replication is the signature for competition of one cell cycle. Methylation of *Cori* by CcrM of alpha-proteobacteria

Table 4.1 Roles of DNA methyltransferases in the physiology of different bacterial species

Bacterial pathogen	Methylase	Recognition sequence	Effect of deletion	Effect of overexpression
Escherichia coli	Dam	GATC	Suppresses virulence, enhances mutation rates	Enhances the mutation rate
Caulobacter crescentus	CcrM	GANTC	Decreases viability	Defects in cell division and morphology
Brucella abortus	CcrM	GANTC	Decreases viability	Inhibits growth
Salmonella enterica	Dam	GATC	Attenuates virulence	
Vibrio cholerae	Dam	GATC	Essential for viability	Inhibits colonization in suckling mouse model
Yersinia pseudotuberculosis	Dam	GATC	Essential for viability	Attenuates virulence in murine model
Nesseria meningitidis	Dam	GATC	Enhances the rate of phase variation	
Haemophilus influenzae	Dam	GATC	Strain specific reduction in invasion and adhesion	
Yersinia enterocolitica	Dam	GATC	Essential in certain strains	Increase in motility, alters lipopolysaccharide O-antigen
Helicobacter pylori strain PG227	M.HpyAVIA	GAGG	Defects in growth	
Helicobacter pylori	M.HpyI	CATG	Alteration in *dnaK* stress-responsive operon expression	
Klebsiella pneumoniae	Dam	GATC	Partial attenuation in mice infection model	
Seromonas hydrophila	Dam	GATC	Necessary for viability	Defects in virulence
Campylobacter jejuni	CJ1461		Reduced motility and reduced invasion	
Haemophilus influenzae	Mod subunit of Type III R-M system (Phase variable)		Affects transcript levels of *dnaK* and several genes coding for surface proteins	
N. gonorrhoeae strain FA1090	ModA13	AGAAA	Five of the differentially regulated genes have roles in virulence; four in oxidative stress and one in antimicrobial resistance	
Edwardsiella tarda	Dam	GATC	Significant attenuation of overall bacterial virulence and altered several stress responses including spontaneous mutation, recovering from UV radiation and H_2O_2 exposure	

provides a signal for the initiation of next round of replication in these bacteria (Wright et al. 1996). Bacterial cells achieve this stringent control by regulating the concentration of CcrM and CtrA (a global regulator). CtrA controls the expression of CcrM and methylation by CcrM controls its own expression and CtrA expression. One of the two promoters of *ctrA* contains a GANTC site near −35 module (Marczynski and Shapiro 2002).

A number of adenine methyltransferases have been identified in *H. pylori* and it has been suggested that they could be playing significant roles in the gene expression regulation. M.*HpyI*, is M.*NlaIII* homolog present in *H. pylori* and is highly conserved among all *H. pylori* strains. It encodes a 329-amino-acid protein which recognizes DNA at sites containing CATG (Xu et al. 2000). The transcript levels of *hpyIM* expression vary with *in vitro* growth phase, with higher expression being noted during exponential growth than during stationary phase. Inactivation of *hpyIM* results in pleiotropic bacterial morphology including alteration in the expression of the *H.pylori dnaK* stress-responsive operon (Donahue et al. 2002). Recently, it has been shown in *H. pylori* strains PG227 and 128, deletion of a N^6 adenine methyltransferase M.HpyAVIA resulted in slow growth (Kumar et al. 2010).

4.3.3 DNA Methylation Affects the Host-Pathogen Interaction

Several reports suggest that various virulence-associated phenotypes are influenced by DNA methylation and DNA methylation targets regulatory processes modulating the composition and function of bacterial surface and thus, affecting the bacterial interaction with host (Heusipp et al. 2006). The involvement of DNA methylation in controlling the virulence was first described for *S. enterica* serovar Typhimurium. Deletion of Dam decreases the efficiency of the strain to colonize mice. *dam*⁻ strains of *Salmonella* are attenuated and present pleiotropic virulence-related defects including membrane instability, leakage of proteins, defect in motility and release of vesicles. *S. enterica* lacking Dam were effective in colonization of mucosal sites but showed several defects in colonization of deeper tissues (Heithoff et al. 1999; Oza et al. 2005; Heusipp et al. 2006). Deletion of *dam* in *H. influenzae* results in similar defects as observed in *S. enterica*, including decrease in invasion of host cells and decrease in virulence (Watson et al. 2004; López-Garrido and Casadesús 2010). In *Klebsiella pneumoniae*, a mutation eliminating Dam methylation has been shown to result in partial attenuation in a mouse infection model (Mehling et al. 2006). Dam mutants of *Yersinia pestis* and *Salmonella* serovar Typhimurium show 2,000-fold and 10,000-fold increase in lethal dose compared to wild type strain in orally infected mice (Heithoff et al. 1999; Julio et al. 2001; Robinson et al. 2005). It has been shown that natural population of *Neisseria* contains two biotype *dam*⁻ and *dam*⁺. Deletion of *dam* in one biotype results in high rates of phase variation in surface associated components and thus, affects its interaction with the host (Table 4.1) (Bucci et al. 1999). *Edwardsiella tarda* is a serious aquaculture pathogen and infects many fishes. It has been shown that deletion of *dam* in *E. tarada* alters

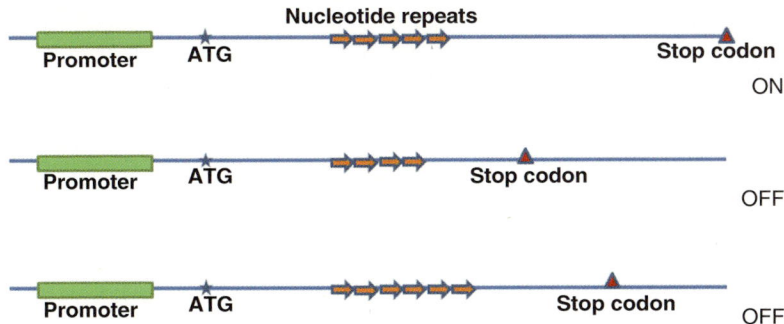

Fig. 4.3 Alteration in the number of repeats results in phase variation

the binding to host mucus (Sun et al. 2010) Deletion of *dam* erases the methylation pattern in the cells thus, affecting the number of protein- DNA interactions. It is not only the deletion of *dam* that affects the bacterial interaction with host but overexpression of Dam in *Vibrio cholera* and *P. multocida* have been shown to decrease virulence (Julio et al. 2001; Chen et al. 2003). Overexpression of Dam alters motility and invasiveness of *Y. enterocolitica* (Julio et al. 2001).

4.3.4 DNA Methylation and Phase Variation

A number of bacterial pathogens, such as *H. pylori*, *H. influenzae*, *Neisseria gonorrhoeae* and *Neisseria meningitidis* have evolved molecular mechanisms that can help them to adapt to the host. These organisms are capable of generating genetic variation, which in turn help them to cope up with the changing host environment and immune response. One common mechanism is phase variation, a reversible switch between an "all-or-none" (ON/OFF) expressing phase (Fig. 4.3), resulting in variation in the level of expression of one or more proteins between individual cells of a clonal population (Weiser et al. 1990; van Ham et al. 1993; Saunders et al. 1998; Hallet 2001). Phase variation results in two sub populations in a clonal population: one lacking or having a decreased level of expression of a phase variable gene and other population with full expression of the gene. The classical view of phase variation and antigenic variation is that its role is to help the bacterium evade the host immune system. This is supported by the fact that most of the genes that are subjected to phase variation are surface antigens such as lipopolysaccharide (LPS) and outer-membrane proteins (OMPs) where they would be exposed to the immune system (Saunders et al. 1998; Jennings et al. 1995; van der Ende et al. 1995) and play a vital role in bacteria and host interaction. Phase variation, therefore helps the pathogen in colonization of the host, adaptation to the dynamic host environment and evasion of immune responses (Moxon et al. 2006). Phase variation can occur through several mechanisms like (i) Short sequence repeats and slipped-strand mispairing (SSM) mechanisms (ii) Homologous (general) recombination

(iii) Site-specific recombination (iv) Environmental regulation and (v) Epigenetic Regulation (Moxon and Thaler 1997; Robertson and Meyer 1992; van der Woude and Baumler 2004; van der Woude 2006). In contrast to the other mechanisms mentioned above, epigenetic regulation of phase variation occurs in the absence of a change in DNA sequence. In epigenetic regulation, the phenotypes are altered but the genotypes are not. DNA methylation plays a critical role in this type of phase variation system, as the methylation state of a target sequence at a specific site in the chromosome affects the DNA binding of a regulatory protein that directly regulates transcription. *E.coli* and *S.entrica* serotype Typhimurium exhibit methylation-dependent phase variation (van der Woude 2006; Broadbent et al. 2010).

Pap phase variation and Ag43 phase variation in *E. coli* are two most studied systems. Pap is pyelonephritis-associated pili (Pap or P pili) and plays important role in virulence in urinary track infection (Braaten et al. 1994). *ag43* gene encodes the outer membrane protein Ag43, which causes autoaggregation, enhances biofilm formation, and may affect phage adsorption (Hallet 2001; Haagmans and van der Woude 2000). Expression of the *pap* operon and *ag43* varies and is dependent on the methylation status of Dam target sequences (GATC) present in their regulatory region (Braaten et al. 1994; Haagmans and van der Woude 2000). Pap expression further requires the global regulator leucine-responsive regulatory protein (Lrp), pap specific regulatory proteins PapI and PapB and the catabolite activator protein (CAP). Methylation status of two Dam sites in the regulatory region controls the binding of the above mentioned regulatory proteins and thus, transcription (Jafri et al. 2002; Casadesu's and Low 2006). *ag43* expression is controlled by the methylation status of three Dam sites present overlapping with the binding site of oxidative stress response protein (OxyR). OxyR is a negative regulator of *ag43* and methylation of GATC sites abrogates the binding of OxyR and results in the "ON" phase. Binding of the OxyR to the regulatory region in the absence of methylation establishes the "OFF" phase (Waldron et al. 2002). Methylation dependent phase variation depends on the concentration of regulatory proteins and Dam to DNA ratio in the cell. Recently it has been shown in *S. enterica*, the product of glycosyltransferase operons (*gtr*) involved in the modifications of the O-antigen that can affect the Serotype is under the control of phase variation. This phase variation occurs by a novel epigenetic mechanism requiring OxyR in conjunction with the DNA methyltransferase Dam (Broadbent et al. 2010). Till now methylation dependent phase variation is only known for Dam, but it could potentially also be mediated by methyltransferases belonging to R-M systems.

4.4 Phase Variable DNA Methyltransferase: A New Dimension

A number of specific roles have been proposed for phase variation. It has long been proposed that phase variation is a mechanism for immune evasion by pathogenic organisms, as the majority of phase-variable genes are predicted to be involved in the biosynthesis of surface structures. Interesting exceptions are genes encoding

R-M enzymes. Phase-variable expression of R-M enzymes has been found in a variety of bacterial pathogens, including, *Mycoplasma pulmonis* (Dybvig et al. 1998), *H. pylori* (Tomb et al. 1997; Alm et al. 1999; de Vries et al. 2002), *Pasteurella haemolytica* (Ryan and Lo 1999) and *H. influenzae* (De Bolle et al. 2000), *N. gonorrhoea* (Adamczyk-Poplawska et al. 2009), *N. meningitides* (Srikhanta et al. 2009, 2010), and *Morexella catarrahalis* (Seib et al. 2002).

Type III R-M systems represent an atypical class of phase variable genes. Type III R-M systems consist of two subunits, Mod (M), which is a functional DNA-methyltransferase, able to recognize and modify the target sequences, and Res (R), which is responsible for DNA cleavage, but functions only in a complex with the Mod subunit. Phase variable Type III R-M systems are present in a variety of pathogenic bacteria. The presence of repeats within a gene is a strong indicator that the gene is phase variable at a frequency determined by the hypermutability of the repeat regions. It has been experimentally proven in organisms, *H. influenzae* (De Bolle et al. 2000; Srikhanta et al. 2005) and *H. pylori* (de Vries et al. 2002) that the presence of repeats in R-M systems results in phase variation. Any alteration in the number of repeats can result in a frameshift mutation and thus, results in the synthesis of a truncated protein (Fig. 4.3). In turn, this results in a phase variable ON/OFF switching of Mod dependent regulons. This is believed to be an adaptive strategy used by bacterial populations facing fluctuating environmental conditions. From sequence analysis, it was predicted that Type III methyltransferases would undergo phase variation in pathogenic organisms like *Pasteurella haemolytica* (Ryan and Lo 1999), *N. meningitidis* (Fox et al. 2007; Adamczyk-Poplawska et al. 2009), *N. gonorrhoeae* (Saunders et al. 2000) and *Moraxella catarrhalis* (Seib et al. 2002). It has been shown that within a single strain of human pathogens, *N. meningitidis, N. gonorrhoeae, H. pylori* and *M. catarrhalis*, multiple phase variable Type III *mod* genes are present. Therefore, it was proposed that DNA restriction is not the only function of phase variable Type III R-M systems (Fox et al. 2007). Phasevarion is a group of genes whose expression is controlled by a phase variable DNA methyltransferase. A recent report suggests that multiple phasevarions exist within the pathogenic *Neisseria*, each regulating a different set of genes (Srikhanta et al. 2009). Both *N. meningitidis* and *N. gonorrhoeae* have two distinct *mod* genes-*modA* and *modB* and these genes switch independently. There are also distinct alleles of *modA* (major alleles include, *modA11, 12, 13*, and minor, *modA4, 15, 18*) and *modB* (*modB1, 2*). These alleles differ only in their DNA recognition domain. *modA11* was only found in *N. meningitidis* and *modA13* only in *N. gonorrhoeae*. ModA13 recognises and methylates 5'-AGAA^{m6}A-3'. When expressed, ModA13 methylates all AGAAA sites in the genome, and thereby controls the gene expression. It was shown that two strains with the same DNA recognition domain (*modA13* allele) regulated the same set of genes, while, *N. meningitides modA11* and *modA12* were found to regulate the expression of different sets of genes, consistent with differences in their DNA recognition domain (Srikhanta et al. 2009). It has been shown that *modA13* ON and OFF strains of *N. gonorrhoeae* have distinct phenotypes in antimicrobial resistance, differential virulence in a primary human cervical epithelial cell model of infection, and in biofilm formation. A number of DNA methyltransferases from Type II class in *H. pylori* have been predicted to exhibit phase variation (Salaun et al. 2004, 2005). Widespread distribution of phase variable R-M systems in host-adapted pathogenic bacteria suggests that this regulated random switching of multiple genes may be a commonly used strategy for generation of distinct, cell types with distinct niche specialization in host adapted bacterial pathogens. Why most of the phase variable methyltransferases found so far belongs to Type III R-M systems, is still an open question. Whether their unique genetic make up, cleavage specificities, or unusual properties of methyltransferases, dictate them to be phase variable, needs to be addressed. By modulating the activity of R-M systems by phase variation bacteria can regulate their ability to take up DNA. OFF phase R-M systems facilitate foreign DNA uptake and thus contributes to genetic variability. Alternately, the ON phase may limit the uptake of DNA from the surrounding. The balance between the two phases is critical for the survival and may be controlled by host and environmental factors. In case of Types I and III R-M systems, phase variation in Mod subunit can turn ON or OFF both the activities of the R-M system that is, methylation and DNA cleavage. On the other hand, in case of Type II R-M system phase variation in methyltransferase can create a lethal phenotype. In the absence of methyltransferase, the cognate restriction enzyme will cleave the DNA. Inactivation of modification components can cause continuous 'bacterial suicides' that may encourage intergenomic recombination and thus help in creating variability in bacteria.

4.5 Molecular Evolution of DNA Methyltransferases

DNA methyltransferases represent a highly diverse group of enzymes. Compared to methyltransferases acting on other substrates like RNA, lipid, protein and small molecules, DNA methyltransferases exhibit sequence permutation, despite similarity in structure. Amino acid sequence alignments of DNA methyltransferases have shown the presence of several conserved motifs. Motifs I-VIII and X are present in most subfamilies with a region of high variability (Malone et al. 1995). The variable region is implicated in recognition of the target sequence and termed as Target recognition domain (TRD). Motifs IV-VIII were involved in catalysis and motifs X and I-III form AdoMet-binding pocket (Schluckebier et al. 1995). Based on the sequential order of the cofactor-binding domain, the catalytic domain, and the recognition domain (TRD), six groups of DNA methyltransferases have been proposed (Fig. 4.1) (Bujnicki 2002). Two models have been proposed to explain the sequence permutation which has resulted in evolution of DNA methyltransferase (Bujnicki and Radlinska 1999; Jeltsch 1999).

Jeltsch proposed that circular permutations play a critical role in the evolution of new methyltransferases (Jeltsch 1999). This model based on the concept that a permutated protein can arise naturally from tandem repeats. This model includes the duplication and in frame fusion of a methyltransferase gene, resulting in a fusion methyltransferase with two catalytic domains and TRDs. Introduction of a start codon in first copy and stop codon in next can result in a circularly permutated variant of a methyltransferase. It was proposed that β sub group of methyltransferase have been

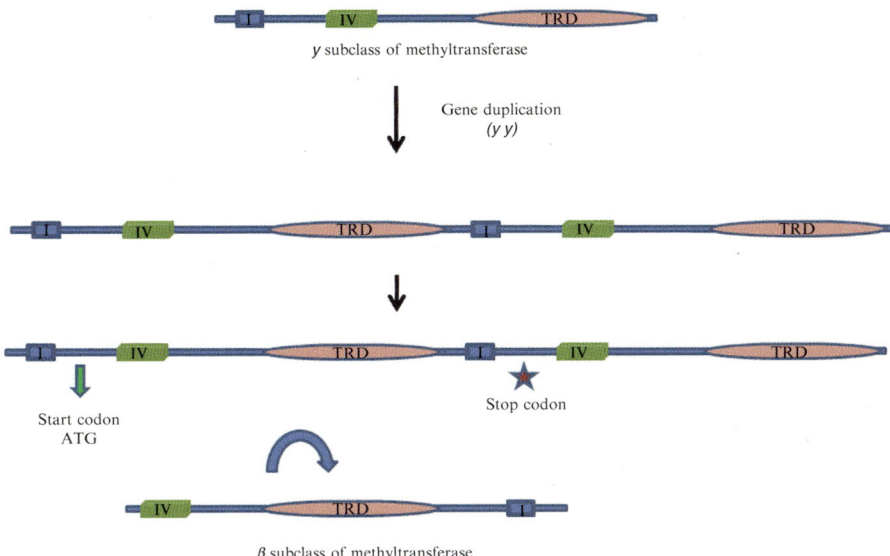

Fig. 4.4 Evolution of β subclass of methyltransferase from γ subclass of methyltransferase by circular permutation

evolved from the duplication of γ subgroup of methyltransferase or vice versa (Fig. 4.4) (Jeltsch 1999). M.FokI represents an interesting example which can be considered as the closest intermediate in circular permutation model. M.FokI contains two complete sets of motifs present in an adenine methyltransferase (Kita et al. 1989; Sugisaki et al. 1989). Recently it has been shown that in *H. pylori*, a single nucleotide insertion between two adjacent methyltransferases can result in a fused methyltransferase with two sets of catalytic motifs, AdoMet binding motifs and TRDs (Kumar and Rao 2011). But this model fails to explain the differences between the TRDs of β and γ sub group of methyltransferase. It is, therefore, unlikely that simple circular permutations from γγ to β or from γγ to β occurred in nature (Bujnicki 2002).

A second model suggests that DNA methyltransferases have evolved by intra-or intergenic rearrangements of gene fragments (Lupas et al. 2001). This model proposes the fragmentation of DNA methyltransferase gene by the action of nuclease and then reassembly of the fragments into a functional form. Reassembly results in shuffling of the motifs. Bujnicki (2002), proposed that M.*MwoI* (δ) has evolved from M.*SfiI* (β) by shuffling of gene fragments and M.*TvoORF1413P* (ζ) evolved from M. *ThaI* (β) sub group of methyltransferase by gene duplication or cut and paste mechanism. It is believed that the two mechanisms are not mutually exclusive and could be possible that both played significant roles in the evolution of permuted methyltransferase.

One of the most intriguing questions is the evolution of R-M enzymes, where the restriction endonuclease and the cognate methylase usually recognize the same DNA sequence. It is straightforward to expect that such enzymes would share at least some elements of the target recognition machinery. Previous attempts to find evidence for such inheritance by pair wise comparison of corresponding restriction enzymes and methyltransferases were unsuccessful. Moreover, common themes of sequence recognition so far have not been found among restriction endonuclease themselves (Wilson and Murray 1991). One possibility is that each component evolved independently from the other and later combined to form R-M systems. Alternatively, due to strong divergence the original domains are barely detectable at the primary structure level.

4.6 Conclusions and Perspectives

The ability to covalently add methyl groups to targeted adenine or cytosine residues in specific DNA sequences without any other changes in the primary DNA structure is a remarkable feature of DNA methyltransferases. Recent evidences suggest that, in addition to protecting the bacterial genome from external DNA, R-M systems have other biological functions (Ando et al. 2010; Jeltsch 2003). The genome analysis of a number of sequenced strains of *H. pylori* showed that this bacterium possesses an unusually high number of strain-specific R-M genes. *H. pylori* is well adapted to the gastric environment, and acquisition of numerous R-M systems might be related to its unique adaptability to acidic environment. Phase variable Type III R-M systems are present in a variety of pathogenic bacteria. Widespread distribution of phase variable R-M systems in host-adapted pathogenic bacteria suggests that this regulated random switching of multiple genes may be a commonly used strategy for bacterial pathogens. Identification of phase variable R-M systems and their role in gene regulation has added another dimension in the field of epigenetic regulation in bacteria. The identification and study of both species-specific and strain-specific phase variable methyltransferases of pathogenic bacteria may therefore, improve the understanding of their pathogenic mechanisms.

The finding that N^6 methyladenine is essential for the virulence and pathogenesis in many bacteria raises the possibility of using an inhibitor of adenine methyltransferase as antimicrobial agents. Decrease in the virulence by the deletion of *dam* in a number of pathogenic bacteria, combined with their capacity to persist at low levels in animals organs make them an interesting target to use as Dam-based live vaccines.

References

Adamczyk-Poplawska M, Lower M, Piekarowicz A (2009) Characterization of the NgoAXP: phase-variable type III restriction-modification system in *Neisseria gonorrhoeae*. FEMS Microbiol Lett 300:25–35

Alm RA, Ling LSL, Moir DT, King BL, Brown ED, Doig PC, Smith DR, Noonan B, Guild BC, deJonge BL, Carmel G, Tummino PJ, Caruso A, Uria-Nickelsen M, Mills DM, Ives C, Gibson R, Merberg D, Mills SD, Jiang Q, Taylor DE, Vovis GF, Trost TJ (1999) Genomic-sequence comparison of two unrelated isolates of the human gastric pathogen *Helicobacter pylori.* Nature 397:176–180

Ando T, Ishiguro K, Watanabe O, Miyake N, Kato T, Hibi S, Mimura S, Nakamura M, Miyahara R, Ohmiya N, Niwa Y, Goto H (2010) Restriction-modification systems may be associated with *Helicobacter pylori* virulence. J Gastroenterol Hepatol 1:S95–S98

Arber W (2000) Genetic variation: molecular mechanisms and impact on microbial evolution. FEMS Microbiol Rev 24:1–7

Arber W (2002) Evolution of prokaryotic genomes. Curr Top Microbiol Immunol 264:1–14

Arber W, Dussoix D (1962) Host specificity of DNA produced by *Escherichia coli.* I. Host controlled modification of bacteriophage lambda. J Mol Biol 5:18–36

Au KG, Welsh K, Modrich P (1992) Initiation of methyl-directed mismatch repair. J Biol Chem 267:12142–12148

Barbeyron T, Kean K, Forterre P (1984) DNA adenine methylation of GATC sequences appeared recently in the *Escherichia coli* lineage. J Bacteriol 160:586–590

Bell DC, Cupples CG (2001) Very-short-patch repair in *Escherichia coli* requires the *dam* adenine methylase. J Bacteriol 183:3631–3635

Bergerat A, Kriebardis A, Guschlbauer W (1989) Preferential site-specific hemimethylation of GATC sites in pBR322 DNA by Dam methyltransferase from *Escherichia coli.* J Biol Chem 264:4064–4070

Bertani G, Weigle JJ (1953) Host controlled variation in bacterial viruses. J Bacteriol 65:113–121

Bestor TH (2000) The DNA methyltransferases of mammals. Hum Mol Genet 9:2395–2402

Bheemanaik S, Reddy YV, Rao DN (2006) Structure, function and mechanism of exocyclic DNA methyltransferases. Biochem J 399:177–190

Bickle TA, Kruger DH (1993) Biology of DNA restriction. Microbiol Rev 57:434–450

Blaisdell BE, Campbell AM, Karlin S (1996) Similarities and dissimilarities of phage genomes. Proc Natl Acad Sci USA 93:5854–5859

Bogan JA, Helmstetter CE (1997) DNA sequestration and transcription in the oriC region of *Escherichia coli.* Mol Microbiol 26:889–896

Braaten BA, Nou X, Kaltenbach LS, Low DA (1994) Methylation patterns in pap regulatory DNA control pyelonephritis-associated pili phase variation in *E. coli.* Cell 76:577–588

Braun RE, O'Day K, Wright A (1985) Autoregulation of the DNA replication gene *dnaA* in *E. coli* K-12. Cell 40:159–169

Broadbent SE, Davies MR, van der Woude MW (2010) Phase variation controls expression of Salmonella lipopolysaccharide modification genes by a DNA methylation-dependent mechanism. Microbiology 77:337–353

Bucci C, Lavitola A, Salvatore P, Del Giudice L, Massardo DR, Bruni CB, Alifano P (1999) Hypermutation in pathogenic bacteria: frequent phase variation in meningococci is a phenotypic trait of a specialized mutator biotype. Mol Cell 3:435–445

Bujnicki JM (2002) Sequence permutations in the molecular evolution of DNA methyltransferases. BMC Evol Biol 2:3

Bujnicki JM, Radlinska M (1999) Molecular evolution of DNA-(–cytosine N4) methyltransferases: evidence for their polyphyletic origin. Nucleic Acids Res 27:4501–4509

Calmann MA, Marinus MG (2003) Regulated expression of the *Escherichia coli dam* gene. J Bacteriol 185:5012–5014

Camacho EM, Casadesus J (2002) Conjugal transfer of the virulence plasmid of *Salmonella enterica* is regulated by the leucine-responsive regulatory protein and DNA adenine methylation. Mol Microbiol 44:1589–1598

Campbell JL, Kleckner N (1990) *E. coli* oriC and the dnaA gene promoter are sequestered from *dam* methyltransferase following the passage of the chromosomal replication fork. Cell 62:967–979

Carlson K, Kosturko LD (1998) Endonuclease II of coliphage T4: a recombinase disguised as a restriction endonuclease? Mol Microbiol 27:671–676

Casadesu's J, Low D (2006) Epigenetic gene regulation in the bacterial world. Microbiol Mol Biol Rev 70:830–856

Chen L, Paulsen DB, Scruggs DW, Banes MM, Reeks BY, Lawrence ML (2003) Alteration of DNA adenine methylase (Dam) activity in *Pasteurella multocida* causes increased spontaneous mutation frequency and attenuation in mice. Microbiology 149:2283–2290

Cheng X (1995) Structure and function of DNA methyltransferases. Annu Rev Biophys Biomol Struct 24:293–318

De Bolle X, Bayliss CD, Field D, van de Ven T, Saunders NJ, Hood DW, Moxon ER (2000) The length of a tetranucleotide repeat tract in *Haemophilus influenzae* determines the phase variation rate of a gene with homology to type III DNA methyltransferases. Mol Microbiol 35:211–222

de Vries N, Duinsbergen D, Kuipers EJ, Pot RG, Wiesenekker P, Penn CW, van Vliet AH, Vandenbroucke-Grauls CM, Kusters JG (2002) Transcriptional phase variation of a type III restriction-modification system in *Helicobacter pylori*. J Bacteriol 184:6615–6623

Dodson KW, Berg DE (1989) Factors affecting transposition activity of IS50 and Tn5 ends. Gene 76:207–213

Donahue JP, Israel DA, Torres VJ, Necheva AS, Miller GG (2002) Inactivation of a *Helicobacter pylori* DNA methyltransferase alters *dnaK* operon expression following host-cell adherence. FEMS Microbiol Lett 208:295–301

Dybvig K, Sitaraman R, French CT (1998) A family of phase-variable restriction enzymes with differing specificities generated by high-frequency gene rearrangements. Proc Natl Acad Sci USA 95:13923–13928

Fa¨lker S, Schmidt MA, Heusipp G (2005) DNA methylation in *Yersinia enterocolitica*: role of the DNA adenine methyltransferase in mismatch repair and regulation of virulence factors. Microbiology 151:2291–2299

Fälker S, Schmidt MA, Heusipp G (2007) DNA adenine methylation and bacterial pathogenesis. Int J Med Microbiol 297:1–7

Fox KL, Srikhanta YN, Jennings MP (2007) Phase variable Type III restriction-modification systems of host-adapted bacterial pathogens. Mol Microbiol 65:1375–1379

Garcia-Del Portillo F, Pucciarelli MG, Casadesus J (1999) DNA adenine methylase mutants of *Salmonella typhimurium* show defects in protein secretion, cell invasion, and M cell cytotoxicity. Proc Natl Acad Sci USA 96:11578–11583

Geier GE, Modrich P (1979) Recognition sequence of the *dam* methylase of *Escherichia coli* K12 and mode of cleavage of Dpn I endonuclease. J Biol Chem 254:1408–1413

Haagmans W, van der Woude M (2000) Phase variation of Ag43 in *Escherichia coli*: dam-dependent methylation abrogates OxyR binding and OxyR-mediated repression of transcription. Mol Microbiol 35:877–887

Hallet B (2001) Playing Dr Jekyll and Mr Hyde: combined mechanisms of phase variation in bacteria. Curr Opin Microbiol 4:570–581

Hattman S, Malygin EG (2004) Bacteriophage T2Dam and T4Dam DNA-[N6-adenine]-methyltransferases. Prog Nucleic Acid Res Mol Biol 77:67–126

Hattman S, Sun W (1997) *Escherichia coli* OxyR modulation of bacteriophage Mu *mom* expression in *dam*+ cells can be attributed to its ability to bind Pmom promoter DNA. Nucleic Acids Res 25:4385–4388

Heithoff D, Sinsheimer RL, Low DA, Mahan MJ (1999) An essential role for DNA adenine methylation in bacterial virulence. Science 284:967–970

Heithoff DM, Enioutina EI, Daynes RA, Sinsheimer RL, Low DA, Mahan MJ (2001) *Salmonella* DNA adenine methylase mutants confer cross-protective immunity. Infect Immun 69:6725–6730

Heitman J (1993) On the origins, structures and functions of restriction-modification enzymes. Genet Eng (NY) 15:57–108

Herman GE, Modrich P (1982) *Escherichia coli dam* methylase. Physical and catalytic properties of the homogeneous enzyme. J Biol Chem 257:2605–2612

Heusipp G, Falker S, Schmidt MA (2006) DNA adenine methylation and bacterial pathogenesis. Int J Med Microbiol 29:1–7

Jafri S, Chen S, Calvo JM (2002) *ilvIH* operon expression in *Escherichia coli* requires Lrp binding to two distinct regions of DNA. J Bacteriol 184:5293–5300

Jeltsch A (1999) Circular permutation in the molecular evolution of DNA methyltransferases. J Mol Evol 49:161–164

Jeltsch A (2002) Beyond Watson and Crick: DNA methylation and molecular enzymology of DNA methyltransferases. Chembiochem 3:274–293

Jeltsch A (2003) Maintenance of species identity and controlling speciation of bacteria: a new function for restriction/modification systems? Gene 317:13–16

Jennings MP, Hood DW, Peak IR, Virji M, Moxon ER (1995) Molecular analysis of a locus for the biosynthesis and phase-variable expression of the lacto-*N*-neotetraose terminal lipopolysaccharide structure in *Neisseria meningitidis*. Mol Microbiol 18:729–740

Jorgensen HF, Bird A (2002) MeCP2 and other methyl-CpG binding proteins. Ment Retard Dev Disabil Res Rev 8:87–93

Julio SM, Heithoff DM, Provenzano D, Klose KE, Sinsheimer RL, Low DA, Mahan MJ (2001) DNA adenine methylase is essential for viability and plays a role in the pathogenesis of *Yersinia pseudotuberculosis* and *Vibrio cholerae*. Infect Immun 69:7610–7615

Julio SM, Heithoff DM, Sinsheimer RL, Low DA, Mahan MJ (2002) DNA adenine methylase overproduction in *Yersinia pseudotuberculosis* alters YopE expression and secretion and host immune responses to infection. Infect Immun 70:1006–1009

Kim JS, Li J, Barnes IH et al (2008) Role of the *Campylobacter jejuni* Cj1461 DNA methyltransferase in regulating virulence characteristics. J Bacteriol 190:6524–6529

Kita K, Kotani H, Sugisaki H, Takanami M (1989) The FokI restriction-modification system I. Organization and nucleotide sequences of the restriction and modification genes. J Biol Chem 264:5751–5756

Klose RJ, Bird AP (2006) Genomic DNA methylation: the mark and its mediators. Trends Biochem Sci 31:89–97

Kobayashi I (2004) Restriction-modification systems as minimal forms of life. In: Pingoud A (ed) Restriction endonucleases. Springer, Berlin, pp 19–62

Kossykh VG, Schlagman SL, Hattman S (1995) Phage T4 DNA [N6-adenine]methyltransferase. Overexpression, purification, and characterization. J Biol Chem 270:14389–14393

Kumar R, Rao DN (2011) A nucleotide insertion between two solitary MTases results in a bifunctional fusion MTase in *H.pylori*. Biochem J 433:487–495

Kumar S, Cheng X, Klimasauskas S, Mi S, Posfai J, Roberts RJ, Wilson GG (1994) The DNA (cytosine-5) methyltransferases. Nucleic Acids Res 22:1–10

Kumar R, Mukhopadhyay AK, Rao DN (2010) Characterization of an N^6 adenine methyltransferase from *H. pylori* strain 26695 which methylates adjacent adenines on the same strand. FEBS J 277:1666–1683

Lewis JD, Meehan RR, Henzel WJ, Maurer-Fogy I, Jeppesen P, Klein F, Bird A (1992) Purification, sequence, and cellular localization of a novel chromosomal protein that binds to methylated DNA. Cell 69:905–914

Lin LF, Posfai J, Roberts RJ, Kong H (2001) Comparative genomics of the restriction-modification systems in *Helicobacter pylori*. Proc Natl Acad Sci USA 98:2740–2745

Løbner-Olesen A, Marinus MG, Hansen FG (2003) Role of SeqA and Dam in *Escherichia coli* gene expression: a global/microarray analysis. Proc Natl Acad Sci USA 100:4672–4677

Lobner-Olesen A, Skovgaard O, Marinus MG (2005) Dam methylation:coordinating cellular processes. Curr Opin Microbiol 8:154–160

López-Garrido J, Casadesús J (2010) Regulation of *Salmonella enterica* pathogenicity island 1 by DNA adenine methylation. Genetics 184:637–649

Low DA, Weyand NJ, Mahan MJ (2001) Roles of DNA adenine methylation in regulating bacterial gene expression and virulence. Infect Immun 69:7197–7204

Lu M, Campbell JL, Boye E, Kleckner N (1994) SeqA: a negative modulator of replication initiation in *E. coli*. Cell 77:413–426

Lupas AN, Pointing CP, Russell RB (2001) On the evolution of protein folds: are similar motifs in different protein folds the result of convergence, insertion, or relics of an ancient peptide world? J Struct Biol 134:191–203

Luria SE, Human ML (1952) A nonhereditary, host-induced variation of bacterial viruses. J Bacteriol 64:557–569

Malone T, Blumenthal RM, Cheng X (1995) Structure-guided analysis reveals nine sequence motifs conserved among DNA amino-methyltransferases, and suggests a catalytic mechanism for these enzymes. J Mol Biol 253:618–632

Marczynski GT, Shapiro L (2002) Control of chromosome replication in *Caulobacter crescentus*. Annu Rev Microbiol 56:625–656

Marinus MG (1996) Methylation of DNA. In: Neidhardt FC, Curtiss R, Ingraham JL, Lin ECC, Low KB, Magasanik B, Reznikoff WS, Riley M, Schaechter M, Umbarger HE (eds) *Escherichia coli* and *Salmonella*: cellular and molecular biology. ASM Press, Washington DC, pp 782–791

Marinus MG, Casadesus J (2009) Roles of DNA adenine methylation in host-pathogen interactions: mismatch repair, transcriptional regulation, and more. FEMS Microbiol Rev 33:488–503

Mashhoon N, Carroll M, Pruss C, Eberhard J, Ishikawa S, Estabrook RA, Reich N (2004) Functional characterization of Escherichia coli DNA adenine methyltransferase, a novel target for antibiotics. J Biol Chem 279:52075–52081

McClain MS, Shaffer CL, Israel DA, Peek RM Jr, Cover TL (2009) Genome sequence analysis of *Helicobacter pylori* strains associated with gastric ulceration and gastric cancer. BMC Genomics 10:3

McCommas SA, Syvanen M (1988) Temporal control of transposition in Tn5. J Bacteriol 170:889–894

McKane M, Milkman R (1995) Transduction, restriction and recombination patterns in *Escherichia coli*. Genetics 139:35–43

Mehling JS, Lavender H, Clegg S (2006) A Dam methylation mutant of *Klebsiella pneumoniae*, is partially attenuated. FEMS Microbiol Lett 268:187–193

Messer W, Noyer-Weidner M (1988) Timing and targeting: the biological functions of Dam methylation in *E. coli*. Cell 54:735–737

Modrich P, Lahue R (1996) Mismatch repair in replication fidelity, genetic recombination, and cancer biology. Annu Rev Biochem 65:101–133

Moxon ER, Thaler DS (1997) Microbial genetics. The tinkerer's evolving tool-box. Nature 387:659–662

Moxon R, Bayliss C, Hood D (2006) Bacterial contingency loci: the role of simple sequence DNA repeats in bacterial adaptation. Annu Rev Genet 40:307–333

Naito T, Kusano K, Kobayashi I (1995) Selfish behavior of restriction-modification systems. Science 267:897–899

Nan X, Cross S, Bird A (1998) Gene silencing by methyl-CpGbinding proteins. Novartis Found Symp 214:6–16

Ogden GB, Pratt MJ, Schaechter M (1988) The replicative origin of the *E. coli* chromosome binds to cell membranes only when hemimethylated. Cell 54:121–135

Oshima T, Wada C, Kawagoe Y, Ara T, Maeda M, Masuda Y, Hiraga S, Mori H (2002) Genome-wide analysis of deoxyadenosine methyltransferase-mediated control of gene expression in *Escherichia coli*. Mol Microbiol 45:673–695

Oza JP, Yeh JB, Reich NO (2005) DNA methylation modulates *Salmonella enterica* serovar Typhimurium virulence in *Caenorhabditis elegans*. FEMS Microbiol Lett 245:53–59

Pingoud A, Fuxreiter M, Pingoud V, Wende W (2005) Type II restriction endonucleases: structure and mechanism. Cell Mol Life Sci 62:685–707

Polaczek P, Kwan K, Campbell L (1998) GATC motifs may alter the conformation of DNA depending on sequence context and N6-adenine methylation status: possible implications for DNA-protein recognition. Mol Gen Genet 258:488–493

Posfai J, Bhagwat AS, Posfai G, Roberts RJ (1989) Predictive motifs derived from cytosine methyltransferases. Nucleic Acids Res 17:2421–2435

Raleigh EA, Brooks JE (1998) Restriction modification systems: where they are and what they do. In: De Bruijn FJ, Lupski JR, Weinstock GM (eds) Bacterial genomes. Chapman and Hall, New York, pp 78–92

Reisenauer A, Shapiro L (2002) DNA methylation affects the cell cycle transcription of the CtrA global regulator in Caulobacter. EMBO J 21:4969–4977

Reisenauer A, Kahng LS, McCollum S, Shapiro L (1999) Bacterial DNA methylation: a cell cycle regulator? J Bacteriol 181:5135–5139

Reznikoff WS (1993) The Tn5 transposon. Annu Rev Microbiol 47:945–963

Roberts D, Hoopes BC, McClure WR, Kleckner N (1985) IS10 transposition is regulated by DNA adenine methylation. Cell 43:117–130

Roberts RJ, Belfort M, Bestor T, Bhagwat AS, Bickle TA, Bitinaite J, Blumenthal RM, Degtyarev S, Dryden DT, Dybvig K et al (2003) A nomenclature for restriction enzymes, DNA methyltransferases, homing endonucleases and their genes. Nucleic Acids Res 31:1805–1812

Roberts RJ, Vincze T, Posfai J, Macelis D (2007) REBASE–enzymes and genes for DNA restriction and modification. Nucleic Acids Res 35:D269–D270

Robertson BD, Meyer TF (1992) Genetic variation in pathogenic bacteria. Trends Genet 8:422–427

Robertson GT, Reisenauer A, Wright R, Jensen RB, Jensen A, Shapiro L, Roop RM II (2000) The *Brucella abortus* CcrM DNA methyltransferase is essential for viability, and its overexpression attenuates intracellular replication in murine macrophages. J Bacteriol 182:3482–3489

Robinson VL, Oyston PC, Titball RW (2005) Oral immunization with a *dam* mutant of *Yersinia pseudotuberrculosis* protects against plague. Microbiology 151:1919–1926

Ryan KA, Lo RY (1999) Characterization of a CACAG pentanucleotide repeat in *Pasteurella haemolytica* and its possible role in modulation of a novel type III restriction-modification system. Nucleic Acids Res 27:1505–1511

Salaun L, Bodo L, Suerbaum S, Saunders NJ (2004) The diversity within an expanded and redefined repertoire of phase-variable genes in *Helicobacter pylori*. Microbiology 150:817–830

Salaun L, Ayraud S, Saunders NJ (2005) Phase variation mediated niche adaptation during prolonged experimental murine infection with *Helicobacter pylori*. Microbiology 151:917–923

Saunders NJ, Peden JF, Hood DW, Moxon ER (1998) Simple sequence repeats in the *Helicobacter pylori* genome. Mol Microbiol 27:1091–1098

Saunders NJ, Jeffries AC, Peden JF, Hood DW, Tettelin H, Rappuoli R, Moxon ER (2000) Repeat-associated phase variable genes in the complete genome sequence of *Neisseria meningitidis* strain MC58. Mol Microbiol 37:207–215

Schlagman SL, Hattman S (1983) Molecular cloning of a functional dam+ gene coding for phage T4 DNA adenine methylase. Gene 22:139–156

Schlagman SL, Miner Z, Feher Z, Hattman S (1988) The DNA [adenine-N6]methyltransferase (Dam) of bacteriophage T4. Gene 73:517–530

Schluckebier G, O'Gara M, Saenger W, Cheng X (1995) Universal catalytic domain structure of AdoMet-dependent methyltransferases. J Mol Biol 247:16–20

Seib KL, Peak IR, Jennings MP (2002) Phase variable restriction-modification systems in *Moraxella catarrhalis*. FEMS Immunol Med Microbiol 32:159–165

Sistla S, Rao DN (2004) S-Adenosyl-L-methionine-dependent restriction enzymes. Crit Rev Biochem Mol Biol 39:1–19

Sohail A, Lieb M, Dar M, Bhagwat AS (1990) A gene required for very short patch repair in E. coli is adjacent to the DNA cytosine methylase gene. J Bacteriol 172:4214–4221

Srikhanta YN, Maguire TL, Stacey KJ, Grimmond SM, Jennings MP (2005) The phasevarion: a genetic system controlling coordinated, random switching of expression of multiple genes. Proc Natl Acad Sci USA 102:5547–5551

Srikhanta YN, Dowideit SJ, Edwards JL, Falsetta ML, Wu H-J, Harrison OB, Fox KL, Seib KL, Maguire TL, Wang AH-J, Maiden MC, Grimmond SM, Apicella MA, Jennings MP (2009) Phasevarions mediate random switching of gene expression in pathogenic Neisseria. PLoS Pathog 5:e1000400

Srikhanta YN, Fox KL, Jennings MP (2010) The phasevarion: phase variation of Type III DNA methyltransferases controls coordinated switching in multiple genes. Nat Rev Microbiol 8:196–206

Sternberg N, Sauer B, Hoess R, Abremski K (1986) Bacteriophage P1 cre gene and its regulatory region. Evidence for multiple promoters and for regulation by Dam methylation. J Mol Biol 187:197–212

Sugisaki H, Kita K, Takanami M (1989) The FokI restriction-modification system II. Presence of two domains in FokI methylase responsible for modification of different strands. J Biol Chem 264:5757–5761

Sun K, Jiao XD, Zhang M, Sun L (2010) DNA adenine methylase is involved in the pathogenesis of *Edwardsiella tarda*. Vet Microbiol 141:149–154

Taylor VL, Titball RW, Oyston PCF (2005) Oral immunization with a *dam* mutant of *Yersinia pseudotuberculosis* protects against plague. Microbiology 151:1919–1926

Tomb JF, White O, Kerlavage AR, Clayton RA, Sutton GG, Fleischmann RD et al (1997) The complete genome sequence of the gastric pathogen *Helicobacter pylori*. Nature 388:539–547

Torreblanca J, Marqués S, Casadesús J (1999) Synthesis of FinP RNA by plasmids F and pSLT is regulated by DNA adenine methylation. Genetics 152:31–45

Urig S, Gowher H, Hermann A, Beck C, Fatemi M, Humeny A, Jeltsch A (2002) The Escherichia coli dam DNA methyltransferase modifies DNA in a highly processive reaction. J Mol Biol 319:1085–1096

van der Ende A, Hopman CT, Zaat S, Essink BB, Berkhout B, Dankert J (1995) Variable expression of class 1 outer membrane protein in Neisseria meningitides is caused by variation in the spacing between the −10 and −35 regions of the promoter. J Bacteriol 177:2475–2480

van der Woude MW (2006) Re-examining the role and random nature of phase variation. FEMS Microbiol Lett 254:190–197

van der Woude MW, Baumler AJ (2004) Phase and antigenic variation in bacteria. Clin Microbiol Rev 17:581–611

Van Etten JL (2003) Unusual life style of giant chlorella viruses. Annu Rev Genet 37:153–195

van Ham SM, van Alphen L, Mooi FR, van Putten JP (1993) Phase variation of *Haemophilus influenzae* fimbriae: transcriptional control of two divergent genes through a variable combined promoter region. Cell 73:1187–1196

Vitkute J, Stankevicius K, Tamulaitiene G, Maneliene Z, Timinskas A, Berg DE, Janulaitis A (2001) Specificities of eleven different DNA methyltransferases of *Helicobacter pylori* strain 26695. J Bacteriol 183:443–450

Waldron DE, Owen P, Dorman CJ (2002) Competitive interaction of the OxyR DNA-binding protein and the Dam methylase at the antigen 43 gene regulatory region in *Escherichia coli*. Mol Microbiol 44:509–520

Watson ME, Jarisch J, Smith AL (2004) Inactivation of deoxyadenosine methyltransferase (*dam*) attenuates *Haemophilus influenzae* virulence. Mol Microbiol 55:651–654

Weiser JN, Williams A, Moxon ER (1990) Phasevariable lipopolysaccharide structures enhance the invasive capacity of *Haemophilus influenzae*. Infect Immun 58:3455–3457

Wilson GG, Murray NE (1991) Restriction and modification systems. Annu Rev Genet 25:585–627

Wion D, Casadesus J (2006) N6-methyl-adenine: an epigenetic signal for DNA-protein interactions. Nat Rev Microbiol 4:183–192

Wright R, Stephens C, Zweiger G, Shapiro L, Alley MR (1996) Caulobacter Lon protease has a critical role in cell-cycle control of DNA methylation. Genes Dev 10:1532–1542

Xu Q, Stickel S, Roberts RJ, Blaser MJ, Morgan RD (2000) Purification of the novel endonuclease, Hpy188I, and cloning of its restriction-modification genes reveal evidence of its horizontal transfer to the *Helicobacter pylori* genome. J Biol Chem 275:17086–17093

Yamaki H, Ohtsubo E, Nagai K, Maeda Y (1988) The *oriC* unwinding by *dam* methylation in *Escherichia coli*. Nucleic Acids Res 16:5067–5073

Yarmolinski MB, Sternberg N (1988) Bacteriophage P1. In: Calendar R (ed) The bacteriophages, vol 1. Plenum Press, New York, pp 782–791

Zweiger G, Marczynski G, Shapiro L (1994) A Caulobacter DNA methyltransferase that functions only in the predivisional cell. J Mol Biol 235:472–485

Part II
Developmental Epigenetics

Chapter 5
Metabolic Aspects of Epigenome: Coupling of *S*-Adenosylmethionine Synthesis and Gene Regulation on Chromatin by SAMIT Module

Kazuhiko Igarashi and Yasutake Katoh

Abstract Histone and DNA methyltransferases utilize *S*-adenosyl-*L*-methionine (SAM), a key intermediate of sulfur amino acid metabolism, as a donor of methyl group. SAM is biosynthesized by methionine adenosyltransferase (MAT) using two substrates, methionine and ATP. Three distinct forms of MAT (MATI, MATII and MATIII), encoded by two distinct genes (*MAT1A* and *MAT2A*), have been identified in mammals. MATII consists of α2 catalytic subunit encoded by *MAT2A* and β regulatory subunit encoded by *MAT2B*, but the physiological function of the β subunit is not clear. MafK is a member of Maf oncoproteins and functions as both transcription activator and repressor by forming diverse heterodimers to bind to DNA elements termed Maf recognition elements. Proteomics analysis of MafK-interactome revealed its interaction with both MATIIα and MATIIβ. They are recruited specifically to MafK target genes and are required for their repression by MafK and its partner Bach1. Because the catalytic activity of MATIIα is required for the MafK target gene repression, MATIIα is suggested to provide SAM locally on chromatin where it is recruited. One of the unexpected features of MATII is that MATIIα interacts with many chromatin-related proteins of diverse functions such as histone modification, chromatin remodeling, transcription regulation, and nucleo-cytoplasmic transport. MATIIα appears to generate multiple, heterogenous regulatory complexes where it provides SAM. Considering their function, the heterooligomer of MATIIα and β is named SAMIT (SAM-integrating transcription) module within their interactome where it serves SAM for nuclear methyltransferases.

K. Igarashi (✉) • Y. Katoh
Department of Biochemistry and Center for Regulatory Epigenome and Diseases,
Tohoku University School of Medicine, Seiryo 2-1, Sendai 980-8575, Japan
e-mail: igarashi@med.tohoku.ac.jp

5.1 Introduction: Overlooked Metabolites in Epigenetics

Post-translational modifications of histone tails especially acetylation and methylation regulate gene expression as a part of the epigenome system. Histone acetyltransferases utilize acetyl-Coenzyme A (Ac-CoA), a molecular hub of carbohydrate, lipid, and energy metabolism. Histone methyltransferases utilize S-adenosyl-L-methionine (SAM or AdoMet), a key intermediate of sulfur amino acid metabolism. DNA methyltransferases also utilize SAM. Viewing these simple facts, it seems reasonable to speculate that cellular metabolism and chromatin regulation are intimately connected. However, a link between chromatin and metabolism has been overlooked. Such a link would be layered at several mechanistic aspects. First, reactions of methylation and acetylation within nuclei may be substrate-limited. Amounts of Ac-CoA or SAM within a cell may fluctuate depending on nutrition and/or energy states, limiting or enhancing histone acetylation and methylation. Consistent with the idea, histone acetylation increases when cells, cultured with low glucose, are supplied with acetate. The acetate can be metabolized to Ac-CoA in the cell (Wellen et al. 2009), indicating that Ac-CoA is limiting for histone acetylation under certain conditions. Therefore, epigenetic modifications may respond to nutritional conditions and metabolic status. Second, enzymes for the synthesis of Ac-CoA and SAM may be compartmentalized within the nuclei. Subcellular localization of Ac-CoA or SAM synthesizing enzymes may affect gene expression. As for example, Ac-CoA synthase 2p in yeast is localized in nuclei (Takahashi et al. 2006), suggesting that local, nuclear synthesis of Ac-CoA may facilitate histone acetylation. In other words, availability and activity of metabolic enzymes related to Ac-CoA or SAM within nuclei may significantly affect histone modification and hence gene expression. Third, these enzymes may be parts of protein complexes that would allow an efficient flow of substrates (i.e., substrate channeling). In this case, assembly and disassembly of such complexes would confer dynamic regulation of epigenome. Despite these interesting possibilities pertaining to the epigenome, little is known about whether SAM is limiting in the nuclei or whether enzymes that synthesize SAM play a role in the nucleus.

Methylation of both histone and DNA is more dynamic than originally thought, as active demethylation of histone and DNA has been discovered (Lan et al. 2008). This raises the possibility that a subtle change in the process of methylation can affect local and global levels of methylation of histone and DNA. In this review, we discuss a connection between the SAM synthesis pathway and histone methylation that has been highlighted by a recent identification of the SAMIT (SAM-integrating transcription) module for chromatin regulation (Katoh et al. 2011).

5.2 S-Adenosylmethionine (SAM) in Epigenetics

5.2.1 SAM as a Hub of Metabolism

SAM was discovered in 1953 (Cantoni 1953), the year marked by the discovery of DNA double helix. SAM is the second most widely used enzyme substrate after ATP (Cantoni 1975). SAM is the major methyl donor among other donors such as folate in enzymatic reactions of methyl transfer. It is involved in the biosynthesis of diverse bioactive substances including hormones and neurotransmitters. SAM is utilized as a substrate by all of the known methyltransferases of histones and DNA. The use of SAM as a methyl donor is because of the strong electrophilic character of its methyl group. This poses a risk: cells need to avoid non-specific reactions between SAM and nucleophiles such as DNA. In vitro SAM acts as an alkylating mutagen (Rydberg and Lindahl 1982). However, its mutagenic activity in *E. coli* cell has been challenged (Posnick and Samson 1999). As the mutagenic potential of SAM in eukaryotes awaits further study, it leaves the pathological significance of SAM as a mutagen unclear. It is therefore reasonable to consider that the evolution of life has selected systems that balance between essential functions and detrimental effects of SAM in the nuclei.

SAM is biosynthesized by methionine adenosyltransferase (MAT) using two substrates, methionine and ATP (Fig. 5.1). Upon transmethylation reaction by

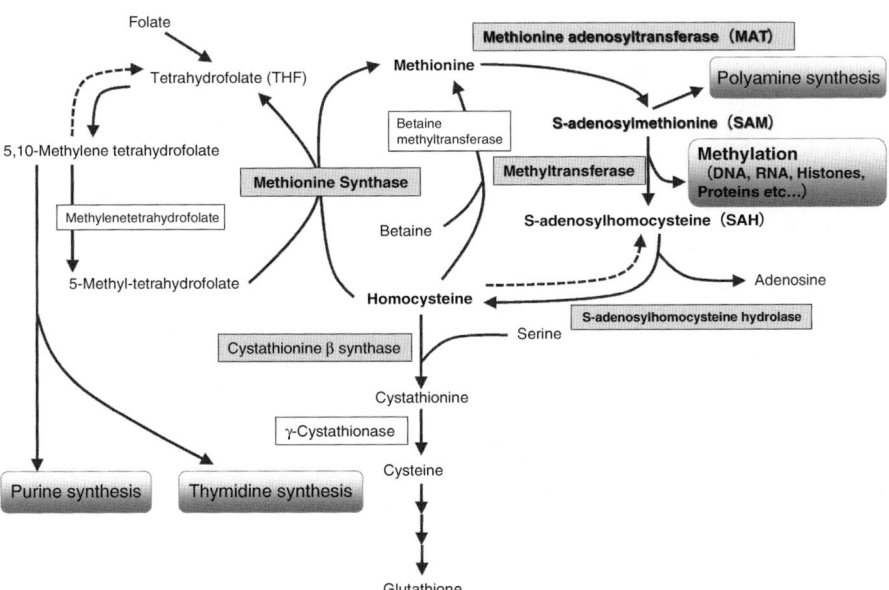

Fig. 5.1 S-adenosylmethionine (SAM) cycle and metabolism. Metabolites and their connections are shown with respective enzymes (*boxed*). Reverse reactions are shown with *dashed lines*

methyltransferases, SAM is converted into S-adenosylhomocysteine (SAH). SAH is then cleaved by SAH hydrolase, generating homosysteine and adenosine. Homocysteine is methylated to regenerate methionine by methionine synthase using the methyl group of 5-methyltetrahydrofolate, connecting the overall reactions into a single methionine cycle. Completion of the reaction cycle occurs only in the liver, since most other tissues lack methionine synthase and/or other enzymes in the cycle (Mato et al. 1997). The metabolites of the methionine cycle are also connected to important pathways other than transmethylation. SAM is the precursor of spermidine and spermine synthesis. SAH provides adenosine for the synthesis of ATP. Homocysteine is metabolized through the transsulfuration pathway to generate cysteine, which is then utilized for synthesis of diverse molecules including protein, glutathione, and coenzyme A. Methyltetrahydrofolate and tetrahydrofolate, a product of methionine regeneration, are parts of the folate cycle which participates in the synthesis of thymidylate (dTMP) and purines.

5.2.2 Enzymes for SAM Synthesis

Three distinct forms of MAT (MATI, MATII and MATIII), encoded by two distinct genes (*MAT1A* and *MAT2A*), have been identified in mammals (Mato et al. 1997; Suma et al. 1983; Mitsui et al. 1988; Sakata et al. 1993; Kotb et al. 1997; Markham and Pajares 2009). MATI and MATIII are a tetramer and a dimer respectively of $\alpha 1$ catalytic subunit, which is encoded by *MAT1A*. MATII consists of the catalytic subunit $\alpha 2$, encoded by *MAT2A* and the regulatory subunit β, encoded by *MAT2B*. The subunit structure is presumably $\alpha 2_2 \beta_1$. The MATII β subunit increases MATII catalytic activity by lowering Km for L-methionine (LeGros et al. 2000), but sensitizes it to SAM-mediated product inhibition by lowering Ki for SAM (Halim et al. 1999). Physiological function of the β subunit in MATII is not clear (see below). MATI has been considered to be liver-specific, whereas MATII is a ubiquitous enzyme. However, it has been reported recently that MATI is expressed in a wider range of tissues in rat including lung and pancreas (Reytor et al. 2009). The mammalian MAT isozymes differ in their Km for methionine, which is ~30 μM for MATII, ~100 μM for MATI, and 1 mM for MATIII (Markham and Pajares 2009). While the physiological significance of the differences in Km values for methionine is not clear at present, they may reflect different microenvironments (i.e., methionine concentrations) where the respective isozymes function. MATII may be suitable for reactions where methionine is relatively sparse. Even yeast cells possess two isozymes of MAT (Chiang and Cantoni 1977), suggesting that their division of labor is fundamental to life.

Several investigators reported intracellular concentrations of SAM. SAM levels in lymphoid cells increase upon activation: it is around 5 μM and 30 μM in resting and activated lymphoid cells, respectively (German et al. 1983). Much higher levels of around 300 μM are reported in human acute myeloid leukemia cell line HL-60 (Chiba et al. 1988). Although these observations may suggest that SAM is not limiting within a cell, little is known about nuclear SAM levels. Subcellular distribution of SAM has been addressed recently (Brown et al. 2010). In their report, the authors measured SAM levels before and after treatment of mice with acetaminophen, a leading cause of drug-induced liver disease and known to reduce SAM levels. Despite of being out of their scope, they report that liver tissue contains roughly 60 nmol of SAM per g tissue, which partition into fractions of soluble cytoplasm, mitochondoria, and nuclei at roughly 20, 10, and 0.2 per g tissue, respectively. Thus, the nuclear compartment contains significantly less SAM compared to other compartments. It remains possible however, that SAM may leak out of nuclei during the biochemical fractionation. It is a fundamental question whether SAM constitutes a single pool in cytoplasmic and nuclear compartments. In an extreme and simplistic view, SAM may be synthesized by cytoplasmic MAT enzyme and can be provided to nuclear methyltransferases by diffusion. Such a system is simple but appears to lack regulatory capability. It may be difficult to balance SAM requirement in nuclei when it is synthesized elsewhere.

5.3 SAM Synthesis in Nuclear Compartment

5.3.1 *MafK-MATII Interaction and SAMIT Module*

MafK is a member of Maf oncoproteins that functions as both transcription activator and repressor. It forms diverse heterodimers that bind to DNA elements termed Maf recognition elements (MAREs) (Fujiwara et al. 1993; Igarashi et al. 1994; Igarashi and Sun 2006; Motohashi and Yamamoto 2007). Many of the MafK target genes such as heme oxygenase-1 (HO-1) and ferritin are oxidative stress-responsive genes. These target genes are repressed and activated by MafK-Bach1 and MafK-Nrf2 heterodimers, respectively (Igarashi and Sun 2006; Motohashi and Yamamoto 2007). Proteomics analysis of MafK-interactome revealed its interaction with both MATIIα and MATIIβ (Katoh et al. 2011). They are recruited specifically to MafK target genes such as HO-1 and globin genes (Fig. 5.2a). They are required for HO-1 gene repression by MafK and Bach1. Because the catalytic activity of MATIIα is required for the MafK target gene repression (Katoh et al. 2011), MATIIα is suggested to provide SAM locally on chromatin where it is recruited. Interestingly, chromatin of the enhancer regions of HO-1 gene contain repressive histone methylation such as histone H3 lysine 9 (H3 K9) and H3 K4 dimethylation which play roles in gene repression (Kim and Buratowski 2009). While exact connection between this particular methylation and MATIIα is not clear, the latter may be involved in writing of the repressive methylation when recruited locally.

One of the unexpected features of MATII is that MATIIα interacts with many chromatin-related proteins of diverse functions such as histone modification, chromatin remodeling, transcription regulation, and nucleo-cytoplasmic transport (Katoh et al. 2011) (Table 5.1). *Drosophila melanogaster* MATIIα (see below) binds to multiple proteins with known and unknown functions in yeast two hybrid assays

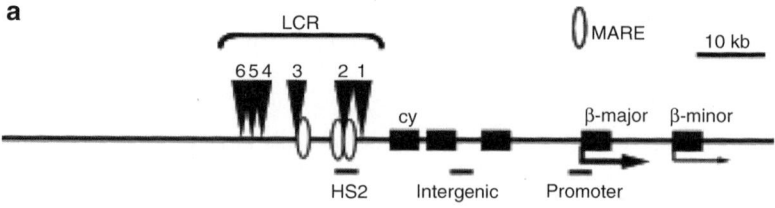
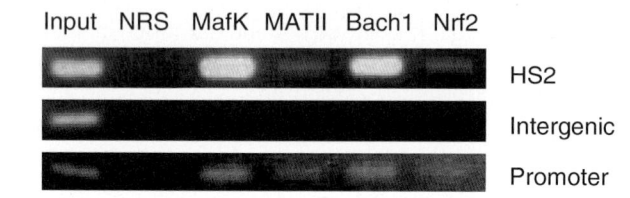
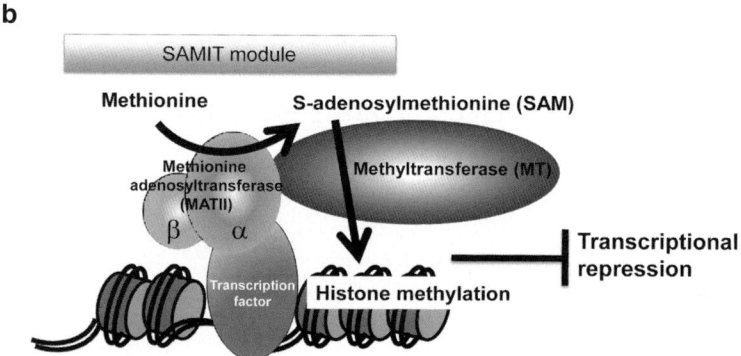

Fig. 5.2 Coupling of SAM synthesis and methylation by SAMIT module. (a) MATII was recruited to β-globin locus. Schematic representation of mouse β-globin locus. PCR primer pairs for ChIP analyses were designed to amplify the LCR hypersensitive sites, intergenic region, and β-major globin promoter. ChIP assays were performed by using anti-MATII, anti-MafK, anti-Bach1, and anti-Nrf2 antibodies and control rabbit IgG (NRS) with MEL cells. Gel images of PCR products of HS2, intergenic and promoter using input and precipitated chromatin as template are shown. NRS, normal rabbit serum. (b) The core of the SAM-dependent regulatory complex, α and β subunits, are named SAMIT module. This module is integrated into a complex with histone H1 and H3 methyltransferase activities. This complex catalyzes both SAM synthesis and histone methylation in vitro. SAMIT is likely to couple these two steps of the reaction by physically interacting with histone methyltransferases

(BioGRID:71491), suggesting its direct interaction with diverse proteins. Because MATIIα sediments in glycerol gradient analysis as several peaks including high molecular weight species, it appears to participate in multiple distinct complexes (Katoh et al. 2011). These observations suggest that the MATIIα interactome is heterogenous and modular in terms of function and structure.

Table 5.1 Shared interactome of MafK and MATIIα

Functional/proteomic cluster	Protein	Feature
(Complex)		
Swi/Snf	Baf53a	Requirement for maximal ATPase activity of BRG1
	Baf57	Remodeling histone H1-containing chromatin at the CD4 silencer
	Baf60b	Bridging interactions between transcription factors and SWI/SNF complexes
	Baf155	p53 target and a tumor suppressor by modulating p21(WAF1/CIP1) expression
	Baf180	Coronary vessel formation
NuRD	CHD4	Regulation of the DNA-damage response (DDR) and G1/S cell-cycle transition
	MBD3	A component of the NuRD co-repressor complex for development of pluripotent cells
	Gatad2b	Mediation of MBD2 and histone interaction
CHRAC	ACF1	Requirement for DNA replication through heterochromatin
PARP	PARP1	The attachment of ADP ribose units to target proteins
	Ku70	One of two subunits of Ku. Ku plays a critical role in the regulation of many cellular processes
	Ku80	Another of two subunits of Ku
	Ssrp1	The smaller of the two subunits of FACT. FACT mediates nucleosome reorganization
	Supt16h	The larger of the two subunits of FACT
(Specific domain)		
Chromodomain	CHD5	A tumor suppressor at human 1p36
	CHD6	A DNA-dependent ATPase and localization at nuclear sites of mRNA synthesis
	CHD7	A critical regulator of important developmental processes in organs affected by human CHARGE syndrome
	CHD8	Regulation of a serum response factor activity and smooth muscle cell apoptosis
	CHD9	A recently identified chromatin remodeler in osteogenic cell differentiation
(Function)		
DNA repair	Rad50	A regulator of cell cycle checkpoints and DNA repair
	Fen1	A member of the Rad2 structure-specific nuclease family, possesses 5'-exonuclease and gap-endonuclease activities
	HSP70	Ubiquitous molecular chaperones
Chromosome	p400	Modulation of cell fate decisions by the regulation of ROS homeostasis
	TopoIIα	The separation of chromosomes for DNA replication
	H3.3B	A replacement histone subtype
Transcription	Bach1	Transcriptional repression of oxidative-stress-response genes

(continued)

Table 5.1 (continued)

Functional/ proteomic cluster	Protein	Feature
Histone acetylation	Myst2	H4-specific histone acetylase, and a coactivator of the DNA replication licensing factor Cdt1
Histone methylation	G9a	Histone methyltransferase for dimethylation of histone H3 K9
	Ehmt1/GLP	Histone methyltransferase for dimethylation of histone H3 K9
	ALL1	Histone methyltransferase for dimethylation of histone H3 K4
	MATIIβ	A regulatory subunit of MATII
Proteolysis	Senp1	Redox sensors and effectors modulating the desumoylation pathway and specific cellular responses to oxidative stress
Transcriptional cofactor	Sin3a	A transcriptional regulatory protein
	PML	A member of the tripartite motif (TRIM) family
	DMAP1	A co-repressor that stimulates DNA methylation globally and locally at sites of double strand break repair
Transport	RanBP2	Direct interaction with the E2 enzyme UBC9 and strongly enhancing SUMO1 transfer
Spindle microtubule	Kif4	Regulation of activity-dependent neuronal survival by suppressing PARP-1 enzymatic activity
Other	Ubqln4	Proteasome-mediated degradation of proteins and interaction with ataxin-1

In the context of transcription regulation by MafK and Bach1, several MATIIα-interactants such as MATIIβ, poly-ADP-ribose polymerase 1 (PARP1), BAF53a, and CHD4 (Katoh et al. 2011) are critical for transcription repression. They form higher molecular weight complexes of roughly 600 kDa, among which at least MATIIβ, BAF53a, and CHD4 are recruited to HO-1. While MATIIα and β participate in MafK-mediated gene repression, the β subunit also plays a role independent of α subunit (see below). Therefore, the core of the SAM-dependent regulatory complex, comprising the α and β subunits, are named SAMIT module for SAM-integrating transcription repression. Because CHD4 is a rather large protein with many functional domains such as ATPase, chromo domain, and PHD finger, it may function as a platform to assemble SAMIT module into a larger operative complex and to organize its function with other proteins including PARP1 and BAF53a.

The fact that MATIIβ is recruited to MafK target genes strongly suggests that it is not a classical regulatory subunit for fine-tuning of its catalytic subunit, as has been considered previously (LeGros et al. 2000; Halim et al. 1999). A structural feature of MATIIβ also supports this idea. Human MATIIβ show 28% homology with a family of bacterial enzymes that catalyze the reduction of TDP-linked sugars such as dTDP-4-dehydrorhamnose reductase and other proteins involved in the production of polysaccharides. Therefore, MATIIβ itself may be an enzyme in

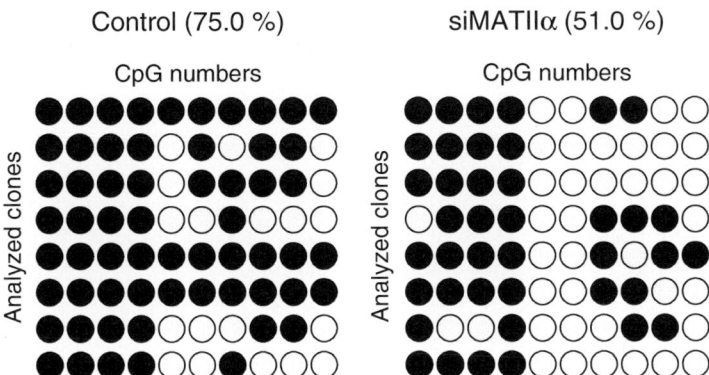

Fig. 5.3 Involvement of MATIIα in DNA methylation. Bisulphite sequencing analysis of LINE L1 locus in Hepa1 cells transfected with control or MATIIα siRNAs. The overall percentages of methylated CpGs (*black circle*) are indicated in parentheses

carbohydrate metabolism. MATIIβ interacts with RNA binding protein HuR. MATIIβ promotes cytoplasmic accumulation of HuR and translation of HuR target mRNAs such as cyclin A (Xia et al. 2010). Furthermore, MATIIβ interacts with diverse nuclear proteins (our unpublished observation) and re-evaluation of its function beyond the regulatory subunit of MATII is necessary.

5.3.2 Coupling of SAM Synthesis and Methylation by SAMIT

SAMIT module is integrated into a complex with histone H1 and H3 methyltransferase activities (Katoh et al. 2011). Furthermore, this complex catalyzes both SAM synthesis and histone methylation in vitro. In the newly developed assay, radio-labeled methionine is used instead of radio-labeled SAM. Methylation of target proteins such as histones can be detected only when a protein complex contains both SAM synthesizing and methyltransferase activities. SAMIT and its interacting proteins mediate efficient methylation of histone H1 and H3 in vitro (Katoh et al. 2011). Thus, SAMIT is likely to couple the two steps of the reaction by physically interacting with histone methyltransferases (Fig. 5.2b). Such a coupling may confer several advantages. First, juxtaposition of SAM synthesis to methylation can avoid genotoxic effects of SAM by lowering overall SAM levels but allowing local methylation reaction. Second, association and dissociation of SAMIT with methyltransferases provide a window for dynamic regulation. Methylation of chromatin may be promoted or inhibited depending on their interaction, conferring signal responsiveness of histone and/or DNA methylation.

Another interesting possibility is in the regulation of DNA methylation. MATIIα is also involved in DNA methylation. Upon its knockdown, DNA methylation at LINE L1 repeats is reduced (Fig. 5.3). Therefore, it will be important to examine interaction of SAMIT module with DNA methyltransferases. Active DNA demethylation has been suggested as a mechanism for conversion of methyl-cytosine to cytosine. Among several candidate enzymes for this reaction, DNA methyltransferase 3A (DNMT3A) is unique, in that it catalyzes both methylation and demethylation. However, demethylation can occur only when SAM concentration is very low. In this sense, SAM levels must fluctuate in a wide range. Considering that SAM is essential for diverse biochemical reactions, it is unlikely that gross SAM levels within cells become very low to allow DNA demethylation by DNMT3A (Wu and Zhang 2010). Coupling of MATII with methyltransferases is expected to resolve this conundrum since such a system allows lowering local SAM levels by dissociating MATII from target genes and/or methyltransferases.

5.3.3 Genetics of MATII in Drosophila melanogaster

In *Drosophila melanogaster*, modifier mutations of position effect variegation and *Polycomb* group (Pc-G) genes have been useful to identify regulators of chromatin structure. MATII was identified in such an analysis. *zeste* encodes a DNA binding protein that recognizes polycomb responsive elements (PREs) and regulatory regions of *white* gene. The neomorphic *zeste[1]* mutation renders the encoded protein extremely sticky and abnormally efficient in white gene repression. *Su(z)5* was reported as a suppressor mutation of *zeste[1]* and a homozygous embryonic lethal mutation (Person 1976), suggesting its involvement in transcription repression. This mutation was followed up further by others, revealing that *Su(z)5* is an enhancer of a polycomb mutation (Larsson et al. 1996), again suggesting its involvement in gene repression. The cloned wild-type allele of *Su(z)5* was found to encode MAT. Fly has two MAT isozyme genes and according to genome data base (Gene ID: 48552), *Su(z)5* encodes MATIIα. Even though *Su(z)5* enhance the *Pc* mutant, *Su(z)5* does not induce ectopic expression of the homeotic genes by itself (Larsson and Rasmuson-Lestander 1998). This finding may suggest that the MATIIα function is at least partially compensated by MATI or other chromatin proteins. The authors suggested that the phenotypes of *Su(z)5* mutation was due to an obstruction of the polyamine biosynthesis (Larsson et al. 1996). However, considering the findings regarding mouse MATIIα, it is more likely due to a collapse in methylation of chromatin.

5.3.4 Nuclear MATI

MATI is mainly cytoplasmic but a portion of it is localized within nuclei (Reytor et al. 2009). It is becoming clear that MATI also functions in the nucleus and regulates gene expression. Overexpression of MATI in CHO cells causes a mild but selective increase in trimethylation of H3 K27, suggesting that MATI is functionally coupled with a specific methyltransferase for this modification. This observation also suggests that SAM is limiting for histone methylation in nuclei. The presence of nuclear localization signal on MATI (Reytor et al. 2009) is consistent with its nuclear function. However, detailed nuclear function and mechanism of MATI are unclear.

5.4 Metabolic View of Epigenome

5.4.1 Nuclear Methionine Cycle?

All of the metabolic substrates involved in the methionine cycle are small and are expected to diffuse freely in a cell. From the viewpoint of biochemistry, however, these enzymatic reactions may be coupled for efficiency and regulation by enzyme compartmentalization, just as the case for the citric acid cycle. Indeed, SAH hydrolase is localized with in the nuclei in Xenopus laevis (Radomski et al. 1999). Because most methyltransferases bind SAH with higher affinity than SAM, they are subject to potent product inhibition (Chiang et al. 1996). Therefore, a nuclear localization of SAH hydrolase may facilitate rapid turnover of SAH, promoting histone methylation by histone and DNA methyltransferases within nuclei. However, it is not known whether nuclear SAH hydrolase is universal among other types of cells or it plays any physiological role within nucleus. Myc induces the expression of SAH hydrolase to promote mRNA Cap methylation (Fernandez-Sanchez et al. 2009), raising the possibility that the induced enzyme functions in nuclei. It will be an important question whether these enzymes are coupled on chromatin to regulate histone and DNA methylation.

5.4.2 Versions on a Theme: Ac-CoA and Deoxyribonucleotides in Nuclei

In the yeast *Saccharomyces cerevisiae*, Ac-CoA synthetase (Acs) catalyzes synthesis of Ac-CoA from acetate. One of its isozyme Acs2p is required for histone acetylation and global gene expression (Takahashi et al. 2006). While Acs2p is normally present in the nuclear compartment, this localization appears irrelevant for its function. A temperature-sensitive mutant of *acs2* can be rescued by prokaryotic Acs tagged with a nuclear export signal (Takahashi et al. 2006). This observation suggests that the nuclear and cytosolic Ac-CoA exists in a single pool. However, this observation should be carefully interpreted because the engineered protein may transit nuclei.

In contrast to yeast, most mammalian cells do not utilize acetate as a source of Ac-CoA synthesis. Ac-CoA outside of mitochondria is synthesized from mitochondoria-derived citrate by the enzyme ATP-citrate lyase (ACL). ACL is present in nuclei and required for histone acetylation (Wellen et al. 2009). However, it has not been tested whether nuclear localization of ACL is critical for its function. This is important to understand whether Ac-CoA is present in a single pool in the nuclear and cytosolic compartments (Takahashi et al. 2006).

In DNA repair and replication, a balanced supply of deoxyribonucleotide is essential. It has been reported that ribonucleotide reductase subunits accumulate very rapidly at a DNA damage site, suggesting that efficient DNA repair is dependent upon localized supply of substrates for DNA synthesis (Niida et al. 2010). This recruitment of ribonucleotide reductase subunits is dependent on their binding to histone acetylase Tip60. Furthermore, binding of these subunits to Tip60 is necessary for efficient DNA repair (Niida et al. 2010). However, not all enzymes for nucleotide synthesis are compartmentalized within nuclei. Purines are essential building blocks for RNA and DNA. De novo synthesis of adenosine and guanosine involves ten chemical reactions that transform phosphoribosyl pyrophosphate to inosine monophosphate. Enzymes involved in these reactions are mainly cytoplasmic and interact with each other when purines are low to form clusters or "purisome" to carry out de novo purine biosynthesis (An et al. 2008). Although the place for the reactions is cytoplasmic, this example reiterates the importance of interactions of enzymes in a pathway for efficient reactions and regulation. Therefore, even when cytoplasmic and nuclear compartments share a common pool of metabolites, localized assembly of enzymes is expected to promote efficient, tunable flow of metabolites and their ultimate incorporation into macromolecules as constituents or modifications.

5.5 Perspectives

While nuclear functions of the enzymes involved in the synthesis of SAM and Ac-CoA are emerging, our understanding is still fragmentary. We do not know much about concentrations of these metabolites in nuclei and how they change during processes such as stress response and differentiation. Because small metabolites would leak out of nuclei during biochemical fractionation, probes for in situ imaging of the relevant metabolites will be required to address this issue.

While MATII interacts with diverse nuclear proteins involved in different processes such as DNA replication and repair (Katoh et al. 2011), nothing is known about physiological and pathological significance of these interactions. It is very likely that localized SAM synthesis is coupled with writing of methyl mark on chromatin during these reactions.

Based on the fact that leukemic cells utilize significantly higher levels of SAM than normal lymphocytes, it has been shown that leukemic cells are more sensitive to reduction of MATIIβ by RNA interference (Attia et al. 2008). Therefore, therapeutic exploitation of the nuclear metabolic enzymes will be an interesting future issue.

In a broader perspective, mapping nuclear protein networks will be critical to understand a cross-talk of metabolism and epigenome. With procedures that allow recovery of weak and transient interaction, in combination with sensitive and accurate mass spectrometry analysis, we will uncover more and more subtle but salient interactions. Such interactions are usually visualized as substoichiometric bands in protein gels and have been discarded because they do not conform to the criteria of "complex". However, nuclear "complexome" appears to be modular in nature and stable hetero-oligomers of proteins (modules) participate in multiple different complexes (Malovannaya et al. 2011). Substoichiometric proteins may reflect protein modules that interact with a sub-fraction of complexes to fulfill specific and/or dynamic function. Therefore, SAMIT module will be a unique bridgehead to initiate our exploration into the modularity of nuclear protein and metabolic networks.

References

An S, Kumar R, Sheets ED, Benkovic SJ (2008) Reversible compartmentalization of de novo purine biosynthetic complexes in living cells. Science 320(5872):103–106

Attia RR, Gardner LA, Mahrous E, Taxman DJ, Legros L, Rowe S, Ting JP, Geller A, Kotb M (2008) Selective targeting of leukemic cell growth in vivo and in vitro using a gene silencing approach to diminish S-adenosylmethionine synthesis. J Biol Chem 283(45):30788–30795

Brown JM, Ball JG, Hogsett A, Williams T, Valentovic M (2010) Temporal study of acetaminophen (APAP) and S-adenosyl-L-methionine (SAMe) effects on subcellular hepatic SAMe levels and methionine adenosyltransferase (MAT) expression and activity. Toxicol Appl Pharmacol 247(1):1–9

Cantoni GL (1953) S-adenosylmethionine; a new intermediate formed enzymatically from L-methionine and adenosinetriphosphate. J Biol Chem 204:403–416

Cantoni GL (1975) Biological methylation: selected aspects. Annu Rev Biochem 44:435–451

Chiang PK, Cantoni GL (1977) Activation of methionine for transmethylation. Purification of the S-adenosylmethionine synthetase of bakers' yeast and its separation into two forms. J Biol Chem 252(13):4506–4513

Chiang PK, Gordon RK, Tal J, Zeng GC, Doctor BP, Pardhasaradhi K, McCann PP (1996) S-Adenosylmethionine and methylation. FASEB J 10(4):471–480

Chiba P, Wallner C, Kaiser E (1988) S-adenosylmethionine metabolism in HL-60 cells: effect of cell cycle and differentiation. Biochim Biophys Acta 971(1):38–45

Fernandez-Sanchez ME, Gonatopoulos-Pournatzis T, Preston G, Lawlor MA, Cowling VH (2009) S-adenosyl homocysteine hydrolase is required for Myc-induced mRNA cap methylation, protein synthesis, and cell proliferation. Mol Cell Biol 29(23):6182–6191

Fujiwara KT, Kataoka K, Nishizawa M (1993) Two new members of the maf oncogene family, mafK and mafF, encode nuclear b-Zip proteins lacking putative trans-activator domain. Oncogene 8(9):2371–2380

German DC, Bloch CA, Kredich NM (1983) Measurements of S-adenosylmethionine and L-homocysteine metabolism in cultured human lymphoid cells. J Biol Chem 258(18):10997–11003

Halim AB, LeGros L, Geller A, Kotb M (1999) Expression and functional interaction of the catalytic and regulatory subunits of human methionine adenosyltransferase in mammalian cells. J Biol Chem 274(42):29720–29725

Igarashi K, Sun J (2006) The heme-Bach1 pathway in the regulation of oxidative stress response and erythroid differentiation. Antioxid Redox Signal 8(1–2):107–118

Igarashi K, Kataoka K, Itoh K, Hayashi N, Nishizawa M, Yamamoto M (1994) Regulation of transcription by dimerization of erythroid factor NF-E2 p45 with small Maf proteins. Nature 367(6463):568–572

Katoh Y, Ikura T, Hoshikawa Y, Tashiro S, Ito T, Ohta M, Kera Y, Noda T, Igarashi K (2011) Methionine adenosyltransferase II serves as a transcriptional corepressor of Maf oncoprotein. Mol Cell 41(5):554–566

Kim T, Buratowski S (2009) Dimethylation of H3K4 by Set1 recruits the Set3 histone deacetylase complex to 5′ transcribed regions. Cell 137(2):259–272

Kotb M, Mudd SH, Mato JM, Geller AM, Kredich NM, Chou JY, Cantoni GL (1997) Consensus nomenclature for the mammalian methionine adenosyltransferase genes and gene products. Trends Genet 13(2):51–52

Lan F, Nottke AC, Shi Y (2008) Mechanisms involved in the regulation of histone lysine demethylases. Curr Opin Cell Biol 20(3):316–325

Larsson J, Rasmuson-Lestander A (1998) Somatic and germline clone analysis in mutants of the S-adenosylmethionine synthetase encoding gene in *Drosophila melanogaster*. FEBS Lett 427(1):119–123

Larsson J, Zhang J, Rasmuson-Lestander A (1996) Mutations in the *Drosophila melanogaster* gene encoding S-adenosylmethionine synthetase [corrected] suppress position-effect variegation. Genetics 143(2):887–896

LeGros HL Jr, Halim AB, Geller AM, Kotb M (2000) Cloning, expression, and functional characterization of the beta regulatory subunit of human methionine adenosyltransferase (MAT II). J Biol Chem 275(4):2359–2366

Malovannaya A, Lanz RB, Jung SY, Bulynko Y, Le NT, Chan DW, Ding C, Shi Y, Yucer N, Krenciute G et al (2011) Analysis of the human endogenous coregulator complexome. Cell 145(5):787–799

Markham GD, Pajares MA (2009) Structure-function relationships in methionine adenosyltransferases. Cell Mol Life Sci 66(4):636–648

Mato JM, Alvarez L, Ortiz P, Pajares MA (1997) S-adenosylmethionine synthesis: molecular mechanisms and clinical implications. Pharmacol Ther 73(3):265–280

Mitsui K, Teraoka H, Tsukada K (1988) Complete purification and immunochemical analysis of S-adenosylmethionine synthetase from bovine brain. J Biol Chem 263(23):11211–11216

Motohashi H, Yamamoto M (2007) Carcinogenesis and transcriptional regulation through Maf recognition elements. Cancer Sci 98(2):135–139

Niida H, Katsuno Y, Sengoku M, Shimada M, Yukawa M, Ikura M, Ikura T, Kohno K, Shima H, Suzuki H et al (2010) Essential role of Tip60-dependent recruitment of ribonucleotide reductase at DNA damage sites in DNA repair during G1 phase. Genes Dev 24(4):333–338

Person K (1976) Modification of the eye colour mutant zeste by suppressor, enhancer and minute genes in *Drosophila melanogaster*. Hereditas 82:111–119

Posnick LM, Samson LD (1999) Influence of S-adenosylmethionine pool size on spontaneous mutation, dam methylation, and cell growth of *Escherichia coli*. J Bacteriol 181(21):6756–6762

Radomski N, Kaufmann C, Dreyer C (1999) Nuclear accumulation of S-adenosylhomocysteine hydrolase in transcriptionally active cells during development of *Xenopus laevis*. Mol Biol Cell 10(12):4283–4298

Reytor E, Perez-Miguelsanz J, Alvarez L, Perez-Sala D, Pajares MA (2009) Conformational signals in the C-terminal domain of methionine adenosyltransferase I/III determine its nucleo-cytoplasmic distribution. FASEB J 23(10):3347–3360

Rydberg B, Lindahl T (1982) Nonenzymatic methylation of DNA by the intracellular methyl group donor S-adenosyl-L-methionine is a potentially mutagenic reaction. EMBO J 1:211–216

Sakata SF, Shelly LL, Ruppert S, Schutz G, Chou JY (1993) Cloning and expression of murine S-adenosylmethionine synthetase. J Biol Chem 268(19):13978–13986

Suma Y, Yamanaka Y, Tsukada K (1983) Synthesis of S-adenosylmethionine synthetase isozymes from rat and mouse livers. Immunochemical studies. Biochim Biophys Acta 755(2):287–292

Takahashi H, McCaffery JM, Irizarry RA, Boeke JD (2006) Nucleocytosolic acetyl-coenzyme a synthetase is required for histone acetylation and global transcription. Mol Cell 23(2):207–217

Wellen KE, Hatzivassiliou G, Sachdeva UM, Bui TV, Cross JR, Thompson CB (2009) ATP-citrate lyase links cellular metabolism to histone acetylation. Science 324(5930):1076–1080

Wu SC, Zhang Y (2010) Active DNA demethylation: many roads lead to Rome. Nat Rev Mol Cell Biol 11:607–620

Xia M, Chen Y, Wang LC, Zandi E, Yang H, Bemanian S, Martinez-Chantar ML, Mato JM, Lu SC (2010) Novel function and intracellular localization of methionine adenosyltransferase 2beta splicing variants. J Biol Chem 285(26):20015–20021

Chapter 6
Epigenetic Regulation of Male Germ Cell Differentiation

Oliver Meikar, Matteo Da Ros, and Noora Kotaja

Abstract Male germ cell differentiation is a complex developmental program that produces highly specialized mature spermatozoa capable of independent movement and fertilization of an egg. Germ cells are unique in their capability to generate new organisms, and extra caution has to be taken to secure the correct inheritance of genetic and epigenetic information. Male germ cells are epigenetically distinct from somatic cells and they undergo several important epigenetic transitions. In primordial germ cells (PGCs), epigenome is reprogrammed by genome-wide resetting of epigenetic marks, including the sex-specific imprinting of certain genes. Postnatal spermatogenesis is characterized by drastic chromatin rearrangements during meiotic recombination, sex chromosome silencing, and compaction of sperm nuclei, which is accomplished by replacing near to all histones by sperm-specific protamines. Small RNAs, including microRNAs (miRNAs), endogenous small interfering RNAs (endo-siRNAs) and PIWI-interacting RNAs (piRNAs) are also involved in the control of male gamete production. The activities of small RNAs in male germ cells are diverse, and include miRNA- and endo-siRNA-mediated posttranscriptional mRNA regulation and piRNA-driven transposon silencing and the control of DNA methylation in PGCs. In this chapter, we give a brief review on the epigenetic processes that govern chromatin organization and germline-specific gene expression in differentiating male germ cells.

O. Meikar • M. Da Ros • N. Kotaja (✉)
Institute of Biomedicine, Department of Physiology, University of Turku,
Kiinamyllynkatu 10, Turku FIN-20520, Finland
e-mail: noora.kotaja@utu.fi

6.1 Introduction to Spermatogenesis

6.1.1 From Primordial Germ Cells to Spermatozoa

In mammalian embryos, germ cells are specified shortly after implantation, emerging from the epiblast that consists of pluripotent cells. Before gastrulation, precursor cells of primordial germ cells (PGCs) are induced within the proximal rim of the epiblast by morphogenetic signals provided from adjacent extra-embryonic ectoderm. During gastrulation, the nascent PGCs start to migrate toward the future gonads and rapidly proliferate (Matsui 2010). After migrating to the gonadal ridge, PGCs become gonocytes within cords that are formed by Sertoli precursors and surrounded by peritubular cells. Gonocytes show a burst of mitotic activity, then arrest in the G_0 phase of cell cycle remaining mitotically quiescent until after birth, when they give rise to spermatogonia (de Rooij and Russell 2000). The postnatal production of mature spermatozoa from spermatogonial stem cells is a complex process that takes place in seminiferous tubules inside the testis (Fig. 6.1). Sertoli cells are the somatic cells in the seminiferous epithelium that control the correct progression of spermatogenesis through cell-cell interactions and communication with the differentiating germ cells. Sertoli cell function itself is regulated by the hypothalamic-pituitary-gonadal axis, and depends on the actions of follicle stimulating hormone (FSH) and testosterone (T, produced by Leydig cells in the testis after stimulation by luteinizing hormone, LH) (Fig. 6.1).

Spermatogenesis includes proliferation, differentiation and morphogenesis of male germ cells (Hess and de Franca 2008). The process begins when diploid spermatogonia multiply by consecutive mitotic divisions and then enter the meiotic program and become spermatocytes. During meiosis, aligned homologous chromosomes pair and the synaptonemal complex is formed. Synapsis permits genetic crossover at sites along the synaptonemal complex known as recombination nodules. Finally the synaptonemal complex disintegrates, bivalent chromosomes align on the metaphase plate and sister chromatids dissociate into two daughter cells. The second meiotic division of secondary spermatocytes results in the production of haploid spermatids. The postmeiotic developmental phase, spermiogenesis, involves the differentiation of spermatids into spermatozoa. This phase includes an enormous morphogenetic transformation involving DNA compaction, cytoplasmic ejection and acrosome and flagellar formation (Kimmins et al. 2004).

Sertoli cells are attached to the basal lamina of the seminiferous tubules and extend the cytoplasm towards the lumen and interact with all the differentiating germ cell types. Spermatogonia are also localized at the level of the basal lamina, while spermatocytes, round spermatids, elongating spermatids and mature spermatozoa are present at successive steps moving from the basal lamina to the lumen. Spermatogenesis is completed when the mature spermatozoa are released from the Sertoli cells and transported to the epididymis. The whole process occurs in an ordered manner, referred to as the spermatogenic cycle, which is divided in a species-specific number of precisely timed stages (i.e. 12 in the mouse), easily

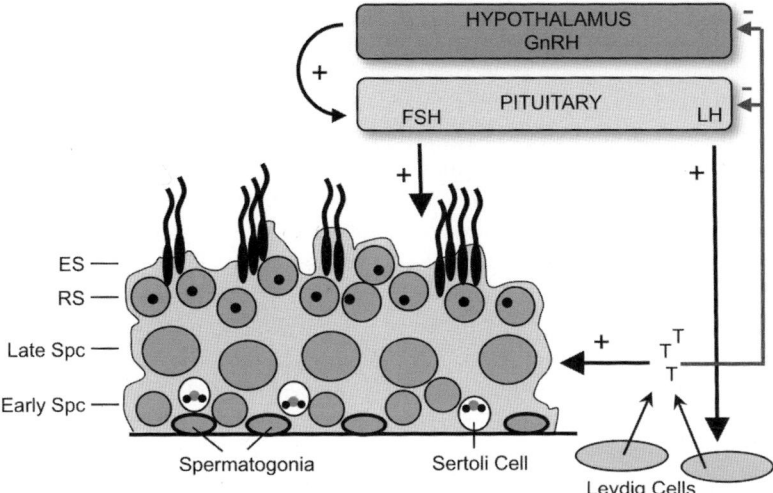

Fig. 6.1 Organization of seminiferous epithelium and hormonal control of spermatogenesis. Inside the seminiferous tubules, germ cells are embedded in the cytoplasmic pockets of somatic Sertoli cells that regulate their functions through direct cell contacts and paracrine signaling. The least differentiated germ cells, spermatogonia, are localized in the basal compartment of the tubule close to the basal lamina. In the course of differentiation, germ cells move towards the luminal compartment so that the next layer from spermatogonia contains meiotic spermatocytes, followed by the postmeiotic round spermatids and finally elongating spermatids. Mature sperm is released into the lumen. Somatic, testosterone-producing Leydig cells are situated in the interstitial space between the tubules. Spermatogenesis is under a direct control of the hypothalamus-pituitary axis. Gonadotropin-releasing hormone (GnRH) secreted by hypothalamus induces the production of follicle-stimulating hormone (FSH) and luteinizing hormone (LH) in pituitary gland. LH stimulates Leydig cells to produce testosterone (T) that controls spermatogenesis. FSH acts on Sertoli cells inside the seminiferous epithelium. *Early Spc* early spermatocytes, *Late Spc* late spermatocytes, *RS* round spermatids, *ES* elongating spermatids

recognizable from the cell associations (spermatogonia, spermatocytes and spermatids) present in a cross section of the seminiferous tubules (Oakberg 1956). This organization ensures the constant production of high number of spermatozoa during the whole course of sexual maturity.

6.1.2 Regulation of Gene Expression During Spermatogenesis

It has been estimated that 60% of the mouse genome is expressed during testis development from birth to adulthood. Since male germ cell differentiation consists of several unique processes and mechanisms, some of these genes are either male germ cell -specific or testis predominant. Indeed, according to the transcriptome analyses, even 4% of the mouse genome seems to be dedicated to the expression of genes unique to spermatogenesis (Lee et al. 2009). Gene expression is governed by highly specific regulatory rules in order to control chromatin organization as well as transcriptional events during spermatogenesis (Kimmins et al. 2004; Kimmins and Sassone-Corsi 2005). Spermatocytes and round spermatids are transcriptionally very active, but during the late steps of spermatogenesis, the compaction of sperm chromatin results in a drastic inhibition of transcriptional activity (Tanaka and Baba 2005). Translational control of mRNAs becomes prominent, and after chromatin condensation, gene control is regulated almost exclusively at the posttranscriptional level (Kleene 2003). This is reflected by a high number of RNA-binding proteins in spermatogenic cells, many of them being testis-specific (Paronetto and Sette 2010). Similar to the germline of many different organisms, mammalian male germ cells are characterized by distinct RNA and protein-rich non-membranous cytoplasmic domains called germ granules (Eddy 1970; Chuma et al. 2009). The most prominent of them are the intermitochondrial cement (IMC) in pachytene spermatocytes and the chromatoid body (CB) in round spermatids (Meikar et al. 2011). Given their distinct features and the known protein and RNA composition, germ granules are likely to have a specialized role in mRNA regulation and small RNA-dependent pathways during male germ cell differentiation (Meikar et al. 2011).

6.2 Chromatin Modifications in Male Germ Cells

6.2.1 Chromatin Modifying Proteins

Epigenetic mechanisms play an important part in the control of gene expression programs in male germline. Furthermore, differentiating male germ cells undergo several important epigenetic transitions that are characterized by extensive chromatin alterations (Fig. 6.2). These include the reprogramming of epigenetic marks in PGCs, meiotic chromatin organization and recombination events, and the compaction of haploid genome inside sperm nuclei (Kota and Feil 2010). Therefore, it is not surprising that several chromatin modifying enzymes are expressed during male germ cell differentiation in temporally regulated manner, and their functions are required for the correct development of a male gamete (Godmann et al. 2009). The methylation of cytosine residues of genomic DNA, which takes place predominantly within CpG dinucleotides, is associated with gene silencing, especially in processes such as transposon and repeat satellite sequence silencing and imprinting. Intriguingly, the sperm genome-wide methylation patterns differ markedly from that of somatic cells, thus highlighting the importance of DNA methylation in the control of male germline –specific processes, including the maintenance of unique chromosomal structures in male germ cells (La Salle et al. 2007; Oakes et al. 2007). Five DNA (cytosine-5)-methyltransferases (DNMTs) are characterized in mammals (Trasler 2009). DNMT1 is critical for maintenance of methylation patterns during replication of DNA. DNMT3A and DNMT3B are *de novo* methyltransferases that are highly expressed during embryonic development. DNMT3L does not have

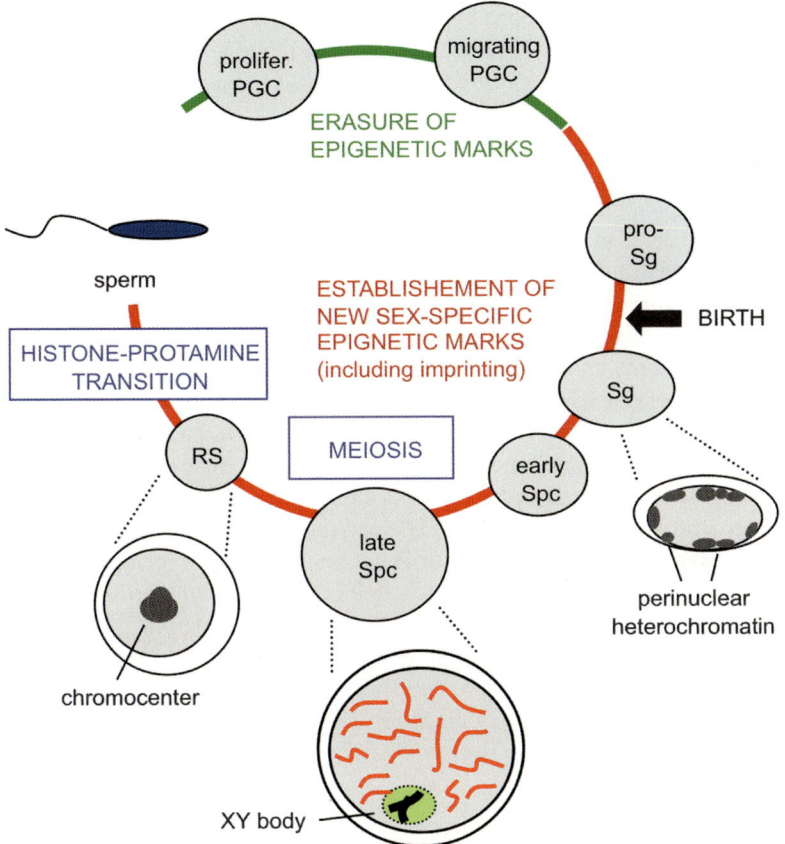

Fig. 6.2 Epigenetic transitions and chromatin dynamics in male germ cells. Germ cell development is accompanied by a massive reprogramming of epigenetic marks. Erasure of the epigenetic marks (indicated with *green line* and *green font*) takes place in PGCs. Novel male specific marks, including the imprinting of certain genes (indicated with *red line* and *red font*) starts in PGCs and continues during the postnatal differentiation. The major epigenetic processes that involve dramatic chromatin dynamics and reorganization, meiosis and chromatin compaction in late spermiogenesis by the histone-protamine transition, are indicated. Special heterochromatin structures involved in gene silencing in male germ cells are highlighted by *drawings*. These structures include the perinuclear heterochromatin in differentiating spermatogonia, the sex body (XY body) in spermatocytes and the chromocenter in spermatids. The details of formation and the purpose of perinuclear heterochromatin in spermatogonia is still obscure. Sex body forms during meiosis when sex chromosomes are condensed and silenced by MSCI. Homologous chromosomes paired by synaptonemal complex are shown in *red* in the nucleus. X and Y chromosomes condensed by heterochromatinization are shown in *black* and sex body is indicated as *green circle*. Chromocenter in the nucleus of round spermatids contains the centromeric heterochromatin of each chromosome, and is thought to serve as an initial organizer of the DNA packing that takes place in the late spermiogenesis

catalytic activity but it functions by interacting with DNMT3A and DNMT3B and modulating their activities. DNMT1, DNMT3A, DNMT3B and DNMT3L are all differentially expressed in germ cell type-specific patterns in both prenatal and postnatal male germ cells (Trasler 2009).

A defined composition of repressive and activating histone modifications ensures proper meiotic and postmeiotic events, but also mediates a highly orchestrated expression of male germ cell –specific genes. Disturbances of the balance between different epigenetic marks usually results in a failure of male germ cell development, as demonstrated by several knockout studies (Godmann et al. 2009). For example, correct methylation patterns on H3 are crucial for normal progression of spermatogenesis, since the deletion H3 methyltransferases has been shown to lead to meiotic problems and male infertility. Double knockout mice for H3 Lysine 9 (H3K9) trimethylases KMT1A and KMT1B (also called SUV39H1 and SUV39H2) and knockout mice for mono/dimethylase KMT1C (also called G9A or EHMT2) show similar meiotic phenotypes even though the mechanisms are presumably different because KMT1A/B are known to catalyze H3K9 methylation associated with heterochromatin, whereas KMT1C acts on euchromatin (Peters et al. 2001; Tachibana et al. 2007). Methylation of H3 at residue K4 is also important, and the deletion of H3K4 histone methyltransferase PRDM9 (also called MEISETZ) results in sterility due to a disrupted spermatogenesis at the meiotic pachytene stage (Hayashi et al. 2005). Likewise, other histone modifications such as acetylation, phosphorylation and ubiquitination, have been demonstrated as essential regulators of spermatogenesis (Kimmins et al. 2007; Kolthur-Seetharam et al. 2009; Roest et al. 1996; Baarends et al. 2003).

6.2.2 *Epigenetic Reprogramming*

In germ cells, the somatic cell state has to be reprogrammed by genome-wide epigenetic resetting. Reprogramming begins during embryonic development at the time when PGCs migrate and colonize genital ridges (Fig. 6.2). The process begins by near to complete erasure of somatic methylation marks, and is followed by the re-establishment of novel sex-specific DNA methylation patterns, including differential imprinting of the genes in the male and female germ cells (Sasaki and Matsui 2008; Ewen and Koopman 2010). *De novo* methyltransferases DNMT3A and DNMT3B have been demonstrated to be critical for gene imprinting and silencing of repeat sequences in the germline (Kaneda et al. 2004; Kato et al. 2007). While other methyltransferases are expressed also in somatic cells, the expression of the catalytically inactive family member, DNMT3L, is restricted to male and female germ cells. DNMT3L deficiency results in defective methylation of retrotransposons, which induces aberrant transposon expression in the germline (La Salle et al. 2007; Bourc'his and Bestor 2004; Webster et al. 2005). These changes results in severe problems in meiotic progression including widespread nonhomologous synapsis at the pachytene stage. Meiotic defects may be derived from inappropriate alignment of unmethylated retrotransposon elements, chromosome breaks induced by aberrant retrotransposition events, or altered gene expression. DNA demethylation in PGCs during epigenetic reprogramming is also linked to changes in nuclear architecture, loss of histone modifications, and widespread histone replacement (Seki et al. 2007; Hajkova et al. 2008). The molecular details of massive DNA demethylation was recently clarified by a study demonstrating that the base excision DNA repair pathway provides a mechanism for the demethylation and the extensive chromatin remodeling in the mouse PGCs (Hajkova et al. 2010). Some epigenetic marks remain intact during the epigenetic reprogramming, and thus provide possible means for transgenerational inheritance of epigenome – it is also clear that the period of epigenetic reprogramming is critical for correct maintenance of germline epigenome, and it is especially vulnerable to environmental agents that may induce changes in the epigenome (discussed below).

6.2.3 Epigenetic Control of Meiosis

A central process in gametogenesis is meiosis, which leads to the conversion of diploid cells to haploid gametes. Meiotic events are controlled by epigenetic mechanisms including DNA and histone modifications and chromatin remodeling. Several factors that control histone methylation are essential for meiotic transitions, particularly in the male germline (Nottke et al. 2009). However, on the basis of current knowledge, it is still unclear whether the meiotic problems are caused by the defects in meiosis or for example by indirect effects due to altered gene expression. During meiosis, homologous chromosomes become aligned and form synapsis, which is a prerequisite for the chromosomal crossover and meiotic recombination. Recombination events guarantee that the new organism generated after fertilization is genetically different from its parents. In mammals, crossovers cluster at preferential sites, which are called recombination hotspots. In addition to specific DNA elements, the initiation sites of recombination are enriched with certain epigenetic modifications such as H3K4 trimethylation and H3K9 acetylation (Buard et al. 2009). Specific sequence elements connected to recombination hotspots are recognized by a meiosis-specific H3 methyltransferase PRDM9 (Baudat et al. 2010). Binding of PRDM9 leads to trimethylation of H3K4, which subsequently triggers SPO11-meidated double strand break formation that is required for crossovers. It is important to prevent illegitimate recombination and crossing over at unpaired sites and between nonhomologous chromosomes. A process called meiotic sex chromosome inactivation (MSCI) is utilized to silence unsynapsed sex chromosomes in a chromosomal domain called sex body, and avert the deleterious recombination events during first meiotic prophase (Burgoyne et al. 2009). MSCI and sex body formation are controlled by epigenetic mechanisms, including incorporation of specific histone variants and specific histone modifications, such as phosphorylation, ubiquitination and sumoylation. ATR kinase-mediated phosphorylation of histone variant H2AX at serine-139 (γH2AX), and the subsequent accumulation of γH2AX is known to be an important signal for triggering MSCI (Burgoyne et al. 2009).

6.2.4 Postmeiotic Histone Replacement and Chromatin Compaction

After the two meiotic divisions, the final step of spermatogenesis, spermiogenesis, takes place during which the maturation of spermatids to fully differentiated spermatozoa is completed. Postmeiotic male germ cells undergo a remarkable chromatin remodeling, which enables the extreme compaction of male germ cell nuclei. Chromatin compaction may help to optimize nuclear shape and hence support the ability of sperm cells to swim across the female reproductive tract, and it may also confer additional protection from the effects of genotoxic factors (Kimmins and Sassone-Corsi 2005; Miller et al. 2010). Despite the fundamental nature of this process, the molecular basis of the mechanisms involved remains largely unknown. The major steps in mammals involve the replacement of most of the nucleosomal histones first by transition proteins (TNPs) and subsequently by protamines (PRMs). The timing of TNP and PRM expression is tightly regulated and involves local chromatin changes, including H3K9 demethylation by KDM3A/JHDM2A (Okada et al. 2007). Despite the fact that TNPs constitute even 90% of all chromatin basic proteins after histone removal and before protamines deposition, the exact role of TNPs in histone replacement and chromatin compaction is still obscure. In double knockout mice lacking both TNP1 and TNP2, histones were found to be displaced normally, demonstrating that TNPs do not play a role in the removal of histones. *Tnp1/Tnp2* double knockout spermatids have, however, problems in chromatin condensation (Zhao et al. 2004a, b).

Protamines are small, highly basic arginine-rich proteins that are evolutionarily related to histone H1. The properties of protamines enable over ten times more efficient packing of paternal DNA than what is achieved by histone-built nucleosomes. Mice and humans express two protamines, PRM1 and PRM2. Protamines are essential for the production of spermatozoa, and the haploinsufficiency of either PRM1 or PRM2 disrupts nuclear formation and normal sperm function (Cho et al. 2001). Even though protamine incorporation covers most of the sperm genome, many regions retain nucleosomal histones, which are enriched in specific epigenetic modifications, such as trimethylated H3K4, trimethylated H3K27, and unmethylated DNA. These loci include key developmental genes, imprinted genes, microRNAs and homeotic genes. The specific epigenetic modifications of these genes potentially contribute to the appropriate gene expression during early embryonic development (Kota and Feil 2010; Hammoud et al. 2009; Brykczynska et al. 2010).

The exact mechanisms for histone replacement are still unclear. As discussed below, the incorporation of histone variants into nucleosomes prior to histone-protamine transition is probably involved in creating less stable nucleosomes. In addition, a genome-wide massive histone hyperacetylation, which takes place in late round spermatids and early elongating spermatids before chromatin compaction, seems to be tightly linked to histone replacement (Hazzouri et al. 2000). A testis-specific bromodomain-containing protein BRDT, which binds acetylated lysines and is implicated in chromatin remodeling, is associated with acetylated histones located in the pericentric regions in elongating spermatids (Govin et al. 2006). Functional BRDT is required for the normal elongation of spermatids, suggesting that it may be involved in acetylation-mediated events during chromatin compaction (Shang et al. 2007). Interestingly, ubiquitination of H2A and H2B by the E3 ubiquitin ligase RNF8 has been demonstrated to be an important signal in promoting H4 hyperacetylation in elongating spermatids (Lu et al. 2010). Disruption of *Rnf8* gene in mice disrupts the H4 hyperacetylation and the mice are deficient in global nucleosome removal (Lu et al. 2010).

6.2.5 Histone Variants

Before the replacement of almost all the histones by protamines, the male germ cell genome is organized by the replacement of canonical histones by a variety of histone variants (Kimmins and Sassone-Corsi 2005; Gaucher et al. 2010; Boussouar et al. 2008). A very high number of both core and linker histone variants are expressed in male germ cells, many of them being testis-specific. The large scale histone replacement starts already very early during embryonic germ cell development, as demonstrated by a study revealing the involvement of histone displacement in the erasure and re-establishment of DNA methylation in PGCs (Hajkova et al. 2008). However, the most active nucleosome re-organization by histone variants takes place during meiotic and postmeiotic differentiation. During meiosis, the formation of sex body in pachytene spermatocytes is accompanied by the incorporation of specific histone variants such as H3.3 and macroH2A. Histone variants are suggested to create less stable nucleosomes, thus a large-scale incorporation of histone variants in haploid cells serves as a potential mechanism for histone replacement by protamines. It has been hypothesized that the incorporated histone variants form chromatin domains with unstable nucleosomes, which may then constitute preferential targets for nucleosome disassembly and histone displacement (Gaucher et al. 2010). A region-specific assembly of histone variants in specific chromosomal domains has been reported in spermatids. Two H2A-like variants H2AL1 and H2AL2 are synthesized in elongating spermatids and become specifically associated with pericentric regions just before and during the assembly of protamines, thus linking these variants to the differential organization of pericentric heterochromatin during mouse spermiogenesis (Govin et al. 2007). A testis-specific linker histone variant H1T2 associates with chromatin domains localized at the apical pole of the nucleus of late round and elongating spermatids, and is thus involved in creating a polarity in the spermatid nucleus (Martianov et al. 2005). Mice deficient for H1T2 show impaired chromatin compaction resulting in reduced fertility (Martianov et al. 2005).

6.3 Small RNAs in Male Germ Cells

6.3.1 Introduction to Male Germline Small RNAs

There are at least three different populations of small RNAs in mammalian testis: microRNA (miRNAs), endogenous small interfering RNAs (endo-siRNAs) and PIWI-interacting RNAs (piRNAs) (Fig. 6.3). The miRNA pathway is a well-understood posttranscriptional silencing mechanism. miRNAs precursors are endogenous long hairpin loop primary transcripts, which are processed in the nucleus by double-stranded RNA endonuclease Drosha into pre-miRNAs. In the cytoplasm, the pre-miRNAs become targets for another endonuclease, Dicer, which cuts them into short (usually 21 nt) double-stranded miRNAs. Eventually one strand

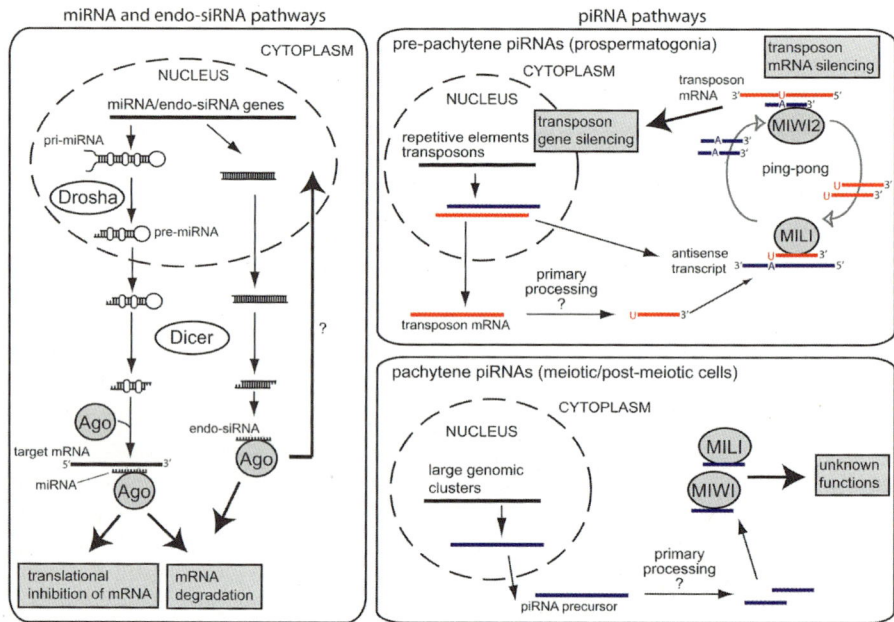

Fig. 6.3 Biosynthesis and functions of small RNAs in mammalian male germline. miRNAs are produced from long imperfect hairpin loops and cut successively by Drosha and Dicer into mature miRNAs. Endo-siRNAs are processed by Dicer from double-stranded RNA precursors. miRNAs and siRNAs form RNA-induced silencing complex (RISC) together with AGO proteins, and regulate the stability or translation of their target mRNAs. Other functions, including the epigenetic chromatin modification in the nucleus, have also been suggested. The production of MILI and MIWI2-bound pre-pachytene piRNAs is explained by the so-called ping-pong mechanism where piRNAs from the opposite strand are promoting each others synthesis in a feed-forward way. Pre-pachytene piRNAs silence transposons posttranscriptionally as well as at the chromatin level through methylation-dependent mechanisms. The primary piRNAs that start the ping-pong cycle are cut out from the transposon mRNAs by an unknown mechanism. Pachytene piRNAs bind MIWI and MILI and are produced in vast quantities presumably from long single-stranded piRNA precursors. The details of their production mechanisms or functions are not yet characterized

of the processed mature miRNA is loaded onto AGO subfamily member of Argonaute proteins in the RNA induced silencing complex (RISC), where it guides the recognition of target mRNAs by imperfect base-pairing leading to mRNA translational inhibition or decay (Lee et al. 2004; Krol et al. 2010). The predominant function of mammalian miRNAs has been reported to be the destabilization of their target mRNAs (Guo et al. 2010). Mature siRNAs are similar to mature miRNAs, but they are fully complementary with their targets and result in the target RNA cleavage. siRNAs derive from long dsRNA precursors and their synthesis is Drosha-independent and requires only Dicer activity. siRNA precursors are usually introduced in cells exogenously, for example by viruses, but as originally described in plants and nematodes, also endogenous siRNAs (endo-siRNAs) are produced and participate in gene silencing (Lau 2010). Moreover, there is increasing evidence that endo-siRNAs can be used as a gene control mechanism in mammals (Lau 2010; Tam et al. 2008; Watanabe et al. 2008; Song et al. 2011).

In contrast to miRNAs that are widely expressed in different tissues, piRNAs are predominantly expressed in the germline and are currently mostly known by their role in the silencing of transposable elements that become activated during DNA demethylation in PGCs (Siomi et al. 2011). piRNAs are named by their direct association with the PIWI proteins (MILI, MIWI and MIWI2 in mice), that belong to evolutionarily conserved Argonaute protein family together with AGO proteins (Cenik and Zamore 2011). piRNAs were first identified either by sequencing the characteristic ~30-nt band from total testis RNA extracts or by sequencing the small RNAs that co-immunoprecipitated with PIWI proteins in mouse testis (Aravin et al. 2006; Girard et al. 2006; Grivna et al. 2006; Watanabe et al. 2006). piRNAs are highly abundant in mammalian testis, and after several deep sequencing experiments, the suggested amount of different piRNAs is in hundreds of thousands (Aravin et al. 2007a). Processing of piRNAs does not involve Drosha or Dicer, and is thus separate from miRNA and siRNA biogenesis. Careful analysis of piRNAs that associate with different PIWI proteins has revealed at least two distinguishable classes of piRNAs in mammalian testis which are named by their expression time as pre-pachytene and pachytene piRNAs. Although similar at the molecular level, they have different tasks and mechanism of action, they associate with different PIWI proteins and derive from different regions of the genome.

6.3.2 Control of Spermatogenesis by miRNAs and Endogenous siRNAs

Components of miRNA machinery are found in the male germline and there are several miRNAs that are expressed either specifically or predominantly in the testis (Ro et al. 2007; Chiang et al. 2010; Gonzalez-Gonzalez et al. 2008; Kotaja et al. 2006; Korhonen et al. 2011). Interestingly, many of the testis-expressed miRNA genes are located in miRNA gene clusters on the X chromosome where they are able to escape from the transcriptional silencing of sex chromatin during meiosis by an unknown mechanism (Song et al. 2009). The importance of Dicer in mouse female and male germ cell maturation has been demonstrated by studies analysing different conditional tissue-specific knockout mouse lines (Korhonen et al. 2011; Murchison et al. 2007; Tang et al. 2007; Hayashi et al. 2008; Maatouk et al. 2008). In mouse testis, Sertoli cell–specific deletion of *Dicer1* revealed its crucial importance for the normal function of these somatic nursing cells in supporting male germ cell differentiation (Papaioannou et al. 2009, 2010). Knockout mice with spermatogonia-specific deletion of *Dicer1* demonstrated the imperative role of Dicer and Dicer-dependent small RNAs in the regulation of postnatal male germ cell development (Korhonen et al. 2011). The most prominent defects in *Dicer1*-depleted spermatids were found during haploid differentiation, especially in the elongation of nuclei and the organization of chromatin. The molecular mechanisms that are used by Dicer-dependent pathways to control spermatogenesis are still unclear. Posttranscriptional control is of central importance during haploid differentiation due to the transcriptional silencing that results from the tight packing of chromatin with protamines (Kimmins and Sassone-Corsi 2005; Gaucher et al. 2010). Since the most striking defects in the differentiation of Dicer-null male germ cells coincide chronologically with chromatin condensation and transcriptional silencing, it can be envisaged that Dicer is involved in this posttranscriptional control of haploid mRNAs.

Dicer is involved also in the production of endo-siRNAs in male germline, which emphasizes the diversity of Dicer-dependent pathways (Song et al. 2011). Endo-siRNA pathways were originally thought to be absent in mammalian cells due to an aggressive innate immune response to intracellular dsRNAs that is indicative of a viral infection (interferon response) (Sen and Sarkar 2007). Germ cells are notable exceptions since they do not activate such a response when injected with dsRNA (Svoboda et al. 2000). Male germ cell endo-siRNAs have been reported to function in posttranscriptional control of a wide variety of protein encoding mRNAs, but other functions, possibly related to chromatin modifications, are also suggested (Song et al. 2011). Evidence of a small RNA–mediated transcriptional gene silencing and regulation of heterochromatin formation and maintenance is emerging, even though the mechanistic aspects still remain unclear (Moazed 2009). Possible nuclear functions of Dicer are supported by its localization in the nucleus and specifically on certain chromosomal domains (Khalil and Driscoll 2010; Sinkkonen et al. 2010), and its involvement in the centromeric repeat transcript silencing in mouse male germ cells (Korhonen et al. 2011) and in mouse embryonic stem cells (Kanellopoulou et al. 2005; Murchison et al. 2005). However, further studies will be required to reveal the mechanistic connection of Dicer and Dicer-dependent small RNAs with heterochromatin formation, regulation of repeat-derived transcripts and control of chromatin organization in differentiating male germ cells.

6.3.3 Pre-Pachytene piRNAs as Regulators of Transposon Expression

Each PIWI protein is expressed at a different timeframe during male germ cell development and binds a different subset of piRNAs. MIWI2 and MILI expression starts in fetal prospermatogonia, which are undergoing the genome-wide reorganization of DNA methylation pattern to generate novel gametic epigenetic marks (Aravin et al. 2008). During that time, everything which is otherwise epigenetically silenced becomes derepressed, so the cells need an alternative mechanism to protect their genome against the invasion of activated transposable elements. When analysing the MILI- and MIWI2-bound piRNAs from the total piRNA pool, a strikingly uniform subclass of piRNAs emerged that originate from repeat sequences related to transposable elements and heterochromatic regions. They were called pre-pachytene piRNAs and they are now the most studied and understood piRNAs, although they represent only a tiny fraction of all mammalian piRNAs. MILI and MIWI2 together with pre-pachytene piRNAs participate in silencing of transposable elements both at epigenetic and posttranscriptional level in fetal and neonatal germ cells (Aravin et al. 2007a, 2008; Carmell et al. 2007; Kuramochi-Miyagawa et al. 2008). In the knock-out mice of either of these proteins the transposons are uncontrollably expressed, causing damage in the genome integrity of the cell, which eventually leads to meiotic arrest and sterility (Carmell et al. 2007; Kuramochi-Miyagawa et al. 2004).

Analysis of the piRNA sequences and timing of their expression has led to the model of piRNA biogenesis mechanism, called the ping-pong cycle, that is mediated by PIWI proteins (Fig. 6.3) (Aravin et al. 2008; Brennecke et al. 2007). By this model, the transposon mRNAs activate the amplification of their target piRNAs by sense-antisense RNA amplification loop. In mouse, the transposon transcripts themselves become the substrates of so-called primary piRNAs by a yet unknown primary piRNA processing pathway. These transposon-derived sense primary piRNAs are bound to MILI and mediate the production of the secondary piRNAs by cleaving the antisense transposon transcripts that are generated genetically from piRNA clusters. The secondary piRNAs pair with MIWI2 and in turn, target the respective transposon mRNAs, which becomes the substrates of next piRNA molecules. This amplification loop is triggered and regulated by the presence of transposon mRNAs. The primary piRNAs share a 5′ U-bias, which corresponds to the bias of 10A in the complementary secondary piRNAs. This footprint is characteristic to the ping-pong cycle in different organisms where it acts as an adaptive immune system that optimizes the piRNA population accordingly to the target elements.

6.3.4 Pachytene piRNAs and the Chromatoid Body

Pachytene piRNAs arise in spermatocytes during meiosis around day 14 post partum, peak in haploid round spermatids and disappear during later steps of spermiogenesis, overlapping the expression of their binding PIWI partners, MILI and MIWI. The amount of pachytene piRNAs per each pachytene spermatocyte or round spermatid is remarkable as it is possible to visualize the characteristic 30-nucleotide band in total testis RNA extract just in ethidium bromide or SybrGold stained polyacrylamide gel (Aravin et al. 2006; Girard et al. 2006; Grivna et al. 2006). The amount of pachytene piRNAs in cells is suggested to be around a million molecules, but the individual copy number is low, which means that piRNA population is very heterogeneous consisting of hundreds of thousands of different piRNAs (Aravin et al. 2006, 2007b). Compared to their pre-pachytene counterparts, the biogenesis and function of pachytene piRNAs is unknown. They share the preference to 5′ U, but lack the 10A preference and do not produce homologous antisense transcripts, which is characteristic to the ping-pong cycle. Pachytene piRNAs are also devoid of sequences relative to active transposons. Instead they map into large sparse clusters, from tens to hundreds of kilobases along the genome with most of the clusters being derived from one of the two genomic strands (Aravin et al. 2007a, b). There is very little conservation of individual piRNA sequences between different mammals, but surprisingly there is a significant conservation of the genomic locations of mammalian piRNA clusters (Betel et al. 2007). Interestingly, pachytene piRNAs in round spermatid are concentrated in chromatoid bodies – male germ cell-specific cytoplasmic ribonucleoprotein germ granules of remarkable size and peculiar features (Meikar et al. 2010, 2011; Kotaja and Sassone-Corsi 2007). As the chromatoid body also concentrates MIWI, RNA binding proteins and helicases in addition to longer polyadenylated RNAs (Kotaja et al. 2006; Meikar et al. 2010), it is tempting to speculate that it serves as a processing centre for pachytene piRNAs and mRNAs.

6.4 Implications to Human Diseases

6.4.1 Epigenetic Transgenerational Inheritance

Epigenetic status of the chromatin is mitotically stable – as the cell undergoes mitosis the epigenome is replicated. Germline cells have the capacity of erasing the epigenetic memory and resetting the epigenome. Certain marks, however, appear to escape the reprogramming and can be transmitted to the offspring, which enables transgenerational epigenetic inheritance. Because the environmentally induced changes in the epigenome of the germline becomes permanently programmed and the altered epigenome and phenotype can be transmitted to subsequent progeny, environmental epigenetics has certainly a critical role in disease etiology (Walker and Gore 2011; Skinner 2011). Environmental factors, such as endocrine disrupting chemicals (EDCs) that interfere with the ability of endocrine systems to maintain homeostasis, are able to influence the epigenome. The critical time for environmental exposure is the period when the germline cell fate is determined and epigenome is reprogrammed in mammalian fetus. If the mother is exposed to environmental agents such as EDCs during this time, the germline epigenome of male offspring can be modified and permanently altered. Studies in mice have demonstrated the transgenerational inheritance of the epigenetic changes in DNA methylation caused by the exposure of embryos to vinclozolin, diethylstilbesterol, bisphenol A and polychlorinated biphenyls that are representative EDCs (Walker and Gore 2011; Skinner 2011). DNA methylation has been proposed as a major mechanism for transgenerational epigenetic effects because of vast changes of DNA methylation patterns during early mammalian development (Lange and Schneider 2010). Since piRNAs in mammalian PGCs have been implicated in the methylation of transposable elements upstream of DNMT, it has been suggested that small RNAs could mediate epigenetic inheritance through homology-dependent silencing systems (Suter and Martin 2010).

6.4.2 Epigenetic Control of Germline Antigens in Cancer

Cancer germline (CG) -antigens are classified as factors normally expressed only by germ cells, but found to be aberrantly expressed in a wide range of human cancers. Epigenetic status of germline cells are programmed to repress somatic genes and support germline differentiation. Cancer cells undergo large-scale epigenome alterations, which may lead to aberrant activation of the CG genes that are normally silenced outside the germline (Wang et al. 2011). Recent evidence suggests epigenetic mechanisms, particularly DNA methylation, as primary regulators of CG gene expression in normal and cancer cells. The mechanisms by which CG antigens promote tumour growth are still unclear, but it has been proposed that the off-context activity of some germline epigenome regulators could reprogram the somatic epigenome toward a malignant state by favouring self-renewal and sustaining cell proliferation, and in this way to support oncogenic properties of the cells (Wang et al. 2011). There has been a search for good tumour antigens in the past 100 years. An ideal cancer antigen for immunotherapy would be specifically expressed in tumours, absent from healthy tissue and critical for the survival of cancer. Importantly, CG-antigens have potentially a high clinical relevance in cancer immunotherapy because they are immunogenic and because of their restricted expression pattern in physiological conditions (Simpson et al. 2005; Akers et al. 2010).

Acknowledgements Work in our laboratory is supported by research grants from the Academy of Finland and Emil Aaltonen Foundation. OM and MDR are supported by the Turku Doctoral Programme of Biomedical Sciences.

References

Akers SN, Odunsi K, Karpf AR (2010) Regulation of cancer germline antigen gene expression: implications for cancer immunotherapy. Future Oncol 6(5):717–732

Aravin A, Gaidatzis D, Pfeffer S, Lagos-Quintana M, Landgraf P, Iovino N, Morris P, Brownstein MJ, Kuramochi-Miyagawa S, Nakano T, Chien M, Russo JJ, Ju J, Sheridan R, Sander C, Zavolan M, Tuschl T (2006) A novel class of small RNAs bind to MILI protein in mouse testes. Nature 442(7099):203–207

Aravin AA, Hannon GJ, Brennecke J (2007a) The Piwi-piRNA pathway provides an adaptive defense in the transposon arms race. Science 318(5851):761–764

Aravin AA, Sachidanandam R, Girard A, Fejes-Toth K, Hannon GJ (2007b) Developmentally regulated piRNA clusters implicate MILI in transposon control. Science 316(5825):744–747

Aravin AA, Sachidanandam R, Bourc'his D, Schaefer C, Pezic D, Toth KF, Bestor T, Hannon GJ (2008) A piRNA pathway primed by individual transposons is linked to de novo DNA methylation in mice. Mol Cell 31(6):785–799

Baarends WM, Wassenaar E, Hoogerbrugge JW, van Cappellen G, Roest HP, Vreeburg J, Ooms M, Hoeijmakers JH, Grootegoed JA (2003) Loss of HR6B ubiquitin-conjugating activity results in damaged synaptonemal complex structure and increased crossing-over frequency during the male meiotic prophase. Mol Cell Biol 23(4):1151–1162

Baudat F, Buard J, Grey C, Fledel-Alon A, Ober C, Przeworski M, Coop G, de Massy B (2010) PRDM9 is a major determinant of meiotic recombination hotspots in humans and mice. Science 327(5967):836–840

Betel D, Sheridan R, Marks DS, Sander C (2007) Computational analysis of mouse piRNA sequence and biogenesis. PLoS Comput Biol 3(11):e222

Bourc'his D, Bestor TH (2004) Meiotic catastrophe and retrotransposon reactivation in male germ cells lacking Dnmt3L. Nature 431(7004):96–99

Boussouar F, Rousseaux S, Khochbin S (2008) A new insight into male genome reprogramming by histone variants and histone code. Cell Cycle 7(22):3499–3502

Brennecke J, Aravin AA, Stark A, Dus M, Kellis M, Sachidanandam R, Hannon GJ (2007) Discrete small RNA-generating loci as master regulators of transposon activity in Drosophila. Cell 128(6):1089–1103

Brykczynska U, Hisano M, Erkek S, Ramos L, Oakeley EJ, Roloff TC, Beisel C, Schubeler D, Stadler MB, Peters AH (2010) Repressive and active histone methylation mark distinct promoters in human and mouse spermatozoa. Nat Struct Mol Biol 17(6):679–687

Buard J, Barthes P, Grey C, de Massy B (2009) Distinct histone modifications define initiation and repair of meiotic recombination in the mouse. EMBO J 28(17):2616–2624

Burgoyne PS, Mahadevaiah SK, Turner JM (2009) The consequences of asynapsis for mammalian meiosis. Nat Rev Genet 10(3):207–216

Carmell MA, Girard A, van de Kant HJ, Bourc'his D, Bestor TH, de Rooij DG, Hannon GJ (2007) MIWI2 is essential for spermatogenesis and repression of transposons in the mouse male germline. Dev Cell 12(4):503–514

Cenik ES, Zamore PD (2011) Argonaute proteins. Curr Biol 21(12):R446–R449

Chiang HR, Schoenfeld LW, Ruby JG, Auyeung VC, Spies N, Baek D, Johnston WK, Russ C, Luo S, Babiarz JE, Blelloch R, Schroth GP, Nusbaum C, Bartel DP (2010) Mammalian microRNAs: experimental evaluation of novel and previously annotated genes. Genes Dev 24(10):992–1009

Cho C, Willis WD, Goulding EH, Jung-Ha H, Choi YC, Hecht NB, Eddy EM (2001) Haploinsufficiency of protamine-1 or –2 causes infertility in mice. Nat Genet 28(1):82–86

Chuma S, Hosokawa M, Tanaka T, Nakatsuji N (2009) Ultrastructural characterization of spermatogenesis and its evolutionary conservation in the germline: germinal granules in mammals. Mol Cell Endocrinol 306(1–2):17–23

de Rooij DG, Russell LD (2000) All you wanted to know about spermatogonia but were afraid to ask. J Androl 21(6):776–798

Eddy EM (1970) Cytochemical observations on the chromatoid body of the male germ cells. Biol Reprod 2(1):114–128

Ewen KA, Koopman P (2010) Mouse germ cell development: from specification to sex determination. Mol Cell Endocrinol 323(1):76–93

Gaucher J, Reynoird N, Montellier E, Boussouar F, Rousseaux S, Khochbin S (2010) From meiosis to postmeiotic events: the secrets of histone disappearance. FEBS J 277(3):599–604

Girard A, Sachidanandam R, Hannon GJ, Carmell MA (2006) A germline-specific class of small RNAs binds mammalian Piwi proteins. Nature 442(7099):199–202

Godmann M, Lambrot R, Kimmins S (2009) The dynamic epigenetic program in male germ cells: its role in spermatogenesis, testis cancer, and its response to the environment. Microsc Res Tech 72(8):603–619

Gonzalez-Gonzalez E, Lopez-Casas PP, del Mazo J (2008) The expression patterns of genes involved in the RNAi pathways are tissue-dependent and differ in the germ and somatic cells of mouse testis. Biochim Biophys Acta 1779(5):306–311

Govin J, Lestrat C, Caron C, Pivot-Pajot C, Rousseaux S, Khochbin S (2006) Histone acetylation-mediated chromatin compaction during mouse spermatogenesis. Ernst Schering Res Found Workshop 57:155–172

Govin J, Escoffier E, Rousseaux S, Kuhn L, Ferro M, Thevenon J, Catena R, Davidson I, Garin J, Khochbin S, Caron C (2007) Pericentric heterochromatin reprogramming by new histone variants during mouse spermiogenesis. J Cell Biol 176(3):283–294

Grivna ST, Beyret E, Wang Z, Lin H (2006) A novel class of small RNAs in mouse spermatogenic cells. Genes Dev 20(13):1709–1714

Guo H, Ingolia NT, Weissman JS, Bartel DP (2010) Mammalian microRNAs predominantly act to decrease target mRNA levels. Nature 466(7308):835–840

Hajkova P, Ancelin K, Waldmann T, Lacoste N, Lange UC, Cesari F, Lee C, Almouzni G, Schneider R, Surani MA (2008) Chromatin dynamics during epigenetic reprogramming in the mouse germ line. Nature 452(7189):877–881

Hajkova P, Jeffries SJ, Lee C, Miller N, Jackson SP, Surani MA (2010) Genome-wide reprogramming in the mouse germ line entails the base excision repair pathway. Science 329(5987):78–82

Hammoud SS, Nix DA, Zhang H, Purwar J, Carrell DT, Cairns BR (2009) Distinctive chromatin in human sperm packages genes for embryo development. Nature 460(7254):473–478

Hayashi K, Yoshida K, Matsui Y (2005) A histone H3 methyltransferase controls epigenetic events required for meiotic prophase. Nature 438(7066):374–378

Hayashi K, de Sousa C, Lopes SM, Kaneda M, Tang F, Hajkova P, Lao K, O'Carroll D, Das PP, Tarakhovsky A, Miska EA, Surani MA (2008) MicroRNA biogenesis is required for mouse primordial germ cell development and spermatogenesis. PLoS One 3(3):e1738

Hazzouri M, Pivot-Pajot C, Faure AK, Usson Y, Pelletier R, Sele B, Khochbin S, Rousseaux S (2000) Regulated hyperacetylation of core histones during mouse spermatogenesis: involvement of histone deacetylases. Eur J Cell Biol 79(12):950–960

Hess RA, de Franca LR (2008) Spermatogenesis and cycle of the seminiferous epithelium. Adv Exp Med Biol 636:1–15

Kaneda M, Okano M, Hata K, Sado T, Tsujimoto N, Li E, Sasaki H (2004) Essential role for de novo DNA methyltransferase Dnmt3a in paternal and maternal imprinting. Nature 429(6994):900–903

Kanellopoulou C, Muljo SA, Kung AL, Ganesan S, Drapkin R, Jenuwein T, Livingston DM, Rajewsky K (2005) Dicer-deficient mouse embryonic stem cells are defective in differentiation and centromeric silencing. Genes Dev 19(4):489–501

Kato Y, Kaneda M, Hata K, Kumaki K, Hisano M, Kohara Y, Okano M, Li E, Nozaki M, Sasaki H (2007) Role of the Dnmt3 family in de novo methylation of imprinted and repetitive sequences during male germ cell development in the mouse. Hum Mol Genet 16(19):2272–2280

Khalil AM, Driscoll DJ (2010) Epigenetic regulation of pericentromeric heterochromatin during mammalian meiosis. Cytogenet Genome Res 129(4):280–289

Kimmins S, Sassone-Corsi P (2005) Chromatin remodelling and epigenetic features of germ cells. Nature 434(7033):583–589

Kimmins S, Kotaja N, Davidson I, Sassone-Corsi P (2004) Testis-specific transcription mechanisms promoting male germ-cell differentiation. Reproduction 128(1):5–12

Kimmins S, Crosio C, Kotaja N, Hirayama J, Monaco L, Hoog C, van Duin M, Gossen JA, Sassone-Corsi P (2007) Differential functions of the Aurora-B and Aurora-C kinases in mammalian spermatogenesis. Mol Endocrinol 21(3):726–739

Kleene KC (2003) Patterns, mechanisms, and functions of translation regulation in mammalian spermatogenic cells. Cytogenet Genome Res 103(3–4):217–224

Kolthur-Seetharam U, Teerds K, de Rooij DG, Wendling O, McBurney M, Sassone-Corsi P, Davidson I (2009) The histone deacetylase SIRT1 controls male fertility in mice through regulation of hypothalamic-pituitary gonadotropin signaling. Biol Reprod 80(2):384–391

Korhonen HM, Meikar O, Yadav RP, Papaioannou MD, Romero Y, Da Ros M, Herrera PL, Toppari J, Nef S, Kotaja N (2011) Dicer is required for haploid male germ cell differentiation in mice. PLoS One 6:e24821

Kota SK, Feil R (2010) Epigenetic transitions in germ cell development and meiosis. Dev Cell 19(5):675–686

Kotaja N, Sassone-Corsi P (2007) The chromatoid body: a germ-cell-specific RNA-processing centre. Nat Rev Mol Cell Biol 8(1):85–90

Kotaja N, Bhattacharyya SN, Jaskiewicz L, Kimmins S, Parvinen M, Filipowicz W, Sassone-Corsi P (2006) The chromatoid body of male germ cells: similarity with processing bodies and presence of Dicer and microRNA pathway components. Proc Natl Acad Sci U S A 103(8):2647–2652

Krol J, Loedige I, Filipowicz W (2010) The widespread regulation of microRNA biogenesis, function and decay. Nat Rev Genet 11(9):597–610

Kuramochi-Miyagawa S, Kimura T, Ijiri TW, Isobe T, Asada N, Fujita Y, Ikawa M, Iwai N, Okabe M, Deng W, Lin H, Matsuda Y, Nakano T (2004) Mili, a mammalian member of piwi family gene, is essential for spermatogenesis. Development 131(4):839–849

Kuramochi-Miyagawa S, Watanabe T, Gotoh K, Totoki Y, Toyoda A, Ikawa M, Asada N, Kojima K, Yamaguchi Y, Ijiri TW, Hata K, Li E, Matsuda Y, Kimura T, Okabe M, Sakaki Y, Sasaki H, Nakano T (2008) DNA methylation of retrotransposon genes is regulated by Piwi family members MILI and MIWI2 in murine fetal testes. Genes Dev 22(7):908–917

La Salle S, Oakes CC, Neaga OR, Bourc'his D, Bestor TH, Trasler JM (2007) Loss of spermatogonia and wide-spread DNA methylation defects in newborn male mice deficient in DNMT3L. BMC Dev Biol 7:104

Lange UC, Schneider R (2010) What an epigenome remembers. Bioessays 32(8):659–668

Lau NC (2010) Small RNAs in the animal gonad: guarding genomes and guiding development. Int J Biochem Cell Biol 42(8):1334–1347

Lee Y, Kim M, Han J, Yeom KH, Lee S, Baek SH, Kim VN (2004) MicroRNA genes are transcribed by RNA polymerase II. EMBO J 23(20):4051–4060

Lee TL, Pang AL, Rennert OM, Chan WY (2009) Genomic landscape of developing male germ cells. Birth Defects Res C Embryo Today 87(1):43–63

Lu LY, Wu J, Ye L, Gavrilina GB, Saunders TL, Yu X (2010) RNF8-dependent histone modifications regulate nucleosome removal during spermatogenesis. Dev Cell 18(3):371–384

Maatouk DM, Loveland KL, McManus MT, Moore K, Harfe BD (2008) Dicer1 is required for differentiation of the mouse male germline. Biol Reprod 79(4):696–703

Martianov I, Brancorsini S, Catena R, Gansmuller A, Kotaja N, Parvinen M, Sassone-Corsi P, Davidson I (2005) Polar nuclear localization of H1T2, a histone H1 variant, required for spermatid elongation and DNA condensation during spermiogenesis. Proc Natl Acad Sci U S A 102(8):2808–2813

Matsui Y (2010) The molecular mechanisms regulating germ cell development and potential. J Androl 31(1):61–65

Meikar O, Da Ros M, Liljenback H, Toppari J, Kotaja N (2010) Accumulation of piRNAs in the chromatoid bodies purified by a novel isolation protocol. Exp Cell Res 316(9):1567–1575

Meikar O, Da Ros M, Korhonen H, Kotaja N (2011) Chromatoid body and small RNAs in male germ cells. Reproduction 142(2):195–209

Miller D, Brinkworth M, Iles D (2010) Paternal DNA packaging in spermatozoa: more than the sum of its parts? DNA, histones, protamines and epigenetics. Reproduction 139(2):287–301

Moazed D (2009) Small RNAs in transcriptional gene silencing and genome defence. Nature 457(7228):413–420

Murchison EP, Partridge JF, Tam OH, Cheloufi S, Hannon GJ (2005) Characterization of Dicer-deficient murine embryonic stem cells. Proc Natl Acad Sci U S A 102(34):12135–12140

Murchison EP, Stein P, Xuan Z, Pan H, Zhang MQ, Schultz RM, Hannon GJ (2007) Critical roles for Dicer in the female germline. Genes Dev 21(6):682–693

Nottke A, Colaiacovo MP, Shi Y (2009) Developmental roles of the histone lysine demethylases. Development 136(6):879–889

Oakberg EF (1956) Duration of spermatogenesis in the mouse and timing of stages of the cycle of the seminiferous epithelium. Am J Anat 99(3):507–516

Oakes CC, La Salle S, Smiraglia DJ, Robaire B, Trasler JM (2007) A unique configuration of genome-wide DNA methylation patterns in the testis. Proc Natl Acad Sci U S A 104(1):228–233

Okada Y, Scott G, Ray MK, Mishina Y, Zhang Y (2007) Histone demethylase JHDM2A is critical for Tnp1 and Prm1 transcription and spermatogenesis. Nature 450(7166):119–123

Papaioannou MD, Pitetti JL, Ro S, Park C, Aubry F, Schaad O, Vejnar CE, Kuhne F, Descombes P, Zdobnov EM, McManus MT, Guillou F, Harfe BD, Yan W, Jegou B, Nef S (2009) Sertoli cell Dicer is essential for spermatogenesis in mice. Dev Biol 326(1):250–259

Papaioannou MD, Lagarrigue M, Vejnar CE, Rolland AD, Kuhne F, Aubry F, Schaad O, Fort A, Descombes P, Neerman-Arbez M, Guillou F, Zdobnov EM, Pineau C, Nef S (2010) Loss of Dicer in Sertoli cells has a major impact on the testicular proteome of mice. Mol Cell Proteomics 10(4):M900587MCP200

Paronetto MP, Sette C (2010) Role of RNA-binding proteins in mammalian spermatogenesis. Int J Androl 33(1):2–12

Peters AH, O'Carroll D, Scherthan H, Mechtler K, Sauer S, Schofer C, Weipoltshammer K, Pagani M, Lachner M, Kohlmaier A, Opravil S, Doyle M, Sibilia M, Jenuwein T (2001) Loss of the Suv39h histone methyltransferases impairs mammalian heterochromatin and genome stability. Cell 107(3):323–337

Ro S, Park C, Sanders KM, McCarrey JR, Yan W (2007) Cloning and expression profiling of testis-expressed microRNAs. Dev Biol 311(2):592–602

Roest HP, van Klaveren J, de Wit J, van Gurp CG, Koken MH, Vermey M, van Roijen JH, Hoogerbrugge JW, Vreeburg JT, Baarends WM, Bootsma D, Grootegoed JA, Hoeijmakers JH (1996) Inactivation of the HR6B ubiquitin-conjugating DNA repair enzyme in mice causes male sterility associated with chromatin modification. Cell 86(5):799–810

Sasaki H, Matsui Y (2008) Epigenetic events in mammalian germ-cell development: reprogramming and beyond. Nat Rev Genet 9(2):129–140

Seki Y, Yamaji M, Yabuta Y, Sano M, Shigeta M, Matsui Y, Saga Y, Tachibana M, Shinkai Y, Saitou M (2007) Cellular dynamics associated with the genome-wide epigenetic reprogramming in migrating primordial germ cells in mice. Development 134(14):2627–2638

Sen GC, Sarkar SN (2007) The interferon-stimulated genes: targets of direct signaling by interferons, double-stranded RNA, and viruses. Curr Top Microbiol Immunol 316:233–250

Shang E, Nickerson HD, Wen D, Wang X, Wolgemuth DJ (2007) The first bromodomain of Brdt, a testis-specific member of the BET sub-family of double-bromodomain-containing proteins, is essential for male germ cell differentiation. Development 134(19):3507–3515

Simpson AJ, Caballero OL, Jungbluth A, Chen YT, Old LJ (2005) Cancer/testis antigens, gametogenesis and cancer. Nat Rev Cancer 5(8):615–625

Sinkkonen L, Hugenschmidt T, Filipowicz W, Svoboda P (2010) Dicer is associated with ribosomal DNA chromatin in mammalian cells. PLoS One 5(8):e12175

Siomi MC, Sato K, Pezic D, Aravin AA (2011) PIWI-interacting small RNAs: the vanguard of genome defence. Nat Rev Mol Cell Biol 12(4):246–258

Skinner MK (2011) Environmental epigenetic transgenerational inheritance and somatic epigenetic mitotic stability. Epigenetics 6(7):838–842

Song R, Ro S, Michaels JD, Park C, McCarrey JR, Yan W (2009) Many X-linked microRNAs escape meiotic sex chromosome inactivation. Nat Genet 41(4):488–493

Song R, Hennig GW, Wu Q, Jose C, Zheng H, Yan W (2011) Male germ cells express abundant endogenous siRNAs. Proc Natl Acad Sci U S A 108(32):13159–13164

Suter CM, Martin DI (2010) Paramutation: the tip of an epigenetic iceberg? Trends Genet 26(1):9–14

Svoboda P, Stein P, Hayashi H, Schultz RM (2000) Selective reduction of dormant maternal mRNAs in mouse oocytes by RNA interference. Development 127(19):4147–4156

Tachibana M, Nozaki M, Takeda N, Shinkai Y (2007) Functional dynamics of H3K9 methylation during meiotic prophase progression. EMBO J 26(14):3346–3359

Tam OH, Aravin AA, Stein P, Girard A, Murchison EP, Cheloufi S, Hodges E, Anger M, Sachidanandam R, Schultz RM, Hannon GJ (2008) Pseudogene-derived small interfering RNAs regulate gene expression in mouse oocytes. Nature 453(7194):534–538

Tanaka H, Baba T (2005) Gene expression in spermiogenesis. Cell Mol Life Sci 62(3):344–354

Tang F, Kaneda M, O'Carroll D, Hajkova P, Barton SC, Sun YA, Lee C, Tarakhovsky A, Lao K, Surani MA (2007) Maternal microRNAs are essential for mouse zygotic development. Genes Dev 21(6):644–648

Trasler JM (2009) Epigenetics in spermatogenesis. Mol Cell Endocrinol 306(1–2):33–36

Walker DM, Gore AC (2011) Transgenerational neuroendocrine disruption of reproduction. Nat Rev Endocrinol 7(4):197–207

Wang J, Emadali A, Le Bescont A, Callanan M, Rousseaux S, Khochbin S (2011) Induced malignant genome reprogramming in somatic cells by testis-specific factors. Biochim Biophys Acta 1809(4–6):221–225

Watanabe T, Takeda A, Tsukiyama T, Mise K, Okuno T, Sasaki H, Minami N, Imai H (2006) Identification and characterization of two novel classes of small RNAs in the mouse germline: retrotransposon-derived siRNAs in oocytes and germline small RNAs in testes. Genes Dev 20(13):1732–1743

Watanabe T, Totoki Y, Toyoda A, Kaneda M, Kuramochi-Miyagawa S, Obata Y, Chiba H, Kohara Y, Kono T, Nakano T, Surani MA, Sakaki Y, Sasaki H (2008) Endogenous siRNAs from naturally formed dsRNAs regulate transcripts in mouse oocytes. Nature 453(7194):539–543

Webster KE, O'Bryan MK, Fletcher S, Crewther PE, Aapola U, Craig J, Harrison DK, Aung H, Phutikanit N, Lyle R, Meachem SJ, Antonarakis SE, de Kretser DM, Hedger MP, Peterson P, Carroll BJ, Scott HS (2005) Meiotic and epigenetic defects in Dnmt3L-knockout mouse spermatogenesis. Proc Natl Acad Sci U S A 102(11):4068–4073

Zhao M, Shirley CR, Hayashi S, Marcon L, Mohapatra B, Suganuma R, Behringer RR, Boissonneault G, Yanagimachi R, Meistrich ML (2004a) Transition nuclear proteins are required for normal chromatin condensation and functional sperm development. Genesis 38(4):200–213

Zhao M, Shirley CR, Mounsey S, Meistrich ML (2004b) Nucleoprotein transitions during spermiogenesis in mice with transition nuclear protein Tnp1 and Tnp2 mutations. Biol Reprod 71(3):1016–1025

Chapter 7
Epigenetic Regulation of Skeletal Muscle Development and Differentiation

Narendra Bharathy, Belinda Mei Tze Ling, and Reshma Taneja

Abstract Skeletal muscle cells have served as a paradigm for understanding mechanisms leading to cellular differentiation. Formation of skeletal muscle involves a series of steps in which cells are commited towards the myogenic lineage, undergo expansion to give rise to myoblasts that differentiate into multinucleated myotubes, and mature to form adult muscle fibers. The commitment, proliferation, and differentiation of progenitor cells involve both genetic and epigenetic changes that culminate in alterations in gene expression. Members of the Myogenic regulatory factor (MRF), as well as the Myocyte Enhancer Factor (MEF2) families control distinct steps of skeletal muscle proliferation and differentiation. In addition, growing evidence indicates that chromatin modifying enzymes and remodeling complexes epigenetically reprogram muscle promoters at various stages that preclude or promote MRF and MEF2 activites. Among these, histone deacetylases (HDACs), histone acetyltransferases (HATs), histone methyltransferases (HMTs) and SWI/SNF complexes alter chromatin structure through post-translational modifications to impact MRF and MEF2 activities. With such new and emerging knowledge, we are beginning to develop a true molecular understanding of the mechanisms by which skeletal muscle development and differentiation is regulated. Elucidation of the mechanisms by which epigenetic regulators control myogenesis will likely provide a new foundation for the development of novel therapeutic drugs for muscle

---

N. Bharathy • B.M.T. Ling
Department of Physiology, Yong Loo Lin School of Medicine, National University of Singapore, Block MD9, 2 Medical Drive, Singapore 117597, Singapore

R. Taneja (✉)
Department of Physiology, Yong Loo Lin School of Medicine, National University of Singapore, Block MD9, 2 Medical Drive, Singapore 117597, Singapore

NUS Graduate School of Integrative Sciences and Engineering, Singapore, Singapore
e-mail: phsrt@nus.edu.sg

dystrophies, ageing-related regeneration defects that occur due to altered proliferation and differentiation, and other malignancies.

7.1 Introduction

7.1.1 Skeletal Myogenesis in the Embryo and the Adult

All skeletal muscle of the vertebrate body and some head muscles are derived from somites, that are laid down on each side of the neural tube and notochord during embryogenesis (Buckingham 2001; Cossu et al. 1996a; Pownall et al. 2002; Tajbakhsh and Cossu 1997). Cells in the dorsal portion of the somite form the dermomyotome, which gives rise to the dermis and most skeletal muscles of the body. Through inductive and repressive signals from surrounding tissues, somites acquire dorsal-ventral, anterior-posterior, and medial-lateral polarity. In response to signals from the notochord such as Wnt and Sonic Hedgehog, the expression of the paired homeobox transcription factors Pax3 and Pax7 are induced in muscle progenitor cells in the dermamyotome resulting in the specification of muscle cells. Pax3 induces the expression of the myogenic regulatory factors Myf5, and consequently MyoD, commiting cells to the myogenic lineage. Myf5 and MyoD establish two distinct populations of cells, that give rise to the epaxial and hypaxial muscles (Ordahl and Le Douarin 1992; Cossu et al. 1996b). Differentiation of muscle cells is subsequently mediated by MyoD, Myogenin and MRF4.

Satellite cells, the local muscle stem cells, arise from Pax3 and Pax7 expressing muscle progenitor cells that originate in the dermomyotome (Relaix et al. 2005; Gros et al. 2005). In the adult muscle, satellite cells are located between the sarcolemma and basal lamina of the muscle fiber, and are the main source of myogenic precursors in the adult. The expression of Pax7 is required for the maintenance of satellite cells and generation of committed progenitors (Relaix et al. 2006; Kuang et al. 2006). Pax7 binds to the Myf5 promoter and recruits the HMT complex Wdr5–Ash2L–MLL2 (McKinnell et al. 2008) that directs trimethylation of histone H3 lysine 4 (H3K4me3). Methylation of H3K4 marks chromatin in a conformation permissive for transcription resulting in the upregulation of Myf5 expression. Thus, Pax7 induces chromatin modifications that stimulate transcriptional activation of Myf5 and thereby regulates commitment into the myogenic developmental programme. Upon muscle injury, quiescent satellite cells are activated and undergo proliferation to give rise to myoblasts. Activated satellite cells expressing Pax7 and MyoD undergo several rounds of proliferation, and subsequently upregulate myogenin and MRF4 to differentiate and form new myofibers repairing the damaged muscles. The onset of differentiation is preceded by down-regulation of Pax7. A distinct population of satellite cells retain Pax7 expression but dowregulate MyoD, reversibly exit the cycle, and relocate to the basal lamina, thereby replenishing the satellite cell pool (Zammit et al. 2004).

7.2 Transcriptional Control of Myogenesis

Differentiation of skeletal myoblasts is regulated by two classes of transcription factors – the Myogenic regulatory factors (MRF) that include MyoD, Myf5, MRF4 and Myogenin; and the Myocyte Enhancer Factor (MEF2) family which includes MEF2-A, -B, -C and -D (Molkentin and Olson 1996; Black and Olson 1998). A landmark discovery that facilitated the understanding of the genetic control of skeletal muscle differentiation was the identification of MyoD as a master regulator of myogenesis through its ability to convert cultured fibroblasts cells into skeletal muscle (Davis et al. 1987). MyoD expression is selectively restricted to skeletal muscle cells. In non-muscle cells MyoD expression is repressed by DNA methylation, an epigenetic modification mediated by DNA methyltransferases (DNMTs). Thus, treatment with the demethylating agent 5-azacytidine derepresses the MyoD promoter and induces the differentiation of fibroblasts into muscle cells (Lassar et al. 1986). Once induced in committed myoblasts, MyoD expression promotes cell cycle exit by regulation of p21 expression (Halevy et al. 1995) as well as expression of myogenin (Fig. 7.1). Terminal differentiation is characterized by the expression

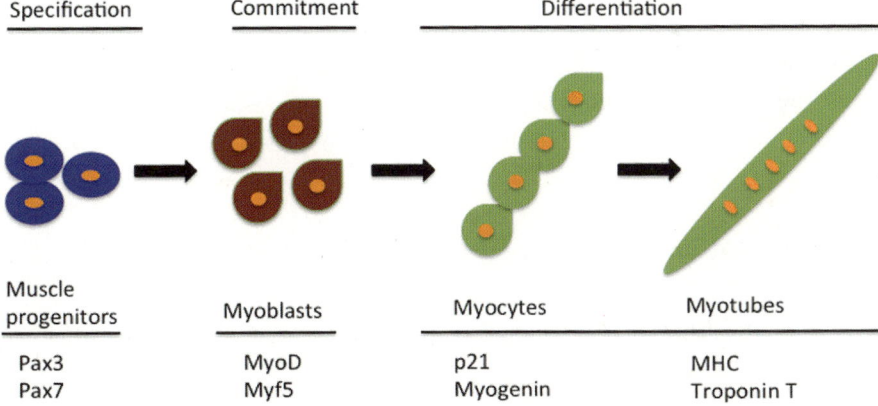

Fig. 7.1 Schematic representation of skeletal muscle differentiation. Pax3/Pax7 positive precursor cells arising from the dermamyotome initiate expression of Myf5 and MyoD and undergo commitment towards the myogenic lineage. MyoD and Myf5 expressing cells undergo expansion and proliferate to give rise to myoblasts. Upon appropriate differentiation cues, myoblasts irreversibly exit the cell cycle and myocytes express early differentiation markers p21 and Myogenin. At later stages, terminally differentiated multinucleated myotubes are characterized by expression of myosin heavy chain (MHC) and Troponin T

of late differentiation markers MHC and Troponin T. Subsequently, three other members of the MRF gene family were identified: Myogenin, Myf5 and MRF4. While all MRFs share the property of converting non-muscle cells to the myogenic lineage, they are expressed at different times during embryonic development. Myf5 is the first MRF to be expressed in the somite followed by MyoD in undifferentiated proliferating myoblasts (Buckingham 1992), whereas Myogenin and MRF4 are activated during differentiation. This led to hypothesis that Myf5 and MyoD are required for determination of myogenic precursor cells whereas myogenin and MRF4 are required for terminal differentiation. Disruption of MyoD or Myf5 alone in mice results in a delay in the formation of hypaxial and epaxial muscles respectively, with no gross defects in muscle differentiation due to functional redundancy between the two genes (Rudnicki et al. 1992; Braun et al. 1992). However, loss of both MyoD and Myf5 in mice results in reduced muscle masses consequent to a defect in the formation of myoblasts (Rudnicki et al. 1993). In the absence of myogenin, early steps of myogenesis occur normally. However the formation of myofibers is impaired and myogenin mutants die perinatally with an absence of differentiated muscles (Hasty et al. 1993; Nabeshima et al. 1993; Venuti et al. 1995). Loss of MRF4 alone results in subtle defects in myogenesis resulting in a slight reduction of muscle specific genes but no overt defect in muscle development (Olson et al. 1996). The phenotype of single and double MRF mutants indicate a genetic pathway of MRF function with some functions that are overlapping between MyoD and Myf5, whereas some are unique. It is unclear whether these differences are due to inherent functional differences, or are mainly caused by different expression patterns and/or transcriptional activation of MRFs. The MEF2 genes are expressed at very low levels in myoblasts and are induced during differentiation by MRFs. Once induced, Myogenin and MEF2C reside in a positive feedback loop enhancing each others transcription, with MEF2C being essential for the correct spatio-temporal expression of myogenin (Dodou et al. 2003; Cserjesi and Olson 1991). Together, MRFs and MEF2 proteins create a positive feedback loop and also regulate transcription of muscle structural genes.

7.3 Regulation of MyoD and MEF2 Activities

All MRFs and MEF2 family members contain DNA-binding and dimerization domains. MRFs dimerize with the ubiquitously expressed bHLH E proteins E12 and E47 (two splice variants of the E2A gene). The MRF-E protein heterodimers bind to their target E-box (CANNTG) site through the basic region to activate expression of downstream targets. The helix-loop-helix (HLH) domain in MRFs is required for dimerization with E proteins. In addition, MRFs interact with MEF2 through their basic domain. All MEF2 factors bind to A/T rich regions in muscle regulatory promoters and share the MADS box domain that mediates dimerization and DNA binding, and a MEF2 domain that mediates co-factor recruitment. Unlike MRFs, MEF2 factors do not possess myogenic activity on their own. However, they are able to potentiate myogenesis induced by MRFs, and increase efficiency of conversion of non-muscle cells (Molkentin et al. 1995).

Since MyoD and MEF2 factors are expressed in myoblasts, several regulatory mechanisms ensure that their activity is tightly regulated until appropriate cues are present that are permissive for differentiation. For instance, MyoD expression and activity in myoblasts is controlled through several mechanisms including the presence of Inhibitor of differentiation (Id) proteins that sequester E proteins from MyoD; as well as other inhibitors such as Twist, Mist1, MyoR, and Sharp-1 that inhibit MyoD transcriptional activity, DNA-binding, and dimerization with E-proteins (Benezra et al. 1990; Spicer et al. 1996; Lu et al. 1999; Lemercier et al. 1998; Azmi et al. 2004). During differentiation, the expression levels of many of these inhibitory molecules including Id1 decline allowing for increased MyoD transcriptional activity. In addition, the transcriptional network that regulates myogenesis, a number of key epigenetic marks at muscle specific loci are altered during the commitment, proliferation and differentiation stages. Post-translational modifications of histone tails mediated by HDACs, HATs and HMTs alter the chromatin configuration allowing for a transcriptional control of MRF and MEF2 activity in undifferentiated and differentiated cells (McKinsey et al. 2001; Guasconi and Puri 2009).

7.4 Chromatin Modifications in Undifferentiated Myoblasts

MyoD is expressed in committed myogenic precursor cells and has the ability to initiate the differentiation program. The premature activation of differentiation is prevented by several epigenetic mechanisms including recruitment of HDACs and HMTs which restrain MyoD and MEF2 activities in myoblasts (Fig. 7.2). Histone deacetylases are divided into three categories based on their homology to yeast proteins Rpd3p (class I), Hda1p (class II), and Sir2p (class III) (de Ruijter et al. 2003; North and Verdin 2004). While class I and II HDACs are sensitive to TSA, class III HDACs are not, and the deacetylase activity of class III enzymes relies on the cofactor NAD. All three classes of HDACs are involved in preventing premature myogenesis and function to regulate MyoD and MEF2 activities, as well as in sensing the redox balance. HDAC1, HDAC4/5 and SirT1 interact with MyoD and MEF2 factors and function not only deacetylate histones but also deacetylate transcription factors. In undifferentiated myoblasts, HDAC1 preferentially associates with MyoD in myoblasts through its bHLH domain. This association results in local deacetylation of histones resulting in a transcriptionally repressive chromatin configuration. Overexpression of HDAC1 inhibits muscle differentiation and is associated with reduced histone acetylation on late muscle promoters MCK and MHC (Puri et al. 2001; Mal et al. 2001). In addition, HDAC1 can deacetylate MyoD in vitro, that may additionally contribute to keeping MyoD inactive (Mal and Harter 2003; Mal et al. 2001). On the other hand, MEF2 activity in myoblasts is negatively regulated by HDAC4 and HDAC5. MEF2 factors interact directly with HDAC 4/5 in resulting in repression of MEF2-dependent transcription (Lu et al. 2000; McKinsey et al. 2001).

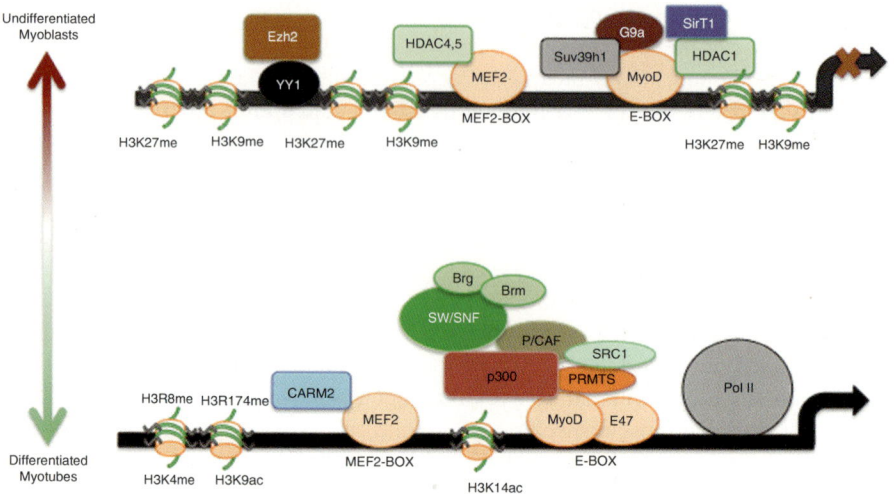

Fig. 7.2 Epigenetic reprogramming at muscle promoters in undifferentiated myoblasts and differentiated myotubes. In undifferentiated myoblasts, MyoD interacts with HDAC1 and SirT1, as well as HMTs Suv39h1 and G9a. Ezh2 is recruited on muscle promoters by YY1. MEF2 interacts with HDAC4/5. HMTs and HDACs result in local repressive chromatin marks such as H3K9me, H3K27me and deacetylation of histone tails. In differentiated cells, recruitment of HATs and chromatin remodelling enzymes by MyoD and MEF2 permits extensive reprograming of muscle promoters resulting H3K9 and H3K14 acetylation and H3K4, H3R8 and H3R17 methylation

Consistently, overexpression of HDAC4 or HDAC5 inhibits differentiation and inhibits both early and late muscle differentiation genes. The Class III HDAC SirT1, whose activity is regulated by the availability of NAD$^+$, forms a complex with MyoD and p300/CBP associated factor (P/CAF). Overexpression of SirT1 inhibits MyoD activity in myoblasts and thereby myogenesis (Fulco et al. 2003). During differentiation, the NAD$^+$/NADH$^+$ ratio decreases resulting in reduced SirT1 activity and allowing P/CAF to acetylate histones and MyoD (see below). Thus SirT1 functions as a redox sensor responding to metabolic changes that occur during differentiation.

In addition to HDACs, which mediate deacetylation of histones in undifferentiated cells, the SET-domain containing HMTs are critical mediators of muscle gene repression. In myoblasts, high levels of Histone H3 lysine 9 methylation (H3K9me) which is associated with gene repression, is apparent on the early myogenic promoters myogenin (Zhang et al. 2002; Mal and Harter 2003). The Su(var)3-9 family molecules were the first H3K9 methyltransferases to be described (Rea et al. 2000) and subsequently many others have been characterized. Almost all of them contain a SET domain and include Suv39h1, 2; G9a, GLP/Eu-HMT1; SETDB1, SETDB2; RIZ1/PRDM2; KYP/SUVH4; and DIM-5. Among these, Suv39h1 is the principal enzyme responsible for accumulation of H3K9me3 and is enriched in heterochromatin. Suv39h1 interacts with MyoD in undifferentiated myoblasts and inhibits its activity and myogenic differentiation (Mal 2006). H3K9me also serves as a platform for recruitment of the heterochromatin protein 1 (HP1) that leads to stable repression via formation of a heterochromatic structure. Similar to Suv39h1, the euchromatic methyltransferase G9a, which is mainly responsible for mono- and dimethylation of H3K9, is also expressed in undifferentiated myoblasts and declines upon myogenic differentiation. Interestingly, G9a mediates H3K9me2 on the myogenin promoter as well as methylates MyoD to control its activity (Ling et al. 2012). Thus it is possible that G9a and Suv39h1 serve to functionally maintain an undifferentiated state by impacting H3K9me on distinct promoters, or may act in sequence to mediate and maintain facultative and stable repression. A second repressive mark seen on late myogenic promoters in undifferentiated cells is H3K27me3 that is mediated by Ezh2, which forms the catalytic unit of Polycomb repressor complex (Caretti et al. 2004). Ezh2 is recruited on late myogenic promoters through the transcription factor YY1 and is found in complexes with HDAC1.

7.5 Chromatin Modifications and Remodelling During Differentiation

One of the early requirements during differentiation is an irreversible exit of myoblasts from the cell cycle. The retinoblastoma protein Rb plays a key role in cell cycle arrest of myoblasts (Huh et al. 2004). Interestingly, HDAC1, which inhibits MyoD in myoblasts, associates with Rb during differentiation. The Rb-HDAC1 complex represses E2F target genes and thereby silences S-phase genes (Puri et al. 2001; Blais et al. 2007). In addition, Suv39h1, which mediates H3K9me, has been shown to silence proliferation genes. These findings are however somewhat at odds with its proposed role as an inhibitor of MyoD activity and muscle differentiation (Ait-Si-Ali et al. 2004; Mal 2006). Finally, Mixed lineage leukemia 5 (MLL5) which is upregulated during quiescence also suppresses S-phase genes including the cyclin A2 promoter (Sebastian et al. 2009). Thus MLL5 may be important in maintaining satellite cell quiescence and expression of muscle determination genes.

The onset of differentiation also requires extensive reprogamming at muscle specific promoters and the replacement of repressive chromatin configuration with those associated with activation. During differentiation, deacetylation of histones H3 and H4, and methylation of H3K9 and H3K27 mediated by HDAC1, HDACII, and HMTs are erased, that correlate with a decline in the expression of Suv39h1, G9a, Ezh2 and HDAC1. Moreover, HDAC4/5 are exported from the nucleus by a CaMK-dependent mechanism during differentiation, resulting in a loss of HDAC4/5 and MEF2 interaction, allowing for a derepression of MEF2 activity (McKinsey et al. 2001).

Conversion of MyoD from a transcriptionally inactive state in myoblasts to an active state during differentiation is contigent upon its activation by several chromatin modifiers and remodeling proteins. These include CBP/p300, P/CAF, the arginine methyltransferase Carm1/Prmt4, Prmt5, and the ATPase dependent SWI/SNF remodeling complexes that are recruited to muscle promoters. PCAF and p300/CBP form a multimeric complex with MyoD and are required for MyoD to promote myogenic differentiation. p300/CBP acetylate histones H3 and H4, followed by acetylation of MyoD by P/CAF (Puri et al. 1997; Sartorelli et al. 1999; Dilworth et al. 2004). Acetylation of MyoD by P/CAF occurs at three lysine residues K99, K102 and K104 in its basic DNA-binding domain, that stimulates MyoD DNA-binding. Activation of MyoD results in myogenin expression, and a replacement of H3K9me3 by the transcriptionally permissive H3K4me3 on the myogenin promoter (Rampalli et al. 2007).

The SWI/SNF chromatin remodelling complexes also play an essential part in activation of the muscle differentiation program. Brg1, the ATPase subunit of SWI/SNF, is required for nucleosome remodeling and correlates with the presence of Pol II holoenzyme on muscle promoters. SWI/SNF recruitment is dependent on p38 activity, thus linking extracellular signals to nucleosome remodelling (Simone et al. 2004; de la Serna et al. 2005; Albini and Puri 2010).

Carm1/Prmt4, a type I methyltransferase interacts with MEF2 proteins and results in dimethylation of H3R17 (H3R17me2), whereas Prmt5, a type II arginine methyltransferase, associates with MyoD and methylates H3R8 (Chen et al. 2002; Dacwag et al. 2009). Both Carm1/Prmt4 and Prmt5 proteins have been demonstrated to be required at distinct steps of skeletal myogenesis. Chromatin IP experiments have revealed that Carm1/Prmt4 and H3R17me2 are present on the promoters of late myogenic regulatory regions such as MCK. On the other hand, Prmt5 and H3R8me2 binding is apparent on both early (myogenin) and late (MCK, dystrophin) myogenic promoters although Prmt5 appears to be dispensable for activation of the late promoters. Loss of Carm1 results in a loss of Brg1 association with myogenic promoter elements indicating that Carm1/Prmt4 binding at late- myogenic promoters facilitates binding of the Brg1 ATP-dependent chromatin-remodeling enzyme and subsequent chromatin remodeling at these regulatory sequences.

These studies collectively demonstrate that epigenetic regulators impact almost every step of myogenesis. DNA methylation by Dnmts regulates skeletal muscle specific MyoD expression, commitment to the myogenic lineage requires Pax7-mediated recruitment of Wdr5–Ash2L–MLL2 that upregulates Myf5 expression, control of cell cycle exit and S-phase genes is mediated by HDAC1, Suv39h1; MLL5 plays a role in quiescence and maintenance of determination gene expression; and MyoD and MEF2 activities during differentiation are controlled through complexes of HDACs, HMTs, HATs and the SWI/SNF remodeling enzymes.

7.6 Targeting Epigenetic Regulators in Muscular Dystrophy

Given the impact of HDACs on inhibition of MyoD and MEF2 activities, treatment of myoblasts with HDAC inhibitors was found to result in increased size of muscles. Interestingly, the inhibition of HDACs resulted in the upregulation of a key target gene, follistatin. Follistatin has no impact on proliferation of myoblasts, but instead mediates increased fusion, resulting in hypernucleated myotubes (Iezzi et al. 2004). Follistatin is also a negative regulator of myostatin/TGFβ signaling which inhibits muscle growth and regeneration. Thus, treatment with various HDAC inhibitors such as trichostatin A (TSA), valproic acid, and phenylbuytrate was found to enhance differentiation of satellite cells. Moreover, intraperitoneal injections with HDAC inhibitors resulted in increased follistatin expression and improved muscle fiber size that exhibit increased resistance to degeneration in mdx mice, a mouse model of muscular dystrophy (Minetti et al. 2006). At the molecular level, these findings reflect a regulatory connection of HDAC2 and follistatin expression. HDAC2 is S-nitrosylated by nitric oxide (NO) and this modification blocks HDAC2-mediated repression of follistatin. In DMD, the loss of the dystrophin results in loss of nitric oxide synthase (nNOS), and thus deregulated NO signaling. Consequently, increased HDAC2 activity leads to constitutive repression of follistatin that is blocked with HDAC inhibitors.

Intriguingly, in mdx hearts, an increase in PCAF and p300 expression, but not that of HDAC1 was reported recently (Colussi et al. 2011). This correlated with increase in N^{ε}-Lysine acetylation of connexin 43 and altered localization. Consistently, treatment with a HAT inhibitor anacardic acid, restored connexin 43 localization and cardiomyopathy in mdx mice. Moreover, the activity of HDAC4 was reduced in dystrophic heart, indicating that the levels of PCAF and HDAC4 may be important in connexin 43 localization. Thus, while HDAC inhibitors may present a viable pharmacological option for therapeutic intervention in muscular dystrophies, the consequences and impact of global HDAC inhibition through systemic treatments with HDAC inhibitors needs further consideration. Contemplation of tissue specific effects is essential in the use of drugs targeting epigenetic regulators for myopathies where systemic delivery is required.

7.7 Conclusion

In this review we have attempted to summarize some of the mechanisms which regulate the commitment and differentiation of cells into the myogenic lineage, and the role of epigenetics in developmental and adult skeletal myogenesis. Many chromatin modifiers and remodeling enzymes play both positive and negative roles at distinct steps to tightly control the initiation and progession of the differentiation program activity. In addition, recent studies have documented striking advances in the possibility of using epigenetic regulators in muscle pathologies. In light of these exciting developments, it is clear that further understanding of various epigenetic modifiers will facilitate and accelerate potential therapeutic strategies in myopathies that result from altered proliferation and differentiation of muscle precursor cells, and continue to push forward discoveries in this important area of research.

References

Ait-Si-Ali S, Guasconi V, Fritsch L, Yahi H, Sekhri R, Naguibneva I, Robin P, Cabon F, Polesskaya A, Harel-Bellan A (2004) A Suv39h-dependent mechanism for silencing S-phase genes in differentiating but not in cycling cells. EMBO J 23:605–615

Albini S, Puri PL (2010) SWI/SNF complexes, chromatin remodeling and skeletal myogenesis: it's time to exchange! Exp Cell Res 316:3073–3080

Azmi S, Ozog A, Taneja R (2004) Sharp-1 inhibits skeletal muscle differentiation through repression of myogenic transcription factors. J Biol Chem 279:52643–52652

Benezra R, Davis RL, Lockshon D, Turner DL, Weintraub H (1990) Cell 61:49–59

Black BL, Olson EN (1998) Transcriptional control of muscle development by myocyte enhancer factor-2 (MEF2) proteins. Annu Rev Cell Dev Biol 14:167–196

Blais A, van Oevelen CJ, Margueron R, Acosta-Alvear D, Dynlacht BD (2007) Retinoblastoma tumor suppressor protein-dependent methylation of histone H3 lysine 27 is associated with irreversible cell cycle exit. J Cell Biol 179:1399–1412

Braun T, Rudnicki MA, Arnold HH, Jaenisch R (1992) Targeted inactivation of the muscle regulatory gene Myf-5 results in abnormal rib development and perinatal death. Cell 7:369–382

Buckingham M (1992) Making muscle in mammals. Trends Genet 8:144–149

Buckingham M (2001) Skeletal muscle formation in vertebrates. Curr Opin Genet Dev 11:440–448

Caretti G, Di Padova M, Micales B, Lyons GE, Sartorelli V (2004) The Polycomb Ezh2 methyltransferase regulates muscle gene expression and skeletal muscle differentiation. Genes Dev 18:2627–2638

Chen SL, Loffler KA, Chen D, Stallcup MR, Muscat GE (2002) The coactivator-associated arginine methyltransferase is necessary for muscle differentiation: CARM1 coactivates myocyte enhancer factor-2. J Biol Chem 277:4324–4333

Colussi C, Rosati J, Straino S, Spallotta F, Berni R, Stilli D, Rossi S, Musso E, Macchi E, Mai A, Sbardella G, Castellano S, Chimenti C, Frustaci A, Nebbioso A, Altucci L, Capogrossi MC, Gaetano C (2011) Nε-lysine acetylation determines dissociation from GAP junctions and lateralization of connexin 43 in normal and dystrophic heart. Proc Natl Acad Sci USA 108:2795–2800

Cossu G, Tajbakhsh S, Buckingham M (1996a) How is myogenesis initiated in the embryo? Trends Genet 12:218–223

Cossu G, Kelly R, Tajbakhsh S, Di Donna S, Vivarelli E, Buckingham M (1996b) Activation of different myogenic pathways: myf-5 is induced by the neural tube and MyoD by the dorsal ectoderm in mouse paraxial mesoderm. Development 122:429–437

Cserjesi P, Olson EN (1991) Myogenin induces the myocyte-specific enhancer binding factor MEF-2 independently of other muscle-specific gene products. Mol Cell Biol 11:4854–4862

Dacwag CS, Bedford MT, Sif S, Imbalzano AN (2009) Distinct protein arginine methyltransferases promote ATP-dependent chromatin remodeling function at different stages of skeletal muscle differentiation. Mol Cell Biol 29:1909–1921

Davis RL, Weintraub H, Lassar AB (1987) Expression of a single transfected cDNA converts fibroblasts to myoblasts. Cell 51:987–1000

de la Serna IL, Ohkawa Y, Berkes CA, Bergstrom DA, Dacwag CS, Tapscott SJ, Imbalzano AN (2005) MyoD targets chromatin remodeling complexes to the myogenin locus prior to forming a stable DNA-bound complex. Mol Cell Biol 25:3997–4009

de Ruijter AJ, van Gennip AH, Caron HN, Kemp S, van Kuilenburg AB (2003) Histone deacetylases (HDACs): characterization of the classical HDAC family. Biochem J 370:737–749

Dilworth FJ, Seaver KJ, Fishburn AL, Htet SL, Tapscott SJ (2004) In vitro transcription system delineates the distinct roles of the coactivators PCAF and p300 during MyoD/E47-dependent transactivation. Proc Natl Acad Sci USA 101:11593–11598

Dodou E, Xu SM, Black BL (2003) mef2c is activated directly by myogenic basic helix–loop–helix proteins during skeletal muscle development *in vivo*. Mech Dev 120:1021–1103

Fulco M, Schiltz RL, Iezzi S, King MT, Zhao P, Kashiwaya Y, Hoffman E, Veech RL, Sartorelli V (2003) Sir2 regulates skeletal muscle differentiation as a potential sensor of the redox state. Mol Cell 12:51–62

Gros J, Manceau M, Thomé V, Marcelle C (2005) A common somitic origin for embryonic muscle progenitors and satellite cells. Nature 435:954–958

Guasconi V, Puri PL (2009) Chromatin: the interface between extrinsic cues and the epigenetic regulation of muscle regeneration. Trends Cell Biol 19:286–294

Halevy O, Novitch BG, Spicer DB, Skapek SX, Rhee J, Hannon GJ, Beach D, Lassar AB (1995) Correlation of terminal cell cycle arrest of skeletal muscle with induction of p21 by MyoD. Science 267:1018–1021

Hasty P, Bradley A, Morris JH, Edmondsnon DG, Venuti JM, Olson EN, Klein WH (1993) Muscle deficiency and neonatal death in mice with a targeted mutation in the *myogenin* gene. Nature 364:501–506

Huh MS, Parker MH, Scimè A, Parks R, Rudnicki MA (2004) Rb is required for progression through myogenic differentiation but not maintenance of terminal differentiation. J Cell Biol 166:865–876

Iezzi S, Di Padova M, Serra C, Caretti G, Simone C, Maklan E, Minetti G, Zhao P, Hoffman EP, Puri PL, Sartorelli V (2004) Deacetylase inhibitors increase muscle cell size by promoting myoblast recruitment and fusion through induction of follistatin. Dev Cell 6:673–684

Kuang S, Chargé SB, Seale P, Huh M, Rudnicki MA (2006) Distinct roles for Pax7 and Pax3 in adult regenerative myogenesis. J Cell Biol 172:103–113

Lassar AB, Paterson BM, Weintraub H (1986) Transfection of a DNA locus that mediates the conversion of 10T1/2 fibroblasts to myoblasts. Cell 47:649–656

Lemercier C, To RQ, Carrasco RA, Konieczny SF (1998) The basic helix-loop-helix transcription factor Mist1 functions as a transcriptional repressor of myoD. EMBO J 17:1412–1422

Ling BM, Bharathy N, Chung TK, Kok WK, Li S, Tan YH, Rao VK, Gopinadhan S, Sartorelli V, Walsh MJ, Taneja R (2012) Lysine methyltransferase G9a methylates the transcription factor MyoD and regulates skeletal muscle differentiation. Proc Natl Acad Sci USA 109:841–846

Lu J, Webb R, Richardson JA, Olson EN (1999) MyoR: a muscle-restricted basic helix-loop-helix transcription factor that antagonizes the actions of MyoD. Proc Natl Acad Sci USA 96:552–557

Lu J, McKinsey TA, Zhang CL, Olson EN (2000) Regulation of skeletal myogenesis by association of the MEF2 transcription factor with class II histone deacetylases. Mol Cell 6:233–244

Mal AK (2006) Histone methyltransferase Suv39h1 represses MyoD-stimulated myogenic differentiation. EMBO J 25:3323–3334

Mal A, Harter ML (2003) MyoD is functionally linked to the silencing of a muscle-specific regulatory gene prior to skeletal myogenesis. Proc Natl Acad Sci USA 100:1735–1739

Mal A, Sturniolo M, Schiltz RL, Ghosh MK, Harter ML (2001) A role for histone deacetylase HDAC1 in modulating the transcriptional activity of MyoD: inhibition of the myogenic program. EMBO J 20:1739–1753

McKinnell IW, Ishibashi J, Le Grand F, Punch VG, Addicks GC, Greenblatt JF, Dilworth FJ, Rudnicki MA (2008) Pax7 activates myogenic genes by recruitment of a histone methyltransferase complex. Nat Cell Biol 10:77–84

McKinsey TA, Zhang CL, Olson EN (2001) Control of muscle development by dueling HATs and HDACs. Curr Opin Genet Dev 11:497–504

Minetti GC, Colussi C, Adami R, Serra C, Mozzetta C, Parente V, Fortuni S, Straino S, Sampaolesi M, Di Padova M, Illi B, Gallinari P, Steinkühler C, Capogrossi MC, Sartorelli V, Bottinelli R, Gaetano C, Puri PL (2006) Functional and morphological recovery of dystrophic muscles in mice treated with deacetylase inhibitors. Nat Med 12:1147–1150

Molkentin JD, Olson EN (1996) Combinatorial control of muscle development by basic helix-loop-helix and MADS-box transcription factors. Proc Natl Acad Sci USA 93:9366–9373

Molkentin JD, Black BL, Martin JF, Olson EN (1995) Cooperative activation of muscle gene expression by MEF2 and myogenic bHLH proteins. Cell 83:1125–1136

Nabeshima Y, Hanaoka K, Hayasaka M, Esumi E, Li S, Nonaka I, Nabeshima Y (1993) Myogenin gene disruption results in perinatal lethality because of severe muscle defect. Nature 364:532–535

North BJ, Verdin E (2004) Sirtuins: Sir2-related NAD-dependent protein deacetylases. Genome Biol 5:224

Olson EN, Arnold HH, Rigby PWJ, Wold BJ (1996) Know your neighbors: three phenotypes in null mutants of the myogenic bHLH gene MRF4. Cell 85:1–4

Ordahl CP, Le Douarin NM (1992) Two myogenic lineages within the developing somite. Development 114:339–353

Pownall ME, Gustafsson MK, Emerson CP Jr (2002) Myogenic regulatory factors and the specification of muscle progenitors in vertebrate embryos. Annu Rev Cell Dev Biol 18:747–783

Puri PL, Sartorelli V, Yang XJ, Hamamori Y, Ogryzko VV, Howard BH, Kedes L, Wang JY, Graessmann A, Nakatani Y, Levrero M (1997) Differential roles of p300 and PCAF acetyltransferases in muscle differentiation. Mol Cell 1:35–45

Puri PL, Iezzi S, Stiegler P, Chen TT, Schiltz RL, Muscat GE, Giordano A, Kedes L, Wang JY, Sartorelli V (2001) Class I histone deacetylases sequentially interact with MyoD and pRb during skeletal myogenesis. Mol Cell 8:885–897

Rampalli S, Li L, Mak E, Ge K, Brand M, Tapscott SJ, Dilworth FJ (2007) p38 MAPK signaling regulates recruitment of Ash2L-containing methyltransferase complexes to specific genes during differentiation. Nat Struct Mol Biol 14:1150–1156

Rea S, Eisenhaber F, O'Carroll D, Strahl BD, Sun ZW, Schmid M, Opravil S, Mechtler K, Ponting CP, Allis CD, Jenuwein T (2000) Regulation of chromatin structure by site-specific histone H3 methyltransferases. Nature 406:593–599

Relaix F, Rocancourt D, Mansouri A, Buckingham M (2005) A Pax3/Pax7-dependent population of skeletal muscle progenitor cells. Nature 435:948–953

Relaix F, Montarras D, Zaffran S, Gayraud-Morel B, Rocancourt D, Tajbakhsh S, Mansouri A, Cumano A, Buckingham M (2006) Pax3 and Pax7 have distinct and overlapping functions in adult muscle progenitor cells. J Cell Biol 172:91–102

Rudnicki MA, Braun T, Hinuma S, Jaenisch R (1992) Inactivation of MyoD in mice leads to upregulation of the myogenic HLH gene Myf-5 and results in apparently normal muscle development. Cell 71:383–390

Rudnicki MA, Schnegelsberg PN, Stead RH, Braun T, Arnold HH, Jaenisch R (1993) MyoD or Myf-5 is required for the formation of skeletal muscle. Cell 75:1351–1359

Sartorelli V, Puri PL, Hamamori Y, Ogryzko V, Chung G, Nakatani Y, Wang JY, Kedes L (1999) Acetylation of MyoD directed by PCAF is necessary for the execution of the muscle program. Mol Cell 4:725–734

Sebastian S, Sreenivas P, Sambasivan R, Cheedipudi S, Kandalla P, Pavlath GK, Dhawan J (2009) MLL5, a trithorax homolog, indirectly regulates H3K4 methylation, represses cyclin A2 expression, and promotes myogenic differentiation. Proc Natl Acad Sci USA 106:4719–4724

Simone C, Forcales SV, Hill DA, Imbalzano AN, Latella L, Puri PL (2004) p38 pathway targets SWI-SNF chromatin-remodeling complex to muscle-specific loci. Nat Genet 36:738–743

Spicer DB, Rhee J, Cheung WL, Lassar AB (1996) Inhibition of myogenic bHLH and MEF2 transcription factors by the bHLH protein twist. Science 272:1476–1480

Tajbakhsh S, Cossu G (1997) Establishing myogenic identity during somitogenesis. Curr Opin Genet Dev 7:634–641

Venuti JM, Morris JH, Vivian JL, Olson EN, Klein WH (1995) Myogenin is required for late but not early aspects of myogenesis during mouse development. J Cell Biol 128:563–576

Zammit PS, Golding JP, Nagata Y, Hudon V, Partridge TA, Beauchamp JR (2004) Muscle satellite cells adopt divergent fates: a mechanism for self-renewal? J Cell Biol 166:347–357

Zhang CL, McKinsey TA, Olson EN (2002) Association of class II histone deacetylases with heterochromatin protein 1: potential role for histone methylation in control of muscle differentiation. Mol Cell Biol 22:7302–7312

Chapter 8
Small Changes, Big Effects: Chromatin Goes Aging

Asmitha Lazarus, Kushal Kr. Banerjee, and Ullas Kolthur-Seetharam*

Abstract Aging is a complex trait and is influenced by multiple factors that are both intrinsic and extrinsic to the organism (Kirkwood et al. 2000; Knight 2000). Efforts to understanding the mechanisms that extend or shorten lifespan have been made since the early twentieth century. Aging is characteristically associated with a progressive decline in the overall fitness of the organism. Several studies have provided valuable information about the molecular events that accompany this process and include accumulation of nuclear and mitochondrial mutations, shortened and dysfunctional telomeres, oxidative damage of protein/DNA, senescence and apoptosis (Muller 2009). Clinical studies and work on model organisms have shown that there is an increased susceptibility to conditions such as neurological disorders, diabetes, cardiovascular diseases, degenerative syndromes and even cancers, with age (Arvanitakis et al. 2006; Lee and Kim 2006; Rodriguez and Fraga 2010).

Investigations into aging mechanisms in unicellular systems, like yeast and *in vitro* cell culture models, have identified several pathways involved in this process. In cells aging is typically associated with a senescent phenotype. Cells are known to have a limited proliferative capacity (Hayflick limit) (Hayflick 1965) and senescence can be defined as a state in which cells cease to proliferate after a finite number of divisions (Adams 2009). Some of the well-known triggers that induce senescence include DNA damage, telomere shortening and redox stress (Rodier and Campisi 2011). From literature, it is evident that in most of these cases, factors/pathways which bring about cell cycle arrest are activated and include p53/p21, and p16/RB

---

*Asmitha Lazarus and Kushal Kr. Banerjee have contributed equally

A. Lazarus • K.K. Banerjee • U. Kolthur-Seetharam (✉)
B-306, Department of Biological Sciences, Tata Institute of Fundamental Research, Dr. Homi Bhabha Road, Colaba, Mumbai 400 005, India
e-mail: ullas@tifr.res.in

pathways (Ben-Porath and Weinberg 2005). However, the cellular/molecular signatures that characterize senescence are typically scored by induction of senescence associated β-galactosidase activity, marks of cell cycle arrest, changes in cellular morphology and/or organization, secretion of numerous proteins including cytokines and chemokines, and DNA damage (Rodier and Campisi 2011). The quest to decipher the molecular events that induce or bring about cellular senescence have unraveled the role of chromatin as an important component of this response (Misteli 2010).

The DNA in every eukaryotic cell exists as a complex with specialized proteins called histones that form chromatin. Chromatin plays a central role in processes that range from gene expression to chromosome dynamics during the cell cycle. Chromatin can be broadly categorized into two types, namely, euchromatin and heterochromatin (Bassett et al. 2009). Euchromatin appears decondensed cytologically and is mostly transcriptionally active. Heterochromatin, on the other hand, is highly compact and mostly contains transcriptionally silenced genes (Bassett et al. 2009; Frenster et al. 1963). The building block of chromatin is a nucleosome that consists of 147 base pairs of DNA wrapped around a protein octamer containing two molecules of each canonical histone H2A, H2B, H3 and H4, and is separated from one another by 10–60 base pairs of linker DNA (Luger et al. 1997). Histones contain a typical histone-fold domain, which is required to form the octamer, and their N-termini protrude out of the nucleosomes (Luger et al. 1997). Most of the residues on these histone tails are subject to posttranslational modifications (Jenuwein and Allis 2001). Some of the most prevalent modifications of histones are phosphorylation, acetylation, methylation, ubiquitination, sumoylation, ADP-ribosylation and biotinylation (Jenuwein and Allis 2001; Margueron et al. 2005). Recent reviews have illustrated biophysical and physiological consequences of such modifications on chromatin structure and function (Li and Reinberg 2011). In addition to histone and non-histone proteins that bind to DNA, modification of DNA (Cytosine methylation in eukaryotes) is also an important component of chromatin (Li and Reinberg 2011). It is interesting to note that there is a dynamic interplay between histone and DNA modifications that determine chromatin structure/function (Bonasio et al. 2010; Li and Reinberg 2011).

As mentioned earlier, the most obvious associations of DNA with aging are increased DNA damage (or reduced repair) (Seviour and Lin 2010) and telomere shortening (Kenyon and Gerson 2007). Increasing evidence in literature indicates that chromatin plays a major role in affecting both these processes (Shin et al. 2011a). In addition to these, the ability of chromatin to affect gene expression patterns in a cell has huge consequences on the ability to maintain homeostasis. Therefore, given the central role of chromatin in affecting various cellular processes, intuitively one would expect it to be a crucial component of cellular aging. In this chapter, we review the recent progress on the role of chromatin in aging. Specifically, we highlight the chromatin changes that have been associated with cellular and/or organismal aging. Importantly, we also highlight the role of histone modifiers in affecting lifespan.

8.1 Chromatin and Aging

8.1.1 DNA Methylation

DNA methylation, one of the most well studied epigenetic marks, involves the methylation of cytosines in CpG dinucleotides and is catalyzed by enzymes termed as DNA methyl transferases (DNMTs: DNMT1, DNMT3a, DNMT3b) (Jurkowska et al. 2011). It is well established that DNA methylation constitutes mechanisms required for both short-term and long-term effects on gene expression (Bonasio et al. 2010; Li and Reinberg 2011). Specifically, alterations in methylation of CpGs at upstream regulatory elements are known to modulate transcription of genes (Li and Reinberg 2011). Due to its ability to control both global and locus specific chromatin functions, DNA methylation is known to play a critical role in cellular physiology. It is important to note that key biological processes such as development, differentiation and cell death are affected by DNA methylation (De Carvalho et al. 2010; Geiman and Muegge 2010; Gibney and Nolan 2010). Its role in aging and/or senescence has been addressed in the recent past, and it is apparent that DNA methylation is one of the key factors involved in cellular and/or organismal aging (Feser and Tyler 2011; Fraga and Esteller 2007; Sedivy et al. 2008; Dimauro and David 2009). Figure 8.1 illustrates the changes in DNA methylation during aging.

An important role for DNA methylation in aging was first evidenced in replicative senescence of primary fibroblasts from mice, hamsters and humans. The study showed that in these cells, levels of 5-methylcytosine markedly declined during senescence (Wilson and Jones 1983). A follow up study demonstrated that accelerated 5-methylcytosine loss (by 5-azacytidine treatment) shortened the *in vitro* lifespan of human diploid fibroblasts (Fairweather et al. 1987). This phenomenon was reconfirmed by various *in vivo* and *in vitro* studies wherein it was observed that DNA methylation levels fell during aging, both at certain specific loci and at a genome wide level (Fairweather et al. 1987; Christensen et al. 2009; Fuke et al. 2004; Singhal et al. 1987). Interestingly, recent reports have indicated that this age-associated decline in total genomic DNA methylation occurs mostly at repetitive DNA sequences (Koch et al. 2011; Romanov and Vanyushin 1981; Singhal et al. 1987; Wilson et al. 1987). These observations have led to the speculation that the decrease in DNA methylation affects constitutive heterochromatin (DePinho 2000). Specifically, it has been suggested that with age de-heterochromatinization of repetitive regions could lead to deleterious recombinations which may cause increased incidences of age-associated diseases such as cancer (DePinho 2000). Further, the importance of DNA methylation in aging is supported by observations that show age-dependent decrease in the expression of the DNA methyltransferase (DNMT1) (see below) (Casillas et al. 2003; Lopatina et al. 2002). Supporting that the gradual loss of DNA methylation could function as a "counting hypothesis" for senescence (Hoal-van Helden and van Helden 1989; Wilson and Jones 1983), CpG methylation was shown to decrease with increased population doublings of normal cells in

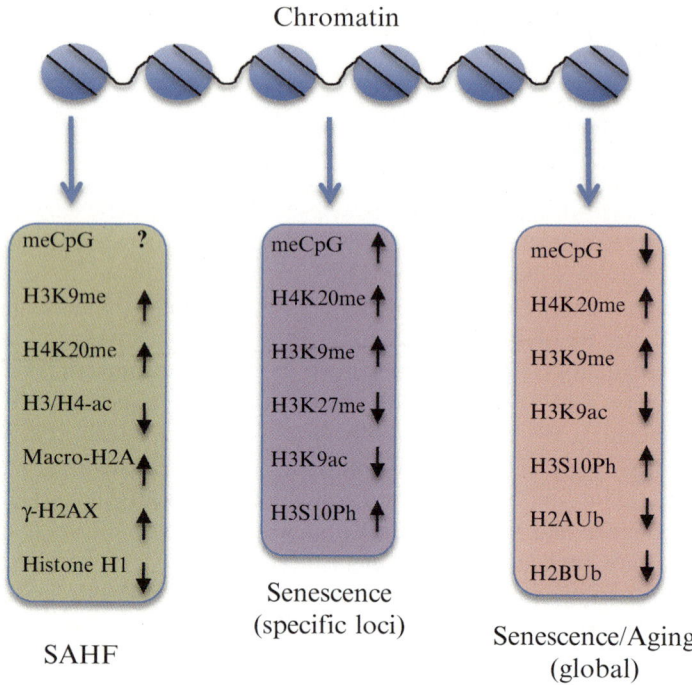

Fig. 8.1 Epigenetic changes during aging. Alterations in epigenetic marks in cultured cells and/or tissues during aging affects chromatin both globally and at specific loci. In general, histone modifications associated with heterochromatin seem to accumulate during aging. The association between DNA methylation and aging is context dependent and often determined by global and locus specific changes

culture (Fairweather et al. 1987; Wilson and Jones 1983) and during organismal aging (Hornsby et al. 1992; Singhal et al. 1987).

Contrary to observations of a decrease in global DNA methylation during aging specific loci and promoter regions of key cell cycle regulatory genes have been shown to be hypermethylated (Fig. 8.1). For example, the Estrogen receptor gene (Issa et al. 1994), *INK4A/ARF/INK4b* locus (which codes for p16, p14 and p15 proteins respectively) (Koch et al. 2011), ribosomal RNA genes (Swisshelm et al. 1990; Oakes et al. 2003) and multiple tumor suppressor or tumour-associated genes like APC and E-cadherin (Bornman et al. 2001; Waki et al. 2003) accumulate DNA methylation during aging. Interestingly, in another study it was observed that DNA methylation levels were maintained in long-term culture of mesenchymal stromal cells (MSE) and MSEs from young and old donors. However, they exhibited differential DNA methylation patterns at specific loci, like in the homeobox genes and genes involved in cell differentiation (Bork et al. 2010). It has been hypothesized that this locus specific hypermethylation in the background of a global reduction of methyl-CpGs could be due to an increase in Dnmt3b expression that has been observed in senescent cells (Casillas et al. 2003; So et al. 2006).

Observations from some studies suggest that hypermethylation depends upon the prevalent density of methyl-cytosines at specific loci with sparsely methylated regions more amenable to hypermethylation (Song et al. 2002; Stirzaker et al. 2004). These observations have led to a 'seeds of methylation' hypothesis based on increasing CpG methylation levels (Rakyan et al. 2010). A recent genome-scale study addressed dynamic changes in the epigenome in normal human aging (Rakyan et al. 2010). This report identified aging-associated differentially methylated regions (aDMRs) that gain methylation with age in different tissues, thus suggesting that aDMR signature is a multi-tissue phenomenon. Further, it was also demonstrated that aging associated DNA hypermethylation occurs predominantly at bivalent chromatin/promoters (Rakyan et al. 2010). These studies point out an interesting aspect of locus specific methylation contributing to aging. In this scenario one would expect an inherent bias in methylation rates at loci that would ultimately (or cumulatively) result in a senescent phenotype.

8.1.2 Histone Modifications

As mentioned in the introduction, histone modifications are one of the most central elements that affect chromatin structure and function. The most common and well-studied histone modifications that are known to impact chromatin are acetylation of lysines, methylation of lysines and arginines, phosphorylation of serine and threonine, and ubiquitination of lysines (Jenuwein and Allis 2001; Margueron et al. 2005). It is evident from literature that interfering with these modifications affects both global and locus specific chromatin, and as a consequence impinges on various cellular processes (Murr 2010). Some recent reviews provide exhaustive information about histone modifications and their role in chromatin structure and function. Figure 8.1 illustrates the histone modifications, which are associated with 'open or closed' chromatin and their associations with aging.

Specific histone modifications undergo distinct changes in profile during aging (Fig. 8.1). The levels of histone H4 lysine-20 tri-methylation (H4K20Me3), a mark of constitutive heterochromatin and which is enriched in differentiated cells, have been found to increase in senescent cells. This has been speculated to cause the accumulation of heterochromatic structures in senescent human fibroblasts (Narita et al. 2003). The total abundance of histone H4K20Me3 has also been reported to increase with age in rat liver and kidney (Kouzarides 2007; Sarg et al. 2002), supporting the notion that heterochromatin may accumulate with tissue aging, at least at some sites. In another study, Bracken et al. observed a loss of histone H3 lysine-27 tri-methylation (H3K27Me3), a mark associated with silent chromatin, at the *INK4b* and *INK4a–ARF* loci in senescent human diploid lung embryonic fibroblast cell line (Bracken et al. 2007). This decrease in H3K27Me3 was accompanied by a decrease in EZH2, the histone methyltransferase responsible for this modification

(Bracken et al. 2007). Several groups have studied changes in histone H3 modifications with age in rat liver. They found that histone H3 lysine-9 acetylation (H3K9Ac) decreased and histone H3 Serine-10 phosphorylation (H3S10Ph) increased with age significantly (Braig et al. 2005; O'Sullivan et al. 2010; Kawakami et al. 2009). These independent observations both in cells in culture and in aged animals clearly establish a positive correlation between heterochromatic marks and aging (Fig. 8.1).

Mono-ubiquitination of histones H2A and H2B is known to alter chromatin dynamics and regulate gene expression. While H2A ubiquitination leads to silencing, ubiquitination of H2B has been implicated in active transcription. Interestingly, these modifications have been associated with aging. The link between histone ubiquitination and aging was first demonstrated by a study which showed that the proportion of ubiquitinated histones was about 30% higher in old mice than in young ones (Morimoto et al. 1993). However, reduced expressions of H2B ubiquitin ligases RNF20/Bre1 have been associated with senescence/aging phenotypes. In yeast, absence of Bre1 results in reduced lifespan during chronological aging due to enhanced apoptotic cell death (Walter et al. 2010). Similarly, depletion of RNF20 has been shown to induce cellular senescence in glioma cells (Gao et al. 2011). Like H2B, ubiquitination of histone H2A has also been implicated in aging. Downregulation of BMI1, a component of the polycomb repressive complex (PRC), which ubiquitinates histone H2A (Cao et al. 2005) has been shown to result in derepression of growth inhibitory genes and putative tumor suppressors. As a consequence these cells display premature senescence and apoptosis (Bommi et al. 2010). Although, these studies suggest that histone ubiquitination is involved in aging, whether these effects are mediated through alterations in global chromatin architecture or transcription of specific genes is still not clear.

8.1.3 Senescence Associated Heterochromatic Foci (SAHF)

Cells grown in culture have provided valuable insights into aging mechanisms. In this regard, most of our understanding of the role of chromatin on aging has come from studies on senescing cells. Not surprisingly, alterations of chromatin structure are associated with the irreversible state of senescent cells (Braig and Schmitt 2006; Narita et al. 2003). Many senescent human cells, when stained with the DNA staining dye 4', 6-diamidino-2-phenylindole (DAPI), show visible punctuate DNA foci known as senescence associated heterochromatic foci (SAHF), a new type of facultative heterochromatin (Narita 2007; Narita et al. 2003). RNA-FISH and *in situ* labeling of nascent RNAs demonstrate that SAHF contain transcriptionally inactive chromatin (Funayama et al. 2006; Narita 2007). In addition to being transcriptionally inactive, these loci are also enriched with typical proteins that are found in heterochromatin. For example, SAHFs in general contain heterochromatin protein-1 (HP1), repressive histone modifications like H3K9 methylation and hypoacetylated histones (Narita et al. 2006). However, these SAHFs do not show some usual marks of condensed chromatin, like the phosphorylation of histone H3 at Serine-10 or

Serine-28, marks of mitotic chromatin or of histone H2B at Serine-14, a mark of apoptotic chromatin (Funayama et al. 2006; Peterson and Laniel 2004).

It is important to note that in addition to changes in histone modifications, histone chaperones, and alterations in chromatin composition have also been implicated in senescence. For example, studies have shown that the formation of SAHF during cellular senescence depends on histone H3 chaperones, ASF1 (anti-silencing function 1) (Zhang et al. 2005) and HIRA (histone cell cycle regulation defective homologue A) (Ye et al. 2007). Interestingly, these loci are also known to contain variants of histones which have been otherwise associated with silenced chromatin. SAHFs are enriched with macro-H2A (histone H2A variant) that is mainly required for inactivation of X-chromosome (Costanzi and Pehrson 1998; Funayama et al. 2006; Zhang et al. 2005). The role of histone variants in the formation of SAHF is also supported by findings that report an increase in γ-H2AX in early neoplastic lesions that contain senescent cells *in vivo* and also in aging tissues. This is thought to contribute to senescence and proliferation arrest of damaged cells (Bartkova et al. 2006; Herbig et al. 2006). Chromatin compaction can also be altered by the recruitment of factors that are known to replace histone H1 and bind to linker DNA. In this regard, the finding which shows that in SAHFs there is a decrease in linker histone H1 occupancy and increased levels of chromatin-bound high mobility group-A proteins (HMGA) becomes relevant (Funayama et al. 2006; Narita et al. 2006). The exact molecular mechanisms of SAHF formation are not very clear, but independent studies have demonstrated that the ectopic expression of either HMGA1 or HMGA2 induces SAHF formation and other senescence phenotypes in normal human fibroblasts. It was also observed that knockdown of HMGA proteins by RNAi prevents SAHF formation, thus indicating that HMGA are essential components for SAHF formation (Funayama et al. 2006; Narita et al. 2006). It has been speculated that the DNA-bending properties of HMG family proteins may help induce SAHF formation by binding and bending linker DNA (Hock et al. 2007; Paull et al. 1993).

Formation of heterochromatin is often facilitated by enzymatic activities that are known to repress transcription. Notably, histone deacetylases and histone methyl transferases that add 'repressive chromatin marks' play essential roles in heterochromatin formation. The Sin3 multiprotein complex is a repressor complex recruited by several sequence specific transcription factors. The repressor activity of the Sin3 complex is brought about by the Sin3A/Sin3B-associated HDAC1 and HDAC2 proteins. A study by Grandinetti et al. has demonstrated that Sin3B-null fibroblasts are resistant to replicative and oncogene-induced senescence (Grandinetti et al. 2009). They also showed that over-expression of Sin3B triggers senescence and the formation of SAHF. However, the role of histone deacetylation in inducing SAHF seems to be HDAC specific. While Sin3 complex via HDAC activity aids in the formation of SAHFs, a study by Huang et al. has suggested that Sirt1, a NAD^+-dependent deacetylase (described below), antagonizes cellular senescence in human diploid fibroblasts (Huang et al. 2008). Their experiments demonstrated that over-expressing Sirt1 led to a reduction of senescence associated biomarkers, which included the formation of SAHFs (Huang et al. 2008).

Although, SAHFs seem to bring about a global change in chromatin architecture, it is not clear if SAHF formation contributes to senescence. In support of SAHF contributing to senescence, evidence show that SAHF formation contributes to stable proliferative arrest by repressing transcription of E2F target genes that are required for G1 to S phase transition (Narita et al. 2003). Chromatin immunoprecipitation analyses have demonstrated that the promoters of E2F target genes become heterochromatic in senescent cells but not in proliferating or quiescent cells. In addition, overexpression of E2F-1 was not able to derepress these genes indicating heterochromatinization mediated transcriptional silencing (Narita et al. 2003). Interestingly, SAHF-dependent silencing of E2F genes requires the retinoblastoma (Rb) protein at these gene promoters (Narita et al. 2003). Further, studies have shown that Rb associates with HP1 and the histone methyltransferase Suv39H1 to facilitate senescence. Specifically, Rb family members have been shown to interact with HDAC1, DNA methyltransferase and polycomb proteins among other transcriptional co-repressors to repress the activity of E2F1 (Trimarchi and Lees 2002; Narita et al. 2003). Prohibitin, a protein implicated in cell cycle control and antiproliferative activities, is found in SAHF and colocalizes with HP1 (Rastogi et al. 2006). This finding suggests that SAHFs might actively contribute to senescence. In this study prohibitin, Suv39H1 and HP1 were detected on E2F target promoters during senescence, and a deletion of prohibitin led to a loss of senescent phenotype (Rastogi et al. 2006).

Although, there is a lot of evidence to suggest that SAHF formation is important for induction of senescence, formation of heterochromatin itself seems to be the most important feature of senescence. A recent study has shown an increase in the abundance of heterochromatin proteins and marks in senescence but without the formation of SAHF (Kosar et al. 2011). Hence, local heterochromatinization, but not global SAHF, may induce senescence-associated proliferation arrest by mediating the silencing of proliferation genes.

8.1.4 microRNAs, Epigenetics and Aging

MicroRNAs are ~22 bases long RNAs, which bind to the 3′UTR of target mRNAs and regulate gene expression post-transcriptionally by translational inhibition or mRNA degradation (He and Hannon 2004). Due to their ability to target multiple mRNAs, they are now considered as major factors that affect cellular physiology (He and Hannon 2004; Sayed and Abdellatif 2011). Originally appreciated for their role in cancer and development, microRNAs have also been shown to be involved in regulating factors or pathways, which impinge on aging. In the recent past, studies have identified many microRNAs that target 'aging' factors and several reviews have highlighted these reports (Bates et al. 2009; Gorospe and Abdelmohsen 2011; Grillari and Grillari-Voglauer 2010). Rather than detailing microRNAs and their targets that have been implicated in aging, we specifically highlight studies that have addressed altered expression of microRNAs during aging.

Intriguingly, global microarray profiling studies suggest that more microRNAs are upregulated rather than downregulated during aging (Li et al. 2011; Maes et al. 2008; Zhang et al. 2010). It is important to note that upregulation of some of these microRNAs have been implicated in regulating the expression of genes, which are known to affect organismal physiology. For example, miR-669c and miR-709 (up-regulated at 18 months with a maximum expression at 33 months), and miR-93 and miR-214 (up-regulated around 33 months) have been shown to target genes associated with detoxification and regenerative capacity of the liver, functions that slowly decline in aged liver (Maes et al. 2008). In another study, Bates et al. profiled microRNAs regulated in the liver of Ames dwarf mice, which display a delayed onset of aging (Steuerwald et al. 2010). They found that miR-27a is upregulated in these dwarf mice at an early age. Their results also suggest that miR-27a regulates two key metabolic proteins ornithine decarboxylase and spermidine synthase. Based on these observations the authors have speculated that miR-dependent regulation of metabolic pathways such as glutathione metabolism, urea cycle, and polyamine biosynthesis maybe important for health span and longevity in these mice (Bates et al. 2010). However, studies which link microRNAs with DNA repair or cell proliferation pathways have raised the possibility that age related alterations in microRNA expressions maybe relevant in mediating the aging process (Chen et al. 2010).

The link between microRNAs and aging has been further strengthened by a study in which reducing the activity of C. elegans linage 4 (*lin-4*) microRNA shortened lifespan and its overexpression led to a longevity phenotype (Boehm and Slack 2005). Another study that looked at senescence in normal human keratinocytes (NHK) found microRNAs miR-137 and miR-668 to be upregulated during replicative senescence (Shin et al. 2011b). Interestingly, induction of senescence by ectopic over-expression of miR-137 and miR-668 was associated with an increase in senescence associated (SA) β-galactosidase activity, p53 and p16INK4A levels. Further, expressions of these microRNAs were also observed to be elevated during organismal aging of normal human oral epithelia (Shin et al. 2011b).

Although, it is increasingly becoming apparent that microRNAs play a vital role in regulating aging, very little is known about epigenetic changes that mediate the expression of such key microRNAs. Recent studies have clearly shown that microRNA expression is regulated by epigenetic marks (Liang et al. 2009). Importantly, their promoters have been shown to exhibit differential DNA methylation and histone modifications, that are reminiscent of modifications on protein coding genes (Lee et al. 2011; Saito and Jones 2006). Lee et al. have demonstrated that inhibition of HDACs triggers cellular senescence by inducing the expression of miR-23a, miR-26a and miR-30a. Interestingly, these microRNAs target and downregulate HMGA2 expression that has been associated with induction of senescence (Lee et al. 2011).

Further studies aimed at profiling microRNAs during aging, and in specific tissues, will aid in appreciating the regulation of pathways that mediate lifespans of organisms. Importantly, investigating the mechanisms that control mircoRNA expression, specifically histone deacetylases and DNA methyltransferases (which have been associated with aging, see below), will highlight the importance of posttranscriptional control of 'aging genes'. In addition, such insights will provide a holistic picture of changes in gene regulation, mediated by chromatin modifiers, in affecting organismal longevity.

8.2 Role of Chromatin Modifiers in Aging

The previous section highlights the importance of chromatin associated changes in aging and cellular senescence. Although, it is clear that these changes are strong correlates of aging, whether they are causal factors or mere consequences of aging remains unclear (Dimauro and David 2009). Also, aging/senescence dependent changes that the enzymes which affect these modifications themselves undergo are less appreciated. As reviewed elsewhere, post-translational modifications of histones are catalyzed by specific enzymatic machineries (Bannister and Kouzarides 2011). Histone acetylation is affected by opposing activities of histone acetyltransferases (HATs) and histone deacetylases (HDACs) (Legube and Trouche 2003). Separate families of enzymes are known to methylate and demethylate lysine/arginine residues in histones (Yoshimi and Kurokawa 2011). Interestingly, unlike these modifications histone phosphorylation and dephosphorylation are brought about by a diverse set of enzymes (Hans and Dimitrov 2001). In this section, we have attempted to review the studies which have given us insights into the role of histone modifiers during aging. Specifically, we will look at the important classes of chromatin modifiers: DNMTs, Histone acetyltransferases Histone deacetylases and Sirtuins (Table 8.1).

Table 8.1 List of chromatin modifiers and their association with aging or senescence

Chromatin modifier	Modification site	Modification	Alteration with age	Role in aging
DNMT1	CpG dinucleotide	Methylation	Decrease	Anti-senescent
DNMT3a	CpG dinucleotide	Methylation	Increase	Locus specific effects
DNMT3b	CpG dinucleotide	Methylation	Increase	Locus specific effects
Mof (HAT)	H4K16	Acetylation	Increase	Pro-senescent
Sas2 (HAT)	H4K16	Acetylation	Increase	Pro-senescent
CBP (HAT)	H3K9, H3K27, H3K56	Acetylation	Decrease	Anti-senescent
P300 (HAT)	H3K9, H3K27, H3K56	Acetylation	Decrease	Anti-senescent
HDAC1	H3K9, H3K56, H4K16	Deacetylation	Increase	Pro-senescent
SIRT1	H3K9, H4K16	Deacetylation	Decrease	Anti-senescent
SIRT6	H3K9	Deacetylation	?	Anti-senescent
EZH2 (HMT)	H3K27	Methylation	Decrease	Pro-senescence

The table illustrates the roles of DNMTs, HMTases, HATs, HDACs and Sirtuins, and depicts changes in their expression during aging

8.2.1 DNA Methyl Transferases (DNMTs)

DNA methyltransferases (DNMTs) catalyze the addition of a methyl group to DNA on cytosines, typically in CpG dinucleotides. DNA methylation has been associated with gene silencing and robust regulation of transcription. DNMT mediated DNA methylation brings about chromatin silencing by inducing the formation of heterochromatin through recruitment of specific proteins like Methyl CpG binding proteins, MeCP2 (Kimura and Shiota 2003) and MBD (Fujita et al. 2003; Villa et al. 2006). It is interesting to note that DNMTs have been shown to be in complex with histone methyl transferases and histone deacetylases. In mammals there are three DNA methyltransferases, namely, Dnmt1, Dnmt3a and Dnmt3b. Dnmt1 is considered as a maintenance methylase since it methylates newly replicated DNA using hemimethylated DNA as a substrate. Dnmt3a and 3b mediate *de novo* methylation, that is, they can methylate previously unmethylated DNA.

The role of DNMTs in aging has been addressed in the recent past because of their ability to affect chromatin/epigenetic modifications. In addition, previous reports have also correlated changes in DNA methylation during aging. Several studies have thrown light upon changes in DNMT levels and activity that may have crucial roles in cellular senescence (Lopatina et al. 2003; Vogt et al. 1998). Consistent with previous observations of a decrease in global DNA methylation in senescence, studies have shown that the levels and activity of the maintenance methylase Dnmt1 decrease in aging fibroblast cells. However, an increase in Dnmt-3a and -3b activity was observed which raises the possibility of a compensatory role for these DNMTs (Casillas et al. 2003; Lopatina et al. 2003). Although, it was long observed that promoter hypermethylation of cell cycle inhibitory genes, $p16^{INK4A}$ and $p21^{CIP1/WAF1}$ tipped the balance between senescence and oncogenesis, a recent report shows the involvement of DNMTs, and their findings have provided further support to the hypothesis that DNMTs are important for inducing senescence. The authors of this study observed that upon inhibition of DNMT1 and DNMT3b in human umbilical cord blood-derived multipotent stem cells (hUCB-MSCs), p16 and p21 expression increased and activated senescence in these cells (So et al. 2011). Not surprisingly, several studies have shown that DNMTs are overexpressed in cancer cell lines resulting in silencing of the expression of p16 (So et al. 2011; Yang et al. 2001).

Calorie/Dietary restriction (CR/DR) is one of the interventions that has been commonly used to understand the molecular factors involved in aging. Although, as previously mentioned it is unclear whether DNMTs play a deterministic role in aging, studies have indicated that Dnmt3a levels in the mouse hippocampus change when the animals are subjected to dietary restriction (Chouliaras et al. 2011a, b). This study also sheds light on the possibility that one of the major mechanisms by which CR/DR mediates organismal aging is by modulating DNA methylation status by regulating expression levels/enzymatic activities of individual DNMTs. However, it is still unclear if DNMTs can by themselves induce senescence or whether other factors that initiate or require chromatin changes affect their expression and/or activities during aging. Nevertheless, these findings clearly indicate that the changes in DNMT expressions can lead to alterations in chromatin structure during aging by influencing both global and gene specific methyl-CpG levels and/or distribution. In spite of these reports that indicate strong associations between DNMT expression/ activities, DNA methylation patterns and aging, it is surprising to find that there are very few attempts to map the changes in DNMT localizations on a genome wide scale. Further, it would be interesting to similarly analyze genome-wide alterations in DNA methylation profiles in various model systems that are known to extend their lifespans in response to dietary interventions.

8.2.2 Histone Acetyl Transferases and Histone Deacetylases

Besides DNA methylation, histone modifications have become the most important determinants of chromatin structure/function involved in various cellular outputs. Specifically, histone acetylation has been one of the hallmarks of active gene transcription and often influences other modifications of histones such as methylation. As previously mentioned, histone acetylation-deacetylation reactions are catalyzed by histone acetyl transferases and histone deacetylases, respectively. These enzymes have been implicated in aging from studies that have attempted to decipher the changes observed in histone modifications during aging and/or map the genetic factors involved in aging. Although, very little is known about the role of HATs, HDACs have been well addressed with regards to their involvement in cellular or organismal aging.

HATs: An important HAT that has been linked to aging is Mof, which mediates acetylation at the H4 lysine 16 (H4K16) residue. Mof has been shown to be important for the maintenance of genome stability and its depletion leads to delayed γ-H2AX foci formation in response to DNA damage and abrogated DNA damage repair. Mof has also been shown to be an important regulator of DNA damage because of its ability to bind to 53BP1 (Krishnan et al. 2011). Mof associates with the nuclear matrix and is a key component of the pre-lamin A complex. In a very recent study, Vaidehi Krishnan et al. have shown that in mice that lack the zinc metalloproteinase (Zmpste 24), Mof localization at the nuclear matrix decreases (Krishnan et al. 2011). This effect has been linked to the accumulation of unprocessed pre-lamin A, which is associated with progeroid symptoms. Incidentally, depletion of Mof has also been shown to exacerbate the senescent phenotype of cell lines that lack Zmpste 24. In support of this, overexpression of Mof has been associated with hyperacetylation of H4K16 and a delay in cellular senescence (Hajji et al. 2010). Another HAT that seems to be an important mediator of aging is Sas2. Studies in *S. cerevisiae* have shown that Sas2 inactivation leads to delayed senescence due to activation of homologous recombination (HR) machinery at telomeric regions, thus delaying senescence by preventing telomere loss (Kozak et al. 2010).

Several studies show that the HATs p300 and CBP are important regulators of senescence (Bandyopadhyay et al. 2002; He et al. 2011; Pedeux et al. 2005; Prieur et al. 2011). The study by Prieur et al. showed that p300 is an important regulator of chromatin dependent mediator of senescence and that this mechanism is independent of p53, p21 and p16 (Prieur et al. 2011).

HDACs: There are four classes of HDACs and specifically, Sirtuins that belong to Class-III HDACs are distinct in their activity because of their dependence on NAD^+. We have described the role of sirtuins in aging in a separate section below. Among the other HDACs, members that belong to Class I have been so far implicated for their potential roles in aging. Several studies have indicated that inhibition of HDAC activity leads to induction of a senescent phenotype. June Munro et al. in their study show that administration of HDAC inhibitors sodium butyrate and trichostatin A (TSA) induces senescence in human fibroblasts. These cells exhibit typical senescent phenotype such as β-galactosidase staining, in addition to an elevation in cyclin-Cdk inhibitors, p21and p16 (Munro et al. 2004). It is interesting to note that HDAC antagonists are potent inhibitors of cancer cell proliferation or tumorigenesis. Suberoylanilide hydroxamic acid (SAHA), is an HDAC inhibitor which is used as an anti-tumor drug. SAHA has been shown to induce polyploidy in human colon cancer cell line HCT116 and human breast cancer cell lines, MCF-7, MDA-MB- 231, and MBA-MD-468, which activates senescence in these cells (Xu et al. 2005). Results that corroborate these findings also show that senescence is accompanied by a decrease in HDAC expression. Contradictory observations regarding the role of HDAC activity and aging have also been made. A recent study showed that HDAC1 overexpression inhibited cell proliferation and induced premature senescence in cervical cancer cells through a pathway that involved the deacetylase Sp1, protein phosphatase A PP2A and retinoblastoma protein Rb (Chuang and Hung 2011).

Contrary to what has been observed in cells in culture, organismal studies have indicated that a reduced/absence of HDAC expression/activity leads to lifespan extension. Rpd3 is an HDAC found in all organisms. A study by Rogina et al. showed that reduction in Rpd3 levels in *Drosophila* renders them a longer lifespan. Moreover, these mutants fail to increase their lifespan any further in response to CR/DR (Rogina and Helfand 2004). It has to be noted that the molecular mechanisms of HDAC-dependent changes in aging and lifespan are not well understood. Interestingly, administering TSA to flies also leads to an extension in lifespan, which is accompanied by an increase in Hsp22 protein levels (Tao et al. 2004; Zhao et al. 2005). However, the global changes in histone acetylation and chromatin architecture that would be associated with an absence or inhibition of HDACs have not been addressed, yet. It is unclear if these results depict physiological differences which are elicited by HDAC inhibition at the cellular and organismal levels. It is also likely that different family members which belong to HDACs have varied roles and involve altered substrate specificities. It has to be noted that HDACs are known to deacetylate and affect non-histone proteins as well (Thevenet et al. 2004; Gregoire et al. 2007). Further analysis of individual HDAC proteins may identify their individual functions in mechanisms that induce senescence.

8.2.3 Sirtuins

Sir2 is the founding member of an evolutionarily conserved family of proteins that was first identified in *S. cerevisiae*. Sir2 has been classified as a Class III HDAC and it has been shown to depend on NAD^+ as a co-substrate for its deacetylase activity (Ghosh et al. 2010; Imai and Guarente 2010; Zhang and Kraus 2010). The role of Sir2 in regulation of chromatin has been attributed to its ability to deacetylate specific residues in histones H3 (lysine-9) and H4 (lysine-16). Sir2 activity is required to silence chromatin at sub-telomeric DNA, mating-type and ribosomal DNA (rDNA) loci (Fig. 8.2) (Ha and Huh 2011; Kaeberlein et al. 1999; Rusche et al. 2003). Further studies in yeast established Sir2 as a key link between chromatin regulation and aging. In *S. cerevisiae*, it is known that recombination at rDNA regions leads to the formation of extra chromosomal rDNA circles, which reduce replicative lifespan (Sinclair and Guarente 1997). The ability of Sir2 to mediate silencing at rDNA is considered paramount for its role as a negative regulator of aging in yeast. Additionally, a study by Dang et al. shows that in replicatively aged yeast cells, a decline in Sir2 levels correlates with an increase in acetylation of H4K16 (Dang et al. 2009).

Studies aimed at deciphering an "anti-aging" function of Sir2 in worms and flies corroborated the findings in yeast. In *C. elegans*, Sir2 is known to extend lifespan and interact with insulin IGF signaling. Similar reports in *D. melanogaster* have shown that the Sir2 ortholog mediates CR/DR dependent lifespan extension (Rogina and Helfand 2004). However, the molecular mechanisms, which are affected by Sir2 in these organisms that regulate lifespan extensions, are still unclear. The chromatin regulatory function of Sir2 orthologs in other organisms (flies and mammals) has also been addressed. From these studies it becomes evident that the role of Sir2 and its orthologs in regulating chromatin-mediated changes is evolutionarily conserved. Sir2 has been identified as a regulator of heterochromatin formation in flies and affects position effect variegation (PEV). Results indicate that the role of Sir2 in PEV is independent of its ability to extend lifespan (Frankel and Rogina 2005; Newman et al. 2002). However, whether its role in mediating locus specific chromatin changes is linked to its role in lifespan extension is still unclear.

In mammals, SIRT1, the homolog of ySir2 has been shown to be a major regulator of chromatin structure and gene expression. SIRT1 is a well-established regulator of transcription by deacetylating a host of transcription factors and co-regulators (Table 8.2) (Brooks and Gu 2008; Deng 2009). It should be noted that some of these transcription factors, such as NF-kB, FOXO and p53, have been implicated in organismal aging. However, the role of SIRT1 in mammalian aging has been difficult to address as most SIRT1 null mice die due to developmental defects. Mammalian SIRT1 regulates chromatin dynamics by mediating deacetylation of H3 lysine 9 (H3K9) and H4 lysine 16 (H4K16). In addition to contributing to histone acetylation changes, SIRT1 has also been shown to impinge on other mediators of chromatin. Importantly, SIRT1 is known to cross-talk with DNMTs (O'Hagan et al. 2008) and histone methyltransferases (Vaquero et al. 2004). SIRT1 has been shown to bind and

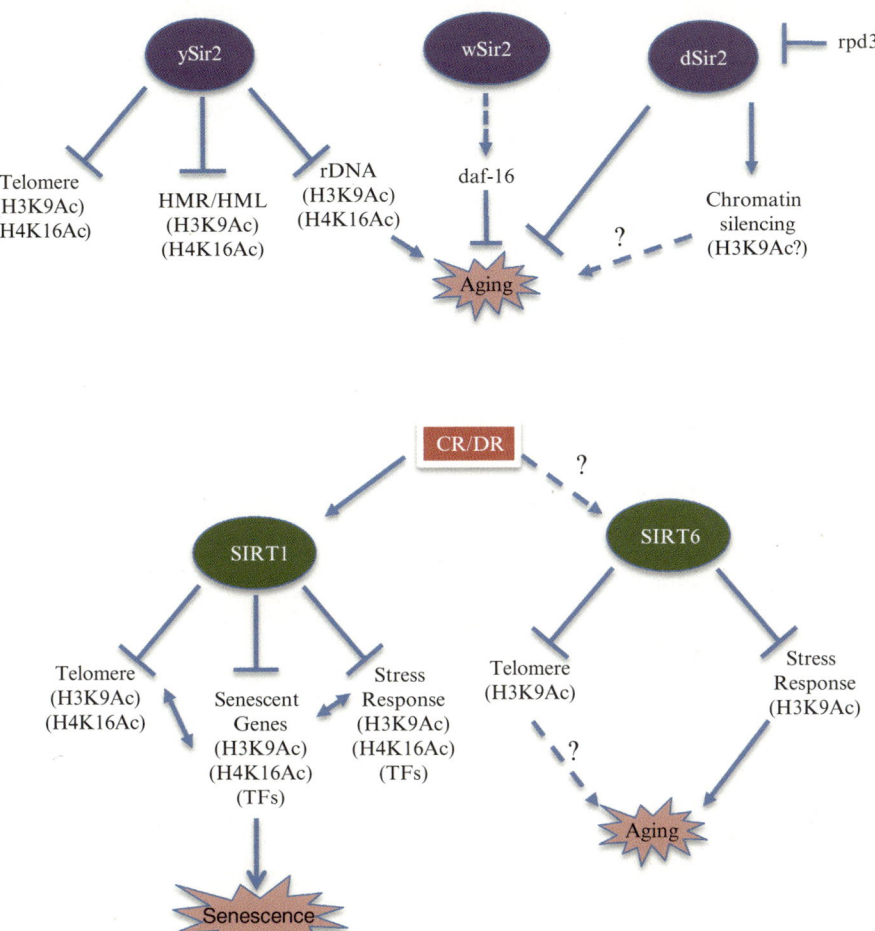

Fig. 8.2 The role of Sirtuins in aging across species. Sir2 and its homologues (including Sirt1 and Sirt6 in mammals) are key players in cellular/organismal aging. Studies in yeast, flies and mammals show that Sir2, Sirt1 and Sirt6 are NAD+-dependent deacetylases and affect chromatin by deacetylating histones (H3K9 or H4K16) as illustrated. Except in yeast, the link between sirtuins and aging is not limited to its role in affecting chromatin since they are known to regulate other pathways/factors. In worms and flies where Sir2 is now known to extend lifespan (and in response to calorie/dietary restriction) the chromatin angle in mediating this effect is still unclear. SIRT1 and SIRT6 are important in regulating the expression of a host of genes that mediate senescence, in addition to their roles at the telomere

deacetylate SUV39H1 which brings about H3K9 trimethylation (Vaquero et al. 2004). It is speculated that SIRT1 mediated histone deacetylation renders the site open for methylation. Additionally, a loss of SIRT1 is associated with a reduction in H3K9me3 levels and a concomitant impairment of heterochromatin protein-1 (HP1)

Table 8.2 Sirt1 deacetylation targets which have been implicated in aging/senescence

Interactor	Biological function
p53	Tumor suppressor and cell cycle regulator
p73	Tumor suppressor and cell cycle regulator
Ezh2	Histone methyl transferase that maintains transcriptionally repressed state
E2F1	A transcription factor important for G1/S transition
PCAF	An acetyltransferase that inhibits apoptosis
RelA/p65	A transcription factor that regulates transcription of NFkB target genes
FOXOs	A family of transcription factors that regulates cell cycle and stress response
SUV39H1	A histone methyl transferase that is important for the maintenance of heterochromatin state

recruitment. Together, these have been proposed to affect heterochromatin formation (Vaquero et al. 2007).

Independent studies have shown that SIRT1 plays a crucial role in cellular senescence. The study by Langley et al. was the first study which showed that SIRT1 negatively regulates cellular aging in mammalian cells (Langley et al. 2002). The authors showed that SIRT1 binds, deacetylates and inhibits p53 transactivation activity leading to its anti-senescent property. Subsequent studies identified that the anti-senescence effects of SIRT1 was a common feature of multiple cell types including human diploid fibroblasts (Huang et al. 2008), human umbilical vein endothelial cell line (Ota et al. 2007) and several cancer cell lines like breast cancer MCF-7, lung cancer H1299 and prostate cancer cells (Jung-Hynes et al. 2009; Ota et al. 2006). Importantly, SIRT1 is known to specifically repress genes involved in cell cycle arrest such as p16 (Huang et al. 2008, p. 21; Rathbone et al. 2008; Yuan et al. 2011, p. 27; Ota et al. 2006).

It is interesting to note that SIRT1 activity and/or levels have been proposed to decrease during aging in cells and mice (Yamakuchi et al. 2008). However, a clear picture that links the chromatin functions of SIRT1 and its role in cellular senescence is still not available. In support of such a role, reports that indicate chromatin relocalization of SIRT1 during aging imply a possible chromatin dependent effect of SIRT1 in aging/senescence (Oberdoerffer et al. 2008). This finding is reminiscent of a similar phenomenon in yeast where the Sir2 redistribution on the genome has been observed in aging yeast cells (Gotta et al. 1997). It is clear that SIRT1 is important for the maintenance of telomeric chromatin in mammalian cell lines (Palacios et al. 2010). However, it is still not known if the functions of SIRT1 at the telomere are important for its role in cellular senescence. Further investigations are required to appreciate the link between Sir2/SIRT1 dependent global and/or locus specific chromatin changes and aging.

SIRT6 another important mammalian sirtuin has been clearly shown to play a major role in aging (Fig. 8.2). Mice deficient for SIRT6 exhibited progeroid symptoms (Kawahara et al. 2011) and results suggest that its ability to regulate DNA damage repair pathways were key to its role in aging (Mostoslavsky et al. 2006). Subsequently, SIRT6 was shown to deacetylate histone H3 at lysine 9 residue (Michishita et al. 2008), which incidentally is also targeted by SIRT1 (Vaquero et al. 2004). Reports that elucidated the ability of SIRT6 to regulate NF-kB dependent transcription showed that its role in aging is mostly determined by its ability to regulate inflammatory responses (Kawahara et al. 2011). It has been suggested that a dynamic relocalization of Sirt6 on chromatin is important for its ability to regulate organismal aging by controlling the expression of essential aging related genes, many of which are NF-kB targets (Kawahara et al. 2011). Further, SIRT6 has been shown to prevent telomere dysfunction in human cells by deacetylating H3K9 at telomeric loci, although, such an effect has not been observed in SIRT6 null mice (Michishita et al. 2008). Put together, it is evident that the functions of SIRT6 in mediating stress responses and at the telomere might have a bearing on aging and is probably dependent on its ability to deacetylate H3K9 residue. It is interesting to note that although both SIRT1 and SIRT6 have been implicated in similar pathways and at telomere functions, it is still unclear if they bring about a coordinated response to regulate aging.

8.3 Progeroid Syndromes and Chromatin

Progeroid syndromes are characterized by symptoms that mimic aging. Two of the most well studied clinical progeroid conditions are Werner syndrome and Hutchinson-Gilford progeria syndrome (HGPS). Werner's is a progeroid syndrome caused by mutations in the *WRN* gene, which encodes a member of the RecQ family of helicases. Intriguingly, some features of this disorder are also present in laminopathies caused by mutant *LMNA* encoding nuclear lamins A/C that causes HGPS. Recent studies suggest that epigenetic modifications in these progeroid genes lead to malignant transformation (Shumaker et al. 2006).

In HGPS, Lamin A gene is mutated resulting in a cryptic splice site in exon-11 causing 150 nucleotide deletion (LAΔ50). It was interesting to find that HGPS was associated with global changes in nuclear and chromatin architecture. Specifically, HGPS fibroblasts exhibit a loss of nuclear peripheral heterochromatin (Dechat et al. 2008), the severity of which depends on the accumulation of the abnormal LAΔ50 protein (Goldman et al. 2004). In addition, in cells derived from older HPGS patients, several heterochromatin marks, such as mono- and tri-methylated H3K9, show a dramatic decrease. A loss of the H3K27 tri-methyl mark was also observed in these cells and was correlated with a nine- to ten-fold decrease in the histone methyltransferase EZH2 expression. Another study, which looked at late-passage HGPS cells, observed an up-regulation of H4K20 tri-methylation (Shumaker et al. 2006). It is important to note that H4K20 tri-methylation has been shown to be elevated in livers of older rats (Sarg et al. 2002) and in SAHFs in cultured cells (described above). The molecular mechanisms that link lamin A to heterochromatin formation are still not very clear. However, studies suggest that retinoblastoma protein (Rb) binds directly to type-A lamins (Ozaki et al. 1994; Johnson et al. 2004). Based on independent observations that Rb regulates histone methylation at H3K27, H3K9 and H4K20 residues, it has been speculated that Rb could be one of the factors that links aberrant histone methylation in HGPS (Blais et al. 2007).

Werner syndrome (WS) provides another example of a gene involved in aging and with tumor suppressor properties. As a result of the mutation in the *WRN* gene, cells from patients with WS show high genomic instability, especially at repetitive loci. But it is still not clear if WRN affects global chromatin that would eventually lead to aging. However, the role of chromatin in tipping the balance between senescence and cell proliferation becomes apparent from studies, which show that *WRN* is frequently repressed by CpG island hypermethylation in many human cancers (Agrelo et al. 2005).

8.4 Conclusion

Epigenetic marks, which are long lasting and inheritable, play a central role in mediating the outputs from the genome, in response to both extrinsic and intrinsic cues, and therefore, known to affect various biological processes. Hence, it is not surprising to find that chromatin is a major player in mediating cellular responses to aging. Although, all the critical chromatin components like DNA methylation, histone modifications and histone variants have been shown to be involved in this process, the mechanistic details that elicit these changes are less understood. Specifically, it will be interesting to address the cross-talk between classical aging pathways/ mechanisms and chromatin signaling. It will be important to address if reversal of any of these chromatin changes would affect the aging process. In this regard, more work needs to be done on the role of chromatin modifiers, which mediate both global and locus specific effects on chromatin structure/function. Studies on proteins like HDACs and Sirtuins have indicated their involvement in the aging process. However, more insights into mechanistic details describing the chromatin effects are needed. Another important aspect that needs to be addressed is the apparent gaps in appreciating the roles of chromatin and chromatin modifiers in cellular and organismal aging.

Since aging is a complex biological process involving multiple factors, interventions aimed at one or more specific pathways/factors (to delay aging) are likely to give limited benefits. Targeting cellular components that would integrate the cues and the responses might turn out to be more beneficial. In this context, interfering with chromatin changes and/or chromatin modifiers that affect aging might become therapeutically relevant. This is crucial since small chemical modulators and/or dietary manipulations have shown promising results with regards to their ability to "delay the aging process" and are often mediated through some of these factors.

References

Adams PD (2009) Healing and hurting: molecular mechanisms, functions, and pathologies of cellular senescence. Mol Cell 36:2–14

Agrelo R, Setien F, Espada J, Artiga MJ, Rodriguez M, Perez-Rosado A, Sanchez-Aguilera A, Fraga MF, Piris MA, Esteller M (2005) Inactivation of the lamin A/C gene by CpG island promoter hypermethylation in hematologic malignancies, and its association with poor survival in nodal diffuse large B-cell lymphoma. J Clin Oncol 23:3940–3947

Arvanitakis Z, Wilson RS, Bennett DA (2006) Diabetes mellitus, dementia, and cognitive function in older persons. J Nutr Health Aging 10:287–291

Bandyopadhyay D, Okan NA, Bales E, Nascimento L, Cole PA, Medrano EE (2002) Down-regulation of p300/CBP histone acetyltransferase activates a senescence checkpoint in human melanocytes. Cancer Res 62:6231–6239

Bannister AJ, Kouzarides T (2011) Regulation of chromatin by histone modifications. Cell Res 21:381–395

Bartkova J, Rezaei N, Liontos M, Karakaidos P, Kletsas D, Issaeva N, Vassiliou LV, Kolettas E, Niforou K, Zoumpourlis VC, Takaoka M, Nakagawa H, Tort F, Fugger K, Johansson F, Sehested M, Andersen CL, Dyrskjot L, Orntoft T, Lukas J, Kittas C, Helleday T, Halazonetis TD, Bartek J, Gorgoulis VG (2006) Oncogene-induced senescence is part of the tumorigenesis barrier imposed by DNA damage checkpoints. Nature 444:633–637

Bassett A, Cooper S, Wu C, Travers A (2009) The folding and unfolding of eukaryotic chromatin. Curr Opin Genet Dev 19:159–165

Bates DJ, Liang R, Li N, Wang E (2009) The impact of noncoding RNA on the biochemical and molecular mechanisms of aging. Biochim Biophys Acta 1790:970–979

Bates DJ, Li N, Liang R, Sarojini H, An J, Masternak MM, Bartke A, Wang E (2010) MicroRNA regulation in Ames dwarf mouse liver may contribute to delayed aging. Aging Cell 9:1–18

Ben-Porath I, Weinberg RA (2005) The signals and pathways activating cellular senescence. Int J Biochem Cell Biol 37:961–976

Blais A, van Oevelen CJ, Margueron R, Acosta-Alvear D, Dynlacht BD (2007) Retinoblastoma tumor suppressor protein-dependent methylation of histone H3 lysine 27 is associated with irreversible cell cycle exit. J Cell Biol 179:1399–1412

Boehm M, Slack F (2005) A developmental timing microRNA and its target regulate life span in C. elegans. Science 310:1954–1957

Bommi PV, Dimri M, Sahasrabuddhe AA, Khandekar J, Dimri GP (2010) The polycomb group protein BMI1 is a transcriptional target of HDAC inhibitors. Cell Cycle 9:2663–2673

Bonasio R, Tu S, Reinberg D (2010) Molecular signals of epigenetic states. Science 330:612–616

Bork S, Pfister S, Witt H, Horn P, Korn B, Ho AD, Wagner W (2010) DNA methylation pattern changes upon long-term culture and aging of human mesenchymal stromal cells. Aging Cell 9:54–63

Bornman DM, Mathew S, Alsruhe J, Herman JG, Gabrielson E (2001) Methylation of the E-cadherin gene in bladder neoplasia and in normal urothelial epithelium from elderly individuals. Am J Pathol 159:831–835

Bracken AP, Kleine-Kohlbrecher D, Dietrich N, Pasini D, Gargiulo G, Beekman C, Theilgaard-Monch K, Minucci S, Porse BT, Marine JC, Hansen KH, Helin K (2007) The Polycomb group proteins bind throughout the INK4A-ARF locus and are disassociated in senescent cells. Genes Dev 21:525–530

Braig M, Schmitt CA (2006) Oncogene-induced senescence: putting the brakes on tumor development. Cancer Res 66:2881–2884

Braig M, Lee S, Loddenkemper C, Rudolph C, Peters AH, Schlegelberger B, Stein H, Dorken B, Jenuwein T, Schmitt CA (2005) Oncogene-induced senescence as an initial barrier in lymphoma development. Nature 436:660–665

Brooks CL, Gu W (2008) p53 Activation: a case against Sir. Cancer Cell 13:377–378

Cao R, Tsukada Y, Zhang Y (2005) Role of Bmi-1 and Ring1A in H2A ubiquitylation and Hox gene silencing. Mol Cell 20:845–854

Casillas MA Jr, Lopatina N, Andrews LG, Tollefsbol TO (2003) Transcriptional control of the DNA methyltransferases is altered in aging and neoplastically-transformed human fibroblasts. Mol Cell Biochem 252:33–43

Chen LH, Chiou GY, Chen YW, Li HY, Chiou SH (2010) MicroRNA and aging: a novel modulator in regulating the aging network. Ageing Res Rev 9(Suppl 1):S59–S66

Chouliaras L, van den Hove DL, Kenis G, Dela Cruz J, Lemmens MA, van Os J, Steinbusch HW, Schmitz C, Rutten BP (2011a) Caloric restriction attenuates age-related changes of DNA methyltransferase 3a in mouse hippocampus. Brain Behav Immun 25:616–623

Chouliaras L, van den Hove DL, Kenis G, Keitel S, Hof PR, van Os J, Steinbusch HW, Schmitz C, Rutten BP (2011b) Prevention of age-related changes in hippocampal levels of 5-methylcytidine by caloric restriction. Neurobiol Aging 33(8):1672–1681

Christensen BC, Houseman EA, Marsit CJ, Zheng S, Wrensch MR, Wiemels JL, Nelson HH, Karagas MR, Padbury JF, Bueno R, Sugarbaker DJ, Yeh RF, Wiencke JK, Kelsey KT (2009) Aging and environmental exposures alter tissue-specific DNA methylation dependent upon CpG island context. PLoS Genet 5:e1000602

Chuang JY, Hung JJ (2011) Overexpression of HDAC1 induces cellular senescence by Sp1/PP2A/pRb pathway. Biochem Biophys Res Commun 407:587–592

Costanzi C, Pehrson JR (1998) Histone macroH2A1 is concentrated in the inactive X chromosome of female mammals. Nature 393:599–601

Dang W, Steffen KK, Perry R, Dorsey JA, Johnson FB, Shilatifard A, Kaeberlein M, Kennedy BK, Berger SL (2009) Histone H4 lysine 16 acetylation regulates cellular lifespan. Nature 459:802–807

De Carvalho DD, You JS, Jones PA (2010) DNA methylation and cellular reprogramming. Trends Cell Biol 20:609–617

Dechat T, Pfleghaar K, Sengupta K, Shimi T, Shumaker DK, Solimando L, Goldman RD (2008) Nuclear lamins: major factors in the structural organization and function of the nucleus and chromatin. Genes Dev 22:832–853

Deng CX (2009) SIRT1, is it a tumor promoter or tumor suppressor? Int J Biol Sci 5:147–152

DePinho RA (2000) The age of cancer. Nature 408:248–254

Dimauro T, David G (2009) Chromatin modifications: the driving force of senescence and aging? Aging (Albany NY) 1:182–190

Fairweather DS, Fox M, Margison GP (1987) The in vitro lifespan of MRC-5 cells is shortened by 5-azacytidine-induced demethylation. Exp Cell Res 168:153–159

Feser J, Tyler J (2011) Chromatin structure as a mediator of aging. FEBS Lett 585:2041–2048

Fraga MF, Esteller M (2007) Epigenetics and aging: the targets and the marks. Trends Genet 23:413–418

Frankel S, Rogina B (2005) Drosophila longevity is not affected by heterochromatin-mediated gene silencing. Aging Cell 4:53–56

Frenster JH, Allfrey VG, Mirsky AE (1963) Repressed and active chromatin isolated from interphase lymphocytes. Proc Natl Acad Sci U S A 50:1026–1032

Fujita N, Watanabe S, Ichimura T, Ohkuma Y, Chiba T, Saya H, Nakao M (2003) MCAF mediates MBD1-dependent transcriptional repression. Mol Cell Biol 23:2834–2843

Fuke C, Shimabukuro M, Petronis A, Sugimoto J, Oda T, Miura K, Miyazaki T, Ogura C, Okazaki Y, Jinno Y (2004) Age related changes in 5-methylcytosine content in human peripheral leukocytes and placentas: an HPLC-based study. Ann Hum Genet 68:196–204

Funayama R, Saito M, Tanobe H, Ishikawa F (2006) Loss of linker histone H1 in cellular senescence. J Cell Biol 175:869–880

Gao Z, Xu MS, Barnett TL, Xu CW (2011) Resveratrol induces cellular senescence with attenuated mono-ubiquitination of histone H2B in glioma cells. Biochem Biophys Res Commun 407:271–276

Geiman TM, Muegge K (2010) DNA methylation in early development. Mol Reprod Dev 77:105–113

Ghosh S, George S, Roy U, Ramachandran D, Kolthur-Seetharam U (2010) NAD: a master regulator of transcription. Biochim Biophys Acta 1799:681–693

Gibney ER, Nolan CM (2010) Epigenetics and gene expression. Heredity 105:4–13

Goldman RD, Shumaker DK, Erdos MR, Eriksson M, Goldman AE, Gordon LB, Gruenbaum Y, Khuon S, Mendez M, Varga R, Collins FS (2004) Accumulation of mutant lamin A causes progressive changes in nuclear architecture in Hutchinson-Gilford progeria syndrome. Proc Natl Acad Sci U S A 101:8963–8968

Gorospe M, Abdelmohsen K (2011) MicroRegulators come of age in senescence. Trends Genet 27:233–241

Gotta M, Strahl-Bolsinger S, Renauld H, Laroche T, Kennedy BK, Grunstein M, Gasser SM (1997) Localization of Sir2p: the nucleolus as a compartment for silent information regulators. EMBO J 16:3243–3255

Grandinetti KB, Jelinic P, DiMauro T, Pellegrino J, Fernandez Rodriguez R, Finnerty PM, Ruoff R, Bardeesy N, Logan SK, David G (2009) Sin3B expression is required for cellular senescence and is up-regulated upon oncogenic stress. Cancer Res 69:6430–6437

Gregoire S, Xiao L, Nie J, Zhang X, Xu M, Li J, Wong J, Seto E, Yang XJ (2007) Histone deacetylase 3 interacts with and deacetylates myocyte enhancer factor 2. Mol Cell Biol 27:1280–1295

Grillari J, Grillari-Voglauer R (2010) Novel modulators of senescence, aging, and longevity: small non-coding RNAs enter the stage. Exp Gerontol 45:302–311

Ha CW, Huh WK (2011) The implication of Sir2 in replicative aging and senescence in *Saccharomyces cerevisiae*. Aging (Albany NY) 3:319–324

Hajji N, Wallenborg K, Vlachos P, Fullgrabe J, Hermanson O, Joseph B (2010) Opposing effects of hMOF and SIRT1 on H4K16 acetylation and the sensitivity to the topoisomerase II inhibitor etoposide. Oncogene 29:2192–2204

Hans F, Dimitrov S (2001) Histone H3 phosphorylation and cell division. Oncogene 20:3021–3027

Hayflick L (1965) The limited in vitro lifetime of human diploid cell strains. Exp Cell Res 37:614–636

He L, Hannon GJ (2004) MicroRNAs: small RNAs with a big role in gene regulation. Nat Rev Genet 5:522–531

He H, Yu FX, Sun C, Luo Y (2011) CBP/p300 and SIRT1 are involved in transcriptional regulation of S-phase specific histone genes. PLoS One 6:e22088

Herbig U, Ferreira M, Condel L, Carey D, Sedivy JM (2006) Cellular senescence in aging primates. Science 311:1257

Hoal-van Helden EG, van Helden PD (1989) Age-related methylation changes in DNA may reflect the proliferative potential of organs. Mutat Res 219:263–266

Hock R, Furusawa T, Ueda T, Bustin M (2007) HMG chromosomal proteins in development and disease. Trends Cell Biol 17:72–79

Hornsby PJ, Yang L, Gunter LE (1992) Demethylation of satellite I DNA during senescence of bovine adrenocortical cells in culture. Mutat Res 275:13–19

Huang J, Gan Q, Han L, Li J, Zhang H, Sun Y, Zhang Z, Tong T (2008) SIRT1 overexpression antagonizes cellular senescence with activated ERK/S6k1 signaling in human diploid fibroblasts. PLoS One 3:e1710

Imai S, Guarente L (2010) Ten years of NAD-dependent SIR2 family deacetylases: implications for metabolic diseases. Trends Pharmacol Sci 31:212–220

Issa JP, Ottaviano YL, Celano P, Hamilton SR, Davidson NE, Baylin SB (1994) Methylation of the oestrogen receptor CpG island links ageing and neoplasia in human colon. Nat Genet 7:536–540

Jenuwein T, Allis CD (2001) Translating the histone code. Science 293:1074–1080

Johnson BR, Nitta RT, Frock RL, Mounkes L, Barbie DA, Stewart CL, Harlow E, Kennedy BK (2004) A-type lamins regulate retinoblastoma protein function by promoting subnuclear localization and preventing proteasomal degradation. Proc Natl Acad Sci U S A 101:9677–9682

Jung-Hynes B, Nihal M, Zhong W, Ahmad N (2009) Role of sirtuin histone deacetylase SIRT1 in prostate cancer. A target for prostate cancer management via its inhibition? J Biol Chem 284:3823–3832

Jurkowska RZ, Jurkowski TP, Jeltsch A (2011) Structure and function of mammalian DNA methyltransferases. Chembiochem 12:206–222

Kaeberlein M, McVey M, Guarente L (1999) The SIR2/3/4 complex and SIR2 alone promote longevity in *Saccharomyces cerevisiae* by two different mechanisms. Genes Dev 13: 2570–2580

Kawahara TL, Rapicavoli NA, Wu AR, Qu K, Quake SR, Chang HY (2011) Dynamic chromatin localization of Sirt6 shapes stress- and aging-related transcriptional networks. PLoS Genet 7:e1002153

Kawakami K, Nakamura A, Ishigami A, Goto S, Takahashi R (2009) Age-related difference of site-specific histone modifications in rat liver. Biogerontology 10:415–421

Kenyon J, Gerson SL (2007) The role of DNA damage repair in aging of adult stem cells. Nucleic Acids Res 35:7557–7565

Kimura H, Shiota K (2003) Methyl-CpG-binding protein, MeCP2, is a target molecule for maintenance DNA methyltransferase, Dnmt1. J Biol Chem 278:4806–4812

Kirkwood TL, Kapahi P, Shanley DP (2000) Evolution, stress, and longevity. J Anat 197(Pt 4): 587–590

Knight JA (2000) The biochemistry of aging. Adv Clin Chem 35:1–62

Koch CM, Suschek CV, Lin Q, Bork S, Goergens M, Joussen S, Pallua N, Ho AD, Zenke M, Wagner W (2011) Specific age-associated DNA methylation changes in human dermal fibroblasts. PLoS One 6:e16679

Kosar M, Bartkova J, Hubackova S, Hodny Z, Lukas J, Bartek J (2011) Senescence-associated heterochromatin foci are dispensable for cellular senescence, occur in a cell type- and insult-dependent manner and follow expression of p16(ink4a). Cell Cycle 10:457–468

Kouzarides T (2007) Chromatin modifications and their function. Cell 128:693–705

Kozak ML, Chavez A, Dang W, Berger SL, Ashok A, Guo X, Johnson FB (2010) Inactivation of the Sas2 histone acetyltransferase delays senescence driven by telomere dysfunction. EMBO J 29:158–170

Krishnan V, Chow MZ, Wang Z, Zhang L, Liu B, Liu X, Zhou Z (2011) Histone H4 lysine 16 hypoacetylation is associated with defective DNA repair and premature senescence in Zmpste24-deficient mice. Proc Natl Acad Sci U S A 108:12325–12330

Langley E, Pearson M, Faretta M, Bauer UM, Frye RA, Minucci S, Pelicci PG, Kouzarides T (2002) Human SIR2 deacetylates p53 and antagonizes PML/p53-induced cellular senescence. EMBO J 21:2383–2396

Lee ST, Kim M (2006) Aging and neurodegeneration. Molecular mechanisms of neuronal loss in Huntington's disease. Mech Ageing Dev 127:432–435

Lee S, Jung JW, Park SB, Roh K, Lee SY, Kim JH, Kang SK, Kang KS (2011) Histone deacetylase regulates high mobility group A2-targeting microRNAs in human cord blood-derived multipotent stem cell aging. Cell Mol Life Sci 68:325–336

Legube G, Trouche D (2003) Regulating histone acetyltransferases and deacetylases. EMBO Rep 4:944–947

Li G, Reinberg D (2011) Chromatin higher-order structures and gene regulation. Curr Opin Genet Dev 21:175–186

Li N, Bates DJ, An J, Terry DA, Wang E (2011) Up-regulation of key microRNAs, and inverse down-regulation of their predicted oxidative phosphorylation target genes, during aging in mouse brain. Neurobiol Aging 32:944–955

Liang R, Bates DJ, Wang E (2009) Epigenetic control of microRNA expression and aging. Curr Genomics 10:184–193

Lopatina N, Haskell JF, Andrews LG, Poole JC, Saldanha S, Tollefsbol T (2002) Differential maintenance and de novo methylating activity by three DNA methyltransferases in aging and immortalized fibroblasts. J Cell Biochem 84:324–334

Lopatina NG, Poole JC, Saldanha SN, Hansen NJ, Key JS, Pita MA, Andrews LG, Tollefsbol TO (2003) Control mechanisms in the regulation of telomerase reverse transcriptase expression in differentiating human teratocarcinoma cells. Biochem Biophys Res Commun 306:650–659

Luger K, Mader AW, Richmond RK, Sargent DF, Richmond TJ (1997) Crystal structure of the nucleosome core particle at 2.8 A resolution. Nature 389:251–260

Maes OC, An J, Sarojini H, Wang E (2008) Murine microRNAs implicated in liver functions and aging process. Mech Ageing Dev 129:534–541

Margueron R, Trojer P, Reinberg D (2005) The key to development: interpreting the histone code? Curr Opin Genet Dev 15:163–176

Michishita E, McCord RA, Berber E, Kioi M, Padilla-Nash H, Damian M, Cheung P, Kusumoto R, Kawahara TL, Barrett JC, Chang HY, Bohr VA, Ried T, Gozani O, Chua KF (2008) SIRT6 is a histone H3 lysine 9 deacetylase that modulates telomeric chromatin. Nature 452:492–496

Misteli T (2010) Higher-order genome organization in human disease. Cold Spring Harb Perspect Biol 2:a000794

Morimoto S, Komatsu S, Takahashi R, Matsuo M, Goto S (1993) Age-related change in the amount of ubiquitinated histones in the mouse brain. Arch Gerontol Geriatr 16:217–224

Mostoslavsky R, Chua KF, Lombard DB, Pang WW, Fischer MR, Gellon L, Liu P, Mostoslavsky G, Franco S, Murphy MM, Mills KD, Patel P, Hsu JT, Hong AL, Ford E, Cheng HL, Kennedy C, Nunez N, Bronson R, Frendewey D, Auerbach W, Valenzuela D, Karow M, Hottiger MO, Hursting S, Barrett JC, Guarente L, Mulligan R, Demple B, Yancopoulos GD, Alt FW (2006) Genomic instability and aging-like phenotype in the absence of mammalian SIRT6. Cell 124:315–329

Muller M (2009) Cellular senescence: molecular mechanisms, in vivo significance, and redox considerations. Antioxid Redox Signal 11:59–98

Munro J, Barr NI, Ireland H, Morrison V, Parkinson EK (2004) Histone deacetylase inhibitors induce a senescence-like state in human cells by a p16-dependent mechanism that is independent of a mitotic clock. Exp Cell Res 295:525–538

Murr R (2010) Interplay between different epigenetic modifications and mechanisms. Adv Genet 70:101–141

Narita M (2007) Cellular senescence and chromatin organisation. Br J Cancer 96:686–691

Narita M, Nunez S, Heard E, Lin AW, Hearn SA, Spector DL, Hannon GJ, Lowe SW (2003) Rb-mediated heterochromatin formation and silencing of E2F target genes during cellular senescence. Cell 113:703–716

Narita M, Krizhanovsky V, Nunez S, Chicas A, Hearn SA, Myers MP, Lowe SW (2006) A novel role for high-mobility group a proteins in cellular senescence and heterochromatin formation. Cell 126:503–514

Newman BL, Lundblad JR, Chen Y, Smolik SM (2002) A Drosophila homologue of Sir2 modifies position-effect variegation but does not affect life span. Genetics 162:1675–1685

O'Hagan HM, Mohammad HP, Baylin SB (2008) Double strand breaks can initiate gene silencing and SIRT1-dependent onset of DNA methylation in an exogenous promoter CpG island. PLoS Genet 4:e1000155

O'Sullivan RJ, Kubicek S, Schreiber SL, Karlseder J (2010) Reduced histone biosynthesis and chromatin changes arising from a damage signal at telomeres. Nat Struct Mol Biol 17:1218–1225

Oakes CC, Smiraglia DJ, Plass C, Trasler JM, Robaire B (2003) Aging results in hypermethylation of ribosomal DNA in sperm and liver of male rats. Proc Natl Acad Sci U S A 100:1775–1780

Oberdoerffer P, Michan S, McVay M, Mostoslavsky R, Vann J, Park SK, Hartlerode A, Stegmuller J, Hafner A, Loerch P, Wright SM, Mills KD, Bonni A, Yankner BA, Scully R, Prolla TA, Alt FW, Sinclair DA (2008) SIRT1 redistribution on chromatin promotes genomic stability but alters gene expression during aging. Cell 135:907–918

Ota H, Tokunaga E, Chang K, Hikasa M, Iijima K, Eto M, Kozaki K, Akishita M, Ouchi Y, Kaneki M (2006) Sirt1 inhibitor, Sirtinol, induces senescence-like growth arrest with attenuated Ras-MAPK signaling in human cancer cells. Oncogene 25:176–185

Ota H, Akishita M, Eto M, Iijima K, Kaneki M, Ouchi Y (2007) Sirt1 modulates premature senescence-like phenotype in human endothelial cells. J Mol Cell Cardiol 43:571–579

Ozaki T, Saijo M, Murakami K, Enomoto H, Taya Y, Sakiyama S (1994) Complex formation between lamin A and the retinoblastoma gene product: identification of the domain on lamin A required for its interaction. Oncogene 9:2649–2653

Palacios JA, Herranz D, De Bonis ML, Velasco S, Serrano M, Blasco MA (2010) SIRT1 contributes to telomere maintenance and augments global homologous recombination. J Cell Biol 191: 1299–1313

Paull TT, Haykinson MJ, Johnson RC (1993) The nonspecific DNA-binding and -bending proteins HMG1 and HMG2 promote the assembly of complex nucleoprotein structures. Genes Dev 7:1521–1534

Pedeux R, Sengupta S, Shen JC, Demidov ON, Saito S, Onogi H, Kumamoto K, Wincovitch S, Garfield SH, McMenamin M, Nagashima M, Grossman SR, Appella E, Harris CC (2005) ING2 regulates the onset of replicative senescence by induction of p300-dependent p53 acetylation. Mol Cell Biol 25:6639–6648

Peterson CL, Laniel MA (2004) Histones and histone modifications. Curr Biol 14:R546–R551

Prieur A, Besnard E, Babled A, Lemaitre JM (2011) p53 and p16(INK4A) independent induction of senescence by chromatin-dependent alteration of S-phase progression. Nat Commun 2:473

Rakyan VK, Down TA, Maslau S, Andrew T, Yang TP, Beyan H, Whittaker P, McCann OT, Finer S, Valdes AM, Leslie RD, Deloukas P, Spector TD (2010) Human aging-associated DNA hypermethylation occurs preferentially at bivalent chromatin domains. Genome Res 20:434–439

Rastogi S, Joshi B, Dasgupta P, Morris M, Wright K, Chellappan S (2006) Prohibitin facilitates cellular senescence by recruiting specific corepressors to inhibit E2F target genes. Mol Cell Biol 26:4161–4171

Rathbone CR, Booth FW, Lees SJ (2008) FoxO3a preferentially induces p27Kip1 expression while impairing muscle precursor cell-cycle progression. Muscle Nerve 37:84–89

Rodier F, Campisi J (2011) Four faces of cellular senescence. J Cell Biol 192:547–556

Rodriguez RM, Fraga MF (2010) Aging and cancer: are sirtuins the link? Future Oncol 6:905–915

Rogina B, Helfand SL (2004) Sir2 mediates longevity in the fly through a pathway related to calorie restriction. Proc Natl Acad Sci U S A 101:15998–16003

Romanov GA, Vanyushin BF (1981) Methylation of reiterated sequences in mammalian DNAs. Effects of the tissue type, age, malignancy and hormonal induction. Biochim Biophys Acta 653:204–218

Rusche LN, Kirchmaier AL, Rine J (2003) The establishment, inheritance, and function of silenced chromatin in *Saccharomyces cerevisiae*. Annu Rev Biochem 72:481–516

Saito Y, Jones PA (2006) Epigenetic activation of tumor suppressor microRNAs in human cancer cells. Cell Cycle 5:2220–2222

Sarg B, Koutzamani E, Helliger W, Rundquist I, Lindner HH (2002) Postsynthetic trimethylation of histone H4 at lysine 20 in mammalian tissues is associated with aging. J Biol Chem 277:39195–39201

Sayed D, Abdellatif M (2011) MicroRNAs in development and disease. Physiol Rev 91:827–887

Sedivy JM, Banumathy G, Adams PD (2008) Aging by epigenetics–a consequence of chromatin damage? Exp Cell Res 314:1909–1917

Seviour EG, Lin SY (2010) The DNA damage response: balancing the scale between cancer and ageing. Aging (Albany NY) 2:900–907

Shin DM, Kucia M, Ratajczak MZ (2011a) Nuclear and chromatin reorganization during cell senescence and aging – a mini-review. Gerontology 57:76–84

Shin KH, Pucar A, Kim RH, Bae SD, Chen W, Kang MK, Park NH (2011b) Identification of senescence-inducing microRNAs in normal human keratinocytes. Int J Oncol 39:1205–1211

Shumaker DK, Dechat T, Kohlmaier A, Adam SA, Bozovsky MR, Erdos MR, Eriksson M, Goldman AE, Khuon S, Collins FS, Jenuwein T, Goldman RD (2006) Mutant nuclear lamin A leads to progressive alterations of epigenetic control in premature aging. Proc Natl Acad Sci U S A 103:8703–8708

Sinclair DA, Guarente L (1997) Extrachromosomal rDNA circles–a cause of aging in yeast. Cell 91:1033–1042

Singhal RP, Mays-Hoopes LL, Eichhorn GL (1987) DNA methylation in aging of mice. Mech Ageing Dev 41:199–210

So K, Tamura G, Honda T, Homma N, Waki T, Togawa N, Nishizuka S, Motoyama T (2006) Multiple tumor suppressor genes are increasingly methylated with age in non-neoplastic gastric epithelia. Cancer Sci 97:1155–1158

So AY, Jung JW, Lee S, Kim HS, Kang KS (2011) DNA methyltransferase controls stem cell aging by regulating BMI1 and EZH2 through microRNAs. PLoS One 6:e19503

Song JZ, Stirzaker C, Harrison J, Melki JR, Clark SJ (2002) Hypermethylation trigger of the glutathione-S-transferase gene (GSTP1) in prostate cancer cells. Oncogene 21:1048–1061

Steuerwald NM, Parsons JC, Bennett K, Bates TC, Bonkovsky HL (2010) Parallel microRNA and mRNA expression profiling of (genotype 1b) human hepatoma cells expressing hepatitis C virus. Liver Int 30:1490–1504

Stirzaker C, Song JZ, Davidson B, Clark SJ (2004) Transcriptional gene silencing promotes DNA hypermethylation through a sequential change in chromatin modifications in cancer cells. Cancer Res 64:3871–3877

Swisshelm K, Disteche CM, Thorvaldsen J, Nelson A, Salk D (1990) Age-related increase in methylation of ribosomal genes and inactivation of chromosome-specific rRNA gene clusters in mouse. Mutat Res 237:131–146

Tao D, Lu J, Sun H, Zhao YM, Yuan ZG, Li XX, Huang BQ (2004) Trichostatin A extends the lifespan of Drosophila melanogaster by elevating hsp22 expression. Acta Biochim Biophys Sin (Shanghai) 36:618–622

Thevenet L, Mejean C, Moniot B, Bonneaud N, Galeotti N, Aldrian-Herrada G, Poulat F, Berta P, Benkirane M, Boizet-Bonhoure B (2004) Regulation of human SRY subcellular distribution by its acetylation/deacetylation. EMBO J 23:3336–3345

Trimarchi JM, Lees JA (2002) Sibling rivalry in the E2F family. Nat Rev Mol Cell Biol 3:11–20

Vaquero A, Scher M, Lee D, Erdjument-Bromage H, Tempst P, Reinberg D (2004) Human SirT1 interacts with histone H1 and promotes formation of facultative heterochromatin. Mol Cell 16:93–105

Vaquero A, Scher M, Erdjument-Bromage H, Tempst P, Serrano L, Reinberg D (2007) SIRT1 regulates the histone methyl-transferase SUV39H1 during heterochromatin formation. Nature 450:440–444

Villa R, Morey L, Raker VA, Buschbeck M, Gutierrez A, De Santis F, Corsaro M, Varas F, Bossi D, Minucci S, Pelicci PG, Di Croce L (2006) The methyl-CpG binding protein MBD1 is required for PML-RARalpha function. Proc Natl Acad Sci U S A 103:1400–1405

Vogt M, Haggblom C, Yeargin J, Christiansen-Weber T, Haas M (1998) Independent induction of senescence by p16INK4a and p21CIP1 in spontaneously immortalized human fibroblasts. Cell Growth Differ 9:139–146

Waki T, Tamura G, Sato M, Motoyama T (2003) Age-related methylation of tumor suppressor and tumor-related genes: an analysis of autopsy samples. Oncogene 22:4128–4133

Walter D, Matter A, Fahrenkrog B (2010) Bre1p-mediated histone H2B ubiquitylation regulates apoptosis in *Saccharomyces cerevisiae*. J Cell Sci 123:1931–1939

Wilson VL, Jones PA (1983) DNA methylation decreases in aging but not in immortal cells. Science 220:1055–1057

Wilson VL, Smith RA, Ma S, Cutler RG (1987) Genomic 5-methyldeoxycytidine decreases with age. J Biol Chem 262:9948–9951

Xu WS, Perez G, Ngo L, Gui CY, Marks PA (2005) Induction of polyploidy by histone deacetylase inhibitor: a pathway for antitumor effects. Cancer Res 65:7832–7839

Yamakuchi M, Ferlito M, Lowenstein CJ (2008) miR-34a repression of SIRT1 regulates apoptosis. Proc Natl Acad Sci U S A 105:13421–13426

Yang X, Phillips DL, Ferguson AT, Nelson WG, Herman JG, Davidson NE (2001) Synergistic activation of functional estrogen receptor (ER)-alpha by DNA methyltransferase and histone deacetylase inhibition in human ER-alpha-negative breast cancer cells. Cancer Res 61:7025–7029

Ye X, Zerlanko B, Zhang R, Somaiah N, Lipinski M, Salomoni P, Adams PD (2007) Definition of pRB- and p53-dependent and -independent steps in HIRA/ASF1a-mediated formation of senescence-associated heterochromatin foci. Mol Cell Biol 27:2452–2465

Yoshimi A, Kurokawa M (2011) Key roles of histone methyltransferase and demethylase in leukemogenesis. J Cell Biochem 112:415–424

Yuan F, Xie Q, Wu J, Bai Y, Mao B, Dong Y, Bi W, Ji G, Tao W, Wang Y, Yuan Z (2011) MST1 promotes apoptosis through regulating Sirt1-dependent p53 deacetylation. J Biol Chem 286:6940–6945

Zhang T, Kraus WL (2010) SIRT1-dependent regulation of chromatin and transcription: linking NAD(+) metabolism and signaling to the control of cellular functions. Biochim Biophys Acta 1804:1666–1675

Zhang R, Poustovoitov MV, Ye X, Santos HA, Chen W, Daganzo SM, Erzberger JP, Serebriiskii IG, Canutescu AA, Dunbrack RL, Pehrson JR, Berger JM, Kaufman PD, Adams PD (2005) Formation of MacroH2A-containing senescence-associated heterochromatin foci and senescence driven by ASF1a and HIRA. Dev Cell 8:19–30

Zhang J, Liu Q, Zhang W, Li J, Li Z, Tang Z, Li Y, Han C, Hall SH, Zhang Y (2010) Comparative profiling of genes and miRNAs expressed in the newborn, young adult, and aged human epididymides. Acta Biochim Biophys Sin (Shanghai) 42:145–153

Zhao Y, Sun H, Lu J, Li X, Chen X, Tao D, Huang W, Huang B (2005) Lifespan extension and elevated hsp gene expression in Drosophila caused by histone deacetylase inhibitors. J Exp Biol 208:697–705

Chapter 9
Homeotic Gene Regulation: A Paradigm for Epigenetic Mechanisms Underlying Organismal Development

Navneet K. Matharu, Vasanthi Dasari, and Rakesh K. Mishra

Abstract The organization of eukaryotic genome into chromatin within the nucleus eventually dictates the cell type specific expression pattern of genes. This higher order of chromatin organization is established during development and dynamically maintained throughout the life span. Developmental mechanisms are conserved in bilaterians and hence they have body plan in common, which is achieved by regulatory networks controlling cell type specific gene expression. Homeotic genes are conserved in metazoans and are crucial for animal development as they specify cell type identity along the anterior-posterior body axis. Hox genes are the best studied in the context of epigenetic regulation that has led to significant understanding of the organismal development. Epigenome specific regulation is brought about by conserved chromatin modulating factors like PcG/trxG proteins during development and differentiation. Here we discuss the conserved epigenetic mechanisms relevant to homeotic gene regulation in metazoans.

9.1 Animal Development

Embryonic development of diversified animals follows a common developmental theme. After fertilization, an embryo undergoes multiple rounds of cleavage to make embryonic cell mass. The ground plan for axis specification is laid down in an embryo early during embryogenesis. The molecular cues for dorso-ventral axis specification are found little later during cleavages. The most important event in embryogenesis in triploblastic organisms is gastrulation, when three germ layers, ectoderm, mesoderm and endoderm are formed. Each layer results in the formation of lineage specific organs and structures of the developing embryo. After crossing this crucial land-

---

N.K. Matharu • V. Dasari • R.K. Mishra (✉)
Centre for Cellular and Molecular Biology, Council of Scientific and Industrial Research (CSIR), Uppal Road, Hyderabad 500 007, India
e-mail: mishra@ccmb.res.in

mark, an embryo starts transforming from the state of totipotency to specifying cell types, which is the concerted work of interconnected regulatory pathways.

Each organism has dorso-ventral, left-right and rostro-caudal axes. The marking of animal body plan onto early embryo is done with the help of few molecular cues known as morphogens, which form the gradient fields in the three-dimensional space of a developing embryo. These morphogens can be the class of transcription factors or the signaling molecules that ultimately set the hierarchy of developmental regulatory networks. In *Drosophila,* maternally deposited mRNA gradients of *bicoid* at the anterior end and *nanos* at the posterior end polarize the embryo and define A-P axis (Dahanukar and Wharton 1996; Driever and Nusslein-Volhard 1988a, b). Levels of these molecules across the A-P axis decide the further downstream activation of regulatory networks in a gradient specific manner. The specification of A-P body axis in vertebrates results from an interplay between the localization of Wnt, Fgf and retinoic acid, that are required for the specification of the posterior neural ectoderm (Iimura et al. 2009; Schier and Talbot 2005). Anterior axis is determined by the presence of an enzyme *cyp26* that degrades retinoic acid and hence specifies the expression of anterior genes forming anterior neural ectoderm (Kudoh et al. 2002). All vertebrate embryos have a convergent stage during embryogenesis wherein the rostral or head part becomes distinct and the neural tube extends towards the posterior end. Further along this axial structure, the segmentation of an embryo starts and the secondary limb bud field is specified from where appendages would arise subsequently, along the A-P axis. The part of the mesoderm, called paraxial mesoderm, extends posteriorly along with segmentation of an embryo into somites which gives rise to the vertebral column. The segmentation of an embryo simultaneously serves as the action ground for homeotic genes that will decide the identity of each segment. These genes are conserved across phyla and are instrumental for the diversification of morphological features along the A-P axis of bilaterians (Gaunt 1994; Holland et al. 1992; Krumlauf 1992).

9.2 Hox Genes

Hox gene products are transcription factors containing the conserved DNA binding homeo domain. This domain is 60 amino acids long and occurs in a conserved helix-turn-helix configuration for providing DNA target site binding specificity (Gehring and Hiromi 1986; Levine and Hoey 1988; Maconochie et al. 1996; Scott et al. 1989). Hox genes are organized in the form of clusters and this feature is conserved among all animals (Fig. 9.1). The Hox cluster of *Drosophila melanogaster* containing eight homeotic genes, is split into two complexes – the Antennapedia Complex (ANT-C) and the bithorax complex (BX-C) (Carroll 1995; Celniker et al. 1989; Maeda and Karch 2006). ANT-C consists of five homeotic genes, *labial (lab), proboscipedia (pb), Deformed (Dfd), Sex combs reduced (Scr)* and *Antennapedia (Antp).* The genes in this complex determine the head and first two thoracic segments (parasegment 1 to parasegment 5 in larval stages). The bithorax complex contains three homeotic

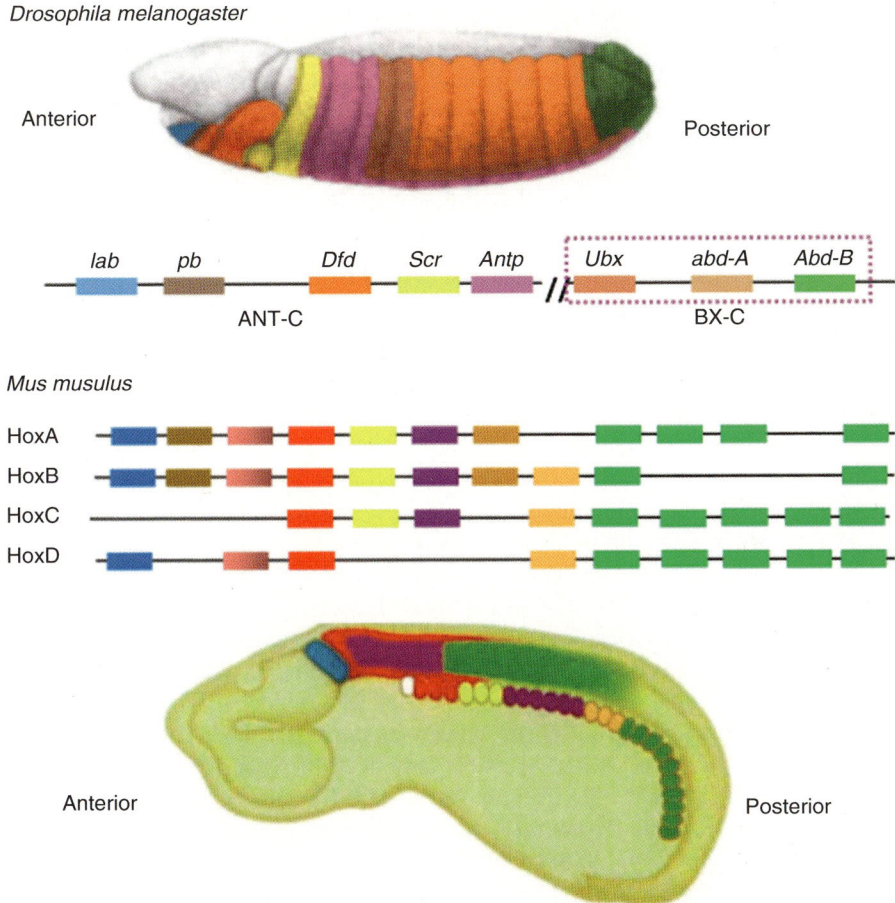

Fig. 9.1 Conservation of organization and expression of hox genes. Hox genes and their expression domains in *Drosophila* embryo (*top*) and mouse embryo (*bottom*) are shown. Flies have split hox cluster while mammals have four hox clusters. Each of the fly gene has one or more homologues in vertebrates. The posterior hox group (*Abd-B* group) in vertebrates has expanded, diversified and co-opted for limb formation. Hox genes and their expression domains are *color-coded* (Adapted from Vasanthi and Mishra 2008)

genes: *Ubx*, *abd-A* and *Abd-B*. These genes are responsible for the identity of parasegment 5 (PS5) to parasegment 14 (PS14) that form the third thoracic and eight abdominal segments in adults (Duncan 1987). There is a considerable level of structural homology between the various *Drosophila* hox genes and vertebrate homeodomain genes. In fact, a careful study showed that each fly hox gene has a corresponding homologue in vertebrates. The vertebrate homeotic complex has a total of 39 genes divided into four unlinked groups namely hox A, B, C and D complex. Each complex has few of the 13 paralogous genes missing in every organism arranged in spatially collinear manner (McGinnis and Krumlauf 1992). As depicted in the Fig. 9.1, *lab* of fly corresponds to hox1, *pb* to hox2, *Dfd* to hox4, *Scr* to hox5, *Antp* to hox6, *Ubx* to hox7, *abd-A* to hox8 and *Abd-B* to hox9-13 paralogous groups in vertebrates (Maconochie et al. 1996). This classification is functionally supported by rescue of fly homeotic phenotypes by mouse hox genes (Lutz et al. 1996). The finding that hox genes are extremely conserved functionally and structurally was remarkable, and led to the theory of co-option of hox genes for conserved body plan.

9.3 Homeotic Transformations: Mutations and Phenotypes

William Bateson, in 1894, in his *'Materials for the Study of Variation, Treated with Especial Regard to Discontinuity in the Origin of Species'*, first used the term 'homeotic' which is derived from the Greek word *'homeosis'*, meaning 'becoming like' to explain the leg emerging in place of antenna of *Drosophila*. These kinds of dramatic changes, where one part of the body becomes like the other, were termed as 'homeotic transformations'. Since hox genes specify the identity of segments along the A-P axis, any mutation in them results in loss of segment identity of the corresponding gene expression. Moreover, any perturbation in the levels of hox gene expression results in the formation of confused segments, consequentially in anterior or posterior transformations. Experimental analysis of vertebrate hox code is done primarily in mice by gene targeting studies. It has been noticed that single hox gene mutant or knockout mice does not necessarily lead to homeotic transformations, suggesting that there is some level of redundancy in hox gene function. Triple homozygous knockout for hoxa10, c10 and d10 in mice show loss of lumbar vertebrae, instead, lumbar vertebrae take over the identity of thoracic segments with skeletal outgrowth as ribs. Similarly homozygous knockout mice for hoxa11, c11 and d11 have sacral vertebrae replaced by lumbar vertebrae. However, single paralogue knockout in mice for hox10 and 11 did not result in homeotic transformations (Wellik and Capecchi 2003). hoxd13 mutations in mice and in human are linked to problems in digit formation and urinogenital system (Fromental-Ramain et al. 1996; Goodman et al. 1998; Kondo et al. 1996; Muragaki et al. 1996). hoxb8 homozygous loss of function mutant mice do not show any transformations in skeletal elements, these mice have altered sensory neuron activity (Holstege et al. 2008). Behavioral analysis of these mice revealed that, they have obsessive-compulsive disorder of pathological grooming (Chen et al. 2010). These mice pull hair from their body and in the process have self-inflicted wounds. Mice transgenic for over expression of hoxb8 have mirror image duplication of posterior skeletal elements of limbs, like duplicated digits, carpals and ulna (van den Akker et al. 2001). Apart from assigning gross skeletal identity, maintenance of hox gene expression is necessary throughout the lifespan of an organism. Homeotic gene products target many downstream processes, which get abrogated upon loss of function or over expression. Upon RA exposure during early embryogenesis, anterior segments take the identity of posterior ones and hence there is loss of posterior segments resulting in shortening of the rostro-caudal skeletal axis. The effects of RA on murine vertebral transformations and hox gene expression have been extensively catalogued (Kessel and Gruss 1991).

9.4 Collinearity of Hox Genes and RA Induction

The remarkable feature of hox genes is that their expression pattern is dependent on the orientation they are arranged on the chromosome. The order of the arrangement of these hox genes on the chromosome exactly follows the order of expression pattern along AP axis of the embryo. Anterior hox genes (3' of the cluster) are expressed in the rostral regions and posterior hox genes (5' of the cluster) in caudal region of an organism. The correspondence of genomic arrangement with the domain specific expression in an embryonic space is called the 'spatial collinearity' (Krumlauf 1994). A distinguishing feature about vertebrate hox complexes is the correlation between the order of genes on the chromosome and time of expression of these genes, anterior genes being expressed early, and posterior genes being expressed later during axis formation and segment specification, this being termed temporal collinearity. The variations in spatio-temporal distribution of hox genes expression along the A-P axis can result in structural novelties in animal kingdom (Dekker et al. 1993; Duboule 1994; Gaunt and Strachan 1996; Kondo et al. 1998). In essence each segment in the body is dependent on 'hox code' for its identity and segment specific variations can occur as a result of subtle variations in this 'hox code'.

During somitogenesis the oscillating gradient of retinoic acid (RA) along with Wnt/Fgf signaling is crucial for segmentation of an embryo. In limb bud formation, RA polarity is necessary for maintaining the zone of polarizing activity (ZPA) and thus activation of region specific induction of separate developmental pathways (Buxton et al. 1997; Stratford et al. 1996). RA has morphogenetic properties; the dose of RA along the A-P axis is crucial in maintaining the hox gene expression in correct order. Retinoic acid induces dose dependent hox gene expression in a collinear fashion. High doses of RA lead to induction of posterior hox genes earlier than specified and thus results in axial malformations. This results in loss of posterior structures of the embryo (Kessel and Gruss 1991). Nevertheless RA acts as morphogen but in higher dosages can lead to severe developmental problems. Within the developing embryo peri-nodal tissue cells surrounding the hensen node, mesodermal floor plate and later on limb buds are endogenous source of retinoids (Horton and Maden 1995; Maden et al. 1988). As the embryo axis elongates, endogenous retinoid production also increases in the embryo (Hernandez et al. 2007).

It is known that anterior hox genes need less RA than posterior hox gene for activation and thus setting up progressively, an increasing front of RA in the elongating embryo (Conlon 1995). Levels and gradients of retinoic acid in an embryo are maintained by its *Raldh2* dependent biosynthesis, and *cyp26* dependent degradation. Embryos deficient in both the pathways result in disturbed axial patterning. Mouse embryos developed in vitamin A deficient mothers show pleiotropic developmental defects, which are also observed upon vitamin A deficiency in vertebrate model systems and are classified as VAD syndrome (Mark et al. 2004, 2006). RA is instructive during hox gene specification of the metameric somites and induces limb bud formation and organogenesis.

9.5 Regulation of Master Regulators

Homeotic genes expression follows a spatio-temporal manner but the understanding of how these genes are regulated to execute the 'hox plan' is still developing. The levels and the kinds of *cis* and *trans* regulatory inputs required by hox genes to bring about the read out in the form of expression pattern in a developing embryo are (may be) enormous and complex. By employing novel strategies of regulating genes such as promoter sharing, promoter competition, enhancer sharing, enhancer selectivity, insulator elements, silencers, repressors, maintenance elements, Retinoic acid response elements (RAREs), long range interactions, higher order chromatin re-organization and non-coding RNAs, homeotic clusters have become the paradigm to understand different ways and complexity of gene regulation in the past three decades.

9.5.1 Features for Tight Regulation of Hox Clusters

The tight association of clustered hox genes is conserved during evolution and necessary to establish hox expression domains in proper space and time during embryo development. It has been shown that a gene from one paralogous hox group can overtake the function of the other, like hoxa1 insertion in place of hoxb1 deletion rescues the hoxb1 phenotype. But when the hox gene position is changed along the cluster such that it resides in non-native regulatory context, its expression is governed by neighbouring regulatory elements. For example, hoxd9-LacZ transgene when placed posterior of the hox cluster, starts expressing in spatio-temporal domain of hoxd13, similar is the case when hoxb1-LacZ transgene is inserted in the posterior locations (Kmita et al. 2000). These experiments clearly illustrate that hox clusters retain the spatio-temporal information for tight regulation in their non-genic region. Any perturbation, for example inversions or duplications or deletions in the organization of the cluster lead to disturbed homeotic gene expression. Many of the homeotic mutations in *Drosophila* mapped to regulatory sequences rather than the coding region of the hox genes (Bender et al. 1983). Homeotic transformations may also result by mutating certain factors, which are *trans*-regulators of homeotic complexes. Expression and regulation of the hox cluster requires both *cis* and *trans* inputs for its initiation, modulation and maintenance in correct space and time.

9.5.2 Cis-Regulatory DNA Elements in Hox Clusters

Eukaryotic genome is dominated by non-coding sequences in its content, while coding (genic) sequence is a small minority, only 2% of the genome. For example in human, to control and modulate this 2% of the genome, sequences of non-coding potential are employed and incorporated in the genome, keeping the coding process itself conserved (Flam 1994). Homeotic genes exist in clusters along with their *cis*-regulatory DNA elements present within or in the flanking regions of the cluster making up the 'homeotic-landscape' (Duboule 2007). Different kinds of *cis*-regulatory modules are juxtaposed in the hox complex to switch on and off, or even to command the level of gene expression in spatially and temporally distinct domains. Most of these *cis* elements are indentified by deletions and are modular in nature as assayed in transgenic context.

9.5.2.1 Enhancers

Several *cis*-elements having enhancer function have been found in vertebrate hox clusters. Apart from enhancing the level of expression of the gene in a tissue specific manner, enhancers in hox clusters have evolved with the need to follow the 'hox clock'. Early expression of hoxc8 is associated with a conserved upstream element having enhancer function. Deletion of the hoxc8 enhancer results in anterior transformations in axial skeleton associated with the delayed expression of hoxc8 (Juan and Ruddle 2003). Deletion of hoxd11 enhancer is responsible for the delayed hoxd11 expression resulting in homeotic transformations in the sacral region of the axial skeleton (Gerard et al. 1997; Herault et al. 1999; Zakany et al. 1997).

hoxb3 and hoxb4 have subset of expression domain, which is overlapping in rhombomere6/7 of hindbrain. This sharing of domain in space is a result of sharing of common enhancer CR3, present in the intergene of hoxb3 and hoxb4. hoxb4 and hoxb5 gene products can auto-regulate and cross-regulate respectively the enhancer by direct binding, unlike hoxb3 protein. This CR3 region in transgenic *Drosophila* can also be activated by *Dfd,* which belongs to hox4 group of genes (Gould et al. 1997; Sharpe et al. 1998). This cross-species conservation in *enhancer sharing* and *auto-regulative* mechanisms highlight the evolutionary constraints imposed upon regulation of hox clusters. hoxb4 and hoxb5 intergene has an atypical shared enhancer element, having neural and limb enhancer modules. In transgenic assay, limb enhancer region can drive only hoxb4 promoter but not hoxb5 promoter, suggesting the *enhancer selectivity*. Another module of the same enhancer has a neural enhancer region, which can drive reporter expression equally from hoxb4 and hoxb5 promoters in transgenic assay. However, when placed in dual reporter construct, the neural enhancer has preference towards hoxb4 promoter. hoxb4 promoter competes for the enhancer, but this *promoter competition* is abolished when distance between hoxb4 promoter and neural enhancer is increased, and neural enhancer starts driving the hoxb5 promoter instead, due to close proximity with the latter (Sharpe et al. 1998). The tight placement of enhancers within the hox cluster is highly context dependent making them more robust and innovative in performance.

Regulation of posterior hoxd genes depends on the large global control region (GCR) identified 240 kb upstream of the HoxD complex having enhancer like function. This GCR is described as a 40 kb region with several regulatory domains. Locus specific control of posterior hoxd genes is governed by GCR, which when deleted, abolishes the expression of posterior hoxd genes in the limb region. Serial deletion and duplication studies on posterior hoxd genes implicate the mechanism of GCR sharing among posterior hoxd genes for 'quantitative collinearity' in limb bud. For example deletion of hoxd13 and hoxd12 genes results in hoxd13 like expression of next posterior gene i.e., hoxd10. This analysis conclusively suggests that global enhancer discriminates its sharing depending upon the location in the posterior HoxD complex and thus forms the quantitative collinear gradient of posterior hoxd genes in the limb region (Kmita et al. 2002a). The long range interaction of enhancer like locus control region is a vertebrate innovation typically to regulate clustered genes.

9.5.2.2 Boundary Elements

On *prima facie,* enhancers in hox clusters are responsible for distinct domain specific expression of hox genes along the body axis, but essentially, enhancers are capable of promiscuity in driving the expression of nearby genes too. Nevertheless, enhancer selectivity and preference has been observed in transgenic assays, activity of enhancers alone cannot justify the tight regulation operating in hox clusters. Certain *cis*-regulatory elements restrict the illegitimate enhancer-promoter crosstalk exist in the genome and thus demarcate functional chromatin domains. Moreover, apposing functional domains within hox clusters demand existence of higher order insulation mechanisms to prevent spread-over of adjacent enhancer action. Boundary elements are modular and can be assayed in transgenic context as blockers of enhancer-promoter interaction. Essentially, boundary or insulators protect genes from flanking chromatin environment but do not by themselves instruct the transcriptional state of a gene.

Hox complex has closely spaced differentially expressed hox genes and boundary elements are required to maintain separate expression units as functional chromatin domains. Dissection of the regulatory landscape of *Drosophila* BX-C complex has provided ample evidence to prove that insulators are critical in defining domain specific hox expression. Segment specific expression pattern of *Ubx, abd-A, Abd-B* is controlled by a large *cis*-regulatory region, ~300 kb (Karch et al. 1985), that is subdivided into nine functionally autonomous *cis*-regulatory domains *(abx/bx, bxd/pbx, iab-2* to *iab-8)*. Each *cis*-regulatory domain directs expression of homeotic genes, *Ubx, abd-A, Abd-B* in a specific parasegment. For example, *iab-5 cis*-regulatory domain controls *Abd-B* expression pattern in parasegment10 (PS10)/A5. Similarly *iab-6, iab-7* and *iab-8* control *Abd-B* expression in PS11/A6, PS12/A7 and PS13/A8 identity respectively (Boulet et al. 1991; Celniker et al. 1989, 1990). Genetic and molecular studies have identified chromatin domain boundaries that demarcate the *cis*-regulatory domains that ensure the functional autonomy of each regulatory domain (Mihaly et al. 1998). For example in PS11, *Fab-7* boundary protects active *iab-6 cis*-regulatory domain from inactive *iab-7* domain restricting inappropriate regulatory interactions between the two domains. Deletion of *Fab-7* boundary allows the next *cis*-regulatory domain, *iab-7* to drive *Abd-B* in PS11 resulting in A6-A7 transformation in adult fly.

In vertebrate hox clusters, the existence of insulator elements is expected, but so far only one such element has been identified. The boundary element of hox clusters that is genetically well dissected is between Evx2-hoxd13. Both Evx2-hoxd13 have a common regulatory network and are under the influence of GCR. Neural expression of posterior hoxd genes is supposedly prevented by Evx2-d13 boundary element (Kmita et al. 2002b). In order to test boundary function effectively, several regions of evx2-d13 intergene were tested by placing it towards the 3' of Evx2 gene tagged with lacZ gene as reporter. A region close to Evx2 promoter harboring boundary activity, when placed downstream to the Evx2 could block neural enhancer from driving Evx2 expression in CNS (Yamagishi et al. 2007). When tested in *Drosophila* and mammalian K562 cells, Evx2-d13 core boundary function could be narrowed down to a 3 kb region harboring binding sites for *Trl*/GAF. The Evx2-d13 boundary needs *Trl*/GAF for its insulation activity (Vasanthi et al. 2010). The *Drosophila* homologue of Evx2 is *even-skipped* that also has promoter associated *Trl* dependent boundary function (Ohtsuki and Levine 1998). The *Trl* dependent insulation mechanism of Evx2-d13 boundary is reminiscent of *Drosophila* hox regulation although there is no sequence similarity whatsoever. Smaller fragments of 1 kb derived from bigger 3 kb Evx2-d13 boundary are capable of enhancer blocking in *Drosophila*, however they are weak blockers. Evx2-d13 boundary model illustrates that boundary function in vertebrate hox cluster is spread-over a few kilobases unlike much smaller boundaries of *Drosophila* BX-C. This may indicate recruitment of additional regulatory motifs in vertebrate boundaries.

The sequence comparison of various boundary elements has failed to identify any significant homology except for small conserved sequence motifs of unknown significance or stretches of AT rich regions (Karch et al. 1994; Vazquez et al. 1993). Chromatin domain boundaries that subdivide the genome into functional units have been isolated using different criteria (West et al. 2002). Interestingly, it has been observed that there is a great degree of conservation of the boundary activity across the species (Chung et al. 1993). This could be because of the conservation of the factors associated with the boundary elements. For example, 5' HS4 boundary element from mouse β *globin* locus acts as strong boundary in the enhancer blocking assay system in *Drosophila*. On the other hand, *Fab-8* boundary element from the *Drosophila* bithorax complex functions as an enhancer blocker in vertebrate cells (Moon et al. 2005).

The mechanism of boundary elements function is largely unknown, though there are several largely speculative models to explain it. (Capelson and Corces 2004; Kuhn and Geyer 2003; Raab and Kamakaka 2010; Valenzuela and Kamakaka 2006). Boundary elements may function in combination with nuclear matrix through SAR/MAR (Scaffold Associated Region/Matrix Associated Region) like elements where the insulating effects are a consequence of this association (Gerasimova and Corces 2001; Pathak et al. 2007). Another way boundary elements might function is by inhibiting the interaction between the promoter and the signal from enhancers (Gaszner and Felsenfeld 2006). One more way in which boundaries might act is by competing with promoter to capture enhancer, namely the promoter decoy model (Dorsett 1999; Gause et al. 2001). Recently, it has been shown in *Drosophila* that the most plausible action of boundary elements is by looping DNA (Gohl et al. 2011). The boundary action of an element is dependent on the kind of other boundary elements flanking the transcriptional unit. Almost all reports of boundary elements discuss possible mechanisms of action but no single model explains fully all the aspects of their action, possibly because more than one mechanism are involved (Iqbal and Mishra 2007; Mishra and Karch 1999).

9.5.2.3 Retinoic Acid Response Elements

Hox gene activation is critically responsive to retinoic acid (RA) within the developing embryo. RA is present in a concentration gradient and thus acts as a morphogen for pattern formation and cell differentiation. Retinoic acid works as a ligand for retinoic acid receptors (RARs) which upon binding to cis-elements called Retinoic acid response elements (RAREs) present in gene promoters brings about changes in transcriptional state of the gene. Exposure of a developing embryo to RA results in pleiotropic effects like perturbations in axial patterning, limb bud formation, segmental identity and numbers of metameric structures along the A-P axis.

In the absence of RA ligand (or in the presence of antagonist ligand) binding domain of RAR/RXR heterodimer recruits co-repressor complex NCoR/SMRT and maintains repression (Leid et al. 1992). Upon RA (agonist ligand) binding, there is a conformational change in the ligand-binding domain at the surface of heterodimer RAR/RXR allowing the binding of coactivators (p300/pCAF complex) with subsequent release of corepressor complexes (Glass and Rosenfeld 2000). Null mutants of these receptors show equally pleiotropic effects like VAD deficiency including disturbed hox gene expression and axial skeletal defects (Kastner et al. 1995; Mark et al. 2009). The release of NCoR/SMRT complex from RARE triggers JMJD3, a H3K27 demethylase, to remove the H3K27me3 repressive mark (Jepsen et al. 2007). RA induced de-repression is accompanied by concomitant release of SUZ12 and erasure of the H3K27me3 mark which essentially puts RA in a more instructive role to remove inactive marks (Gillespie and Gudas 2007). Upon removal of RA treatment SUZ12 re-associates with RARE. RA mediated chromatin restructuring to lift PcG repression from hox genes is shown via another H3K27me3 demethtylase UTX, which even dislodges EZH2 from repressive domain (Agger et al. 2007; Lee et al. 2007). UTX mediated demethylation is thought to ensure quick removal of repressive mark and hence activation in a precise spatio-temporal window. Removal of the H3K27me3 mark from hoxa1 RARE that is 4.6 kb away from hoxa1 promoter, results in increased occupancy of RNAPolII at promoter region. Concomitant increase in H3K4me3 mark in hoxa1 and hoxb1 RARE upon RA exposure does not increase the MLL complex occupancy at these loci pointing to an MLL independent mechanism operating for assignment of activated state (Kashyap et al. 2011).

9.6 PREs – Polycomb Response Elements

With an identical genome, metazoans know the art of making different cell types. Hox genes define and specify the transit of plain embryo to a working ensemble of specified cells and structures. These gene products are quite authoritative in specifying cells to differentiate and follow a particular lineage and this specification of cellular identity has to be memorized by the cell type for the remaining lifetime of an animal. Maintenance of the expression states of crucial cell-identity specifying genes is controlled by robust cellular system comprising conserved *Polycomb/trithorax* group (PcG/trxG) of proteins. These proteins are in-fact so important throughout the lifespan of an animal that loss of PcG/trxG system any time creates total chaos in otherwise well disciplined cell type specific gene expression. Expression of hox genes is maintained by two antagonistic sets of these genes: *Polycomb* group (PcG) and *trithorax* group (trxG) of genes (Kennison 1995). The PcG and trxG of proteins were first identified as *trans*-acting regulators of homeotic genes. The phenotypes observed in case of these genes are similar to the homeotic phenotypes. Suppressors of the PcG phenotype have been identified that fall into *trxG* genes. Several *PcG* and *trxG* genes have been identified in *Drosophila*. Molecular analyses of these proteins showed that they might act in large complexes and modify the local properties of chromatin to maintain transcriptionally repressed or active state of the target genes. PcG proteins target several genes involved in signaling pathways that are conserved, suggesting that these proteins might act as regulators to coordinate a variety of developmental and differentiation processes; consistent with recent evidences suggesting that the function of the conserved PcG/trxG complexes is dynamically modulated in time and space (Klebes et al. 2005; Lee et al. 2005; Maurange et al. 2006). PcG/trxG proteins interact with specific *cis* regulatory elements called <u>P</u>olycomb <u>R</u>esponse <u>E</u>lements (PREs) (Chan et al. 1994; Christen and Bienz 1994; Simon et al. 1993) and are conserved across the species (Ringrose and Paro 2004).

Three crucial steps required to confer Polycomb dependent cellular memory module are PcG targeting to specific DNA, chromatin re-structuring corresponding to the expression state of loci and memorizing it through cell divisions. These steps are the hallmarks of an element responsive to PcG/trxG regulation and thus several assays to identify PRE (Polycomb response element) are designed accordingly.

9.6.1 Targeting of PcG/trxG

Cis-regulatory elements responsive to PcG/trxG systems are called Polycomb Response Element (PRE)/Trithorax Response Element (TRE), having the repressor and activator function respectively. However due to the overlapping nature of existence of these two elements it is sometimes tedious to functionally separate them out.

Polycomb mediated silencing is the action of three repressive complexes, 3MDa Polycomb Repressive Complex 1, PRC1 and 600 kDa, Polycomb Repressive Complex 2, PRC2, which has H3K27 tri-methyltransferase activity and PhoRC (PHO dependent repressive complex) (Table 9.1). Targeting of these complexes to PRE is known by PHO, PHO-like, Trl-GAGA, Pipsqueak, DSP1 (Protein dorsal switch 1), SP1 (specificity protein 1), Grainy head and ZESTE in *Drosophila*. In vertebrates, YY1 (Yin-Yang 1, homologue of PHO) and ThPOK (homologue of Trl-GAGA) (Matharu et al. 2010), are involved in PcG targeting, while other

Table 9.1 PcG complexes: fly genes and vertebrate homologues

Complex	Fly	Vertebrate homologues
PRC1	PC	CBX2, CBX4, CBX6, CBX7, CBX8
	PSC	PGF6 (MBLR), PGF (BMI1), PGF5, PGF3, PGF2 (MEL18), PGF1 (NSPc1)
	PH	PHC1, PHC2, PHC3
	SCE	RING1, RING2
	E(Z)	EZH1,EZH2
	ESC/ESCL	EED1, EED2, EED3, EED4
	SU(Z)12	Suz12
	NURF55	RBAP48, RBAP46
	PCL	PCL1 (PHF1), PCL2 (MTF1), PCL3 (PHF19)

homologues are not yet identified (Beisel and Paro 2011). Mutations in these proteins result in loss of PcG silencing, which suggests that sequence specific targeting of PcG is crucial for further PcG repression. YY1 knockdown in mouse myoblasts prevents binding of PRC2 member EZH2 (enhancer of zeste homologue 2) and thus loss of H3K27 tri-methylation (Caretti et al. 2004). It is difficult to predict PRE due to lack of sequence homology, therefore many bioinformatics tools have been designed to predict the binding motifs for these DNA targeting proteins (Ringrose and Paro 2004; Ringrose et al. 2003). Several ChIP based studies for PcG targets in mammals revealed close association to the target promoters (Boyer et al. 2006; Bracken et al. 2006; Endoh et al. 2008), while in ES cells this targeting is restricted to promoters with high CpG content (Ku et al. 2008). There are only three PREs so far identified in higher vertebrates. In mouse PRE-kr is identified for regulating *MafB* gene, which has YY1 and GAGA binding sites (Sing et al. 2009). Another mammalian PRE identified so far is HoxD11.12 PRE, which again has YY1 binding motifs (Woo et al. 2010). Up to 25 kb upstream to mouse HoxD cluster a fragment initially identified as a repressor element (Kondo and Duboule 1999) by deletion analysis was later shown to recruit EED (PRC2 member) in a temporal fashion for silencing mechanisms (Mishra et al. 2007). This fragment is shown to have repressor function in transgenic *Drosophila* along with eye color variegation and PC recruitment. Targeting of PcG/trxG to PRE thus satisfies one of the criteria for a functional PRE.

Recruitment of trxG is somewhat more complex and less understood than PcG complexes. At least four different complexes containing trxG proteins have been identified from *Drosophila* embryos: SWI/SNF complex (has Brm, Osa, Moira, Snr1 that catalyses ATP dependent chromatin remodeling), NURF complex (has Iswi, N38, N301, N55 that catalyses ATP dependent chromatin remodeling), TAC1 complex (has Trx, dCBP, Sbf1 has SET domain proteins involved in methylation of histones) and Ash1 complex (Ash1, dCBP has SET domain proteins involved in methylation of histones). Many *trithorax* homologues have been discovered in vertebrates, having conserved function as in *Drosophila*. Binding of such complexes has been shown at hox genes but the mechanism of activation is not fully understood (Hsieh et al. 2003; Hughes et al. 2004).

9.6.2 PcG/trxG Dependent Chromatin Modulation

Chromatin regulators such as PcG/trxG directly command the gross restructuring of the chromatin and have an effect on accessibility of the transcriptional machinery. Histone methyl-transferase (HMT) activity of the PRC2 complex resides with E(Z) of flies and EZH2 of mammals. It needs the assembly of core PRC2 complex members, Suz12, and ESC (in flies)/EED (mammals) for its catalytic activity. Tri-methylation of histone 3 at lysine27 strictly tracks with the occupancy of PcG complex while H2K27 mono and di-methyl marks are widespread in the genome but only the former being involved in the PcG dependent silencing of chromatin. The complexity of the PRC2 complex is not yet properly understood. Some studies suggest that certain isoforms of EED (homologue of fly ESC) can methylate H1 linker also but whether this methylation is necessary for PcG silencing is not known (Martin et al. 2006). Another homologue of E(Z) in mammals, EZH1, is expressed in adult non-dividing tissues having less HMTase activity than EZH2 which is crucial for embryogenesis and cell differentiation (Martin et al. 2006). PRC1 complex recognizes the H3K27 modification for PRE targeting, loss of PRC2 leads to loss of PRC1 targeting also. Another complex PC-like (PCL) in flies and PHF1 in mammals is required to enhance the HMTase activity of PRC2 complex. In flies PCL recruits PRC2 complex (Savla et al. 2008) and thus loss of PCL leads to de-repression of some of the target loci but not all. However, this sequence of PRC2 and PRC1 targeting is still debatable because of a line of evidence cited which show PRC1 occupancy and H3K27 tri-methylation to be exclusive at certain loci in *Drosophila* cells (Schwartz et al. 2006). Studies like PHO mediated targeting of the PRC1 complex (Mohd-Sarip et al. 2006) and retention of PRC1 upon reduction of H3K27 tri-methylation at few of the target loci support the targeting of PRC1 independent of H3K27 tri-methylation (Schoeftner et al. 2006). But at many developmentally important loci studied so far PRC1 occupancy coincides with the H3K27 mark and further leads to chromatin modulation such as to impede RNA PolII machinery to establish silencing (Chopra et al. 2009). However recent studies are pointing towards PC mediated long-range interaction of distant Pc sites to assemble in one compartment and thus facilitate the spreading of silencing machinery over large loci (Bantignies et al. 2011; Lanzuolo et al. 2007; Tiwari et al. 2008). This recent school of thought is developing to address the recruitment of PRC1 complex independent of PRC2 for silencing during X-inactivation also (Casanova et al. 2011; Tian et al. 2010; Zhao et al. 2010). The mechanisms of PcG dependent silencing looks simple in *Drosophila* which has limited complexity in PRC1 and PRC2 members, unlike in the case of mammals where the repertoire of Pc family itself is expanded (Table 9.1).

9.6.3 Bivalent Chromatin Domains: PcG/trxG

Since PRE/TRE exist in union many a time, the chromatin domains flanking this common fragment have differential transcriptional information (Fig. 9.2). The mechanisms of PcG silencing and trxG activation are interdependent and their

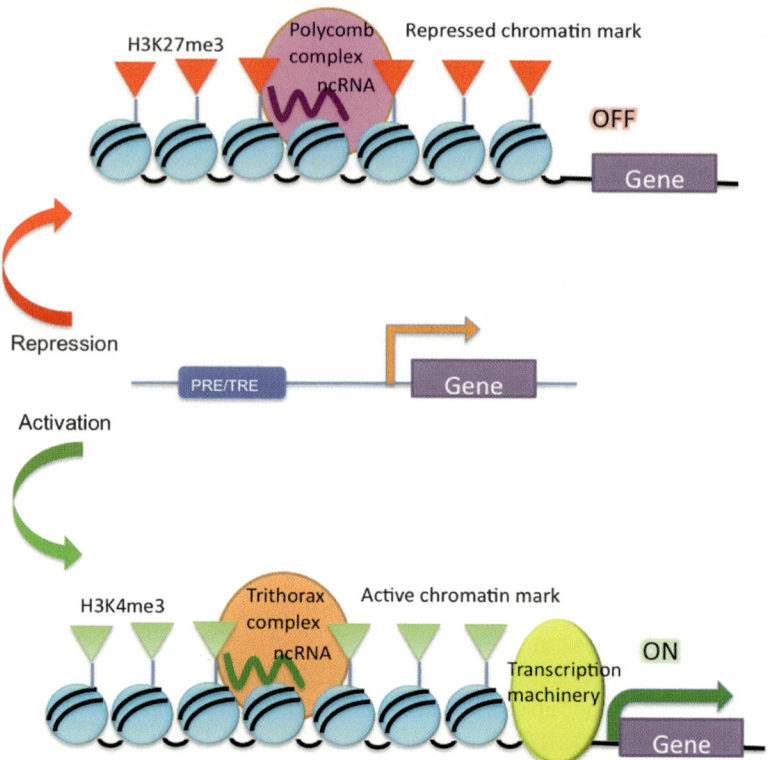

Fig. 9.2 **PcG/trxG complexes and chromatin modification.** Schematic representation of PcG and trxG complexes associated on PRE/TRE. PcG and trxG complexes deposit histone marks that play complementary roles in silencing and activation of their target chromatin. A developmental signal determines whether the PRE/TRE mediates gene activation or gene repression, which is accompanied by histone H3K27me3 and H3K4 demethylation or H3K4me3 and H3K27 demethylation, respectively. PcG/trxG complex can form 'bivalent chromatin domains' by having adjacent active and silenced chromatin mark. PcG/trxG complexes interact with non-coding RNAs, which helps in recruitment. PRC complex leads to histone modification H3K27me3 causing local silenced chromatin to be silenced. On the contrary for active chromatin, trxG complexes cause H3K4me3 modification and thus decides recruitment of transcription machinery for expression of genes

assembly on chromatin is also concerted. *Drosophila* PREs are known to have distinct TRE fragment, for example *bxd*-PRE regulates *Ubx* and has dissectable TRE (Tillib et al. 1999). This makes sense because the Ubx gene has to be switched on and off in a spatial manner during development. PcG takes care in the domain where the *Ubx* expression has to be silenced and trxG where it has to switch on. This PRE/TRE switch is common feature in *Drosophila* hox complex while in mammalian development this possibility has not yet been explored. However, a study in mice showed antagonist roles of Bmi1 (PcG member) and Mll (trxG member) during regulation of hox genes. *Bmi1* mutant showed posterior transformations of cervical vertebrae, which got corrected to certain extent in the background of *Mll* heterozygous mutation. HoxC in the same study was analyzed as target locus for Bmi1 and Mll. Loss of Bmi1 results in up-regulation and loss of Mll results in down-regulation of hoxc8 in mice (Hanson et al. 1999).

Histone modifications studied across developmentally important loci in ES cells revealed 'bivalent domains' having H3K27me3 and H3K4me3 at promoters (Bernstein et al. 2006) (Fig. 9.2). Genome-wide co-occupancy of H3K27 tri-methylation and H3K4 tri-methylation in ES cells found at many promoters is either bound by both PRC1 and PRC2 or only PRC2 (Ku et al. 2008). However PRC1 occupied bivalent domains are more stable and retain silenced information after differentiation. 'Bivalent chromatin domains' occupied by PRC1 and PRC2 have transcriptionally poised Ser-5 phosphorylated RNA PolII. Loss of any PRC1 or PRC2 member results in firing of transcription by RNA PolII at these promoters. H3K4 demethylase, Jarid2, associates with PRC2 components and necessary for poising of RNA PolII. (Landeira et al. 2010; Li et al. 2010; Pasini et al. 2010). Intriguingly, Jarid2 depletion results in loss of poised RNA polII with lower transcript levels as compared to even wild type ES cells and hence paused differentiation (Landeira et al. 2010). Another demethylase, RBP2 (retinoblastoma binding protein 2-H3K4 demethylase) has been shown to localize with PRC2 complex suggesting robust silencing by PRC2 targets. Although the interaction is transient, it is functionally relevant. During hox gene activation RBP2 dislodges from transcriptionally poised promoters, and again is required to silence it after differentiation (Pasini et al. 2008).

Another class of demethylases for H3K27 is also implicated in ES cells differentiation and embryonic development. UTX and Jmjd3 associate with Mll2/3 trithorax complex and removes H3K27me3 repressive mark. UTX acts on many hox promoters to remove the repressive mark, which are subsequently marked with H3K4me3 active mark upon RA signaling (Lee et al. 2007). Knock down of UTX in zebrafish embryos shows anteriorization phenotype with posterior shift of hoxc8 anterior expression boundary and reduced transcription of posterior hox genes (Lan et al. 2007).

Zygotic genome remains inactive till maternal-zygotic transition. Out of histone3 lysine modifications, H3K27me3 and H3K4me3 marks are detected only after maternal-zygotic transition in zebrafish embryos (Vastenhouw et al. 2010). The embryonic pluripotency coincides with the arrival of these two marks in the zygotic epigenome. The existence of bivalent domains corroborating the ES cells studies is suggested during zebrafish embryogenesis.

9.6.4 H2A Ubiquitination and PcG Silencing

Another covalent histone modification found to be involved in PcG mediated silencing is H2AK119 ubiquitination (H2Aub) (Shilatifard 2006). Loss of H2A ubiquitination from PcG target promoters dislodges PRC1 complex but H3K27me3 mark remains. RING1 and BMI1 (PSC in flies) members of PRC1 are part of this H2Aub complex named PRC1-L (PRC1-like)/dRAF (dRing associated factors) in flies/BCOR (BCL6 CoRepressor) complex in mammals (Table 9.1). PRC1-L complex can ubiquitylate

H2A at lysine 119. Loss of dRING in *Drosophila* leads to PcG mediated derepression of *Ubx* concomitant with the loss of H2Aub (Wang et al. 2004). Interestingly PRC1-L complexes contain KDM2 (a histone lysine demethylase) as one of the component, which has been shown to have high ubiquitinating activity at PcG targets. KDM2 demethylates H3K36me2, an active mark, and facilitates H2AK119 ubiquitination. *Drosophila* KDM2 is found to be the enhancer of PcG phenotype and suppressor of trxG phenotype (Lagarou et al. 2008). Loss of H3K27me3 demethyalse, UTX, results in increased ubiquitination of H2AK119 at several hox promoters, which reinstates the role of H2Aub in PcG silencing. H2A de-ubiquitinating complex (PR-DUB-Polycomb repressive deubiquitinase) contains *Calypso* and ASX, mutants of which show de-repression of PcG targets (Scheuermann et al. 2010). Interestingly these mutations also show slight increase in H2B ubiquitination along with H3K4 methylation, both being associated with active mark. H2Bub is shown to be important for sub-telomeric anti-silencing (Wan et al. 2010), however its role in PcG silencing or trxG activation is yet to be deciphered. ZRF1 (zuotin-related factor 1) in mammals is recruited to H2Aub and displaces PRC1 complex during differentiation (Richly et al. 2010). ZRF1 is required for retinoic acid dependent transcriptional activation and H2A de-ubiquitination of hoxA promoters. Ubp-M (Usp12) has the activity of H2A specific de-ubiquitination, abrogation in Ubp-M function results in repressed hoxd10 and abnormal posterior development of *Xenopus* (Joo et al. 2007). H3K27me3 is required to target PRC1 but considering another layer of histone modifications operating to stabilize PRC1 suggest that H3K27me3 mark is surely not enough to hold PRC1.

9.6.5 Elongation Block and PcG Silencing

PcG/trxG co-existence assures the cell to execute developmental processes in time and in an energy efficient manner. At the core of gene regulation the transcriptional machinery has to switch ON/OFF. PcG mediated chromatin marks were thought to be permanent previously but in the light of new studies it seems to be more dynamic. PcG silencing correlated for 'transcriptionally inactive chromatin' is now taken as 'transcriptionally poised chromatin'. It has been shown that *Ubx* and *AbdB* contain paused RNA polymerase when silenced. Co-occupancy of stalled RNA PolII along with PcG members at hox promoters (Chopra et al. 2009) is explained as the cellular mechanism to tilt the balance to ON/OFF expression state whenever required. Many 'bivalent' domains in gene promoters show the presence of stalled RNA PolII (has phosphorylated at Ser5) and these genes also contain H2Aub marks. Depletion of RING1 releases this elongation block by having RNA PolII phosphorylated at Ser2 (Stock et al. 2007). Another line of study about H2A ubiquitination suggests that NcoR (nucleosomal co repressor factor) dependent H2Aub activity is necessary for stalled RNA PolII, which prevents the recruitment of FACT complex and hence blocks elongation (Zhou et al. 2008). Conversely H2B ubiquitination associates with FACT complex on actively transcribing promoters (Minsky et al. 2008).

Studies independent of PcG mechanisms also point towards elongation switch for regulation of gene expression. Promoters of developmentally important genes in ES cells having actively marked nucleosomes, H3K4me3/H3K9Ac along with RNA PolII undergo transcriptional initiation, but only subset of genes are fired for elongation (Guenther et al. 2007). Trithorax members are known facilitators for this elongation to overcome PcG repression.

9.6.6 Non-Coding RNA in PcG/trxG Regulation

Nearly 98% of the eukaryotic genome is non-coding, and new techniques for genome level sequencing have shown that most of the non-coding genome is being transcribed. Emerging roles of non-coding RNAs in the genome suggest that they constitute another layer of regulatory networks operating in the genome. Targeting of PcG/trxG has been shown to be dependent on either *cis* or *trans* non-coding RNA transcription. Both in flies as well as in mammals non-coding RNAs, sometimes in both the orientations are detected through PRE/TRE, which can also directly bind to PcG/trxG complex (Fig. 9.2). These can be few nucleotides (100–200 bp) to few kilobases in length. A non-coding RNA of 26 kb length has been detected through fly *bxd*-PRE, which is further processed (Lipshitz et al. 1987). There is no general rule as of expression domain of these non-coding RNA with that of its cognate hox gene expression state. However these ncRNAs are expressed before the appearance of hox gene mRNA, setting up the stage for 'PcG or trxG recruitment' depending upon the regulation required in spatio-temporal context. This kind of PcG/trxG recruitment to PRE/TRE by ncRNAs is thought to be sequence dependent for exact targeting.

Bivalent chromatin domains at promoters are associated with transcription of ncRNA around few hundred nucleotides of (TSS) transcription start site, which may explain the retention of H3K4me3 mark even on 'poised promoters'. These ncRNAs in turn recruit PRC2 in *cis* for repression state. However certain line of evidence point that TSS ncRNA runs through and dislodges repressive complex and facilitates recruitment of ASH1 (trxG) complex by its direct binding, which further results in release of elongation block for RNA PolII. The PcG/trxG switch needs transcription through corresponding PRE/TRE to overcome silencing (Schmitt et al. 2005). In vertebrates PRE/TRE are not defined, however RA induced transcription of ncRNA through PcG sites has been observed with concomitant loss of SUZ12 and EZH2 from target regions (Sessa et al. 2007) to switch on active state. Like in flies this ncRNA transcription precedes hox expression but ectopically increasing the ncRNA does not have any effect on hox gene expression, suggesting that transcription *per se* is important rather than the stoichiometry.

More than 200 long non-coding RNA (lncRNA) are transcribed through four human hox clusters. One of the lncRNA transcribed from HoxC cluster is 2.2 kb long, named as HOTAIR and targets HoxD cluster (Rinn et al. 2007; Woo and Kingston 2007). Depletion of HOTAIR results in loss of H3K27me3 in HoxD cluster and hence de-repression of many hoxd genes while there is no effect on HoxC genes.

HOTAIR directly binds to PRC2 complex and hence acts as a structural platform to target silencing machinery on HoxD (Tsai et al. 2010). This kind of trans-regulation by ncRNA raises more questions than answers got from ncRNA for cis-regulation where the transcription of ncRNA coincides with its PcG/trxG recruitment. It is not yet known whether recognition of ncRNA by PcG is dependent on primary sequence or on the structural information of ncRNA. Recently a study has shown long non-coding RNAs to be identified having enhancer like activation function (Orom et al. 2010). Some of these long non-coding RNAs have typically H3K4me3 at their start site while others have H3K4me1, both having H3K36 methylation downstream, which is hallmark of transcriptional elongation. Since ectopic expression of these RNAs up-regulates gene expression, it is speculated that TSS of these RNAs can serve as *cis*-enhancers or PRE/TRE domain for the corresponding coding gene.

9.6.7 PcG/trxG Memory

The expression state decided by PcG/trxG is transmitted through DNA replication and cell divisions. The transmission of this epigenetic memory information resides within the chromatin structure itself. That is why several kinds of epigenomes exist in a single animal body with liver cells always giving rise to cells having liver epigenomes and skin cells to cells with skin epigenome after every cell division. The maintenance of cellular memory module is done by PcG/trxG complexes and therefore crucial for differentiation and embryogenesis.

Evidences support the idea that members of PcG/trxG sustain through DNA replication and as a matter of fact their mark do not get 'erased'. PcG bound chromatin even after replication *in vitro* retains binding. trxG proteins like MLL1 and Trl-GAGA have been shown to be associated with mitotic condensed chromosomes and thus enables activation of a wave of transcription quickly upon mitotic exit (Bhat et al. 1996; Blobel et al. 2009). This model summarizes that H3K27me3 mark can also recruit PRC2 during G1 phase in proliferating cells and further methylates daughter strand and adjacent regions to recruit PRC1 silencing complex (Margueron et al. 2009).

One of the mechanisms of PcG silencing is via long-range chromatin looping to maintain transcriptionally poised domains (Bantignies et al. 2011; Lanzuolo et al. 2007). Loss of PcG disrupts this looping and thus leads to de-repression. A recent study shows that the loss of looping can also be due to loss of DNA methyltransferases (DNMT1/3A) (Tiwari et al. 2008). Epigenetically silenced CpG-hypermethylated promoters are known to be overlapping with domain of PcG binding in mammalian genome. It has been shown that promoters having histone H3 lysine 27 methylation are more frequently *de novo* methylated than other promoters (Mohn et al. 2008). On the contrary Dnmt3a dependent non-promoter methylation is shown necessary to counteract PcG silencing and thus maintains the active state of gene expression (Wu et al. 2010). However it is not clear whether DNA methylation that can be transmitted can serve as memory indicator for PcG silencing.

9.7 Epigenomic Constraints and Their Co-Option with Hox Genes for Complexity

Coding regions need appropriate epigenetic environment to perform their functions in a defined cell type. Upgradation in the number of germ layers from single layered poriferans to diploblastic cnidarians to triploblastic organisms is one of the prerequisite mechanisms in specifying cell type specific epigenomes. Increase in number of cell type specific epigenomes is the complexity incorporated in metazoan evolutionary inventory. Lower bilaterians have around 30 cell types, and as the complexity increased during evolution number of cell types also increased from 150 in amphibians to around 200 cell types in mammals irrespective of genome size (Denis and Lacroix 1993; Milinkovitch et al. 2010; Sarras et al. 2002). Certainly, restriction on number of cell types in lower organisms, poses **cell type epigenome constraint** for their evolvability. Increasing cell types effectively increased the number of epigenetic modules not only at the cellular level but also increased the inter-cellular interactions between different cell types. Several cell types have to be networked for a single functional module. For example removal of six somites in vertebrate embryo that give rise to forelimb, leads to abrogation in developing muscular connections and hence defective limb formation (Chevallier et al. 1978; Lee 1992; Lee and Chan 1991).

Many chromatin-mediated factors have duplicated and diverged along with the cell type complexity. Epigenetic regulators of homeotic genes Polycomb/trithorax is one such group that has expanded in vertebrate lineage (Senthilkumar and Mishra 2009) and so have the cell types. Homeotic selector genes specify cell identity not as standalone rather need chromatin reorganization done by the cascade of upstream events. Since in vertebrates, the axis elongation and segment specifications go hand in hand, the role of PcG/trxG in conjunction with homeotic genes is more important. Similarly the members of the family of Wnts, Fgfs, BMPs, RARs (Dale and Jones 1999; Dickinson and McMahon 1992; Hogan 1996; Katsube et al. 2009; Moon et al. 1997; Slack 1990; Yamaguchi and Rossant 1995) and many other important development regulators have increased and diversified in more complex organisms.

9.8 Chromatin Organization of Hox Complex in Retrospect

Primary sequence of DNA imparts information to carry out much of the basic cellular functions, such that 'naked DNA' is adequate for basic transcription machinery along with some co-factors to transcribe it into RNA. Only 'transcription' of coding region cannot suffice for the sustenance of cellular system that needs higher order of regulated expression of its coding region. The same DNA sequence can project variety of functionalities depending upon how it is packaged, in other words how it is 'chromatinized'. Such is the profound effect of the chromatin on gene expression that even neighboring state of chromatin matters. Before gene expression to take place it is important to prepare its 'chromatin state' and at the same time to ward it off from the neighboring not so congenial chromatin environment in a most efficient way possible. For example an active gene can be repressed by positioning it near to the inactive chromatin (heterochromatin).

147 bp of linear DNA is wrapped around per nucelosome/histone octamer maintaining at least 14 distinct contact points (Ebralidse et al. 1988; Lambert and Thomas 1986). Despite being dynamic with high turnover, this packaging is energetically most favored and most stable DNA-protein interaction in cellular conditions. The degree of compaction of a nucleosome decorated DNA is a function of histone modifications it carries. There are several post-translational modifications of protruding histone tails that are known to effect the compaction of DNA and hence associated with transcriptionally *active* or *inactive* chromatin marks. Most of these modifications are localized upstream of the coding sequence directly modulating the gene expression. In co-expressed genes histone acetylations (active mark) in one gene are speculated to trigger or spread to the effector genes. However histone methylations are meticulously regulated requiring specific DNA sequence and specific catalytic enzymes resulting in critical structural changes in the chromatin by further recruitment of chromatin complexes (Suganuma and Workman 2008). Acetylated histones relax the DNA compaction and hence facilitate recruitment of co-activators for transcription. The RNA PolII initiation and elongation phase are marked by its Ser-5-CTD and Ser-2-CTD phosphorylation respectively. PAF complex is loaded onto Ser-5 CTD and is central to the recruitment of many Ser-5-CTD associated complexes. PAF is essential for H2B ubiquitination, which is prior to the recruitment of H3K4 di-tri-methyltransferase, Set1 required for elongation (Ng et al. 2003). MLL complex in mammals is essential for converting H3K4me2 into H3K4me3, the mark enriched in the promoter regions of the actively transcribing genes (Wysocka et al. 2005). RNAPolII can run through the nucleosome-DNA complex if at least one H2A/H2B dimer is evicted. ATP dependent remodeling complex RSC, which has multiple bromodomain factors facilitates the eviction of H2A/H2B dimer from the paused RNAPolII sites and FACT histone chaperone complex takes care of the assembly and disassembly of histones during elongation (Belotserkovskaya et al. 2003; Saunders et al. 2003; Workman and Kingston 1998).

Chromatin needs to be removed off the repressive marks and consecutively marked with active histone modifications to come closer to the possibility of being transcribed, as it has been described above that several other factors associated directly with the RNAPolII machinery have to fall in place for productive gene expression. 'Opening up' of HoxD cluster is preceded by the removal of H3K27me3 mark in a temporally collinear fashion and replacement by acetylated histone3 (Soshnikova and Duboule 2009). This temporal opening of HoxD cluster gets disturbed upon gross genomic rearrangement of the hox locus (Fig. 9.3). The de-condensation and opening up of hoxb1 gene upon induction (RA) allows it to 'break free' from its repressive environment to get expressed (Chambeyron and Bickmore 2004a; Chambeyron et al. 2005; Morey et al. 2007, 2008), while hoxb9 is still retained in the inactive region of the chromosomal territory. The looping out of co-expressed genes has been documented in beta-globin clusters also, where LCR induces the looping out of the genes followed by their expression (Chambeyron and Bickmore 2004b). Relocating beta-globin LCR on another chromosome leads to high expression of genes in the vicinity along with the GFP reporter targeting them to so called RNAPolII speckles or transcription factories (Ragoczy et al. 2006). With the emerging idea of the nucleus being functionally compartmentalized, it has been shown that elongating RNAPolII foci are discrete and distinct than poised RNApoII foci. Within the cluster the looping out of genes is thought to target them to different RNAPolII foci and thus decides the transcription state of the gene (Sutherland and Bickmore 2009).

After induction of hox genes how is chromatin maintained? The answer to this question is still poorly understood. However, 3C based study of human testicular cell lines revealed that resting hox clusters are maintained in distinct loops having contact points within the cluster (Ferraiuolo et al. 2010). Similar looping within the cluster is also observed in tissues expressing beta-globin locus (Tolhuis et al. 2002). Few DNA-FISH studies in *Drosophila* have given an idea that long-range interaction within BX-C is necessary for correct expression of hox genes. Distinct loop interactions were found in different anterior-posterior compartments in the same embryo (Lanzuolo et al. 2007), where expression of hox gene requires it to leave the repressive polycomb bodies. In the head region where both *Ubx* and *Abd-B* repressed, bxd-PRE and Fab7-PRE colocalize with PC body, while in abdomen region they do not. These interactions are abrogated in PC mutations suggesting the requirement of PcG for 'loop maintenance' (Bantignies et al. 2003). Another study reported that insulators of *Mcp* and *Fab7* are required rather than PRE for this long-range interaction in transgenic context (Li et al. 2011).

9.9 Summary and Future Directions

Since past few years understanding of the mechanism of coordinated gene expression of clustered genes is developing in terms of chromatin structure. Emerging scenario for the regulation of homeotic complexes portrays the modular entities as having interdependence for their function. Modular elements like enhancers, boundaries, PRE/TRE, RARE are identified separately in transgenic or deletion assays, but when sewn together, in a typical fashion modulate chromatin in the most desirable manner. It is not necessary that all of these regulatory elements have to be functional along A-P axis, but one's functionality may ensure how the other element would behave in a segment specific manner (Fig. 9.3). The initial opening up of the vertebrate hox cluster is dependent on global mechanism while fine-tuning of the cluster and maintenance may need several modules apposed together. Since vertebrate embryo grows in time and space, the initiation/establishment and maintenance of the complex takes place simultaneously. Contribution of epigenetic mechanisms has emerged as the key factor in this complex regulatory process.

Homeotic gene cluster, specially the one in *Drosophila melanogaster*, has served as an excellent model system for genetic screens and discovery of regulatory mechanisms involved in the genetic basis of animal body plan (Lewis 1978). This and similar, but more extensive genetic screens that followed soon after (Nusslein-Volhard and Wieschaus 1980) have led to the discovery of molecular

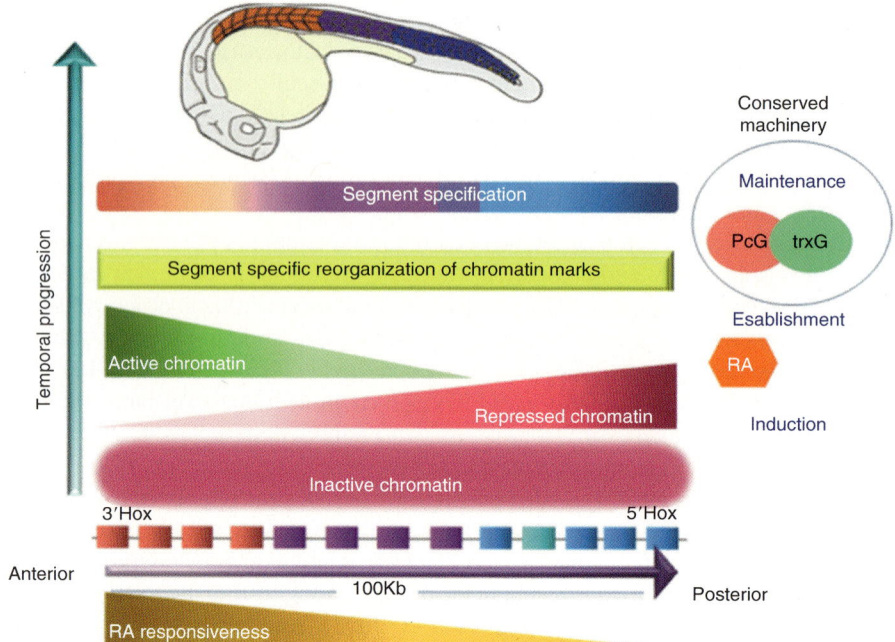

Fig. 9.3 Chromatin organization of hox genes during development. Schematic representation of vertebrate Hox complex is shown. Initially during embryonic stages hox locus is silenced and all the hox genes are in repressed condition. As development proceeds (from *top* to *bottom* of the figure), the genes get expressed one after the other in the order of chromosomal location. 3′ hox genes in the anterior part of the body have active chromatin marks and 5′ hox genes have repressed chromatin. Induction of hox expression is controlled by retinoic acid signaling in vertebrates. Anterior genes are highly responsive or sensitive to retinoic acid induction than posterior genes and hence are expressed earlier even at lower doses of retinoic acid. Once the hox gene expression is induced, it is established and maintained by chromatin reorganization controlled by PcG/trxG proteins. The segment specific reorganization of chromatin marks determines the correct body patterning along A-P axis

components of embryonic development that have turned out to be amazingly conserved from flies to human. These historic screens and analyses paved the way to understand the molecular basis of embryonic development and led us to the point where the question that stares at us is – how the identical genome gives hundreds of cell type during embryonic development? Can we have 'epigenetic screens' to dissect out epigenetic details? One needs a way to isolate factors in cell type and temporally specified context as against the genetic screens that lead to the final phenotypes in an organism. While this appears to be an intractable problem, one way forward is to map the epigenetic profile of a cell type, in a temporally defined context. This and a 3D organization of the genome within the nuclear space of specific cells in a temporal context can give us the ultimate set of information to understand how the single genome gives multiple cell types.

Epigenetic mechanisms, by definition, act differentially on the identical genome in distinct cell types. This also means that cis-elements that recruit or guide epigenetic modifications of specific loci are interpreted differently by recruitment factors in different cell types. Identification of such 'Epi-cis-elements' of the genome will be a step forward to understand how this regulation of highest dimension and dynamics has evolved to take genetic information to the level of sophistication and complexity as is seen in organisms of such a vast variety.

References

Agger K, Cloos PA, Christensen J, Pasini D, Rose S, Rappsilber J, Issaeva I, Canaani E, Salcini AE, Helin K (2007) UTX and JMJD3 are histone H3K27 demethylases involved in HOX gene regulation and development. Nature 449:731–734

Bantignies F, Grimaud C, Lavrov S, Gabut M, Cavalli G (2003) Inheritance of Polycomb-dependent chromosomal interactions in Drosophila. Genes Dev 17:2406–2420

Bantignies F, Roure V, Comet I, Leblanc B, Schuettengruber B, Bonnet J, Tixier V, Mas A, Cavalli G (2011) Polycomb-dependent regulatory contacts between distant Hox loci in Drosophila. Cell 144:214–226

Beisel C, Paro R (2011) Silencing chromatin: comparing modes and mechanisms. Nat Rev Genet 12:123–135

Belotserkovskaya R, Oh S, Bondarenko VA, Orphanides G, Studitsky VM, Reinberg D (2003) FACT facilitates transcription-dependent nucleosome alteration. Science 301:1090–1093

Bender W, Akam M, Karch F, Beachy PA, Peifer M, Spierer P, Lewis EB, Hogness DS (1983) Molecular genetics of the bithorax complex in Drosophila melanogaster. Science 221:23–29

Bernstein BE, Mikkelsen TS, Xie X, Kamal M, Huebert DJ, Cuff J, Fry B, Meissner A, Wernig M, Plath K, Jaenisch R, Wagschal A, Feil R, Schreiber SL, Lander ES (2006) A bivalent chromatin structure marks key developmental genes in embryonic stem cells. Cell 125:315–326

Bhat KM, Farkas G, Karch F, Gyurkovics H, Gausz J, Schedl P (1996) The GAGA factor is required in the early Drosophila embryo not only for transcriptional regulation but also for nuclear division. Development 122:1113–1124

Blobel GA, Kadauke S, Wang E, Lau AW, Zuber J, Chou MM, Vakoc CR (2009) A reconfigured pattern of MLL occupancy within mitotic chromatin promotes rapid transcriptional reactivation following mitotic exit. Mol Cell 36:970–983

Boulet AM, Lloyd A, Sakonju S (1991) Molecular definition of the morphogenetic and regulatory functions and the cis-regulatory elements of the Drosophila Abd-B homeotic gene. Development 111:393–405

Boyer LA, Plath K, Zeitlinger J, Brambrink T, Medeiros LA, Lee TI, Levine SS, Wernig M, Tajonar A, Ray MK, Bell GW, Otte AP, Vidal M, Gifford DK, Young RA, Jaenisch R (2006) Polycomb complexes repress developmental regulators in murine embryonic stem cells. Nature 441:349–353

Bracken AP, Dietrich N, Pasini D, Hansen KH, Helin K (2006) Genome-wide mapping of Polycomb target genes unravels their roles in cell fate transitions. Genes Dev 20:1123–1136

Buxton PG, Kostakopoulou K, Brickell P, Thorogood P, Ferretti P (1997) Expression of the transcription factor slug correlates with growth of the limb bud and is regulated by FGF-4 and retinoic acid. Int J Dev Biol 41:559–568

Capelson M, Corces VG (2004) Boundary elements and nuclear organization. Biol Cell 96:617–629

Caretti G, Di Padova M, Micales B, Lyons GE, Sartorelli V (2004) The Polycomb Ezh2 methyltransferase regulates muscle gene expression and skeletal muscle differentiation. Genes Dev 18:2627–2638

Carroll SB (1995) Homeotic genes and the evolution of arthropods and chordates. Nature 376:479
Casanova M, Preissner T, Cerase A, Poot R, Yamada D, Li X, Appanah R, Bezstarosti K, Demmers J, Koseki H, Brockdorff N (2011) Polycomblike 2 facilitates the recruitment of PRC2 Polycomb group complexes to the inactive X chromosome and to target loci in embryonic stem cells. Development 138:1471–1482
Celniker SE, Keelan DJ, Lewis EB (1989) The molecular genetics of the bithorax complex of Drosophila: characterization of the products of the Abdominal-B domain. Genes Dev 3:1424–1436
Celniker SE, Sharma S, Keelan DJ, Lewis EB (1990) The molecular genetics of the bithorax complex of Drosophila: cis-regulation in the Abdominal-B domain. EMBO J 9:4277–4286
Chambeyron S, Bickmore WA (2004a) Chromatin decondensation and nuclear reorganization of the HoxB locus upon induction of transcription. Genes Dev 18:1119–1130
Chambeyron S, Bickmore WA (2004b) Does looping and clustering in the nucleus regulate gene expression? Curr Opin Cell Biol 16:256–262
Chambeyron S, Da Silva NR, Lawson KA, Bickmore WA (2005) Nuclear re-organisation of the Hoxb complex during mouse embryonic development. Development 132:2215–2223
Chan CS, Rastelli L, Pirrotta V (1994) A Polycomb response element in the Ubx gene that determines an epigenetically inherited state of repression. EMBO J 13:2553–2564
Chen SK, Tvrdik P, Peden E, Cho S, Wu S, Spangrude G, Capecchi MR (2010) Hematopoietic origin of pathological grooming in Hoxb8 mutant mice. Cell 141:775–785
Chevallier A, Kieny M, Mauger A (1978) Limb-somite relationship: effect of removal of somitic mesoderm on the wing musculature. J Embryol Exp Morphol 43:263–278
Chopra VS, Hong JW, Levine M (2009) Regulation of Hox gene activity by transcriptional elongation in Drosophila. Curr Biol 19:688–693
Christen B, Bienz M (1994) Imaginal disc silencers from Ultrabithorax: evidence for Polycomb response elements. Mech Dev 48:255–266
Chung JH, Whiteley M, Felsenfeld G (1993) A 5′ element of the chicken beta-globin domain serves as an insulator in human erythroid cells and protects against position effect in Drosophila. Cell 74:505–514
Conlon RA (1995) Retinoic acid and pattern formation in vertebrates. Trends Genet 11:314–319
Dahanukar A, Wharton RP (1996) The Nanos gradient in Drosophila embryos is generated by translational regulation. Genes Dev 10:2610–2620
Dale L, Jones CM (1999) BMP signalling in early Xenopus development. Bioessays 21:751–760
Dekker EJ, Pannese M, Houtzager E, Boncinelli E, Durston A (1993) Colinearity in the Xenopus laevis Hox-2 complex. Mech Dev 40:3–12
Denis H, Lacroix JC (1993) The dichotomy between germ line and somatic line, and the origin of cell mortality. Trends Genet 9:7–11
Dickinson ME, McMahon AP (1992) The role of Wnt genes in vertebrate development. Curr Opin Genet Dev 2:562–566
Dorsett D (1999) Distant liaisons: long-range enhancer-promoter interactions in Drosophila. Curr Opin Genet Dev 9:505–514
Driever W, Nusslein-Volhard C (1988a) A gradient of bicoid protein in Drosophila embryos. Cell 54:83–93
Driever W, Nusslein-Volhard C (1988b) The bicoid protein determines position in the Drosophila embryo in a concentration-dependent manner. Cell 54:95–104
Duboule D (1994) Temporal colinearity and the phylotypic progression: a basis for the stability of a vertebrate Bauplan and the evolution of morphologies through heterochrony. Development Suppl:135–142
Duboule D (2007) The rise and fall of Hox gene clusters. Development 134:2549–2560
Duncan I (1987) The bithorax complex. Annu Rev Genet 21:285–319
Ebralidse KK, Grachev SA, Mirzabekov AD (1988) A highly basic histone H4 domain bound to the sharply bent region of nucleosomal DNA. Nature 331:365–367

Endoh M, Endo TA, Endoh T, Fujimura Y, Ohara O, Toyoda T, Otte AP, Okano M, Brockdorff N, Vidal M, Koseki H (2008) Polycomb group proteins Ring1A/B are functionally linked to the core transcriptional regulatory circuitry to maintain ES cell identity. Development 135:1513–1524

Ferraiuolo MA, Rousseau M, Miyamoto C, Shenker S, Wang XQ, Nadler M, Blanchette M, Dostie J (2010) The three-dimensional architecture of Hox cluster silencing. Nucleic Acids Res 38:7472–7484

Flam F (1994) Hints of a language in junk DNA. Science 266:1320

Fromental-Ramain C, Warot X, Messadecq N, LeMeur M, Dolle P, Chambon P (1996) Hoxa-13 and Hoxd-13 play a crucial role in the patterning of the limb autopod. Development 122:2997–3011

Gaszner M, Felsenfeld G (2006) Insulators: exploiting transcriptional and epigenetic mechanisms. Nat Rev Genet 7:703–713

Gaunt SJ (1994) Conservation in the Hox code during morphological evolution. Int J Dev Biol 38:549–552

Gaunt SJ, Strachan L (1996) Temporal colinearity in expression of anterior Hox genes in developing chick embryos. Dev Dyn 207:270–280

Gause M, Morcillo P, Dorsett D (2001) Insulation of enhancer-promoter communication by a gypsy transposon insert in the Drosophila cut gene: cooperation between suppressor of hairy-wing and modifier of mdg4 proteins. Mol Cell Biol 21:4807–4817

Gehring WJ, Hiromi Y (1986) Homeotic genes and the homeobox. Annu Rev Genet 20:147–173

Gerard M, Zakany J, Duboule D (1997) Interspecies exchange of a Hoxd enhancer in vivo induces premature transcription and anterior shift of the sacrum. Dev Biol 190:32–40

Gerasimova TI, Corces VG (2001) Chromatin insulators and boundaries: effects on transcription and nuclear organization. Annu Rev Genet 35:193–208

Gillespie RF, Gudas LJ (2007) Retinoid regulated association of transcriptional co-regulators and the polycomb group protein SUZ12 with the retinoic acid response elements of Hoxa1, RARbeta(2), and Cyp26A1 in F9 embryonal carcinoma cells. J Mol Biol 372:298–316

Glass CK, Rosenfeld MG (2000) The coregulator exchange in transcriptional functions of nuclear receptors. Genes Dev 14:121–141

Gohl D, Aoki T, Blanton J, Shanower G, Kappes G, Schedl P (2011) Mechanism of chromosomal boundary action: roadblock, sink, or loop? Genetics 187:731–748

Goodman F, Giovannucci-Uzielli ML, Hall C, Reardon W, Winter R, Scambler P (1998) Deletions in HOXD13 segregate with an identical, novel foot malformation in two unrelated families. Am J Hum Genet 63:992–1000

Gould A, Morrison A, Sproat G, White RA, Krumlauf R (1997) Positive cross-regulation and enhancer sharing: two mechanisms for specifying overlapping Hox expression patterns. Genes Dev 11:900–913

Guenther MG, Levine SS, Boyer LA, Jaenisch R, Young RA (2007) A chromatin landmark and transcription initiation at most promoters in human cells. Cell 130:77–88

Hanson RD, Hess JL, Yu BD, Ernst P, van Lohuizen M, Berns A, van der Lugt NM, Shashikant CS, Ruddle FH, Seto M, Korsmeyer SJ (1999) Mammalian Trithorax and polycomb-group homologues are antagonistic regulators of homeotic development. Proc Natl Acad Sci USA 96:14372–14377

Herault Y, Beckers J, Gerard M, Duboule D (1999) Hox gene expression in limbs: colinearity by opposite regulatory controls. Dev Biol 208:157–165

Hernandez RE, Putzke AP, Myers JP, Margaretha L, Moens CB (2007) Cyp26 enzymes generate the retinoic acid response pattern necessary for hindbrain development. Development 134:177–187

Hogan BL (1996) Bone morphogenetic proteins in development. Curr Opin Genet Dev 6:432–438

Holland PW, Holland LZ, Williams NA, Holland ND (1992) An amphioxus homeobox gene: sequence conservation, spatial expression during development and insights into vertebrate evolution. Development 116:653–661

Holstege JC, de Graaff W, Hossaini M, Cano SC, Jaarsma D, van den Akker E, Deschamps J (2008) Loss of Hoxb8 alters spinal dorsal laminae and sensory responses in mice. Proc Natl Acad Sci U S A 105:6338–6343

Horton C, Maden M (1995) Endogenous distribution of retinoids during normal development and teratogenesis in the mouse embryo. Dev Dyn 202:312–323

Hsieh JJ, Cheng EH, Korsmeyer SJ (2003) Taspase1: a threonine aspartase required for cleavage of MLL and proper HOX gene expression. Cell 115:293–303

Hughes CM, Rozenblatt-Rosen O, Milne TA, Copeland TD, Levine SS, Lee JC, Hayes DN, Shanmugam KS, Bhattacharjee A, Biondi CA, Kay GF, Hayward NK, Hess JL, Meyerson M (2004) Menin associates with a trithorax family histone methyltransferase complex and with the hoxc8 locus. Mol Cell 13:587–597

Iimura T, Denans N, Pourquie O (2009) Establishment of Hox vertebral identities in the embryonic spine precursors. Curr Top Dev Biol 88:201–234

Iqbal H, Mishra R (2007) Chromatin domain boundaries: defining the functional domains in genome. Proc Indian Natl Sci Acad 73:239–253

Jepsen K, Solum D, Zhou T, McEvilly RJ, Kim HJ, Glass CK, Hermanson O, Rosenfeld MG (2007) SMRT-mediated repression of an H3K27 demethylase in progression from neural stem cell to neuron. Nature 450:415–419

Joo HY, Zhai L, Yang C, Nie S, Erdjument-Bromage H, Tempst P, Chang C, Wang H (2007) Regulation of cell cycle progression and gene expression by H2A deubiquitination. Nature 449:1068–1072

Juan AH, Ruddle FH (2003) Enhancer timing of Hox gene expression: deletion of the endogenous Hoxc8 early enhancer. Development 130:4823–4834

Karch F, Weiffenbach B, Peifer M, Bender W, Duncan I, Celniker S, Crosby M, Lewis EB (1985) The abdominal region of the bithorax complex. Cell 43:81–96

Karch F, Galloni M, Sipos L, Gausz J, Gyurkovics H, Schedl P (1994) Mcp and Fab-7: molecular analysis of putative boundaries of cis-regulatory domains in the bithorax complex of *Drosophila melanogaster*. Nucleic Acids Res 22:3138–3146

Kashyap V, Gudas LJ, Brenet F, Funk P, Viale A, Scandura JM (2011) Epigenomic reorganization of the clustered Hox genes in embryonic stem cells induced by retinoic acid. J Biol Chem 286:3250–3260

Kastner P, Mark M, Chambon P (1995) Nonsteroid nuclear receptors: what are genetic studies telling us about their role in real life? Cell 83:859–869

Katsube K, Sakamoto K, Tamamura Y, Yamaguchi A (2009) Role of CCN, a vertebrate specific gene family, in development. Dev Growth Differ 51:55–67

Kennison JA (1995) The Polycomb and trithorax group proteins of Drosophila: trans-regulators of homeotic gene function. Annu Rev Genet 29:289–303

Kessel M, Gruss P (1991) Homeotic transformations of murine vertebrae and concomitant alteration of Hox codes induced by retinoic acid. Cell 67:89–104

Klebes A, Sustar A, Kechris K, Li H, Schubiger G, Kornberg TB (2005) Regulation of cellular plasticity in Drosophila imaginal disc cells by the Polycomb group, trithorax group and lama genes. Development 132:3753–3765

Kmita M, van Der Hoeven F, Zakany J, Krumlauf R, Duboule D (2000) Mechanisms of Hox gene colinearity: transposition of the anterior Hoxb1 gene into the posterior HoxD complex. Genes Dev 14:198–211

Kmita M, Fraudeau N, Herault Y, Duboule D (2002a) Serial deletions and duplications suggest a mechanism for the collinearity of Hoxd genes in limbs. Nature 420:145–150

Kmita M, Tarchini B, Duboule D, Herault Y (2002b) Evolutionary conserved sequences are required for the insulation of the vertebrate Hoxd complex in neural cells. Development 129:5521–5528

Kondo T, Duboule D (1999) Breaking colinearity in the mouse HoxD complex. Cell 97:407–417

Kondo T, Dolle P, Zakany J, Duboule D (1996) Function of posterior HoxD genes in the morphogenesis of the anal sphincter. Development 122:2651–2659

Kondo T, Zakany J, Duboule D (1998) Control of colinearity in AbdB genes of the mouse HoxD complex. Mol Cell 1:289–300

Krumlauf R (1992) Evolution of the vertebrate Hox homeobox genes. Bioessays 14:245–252
Krumlauf R (1994) Hox genes in vertebrate development. Cell 78:191–201
Ku M, Koche RP, Rheinbay E, Mendenhall EM, Endoh M, Mikkelsen TS, Presser A, Nusbaum C, Xie X, Chi AS, Adli M, Kasif S, Ptaszek LM, Cowan CA, Lander ES, Koseki H, Bernstein BE (2008) Genomewide analysis of PRC1 and PRC2 occupancy identifies two classes of bivalent domains. PLoS Genet 4:e1000242
Kudoh T, Wilson SW, Dawid IB (2002) Distinct roles for Fgf, Wnt and retinoic acid in posteriorizing the neural ectoderm. Development 129:4335–4346
Kuhn EJ, Geyer PK (2003) Genomic insulators: connecting properties to mechanism. Curr Opin Cell Biol 15:259–265
Lagarou A, Mohd-Sarip A, Moshkin YM, Chalkley GE, Bezstarosti K, Demmers JA, Verrijzer CP (2008) dKDM2 couples histone H2A ubiquitylation to histone H3 demethylation during Polycomb group silencing. Genes Dev 22:2799–2810
Lambert SF, Thomas JO (1986) Lysine-containing DNA-binding regions on the surface of the histone octamer in the nucleosome core particle. Eur J Biochem 160:191–201
Lan F, Bayliss PE, Rinn JL, Whetstine JR, Wang JK, Chen S, Iwase S, Alpatov R, Issaeva I, Canaani E, Roberts TM, Chang HY, Shi Y (2007) A histone H3 lysine 27 demethylase regulates animal posterior development. Nature 449:689–694
Landeira D, Sauer S, Poot R, Dvorkina M, Mazzarella L, Jorgensen HF, Pereira CF, Leleu M, Piccolo FM, Spivakov M, Brookes E, Pombo A, Fisher C, Skarnes WC, Snoek T, Bezstarosti K, Demmers J, Klose RJ, Casanova M, Tavares L, Brockdorff N, Merkenschlager M, Fisher AG (2010) Jarid2 is a PRC2 component in embryonic stem cells required for multi-lineage differentiation and recruitment of PRC1 and RNA Polymerase II to developmental regulators. Nat Cell Biol 12:618–624
Lanzuolo C, Roure V, Dekker J, Bantignies F, Orlando V (2007) Polycomb response elements mediate the formation of chromosome higher-order structures in the bithorax complex. Nat Cell Biol 9:1167–1174
Lee KK (1992) The regulative potential of the limb region in 11.5-day rat embryos following the amputation of the fore-limb bud. Anat Embryol (Berl) 186:67–74
Lee KK, Chan WY (1991) A study on the regenerative potential of partially excised mouse embryonic fore-limb bud. Anat Embryol (Berl) 184:153–157
Lee N, Maurange C, Ringrose L, Paro R (2005) Suppression of Polycomb group proteins by JNK signalling induces transdetermination in Drosophila imaginal discs. Nature 438:234–237
Lee MG, Villa R, Trojer P, Norman J, Yan KP, Reinberg D, Di Croce L, Shiekhattar R (2007) Demethylation of H3K27 regulates polycomb recruitment and H2A ubiquitination. Science 318:447–450
Leid M, Kastner P, Chambon P (1992) Multiplicity generates diversity in the retinoic acid signalling pathways. Trends Biochem Sci 17:427–433
Levine M, Hoey T (1988) Homeobox proteins as sequence-specific transcription factors. Cell 55:537–540
Lewis EB (1978) A gene complex controlling segmentation in Drosophila. Nature 276:565–570
Li G, Margueron R, Ku M, Chambon P, Bernstein BE, Reinberg D (2010) Jarid2 and PRC2, partners in regulating gene expression. Genes Dev 24:368–380
Li HB, Muller M, Bahechar IA, Kyrchanova O, Ohno K, Georgiev P, Pirrotta V (2011) Insulators, not Polycomb response elements, are required for long-range interactions between Polycomb targets in *Drosophila melanogaster*. Mol Cell Biol 31:616–625
Lipshitz HD, Peattie DA, Hogness DS (1987) Novel transcripts from the Ultrabithorax domain of the bithorax complex. Genes Dev 1:307–322
Lutz B, Lu HC, Eichele G, Miller D, Kaufman TC (1996) Rescue of Drosophila labial null mutant by the chicken ortholog Hoxb-1 demonstrates that the function of Hox genes is phylogenetically conserved. Genes Dev 10:176–184
Maconochie M, Nonchev S, Morrison A, Krumlauf R (1996) Paralogous Hox genes: function and regulation. Annu Rev Genet 30:529–556
Maden M, Ong DE, Summerbell D, Chytil F (1988) Spatial distribution of cellular protein binding to retinoic acid in the chick limb bud. Nature 335:733–735

Maeda RK, Karch F (2006) The ABC of the BX-C: the bithorax complex explained. Development 133:1413

Margueron R, Justin N, Ohno K, Sharpe ML, Son J, Drury WJ 3rd, Voigt P, Martin SR, Taylor WR, De Marco V, Pirrotta V, Reinberg D, Gamblin SJ (2009) Role of the polycomb protein EED in the propagation of repressive histone marks. Nature 461:762–767

Mark M, Ghyselinck NB, Chambon P (2004) Retinoic acid signalling in the development of branchial arches. Curr Opin Genet Dev 14:591–598

Mark M, Ghyselinck NB, Chambon P (2006) Function of retinoid nuclear receptors: lessons from genetic and pharmacological dissections of the retinoic acid signaling pathway during mouse embryogenesis. Annu Rev Pharmacol Toxicol 46:451–480

Mark M, Ghyselinck NB, Chambon P (2009) Function of retinoic acid receptors during embryonic development. Nucl Recept Signal 7:e002

Martin C, Cao R, Zhang Y (2006) Substrate preferences of the EZH2 histone methyltransferase complex. J Biol Chem 281:8365–8370

Matharu NK, Hussain T, Sankaranarayanan R, Mishra RK (2010) Vertebrate homologue of Drosophila GAGA factor. J Mol Biol 400:434–447

Maurange C, Lee N, Paro R (2006) Signaling meets chromatin during tissue regeneration in Drosophila. Curr Opin Genet Dev 16:485–489

McGinnis W, Krumlauf R (1992) Homeobox genes and axial patterning. Cell 68:283–302

Mihaly J, Hogga I, Barges S, Galloni M, Mishra RK, Hagstrom K, Muller M, Schedl P, Sipos L, Gausz J, Gyurkovics H, Karch F (1998) Chromatin domain boundaries in the Bithorax complex. Cell Mol Life Sci 54:60–70

Milinkovitch MC, Helaers R, Tzika AC (2010) Historical constraints on vertebrate genome evolution. Genome Biol Evol 2:13–18

Minsky N, Shema E, Field Y, Schuster M, Segal E, Oren M (2008) Monoubiquitinated H2B is associated with the transcribed region of highly expressed genes in human cells. Nat Cell Biol 10:483–488

Mishra RK, Karch F (1999) Boundaries that demarcate structural and functional domains of chromatin. J Biosci 24:377–399

Mishra RK, Yamagishi T, Vasanthi D, Ohtsuka C, Kondo T (2007) Involvement of polycomb-group genes in establishing HoxD temporal colinearity. Genesis 45:570–576

Mohd-Sarip A, van der Knaap JA, Wyman C, Kanaar R, Schedl P, Verrijzer CP (2006) Architecture of a polycomb nucleoprotein complex. Mol Cell 24:91–100

Mohn F, Weber M, Rebhan M, Roloff TC, Richter J, Stadler MB, Bibel M, Schubeler D (2008) Lineage-specific polycomb targets and de novo DNA methylation define restriction and potential of neuronal progenitors. Mol Cell 30:755–766

Moon RT, Brown JD, Torres M (1997) WNTs modulate cell fate and behavior during vertebrate development. Trends Genet 13:157–162

Moon H, Filippova G, Loukinov D, Pugacheva E, Chen Q, Smith ST, Munhall A, Grewe B, Bartkuhn M, Arnold R, Burke LJ, Renkawitz-Pohl R, Ohlsson R, Zhou J, Renkawitz R, Lobanenkov V (2005) CTCF is conserved from Drosophila to humans and confers enhancer blocking of the Fab-8 insulator. EMBO Rep 6:165–170

Morey C, Da Silva NR, Perry P, Bickmore WA (2007) Nuclear reorganisation and chromatin decondensation are conserved, but distinct, mechanisms linked to Hox gene activation. Development 134:909–919

Morey C, Da Silva NR, Kmita M, Duboule D, Bickmore WA (2008) Ectopic nuclear reorganisation driven by a Hoxb1 transgene transposed into Hoxd. J Cell Sci 121:571–577

Muragaki Y, Mundlos S, Upton J, Olsen BR (1996) Altered growth and branching patterns in synpolydactyly caused by mutations in HOXD13. Science 272:548–551

Ng HH, Robert F, Young RA, Struhl K (2003) Targeted recruitment of Set1 histone methylase by elongating Pol II provides a localized mark and memory of recent transcriptional activity. Mol Cell 11:709–719

Nusslein-Volhard C, Wieschaus E (1980) Mutations affecting segment number and polarity in Drosophila. Nature 287:795–801

Ohtsuki S, Levine M (1998) GAGA mediates the enhancer blocking activity of the eve promoter in the Drosophila embryo. Genes Dev 12:3325–3330

Orom UA, Derrien T, Beringer M, Gumireddy K, Gardini A, Bussotti G, Lai F, Zytnicki M, Notredame C, Huang Q, Guigo R, Shiekhattar R (2010) Long noncoding RNAs with enhancer-like function in human cells. Cell 143:46–58

Pasini D, Hansen KH, Christensen J, Agger K, Cloos PA, Helin K (2008) Coordinated regulation of transcriptional repression by the RBP2 H3K4 demethylase and polycomb-repressive complex 2. Genes Dev 22:1345–1355

Pasini D, Cloos PA, Walfridsson J, Olsson L, Bukowski JP, Johansen JV, Bak M, Tommerup N, Rappsilber J, Helin K (2010) JARID2 regulates binding of the Polycomb repressive complex 2 to target genes in ES cells. Nature 464:306–310

Pathak RU, Rangaraj N, Kallappagoudar S, Mishra K, Mishra RK (2007) Boundary element-associated factor 32B connects chromatin domains to the nuclear matrix. Mol Cell Biol 27:4796–4806

Raab JR, Kamakaka RT (2010) Insulators and promoters: closer than we think. Nat Rev Genet 11:439–446

Ragoczy T, Bender MA, Telling A, Byron R, Groudine M (2006) The locus control region is required for association of the murine beta-globin locus with engaged transcription factories during erythroid maturation. Genes Dev 20:1447–1457

Richly H, Rocha-Viegas L, Ribeiro JD, Demajo S, Gundem G, Lopez-Bigas N, Nakagawa T, Rospert S, Ito T, Di Croce L (2010) Transcriptional activation of polycomb-repressed genes by ZRF1. Nature 468:1124–1128

Ringrose L, Paro R (2004) Epigenetic regulation of cellular memory by the Polycomb and Trithorax group proteins. Annu Rev Genet 38:413–443

Ringrose L, Rehmsmeier M, Dura JM, Paro R (2003) Genome-wide prediction of Polycomb/Trithorax response elements in *Drosophila melanogaster*. Dev Cell 5:759–771

Rinn JL, Kertesz M, Wang JK, Squazzo SL, Xu X, Brugmann SA, Goodnough LH, Helms JA, Farnham PJ, Segal E, Chang HY (2007) Functional demarcation of active and silent chromatin domains in human HOX loci by noncoding RNAs. Cell 129:1311–1323

Sarras MP Jr, Yan L, Leontovich A, Zhang JS (2002) Structure, expression, and developmental function of early divergent forms of metalloproteinases in hydra. Cell Res 12:163–176

Saunders A, Werner J, Andrulis ED, Nakayama T, Hirose S, Reinberg D, Lis JT (2003) Tracking FACT and the RNA polymerase II elongation complex through chromatin in vivo. Science 301:1094–1096

Savla U, Benes J, Zhang J, Jones RS (2008) Recruitment of Drosophila Polycomb-group proteins by Polycomblike, a component of a novel protein complex in larvae. Development 135:813–817

Scheuermann JC, de Ayala Alonso AG, Oktaba K, Ly-Hartig N, McGinty RK, Fraterman S, Wilm M, Muir TW, Muller J (2010) Histone H2A deubiquitinase activity of the Polycomb repressive complex PR-DUB. Nature 465:243–247

Schier AF, Talbot WS (2005) Molecular genetics of axis formation in zebrafish. Annu Rev Genet 39:561–613

Schmitt S, Prestel M, Paro R (2005) Intergenic transcription through a polycomb group response element counteracts silencing. Genes Dev 19:697–708

Schoeftner S, Sengupta AK, Kubicek S, Mechtler K, Spahn L, Koseki H, Jenuwein T, Wutz A (2006) Recruitment of PRC1 function at the initiation of X inactivation independent of PRC2 and silencing. EMBO J 25:3110–3122

Schwartz YB, Kahn TG, Nix DA, Li XY, Bourgon R, Biggin M, Pirrotta V (2006) Genome-wide analysis of Polycomb targets in *Drosophila melanogaster*. Nat Genet 38:700–705

Scott MP, Tamkun JW, Hartzell GW 3rd (1989) The structure and function of the homeodomain. Biochim Biophys Acta 989:25–48

Senthilkumar R, Mishra RK (2009) Novel motifs distinguish multiple homologues of Polycomb in vertebrates: expansion and diversification of the epigenetic toolkit. BMC Genomics 10:549

Sessa L, Breiling A, Lavorgna G, Silvestri L, Casari G, Orlando V (2007) Noncoding RNA synthesis and loss of Polycomb group repression accompanies the colinear activation of the human HOXA cluster. RNA 13:223–239

Sharpe J, Nonchev S, Gould A, Whiting J, Krumlauf R (1998) Selectivity, sharing and competitive interactions in the regulation of Hoxb genes. EMBO J 17:1788–1798

Shilatifard A (2006) Chromatin modifications by methylation and ubiquitination: implications in the regulation of gene expression. Annu Rev Biochem 75:243–269

Simon J, Chiang A, Bender W, Shimell MJ, O'Connor M (1993) Elements of the Drosophila bithorax complex that mediate repression by Polycomb group products. Dev Biol 158:131–144

Sing A, Pannell D, Karaiskakis A, Sturgeon K, Djabali M, Ellis J, Lipshitz HD, Cordes SP (2009) A vertebrate Polycomb response element governs segmentation of the posterior hindbrain. Cell 138:885–897

Slack JM (1990) Growth factors as inducing agents in early Xenopus development. J Cell Sci 13(Suppl):119–130

Soshnikova N, Duboule D (2009) Epigenetic temporal control of mouse Hox genes in vivo. Science 324:1320–1323

Stock JK, Giadrossi S, Casanova M, Brookes E, Vidal M, Koseki H, Brockdorff N, Fisher AG, Pombo A (2007) Ring1-mediated ubiquitination of H2A restrains poised RNA polymerase II at bivalent genes in mouse ES cells. Nat Cell Biol 9:1428–1435

Stratford T, Horton C, Maden M (1996) Retinoic acid is required for the initiation of outgrowth in the chick limb bud. Curr Biol 6:1124–1133

Suganuma T, Workman JL (2008) Crosstalk among histone modifications. Cell 135:604–607

Sutherland H, Bickmore WA (2009) Transcription factories: gene expression in unions? Nat Rev Genet 10:457–466

Tian D, Sun S, Lee JT (2010) The long noncoding RNA, Jpx, is a molecular switch for X chromosome inactivation. Cell 143:390–403

Tillib S, Petruk S, Sedkov Y, Kuzin A, Fujioka M, Goto T, Mazo A (1999) Trithorax- and Polycomb-group response elements within an Ultrabithorax transcription maintenance unit consist of closely situated but separable sequences. Mol Cell Biol 19:5189–5202

Tiwari VK, McGarvey KM, Licchesi JD, Ohm JE, Herman JG, Schubeler D, Baylin SB (2008) PcG proteins, DNA methylation, and gene repression by chromatin looping. PLoS Biol 6:2911–2927

Tolhuis B, Palstra RJ, Splinter E, Grosveld F, de Laat W (2002) Looping and interaction between hypersensitive sites in the active beta-globin locus. Mol Cell 10:1453–1465

Tsai MC, Manor O, Wan Y, Mosammaparast N, Wang JK, Lan F, Shi Y, Segal E, Chang HY (2010) Long noncoding RNA as modular scaffold of histone modification complexes. Science 329:689–693

Valenzuela L, Kamakaka RT (2006) Chromatin insulators. Annu Rev Genet 40:107–138

van den Akker E, Fromental-Ramain C, de Graaff W, Le Mouellic H, Brulet P, Chambon P, Deschamps J (2001) Axial skeletal patterning in mice lacking all paralogous group 8 Hox genes. Development 128:1911–1921

Vasanthi D, Mishra RK (2008) Epigenetic regulation of genes during development: a conserved theme from flies to mammals. J Genet Genomics 35:413–429

Vasanthi D, Anant M, Srivastava S, Mishra RK (2010) A functionally conserved boundary element from the mouse HoxD locus requires GAGA factor in Drosophila. Development 137:4239–4247

Vastenhouw NL, Zhang Y, Woods IG, Imam F, Regev A, Liu XS, Rinn J, Schier AF (2010) Chromatin signature of embryonic pluripotency is established during genome activation. Nature 464:922–926

Vazquez J, Farkas G, Gaszner M, Udvardy A, Muller M, Hagstrom K, Gyurkovics H, Sipos L, Gausz J, Galloni M et al (1993) Genetic and molecular analysis of chromatin domains. Cold Spring Harb Symp Quant Biol 58:45–54

Wan Y, Chiang JH, Lin CH, Arens CE, Saleem RA, Smith JJ, Aitchison JD (2010) Histone chaperone Chz1p regulates H2B ubiquitination and subtelomeric anti-silencing. Nucleic Acids Res 38:1431–1440

Wang H, Wang L, Erdjument-Bromage H, Vidal M, Tempst P, Jones RS, Zhang Y (2004) Role of histone H2A ubiquitination in Polycomb silencing. Nature 431:873–878

Wellik DM, Capecchi MR (2003) Hox10 and Hox11 genes are required to globally pattern the mammalian skeleton. Science 301:363–367

West AG, Gaszner M, Felsenfeld G (2002) Insulators: many functions, many mechanisms. Genes Dev 16:271–288

Woo CJ, Kingston RE (2007) HOTAIR lifts noncoding RNAs to new levels. Cell 129:1257–1259

Woo CJ, Kharchenko PV, Daheron L, Park PJ, Kingston RE (2010) A region of the human HOXD cluster that confers polycomb-group responsiveness. Cell 140:99–110

Workman JL, Kingston RE (1998) Alteration of nucleosome structure as a mechanism of transcriptional regulation. Annu Rev Biochem 67:545–579

Wu H, Coskun V, Tao J, Xie W, Ge W, Yoshikawa K, Li E, Zhang Y, Sun YE (2010) Dnmt3a-dependent nonpromoter DNA methylation facilitates transcription of neurogenic genes. Science 329:444–448

Wysocka J, Milne TA, Allis CD (2005) Taking LSD 1 to a new high. Cell 122:654–658

Yamagishi T, Ozawa M, Ohtsuka C, Ohyama-Goto R, Kondo T (2007) Evx2-Hoxd13 intergenic region restricts enhancer association to Hoxd13 promoter. PLoS One 2:e175

Yamaguchi TP, Rossant J (1995) Fibroblast growth factors in mammalian development. Curr Opin Genet Dev 5:485–491

Zakany J, Gerard M, Favier B, Duboule D (1997) Deletion of a HoxD enhancer induces transcriptional heterochrony leading to transposition of the sacrum. EMBO J 16:4393–4402

Zhao J, Ohsumi TK, Kung JT, Ogawa Y, Grau DJ, Sarma K, Song JJ, Kingston RE, Borowsky M, Lee JT (2010) Genome-wide identification of polycomb-associated RNAs by RIP-seq. Mol Cell 40:939–953

Zhou W, Zhu P, Wang J, Pascual G, Ohgi KA, Lozach J, Glass CK, Rosenfeld MG (2008) Histone H2A monoubiquitination represses transcription by inhibiting RNA polymerase II transcriptional elongation. Mol Cell 29:69–80

Part III
Epigenetics and Transcription Regulation

Chapter 10
Basic Mechanisms in RNA Polymerase I Transcription of the Ribosomal RNA Genes

Sarah J. Goodfellow and Joost C.B.M. Zomerdijk

Abstract RNA Polymerase (Pol) I produces ribosomal (r)RNA, an essential component of the cellular protein synthetic machinery that drives cell growth, underlying many fundamental cellular processes. Extensive research into the mechanisms governing transcription by Pol I has revealed an intricate set of control mechanisms impinging upon rRNA production. Pol I-specific transcription factors guide Pol I to the rDNA promoter and contribute to multiple rounds of transcription initiation, promoter escape, elongation and termination. In addition, many accessory factors are now known to assist at each stage of this transcription cycle, some of which allow the integration of transcriptional activity with metabolic demands. The organisation and accessibility of rDNA chromatin also impinge upon Pol I output, and complex mechanisms ensure the appropriate maintenance of the epigenetic state of the nucleolar genome and its effective transcription by Pol I. The following review presents our current understanding of the components of the Pol I transcription machinery, their functions and regulation by associated factors, and the mechanisms operating to ensure the proper transcription of rDNA chromatin. The importance of such stringent control is demonstrated by the fact that deregulated Pol I transcription is a feature of cancer and other disorders characterised by abnormal translational capacity.

10.1 Introduction

In eukaryotic cells, the task of transcribing nuclear genes is shared by Pols I, II and III. Each of these polymerases is dedicated to the transcription of a different set of genes, known as class I, II or III genes, accordingly. Pol II produces messenger (m) RNAs, which code for cellular proteins, and many small nuclear (sn)RNAs, which

---

S.J. Goodfellow • J.C.B.M. Zomerdijk (✉)
Wellcome Trust Centre for Gene Regulation and Expression, College of Life Sciences, University of Dundee, Dundee DD1 5EH, UK
e-mail: j.zomerdijk@dundee.ac.uk

are involved in mRNA processing. Pol III synthesises the transfer (t)RNAs, the 5S rRNA and a variety of other small, untranslated RNAs with essential roles in metabolism (White 2008). Unlike Pols II and III, which transcribe a variety of different genes, Pol I is dedicated to the synthesis of rRNA, and this accounts for up to 60% of transcriptional activity in a eukaryotic cell (Moss and Stefanovsky 2002). In mammals, the 47S pre-rRNA produced by Pol I is processed into mature 18S, 5.8S and 28S rRNAs, which are essential structural and catalytic components of ribosomes (Moore and Steitz 2002). Ribosomes constitute the core of the protein synthetic machinery; consequently, ribosome content is a critical determinant of protein accumulation and hence cell growth and division (Camacho et al. 1990; Siehl et al. 1985; Zetterberg and Killander 1965). The abundance of ribosomes within a cell depends upon the availability of rRNA (Liebhaber et al. 1978). Therefore, given the high metabolic burden of rRNA production and its direct influence on protein synthetic capacity, a tightly regulated transcription system has evolved, devoted to ensuring that rRNA synthesis is closely coupled with cellular growth demands.

10.2 The Ribosomal DNA

Ribosomal RNA synthesis by Pol I occurs in the nucleolus. This nuclear structure is also the site of ribosome assembly, which involves the incorporation of the rRNAs produced by Pol I along with Pol III-produced 5S rRNA and many ribosomal proteins (for reviews see Boisvert et al. 2007; Tschochner and Hurt 2003). The nucleolus forms around the nucleolar organiser regions (NORs) containing hundreds of rRNA genes, the majority (although not all) of which are organised head-to-tail into tandem arrays (Caburet et al. 2005; Nemeth and Langst 2011). Recent genome-wide mapping studies aimed at characterising the nucleolar genome have discovered that specific chromatin regions unrelated to the rDNA, and localised to distinct chromosomes, are also associated with nucleoli (Nemeth et al. 2010; van Koningsbruggen et al. 2010). These studies point towards a role for the nucleolus in the general organisation of chromosomes in the nucleus and imply a correlation between tethering to the nucleolar periphery and transcriptional silencing.

Approximately 150–200 rDNA repeats are found in the yeast genome, whereas human diploid cells have approximately 400 repeats (Birch and Zomerdijk 2008). However, only a subset of these genes (~50%) is transcribed at any given time. A recent study proposed that allelic inactivation of mammalian rRNA genes, occurring early in development, could account for this (Schlesinger et al. 2009). Active and inactive rDNA repeats are distinguished by distinct chromatin states, epigenetic marks, topological organisation and sub-nucleolar localisation (for reviews see Birch and Zomerdijk 2008; McStay and Grummt 2008; Nemeth and Langst 2011; Sanij and Hannan 2008). Transcriptionally inactive rDNA is found in a tightly packaged, heterochromatic state characterised by methylation of the DNA and repressive histone modifications, and is localised outwith the nucleolar fibrillar centre/fibrillar centre-dense fibrillar component border regions where rDNA transcription takes place. On the other hand, active rRNA genes are found in a more open chromatin state characterised by hypomethylated DNA, acetylated histones and, in mammals, an enrichment of the Pol I activator upstream binding factor (UBF). A related protein, Hmo1, specifically associates with active rDNA repeats in yeast (Merz et al. 2008). UBF binds DNA as a dimer through its high mobility group (HMG) domains, inducing substantial topological changes in DNA (Bazett-Jones et al. 1994; Jantzen et al. 1990). UBF binding is thought to be important for maintaining euchromatic rDNA, partly through its displacement of the repressive histone H1 (Kermekchiev et al. 1997; Nemeth and Langst 2011; Sanij et al. 2008) and is critical for nucleolar architecture, underpinning the structural organisation of rDNA into NORs (Mais et al. 2005). UBF has recently been shown to interact with the transcriptional regulator CTCF (CCCTC binding factor) (van de Nobelen et al. 2010), which has been implicated in the regional organisation of nucleolar rDNA (Guerrero and Maggert 2011). Extensive DNA looping is postulated to occur specifically in the active rDNA repeats, juxtaposing sites of transcription initiation and termination, mediated by Pol I-specific transcription factors and the proto-oncogene c-Myc (Denissov et al. 2011; Nemeth et al. 2008; Shiue et al. 2009). This higher order chromatin conformation of active rRNA genes further demonstrates the complexity of genome organisation within the nucleolus, the regulation of which is only just beginning to be understood.

In mammals, each rDNA repeat is approximately 43 kb and contains regulatory elements including promoters, repetitive enhancers and terminators within an intergenic spacer (IGS) of approximately 30 kb, and a single transcribed region of approximately 13 kb containing the 47S coding region (Fig. 10.1) (reviewed by McStay and Grummt 2008). In *S. cerevisiae*, each rDNA repeat is approximately 9.1 kb and contains a 6.9 kb 35S pre-rRNA coding region and a comparatively short IGS (Albert et al. 2011; French et al. 2003). Although described as rDNA 'repeats', recent evidence indicates that the multiple rRNA genes are not simply identical copies of the same transcription unit but that, in fact, several rDNA variants exist and these can be differentially expressed and regulated (Santoro et al. 2010; Tseng et al. 2008).

Eukaryotic rRNA gene promoters contain two regulatory elements important for directing accurate and efficient transcription initiation: the core promoter and the upstream control element (UCE; functionally analogous to the yeast upstream promoter element (UPE)) (Fig. 10.1). The core promoter is sufficient for basal transcription by Pol I in most species (Paule and White 2000). The UCE lies further upstream (−156 to −107 relative to the transcription start site of human rRNA genes) and is important for stimulating transcription from the core promoter (Paule and White 2000; Russell and Zomerdijk 2005). Although the general layout of the rDNA promoter is conserved from yeast to humans, with the spacing and orientation of the core promoter and UPE/UCE being critical, there is little sequence similarity between elements and, as a result, Pol I transcription is highly species specific (for reviews see Grummt 2003; Heix and Grummt 1995). In addition to the main rRNA gene promoter, which directs pre-rRNA synthesis, related sequence elements known as spacer promoters have been identified within the IGS from several species

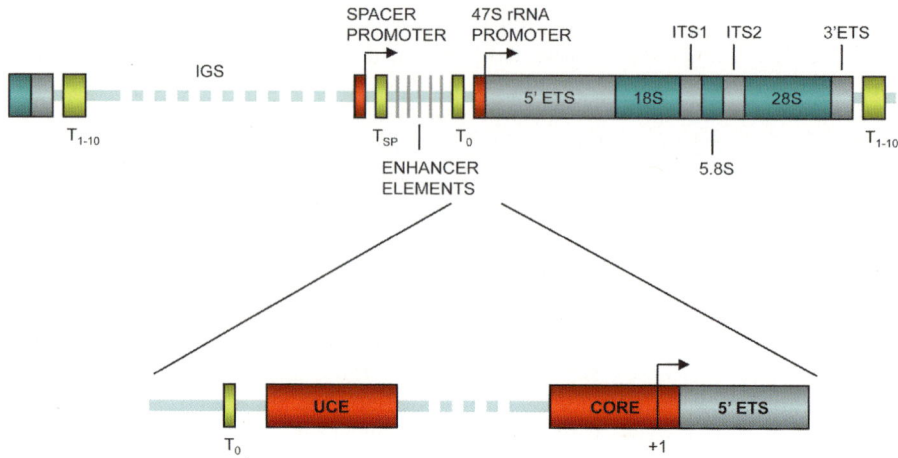

Fig. 10.1 Mammalian rDNA repeat and 47S rRNA promoter. The *top* panel illustrates key elements and the general organisation of a mammalian rDNA repeat. The IGS includes the spacer and 47S rRNA promoters, enhancer repeats and the TTF-I binding sites T_0 and T_{sp}. *Arrows* indicate start sites and direction of transcription. The coding region contains 5′ and 3′ external transcribed spacer (ETS) and two internal transcribed spacer (ITS) regions, along with regions encoding 18S, 5.8S and 28S rRNAs. Terminator elements (T_{1-10}) downstream of the 47S rRNA gene are also indicated. The *lower* panel illustrates the layout of the 47S rRNA promoter, which directs the assembly of the Pol I PIC and consists of an upstream control element, and a core promoter element overlapping the transcription start site

(De Winter and Moss 1986; Grimaldi and Di Nocera 1988; Kuhn and Grummt 1987; Labhart and Reeder 1984) (Fig. 10.1). Studies using mouse cells suggest that transcripts produced by Pol I from these promoters (known as promoter RNA (pRNA)) are involved in transcriptional silencing of the rRNA genes (Mayer et al. 2006).

In plants, IGS transcripts of as yet undefined origin serve as precursors for the RNA-dependent RNA Polymerase 2 (RDR2). RDR2 works together with, among other enzymes, the plant-specific Pols IV and V in the production of siRNAs which also mediate class I gene silencing through epigenetic mechanisms (Lawrence et al. 2004; Pontes et al. 2006; Preuss et al. 2008).

10.3 Transcription by Pol I

Using the rDNA repeats as a template, Pol I catalyses the synthesis of rRNA. However, in order to do this accurately and efficiently, Pol I requires a number of accessory factors, which facilitate polymerase recruitment, initiation, promoter escape, elongation, termination and re-initiation, as discussed below.

10.3.1 Pre-initiation Complex (PIC) Assembly

RNA polymerases themselves have little affinity for promoter sequence elements and so rely upon specific transcription factors for accurate recruitment. Therefore, as with Pol II- and Pol III-driven transcription, the first stage of transcription by Pol I is the formation of a PIC at the gene promoter. A common feature of the basal transcription machinery used by Pols I, II and III is the requirement for a TBP-containing transcription factor complex. However, in each case, the combination of TBP-associated factors (TAFs) is polymerase-specific. Transcription by Pol I in mammalian cells is dependent upon selectivity factor 1 (SL1, termed TIF1B in mouse), which is a complex of TBP and at least four Pol I-specific TAFs: TAF_I110 (TAF1C), TAF_I63 (TAF1B), TAF_I48 (TAF1A) and TAF_I41 (TAF1D) (Comai et al. 1992, 1994; Eberhard et al. 1993; Gorski et al. 2007; Heix et al. 1997; Zomerdijk et al. 1994). An additional TAF, TAF12, which was originally described as a factor involved in the transcription of class II genes, has also been implicated as a component of the mammalian SL1 complex (Denissov et al. 2007).

SL1 is pivotal to PIC formation. It confers promoter specificity by recognising and binding the core promoter element in the rDNA repeat, it is essential for Pol I recruitment to the transcription start site, and it promotes a stable interaction between UBF and the rDNA promoter (Beckmann et al. 1995; Cavanaugh et al. 2002; Friedrich et al. 2005; Miller et al. 2001; Rudloff et al. 1994). Recently, SL1 has been shown to also have an essential post-polymerase recruitment role, operating through TAF1B (Naidu et al. 2011). TAF1B and the yeast orthologue Rrn7 are structurally and functionally related to TFIIB and the Brf proteins, which are involved in Pol II and Pol III transcription, respectively, thus extending and underscoring the parallels between the eukaryotic transcription machineries (Knutson and Hahn 2011; Naidu et al. 2011).

Furthermore, a role for SL1 in maintaining the promoters of active rRNA genes in a hypomethylated state has been proposed. This involves the TAF12-mediated recruitment of GADD45a (growth arrest and DNA damage inducible protein 45 alpha) and various components of the nucleotide excision repair (NER) machinery (Schmitz et al. 2009). In addition, SL1 is thought to contribute to the structural organisation of actively transcribed rRNA genes through interactions with promoter, upstream enhancer and terminator elements (Denissov et al. 2011). Transcription by Pol I in yeast also requires TBP and the core promoter binding complex core factor (CF), which is composed of three associated proteins, RRN6, 7 and 11 (Reeder 1999). RRN6, 7 and 11 are only distantly related to the mammalian SL1 subunits TAF_I110, TAF_I63 and TAF_I48, respectively, with limited sequence homologies (Boukhgalter et al. 2002; unpublished observations), highlighting the divergent nature of the rDNA promoter elements.

Core promoter selection and binding by SL1 is solely mediated by the TAFs, with TAF_I110, TAF_I63 and TAF_I48 being reported to make direct contacts with the DNA (Beckmann et al. 1995; Rudloff et al. 1994). SL1 TAFs are also crucial for the recruitment of Pol I. Pol I has 14 polypeptide subunits in yeast, homologues for 13

Table 10.1 Eukaryotic RNA polymerase subunits

S. cerevisiae Pol I subunits	Human Pol I subunits	Homologues in Pols II/III [or associated factors]
Shared subunits		
RPB5 (ABC27, POLR2E)	hRPB5	shared
RPB6 (ABC23, POLR2F)	hRPB6	shared
RPB8 (ABC14.5, POLR2H)	hRPB8	shared
RPB10 (ABC10β, POLR2L)	hRPB10	shared
RPB12 (ABC10α, POLR2K)	hRPB12	shared
RPA40 (AC40, POLR1C)	hRPA40	RPB3/shared
RPA19 (AC19, POLR1D)	hRPA19	RPB11/shared
Homologous subunits		
RPA190 (A190, POLR1A)	hRPA190	RPB1/RPC160
RPA135 (A135, POLR1B)	hRPA135	RPB2/RPC128
RPA43 (A43, POLR1F)	hRPA43	RPB7/RPC25
RPA14 (A14)	*	RPB4/RPC17
RPA12 (A12.2, POLR1H)	hRPA12	RPB9/RPC11
RPA49 (A49, POLR1E)	PAF53	[TFIIF (Rap74 subunit) & TFIIE-β/RPC37 & RPC34]
RPA34.5 (A34.5, POLR1G)	CAST (PAF49)	[TFIIF (Rap30 subunit)/RPC53]

*Human counterpart not yet identified

of which have been identified in mammals (Table 10.1). A catalytic core is formed by ten of these subunits, which are shared with or homologous to subunits found in Pols II and III (Kuhn et al. 2007; Werner et al. 2009). At the periphery, the A14 and A43 Pol I subunits associate as a heterodimer (Kuhn et al. 2007). Similar heterodimeric structures are also found in Pols II and III (see Table 10.1 for homologous subunits) (Werner et al. 2009). The remaining two Pol I-specific subunits (*S. cerevisiae* A49 and A34.5; mammalian PAF53 and CAST/PAF49) form a heterodimeric subcomplex that can dissociate from Pol I (Hanada et al. 1996; Huet et al. 1975; Kuhn et al. 2007; Yamamoto et al. 2004). Studies in yeast indicate that this subcomplex is structurally and functionally related to the TFIIE and TFIIF initiation factors used by Pol II (as indicated in Table 10.1), and can bind DNA and promote RNA cleavage (Geiger et al. 2010; Kuhn et al. 2007). These specific Pol I subunits function at multiple stages in the Pol I transcription cycle, playing important roles in polymerase recruitment, promoter escape and elongation (Albert et al. 2011; Beckouet et al. 2008; Kuhn et al. 2007; Panov et al. 2006a, b).

The multisubunit Pol I complex exists as at least two distinct subpopulations (Milkereit and Tschochner 1998; Miller et al. 2001), known as Pol Iα and Pol Iβ in mammalian cells (Miller et al. 2001). Both forms of Pol I are active and can catalyse the synthesis of RNA, but only Pol Iβ, which represents less than 10% of the total Pol I in a cell, can be incorporated into PICs and initiate accurate, promoter-specific transcription (Milkereit and Tschochner 1998; Miller et al. 2001). This is due, at least in part, to the association of Pol Iβ with RRN3 (Milkereit and Tschochner 1998; Miller et al. 2001) (murine TIF1A (Bodem et al. 2000)). RRN3 interacts directly with Pol I, through its A43 subunit (Cavanaugh et al. 2002; Peyroche et al. 2000). In addition, the Pol I-specific A49/A34.5 subcomplex is important for the association of RRN3 with Pol I, although it is unclear whether this is mediated by direct interactions between RRN3 and these subunits (Beckouet et al. 2008). RRN3 also binds the CF subunit RRN6 in yeast, and SL1 subunits TAF_I110, TAF_I63 and TAF_I41 in mammals (Cavanaugh et al. 2002; Gorski et al. 2007; Miller et al. 2001; Peyroche et al. 2000). Therefore, RRN3 plays an essential, evolutionarily-conserved role in mediating specific transcription initiation at class I genes by connecting Pol I with an essential promoter-binding factor, and thus facilitating polymerase recruitment to the PIC at the rDNA promoter.

In addition to RRN3, various other proteins have been found specifically associated with Pol Iβ. For example, the serine/threonine kinase CK2 is present in Pol Iβ but not Pol Iα complexes and is found at the rDNA promoter in cells (Lin et al. 2006; Panova et al. 2006). Various roles have been proposed for this kinase in the regulation of transcription by Pol I (Bierhoff et al. 2008; Lin et al. 2006; Panova et al. 2006; Voit et al. 1992). Reports suggest that CK2 targets TAF_I110 and UBF and in this way regulates PIC assembly and stability, although the precise mechanistic details of this remain unclear (Lin et al. 2006; Panova et al. 2006). CK2 also phosphorylates the essential initiation factor TIF1A (the mouse counterpart of RRN3) (Bierhoff et al. 2008). However, rather than influencing PIC assembly, this modification seems important for the release of Pol I from promoter-bound initiation factors and thus elongation (Bierhoff et al. 2008). Another protein found specifically associated with the initiation-competent Pol Iβ complex is topoisomerase IIα (Panova et al. 2006). Interestingly, this topoisomerase IIα was found to be targeted by Pol Iβ-associated CK2. However, the significance of these observations to the regulation of transcription by Pol I have yet to be elucidated.

SL1 and Pol Iβ alone are sufficient to support basal levels of Pol I transcription *in vitro*. However, to achieve activated transcription, UBF must also be incorporated into the Pol I PIC. As discussed above, UBF binds throughout the rDNA in cells (O'Sullivan et al. 2002), playing critical roles as a nucleolar scaffold protein and in promoting decondensation of rDNA chromatin (Chen et al. 2004; Mais et al. 2005; Sanij et al. 2008). Crucially, UBF can also activate promoter-specific transcription by Pol I. UBF interacts cooperatively with SL1 at the rDNA promoter, with SL1 binding the highly acidic C-terminus of UBF through its TAF_I48 and TBP subunits (Beckmann et al. 1995; Bell et al. 1988; Hempel et al. 1996; Jantzen et al. 1992; Kihm et al. 1998; Tuan et al. 1999). This stabilises the association of UBF with the PIC (Friedrich et al. 2005), hence facilitating promoter-specific transcriptional activation. In addition to SL1, UBF also interacts with the PAF53 and PAF49/CAST subunits of Pol I (Hanada et al. 1996; Panov et al. 2006a, b; Seither et al. 1997; Whitehead et al. 1997). The HMG-box protein Hmo1 is involved in rDNA transcription in yeast and is, perhaps, the functional analogue of mammalian UBF (Gadal et al. 2002). Like UBF, Hmo1 binds throughout the rDNA repeat and acts synergistically

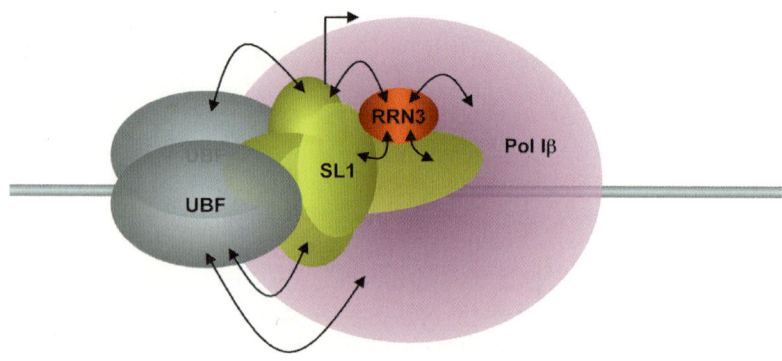

Fig. 10.2 The mammalian Pol I pre-initiation complex. Activated transcription by Pol I requires the assembly of Pol I-specific transcription factors SL1 and UBF at the rRNA promoter. In addition to contacts made between these transcription factors and the rDNA, several protein-protein interactions are also known to facilitate PIC assembly, as indicated by double-headed *arrows* (described in the text). A multitude of other factors cooperate with this transcription machinery to enhance PIC assembly and promote efficient rRNA synthesis by Pol I *in vivo*

with the Pol I-subunit A49 to activate transcription (Gadal et al. 2002; Hall et al. 2006; Kasahara et al. 2007). Interestingly, Albert et al. (2011) recently demonstrated the importance of the yeast A34.5-A49 subcomplex to nucleolar architecture. Given the fundamental role played by UBF as a nucleolar scaffold, it will be interesting to ascertain whether this property of these Pol I subunits is influenced by their interaction with Hmo1/UBF.

In summary, a series of cooperative protein-protein and protein-DNA interactions involving SL1, UBF (or their functional equivalents) and specific promoter elements are required for the recruitment of polymerase poised for the activated transcription of rRNA genes, as depicted in Fig. 10.2. Live cell imaging coupled with computational kinetic modelling has demonstrated a direct correlation between the efficiency of PIC assembly and transcriptional output in cells (Gorski et al. 2008).

10.3.2 Initiation and Promoter Escape

Following the assembly of a productive PIC at the rDNA promoter, promoter opening and transcription initiation by Pol I can commence, defined by the incorporation of the first ribonucleotides of the RNA chain. However, for productive RNA synthesis to ensue, Pol I must dissociate from the promoter-bound initiation factors in a process known as promoter escape (Panov et al. 2006a, b; Russell and Zomerdijk 2005). This post-PIC assembly event is rate-limiting for rRNA synthesis *in vitro* (Panov et al. 2001).

Promoter escape following transcription initiation coincides with the release of RRN3 from polymerase (Aprikian et al. 2001; Hirschler-Laszkiewicz et al. 2003; Milkereit and Tschochner 1998). In mouse cells, covalent attachment of RRN3 to the A43 Pol I subunit, with which RRN3 interacts, impairs rDNA transcription and cell cycle progression (Bierhoff et al. 2008). However, a similar approach pioneered in yeast strains lacking RRN3 and A43, but instead expressing a non-dissociable Pol I-RRN3 complex, did not detect any defects in rRNA synthesis or growth (Laferte et al. 2006). These studies suggest potential species-specific differences in the relative importance of RRN3 dissociation to the transcription cycle. The interaction between RRN3 and Pol I is controlled at least in part by phosphorylation. However, regulatory phosphorylation events also appear to vary from yeast to mammals: in yeast, Pol I phosphorylation apparently regulates this interaction, whereas in mammals, RRN3 phosphorylation seems important (Bierhoff et al. 2008; Cavanaugh et al. 2002; Fath et al. 2001). Bierhoff et al. (2008) looked specifically at the phosphorylation events regulating the dissociation of Pol I from RRN3 during promoter escape, and demonstrated that phosphorylation of two specific serine residues in mouse RRN3 (TIF1A) by CK2 promotes its release from polymerase.

The mammalian activator of Pol I transcription, UBF, also plays an important role in stimulating promoter escape (Panov et al. 2006a). UBF interacts with the Pol I-specific heterodimer PAF49/CAST-PAF53 (Hanada et al. 1996; Panov et al. 2006b), and this is important for transcriptional activation by UBF, which occurs subsequent to PIC assembly (Panov et al. 2006b). However, the mechanisms underlying this are unclear, although changes in DNA and/or polymerase conformation have been proposed (Panov et al. 2006b). In yeast, the homologues of these Pol I subunits (A34.5-A49) interact structurally and functionally with the probable yeast counterpart of UBF, Hmo1, and play an important role in promoter escape by promoting the release of RRN3 from elongating polymerase (Beckouet et al. 2008; Gadal et al. 2002; Schnapp et al. 1994). Conceivably, a network of interactions involving these factors could induce conformational changes in the PIC, triggering any post-translational modifications and the release of RRN3, converting initiation-competent Pol I into an elongating form.

10.3.3 Elongation

A mammalian cell requires approximately 8–10 million rRNA transcripts every 24 h to sustain adequate levels of ribosome biogenesis (Lewis and Tollervey 2000). Accordingly, transcription elongation by Pol I is highly efficient with, on average, 100 polymerases transcribing each active gene at a rate of approximately 95 nucleotides per second (Dundr et al. 2002). Similar elongation rates have also been observed for yeast Pol I (French et al. 2003). This impressive transcriptional output is achieved through the intrinsic processivity of Pol I and its cooperation with a multitude of other proteins.

Factors TFIIF and TFIIS are involved in elongation by Pol II (Saunders et al. 2006). Recent work using yeast has demonstrated that the A34.5-A49 subcomplex of Pol I is structurally and functionally analogous to TFIIF and is important for Pol I processivity (Geiger et al. 2010; Kuhn et al. 2007). Furthermore, the A12.2 subunit of Pol I stimulates the intrinsic RNA cleavage activity of Pol I and shares functional and structural homology with TFIIS, which enhances the weak 3'-RNA cleavage activity of Pol II (Haag and Pikaard 2007; Kuhn et al. 2007). This cleavage activity might be required for RNA proofreading and to stimulate elongation by creating a new and correctly aligned 3'OH in the polymerase active site after stalling and backtracking of polymerase. A role for the A34.5-A49 subcomplex in permitting contact between adjacent Pol I molecules on the same rDNA template, which might contribute to efficient transcription elongation, has also been proposed (Albert et al. 2011).

Pol I-specific transcription factors are also thought to play a role in elongation. UBF is distributed throughout the rDNA repeats and has been reported to regulate Pol I elongation by phosphorylation-dependent remodelling of the rDNA chromatin (O'Sullivan et al. 2002; Stefanovsky et al. 2006). More recently, SL1 has been proposed to assist elongating Pol I via its role in anchoring the core promoter, upstream region and terminator, which provides a spatial arrangement favourable for productive rRNA synthesis (Denissov et al. 2011).

Furthermore, several additional factors are crucial for transcription elongation by Pol I in cells, allowing the polymerase to negotiate rDNA in the context of chromatin. For example, the histone chaperones nucleolin, nucleophosmin and FACT assist in rDNA transcription in mammalian cells (Birch et al. 2009; Murano et al. 2008; Rickards et al. 2007) and in yeast, Spt4/5 and Paf1C are important (Schneider et al. 2006, 2007; Zhang et al. 2009, 2010). Moreover, various chromatin remodelling and modifying activities have been shown to promote transcription by Pol I *in vivo* including Chd1p, Isw1p and Isw2p in yeast (Jones et al. 2007) and tip60, Williams syndrome transcription Factor (WSTF)-SNF2h and the histone methyltransferase G9a in mammalian cells (Halkidou et al. 2004; Percipalle et al. 2006; Yuan et al. 2007). Nuclear actin and myosin I drive transcription elongation by Pol I and this might, in part, be attributed to their interaction with the chromatin remodeler WSTF at the rDNA (Percipalle et al. 2006; Ye et al. 2008).

Given the high loading density of Pol I on rRNA genes, it is important that any other physical impediments encountered by a transcribing polymerase, caused by topological changes in the rDNA or DNA damage, for example, are efficiently resolved. Consequently, mechanisms ensure that such constraints are minimised. For instance, topoisomerases promote transcriptional elongation in yeast by relieving the positive and negative supercoiling that occurs ahead of and behind transcribing Pol I, respectively (Brill et al. 1987; El Hage et al. 2010; French et al. 2011; Schultz et al. 1992). Topoisomerase IIα is a component of Pol Iβ in human cells, as discussed above (Panova et al. 2006). Furthermore, topoisomerase I was found associated with Pol I complexes in mouse cells (Hannan et al. 1999; Rose et al. 1988), and has been proposed to assist transcriptional elongation by Pol I in human cells (Zhang et al. 1988). Therefore, this function of topoisomerases in relieving torsional strain during transcriptional elongation by Pol I may be evolutionarily conserved.

Signalling pathways invoked by DNA damage lead to a transient repression of rRNA synthesis, partly through the ATM-mediated displacement of elongating polymerase (Kruhlak et al. 2007). Resumption of Pol I transcription is dependent upon functional DNA repair mechanisms (Kruhlak et al. 2007). Transcription-coupled DNA repair occurs at rDNA genes (Conconi et al. 2002) and various DNA repair proteins have been found in Pol I complexes, including TFIIH, Cockayne syndrome B protein (CSB), Werner's syndrome helicase (WRN), Ku70/80 and several components of the NER machinery (Bradsher et al. 2002; Hannan et al. 1999; Iben et al. 2002; Schmitz et al. 2009; Shiratori et al. 2002). In many cases, these interactions have been shown to promote transcription by Pol I. However, a direct role for these factors in the transcription-coupled repair of rDNA has yet to be demonstrated.

As elongation by Pol I proceeds, the nascent pre-rRNA associates with components of the processing machinery, allowing co-transcriptional maturation of the rRNA and assembly of ribosomal particles (reviewed by Granneman and Baserga 2005). As a result, pre-rRNA synthesis and processing are closely coordinated, such that defective transcription by Pol I impairs pre-rRNA processing and vice versa (Granneman and Baserga 2005; Schneider et al. 2006, 2007). Although the mechanisms responsible for this coupling are incompletely defined, factors implicated in yeast include Spt4 and Spt5, which interact both with elongating Pol I and components of the pre-rRNA processing machinery (Leporé and Lafontaine 2011; Schneider et al. 2006, 2007). Such rigorous coordination likely contributes to the highly efficient and tightly regulated production of ribosomes.

10.3.4 Termination

Transcription termination by Pol I is a multistep process involving specific DNA sequence elements and regulatory proteins. In mammals, transcription termination factor TTF-I binds terminator elements downstream of the rRNA gene (T_1-T_{10}; Fig. 10.1), causing polymerase pausing. Dissociation of the paused transcription complex is then mediated by Pol I and transcript release factor PTRF. A similar mechanism is thought to operate in yeast, involving the TTF-I homologue Reb1p (Jansa and Grummt 1999). However, recent studies using yeast have uncovered further complexity in the control of transcription termination by Pol I, by demonstrating the existence of a 'torpedo' mechanism, similar to that employed for the termination of transcription by Pol II. This process begins with cleavage of the nascent pre-rRNA by the endonuclease Rnt1, followed by the progressive digestion of the resulting Pol I-associated RNA cleavage product mediated by the cooperative actions of the 5' to 3' exonuclease Rat1 (mammalian Xrn2) and the RNA helicase Sen1 (Braglia et al. 2010, 2011; El Hage et al. 2008; Kawauchi et al. 2008). Recognition of the Rnt1-cleaved pre-rRNA by Rat1 is thought to be controlled by phosphorylation of the 5'end of the RNA by the polynucleotide kinase Grc3 (Braglia et al. 2010). Once Rat1 reaches elongating Pol I, the transcription complex becomes unstable and dissociates from DNA, thus resulting in transcription termination.

In addition, the smallest Pol I subunit A12.2/RPA12 is critical for effective transcription termination, potentially mediated by its stimulation of the intrinsic 3′-end RNA cleavage activity of Pol I (Haag and Pikaard 2007; Kuhn et al. 2007; Prescott et al. 2004). The Pol II and III homologues of RPA12, RPB9 and RPC11, are also important for relief of polymerase pausing and termination. Furthermore, each of these polymerase subunits shares homology with the RNA cleavage enhancing factor TFIIS (Prescott et al. 2004). It is possible that several mechanisms co-exist to ensure accurate, efficient termination of transcription by Pol I and thus cell viability (Braglia et al. 2011).

10.3.5 Re-initiation

Correct termination of transcription and release of the nascent rRNA is required for re-initiation by Pol I. Once a gene is activated, the rate of re-initiation will contribute to the overall level of transcripts produced. In addition to terminator elements located downstream of the rRNA coding region, TTF-I binding sites are also found immediately upstream of the rDNA promoter (termed T_0) and downstream of the spacer promoter (T_{sp}) (Fig. 10.1) (Nemeth et al. 2008). Interactions between TTF-1 and its binding sites are thought to be important for epigenetic and topological regulation of the rDNA (for reviews see McStay and Grummt 2008; Nemeth and Langst 2011). In cells, efficient recycling and re-initiation by Pol I might be facilitated by TTF-1-mediated juxtaposition of the terminator and promoter elements, which results in the formation of DNA loops (Nemeth et al. 2008; Nemeth and Langst 2011; Shiue et al. 2009). In addition to TTF-I, SL1 and c-Myc have been implicated in the formation of such DNA loops (Denissov et al. 2011; Nemeth and Langst 2011; Shiue et al. 2009).

Another important event in transcription re-initiation by Pol I is the re-association of the essential initiation factor RRN3 with polymerase, allowing Pol I to re-assemble with SL1 and UBF, which remain promoter bound following escape of elongating polymerase (Lin et al. 2006; Panov et al. 2001). This reversible interaction between Pol I and RRN3 underpins critical transitions in the Pol I transcription cycle, and is evolutionarily conserved. However, the mechanisms underlying this are not fully elucidated. Such a dynamic partnership is likely controlled through reversible post-translational modifications (Bierhoff et al. 2008; Cavanaugh et al. 2002; Fath et al. 2001). In support of this, Bierhoff et al. (2008) demonstrated that dephosphorylation of the CK2 target sites (serines 170 and 172) in TIF-1A (mouse RRN3) by the protein phosphatase FCP1 is required for the re-association of TIF-1A with Pol I, and hence re-initiation. In contrast, these particular residues are not conserved in yeast RRN3 and phosphorylation of Pol I itself is thought to regulate Pol I-RRN3 complex formation (Bierhoff et al. 2008; Fath et al. 2001). FCP1 has been implicated in the regulation of Pol I transcription in yeast, but rather than functioning in re-initiation, this phosphatase appears to promote chain elongation (Fath et al. 2004). Whether additional kinases and/or phosphatases targeting RRN3 and/or Pol I are involved in a conserved mechanism for promoting multiple rounds of transcription remains to be determined.

10.4 Regulation of rDNA Transcription

Stringent regulatory mechanisms operate to ensure a precise balance between the requirement for and availability of rRNA, allowing cells to control their capacity for protein synthesis in response to changing metabolic needs. For example, transcription by Pol I is low when nutrients or mitogens are limiting, but upregulated when the availability of these growth stimuli increase. In addition, Pol I transcription is regulated in response to a range of cellular stresses and during many fundamental cellular processes, including cellular differentiation and throughout the cell cycle (for reviews see Drygin et al. 2010; Grummt 2003; Grummt and Voit 2010; Mayer and Grummt 2005; Moss 2004; Russell and Zomerdijk 2005). The level of cellular rRNA is determined by the rate at which active rRNA genes are transcribed, and also by the number of active genes. Some of the mechanisms influencing these different aspects of class I gene expression are discussed below.

10.4.1 Regulation of the Pol I Transcription Machinery

In yeast, growth-dependent changes in rRNA synthesis can be achieved both through altering the proportion of active genes, and the rate of transcription from already active loci (Grummt and Pikaard 2003; Russell and Zomerdijk 2005). However, in mammalian cells, changes in rRNA production in response to growth signals seem to be mediated mainly by the latter mechanism. A plethora of cellular control pathways have been shown to mediate such acute changes in rRNA synthesis by directly regulating the activity of the Pol I transcription machinery, with positive regulators of growth activating transcription and negative regulators of growth having repressive effects (Drygin et al. 2010; Grummt 2003; Moss 2004; Russell and Zomerdijk 2005). Some of these regulatory proteins, and their effects on the Pol I transcription machinery, are listed in Table 10.2.

10.4.2 Regulation of rDNA Chromatin

In addition to these mechanisms that modulate the activity of the Pol I transcription machinery, epigenetic regulation of the rDNA chromatin, which determines the number of active rRNA genes, can also influence the level of rRNA produced. Such epigenetic regulation is thought to be stably propagated throughout cell divisions to ensure that an appropriate proportion of active and inactive rDNA repeats is maintained. The importance of this to nucleolar integrity, genomic stability and the global regulation of gene expression has been proposed (Espada et al. 2007; Guetg et al. 2010; Ide et al. 2010; McStay and Grummt 2008; Paredes and Maggert 2009).

Table 10.2 Positive and negative regulators of cell growth target the Pol I transcription machinery

Regulatory factor	Targets in Pol I transcription machinery	References
Activators		
G1-specific cyclin/CDKs	UBF	Voit et al. (1999) and Voit and Grummt (2001)
ERK	RRN3, UBF	Stefanovsky et al. (2001) and Zhao et al. (2003)
RSK	RRN3	Zhao et al. (2003)
CK2	UBF, RRN3	Lin et al. (2006), Panova et al. (2006), and Bierhoff et al. (2008)
mTOR	RRN3, UBF	Hannan et al. (2003), Claypool et al. (2004), and Mayer et al. (2004)
CBP	UBF	Pelletier et al. (2000)
PCAF	SL1	Muth et al. (2001)
TIP60	UBF	Halkidou et al. (2004)
c-Myc	SL1	Arabi et al. (2005) and Grandori et al. (2005)
RasL11a	UBF	Pistoni et al. (2010)
Repressors		
p53	SL1	Zhai and Comai (2000)
RB/p130	UBF	Cavanaugh et al. (1995), Voit et al. (1997), and Hannan et al. (2000)
CK2	SL1	Panova et al. (2006)
PTEN	SL1	Zhang et al. (2005)
p14ARF	UBF, TTF-I	Ayrault et al. (2006) and Lessard et al. (2010)
GSK3β	SL1	Vincent et al. (2008)
AMPK	RRN3	Hoppe et al. (2009)
JNK2	RRN3	Mayer et al. (2005)

One of the key factors in establishing the epigenetic state of rRNA genes in mammals is TTF-I, which, in addition to its role as a transcription terminator and potential regulator of rDNA topology, can define active or inactive rDNA conformations through its association with chromatin remodelling complexes (reviewed by McStay and Grummt 2008). Epigenetic silencing by TTF-I is mediated by its recruitment of the nucleolar remodelling complex (NoRC) to the promoter-proximal T_0 element. The NoRC subunit TIP5 interacts with TTF-I, and NoRC in turn recruits DNA methyltransferases DNMT1 and DNMT3, and the histone deacetylase-containing Sin3 complex, which mediate transcriptional repression (Santoro and Grummt 2005; Santoro et al. 2002; Zhou et al. 2002). Methylation of a single CpG dinucleotide at the mouse rDNA promoter seems particularly important for transcriptional silencing, as this diminishes binding of the activator UBF to the rDNA (Santoro and Grummt 2001, 2005). NoRC function is dependent on the association of TIP5 with pRNAs. These 150–300 nucleotide transcripts are derived from the IGS by Pol I-driven transcription from the spacer promoters of a subset of hypomethylated rRNA genes (Mayer et al. 2006, 2008 ; Santoro et al. 2010). Such transcripts are essential for epigenetic silencing of rDNA. Furthermore, a recent study has shown that pRNA can induce *de novo* methylation of rDNA and transcriptional silencing independently of TTF-I and NoRC, by interacting directly with the T_0 element forming a DNA-RNA triplex which is recognised by DNMT3b (Schmitz et al. 2010). In fact, binding of TTF-I and pRNA to T_0 are mutually exclusive (Schmitz et al. 2010). The levels of pRNA and its association with T_0 vary during S phase progression, suggesting a potential link between pRNA and the transmission of epigenetic rDNA silencing between cell divisions (Santoro et al. 2010; Schmitz et al. 2010). Studies of the molecular basis of nucleolar dominance in plant hybrids, whereby NORs from one parental species are dominant over the other, also suggest an involvement of non-coding RNAs derived from the rDNA IGS in determining a repressive pattern of DNA methylation and histone deacetylation (reviewed by Tucker et al. 2010). Therefore, the involvement of such RNAs in selecting the proportion of transcriptionally silenced rDNA repeats, through epigenetic mechanisms, appears to be evolutionarily conserved.

Mechanisms also exist to maintain a proportion of rDNA repeats in an active chromatin conformation. For instance, binding of TTF-I to the T_0 elements of certain rDNA repeats in mouse cells induces chromatin remodelling and transcriptional activation (Langst et al. 1997). This is mediated by an interaction between TTF-I and the chromatin remodeler CSB (Yuan et al. 2007). Transcriptional activation by CSB is dependent on its intrinsic ATPase activity, and also its association with the histone methyltransferase G9a (Yuan et al. 2007). Therefore, TTF-I is integral in determining whether rDNA repeats adopt an active or an inactive epigenetic conformation. However, it is unclear how TTF-I interacts differentially with these positive and negative regulators of transcription to achieve a precise balance between these alternative chromatin states.

Other factors proposed to maintain rDNA in a euchromatic state include the methyl-CpG binding domain protein MBD3, TAF12-recruited GADD45a and the putative chromatin remodeler CHD7 (chromodomain helicase DNA-binding protein 7). These proteins prevent repressive methylation of the rDNA and/or promote the active demethylation of this region (Brown and Szyf 2007; Schmitz et al. 2009; Zentner et al. 2010). In addition to DNA methylation, the methylation state of the rDNA-associated histones also correlates with transcriptional activity: di- and tri-methylation of Lys4 and mono- and di-methylation of Lys36 of histone H3 mark active repeats, whereas di-methylation of Lys9 of histone H3 is associated with silenced rDNA chromatin. The JmjC domain-containing lysine demethylases JHDM1B, KDM2A and PHF8 associate with rDNA and influence this histone methylation pattern (Feng et al. 2010; Frescas et al. 2007; Tanaka et al. 2010).

UBF is also involved in determining the number of active rDNA repeats (Sanij et al. 2008). However, this does not appear to involve epigenetic modifications of the chromatin, but instead occurs through the ability of UBF to displace histone H1, thus preventing H1-induced chromatin condensation (Sanij et al. 2008). In yeast, the UBF-related protein Hmo1 localises to active rDNA repeats, and a recent study has demonstrated the importance of this to the maintenance of a transcriptionally-competent chromatin state, established following DNA replication through the Pol I transcription-dependent (and potentially histone chaperone-dependent) displacement of nucleosomes (Wittner et al. 2011).

Although mechanisms directly influencing the activity of the Pol I transcription machinery are important for the modulation of rRNA production in response to growth signals, alterations in the rDNA chromatin are also likely to contribute. For instance, Murayama et al. (2008) have described a complex known as eNoSC (energy-dependent nucleolar silencing complex), which mediates the epigenetic repression of rRNA genes in response to energy deprivation. Furthermore, the lysine demethylase KDM2A targets mono- and di-methylated Lys36 of histone H3 and in this way represses transcription by Pol I in response to starvation (Tanaka et al. 2010). In addition to these dynamic alterations in rDNA chromatin in response to changing metabolic conditions, reducing the number of active rDNA repeats has been proposed to contribute to the down-regulation of Pol I transcription that accompanies differentiation (Sanij and Hannan 2008). Moreover, the proportion of active rDNA repeats varies depending on the developmental stage and cell type, indicating that lineage-specific regulation of the number of actively transcribed rRNA genes could be important for vertebrate development (Haaf et al. 1991; Schlesinger et al. 2009).

10.4.3 Deregulated Transcription by Pol I and Disease

The regulation of Pol I transcription is clearly a crucial feature of normal cellular growth and proliferation. The importance of such stringent control is highlighted by the fact that transcription by Pol I is deregulated in various disease states. Most notably, pre-rRNA levels are elevated in a wide range of tumour types and this is thought to be a general feature of human cancers (reviewed by Ruggero and Pandolfi 2003; White 2008). Inactivation of tumour suppressors, aberrant activation of oncogenes (many of which target the Pol I transcription machinery directly as outlined in Table 10.2) and loss of rDNA methylation, are all thought to play a role in this abnormal activation of rDNA expression, contributing to the uncontrolled cell growth and division that is characteristic of tumour cells. In addition, elevated transcription by Pol I underlies the hypertrophic growth of cardiomyocytes, which is a characteristic feature of various cardiovascular disorders (Brandenburger et al. 2001). In contrast, decreased rRNA production, as a result of rDNA promoter hypermethylation in the cerebral cortex, has been described as a feature of the neurodegeneration that accompanies Alzheimer's disease (Pietrzak et al. 2011). Furthermore, the demethylase PHF8 might link dynamic histone methylation at rDNA to mental retardation with cleft lip and palate (Feng et al. 2010).

The consequences of deregulated Pol I transcription during development are highlighted by recent findings regarding the genetic basis of Treacher Collins Syndrome, which can be caused by mutations in Pol I/III subunits or the UBF-interacting protein Treacle (Dauwerse et al. 2011; Valdez et al. 2004). This craniofacial autosomal-dominant disorder is characterised by a deficiency in neural crest cells, resulting from inadequate ribosome production during development. Abnormalities in ribosome biogenesis give rise to a variety of other congenital disorders, emphasizing the importance of understanding the mechanisms impinging upon ribosome production (Narla and Ebert 2010).

10.5 Conclusions

Transcription by Pol I underlies fundamental cellular functions. Our understanding of this process has grown in recent years, revealing unanticipated complexity. An increasing number of functions are being attributed to the Pol I-specific transcription factors, unravelling the means by which they facilitate rRNA synthesis. Furthermore, many additional factors are now known to bind and regulate this machinery. Recent insights into the intricacies of the structural and topological organisation of rDNA further demonstrate the multifaceted cellular control mechanisms which impinge upon rRNA production. Such elaborate regulation highlights the importance for precise control of transcription by Pol I, a point further emphasized by the apparently universal deregulation of rRNA expression in human tumours. Despite the ever-expanding list of factors and epigenetic control mechanisms that influence transcription by Pol I, the full implications of many of these discoveries have yet to be established. Therefore, further research directed towards resolving the many unanswered questions regarding the complex interplay between regulatory mechanisms targeting the rDNA chromatin and the Pol I transcription machinery is essential, not only to enhance our understanding of cellular growth controls, but also to enable the development of prognostic tools and therapeutic strategies for disease.

Acknowledgements We thank Dr Jackie Russell for critical reading and helpful comments. We thank the Wellcome Trust for supporting our research through a Wellcome Trust Programme Grant (085441/Z/08/Z) awarded to JCBMZ.

References

Albert B, Leger-Silvestre I, Normand C, Ostermaier MK, Perez-Fernandez J, Panov KI, Zomerdijk JC, Schultz P, Gadal O (2011) RNA polymerase I-specific subunits promote polymerase clustering to enhance the rRNA gene transcription cycle. J Cell Biol 192:277–293

Aprikian P, Moorefield B, Reeder RH (2001) New model for the yeast RNA polymerase I transcription cycle. Mol Cell Biol 21:4847–4855

Arabi A, Wu S, Ridderstrale K, Bierhoff H, Shiue C, Fatyol K, Fahlen S, Hydbring P, Soderberg O, Grummt I, Larsson LG, Wright AP (2005) c-Myc associates with ribosomal DNA and activates RNA polymerase I transcription. Nat Cell Biol 7:303–310

Ayrault O, Andrique L, Fauvin D, Eymin B, Gazzeri S, Seite P (2006) Human tumor suppressor p14ARF negatively regulates rRNA transcription and inhibits UBF1 transcription factor phosphorylation. Oncogene 25:7577–7586

Bazett-Jones DP, Leblanc B, Herfort M, Moss T (1994) Short-range DNA looping by the Xenopus HMG-box transcription factor, xUBF. Science 264:1134–1137

Beckmann H, Chen JL, O'Brien T, Tjian R (1995) Coactivator and promoter-selective properties of RNA polymerase I TAFs. Science 270:1506–1509

Beckouet F, Labarre-Mariotte S, Albert B, Imazawa Y, Werner M, Gadal O, Nogi Y, Thuriaux P (2008) Two RNA polymerase I subunits control the binding and release of Rrn3 during transcription. Mol Cell Biol 28:1596–1605

Bell SP, Learned RM, Jantzen HM, Tjian R (1988) Functional cooperativity between transcription factors UBF1 and SL1 mediates human ribosomal RNA synthesis. Science 241:1192–1197

Bierhoff H, Dundr M, Michels AA, Grummt I (2008) Phosphorylation by casein kinase 2 facilitates rRNA gene transcription by promoting dissociation of TIF-IA from elongating RNA polymerase I. Mol Cell Biol 28:4988–4998

Birch JL, Zomerdijk JC (2008) Structure and function of ribosomal RNA gene chromatin. Biochem Soc Trans 36:619–624

Birch JL, Tan BC, Panov KI, Panova TB, Andersen JS, Owen-Hughes TA, Russell J, Lee SC, Zomerdijk JC (2009) FACT facilitates chromatin transcription by RNA polymerases I and III. EMBO J 28:854–865

Bodem J, Dobreva G, Hoffmann-Rohrer U, Iben S, Zentgraf H, Delius H, Vingron M, Grummt I (2000) TIF-IA, the factor mediating growth-dependent control of ribosomal RNA synthesis, is the mammalian homolog of yeast Rrn3p. EMBO Rep 1:171–175

Boisvert FM, van Koningsbruggen S, Navascues J, Lamond AI (2007) The multifunctional nucleolus. Nat Rev Mol Cell Biol 8:574–585

Boukhgalter B, Liu M, Guo A, Tripp M, Tran K, Huynh C, Pape L (2002) Characterization of a fission yeast subunit of an RNA polymerase I essential transcription initiation factor, SpRrn7h/TAF(I)68, that bridges yeast and mammals: association with SpRrn11h and the core ribosomal RNA gene promoter. Gene 291:187–201

Bradsher J, Auriol J, Proietti de Santis L, Iben S, Vonesch JL, Grummt I, Egly JM (2002) CSB is a component of RNA pol I transcription. Mol Cell 10:819–829

Braglia P, Heindl K, Schleiffer A, Martinez J, Proudfoot NJ (2010) Role of the RNA/DNA kinase Grc3 in transcription termination by RNA polymerase I. EMBO Rep 11:758–764

Braglia P, Kawauchi J, Proudfoot NJ (2011) Co-transcriptional RNA cleavage provides a failsafe termination mechanism for yeast RNA polymerase I. Nucleic Acids Res 39:1439–1448

Brandenburger Y, Jenkins A, Autelitano DJ, Hannan RD (2001) Increased expression of UBF is a critical determinant for rRNA synthesis and hypertrophic growth of cardiac myocytes. FASEB J 15:2051–2053

Brill SJ, DiNardo S, Voelkel-Meiman K, Sternglanz R (1987) Need for DNA topoisomerase activity as a swivel for DNA replication for transcription of ribosomal RNA. Nature 326:414–416

Brown SE, Szyf M (2007) Epigenetic programming of the rRNA promoter by MBD3. Mol Cell Biol 27:4938–4952

Caburet S, Conti C, Schurra C, Lebofsky R, Edelstein SJ, Bensimon A (2005) Human ribosomal RNA gene arrays display a broad range of palindromic structures. Genome Res 15:1079–1085

Camacho JA, Peterson CJ, White GJ, Morgan HE (1990) Accelerated ribosome formation and growth in neonatal pig hearts. Am J Physiol 258:C86–91

Cavanaugh AH, Hempel WM, Taylor LJ, Rogalsky V, Todorov G, Rothblum LI (1995) Activity of RNA polymerase I transcription factor UBF blocked by Rb gene product. Nature 374:177–180

Cavanaugh AH, Hirschler-Laszkiewicz I, Hu Q, Dundr M, Smink T, Misteli T, Rothblum LI (2002) Rrn3 phosphorylation is a regulatory checkpoint for ribosome biogenesis. J Biol Chem 277:27423–27432

Chen D, Belmont AS, Huang S (2004) Upstream binding factor association induces large-scale chromatin decondensation. Proc Natl Acad Sci USA 101:15106–15111

Claypool JA, French SL, Johzuka K, Eliason K, Vu L, Dodd JA, Beyer AL, Nomura M (2004) Tor pathway regulates Rrn3p-dependent recruitment of yeast RNA polymerase I to the promoter but does not participate in alteration of the number of active genes. Mol Biol Cell 15:946–956

Comai L, Tanese N, Tjian R (1992) The TATA-binding protein and associated factors are integral components of the RNA polymerase I transcription factor, SL1. Cell 68:965–976

Comai L, Zomerdijk JC, Beckmann H, Zhou S, Admon A, Tjian R (1994) Reconstitution of transcription factor SL1: exclusive binding of TBP by SL1 or TFIID subunits. Science 266:1966–1972

Conconi A, Bespalov VA, Smerdon MJ (2002) Transcription-coupled repair in RNA polymerase I-transcribed genes of yeast. Proc Natl Acad Sci USA 99:649–654

Dauwerse JG, Dixon J, Seland S, Ruivenkamp CA, van Haeringen A, Hoefsloot LH, Peters DJ, Boers AC, Daumer-Haas C, Maiwald R, Zweier C, Kerr B, Cobo AM, Toral JF, Hoogeboom AJ, Lohmann DR, Hehr U, Dixon MJ, Breuning MH, Wieczorek D (2011) Mutations in genes encoding subunits of RNA polymerases I and III cause Treacher Collins syndrome. Nat Genet 43:20–22

De Winter RF, Moss T (1986) Spacer promoters are essential for efficient enhancement of *X. laevis* ribosomal transcription. Cell 44:313–318

Denissov S, van Driel M, Voit R, Hekkelman M, Hulsen T, Hernandez N, Grummt I, Wehrens R, Stunnenberg H (2007) Identification of novel functional TBP-binding sites and general factor repertoires. EMBO J 26:944–954

Denissov S, Lessard F, Mayer C, Stefanovsky V, van Driel M, Grummt I, Moss T, Stunnenberg HG (2011) A model for the topology of active ribosomal RNA genes. EMBO Rep 12:231–237

Drygin D, Rice WG, Grummt I (2010) The RNA polymerase I transcription machinery: an emerging target for the treatment of cancer. Annu Rev Pharmacol Toxicol 50:131–156

Dundr M, Hoffmann-Rohrer U, Hu Q, Grummt I, Rothblum LI, Phair RD, Misteli T (2002) A kinetic framework for a mammalian RNA polymerase in vivo. Science 298:1623–1626

Eberhard D, Tora L, Egly JM, Grummt I (1993) A TBP-containing multiprotein complex (TIF-IB) mediates transcription specificity of murine RNA polymerase I. Nucleic Acids Res 21:4180–4186

El Hage A, Koper M, Kufel J, Tollervey D (2008) Efficient termination of transcription by RNA polymerase I requires the 5′ exonuclease Rat1 in yeast. Genes Dev 22:1069–1081

El Hage A, French SL, Beyer AL, Tollervey D (2010) Loss of Topoisomerase I leads to R-loop-mediated transcriptional blocks during ribosomal RNA synthesis. Genes Dev 24:1546–1558

Espada J, Ballestar E, Santoro R, Fraga MF, Villar-Garea A, Nemeth A, Lopez-Serra L, Ropero S, Aranda A, Orozco H, Moreno V, Juarranz A, Stockert JC, Langst G, Grummt I, Bickmore W, Esteller M (2007) Epigenetic disruption of ribosomal RNA genes and nucleolar architecture in DNA methyltransferase 1 (Dnmt1) deficient cells. Nucleic Acids Res 35:2191–2198

Fath S, Milkereit P, Peyroche G, Riva M, Carles C, Tschochner H (2001) Differential roles of phosphorylation in the formation of transcriptional active RNA polymerase I. Proc Natl Acad Sci USA 98:14334–14339

Fath S, Kobor MS, Philippi A, Greenblatt J, Tschochner H (2004) Dephosphorylation of RNA polymerase I by Fcp1p is required for efficient rRNA synthesis. J Biol Chem 279: 25251–25259

Feng W, Yonezawa M, Ye J, Jenuwein T, Grummt I (2010) PHF8 activates transcription of rRNA genes through H3K4me3 binding and H3K9me1/2 demethylation. Nat Struct Mol Biol 17(4):445–450

French SL, Osheim YN, Cioci F, Nomura M, Beyer AL (2003) In exponentially growing *Saccharomyces cerevisiae* cells, rRNA synthesis is determined by the summed RNA polymerase I loading rate rather than by the number of active genes. Mol Cell Biol 23:1558–1568

French SL, Sikes ML, Hontz RD, Osheim YN, Lambert TE, El Hage A, Smith MM, Tollervey D, Smith JS, Beyer AL (2011) Distinguishing the roles of Topoisomerases I and II in relief of transcription-induced torsional stress in yeast rRNA genes. Mol Cell Biol 31:482–494

Frescas D, Guardavaccaro D, Bassermann F, Koyama-Nasu R, Pagano M (2007) JHDM1B/FBXL10 is a nucleolar protein that represses transcription of ribosomal RNA genes. Nature 450:309–313

Friedrich JK, Panov KI, Cabart P, Russell J, Zomerdijk JC (2005) TBP-TAF complex SL1 directs RNA polymerase I pre-initiation complex formation and stabilizes upstream binding factor at the rDNA promoter. J Biol Chem 280:29551–29558

Gadal O, Labarre S, Boschiero C, Thuriaux P (2002) Hmo1, an HMG-box protein, belongs to the yeast ribosomal DNA transcription system. EMBO J 21:5498–5507

Geiger SR, Lorenzen K, Schreieck A, Hanecker P, Kostrewa D, Heck AJ, Cramer P (2010) RNA polymerase I contains a TFIIF-related DNA-binding subcomplex. Mol Cell 39:583–594

Gorski JJ, Pathak S, Panov K, Kasciukovic T, Panova T, Russell J, Zomerdijk JC (2007) A novel TBP-associated factor of SL1 functions in RNA polymerase I transcription. EMBO J 26:1560–1568

Gorski SA, Snyder SK, John S, Grummt I, Misteli T (2008) Modulation of RNA polymerase assembly dynamics in transcriptional regulation. Mol Cell 30:486–497

Grandori C, Gomez-Roman N, Felton-Edkins ZA, Ngouenet C, Galloway DA, Eisenman RN, White RJ (2005) c-Myc binds to human ribosomal DNA and stimulates transcription of rRNA genes by RNA polymerase I. Nat Cell Biol 7:311–318

Granneman S, Baserga SJ (2005) Crosstalk in gene expression: coupling and co-regulation of rDNA transcription, pre-ribosome assembly and pre-rRNA processing. Curr Opin Cell Biol 17:281–286

Grimaldi G, Di Nocera PP (1988) Multiple repeated units in *Drosophila melanogaster* ribosomal DNA spacer stimulate rRNA precursor transcription. Proc Natl Acad Sci USA 85:5502–5506

Grummt I (2003) Life on a planet of its own: regulation of RNA polymerase I transcription in the nucleolus. Genes Dev 17:1691–1702

Grummt I, Pikaard CS (2003) Epigenetic silencing of RNA polymerase I transcription. Nat Rev Mol Cell Biol 4:641–649

Grummt I, Voit R (2010) Linking rDNA transcription to the cellular energy supply. Cell Cycle 9:225–226

Guerrero PA, Maggert KA (2011) The CCCTC-binding factor (CTCF) of Drosophila contributes to the regulation of the ribosomal DNA and nucleolar stability. PLoS One 6:e16401

Guetg C, Lienemann P, Sirri V, Grummt I, Hernandez-Verdun D, Hottiger MO, Fussenegger M, Santoro R (2010) The NoRC complex mediates the heterochromatin formation and stability of silent rRNA genes and centromeric repeats. EMBO J 29:2135–2146

Haaf T, Hayman DL, Schmid M (1991) Quantitative determination of rDNA transcription units in vertebrate cells. Exp Cell Res 193:78–86

Haag JR, Pikaard CS (2007) RNA polymerase I: a multifunctional molecular machine. Cell 131:1224–1225

Halkidou K, Logan IR, Cook S, Neal DE, Robson CN (2004) Putative involvement of the histone acetyltransferase Tip60 in ribosomal gene transcription. Nucleic Acids Res 32:1654–1665

Hall DB, Wade JT, Struhl K (2006) An HMG protein, Hmo1, associates with promoters of many ribosomal protein genes and throughout the rRNA gene locus in *Saccharomyces cerevisiae*. Mol Cell Biol 26:3672–3679

Hanada K, Song CZ, Yamamoto K, Yano K, Maeda Y, Yamaguchi K, Muramatsu M (1996) RNA polymerase I associated factor 53 binds to the nucleolar transcription factor UBF and functions in specific rDNA transcription. EMBO J 15:2217–2226

Hannan RD, Cavanaugh A, Hempel WM, Moss T, Rothblum L (1999) Identification of a mammalian RNA polymerase I holoenzyme containing components of the DNA repair/replication system. Nucleic Acids Res 27:3720–3727

Hannan KM, Hannan RD, Smith SD, Jefferson LS, Lun M, Rothblum LI (2000) Rb and p130 regulate RNA polymerase I transcription: Rb disrupts the interaction between UBF and SL-1. Oncogene 19:4988–4999

Hannan KM, Brandenburger Y, Jenkins A, Sharkey K, Cavanaugh A, Rothblum L, Moss T, Poortinga G, McArthur GA, Pearson RB, Hannan RD (2003) mTOR-dependent regulation of ribosomal gene transcription requires S6K1 and is mediated by phosphorylation of the carboxy-terminal activation domain of the nucleolar transcription factor UBF. Mol Cell Biol 23:8862–8877

Heix J, Grummt I (1995) Species specificity of transcription by RNA polymerase I. Curr Opin Genet Dev 5:652–656

Heix J, Zomerdijk JC, Ravanpay A, Tjian R, Grummt I (1997) Cloning of murine RNA polymerase I-specific TAF factors: conserved interactions between the subunits of the species-specific transcription initiation factor TIF-IB/SL1. Proc Natl Acad Sci USA 94:1733–1738

Hempel WM, Cavanaugh AH, Hannan RD, Taylor L, Rothblum LI (1996) The species-specific RNA polymerase I transcription factor SL-1 binds to upstream binding factor. Mol Cell Biol 16:557–563

Hirschler-Laszkiewicz I, Cavanaugh AH, Mirza A, Lun M, Hu Q, Smink T, Rothblum LI (2003) Rrn3 becomes inactivated in the process of ribosomal DNA transcription. J Biol Chem 278:18953–18959

Hoppe S, Bierhoff H, Cado I, Weber A, Tiebe M, Grummt I, Voit R (2009) AMP-activated protein kinase adapts rRNA synthesis to cellular energy supply. Proc Natl Acad Sci USA 106:17781–17786

Huet J, Buhler JM, Sentenac A, Fromageot P (1975) Dissociation of two polypeptide chains from yeast RNA polymerase A. Proc Natl Acad Sci USA 72:3034–3038

Iben S, Tschochner H, Bier M, Hoogstraten D, Hozak P, Egly JM, Grummt I (2002) TFIIH plays an essential role in RNA polymerase I transcription. Cell 109:297–306

Ide S, Miyazaki T, Maki H, Kobayashi T (2010) Abundance of ribosomal RNA gene copies maintains genome integrity. Science 327:693–696

Jansa P, Grummt I (1999) Mechanism of transcription termination: PTRF interacts with the largest subunit of RNA polymerase I and dissociates paused transcription complexes from yeast and mouse. Mol Gen Genet 262:508–514

Jantzen HM, Admon A, Bell SP, Tjian R (1990) Nucleolar transcription factor hUBF contains a DNA-binding motif with homology to HMG proteins. Nature 344:830–836

Jantzen HM, Chow AM, King DS, Tjian R (1992) Multiple domains of the RNA polymerase I activator hUBF interact with the TATA-binding protein complex hSL1 to mediate transcription. Genes Dev 6:1950–1963

Jones HS, Kawauchi J, Braglia P, Alen CM, Kent NA, Proudfoot NJ (2007) RNA polymerase I in yeast transcribes dynamic nucleosomal rDNA. Nat Struct Mol Biol 14:123–130

Kasahara K, Ohtsuki K, Ki S, Aoyama K, Takahashi H, Kobayashi T, Shirahige K, Kokubo T (2007) Assembly of regulatory factors on rRNA and ribosomal protein genes in *Saccharomyces cerevisiae*. Mol Cell Biol 27:6686–6705

Kawauchi J, Mischo H, Braglia P, Rondon A, Proudfoot NJ (2008) Budding yeast RNA polymerases I and II employ parallel mechanisms of transcriptional termination. Genes Dev 22:1082–1092

Kermekchiev M, Workman JL, Pikaard CS (1997) Nucleosome binding by the polymerase I transactivator upstream binding factor displaces linker histone H1. Mol Cell Biol 17:5833–5842

Kihm AJ, Hershey JC, Haystead TA, Madsen CS, Owens GK (1998) Phosphorylation of the rRNA transcription factor upstream binding factor promotes its association with TATA binding protein. Proc Natl Acad Sci USA 95:14816–14820

Knutson BA, Hahn S (2011) Yeast Rrn7 and human TAF1B are TFIIB-related RNA polymerase I general transcription factors. Science 333:1637–1640

Kruhlak M, Crouch EE, Orlov M, Montano C, Gorski SA, Nussenzweig A, Misteli T, Phair RD, Casellas R (2007) The ATM repair pathway inhibits RNA polymerase I transcription in response to chromosome breaks. Nature 447:730–734

Kuhn A, Grummt I (1987) A novel promoter in the mouse rDNA spacer is active in vivo and in vitro. EMBO J 6:3487–3492

Kuhn CD, Geiger SR, Baumli S, Gartmann M, Gerber J, Jennebach S, Mielke T, Tschochner H, Beckmann R, Cramer P (2007) Functional architecture of RNA polymerase I. Cell 131:1260–1272

Labhart P, Reeder RH (1984) Enhancer-like properties of the 60/81 bp elements in the ribosomal gene spacer of *Xenopus laevis*. Cell 37:285–289

Laferte A, Favry E, Sentenac A, Riva M, Carles C, Chedin S (2006) The transcriptional activity of RNA polymerase I is a key determinant for the level of all ribosome components. Genes Dev 20:2030–2040

Langst G, Blank TA, Becker PB, Grummt I (1997) RNA polymerase I transcription on nucleosomal templates: the transcription termination factor TTF-I induces chromatin remodeling and relieves transcriptional repression. EMBO J 16:760–768

Lawrence RJ, Earley K, Pontes O, Silva M, Chen ZJ, Neves N, Viegas W, Pikaard CS (2004) A concerted DNA methylation/histone methylation switch regulates rRNA gene dosage control and nucleolar dominance. Mol Cell 13:599–609

Leporé N, Lafontaine DLJ (2011) A functional interface at the rDNA connects rRNA synthesis, pre-rRNA processing and nucleolar surveillance in budding yeast. PLoS One 6:e24962

Lessard F, Morin F, Ivanchuk S, Langlois F, Stefanovsky V, Rutka J, Moss T (2010) The ARF tumor suppressor controls ribosome biogenesis by regulating the RNA polymerase I transcription factor TTF-I. Mol Cell 38:539–550

Lewis JD, Tollervey D (2000) Like attracts like: getting RNA processing together in the nucleus. Science 288:1385–1389

Liebhaber SA, Wolf S, Schlessinger D (1978) Differences in rRNA metabolism of primary and SV40-transformed human fibroblasts. Cell 13:121–127

Lin CY, Navarro S, Reddy S, Comai L (2006) CK2-mediated stimulation of Pol I transcription by stabilization of UBF-SL1 interaction. Nucleic Acids Res 34:4752–4766

Mais C, Wright JE, Prieto JL, Raggett SL, McStay B (2005) UBF-binding site arrays form pseudo-NORs and sequester the RNA polymerase I transcription machinery. Genes Dev 19:50–64

Mayer C, Grummt I (2005) Cellular stress and nucleolar function. Cell Cycle 4:1036–1038

Mayer C, Zhao J, Yuan X, Grummt I (2004) mTOR-dependent activation of the transcription factor TIF-IA links rRNA synthesis to nutrient availability. Genes Dev 18:423–434

Mayer C, Bierhoff H, Grummt I (2005) The nucleolus as a stress sensor: JNK2 inactivates the transcription factor TIF-IA and down-regulates rRNA synthesis. Genes Dev 19:933–941

Mayer C, Schmitz KM, Li J, Grummt I, Santoro R (2006) Intergenic transcripts regulate the epigenetic state of rRNA genes. Mol Cell 22:351–361

Mayer C, Neubert M, Grummt I (2008) The structure of NoRC-associated RNA is crucial for targeting the chromatin remodelling complex NoRC to the nucleolus. EMBO Rep 9:774–780

McStay B, Grummt I (2008) The epigenetics of rRNA genes: from molecular to chromosome biology. Annu Rev Cell Dev Biol 24:131–157

Merz K, Hondele M, Goetze H, Gmelch K, Stoeckl U, Griesenbeck J (2008) Actively transcribed rRNA genes in *S. cerevisiae* are organized in a specialized chromatin associated with the high-mobility group protein Hmo1 and are largely devoid of histone molecules. Genes Dev 22:1190–1204

Milkereit P, Tschochner H (1998) A specialized form of RNA polymerase I, essential for initiation and growth-dependent regulation of rRNA synthesis, is disrupted during transcription. EMBO J 17:3692–3703

Miller G, Panov KI, Friedrich JK, Trinkle-Mulcahy L, Lamond AI, Zomerdijk JC (2001) hRRN3 is essential in the SL1-mediated recruitment of RNA polymerase I to rRNA gene promoters. EMBO J 20:1373–1382

Moore PB, Steitz TA (2002) The involvement of RNA in ribosome function. Nature 418:229–235

Moss T (2004) At the crossroads of growth control; making ribosomal RNA. Curr Opin Genet Dev 14:210–217

Moss T, Stefanovsky VY (2002) At the center of eukaryotic life. Cell 109:545–548

Murano K, Okuwaki M, Hisaoka M, Nagata K (2008) Transcription regulation of the rRNA gene by a multifunctional nucleolar protein, B23/nucleophosmin, through its histone chaperone activity. Mol Cell Biol 28:3114–3126

Murayama A, Ohmori K, Fujimura A, Minami H, Yasuzawa-Tanaka K, Kuroda T, Oie S, Daitoku H, Okuwaki M, Nagata K, Fukamizu A, Kimura K, Shimizu T, Yanagisawa J (2008) Epigenetic control of rDNA loci in response to intracellular energy status. Cell 133:627–639

Muth V, Nadaud S, Grummt I, Voit R (2001) Acetylation of TAF(I)68, a subunit of TIF-IB/SL1, activates RNA polymerase I transcription. EMBO J 20:1353–1362

Naidu S, Friedrich JK, Russell J, Zomerdijk JC (2011) TAF1B is a TFIIB-like component of the basal transcription machinery for RNA polymerase I. Science 333:1640–1642

Narla A, Ebert BL (2010) Ribosomopathies: human disorders of ribosome dysfunction. Blood 115:3196–3205

Nemeth A, Langst G (2011) Genome organization in and around the nucleolus. Trends Genet 27:149–156

Nemeth A, Guibert S, Tiwari VK, Ohlsson R, Langst G (2008) Epigenetic regulation of TTF-I-mediated promoter-terminator interactions of rRNA genes. EMBO J 27:1255–1265

Nemeth A, Conesa A, Santoyo-Lopez J, Medina I, Montaner D, Peterfia B, Solovei I, Cremer T, Dopazo J, Langst G (2010) Initial genomics of the human nucleolus. PLoS Genet 6:e1000889

O'Sullivan AC, Sullivan GJ, McStay B (2002) UBF binding in vivo is not restricted to regulatory sequences within the vertebrate ribosomal DNA repeat. Mol Cell Biol 22:657–668

Panov KI, Friedrich JK, Zomerdijk JC (2001) A step subsequent to preinitiation complex assembly at the ribosomal RNA gene promoter is rate limiting for human RNA polymerase I-dependent transcription. Mol Cell Biol 21:2641–2649

Panov KI, Friedrich JK, Russell J, Zomerdijk JC (2006a) UBF activates RNA polymerase I transcription by stimulating promoter escape. EMBO J 25:3310–3322

Panov KI, Panova TB, Gadal O, Nishiyama K, Saito T, Russell J, Zomerdijk JC (2006b) RNA polymerase I-specific subunit CAST/hPAF49 has a role in the activation of transcription by upstream binding factor. Mol Cell Biol 26:5436–5448

Panova TB, Panov KI, Russell J, Zomerdijk JC (2006) Casein kinase 2 associates with initiation-competent RNA polymerase I and has multiple roles in ribosomal DNA transcription. Mol Cell Biol 26:5957–5968

Paredes S, Maggert KA (2009) Ribosomal DNA contributes to global chromatin regulation. Proc Natl Acad Sci USA 106:17829–17834

Paule MR, White RJ (2000) Survey and summary: transcription by RNA polymerases I and III. Nucleic Acids Res 28:1283–1298

Pelletier G, Stefanovsky VY, Faubladier M, Hirschler-Laszkiewicz I, Savard J, Rothblum LI, Cote J, Moss T (2000) Competitive recruitment of CBP and Rb-HDAC regulates UBF acetylation and ribosomal transcription. Mol Cell 6:1059–1066

Percipalle P, Fomproix N, Cavellan E, Voit R, Reimer G, Kruger T, Thyberg J, Scheer U, Grummt I, Farrants AK (2006) The chromatin remodelling complex WSTF-SNF2h interacts with nuclear myosin 1 and has a role in RNA polymerase I transcription. EMBO Rep 7:525–530

Peyroche G, Milkereit P, Bischler N, Tschochner H, Schultz P, Sentenac A, Carles C, Riva M (2000) The recruitment of RNA polymerase I on rDNA is mediated by the interaction of the A43 subunit with Rrn3. EMBO J 19:5473–5482

Pietrzak M, Rempala G, Nelson PT, Zheng JJ, Hetman M (2011) Epigenetic silencing of nucleolar rRNA genes in Alzheimer's disease. PLoS One 6:e22585

Pistoni M, Verrecchia A, Doni M, Guccione E, Amati B (2010) Chromatin association and regulation of rDNA transcription by the Ras-family protein RasL11a. EMBO J 29:1215–1224

Pontes O, Li CF, Costa Nunes P, Haag J, Ream T, Vitins A, Jacobsen SE, Pikaard CS (2006) The Arabidopsis chromatin-modifying nuclear siRNA pathway involves a nucleolar RNA processing center. Cell 126:79–92

Prescott EM, Osheim YN, Jones HS, Alen CM, Roan JG, Reeder RH, Beyer AL, Proudfoot NJ (2004) Transcriptional termination by RNA polymerase I requires the small subunit Rpa12p. Proc Natl Acad Sci USA 101:6068–6073

Preuss SB, Costa-Nunes P, Tucker S, Pontes O, Lawrence RJ, Mosher R, Kasschau KD, Carrington JC, Baulcombe DC, Viegas W, Pikaard CS (2008) Multimegabase silencing in nucleolar dominance involves siRNA-directed DNA methylation and specific methylcytosine-binding proteins. Mol Cell 32:673–684

Reeder RH (1999) Regulation of RNA polymerase I transcription in yeast and vertebrates. Prog Nucleic Acid Res Mol Biol 62:293–327

Rickards B, Flint SJ, Cole MD, LeRoy G (2007) Nucleolin is required for RNA polymerase I transcription in vivo. Mol Cell Biol 27:937–948

Rose KM, Szopa J, Han FS, Cheng YC, Richter A, Scheer U (1988) Association of DNA topoisomerase I and RNA polymerase I: a possible role for topoisomerase I in ribosomal gene transcription. Chromosoma 96:411–416

Rudloff U, Eberhard D, Tora L, Stunnenberg H, Grummt I (1994) TBP-associated factors interact with DNA and govern species specificity of RNA polymerase I transcription. EMBO J 13:2611–2616

Ruggero D, Pandolfi PP (2003) Does the ribosome translate cancer? Nat Rev Cancer 3:179–192

Russell J, Zomerdijk JC (2005) RNA-polymerase-I-directed rDNA transcription, life and works. Trends Biochem Sci 30:87–96

Sanij E, Hannan RD (2008) Chromatin organization and expression. Genome Biol 9:305

Sanij E, Poortinga G, Sharkey K, Hung S, Holloway TP, Quin J, Robb E, Wong LH, Thomas WG, Stefanovsky V, Moss T, Rothblum L, Hannan KM, McArthur GA, Pearson RB, Hannan RD (2008) UBF levels determine the number of active ribosomal RNA genes in mammals. J Cell Biol 183:1259–1274

Santoro R, Grummt I (2001) Molecular mechanisms mediating methylation-dependent silencing of ribosomal gene transcription. Mol Cell 8:719–725

Santoro R, Grummt I (2005) Epigenetic mechanism of rRNA gene silencing: temporal order of NoRC-mediated histone modification, chromatin remodeling, and DNA methylation. Mol Cell Biol 25:2539–2546

Santoro R, Li J, Grummt I (2002) The nucleolar remodeling complex NoRC mediates heterochromatin formation and silencing of ribosomal gene transcription. Nat Genet 32:393–396

Santoro R, Schmitz KM, Sandoval J, Grummt I (2010) Intergenic transcripts originating from a subclass of ribosomal DNA repeats silence ribosomal RNA genes in trans. EMBO Rep 11:52–58

Saunders A, Core LJ, Lis JT (2006) Breaking barriers to transcription elongation. Nat Rev Mol Cell Biol 7:557–567

Schlesinger S, Selig S, Bergman Y, Cedar H (2009) Allelic inactivation of rDNA loci. Genes Dev 23:2437–2447

Schmitz KM, Schmitt N, Hoffmann-Rohrer U, Schafer A, Grummt I, Mayer C (2009) TAF12 recruits Gadd45a and the nucleotide excision repair complex to the promoter of rRNA genes leading to active DNA demethylation. Mol Cell 33:344–353

Schmitz KM, Mayer C, Postepska A, Grummt I (2010) Interaction of noncoding RNA with the rDNA promoter mediates recruitment of DNMT3b and silencing of rRNA genes. Genes Dev 24:2264–2269

Schnapp G, Santori F, Carles C, Riva M, Grummt I (1994) The HMG box-containing nucleolar transcription factor UBF interacts with a specific subunit of RNA polymerase I. EMBO J 13:190–199

Schneider DA, French SL, Osheim YN, Bailey AO, Vu L, Dodd J, Yates JR, Beyer AL, Nomura M (2006) RNA polymerase II elongation factors Spt4p and Spt5p play roles in transcription elongation by RNA polymerase I and rRNA processing. Proc Natl Acad Sci USA 103:12707–12712

Schneider DA, Michel A, Sikes ML, Vu L, Dodd JA, Salgia S, Osheim YN, Beyer AL, Nomura M (2007) Transcription elongation by RNA polymerase I is linked to efficient rRNA processing and ribosome assembly. Mol Cell 26:217–229

Schultz MC, Brill SJ, Ju Q, Sternglanz R, Reeder RH (1992) Topoisomerases and yeast rRNA transcription: negative supercoiling stimulates initiation and topoisomerase activity is required for elongation. Genes Dev 6:1332–1341

Seither P, Zatsepina O, Hoffmann M, Grummt I (1997) Constitutive and strong association of PAF53 with RNA polymerase I. Chromosoma 106:216–225

Shiratori M, Suzuki T, Itoh C, Goto M, Furuichi Y, Matsumoto T (2002) WRN helicase accelerates the transcription of ribosomal RNA as a component of an RNA polymerase I-associated complex. Oncogene 21:2447–2454

Shiue CN, Berkson RG, Wright AP (2009) c-Myc induces changes in higher order rDNA structure on stimulation of quiescent cells. Oncogene 28:1833–1842

Siehl D, Chua BH, Lautensack-Belser N, Morgan HE (1985) Faster protein and ribosome synthesis in thyroxine-induced hypertrophy of rat heart. Am J Physiol 248:C309–319

Stefanovsky VY, Pelletier G, Hannan R, Gagnon-Kugler T, Rothblum LI, Moss T (2001) An immediate response of ribosomal transcription to growth factor stimulation in mammals is mediated by ERK phosphorylation of UBF. Mol Cell 8:1063–1073

Stefanovsky V, Langlois F, Gagnon-Kugler T, Rothblum LI, Moss T (2006) Growth factor signaling regulates elongation of RNA polymerase I transcription in mammals via UBF phosphorylation and r-chromatin remodeling. Mol Cell 21:629–639

Tanaka Y, Okamoto K, Teye K, Umata T, Yamagiwa N, Suto Y, Zhang Y, Tsuneoka M (2010) JmjC enzyme KDM2A is a regulator of rRNA transcription in response to starvation. EMBO J 29:1510–1522

Tschochner H, Hurt E (2003) Pre-ribosomes on the road from the nucleolus to the cytoplasm. Trends Cell Biol 13:255–263

Tseng H, Chou W, Wang J, Zhang X, Zhang S, Schultz RM (2008) Mouse ribosomal RNA genes contain multiple differentially regulated variants. PLoS One 3:e1843

Tuan JC, Zhai W, Comai L (1999) Recruitment of TATA-binding protein-TAFI complex SL1 to the human ribosomal DNA promoter is mediated by the carboxy-terminal activation domain of upstream binding factor (UBF) and is regulated by UBF phosphorylation. Mol Cell Biol 19:2872–2879

Tucker S, Vitins A, Pikaard CS (2010) Nucleolar dominance and ribosomal RNA gene silencing. Curr Opin Cell Biol 22:351–356

Valdez BC, Henning D, So RB, Dixon J, Dixon MJ (2004) The Treacher Collins syndrome (TCOF1) gene product is involved in ribosomal DNA gene transcription by interacting with upstream binding factor. Proc Natl Acad Sci USA 101:10709–10714

van de Nobelen S, Rosa-Garrido M, Leers J, Heath H, Soochit W, Joosen L, Jonkers I, Demmers J, van der Reijden M, Torrano V, Grosveld F, Delgado MD, Renkawitz R, Galjart N, Sleutels F (2010) CTCF regulates the local epigenetic state of ribosomal DNA repeats. Epigenetics Chromatin 3:19

van Koningsbruggen S, Gierlinski M, Schofield P, Martin D, Barton GJ, Ariyurek Y, den Dunnen JT, Lamond AI (2010) High-resolution whole-genome sequencing reveals that specific chromatin domains from most human chromosomes associate with nucleoli. Mol Biol Cell 21:3735–3748

Vincent T, Kukalev A, Andang M, Pettersson R, Percipalle P (2008) The glycogen synthase kinase (GSK) 3beta represses RNA polymerase I transcription. Oncogene 27:5254–5259

Voit R, Grummt I (2001) Phosphorylation of UBF at serine 388 is required for interaction with RNA polymerase I and activation of rDNA transcription. Proc Natl Acad Sci USA 98:13631–13636

Voit R, Schnapp A, Kuhn A, Rosenbauer H, Hirschmann P, Stunnenberg HG, Grummt I (1992) The nucleolar transcription factor mUBF is phosphorylated by casein kinase II in the C-terminal hyperacidic tail which is essential for transactivation. EMBO J 11:2211–2218

Voit R, Schafer K, Grummt I (1997) Mechanism of repression of RNA polymerase I transcription by the retinoblastoma protein. Mol Cell Biol 17:4230–4237

Voit R, Hoffmann M, Grummt I (1999) Phosphorylation by G1-specific cdk-cyclin complexes activates the nucleolar transcription factor UBF. EMBO J 18:1891–1899

Werner M, Thuriaux P, Soutourina J (2009) Structure-function analysis of RNA polymerases I and III. Curr Opin Struct Biol 19:740–745

White RJ (2008) RNA polymerases I and III, non-coding RNAs and cancer. Trends Genet 24:622–629

Whitehead CM, Winkfein RJ, Fritzler MJ, Rattner JB (1997) ASE-1: a novel protein of the fibrillar centres of the nucleolus and nucleolus organizer region of mitotic chromosomes. Chromosoma 106:493–502

Wittner M, Hamperl S, Stockl U, Seufert W, Tschochner H, Milkereit P, Griesenbeck J (2011) Establishment and maintenance of alternative chromatin states at a multicopy gene locus. Cell 145:543–554

Yamamoto K, Yamamoto M, Hanada K, Nogi Y, Matsuyama T, Muramatsu M (2004) Multiple protein-protein interactions by RNA polymerase I-associated factor PAF49 and role of PAF49 in rRNA transcription. Mol Cell Biol 24:6338–6349

Ye J, Zhao J, Hoffmann-Rohrer U, Grummt I (2008) Nuclear myosin I acts in concert with polymeric actin to drive RNA polymerase I transcription. Genes Dev 22:322–330

Yuan X, Feng W, Imhof A, Grummt I, Zhou Y (2007) Activation of RNA polymerase I transcription by cockayne syndrome group B protein and histone methyltransferase G9a. Mol Cell 27:585–595

Zentner GE, Hurd EA, Schnetz MP, Handoko L, Wang C, Wang Z, Wei C, Tesar PJ, Hatzoglou M, Martin DM, Scacheri PC (2010) CHD7 functions in the nucleolus as a positive regulator of ribosomal RNA biogenesis. Hum Mol Genet 19:3491–3501

Zetterberg A, Killander D (1965) Quantitative cytophotometric and autoradiographic studies on the rate of protein synthesis during interphase in mouse fibroblasts in vitro. Exp Cell Res 40:1–11

Zhai W, Comai L (2000) Repression of RNA polymerase I transcription by the tumor suppressor p53. Mol Cell Biol 20:5930–5938

Zhang H, Wang JC, Liu LF (1988) Involvement of DNA topoisomerase I in transcription of human ribosomal RNA genes. Proc Natl Acad Sci USA 85:1060–1064

Zhang C, Comai L, Johnson DL (2005) PTEN represses RNA Polymerase I transcription by disrupting the SL1 complex. Mol Cell Biol 25:6899–6911

Zhang Y, Sikes ML, Beyer AL, Schneider DA (2009) The Paf1 complex is required for efficient transcription elongation by RNA polymerase I. Proc Natl Acad Sci USA 106:2153–2158

Zhang Y, Smith AD, Renfrow MB, Schneider DA (2010) The RNA polymerase-associated factor 1 complex (Paf1C) directly increases the elongation rate of RNA polymerase I and is required for efficient regulation of rRNA synthesis. J Biol Chem 285:14152–14159

Zhao J, Yuan X, Frodin M, Grummt I (2003) ERK-dependent phosphorylation of the transcription initiation factor TIF-IA is required for RNA polymerase I transcription and cell growth. Mol Cell 11:405–413

Zhou Y, Santoro R, Grummt I (2002) The chromatin remodeling complex NoRC targets HDAC1 to the ribosomal gene promoter and represses RNA polymerase I transcription. EMBO J 21:4632–4640

Zomerdijk JC, Beckmann H, Comai L, Tjian R (1994) Assembly of transcriptionally active RNA polymerase I initiation factor SL1 from recombinant subunits. Science 266:2015–2018

Chapter 11
The RNA Polymerase II Transcriptional Machinery and Its Epigenetic Context

Maria J. Barrero and Sohail Malik

Abstract RNA polymerase II (Pol II) is the main engine that drives transcription of protein-encoding genes in eukaryotes. Despite its intrinsic subunit complexity, Pol II is subject to a host of factors that regulate the multistep transcription process. Indeed, the hallmark of the transcription cycle is the dynamic association of Pol II with initiation, elongation and other factors. In addition, Pol II transcription is regulated by a series of cofactors (coactivators and corepressors). Among these, the Mediator has emerged as one of the key regulatory factors for Pol II. Transcription by Pol II takes place in the context of chromatin, which is subject to numerous epigenetic modifications. This chapter mainly summarizes the various biochemical mechanisms that determine formation and function of a Pol II preinitiation complex (PIC) and those that affect its progress along the gene body (elongation). It further examines the various epigenetic modifications that the Pol II machinery encounters, especially in certain developmental contexts, and highlights newer evidence pointing to a likely close interplay between this machinery and factors responsible for the chromatin modifications.

11.1 Introduction

The general transcription machinery of a eukaryotic cell constitutes the nuts and bolts through which gene regulation is largely achieved. In recent years, epigenetic modifications that are acquired by a given gene have come to be seen as providing

---

M.J. Barrero
Center for Regenerative Medicine, Dr Aiguader 88, Barcelona, Spain
e-mail: mjbarrero@cmrb.eu

S. Malik (✉)
Laboratory of Biochemistry and Molecular Biology, Rockefeller University,
1230 York Avenue, New York, NY 10465, USA
e-mail: maliks@rockefeller.edu

the major blueprint for how its transcriptional program will be executed over the longer term. However, the actual translation of the program into actionable output remains the task of the general transcriptional machinery.

In addition to RNA polymerases, this machinery consists of a series of initiation factors that both confer specificity and help launch the polymerase so that it can actually transcribe the body of the gene. During this phase the polymerase is associated with a set of elongation factors that, in addition to ensuring processivity, regulate polymerase progress across the gene through various checkpoints and hurdles that include nucleosomes. The overall process of transcription is thus characterized by a multitude of factors that dynamically interact with the polymerases.

Yet the machineries responsible for the apparently distinct processes underlying transcription on the one hand and epigenetic modifications on the other might intersect, both structurally and functionally. This chapter will mainly describe the workings of the general transcription machinery. It will also briefly examine some selected examples that illustrate the potential for interactions of this machinery with epigenetically modified chromatin and factors responsible for these modifications. Finally, it will touch upon some important biological consequences of such interactions.

11.2 Transcription Is a Multistep Process

Given the complexity of their genomes, eukaryotic cells possess three RNA polymerases, each dedicated to the transcription of a specific class of genes (Roeder and Rutter 1969). Two additional variant polymerases have also been identified in plant cells (Pikaard et al. 2008). RNA polymerase II (Pol II), the enzyme responsible for transcription of genes encoding proteins and some small RNAs, will constitute the subject of this chapter. However, it should be kept in mind that the mechanisms by which the various polymerases assemble into initiation–competent complexes at their promoters are highly analogous and are mediated by functionally equivalent, as well as shared, factors. Thus, much of what we currently know about Pol II would potentially be of relevance to the other polymerases, each of which likely operates within its own particular epigenetic environment.

Pol II genes are under the control of numerous transcription factors that carry regulatory information emanating for example, from developmental, hormonal or pharmacological cues (Brivanlou and Darnell 2002). Typically these factors bind within the vicinity of the gene promoter and through direct or indirect interactions with the factors associated with Pol II (see further below) they parlay the signals into appropriate transcriptional responses. Interestingly, and relevant to the present topic, chromatinization of the template dictates that the activation process take place, broadly, in two steps (Roeder 1998, 2005). In the first step (Fig. 11.1), the transcription factors, in concert with chromatin factors (discussed below), act to breach the chromatin barrier. It is only in the second step that Pol II gains access to its cognate promoter. It has been argued that in contrast to the relatively straightforward transcriptional activation pathways in prokaryotes, chromatin has imparted a distinct logic to transcriptional regulation in eukaryotes (Struhl 1999). The latter have thus both evolved strategies to deal with the chromatin and developed mechanisms that couple chromatin remodeling with the transcription process *per se*. Otherwise, the core transcriptional machineries of prokaryotes and eukaryotes display a remarkable degree of evolutionary conservation (Cramer 2002; Ebright 2000).

11.3 Pol II and the Preinitiation Complex

Pol II is a 12-subunit enzyme whose central mass consists of the two largest subunits (RPB1 and RPB2) that are orthologs of prokaryotic RNA polymerase components β and β'. Together, they generate a jaw-like structure with a cleft running along its length (Cramer et al. 2008). RPB3 and RPB11, the orthologs of the two prokaryotic α subunits, are located distal to the cleft. Overall, the central core of Pol II is virtually indistinguishable from the corresponding prokaryotic structure (Cramer et al. 2000; Zhang et al. 1999). The remaining (relatively small) subunits, which are mostly eukaryote-specific but include RPB6, the ortholog of prokaryotic ω subunit (Minakhin et al. 2001), are dispersed around this core. The active site, which is responsible for templated incorporation of nucleotides into the growing RNA chain, consists of highly conserved residues and catalytic magnesium ions buried deep within the cleft (Cramer 2002).

Although catalytically competent, Pol II, like prokaryotic RNA polymerases that are dependent on cognate sigma factors, is incapable of accurately initiating from promoter-directed transcription start sites (TSS) in the absence of general transcription factors (GTFs) that include TFIIA, TFIIB, TFIID, TFIIE, TFIIF, and TFIIH (Hahn 2004; Orphanides et al. 1996; Roeder 1996; Thomas and Chiang 2006). The primary role of these GTFs is thus to accurately position and orient Pol II on the promoter and to facilitate initial access of its catalytic site to the transcribed strand of the template. Recall that promoters essentially consist of some combination of a series of defined DNA elements in the vicinity of the TSS (Juven-Gershon et al. 2008). These elements include the TATA box (located at −30 relative to the TSS) and the initiator (+1), as well as more recently defined motifs such as MTE, DPE, and DCE that are mostly located downstream of the TSS. Together with Pol II, the GTFs assemble into a preinitiation complex (PIC), a key intermediate in the transcription activation pathway. To a very large extent, it is the formation and function of the PIC that controls gene activity (Roeder 1998, 2005).

A high-resolution structure of a complete Pol II PIC has not yet been described (Kornberg 2007). However, multiple lines of evidence, including gel mobility shift studies (Buratowski et al. 1989; Flores et al. 1992) and chemical cross-linking (Chen and Hahn 2004; Eichner et al. 2010) have provided detailed insights into how it is assembled and organized, at least on TATA-containing core promoters. PIC assembly begins with the binding of the TBP subunit of TFIID to the TATA box via minor groove interactions (Burley and Roeder 1996). On TATA-less promoters, the TAF components of TFIID are thought to contribute via interactions with any of the

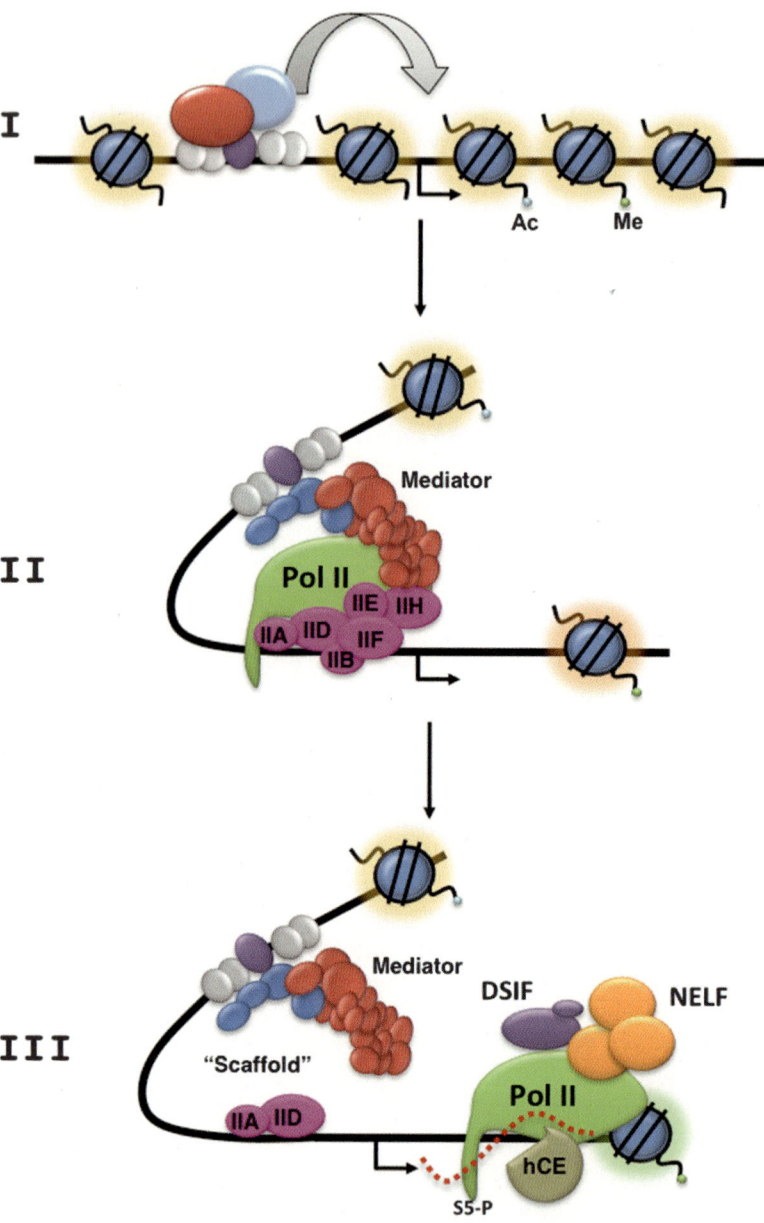

Fig. 11.1 Transcription by Pol II is a multi-step process. The figure depicts a simplified transcription activation pathway that commences when transcriptional activators bind to their cognate sites in the regulatory region of the gene that may be buried in chromatin. These factors recruit a series of chromatin coactivators that can both covalently modify nucleosomes at specific histone residues and mobilize the nucleosomes via ATP-requiring reactions. The resulting intermediate (I) contains chromatin that is characterized by distinct covalent modifications such as acetylation (Ac) and methylation (Me); some nucleosomes may be "evicted". The activators then recruit Mediator (intermediate II). Although the intact Mediator consisting of the core and the kinase module might

other motifs that might be present in the promoter (Burke and Kadonaga 1997; Martinez et al. 1995). In some cases, as-yet-unidentified factors might additionally be required (Martinez et al. 1998). TBP binding to promoter DNA leads to an unusual distortion and sharp bending of the template, which potentially contributes to the final topology of the PIC. TFIIA, which binds upstream of the TATA box, greatly facilitates this interaction. Additional stabilization comes from the interaction of the C-terminal core domains of TFIIB, which makes contacts both with the underside of TBP (Nikolov et al. 1995) and with DNA upstream and downstream of the TATA box, which constitutes the B-recognition element (BRE) (Lagrange et al. 1998). The N-terminal domain of TFIIB makes intimate contacts with Pol II (Kostrewa et al. 2009; Liu et al. 2010), which is recruited to the PIC in conjunction with TFIIF (Flores et al. 1992). In this way, Pol II is positioned accurately and in the correct orientation. TFIIE and TFIIH are the last to enter the PIC and play active roles in the firing of the PIC (Buratowski et al. 1989; Flores et al. 1992; Ohkuma and Roeder 1994).

As to the generality of action of the GTFs, data from genome-wide analyses are not comprehensive enough at this time to conclude whether all the GTFs are required at all genes. Nonetheless, studies of TBP paralogs (TRFs) have revealed that in certain developmental contexts, canonical TFIID might be replaced with a pared down counterpart consisting of TRF3 and at most one TAF (TAF3) (Deato et al. 2008; Deato and Tjian 2007).

11.4 PIC Function

PIC assembly is followed by a series of well-orchestrated events that ultimately lead to RNA synthesis in earnest. The PIC can be thought of as being the counterpart of the prokaryotic "closed complex". Therefore, promoter melting, the process in which the double-stranded DNA around the TSS is partially unwound, is a prerequisite for the onset of RNA synthesis. Of the two ATPases/helicases present in TFIIH, ERCC3 (XPB) has been implicated in promoter melting (Lin et al. 2005; Tirode et al. 1999). However, unlike conventional helicases, TFIIH seems to unwind promoter DNA through a novel torsional mechanism that entails ATP hydrolysis (Kim et al. 2000). TFIIB and TFIIE also contribute to promoter melting,

Fig. 11.1 (contiuned) be recruited with subsequent loss of the kinase module as the PIC matures, only the core Mediator is shown. PIC assembly entails entry of the various GTFs (TFIIA, TFIIB, TFIID, TFIIE, TFIIF, and TFIIH) and Pol II. After transcription initiation (abortive initiation, which accompanies this step is not illustrated), Pol II clears the promoter. Prior to fully entering the elongation phase (see Fig. 11.2), Pol II, which by now is phosphorylated at Ser5, may pass through a capping checkpoint. The scaffold complex (intermediate III) containing a subset of GTFs and Mediator remains behind at the promoter and can contribute to subsequent rounds of transcription. At the capping checkpoint, Pol II becomes associated with elongation factors including DSIF and NELF. The capping enzyme (hCE) modifies (7MeG) the nascent RNA. Pol II is released from this pause through recruitment of P-TEFb (see Fig. 11.2 legend)

the former most likely through stabilization of the melted state via interactions with the non-template strand (Kostrewa et al. 2009). Melting takes place in multiple steps. Initially, a region extending from −9 to +2 is unwound (Holstege et al. 1997). Subsequently, the melted region ("bubble") is extended, first to +4, and later further downstream, to allow templated phosphodiester bond formation to occur. The discontinuous nature of the bubble extension and early transcription events is particularly noteworthy. Thus, as in prokaryotic systems, the nascent RNA is not extended right away. Rather, abortively synthesized short RNAs are first generated in a stoichiometric excess. It is believed that this is a result of the strains that the PIC is subjected to in this step (Pal et al. 2005). As the strain is periodically relieved through bubble collapse, transcription is aborted leading to release of the short RNAs. At the structural level, this can also be attributed to the N-terminal B-finger/B-reader of TFIIB that reaches into the active site of Pol II, where it can sterically clash with the nascent RNA (Bushnell et al. 2004; Kostrewa et al. 2009; Pal et al. 2005). Because of the divergent demands on the system (bubble collapse versus RNA chain extension), multiple attempts are needed for ensuring that long RNA chain synthesis will eventually win out in so far as TFIIB is involved. It therefore seems also to play an important role in ensuring an orderly transition into the productive phase of transcription by overseeing the process of promoter escape.

Strictly speaking, promoter escape refers to the process whereby Pol II relinquishes its initial contacts within the PIC and begins to acquire a conformation that is more typical of the elongation phase. This process, too, entails multiple substeps. As the RNA is extended past circa +9, the PIC begins to undergo extensive rearrangements (Holstege et al. 1997). In addition to Pol II movements and associated changes in the template structure, the PIC loses a subset of GTFs. While there is some uncertainty as to which subunits are lost, the most convincing data come from studies in yeast and indicate that following Pol II escape a "scaffold" complex consisting of TFIIA, TFIID, TFIIE, TFIIH, and Mediator (below) is left behind; TFIIB is likely shed while TFIIF travels with Pol II (Yudkovsky et al. 2000). The resulting scaffold can serve as an efficient platform for assembly of complete PICs in subsequent rounds of transcription requiring only TFIIB and TFIIF. With regard to the mechanisms underlying promoter escape, in addition to the above-mentioned contribution of TFIIB, the TFIIH-associated ATPase activity likely plays a dominant role in the escape (Lin et al. 2005). However, the precise mechanisms remain uncharacterized.

A hallmark of the escaping Pol II is that it undergoes phosphorylation at the Serine 5 residues of an amino acid sequence ($YS_2PTS_5PS_7$) that is present in highly repeated copies (26 in yeast; 52 in mammals) at the C-terminal end of the RPB1 subunit (CTD) (Buratowski 2009; Egloff and Murphy 2008). Ser5 phosphorylation is carried out by another enzymatic subunit of TFIIH, CDK7. However, it is important to note that, at least in experimental systems reconstituted from purified components, this modification is not likely to be a cause of Pol II escape (Akoulitchev et al. 1995; Gu et al. 1999; Li and Kornberg 1994). Nonetheless, in vivo, the phosphorylation could contribute to this step, for example, by weakening some protein-protein interactions that anchor Pol II in the PIC (Naar et al. 2002). As discussed below, it is more likely that this modification represents a "mark" of a Pol II molecule that has successfully completed the process of initiation and is now competent to respond to factors involved in transcription elongation and related processes (below). Indeed, highly phosphorylated Pol II is incapable of entering the PIC necessitating dephosphorylation by, among others, Rtr1 (Mosley et al. 2009) and the TFIIF-dependent FCP1 phosphatases (Kobor et al. 1999).

11.5 Factors Affecting Pol II Transcription Post-initiation

After Pol II clears the promoter, it remains subject to numerous additional controls, this time through general transcription elongation factors (Saunders et al. 2006; Sims et al. 2004). The so-called capping checkpoint is the first major regulatory site encountered by Pol II. Most eukaryotic mRNAs bear a 7-methyl guanosine modification at their 5′ ends, which is appended by the capping enzyme at this point as the nascent RNA begins to emerge from the Pol II RNA exit channel. This step coincides with a transient pause that Pol II undergoes following association with the elongation factors DSIF and NELF. Pol II resumes transcription when the negative effects of DSIF and NELF are reversed through the action of a positively acting elongation factor P-TEFb, which may be recruited via an activator (Rahl et al. 2010) and as part of a much larger complex, the "super elongation complex" (SEC), which contains additional elongation factors (Smith et al. 2011). The CDK9 kinase subunit of P-TEFb phosphorylates multiple factors in the paused elongation complex. Its targets include both NELF and DSIF. Phosphorylation of NELF leads to its release whereas phosphorylation of the Spt5 subunit of DSIF transforms it into a positively acting elongation factor that goes on to travel with Pol II further downstream (Fig. 11.2). P-TEFb also phosphorylates the Ser2 residue of the RPB1 CTD, a modification that also is important in coupling transcription elongation to post-transcriptional processes.

Perhaps the most obvious impediment faced by the elongating Pol II is the nucleosome (see further below). Even though the intrinsic forces generated by the advancing Pol II allow it to make some headway into the chromatin (Hodges et al. 2009), the relatively tight wrapping of template DNA around the nucleosome necessitates involvement of additional elongation factors (Sims et al. 2004). These factors use diverse mechanisms that range from mobilization of the nucleosome to alterations of the catalytic properties of Pol II. Among the factors representing the former mechanism are the numerous ATP-dependent chromatin-remodeling factors (such as the SWI/SNF and I-SWI class complexes) that physically displace the nucleosomes (Mohrmann and Verrijzer 2005). Additionally, chaperones, of which FACT is the best characterized, also disrupt the nucleosome by acting as acceptors for displaced histones (Winkler and Luger 2011). Representing the latter mechanism for facilitating transcription through a nucleosome is SII (Guermah et al. 2006; Kireeva et al. 2005), which stimulates the intrinsic RNA cleavage activity of Pol II (Kettenberger et al. 2003). Pol II stalled within nucleosomes tends to backtrack and lose its register; RNA

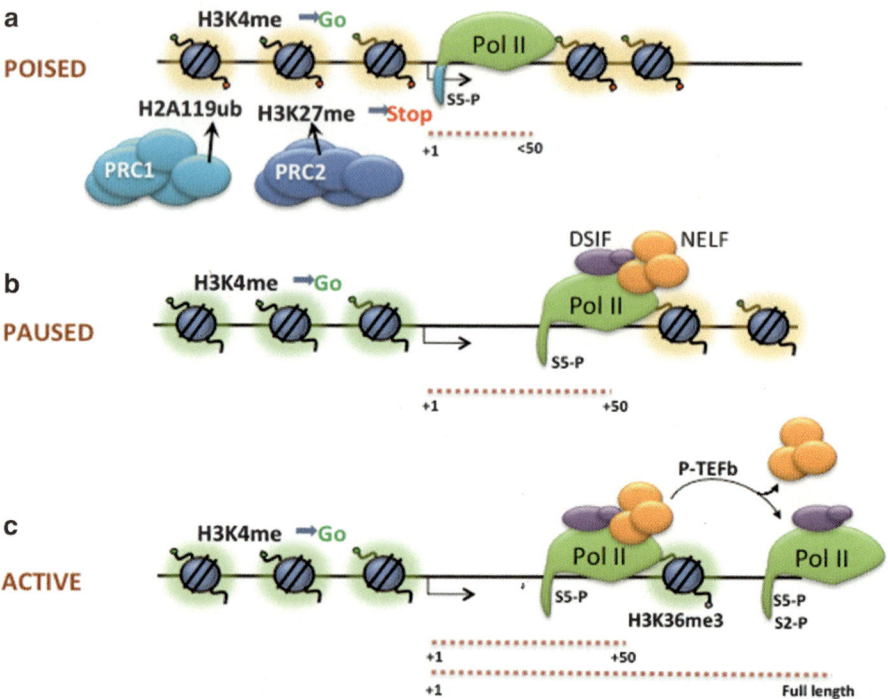

Fig. 11.2 Control of Pol II elongation. (**a**) Genes such as those in ES cells that contain bivalent domains, which are marked by simultaneous methylation of histone H3 at lysine 4 and lysine 27, are characterized by a *poised* Pol II. The poised Pol II is located very close to the TSS (<+50) and is phosphorylated at Ser5 of the CTD, but not Ser2. It is in a conformation that is refractory to detection by antibodies against unmodified CTD. Targets of PRC1 also present ubiquitinated H2A, which through unknown mechanisms has been suggested to restrain Pol II in a poised configuration at these genes. (**b**) Other genes, which might be functionally inactive, carry a *paused* Pol II. While this Pol II is also at a promoter-proximal location, this location is typically somewhat further downstream (circa +50) relative to the poised Pol II in (**a**). The paused Pol II is phosphorylated at Ser5 and may be associated with the elongation factors DSIF and NELF. (**c**) At highly transcribed genes, the pause is released through the phosphorylation of NELF and DSIF by P-TEFb, which leads to NELF release whereas phosphorylated DSIF becomes a positively-acting elongation factor that travels with Pol II. The actively elongating Pol II is further modified at Ser2 and recruits the HMT Set2 that in turn methylates histone H3 at lysine 36 during the elongation phase

cleavage provides the enzyme another chance to go through the blockage. Interestingly, SII appears also to enter the promoter-bound PIC suggesting that it might have additional as-yet-uncharacterized roles in transcription initiation (Kim et al. 2007). At least one other multisubunit elongation factor (PAF1), which has been shown to function in cooperation with SII (Kim et al. 2010), appears to act at the elongation stage both by modulating Pol II elongation rates as well as a device for coupling elongation to the specific epigenetic environment of the gene as discussed below. Although multiple factors that facilitate Pol II transit across this barrier have already been identified, it is likely that given the rate limiting nature of this step – and the potential for regulation – other factors remain to be discovered.

More recently, genome-wide analyses in metazoans have revealed another important post-initiation checkpoint of sorts. These studies show that regardless of whether or not a full-length RNA is transcribed from them, an apparently "paused" Pol II is detectable at a promoter-proximal location (circa +50) at a large number of genes (Margaritis and Holstege 2008; Muse et al. 2007; Zeitlinger et al. 2007). The current interpretation of this observation is that, analogous to what was originally observed for heat-shock genes (Lis 1998), the paused Pol II molecules are likely to reflect species that have initiated but are stalled, potentially in anticipation of a stimulatory signal that would allow them to transition into a full-fledged transcription elongation complex (Nechaev and Adelman 2008). Whereas the precise factorology for this phenomenon has not yet been worked out, nor has it become clear how it relates to the capping checkpoint, there are indications that mechanisms entailing NELF-DSIF-P-TEFb interplay as well as potentially novel pathways might be involved (Muse et al. 2007).

As evident from the preceding discussion, a hallmark of the Pol II transcription cycle is the dynamic association of distinct classes of factors with Pol II. To a significant extent, this is determined by the CTD phosphorylation status of Pol II. It was noted above that Pol II that is competent for PIC formation is essentially in an unphosphorylated state. By the time it arrives at the capping checkpoint it has already acquired the Ser5 modification, which in fact is a prerequisite for interaction with the capping enzyme (Fabrega et al. 2003). Similarly, the actively elongating Pol II is further modified at Ser2. This modification mediates interactions with additional elongation factors (including Spt6 (Sun et al. 2010) and Set2 (Kizer et al. 2005)), as well as RNA processing factors, including those responsible for polyadenylation and termination (Meinhart et al. 2005). In this way, the CTD also is a device for coupling transcription to post-transcriptional processes. Ser7 phosphorylation has also been described and serves to recruit the Integrator complex (a potential transcription cofactor whose precise functions remain uncharacterized) to the atypical U1 and U2 genes (Egloff et al. 2007). In addition to serine phosphorylation, tyrosine phosphorylation (Baskaran et al. 1993), proline isomerization (Xu and Manley 2004) and serine and threonine glycosylation (Kelly et al. 1993) can also occur. Most recently, arginine methylation at one of the non-consensus CTD repeats has also been described (Sims et al. 2011). Together, these modifications have been construed as delineating an underlying "CTD code" that provides readable instructions to the Pol II-interacting factors (Egloff and Murphy 2008).

11.6 Involvement of General Cofactors in Pol II Transcription: Central Role of the Mediator Complex

The general transcription machinery discussed above can function autonomously, especially in what is referred to as "basal" transcription in in vitro studies of transcription (Roeder1996, 1998). However, in the cellular context the machinery is designed to be responsive to regulatory transcription factors. Despite the potential for direct interactions with components of the general machinery, transcriptional activators typically function through intermediary factors, or coactivators (Roeder 1998). As mentioned, given the chromatin environment, it is not surprising that many coactivators are actually chromatin remodeling factors. Importantly, there is a class of coactivators that by virtue of their close association with the general transcriptional machinery function at the level of PIC. Of these, the 30-subunit Mediator is perhaps the most prominent (Malik and Roeder 2010).

Multiple lines of investigation have converged to establish that Mediator is the major point of control for the Pol II PIC. Originally isolated in yeast (Kornberg 2005; Lee and Young 2000) as a Pol II-interacting cofactor, it has also been shown to interact with numerous transcriptional activators over the years, especially in extensive studies done in metazoan systems (Blazek et al. 2005; Boyer et al. 1999; Fondell et al. 1996; Naar et al. 1999). The earliest models for its function therefore postulated that it interacts as an interface between activators bound to their cognate elements and the PIC assembled at the core promoter (Malik and Roeder 2000). In this capacity its main role has been seen as facilitating recruitment of Pol II to the PIC (Fig. 11.1). However, as argued below, the multifarious properties that have thus far been ascribed to the Mediator favor an alternative view in which it acts as a hub for processing regulatory information. Further, in this view, rather than acting as a binary switch for turning transcription on (or off) it acts to deliver finely calibrated output to the PIC (Malik and Roeder 2010).

Structurally, the Mediator, many of whose individual subunits have been conserved from yeast to human (Bourbon 2008), displays a modular organization. Three main modules (head, middle, tail) constitute the major form of the Mediator. A fourth module, containing the CDK8-cyclin C pair, as well as two other large subunits, reversibly associates with the bulk Mediator. Association of this "kinase" module, confers generally negative properties on the Mediator although recent reports have suggested dual (i.e., both positive and negative) functions for it (Donner et al. 2007; Furumoto et al. 2007).

Broadly, whereas some of the resident subunits in the head and tail have been implicated in Pol II (and associated) interactions (Cai et al. 2009; Soutourina et al. 2011; Takagi et al. 2006), subunits in the tail are typically responsible for interactions with transcriptional activators (Malik and Roeder 2010). Through such interactions, specific Mediator subunits have been implicated in distinct cellular pathways. Most prominently, the MED1 subunit (its precise modular location is not yet mapped, but it is likely to lie close to the tail (Malik and Roeder 2010)) serves as the target for several nuclear receptors (Ito et al. 2000), which because of their responsiveness to hormonal signals are major regulators of animal physiology. In this way, the MED1 subunit channels transcription signals borne by the receptors to the PICs formed on their target genes. Ultimately, this may be translated into the appropriate cellular response. The classic illustration of this high degree of specificity in Mediator action comes from analyses of a model developmental system comprised of mouse embryo fibroblasts (MEFs) isolated from embryos of mice in which the gene encoding MED1 has been ablated (Ge et al. 2002; Ito et al. 2000). Although $MED1^{-/-}$ mice are embryonic lethal (reflecting the importance of the subunit in the overall development of the animal), MEFs isolated from such embryos just prior to death can be maintained in culture (reflecting the subunit's dispensability for housekeeping gene expression). Importantly, while wild-type MEFs can be differentiated into adipocytes through stimulation of PPARγ receptor-dependent pathways, *MED1*[-/-] MEFs fail to undergo such differentiation. Other Mediator subunits, notably MED15 (Yang et al. 2006), MED23 (Stevens et al. 2002) and MED12, a component of the kinase module (Carrera et al. 2008), also control discrete cellular pathways.

Although the coactivator role of Mediator was initially emphasized, it has become clear that it also affects basal (i.e., activator-independent) transcription (Baek et al. 2006; Mittler et al. 2001; Takagi and Kornberg 2006). Indeed, genome-wide analyses have revealed a near-universal requirement for Mediator at all genes, at least in yeast (Holstege et al. 1998). Interestingly, in this system and in contrast to the conditional requirement for MED1 mentioned above, deletion of a core (head) subunit is as deleterious as deletion of a Pol II subunit such that transcription of a vast majority of genes in the cell is affected. Also of note, multiple physical and functional interactions between Mediator and GTFs have been documented. These include Mediator's ability not only to recruit TFIIH to the PIC (Esnault et al. 2008; Pavri et al. 2005) but also its effects on the CTD phosphorylation function of TFIIH. Mediator effects on TFIIH can be both positive (Kim et al. 1994) and negative, the latter through phosphorylation of cyclin H in TFIIH by the CDK8 kinase (Akoulitchev et al. 2000). Further, efficient TFIIB recruitment to the PIC is also dependent on prior Mediator association (Baek et al. 2006). Mediator and TFIID also physically interact (Johnson et al. 2002). Thus, in the updated view, Mediator may even be regarded as a bona fide GTF.

Even more intriguingly, Mediator's role in PIC function might also potentially extend to post-initiation steps. Genetic data in yeast originally provided evidence for common roles of the MED31 subunit and SII (Guglielmi et al. 2007) as well as for links between this subunit and the Set2 complex, which can also function as an elongation factor (Krogan et al. 2003b). Functional interplay between Mediator and the elongation factor DSIF was also reported (Malik et al. 2007). Similarly, there are indications that Mediator interactions with P-TEFb and other SEC components might affect transcription elongation (Donner et al. 2010; Takahashi et al. 2011). Thus, Mediator might be uniquely situated to tie together the PIC formation, initiation, and elongation phases of transcription. In this light, an outstanding question is whether (and how) Mediator might also regulate the promoter-proximal paused Pol II.

In addition to its effects on the PIC, Mediator also operates within the context of multiple coactivators that impinge on the control regions of the gene. The emerging theme is that Mediator might coordinate the function of these diverse factors (Malik and Roeder 2010). For the most part, these are chromatin coactivators that help to "loosen" up the template prior to PIC assembly. Recall that the nucleosome, the fundamental unit of chromatin, is composed of two copies each of four core histones, H2A, H2B, H3, and H4, around which 146 bp of DNA is wrapped (Luger et al. 1997). The N-terminal tails of histones are relatively accessible to enzymatic modifications such as acetylation, methylation, phosphorylation, ubiquitination and sumoylation (Kouzarides 2007). Of all known histone modifications, acetylation is the only modification that directly causes the structural relaxation of chromatin by introducing a negative charge, favoring the recruitment of transcription factors as well as the process of transcription. Other histone modifications mediate the recruitment of activating and repressing effector complexes that ultimately mediate: (a) acetylation or deacetylation of histones; (b) nucleosome mobility, or even nucleosome eviction, to facilitate the compaction or relaxation of chromatin typically via mechanisms entailing ATP hydrolysis; and (c) recruitment of structural non-histone proteins that participate in chromatin compaction and assembly of higher order structures that are typical of heterochromatin.

Among the best-characterized chromatin cofactors that participate in short-term activation of genes are the various ATP-dependent remodelers like the p300 and GCN5 acetyltransferases. GCN5 is part of a larger complex (SAGA) that also includes certain TAFs and might function as a core promoter factor in lieu of TFIID at some promoters (Rodriguez-Navarro 2009). Additional factors (mostly methyl transferases and demethylases) whose actions lead to changes of a more long-term nature ("epigenetic") are discussed further below. The mechanisms of action of p300 and GCN5 nicely exemplify how histone modification is closely tied to Mediator and indeed Mediator might be involved in facilitating transitions between the chromatin machinery and the PIC (Black et al. 2006; Liu et al. 2008; Wallberg et al. 2003). Thus, although our simplified models for transcription activation describe it as a two-step process, Mediator likely provides continuity between them.

11.7 Negative Control of the Pol II Machinery

Although activation mechanisms have been emphasized here, it must be remembered that negative control is an integral component of the logic of gene regulation in eukaryotes. Indeed, "anti-repression" can add an additional layer of control to the overall activation of any given gene (Roeder 1998). Most obvious is the negative control exerted at the level of the chromatin structure, both through the nucleosomes themselves and through additional interacting proteins (e.g., linker histones) that further compact and condense the chromatin. Additional factors that contribute at this level include chromatin-modifying enzymes, especially histone deacetylases, or HDACs, and other co-repressors, which may be recruited in a gene-specific manner to distinct loci (Smith and Workman 2009). Here we point out some of the more salient mechanisms that most directly impact on the Pol II machinery and include a series of negative cofactors that target distinct GTFs. The dual (positive and negative) features of the Mediator have already been alluded to above. Other factors include NC2 and Mot1, which impact on TFIID. NC2 can interact with TBP and sterically occlude TFIIA and TFIIB from entering the PIC (Kamada et al. 2001). Mot1, on the other hand, is an ATPase that actively displaces TBP from promoters (Auble et al. 1994). Surprisingly, at some loci, these factors manifest as positive factors. In the case of NC2, it stimulates transcription from DPE-containing promoters at the expense of TATA-containing promoters (Willy et al. 2000). It has been suggested that NC2 affects the conformation of the TBP-TATA complex and thus mobilizes TBP towards weaker DPE-containing promoters (Schluesche et al. 2007). Similarly, Mot1 might also act to displace TBP from cryptic promoters and make it available to weak, but bona fide, promoters (Sprouse et al. 2008).

11.8 RNA Pol II Function Is Linked to Histone Modifications

As is evident from the above discussion of how Pol II and its associated factors function, the presence of densely packed nucleosomes offers a physical barrier for the efficient recruitment and processivity of Pol II. This supports the notion that genes immersed in highly packed areas of chromatin are less likely to engage RNA polymerase into productive rounds of transcription. Yet, more recent discoveries intriguingly suggest that histone modifications, especially those associated with epigenetic effects, can also play a direct role in modulating Pol II recruitment and processivity.

Most convincingly, genome-wide occupancy studies have revealed striking correlations between the presence of Pol II and certain histone modifications. Thus, transcriptional initiation strongly correlates with histone H3 trimethylated at lysine 4 (H3K4me3) and acetylation of histone H3 at lysine 9 and 14 (H3K9,14 ac) at nucleosomes near the TSS (Bernstein et al. 2002; Guenther et al. 2007; Santos-Rosa et al. 2002; Schubeler et al. 2004). Transcription elongation correlates with the presence of H3 trimethylated at lysine 36 (H3K36me3) at nucleosomes located in the body of the gene (Bannister et al. 2005; Kharchenko et al. 2010). Whether the presence of these marks is a consequence of transcription, or whether they play an active role in Pol II recruitment and processivity is presently unclear. Nonetheless, the Set1 complex, which catalyzes the H3K4me3 modification in yeast, is recruited to actively transcribed genes by interacting with the RNA Pol II-associated PAF1 complex (Krogan et al. 2003a). Experiments in yeast further suggest that H3K4me3 modification is deposited after Pol II recruitment and phosphorylation of Ser5 at the CTD. Accordingly, the recruitment to chromatin of the MLL1 complex, a mammalian homolog of the yeast Set1 complex, might in fact occur through the direct interaction of MLL components with the CTD of Pol II that has been phosphorylated at Ser5 (Hughes et al. 2004). This not only suggests that the establishment of the H3K4me3 mark is a consequence of Pol II recruitment but that Pol II is an active player in the process. On the other hand, specific TAFs in TFIID can bind directly to the H3K4me3 mark and this interaction seems potentiated by H3K9 and K14 acetylation (Vermeulen et al. 2007). This therefore suggests that H3K4me3 could also actively participate in PIC formation, and hence Pol II recruitment.

The multifunctional Mediator complex has also been implicated in establishing the epigenetically silenced status of certain neuronal genes in extraneuronal cells. In this system, the MED12 subunit of the generally repressive Mediator kinase module contributes to long-term repression of the target gene through interactions with multiple factors that include the G9a methyl transferase, which methylates H3K9 (Ding et al.

2008). This modification (H3K9me2) is known to attract heterochromatin protein 1 (HP1), which in turn is recognized by the DNA-methylating enzyme DNMT1. Together with other repression mechanisms that also operate at these loci (Ooi and Wood 2007), repression of the target genes is thus ensured through subsequent generations. Conversely, and reflecting the dual modes of function of this module, the MED12 and MED13 subunits of the Mediator interact with the Pygopus factor in Drosophila, which is a PHD finger-domain-containing protein that recognizes H3K4me3, the mark associated with transcriptional activation (Carrera et al. 2008; Fiedler et al. 2008).

As for histone modifications near the TSS, transcriptional elongation by Pol II also may influence epigenetic modifications in the body of the gene. For example, the deposition of H3K36me3 along the gene body in actively transcribed genes is a consequence of the recruitment of the histone methyltransferase (HMT) Set2 by the elongating Pol II. It is believed that methylated H3K36 acts in part to prevent cryptic initiation within genes through the recruitment of histone-deacetylase activities (Lee and Shilatifard 2007).

11.9 Epigenetic and Transcriptional Regulation in ES Cells: A Case Study

Transcription regulation in embryonic stem (ES) cells, with their unusual potential for both self renewal and pluripotency, illustrates how an intricate choreography entailing the Pol II machinery, on the one hand, and the epigenetic machinery, on the other, can lead to dramatic biological outcomes.

Chromatin immunoprecipitation (ChIP) coupled to DNA microarray analyses have revealed that in human ES cells, the phenomenon of promoter-proximal pausing of Pol II is widespread. Thus, 75% of all annotated promoters are enriched for Pol II and the H3K4me3, H3K9,14 ac epigenetic marks. However, only about half of these produce detectable transcripts and show elongation-associated H3K36me3 (Guenther et al. 2007), suggesting that successful elongation is a rate-limiting step for productive transcription. Further, a recent study using the new technique of GRO-seq (Min et al. 2011), in which nascent transcripts are mapped, confirms that promoter-proximal pausing of Pol II is a potential rate-limiting step for at least 40% of the genes in ES cells (as well as in MEFs, another cell type that was analyzed). Even at highly transcribed genes, Pol II occupancy is markedly higher near the TSS compared to the coding region. However, relative Pol II occupancy at TSS versus the coding region decreases with increasing gene activity, suggesting that overcoming the paused configuration of Pol II contributes towards overall activation of the gene. Comparison of GRO-seq signals between ES and MEFs also revealed that both Pol II recruitment (PIC formation) and transcriptional elongation contribute to the differential gene expression seen in these cell types (Min et al. 2011).

In ES cells, a prominent group of genes that recruit Pol II, but are not actively transcribed, are those encoding developmental regulators (Min et al. 2011; Stock et al. 2007). These genes display peculiar modification patterns, called bivalent domains, which are characterized by the presence of large stretches of chromatin containing H3K27me3 that in turn harbour smaller regions of H3K4me3 around the TSS (Fig. 11.2). The coexistence of these two antagonistic marks, one (H3K27me3) associated with transcriptional repression and the other (H3K4me3) with activation, has been suggested to play a role in silencing developmental genes in ES cells while keeping them poised for activation upon initiation of specific developmental pathways (Azuara et al. 2006; Bernstein et al. 2006). The K27me3 modification is catalyzed by the Polycomb group of proteins (PcG), which, in mammals, are found in at least two distinct complexes. The Polycomb repressive complex 2 (PRC2) contains four core components including the SET domain-containing EZH2 or EZH1 subunits, which are able to trimethylate H3K27. This mark is specifically recognized by the chromodomain of CBX proteins present in the Polycomb repressive complex 1 (PRC1), which also contains, among other subunits, a RING domain protein that mediates the ubiquitylation of histone H2A (Sauvageau and Sauvageau 2010).

Bivalent domains can be classified into two types based on whether they contain both PRC1 and PRC2, or only PRC2 (Ku et al. 2008). Genes occupied by PRC1 in ES cells usually encode for transcription factors and morphogenesis molecules that play key roles in development. GRO-seq data confirms that genes with bivalent domains are transcribed at very low levels. However, and especially in the case of genes bound by PRC2 only, notable levels of transcriptionally engaged Pol II are observed near their TSS, compared to the greatly reduced levels of productively elongating Pol II (Min et al. 2011). Genome-wide studies in Drosophila melanogaster also indicate that many genes that contain Polycomb responsive elements (PRE) proximal to the TSS produce short transcripts suggesting that transcription starts but elongation fails (Kharchenko et al. 2010). Interestingly, genes encoding regulatory and developmental functions are overrepresented in this class and they are further marked with H3K27me3 and either H3K4me1 or H3K4me2 but not H3K4me3. However, GRO-seq data in ES cells suggest that Pol II at bivalent gene promoters is confined to extremely 5′ proximal regions (Min et al. 2011). Whereas this again suggests that Pol II is stalled at these promoters, it appears that it is in a conformation that is different from that of the paused Pol II observed on actively transcribed genes (Stock et al. 2007). It can therefore be referred to as being in a "poised" conformation although precise structural features remain unidentified (Fig. 11.2). This poised Pol II is phosphorylated at Ser5 of the CTD but not at Ser2. Indeed, it cannot be detected with standard antibodies against CTD, suggesting that its conformation is different from that of the paused Pol II that is more widely observed. Further, ablation of Ring1B, the PRC1 subunit that mediates ubiquitination of histone H2A, results in loss of ubiquitinated H2A at bivalent genes and de-repression of these genes without changes in H3K27me3. Interestingly, the levels of Ser5 and Ser2 phosphorylation of Pol II at these genes does not change, but Pol II is now detectable with antibodies against CTD, suggesting that a conformational change has released Pol II from the poised state (Stock et al. 2007). Therefore, Ring 1-mediated ubiquitination of H2A seems to restrain poised Pol II at bivalent genes. The precise role of the H3K4me3 mark at bivalent genes, especially in relation to Pol II recruitment, remains unclear. Knock-down of selected H3K4 demethylases in

ES cells causes an increase in the H3K4me3 levels and the de-repression of certain bivalent genes (Adamo et al. 2011; Pasini et al. 2008). In contrast, ES cells null for Jarid2, a subunit of PRC1 involved in its recruitment to target genes, fail to recruit Pol II to bivalent genes while levels of H3K4me3 are unchanged (Landeira et al. 2010).

In summary, several histone modifications seem to be directly correlated with the presence of Pol II at the TSS or coding regions of genes notably, although not exclusively, in developmental systems. While the details of the interplay between these modifications and Pol II function are still not clear, the recent findings discussed above show that histone modifications play direct roles in regulating the transcriptional machinery.

11.10 Conclusions

The past two decades have witnessed a major expansion in our knowledge of how the Pol II machinery functions. Numerous factors that allow it to assemble into an active PIC at gene promoters as well as factors that act as it is navigating the gene body have been identified and characterized. In many cases, structural and biochemical studies have combined to reveal, at an unprecedented level of detail, how these factors impact Pol II function. With more recent studies linking Pol II function with specific epigenetic marks in the genome, the latest challenge in the field now is to describe in equivalent detail how the Pol II machinery plays out in this context. Epigenetic modifications have also been described in terms of a "histone code", which is laid down by so-called "writers" and subsequently interpreted by "readers" (Jenuwein and Allis 2001; Strahl and Allis 2000). As discussed here, the Pol II machinery seems to contain both writers and readers. Future work is sure to reveal further how other individual components of this elaborate machinery are affected by and, in turn, influence the epigenetic landscape, and ultimately lead to interesting biological outcomes.

Acknowledgments Given the scope of this chapter, citations have been limited to review articles in many cases. We therefore apologize to colleagues whose original publications we could not directly cite. M.J.B is partially supported by the Ramón y Cajal program and by grants RYC-2007-01510 and SAF2009-08588 from the Ministerio de Ciencia e Innovación of Spain. S.M. is partially supported by 1RC1GM09029. M.J.B and S.M. would also like to thank Dr Robert G. Roeder for his support and encouragement.

References

Adamo A, Sese B, Boue S, Castano J, Paramonov I, Barrero MJ, Izpisua Belmonte JC (2011) LSD1 regulates the balance between self-renewal and differentiation in human embryonic stem cells. Nat Cell Biol 13:652–659

Akoulitchev S, Makela TP, Weinberg RA, Reinberg D (1995) Requirement for TFIIH kinase activity in transcription by RNA polymerase II. Nature 377:557–560

Akoulitchev S, Chuikov S, Reinberg D (2000) TFIIH is negatively regulated by cdk8-containing mediator complexes. Nature 407:102–106

Auble DT, Hansen KE, Mueller CG, Lane WS, Thorner J, Hahn S (1994) Mot1, a global repressor of RNA polymerase II transcription, inhibits TBP binding to DNA by an ATP-dependent mechanism. Genes Dev 8:1920–1934

Azuara V, Perry P, Sauer S, Spivakov M, Jorgensen HF, John RM, Gouti M, Casanova M, Warnes G, Merkenschlager M, Fisher AG (2006) Chromatin signatures of pluripotent cell lines. Nat Cell Biol 8:532–538

Baek HJ, Kang YK, Roeder RG (2006) Human mediator enhances basal transcription by facilitating recruitment of transcription factor IIB during preinitiation complex assembly. J Biol Chem 281:15172–15181

Bannister AJ, Schneider R, Myers FA, Thorne AW, Crane-Robinson C, Kouzarides T (2005) Spatial distribution of di- and tri-methyl lysine 36 of histone H3 at active genes. J Biol Chem 280:17732–17736

Baskaran R, Dahmus ME, Wang JY (1993) Tyrosine phosphorylation of mammalian RNA polymerase II carboxyl-terminal domain. Proc Natl Acad Sci USA 90:11167–11171

Bernstein BE, Humphrey EL, Erlich RL, Schneider R, Bouman P, Liu JS, Kouzarides T, Schreiber SL (2002) Methylation of histone H3 Lys 4 in coding regions of active genes. Proc Natl Acad Sci USA 99:8695–8700

Bernstein BE, Mikkelsen TS, Xie X, Kamal M, Huebert DJ, Cuff J, Fry B, Meissner A, Wernig M, Plath K, Jaenisch R, Wagschal A, Feil R, Schreiber SL, Lander ES (2006) A bivalent chromatin structure marks key developmental genes in embryonic stem cells. Cell 125:315–326

Black JC, Choi JE, Lombardo SR, Carey M (2006) A mechanism for coordinating chromatin modification and preinitiation complex assembly. Mol Cell 23:809–818

Blazek E, Mittler G, Meisterernst M (2005) The mediator of RNA polymerase II. Chromosoma 113:399–408

Bourbon HM (2008) Comparative genomics supports a deep evolutionary origin for the large, four-module transcriptional mediator complex. Nucleic Acids Res 36:3993–4008

Boyer TG, Martin ME, Lees E, Ricciardi RP, Berk AJ (1999) Mammalian Srb/Mediator complex is targeted by adenovirus E1A protein. Nature 399:276–279

Brivanlou AH, Darnell JE Jr (2002) Signal transduction and the control of gene expression. Science 295:813–818

Buratowski S (2009) Progression through the RNA polymerase II CTD cycle. Mol Cell 36:541–546

Buratowski S, Hahn S, Guarente L, Sharp PA (1989) Five intermediate complexes in transcription initiation by RNA polymerase II. Cell 56:549–561

Burke TW, Kadonaga JT (1997) The downstream core promoter element, DPE, is conserved from Drosophila to humans and is recognized by TAFII60 of Drosophila. Genes Dev 11:3020–3031

Burley SK, Roeder RG (1996) Biochemistry and structural biology of transcription factor IID (TFIID). Annu Rev Biochem 65:769–799

Bushnell DA, Westover KD, Davis RE, Kornberg RD (2004) Structural basis of transcription: an RNA polymerase II-TFIIB cocrystal at 4.5 Angstroms. Science 303:983–988

Cai G, Imasaki T, Takagi Y, Asturias FJ (2009) Mediator structural conservation and implications for the regulation mechanism. Structure 17:559–567

Carrera I, Janody F, Leeds N, Duveau F, Treisman JE (2008) Pygopus activates Wingless target gene transcription through the mediator complex subunits Med12 and Med13. Proc Natl Acad Sci USA 105:6644–6649

Chen HT, Hahn S (2004) Mapping the location of TFIIB within the RNA polymerase II transcription preinitiation complex: a model for the structure of the PIC. Cell 119:169–180

Cramer P (2002) Multisubunit RNA polymerases. Curr Opin Struct Biol 12:89–97

Cramer P, Bushnell DA, Fu J, Gnatt AL, Maier-Davis B, Thompson NE, Burgess RR, Edwards AM, David PR, Kornberg RD (2000) Architecture of RNA polymerase II and implications for the transcription mechanism. Science 288:640–649

Cramer P, Armache KJ, Baumli S, Benkert S, Brueckner F, Buchen C, Damsma GE, Dengl S, Geiger SR, Jasiak AJ, Jawhari A, Jennebach S, Kamenski T, Kettenberger H, Kuhn CD,

Lehmann E, Leike K, Sydow JF, Vannini A (2008) Structure of eukaryotic RNA polymerases. Annu Rev Biophys 37:337–352

Deato MD, Tjian R (2007) Switching of the core transcription machinery during myogenesis. Genes Dev 21:2137–2149

Deato MD, Marr MT, Sottero T, Inouye C, Hu P, Tjian R (2008) MyoD targets TAF3/TRF3 to activate myogenin transcription. Mol Cell 32:96–105

Ding N, Zhou H, Esteve PO, Chin HG, Kim S, Xu X, Joseph SM, Friez MJ, Schwartz CE, Pradhan S, Boyer TG (2008) Mediator links epigenetic silencing of neuronal gene expression with x-linked mental retardation. Mol Cell 31:347–359

Donner AJ, Szostek S, Hoover JM, Espinosa JM (2007) CDK8 is a stimulus-specific positive coregulator of p53 target genes. Mol Cell 27:121–133

Donner AJ, Ebmeier CC, Taatjes DJ, Espinosa JM (2010) CDK8 is a positive regulator of transcriptional elongation within the serum response network. Nat Struct Mol Biol 17:194–201

Ebright RH (2000) RNA polymerase: structural similarities between bacterial RNA polymerase and eukaryotic RNA polymerase II. J Mol Biol 304:687–698

Egloff S, Murphy S (2008) Cracking the RNA polymerase II CTD code. Trends Genet 24:280–288

Egloff S, O'Reilly D, Chapman RD, Taylor A, Tanzhaus K, Pitts L, Eick D, Murphy S (2007) Serine-7 of the RNA polymerase II CTD is specifically required for snRNA gene expression. Science 318:1777–1779

Eichner J, Chen HT, Warfield L, Hahn S (2010) Position of the general transcription factor TFIIF within the RNA polymerase II transcription preinitiation complex. EMBO J 29:706–716

Esnault C, Ghavi-Helm Y, Brun S, Soutourina J, Van Berkum N, Boschiero C, Holstege F, Werner M (2008) Mediator-dependent recruitment of TFIIH modules in preinitiation complex. Mol Cell 31:337–346

Fabrega C, Shen V, Shuman S, Lima CD (2003) Structure of an mRNA capping enzyme bound to the phosphorylated carboxy-terminal domain of RNA polymerase II. Mol Cell 11:1549–1561

Fiedler M, Sanchez-Barrena MJ, Nekrasov M, Mieszczanek J, Rybin V, Muller J, Evans P, Bienz M (2008) Decoding of methylated histone H3 tail by the Pygo-BCL9 Wnt signaling complex. Mol Cell 30:507–518

Flores O, Lu H, Reinberg D (1992) Factors involved in specific transcription by mammalian RNA polymerase II. Identification and characterization of factor IIH. J Biol Chem 267:2786–2793

Fondell JD, Ge H, Roeder RG (1996) Ligand induction of a transcriptionally active thyroid hormone receptor coactivator complex. Proc Natl Acad Sci USA 93:8329–8333

Furumoto T, Tanaka A, Ito M, Malik S, Hirose Y, Hanaoka F, Ohkuma Y (2007) A kinase subunit of the human mediator complex, CDK8, positively regulates transcriptional activation. Genes Cells 12:119–132

Ge K, Guermah M, Yuan CX, Ito M, Wallberg AE, Spiegelman BM, Roeder RG (2002) Transcription coactivator TRAP220 is required for PPAR gamma 2-stimulated adipogenesis. Nature 417:563–567

Gu W, Malik S, Ito M, Yuan CX, Fondell JD, Zhang X, Martinez E, Qin J, Roeder RG (1999) A novel human SRB/MED-containing cofactor complex, SMCC, involved in transcription regulation. Mol Cell 3:97–108

Guenther MG, Levine SS, Boyer LA, Jaenisch R, Young RA (2007) A chromatin landmark and transcription initiation at most promoters in human cells. Cell 130:77–88

Guermah M, Palhan VB, Tackett AJ, Chait BT, Roeder RG (2006) Synergistic functions of SII and p300 in productive activator-dependent transcription of chromatin templates. Cell 125:275–286

Guglielmi B, Soutourina J, Esnault C, Werner M (2007) TFIIS elongation factor and Mediator act in conjunction during transcription initiation in vivo. Proc Natl Acad Sci USA 104:16062–16067

Hahn S (2004) Structure and mechanism of the RNA polymerase II transcription machinery. Nat Struct Mol Biol 11:394–403

Hodges C, Bintu L, Lubkowska L, Kashlev M, Bustamante C (2009) Nucleosomal fluctuations govern the transcription dynamics of RNA polymerase II. Science 325:626–628

Holstege FC, Fiedler U, Timmers HT (1997) Three transitions in the RNA polymerase II transcription complex during initiation. EMBO J 16:7468–7480

Holstege FC, Jennings EG, Wyrick JJ, Lee TI, Hengartner CJ, Green MR, Golub TR, Lander ES, Young RA (1998) Dissecting the regulatory circuitry of a eukaryotic genome. Cell 95: 717–728

Hughes CM, Rozenblatt-Rosen O, Milne TA, Copeland TD, Levine SS, Lee JC, Hayes DN, Shanmugam KS, Bhattacharjee A, Biondi CA, Kay GF, Hayward NK, Hess JL, Meyerson M (2004) Menin associates with a trithorax family histone methyltransferase complex and with the hoxc8 locus. Mol Cell 13:587–597

Ito M, Yuan CX, Okano HJ, Darnell RB, Roeder RG (2000) Involvement of the TRAP220 component of the TRAP/SMCC coactivator complex in embryonic development and thyroid hormone action. Mol Cell 5:683–693

Jenuwein T, Allis CD (2001) Translating the histone code. Science 293:1074–1080

Johnson KM, Wang J, Smallwood A, Arayata C, Carey M (2002) TFIID and human mediator coactivator complexes assemble cooperatively on promoter DNA. Genes Dev 16:1852–1863

Juven-Gershon T, Hsu JY, Theisen JW, Kadonaga JT (2008) The RNA polymerase II core promoter – the gateway to transcription. Curr Opin Cell Biol 20:253–259

Kamada K, Shu F, Chen H, Malik S, Stelzer G, Roeder RG, Meisterernst M, Burley SK (2001) Crystal structure of negative cofactor 2 recognizing the TBP-DNA transcription complex. Cell 106:71–81

Kelly WG, Dahmus ME, Hart GW (1993) RNA polymerase II is a glycoprotein. Modification of the COOH-terminal domain by O-GlcNAc. J Biol Chem 268:10416–10424

Kettenberger H, Armache KJ, Cramer P (2003) Architecture of the RNA polymerase II-TFIIS complex and implications for mRNA cleavage. Cell 114:347–357

Kharchenko PV, Alekseyenko AA, Schwartz YB, Minoda A, Riddle NC, Ernst J, Sabo PJ, Larschan E, Gorchakov AA, Gu T, Linder-Basso D, Plachetka A, Shanower G, Tolstorukov MY, Luquette LJ, Xi R, Jung YL, Park RW, Bishop EP, Canfield TK, Sandstrom R, Thurman RE, MacAlpine DM, Stamatoyannopoulos JA, Kellis M, Elgin SC, Kuroda MI, Pirrotta V, Karpen GH, Park PJ (2010) Comprehensive analysis of the chromatin landscape in *Drosophila melanogaster*. Nature 471:480–485

Kim YJ, Bjorklund S, Li Y, Sayre MH, Kornberg RD (1994) A multiprotein mediator of transcriptional activation and its interaction with the C-terminal repeat domain of RNA polymerase II. Cell 77:599–608

Kim TK, Ebright RH, Reinberg D (2000) Mechanism of ATP-dependent promoter melting by transcription factor IIH. Science 288:1418–1422

Kim B, Nesvizhskii AI, Rani PG, Hahn S, Aebersold R, Ranish JA (2007) The transcription elongation factor TFIIS is a component of RNA polymerase II preinitiation complexes. Proc Natl Acad Sci USA 104:16068–16073

Kim J, Guermah M, Roeder RG (2010) The human PAF1 complex acts in chromatin transcription elongation both independently and cooperatively with SII/TFIIS. Cell 140:491–503

Kireeva ML, Hancock B, Cremona GH, Walter W, Studitsky VM, Kashlev M (2005) Nature of the nucleosomal barrier to RNA polymerase II. Mol Cell 18:97–108

Kizer KO, Phatnani HP, Shibata Y, Hall H, Greenleaf AL, Strahl BD (2005) A novel domain in Set2 mediates RNA polymerase II interaction and couples histone H3 K36 methylation with transcript elongation. Mol Cell Biol 25:3305–3316

Kobor MS, Archambault J, Lester W, Holstege FC, Gileadi O, Jansma DB, Jennings EG, Kouyoumdjian F, Davidson AR, Young RA, Greenblatt J (1999) An unusual eukaryotic protein phosphatase required for transcription by RNA polymerase II and CTD dephosphorylation in *S. cerevisiae*. Mol Cell 4:55–62

Kornberg RD (2005) Mediator and the mechanism of transcriptional activation. Trends Biochem Sci 30:235–239

Kornberg RD (2007) The molecular basis of eukaryotic transcription. Proc Natl Acad Sci USA 104:12955–12961

Kostrewa D, Zeller ME, Armache KJ, Seizl M, Leike K, Thomm M, Cramer P (2009) RNA polymerase II-TFIIB structure and mechanism of transcription initiation. Nature 462:323–330

Kouzarides T (2007) Chromatin modifications and their function. Cell 128:693–705

Krogan NJ, Dover J, Wood A, Schneider J, Heidt J, Boateng MA, Dean K, Ryan OW, Golshani A, Johnston M, Greenblatt JF, Shilatifard A (2003a) The Paf1 complex is required for histone H3 methylation by COMPASS and Dot1p: linking transcriptional elongation to histone methylation. Mol Cell 11:721–729

Krogan NJ, Kim M, Tong A, Golshani A, Cagney G, Canadien V, Richards DP, Beattie BK, Emili A, Boone C, Shilatifard A, Buratowski S, Greenblatt J (2003b) Methylation of histone H3 by Set2 in *Saccharomyces cerevisiae* is linked to transcriptional elongation by RNA polymerase II. Mol Cell Biol 23:4207–4218

Ku M, Koche RP, Rheinbay E, Mendenhall EM, Endoh M, Mikkelsen TS, Presser A, Nusbaum C, Xie X, Chi AS, Adli M, Kasif S, Ptaszek LM, Cowan CA, Lander ES, Koseki H, Bernstein BE (2008) Genomewide analysis of PRC1 and PRC2 occupancy identifies two classes of bivalent domains. PLoS Genet 4:e1000242

Lagrange T, Kapanidis AN, Tang H, Reinberg D, Ebright RH (1998) New core promoter element in RNA polymerase II-dependent transcription: sequence-specific DNA binding by transcription factor IIB. Genes Dev 12:34–44

Landeira D, Sauer S, Poot R, Dvorkina M, Mazzarella L, Jorgensen HF, Pereira CF, Leleu M, Piccolo FM, Spivakov M, Brookes E, Pombo A, Fisher C, Skarnes WC, Snoek T, Bezstarosti K, Demmers J, Klose RJ, Casanova M, Tavares L, Brockdorff N, Merkenschlager M, Fisher AG (2010) Jarid2 is a PRC2 component in embryonic stem cells required for multi-lineage differentiation and recruitment of PRC1 and RNA Polymerase II to developmental regulators. Nat Cell Biol 12:618–624

Lee JS, Shilatifard A (2007) A site to remember: H3K36 methylation a mark for histone deacetylation. Mutat Res 618:130–134

Lee TI, Young RA (2000) Transcription of eukaryotic protein-coding genes. Annu Rev Genet 34:77–137

Li Y, Kornberg RD (1994) Interplay of positive and negative effectors in function of the C-terminal repeat domain of RNA polymerase II. Proc Natl Acad Sci USA 91:2362–2366

Lin YC, Choi WS, Gralla JD (2005) TFIIH XPB mutants suggest a unified bacterial-like mechanism for promoter opening but not escape. Nat Struct Mol Biol 12:603–607

Lis J (1998) Promoter-associated pausing in promoter architecture and postinitiation transcriptional regulation. Cold Spring Harb Symp Quant Biol 63:347–356

Liu X, Vorontchikhina M, Wang YL, Faiola F, Martinez E (2008) STAGA recruits Mediator to the MYC oncoprotein to stimulate transcription and cell proliferation. Mol Cell Biol 28:108–121

Liu X, Bushnell DA, Wang D, Calero G, Kornberg RD (2010) Structure of an RNA polymerase II-TFIIB complex and the transcription initiation mechanism. Science 327:206–209

Luger K, Mader AW, Richmond RK, Sargent DF, Richmond TJ (1997) Crystal structure of the nucleosome core particle at 2.8 A resolution. Nature 389:251–260

Malik S, Roeder RG (2000) Transcriptional regulation through Mediator-like coactivators in yeast and metazoan cells. Trends Biochem Sci 25:277–283

Malik S, Roeder RG (2010) The metazoan Mediator co-activator complex as an integrative hub for transcriptional regulation. Nat Rev Genet 11:761–772

Malik S, Barrero MJ, Jones T (2007) Identification of a regulator of transcription elongation as an accessory factor for the human Mediator coactivator. Proc Natl Acad Sci USA 104:6182–6187

Margaritis T, Holstege FC (2008) Poised RNA polymerase II gives pause for thought. Cell 133:581–584

Martinez E, Zhou Q, L'Etoile ND, Oelgeschlager T, Berk AJ, Roeder RG (1995) Core promoter-specific function of a mutant transcription factor TFIID defective in TATA-box binding. Proc Natl Acad Sci USA 92:11864–11868

Martinez E, Ge H, Tao Y, Yuan CX, Palhan V, Roeder RG (1998) Novel cofactors and TFIIA mediate functional core promoter selectivity by the human TAFII150-containing TFIID complex. Mol Cell Biol 18:6571–6583

Meinhart A, Kamenski T, Hoeppner S, Baumli S, Cramer P (2005) A structural perspective of CTD function. Genes Dev 19:1401–1415

Min IM, Waterfall JJ, Core LJ, Munroe RJ, Schimenti J, Lis JT (2011) Regulating RNA polymerase pausing and transcription elongation in embryonic stem cells. Genes Dev 25:742–754

Minakhin L, Bhagat S, Brunning A, Campbell EA, Darst SA, Ebright RH, Severinov K (2001) Bacterial RNA polymerase subunit omega and eukaryotic RNA polymerase subunit RPB6 are sequence, structural, and functional homologs and promote RNA polymerase assembly. Proc Natl Acad Sci USA 98:892–897

Mittler G, Kremmer E, Timmers HT, Meisterernst M (2001) Novel critical role of a human Mediator complex for basal RNA polymerase II transcription. EMBO Rep 2:808–813

Mohrmann L, Verrijzer CP (2005) Composition and functional specificity of SWI2/SNF2 class chromatin remodeling complexes. Biochim Biophys Acta 1681:59–73

Mosley AL, Pattenden SG, Carey M, Venkatesh S, Gilmore JM, Florens L, Workman JL, Washburn MP (2009) Rtr1 is a CTD phosphatase that regulates RNA polymerase II during the transition from serine 5 to serine 2 phosphorylation. Mol Cell 34:168–178

Muse GW, Gilchrist DA, Nechaev S, Shah R, Parker JS, Grissom SF, Zeitlinger J, Adelman K (2007) RNA polymerase is poised for activation across the genome. Nat Genet 39:1507–1511

Naar AM, Beaurang PA, Zhou S, Abraham S, Solomon W, Tjian R (1999) Composite co-activator ARC mediates chromatin-directed transcriptional activation. Nature 398:828–832

Naar AM, Taatjes DJ, Zhai W, Nogales E, Tjian R (2002) Human CRSP interacts with RNA polymerase II CTD and adopts a specific CTD-bound conformation. Genes Dev 16:1339–1344

Nechaev S, Adelman K (2008) Promoter-proximal Pol II: when stalling speeds things up. Cell Cycle 7:1539–1544

Nikolov DB, Chen H, Halay ED, Usheva AA, Hisatake K, Lee DK, Roeder RG, Burley SK (1995) Crystal structure of a TFIIB-TBP-TATA-element ternary complex. Nature 377:119–128

Ohkuma Y, Roeder RG (1994) Regulation of TFIIH ATPase and kinase activities by TFIIE during active initiation complex formation. Nature 368:160–163

Ooi L, Wood IC (2007) Chromatin crosstalk in development and disease: lessons from REST. Nat Rev Genet 8:544–554

Orphanides G, Lagrange T, Reinberg D (1996) The general transcription factors of RNA polymerase II. Genes Dev 10:2657–2683

Pal M, Ponticelli AS, Luse DS (2005) The role of the transcription bubble and TFIIB in promoter clearance by RNA polymerase II. Mol Cell 19:101–110

Pasini D, Hansen KH, Christensen J, Agger K, Cloos PA, Helin K (2008) Coordinated regulation of transcriptional repression by the RBP2 H3K4 demethylase and Polycomb-Repressive Complex 2. Genes Dev 22:1345–1355

Pavri R, Lewis B, Kim TK, Dilworth FJ, Erdjument-Bromage H, Tempst P, de Murcia G, Evans R, Chambon P, Reinberg D (2005) PARP-1 determines specificity in a retinoid signaling pathway via direct modulation of mediator. Mol Cell 18:83–96

Pikaard CS, Haag JR, Ream T, Wierzbicki AT (2008) Roles of RNA polymerase IV in gene silencing. Trends Plant Sci 13:390–397

Rahl PB, Lin CY, Seila AC, Flynn RA, McCuine S, Burge CB, Sharp PA, Young RA (2010) c-Myc regulates transcriptional pause release. Cell 141:432–445

Rodriguez-Navarro S (2009) Insights into SAGA function during gene expression. EMBO Rep 10:843–850

Roeder RG (1996) The role of general initiation factors in transcription by RNA polymerase II. Trends Biochem Sci 21:327–335

Roeder RG (1998) Role of general and gene-specific cofactors in the regulation of eukaryotic transcription. Cold Spring Harb Symp Quant Biol 63:201–218

Roeder RG (2005) Transcriptional regulation and the role of diverse coactivators in animal cells. FEBS Lett 579:909–915

Roeder RG, Rutter WJ (1969) Multiple forms of DNA-dependent RNA polymerase in eukaryotic organisms. Nature 224:234–237

Santos-Rosa H, Schneider R, Bannister AJ, Sherriff J, Bernstein BE, Emre NC, Schreiber SL, Mellor J, Kouzarides T (2002) Active genes are tri-methylated at K4 of histone H3. Nature 419:407–411

Saunders A, Core LJ, Lis JT (2006) Breaking barriers to transcription elongation. Nat Rev Mol Cell Biol 7:557–567

Sauvageau M, Sauvageau G (2010) Polycomb group proteins: multi-faceted regulators of somatic stem cells and cancer. Cell Stem Cell 7:299–313

Schluesche P, Stelzer G, Piaia E, Lamb DC, Meisterernst M (2007) NC2 mobilizes TBP on core promoter TATA boxes. Nat Struct Mol Biol 14:1196–1201

Schubeler D, MacAlpine DM, Scalzo D, Wirbelauer C, Kooperberg C, van Leeuwen F, Gottschling DE, O'Neill LP, Turner BM, Delrow J, Bell SP, Groudine M (2004) The histone modification pattern of active genes revealed through genome-wide chromatin analysis of a higher eukaryote. Genes Dev 18:1263–1271

Sims RJ 3rd, Belotserkovskaya R, Reinberg D (2004) Elongation by RNA polymerase II: the short and long of it. Genes Dev 18:2437–2468

Sims RJ 3rd, Rojas LA, Beck D, Bonasio R, Schuller R, Drury WJ 3rd, Eick D, Reinberg D (2011) The C-terminal domain of RNA polymerase II is modified by site-specific methylation. Science 332:99–103

Smith KT, Workman JL (2009) Histone deacetylase inhibitors: anticancer compounds. Int J Biochem Cell Biol 41:21–25

Smith E, Lin C, Shilatifard A (2011) The super elongation complex (SEC) and MLL in development and disease. Genes Dev 25:661–672

Soutourina J, Wydau S, Ambroise Y, Boschiero C, Werner M (2011) Direct interaction of RNA polymerase II and mediator required for transcription in vivo. Science 331:1451–1454

Sprouse RO, Shcherbakova I, Cheng H, Jamison E, Brenowitz M, Auble DT (2008) Function and structural organization of Mot1 bound to a natural target promoter. J Biol Chem 283:24935–24948

Stevens JL, Cantin GT, Wang G, Shevchenko A, Berk AJ (2002) Transcription control by E1A and MAP kinase pathway via Sur2 mediator subunit. Science 296:755–758

Stock JK, Giadrossi S, Casanova M, Brookes E, Vidal M, Koseki H, Brockdorff N, Fisher AG, Pombo A (2007) Ring1-mediated ubiquitination of H2A restrains poised RNA polymerase II at bivalent genes in mouse ES cells. Nat Cell Biol 9:1428–1435

Strahl BD, Allis CD (2000) The language of covalent histone modifications. Nature 403:41–45

Struhl K (1999) Fundamentally different logic of gene regulation in eukaryotes and prokaryotes. Cell 98:1–4

Sun M, Lariviere L, Dengl S, Mayer A, Cramer P (2010) A tandem SH2 domain in transcription elongation factor Spt6 binds the phosphorylated RNA polymerase II C-terminal repeat domain (CTD). J Biol Chem 285:41597–41603

Takagi Y, Kornberg RD (2006) Mediator as a general transcription factor. J Biol Chem 281:80–89

Takagi Y, Calero G, Komori H, Brown JA, Ehrensberger AH, Hudmon A, Asturias F, Kornberg RD (2006) Head module control of mediator interactions. Mol Cell 23:355–364

Takahashi H, Parmely TJ, Sato S, Tomomori-Sato C, Banks CA, Kong SE, Szutorisz H, Swanson SK, Martin-Brown S, Washburn MP, Florens L, Seidel CW, Lin C, Smith ER, Shilatifard A, Conaway RC, Conaway JW (2011) Human mediator subunit MED26 functions as a docking site for transcription elongation factors. Cell 146:92–104

Thomas MC, Chiang CM (2006) The general transcription machinery and general cofactors. Crit Rev Biochem Mol Biol 41:105–178

Tirode F, Busso D, Coin F, Egly JM (1999) Reconstitution of the transcription factor TFIIH: assignment of functions for the three enzymatic subunits, XPB, XPD, and cdk7. Mol Cell 3:87–95

Vermeulen M, Mulder KW, Denissov S, Pijnappel WW, van Schaik FM, Varier RA, Baltissen MP, Stunnenberg HG, Mann M, Timmers HT (2007) Selective anchoring of TFIID to nucleosomes by trimethylation of histone H3 lysine 4. Cell 131:58–69

Wallberg AE, Yamamura S, Malik S, Spiegelman BM, Roeder RG (2003) Coordination of p300-mediated chromatin remodeling and TRAP/mediator function through coactivator PGC-1alpha. Mol Cell 12:1137–1149

Willy PJ, Kobayashi R, Kadonaga JT (2000) A basal transcription factor that activates or represses transcription. Science 290:982–985

Winkler DD, Luger K (2011) The histone chaperone FACT: structural insights and mechanisms for nucleosome reorganization. J Biol Chem 286:18369–18374

Xu YX, Manley JL (2004) Pinning down transcription: regulation of RNA polymerase II activity during the cell cycle. Cell Cycle 3:432–435

Yang F, Vought BW, Satterlee JS, Walker AK, Jim Sun ZY, Watts JL, DeBeaumont R, Saito RM, Hyberts SG, Yang S, Macol C, Iyer L, Tjian R, van den Heuvel S, Hart AC, Wagner G, Naar AM (2006) An ARC/Mediator subunit required for SREBP control of cholesterol and lipid homeostasis. Nature 442:700–704

Yudkovsky N, Ranish JA, Hahn S (2000) A transcription reinitiation intermediate that is stabilized by activator. Nature 408:225–229

Zeitlinger J, Stark A, Kellis M, Hong JW, Nechaev S, Adelman K, Levine M, Young RA (2007) RNA polymerase stalling at developmental control genes in the *Drosophila melanogaster* embryo. Nat Genet 39:1512–1516

Zhang G, Campbell EA, Minakhin L, Richter C, Severinov K, Darst SA (1999) Crystal structure of *Thermus aquaticus* core RNA polymerase at 3.3 Å resolution. Cell 98:811–824

Chapter 12
RNA Polymerase III Transcription – Regulated by Chromatin Structure and Regulator of Nuclear Chromatin Organization

Chiara Pascali and Martin Teichmann

Abstract RNA polymerase III (Pol III) transcription is regulated by modifications of the chromatin. DNA methylation and post-translational modifications of histones, such as acetylation, phosphorylation and methylation have been linked to Pol III transcriptional activity. In addition to being regulated by modifications of DNA and histones, Pol III genes and its transcription factors have been implicated in the organization of nuclear chromatin in several organisms. In yeast, the ability of the Pol III transcription system to contribute to nuclear organization seems to be dependent on direct interactions of Pol III genes and/or its transcription factors TFIIIC and TFIIIB with the structural maintenance of chromatin (SMC) protein-containing complexes cohesin and condensin. In human cells, Pol III genes and transcription factors have also been shown to colocalize with cohesin and the transcription regulator and genome organizer CCCTC-binding factor (CTCF). Furthermore, chromosomal sites have been identified in yeast and humans that are bound by partial Pol III machineries (extra TFIIIC sites – ETC; chromosome organizing clamps – COC). These ETCs/COC as well as Pol III genes possess the ability to act as boundary elements that restrict spreading of heterochromatin.

12.1 Introduction

Transcription in eukaryotes is carried out by multiple DNA-dependent RNA polymerases with specialized functions (Pol I, Pol II and Pol III in animal cells and in addition Pol IV and Pol V in plant cells). Being composed of 17 subunits, RNA

C. Pascali • M. Teichmann (✉)
Institut Européen de Chimie et Biologie (IECB), Université Bordeaux
Segalen / INSERM U869, 2, rue Robert Escarpit, 33607 Pessac, France
e-mail: Martin.Teichmann@inserm.fr

polymerase III (Pol III) displays the most complex protein composition of these RNA polymerases. Pol III transcribes genes encoding small untranslated RNAs that are characterized by a limited number of promoters (three major types in mammals) and are accordingly recognized by general transcription factors (reviewed in Geiduschek and Kassavetis 2001; Huang and Maraia 2001; Schramm and Hernandez 2002; Roeder 2003; Dumay-Odelot et al. 2010). Many of these genes are conserved from unicellular to multicellular organisms (5S RNA; tRNAs; U6 RNA; RNAse P and RNAse MRP RNAs; 7SL RNA). These RNAs participate in the regulation of different aspects of gene expression, including transcription, RNA processing, translation and protein translocation to the endoplasmic reticulum. Other genes are transcribed by Pol III in a variety of eukaryotes from protozoa to metazoa, including the genes coding for vault RNAs, Y RNAs and the 7SK RNA (Dieci et al. 2007). Higher eukaryotes contain in addition a variable number of Pol III-transcribed retrotransposed elements (small interspersed nuclear elements -SINEs) that constitute a substantial fraction of the DNA of a given genome (about 11% of the human genome; Cordaux and Batzer 2009). Besides these Pol III-transcribed genes encoded by eukaryotic genomes, several virus-encoded Pol III transcription units have been described that require the transcription apparatus of the host to be expressed (adenovirus-associated VA genes and Epstein-Barr virus-associated EBER genes). In yeast, Pol III promoters are essentially located downstream of the transcription start site (TSS), although sequences 5′ the TSS have been described that influence transcription efficiency and even start site selection *in vitro* and *in vivo* (Eschenlauer et al. 1993; Giuliodori et al. 2003). In mammals, type 1 (5S gene) and type 2 promoters are internal to the gene (e.g. tRNA; VA genes), whereas type 3 promoters (e.g. U6; 7SK; RNAse P; RNAse MRP genes) are entirely located upstream the TSS (Dieci et al. 2007; Orioli et al. 2012). Transcription factors recognizing these genes and being required for their expression have extensively been reviewed (Dumay-Odelot et al. 2010; Geiduschek and Kassavetis 2001; Huang and Maraia 2001; Schramm and Hernandez 2002) and they are schematically represented in Figs. 12.1 and 12.2 for human type 2 and type 3 promoters.

For a long time, Pol III transcription has been known to be essential for the survival of cells, but data supporting the participation of Pol III transcription in the regulation of cell fates or disease have only emerged recently (reviewed in Dumay-Odelot et al. 2010; Marshall and White 2008; White 2008). Some of these functions appear to be intimately connected to the regulation of chromatin dynamics and nuclear architecture. Here, we will describe our current knowledge of the genes and proteins that are involved in these processes.

In general, two functional connections of Pol III transcription with chromatin configuration and nuclear architecture can be discerned. First, the regulatory function of chromatin structure on Pol III transcription activity. Second, the contribution of Pol III genes or Pol III promoter elements to the organization of the genome within the nucleus.

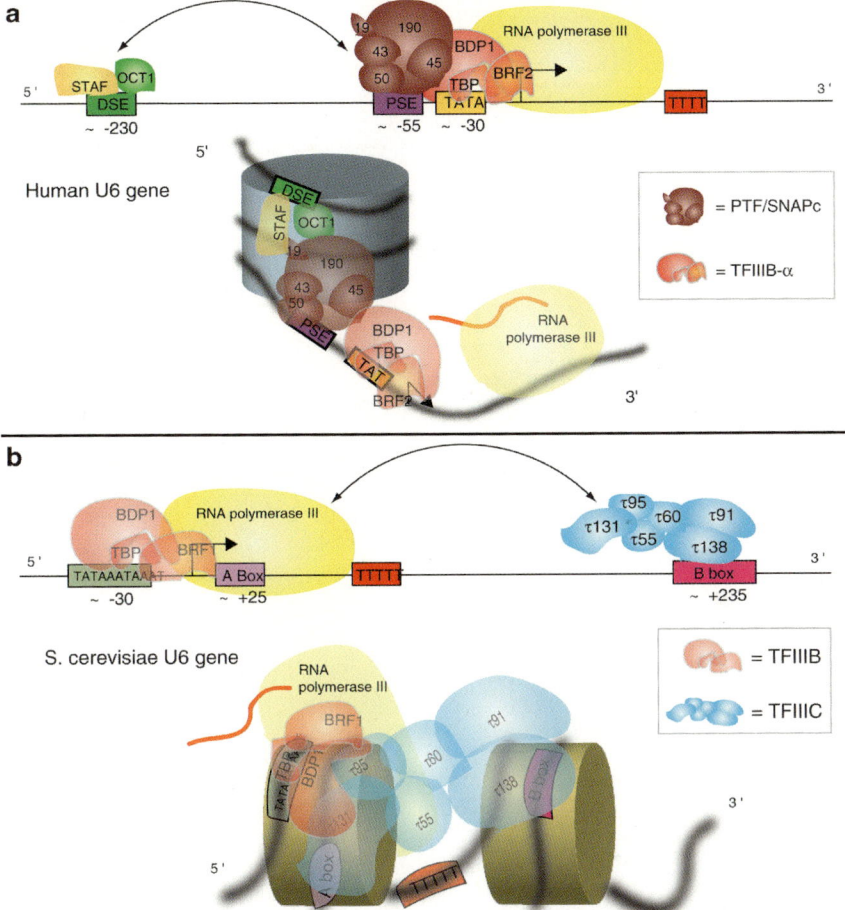

Fig. 12.1 Transcription of the *S. cerevisiae* and of the human U6 gene is facilitated by nucleosomal organization of their promoter/enhancer elements. (**a**) The human U6 gene regulatory sequences are composed of the distal and proximal sequence elements (DSE and PSE), as well as a TATA-box. Factors required for the transcription of the U6 gene are STAF, OCT1, PTF/SNAPc (subunits of 190, 50, 45, 43 and 19 kDa) and TFIIIB-α (subunits BDP1, TBP and BRF2; reviewed in 1). The *upper panel* shows a symbolized representation of U6 gene regulatory elements and transcription factors on linear DNA, indicating that contacts of DSE-bound STAF/OCT1 transcription factors with the PSE/TATA-bound basal Pol III transcription machinery (PTF/TFIIIB-α, RNA polymerase III) cannot be established without physically approaching these elements/factors. The *lower panel* shows a model of a nucleosome positioned in between DSE and PSE (Stünkel et al. 1997; Zhao et al. 2001), permitting the establishment of direct contacts of OCT1/STAF with the basal Pol III machinery and transcription initiation. Nascent RNA is symbolized by a *red line*. (**b**) The *S. cerevisiae* U6 gene regulatory elements are composed of a TATA-like box 5′ of the transcription initiation site, a gene-internal *A Box* and a *B Box* downstream the transcription termination site (Burnol et al. 1993a, b; Eschenlauer et al. 1993; reviewed in Teichmann et al. 2010). RNA polymerase III is recruited to the *S. cerevisiae* U6 gene by TFIIIC (composed by the tau subunits 138, 131, 95, 91, 60 and 55) and TFIIIB (composed by TBP, BDP1 and BRF1). The *upper panel* indicates that TFIIIC-TFIIIB interactions on a linear U6 template are limited due to the distance in between A and B boxes (symbolized by a *line* with two *arrowheads*). The *lower panel* shows a model of a positioned nucleosome that approaches A- and B-Boxes, thereby facilitating contacts of TFIIIC with both elements leading to the recruitment of TFIIIB and productive transcription (Burnol et al. 1993a, b). Nascent RNA is symbolized by a *red line*

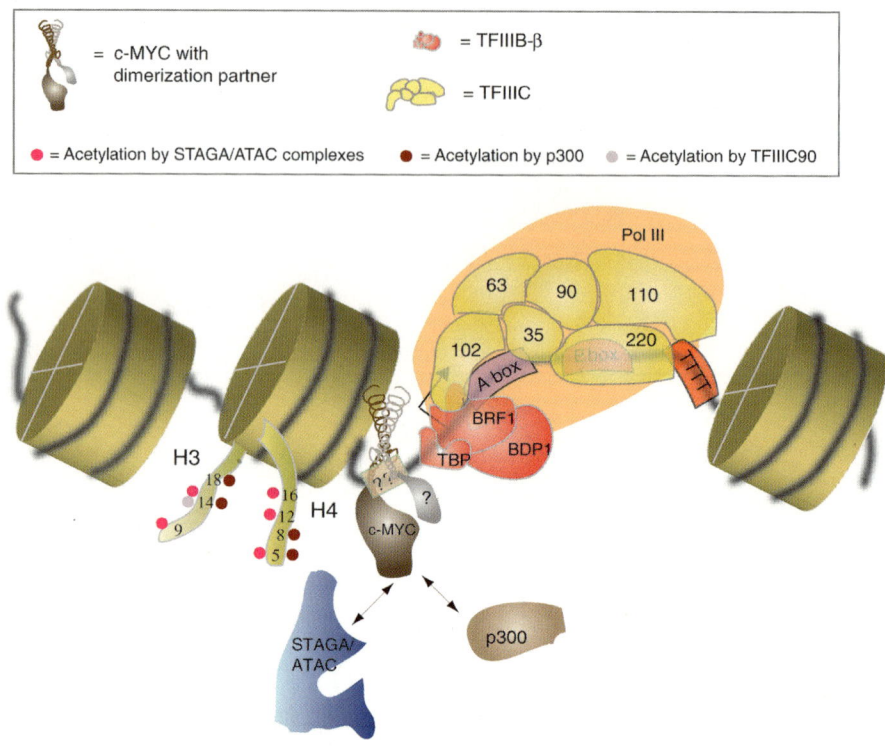

Fig. 12.2 Model of the activation of type 2 promoter (tRNA) transcription by c-MYC in the context of nucleosomal organization of promoter elements. Recognition of an unidentified DNA sequence close to Pol III genes (E box?) by c-MYC, probably heterodimerized with a yet unknown transcription factor (MAX?) leads to the recruitment of the histone acetyltransferases (HATs) GCN5 (in SAGA or ATAC complexes) and/or p300 (Gomez-Roman et al. 2003; Kenneth et al. 2007; Mertens and Roeder 2008; reviewed in Dumay-Odelot et al. 2010). These HATs acetylate the appropriately indicated lysine residues in histone tails (for simplicity, only one tail is shown for histones H3 and H4, respectively). The basal Pol III transcription machinery, composed of TFIIIC (subunits 220, 110, 102, 90, 63 and 35), TFIIIB-β (components BDP1, TBP and BRF1) and Pol III itself. TFIIIC90 contributes to acetylation of histone H3K14 (Hsieh et al. 1999)

12.2 Regulation of Pol III Transcription in Chromatin

Transcription by RNA polymerase III has best been studied in the yeast *S. cerevisiae* and in human cells. All proteins that are essential for the transcription of Pol III genes *in vitro* in these organisms have been identified and the corresponding cDNAs have been cloned. Thus, *in vitro* transcription systems on 'naked' DNA have been described that are reconstituted from highly purified recombinant or epitope-tagged transcription factors and RNA polymerase III itself. In *S. cerevisiae*, it has been shown that tRNA genes can be transcribed by recombinant TFIIIC (six subunits), recombinant TFIIIB (three subunits) and recombinant Sub1 (orthologous to mammalian PC4), together with affinity-purified Pol III (Ducrot et al. 2006; Tavenet et al. 2009). A similar *in vitro* system has been described for transcription of the adenoviral VA1 gene with human factors. In this system, recombinant TFIIIC has been replaced by affinity-purified TFIIIC (Dumay-Odelot et al. 2007; Mertens and Roeder 2008; Haurie et al. 2010; transcription factors required for the expression of type 2 promoters are schematically shown in Fig. 12.2; PC4 is omitted from this Figure since its mode of action is not known). These yeast and human *in vitro* transcription systems suggest that the essential basal Pol III transcription factors have been identified.

In the past few years and based on the knowledge of the basal Pol III machinery, research of gene expression by Pol III has developed into new directions. First, the influence of the chromatin and of modulators thereof on the expression of Pol III genes has been analyzed. Second, genome-wide analyses have been undertaken for determining the presence of basal factors and of chromatin regulators at Pol III genes in living cells.

12.2.1 The Impact of Chromatin on Pol III Transcription

Pioneering experiments demonstrating the influence of chromatin on the regulation of Pol III transcription were carried out shortly after the discovery of three nuclear DNA-dependent RNA polymerases (Roeder and Rutter 1969). It was shown that isolated nuclei (Weinmann and Roeder 1974) or chromatin prepared from mouse myeloma cells (Marzluff and Huang 1975) retained the ability of faithfully transcribing 5S and tRNA genes. Chromatin-derived transcription of the 5S gene could be specifically enhanced by exogenously added Pol III, indicating that the chromatin contained factors that were required for transcription of the gene and that allowed the recruitment of Pol III (Parker and Roeder 1977). Assembly of nucleosomes with purified histones and DNA by employing a salt dialysis method showed that these basal units of chromatin inhibited transcription (by Pol I and Pol II), even in the absence of linker histones (Wasylyk et al. 1979). Preassembly of the Pol III preinitiation complex in such a chromatin assembly system, however, prevented repressive effects of chromatin (Gottesfeld and Bloomer 1982; Felts et al. 1990; Tremethick et al. 1990). Minichromosomes isolated by sucrose gradient centrifugation confirmed the repressive effect of chromatin on 5S RNA transcription, but also showed that a similarly prepared minichromosome carrying the $tRNA_i^{Met}$ gene remained accessible for transcription (Lassar et al. 1985).

The importance of promoter recognition by Pol III transcription factors with respect to transcription in the context of chromatin *in vivo* was first demonstrated for the *S. cerevisiae* SNR6 (U6 RNA) gene. TFIIIC-binding to a B Box element downstream the transcription termination site was essential for transcription of the U6 gene in reconstituted chromatin *in vitro* as well as *in vivo* (Burnol et al. 1993a, b; Fig. 12.1a). Although the promoter context of the U6 gene in humans is distinct from that found in *S. cerevisiae* (Teichmann et al. 2010) and transcription of this gene does not require TFIIIC (Waldschmidt et al. 1991), it was also shown to depend on a specific nuclosomal configuration. A nucleosome specifically positioned in between distal and proximal sequence elements was demonstrated to be a central determinant for the expression of not only the U6 (Stünkel et al. 1997; Zhao et al. 2001; Fig. 12.1b), but also the 7SK gene (Boyd et al. 2000).

In contrast, the gene-internal tRNA promoters were shown to be nucleosome-free or able to prevent nucleosome assembly if being actively transcribed. By employing nucleosome positioning signals, artificially directing a nucleosome over the A Box of a tRNA gene *in vivo* it was demonstrated that ongoing Pol III transcription prohibited nucleosome delivery over the tRNA promoter (Morse et al. 1992). Later on, by genome-wide localization, it was shown that most tRNA genes reside in nucleosome-free regions in *S. cerevisiae* (Albert et al. 2007) and *P. falciparum* (Westenberger et al. 2009). The histone variant H2A.Z was shown to be present at the 5' end of genes transcribed by Pol II or Pol III and to regulate the formation of a nucleosome-free region over the transcription start site (Zhang et al. 2005; Albert et al. 2007; Marques et al. 2010; Mahapatra et al. 2011). In the case of the tRNA SUP4 gene, the histone chaperone FACT, together with the HMG-box protein NHP6 and chromatin remodeling complex RSC were shown to be involved in depositing and removing H2A.Z from gene-flanking nucleosomes (Mahapatra et al. 2011). It was furthermore demonstrated that chromatin structure affects transcription of the ZOD1 gene without altering the occupancy of this gene by Pol III transcription factors (Guffanti et al. 2006). Taking into account that NHP6 is involved in establishing nucleosomal configurations that are permissive for Pol III transcription (Mahapatra et al. 2011) and that NHP6 requirement for transcription increases if Pol III gene-flanking regions comprise suboptimal TFIIIB binding sites (Braglia et al. 2007), it may well be that nucleosomal reorganization at the ZOD1 gene is dependent on NHP6. Such a possible action of NHP6 at the ZOD1 promoter may remove a nucleosome positioned over the 5' region of the gene, resulting in productive transcription. The importance of nucleosome positioning at tRNA genes was put into a distinct perspective when it was uncovered that tRNA genes do not only represent sites for transcription, but also serve as boundary elements, at least in the yeasts *S. cerevisiae* and *S. pombe*, separating euchromatin from heterochromatin (see also Sect. 12.5).

12.2.2 Regulatory Proteins That Influence Pol III Activity in Chromatin

12.2.2.1 MYC

Robert J. White and colleagues described that the protein encoded by the proto-oncogene *c-MYC* is able of stimulating mammalian Pol III transcription. The activation of Pol III transcription was attributed to direct interactions of c-MYC with the basal transcription factor TFIIIB (Gomez-Roman et al. 2003). However, c-MYC plays several roles in the activation of transcription by RNA polymerase II, including the recruitment of chromatin modifying complexes, most notably those containing the histone acetyltransferase GCN5 (KAT2A; Flinn et al. 2002). GCN5, together with its essential cofactor TRRAP has been shown to activate Pol III transcription (Kenneth et al. 2007; Fig. 12.2) in a process that is regulated by the ribosomal protein L11 (Dai et al. 2010). Furthermore, c-MYC-dependent depositioning of histone H2A.Z was demonstrated to influence Pol II transcription (Martinato et al. 2008) and a similar function of c-MYC in Pol III transcription cannot be excluded. Thus, c-MYC may contribute to the regulation of Pol III transcription in multiple ways, by affecting the chromatin structure of Pol III-transcribed genes and by the recruitment of the basal machinery to Pol III genes. These actions may critically contribute to c-MYC's ability of participating in cellular transformation (Eilers and Eisenman 2008). Indeed, it was shown that ongoing Pol III activity is required for c-MYC's contribution to transforming primary cells into tumor cells (Johnson et al. 2008). Thus, c-MYC-dependent modulation of chromatin modification and conformation at Pol III gene loci may turn out critical for cellular transformation.

12.2.2.2 Histone (Lysine) Acetyl Transferases (KATs)

Besides the lysine acetyl transferase GCN5 (KAT2A) mentioned above, also p300 (KAT3B) was shown to enhance Pol III transcription (Mertens and Roeder 2008). The ability of KAT3B to stimulate Pol III transcription on *in vitro* assembled chromatin templates was dependent on KAT3B acetyltransferase activity. In addition, KAT3B activated Pol III transcription of a naked $tDNA_i^{Met}$ template as well as TFIIIC-promoter-binding in the absence of Acetyl-CoA. Thus KAT3B may regulate Pol III transcription during the process of chromatin opening and subsequently during preinitiation complex formation and transcription initiation. In addition to KAT2A and KAT3B that both have been best described for their functions in Pol II transcription (Sterner and Berger 2000; Nagy and Tora 2007), a Pol III-specific KAT activity intrinsic to human transcription factor TFIIIC was identified. The TFIIIC90 subunit was shown to specifically acetylate lysine 14 of histone H3, whereas the entire, affinity-purified TFIIIC complex preferentially exhibited H4 acetyltransferase activity (Kundu et al. 1999; Hsieh et al. 1999; Fig. 12.2). However, the H4 acetyltransferase activity could not yet be attributed to a specific TFIIIC subunit and may either be due to altered specificity of TFIIIC in the nucleosomal context or may have been conferred by a protein co-purifying with TFIIIC. Recently, it was furthermore shown that histone H3S28 phosphorylation enhances Pol III transcription in chromatin (Zhang et al. 2011).

12.3 New Ideas from ChIP-Seq

ChIP sequencing (ChIP-seq) performed in several distinct human cell lines provided important information about the occupancy of Pol III loci by RNA polymerase III, suggestive of their expression and also of cell type-specific differences in their occupation (Barski et al. 2010; Canella et al. 2010; Moqtaderi et al. 2010; Oler et al. 2010; Raha et al. 2010). ChIP-seq data suggest that nucleosomes are depleted over tRNA genes (Moqtaderi et al. 2010), similar to the observations made in yeast (Lee et al. 2007; Mavrich et al. 2008). Histone modifications typical for Pol II transcription start sites (TSS) or Pol II enhancers (H3K4me1; H3K4me2; H3K4me3; H3K4ac; H3K9ac; H3K27ac; H3K36ac; H4K5ac; H4K8ac; H4K91ac; H2B5ac; H2B12ac; H2B120ac; Wang et al. 2008; Zhou et al. 2011) are enriched in the vicinity of Pol III expressed genes (Barski et al. 2010; Oler et al. 2010). Thus, enzymes adding or removing these marks (see Table 12.1) may contribute to the regulation of Pol III transcription. In addition, the histone variants H2A.Z and H3.3 that are associated with actively transcribed Pol II gene regulatory regions (Jin et al. 2009) are also enriched at Pol III genes (Barski et al. 2010; Oler et al. 2010). In agreement with an earlier study (Listerman et al. 2007), Pol II itself also specifically crosslinks close to Pol III TSS. Furthermore, several Pol II transcription activators (STAT1; ETS1; c-JUN; JUN-D; c-FOS; c-MYC) and co-activators (CBP) were crosslinked nearby Pol III TSS (Oler et al. 2010; Raha et al. 2010). In contrast, Pol III genes that were not occupied by its transcription machinery were enriched in histone H3K27me3 (Barski et al. 2010; Oler et al. 2010), a histone mark typically associated with inactive genes (Zhou et al. 2011).

The number of TFIIIC binding sites largely exceeds that of the sites occupied by Pol III (Moqtaderi and Struhl 2004; Moqtaderi et al. 2010). Two thirds of these extra TFIIIC (ETC) sites are found close to Pol II genes (<1 kb). The ETC loci are also associated with histone modifications that are typically found at active Pol II genes. In contrast to transcribed Pol III genes, however, the highest density of these histone modifications is found directly next to the sites of TFIIIC-binding and diminishes with the distance from these sites. This observation indicates that these histone modifications may be introduced in a TFIIIC-dependent manner. Furthermore, TFIIIC co-localizes with CTCF at ETCs and to a lesser extent at transcribed Pol III genes (Moqtaderi et al. 2010; Oler et al. 2010), indicating a role for TFIIIC in the organization of the chromosomes within the nucleus (Fig. 12.3; see also Sect. 12.4.2).

Taken together, the ChIP-seq studies from a variety of human cell lines demonstrated not only that Pol III enrichment at its target promoters, but also that TFIIIC-binding to ETCs coincides with nucleosomal environments that are favourable for transcription, including typical histone modifications and the presence of specific histone variants.

12.4 Nuclear Organization by Pol III

Increasing evidence suggests that genes transcribed by Pol III are clustered within the nucleus and may be involved in the spatial organization of the nucleus. Most data on Pol III gene clustering have been obtained in the yeasts *S. cerevisiae* and

Table 12.1 Histone modifications being associated with (**A**) or excluded from (**B**) genes in contact with the Pol III transcription machinery (Barski et al. 2010). Enzymaes that have been shown to be involved in the respective modifications have been indicated to the right. With the exception of GCN5 (KAT2A; Kenneth et al. 2007) and p300 (KAT3B; Mertens and Roeder 2008), the involvement of these enzymes in the regulation of Pol III transcription has not been documented, but they may represent possible regulators of this transcription system

Modification	Addition by:	Removal by:
(A)		
H3K4me1	NSD3	LSD1; KDM1A
	SET1A, B; KMT2F, G	AOF1; KDM1B
	MLL1-5; KMT2A-E	JARID1B; KDM5B
	Nimura et al. (2010) and Black and Whestine (2011)	Black and Whestine (2011)
H3K4me2	NSD3	LSD1; KDM1A
	SET1A, B; KMT2F, G	AOF1; KDM1B
	MLL1-5; KMT2A-E	JARID1A-D; KDM5A-D
	SET7; KMT7	Black and Whestine (2011)
	SMYD3	
	Nimura et al. (2010) and Black and Whestine (2011)	
H3K4me3	SET1A, B; KMT2F, G	JHDM1B; KDM2B
	MLL1-5; KMT2A-E	JARID1A-D; KDM5A-D
	SMYD3	Black and Whestine (2011)
	PRDM7	
	Nimura et al. (2010), Black and Whestine (2011), and Scharf and Imhof (2011)	
H3K4ac	Mst1; KAT5	HDAC3
	Xhemalce and Kouzarides (2010)	Eot-Houllier et al. (2008)
H3K9ac	GCN5A; KAT2A	SIRT6
	PCAF; KAT2B	Michishita et al. (2008)
	Nagy et al. (2010)	
H3K27ac	CBP; KAT3A	NURD (Reynolds et al. 2011)
	Tie et al. (2009)	
H3K36ac	GCN5; KAT2A	?
	Morris et al. (2007)	
H4K5ac	GCN5A; KAT2A	HDAC3 (Hartman et al. 2005; Demmerle et al. 2012)
	PCAF; KAT2B	
	CBP; KAT3A	
	p300; KAT3B	
	Anamika et al. (2010), Altaf et al. (2010)	
H4K8ac	GCN5A; KAT2A	?
	PCAF; KAT2B	
	MOF; KAT8	
	CBP; KAT3A	
	p300; KAT3B	
	Anamika et al. (2010), Altaf et al. (2010)	

(continued)

Table 12.1 (continued)

Modification	Addition by:	Removal by:
H4K91ac	HAT4 (Yang et al. (2011)	?
H2BK5ac	CBP; KAT3A p300; KAT3B Anamika et al. (2010)	?
H2BK12ac	CBP; KAT3A p300; KAT3B Anamika et al. (2010)	?
H2BK120ac	CBP; KAT3A p300; KAT3B Gatta et al. (2010)	?
(B)		
H3K9me3	Ash1; KMT2H Suv39H1,2; KMT1A,B RIZ1; KMT8 SETDB1; KMT1E Nimura et al. (2010), Black and Whestine (2011), and Scharf and Imhof (2011)	JMJD2; KDM4 Nimura et al. (2010), Black and Whestine (2011), and Scharf and Imhof (2011)
H3K27me3	EZH2; KMT6 NSD3 Nimura et al. (2010), Black and Whestine (2011), and Scharf and Imhof (2011)	JMJD3; KDM6A UTX; KDM6B Nimura et al. (2010), Black and Whestine (2011), and Scharf and Imhof (2011)

S. pombe, but recent data also support Pol III gene-mediated genome organization in human cells. A large proportion of the 274 tRNA genes in the yeast *S. cerevisiae* have been localized to the nucleolus, regardless of their primary genomic localization (Thompson et al. 2003; Wang et al. 2005). This spatial relocation brings them into vicinity of 5S RNA genes that are embedded in the 100–200 rDNA repeats and thus found in the nucleolus. In the yeast *S. pombe*, tRNA genes have been found close to the nuclear periphery, in regions that also contain centromeric DNA. Some of the *S. pombe* tRNA genes are close to centromeres on linear maps of the genome (52 of the 174 fission yeast tRNA genes; Takahashi et al. 1991), but others that are dispersed throughout the three chromosomes have also been found associated with centromeres (Iwasaki et al. 2010). In both cases, *S. cerevisiae* and *S. pombe*, through direct genomic association or through intranuclear relocalization to specific environments, tRNA genes are found in the vicinity of 5S RNA genes. Thus, the nuclear organization of Pol III-transcribed genes may facilitate coordinated expression of tRNA and 5S RNA genes and it may be important for the co-regulation of Pol I and Pol III transcription (Briand et al. 2001).

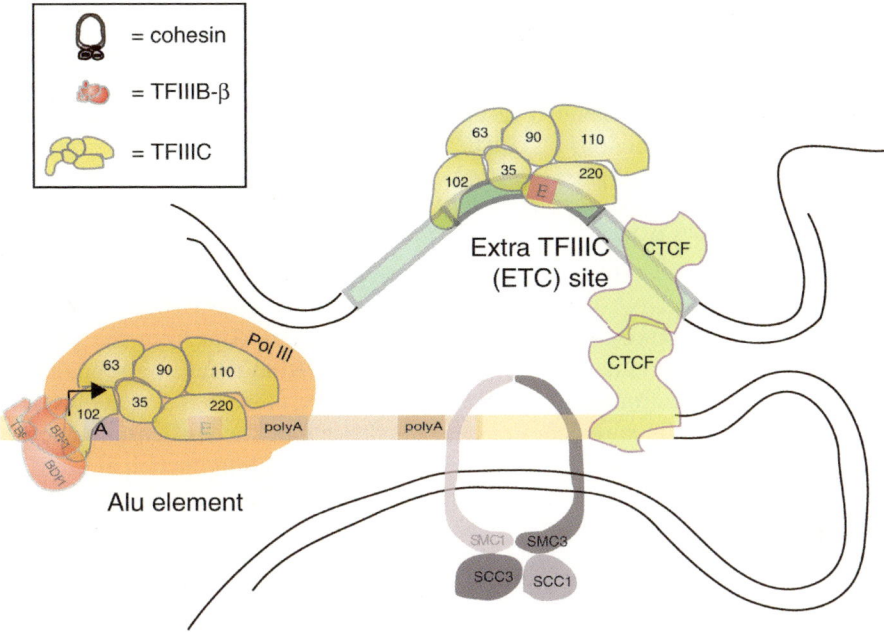

Fig. 12.3 Hypothetical model of a possible contribution of Pol III genes and promoter elements in the establishment of intra- and interchromosomal interactions in mammalian cells. An Alu element is associated with cohesin (Hakimi et al. 2002), which may enable the formation of intrachromosomal loops (Wood et al. 2010). Cohesin associates with CTCF (Parelho et al. 2008; Wendt et al. 2008) and CTCF may form homodimeris or multimers (Phillips and Corces 2009) with CTCF bound to other chromosomal sites (e.g. Extra TFIIIC (ETC) locus (Moqtaderi et al. 2010)). Such multimerizations of CTCF bound to distinct sites may permit the formation of interchromosomal contacts and help to structure nuclear chromatin, possibly in a cell type-specific manner. Subunits of TFIIIC and TFIIIB-β (see Fig. 12.2), as well as of cohesin (SMC1, 3 and SCC 1,3) are appropriately indicated. The ETC is composed of a B box or an unrelated DNA sequence and is bound by TFIIIC, whereas little or no TFIIIB or Pol III are present at this site (Moqtaderi et al. 2010)

12.4.1 Condensin

Subnuclear localization of tRNA genes in the yeast *S. cerevisiae* and *S. pombe* involves the condensin complex (D'Ambrosio et al. 2008; Haeusler et al. 2008; Iwasaki et al. 2010). The condensin complex in budding yeast is composed of two "structural maintenance of chromosomes – SMC" subunits (SMC2 and SMC4) and three non-SMC subunits (Ycs4, Ycs5/Ycg1 and Brn1; Hirano 2005). Vertebrates contain two paralogous condensin complexes composed of five subunits each (the ATPases SMC2 and 4 and auxillary subunits CAPG/G2, CAP-D2/D3 and CAP-H/H2; Hirano 2005; Hudson et al. 2009). In yeast, condensin binding sites are found close to TFIIIC-binding sites and to the "sister chromatid cohesion" 2 and 4 subunits (SCC2/4) of the cohesin loading complex. SCC2/4 and TFIIIC are thought to assist condensin association with chromosomes (D'Ambrosio et al. 2008). Mutation of condensin subunits resulted in the loss of tRNA clustering at the nucleolus. Direct interactions of TFIIIC, TFIIIB and condensin were suggested by co-immunoprecipitation (Haeusler et al. 2008). The co-localization of condensin with TFIIIC was not restricted to Pol III-transcribed genes, but also occurred at extra TFIIIC (ETC) B Box sequences and even at genomic sites that did not contain clear B Box motifs. In addition, a mutation in the TFIIIC subunit Tfc3 (τ138) resulted in reduced DNA-recruitment of the condensin subunit BRN1 and of its loader SCC4 (D'Ambrosio et al. 2008). Complementary data from the yeast *S. pombe* showed that a mutation in SMC4 (*cut3-477*) led to reduced association of tRNA genes with centromeres (which are often located close to the nucleolus) and at the same time to impaired chromosome condensation. The condensation defect and the tRNA relocation could be rescued by simultaneous introduction of a mutant TFIIIC subunit (orthologous to τ138; *sfc3-1*) that negatively impacts on Pol III transcription (Iwasaki et al. 2010). Thus, in *S. pombe*, reduced Pol III transcription favours the centromeric location of Pol III genes and their association with condensin. Together the data from *S. cerevisiae* and *S. pombe* demonstrate that TFIIIC assists in proper loading of condensin to chromosomes. This function of TFIIIC is also reflected by the presence of condensin at all 274 nuclear *S. cerevisiae* tRNA genes (D'Ambrosio et al. 2008). TFIIIC and possibly TFIIIB, by interacting with condensin and with Pol III genes may play a role in structuring chromatin within the nucleus in yeasts. The clustered nucleolar (*S. cerevisiae*) or centromeric (*S. pombe*) organization of tRNA genes probably involves functional or physical contacts of TFIIIC, SCC2/4 and condensin present at two or more distinct Pol III genes. These contacts may possibly also be established in between tRNA genes and ETC sequences of the same or different chromosomes. According to such a model, the nucleolar localization of tRNA genes in *S. cerevisiae* could be dependent on condensin- interactions with TFIIIC/TFIIIB-complexes that are present on tRNA genes and on condensin being bound in the vicinity of 35S rRNA genes (Lavoie et al. 2004; Wang et al. 2005). Such a hypothetical condensin-dependent interaction of tRNA genes and 35S rRNA genes may, at least in parts, explain why the tRNA genes are often found within the nucleolus of *S. cerevisiae*. Likewise, interactions of centromer-bound and tRNA-bound condensin may help tethering these tRNA genes to centromeres in *S. pombe*.

A similar coupling of condensin and TFIIIC functions in higher eukaryotes has not yet been described. However, several data are not in favour with an absolute evolutionary conservation of these concerted functions of condensin and TFIIIC. For example, yeast do not contain subunits orthologous to vertebrate condensin II, suggesting that condesin II may not be involved in Pol III gene-dependent clustering. In addition, vertebrate condensin I, which is evolutionarily related to yeast condensin is excluded from the nucleus during interphase (Hudson et al. 2009), making it unlikely that it could be involved in similar processes of nuclear organization as those having been discovered in yeast, which comprise condensin functions in tethering tRNA genes to centromeres during interphase (Iwasaki et al. 2010). Accordingly, clustering of tRNA genes has not yet been reported in vertebrates (Pombo et al. 1999). Although an involvement of condensin in establishing Pol III gene-dependent subnuclear structures in higher eukaryotes cannot firmly be excluded, the above cited data point to alternative mechanisms of nuclear organization in higher eukaryotes, probably relying on genome organizing molecules such as CTCF (see Sect. 12.4.2).

12.4.2 Cohesin

The ring shaped tetrameric cohesin complex is composed of SMC1/3 and SCC 1/3 subunits (Bose and Gerton 2010). As for condensin (D'Ambrosio et al. 2008), loading of cohesin onto chromosomal DNA requires the SCC2/4 loading complex (Ciosk et al. 2000). Accessory proteins including Pds5, Rad61 and Eco1 are involved in cohesin function (Bose and Gerton 2010). In the yeasts *S. cerevisiae* and *S. pombe*, cohesin is often found close to convergently transcribed genes, suggesting that it is translocated to these sites after having been loaded onto chromatin by the SCC2/4 loading complex (Lengronne et al. 2004). Cohesin has functionally been connected to Pol III transcription (Donze et al. 1999; Dubey and Gartenberg 2007). Mutation of *SMC1* or *SMC3* genes led to the loss of tDNA boundary activity at the transcriptionally repressed HMR mating type locus (Donze et al. 1999) (see also Sect. 12.5). Boundary element activity was shown to depend on an actively transcribed tRNA^{Thr} gene, thus linking cohesin function to Pol III transcription (Donze and Kamakaka 2001). Deletion of silent information regulator 3 (SIR3) led to loss of SCC1 association with the HMR locus (Chang et al. 2005). SIR2-4 proteins form a complex that, together with SIR1, is required for the establishment of heterochromatin at the silent mating-type loci. SIR2 is a NAD-dependent histone H4K16ac deacetylase and SIR3 binds to deacetylated H4K16. SIR 3 and SIR 4 are required for heterochromatin spreading (Johnson et al. 2009). Furthermore, deletion of the tDNA, mutation of its B Box or mutation of the BRF1 subunit of TFIIIB reduced cohesion of the HMR locus (Dubey and Gartenberg 2007). Thus, the binding of cohesin to HMR is dependent on a component of the SIR heterochromatin establishment complex and on the recruitment of TFIIIB to the tRNA^{Thr} gene. Moreover, mutation of the chromatin remodelling complex subunit RSC2 also led to reduced cohesion at HMR. In this context, it should be mentioned that a genome-wide analysis demonstrated that RSC is recruited to numerous tRNA genes in *S. cerevisiae* (Ng et al. 2002) and interacts with RNA polymerase III (Soutourina et al. 2006). Interestingly, mutation of the ISW2 subunit of the ISWI complex that is required for tDNA-dependent barrier function at the HMR (Gelbart et al. 2005; Oki and Kamakaka 2005; Tackett et al. 2005) did not result in alteration of cohesion at this locus. Thus, tDNA barrier activity requiring ISWI complex function and cohesin loading depend on distinct chromatin modifying complexes. In summary, the following model may be proposed: Rsc, recruited by tRNA-bound TFIIIB/Pol III is involved in assisting cohesin loading, which is required for ISWI-dependent tDNA boundary function.

Although tRNAs in mammals have not been shown to co-localize at specific sites within the nucleus and neither condensin nor cohesin association with these genes has hitherto been described, related, but distinct connections of cohesin and Pol III-transcribed genes have been reported. It was shown that human Rad21 (orthologous to the *S. cerevisiae* cohesin subunit SCC1) directly interacts with the Snf2h subunit of the ISWI complex at Pol III-transcribed Alu elements *in vivo* (Hakimi et al. 2002). Furthermore, cohesin was found to co-localize with the CCCTC-binding factor (CTCF) which may be involved in positioning cohesin complexes after having been loaded onto DNA (Nasmyth and Haering 2009). CTCF has been described as a regulator involved in insulation and also in transcriptional activation or repression. It has been shown that CTCF contributes to intra- and interchromosomal interactions, thereby helping to establish higher order chromatin structures (Ling et al. 2006; Kurukuti et al. 2006; Phillips and Corces 2009). CTCF has also been localized close to ETC sites in human K562 cells (Moqtaderi et al. 2010). Thus, cohesin and CTCF have both been linked to human Pol III promoter sequences and they have been shown to colocalize at insulator regions (Parelho et al. 2008; Wendt et al. 2008). These data suggest that some of the underlying mechanisms of establishing subnuclear architectures involving cohesin and RNA polymerase III transcription components have been evolutionary conserved, even if the participation of tRNA genes in this process may have become less evident (a hypothetical model is shown in Fig. 12.3). It should be taken into consideration that yeast and human genomes differ in size (about 3×10^9 base pairs in humans (International Human Genome Sequencing Consortium 2004) versus 1.2×10^7 base pairs in *S. cerevisiae* (Goffeau et al. 1996)). This 250-fold difference in size faces a comparatively minor difference in the number of tRNA genes being in contact with the Pol III transcription machinery (274 in *S. cerevisiae* (Harismendy et al. 2003; Roberts et al. 2003; Moqtaderi and Struhl 2004)) versus about 300 in humans (Barski et al. 2010; Canella et al. 2010; Moqtaderi et al. 2010; Oler et al. 2010; Raha et al. 2010). As a consequence, tRNA genes genes are much more 'diluted' within the human genome (on average about 1 transcribed tRNA gene per 10^7 nucleotides) than in *S. cerevisiae* (one tRNA gene per 5×10^4 base pairs). In view of this difference, it is difficult to imagine that human tRNA genes are implicated in the same condensin- and cohesin-dependent functional structuring of the genome as has been shown in yeast. If ever it may turn out that Pol III genes participate in the organization of the genome in higher eukaryotes including humans, it may well be that repetitive elements such as SINEs (see also below) contribute to this function. The finding that cohesin colocalizes with Alu sequences (Hakimi et al. 2002) supports speculations about such a model.

Mutations in cohesin have been shown to be the underlying cause for genetic diseases. The Cornelia de Lange syndrome (CdLS) and the Roberts Syndrome (RS) are attributed to mutations in proteins affecting cohesin function. Mutations in *SMC1*, *SMC3* and *SCC2* have been linked to CdLS and mutations in *ESCO2* (orthologue of *S. cerevisiae ECO1*; acetyltransferase of SMC3) to RS. CdLS and RS are characterized by mental and growth retardation, including craniofacial anomalies and the RS in particular by symetric hypomelia (Liu and Krantz 2008). In order to understand the molecular mechanisms leading to CdLS and RD, disease-related mutations in *SCC2* and *ESCO2* have been introduced into the orthologous *S. cerevisiae* proteins (scc2-D730V and eco1-W216G). These mutations lead to chromosome decondensation and changes in nuclear architecture. Importantly, colocalization of tRNA genes and the Pol II-transcribed GAL2 gene was impaired, whereas no apparent phenotype on cohesin binding to and distribution on chromatin, as well as chromatid cohesion could be observed (Gard et al. 2009). The loss of GAL2-tRNA colocalization upon mutation of *SCC2* or *ECO1* may be explained by an impaired ability of mutated cohesin to establish intrachromosomal loops that otherwise may approach the GAL2 gene and tRNA genes and subsequently lead to their nucleolar location (similar to the model of an intrachromosomal loop established by cohesin shown in Fig. 12.3). These data indicate that the development of cohesinopathies may depend on defects in gene expression, possibly involving alterations in Pol III gene-dependent establishment of nuclear architecture.

12.5 tDNA Boundary Function

tRNA genes were shown to act as insulator elements. The term 'insulator element' is employed with regard to describing two distinct mechanisms: First, the insulator element may separate enhancer (or repressor) elements from a basal promoter, thereby hindering enhancer-bound activators (or repressors) from activating (repressing) transcription. A second function of an insulator element is its ability to form a boundary in between hetero- and euchromatin. This mode of action leads to the separation of repressive chromatin structures from those that are permissive for transcription. Both mechanisms have been suggested to operate in the case of tRNA genes.

Concerning the former mechanism, tRNA genes have been shown to influence the expression of close-by located Pol II-transcribed genes. Several lines of evidence have been reported in support of such a Pol III gene-mediated influence on the expression of Pol II-transcribed genes. It has been shown that active transcription of tRNA genes represses Pol II transcription (Kinsey and Sandmeyer 1991; Hull et al. 1994; Kendall et al. 2000). The tRNA mediated gene silencing (tgm) was dependent on nucleolar localization of tRNA genes, since mutants that disrupted nucleolar integrity lead to scattered nuclear tRNA gene distribution, accompanied by the loss of tgm (Wang et al. 2005; Haeusler et al. 2008). Thus, nuclear organization and Pol III transcriptional activity were both demonstrated crucial for the inhibition of Pol II transcription by tRNA genes.

With respect to the second mechanism, it was shown that a tRNA^{Thr} gene restricts the spread of heterochromatin at the HMR mating type locus in *S. cerevisiae* (Donze et al. 1999; Donze and Kamakaka 2001). Similarly, STE6 α2 operator-mediated repression of CBT1 expression was blocked by a tRNA^{Thr} gene in Matα cells

(Simms et al. 2004). In the case of the HMR-E silencer, it was shown that tRNA-mediated silencing depends on several features of the Pol III-transcribed gene. First, the tRNA^{Thr} gene could not be replaced by certain other Pol III-transcribed genes, such as U6 or 5S RNA genes. Second, the spacing of A- and B-Boxes turned out crucial, since an intron-containing tRNA gene could only replace the tRNA^{Thr} gene after deletion of its intron. Third, sequences surrounding the tRNA gene were likewise important for barrier function. Fourth, mutations affecting Pol III transcription abolished heterochromatin barrier. In addition to DNA sequence requirements, mutations in tfc3 (τ138; largest subunit of TFIIIC) or in BRF1 (TFIIIB subunit) impaired barrier activity, whereas a transcription initiation incompetent mutation of Pol III (*rpc31-236*) did not affect insulator function. Interestingly, mutation of GCN5 or SAS2 histone acetyltransferases (HAT) (Donze and Kamakaka 2001) or combined deletion of the tRNA^{Thr} gene and mutation of the EAF3 HAT (Oki and Kamakaka 2005) demonstrated that these HATs cooperate with the tRNA^{Thr} gene to establish the heterochromatin barrier. Together, these data point to a complex barrier mechanism that depends on the tRNA promoter and surrounding sequences, on transcription factor binding and HAT activity. However, mechanisms for establishing barrier activity may vary in between different tRNA genes. In the case of the CBT1 gene, it was shown that histone acetylation at this locus did not change upon deletion of the tRNA gene barrier. Either the tRNA gene barrier at the STE6 α2 operator functions different from that at the HMR locus or the contribution of GCN5 and SAS2 HATs to barrier function of the latter is indirect, possibly involving chromatin remodelling and nucleosome positioning. Indeed, it was demonstrated that an actively transcribed tRNA gene affects nucleosome positioning *in vivo* (Morse et al. 1992). Furthermore, components of Pol III pre-initiation complexes have been shown to interact and to recruit subunits of chromatin remodelling complexes to Pol III genes. The TFIIIB component BDP1 recruits ISW2 to about 50% of the *S. cerevisiae* tRNA genes (Bachman et al. 2005; Gelbart et al. 2005). ISW2 is involved in nucleosomal spacing and in transcription repression (Mellor and Morillon 2004). However, it seems as if ISW2 recruitment does not affect Pol III transcription, but rather impacts on Ty integration upstream the tRNA genes (Bachman et al. 2005). Rsc has also been shown to be recruited to Pol III promoters, independently of their transcriptional activity or the presence of TFIIIB (Ng et al. 2002; Soutourina et al. 2006). Importantly, Rsc together with the HAT Rtt109 (being structurally related to p300; Tang et al. 2008) were shown to be required for nuleosome eviction at the HMR silent mating type locus, creating an about 700 nucleotide nucleosome-free region surrounding the tRNA^{Thr} gene (Dhillon et al. 2009). Full barrier activity at HMR was furthermore dependent on the HMG-box containing protein NHP6 (Braglia et al. 2007). Taken together, the documented recruitment of HAT and chromatin remodelling activities to tRNA genes and their requirement for the establishment of barrier function suggest that basal Pol III gene-bound transcription factors, by recruiting chromatin modifying activities, participate in the remodelling of chromatin leading to the reconfiguration of local chromosome structures.

Not only subnuclear repositioning of tRNA genes, but also barrier functions of tRNA genes have been conserved from *S. cerevisiae* to *S. pombe*. According to its genomic localization close to the centromere of chromosome 1, cen1, a tRNA^{Ala} gene has been shown to block the propagation of pericentromeric heterochromatin into the flanking euchromatin (Scott et al. 2006). Also in this case, transcriptional activity of the tRNA gene was required for barrier function. However, a transcription-independent, but TFIIIC-dependent mode of barrier function has in addition been identified in both fission and budding yeast. It relies on the assembly of partial Pol III pre-initiation complexes that do not result in transcription of associated genomic sequences. These sites, referred to as 'extra TFIIIC' sites (ETC; Moqtaderi and Struhl 2004) in budding yeast or 'chromosome organizing clamps' (COC; Noma et al. 2006) in fission yeast recruit TFIIIC only or TFIIIC and TFIIIB (Noma et al. 2006; Simms et al. 2008; Valenzuela et al. 2009). Such ETCs seem to be evolutionarily conserved and have likewise been identified by ChIP-Seq experiments in human cells (Moqtaderi et al. 2010; Oler et al. 2010), often being located close to CTCF binding sites.

12.6 Alu-SINEs

Small interspersed nuclear elements (SINEs) belong to transposable elements (TEs). TEs can be subdivided into DNA transposons and retrotransposons. Retrotransposons can be further distinguished into long terminal repeat (LTR)-containing and non-LTR retrotransposons, the latter of which include Alu SINEs. The about one million Alu elements in humans collectively account for ~11% of the genome (International human genome sequencing consortium 2001; Cordaux and Batzer 2009). Alu RNAs represent fusion products of left and right monomers, each of which have been derived from the SRP (7SL) RNA. The left arm of Alu elements carries an internal type 2 promoter composed of an A- and a B Box that, together with 5' and 3'-flanking sequences, directs transcription (Shaikh et al. 1997; Alemán et al. 2000; Berger and Strub 2011). Alu RNA has been shown to be involved in the regulation of Pol II transcription, alternative splicing, mRNA stability and of translation (reviewed in Häsler et al. 2007; Ponicsan et al. 2010; Berger and Strub 2011). Under normal physiologic conditions, most Alu elements are thought to be transcriptionally silent (Liu et al. 1994; Schmid 1998). Repression of Alu transcription is conferred, at least in parts, by the presence of positioned nucleosomes over the transcribed region (Englander et al. 1993; Englander and Howard 1995; Russanova et al. 1995; Tanaka et al. 2010). Various cellular stresses activate Alu transcription, possibly due to alterations in chromatin structure at these sites (Russanova et al. 1995; Li et al. 2000; Kim et al. 2001). DNA methylation is also involved in the repression of Alu element transcription and treatment with 5-azacytidine coincides with hypomethylation of these elements and increased abundance of Alu RNA (Liu and Schmid 1993; Liu et al. 1994).

Several genetic diseases have been linked to Alu elements. Retrotransposition of transcriptionally active young Alu elements into introns or exons of regulatory genes has been shown to be the underlying cause of diseases like hemophilia A (factor VIII) and B (factor IX), neuofibromatosis (NF1), Huntington disease (ADD1), Apert syndrome (FGFR2) or also breast cancer (BRCA2) (reviewed in Ostertag et al. 2003; Callinan and Batzer 2006). In addition to Alu retrotransposition-mediated diseases, there are other examples that link Alu elements as the source of microsatellites to disease development. Alu element-dependent expansion of a GAA triplet repeat within the first intron of the frataxin gene has been associated with Friedreich ataxia (Justice et al. 2001; Clark et al. 2004). In addition, Spinocerebellar ataxia type 10 was shown to be caused by the Alu-mediated acquisition of unstable pentanucleotide repeats within intron 9 of the human ataxin 10 (ATXN10) gene (Matsuura et al. 2000; Kurosaki et al. 2006, 2009). With respect to trinucleotide expansion diseases, it is noteworthy that CTCF, which colocalizes with Pol III promoter elements (Moqtaderi et al. 2010) has likewise been implicated in the origin of generating fragile sites within trinucleotide repeats (Libby et al. 2008). Alu elements are furthermore sites at which genomic duplications or recombination-mediated deletions occur, which have been shown to result in genetic diseases, including cancer (Deininger and Batzer 1999).

Alu elements do not only contribute to the modification of primate genomes by causing insertions or deletions, but they also affect the expression of nearby genes through intrinsic boundary element activity. The Alu2 element that flanks the human keratin 18 (K18) gene permits position-independent expression of the K18 gene (or of reporter genes) in transgenic mice. Transcription interference with neighbouring genes was dependent on Pol III activity, but boundary element insulator activity was not (Neznanov and Oshima 1993; Willoughby et al. 2000). In addition, a B2 SINE was implicated in the stage-specific activation of the growth hormone gene in mice. B2 SINEs are retrotransposons derived from tRNA genes. At embryonic stage 17.5, the chromatin structure at the GH promoter and at a site 10–14 kb upstream the transcription start site (TSS) changed from hetero- to euchromatin. The site 10–14 kb upstream the TSS was shown to contain a B2 element and both Pol III sense as well as Pol II antisense transcripts could be detected. Boundary activity was dependent on Pol II transcription only (Lunyak et al. 2007). Recently, it was shown that Pol II and Pol III transcription from the same DNA strand of an Alu-like mouse B1 SINE (B1-X35S) was required for insulator activity (Román et al. 2011). Interestingly, the B1-X35S was bound by CTCF which has also been shown to colocalize with cohesin (Nasmyth and Haering 2009). As mentioned before, cohesin is found at human Alu elements, together with the ISWI chromatin remodelling complex (Hakimi et al. 2002; Fig. 12.3).

12.7 Perinucleolar Compartments (PNCs)

The perinucleolar compartment (PNC) is a dynamic structure that is localized close to the nucleolus (Matera et al. 1995). Its formation as a visible subnuclear body has been correlated with cell transformation and particularly with the capacity of cells to form metastases. The PNC is enriched in RNA-binding proteins, such as CUG-BP,

KSRP, nucleolin, PTB, raver 1/2 and ROD1, which contribute to RNA processing and splicing. In addition to these proteins, Pol III-transcribed RNAs, including RNAse P RNA, RNAse MRP RNA, SRP (7SL) RNA, hY RNAs and Alu RNAs have been identified in PNCs (Pollock and Huang 2010). PNCs seem to be associated with DNA, since treatment of cells with histone deacetylase inhibitors alters the morphology of PNCs and since DNA replication uncoupled from cell division increases the number of PNCs (Norton et al. 2009). However, the DNA loci do not contain Pol III genes. Nevertheless, the establishment of PNCs is dependent on ongoing Pol III transcription, which can partially be replaced by expression of RNAse MRP RNA from an ectopic Pol II promoter (Wang et al. 2003). Thus, Pol III RNAs are involved in the formation of a discrete subnuclear structure, the PNC. Analysis of more than 50 non-transformed to fully transformed cell lines derived from mouse and human stromal, endothelial or hematopoietic cells and also from embryonic stem cells showed that PNCs specifically form in cells from solid tumor tissues (Norton et al. 2008) and that their prevalence reaches near 100% in metastatic breast cancer (Kamath et al. 2005). Interestingly, PNCs were not detectable in mouse or human WA07 embryonic stem cells, but in murine F9 teratocarcinoma cells. The identification of an isoform of human RNA polymerase III (Pol IIIα) that is expressed in embryonic H1 stem cells and in cells transformed by defined genetic elements (Haurie et al. 2010) suggests that this isoform alone cannot be responsible for the establishment of PNCs, but may well participate or even be essential for their formation.

12.8 Conclusions

RNA polymerase III transcription is regulated by chromatin structure. In turn, Pol III genes or promoter elements, together with their partial or complete transcription machinery, in the absence or presence of ongoing transcription and depending on the genomic context, contribute to nuclear chromatin organization and the separation of chromatin domains. Thus, Pol III transcription and regulation of chromatin accessibility are interwoven by multiple mechanisms which may be linked to each other by feedback-loops, where chromatin accessibility allows for Pol III transcription, permitting the establishment of chromatin barriers, leading to nucleosome-free regions that, as a consequence, permit transcription and so on. The exciting finding that Pol III genes are occupied by proteins, such as cohesin, condensin or CTCF that contribute to establishing higher order chromatin structures leads to the assumption that the Pol III transcription machinery may contribute to cellular homeostasis beyond the traditionally appreciated role in the production of essential RNAs. In particular, the interaction of ETCs with CTCF or of cohesin with Alu elements, as well as the knowledge of cell type-specific interactions of CTCF with DNA and of cell type-specific methylation of Alu elements raises the possibility that Pol III-dependent establishment of higher chromatin structures may contribute to the regulation of gene expression by Pol II in a tissue- or cell type-specific manner.

Future experiments will show to which extent novel, chromatin-associated functions of Pol III transcription are important for the regulation of developmental processes, of cellular differentiation or, more in general, of health and disease.

Acknowledgements The authors' laboratory is funded by the national league against cancer (Equipe labellisée par la Ligue National Contre le Cancer). In addition it has received grants from the French national institutes of health (INSERM), the national research agency (ANR), the national institute of cancer (INCa), the regional government of Aquitaine and the French-Italian University.

References

Albert I, Mavrich TN, Tomsho LP, Qi J, Zanton SJ, Schuster SC, Pugh BF (2007) Translational and rotational settings of H2A.Z nucleosomes across the *Saccharomyces cerevisiae* genome. Nature 446(7135):572–576

Alemán C, Roy-Engel AM, Shaikh TH, Deininger PL (2000) Cis-acting influences on Alu RNA levels. Nucleic Acids Res 28(23):4755–4761

Altaf M, Auger A, Monnet-Saksouk J, Brodeur J, Piquet S, Cramet M, Bouchard N, Lacoste N, Utley RT, Gaudreau L, Côté J (2010) NuA4-dependent acetylation of nucleosomal histones H4 and H2A directly stimulates incorporation of H2A.Z by the SWR1 complex. J Biol Chem 285(21):15966–15977

Anamika K, Krebs AR, Thompson J, Poch O, Devys D, Tora L (2010) Lessons from genome-wide studies: an integrated definition of the coactivator function of histone acetyl transferases. Epigenetics Chromatin 3(1):18

Bachman N, Gelbart ME, Tsukiyama T, Boeke JD (2005) TFIIIB subunit Bdp1p is required for periodic integration of the Ty1 retrotransposon and targeting of Isw2p to *S. cerevisiae* tDNAs. Genes Dev 19(8):955–964

Bannister AJ, Kouzarides T (2011) Regulation of chromatin by histone modifications. Cell Res 21(3):381–395

Barski A, Chepelev I, Liko D, Cuddapah S, Fleming AB, Birch J, Cui K, White RJ, Zhao K (2010) Pol II and its associated epigenetic marks are present at Pol III-transcribed noncoding RNA genes. Nat Struct Mol Biol 17(5):629–634

Berger A, Strub K (2011) Multiple roles of Alu-related noncoding RNAs. Prog Mol Subcell Biol 51:119–146

Black JC, Whetstine JR (2011) Chromatin landscape: methylation beyond transcription. Epigenetics 6(1):9–15

Bose T, Gerton JL (2010) Cohesinopathies, gene expression, and chromatin organization. J Cell Biol 189(2):201–210

Boyd DC, Greger IH, Murphy S (2000) In vivo footprinting studies suggest a role for chromatin in transcription of the human 7SK gene. Gene 247(1–2):33–44

Braglia P, Dugas SL, Donze D, Dieci G (2007) Requirement of Nhp6 proteins for transcription of a subset of tRNA genes and heterochromatin barrier function in *Saccharomyces cerevisiae*. Mol Cell Biol 27(5):1545–1557

Briand JF, Navarro F, Gadal O, Thuriaux P (2001) Cross talk between tRNA and rRNA synthesis in *Saccharomyces cerevisiae*. Mol Cell Biol 21(1):189–195

Burnol AF, Margottin F, Huet J, Almouzni G, Prioleau MN, Méchali M, Sentenac A (1993a) TFIIIC relieves repression of U6 snRNA transcription by chromatin. Nature 362(6419):475–477

Burnol AF, Margottin F, Schultz P, Marsolier MC, Oudet P, Sentenac A (1993b) Basal promoter and enhancer element of yeast U6 snRNA gene. J Mol Biol 233(4):644–658

Callinan PA, Batzer MA (2006) Retrotransposable elements and human disease. Genome Dyn 1:104–115

Canella D, Praz V, Reina JH, Cousin P, Hernandez N (2010) Defining the RNA polymerase III transcriptome: genome-wide localization of the RNA polymerase III transcription machinery in human cells. Genome Res 20(6):710–721

Chang CR, Wu CS, Hom Y, Gartenberg MR (2005) Targeting of cohesin by transcriptionally silent chromatin. Genes Dev 19(24):3031–3042

Ciosk R, Shirayama M, Shevchenko A, Tanaka T, Toth A, Shevchenko A, Nasmyth K (2000) Cohesin's binding to chromosomes depends on a separate complex consisting of Scc2 and Scc4 proteins. Mol Cell 5(2):243–254

Clark RM, Dalgliesh GL, Endres D, Gomez M, Taylor J, Bidichandani SI (2004) Expansion of GAA triplet repeats in the human genome: unique origin of the FRDA mutation at the center of an Alu. Genomics 83(3):373–383

Cordaux R, Batzer MA (2009) The impact of retrotransposons on human genome evolution. Nat Rev Genet 10(10):691–703

D'Ambrosio C, Schmidt CK, Katou Y, Kelly G, Itoh T, Shirahige K, Uhlmann F (2008) Identification of cis-acting sites for condensin loading onto budding yeast chromosomes. Genes Dev 22(16):2215–2227

Dai MS, Sun XX, Lu H (2010) Ribosomal protein L11 associates with c-Myc at 5 S rRNA and tRNA genes and regulates their expression. J Biol Chem 285(17):12587–12594

Deininger PL, Batzer MA (1999) Alu repeats and human disease. Mol Genet Metab 67(3):183–193

Demmerle J, Koch AJ, Holaska JM (2012) The nuclear envelope protein emerin binds directly to histone deacetylase 3 (HDAC3) and activates HDAC3 activity. J Biol Chem [Epub ahead of print]

Dhillon N, Raab J, Guzzo J, Szyjka SJ, Gangadharan S, Aparicio OM, Andrews B, Kamakaka RT (2009) DNA polymerase epsilon, acetylases and remodellers cooperate to form a specialized chromatin structure at a tRNA insulator. EMBO J 28(17):2583–2600

Dieci G, Fiorino G, Castelnuovo M, Teichmann M, Pagano A (2007) The expanding RNA polymerase III transcriptome. Trends Genet 23(12):614–622

Donze D, Kamakaka RT (2001) RNA polymerase III and RNA polymerase II promoter complexes are heterochromatin barriers in *Saccharomyces cerevisiae*. EMBO J 20(3):520–531

Donze D, Adams CR, Rine J, Kamakaka RT (1999) The boundaries of the silenced HMR domain in *Saccharomyces cerevisiae*. Genes Dev 13(6):698–708

Dubey RN, Gartenberg MR (2007) A tDNA establishes cohesion of a neighboring silent chromatin domain. Genes Dev 21(17):2150–2160

Ducrot C, Lefebvre O, Landrieux E, Guirouilh-Barbat J, Sentenac A, Acker J (2006) Reconstitution of the yeast RNA polymerase III transcription system with all recombinant factors. J Biol Chem 281(17):11685–11692

Dumay-Odelot H, Marck C, Durrieu-Gaillard S, Lefebvre O, Jourdain S, Prochazkova M, Pflieger A, Teichmann M (2007) Identification, molecular cloning, and characterization of the sixth subunit of human transcription factor TFIIIC. J Biol Chem 282(23):17179–17189

Dumay-Odelot H, Durrieu-Gaillard S, Da Silva D, Roeder RG, Teichmann M (2010) Cell growth- and differentiation-dependent regulation of RNA polymerase III transcription. Cell Cycle 9(18):3687–3699

Eilers M, Eisenman RN (2008) Myc's broad reach. Genes Dev 22(20):2755–2766

Englander EW, Howard BH (1995) Nucleosome positioning by human Alu elements in chromatin. J Biol Chem 270(17):10091–10096

Englander EW, Wolffe AP, Howard BH (1993) Nucleosome interactions with a human Alu element. Transcriptional repression and effects of template methylation. J Biol Chem 268(26):19565–19573

Eot-Houllier G, Fulcrand G, Watanabe Y, Magnaghi-Jaulin L, Jaulin C (2008) Histonedeacetylase 3 is required for centromeric H3K4 deacetylation and sister chromatid cohesion. Genes Dev 22(19):2639–2644

Eschenlauer JB, Kaiser MW, Gerlach VL, Brow DA (1993) Architecture of a yeast U6 RNA gene promoter. Mol Cell Biol 13(5):3015–3026

Felts SJ, Weil PA, Chalkley R (1990) Transcription factor requirements for in vitro formation of transcriptionally competent 5S rRNA gene chromatin. Mol Cell Biol 10(5):2390–2401

Flinn EM, Wallberg AE, Hermann S, Grant PA, Workman JL, Wright AP (2002) Recruitment of Gcn5-containing complexes during c-Myc-dependent gene activation. Structure and function aspects. J Biol Chem 277(26):23399–23406

Gard S, Light W, Xiong B, Bose T, McNairn AJ, Harris B, Fleharty B, Seidel C, Brickner JH, Gerton JL (2009) Cohesinopathy mutations disrupt the subnuclear organization of chromatin. J Cell Biol 187(4):455–462

Gatta R, Mantovani R (2010) Single nucleosome ChIPs identify an extensive switch of acetyl marks on cell cycle promoters. Cell Cycle 9(11):2149–2159

Geiduschek EP, Kassavetis GA (2001) The RNA polymerase III transcription apparatus. J Mol Biol 310(1):1–26

Gelbart ME, Bachman N, Delrow J, Boeke JD, Tsukiyama T (2005) Genome-wide identification of Isw2 chromatin-remodeling targets by localization of a catalytically inactive mutant. Genes Dev 19(8):942–954

Giuliodori S, Percudani R, Braglia P, Ferrari R, Guffanti E, Ottonello S, Dieci G (2003) A composite upstream sequence motif potentiates tRNA gene transcription in yeast. J Mol Biol 333(1):1–20

Goffeau A, Barrell BG, Bussey H, Davis RW, Dujon B, Feldmann H, Galibert F, Hoheisel JD, Jacq C, Johnston M, Louis EJ, Mewes HW, Murakami Y, Philippsen P, Tettelin H, Oliver SG (1996) Life with 6000 genes. Science 274(5287):546, 563–567

Gomez-Roman N, Grandori C, Eisenman RN, White RJ (2003) Direct activation of RNA polymerase III transcription by c-Myc. Nature 421(6920):290–294

Gottesfeld J, Bloomer LS (1982) Assembly of transcriptionally active 5S RNA gene chromatin in vitro. Cell 28(4):781–791

Guffanti E, Percudani R, Harismendy O, Soutourina J, Werner M, Iacovella MG, Negri R, Dieci G (2006) Nucleosome depletion activates poised RNA polymerase III at unconventional transcription sites in *Saccharomyces cerevisiae*. J Biol Chem 281(39):29155–29164

Haeusler RA, Pratt-Hyatt M, Good PD, Gipson TA, Engelke DR (2008) Clustering of yeast tRNA genes is mediated by specific association of condensin with tRNA gene transcription complexes. Genes Dev 22(16):2204–2214

Hakimi MA, Bochar DA, Schmiesing JA, Dong Y, Barak OG, Speicher DW, Yokomori K, Shiekhattar R (2002) A chromatin remodelling complex that loads cohesin ontohuman chromosomes. Nature 418(6901):994–998

Harismendy O, Gendrel CG, Soularue P, Gidrol X, Sentenac A, Werner M, Lefebvre O (2003) Genome-wide location of yeast RNA polymerase III transcription machinery. EMBO J 22(18):4738–4747

Hartman HB, Yu J, Alenghat T, Ishizuka T, Lazar MA (2005) The histone-binding code of nuclear receptor co-repressors matches the substrate specificity of histone deacetylase 3. EMBO Rep 6(5):445–451

Häsler J, Samuelsson T, Strub K (2007) Useful 'junk': Alu RNAs in the human transcriptome. Cell Mol Life Sci 64(14):1793–1800

Haurie V, Durrieu-Gaillard S, Dumay-Odelot H, Da Silva D, Rey C, Prochazkova M, Roeder RG, Besser D, Teichmann M (2010) Two isoforms of human RNA polymerase III with specific functions in cell growth and transformation. Proc Natl Acad Sci USA 107(9):4176–4181

Hirano T (2005) Condensins: organizing and segregating the genome. Curr Biol 15(7):R265–R275

Hsieh YJ, Kundu TK, Wang Z, Kovelman R, Roeder RG (1999) The TFIIIC90 subunit of TFIIIC interacts with multiple components of the RNA polymerase III machinery and contains a histone-specific acetyltransferase activity. Mol Cell Biol 19(11):7697–7704

Huang Y, Maraia RJ (2001) Comparison of the RNA polymerase III transcription machinery in *Schizosaccharomyces pombe*, *Saccharomyces cerevisiae* and human. Nucleic Acids Res 29(13):2675–2690

Hudson DF, Marshall KM, Earnshaw WC (2009) Condensin: architect of mitotic chromosomes. Chromosome Res 17(2):131–144

Hull MW, Erickson J, Johnston M, Engelke DR (1994) tRNA genes as transcriptional repressor elements. Mol Cell Biol 14(2):1266–1277

International Human Genome Sequencing Consortium (2001) Initial sequencing and analysis of the human genome. Nature 409(6822):860–921

International Human Genome Sequencing Consortium (2004) Finishing the euchromatic sequence of the human genome. Nature 431(7011):931–945

Iwasaki O, Tanaka A, Tanizawa H, Grewal SI, Noma K (2010) Centromeric localization of dispersed Pol III genes in fission yeast. Mol Biol Cell 21(2):254–265

Jin C, Zang C, Wei G, Cui K, Peng W, Zhao K, Felsenfeld G (2009) H3.3/H2A.Z double variant-containing nucleosomes mark 'nucleosome-free regions' of active promoters and other regulatory regions. Nat Genet 41(8):941–945

Johnson SA, Dubeau L, Johnson DL (2008) Enhanced RNA polymerase III-dependent transcription is required for oncogenic transformation. J Biol Chem 283(28):19184–19191

Johnson A, Li G, Sikorski TW, Buratowski S, Woodcock CL, Moazed D (2009) Reconstitution of heterochromatin-dependent transcriptional gene silencing. Mol Cell 35(6):769–781

Justice CM, Den Z, Nguyen SV, Stoneking M, Deininger PL, Batzer MA, Keats BJ (2001) Phylogenetic analysis of the Friedreich ataxia GAA trinucleotide repeat. J Mol Evol 52(3):232–238

Kamath RV, Thor AD, Wang C, Edgerton SM, Slusarczyk A, Leary DJ, Wang J, Wiley EL, Jovanovic B, Wu Q, Nayar R, Kovarik P, Shi F, Huang S (2005) Perinucleolar compartment prevalence has an independent prognostic value for breast cancer. Cancer Res 65(1):246–253

Kendall A, Hull MW, Bertrand E, Good PD, Singer RH, Engelke DR (2000) A CBF5 mutation that disrupts nucleolar localization of early tRNA biosynthesis in yeast also suppresses tRNA gene-mediated transcriptional silencing. Proc Natl Acad Sci USA 97(24):13108–13113

Kenneth NS, Ramsbottom BA, Gomez-Roman N, Marshall L, Cole PA, White RJ (2007) TRRAP and GCN5 are used by c-Myc to activate RNA polymerase III transcription. Proc Natl Acad Sci USA 104(38):14917–14922

Kim C, Rubin CM, Schmid CW (2001) Genome-wide chromatin remodeling modulates the Alu heat shock response. Gene 276(1–2):127–133

Kinsey PT, Sandmeyer SB (1991) Adjacent pol II and pol III promoters: transcription of the yeast retrotransposon Ty3 and a target tRNA gene. Nucleic Acids Res 19(6):1317–1324

Kundu TK, Wang Z, Roeder RG (1999) Human TFIIIC relieves chromatin-mediated repression of RNA polymerase III transcription and contains an intrinsic histone acetyltransferase activity. Mol Cell Biol 19(2):1605–1615

Kurosaki T, Ninokata A, Wang L, Ueda S (2006) Evolutionary scenario for acquisition of CAG repeats in human SCA1 gene. Gene 373:23–27

Kurosaki T, Matsuura T, Ohno K, Ueda S (2009) Alu-mediated acquisition of unstable ATTCT pentanucleotide repeats in the human ATXN10 gene. Mol Biol Evol 26(11):2573–2579

Kurukuti S, Tiwari VK, Tavoosidana G, Pugacheva E, Murrell A, Zhao Z, Lobanenkov V, Reik W, Ohlsson R (2006) CTCF binding at the H19 imprinting control region mediates maternally inherited higher-order chromatin conformation to restrict enhancer access to Igf2. Proc Natl Acad Sci USA 103(28):10684–10689

Lassar AB, Hamer DH, Roeder RG (1985) Stable transcription complex on a class III gene in a minichromosome. Mol Cell Biol 5(1):40–45

Lavoie BD, Hogan E, Koshland D (2004) In vivo requirements for rDNA chromosome condensation reveal two cell-cycle-regulated pathways for mitotic chromosome folding. Genes Dev 18(1):76–87

Lee W, Tillo D, Bray N, Morse RH, Davis RW, Hughes TR, Nislow C (2007) A high-resolution atlas of nucleosome occupancy in yeast. Nat Genet 39(10):1235–1244

Lengronne A, Katou Y, Mori S, Yokobayashi S, Kelly GP, Itoh T, Watanabe Y, Shirahige K, Uhlmann F (2004) Cohesin relocation from sites of chromosomal loading to places of convergent transcription. Nature 430(6999):573–578

Li TH, Kim C, Rubin CM, Schmid CW (2000) K562 cells implicate increased chromatin accessibility in Alu transcriptional activation. Nucleic Acids Res 28(16):3031–3039

Libby RT, Hagerman KA, Pineda VV, Lau R, Cho DH, Baccam SL, Axford MM, Cleary JD, Moore JM, Sopher BL, Tapscott SJ, Filippova GN, Pearson CE, La Spada AR (2008) CTCF cis-regulates trinucleotide repeat instability in an epigenetic manner: a novel basis for mutational hot spot determination. PLoS Genet 4(11):e1000257

Ling JQ, Li T, Hu JF, Vu TH, Chen HL, Qiu XW, Cherry AM, Hoffman AR (2006) CTCF mediates interchromosomal colocalization between Igf2/H19 and Wsb1/Nf1. Science 312(5771): 269–272

Listerman I, Bledau AS, Grishina I, Neugebauer KM (2007) Extragenic accumulation of RNA polymerase II enhances transcription by RNA polymerase III. PLoS Genet 3(11):e212

Liu J, Krantz ID (2008) Cohesin and human disease. Annu Rev Genomics Hum Genet 9:303–320

Liu WM, Schmid CW (1993) Proposed roles for DNA methylation in Alu transcriptional and mutational inactivation. Nucleic Acids Res 21(6):1351–1359

Liu WM, Maraia RJ, Rubin CM, Schmid CW (1994) Alu transcripts: cytoplasmic localisation and regulation by DNA methylation. Nucleic Acids Res 22(6):1087–1095

Lunyak VV, Prefontaine GG, Núñez E, Cramer T, Ju BG, Ohgi KA, Hutt K, Roy R, García-Díaz A, Zhu X, Yung Y, Montoliu L, Glass CK, Rosenfeld MG (2007) Developmentally regulated activation of a SINE B2 repeat as a domain boundary in organogenesis. Science 317(5835):248–251

Mahapatra S, Dewari PS, Bhardwaj A, Bhargava P (2011) Yeast H2A.Z, FACT complex and RSC regulate transcription of tRNA gene through differential dynamics of flanking nucleosomes. Nucleic Acids Res 39(10):4023–4034

Marques M, Laflamme L, Gervais AL, Gaudreau L (2010) Reconciling the positive and negative roles of histone H2A.Z in gene transcription. Epigenetics 5(4):267–272

Marshall L, White RJ (2008) Non-coding RNA production by RNA polymerase III is implicated in cancer. Nat Rev Cancer 8(12):911–914

Martinato F, Cesaroni M, Amati B, Guccione E (2008) Analysis of Myc-induced histone modifications on target chromatin. PLoS One 3(11):e3650

Marzluff WF Jr, Huang RC (1975) Chromatin directed transcription of 5S and tRNA genes. Proc Natl Acad Sci USA 72(3):1082–1086

Matera AG, Frey MR, Margelot K, Wolin SL (1995) A perinucleolar compartment contains several RNA polymerase III transcripts as well as the polypyrimidine tract-binding protein, hnRNP I. J Cell Biol 129(5):1181–1193

Matsuura T, Yamagata T, Burgess DL, Rasmussen A, Grewal RP, Watase K, Khajavi M, McCall AE, Davis CF, Zu L, Achari M, Pulst SM, Alonso E, Noebels JL, Nelson DL, Zoghbi HY, Ashizawa T (2000) Large expansion of the ATTCT pentanucleotide repeat in spinocerebellar ataxia type 10. Nat Genet 26(2):191–194

Mavrich TN, Ioshikhes IP, Venters BJ, Jiang C, Tomsho LP, Qi J, Schuster SC, Albert I, Pugh BF (2008) A barrier nucleosome model for statistical positioning of nucleosomes throughout the yeast genome. Genome Res 18(7):1073–1083

Mellor J, Morillon A (2004) ISWI complexes in *Saccharomyces cerevisiae*. Biochim Biophys Acta 1677(1–3):100–112

Mertens C, Roeder RG (2008) Different functional modes of p300 in activation of RNA polymerase III transcription from chromatin templates. Mol Cell Biol 28(18):5764–5776

Michishita E, McCord RA, Berber E, Kioi M, Padilla-Nash H, Damian M, Cheung P, Kusumoto R, Kawahara TL, Barrett JC, Chang HY, Bohr VA, Ried T, Gozani O, Chua KF (2008) SIRT6 is a histone H3 lysine 9 deacetylase that modulates telomeric chromatin. Nature 452(7186):492–496

Moqtaderi Z, Struhl K (2004) Genome-wide occupancy profile of the RNA polymerase III machinery in *Saccharomyces cerevisiae* reveals loci with incomplete transcription complexes. Mol Cell Biol 24(10):4118–4127

Moqtaderi Z, Wang J, Raha D, White RJ, Snyder M, Weng Z, Struhl K (2010) Genomic binding profiles of functionally distinct RNA polymerase III transcription complexes in human cells. Nat Struct Mol Biol 17(5):635–640

Morris SA, Rao B, Garcia BA, Hake SB, Diaz RL, Shabanowitz J, Hunt DF, Allis CD, Lieb JD, Strahl BD (2007) Identification of histone H3 lysine 36 acetylation as a highly conserved histone modification. J Biol Chem 282(10):7632–7640

Morse RH, Roth SY, Simpson RT (1992) A transcriptionally active tRNA gene interferes with nucleosome positioning in vivo. Mol Cell Biol 12(9):4015–4025

Nagy Z, Tora L (2007) Distinct GCN5/PCAF-containing complexes function as co-activators and are involved in transcription factor and global histone acetylation. Oncogene 26(37):5341–5357

Nagy Z, Riss A, Fujiyama S, Krebs A, Orpinell M, Jansen P, Cohen A, Stunnenberg HG, Kato S, Tora L (2010) The metazoan ATAC and SAGA coactivator HAT complexes regulate different sets of inducible target genes. Cell Mol Life Sci 67(4):611–628

Nasmyth K, Haering CH (2009) Cohesin: its roles and mechanisms. Annu Rev Genet 43:525–558

Neznanov NS, Oshima RG (1993) Cis regulation of the keratin 18 gene in transgenic mice. Mol Cell Biol 13(3):1815–1823

Ng HH, Robert F, Young RA, Struhl K (2002) Genome-wide location and regulated recruitment of the RSC nucleosome-remodeling complex. Genes Dev 16(7):806–819

Nimura K, Ura K, Kaneda Y (2010) Histone methyltransferases: regulation of transcription and contribution to human disease. J Mol Med 88(12):1213–1220

Noma K, Cam HP, Maraia RJ, Grewal SI (2006) A role for TFIIIC transcription factor complex in genome organization. Cell 125(5):859–872

Norton JT, Pollock CB, Wang C, Schink JC, Kim JJ, Huang S (2008) Perinucleolar compartment prevalence is a phenotypic pancancer marker of malignancy. Cancer 113(4):861–869

Norton JT, Wang C, Gjidoda A, Henry RW, Huang S (2009) The perinucleolar compartment is directly associated with DNA. J Biol Chem 284(7):4090–4101

Oki M, Kamakaka RT (2005) Barrier function at HMR. Mol Cell 19(5):707–716

Oler AJ, Alla RK, Roberts DN, Wong A, Hollenhorst PC, Chandler KJ, Cassiday PA, Nelson CA, Hagedorn CH, Graves BJ, Cairns BR (2010) Human RNA polymerase III transcriptomes and relationships to Pol II promoter chromatin and enhancer-binding factors. Nat Struct Mol Biol 17(5):620–628

Orioli A, Pascali C, Pagano A, Teichmann M, Dieci G (2012) RNA polymerase III transcription control elements: themes and variations. Gene 493(2):185–194

Ostertag EM, Goodier JL, Zhang Y, Kazazian HH Jr (2003) SVA elements are nonautonomous retrotransposons that cause disease in humans. Am J Hum Genet 73(6):1444–1451

Parelho V, Hadjur S, Spivakov M, Leleu M, Sauer S, Gregson HC, Jarmuz A, Canzonetta C, Webster Z, Nesterova T, Cobb BS, Yokomori K, Dillon N, Aragon L, Fisher AG, Merkenschlager M (2008) Cohesins functionally associate with CTCF on mammalian chromosome arms. Cell 132(3):422–433

Parker CS, Roeder RG (1977) Selective and accurate transcription of the Xenopus laevis 5S RNA genes in isolated chromatin by purified RNA polymerase III. Proc Natl Acad Sci USA 74(1):44–48

Phillips JE, Corces VG (2009) CTCF: master weaver of the genome. Cell 137(7):1194–1211

Pollock C, Huang S (2010) The perinucleolar compartment. Cold Spring Harb Perspect Biol 2(2):a000679

Pombo A, Jackson DA, Hollinshead M, Wang Z, Roeder RG, Cook PR (1999) Regional specialization in human nuclei: visualization of discrete sites of transcription by RNA polymerase III. EMBO J 18(8):2241–2253

Ponicsan SL, Kugel JF, Goodrich JA (2010) Genomic gems: SINE RNAs regulate mRNA production. Curr Opin Genet Dev 20(2):149–155

Raha D, Wang Z, Moqtaderi Z, Wu L, Zhong G, Gerstein M, Struhl K, Snyder M (2010) Close association of RNA polymerase II and many transcription factors with Pol III genes. Proc Natl Acad Sci USA 107(8):3639–3644

Reynolds N, Salmon-Divon M, Dvinge H, Hynes-Allen A, Balasooriya G, Leaford D, Behrens A, Bertone P, Hendrich B (2011) NuRD-mediated deacetylation of H3K27 facilitates recruitment of Polycomb Repressive Complex 2 to direct gene repression. EMBO J 31(3):593–605

Roberts DN, Stewart AJ, Huff JT, Cairns BR (2003) The RNA polymerase III transcriptome revealed by genome-wide localization and activity-occupancy relationships. Proc Natl Acad Sci USA 100(25):4695–4700

Roeder RG (2003) Lasker Basic Medical Research Award. The eukaryotic transcriptional machinery: complexities and mechanisms unforeseen. Nat Med 9:1239–1244

Roeder RG, Rutter WJ (1969) Multiple forms of DNA-dependent RNA polymerase in eukaryotic organisms. Nature 224(5216):234–237

Román AC, González-Rico FJ, Moltó E, Hernando H, Neto A, Vicente-Garcia C, Ballestar E, Gómez-Skarmeta JL, Vavrova-Anderson J, White RJ, Montoliu L, Fernández-Salguero PM (2011) Dioxin receptor and SLUG transcription factors regulate the insulator activity of B1 SINE retrotransposons via an RNA polymerase switch. Genome Res 21(3):422–432

Russanova VR, Driscoll CT, Howard BH (1995) Adenovirus type 2 preferentially stimulates polymerase III transcription of Alu elements by relieving repression: a potential role for chromatin. Mol Cell Biol 15(8):4282–4290

Scharf AN, Imhof A (2011) Every methyl counts – epigenetic calculus. FEBS Lett 585(13): 2001–2007

Schmid CW (1998) Does SINE evolution preclude Alu function? Nucleic Acids Res 26(20): 4541–4550

Schramm L, Hernandez N (2002) Recruitment of RNA polymerase III to its target promoters. Genes Dev 16(20):2593–2620

Scott KC, Merrett SL, Willard HF (2006) A heterochromatin barrier partitions the fission yeast centromere into discrete chromatin domains. Curr Biol 16(2):119–129

Shaikh TH, Roy AM, Kim J, Batzer MA, Deininger PL (1997) cDNAs derived from primary and small cytoplasmic Alu (scAlu) transcripts. J Mol Biol 271(2):222–234

Simms TA, Miller EC, Buisson NP, Jambunathan N, Donze D (2004) The *Saccharomyces cerevisiae* TRT2 tRNAThr gene upstream of STE6 is a barrier to repression in MATalpha cells and exerts a potential tRNA position effect in MATa cells. Nucleic Acids Res 32(17):5206–5213

Simms TA, Dugas SL, Gremillion JC, Ibos ME, Dandurand MN, Toliver TT, Edwards DJ, Donze D (2008) TFIIIC binding sites function as both heterochromatin barriers and chromatin insulators in *Saccharomyces cerevisiae*. Eukaryot Cell 7(12):2078–2086

Soutourina J, Bordas-Le Floch V, Gendrel G, Flores A, Ducrot C, Dumay-Odelot H, Soularue P, Navarron F, Cairns BR, Lefebvre O, Werner M (2006) Rsc4 connects the chromatin remodeler RSC to RNA polymerases. Mol Cell Biol 26(13):4920–4933

Sterner DE, Berger SL (2000) Acetylation of histones and transcription-related factors. Microbiol Mol Biol Rev 64(2):435–459

Stünkel W, Kober I, Seifart KH (1997) A nucleosome positioned in the distal promoter region activates transcription of the human U6 gene. Mol Cell Biol 17(8):4397–4405

Tackett AJ, Dilworth DJ, Davey MJ, O'Donnell M, Aitchison JD, Rout MP, Chait BT (2005) Proteomic and genomic characterization of chromatin complexes at a boundary. J Cell Biol 169(1):35–47

Takahashi K, Murakami S, Chikashige Y, Niwa O, Yanagida M (1991) A large number of tRNA genes are symmetrically located in fission yeast centromeres. J Mol Biol 218(1):13–17

Tanaka Y, Yamashita R, Suzuki Y, Nakai K (2010) Effects of Alu elements on global nucleosome positioning in the human genome. BMC Genomics 11:309

Tang Y, Holbert MA, Wurtele H, Meeth K, Rocha W, Gharib M, Jiang E, Thibault P, Verreault A, Cole PA, Marmorstein R (2008) Fungal Rtt109 histone acetyltransferase is an unexpected structural homolog of metazoan p300/CBP. Nat Struct Mol Biol 15(9):998

Tavenet A, Suleau A, Dubreuil G, Ferrari R, Ducrot C, Michaut M, Aude JC, Dieci G, Lefebvre O, Conesa C, Acker J (2009) Genome-wide location analysis reveals a role for Sub1 in RNA polymerase III transcription. Proc Natl Acad Sci USA 106(34):14265–14270

Teichmann M, Dieci G, Pascali C, Boldina G (2010) General transcription factors and subunits of RNA polymerase III: paralogs for promoter- and cell type-specific transcription in multicellular eukaryotes. Transcription 1(3):130–135

Thompson M, Haeusler RA, Good PD, Engelke DR (2003) Nucleolar clustering of dispersed tRNA genes. Science 302(5649):1399–1401

Tie F, Banerjee R, Stratton CA, Prasad-Sinha J, Stepanik V, Zlobin A, Diaz MO, Scacheri PC, Harte PJ (2009) CBP-mediated acetylation of histone H3 lysine 27 antagonizes Drosophila Polycomb silencing. Development 136(18):3131–3141

Tremethick D, Zucker K, Worcel A (1990) The transcription complex of the 5S RNA gene, but not transcription factor IIIA alone, prevents nucleosomal repression of transcription. J Biol Chem 265(9):5014–5023

Valenzuela L, Dhillon N, Kamakaka RT (2009) Transcription independent insulation at TFIIIC-dependent insulators. Genetics 183(1):131–148

Waldschmidt R, Wanandi I, Seifart KH (1991) Identification of transcription factors required for the expression of mammalian U6 genes in vitro. EMBO J 10(9):2595–2603

Wang C, Politz JC, Pederson T, Huang S (2003) RNA polymerase III transcripts and the PTB protein are essential for the integrity of the perinucleolar compartment. Mol Biol Cell 14(6):2425–2435

Wang L, Haeusler RA, Good PD, Thompson M, Nagar S, Engelke DR (2005) Silencing near tRNA genes requires nucleolar localization. J Biol Chem 280(10):8637–8639

Wang Z, Zang C, Rosenfeld JA, Schones DE, Barski A, Cuddapah S, Cui K, Roh TY, Peng W, Zhang MQ, Zhao K (2008) Combinatorial patterns of histone acetylations and methylations in the human genome. Nat Genet 40(7):897–903

Wasylyk B, Oudet P, Chambon P (1979) Preferential in vitro assembly of nucleosome cores on some AT-rich regions of SV40 DNA. Nucleic Acids Res 7(3):705–713

Weinmann R, Roeder RG (1974) Role of DNA-dependent RNA polymerase 3 in the transcription of the tRNA and 5S RNA genes. Proc Natl Acad Sci USA 71(5):1790–1794

Wendt KS, Yoshida K, Itoh T, Bando M, Koch B, Schirghuber E, Tsutsumi S, Nagae G, Ishihara K, Mishiro T, Yahata K, Imamoto F, Aburatani H, Nakao M, Imamoto N, Maeshima K, Shirahige K, Peters JM (2008) Cohesin mediates transcriptional insulation by CCCTC-binding factor. Nature 451(7180):796–801

Westenberger SJ, Cui L, Dharia N, Winzeler E, Cui L (2009) Genome-wide nucleosome mapping of Plasmodium falciparum reveals histone-rich coding and histone-poor intergenic regions and chromatin remodeling of core and subtelomeric genes. BMC Genomics 10:610

White RJ (2008) RNA polymerases I and III, non-coding RNAs and cancer. Trends Genet 24(12):622–629

Willoughby DA, Vilalta A, Oshima RG (2000) An Alu element from the K18 gene confers position-independent expression in transgenic mice. J Biol Chem 275(2):759–768

Wood AJ, Severson AF, Meyer BJ (2010) Condensin and cohesin complexity: the expanding repertoire of functions. Nat Rev Genet 11(6):391–404

Xhemalce B, Kouzarides T (2010) A chromodomain switch mediated by histone H3 Lys 4 acetylation regulates heterochromatin assembly. Genes Dev 24(7):647–652

Yang X, Yu W, Shi L, Sun L, Liang J, Yi X, Li Q, Zhang Y, Yang F, Han X, Zhang D, Yang J, Yao Z, Shang Y (2011) HAT4, a Golgi apparatus-anchored B-type histone acetyltransferase, acetylates free histone H4 and facilitates chromatin assembly. Mol Cell 44(1):39–50

Zhang H, Roberts DN, Cairns BR (2005) Genome-wide dynamics of Htz1, a histone H2A variant that poises repressed/basal promoters for activation through histone loss. Cell 123(2):219–231

Zhang Q, Zhong Q, Evans AG, Levy D, Zhong S (2011) Phosphorylation of histone H3 serine 28 modulates RNA polymerase III-dependent transcription. Oncogene 30(37):3943–3952

Zhao X, Pendergrast PS, Hernandez N (2001) A positioned nucleosome on the human U6 promoter allows recruitment of SNAPc by the Oct-1 POU domain. Mol Cell 7(3):539–549

Zhou VW, Goren A, Bernstein BE (2011) Charting histone modifications and the functional organization of mammalian genomes. Nat Rev Genet 12(1):7–18

Chapter 13
The Role of DNA Methylation and Histone Modifications in Transcriptional Regulation in Humans

Jaime L. Miller and Patrick A. Grant

Abstract Although the field of genetics has grown by leaps and bounds within the last decade due to the completion and availability of the human genome sequence, transcriptional regulation still cannot be explained solely by an individual's DNA sequence. Complex coordination and communication between a plethora of well-conserved chromatin modifying factors are essential for all organisms. Regulation of gene expression depends on histone post translational modifications (HPTMs), DNA methylation, histone variants, remodeling enzymes, and effector proteins that influence the structure and function of chromatin, which affects a broad spectrum of cellular processes such as DNA repair, DNA replication, growth, and proliferation. If mutated or deleted, many of these factors can result in human disease at the level of transcriptional regulation. The common goal of recent studies is to understand disease states at the stage of altered gene expression. Utilizing information gained from new high-throughput techniques and analyses will aid biomedical research in the development of treatments that work at one of the most basic levels of gene expression, chromatin. This chapter will discuss the effects of and mechanism by which histone modifications and DNA methylation affect transcriptional regulation.

13.1 DNA Methylation

13.1.1 CpG Islands

With respect to epigenetic research and a causal relationship to human disease, DNA methylation is the most characterized modification. The enzymatic addition of a methyl group to DNA is performed by DNA methyltransferase (DNMT) on the

---

J.L. Miller • P.A. Grant (✉)
Department of Biochemistry and Molecular Genetics, University of Virginia
School of Medicine, Charlottesville, VA 22908, USA
e-mail: pag9n@virginia.edu

5′-carbon of the pyrimidine ring in cytosine. Four human DNMTs have been characterized: DNMT1 (Bestor et al. 1988), DNMT2 (Yoder and Bestor 1998), DNMT3a and DNMT3b (Okano et al. 1999). *De novo* DNA methylation patterns are established early in development by DNMT3a and DNMT3b and maintained by DNMT1, which prefers to methylate hemi-methylated templates during DNA replication through its recruitment by proliferating cell nuclear antigen (PCNA). Around 3% of cytosines are methylated in the human genome almost exclusively in the context of the dinucleotide, CpG. 5-methylcytosine (5-mC) is also found in very low abundance at the trinucleotide, CpNpG (Clark et al. 1995; Lee et al. 2010).

CpG dinucleotides are rarer than expected in the human genome (~1%) (Josse et al. 1961; Swartz et al. 1962) as a result of 5-mC deamination and subsequent mutation to thymine (Scarano et al. 1967). 70–80% of CpG dinucleotides are methylated and those dinucleotides that are unmethylated tend to cluster in islands (Ehrlich et al. 1982). Regions containing the normal expected density of CpG dinucleotides are called CpG islands (CGI), which are regions no smaller than 200 bp that contain a GC content of more than 55% and an expected GC content to observed GC content ratio greater than 0.65 (Takai and Jones 2002).

Approximately 60% of human gene promoters and first exons are associated with CGIs (Bird 2002). CGIs at promoters are frequently hypomethylated corresponding to a permissive chromatin structure in order to poise genes for a transcriptional activation (Larsen et al. 1992; Antequera and Bird 1993) while some are hypermethylated during development, which stably silences the promoter (Fig. 13.1a) (Straussman et al. 2009). Such programmed CGI methylation is important for genomic imprinting, which results in monoallelic expression through the silencing of a parental allele (Kacem and Feil 2009) and gene dosage compensation such as X-chromosome inactivation in females (Reik and Lewis 2005). Recently, Doi et al. has shown that limited gene expression in differing tissue types is caused by differential methylation of CpG island shores (2009), which are located within 2.0 kb of CGIs (Fig. 13.1b) (Saxonov et al. 2006). Still, a fraction of CGIs are prone to methylation in some tissues due to aging, in promoters of tumor suppressor genes in cancer cells (Issa 2000) and committed cell lines (Jones et al. 1990). The remaining 40% of CGIs are located intra- and intergenically. Intragenically located CGIs within the coding region of genes are methylated at trinucleotides CpXpG (Lister et al. 2009) and are commonly found in highly expressed, constitutively active genes (Fig. 13.1c) (Zhang et al. 2006) while intergenic CGIs may be used for transcription of non-coding RNAs (Illingworth et al. 2008).

13.1.2 Transcriptional Regulation

More often than not, DNA methylation is usually associated with gene silencing due to (1) the occlusion of DNA binding proteins that act as or recruit transcriptional activators or (2) the recruitment of methyl-binding proteins (MBPs), which recruit transcriptional corepressor complexes (Fig. 13.1a). Transcriptional activators and

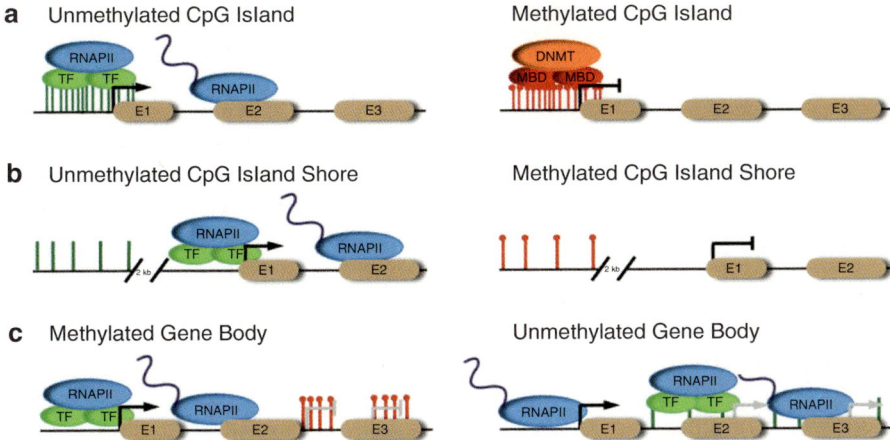

Fig. 13.1 Various sites and effects of DNA methylation throughout the genome. DNA methylation is found at inter- and intragenic regions throughout the genome. DNA methylation dependent transcriptional activity is contingent on CpG dinucleotide genic location and density. Normal methylation events and subsequent effects are shown on the left. (**a**) CpG islands at promoters are normally unmethylated resulting in gene expression. However, aberrant hypermethylation at the same promoter results in corepressor complex recruitment and subsequent gene repression. (**b**) Intragenic regions characterized by scattered CpG dinucleotides located 2 kb upstream of the promoter called CpG island shores are regulated in the same manner as (**a**). (**c**) DNA methylation within the gene body prevents initiation of transcription from spurious sites in the gene. If unmethylated, these sites become transcriptional start sites resulting in an incorrect product (Portela and Esteller 2010)

repressors recruit histone modifying and chromatin remodeling complexes that can remodel chromatin, which ultimately changes the transcriptional activity of a gene. Modifications made by such complexes and subsequent effects on transcription will be discussed later.

Even previous to DNA methylation, DNMTs can be recruited to DNA via DNA binding transcription factors, which results in specific promoter DNA methylation and regulatory gene repression. For example, studies (Di Croce et al. 2002) showed that DNMTs interact with the oncogenic transcription factor formed by the fusion of promyelocytic leukemia protein and retinoic acid receptor (PML-RAR), found in acute promyelocytic leukemia. DNMT recruitment to the *RARβ2* gene promoter by PML-RAR results in promoter hypermethylation and subsequent gene silencing (Di Croce et al. 2002). A similar mechanism has been described for Myc, a DNA binding transcription factor. Myc interacts with DNMT3a and is recruited to the p21 gene promoter resulting in subsequent DNA methylation and p21 gene repression (Brenner et al. 2005). In addition, p53 also interacts with DNMT3a and represses p53′s transactivator function at the p21 gene promoter but in a DNA methylation independent manner (Wang et al. 2005). Both mechanisms elucidate cancer promoting pathways that intersect with DNA methylation and cause repression of p21 expression, a cyclin dependent kinase inhibitor. Moreover, one study (Hervouet et al. 2009) showed that DNMT3a/b interacts with 79 different DNA binding transcription factors. Some interactions were exclusive to each DNMT while some were shared between both (Hervouet et al. 2009). The diversity of interactions further illustrates the importance of DNA methylation on gene expression regulation through DNMT recruitment via DNA binding transcription factors.

Once DNA is methylated, DNA methyl-binding proteins (MBP) can bind to DNA and recruit transcriptional corepressors such as histone deacetylase (HDAC) complexes, polycomb proteins, and chromatin remodeling complexes. One family consists of MBPs, which possess a conserved methyl-CpG-binding-domain (MBD) and includes MBD1, MBD2, MBD3, MBD4, and MeCP2. MeCP2 is the founding member of the MBD family and contains a MBD in addition to an adjacent transcriptional repressor domain (TRD) (Klose and Bird 2006). The TRD interacts with Sin3 corepressor complex containing HDAC1 and 2 (Nan et al. 1998). MBD1 also contains three zinc-binding domains (CxxC), which has been shown to be responsible for its ability to bind unmethylated CpG sites (Jorgensen et al. 2004). MBD1 and 2 both contain a TRD that recruits different transcriptional corepressor complexes containing HDACs. MBD3 contains a MBD but does not bind methylated DNA due to two amino acid substitutions (Hendrich and Tweedie 2003) but is associated with the nucleosome remodeling and histone deacetylase (NuRD) corepressor complex, which contains HDACs necessary for transcriptional silencing. MBD4 is a thymidine glycosylase DNA repair enzyme that excises mismatched thymines that have resulted from 5-methylcytosine deamination in the context of CpG dinucleotides (Hendrich et al. 1999).

The second family of MBPs includes Kaiso, zinc finger and BTB (for BR-C, ttk, and bab) domain containing (ZBTB) 4 and ZBTB38 (Zollman et al. 1994). These are atypical MBPs, because they depend on a zinc-finger domain to recognize methylated DNA and a POZ (for Pox virus and Zinc finger) (Bardwell and Treisman 1994)/BTB domain to repress transcription through its interaction with nuclear receptor co-repressor-1 (N-CoR) (Prokhortchouk and Defossez 2008). Another study (Iioka et al. 2009) showed that Kaiso can regulate transcription factor activity by modulating the interaction between β-catenin and HDAC1 activity. The third family of MBPs includes ubiquitin-like plant homeodomain and RING finger (UHRF)-domain containing protein 1 and 2. Both contain SET and RING associated (SRA) domains, which preferentially bind to DNMT1's substrate, hemi-methylated DNA (Bostick et al. 2007). Furthermore, UHRF1 has been shown to colocalize with DMNT1, which suggests that this family of MBPs may help target DMNT1 to DNA (Bostick et al. 2007).

DNA methylation is usually associated with transcriptional silencing, and one of the most well known cases where differential DNA methylation induces and suppresses expression is genomic imprinting at the *H19/IGF2* locus. Genomic imprinting is a form of gene regulation in which an allele is expressed from one of the two parental homologous genes. *H19* and *IGF2* are reciprocally imprinted so that *H19* is expressed from the maternally inherited allele and *IGF2* from the paternally inherited allele (Bell and Felsenfeld 2000). Transcriptional regulation of these genes is dependent on a differentially methylated DNA domain (DMD) or imprinting control region (ICR) located upstream of *H19* and downstream of *IGF2*. The DMD/ICR is methylated on the paternal allele but not the maternal allele (Bell and Felsenfeld 2000; Hark et al. 2000; Szabo et al. 2000; Kanduri et al. 2000). CCCTC-binding factor (CTCF) binds to the unmethylated ICR of the maternal allele, which blocks an enhancer region located downstream of *H19* from activating transcription of *IGF2* (Hark et al. 2000). CTCF binding also protects against *de novo* methylation and subsequent repression at the *H19* locus on the maternal allele (Rand et al. 2004). This is one of the most basic examples of how differentially methylated regions can determine levels of gene expression. Mutations or deletions in the *H19* promoter, ICR, or enhancer can lead to growth defects such as Beckwith-Wiedemann Syndrome or Silver-Russell dwarfism (Delaval et al. 2006).

With the advent of microarrays and high-throughput technologies, an explosion of gene expression profile comparisons in normal and diseased cells has occurred. Many studies have pursued genes of interest by comparing the DNA methylation status of a gene's 5' promoter region (Weber et al. 2005; Hatada et al. 2006), and presently, more comprehensive results are available as more direct solutions to discovering gene expression controlled by DNA methylation are established. Using *Arabidopsis thaliana* as a model system, Zhang et al. analyzed and compared whole genome methylome tiling arrays gathered from immunoprecipitating 5-mC or chromatin crosslinked MBPs in normal and mutant cells (Zhang et al. 2006). Another study (Javierre et al. 2010) compared the DNA methylome of monozygotic twins who were differently affected by the disease, systemic lupus erythematosus (SLE) (Javierre et al. 2010). In comparison to the healthy twin, the twin affected by SLE had a decrease in promoter DNA methylation for many genes involved in immune system function including *IFNGR2, MMP14, LCN2, CSF3R, PECAM1, CD9, AIM2,* and *PDX1*. These genes had also previously been shown to participate in the development of SLE (Javierre et al. 2010).

13.1.3 5-Hydroxymethylcytosine

In the previous sections, 5-methylcytosine (5-mC) was discussed extensively. 5-mC can be converted to 5-hydroxymethylcytosine (5-hmC) by an oxidation reaction carried out the ten-eleven-translocation (TET) family of proteins (Tahiliani et al. 2009). 5-hmC was first discovered in bacteriophage DNA in 1952 (Wyatt and Cohen 1952; Warren 1980) and has since been found to be enriched in mouse brain (Kriaucionis and Heintz 2009), embryonic stem cells (Tahiliani et al. 2009), and human tissues (Li and Liu 2011).

Levels of 5-hmC are dynamically regulated by TET1-3 in stem cells and seem to be higher in pluripotent cells. Knockdown of *TET1* and *TET2* causes a decrease in 5-hmC levels and an increase in 5-mC at stem cell related gene promoters (Ficz et al. 2011). These genes are subsequently silenced. *TET3* is highly expressed in zygotes and oocytes (Wossidlo et al. 2011) and a recent study (Iqbal et al. 2011) has shown that after fertilization, 5-mC is converted to 5-hmC in the male but not the female pronucleus. This data (Iqbal et al. 2011) suggests an alternative to the global demethylation theory during cellular dedifferentiation where genome-wide 5-mC may be converted to 5-hmC by TET3 and differentiation is promoted by a decrease in TET3 and an increase in TET1 and 2 (Koh et al. 2011; Walter 2011). The mechanisms behind 5-hmC's role in cellular differentiation (Ito et al. 2010), carcinogenesis, (Li and Liu 2011) and association with actively transcribed genes is a mystery (Ficz et al. 2011). One clue provided is that 5-hmC prevents the binding of MBDs (Valinluck et al. 2004) and DNMTs (Valinluck and Sowers 2007).

13.2 Histone Modifications

13.2.1 Types of Modifications

As mentioned in the previous section, methylated DNA can recruit different transcriptional activator and repressor complexes. In most cases, these complexes contain histone modifying and chromatin remodeling enzymes that regulate chromatin structure, which ultimately changes the transcriptional activity of a gene. Such complexes are not just recruited by DNA methylation but also by various post-translational modifications (PTMs) of the proteins that make up chromatin. In this section, the effects of histone modifications and chromatin remodeling on gene expression will be discussed.

Chromatin is the organization of the eukaryotic genome into a condensed form due to the function of many proteins and RNAs. The fundamental unit of the highly ordered chromatic fiber is the nucleosome, which consists of 146 base pairs of DNA wrapped around an octamer of core histones that contains two of each histone H2A, H2B, H3, and H4. A linker histone, H1, binds to DNA as it enters and exits its 1.65 turns around the nucleosome (Luger et al. 1997). Naturally, the condensed structure forms a barrier to cell processes that require accessibility to DNA such as DNA replication, damage repair, and transcription (Workman and Kingston 1998). Covalent HPTMs such as acetylation, methylation, phosphorylation, ubiquitination, SUMOylation, ADP-ribosylation, deimination, and the non-covalent proline isomerization (Kouzarides 2007) can affect the condensation of chromatin as to organize the genome into transcriptionally active and inactive regions termed euchromatin and heterochromatin (Heitz 1929) respectively and recruit effector proteins (Jenuwein 2001).

The core histones are highly conserved basic proteins composed of a globular domain and highly flexible N-terminal tails that protrude from the DNA wrapped nucleosome (Luger et al. 1997). All histone N-terminal tails and globular domains are subject to modification and more is known about the smaller covalent modifications methylation, acetylation, and phosphorylation. Lysine residues can be mono-, di-, and trimethylated while arginine residues can only be mono- or symmetrically or asymmetrically dimethylated (Bannister and Kouzarides 2011). The interactions between chromatin associated proteins that bind HPTMs can act

13 The Role of DNA Methylation and Histone Modifications...

Table 13.1 Transcriptional and cellular role of histone modifications

Modification	Histone residues modified	Role in cell activity and transcription	Histone modification readers
Acetylated lysine (Kac)	H3 (K4,9,14,18,23, 27,36,56) H4 (K5,8,12,16,19)	Activation	Bromodomain
	H2A (K5,9) H2B (K5,6,7,12,16, 17,20,120)	DNA damage repair	Tandem PHD
Phosphorylated serine/	H3 (S10,28 and T3,11,45)	Apoptosis	14-3-3 Domain
	H4 (S1,47)	Activation	
Threonine (S/Tph)	H2A (S1)	Mitosis (Baker et al. 2010)	
Methylated lysine (Kme)	H3 (K4,23,36,79)	Activation	MBT PHD
	H3 (K9, 27) and H4 (K20)	Repression	Tudor Chromodomain WD40
Methylated arginine (Rme)	H3 (R2,17,26) and H4 (R3)	Activation	Tudor (Yang et al. 2010; Chen et al. 2011)
	H3 (R8)	Repression	ADD (Zhao et al. 2009)
Ubiquitylated lysine (Kub)	H2A (K119) H2B (K120)	Repression Activation (Zhu et al. 2005)	Cps35 (Lee et al. 2007a; Zheng et al. 2010)
Sumoylated lysine (Ksu)	H4 (?)	Repression (Shiio and Eisenman 2003)	

synergistically or antagonistically with one another resulting in various gradients of transcriptional activation and repression across the genome. The term "histone code" was coined in order to convey that chromatin modifying proteins ultimately determine phenotype rather than simple possession of a certain genetic code (Strahl and Allis 2000; Jenuwein and Allis 2001). HPTMs specific roles in gene expression and cellular activities are shown in Table 13.1 (adapted from Kouzarides and Berger 2007; Wang et al. 2008).

Euchromatin is characterized by high levels of acetylation and high levels of H3K4me1/2/3, H3K36me3 and H3K79me1/2/3. On the other hand, heterochromatin is characterized by low levels of acetylation and high levels of H3K9me2/3, H3K27me2/3 and H4K20me3 (Table 13.1) (Li et al. 2007a). More recently, a group (Wang et al. 2008) performed chromatin immunoprecipitation sequencing (ChIP-seq) on 39 different core histone acetylations and methylations at 3,286 promoter regions. As shown in previous studies (Turner et al. 1992), acetylated histones consistently correlate with increased gene transcription. However, certain modifications localized to specific gene regions rather than just at transcriptional start sites (TSS). H2AK9ac, H2BK5ac, H3K9ac, H3K18ac, H3K27ac, H3K36ac and H4K91ac were mainly located in the region surrounding the TSS, whereas

H2BK12ac, H2BK20ac, H2BK120ac, H3K4ac, H4K5ac, H4K8ac, H4K12ac and H4K16ac were prominent in the promoter and transcribed regions of active genes (Wang et al. 2008).

Another group (Karlic et al. 2010) analyzed ChIP-seq data produced by the Zhao lab in order to create a model that could predict levels of gene expression based on HPTM levels present at promoters. They found that actively transcribed genes are characterized by high levels of H3K4me3, H3K27ac, H2BK5ac and H4K20me1 in the promoter and H3K79me1 and H4K20me1 along the gene body. Moreover, they found high levels of H4K20me1 and H3K27ac at promoters that contained high CpG content and H3K4me3 and H3K79me1 at promoters with low CpG content (Karlic et al. 2010). Although there is no model that explains the HPTM difference at the two types of promoters, one can guess that the difference is caused by different regulatory mechanisms and possibly, changes in DNA methylation. In agreement with this theory, a recent paper (Ernst and Kellis 2010) used previous ChIP-seq data for HPTMs, CTCF, RNA Polymerase II (RNAPII), and the histone variant, H2A.Z, to describe 51 distinct chromatin states. Each state is described by the enrichment of different HPTMs and chromatin associated proteins across the genome. Moreover, biological states of cells (cell cycle, developmental, T-cell activation, *etc.*) were predicted using the 51 epigenetic states (Ernst and Kellis 2010). Another interesting study (Mikkelsen et al. 2007) showed that embryonic stem cells contained a bivalent pattern of HPTMs at promoters of genes that regulate development. Surprisingly, they found H3K4me3, an activation mark, and H3K27me3, a repressive mark, co-localizing at these promoters in stem cells (Mikkelsen et al. 2007). These bivalent domains can resolve into four different chromatin states: (1) marked with both H3K4me3 and H3K27me3; (2) marked with neither H3K4me3 nor H3K27me3; (3) marked with H3K4me3 alone; (4) marked with H3K27me3 alone. Maintenance or loss of both marks results in a poised transcriptional state while preservation of H3K27me3 alone or H3K4me3 alone results in inactive and active transcription respectively (Cui et al. 2009). This data (Mikkelsen et al. 2007) suggests that HPTM bivalency at promoters allows for plasticity during cellular differentiation and development (Bernstein et al. 2006).

13.2.2 Transcriptional Regulation

As presented in the previous section, various HPTMs correlate with gene expression and repression (Fig. 13.2). Until recently, elucidating the mechanisms by which HPTMs interact with one another to control transcriptional activity has been complicated due to the layered complexity of combinatorial HPTMs and HPTM crosstalk. However, analysis of recently acquired ChIP-seq data and associated gene expression profiles has speedily facilitated decipherment of the histone code and its effect on transcriptional activity (Figs. 13.2 and 13.3) (Barski et al. 2007; Wang et al. 2008; Heintzman et al. 2007; Mikkelsen et al. 2007). Three broad effects on transcription can be attributed to HPTMs: (1) HPTMs can prevent certain chromatin

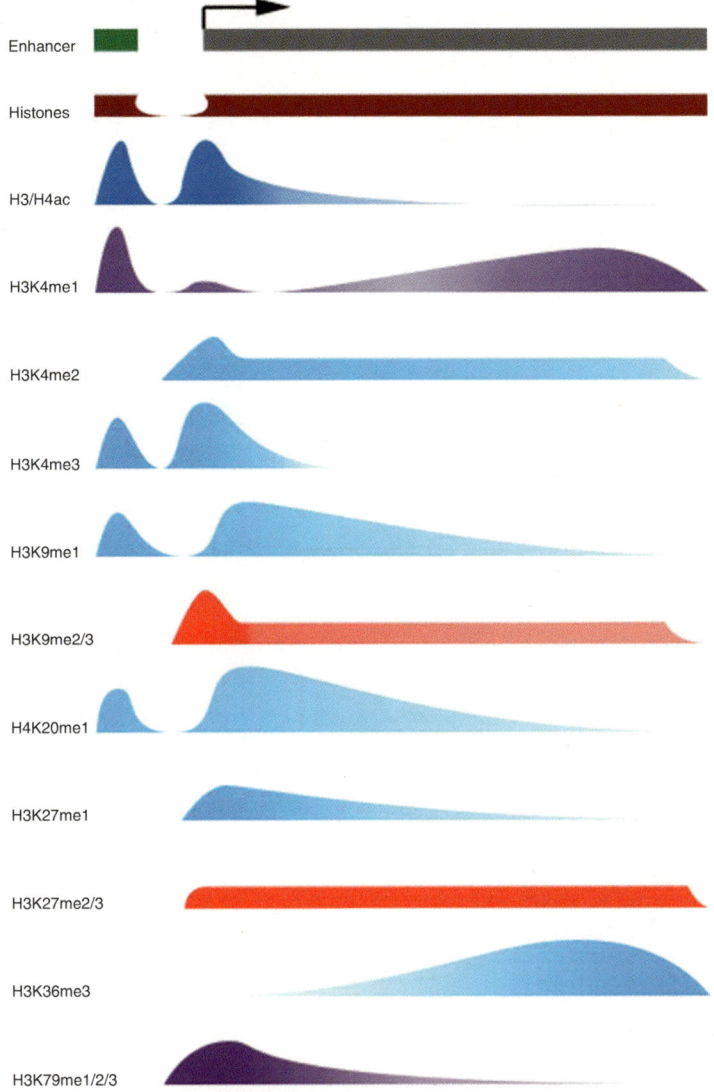

Fig. 13.2 Localization of histone modifications across genes as it relates to transcriptional regulation. Patterns of histone modification enrichment are shown across an arbitrary enhancer and gene. The enhancer is shown as the smaller region succeeded by a gap denoting a nucleosome-free region and transcriptional start site as shown by the *arrow*. Data used to compile the profiles are from GWAS on histone modifications. The correlative effects of the modifications on gene expression are indicated by different colors: *blue* for expression, *red* for repression, and *purple* for enrichment of the mark in both repressed and expressed genes (Wang et al. 2008; Barski et al. 2007; Li et al. 2007a)

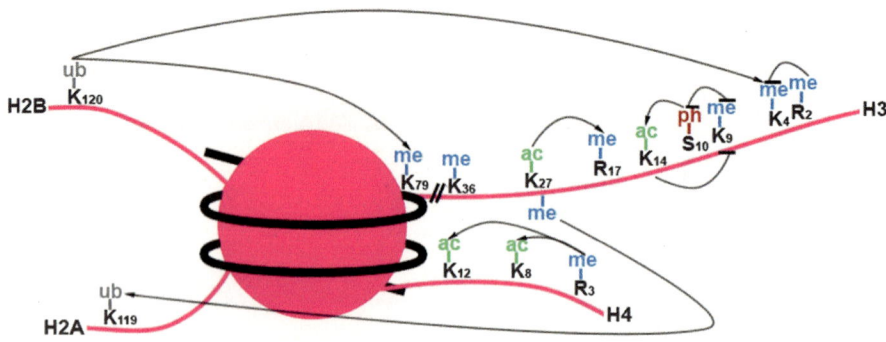

Fig. 13.3 Histone modification crosstalk. Various post-translational histone modifications affect the binding of certain domains and catalysis of other HPTMs. *Arrows* indicate a positive effect and *bars* indicate inhibitory effects on other HPTMs (Bannister and Kouzarides 2011)

binding proteins from binding. For example, H3S10ph prevents heterochromatin protein 1 (HP1) from binding H3K9me3 (Kouzarides and Berger 2007); (2) HPTMs can recruit certain chromatin binding proteins, which can enhance or inhibit gene activation. For example, H3K9me3, a marker for mammalian heterochromatin, is bound by the chromodomain of HP1 resulting in chromatin condensation and occlusion of DNA and nucleosomal binding sites utilized by coactivators, transcription factors, and RNAPII (Kouzarides and Berger 2007); (3) HPTMs can act *in cis* by affecting transcription through alteration of chromatin structure. For example, H4K16ac alone prevents the formation of a higher ordered compacted chromatin structure resulting in chromatin decondensation and increased transcriptional activity (Shogren-Knaak et al. 2006).

The following sections will focus on the effects of histone (de)acetylation and methylation on gene expression. It should be noted that much of the mechanistic research done on transcriptional regulation and HPTMs is pioneered through the use of yeast model systems because genetic manipulation and high-yield results have been easier to obtain as compared to humans. Importantly, many yeast proteins have correlative homologs that serve in the same manner as they do in mammals. However, there are some differences between the two eukaryotic organisms. For example, yeast do not possess the repressive mark H3K27me and in some cases, homologous complexes may contain different chromatin targeting proteins.

13.2.3 Histone Acetylation and Deacetylation

Histone acetylation at conserved lysine residues is the most intensely studied HPTM and was the first modification linked to transcriptional activity (Hebbes et al. 1988). It was not until 1996, that a direct molecular link was made between acetylation and transcription. The first nuclear histone acetyltransferase (HAT) discovered, p55, was orthologous to a previously isolated transcriptional coactivator in *Saccharomyces cerevisiae*, Gcn5 (Brownell et al. 1996). HATs catalyze the addition of acetyl-coA to the ε-amino group on lysine side chains resulting in charge neutralization and affinity reduction between negatively charged DNA and basic histones. Acetylation ultimately creates an "open" chromatin structure (Shogren-Knaak et al. 2006) poised for active transcription through exposure of DNA-binding sites (Vettese-Dadey et al. 1996). There are two types of HATs: type-A (nuclear) and type-B (cytoplasmic). This discussion will only focus on type-A as they catalyze reactions related to active transcription (Bannister and Kouzarides 2011).

Type-A HATs are further divided into five families including the GCN5-related *N*-acetyltransferases (GNATs); the MOZ, Ybf2/Sas3, Sas2 and Tip60 (MYST)-related HATs; p300/CREB-binding protein (CBP) HATs; the general transcription factor HATs including the TFIID subunit TBP-associated factor-1 (TAF1); and the nuclear hormone-related HATs SRC1 and ACTR (SRC3) (Nagy and Tora 2007). They are often part of larger protein complexes and are recruited by DNA binding activators. For instance, in yeast, Gcn5 is part of the Spt-Ada-Gcn5-Acetyltransferase (SAGA) and Adaptor (ADA) complexes (Grant et al. 1997). In SAGA, Gcn5 is associated with three protein families known to be involved in gene expression: Spt, Ada, and a subset of TAFs (Grant et al. 1998). SAGA is recruited to active promoters via the SAGA subunit, Tra1's interaction with acidic activator domains of transcriptional activators and subsequent recruitment of the TATA-binding protein (TBP) by the subunit, Spt3 (Grant et al. 1998; Larschan and Winston 2001; Brown et al. 2001; Reeves and Hahn 2005). Similar complex subunits have been found to be associated with Gcn5 human homologs, p300/CBP associated factor (P/CAF) and hGcn5 (Ogryzko et al. 1998; Martinez et al. 1998; Nagy and Tora 2007). Human Gcn5 is found in the SAGA complex homolog Spt3-Taf9-Gcn5-Acetyltransferase (STAGA) complex and is recruited to promoters by the Tra1 human homolog, Transactivation/transformation domain associated protein (TRRAP) via its interaction with the transactivation domain of c-Myc (McMahon et al. 2000; Liu et al. 2003).

Furthermore, Gcn5, P/CAF, and p300 contain a bromodomain that bind acetyl-lysine. Taf1 contains two bromodomains (Jacobson et al. 2000). The exact function of bromodomains has yet to be elucidated. However, it is speculated that once HAT complexes are targeted to the promoter and perform acetylation, subsequent coactivators can stably bind to acetylated histone rich promoter regions via bromodomains, which would facilitate an acetylation cascade. Consistent with this hypothesis, SAGA requires the functional bromodomains of Gcn5 and the remodeling complex proteins Swi2/Snf2 for stable promoter occupancy, efficient HAT activity, and increase in gene expression resulting from an "open" chromatin conformation, and subsequent gene activation (Hassan et al. 2002). It should also be noted that HATs also acetylate non-histone proteins including the tumor suppressor p53 and various transcription factors (Glozak et al. 2005), which ultimately regulates gene expression (Sterner and Berger 2000).

Like many bromodomains, DPF3b, a novel acetyl-lysine reader and BAF remodeling complex associated subunit also binds ambiguously to acetylated H3 and H4

(Lange et al. 2008) via its tandem plant homeodomain (PHD) fingers (Zeng et al. 2010). One PHD finger only has affinity for acetylated H3K14, which increases full-length DPF3b's affinity for acetylated H3 and H4 (Zeng et al. 2010). Loss of DPF3b affects both skeletal and heart muscle development through transcriptional deregulation of other transcriptional factors (Lange et al. 2008).

On the other hand, histone deacetylases (HDACs) reverse the reaction carried out by HATs by removing acetyl marks on lysine to restore the positive charge. They fall into four classes: Class I (HDAC1, 2, 3, and 8), II (HDAC4, 5, 6, 7, 9, 10), III or Sir2-related enzymes, and class IV, which contains one member, HDAC11. Class III HDACs require the cofactor NAD^+ for its activity (Yang and Seto 2007). HDAC1 and 2 are found in the mammalian complexes Sin3A/B, NuRD, and corepressor for RE1 silencing transcription factor/neural-restrictive silencing factor (CoREST) while HDAC3 is found in nuclear receptor corepressor/silencing mediator for retinoid and thyroid hormone receptors (N-CoR/SMRT) (Yang and Seto 2008). Some of these corepressor complexes contain methyl-lysine binders that help target complexes to specific site on chromatin. For instance, a subunit of the Sin3a complex, ING2, contains a PHD finger domain that binds H3K4me3 (Champagne and Kutateladze 2009) in response to DNA damage. Once Sin3a is recruited, HDAC1 activity is stimulated, which stabilizes nucleosomes resulting in the repression of cell proliferation genes as a response to genotoxic events (Shi et al. 2006).

13.2.4 Histone Methylation

Histone methylation is performed on the residues lysine and arginine by histone methyltransferase (HMT) enzymes. Lysines can be mono-, di-, and trimethylated while arginines can be mono- and symmetrically or asymmetrically dimethylated. There are over 20 sites of methylation that have been identified on the core histones. Given all the possible combinations of histone methylation, it is one of the most complex HPTMs to study in a static model. The modifications most relevant to transcriptional regulation have been listed in Table 13.1 and a few of the most studied histone methylations will be discussed in this section. Fig. 13.2 summarizes the transcriptional effects and genomic enrichment of the HPTMs discussed below.

13.2.4.1 H3K4

H3K4 methylation is usually enriched at the enhancers and promoters of actively transcribed genes (Wang et al. 2008; Santos-Rosa et al. 2002). H3K4me1 is highly enriched at enhancers (Wang et al. 2005). H3K4me2 is commonly found in the body of active genes while H3K4me3 is largely observed at the 5′ ORF of genes (Pokholok et al. 2005). Methylation of H3K4 results from the recruitment of various H3K4 HMT enzymes by transcriptional machinery, specifically RNAPII. Once RNAPII is poised for active transcription through phosphorylation of serine-5 of the carboxy-terminal domain (CTD) by TFIIH (Phatnani and Greenleaf 2006), the Set1 containing H3K4 HMT complex, COMPASS, is recruited by the PAF complex (Ng et al. 2003; Wood et al. 2003). RNAPII is released into an early elongating complex where H2BK120 (K123 in yeast) is ubiquitylated, which is required for further Set1 activity. Sometime during elongation, RNAPII is phosphorylated at serine-2 resulting in the release of Set1 (reviewed in Martin and Zhang 2005).

Furthermore PAF also interacts with chromodomain containing protein Chd1 (Simic et al. 2003). Proteins possessing methyl-binding domains, called chromodomains, are recruited to the H3K4me3 enriched promoter. SAGA also interacts with Chd1, which has two chromodomains, one which helps recruit SAGA to sites of H3K4me2/3 (Pray-Grant et al. 2005). As discussed earlier, SAGA recruitment results in an acetylation cascade that further promotes transcriptional activation. In humans, the HMT containing mixed-lineage-leukemia (MLL) complex is recruited by the H3K4me2 binding domain, WDR5. WDR5 interacts preferentially with H3K4me2 through its WD40-repeat domain (Wysocka et al. 2005). MLL can then convert H3K4me2 to H3K4me3.

13.2.4.2 H3K36

Unlike the 5' localization of H3K4 methylation, H3K36 methylation is highly enriched in the coding region and 3' ORF of genes (Kolasinska-Zwierz et al. 2009). As mentioned in the previous section, once the CTD of RNAPII is phosphorylated at Serine-2 by Ctk1 and Bur1 kinases (Keogh et al. 2003; Qiu et al. 2009), Set1 is released and chromatin is primed for transcriptional elongation through recruitment of Set2 (Xiao et al. 2003; Krogan et al. 2003). Set2 HMT catalyzes H3K36 methylation and specifically binds to phosphorylated Serine-2 of RNAPII's CTD (Hampsey and Reinberg 2003). This form of RNAPII is found in the transcribed regions of genes and the 3' end of genes, which correlates with H3K36me2/3 localization (Xiao et al. 2003; Krogan et al. 2003; Hampsey and Reinberg 2003; Li et al. 2003). The passage of RNAPII during transcriptional elongation results in histone displacement and positioning behind RNAPII. These histones are hyperacetylated and subsequently methylated by Set2 (Hampsey and Reinberg 2003; Carrozza et al. 2005; Joshi and Struhl 2005; Keogh et al. 2005).

H3K36me2 is recognized by the chromodomain of Eaf3 and PHD finger of Rco1, which are subunits of the Rpd3S HDAC complex (Joshi and Struhl 2005; Govind et al. 2010). During transcriptional elongation, Rpd3S is recruited via the serine-2/serine-5-diphosphorylated CTD repeats followed by H3K36me2 binding by Eaf3 and Rco1 (Keogh et al. 2005; Govind et al. 2010). Once Eaf3 and Rci1 are recruited by H3K36me2, Rpd3 is transferred from the phosphorylated CTD to H3 where its HDAC activity creates a hypoacetylated environment within gene bodies and at the 3' end. (Li et al. 2007b; Govind et al. 2010). Deletion of Rco1 or Eaf3 results in hyperacetylation of ORFs and the production of aberrant transcripts that are presumably initiated from cryptic promoters that are usually silenced by the Set2-Rpd3 pathway after RNAPII progression (Carrozza et al. 2005; Joshi and Struhl 2005; Keogh et al. 2005).

13.2.4.3 H3K79

Unlike the previously discussed HPTMs, H3K79 methylation occurs in the globular domain of H3 and within the core of the nucleosome. It is found within the coding regions of genes and is usually associated with active chromatin. H3K79 methylation is catalyzed by the HMT, Dot1. Dot1 is the first lysine HMT that has been identified that's lacks an identifiable SET domain (Feng et al. 2002). Dot1 is required to prevent the spread of HDACs into active chromosomal regions (van Leeuwen et al. 2002). There is no protein that links H3K79 methylation to transcriptional regulation. However, mammalian hDot1L has been implicated in mediating the leukemogenic fusion protein MLL (for mixed lineage leukemia)-AF10. It was found that hDot1L is recruited to MLL–AF10 target genes, such as *HOXA9*, through an interaction between hDot1L and AF10 (Okada et al. 2005). Upregulation of *HOXA9* expression results in defective hematopoiesis and leukemogenic transformation making regulation of H3K79 methylation a possible therapeutic target. Also, mammalian protein 53BP1 interacts with H3K79me3 through a tudor domain at sites of DNA damage (Huyen et al. 2004).

13.2.4.4 H3K27

In mammals, H3K27 methylation is a repressive mark catalyzed by the Polycomb Repressor Complex 2 (PRC2), which contains the SET-domain containing lysine HMT, Enhancer of Zester 2 (EZH2). H3K27me3 serves as a repressive mark at homeotic genes, the inactive X-chromosome, and imprinted genes while H3K27me1 is enriched at pericentric heterochromatin (Martin and Zhang 2005). PRC2 is made up of four core components: EZH2, embryonic ectoderm development (EED), suppressor of zeste 12 homolog (SUZ12), and histone-binding protein retinoblastoma-binding protein p48/46 (RbAp48/46). Both EED and SUZ12 are necessary for EZH2 HMT activity (Simon and Kingston 2009). EED contains repeats of WD40 domains that bind H3K27me3 and promote PRC2 propagation (Margueron et al. 2009) and SUZ12 contains C_2-H_2 zinc finger and VEFS domain. RbAp48/46 contains six WD40 domains and is a core histone binding subunit.

PCR2 also interacts with AEBP2, PCLs and JARID2. AEBP2 contains three zinc-fingers that may play a role in DNA binding (Kim et al. 2009a). PCL1, PCL2 and PCL3 (also known as PHF1, MTF2 and PHF19, respectively) contain a tudor domain and two PHD finger proteins, a PCL extended domain and a carboxy-terminal domain tail (Wang et al. 2004a). PCL proteins interact with PRC2 through EZH2, and to some extent through SUZ12 and the histone chaperones RbAp46/48 (Nekrasov et al. 2007). JARID2 is the founding member of the Jumonji family of proteins that catalyses the demethylation of histone proteins. However, it lacks demethylase activity. JARID2 contains JmjC and JmjN domains and two potential DNA binding domains, ARID and a zinc finger (Margueron and Reinberg 2011). The core components of PRC2 and its associated proteins discussed above are all necessary for EZH2 optimal function.

The targeting of PRC2 in *D. melanogaster* is a well understood mechanism compared to humans. In *D. melanogaster*, transcription factors, such as Pho and PhoL, bind to the Polycomb responsive element and recruit EZ of PRC2. Only now is the mammalian mechanism coming to light with the recent discovery of long non-coding RNA (lncRNA) dependent PRC2 recruitment. The lncRNA, *HOTAIR*, is transcribed from the HOXC locus, binds PRC2, and targets the complex to the HOXD locus where several genes are repressed (Rinn et al. 2007). Also, the lncRNA *Xist* and a short internal transcript *RepA* have been to shown target PRC2 to the inactivated female X-chromosome, which subsequently is repressed and enriched with H3K27me3. In contrast the lncRNA and antagonist to *Xist*, *Tsix*, also interacts with PRC2 suggesting an inhibitory mechanism to X-chromosome inactivation (Zhao et al. 2008).

13.2.4.5 H3K9

H3K9 methylation is one of the most intensely studied histone modifications to date. H3K9me1 is catalyzed by methyltransferases HMT1C/G9a or demethylases KDM3A/JMJD1A and KDM4D/JMJD2D (Shi and Whetstine 2007). The mark is enriched at the 5′ UTR and found minimally in non-genic regions (Barski et al. 2007; Rosenfeld et al. 2009). Although no function has been ascribed to H3K9me1, its proposed mechanism of action may be to act as an intermediary between gene activation and repression through rapid methylation or demethylation (Black and Whetstine 2011). Most studies have focused on H3K9me2/3 as a heterochromatin mark catalyzed by the lysine HMT SUV39H1/2 and recognized by the chromodomain of heterochromatin protein-1 (HP1), which dictates the compaction of heterochromatin. H3K9me2/3 is enriched in pericentromeric, subtelomeric, and gene desert regions (Rice et al. 2003). Gene deserts are megabase sized regions devoid of coding genes, and unlike H3K9me3, H3K9me2 is rarely found in at individual active or silenced genes (Rosenfeld et al. 2009). In support of H3K9me2′s function as a repressive mark, it has been shown to associate with Lamin B1, a protein localized to the nuclear periphery and part of the nuclear lamina, which is commonly associated with inactive genes. Lamin B1 associated regions are also devoid of the activating mark, H3K4me3, and RNAPII further suggesting H3K9me2 is most likely a repressive mark that facilitates separation of active and inactive genes through chromosomal localization within the nuclear architecture (Guelen et al. 2008).

H3K9me3 is commonly found at heterochromatin and repressed promoters, and unlike H3K9me2, H3K9me3 is also localized to centromeres, subtelomeric regions, and in some cases, the coding region of genes (Vakoc et al. 2006; Mikkelsen et al. 2007). H3K9me3 is usually associated with H3K20me3 at heterochromatic locations such as pericentromeric chromatin, but this bivalent mark is absent at subtelomeric regions and gene deserts suggesting different silencing mechanisms at these different heterochromatic regions (Rosenfeld et al. 2009). In addition to its heterochromatin formation function, H3K9me2/3 is implicated in the silencing of euchromatic genes. RB and KAP1 corepressor complexes recruit lysine HMTs SUV39H1 and ESET/SETDB1 respectively to promoters of active genes. HP1 is recruited to sites of

H3K9 methylation but is restricted to the promoter region of genes and does not spread (Kouzarides and Berger 2007). The role of H3K9me3 in the coding region of genes has not been elucidated, but enrichment of H3K9me3 at the 3′ ORF increases and co-localizes with the elongating form of RNAPII during active transcription. Moreover, despite the accepted dogma that HP1 is thought to always be repressive, a γ-isoform of HP1 has been found to also be enriched in the coding regions of active genes (Vakoc et al. 2005). During transcriptional activation, promoter repression by HP1β is replaced by HP1γ, which seems to facilitate RNAPII processivity through the coding region of the gene in addition to an increase in H3K9me3 (Mateescu et al. 2008).

13.2.4.6 H3K20

In addition to H3K9me2/3, H4K20me3 is also indicative of silenced chromatin. H4K20 methylation is catalyzed by two SET-domain containing lysine HMTs, SUV4-20H1 and SUV4-20H2. Interestingly, both of these HMTs have been shown to interact with the repressive HP1 isoforms, α and β, indicating a possible upstream function for H3K9 methylation and subsequent H4K20 methylation (Schotta et al. 2004). This idea is further illustrated by the dual enrichment of H3K9me3 and H4K20me3 at constitutively repressed regions such as transposons, satellite and long terminal repeats (LTRs), and pericentromeric chromatin, a region rich with repetitive satellite elements and interspersed with long and short interspersed nuclear elements (LINEs and SINEs). As discussed in the previous section, gene deserts are enriched with H3K9me2/3 but not H4K20me3. Interestingly, neither mark is found at telomeric and subtelomeric regions, which suggests a different mechanism of repression mediates constitutive heterochromatin at telomeres (Rosenfeld et al. 2009).

In contrast to H4K20me3, H4K20me1 is associated with highly expressed genes and is enriched at the 5′ coding region along with H2BK5me1, H3K4me1/2/3, H3K9me1, H3K27me1, and H3K79me1/2/3 (Wang et al. 2008). As previously discussed, H3K36me3 is located at the 3′ end of the coding region and marks transcriptionally active genes. Studies have shown that H4K20me1, H3K36me3, and H3K79me1/2/3 facilitate transcriptional elongation as all three marks fluctuate in a similar temporal manner during gene activation and subsequent transcription (Vakoc et al. 2006). H4K20me2 also seems to be required for checkpoint function and cell survival after DNA damage through the recruitment of Tudor-domain containing protein Crb2 (Greeson et al. 2008).

13.2.5 Histone Demethylases

Reversal of histone methylation was thought to be impossible due to the stable nature of the modification until the discovery of lysine-specific demethylase 1 (LSD1). LSD1 is a FAD dependent amine oxidase that catalyzes lysine demethylation and releases the product hydrogen peroxide (Shi et al. 2004). Protein arginine deiminase 4 (PADI4) converts methyl-arginine to citrulline rather than an unmodified arginine. PADI4 does not complete full demethylation and therefore requires processing by histone replacement or aminotransferases for complete arginine demethylation (Bannister et al. 2002). Lastly, the JumonjiC-domain containing histone demethylases (JHDMs) are Fe^{2+} and α-ketoglutarate dependent histone demethylases that release the product formaldehyde (Tsukada et al. 2006). Specifics about individual enzymes, mechanisms, specificity, and transcriptional activity can be found in Table 13.2.

13.2.6 Histone Proteolysis

In addition to demethylation and deacetylation, previous reports of H3 N-terminal tail proteolytic cleavage have also been described as a mechanism that facilitates the removal of HPTMs (Allis et al. 1980). Recently, H3 tail cleavage by Cathepsin L has been linked to transcriptional activation and induction of differentiation in embryonic stem cells. N-terminal tail cleavage is also regulated by the HPTMs present on the tail (Duncan et al. 2008). Studies have shown (Santos-Rosa et al. 2009) that cleavage is inhibited by the activation mark H3K4me3 and facilitated by the repressive mark H3R2me2 suggesting that tail clipping is a rapid way to void promoters of repressive marks and complexes during the regulation of gene expression. Moreover, tail clipping directly precedes histone eviction at promoters, which provides strong evidence that H3 tail cleavage is a gene activating event (Santos-Rosa et al. 2009). A major challenge in the chromatin field remains in understanding how patterns of modifications are generated and interpreted by nuclear machinery.

13.2.7 Histone Crosstalk

Given all the histone modifications discussed in the previous sections, regulation of chromatin structure and transcriptional activity can be tightly controlled through the use of combinatorial modifications (Zhang and Reinberg 2001). Histone modifications can affect the stimulation or inhibition of multiple cellular processes, which subsequently affects the capacity for the creation or erasure of other HPTMs (Fig. 13.3) (adapted from Bannister and Kouzarides 2011). Some modifications can inhibit the targeting of other modifications as seen with H3K27, which can be exclusively methylated or acetylated. Various modifications are also dependent on one another. For example, H2B120 ubiquitylation is necessary for H3K79 methylation in both yeast and humans (Lee et al. 2007a; Kim et al. 2009b). Modifications can also prevent the binding of certain effector proteins as is the case with the inhibition of HP1's targeting to H3K9me2/3 by H3S10 phosphorylation (Fischle et al. 2005). Some marks can also facilitate the binding of effector proteins that in turn perform

Table 13.2 Histone demethylases

Enzymatic family	Subfamily	Enzymes	Specific residue activity	Transcriptional activity	References
PADI		PAD4	H3R2me1 H3R8me1 H3R17me1 H3R26me1 H4R3me1	Derepressors	Bannister et al. (2002), Wang et al. (2004b), and Cuthbert et al. (2004)
Amine oxidase	LSD1		H3K4me1/2	Repressors: CoREST, NuRD Activator: AR/ERα	Lee et al. (2005), Shi et al. (2005), Wang et al. (2009), Metzger et al. (2005), and Garcia-Bassets et al. (2007)
JmjC	JHDM1	JHDM1A JHDM1B	H3K9me1/2 H3K36me1/2		Tsukada et al. (2006)
	JHDM3/JMJD2	JMJD2/JHDM3A JMJD2B JMJD2C/GASC1 JMJD2D	H3K9me2/3 H3K36me2/3		Whetstine et al. (2006), Klose et al. (2006), Cloos et al. (2006), and Fodor et al. (2006)
	JARID	JARID1A JARID1B JARID1C JARID1D	H3K4me2/3	Repressor of growth inhibitors	Iwase et al. (2007), Klose et al. (2007), Lee et al. (2007b), and Yamane et al. (2007)
	UTX/UTY	JMJD3 UTX	H3K27me2/3	Activator: MLL	Agger et al. (2007) and Issaeva et al. (2007)
	JHDM2	JHDM2A JHDM2B JHDM2C	H3K9me1/2	Activator: AR	Yamane et al. (2006)

Acronyms: *PADI* Peptidyl arginine deiminase, *LSD* Lysine specific demethylase, *JmjC* Jumonji C, *JHDM* JmjC-domain-containing histone demethylase, *AR* Androgen receptor, *ER* Estrogen receptor, *CoREST* Corepressor for RE1 silencing transcription factor/neural-restrictive silencing factor, *NuRD* Nucleosome remodeling and histone deacetylase

other modifications. As mentioned above, ING2 contains a PHD finger domain that binds H3K4me3 (Champagne and Kutateladze 2009) in response to DNA damage. Once Sin3a is recruited, HDAC1 activity is stimulated to deacetylate histones and reduce transcriptional activity of genes that promote cell growth and division (Shi et al. 2006).

13.3 Epigenetics and Human Disease

The human body is comprised of trillions of cells, each of which concurrently performs a specific function in order to form a functional human being. The function that one cell serves may be drastically different from another, yet each cell contains identical genetic information. Such phenotypic diversity is a result of a cell's distinctive gene expression profile. Gene expression is directly influenced by various factors including histone modifications, DNA methylation, histone variants, and availability of functional chromatin modifying complexes. Occasionally, DNA sequences targeted for modifications are expanded or contracted, or the enzymes that catalyze the addition or removal of modifications are lost or mutated. Respectively, these events cause a redistribution of DNA methylation and histone modification patterns. Alteration in the localization of these marks at sites such as promoters, repeat elements, and constitutive heterochromatin ultimately result in diseased states due to dysregulated gene expression (Kaufman and Rando 2010).

The idea that influences beyond the genetic code could determine phenotype is not by any means novel. In 1942, C.H. Waddington coined the phrase "epigenetic landscape" to denote changes in phenotype during development despite an identical genotype (Waddington 1957). To date, the epigenetic landscape portrayed by Waddington could be described by two important areas of chromatin research: the elaborate patterns of histone modifications and histone variant substitutions coined, "the histone code" (Jenuwein and Allis 2001) and DNA methylation patterns (Bird and Wolffe 1999). Through its direct effects on transcriptional regulation, histone modifications and DNA methylation affect many essential cellular processes such as embryogenesis, genomic imprinting, DNA replication, microRNA expression, and X-chromosomal inactivation.

Evidence that some human diseases are caused by something other than just the genes you possess is seen in cancer (Esteller 2007), autoimmune disorders (Javierre et al. 2010), and health related issues such as type 2 diabetes (Miao et al. 2008), coronary artery disease (Ordovas and Smith 2010), and obesity (Campion et al. 2009), to name a few. The role of epigenetics in the development of disease is further illustrated by the discordance of disease and trait development in monozyotic twins. Based on this study, environmental factors seem to play a significant role in disease susceptibility and dictating an individual's epigenetic landscape (Fraga et al. 2005). Ultimately, an increase in disease susceptibility can be attributed to environmentally influenced differences in DNA methylation and histone modification patterns that affect levels of gene expression.

With so many new advents in biomedical research, using human epigenetic profiling for understanding disease and even developing medical treatments has never seemed so tangible. Genome-wide association studies (GWAS) and high-throughput sequencing has allowed for high resolution comparison of modifications and gene expression in various organisms. With a future understanding of the basic functional roles these modifications play as transcriptional regulators in the cell, development of targeted treatments resulting in artificial epigenetic landscaping can potentially be established.

13.4 Summary

Despite the rapid progression of discoveries in the epigenetics fields, there still remain many obstacles and questions left unanswered. Several HPTMs have been discovered without finding the enzyme or complex that performs the covalent addition onto histones or its' removal. Some chromatin modifications are scarce enough that studying them would be impossible without new nanotechnologies such as ChIP-seq, RNA-seq, and MeDIP-seq. However, a problem that many GWAS run into is that many modifications are context dependent. Frequently, experiments performed to locate modifications and their effects on transcription produce results that represent a static state for a specific cell type. Both the cell type and time point at which the data was collected also affects what genes are expressed. Moreover, as in the case of H3K9 methylation, some modifications have the ability to alter a gene's 3D spatial positioning within the nucleus (Guelen et al. 2008). Therefore, in addition to the direct effects that chromatin modifying complexes and covalent histone modifications have on promoters, another layer of complexity is added to transcriptional regulation by the way of a gene's spatiotemporal positioning and organization within the nuclear architecture.

In this chapter, several classes of chromatin modifications and their subsequent effects on transcription have been described. There are many other mechanisms of transcriptional regulation that were mentioned but not discussed including arginine methylation, ubiquitylation, deiminination, and sumoylation. Although the enzymes that catalyze many of these and previously discussed reactions have been discovered, the mechanisms by which they control transcription, are established during development, and are stably maintained in somatic cells are still unclear. Albeit, many modifications have been characterized by their individual effects on gene transcription, developing a more complete picture of the complex orchestration between the enzymes that catalyze the reactions of chromatin modifications will lead to a better understanding of transcriptional regulation. Elucidating the code behind the interplay between chromatin modifying complexes and HPTMs provides exciting new prospects for development of medical treatments in the future that will target chromatin modifying enzymes.

References

Agger K, Cloos PA, Christensen J, Pasini D, Rose S, Rappsilber J et al (2007) UTX and JMJD3 are histone H3K27 demethylases involved in HOX gene regulation and development. Nature 449(7163):731–734

Allis CD, Bowen JK, Abraham GN, Glover CV, Gorovsky MA (1980) Proteolytic processing of histone H3 in chromatin: a physiologically regulated event in tetrahymena micronuclei. Cell 20(1):55–64

Antequera F, Bird A (1993) Number of CpG islands and genes in human and mouse. Proc Natl Acad Sci USA 90(24):11995–11999

Baker SP, Phillips J, Anderson S, Qiu Q, Shabanowitz J, Smith MM et al (2010) Histone H3 thr 45 phosphorylation is a replication-associated post-translational modification in *S. cerevisiae*. Nat Cell Biol 12(3):294–298

Bannister AJ, Kouzarides T (2011) Regulation of chromatin by histone modifications. Cell Res 21(3):381–395

Bannister AJ, Schneider R, Kouzarides T (2002) Histone methylation: dynamic or static? Cell 109(7):801–806

Bardwell VJ, Treisman R (1994) The POZ domain: a conserved protein-protein interaction motif. Genes Dev 8(14):1664–1677

Barski A, Cuddapah S, Cui K, Roh TY, Schones DE, Wang Z et al (2007) High-resolution profiling of histone methylations in the human genome. Cell 129(4):823–837

Bell AC, Felsenfeld G (2000) Methylation of a CTCF-dependent boundary controls imprinted expression of the Igf2 gene. Nature 405(6785):482–485

Bernstein BE, Mikkelsen TS, Xie X, Kamal M, Huebert DJ, Cuff J et al (2006) A bivalent chromatin structure marks key developmental genes in embryonic stem cells. Cell 125(2):315–326

Bestor T, Laudano A, Mattaliano R, Ingram V (1988) Cloning and sequencing of a cDNA encoding DNA methyltransferase of mouse cells. The carboxyl-terminal domain of the mammalian enzymes is related to bacterial restriction methyltransferases. J Mol Biol 203(4):971–983

Bird A (2002) DNA methylation patterns and epigenetic memory. Genes Dev 16(1):6–21

Bird AP, Wolffe AP (1999) Methylation-induced repression–belts, braces, and chromatin. Cell 99(5):451–454

Black JC, Whetstine JR (2011) Chromatin landscape: methylation beyond transcription. Epigenetics Off J DNA Methylation Soc 6(1):9–15

Bostick M, Kim JK, Esteve PO, Clark A, Pradhan S, Jacobsen SE (2007) UHRF1 plays a role in maintaining DNA methylation in mammalian cells. Science (New York, NY) 317(5845):1760–1764

Brenner C, Deplus R, Didelot C, Loriot A, Vire E, De Smet C et al (2005) Myc represses transcription through recruitment of DNA methyltransferase corepressor. EMBO J 24(2):336–346

Brown CE, Howe L, Sousa K, Alley SC, Carrozza MJ, Tan S et al (2001) Recruitment of HAT complexes by direct activator interactions with the ATM-related Tra1 subunit. Science (New York, NY) 292(5525):2333–2337

Brownell JE, Zhou J, Ranalli T, Kobayashi R, Edmondson DG, Roth SY et al (1996) Tetrahymena histone acetyltransferase A: a homolog to yeast Gcn5p linking histone acetylation to gene activation. Cell 84(6):843–851

Campion J, Milagro FI, Martinez JA (2009) Individuality and epigenetics in obesity. Obes Rev Off J Int Assoc Study Obes 10(4):383–392

Carrozza MJ, Li B, Florens L, Suganuma T, Swanson SK, Lee KK et al (2005) Histone H3 methylation by Set2 directs deacetylation of coding regions by Rpd3S to suppress spurious intragenic transcription. Cell 123(4):581–592

Champagne KS, Kutateladze TG (2009) Structural insight into histone recognition by the ING PHD fingers. Curr Drug Targets 10(5):432–441

Chen C, Nott TJ, Jin J, Pawson T (2011) Deciphering arginine methylation: tudor tells the tale. Nat Rev Mol Cell Biol 12(10):629–642

Clark SJ, Harrison J, Frommer M (1995) CpNpG methylation in mammalian cells. Nat Genet 10(1):20–27

Cloos PA, Christensen J, Agger K, Maiolica A, Rappsilber J, Antal T et al (2006) The putative oncogene GASC1 demethylates tri- and dimethylated lysine 9 on histone H3. Nature 442(7100):307–311

Cui K, Zang C, Roh TY, Schones DE, Childs RW, Peng W et al (2009) Chromatin signatures in multipotent human hematopoietic stem cells indicate the fate of bivalent genes during differentiation. Cell Stem Cell 4(1):80–93

Cuthbert GL, Daujat S, Snowden AW, Erdjument-Bromage H, Hagiwara T, Yamada M et al (2004) Histone deimination antagonizes arginine methylation. Cell 118(5):545–553

Delaval K, Wagschal A, Feil R (2006) Epigenetic deregulation of imprinting in congenital diseases of aberrant growth. BioEssays News Rev Mol Cell Dev Biol 28(5):453–459

Di Croce L, Raker VA, Corsaro M, Fazi F, Fanelli M, Faretta M et al (2002) Methyltransferase recruitment and DNA hypermethylation of target promoters by an oncogenic transcription factor. Science (New York, NY) 295(5557):1079–1082

Doi A, Park IH, Wen B, Murakami P, Aryee MJ, Irizarry R et al (2009) Differential methylation of tissue- and cancer-specific CpG island shores distinguishes human induced pluripotent stem cells, embryonic stem cells and fibroblasts. Nat Genet 41(12):1350–1353

Duncan EM, Muratore-Schroeder TL, Cook RG, Garcia BA, Shabanowitz J, Hunt DF et al (2008) Cathepsin L proteolytically processes histone H3 during mouse embryonic stem cell differentiation. Cell 135(2):284–294

Ehrlich M, Gama-Sosa MA, Huang LH, Midgett RM, Kuo KC, McCune RA et al (1982) Amount and distribution of 5-methylcytosine in human DNA from different types of tissues of cells. Nucleic Acids Res 10(8):2709–2721

Ernst J, Kellis M (2010) Discovery and characterization of chromatin states for systematic annotation of the human genome. Nat Biotechnol 28(8):817–825

Esteller M (2007) Cancer epigenomics: DNA methylomes and histone-modification maps. Nat Rev Genet 8(4):286–298

Feng Q, Wang H, Ng HH, Erdjument-Bromage H, Tempst P, Struhl K et al (2002) Methylation of H3-lysine 79 is mediated by a new family of HMTases without a SET domain. Curr Biol (CB) 12(12):1052–1058

Ficz G, Branco MR, Seisenberger S, Santos F, Krueger F, Hore TA et al (2011) Dynamic regulation of 5-hydroxymethylcytosine in mouse ES cells and during differentiation. Nature 473(7347):398–402

Fischle W, Tseng BS, Dormann HL, Ueberheide BM, Garcia BA, Shabanowitz J et al (2005) Regulation of HP1-chromatin binding by histone H3 methylation and phosphorylation. Nature 438(7071):1116–1122

Fodor BD, Kubicek S, Yonezawa M, O'Sullivan RJ, Sengupta R, Perez-Burgos L et al (2006) Jmjd2b antagonizes H3K9 trimethylation at pericentric heterochromatin in mammalian cells. Genes Dev 20(12):1557–1562

Fraga MF, Ballestar E, Paz MF, Ropero S, Setien F, Ballestar ML et al (2005) Epigenetic differences arise during the lifetime of monozygotic twins. Proc Natl Acad Sci USA 102(30): 10604–10609

Garcia-Bassets I, Kwon YS, Telese F, Prefontaine GG, Hutt KR, Cheng CS et al (2007) Histone methylation-dependent mechanisms impose ligand dependency for gene activation by nuclear receptors. Cell 128(3):505–518

Glozak MA, Sengupta N, Zhang X, Seto E (2005) Acetylation and deacetylation of non-histone proteins. Gene 363:15–23

Govind CK, Qiu H, Ginsburg DS, Ruan C, Hofmeyer K, Hu C et al (2010) Phosphorylated Pol II CTD recruits multiple HDACs, including Rpd3C(S), for methylation-dependent deacetylation of ORF nucleosomes. Mol Cell 39(2):234–246

Grant PA, Duggan L, Cote J, Roberts SM, Brownell JE, Candau R et al (1997) Yeast Gcn5 functions in two multisubunit complexes to acetylate nucleosomal histones: characterization of an ada complex and the SAGA (Spt/Ada) complex. Genes Dev 11(13):1640–1650

Grant PA, Schieltz D, Pray-Grant MG, Steger DJ, Reese JC, Yates JR 3rd et al (1998) A subset of TAF(II)s are integral components of the SAGA complex required for nucleosome acetylation and transcriptional stimulation. Cell 94(1):45–53

Greeson NT, Sengupta R, Arida AR, Jenuwein T, Sanders SL (2008) Di-methyl H4 lysine 20 targets the checkpoint protein Crb2 to sites of DNA damage. J Biol Chem 283(48): 33168–33174

Guelen L, Pagie L, Brasset E, Meuleman W, Faza MB, Talhout W et al (2008) Domain organization of human chromosomes revealed by mapping of nuclear lamina interactions. Nature 453(7197):948–951

Hampsey M, Reinberg D (2003) Tails of intrigue: phosphorylation of RNA polymerase II mediates histone methylation. Cell 113(4):429–432

Hark AT, Schoenherr CJ, Katz DJ, Ingram RS, Levorse JM, Tilghman SM (2000) CTCF mediates methylation-sensitive enhancer-blocking activity at the H19/Igf2 locus. Nature 405(6785):486–489

Hassan AH, Prochasson P, Neely KE, Galasinski SC, Chandy M, Carrozza MJ et al (2002) Function and selectivity of bromodomains in anchoring chromatin-modifying complexes to promoter nucleosomes. Cell 111(3):369–379

Hatada I, Fukasawa M, Kimura M, Morita S, Yamada K, Yoshikawa T et al (2006) Genome-wide profiling of promoter methylation in human. Oncogene 25(21):3059–3064

Hebbes TR, Thorne AW, Crane-Robinson C (1988) A direct link between core histone acetylation and transcriptionally active chromatin. EMBO J 7(5):1395–1402

Heintzman ND, Stuart RK, Hon G, Fu Y, Ching CW, Hawkins RD et al (2007) Distinct and predictive chromatin signatures of transcriptional promoters and enhancers in the human genome. Nat Genet 39(3):311–318

Heitz E (1929) Heterochromatin, chromocentren, chromomenen. Berichte der Deutschen Botanischen Gesellschaft 47:274–284

Hendrich B, Tweedie S (2003) The methyl-CpG binding domain and the evolving role of DNA methylation in animals. Trends Genet (TIG) 19(5):269–277

Hendrich B, Hardeland U, Ng HH, Jiricny J, Bird A (1999) The thymine glycosylase MBD4 can bind to the product of deamination at methylated CpG sites. Nature 401(6750):301–304

Hervouet E, Vallette FM, Cartron PF (2009) Dnmt3/transcription factor interactions as crucial players in targeted DNA methylation. Epigenet Off J DNA Methyl Soc 4(7):487–499

Huyen Y, Zgheib O, Ditullio RA Jr, Gorgoulis VG, Zacharatos P, Petty TJ et al (2004) Methylated lysine 79 of histone H3 targets 53BP1 to DNA double-strand breaks. Nature 432(7015):406–411

Iioka H, Doerner SK, Tamai K (2009) Kaiso is a bimodal modulator for Wnt/beta-catenin signaling. FEBS Lett 583(4):627–632

Illingworth R, Kerr A, Desousa D, Jorgensen H, Ellis P, Stalker J et al (2008) A novel CpG island set identifies tissue-specific methylation at developmental gene loci. PLoS Biol 6(1):e22

Iqbal K, Jin SG, Pfeifer GP, Szabo PE (2011) Reprogramming of the paternal genome upon fertilization involves genome-wide oxidation of 5-methylcytosine. Proc Natl Acad Sci USA 108(9):3642–3647

Issa JP (2000) CpG-island methylation in aging and cancer. Curr Top Microbiol Immunol 249:101–118

Issaeva I, Zonis Y, Rozovskaia T, Orlovsky K, Croce CM, Nakamura T et al (2007) Knockdown of ALR (MLL2) reveals ALR target genes and leads to alterations in cell adhesion and growth. Mol Cell Biol 27(5):1889–1903

Ito S, D'Alessio AC, Taranova OV, Hong K, Sowers LC, Zhang Y (2010) Role of tet proteins in 5mC to 5hmC conversion, ES-cell self-renewal and inner cell mass specification. Nature 466(7310):1129–1133

Iwase S, Lan F, Bayliss P, de la Torre-Ubieta L, Huarte M, Qi HH et al (2007) The X-linked mental retardation gene SMCX/JARID1C defines a family of histone H3 lysine 4 demethylases. Cell 128(6):1077–1088

Jacobson RH, Ladurner AG, King DS, Tjian R (2000) Structure and function of a human TAFII250 double bromodomain module. Science (New York, NY) 288(5470):1422–1425

Javierre BM, Fernandez AF, Richter J, Al-Shahrour F, Martin-Subero JI, Rodriguez-Ubreva J et al (2010) Changes in the pattern of DNA methylation associate with twin discordance in systemic lupus erythematosus. Genome Res 20(2):170–179

Jenuwein T (2001) Re-SET-ting heterochromatin by histone methyltransferases. Trends Cell Biol 11(6):266–273

Jenuwein T, Allis CD (2001) Translating the histone code. Science (New York, NY) 293(5532): 1074–1080

Jones PA, Wolkowicz MJ, Rideout WM 3rd, Gonzales FA, Marziasz CM, Coetzee GA et al (1990) De novo methylation of the MyoD1 CpG island during the establishment of immortal cell lines. Proc Natl Acad Sci USA 87(16):6117–6121

Jorgensen HF, Ben-Porath I, Bird AP (2004) Mbd1 is recruited to both methylated and nonmethylated CpGs via distinct DNA binding domains. Mol Cell Biol 24(8):3387–3395

Joshi AA, Struhl K (2005) Eaf3 chromodomain interaction with methylated H3-K36 links histone deacetylation to Pol II elongation. Mol Cell 20(6):971–978

Josse J, Kaiser AD, Kornberg A (1961) Enzymatic synthesis of deoxyribonucleic acid. VIII. Frequencies of nearest neighbor base sequences in deoxyribonucleic acid. J Biol Chem 236:864–875

Kacem S, Feil R (2009) Chromatin mechanisms in genomic imprinting. Mamm Genome Off J Int Mamm Genome Soc 20(9–10):544–556

Kanduri C, Pant V, Loukinov D, Pugacheva E, Qi CF, Wolffe A et al (2000) Functional association of CTCF with the insulator upstream of the H19 gene is parent of origin-specific and methylation-sensitive. Curr Biol (CB) 10(14):853–856

Karlic R, Chung HR, Lasserre J, Vlahovicek K, Vingron M (2010) Histone modification levels are predictive for gene expression. Proc Natl Acad Sci USA 107(7):2926–2931

Kaufman PD, Rando OJ (2010) Chromatin as a potential carrier of heritable information. Curr Opin Cell Biol 22(3):284–290

Keogh MC, Podolny V, Buratowski S (2003) Bur1 kinase is required for efficient transcription elongation by RNA polymerase II. Mol Cell Biol 23(19):7005–7018

Keogh MC, Kurdistani SK, Morris SA, Ahn SH, Podolny V, Collins SR et al (2005) Cotranscriptional set2 methylation of histone H3 lysine 36 recruits a repressive Rpd3 complex. Cell 123(4):593–605

Kim H, Kang K, Kim J (2009a) AEBP2 as a potential targeting protein for polycomb repression complex PRC2. Nucleic Acids Res 37(9):2940–2950

Kim J, Guermah M, McGinty RK, Lee JS, Tang Z, Milne TA et al (2009b) RAD6-mediated transcription-coupled H2B ubiquitylation directly stimulates H3K4 methylation in human cells. Cell 137(3):459–471

Klose RJ, Bird AP (2006) Genomic DNA methylation: the mark and its mediators. Trends Biochem Sci 31(2):89–97

Klose RJ, Yamane K, Bae Y, Zhang D, Erdjument-Bromage H, Tempst P et al (2006) The transcriptional repressor JHDM3A demethylates trimethyl histone H3 lysine 9 and lysine 36. Nature 442(7100):312–316

Klose RJ, Yan Q, Tothova Z, Yamane K, Erdjument-Bromage H, Tempst P et al (2007) The retinoblastoma binding protein RBP2 is an H3K4 demethylase. Cell 128(5):889–900

Koh KP, Yabuuchi A, Rao S, Huang Y, Cunniff K, Nardone J et al (2011) Tet1 and Tet2 regulate 5-hydroxymethylcytosine production and cell lineage specification in mouse embryonic stem cells. Cell Stem Cell 8(2):200–213

Kolasinska-Zwierz P, Down T, Latorre I, Liu T, Liu XS, Ahringer J (2009) Differential chromatin marking of introns and expressed exons by H3K36me3. Nat Genet 41(3):376–381

Kouzarides T (2007) Chromatin modifications and their function. Cell 128(4):693–705

Kouzarides T, Berger S (2007) Chromatin modifications and their mechanisms of action. In: Allis CD, Jenuwein T, Reinberg D (eds) Epigenetics. Cold Spring Harbor Laboratory Press, Plainview, pp 191–206

Kriaucionis S, Heintz N (2009) The nuclear DNA base 5-hydroxymethylcytosine is present in Purkinje neurons and the brain. Science (New York, NY) 324(5929):929–930

Krogan NJ, Kim M, Tong A, Golshani A, Cagney G, Canadien V et al (2003) Methylation of histone H3 by Set2 in *Saccharomyces cerevisiae* is linked to transcriptional elongation by RNA polymerase II. Mol Cell Biol 23(12):4207–4218

Lange M, Kaynak B, Forster UB, Tonjes M, Fischer JJ, Grimm C et al (2008) Regulation of muscle development by DPF3, a novel histone acetylation and methylation reader of the BAF chromatin remodeling complex. Genes Dev 22(17):2370–2384

Larschan E, Winston F (2001) The *S. cerevisiae* SAGA complex functions in vivo as a coactivator for transcriptional activation by Gal4. Genes Dev 15(15):1946–1956

Larsen F, Gundersen G, Lopez R, Prydz H (1992) CpG islands as gene markers in the human genome. Genomics 13(4):1095–1107

Lee MG, Wynder C, Cooch N, Shiekhattar R (2005) An essential role for CoREST in nucleosomal histone 3 lysine 4 demethylation. Nature 437(7057):432–435

Lee JS, Shukla A, Schneider J, Swanson SK, Washburn MP, Florens L et al (2007a) Histone cross-talk between H2B monoubiquitination and H3 methylation mediated by COMPASS. Cell 131(6):1084–1096

Lee MG, Norman J, Shilatifard A, Shiekhattar R (2007b) Physical and functional association of a trimethyl H3K4 demethylase and Ring6a/MBLR, a polycomb-like protein. Cell 128(5):877–887

Lee J, Jang SJ, Benoit N, Hoque MO, Califano JA, Trink B et al (2010) Presence of 5-methylcytosine in CpNpG trinucleotides in the human genome. Genomics 96(2):67–72

Li W, Liu M (2011) Distribution of 5-hydroxymethylcytosine in different human tissues. J Nucleic Acids 2011:870726

Li B, Howe L, Anderson S, Yates JR 3rd, Workman JL (2003) The Set2 histone methyltransferase functions through the phosphorylated carboxyl-terminal domain of RNA polymerase II. J Biol Chem 278(11):8897–8903

Li B, Carey M, Workman JL (2007a) The role of chromatin during transcription. Cell 128(4):707–719

Li B, Gogol M, Carey M, Lee D, Seidel C, Workman JL (2007b) Combined action of PHD and chromo domains directs the Rpd3S HDAC to transcribed chromatin. Science (New York, NY) 316(5827):1050–1054

Lister R, Pelizzola M, Dowen RH, Hawkins RD, Hon G, Tonti-Filippini J et al (2009) Human DNA methylomes at base resolution show widespread epigenomic differences. Nature 462(7271):315–322

Liu X, Tesfai J, Evrard YA, Dent SY, Martinez E (2003) c-myc transformation domain recruits the human STAGA complex and requires TRRAP and GCN5 acetylase activity for transcription activation. J Biol Chem 278(22):20405–20412

Luger K, Mader AW, Richmond RK, Sargent DF, Richmond TJ (1997) Crystal structure of the nucleosome core particle at 2.8 A resolution. Nature 389(6648):251–260

Margueron R, Reinberg D (2011) The polycomb complex PRC2 and its mark in life. Nature 469(7330):343–349

Margueron R, Justin N, Ohno K, Sharpe ML, Son J, Drury WJ 3rd et al (2009) Role of the polycomb protein EED in the propagation of repressive histone marks. Nature 461(7265):762–767

Martin C, Zhang Y (2005) The diverse functions of histone lysine methylation. Nat Rev Mol Cell Biol 6(11):838–849

Martinez E, Kundu TK, Fu J, Roeder RG (1998) A human SPT3-TAFII31-GCN5-L acetylase complex distinct from transcription factor IID. J Biol Chem 273(37):23781–23785

Mateescu B, Bourachot B, Rachez C, Ogryzko V, Muchardt C (2008) Regulation of an inducible promoter by an HP1beta-HP1gamma switch. EMBO Rep 9(3):267–272

McMahon SB, Wood MA, Cole MD (2000) The essential cofactor TRRAP recruits the histone acetyltransferase hGCN5 to c-myc. Mol Cell Biol 20(2):556–562

Metzger E, Wissmann M, Yin N, Muller JM, Schneider R, Peters AH et al (2005) LSD1 demethylates repressive histone marks to promote androgen-receptor-dependent transcription. Nature 437(7057):436–439

Miao F, Smith DD, Zhang L, Min A, Feng W, Natarajan R (2008) Lymphocytes from patients with type 1 diabetes display a distinct profile of chromatin histone H3 lysine 9 dimethylation: an epigenetic study in diabetes. Diabetes 57(12):3189–3198

Mikkelsen TS, Ku M, Jaffe DB, Issac B, Lieberman E, Giannoukos G et al (2007) Genome-wide maps of chromatin state in pluripotent and lineage-committed cells. Nature 448(7153):553–560

Nagy Z, Tora L (2007) Distinct GCN5/PCAF-containing complexes function as co-activators and are involved in transcription factor and global histone acetylation. Oncogene 26(37):5341–5357

Nan X, Ng HH, Johnson CA, Laherty CD, Turner BM, Eisenman RN et al (1998) Transcriptional repression by the methyl-CpG-binding protein MeCP2 involves a histone deacetylase complex. Nature 393(6683):386–389

Nekrasov M, Klymenko T, Fraterman S, Papp B, Oktaba K, Kocher T et al (2007) Pcl-PRC2 is needed to generate high levels of H3-K27 trimethylation at polycomb target genes. EMBO J 26(18):4078–4088

Ng HH, Robert F, Young RA, Struhl K (2003) Targeted recruitment of Set1 histone methylase by elongating Pol II provides a localized mark and memory of recent transcriptional activity. Mol Cell 11(3):709–719

Ogryzko VV, Kotani T, Zhang X, Schiltz RL, Howard T, Yang XJ et al (1998) Histone-like TAFs within the PCAF histone acetylase complex. Cell 94(1):35–44

Okada Y, Feng Q, Lin Y, Jiang Q, Li Y, Coffield VM et al (2005) hDOT1L links histone methylation to leukemogenesis. Cell 121(2):167–178

Okano M, Bell DW, Haber DA, Li E (1999) DNA methyltransferases Dnmt3a and Dnmt3b are essential for de novo methylation and mammalian development. Cell 99(3):247–257

Ordovas JM, Smith CE (2010) Epigenetics and cardiovascular disease. Nat Rev Cardiol 7(9):510–519

Phatnani HP, Greenleaf AL (2006) Phosphorylation and functions of the RNA polymerase II CTD. Genes Dev 20(21):2922–2936

Pokholok DK, Harbison CT, Levine S, Cole M, Hannett NM, Lee TI et al (2005) Genome-wide map of nucleosome acetylation and methylation in yeast. Cell 122(4):517–527

Portela A, Esteller M (2010) Epigenetic modifications and human disease. Nat Biotechnol 28(10):1057–1068

Pray-Grant MG, Daniel JA, Schieltz D, Yates JR 3rd, Grant PA (2005) Chd1 chromodomain links histone H3 methylation with SAGA- and SLIK-dependent acetylation. Nature 433(7024):434–438

Prokhortchouk E, Defossez PA (2008) The cell biology of DNA methylation in mammals. Biochimica Et Biophysica Acta 1783(11):2167–2173

Qiu H, Hu C, Hinnebusch AG (2009) Phosphorylation of the Pol II CTD by KIN28 enhances BUR1/BUR2 recruitment and Ser2 CTD phosphorylation near promoters. Mol Cell 33(6):752–762

Rand E, Ben-Porath I, Keshet I, Cedar H (2004) CTCF elements direct allele-specific undermethylation at the imprinted H19 locus. Curr Biol (CB) 14(11):1007–1012

Reeves WM, Hahn S (2005) Targets of the Gal4 transcription activator in functional transcription complexes. Mol Cell Biol 25(20):9092–9102

Reik W, Lewis A (2005) Co-evolution of X-chromosome inactivation and imprinting in mammals. Nat Rev Genet 6(5):403–410

Rice JC, Briggs SD, Ueberheide B, Barber CM, Shabanowitz J, Hunt DF et al (2003) Histone methyltransferases direct different degrees of methylation to define distinct chromatin domains. Mol Cell 12(6):1591–1598

Rinn JL, Kertesz M, Wang JK, Squazzo SL, Xu X, Brugmann SA et al (2007) Functional demarcation of active and silent chromatin domains in human HOX loci by noncoding RNAs. Cell 129(7):1311–1323

Rosenfeld JA, Wang Z, Schones DE, Zhao K, DeSalle R, Zhang MQ (2009) Determination of enriched histone modifications in non-genic portions of the human genome. BMC Genomics 10:143

Santos-Rosa H, Schneider R, Bannister AJ, Sherriff J, Bernstein BE, Emre NC et al (2002) Active genes are tri-methylated at K4 of histone H3. Nature 419(6905):407–411

Santos-Rosa H, Kirmizis A, Nelson C, Bartke T, Saksouk N, Cote J et al (2009) Histone H3 tail clipping regulates gene expression. Nat Struct Mol Biol 16(1):17–22

Saxonov S, Berg P, Brutlag DL (2006) A genome-wide analysis of CpG dinucleotides in the human genome distinguishes two distinct classes of promoters. Proc Natl Acad Sci USA 103(5):1412–1417

Scarano E, Iaccarino M, Grippo P, Parisi E (1967) The heterogeneity of thymine methyl group origin in DNA pyrimidine isostichs of developing sea urchin embryos. Proc Natl Acad Sci USA 57(5):1394–1400

Schotta G, Lachner M, Sarma K, Ebert A, Sengupta R, Reuter G et al (2004) A silencing pathway to induce H3-K9 and H4-K20 trimethylation at constitutive heterochromatin. Genes Dev 18(11):1251–1262

Shi Y, Whetstine JR (2007) Dynamic regulation of histone lysine methylation by demethylases. Mol Cell 25(1):1–14

Shi Y, Lan F, Matson C, Mulligan P, Whetstine JR, Cole PA et al (2004) Histone demethylation mediated by the nuclear amine oxidase homolog LSD1. Cell 119(7):941–953

Shi YJ, Matson C, Lan F, Iwase S, Baba T, Shi Y (2005) Regulation of LSD1 histone demethylase activity by its associated factors. Mol Cell 19(6):857–864

Shi X, Hong T, Walter KL, Ewalt M, Michishita E, Hung T et al (2006) ING2 PHD domain links histone H3 lysine 4 methylation to active gene repression. Nature 442(7098):96–99

Shiio Y, Eisenman RN (2003) Histone sumoylation is associated with transcriptional repression. Proc Natl Acad Sci USA 100(23):13225–13230

Shogren-Knaak M, Ishii H, Sun JM, Pazin MJ, Davie JR, Peterson CL (2006) Histone H4-K16 acetylation controls chromatin structure and protein interactions. Science (New York, NY) 311(5762):844–847

Simic R, Lindstrom DL, Tran HG, Roinick KL, Costa PJ, Johnson AD et al (2003) Chromatin remodeling protein Chd1 interacts with transcription elongation factors and localizes to transcribed genes. EMBO J 22(8):1846–1856

Simon JA, Kingston RE (2009) Mechanisms of polycomb gene silencing: knowns and unknowns. Nat Rev Mol Cell Biol 10(10):697–708

Sterner DE, Berger SL (2000) Acetylation of histones and transcription-related factors. Microbiol Mol Biol Rev (MMBR) 64(2):435–459

Strahl BD, Allis CD (2000) The language of covalent histone modifications. Nature 403(6765):41–45

Straussman R, Nejman D, Roberts D, Steinfeld I, Blum B, Benvenisty N et al (2009) Developmental programming of CpG island methylation profiles in the human genome. Nat Struct Mol Biol 16(5):564–571

Swartz MN, Trautner TA, Kornberg A (1962) Enzymatic synthesis of deoxyribonucleic acid. XI. Further studies on nearest neighbor base sequences in deoxyribonucleic acids. J Biol Chem 237:1961–1967

Szabo P, Tang SH, Rentsendorj A, Pfeifer GP, Mann JR (2000) Maternal-specific footprints at putative CTCF sites in the H19 imprinting control region give evidence for insulator function. Curr Biol (CB) 10(10):607–610

Tahiliani M, Koh KP, Shen Y, Pastor WA, Bandukwala H, Brudno Y et al (2009) Conversion of 5-methylcytosine to 5-hydroxymethylcytosine in mammalian DNA by MLL partner TET1. Science (New York, NY) 324(5929):930–935

Takai D, Jones PA (2002) Comprehensive analysis of CpG islands in human chromosomes 21 and 22. Proc Natl Acad Sci USA 99(6):3740–3745

Tsukada Y, Fang J, Erdjument-Bromage H, Warren ME, Borchers CH, Tempst P et al (2006) Histone demethylation by a family of JmjC domain-containing proteins. Nature 439(7078): 811–816

Turner BM, Birley AJ, Lavender J (1992) Histone H4 isoforms acetylated at specific lysine residues define individual chromosomes and chromatin domains in drosophila polytene nuclei. Cell 69(2):375–384

Vakoc CR, Mandat SA, Olenchock BA, Blobel GA (2005) Histone H3 lysine 9 methylation and HP1gamma are associated with transcription elongation through mammalian chromatin. Mol Cell 19(3):381–391

Vakoc CR, Sachdeva MM, Wang H, Blobel GA (2006) Profile of histone lysine methylation across transcribed mammalian chromatin. Mol Cell Biol 26(24):9185–9195

Valinluck V, Sowers LC (2007) Endogenous cytosine damage products alter the site selectivity of human DNA maintenance methyltransferase DNMT1. Cancer Res 67(3):946–950

Valinluck V, Tsai HH, Rogstad DK, Burdzy A, Bird A, Sowers LC (2004) Oxidative damage to methyl-CpG sequences inhibits the binding of the methyl-CpG binding domain (MBD) of methyl-CpG binding protein 2 (MeCP2). Nucleic Acids Res 32(14):4100–4108

van Leeuwen F, Gafken PR, Gottschling DE (2002) Dot1p modulates silencing in yeast by methylation of the nucleosome core. Cell 109(6):745–756

Vettese-Dadey M, Grant PA, Hebbes TR, Crane-Robinson C, Allis CD, Workman JL (1996) Acetylation of histone H4 plays a primary role in enhancing transcription factor binding to nucleosomal DNA in vitro. EMBO J 15(10):2508–2518

Waddington CH (1957) The strategy of the genes; a discussion of some aspects of theoretical biology. Allen & Unwin, London

Walter J (2011) An epigenetic tet a tet with pluripotency. Cell Stem Cell 8(2):121–122

Wang S, Robertson GP, Zhu J (2004a) A novel human homologue of drosophila polycomblike gene is up-regulated in multiple cancers. Gene 343(1):69–78

Wang Y, Wysocka J, Sayegh J, Lee YH, Perlin JR, Leonelli L et al (2004b) Human PAD4 regulates histone arginine methylation levels via demethylimination. Science (New York, NY) 306(5694):279–283

Wang YA, Kamarova Y, Shen KC, Jiang Z, Hahn MJ, Wang Y et al (2005) DNA methyltransferase-3a interacts with p53 and represses p53-mediated gene expression. Cancer Biol Ther 4(10):1138–1143

Wang Z, Zang C, Rosenfeld JA, Schones DE, Barski A, Cuddapah S et al (2008) Combinatorial patterns of histone acetylations and methylations in the human genome. Nat Genet 40(7): 897–903

Wang Y, Zhang H, Chen Y, Sun Y, Yang F, Yu W et al (2009) LSD1 is a subunit of the NuRD complex and targets the metastasis programs in breast cancer. Cell 138(4):660–672

Warren RA (1980) Modified bases in bacteriophage DNAs. Annu Rev Microbiol 34:137–158

Weber M, Davies JJ, Wittig D, Oakeley EJ, Haase M, Lam WL et al (2005) Chromosome-wide and promoter-specific analyses identify sites of differential DNA methylation in normal and transformed human cells. Nat Genet 37(8):853–862

Whetstine JR, Nottke A, Lan F, Huarte M, Smolikov S, Chen Z et al (2006) Reversal of histone lysine trimethylation by the JMJD2 family of histone demethylases. Cell 125(3):467–481

Wood A, Schneider J, Dover J, Johnston M, Shilatifard A (2003) The Paf1 complex is essential for histone monoubiquitination by the Rad6-Bre1 complex, which signals for histone methylation by COMPASS and Dot1p. J Biol Chem 278(37):34739–34742

Workman JL, Kingston RE (1998) Alteration of nucleosome structure as a mechanism of transcriptional regulation. Annu Rev Biochem 67:545–579

Wossidlo M, Nakamura T, Lepikhov K, Marques CJ, Zakhartchenko V, Boiani M et al (2011) 5-Hydroxymethylcytosine in the mammalian zygote is linked with epigenetic reprogramming. Nat Commun 2:241

Wyatt GR, Cohen SS (1952) A new pyrimidine base from bacteriophage nucleic acids. Nature 170(4338):1072–1073

Wysocka J, Swigut T, Milne TA, Dou Y, Zhang X, Burlingame AL et al (2005) WDR5 associates with histone H3 methylated at K4 and is essential for H3 K4 methylation and vertebrate development. Cell 121(6):859–872

Xiao T, Hall H, Kizer KO, Shibata Y, Hall MC, Borchers CH et al (2003) Phosphorylation of RNA polymerase II CTD regulates H3 methylation in yeast. Genes Dev 17(5):654–663

Yamane K, Toumazou C, Tsukada Y, Erdjument-Bromage H, Tempst P, Wong J et al (2006) JHDM2A, a JmjC-containing H3K9 demethylase, facilitates transcription activation by androgen receptor. Cell 125(3):483–495

Yamane K, Tateishi K, Klose RJ, Fang J, Fabrizio LA, Erdjument-Bromage H et al (2007) PLU-1 is an H3K4 demethylase involved in transcriptional repression and breast cancer cell proliferation. Mol Cell 25(6):801–812

Yang XJ, Seto E (2007) HATs and HDACs: from structure, function and regulation to novel strategies for therapy and prevention. Oncogene 26(37):5310–5318

Yang XJ, Seto E (2008) The Rpd3/Hda1 family of lysine deacetylases: from bacteria and yeast to mice and men. Nat Rev Mol Cell Biol 9(3):206–218

Yang Y, Lu Y, Espejo A, Wu J, Xu W, Liang S et al (2010) TDRD3 is an effector molecule for arginine-methylated histone marks. Mol Cell 40(6):1016–1023

Yoder JA, Bestor TH (1998) A candidate mammalian DNA methyltransferase related to pmt1p of fission yeast. Hum Mol Genet 7(2):279–284

Zeng L, Zhang Q, Li S, Plotnikov AN, Walsh MJ, Zhou MM (2010) Mechanism and regulation of acetylated histone binding by the tandem PHD finger of DPF3b. Nature 466(7303):258–262

Zhang Y, Reinberg D (2001) Transcription regulation by histone methylation: interplay between different covalent modifications of the core histone tails. Genes Dev 15(18):2343–2360

Zhang X, Yazaki J, Sundaresan A, Cokus S, Chan SW, Chen H et al (2006) Genome-wide high-resolution mapping and functional analysis of DNA methylation in arabidopsis. Cell 126(6):1189–1201

Zhao J, Sun BK, Erwin JA, Song JJ, Lee JT (2008) Polycomb proteins targeted by a short repeat RNA to the mouse X chromosome. Science (New York, NY) 322(5902):750–756

Zhao Q, Rank G, Tan YT, Li H, Moritz RL, Simpson RJ et al (2009) PRMT5-mediated methylation of histone H4R3 recruits DNMT3A, coupling histone and DNA methylation in gene silencing. Nat Struct Mol Biol 16(3):304–311

Zheng S, Wyrick JJ, Reese JC (2010) Novel trans-tail regulation of H2B ubiquitylation and H3K4 methylation by the N terminus of histone H2A. Mol Cell Biol 30(14):3635–3645

Zhu B, Zheng Y, Pham AD, Mandal SS, Erdjument-Bromage H, Tempst P et al (2005) Monoubiquitination of human histone H2B: the factors involved and their roles in HOX gene regulation. Mol Cell 20(4):601–611

Zollman S, Godt D, Prive GG, Couderc JL, Laski FA (1994) The BTB domain, found primarily in zinc finger proteins, defines an evolutionarily conserved family that includes several developmentally regulated genes in drosophila. Proc Natl Acad Sci USA 91(22):10717–10721

Chapter 14
Histone Variants and Transcription Regulation

Cindy Law and Peter Cheung

Abstract Histones are the protein components of chromatin and are important for its organization and compaction. Although core histones are exclusively expressed during S phase of the cell cycle, there exist variants of canonical histones that are expressed throughout the cell cycle. These histone variants are often deposited at defined regions of the genome and they play important roles in a variety of cellular processes, such as transcription regulation, heterochromatin formation and DNA repair. In this chapter, we will focus on several histone variants that have been linked to transcription regulation, and highlight their physical and functional features that facilitate their activities in this context.

14.1   Overview of the Biology of Histones

14.1.1   Formation of Chromatin

DNA is the genetic material for all living organisms. When stretched out, the DNA in a human cell is approximately 2 m long. Therefore, a sophisticated mechanism is required to organize the DNA so that it can be stored in a nucleus that is only a few microns in diameter, but also still be accessible for biological functions. In eukaryotic cells, chromatin is formed through the physical association of DNA and histone proteins, and is organized through multiple levels of compaction. At the lowest compaction level, 147 bp of DNA is wrapped around two copies-each of the four core histones, H2A, H2B, H3 and H4, to form the nucleosome core particle

---

C. Law • P. Cheung (✉)
Ontario Cancer Institute, 610 University Avenue, Toronto, ON M5G 2M9, Canada

Department of Medical Biophysics, University of Toronto, Toronto, ON M5G 2M9, Canada
e-mail: pcheung@uhnres.utoronto.ca

(Kornberg 1974). Adjacent nucleosomes are linked together by short stretches of linker DNA to form the "beads on a string" structure, which is also referred to as the 10 nm fibre. With the addition of the linker histone H1, chromatin is further assembled into the 30 nm fibre under physiological conditions. The 30 nm fibre can coil to form solenoid structures, and upon further higher order folding, the most condensed form of chromatin is the metaphase chromosome (Alberts 2002).

14.1.2 Canonical Histones vs. Histone Variants

Histones are amongst the most conserved proteins in all eukaryotes (Pusarla and Bhargava 2005). Core histones (also known as canonical histones) are encoded by multicopy genes that are intronless, and are transcribed into mRNA that are not polyadenylated (Albig and Doenecke 1997). Instead of a poly-A tail, canonical histone transcripts contain a conserved 3' stemloop that is responsible for their restricted expression at S phase, a time when additional histones are needed to associate with the newly synthesized DNA (Harris et al. 1991; Dominski and Marzluff 1999). Histone variants are non-allelic isoforms of core histones that differ in parts of their amino acid sequences compared to their core counterparts. Unlike core histones, histone variants are encoded by single-copy genes that contain introns. Also, histone variant mRNAs are polyadenylated and are constitutively transcribed at all stages of the cell cycle. This last feature suggests that they have additional functions beyond DNA compaction (Kamakaka 2005). While core histone deposition is replication-dependent, most histone variants are deposited in a replication-independent manner. All core histones have variant counterparts: for example, a variant form of H4 was very recently identified in human adipocytes (Banaszynski et al. 2010; Jufvas et al. 2011). Most H2B variants are also tissue-specific and have specialized functions in the testes of vertebrates and invertebrates (Ausió 2006). Finally, variants of H3 and H2A are more ubiquitously found and are involved in diverse functions in the cell (Ausió 2006).

14.1.3 Histone Variants and PTMs: Formation of Chromatin Domains

Higher-order chromatin folding reduces accessibility of DNA, yet local access of defined DNA sequences by nuclear factors is essential for transcription, replication and DNA repair. Therefore, the mechanisms that efficiently compact chromatin must also allow for dynamic decondensation at functional regions of the genome. Two general ways to regulate DNA accessibility are by exploiting changes in histone-DNA and histone-histone interactions through post-translational modifications (PTMs) on histones, as well as by incorporation of histone variants (Strahl and Allis 2000;

Henikoff and Ahmad 2005). Additionally, changes to chromatin can also be mediated through physical sliding or removal of nucleosomes by the actions of ATP-dependent chromatin remodelling complexes (Lusser and Kadonaga 2003).

Histones are subjected to a variety of PTMs, including acetylation, methylation, phosphorylation and ubiquitylation. These modifications primarily occur on the N-terminal tails of histones that protrude out from the nucleosome, but can also occur at sites within the histone fold domains, as well as the C-terminal tails of H2A and H2B (Kouzarides 2007). Histone modifications are involved in diverse functions such as epigenetic inheritance, gene expression, heterochromatinization, and the formation of specialized regions such as telomeres and centromeres (Talbert and Henikoff 2010). The addition of different PTMs on histones can alter the interactions between nucleosome and DNA, or between nucleosomes (Kouzarides 2007). They can also create binding sites for nuclear factors and regulate recruitment of protein complexes that mediate downstream functions (Taverna et al. 2007).

Histone variants can replace core histones at defined regions of the genome to confer unique functions. Since histone variants differ in parts of their amino acid sequences compared to core histones, they may also have distinct modifications that recruit different proteins to these regions (Taverna et al. 2007). Together, the incorporation of histone variants and histone PTMs can create structurally and functionally distinct chromatin domains. Chromatin is often categorized into two main compaction states: euchromatin and heterochromatin. Euchromatin has an open and decondensed structure, and are enriched for active genes. Consistent with that, histone variants and modifications that are associated with transcriptional activation are often found in the euchromatic regions of the genome (Kouzarides 2007). Conversely, regions of heterochromatin are compact and condensed in structure. These regions contain inactive genes, and are enriched for different sets of histone variants and modifications that establish a transcriptionally repressive state of chromatin (Kouzarides 2007). As variants of H3 and H2A are most often associated with distinct transcriptional states of chromatin, we will focus on these variants in this chapter.

14.2 The Role of H3.3 in Transcription Regulation

14.2.1 Overview of H3 Variants

Histone H3 has two canonical variants, H3.2 and the mammalian-specific H3.1 (Marzluff et al. 2002). The expression and deposition of these core H3 variants are replication-dependent (Osley 1991). Since H3.1 and H3.2 only differ by one amino acid, most studies tend to group these variants together. For this reason, and for simplicity, we will use the general term H3 to refer to both of these variants in this section. In addition of canonical H3, there are four replacement H3 variants: the centromere-specific variant, CENP-A; two testis-specific variants, H3t and H3.5; and the transcription-linked H3.3. CENP-A is the variant that substitutes for H3 at

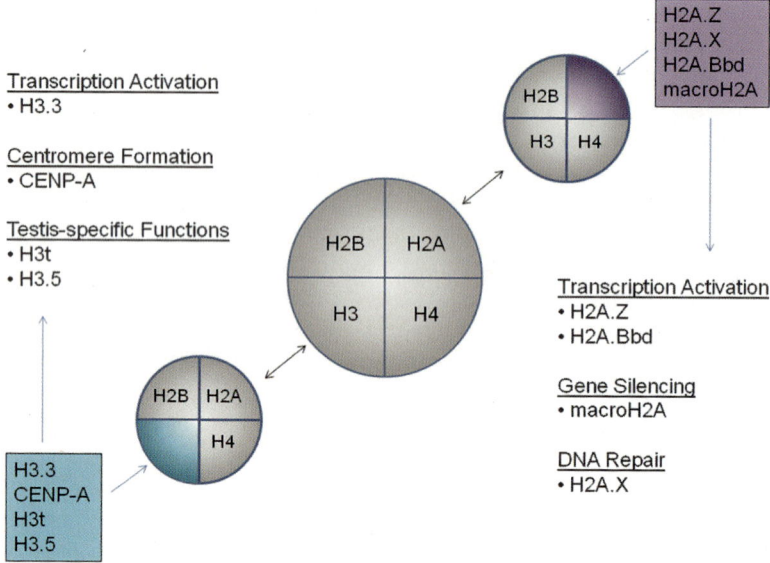

Fig. 14.1 Incorporation of histone variants mediates various cellular processes. When histone variants replace canonical histones in the nucleosome, they may impart distinct functions. The functions of various histone H3 and H2A variants are highlighted in this figure

centromeric nucleosomes. It is required for the recruitment of proteins essential for kinetochore formation and for chromosome segregation (Sullivan et al. 1994). H3t is only expressed in testes and has four additional amino acids compared to H3.1 (Hake and Allis 2006). H3.5 is also a testis-specific variant. It has a ~96% sequence identity with H3.3 and it could rescue H3.3 deficiency in knockdown experiments; however, whether it specifically compensates H3.3's function in transcription is unknown (Schenk et al. 2011). Finally, amongst all the H3 replacement variants, H3.3 is the best studied, and the most clearly linked to transcription regulation (Fig. 14.1).

14.2.2 H3.3 vs. Canonical H3

H3.3 differs from H3.2 at amino acid residues 31, 87, 89 and 90, and it differs from H3.1 at an additional residue at position 96 (Akhmanova et al. 1995; Szenker et al. 2011). This variant is deposited in a replication-independent manner and it is the predominant form of H3 expressed in quiescent, G1 and G2 cells (Frank et al. 2003). The difference in sequence at residues 87–90 between H3 and H3.3 is responsible for this replication-independent deposition. In *Drosophila*, any amino acid substitution of H3 towards the H3.3 residues at these positions allows some H3 deposition in a replication-independent manner (Ahmad and Henikoff 2002).

Table 14.1 Chaperones and genomic location of H3.3, H2A.Z, H2A.Bbd and MacroH2A

Histone variant	Chaperone	Genomic location	Variant/ Chaperone knockout phenotype	References
H3.3	HIRA	Euchromatin, Promoter and gene bodies	HIRA: Embryonic lethality	Roberts et al. (2002)
H3.3	Daxx-ATRX	Telomeres, Pericentric heterochromatin	Daxx-ATRX: Embryonic lethality	Michaelson et al. (1999) Garrick et al. (2006)
H2A.Z	Chz1, Swr1	Promoter, Boundary elements, Pericentric heterochromatin	H2A.Z: Embryonic lethality Swr1: Lethality	Faast et al. (2001) Updike and Mango (2006)
H2A.Bbd	Unknown	Unknown	Unknown	–
MacroH2A	Unknown	Inactive X chromosome, Promoter	MacroH2A: Malformations in the brain and body (zebrafish)	Buschbeck et al. (2009)

This region may also be important for specific chaperone recognition since H3 and H3.3 are deposited into chromatin by different chaperones. HIRA, a nucleosome assembly factor, is found to specifically deposit H3.3 at the promoter and coding regions of active genes in a replication-independent manner, whereas H3.1 is exclusively deposited by CAF-1 during DNA replication (Tagami et al. 2004; Goldberg et al. 2010). More recently, another chaperone, Daxx, in a complex with ATRX, was found to deposit H3.3 at telomeres in murine embryonic stem cells and at pericentric heterochromatin in mouse embryonic fibroblasts (Lewis et al. 2010). For now, the function of H3.3 at telomeres is not known, nor is it clear whether there are distinguishing characteristics on the H3.3 pools associated with the different chaperone complexes (Table 14.1).

14.2.3 Evidence Linking H3.3 to Transcriptional Activation

Genome-wide mapping of H3.3 in *Drosophila, C. elegans* and mammalian cells revealed that H3.3 is mostly enriched at actively transcribed genes (Mito et al. 2005; Ooi et al. 2010; Delbarre et al. 2010). While some studies showed that H3.3 is only found at the promoters of genes, others found enrichment of H3.3 at both the promoter and coding region (Chow et al. 2005; Daury et al. 2006). ChIP-chip experiments showed that the peaks of H3.3 localization overlap with the genomic regions enriched for K4-methylated H3 and RNA Pol II, which are both marks of active transcription (Mito et al. 2005). Indeed, biochemical and mass spectrometry analyses showed that this variant is enriched for transcription-associated PTMs, such as methylation on K4, K36, and K79 and acetylation on K9, K14, K27 and K79 (McKittrick et al. 2004; Hake et al. 2006; Loyola et al. 2006). All evidence together suggest that the majority of H3.3 is physically associated with active genes.

14.2.4 H3.3 and Transcription Regulation

To determine H3.3's function, various labs have examined the dynamics of H3.3 incorporation during transcriptional activation. For example, using inducible gene expression systems, it was found that transcriptional activation is accompanied by displacement of H3 and selective re-deposition of H3.3 at these genes (Schwartz and Ahmad 2005; Wirbelauer et al. 2005; Sutcliffe et al. 2009). In spite of the steady-state distribution of H3.3 at promoters and coding regions, Tamura et al. found that transcription-induced incorporation of H3.3 is greatest at the distal end of the coding region of interferon-stimulated genes upon gene induction, suggesting that H3.3 replaces the evicted canonical H3 during transcription (Tamura et al. 2009). In addition, knockdown of H3.3 or HIRA, the main H3.3 chaperone, inhibited induction of H3.3-target genes, indicating that H3.3 is required for gene activation (Placek et al. 2009; Tamura et al. 2009). Together, these results suggest that H3.3 incorporation is necessary for the rapid exchange of H3 components and possibly for depositing transcription-linked PTM signatures on the H3 at induced genes.

14.2.5 Properties of H3.3 Nucleosomes

Incorporation of H3.3 can alter the biophysical properties of the nucleosome, either by nucleosome destabilization, which would facilitate its rapid removal, or by changing the interactions of the nucleosome with DNA or other nucleosomes, which would antagonize formation of higher-order chromatin structures. Enrichment of H3.3 at transcription start sites (TSS) is inversely correlated with the deposition of histone H1, a linker histone involved in chromatin compaction. Knockdown of H3.3 in *Drosophila* cells by siRNA results in increased H1 binding at TSS, as well as increased nucleosome repeat length, suggesting that H3.3 nucleosomes inhibit further compaction of chromatin (Braunschweig et al. 2009). H3.3-nucleosomes are also more sensitive to salt-dependent disruptions and, compared to H3-nucleosomes, are more prone to losing their H2A/H2B dimers. Such increased nucleosome mobility would allow the H3.3-nucleosomes to be remodelled rapidly during transcriptional activation (Jin and Felsenfeld 2007).

Although much has been learnt about H3.3 over the last decade, there are also interesting questions regarding this variant that remain unresolved. For example, with only four/five amino acids that differ between canonical H3 and H3.3, how do histone chaperones and modifying enzymes differentiate between them? Given that H3.3, H3.2 and H3.1 have distinct PTM signatures, is this due to preferential

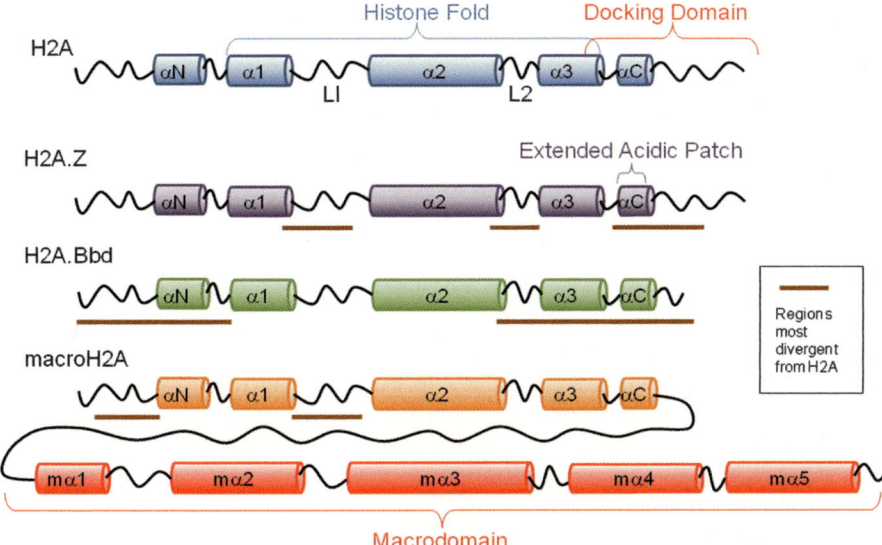

Fig.14.2 Schematic diagram of the secondary structures of histone H2A and H2A variants. The α-helices are represented by *cylinders*. The histone fold, docking domain and the L1 and L2 loops are labelled on the H2A schematic. The *bars* below the H2A variant schematic highlight regions that are most divergent from H2A. H2A.Z has significant differences in the L1 loop region that is important for interactions between the two H2A/H2A.Z-H2B dimers, as well as in the docking domain that is important for interactions between H2A.Z and the H3-H4 tetramer. H2A.Z also contains an extended acidic patch that is important for interactions with other nucleosomes. H2A.Bbd is most divergent from H2A in the N- and C-terminal regions. H2A.Bbd lack regions corresponding to the docking domain and the acidic patch of canonical H2A. MacroH2A differs from H2A in the L1 loop region and also contain a macrodomain that encompass two-third of the size of this histone

recognition by distinct histone modifying enzymes or due to the differential positioning of the variants at different regions of the genome (Hake et al. 2005)? Finally, new evidence shows that H3.3 is also found at intergenic regions, telomeres and centromeres (Mito et al. 2005; Wong et al. 2009; Drané et al. 2010; Santenard et al. 2010). So what are the functions of H3.3 at these regions that are often thought to be transcriptionally silent? These remaining questions will spur on the current and future research on H3.3.

14.3 Variants of Histone H2A

The Histone H2A family of variants consists of H2A.X, H2A.Z, macroH2A and H2A.Bbd. Whereas H2A.Z is conserved throughout evolution, the other variants, such as macroH2A and H2A.Bbd, are only found in vertebrates or mammals (Ausió 2006). H2A.X has a critical role in the process of DNA repair. Phosphorylation of H2A.X marks the site of DNA damage on chromatin and physically recruits DNA repair factors to the DNA breaks (Rogakou et al. 1998). Since the function of H2A.X in DNA repair is well documented, we will refer interested readers to the many excellent reviews already published (Foster and Downs 2005; Thiriet and Hayes 2005).

14.3.1 H2A.Z – The Essential H2A Variant

H2A.Z is one of the best-studied H2A variants and represents about 5–10% of total cellular H2A. H2A.Z is highly conserved throughout evolution with a sequence identity of ~90% between species, whereas its sequence identity to canonical H2A is only about ~60% (Iouzalen et al. 1996; Jackson and Gorovsky 2000) (Fig. 14.2). H2A.Z has an essential function in complex eukaryotes since mutants that are genetically null for H2A.Z in *Tetrahymena*, *Drosophila* and mice are all inviable (van Daal and Elgin 1992; Gorovsky 1996; Faast et al. 2001). H2A.Z has been implicated in a variety of biological processes, including transcriptional activation, heterochromatin formation, chromosome segregation and regulation of cell cycle progression (Zlatanova and Thakar 2008). At present, which of these putative functions of H2A.Z is essential to cell survival is still unclear.

14.3.1.1 Evidence Linking H2A.Z to Transcriptional Activation

H2A.Z has long been thought to have a transcriptional activation function since initial observations showed that it is only found in the transcriptionally active macronucleus of *Tetrahymena* (Allis et al. 1980). In yeast, genome-wide analyses from different groups showed that H2A.Z is enriched at promoters and often flank nucleosome-free regions adjacent to the transcription start site of many genes (Zhang et al. 2005a; Guillemette et al. 2005; Li et al. 2005; Millar et al. 2006). Correlation of the genome-wide localization studies with gene expression data showed that H2A.Z in *S. cerevisiae* is mostly enriched at inactive or repressed genes. However, loss of H2A.Z does not de-repress these genes, but disrupt their inducibility. Therefore, H2A.Z in yeast, does not have a direct transcriptional repression function, but may be required to poise inducible genes for activation (Santisteban et al. 2000; Adam et al. 2001). In contrast to the yeast studies, ChIP-seq analyses by the Zhao lab using human T cells showed that there is an enrichment of H2A.Z at the promoters of actively transcribing genes (Barski et al. 2007). The contrasting observations in yeast and human cells have not been reconciled at present.

Early studies in yeast showed that when genes are activated, their promoters are remodelled to lose H2A.Z. Htz1 (the H2A.Z gene in *S. cerevisiae*) deletion mutants are viable but defective in the activation of H2A.Z-target genes and they exhibit a slow growth phenotype (Santisteban et al. 2000; Adam et al. 2001). In mammalian cells, a number of studies showed that H2A.Z is already present at the regulatory regions of some inducible genes (Farris et al. 2005; Gévry et al. 2007; Sutcliffe et al. 2009; Draker et al. 2011), but gene activation-dependent recruitment of H2A.Z has also been reported (Hardy et al. 2009; Gévry et al. 2009). Similar to yeast studies, a net loss of H2A.Z is often observed after transcriptional induction of mammalian genes (Draker et al. 2011; Hardy et al. 2009; Gévry et al. 2009; Amat and Gudas 2011). Finally, shRNA-mediated knockdown of H2A.Z in different cell lines showed that this variant is required for gene activation (Draker et al. 2011; Gévry et al. 2009; Cuadrado et al. 2010). At present, a consensus model of how deposition and eviction of H2A.Z is involved in transcription regulation has remained elusive; however, the loss of function studies clearly showed that H2A.Z is required for the activation of many genes in both yeast and mammalian cells. In support of this, deletion or knockdown of SRCAP (SWR1 in yeast), an ATPase responsible for H2A.Z deposition, causes decreased H2A.Z deposition as well as transcriptional defects, confirming that H2A.Z has an critical role in transcription regulation (Santisteban et al. 2000; Kobor et al. 2004; Wong et al. 2007; Slupianek et al. 2010).

14.3.1.2 Mechanism of H2A.Z Function: Poising Promoters for Transcription

Structural analyses of the H2A.Z nucleosome showed that it differs from the canonical nucleosome at the H3/H4 docking domain, and this change is thought to destabilize the H2A.Z/H3 interaction (Suto et al. 2000). Sedimentation analyses under changing ionic strength showed a substantial instability of the H2A.Z core particle, indicating a less tight binding of the H2A.Z-H2B dimer to the rest of the octamer (Abbott et al. 2001). Consistent with these stability assays, additional studies showed that incorporation of H2A.Z increases the mobility of the nucleosome. For example, studies in yeast showed that H2A.Z nucleosomes have higher turnover rates (Dion et al. 2007). Also, *in vitro* thermal mobility assays showed that H2A.Z nucleosomes are more mobile at increasing temperatures as compared to H2A nucleosomes, suggesting that H2A.Z nucleosomes may be more easily remodelled (Flaus et al. 2004). Since H2A.Z is preferentially found at the promoters of repressed genes and is lost upon gene activation, the increased mobility of H2A.Z nucleosomes may be important for maintaining genes in a poised state and allow rapid nucleosome remodelling upon induction.

Biophysical studies of chromatin fibres containing H2A.Z showed that it resists condensation and assumes a more relaxed conformation as compared to ones containing the H2A counterpart (Fan et al. 2002). Sedimentation velocity experiments showed that H2A.Z facilitates intramolecular folding of nucleosomal arrays, as in the formation of 30 nm fibre (Fan et al. 2002). The extended acidic patch on H2A.Z is important for its interactions with the N-terminal tail of H4 from a neighbouring nucleosome, which may allow H2A.Z to facilitate the folding of the DNA (Suto et al. 2000). At higher salt concentrations, H2A.Z nucleosomal arrays sedimented more slowly when compared to control arrays. This indicates that H2A.Z inhibits the oligomerization of the array, and suggests that H2A.Z inhibits the formation of highly condensed structures that would result from intermolecular interactions (Fan et al. 2002). The incorporation of H2A.Z is thought to resist permanent silencing of genes by inhibiting heterochromatin formation and, thus, keep genes poised for transcription (Meneghini et al. 2003). Indeed, deletion studies in yeast showed that loss of H2A.Z results in spreading of the telomeric heterochromatin into the adjacent euchromatic regions, suggesting that it has an anti-silencing function. In addition, H2A.Z also functionally antagonizes DNA methylation in *Arabidopsis*, so it could protect promoter regions from being silenced by DNA methylation (Zilberman et al. 2008). In line with H2A.Z's function in keeping promoters poised, H2A.Z can also affect nucleosome positioning at the promoters of target genes (Guillemette et al. 2005). Nucleosome mapping of the TFF1 promoter showed that knock-down of H2A.Z impairs stabilization of nucleosomes at the promoter. The adoption of preferential sites by nucleosomes at the TFF1 promoter allows ERα and its partners to interact with their binding sites (Gévry et al. 2009). Therefore, incorporation of H2A.Z may mediate nucleosome positioning that sets up a chromatin environment that favours binding of transcription-associated proteins.

14.3.1.3 Mechanism of H2A.Z Function: Recruitment of Transcription Machinery

H2A.Z's role in transcription regulation is not only limited to setting up the chromatin environment and for rapid nucleosome exchange, but it may also be important for direct recruitment of components of the transcription machinery. In yeast, loss of H2A.Z results in defective recruitment of RNA Pol II and TBP to the GAL1 promoter. Physical interaction between RNA Pol II and the C-terminal tail of H2A.Z, shown through biochemical analyses, further suggests a direct recruitment function of H2A.Z in transcriptional initiation (Adam et al. 2001). Recent studies by Santisteban et al. showed that H2A.Z is also important for transcriptional elongation. They found that nucleosome remodelling over the coding region during transcription requires H2A.Z. In the absence of H2A.Z, RNA Pol II elongation rate is 24% slower and RNA Pol II phos-Ser2 levels are reduced as compared to WT strains (Santisteban et al. 2011). H2A.Z has also been linked to the recruitment of RNA Pol II to the promoter in mammalian cells since knock-down of H2A.Z impaired RNA Pol II recruitment to the IL8 promoter upon gene induction (Hardy et al. 2009). Therefore, H2A.Z may regulate gene expression at multiple steps, including promoter poising, transcriptional initiation and elongation.

14.3.1.4 Relationship Between H3.3 and H2A.Z

Several recent studies have suggested an interesting relationship between H2A.Z and H3.3. H3.3 and H2A.Z are both found at promoters and are both involved in nucleosome positioning at these regions (Jin et al. 2009; Thakar et al. 2009). When studied individually, both variants are involved in transcriptional activation and their incorporation destabilizes the nucleosome. Genome-wide studies showed that both H3.3 and H2A.Z are enriched at enhancers and insulator regions (Mito et al. 2005; Barski et al. 2007). Moreover, salt extraction studies showed that nucleosomes containing both H3.3 and H2A.Z are less stable than nucleosomes with just H3.3 and H2A (Jin and Felsenfeld 2007). Therefore, the two variants may cooperate and synergize at specific promoters to allow rapid remodelling and activation.

14.3.1.5 Heterochromatin Formation and PTMs of H2A.Z

Although most evidence supports H2A.Z's role in transcriptional activation, there are also studies that linked H2A.Z to heterochromatin formation and gene silencing. For example, H2A.Z at pericentric heterochromatin colocalizes with heterochromatin protein 1 α (HP1α) during early mouse development, and *in vitro* studies showed that it promotes the folding of chromatin fiber through an interaction with HP1α (Rangasamy et al. 2003; Fan et al. 2004). Genome-wide analyses also found an enrichment of H2A.Z at facultative heterochromatin (Creyghton et al. 2008; Hardy et al. 2009). The contrasting findings showing H2A.Z's association with both euchromatin and heterochromatin have not yet been clearly resolved. However, approximately 25% of total H2A.Z in mammalian cells is mono-ubiquitylated at K120 or K121 (Sarcinella et al. 2007). Moreover, the mono-ubiquitylated form of H2A.Z is enriched on the epigenetically-silenced inactive X chromosome in human female cells. Therefore, PTMs on H2A.Z may distinguish its association with the different chromatin states.

In addition to mono-ubiquitylation, human H2A.Z is also known to be acetylated on Lys 4, 7 and 11 (Beck et al. 2006). Genome-wide analyses in yeast demonstrated that, whereas H2A.Z is enriched at repressed genes, AcH2A.Z is found at the promoters of active genes (Millar et al. 2006). ChIP analyses in chicken cells also found that AcH2A.Z is associated with active genes (Bruce et al. 2005). Insofar as acetylation of H2A.Z is associated with active transcription, whereas ubiquitylation of H2A.Z is associated with gene silencing, differential modifications by these PTMs may specify the contrasting downstream functions associated with H2A.Z (Draker and Cheung 2009).

14.3.1.6 H2A.Z's Roles in Hormone-Responsive Gene Activation and Links to Cancer

Given H2A.Z's links to transcription regulation, several studies have focused on studying its function in the activation of hormone-responsive genes. Depletion of H2A.Z or of H2A.Z incorporation impedes both estrogen and androgen signalling, and blocks efficient activation of ER- and AR-regulated genes (Gévry et al. 2009; Draker et al. 2011). Moreover, de-ubiquitylation of H2A and H2A.Z correlates with hormone activation of AR-regulated genes, supporting the idea that modulating the PTMs on H2A.Z could function as a regulatory step in the expression of inducible genes (Draker et al. 2011). Deregulation of ER and AR-activated genes is often linked to the development of breast and prostate cancer respectively. Consistent with the fact that H2A.Z plays important roles in controlling expression of these genes, dysregulation of H2A.Z has also been linked to tumourigenesis. For example, recent studies show that H2A.Z is overexpressed in different cancers and its overexpression correlates with metastatic and undifferentiated cancer types (Zucchi et al. 2004; Hua et al. 2008; Svotelis et al. 2010). These types of cancer are more aggressive and are associated with poorer patient outcome (Rhodes et al. 2004). Therefore, further studies investigating the detailed functions of H2A.Z in gene expression regulation and further identification of H2A.Z-regulated genes will potentially shed light on its role in tumourigenesis.

14.3.2 H2A.Bbd: Variant of Active Transcription

H2A.Bbd is the smallest H2A variant and the most divergent compared to the canonical H2A (only 48% sequence identity) (Fig. 14.2). Its name, Barr body deficient (Bbd), originated from the initial observation that it is excluded from the inactive X chromosome or the Barr body in female mammalian cells. As part of the dosage compensation phenomenon, one of the two X chromosomes in each cell of female mammals is randomly silenced (inactivated) by chromatin compaction and epigenetic mechanisms (see Ng et al. 2007 for more thorough review). Therefore, the conspicuous absence of H2A.Bbd from the silenced inactive X chromosome suggests that it is physically incompatible with silenced or compacted chromatin. Moreover, its preferential colocalization with acetylated H4, a mark of active transcription, further suggests that it has a positive role in transcription regulation (Chadwick and Willard 2001a).

14.3.2.1 H2A.Bbd Lacks the Acidic Patch and the C-terminal Docking Domain of H2A

Structural comparison between H2A and H2A.Bbd showed that the acidic patch of H2A, which is extended in H2A.Z, is absent from H2A.Bbd. Since the acidic patch on H2A is important for the folding of chromatin into the 30 nm fibre, the lack of this region on H2A.Bbd suggests that it may have weaker intermolecular interactions and adopt a more relaxed chromatin structure (Zhou et al. 2007). H2A.Bbd nucleosomal arrays also lack H1 linker histones that are necessary for the formation of higher-order chromatin structures, further suggesting that H2A.Bbd physically antagonizes chromatin compaction, which could serve to facilitate transcriptional activation (Shukla et al. 2010).

Beside the acidic patch, H2A.Bbd also lacks the region corresponding to the H2A C-terminus and part of the docking domain responsible for the interactions between the H2A/H2B dimer and the H3/H4 tetramer (Chadwick and Willard 2001a). H2A. Bbd is less tightly bound to chromatin and the difference in the docking domain is largely responsible for this altered nucleosomal association (Bao et al. 2004; Doyen et al. 2006b). H2A.Bbd nucleosomes are exchanged more rapidly than canonical H2A nucleosomes *in vivo* (Gautier et al. 2004). Recombinant H2A.Bbd histones do not form stable octamers in the presence of the other core histones *in vitro*, but can form nucleosome particles in the presence of DNA (Bao et al. 2004). These biophysical properties alter the stability of H2A.Bbd-containing nucleosomes, and change the amount of DNA wrapped around each nucleosome from 147 bp to ~130 bp (Doyen et al. 2006b). Restriction enzyme analyses also showed increased accessibility in H2A.Bbd nucleosome-containing arrays (Shukla et al. 2010). Therefore, the overall destabilization of the H2A.Bbd nucleosome and nucleosomal arrays is thought to render the chromatin more permissive to active transcription (Zhou et al. 2007). Finally, it is interesting to note that H2A.Bbd only contains 1 lysine compared to the 14 lysine residues in H2A. This unique amino acid composition not only results in a less basic protein, but it also means that H2A.Bbd lacks most of the residues subjected to PTMs on H2A (González-Romero et al. 2008). The significance of this physical feature on the function of this variant remains to be determined.

14.3.2.2 H2A.Bbd: Positive Regulator of Transcription

As a recently discovered variant of H2A, still very little is known about H2A. Bbd. In addition to the previously mentioned studies that link this variant to more accessible chromatin, Angelov et al. found that H2A.Bbd nucleosomes are better substrates for p300-mediated acetylation of histone tails *in vitro*. This suggests that incorporation of H2A.Bbd can enhance acetylation of other histones in the nucleosome context, and further support the prevailing model that H2A.Bbd functions as a positive regulator of transcription (Angelov et al. 2004). As more details of this variant are uncovered in the future, it would be interesting to determine whether H2A.Bbd, like some of the other H2A variants, is also involved in human diseases.

14.3.3 MacroH2A: A Repressive Histone Variant

MacroH2A is a histone H2A variant best known to be associated with gene repression. Directly opposite to H2A.Bbd, immunofluorescence studies showed that macroH2A is enriched at the inactive X chromosome in female mammalian cells (Costanzi and Pehrson 1998). To date, three macroH2A variants have been identified in mammals. MacroH2A1 and macroH2A2 are encoded by two separate genes and the macroH2A1 transcript is alternatively spliced to give rise to two splice variants: macroH2A1.1 and macroH2A1.2. Although there are subtle differences in the localization of the three variants at autosomes, all three variants are enriched on the inactive X chromosome, indicating they have similar functions (Chadwick and Willard 2001b; Costanzi and Pehrson 2001). Most studies have focused on the more abundant macroH2A1, and unless otherwise indicated, the studies presented below mostly refer to this macroH2A variant.

MacroH2A evolved comparatively recently and is significantly divergent from the canonical H2A. It has an N-terminal histone domain that is only 64% in sequence identity with H2A and it also has an extended 25 kDa non-histone region at the C-terminus (Fig. 14.2). The non-histone region, known as the macrodomain, comprises two thirds of the protein's molecular mass (Pehrson and Fried 1992). The macrodomain of macroH2A has been shown to bind and inhibit the enzymatic activity of poly-ADP-ribose-polymerase 1 (PARP-1) (Ouararhni et al. 2006). Catalytically inactive PARP-1 is involved in the maintenance of heterochromatic regions in *Drosophila*. Therefore, macroH2A and PARP-1 together may be involved in the maintenance of the inactive X chromosome and the silencing of some autosomal genes (Tulin et al. 2002; Nusinow et al. 2007).

14.3.3.1 MacroH2A Is Involved in Gene Silencing

Various localization studies showed that macroH2A is associated with transcriptionally silenced regions of the genome. First, it is enriched on the inactive X chromosome and on the inactive alleles of imprinted genes, all of which are associated with facultative heterochromatin (Costanzi and Pehrson 1998; Choo et al. 2007). Second, macroH2A occupancy at genes negatively correlates with gene expression (Buschbeck et al. 2009; Gamble et al. 2010; Changolkar et al. 2010). Third, genome-wide analyses showed that macroH2A is localized to transcriptionally silenced regions marked by H3K27me3 and PRC2-binding (Buschbeck et al. 2009; Araya et al. 2010). Finally, macroH2A accumulates at senescence-associated heterochromatin foci, which are domains of repressed chromatin associated with cellular aging (Zhang et al. 2005b). All together, these findings show that macroH2A is almost exclusively found at heterochromatic regions of the genome.

In addition to its function in the silencing of the inactive X chromosome, macroH2A also has a role in the silencing of autosomal genes. Not only is macroH2A found at the promoter of the IL-8 gene when this gene is transcriptionally silenced, but knockdown of macroH2A derepressed IL-8 expression, demonstrating that macroH2A is required for gene silencing (Agelopoulos and Thanos 2006). Mechanistically, *in vitro* transcription assays demonstrated that macroH2A represses p300- and Gal4-VP16-dependent transcription. Similar repression is seen when the assay was performed using recombinant H2A histones fused to the macrodomain of macroH2A, suggesting that this defined domain is sufficient, and likely responsible, for the repressive function of macroH2A (Doyen et al. 2006a).

14.3.3.2 Mechanism of MacroH2A-Mediated Repression

Incorporation of macroH2A into the nucleosome confers a unique conformation as indicated by a lower sedimentation coefficient compared to nucleosomes containing core H2A (Changolkar and Pehrson 2002). MacroH2A differs from the canonical H2A in the L1 loop, the region of H2A responsible for the interaction between the two H2A/H2B dimers within a nucleosome (Chakravarthy et al. 2005). Therefore, the distinct L1 loop on macroH2A may alter the stability of the nucleosome. Consistent with that, macroH2A nucleosomes are more tightly bound to chromatin and the core DNA of those nucleosomes are more resistant to DNase I digestion. Such findings suggest that macroH2A is associated with closed chromatin conformation, which is consistent with the repressive role for this variant in transcription regulation (Changolkar and Pehrson 2002; Abbott et al. 2004).

Additional mechanistic studies suggested that macroH2A inhibits binding of transcription factors or chromatin remodelling complexes through steric hindrance of the macrodomain, or by reducing the accessibility of the DNA through alterations of the DNA-nucleosome contacts. The presence of macroH2A in a positioned nucleosome interferes with the binding of the transcription factor NF-kB (Angelov et al. 2003). Studies from Narlikar and colleagues showed that macroH2A is permissive to remodelling by ACF, an ISWI complex implicated in gene repression, but is refractory to chromatin remodelling by SWI/SNF, which is involved in gene activation (Chang et al. 2008). The unique structure of the macroH2A nucleosome may allow direct recognition by chromatin remodelling complexes that are involved in gene repression. Finally, macroH2A binds to HDAC1 and 2 and, thus, could promote transcriptional repression through de-acetylation of the other histones associated with macroH2A (Chakravarthy et al. 2005; Doyen et al. 2006a).

ChIP-chip analyses using a testicular cancer cell line showed that macroH2A colocalizes with repressive marks at the promoters of key developmental genes. *In vivo* studies showed that knockdown of macroH2A in zebrafish embryos results in obvious developmental defects of body formation (Buschbeck et al. 2009). On the other hand, ChIP-chip experiments in primary human fibroblasts have shown that a subset of autosomal genes that are marked by macroH2A are transcriptionally active (Gamble et al. 2010). Therefore, this variant may also have a non-repressive transcriptional function in mammalian cells that remains to be clarified.

In addition to its role in X chromosome inactivation, macroH2A is also important for regulating many other genes. Recent studies further suggested that macroH2A is involved in cancer. Bernstein and colleagues discovered that the loss of macroH2A correlates with a more malignant phenotype in melanoma cells (Kapoor et al. 2010). Mechanistically, macroH2A suppresses tumour progression of malignant melanoma through down regulation of Cdk8. Finally, other studies have found that perturbed ratios of macroH2A variants can be a predictor of lung cancer recurrence (Sporn et al. 2009). All together, these studies provide direct and indirect evidence linking macroH2A to oncogenesis. Not only can macroH2A levels be potentially used as prognostic marker in cancer progression, but it could potentially be directly targeted for therapy.

14.4 Conclusions

Many studies over the last decade have revealed that histone variants function as key regulators of a variety of cellular processes. The mechanisms associated with these variants are only beginning to be elucidated and new discoveries will certainly add to the complexities of their functions. Even new variants are still being discovered very recently, such as the new H4 variant found in human adipocytes, and the additional H2A.Z isoform, H2A.Z-2, which differ from the original H2A.Z by three amino acids (Coon et al. 2005; Jufvas et al. 2011). In addition, some of the well studied variants may also have surprising functions that are less well understood. For example, although CENP-A is better known for its role in kinetochore formation, it may have additional roles in defining a special chromatin environment at the centromeres that is distinct from euchromatin and heterochromatin. Most of the centromere is thought to be transcriptionally silent, but studies by the Jiang group have shown that interspersed between the silent domains lies active genes in rice centromeres (Nagaki et al. 2004). They also found that in addition to the CENP-A variant, the rice centromere also contains active marks of transcription such as H3K4me2 and H4 acetylation (Yan et al. 2006). Similarly, the Earnshaw group showed that H3K4me2 and transcription may be required for proper CENP-A deposition in humans (Bergmann et al. 2011). At present, whether CENP-A has a direct role in transcription is unclear; however, the overall PTM-signatures on CENP-A and associated histones in the nucleosome context likely dictate its specific functions.

While most studies focus on individual histone variants, it is important to recognize that these variants function in the context of nucleosomes. For example, studies by the Felsenfeld group showed that nucleosomes containing both H2A.Z and H3.3 are prone to displacement and may be important for the formation of nucleosome free regions at active promoters (Jin et al. 2009). Analogously, other distinct combinations of histone variants may also exist for specific functions. For example, given that H2A.Bbd and H3.3 are both associated with active transcription, it would be interesting to determine whether distinct types of genes are associated with nucleosomes that have this specific pairing. Similarly, it would be intriguing to test whether histone variants of disparate functions, such as macroH2A and H3.3, can co-exist in the nucleosome context and yield new functional features. Finally, the complexities of these combinations are even more complicated when one takes into account that each nucleosome has two copies each of the four histone types. For example, it is now apparent that H2A.Z not only can pair up with another H2A.Z to form homotypic nucleosomes, but it can also pair with the canonical H2A to form heterotypic nucleosomes (Fig. 14.3). Although the pathways leading to the formation of these homotypic and heterotypic H2A.Z-containing nucleosomes have been delineated in yeast studies (Luk et al. 2010), the functional differences between such nucleosomes are not yet known. Studies by the Henikoff lab have shown that homotypic H2A.Z nucleosomes are enriched downstream of active promoters in *Drosophila*, while heterotypic H2A.Z nucleosomes are depleted at that region

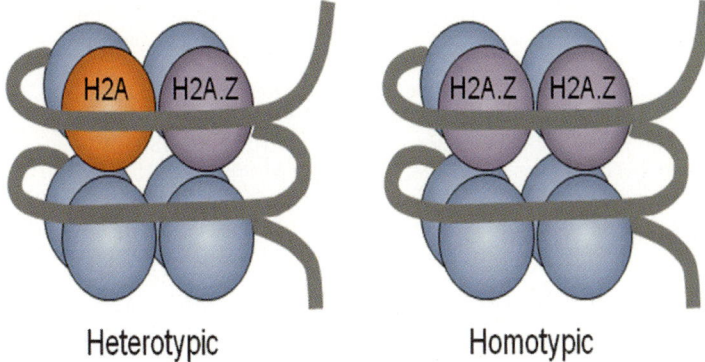

Fig. 14.3 Homotypic and heterotypic nucleosomes. Nucleosomes containing histone variants can exist as homotypic or heterotypic. A homotypic nucleosome contains two copies of the variant histone, in this case two copies of histone variant H2A.Z. A heterotypic nucleosome contains one variant histone, in this case H2A.Z paired with canonical H2A

(Weber et al. 2010), suggesting that they may be functionally distinct. As an extension to the well described "histone code", there may well be a "nucleosome code" or a "histone variant code" that remains to be solved by future studies.

References

Abbott DW, Ivanova VS, Wang X, Bonner WM, Ausió J (2001) Characterization of the stability and folding of H2A.Z chromatin particles: implications for transcriptional activation. J Biol Chem 276:41945–41949

Abbott DW, Laszczak M, Lewis JD, Su H, Moore SC, Hills M, Dimitrov S, Ausió J (2004) Structural characterization of macroH2A containing chromatin. Biochemistry 43:1352–1359

Adam M, Robert F, Larochelle M, Gaudreau L (2001) H2A.Z is required for global chromatin integrity and for recruitment of RNA polymerase II under specific conditions. Mol Cell Biol 21:6270–6279

Agelopoulos M, Thanos D (2006) Epigenetic determination of a cell-specific gene expression program by ATF-2 and the histone variant macroH2A. EMBO J 25:4843–4853

Ahmad K, Henikoff S (2002) The histone variant H3.3 marks active chromatin by replication-independent nucleosome assembly. Mol Cell 9:1191–1200

Akhmanova AS, Bindels PC, Xu J, Miedema K, Kremer H, Hennig W (1995) Structure and expression of histone H3.3 genes in *Drosophila melanogaster* and *Drosophila hydei*. Genome 38:586–600

Alberts B (2002) Molecular biology of the cell, 4th edn. Garland Science, New York

Albig W, Doenecke D (1997) The human histone gene cluster at the D6S105 locus. Hum Genet 101:284–294

Allis CD, Glover CV, Bowen JK, Gorovsky MA (1980) Histone variants specific to the transcriptionally active, amitotically dividing macronucleus of the unicellular eucaryote, *Tetrahymena thermophila*. Cell 20:609–617

Amat R, Gudas LJ (2011) RARγ is required for correct deposition and removal of Suz12 and H2A.Z in embryonic stem cells. J Cell Physiol 226:293–298

Angelov D, Molla A, Perche P-Y, Hans F, Côté J, Khochbin S, Bouvet P, Dimitrov S (2003) The histone variant macroH2A interferes with transcription factor binding and SWI/SNF nucleosome remodeling. Mol Cell 11:1033–1041

Angelov D, Verdel A, An W, Bondarenko V, Hans F, Doyen CM, Studitsky VM, Hamiche A, Roeder RG, Bouvet P, Dimitrov S (2004) SWI/SNF remodeling and p300-dependent transcription of histone variant H2ABbd nucleosomal arrays. EMBO J 23:3815–3824

Araya I, Nardocci G, Morales J, Vera M, Molina A, Alvarez M (2010) MacroH2A subtypes contribute antagonistically to the transcriptional regulation of the ribosomal cistron during seasonal acclimatization of the carp fish. Epigenetics Chromatin 3:14

Ausió J (2006) Histone variants—the structure behind the function. Brief Funct Genomic Proteomic 5:228

Banaszynski LA, Allis CD, Lewis PW (2010) Histone variants in metazoan development. Dev Cell 19:662–674

Bao Y, Konesky K, Park Y-J, Rosu S, Dyer PN, Rangasamy D, Tremethick DJ, Laybourn PJ, Luger K (2004) Nucleosomes containing the histone variant H2A.Bbd organize only 118 base pairs of DNA. EMBO J 23:3314–3324

Barski A, Cuddapah S, Cui K, Roh T-Y, Schones DE, Wang Z, Wei G, Chepelev I, Zhao K (2007) High-resolution profiling of histone methylations in the human genome. Cell 129:823–837

Beck HC, Nielsen EC, Matthiesen R, Jensen LH, Sehested M, Finn P, Grauslund M, Hansen AM, Jensen ON (2006) Quantitative proteomic analysis of post-translational modifications of human histones. Mol Cell Proteomics 5:1314–1325

Bergmann JH, Rodríguez MG, Martins NMC, Kimura H, Kelly DA, Masumoto H, Larionov V, Jansen LET, Earnshaw WC (2011) Epigenetic engineering shows H3K4me2 is required for HJURP targeting and CENP-A assembly on a synthetic human kinetochore. EMBO J 30:328–340

Braunschweig U, Hogan GJ, Pagie L, van Steensel B (2009) Histone H1 binding is inhibited by histone variant H3.3. EMBO J 28:3635–3645

Bruce K, Myers FA, Mantouvalou E, Lefevre P, Greaves I, Bonifer C, Tremethick DJ, Thorne AW, Crane-Robinson C (2005) The replacement histone H2A.Z in a hyperacetylated form is a feature of active genes in the chicken. Nucleic Acids Res 33:5633–5639

Buschbeck M, Uribesalgo I, Wibowo I, Rué P, Martin D, Gutierrez A, Morey L, Guigó R, López-Schier H, Di Croce L (2009) The histone variant macroH2A is an epigenetic regulator of key developmental genes. Nat Struct Mol Biol 16:1074–1079

Chadwick BP, Willard HF (2001a) A novel chromatin protein, distantly related to histone H2A, is largely excluded from the inactive X chromosome. J Cell Biol 152:375–384

Chadwick BP, Willard HF (2001b) Histone H2A variants and the inactive X chromosome: identification of a second macroH2A variant. Hum Mol Genet 10:1101–1113

Chakravarthy S, Gundimella SKY, Caron C, Perche P-Y, Pehrson JR, Khochbin S, Luger K (2005) Structural characterization of the histone variant macroH2A. Mol Cell Biol 25:7616–7624

Chang EY, Ferreira H, Somers J, Nusinow DA, Owen-Hughes T, Narlikar GJ (2008) MacroH2A allows ATP-dependent chromatin remodeling by SWI/SNF and ACF complexes but specifically reduces recruitment of SWI/SNF. Biochemistry 47:13726–13732

Changolkar LN, Pehrson JR (2002) Reconstitution of nucleosomes with histone macroH2A1.2. Biochemistry 41:179–184

Changolkar LN, Singh G, Cui K, Berletch JB, Zhao K, Disteche CM, Pehrson JR (2010) Genome-wide distribution of macroH2A1 histone variants in mouse liver chromatin. Mol Cell Biol 30:5473–5483

Choo JH, Kim JD, Kim J (2007) MacroH2A1 knockdown effects on the Peg3 imprinted domain. BMC Genomics 8:479

Chow C-M, Georgiou A, Szutorisz H, Maia e Silva A, Pombo A, Barahona I, Dargelos E, Canzonetta C, Dillon N (2005) Variant histone H3.3 marks promoters of transcriptionally active genes during mammalian cell division. EMBO Rep 6:354–360

Coon JJ, Ueberheide B, Syka JEP, Dryhurst DD, Ausio J, Shabanowitz J, Hunt DF (2005) Protein identification using sequential ion/ion reactions and tandem mass spectrometry. Proc Natl Acad Sci USA 102:9463–9468

Costanzi C, Pehrson JR (1998) Histone macroH2A1 is concentrated in the inactive X chromosome of female mammals. Nature 393:599–601

Costanzi C, Pehrson JR (2001) MACROH2A2, a new member of the MARCOH2A core histone family. J Biol Chem 276:21776–21784

Creyghton MP, Markoulaki S, Levine SS, Hanna J, Lodato MA, Sha K, Young RA, Jaenisch R, Boyer LA (2008) H2AZ is enriched at polycomb complex target genes in ES cells and is necessary for lineage commitment. Cell 135(4):649–661

Cuadrado A, Corrado N, Perdiguero E, Lafarga V, Muñoz-Canoves P, Nebreda AR (2010) Essential role of p18Hamlet/SRCAP-mediated histone H2A.Z chromatin incorporation in muscle differentiation. EMBO J 29:2014–2025

Daury L, Chailleux C, Bonvallet J, Trouche D (2006) Histone H3.3 deposition at E2F-regulated genes is linked to transcription. EMBO Rep 7:66–71

Delbarre E, Jacobsen BM, Reiner AH, Sorensen AL, Kuntziger T, Collas P (2010) Chromatin environment of histone variant H3. 3 revealed by quantitative imaging and genome-scale chromatin and DNA immunoprecipitation. Mol Biol Cell 21:1872

Dion MF, Kaplan T, Kim M, Buratowski S, Friedman N, Rando OJ (2007) Dynamics of replication-independent histone turnover in budding yeast. Science 315:1405–1408

Dominski Z, Marzluff WF (1999) Formation of the 3′end of histone mRNA. Gene 239:1–14

Doyen C-M, An W, Angelov D, Bondarenko V, Mietton F, Studitsky VM, Hamiche A, Roeder RG, Bouvet P, Dimitrov S (2006a) Mechanism of polymerase II transcription repression by the histone variant macroH2A. Mol Cell Biol 26:1156–1164

Doyen C-M, Montel F, Gautier T, Menoni H, Claudet C, Delacour-Larose M, Angelov D, Hamiche A, Bednar J, Faivre-Moskalenko C, Bouvet P, Dimitrov S (2006b) Dissection of the unusual structural and functional properties of the variant H2A.Bbd nucleosome. EMBO J 25:4234–4244

Draker R, Cheung P (2009) Transcriptional and epigenetic functions of histone variant H2A.Z. Biochem Cell Biol 87:19–25

Draker R, Sarcinella E, Cheung P (2011) USP10 deubiquitylates the histone variant H2A.Z and both are required for androgen receptor-mediated gene activation. Nucleic Acids Res. doi:10.1093/nar/gkq1352

Drané P, Ouararhni K, Depaux A, Shuaib M, Hamiche A (2010) The death-associated protein DAXX is a novel histone chaperone involved in the replication-independent deposition of H3.3. Genes Dev 24:1253–1265

Faast R, Thonglairoam V, Schulz TC, Beall J, Wells JRE, Taylor H, Matthaei K, Rathjen PD, Tremethick DJ, Lyons I (2001) Histone variant H2A.Z is required for early mammalian development. Curr Biol 11:1183–1187

Fan JY, Gordon F, Luger K, Hansen JC, Tremethick DJ (2002) The essential histone variant H2A.Z regulates the equilibrium between different chromatin conformational states. Nat Struct Biol 9:172–176

Fan JY, Rangasamy D, Luger K, Tremethick DJ (2004) H2A.Z alters the nucleosome surface to promote HP1 [alpha]-mediated chromatin fiber folding. Mol Cell 16:655–661

Farris SD, Rubio ED, Moon JJ, Gombert WM, Nelson BH, Krumm A (2005) Transcription-induced chromatin remodeling at the c-myc gene involves the local exchange of histone H2A.Z. J Biol Chem 280:25298–25303

Flaus A, Rencurel C, Ferreira H, Wiechens N, Owen-Hughes T (2004) Sin mutations alter inherent nucleosome mobility. EMBO J 23:343–353

Foster ER, Downs JA (2005) Histone H2A phosphorylation in DNA double-strand break repair. FEBS J 272:3231–3240

Frank D, Doenecke D, Albig W (2003) Differential expression of human replacement and cell cycle dependent H3 histone genes. Gene 312:135–143

Gamble MJ, Frizzell KM, Yang C, Krishnakumar R, Kraus WL (2010) The histone variant macroH2A1 marks repressed autosomal chromatin, but protects a subset of its target genes from silencing. Genes Dev 24:21–32

Garrick D, Sharpe JA, Arkell R, Dobbie L, Smith AJ, Wood WG, Higgs DR, Gibbons RJ (2006) Loss of Atrx affects trophoblast development and the pattern of X-inactivation in extraembryonic tissues. PLoS Genet 2(4):e58

Gautier T, Abbott DW, Molla A, Verdel A, Ausio J, Dimitrov S (2004) Histone variant H2ABbd confers lower stability to the nucleosome. EMBO Rep 5:715–720

Gévry N, Chan HM, Laflamme L, Livingston DM, Gaudreau L (2007) p21 transcription is regulated by differential localization of histone H2A.Z. Genes Dev 21:1869–1881

Gévry N, Hardy S, Jacques P-E, Laflamme L, Svotelis A, Robert F, Gaudreau L (2009) Histone H2A.Z is essential for estrogen receptor signaling. Genes Dev 23:1522–1533

Goldberg AD, Banaszynski LA, Noh K-M, Lewis PW, Elsaesser SJ, Stadler S, Dewell S, Law M, Guo X, Li X (2010) Distinct factors control histone variant H3.3 localization at specific genomic regions. Cell 140:678–691

González-Romero R, Méndez J, Ausió J, Eirín-López JM (2008) Quickly evolving histones, nucleosome stability and chromatin folding: all about histone H2A.Bbd. Gene 413:1–7

Gorovsky MA (1996) Essential and nonessential histone H2A variants in *Tetrahymena thermophila*. Mol Cell Biol 16:4305–4311

Guillemette B, Bataille AR, Gévry N, Adam M, Blanchette M, Robert F, Gaudreau L (2005) Variant histone H2A.Z is globally localized to the promoters of inactive yeast genes and regulates nucleosome positioning. PLoS Biol 3:e384

Hake SB, Allis CD (2006) Histone H3 variants and their potential role in indexing mammalian genomes: the "H3 barcode hypothesis". Proc Natl Acad Sci USA 103:6428–6435

Hake SB, Garcia BA, Duncan EM, Kauer M, Dellaire G, Shabanowitz J, Bazett-Jones DP, Allis CD, Hunt DF (2005) Expression patterns and post-translational modifications associated with mammalian histone H3 variants. J Biol Chem 281:559–568

Hake SB, Garcia BA, Duncan EM, Kauer M, Dellaire G, Shabanowitz J, Bazett-Jones DP, Allis CD, Hunt DF (2006) Expression patterns and post-translational modifications associated with mammalian histone H3 variants. J Biol Chem 281:559–568

Hardy S, Jacques P-E, Gévry N, Forest A, Fortin M-E, Laflamme L, Gaudreau L, Robert F (2009) The euchromatic and heterochromatic landscapes are shaped by antagonizing effects of transcription on H2A.Z deposition. PLoS Genet 5:e1000687

Harris ME, Bohni R, Schneiderman MH, Ramamurthy L, Schumperli D, Marzluff WF (1991) Regulation of histone mRNA in the unperturbed cell cycle: evidence suggesting control at two posttranscriptional steps. Mol Cell Biol 11:2416

Henikoff S, Ahmad K (2005) Assembly of variant histones into chromatin. Annu Rev Cell Dev Biol 21:133–153

Hua S, Kallen CB, Dhar R, Baquero MT, Mason CE, Russell BA, Shah PK, Liu J, Khramtsov A, Tretiakova MS, Krausz TN, Olopade OI, Rimm DL, White KP (2008) Genomic analysis of estrogen cascade reveals histone variant H2A.Z associated with breast cancer progression. Mol Syst Biol 4:188

Iouzalen N, Moreau J, Méchali M (1996) H2A.ZI, a new variant histone expressed during Xenopus early development exhibits several distinct features from the core histone H2A. Nucleic Acids Res 24:3947–3952

Jackson JD, Gorovsky MA (2000) Histone H2A.Z has a conserved function that is distinct from that of the major H2A sequence variants. Nucleic Acids Res 28:3811–3816

Jin C, Felsenfeld G (2007) Nucleosome stability mediated by histone variants H3.3 and H2A.Z. Genes Dev 21:1519–1529

Jin C, Zang C, Wei G, Cui K, Peng W, Zhao K, Felsenfeld G (2009) H3.3/H2A.Z double variant-containing nucleosomes mark "nucleosome-free regions" of active promoters and other regulatory regions. Nat Genet 41:941–945

Jufvas Å, Strålfors P, Vener AV (2011) Histone variants and their post-translational modifications in primary human fat cells. PLoS One 6:e15960

Kamakaka RT (2005) Histone variants: deviants? Genes Dev 19:295–316

Kapoor A, Goldberg MS, Cumberland LK, Ratnakumar K, Segura MF, Emanuel PO, Menendez S, Vardabasso C, Leroy G, Vidal CI, Polsky D, Osman I, Garcia BA, Hernando E, Bernstein E (2010) The histone variant macroH2A suppresses melanoma progression through regulation of CDK8. Nature 468:1105–1109

Kobor MS, Venkatasubrahmanyam S, Meneghini MD, Gin JW, Jennings JL, Link AJ, Madhani HD, Rine J (2004) A protein complex containing the conserved Swi2/Snf2-related ATPase Swr1p deposits histone variant H2A.Z into euchromatin. PLoS Biol 2:E131

Kornberg RD (1974) Chromatin structure: a repeating unit of histones and DNA. Science 184:868–871

Kouzarides T (2007) Chromatin modifications and their function. Cell 128:693–705

Lewis PW, Elsaesser SJ, Noh K-M, Stadler SC, Allis CD (2010) Daxx is an H3.3-specific histone chaperone and cooperates with ATRX in replication-independent chromatin assembly at telomeres. Proc Natl Acad Sci USA 107:14075–14080

Li B, Pattenden SG, Lee D, Gutiérrez J, Chen J, Seidel C, Gerton J, Workman JL (2005) Preferential occupancy of histone variant H2AZ at inactive promoters influences local histone modifications and chromatin remodeling. Proc Natl Acad Sci USA 102:18385–18390

Loyola A, Bonaldi T, Roche D, Imhof A, Almouzni G (2006) PTMs on H3 variants before chromatin assembly potentiate their final epigenetic state. Mol Cell 24:309–316

Luk E, Ranjan A, FitzGerald PC, Mizuguchi G, Huang Y, Wei D, Wu C (2010) Stepwise histone replacement by SWR1 requires dual activation with histone H2A.Z and canonical nucleosome. Cell 143:725–736

Lusser A, Kadonaga JT (2003) Chromatin remodeling by ATP-dependent molecular machines. Bioessays 25:1192–1200

Marzluff WF, Gongidi P, Woods KR, Jin J, Maltais LJ (2002) The human and mouse replication-dependent histone genes. Genomics 80:487–498

McKittrick E, Gafken PR, Ahmad K, Henikoff S (2004) Histone H3.3 is enriched in covalent modifications associated with active chromatin. Proc Natl Acad Sci USA 101:1525–1530

Meneghini MD, Wu M, Madhani HD (2003) Conserved histone variant H2A.Z protects euchromatin from the ectopic spread of silent heterochromatin. Cell 112:725–736

Michaelson JS, Bader D, Kuo F, Kozak C, Leder P (1999) Loss of Daxx, a promiscuously interacting protein, results in extensive apoptosis in early mouse development. Genes Dev 13:1918–1923

Millar CB, Xu F, Zhang K, Grunstein M (2006) Acetylation of H2AZ Lys 14 is associated with genome-wide gene activity in yeast. Genes Dev 20:711–722

Mito Y, Henikoff JG, Henikoff S (2005) Genome-scale profiling of histone H3.3 replacement patterns. Nat Genet 37:1090–1097

Nagaki K, Cheng Z, Ouyang S, Talbert PB, Kim M, Jones KM, Henikoff S, Buell CR, Jiang J (2004) Sequencing of a rice centromere uncovers active genes. Nat Genet 36:138–145

Ng K, Pullirsch D, Leeb M, Wutz A (2007) Xist and the order of silencing. EMBO Rep 8:34–39

Nusinow DA, Hernandez-Munoz I, Fazzio TG, Shah GM, Kraus WL, Panning B (2007) Poly(ADP-ribose) polymerase 1 is inhibited by a histone H2A variant, macroH2A, and contributes to silencing of the inactive X chromosome. J Biol Chem 282:12851–12859

Ooi SL, Henikoff JG, Henikoff S (2010) A native chromatin purification system for epigenomic profiling in *Caenorhabditis elegans*. Nucleic Acids Res 38:e26

Osley MA (1991) The regulation of histone synthesis in the cell cycle. Annu Rev Biochem 60:827–861

Ouararhni K, Hadj-Slimane R, Ait-Si-Ali S, Robin P, Mietton F, Harel-Bellan A, Dimitrov S, Hamiche A (2006) The histone variant mH2A1.1 interferes with transcription by down-regulating PARP-1 enzymatic activity. Genes Dev 20:3324–3336

Pehrson JR, Fried VA (1992) MacroH2A, a core histone containing a large nonhistone region. Science 257:1398–1400

Placek BJ, Huang J, Kent JR, Dorsey J, Rice L, Fraser NW, Berger SL (2009) The histone variant H3.3 regulates gene expression during lytic infection with herpes simplex virus type 1. J Virol 83:1416–1421

Pusarla RH, Bhargava P (2005) Histones in functional diversification. FEBS J 272:5149–5168

Rangasamy D, Berven L, Ridgway P, Tremethick DJ (2003) Pericentric heterochromatin becomes enriched with H2A.Z during early mammalian development. EMBO J 22:1599–1607

Rhodes DR, Yu J, Shanker K, Deshpande N, Varambally R, Ghosh D, Barrette T, Pandey A, Chinnaiyan AM (2004) Large-scale meta-analysis of cancer microarray data identifies common transcriptional profiles of neoplastic transformation and progression. Proc Natl Acad Sci USA 101:9309–9314

Roberts C, Sutherland HF, Farmer H, Kimber W, Halford S, Carey A, Brickman JM, Wynshaw-Boris A, Scambler PJ (2002) Targeted mutagenesis of the Hira gene results in gastrulation defects and patterning abnormalities of mesoendodermal derivatives prior to early embryonic lethality. Mol Cell Biol 22:2318–2328

Rogakou EP, Pilch DR, Orr AH, Ivanova VS, Bonner WM (1998) DNA double-stranded breaks induce histone H2AX phosphorylation on serine 139. J Biol Chem 273:5858

Santenard A, Ziegler-Birling C, Koch M, Tora L, Bannister AJ, Torres-Padilla M-E (2010) Heterochromatin formation in the mouse embryo requires critical residues of the histone variant H3.3. Nat Cell Biol 12:853–862

Santisteban MS, Kalashnikova T, Smith MM (2000) Histone H2A.Z regulates transcription and is partially redundant with nucleosome remodeling complexes. Cell 103:411–422

Santisteban MS, Hang M, Smith MM (2011) Histone variant H2A.Z and RNA polymerase II transcription elongation. Mol Cell Biol 31:1848–1860

Sarcinella E, Zuzarte PC, Lau PNI, Draker R, Cheung P (2007) Monoubiquitylation of H2A.Z distinguishes its association with euchromatin or facultative heterochromatin. Mol Cell Biol 27:6457–6468

Schenk R, Jenke A, Zilbauer M, Wirth S, Postberg J (2011) H3.5 is a novel hominid-specific histone H3 variant that is specifically expressed in the seminiferous tubules of human testes. Chromosoma. doi:10.1007/s00412-011-0310-4

Schwartz BE, Ahmad K (2005) Transcriptional activation triggers deposition and removal of the histone variant H3.3. Genes Dev 19:804–814

Shukla MS, Syed SH, Goutte-Gattat D, Richard JLC, Montel F, Hamiche A, Travers A, Faivre-Moskalenko C, Bednar J, Hayes JJ, Angelov D, Dimitrov S (2010) The docking domain of histone H2A is required for H1 binding and RSC-mediated nucleosome remodeling. Nucleic Acids Res 39:2559–2570

Slupianek A, Yerrum S, Safadi FF, Monroy MA (2010) The chromatin remodeling factor SRCAP modulates expression of prostate specific antigen and cellular proliferation in prostate cancer cells. J Cell Physiol 224:369–375

Sporn JC, Kustatscher G, Hothorn T, Collado M, Serrano M, Muley T, Schnabel P, Ladurner AG (2009) Histone macroH2A isoforms predict the risk of lung cancer recurrence. Oncogene 28:3423–3428

Strahl BD, Allis CD (2000) The language of covalent histone modifications. Nature 403:41–45

Sullivan KF, Hechenberger M, Masri K (1994) Human CENP-A contains a histone H3 related histone fold domain that is required for targeting to the centromere. J Cell Biol 127:581

Sutcliffe EL, Parish IA, He YQ, Juelich T, Tierney ML, Rangasamy D, Milburn PJ, Parish CR, Tremethick DJ, Rao S (2009) Dynamic histone variant exchange accompanies gene induction in T cells. Mol Cell Biol 29:1972–1986

Suto RK, Clarkson MJ, Tremethick DJ, Luger K (2000) Crystal structure of a nucleosome core particle containing the variant histone H2A.Z. Nat Struct Biol 7:1121–1124

Svotelis A, Gévry N, Grondin G, Gaudreau L (2010) H2A.Z overexpression promotes cellular proliferation of breast cancer cells. Cell Cycle 9:364–370

Szenker E, Ray-Gallet D, Almouzni G (2011) The double face of the histone variant H3.3. Cell Res 21:421–434

Tagami H, Ray-Gallet D, Almouzni G, Nakatani Y (2004) Histone H3. 1 and H3. 3 complexes mediate nucleosome assembly pathways dependent or independent of DNA synthesis. Cell 116:51–61

Talbert PB, Henikoff S (2010) Histone variants–ancient wrap artists of the epigenome. Nat Rev Mol Cell Biol 11:264–275

Tamura T, Smith M, Kanno T, Dasenbrock H, Nishiyama A, Ozato K (2009) Inducible deposition of the histone variant H3.3 in interferon-stimulated genes. J Biol Chem 284:12217–12225

Taverna SD, Li H, Ruthenburg AJ, Allis CD, Patel DJ (2007) How chromatin-binding modules interpret histone modifications: lessons from professional pocket pickers. Nat Struct Mol Biol 14:1025–1040

Thakar A, Gupta P, Ishibashi T, Finn R, Silva-Moreno B, Uchiyama S, Fukui K, Tomschik M, Ausio J, Zlatanova J (2009) H2A.Z and H3.3 histone variants affect nucleosome structure: biochemical and biophysical studies. Biochemistry 48:10852–10857

Thiriet C, Hayes JJ (2005) Chromatin in need of a fix: phosphorylation of H2AX connects chromatin to DNA repair. Mol Cell 18:617–622

Tulin A, Stewart D, Spradling AC (2002) The Drosophila heterochromatic gene encoding poly(ADP-ribose) polymerase (PARP) is required to modulate chromatin structure during development. Genes Dev 16:2108–2119

Updike DL, Mango SE (2006) Temporal regulation of foregut development by HTZ-1/H2A.Z and PHA-4/FoxA. PLoS Genet 2:e161

van Daal A, Elgin SC (1992) A histone variant, H2AvD, is essential in *Drosophila melanogaster*. Mol Biol Cell 3:593–602

Weber CM, Henikoff JG, Henikoff S (2010) H2A.Z nucleosomes enriched over active genes are homotypic. Nat Struct Mol Biol 17:1500–1507

Wirbelauer C, Bell O, Schübeler D (2005) Variant histone H3.3 is deposited at sites of nucleosomal displacement throughout transcribed genes while active histone modifications show a promoter-proximal bias. Genes Dev 19:1761–1766

Wong MM, Cox LK, Chrivia JC (2007) The chromatin remodeling protein, SRCAP, is critical for deposition of the histone variant H2A.Z at promoters. J Biol Chem 282:26132–26139

Wong LH, Ren H, Williams E, McGhie J, Ahn S, Sim M, Tam A, Earle E, Anderson MA, Mann J, Choo KHA (2009) Histone H3.3 incorporation provides a unique and functionally essential telomeric chromatin in embryonic stem cells. Genome Res 19:404–414

Yan H, Ito H, Nobuta K, Ouyang S, Jin W, Tian S, Lu C, Venu RC, Wang G-L, Green PJ, Wing RA, Buell CR, Meyers BC, Jiang J (2006) Genomic and genetic characterization of rice Cen3 reveals extensive transcription and evolutionary implications of a complex centromere. Plant Cell 18:2123–2133

Zhang H, Roberts DN, Cairns BR (2005a) Genome-wide dynamics of Htz1, a histone H2A variant that poises repressed/basal promoters for activation through histone loss. Cell 123:219–231

Zhang R, Poustovoitov MV, Ye X, Santos HA, Chen W, Daganzo SM, Erzberger JP, Serebriiskii IG, Canutescu AA, Dunbrack RL (2005b) Formation of MacroH2A-containing senescence-associated heterochromatin foci and senescence driven by ASF1a and HIRA. Dev Cell 8:19–30

Zhou J, Fan JY, Rangasamy D, Tremethick DJ (2007) The nucleosome surface regulates chromatin compaction and couples it with transcriptional repression. Nat Struct Mol Biol 14:1070–1076

Zilberman D, Coleman-Derr D, Ballinger T, Henikoff S (2008) Histone H2A.Z and DNA methylation are mutually antagonistic chromatin marks. Nature 456:125–129

Zlatanova J, Thakar A (2008) H2A.Z: view from the top. Structure 16:166–179

Zucchi I, Mento E, Kuznetsov VA, Scotti M, Valsecchi V, Simionati B, Vicinanza E, Valle G, Pilotti S, Reinbold R, Vezzoni P, Albertini A, Dulbecco R (2004) Gene expression profiles of epithelial cells microscopically isolated from a breast-invasive ductal carcinoma and a nodal metastasis. Proc Natl Acad Sci USA 101:18147–18152

Chapter 15
Noncoding RNAs in Chromatin Organization and Transcription Regulation: An Epigenetic View

Karthigeyan Dhanasekaran, Sujata Kumari, and Chandrasekhar Kanduri

Abstract The Genome of a eukaryotic cell harbors genetic material in the form of DNA which carries the hereditary information encoded in their bases. Nucleotide bases of DNA are transcribed into complimentary RNA bases which are further translated into protein, performing defined set of functions. The central dogma of life ensures sequential flow of genetic information among these biopolymers. Noncoding RNAs (ncRNAs) serve as exceptions for this principle as they do not code for any protein. Nevertheless, a major portion of the human transcriptome comprises noncoding RNAs. These RNAs vary in size, as well as they vary in the spatio-temporal distribution. These ncRnAs are functional and are shown to be involved in diverse cellular activities. Precise location and expression of ncRNA is essential for the cellular homeostasis. Failures of these events ultimately results in numerous disease conditions including cancer. The present review lists out the various classes of ncRNAs with a special emphasis on their role in chromatin organization and transcription regulation.

K. Dhanasekaran • S. Kumari
Transcription and Disease Laboratory, Molecular Biology
and Genetics Unit, Jawaharlal Nehru Centre for Advanced
Scientific Research, Jakkur P.O., Bangalore 560064, India

C. Kanduri (✉)
Science for Life Laboratory, Department of Immunology,
Genetics and Pathology, Rudbeck Laboratory, Uppsala University,
Dag HammarskjöldsVäg 20, S75185 Uppsala, Sweden

Department of Medical and Clinical genetics, Institute of Biomedicine,
The Sahlgrenska Academy, University of Gothenburg, Gothenberg, Sweden
e-mail: kanduri.chandrasekhar@igp.uu.se

15.1 Overview of ncRNA

Noncoding RNAs (ncRNAs) are a large group of functional RNA transcribed by RNA polymerases, but never translated into protein. A decade before the discovery of these noncoding transcripts, most of the sequences in our genome were believed to be part of "Junk DNA". Soon the genome sequencing projects revealed that only a mere 2% of the human genome codes for protein coding genes and almost 98% of the human transcriptome represents the ncRNA. They include well-characterized transfer RNAs and ribosomal RNAs involved in the process of translation, as well as a huge class of other regulatory ncRNAs which have been shown to play a crucial role in gene regulation. ncRNAs in general function as adaptors for the recognition of a particular nucleotide sequence in the target which is later positioned into the enzymatic molecule associated with the specific class of ncRNA. These functional ncRNA are involved in key cellular processes including transcriptional regulation, RNA processing and modification, protein trafficking, genome stability, mRNA stability, and even protein degradation (Hüttenhofer et al. 2005).

Terminologies like "Junk DNA" and "Transcriptional noise" have been challenged since the discovery of regulatory RNAs that are transcribed by both Pol II and Pol III. The list of known ncRNAs is growing larger in numbers since the completion of whole transcriptome analyses like the ENCODE Pilot Project (Birney et al. 2007), the mouse cDNA project FANTOM, and a series of other large scale transcriptome studies performed to fish these transcribed fragments ('transfrags') using various forms of high throughput tiling arrays, ESTs, SAGE tags and RACE techniques. Ultimately the present scenario of the transcriptome is that ncRNA are vast in number covering a huge proportion of the genome consisting of overlapping, bi-directional transcripts. The major obstacle in the identification of these transcripts is that majority of them are expressed in a short spatio-temporal frame, thus it is difficult to recognize such transcripts even after employing the most advanced deep sequencing techniques. Though the ncRNAs have coexisted with protein coding genes they were kept under curtain mainly because of the model organism taken for such studies. Also the mutations that were considered were often expected to have a major impact on the phenotypic outcome of such genetic screens. Mutation in a protein-coding gene can have severe effects on the structure and function of the protein which ultimately shows an altered phenotype, which is often visible in the regular genetic screening due to the high penetrance. On the other hand recessive mutation phenotypes and single-base mutations are harder to identify in comparison to insertions/deletions (Eddy 2001; Kavanaugh and Dietrich 2009). The reason for a shift in the focus of our research towards protein coding genes is that most of the genetic screening techniques and the methodologies followed during the early era of genomics had an inherent bias toward the scanning of known exons and the flanking sequences in that region. Most of the bioinformatic search tools were based on the signatures of protein coding genic regions which does not hold good for the ncRNA prediction especially when they are transcribed from the intergenic deserts. The other reason is that comparative studies on the sequences of ncRNA have failed to show any healthy conservation among the known transcripts reported till date. However, studies on a handful of functional ncRNAs indicate that they carryout common functions via conserved secondary structures, indicating that despite having no sequence similarity they seem to harbor conserved secondary structures. Even though a large number of genomes have been sequenced, the number and diversity of ncRNA-encoding genes is largely unknown, especially due to the incompleteness of the list of various ncRNAs (van Bakel et al. 2010; Farh et al. 2005).

15.2 Classification and Evolution of ncRNA

The existing classes of ncRNAs are in general transcribed by all three possible modes of transcription. The pre-rRNA (28S, 18S, 5.8S) are transcribed by Pol I, and some of the snRNAs and LINEs are transcribed by Pol II where as SINEs, snRNA, 7SL RNA, etc. are transcribed by Pol III transcription machinery. Broadly, ncRNAs can be classified into "housekeeping" and "regulatory" ncRNAs (Morey and Avner 2004). Housekeeping ncRNAs are constitutively expressed and involved in processes like translation, RNA processing, RNA modifications, protein trafficking and genomic stability required for normal cell viability, whereas the regulatory ncRNAs, often expressed in certain specified tissues during different stages of development or in response to an external stimuli and they are comprised of RNAs involved in the process of gene expression/regulation and chromatin organization. The other way to classify ncRNA is based on the size of the functional transcripts as long (9,999–10,000nt), medium (200–999nt), small (24–199nt) and micro (18–31nt) ncRNAs. Noncoding transcripts can also be classified based on the sequence origin as sense or antisense transcripts from the genic, intronic and intergenic region of the genome.

The human genome has approximately 27,161 genes (Flicek et al. 2008) in total, of which about 4,421 are ncRNA genes. Literature on the existing ncRNAs indicates an unequivocal correlation between the rise in the number of noncoding transcripts and the complexity of an organism (Amaral and Mattick 2008). Events responsible for such evolution are by gene duplication, mutation, horizontal transfer and integration of genetic material between different pathogen and host across various phases of evolution. Especially the non-genic deserts are more prone towards such events since the pressure to preserve the functionality of the protein coding genes does not apply to ncRNA coding intergenic regions. Also the drastic mutations can be well tolerated since the constraint for most of the ncRNA is to maintain its secondary structure. Moreover the base change in one strand is often compensated by a complementary mutation across the paired strand. Though ncRNAs in general are rapidly evolving they are earmarked by conservation in their secondary structure and are often found associated with regions spanning promoters, splice junctions, and other regions with specific chromatin signatures in relation to the spatiotemporal expression and subcellular localization pattern (Pang et al. 2006; Bradley et al. 2009).

15.3 Housekeeping ncRNA

rRNAs: Generally eukaryotes have many copies of the rRNA genes organized in tandem repeats; in humans approximately 300–400 rDNA repeats are present in five clusters. All mammalian cells possess two mitochondrial (12S and 16S) and four cytoplasmic rRNA (the 28S, 5.8S, 18S, and 5S subunits) transcribed by RNA polymerase I except 5S rRNA which is transcribed by RNA polymerase III. Most of these rRNAs constitute the active site of ribosomes and also aids in maintaining the fidelity of translation.

tRNAs: tRNAs are the adapter molecules that aid in sequence specific incorporation of various amino acids according to the code present in an mRNA. According to the tRNADB, which is a curated database of tRNA, there are 22 known mitochondrial tRNA genes and 497 nuclear tRNA genes known in humans but the number varies a lot in different organisms (Abe et al. 2011). These genes are found on all chromosomes, except chromosome number 22 and Y chromosome of humans (Lander et al. 2001). tRNAs are transcribed by RNA polymerase III as pre-tRNAs in the nucleus (Dieci et al. 2007) which undergo extensive posttranscriptional modifications. The adaptor function of tRNA lies in its three-dimensional structure wherein one end of the tRNA carries the anticodon that serves as a genetic code to recognize the codon in mRNA during protein biosynthesis. Transfer RNA-like structures (tRNA-like structures) are a separate class of RNA sequences transcribed from the genome of many plant RNA viruses, which have a tRNA like tertiary structure (Crick 1968). These tRNA-like structures mimic some tRNA functions, such as aminoacylation, but only three aminoacylation specificities, valine, histidine and tyrosine have been reported till date (Dreher 2009). Such tRNA-like structures are also known to increase the stability of RNA viruses by encapsulating its RNA genome (Mans et al. 1991). In addition, they act as 3′-translational enhancers (Matsuda and Dreher 2004) and regulators of minus strand synthesis.

tel-sRNAs: Telomere specific small RNAs called as tel-sRNA are found exclusively in the telomeric region of the genome. These small Π-like RNAs are associated asymmetrically to the G-rich strand of telomers which are Dicer-independent, 2′-O-methylated at the 3′ terminus, and conserved from protozoa to mammalian cells. tel-sRNAs were shown for the first time in mouse genome where they aid in the establishment and maintenance of heterochromatin in the telomeric loci (Cao et al. 2009).

tmRNA: The bacterial *tmRNA* has both tRNA-like and mRNA-like function e.g., 10Sa RNA or SsrA. *tmRNA* engages the problematic messenger RNAs and recycles the 70S ribosomes ultimately that incorporates a series of alanine residues which are earmarked for the degradation of those incomplete peptides (Gillet and Felden 2001). For more information about tmRNAs refer to tmRDB, an exclusive database for tmRNAs.

SRP RNA: The RNA component of the signal recognition particle (SRP) ribonucleoprotein complex also known as 7SL, 6S, ffs, or 4.5S RNA, is a universally conserved ncRNA (Rosenblad et al. 2009) that directs the newly synthesized proteins within a cell to the endoplasmic reticulum either co-translationally or post-translationally thereby allowing them to be secreted.

snRNAs: small nuclear RNAs can be broadly classified into two. Firstly, the Sm-class of snRNA that possess a 5′-trimethylguanosine cap, 3′stem-loop and heteroheptameric ring structure that binds to sm-proteins. These non-polyadenylated snRNAs are transcribed by Pol II and processed by integrator. The processed mature snRNAs finally aid in splicing out introns from the pre-mRNA. Secondly, Lsm-class RNAs that possess a monomethylphosphate cap and a 3′stem-loop with uridine rich heteroheptameric ring that binds Lsm-proteins. Pol III transcribes such Lsm-snRNAs using external promoters and Uridine stretch as terminator (Segref et al. 2001). Almost all Lsm-snRNPs are assembled in the nucleus within the cajal body for a brief period after which they diffuse out in the nucleoplasm till they reach their specific nuclear domains like, perichromatin fibrils and interchromatin granule clusters. snRNPs containing such ncRNAs form the core of the spliceosome which are the catalytic centers for splicing introns from pre-mRNA (Matera et al. 2007). Among the snRNAs U7 snRNA needs a special mention which is involved in the processing of 3′ end of histone genes of eukaryotes which possess a unique stem-loop structure instead of a poly-A tail. However, snRNAs are not just restricted to splicing events alone as they have been shown to regulate transcription, independent of their splicing function. 7SK is one such snRNA which mediates Pol II transcriptional inhibition via its interaction with P-TEFb. Apart from the above mentioned snRNAs there are numerous other snRNAs which carryout important biological functions like, RNA Pol III transcribed snaR-A RNA, Intergenic spacer RNA (IGS RNAs) etc., are discussed under regulatory ncRNAs section.

SmY-RNA: These ncRNAs belong to a Small nuclear class of ncRNAs in nematodes SmY-RNA were disovered in *Ascaris lumbricoides* during the year 1996 (Maroney et al. 1996). Based on the evidence obtained from the studies carried out in a related species i.e., *C. elegans* SmY-RNA is believed to be in complex with the spliced-leader RNA and involved in mRNA trans-splicing (MacMorris et al. 2007).

snoRNAs: Small nucleolar RNAs, as the name implies, are retained within the nucleolus and aid as guide strands for incorporating the specific modification like methylations and pseudouridylations, onto other RNA molecules like tRNA, rRNA, snRNA etc. snoRNAs can be further classified into C/D Box RNAs, H/ACA Box RNAs, composite C/D Box and H/ACA Box RNAs and Orphan snoRNAs (Bachellerie et al. 2002; Samarsky et al. 1998). In general C/D box members guide 2′O-ribose-methylations and H/ACA members guide pseudouridylation. snoRNAs are defined by the characteristic secondary structure formed by the signature sequences which varies slightly in the composite snoRNAs. The composite snoRNA contains both C/D and H/ACA box and are retained in the cajal bodies and hence, named as "scaRNAs" (Jády and Kiss 2001). U85 a typical example of composite snoRNA, functions in both 2′-O-ribose methylation and pseudouridylation of snRNA. On the contrary, there are snoRNAs with unidentified substrates that are grouped under the Orphan snoRNAs. Apart from their function in guiding modifications for maintaining a stable pool of ncRNA, some members are even known to act like miRNAs with exclusive regulatory functions and hence they are discussed under the regulatory RNAs section.

15.4 Regulatory RNAs

Regulatory RNAs comprise a subset of both long and small mRNAs having gene expression regulatory function. Regulatory ncRNAs especially the small ncRNAs in general, form base pairs with other RNA or DNA and constitute RNA:RNA or RNA:DNA duplexes. These duplexes are recognised by different complexes like RNA induced silencing complex (RISC), RNA induced transcriptional silencing (RITS) or RNA editing enzymes which act to decipher downstream consequences. These cis-acting regulatory sequences are generally found in non-coding regions of mRNAs and pre-mRNAs. Untranslated regions (UTRs) of mRNA generally act as binding sites for some trans acting regulatory RNAs, though they are also known to form secondary structures facilitating binding of regulatory proteins that in turn control stability, function or localization of mRNAs (Gebauer and Hentze 2004; Moore 2005). Splice junctions provide yet another cis regulatory sequences which along with the aid of spliceosomal snRNAs and other components of spliceosome, a ribonucleoprotein (RNP) complex that controls splicing of the primary transcript (Nilsen 2003; Valadkhan et al. 2007). Regulatory ncRNAs are generally categorized in two classes namely the small (<200 nucleotide) and large (>200 nucleotide) regulatory ncRNAs.

15.4.1 Small Regulatory RNA

There are numerous regulatory RNAs that are <200nt long and show unique spatio-temporal expression in comparison to housekeeping ncRNAs. Some of the well charactreised small regulatory RNAs are discussed in this section.

Regulatory Small nucleolar RNAs (SnoRNAs): Some SnoRNAs show tissue specific expression, like tandemly arranged repeated intron-encoded C/D snoRNA genes in the region downstream from the GTL2 gene at 14q32 show brain specificity. These snoRNA genes associate with human imprinted 14q32 domain suggesting their regulatory role in epigenetic imprinting process (Cavaillé et al. 2002). Two other brain specific snoRNAs, HBII-52 and HBII-85 were reported to be absent from the cortex of a patient with Prader-Willi syndrome (PWS), which is a neuro-genetic disease resulting from a deficiency of paternal gene expression, indicating their role in the etiology of PWS (Cavaillé et al. 2000). Further, it was shown that the snoRNA HBII-52 regulates alternative splicing of the Serotonin Receptor 2C. Lack of HBII-52 in PWS patients generate different messenger RNA (mRNA) isoforms which leads to the loss of high-efficacy serotonin receptor, which could contribute to the disease (Kishore and Stamm 2006).

Apart from regulating mRNA transcription which is central for the regulation of gene expression, other biological reactions comprehending gene expression like mRNA turnover, gene silencing and translation are also controlled by ncRNAs like miRNAs and short interfering RNAs. miRNAs and siRNAs are 21–25 nt long RNAs derived from double stranded RNA precursors. Origin of miRNA is endogenous from short hairpin precursor RNAs whereas siRNA are mostly exogenous from double stranded RNAs or long hairpins. These small RNAs regulate gene expression through translational suppression (post transcriptional) and/or mRNA degradation (transcriptional) by perfect/non-perfect match formed between miRNA and target mRNA (Mattick and Makunin 2005; Yekta et al. 2004; Mansfield et al. 2004). siRNA is also known to regulate gene expression by modulating chromatin structure (discussed in next section). MicroRNA genes are generally transcribed by RNA polymerase II generating primary miRNA (pri-miRNA). Pri-miRNAs are several kilobases long and possess stemloop structure. Pri-miRNAs are cleaved by RNase III, enzyme Drosha, containing multiprotein complex, producing ~70-nt hairpin precursor miRNA (pre-miRNA). Pre-miRNA is exported to cytoplasm where it is processed into ~22 nt miRNA duplex by another RNase III enzyme, Dicer (Bushati and Cohen 2007 for review). Dicer along with protein argonaute form a complex triggering the assembly of ribonucleoprotein complex called as RNA-induced silencing complex (RISC). One strand of miRNA gets incorporated into RISC and guides the complex to target RNA for base pairing. In case of perfect match with target RNA it is cleaved and if base pairing is imperfect and the binding is strong enough to hold, then the translation is repressed. Major mode of action of animal miRNA involves translational repression rather than RNA degradation unlike the plant miRNAs (Millar and Waterhouse 2005a). Target recognition of miRNA mainly depends on the stringency of base pair match at the 5′ end of miRNA called as the "seed region". Nevertheless, when the 5′ sites are dominant, it can function with or without 3′pairing support. In case of insufficient 5′ pairing in some miRNAs, the 3′ compensatory sites play their part by strong pairing with the seed region sequence.

esiRNAs: Initially endogenous siRNAs (esiRNAs) have been detected only in organisms that possess RNA-dependent RNA polymerases (RDRPs) and absent in others which lack endogenous dsRNA (Millar and Waterhouse 2005a, b). However, other sources of dsRNAs including long hairpin structures generated from the palindromic sequences and dsRNAs generated by the annealing of complementary RNAs that are synthesized by two opposing transcription units in the same loci. Such dsRNAs have now proven to be the source of esiRNAs in both *D. melanogaster* and mice (Watanabe et al. 2008; Tam et al. 2008). In Drosophila esiRNAs have been shown to play important role in the formation of heterochromatin within the somatic tissues (Fagegaltier et al. 2009). esiRNAs have also been implicated in suppressing the expression of mobile genetic elements. Mice deficient for Dicer showed elevated expression of only certain transposable elements which are believed to be affected by the esiRNA pathway but the exact mechanism is yet to be discovered (Nilsen 2008).

Viral miRNAs: These are the viral transcripts that are generally employed in processes like immune recognition, cell survival, angiogenesis, proliferation and cell differentiation upon infection of the host cells (Pfeffer et al. 2004; Gottwein et al. 2007; Grey et al. 2010). A recent review on viral miRNAs has listed the known viral miRNAs from different viral species (Plaisance-Bonstaff and Renne 2011). miRNAs in general show a higher degree of conservation but viral miRNAs on the other hand shows very poor sequence homology between viruses. Viral miRNAs targets only a small sub population of viral transcripts and obviously they target the majority of host mRNA transcripts thereby regulating their expression to substantial level to create a conducive, microenvironment for their survival and proliferation of viruses (Grey et al. 2010). Virus encoded miRNAs are known to act as suppressors of RNAi, modulating the host miRNAs and also incorporates epigenetic changes in the host which may aid in the viral oncogenesis (Scaria and Jadhav 2007).

Y RNAs: These are small noncoding RNAs that function as integral part of the Ro RNP. The Ro RNP was discovered by Lerner et al. in systemic lupus erythematosus patients. So far four Y RNA species have been discovered in humans namely hY1 (hY2 is a truncated form of hY1), hY3, hY4, and hY5 RNAs ranging in size from 83 to 112 nucleotides (Hendrick et al. 1981). Y RNAs are expressed in all vertebrate species studied (Perreault et al. 2007). Among the invertebrates Y RNA orthologues have been reported in *Caenorhabditis elegans* (Van Horn et al. 1995; Boria et al. 2010) and *Deinococcus radiodurans* (Chen et al. 2000), but no orthologues in yeasts, plants, or insects. In Deinococcus, Y RNAs are reported to be involved in 23S rRNA maturation (Chen et al. 2007). while the human Y RNAs (hY RNAs) aid in the process of chromosomal DNA replication which ultimately ensures a completely semi-conservative mode of replication throughout the genome. They have been implicated in either the initiation steps to establish an active replication forks or for elongation steps during DNA replication fork progression (Christov et al. 2006). Recently hY RNAs were shown to be even overexpressed in solid tumours, that aids in cell proliferation (Christov et al. 2008). Nevrethless the cause and consequences are not yet completely deciphered.

TSSa-RNAs: Transcription start site–associated RNAs as their name suggests are transcribed either as sense or antisense transcripts from region flanking the active promoters, with peaks of antisense and sense short RNAs peaking between nucleotides −100 and −300 nucleotides upstream and 0 to +50 nucleotides downstream of TSS, respectively. In yeast such TSSa-RNAs are called as cryptic unstable transcripts (CUTs) and stable unannotated transcripts (SUTs) (Neil et al. 2009; Wyers et al. 2005; Xu et al. 2009). TSSa-RNAs are 20–90 nt (Seila et al. 2008) in length and have been proposed to aid in maintaining poised chromatin state at the promoter regions for downstream transcriptional regulatory steps. The transcription initiation factors, RNAPII and the K4-trimethylated histone H3, occupy the same position over the chromatin where TSSa-RNA; whereas, K79-dimethylated histone H3, is located downstream of TSSs. Recently, a long promoter associated ncRNA (pncRNA) has been identified which repress the protein coding transcripts in cis via an RNA binding protein called TLS (Translocated in liposarcoma) that mediates transcription repression through HAT inhibition. Refer Sect. 15.4.3 for more details.

vRNAs: vault RNAs are integral part of the vault particles that were discovered as a vault ribonucleoprotein complex implicated in multidrug resistance and intracellular transport. Generally these are 100 bases long and transcribed by Pol III. vRNAs via a DICER mechanism generate small vault RNAs (svRNAs) that act like miRNAs in downregulating the expression of CYP3A4, an enzyme essential for drug metabolism (Persson et al. 2009).

15.4.2 Long Regulatory ncRNA

Long Regulatory ncRNA (lncRNAs) as mentioned earlier are greater than 200 nt long and both polyadenylated, and nonpolyadenylated transcripts have been reported. Apart from intronic and intergenic (linc RNAs) lncRNAs, they are also encoded from genomic regions enriched with repetitive elements, such as telomeric repeats (TelRNAs), long terminal repeat retrotransposon elements (LINE RNAs), and short interspersed nuclear elements (B2 RNA). lncRNAs often overlap with, or intersperse between the protein-coding and noncoding transcripts. Promoter-associated transcripts, such as promoter-associated long RNAs (PALRs) and promoter upstream transcripts (PROMPTs) have been recently added to the growing list of lncRNAs. Often PROMPTs overlap with PALRs in terms of the size and the distance from promoter. Also they resemble the cryptic unstable transcripts (CUTs) seen in yeasts (Neil et al. 2009). The major roles of most of the lncRNAs are implicated in transcription regulation by altering the enhancers, promoters and other regulatory regions of a gene. This is achieved either by modulating the chromatin structure around these loci or by directly binding to the transcription factors associated to these elements. In general most of these lncRNAs seem to act in a gene-specific manner and recent evidence that lncRNAs themselves may have enhancer activity was suggested by a handful of studies which still remains open for further investigation (Mondal et al. 2010; Ørom et al. 2010).

lincRNAs: The large intergenic non-coding RNAs (lincRNAs) are one among the largest members of lncRNAs which are evolutionarily highly conserved (Guttman et al. 2009). HOTAIR, was the the first lincRNA, identified by Rinn et al. (2007), showing that HOTAIR could influence gene expression in trans by binding PRC2 and targeting it to the HOXD cluster, thereby silencing target genes in HOXD cluster (Rinn et al. 2008). More than 8,000 lincRNAs are known to exist and are well conserved across mammals (Rinn et al. 2008). They are involved in diverse biological processes, like cell-cycle regulation, immune surveillance and in the maintenance of stem cell pluripotency. Often lincRNAs associate with repressive chromatin modifying complexes hence, act as repressors in transcriptional regulatory networks. The typical example being the p53 mediated global gene repression via the lincRNA-p21 triggering apoptosis by recruiting the hnRNP-K on to the defined set of p53 responsive genes (Huarte et al. 2010). Sabine Loewer et al. later described the role of lincRNA in reprogramming events during derivation of human iPSCs which is presently being described as lincRNA-RoR for 'regulator of reprogramming' (Loewer et al. 2010). These observations indicate that ncRNA has wide reach in regulation of various biological functions.

Totally intronic ncRNAs (TIN): E.M. Reis et al. identified the transcribed intronic ncRNAs (Reis et al. 2005) which are lncRNAs of approximately 0.6–2 kb in length. Later Helder I Nakaya et al. based on the *in silico* predictions available on data sets in different ncRNA databases and using the combined intron/exon oligoarrays they were able to point the intronic regions as key sources of potentially regulatory ncRNAs (Nakaya et al. 2007). They showed that TINs have tissue-specific expression signatures for human liver, prostate and kidney. The antisense TIN RNAs were transcribed from introns of protein-coding genes which are reported to be enriched in the 'Regulation of transcription' Gene Ontology category. Intronic RNAs are believed to regulate the abundance or the pattern of exon usage in protein-coding mRNAs. It has been proposed that TINs regulate the corresponding protein coding genes through transcriptional interference at promoters or through the epigenetic modulation of the chromatin architecture (Louro et al. 2009).

T-UCRs: David Haussler et al. (Bejerano et al. 2004) discovered a group of highly conserved transcripts called T-UCRs (Transcribed Ultra Conserved Regions) which do not code for any protein. There are about 481 such transcripts longer than 200 (bp) with 100% identity between the orthologous regions of the human, rat, and mouse genomes. Since these UCRs are often located at fragile sites in the chromatin and also associated to the genomic regions involved in cancers it is not surprising to link T-UCRs with tumorigenesis. It is also known that some of the UCRs' expression is regulated by microRNAs abnormally expressed in human chronic lymphocytic leukemia, and the inhibition of UCR which is overexpressed in colon cancer could even induces apoptosis (Calin et al. 2007). T-UCR expression landscape in neuroblastoma suggests widespread T-UCR involvement in diverse cellular processes that are deregulated in the process of tumourigenesis.

PROMPTs: In mammals certain long, unstable promoter upstream transcripts (PROMPTs) initiate bidirectionally ~0.5–2.5 kb upstream of transcription start sites that are longer than the TSSa-RNAs (Preker et al. 2008). This class of RNA often overlaps with another class of bidirectional promoter-associated long RNAs known as PALRs which are longer than 200 nucleotides (Kapranov et al. 2007) and are distinct from PROMTs. Interestingly, siRNA targeted to promoter upstream regions often resulted in transcriptional gene silencing. Given that promoter upstream regions associated with bidirectional transcripts, siRNA could have mediated transcriptional silencing via promoter associated transcripts targeting to RNAi pathway (Han et al. 2007). However, the functional link between the expression of PROMTs and PALRs with cognate genes is not yet clear.

GRC-RNAs: A polypurine triplet repeat-rich lncRNAs, designated as GAA repeat-containing RNAs, are ~1.5 to ~4 kb long and localize to numerous intra nuclear punctate foci that associate with GAA.TTC-repeat containing genomic regions. These foci drop in number with more differentiation of the cell type. GRC-RNAs are components of the nuclear matrix and interact with various nuclear matrix-associated proteins. In mitotic cells, GRC-RNAs localize to the midbody. The interesting part of GRC-RNA foci is that the number increases during cellular transformation (Zheng et al. 2010).

eRNAs: RNA polymerase II binding was noticed over 25% of the gene enhancers which later turned out that these occupancies were not mere landing pads rather they were more of transcription foci for the novel class of ncRNAs without polyadenylation called the Enhancer RNAs (eRNAs) (Kim et al. 2010; Ren 2010). eRNAs synthesis requires a functional promoter but the requirement of other general transcription factors or the mediator complex proteins is yet to be identified. The expression of eRNAs in the enhancer regions generally correlate with the gene activity of neighboring promoters, indicating that these transcripts may be necessary to activate the nearby promoters either by facilitating the formation of more open chromatin or via promoting enhancer promoter communications. Currently RNAi strategies are employed to decipher the precise mechanism of this class of regulatory ncRNAs.

mlncRNA: The mRNA-like ncRNAs are transcribed by Pol II and poladenylated at 3' and capped at 5' ends. Most of the members are known to be dysregulated in expression during the pathogenesis of multiple human diseases but their functional roles are yet to be assigned. Studies done so far strongly suggest that their expression is tightly regulated to specific subcellular compartments of specific tissues like brain but the exact role of these RNAs are not known (Inagaki et al. 2005; Jiang et al. 2011).

15.4.3 Small and Long Noncoding RNAs in Transcription Regulation

ncRNAs modulating transcription are abundant and were first to be discovered. Noncoding RNAs as transcriptional regulators target different components of transcription. Mostly such RNAs act in cis or trans and target general transcription factors, RNA polymerase, transcriptional activators or repressors. Here we are providing a few examples of ncRNA which regulate different steps of transcriptional process:

Bacterial 6S RNA: The *E. Coli* 6S RNA is one among the first ncRNAs to be discovered. About four decades ago 6S RNA was sequenced. It is 184 nucleotide long RNA having a conserved secondary structure containing largely double stranded and a central single stranded bulge. 6S RNA forms a stable complex with active polymerase tangled with promoter specificity factor σ^{70}. *E. Coli* 6S RNA was shown to interact with RNAP-σ^{70} complex but not with free σ^{70}, thereby suppressing transcription (Trotochaud and Wassarman 2005). Interestingly, this repression of transcription was true for only a subset of promoters, as 6S RNA can activate transcription at promoters requiring Enzyme- σ^S complex (E- σ^S is required for survival during stationary phase), indicating that 6S RNA regulates transcriptional process at multiple levels. Secondary structure of 6S RNA is essential for its activity and notably, single stranded bulge region was found to be critical for its RNAP binding and transcription modulation activity. Furthermore, 6S RNA structure mimic open promoter complex structure seen during transcriptional initiation (as shown in Fig. 15.1a) and thus proposed to inhibit transcription incorporating competition between promoter DNA and the E- σ^{70} (Barrick et al. 2005).

Mouse B2 RNA: B2 RNA is RNAP III encoded transcript, which is transcribed from short interspersed elements (SINE) of mouse genome and it represses RNAP II transcription in response to heat shock (Allen et al. 2004). B2 RNA is 178 nucleotide long and its expression increases many fold upon heat shock. B2 RNA interacts with a RNA docking site on RNAP II and assembles into the preinitiation complex at the promoter disrupting critical contacts between RNAP II and the promoter DNA, thereby inhibiting initiation of transcription (Espinoza et al. 2004). B2 RNA mediated RNAP II transcription repression shows promoter specificity. Recent investigations have explored the mechanisms underlying the B2 RNA mediated repression of RNAPII dependent transcription and found that B2 RNA targets early steps of transcription initiation like the Ser 5 phosphorylation by TFIIH (Espinoza et al. 2007). B2 RNA blocks CTD phosphorylation by TFIIH, only when RNAP II is in a transcriptionally repressed complex over the promoter DNA in an open state (Fig. 15.1b shown in green) prior to the formation of closed (Fig. 15.1b shown in yellow) complex (Yakovchuk et al. 2011).

7SK RNA: The human 7SK RNA is an abundant (2×10^5 copies/cell) evolutionarily conserved nuclear RNA of 331 nucleotides and is transcribed by RNAP III (Murphy et al. 1987 and Zieve et al. 1977). 7SK RNA controls RNAP II elongation by modulating the activity of transcription elongation factor P-TEFb (Nguyen et al. 2001). P-TEFb activates transcriptional elongation by phosphorylating C-Terminal Domain (CTD) of RNAPII. P-TEFb is a heterodimer comprising CDK9 and cyclin T1. In addition to general elongation factor, P-TEFb also functions as an HIV-1 Tat-specific transcription factor. P-TEFb interacts with Tat and the transactivating responsive (TAR) RNA structure located at the 5' end of the nascent viral transcript thus stimulating HIV-1 transcription. 7SK RNA binds to P-TEFb and represses transcription by abrogating its kinase activity. Association of P-TEFb and 7SK RNA is found to be reversible as ultraviolet irradiation and actinomycin D treatment disrupted P-TEFb/7SK RNA complex which can restore transcription (Yang et al. 2001). Further studies showed that inactivation of P-TEFb by 7SK RNA requires their association with other proteins namely MAQ1/HEXIM1 (hexamethylene bisacetamide-induced protein 1) which form the essential components of 7SK RNP. HEXIM1 was shown to inhibit P-TEFb in a 7SK-dependent manner while 7SK serves as a scaffold to mediate the HEXIM1:P-TEFb interaction (Fig. 15.2b) (Yik et al. 2003; Michels et al. 2003, 2004). A recent investigation has demonstrated that 7SK interacts with chromatin with high affinity (Mondal et al. 2010). The latter observation is consistent with the suggestion that 7SK by interacting with the chromatin serves as a scaffold for recruiting HEXIMI:P-TEFb proteins thereby inhibiting transcriptional elongation.

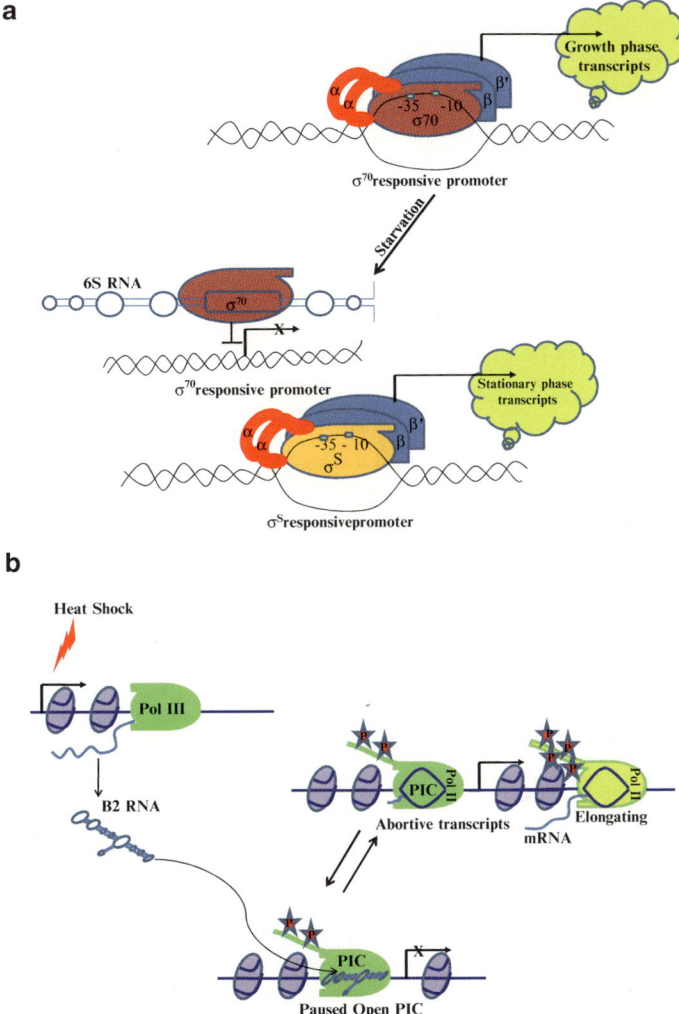

Fig. 15.1 (a) 6S RNA mimics open promoter complex. 6S RNA targets the specificity factor σ70. in E. Coli during stationary phase and sequesters from the active polymerase complex and but not free σ70 and hence, blocking transcription during stationary phase. On the other hand during stationary phase 6S RNA activates transcription at promoters requiring Enzyme- σS complex essential for the survival of bacteria. (**b**) B2RNA docks with RNAP II Preinitiation complex and blocks transcription initiation. In response to heat shock, B2 RNA is transcribed by RNAP III which binds RNA docking site of RNAP II within the paused open preinitiation complex over the promoter prior to the formation of closed complexes. This event blocks the critical contacts between RNAP II and the promoter DNA, and also represses the CTD phosphorylation (depicted in *red* stars: 2 stars and 4 stars:) by TFIIH thereby inhibiting the initiation of transcription by RNAP II

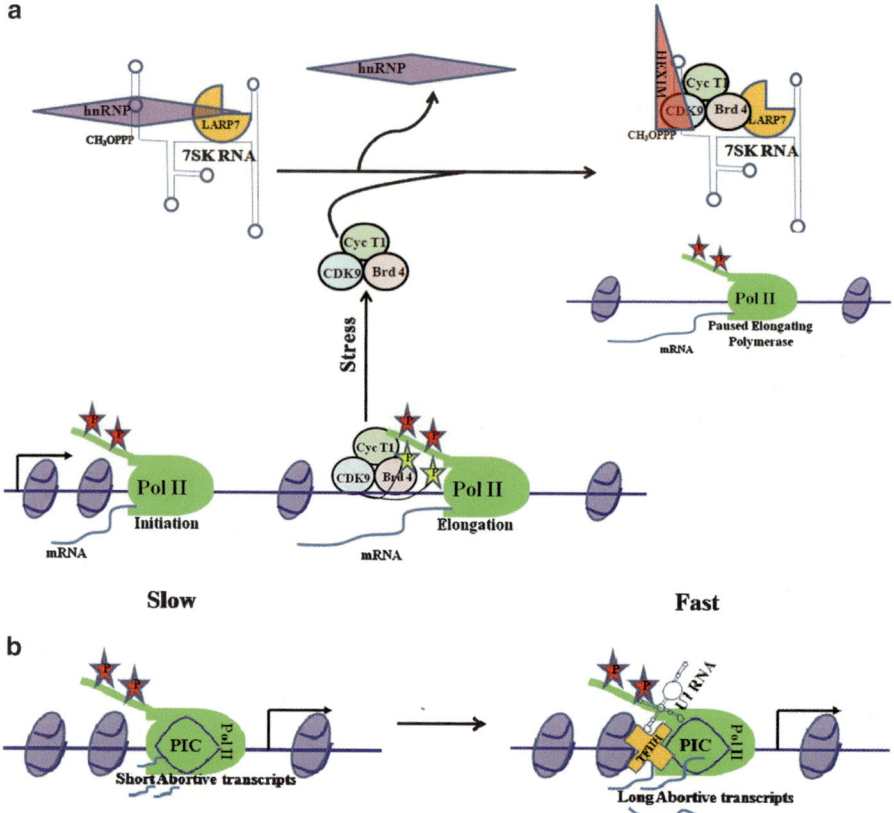

Fig. 15.2 (a) 7SK RNA facilitates HEXIM mediated inhibition of P-TEFb. P-TEFb activates transcriptional elongation by phosphorylating (depicted in *yellow* stars) C-Terminal Domain (CTD) of RNAP II. P-TEFb consists of a kinase CDK9 and cyclin T1 heterodimer along with Brd4. Upon stress, the 7SK snRNP is released from hnRNP complex and binds to P-TEFb thereby abrogating its kinase activity and repression of transcription elongation. This inactivation of P-TEFb by 7SK RNA requires their association with other proteins namely HEXIM1 (hexamethylene bisacetamide-induced protein 1) and LARP7 (La ribonucleoprotein domain family, member 7) which form the essential components of 7SK snRNP upon stress. 7SK acts as a scaffold to mediate the HEXIM1:P-TEFb interaction that in turn blocks transcription elongation. (b) U1snRNA associates with TFIIH and enhances the transcription initiation rate. U1snRNA binds directly to the cyclin-H subunit of TFIIH and stimulates the kinase activity of TFIIH to phosphorylate C-terminal domain (CTD) of RNAP II, thereby stimulating the rate of initiation

U1snRNA: U1snRNA is approximately 160 nucleotide long ncRNA, transcribed by RNAP II. U1snRNA is one among the five small nuclear RNAs (snRNAs) U1-U6 that exist in snRNPs. These snRNPs facilitate splicing by forming the spliceosome together with many other proteins (Kramer 1996; Burge et al. 1999, for reviews). U1 snRNA has been shown to be associated with one of the general transcription factors TFIIH, thereby influencing transcriptional initiation, a critical regulatory stage of gene expression. Specifically it binds directly to cyclin-H subunit of TFIIH and stimulates kinase activity of TFIIH that phosphorylates C-terminal domain (CTD) of RNAP II. Association of TFIIH with U1snRNA stimulates the rate of initiation (rate of formation of first phosphodiester bond) by RNAP II (Fig. 15.2b). Addition of 5′ splice site adjacent to promoter stimulates reinitiation of transcription in TFIIH dependent manner indicating an important role for U1snRNA in transcriptional regulation by RNAP II apart from its well established role in RNA processing (Kwek et al. 2002).

SRA RNA: The steroid receptor RNA activator (SRA) is approximately 700 nucleotide long natural ncRNA. It exists in ribonucleoprotein complexes and functions as transcriptional coactivators of several steroid-hormone receptors (Lanz et al. 1999). Characterization of distinct RNA substructures within the SRA molecule reveals six RNA motifs critical for coactivation (Lanz et al. 2002). It is not clear whether RNA motifs execute transactivation at the RNA level or in cooperation with RNA binding proteins.

HSR1: Heat-shock RNA-1 (HSR1) is a ncRNA which modulates the activity of heat-shock transcription factor 1 (HSF1) upon heat shock response. In response to heat-shock, HSF1 induces the expression of heat shock proteins. In unstressed conditions, HSF1 exist in an inactive monomeric form and upon activation they acquire trimer formation ability and DNA binding properties. HSR1 and translation elongation factor eEF1A (present as ribonucleoprotein complex) are required for HSF1 activation (Shamovsky et al. 2006). eEF1A when free, is available for interaction with HSR1 and HSF1 which as a complex can initiate the heat-shock response. HSR1–eEF1A complexes when formed would capture HSF1 released from the HSP90 complex and assist its assembly into trimers and/or increase the stability of HSF1 trimers which is considered as the active form, which triggers the transcription of heat shock responsive genes (Fig. 15.3a).

NRON RNA: An RNAi based strategy employed to fish out ncRNAs modulating the activity of nuclear factor of activated T cells (NFAT) led to the identification of NRON RNA (Willingham et al. 2005). The nuclear factor of activated T cells (NFAT) refers to a family of transcription factors important in immune responses. These factors are sensitive to calcium signalling and upon activation calcineurin dephosphorylates NFAT resulting in its nuclear import essential for activating transcription. NRON size ranges from 0.8 to 4 kb based on alternative splicing. NRON represses NFAT activity by regulating its nuclear trafficking probably with aid of various transport factors (Fig. 15.3b). Thus, NRON ncRNA provides example of transcriptional regulation not via RNA-protein interactions or activity modulation of activator but through altering subcellular localisation of the latter.

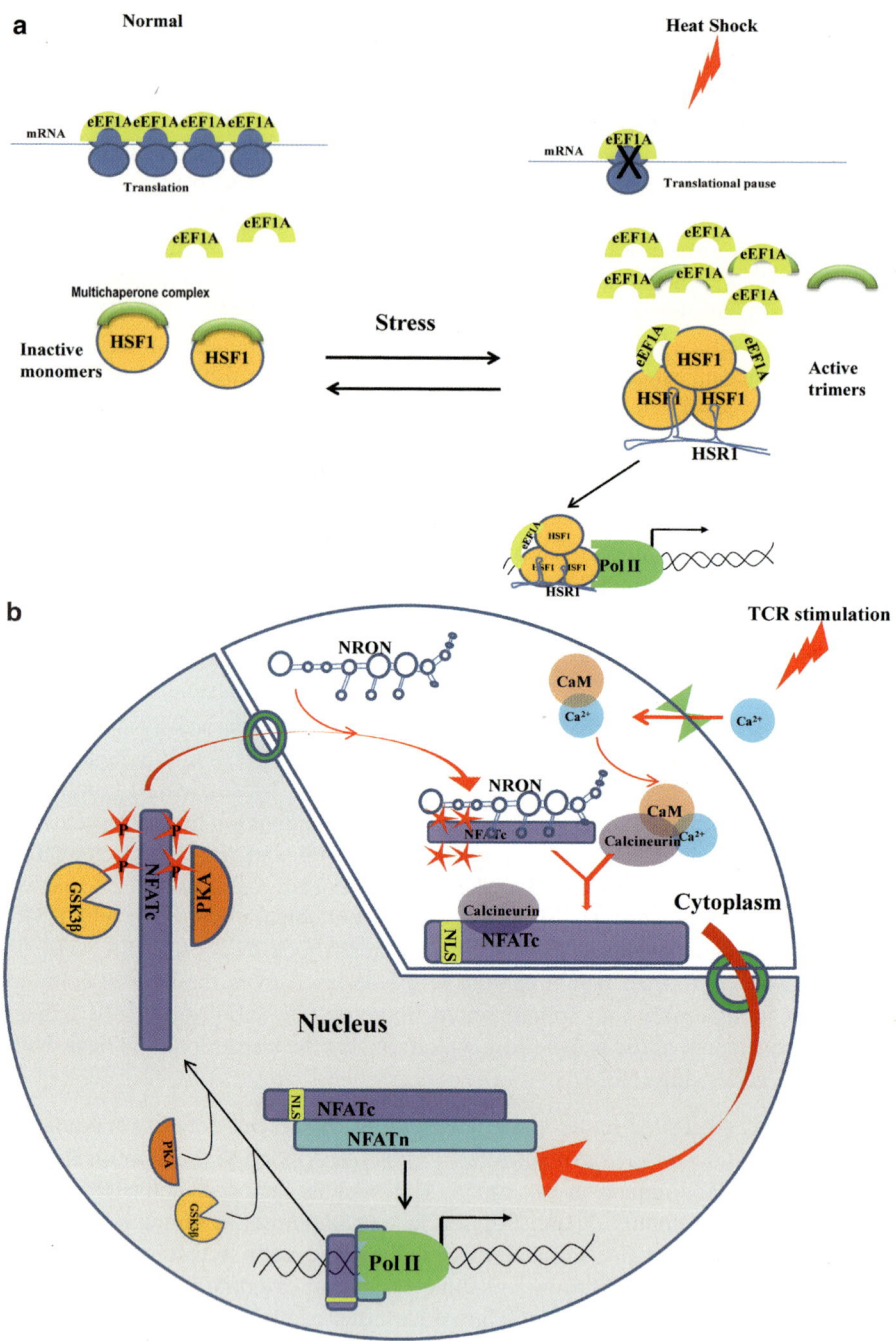

pncRNA: Cyclin D1 (CCND1) promoter is associated with lncRNAs (range in size between 200 and 400nt) which are induced in response to genotoxic factors like ionizing radiation (Wang et al. 2008). The CCND1 pncRNA interacts with an RNA binding protein TLS (Translocated in Liposarcoma) and allosterically modify its activity such that this RNA-Protein interactions exert transcriptional repression by blocking the histone acetyl transferase (HAT) activity of CPB/p300 at the repressed CCND1 promoter.

NRSE dsRNA: Neuron-restrictive silencer element double-stranded RNA (NRSE dsRNA) shares sequence complementarity to promoter element that is bound by NRSF/REST (neuron-restrictive silencing factor/RE-1-silencing transcription factor). NRSF/REST is a repressor protein known to silence neuronal genes in non neuronal cells and restricts neuronal gene expression to neurons. NRSE is a small 20 bp double stranded RNA found to activate neural gene expression thus directing neuronal lineage in stem cells (Kuwabara et al. 2004). Interestingly, activation function of NRSE dsRNA is not via base pairing to promoter element with which it shares sequence homology. Rather, it interacts with NRSF/REST and converts this repressor into transcriptional activator. It is proposed that this RNA:protein interaction might prevent association of NRSF/REST with other corepressor proteins thereby switching neuronal gene expression from repressed state in stem cells to activated state in differentiating cells.

piRNAs: Piwi-interacting RNAs (piRNAs) (24–30 nt) are yet another class of small regulatory RNAs whose functions are not fully understood. Piwi family proteins are a subtype of Argonaute proteins and forms RNA protein complex with piRNA. piRNA are found in both vertebrate and invertebrate class of animal kingdom. The best studied function of the piRNA pathway is shown in germline cells where it is involved in transcriptional silencing of retrotransposons (Aravin et al. 2007). Unlike miRNAs and siRNAs, piRNA biogenesis does not involve Dicer or RISC. Not much is known about piRNA biogenesis however, recently it has been shown that a conserved primary piRNA biogenesis pathway that acts selectively on the 3′ UTRs of messenger RNAs having a functional role in gonadal and germline development (Robine et al. 2009).

◀

Fig. 15.3 (**a**) HSR/HSF1/eEF1A trio complex induces transcription of heat shock responsive genes. In normal unstressed condition, HSF-1 exist in an inactive monomeric form along with the multichaperone complex, while the translation elongation factor eEF1A (present in ribonucleoprotein complex) aid in translation process. Upon heat shock, the eEF1A is no more engaged in translation and so they are free to interact with the HSF1 pool and the HSR1-eEF1A complex could assist its assembly into trimers. The ncRNA HSR1 interact with eEF1A-HSF1 trimers to increase their stability and induce the expression of the downstream heat shock responsive genes. (**b**) NRON blocks NFAT shuttling and inhibits NFAT mediated transcription. In normal resting condition NFAT (nuclear factor of activated T cells) remains phosphorylated and associated with the ncRNA NRON as a complex. In response to TCR stimulation the calcium ion entry activates the phosphatase calcineurin. Calcineurin further dephosphorylates NFAT and exposes the NLS, resulting in its nuclear import essential for activating transcription. Further the cytoplasmic pool is restored upon phosphorylation by kinases like GSK3β and PKA

rasiRNA : Repeat associated small interfering RNA (rasiRNA) is considered to be a subclass of piRNA and associate with both the Ago and Piwi Argonaute protein subfamily unlike piRNA which associates only with the Piwi Argonaute subfamily (Girard et al. 2006; Faehnle and Joshua-Tor 2007). Like piRNAs, rasiRNAs are abundant in germline cells and function in silencing transposons and retrotransposons as well as maintaining heterochromatin structure by controlling repeat sequences transcription (Matzke et al. 2004; Lippman and Martienssen 2004; Aravin and Tuschl 2005).

NanoRNA: NanoRNAs are one among the most recently discovered class of functional small RNAs that are believed to affect gene expression through direct incorporation into a target RNA transcript rather than through a traditional antisense-based mechanism. These nanoRNAs were discovered in *Pseudomonas aeruginosa* as 2–4 nt long oligonucleotides that function as primers for initiating transcription from a set of promoters (Goldman et al. 2011). Still the exact molecular events of gene expression, regulatory role remains open for investigation.

15.4.4 lncRNAs in Genomic Imprinting

Genomic imprinting is an epigenetic phenomenon which restricts expression of some genes to one of the two parental chromosomes. So far more than 100 imprinted genes have been identified and most of them are clustered in large chromosomal domains. The allelic expression of imprinted genes is controlled by imprint control element (ICE). ICE is epigenetically modified by DNA methylation and histone modification to regulate the expression of imprinted genes. Only unmethylated ICE is active in inducing repression of flanking genes. ICE attains methylation during gametogenesis and this germline DNA methylation is established by de novo DNA methyltransferases DNMT3A/DNAMT3L (Bourc'his and Proudhon 2008). Subsequent maintenance of methylation at ICE requires maintenance DNA methylatransferase DNMT1 (Hirasawa et al. 2008). In addition, other protein factors (specific for each ICE) also contribute to the establishment and maintenance of ICE methylation (Li et al. 2008). Histone modifications for methylated and unmethylated ICEs are found different. In general, repressive marks like H3K9Me3, H4K20me3 are associated with DNA-methylated ICE and active marks like H3K4me and H3/H4 acetylation with those of unmethylated ICE.

The mechanism by which ICE is proposed to function is either by constituting an insulator region that prevents promoter enhancer interaction or by activating ncRNA transcription. As seen in the *Igf2* imprinted cluster, a methylation sensitive insulator in the ICE regulates its expression. The chromatin insulator protein CTCF (11-zinc finger protein or CCCTC-binding factor) binds to unmethylated the ICE and prevents the communication between the enhancers downstream of the H19 gene and *Igf2* promoters (Kanduri et al. 2000a, b; Bell and Felsenfeld 2000; Hark et al. 2000). DNA methylation of the ICE prevents CTCF binding and allows the enhancer-*Igf2* promoter communication to facilitate its transcription (Kanduri et al. 2001).

In most of the imprinted gene clusters there is at least one macro ncRNA gene. Some of the tested imprinted macro ncRNA have been shown to be indispensable for the imprinted expression of whole cluster (Pauler et al. 2007; Braidotti et al. 2004). Macro ncRNAs are transcribed from unmethylated ICE. These RNAs possess some unusual features such as low intron/exon ratio i.e. reduced splicing potential, nuclear retention and accumulation at the site of transcription (Pandey et al. 2008; Braidotti et al. 2004; Terranova et al. 2008; Nagano et al. 2008). ncRNA mode of regulation is seen more common in imprinted gene cluster expression in contrast to CTCF dependent chromatin insulation mechanism. *Igf2r* and *Kcnq1* imprinted clusters have been used extensively to investigate the role of macro ncRNAs in genomic imprinting. *Igf2r* cluster harbours four imprinted genes in about 500 kb region on chromosome 17: one macro ncRNA *Airn* is exclusively expressed on the paternal chromosome and three neighboring protein coding imprinted genes, *Ig2r*, *Slc22a2* and *Slc22a3* expressed only from the maternal chromosome (Brandeis et al. 1993; Stoger et al. 1993; Lucifero et al. 2002). The unmethylated ICE on the paternal chromosome serves as promoter for paternally expressed ncRNA, *Airn* (Antisense Igf2r RNA) that overlaps *Igf2r* in antisense orientation. *Airn* ncRNA is about 108 kb long, unspliced and polyadenylated transcript. Targeted deletion of ICE, comprising *Airn* promoter, resulted in loss of silencing of all three neighboring genes on the paternal chromosome, indicating that *Airn* ncRNA plays important role in gene silencing (Wutz et al. 1997).

Kcnq1 domain is a one mega-base imprinted domain containing 8–10 imprinted protein coding genes, which are exclusively expressed from the maternal chromosome, and one lncRNA *Kcnq1ot1* expressed from the paternal chromosome. Expression of *Kcnq1ot1* on the paternal chromosome is linked to silencing of the imprinted protein coding genes (Fitzpatrick et al. 2002; Kanduri et al. 2006; DiNardo et al. 2006). However, on the maternal chromosome the imprinted protein coding genes are expressed due to silencing of *Kcnq1ot1* ncRNA promoter by CpG methylation. It has been shown that *Kcnq1ot1* itself mediates transcriptional gene silencing through interacting with chromatin remodeling machinery such as PRC2 complex members and G9a. Furthermore they are targeted specifically to imprinted gene promoters in a tissue-specific fashion thereby organizing higher order chromatin structure devoid of RNAP II (Pandey et al. 2008; Terranova et al. 2008).

Several recent studies have linked differential ncRNA expression to developmental and tissue specific expression of imprinted genes. One such study reveals that neurons do not show imprinted *Igf2r* expression due to lack of *Airn* ncRNA whereas, glial cells which express *Airn* ncRNA shows imprinting of *Igf2r* expression (Yamasaki et al. 2005). Placenta is another example of tissue specific imprinted expression. Several studies indicate the direct involvement of *Airn* and *Kcnq1ot1* macro ncRNAs in placental genes silencing. *Kcnq1ot1* physically localise to several silent genes lying away from promoter (Pandey et al. 2008). It also interacts with polycomb group proteins and establishes repressive marks. Similarly *Airn* ncRNA bind to H3K9 methyltransferase and lies in close proximity to silent *Slc22a3* promoter of *Igf2r* cluster. Deletion experiments involving *G9a* and polycomb group proteins *EZH2* and *RNF2* shows loss of placental tissue specific imprinted expression in these clusters (Nagano et al. 2008; Terranova et al. 2008; Wagschal et al. 2008).

15.4.5 lncRNAs and X-chromosome Inactivation

A best known phenomenon involving the lncRNA is X-chromosome inactivation (XCI). XCI occurs in mammalian females to ensure equal X-linked gene products between two sexes. Inactivated X chromosome expresses a ncRNA called the inactive X-specific transcript (*Xist*) that localizes and coats one of the X chromosome in cis and bring about gene silencing by establishing a higher order heterochromatic compartment. Recent studies have shown that *Xist* interacts with polycomb group proteins like *EZH2* which induces repressive marks like H3K27me and aid in gene silencing (Silva et al. 2003; Plath et al. 2003). The mechanism by which these repressive chromatin modifiers are recruited to inactive X-chromosome is unknown. On the active X-chromosome, *Xist* is repressed and its repression is carried out by a long ncRNA, *Tsix* which overlaps *Xist* in antisense orientation (Wutz and Gribnau 2007). *Tsix*, unlike Xist, silences only the *Xist* promoter on the active X chromosome. However, the mechanisms by which *Tsix* specifically regulates *Xist* repression is currently not clear. *Tsix* has also been shown to interact with epigenetic regulators such as polycomb proteins (Zhao et al. 2008) and DNA methyltransferases (Sun et al. 2006) and this interaction has been suggested to be crucial for the *Xist* repression on the active X chromosome.

15.5 ncRNA in Disease

15.5.1 An Overview

A wide variety of diseases have been discovered with altered expression or function of ncRNAs. Dyskeratosis congenita, Spinal muscular dystrophy, Autism, Alzheimer's, miR96 associated Hearing loss and Prader-Willi syndrome are some of the diseases where the small RNPs like snRNAs, miRNAs and snoRNAs are altered. The Sm-class snRNPs are not properly assembled in spinal muscular dystrophy (Selenko et al. 2001), and in dyskeratosis congenita mutations occur in telomerase RNA (Vulliamy et al. 2001). Duplication of snRNA SNORD115 is associated with Autism. In Alzheimer's disease. an antisense lncRNA (BACE1–AS) is implicated in increasing the steady state levels of its sense counterpart beta-secretase (BACE1) gene by enhancing its stability via masking certain crucial regulatory elements through sense and antisense interactions (Faghihi et al. 2008). This results in increased cleavage of amyloid precursor protein into amyloid beta1-42 which is a critical component in Alzheimer's disease. In case of Prader-Willi syndrome the paternal copies of the imprinted *SNRPN* and *Necdin* genes along with a cluster of 48 *SNORD116* coding region are deleted (Cavaillé et al. 2000; Skryabin et al. 2007; Ding et al. 2008). One other disorder where ncRNAs are implicated in the disease etiology is a rare forms of hearing disorder where the miRNA, miR-96 is aberrantly expressed (Lewis et al. 2009).

ncRNAs also mediate changes at an epigenetic level that ultimately contribute to certain disease etiology. In a rare form of β-thalassemia, a translocation juxtaposes distantly located *LUC57L* in close proximity to the α–globin gene *HBA2*. This results in transcriptional read through from the truncated *LUC57L* transcription unit and specific methylation of *HBA2* gene thus causing transcriptional silencing of *HBA2* gene (Tufarelli et al. 2003). BC1/BC200 an mRNA like ncRNA is known to be altered in the fragile X syndrome, where the loss of function of FMRP (fragile X mental retardation protein) occurs due to the absence of BC200 binding where the subsequent loss of translational repression of mRNAs in the post synaptic area of such patients (Zalfa et al. 2005). Another related ncRNA which has ancestral similarity towards BC200 called as Psoriasis-related ncRNA (PRINS) (Sonkoly et al. 2005) that like BC200 possess two Alu repetitive sequences and was implicated in Psoriasis via the down-regulation of G1P3 (Szegedi et al. 2010) but the exact mechanism is still unkown. Recent reports have shown some SNPs within the non-coding regions associated with certain disease conditions but the complex patterns of ncRNA expression makes it particularly difficult to screen such SNPs (Mattick 2009a, b).

15.5.2 ncRNA and Cancer

In the recent past there is an increasing appreciation in exploring the functional link between ncRNA expression profiles and cancer. Genome wide association studies (GWAS) have now shifted their focus towards miRNAs and lncRNAs' expression patterns in various cancers. The evidences of altered ncRNAs are often correlated well with cancers to a great extent due to the statistically valid observations made from different geographic locations and gene pools. For some of the cancers, these ncRNAs presently, serve as markers for the diagnosis and scoring the treatment regime. snoRNAs, UCRs and miRNAs are some of the commonly reported class of ncRNAs used for such purposes in cancers (Galasso et al. 2010a, b, c). Numerous lncRNAs have been shown to be altered in multiple cancers. For further reference on the list of lncRNAs and the associated cancer type refer to table 3 in ref (recently reviewed by Gibb et al. 2011) and table 1 in a review by Mattick (2009a, b). T-UCRs (Transcribe ultraconserved Regions) are a class of ncRNA that have been reported to be altered in cancers like adult CLL, colorectal carcinoma, hepatocelluar carcinoma and few neuroblastomas where these RNAs are currently being used to predict the patient prognosis with greater confidence (Braconi et al. 2011).

15.5.3 ncRNA and Therapeutics

As mentioned above, numerous ncRNAs have been implicated in the molecular pathogenesis of various human diseases, especially in cancer a special set of miRNAs possess ongenic properties which are named as "OncomiRs". Targeting these ncRNAs has always been a valid approach to contain such disorders. Unfortunately the existing information on the functional mechanism involving these ncRNAs is incomplete. The major obstacle for this lack of information is the technical difficulties faced by researchers while performing knock-down of the very few ncRNAs that have been distinctly correlated to a disease state using rigorous screening procedures. siRNA based knock-down does not hold good for ncRNAs but the LNA and PNA based AntagomiRs, and the recently developed synthetic ribozyme based enzymes that cleave specific ncRNA population are showing encouraging results. Unfortunately the efficiency of such molecules is poor. Also the delivery of these antagomiRs poses another level of complexity. Currently people are trying to solve the delivery issues using various vehicles like liposome conjugation, cholesterol conjugation, viral vector based infection and other transgenic and nanomaterial approaches (Galasso et al. 2010a, b, c).

15.6 Outlook

The last decade has been a fruitful year for the investigations on noncoding portion of genome, which previously thought to represents a junk portion of the genome. With the development of several high throughput applications such as microarrays and massive parallel sequencing, it is realized that the majority of the noncoding portion of the genome is pervasively transcribed to encode several thousands of small and long transcripts. Though there is a discrepancy as to the extent of transcription across noncoding portion of the genome, the evidence from several independent investigations provides support to the fact that noncoding transcripts are present in several thousands. Early estimates suggest that existence of about 28,000 lncRNAs and their number could grow well beyond the suggested number. Especially when we consider intronic, antisense and promoter associated transcripts. One of the major challenges associated with this huge number is that detailed physical, structural and functional characterization of each transcript. This will enable us to know the extent of transcriptional noise versus functional noncoding transcripts. Unlike protein coding RNA, lncRNA are expressed at very low level, thus posing a problem in functional annotation of lncRNAs. Hence there is a need for technologies to annotate lncRNAs expressed at low levels. Unlike small RNA mediated silencing pathways, lncRNA mediated silencing and activation pathways are ill defined. Base pair interactions primarily define the specificity of small ncRNAs. Given the absence of sequence similarity between lncRNAs and their targets, it is not clear how lncRNAs specifically activate or silence target genes. This is one of the outstanding questions that remain to be investigated. In the recent past, expression profiles of lncRNA in various cancers have been explored to identify potential prognostic and/or disgnostic markers. Like, small RNAs, lncRNAs show distinct expression profiles in various cancers. However, there is not much progress in the treatment of cancers using ncRNAs as targets. Moreover, the molecular pathways by which lncRNAs induce pathogenesis are not well investigated. Hence the molecular pathways that are affected in response to aberrant expression of lncRNAs need to be well investigated in order to devise better intervention strategies using ncRNAs as targets. Detailed functional annotation of ncRNA transcription across the genome is required in order to realize the potential of ncRNAs in mammalian development and disease.

Acknowledgments The authors would like to thank Prof. Tapas K Kundu, for his scientific and technical inputs during the preparation of this chapter. We thank Department of Science and Technology, Govt. of India, and Jawaharlal Nehru Centre for Advanced Scientific Research and Programme support, DBT, Govt. of India, for financial support. DK and SK are Senior research fellows of Council of Scientific and Industrial Research, India.

References

Abe T, Ikemura T, Sugahara J, Kanai A, Ohara Y, Uehara H, Kinouchi M, Kanaya S, Yamada Y, Muto A, Inokuchi H (2011) tRNADB-CE 2011: tRNA gene database curated manually by experts. Nucleic Acids Res 39:D210–D213

Allen TA, Von Kaenel S, Goodrich JA, Kugel JF (2004) The SINE-encoded mouse B2 RNA represses mRNA transcription in response to heat shock. Nat Struct Mol Biol 11:816–821

Amaral PP, Mattick JS (2008) Noncoding RNA in development. Mamm Genome 19:454–492

Aravin A, Tuschl T (2005) Identification and characterization of small RNAs involved in RNA silencing. FEBS Lett 579:5830–5840

Aravin AA, Hannon GJ, Brennecke J (2007) The Piwi-piRNA pathway provides an adaptive defense in the transposon arms race. Science 318:761–764

Bachellerie JP, Cavaillé J, Hüttenhofer A (2002) The expanding snoRNA world. Biochimie 84:775–790

Barrick JE, Sudarsan N, Weinberg Z, Ruzzo WL, Breaker RR (2005) 6S RNA is a widespread regulator of eubacterial RNA polymerase that resembles an open promoter. RNA 11:774–784

Bejerano G, Pheasant M, Makunin I, Stephen S, Kent WJ, Mattick JS, Haussler D (2004) Ultraconserved elements in the human genome. Science 304:1321–1325

Bell AC, Felsenfeld G (2000) Methylation of a CTCF-dependent boundary controls imprinted expression of the Igf2 gene. Nature 405:482–485

Boria I, Gruber AR, Tanzer A, Bernhart SH, Lorenz R, Mueller MM, Hofacker IL, Stadler PF (2010) Nematode sbRNAs: homologs of vertebrate Y RNAs. J Mol Evol 70:346–358

Bourc'his D, Proudhon C (2008) Sexual dimorphism in parental imprint ontogeny and contribution to embryonic development. Mol Cell Endocrinol 282:87–94

Braconi C, Kogure T, Valeri N, Huang N, Nuovo G, Costinean S, Negrini M, Miotto E, Croce CM, Patel T (2011) microRNA-29 can regulate expression of the long non-coding RNA gene MEG3 in hepatocellular cancer. Oncogene. doi:10.1038/onc.2011.193

Bradley RK, Uzilov AV, Skinner ME, Bendaña YR, Barquist L, Holmes I (2009) Evolutionary modeling and prediction of non-coding RNAs in Drosophila. PLoS One 4:e6478

Braidotti G, Baubec T, Pauler F, Seidl C, Smrzka O, Stricker S, Yotova I, Barlow DP (2004) The air noncoding RNA: an imprinted cis-silencing transcript. Cold Spring Harb Symp Quant Biol 69:55–66

Brandeis M, Kafri T, Ariel M, Chaillet JR, McCarrey J, Razin A, Cedar H (1993) The ontogeny of allele-specific methylation associated with imprinted genes in the mouse. EMBO J 12:3669–3677

Burge CB, Tuschl T, Sharp PA (1999) Splicing of precursors to mRNAs by the spliceosome. In: Gesteland RF, Cech TR, Atkins JF (eds) The RNA world, 2nd edn. Cold Spring Harbor Laboratory Press, Cold Spring Harbor, NY, pp 525–560

Bushati N, Cohen SM (2007) MicroRNA functions. Annu Rev Cell Dev Biol 23:175–205

Calin GA, Liu CG, Ferracin M, Hyslop T, Spizzo R, Sevignani C, Fabbri M, Cimmino A, Lee EJ, Wojcik SE, Shimizu M, Tili E, Rossi S, Taccioli C, Pichiorri F, Liu X, Zupo S, Herlea V, Gramantieri L, Lanza G, Alder H, Rassenti L, Volinia S, Schmittgen TD, Kipps TJ, Negrini M, Croce CM (2007) Ultraconserved regions encoding ncRNAs are altered in human leukemias and carcinomas. Cancer Cell 12:215–229

Cao F, Li X, Hiew S, Brady H, Liu Y, Dou Y (2009) Dicer independent small RNAs associate with telomeric heterochromatin. RNA 15:1274–1281

Cavaillé J, Buiting K, Kiefmann M, Lalande M, Brannan CI, Horsthemke B, Bachellerie JP, Brosius J, Hüttenhofer A (2000) Identification of brain-specific and imprinted small nucleolar RNA genes exhibiting an unusual genomic organization. Proc Natl Acad Sci U S A 97:14311–14316

Cavaillé J, Seitz H, Paulsen M, Ferguson-Smith AC, Bachellerie JP (2002) Identification of tandemly-repeated C/D snoRNA genes at the imprinted human 14q32 domain reminiscent of those at the Prader-Willi/Angelman syndrome region. Hum Mol Genet 11:1527–1538

Chen X, Quinn AM, Wolin SL (2000) Ro ribonucleoproteins contribute to the resistance of *Deinococcus radiodurans* to ultraviolet irradiation. Genes Dev 14:777–782

Chen X, Wurtmann EJ, Van Batavia J, Zybailov B, Washburn MP, Wolin SL (2007) An ortholog of the Ro autoantigen functions in 23S rRNA maturation in *D. radiodurans*. Genes Dev 21:1328–1339

Christov CP, Gardiner TJ, Szüts D, Krude T (2006) Functional requirement of noncoding Y RNAs for human chromosomal DNA replication. Mol Cell Biol 26:6993–7004

Christov CP, Trivier E, Krude T (2008) Noncoding human Y RNAs are overexpressed in tumours and required for cell proliferation. Br J Cancer 98:981–988

Crick FH (1968) The origin of the genetic code. J Mol Biol 38:367–379

Dieci G, Fiorino G, Castelnuovo M, Teichmann M, Pagano A (2007) The expanding RNA polymerase III transcriptome. Trends Genet 23:614–622

Ding F, Li HH, Zhang S, Solomon NM, Camper SA, Cohen P, Francke U (2008) SnoRNA Snord116 (Pwcr1/MBII-85) deletion causes growth deficiency and hyperphagia in mice. PLoS One 3:e1709

Dreher TW (2009) Role of tRNA-like structures in controlling plant virus replication. Virus Res 139:217–229

Eddy SR (2001) Non-coding RNA genes and the modern RNA world. Nat Rev Genet 2:919–929

ENCODE Project Consortium, Birney E, Stamatoyannopoulos JA, Dutta A et al (2007) Identification and analysis of functional elements in 1% of the human genome by the ENCODE pilot project. Nature 447:799–816

Espinoza CA, Allen TA, Hieb AR, Kugel JF, Goodrich JA (2004) B2 RNA binds directly to RNA polymerase II to repress transcript synthesis. Nat Struct Mol Biol 11:822–829

Espinoza CA, Goodrich JA, Kugel JF (2007) Characterization of the structure, function, and mechanism of B2 RNA, an ncRNA repressor of RNA polymerase II transcription. RNA 13:583–596

Faehnle CR, Joshua-Tor L (2007) Argonautes confront new small RNAs. Curr Opin Chem Biol 11:569–577

Fagegaltier D, Bougé AL, Berry B, Poisot E, Sismeiro O, Coppée JY, Théodore L, Voinnet O, Antoniewski C (2009) The endogenous siRNA pathway is involved in heterochromatin formation in Drosophila. Proc Natl Acad Sci U S A 106:21258–21263

Faghihi MA, Modarresi F, Khalil AM, Wood DE, Sahagan BG, Morgan TE, Finch CE, St Laurent G 3rd, Kenny PJ, Wahlestedt C (2008) Expression of a noncoding RNA is elevated in Alzheimer's disease and drives rapid feed-forward regulation of beta-secretase. Nat Med 14:723–730

Farh KK, Grimson A, Jan C, Lewis BP, Johnston WK, Lim LP, Burge CB, Bartel DP (2005) The widespread impact of mammalian MicroRNAs on mRNA repression and evolution. Science 310:1817–1821

Fitzpatrick GV, Soloway PD, Higgins MJ (2002) Regional loss of imprinting and growth deficiency in mice with a targeted deletion of KvDMR1. Nat Genet 32:426–431

Flicek P, Aken BL, Beal K et al (2008) Ensembl 2008. Nucleic Acids Res 36:D707–D714

Galasso D, Carnuccio A, Larghi A (2010a) Pancreatic cancer: diagnosis and endoscopic staging. Eur Rev Med Pharmacol Sci 14:375–385

Galasso F, Giannella R, Bruni P, Giulivo R, Barbini VR, Disanto V, Leonardi R, Pansadoro V, Sepe G (2010b) PCA3: a new tool to diagnose prostate cancer (PCa) and a guidance in biopsy decisions. Preliminary report of the UrOP study. Arch Ital Urol Androl 82:5–9

Galasso M, Elena Sana M, Volinia S (2010c) Non-coding RNAs: a key to future personalized molecular therapy? Genome Med 2:12–22

Gebauer F, Hentze MW (2004) Molecular mechanisms of translational control. Nat Rev Mol Cell Biol 5:827–835

Gibb EA, Brown CJ, Lam WL (2011) The functional role of long non-coding RNA in human carcinomas. Mol Cancer 10:38

Gillet R, Felden B (2001) Emerging views on tmRNA-mediated protein tagging and ribosome rescue. Mol Microbiol 42:879–885

Girard A, Sachidanandam R, Hannon GJ, Carmell MA (2006) A germline-specific class of small RNAs binds mammalian Piwi proteins. Nature 442:199–202

Goldman SR, Sharp JS, Vvedenskaya IO, Livny J, Dove SL, Nickels BE (2011) NanoRNAs prime transcription initiation in vivo. Mol Cell 42:817–825

Gottwein E, Mukherjee N, Sachse C, Frenzel C, Majoros WH, Chi JT, Braich R, Manoharan M, Soutschek J, Ohler U, Cullen BR (2007) A viral microRNA functions as an orthologue of cellular miR-155. Nature 450:1096–1099

Grey F, Tirabassi R, Meyers H, Wu G, McWeeney S, Hook L, Nelson JA (2010) A viral microRNA down-regulates multiple cell cycle genes through mRNA 5' UTRs. PLoS Pathog 6:e1000967

Guttman M, Amit I, Garber M, French C, Lin MF, Feldser D, Huarte M, Zuk O, Carey BW, Cassady JP, Cabili MN, Jaenisch R, Mikkelsen TS, Jacks T, Hacohen N, Bernstein BE, Kellis M, Regev A, Rinn JL, Lander ES (2009) Chromatin signature reveals over a thousand highly conserved large non-coding RNAs in mammals. Nature 458:223–227

Han J, Kim D, Morris KV (2007) Promoter-associated RNA is required for RNA-directed transcriptional gene silencing in human cells. Proc Natl Acad Sci U S A 104:12422–12427

Hark AT, Schoenherr CJ, Katz DJ, Ingram RS, Levorse JM, Tilghman SM (2000) CTCF mediates methylation-sensitive enhancer-blocking activity at the H19/Igf2 locus. Nature 405:486–489

Hendrick JP, Wolin SL, Rinke J, Lerner MR, Steitz JA (1981) Ro small cytoplasmic ribonucleoproteins are a subclass of La ribonucleoproteins: further characterization of the Ro and La small ribonucleoproteins from uninfected mammalian cells. Mol Cell Biol 1:1138–1149

Hirasawa R, Chiba H, Kaneda M, Tajima S, Li E, Jaenisch R, Sasaki H (2008) Maternal and zygotic Dnmt1 are necessary and sufficient for the maintenance of DNA methylation imprints during preimplantation development. Genes Dev 22:1607–1616

Huarte M, Guttman M, Feldser D, Garber M, Koziol MJ, Kenzelmann-Broz D, Khalil AM, Zuk O, Amit I, Rabani M, Attardi LD, Regev A, Lander ES, Jacks T, Rinn JL (2010) A large intergenic noncoding RNA induced by p53 mediates global gene repression in the p53 response. Cell 142:409–419

Hüttenhofer A, Schattner P, Polacek N (2005) Non-coding RNAs: hope or hype? Trends Genet 21:289–297

Inagaki S, Numata K, Kondo T, Tomita M, Yasuda K, Kanai A, Kageyama Y (2005) Identification and expression analysis of putative mRNA-like non-coding RNA in Drosophila. Genes Cells 10:1163–1173

Jády BE, Kiss T (2001) A small nucleolar guide RNA functions both in 2'-O-ribose methylation and pseudouridylation of the U5 spliceosomal RNA. EMBO J 20:541–551

Jiang ZF, Croshaw DA, Wang Y, Hey J, Machado CA (2011) Enrichment of mRNA-like noncoding RNAs in the divergence of Drosophila males. Mol Biol Evol 28:1339–1348

Kanduri C (2001) Restriction enzyme BstZ17I is sensitive to cytosine methylation. FEMS Microbiol Lett 200:191–193

Kanduri C, Holmgren C, Pilartz M, Franklin G, Kanduri M, Liu L, Ginjala V, Ulleräs E, Mattsson R, Ohlsson R (2000a) The 5′ flank of mouse H19 in an unusual chromatin conformation unidirectionally blocks enhancer-promoter communication. Curr Biol 10:449–457

Kanduri C, Pant V, Loukinov D, Pugacheva E, Qi CF, Wolffe A, Ohlsson R, Lobanenkov VV (2000b) Functional association of CTCF with the insulator upstream of the H19 gene is parent of origin-specific and methylation-sensitive. Curr Biol 10:853–856

Kanduri C, Thakur N, Pandey RR (2006) The length of the transcript encoded from the Kcnq1ot1 antisense promoter determines the degree of silencing. EMBO J 25:2096–2106

Kapranov P, Willingham AT, Gingeras TR (2007) Genome-wide transcription and the implications for genomic organization. Nat Rev Genet 8:413–423

Kavanaugh LA, Dietrich FS (2009) Non-coding RNA prediction and verification in Saccharomyces cerevisiae. PLoS Genet 5:e1000321

Kim TK, Hemberg M, Gray JM, Costa AM, Bear DM, Wu J, Harmin DA, Laptewicz M, Barbara-Haley K, Kuersten S, Markenscoff-Papadimitriou E, Kuhl D, Bito H, Worley PF, Kreiman G, Greenberg ME (2010) Widespread transcription at neuronal activity-regulated enhancers. Nature 465:182–187

Kishore S, Stamm S (2006) The snoRNA HBII-52 regulates alternative splicing of the serotonin receptor 2C. Science 311:230–232

Krämer A (1996) The structure and function of proteins involved in mammalian pre-mRNA splicing. Annu Rev Biochem 65:367–409

Kuwabara T, Hsieh J, Nakashima K, Taira K, Gage FH (2004) A small modulatory dsRNA specifies the fate of adult neural stem cells. Cell 116:779–793

Kwek KY, Murphy S, Furger A, Thomas B, O'Gorman W, Kimura H, Proudfoot NJ, Akoulitchev A (2002) U1 snRNA associates with TFIIH and regulates transcriptional initiation. Nat Struct Biol 9:800–805

Lander ES, Linton LM, Birren B et al (2001) Initial sequencing and analysis of the human genome. Nature 409:860–921

Lanz RB, McKenna NJ, Onate SA, Albrecht U, Wong J, Tsai SY, Tsai MJ, O'Malley BW (1999) A steroid receptor coactivator, SRA, functions as an RNA and is present in an SRC-1 complex. Cell 97:17–27

Lanz RB, Razani B, Goldberg AD, O'Malley BW (2002) Distinct RNA motifs are important for coactivation of steroid hormone receptors by steroid receptor RNA activator (SRA). Proc Natl Acad Sci U S A 99:16081–16086

Lewis MA, Quint E, Glazier AM, Fuchs H, De Angelis MH, Langford C, van Dongen S, Abreu-Goodger C, Piipari M, Redshaw N, Dalmay T, Moreno-Pelayo MA, Enright AJ, Steel KP (2009) An ENU-induced mutation of miR-96 associated with progressive hearing loss in mice. Nat Genet 41:614–618

Li X, Ito M, Zhou F, Youngson N, Zuo X, Leder P, Ferguson-Smith AC (2008) A maternal-zygotic effect gene, Zfp57, maintains both maternal and paternal imprints. Dev Cell 15:547–557

Lippman Z, Martienssen R (2004) The role of RNA interference in heterochromatic silencing. Nature 431:364–370

Loewer S, Cabili MN, Guttman M, Loh YH, Thomas K, Park IH, Garber M, Curran M, Onder T, Agarwal S, Manos PD, Datta S, Lander ES, Schlaeger TM, Daley GQ, Rinn JL (2010) Large intergenic non-coding RNA-RoR modulates reprogramming of human induced pluripotent stem cells. Nat Genet 42:1113–1117

Louro R, Smirnova AS, Verjovski-Almeida S (2009) Long intronic noncoding RNA transcription: expression noise or expression choice? Genomics 93:291–298

Lucifero D, Mertineit C, Clarke HJ, Bestor TH, Trasler JM (2002) Methylation dynamics of imprinted genes in mouse germ cells. Genomics 79:530–538

MacMorris M, Kumar M, Lasda E, Larsen A, Kraemer B, Blumenthal T (2007) A novel family of C. elegans snRNPs contains proteins associated with trans-splicing. RNA 13:511–520

Mancini-Dinardo D, Steele SJ, Levorse JM, Ingram RS, Tilghman SM (2006) Elongation of the Kcnq1ot1 transcript is required for genomic imprinting of neighboring genes. Genes Dev 20:1268–1282

Mans RM, Pleij CW, Bosch L (1991) tRNA-like structures. Structure, function and evolutionary significance. Eur J Biochem 201:303–324

Mansfield JH, Harfe BD, Nissen R, Obenauer J, Srineel J, Chaudhuri A, Farzan-Kashani R, Zuker M, Pasquinelli AE, Ruvkun G, Sharp PA, Tabin CJ, McManus MT (2004) MicroRNA-responsive 'sensor' transgenes uncover Hox-like and other developmentally regulated patterns of vertebrate microRNA expression. Nat Genet 36:1079–1083

Maroney PA, Yu YT, Jankowska M, Nilsen TW (1996) Direct analysis of nematode cis- and trans-spliceosomes: a functional role for U5 snRNA in spliced leader addition trans-splicing and the identification of novel Sm snRNPs. RNA 2:735–745

Matera AG, Terns RM, Terns MP (2007) Non-coding RNAs: lessons from the small nuclear and small nucleolar RNAs. Nat Rev Mol Cell Biol 8:209–220

Matsuda D, Dreher TW (2004) The tRNA-like structure of Turnip yellow mosaic virus RNA is a 3′-translational enhancer. Virology 321:36–46

Mattick JS (2009a) Has evolution learnt how to learn? EMBO Rep 10:665

Mattick JS (2009b) The genetic signatures of noncoding RNAs. PLoS Genet 5:e1000459

Mattick JS, Makunin IV (2005) Small regulatory RNAs in mammals. Hum Mol Genet 14(Spec No 1):R121-32

Matzke M, Aufsatz W, Kanno T, Daxinger L, Papp I, Mette MF, Matzke AJ (2004) Genetic analysis of RNA-mediated transcriptional gene silencing. Biochim Biophys Acta 1677:129–141

Michels AA, Nguyen VT, Fraldi A, Labas V, Edwards M, Bonnet F, Lania L, Bensaude O (2003) MAQ1 and 7SK RNA interact with CDK9/cyclin T complexes in a transcription-dependent manner. Mol Cell Biol 23:4859–4869

Michels AA, Fraldi A, Li Q, Adamson TE, Bonnet F, Nguyen VT, Sedore SC, Price JP, Price DH, Lania L, Bensaude O (2004) Binding of the 7SK snRNA turns the HEXIM1 protein into a P-TEFb (CDK9/cyclin T) inhibitor. EMBO J 23:2608–2619

Millar AA, Waterhouse PM (2005a) Plant and animal microRNAs: similarities and differences. Funct Integr Genomics 5:129–135

Millar AA, Waterhouse PM (2005b) Small RNAs: endo-siRNAs truly endogenous. Nat Rev Mol Cell Biol 9:426–427

Mondal T, Rasmussen M, Pandey GK, Isaksson A, Kanduri C (2010) Characterization of the RNA content of chromatin. Genome Res 20:899–907

Moore MJ (2005) From birth to death: the complex lives of eukaryotic mRNAs. Science 309:1514–1518

Morey C, Avner P (2004) Employment opportunities for non-coding RNAs. FEBS Lett 567:27–34

Murphy S, Di Liegro C, Melli M (1987) The in vitro transcription of the 7SK RNA gene by RNA polymerase III is dependent only on the presence of an upstream promoter. Cell 51:81–87

Nagano T, Mitchell JA, Sanz LA, Pauler FM, Ferguson-Smith AC, Feil R, Fraser P (2008) The Air noncoding RNA epigenetically silences transcription by targeting G9a to chromatin. Science 322:1717–1720

Nakaya HI, Amaral PP, Louro R, Lopes A, Fachel AA, Moreira YB, El-Jundi TA, da Silva AM, Reis EM, Verjovski-Almeida S (2007) Genome mapping and expression analyses of human intronic noncoding RNAs reveal tissue-specific patterns and enrichment in genes related to regulation of transcription. Genome Biol 8:R43

Neil H, Malabat C, d'Aubenton-Carafa Y, Xu Z, Steinmetz LM, Jacquier A (2009) Widespread bidirectional promoters are the major source of cryptic transcripts in yeast. Nature 457:1038–1042

Nguyen VT, Kiss T, Michels AA, Bensaude O (2001) 7SK small nuclear RNA binds to and inhibits the activity of CDK9/cyclin T complexes. Nature 414:322–325

Nilsen TW (2003) The spliceosome: the most complex macromolecular machine in the cell? Bioessays 25:1147–1149

Nilsen TW (2008) Endo-siRNAs: yet another layer of complexity in RNA silencing. Nat Struct Mol Biol 15:546–548

Ørom UA, Derrien T, Beringer M, Gumireddy K, Gardini A, Bussotti G, Lai F, Zytnicki M, Notredame C, Huang Q, Guigo R, Shiekhattar R (2010) Long noncoding RNAs with enhancer-like function in human cells. Cell 143:46–58

Pandey RR, Mondal T, Mohammad F, Enroth S, Redrup L, Komorowski J, Nagano T, Mancini-Dinardo D, Kanduri C (2008) Kcnq1ot1 antisense noncoding RNA mediates lineage-specific transcriptional silencing through chromatin-level regulation. Mol Cell 32:232–246

Pang KC, Frith MC, Mattick JS (2006) Rapid evolution of noncoding RNAs: lack of conservation does not mean lack of function. Trends Genet 22:1–5

Pauler FM, Koerner MV, Barlow DP (2007) Silencing by imprinted noncoding RNAs: is transcription the answer? Trends Genet 23:284–292

Perreault J, Perreault JP, Boire G (2007) Ro-associated Y RNAs in metazoans: evolution and diversification. Mol Biol Evol 24:1678–1689

Persson H, Kvist A, Vallon-Christersson J, Medstrand P, Borg A, Rovira C (2009) The non-coding RNA of the multidrug resistance-linked vault particle encodes multiple regulatory small RNAs. Nat Cell Biol 11:1268–1271

Pfeffer S, Zavolan M, Grässer FA, Chien M, Russo JJ, Ju J, John B, Enright AJ, Marks D, Sander C, Tuschl T (2004) Identification of virus-encoded microRNAs. Science 304:734–736

Plaisance-Bonstaff K, Renne R (2011) Viral miRNAs. Methods Mol Biol 721:43–66

Plath K, Fang J, Mlynarczyk-Evans SK, Cao R, Worringer KA, Wang H, de la Cruz CC, Otte AP, Panning B, Zhang Y (2003) Role of histone H3 lysine 27 methylation in X inactivation. Science 300:131–135

Preker P, Nielsen J, Kammler S, Lykke-Andersen S, Christensen MS, Mapendano CK, Schierup MH, Jensen TH (2008) RNA exosome depletion reveals transcription upstream of active human promoters. Science 322:1851–1854

Reis EM, Louro R, Nakaya HI, Verjovski-Almeida S (2005) As antisense RNA gets intronic. OMICS 9:2–12

Ren B (2010) Transcription: enhancers make non-coding RNA. Nature 465:173–174

Rinn JL, Kertesz M, Wang JK, Squazzo SL, Xu X, Brugmann SA, Goodnough LH, Helms JA, Farnham PJ, Segal E, Chang HY (2007) Functional demarcation of active and silent chromatin domains in human HOX loci by noncoding RNAs. Cell 129:1311–1323

Rinn JL, Wang JK, Allen N, Brugmann SA, Mikels AJ, Liu H, Ridky TW, Stadler HS, Nusse R, Helms JA, Chang HY (2008) A dermal HOX transcriptional program regulates site-specific epidermal fate. Genes Dev 22:303–307

Robine N, Lau NC, Balla S, Jin Z, Okamura K, Kuramochi-Miyagawa S, Blower MD, Lai EC (2009) A broadly conserved pathway generates 3'UTR-directed primary piRNAs. Curr Biol 19:2066–2076

Rosenblad MA, Larsen N, Samuelsson T, Zwieb C (2009) Kinship in the SRP RNA family. RNA Biol 6:508–516

Samarsky DA, Fournier MJ, Singer RH, Bertrand E (1998) The snoRNA box C/D motif directs nucleolar targeting and also couples snoRNA synthesis and localization. EMBO J 17:3747–3757

Scaria V, Jadhav V (2007) microRNAs in viral oncogenesis. Retrovirology 4:82

Segref A, Mattaj IW, Ohno M (2001) The evolutionarily conserved region of the U snRNA export mediator PHAX is a novel RNA-binding domain that is essential for U snRNA export. RNA 7:351–360

Seila AC, Calabrese JM, Levine SS, Yeo GW, Rahl PB, Flynn RA, Young RA, Sharp PA (2008) Divergent transcription from active promoters. Science 322:1849–1851

Selenko P, Sprangers R, Stier G, Bühler D, Fischer U, Sattler M (2001) SMN tudor domain structure and its interaction with the Sm proteins. Nat Struct Biol 8:27–31

Shamovsky I, Ivannikov M, Kandel ES, Gershon D, Nudler E (2006) RNA-mediated response to heat shock in mammalian cells. Nature 440:556–560

Silva J, Mak W, Zvetkova I, Appanah R, Nesterova TB, Webster Z, Peters AH, Jenuwein T, Otte AP, Brockdorff N (2003) Establishment of histone h3 methylation on the inactive X chromosome requires transient recruitment of Eed-Enx1 polycomb group complexes. Dev Cell 4:481–495

Skryabin BV, Gubar LV, Seeger B, Pfeiffer J, Handel S, Robeck T, Karpova E, Rozhdestvensky TS, Brosius J (2007) Deletion of the MBII-85 snoRNA gene cluster in mice results in postnatal growth retardation. PLoS Genet 3:e235

Sonkoly E, Bata-Csorgo Z, Pivarcsi A, Polyanka H, Kenderessy-Szabo A, Molnar G, Szentpali K, Bari L, Megyeri K, Mandi Y, Dobozy A, Kemeny L, Szell M (2005) Identification and characterization of a novel, psoriasis susceptibility-related noncoding RNA gene, PRINS. J Biol Chem 280:24159–24167

Stöger R, Kubicka P, Liu CG, Kafri T, Razin A, Cedar H, Barlow DP (1993) Maternal-specific methylation of the imprinted mouse Igf2r locus identifies the expressed locus as carrying the imprinting signal. Cell 73:61–71

Sun BK, Deaton AM, Lee JT (2006) A transient heterochromatic state in Xist preempts X inactivation choice without RNA stabilization. Mol Cell 21:617–628

Szegedi K, Sonkoly E, Nagy N, Németh IB, Bata-Csörgo Z, Kemény L, Dobozy A, Széll M (2010) The anti-apoptotic protein G1P3 is overexpressed in psoriasis and regulated by the non-coding RNA, PRINS. Exp Dermatol 19:269–278

Tam OH, Aravin AA, Stein P, Girard A, Murchison EP, Cheloufi S, Hodges E, Anger M, Sachidanandam R, Schultz RM, Hannon GJ (2008) Pseudogene-derived small interfering RNAs regulate gene expression in mouse oocytes. Nature 453:534–538

Terranova R, Yokobayashi S, Stadler MB, Otte AP, van Lohuizen M, Orkin SH, Peters AH (2008) Polycomb group proteins Ezh2 and Rnf2 direct genomic contraction and imprinted repression in early mouse embryos. Dev Cell 15:668–679

Trotochaud AE, Wassarman KM (2005) A highly conserved 6S RNA structure is required for regulation of transcription. Nat Struct Mol Biol 12:313–319

Tufarelli C, Stanley JA, Garrick D, Sharpe JA, Ayyub H, Wood WG, Higgs DR (2003) Transcription of antisense RNA leading to gene silencing and methylation as a novel cause of human genetic disease. Nat Genet 34:157–165

Valadkhan S, Mohammadi A, Wachtel C, Manley JL (2007) Protein-free spliceosomal snRNAs catalyze a reaction that resembles the first step of splicing. RNA 13:2300–2311

van Bakel H, Nislow C, Blencowe BJ, Hughes TR (2010) Most "dark matter" transcripts are associated with known genes. PLoS Biol 8:e1000371

Van Horn DJ, Eisenberg D, O'Brien CA, Wolin SL (1995) Caenorhabditis elegans embryos contain only one major species of Ro RNP. RNA 1:293–303

Vulliamy T, Marrone A, Goldman F, Dearlove A, Bessler M, Mason PJ, Dokal I (2001) The RNA component of telomerase is mutated in autosomal dominant dyskeratosis congenita. Nature 413:432–435

Wagschal A, Sutherland HG, Woodfine K, Henckel A, Chebli K, Schulz R, Oakey RJ, Bickmore WA, Feil R (2008) G9a histone methyltransferase contributes to imprinting in the mouse placenta. Mol Cell Biol 28:1104–1113

Wang X, Arai S, Song X, Reichart D, Du K, Pascual G, Tempst P, Rosenfeld MG, Glass CK, Kurokawa R (2008) Induced ncRNAs allosterically modify RNA-binding proteins in cis to inhibit transcription. Nature 454:126–130

Watanabe T, Totoki Y, Toyoda A, Kaneda M, Kuramochi-Miyagawa S, Obata Y, Chiba H, Kohara Y, Kono T, Nakano T, Surani MA, Sakaki Y, Sasaki H (2008) Endogenous siRNAs from naturally formed dsRNAs regulate transcripts in mouse oocytes. Nature 453:539–543

Willingham AT, Orth AP, Batalov S, Peters EC, Wen BG, Aza-Blanc P, Hogenesch JB, Schultz PG (2005) A strategy for probing the function of noncoding RNAs finds a repressor of NFAT. Science 309:1570–1573

Wutz A, Gribnau J (2007) X inactivation Xplained. Curr Opin Genet Dev 17:387–393

Wutz A, Smrzka OW, Schweifer N, Schellander K, Wagner EF, Barlow DP (1997) Imprinted expression of the Igf2r gene depends on an intronic CpG island. Nature 389:745–749

Wyers F, Rougemaille M, Badis G, Rousselle JC, Dufour ME, Boulay J, Régnault B, Devaux F, Namane A, Séraphin B, Libri D, Jacquier A (2005) Cryptic pol II transcripts are degraded by a nuclear quality control pathway involving a new poly(A) polymerase. Cell 121:725–737

Xu Z, Wei W, Gagneur J, Perocchi F, Clauder-Münster S, Camblong J, Guffanti E, Stutz F, Huber W, Steinmetz LM (2009) Bidirectional promoters generate pervasive transcription in yeast. Nature 457:1033–1037

Yakovchuk P, Goodrich JA, Kugel JF (2011) B2 RNA represses TFIIH phosphorylation of RNA polymerase II. Transcription 2:45–49

Yamasaki Y, Kayashima T, Soejima H, Kinoshita A, Yoshiura K, Matsumoto N, Ohta T, Urano T, Masuzaki H, Ishimaru T, Mukai T, Niikawa N, Kishino T (2005) Neuron-specific relaxation of

Igf2r imprinting is associated with neuron-specific histone modifications and lack of its antisense transcript air. Hum Mol Genet 14:2511–2520

Yang Z, Zhu Q, Luo K, Zhou Q (2001) The 7SK small nuclear RNA inhibits the CDK9/cyclin T1 kinase to control transcription. Nature 414:317–322

Yekta S, Shih IH, Bartel DP (2004) MicroRNA-directed cleavage of HOXB8 mRNA. Science 304:594–596

Yik JH, Chen R, Nishimura R, Jennings JL, Link AJ, Zhou Q (2003) Inhibition of P-TEFb (CDK9/ Cyclin T) kinase and RNA polymerase II transcription by the coordinated actions of HEXIM1 and 7SK snRNA. Mol Cell 12:971–982

Zalfa F, Adinolfi S, Napoli I, Kühn-Hölsken E, Urlaub H, Achsel T, Pastore A, Bagni C (2005) Fragile X mental retardation protein (FMRP) binds specifically to the brain cytoplasmic RNAs BC1/BC200 via a novel RNA-binding motif. J Biol Chem 280:33403–33410

Zhao J, Sun BK, Erwin JA, Song JJ, Lee JT (2008) Polycomb proteins targeted by a short repeat RNA to the mouse X chromosome. Science 322:750–756

Zheng R, Shen Z, Tripathi V, Xuan Z, Freier SM, Bennett CF, Prasanth SG, Prasanth KV (2010) Polypurine-repeat-containing RNAs: a novel class of long non-coding RNA in mammalian cells. J Cell Sci 123:3734–3744

Zieve G, Benecke BJ, Penman S (1977) Synthesis of two classes of small RNA species in vivo and in vitro. Biochemistry 16:4520–4525

Chapter 16
Chromatin Structure and Organization: The Relation with Gene Expression During Development and Disease

Benoît Moindrot, Philippe Bouvet, and Fabien Mongelard

Abstract The elementary level of chromatin fiber, namely the nucleofilament, is known to undergo a hierarchical compaction leading to local chromatin loops, then chromatin domains and ultimately chromosome territories. These successive folding levels rely on the formation of chromatin loops ranging from few kb to some Mb. In addition to a packaging and structural role, the high-order organization of genomes functionally impacts on gene expression program. This review summarises to which extent each level of chromatin compaction does affect gene regulation. In addition, we point out the structural and functional changes observed in diseases. Emphasis will be mainly placed on the large-scale organization of the chromatin.

The genetic material contained in the nucleus has to be transcribed, replicated and repaired all along the cell life, making unlikely the DNA molecules to be haphazardly crammed inside the nucleus. For several decades, evidence has been accumulated showing that the nucleus is highly organized and contains specialized sub-compartments. Several levels of DNA compaction/organization have been depicted: the first and most described one consists in the wrapping of 147 bp around a histone octamer, forming the "bead on a string" or "10 nm" fiber. This elemental level of organization is compacted again in a hierarchical succession of foldings which are all relevant for the understanding of gene expression. In this review, we will mainly focus on the large-scale organization of the chromatin.

B. Moindrot • P. Bouvet • F. Mongelard (✉)
Laboratoire Joliot-Curie, Centre National de la Recherche Scientifique
(CNRS)/Ecole Normale Supérieure de Lyon, Université de Lyon,
46 allée d'Italie, 69007 Lyon, France
e-mail: fabien.mongelard@ens-lyon.fr

16.1 Local Chromatin Folding and Its Impact on Specific Locus Expression

16.1.1 Local Chromatin Loops

Regulatory elements are crucial for the fine regulation of target genes that can sometimes be located several tens of kb away. To explain this long-range effect above considerable distances, looping out of intervening sequences to bring the regulatory elements in close proximity to the target gene has been proposed to be a common phenomenon.

For instance, looping seems to be a common way to allow enhancer-controlled promoter activation. The best example of such looping events was characterised on beta-globin locus by two independent approaches: Chromosome Conformation Capture (3C) (Tolhuis et al. 2002) and RNA-TRAP (Carter et al. 2002). Both methods (for review on 3C and 3C-like methods see (Simonis et al. 2007)) concluded that the Locus Control Region –LCR– of beta-globin locus contacts the beta-globin gene (Fig. 16.1a). This process is highly specific and dynamic since non globin-expressing cells do not show LCR-gene contact, and since the contacted gene changes according to the fetal-to-adult globin expression switch (Palstra et al. 2003). A promoter-enhancer interaction was also depicted in TH2-cytokine locus (Fig. 16.1b) where the LCR is found in close proximity to IL4/5/13 genes promoter (Spilianakis and Flavell 2004). These interactions are specific of the T-cell lineage and are reinforced when cytokines genes become expressed upon differentiation in TH2-lymphocytes. On this locus, SATB1, STAT6 and GATA3 are essential for both the establishment of the 3D-conformation and the cytokine expression (Spilianakis and Flavell 2004; Cai et al. 2006).

Enhancers and promoters are not the only sequences showing specific contacts. In erythroid progenitors, most of the hypersensitive sites (HS) of the beta-globin locus are aggregated (Fig. 16.1a). It has been proposed that bringing together the LCR and the distal HS creates a local microenvironment called "Active Chromatin Hub" (ACH) in which the expressed globin-gene is recruited (Palstra et al. 2003; Patrinos et al. 2004; de Laat and Grosveld 2003). Many nuclear factors are implicated in the maintenance and the formation of the beta-globin ACH, including EKLF (Drissen et al. 2004), GATA1 and FOG1 (Vakoc et al. 2005), as well as Ikaros (Keys et al. 2008). Such a clustering of regulatory elements has also been described on imprinted Igf2-H19 locus. On this locus spanning over ~100 kb, the gathering of the Differentially Methylated Regions (DMR) leads to two different chromatin configurations on both alleles, keeping the enhancer away from Igf2 promoter on the maternal allele (Fig. 16.1c) (Murrell et al. 2004; Qiu et al. 2008). As seen for the beta-globin locus, several proteins, including CTCF (Kurukuti et al. 2006) and cohesin (Nativio et al. 2009) are required to establish this chromatin conformation.

In addition to their role in gene transcription, chromatin loops have been implicated in the DNA rearrangements occurring in the loci belonging to the immunoglobulin superfamily. The V(D)J rearrangements observed on heavy chain locus, kappa light

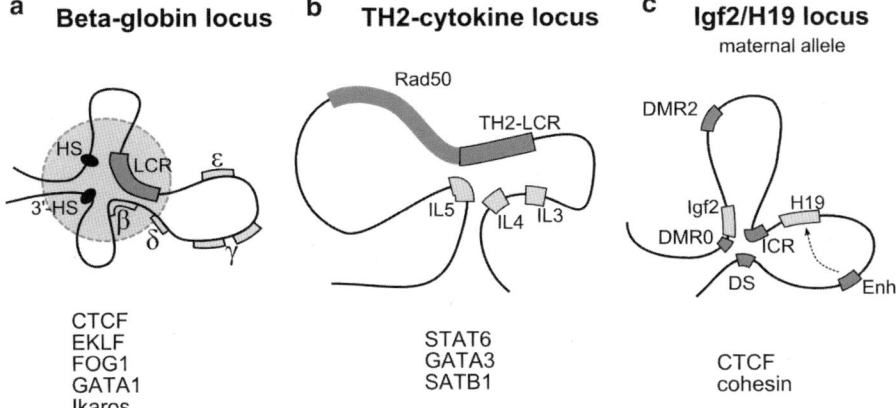

Fig. 16.1 3D-conformation of three loci. Some of the proteins involved in the stabilization/formation of the depicted 3D-conformation are indicated below the drawings. Genes are *light grey* boxes, and regulatory elements are *darker grey* boxes. (**a**) Conformation of beta-globin locus in beta-globin expressing cells. In these cells, the LCR (Locus control region) and the hypersensitive sites (HS) gather, creating a local microenvironment in which the expressed beta-globin gene is recruited (here the β gene). This local microenvironment has been called "active chromatin hub" (depicted in *light grey dashed circle*). Diagram adapted form Patrinos et al. (2004). (**b**) Conformation of the TH2-cytokine locus in the TH2 lymphocytes. The 3' end of *Rad50* corresponds to a locus control region (LCR) promoting the expression of the cytokine-encoding genes (IL3, IL4, IL5). Diagram adapted from Spilianakis and Flavell (2004). (**c**) Conformation of the maternal allele of Igf2/H19 locus. On the maternal allele, the differentially methylated region 0 (DMR0), the imprinting control region (ICR) and the CTCF binding downstream site (DS) gather. The created loops allow the enhancer (Enh) to activate the expression of H19 gene. On maternal allele, the DMR0, DMR2, ICR and DS are unmethylated (Diagram adapted from Nativio et al. (2009))

chain locus and T-Cell-Receptor locus seem to rely on a local chromatin opening (Oestreich et al. 2006; Cobb et al. 2006) and on the formation of specific chromatin loops facilitating the genomic recombination (Sayegh et al. 2005; Oestreich et al. 2006; Skok et al. 2007).

To sum up, loops of chromatin have been observed on numerous loci and are of functional importance. They maintain in close proximity regulatory elements/gene/sequences ongoing recombination through the binding of specific proteins. This functional organization has been also extensively reviewed elsewhere (Fraser 2006; Schoenfelder et al. 2010a; Sexton et al. 2009; Zlatanova and Caiafa 2009; Kadauke and Blobel 2009).

16.1.2 Targeting to the Nuclear Scaffold

The formation and the maintenance of the loops bring the question of the structural component that could provide a stable framework for the support of the chromatin loops. The nuclear matrix is operationally defined as the residual material remaining from the nucleus after a high-salt extraction that removes large amount of nuclear proteins. In those high-salt extracted nuclei, the remaining DNA after DNAseI treatment is thought to be strongly linked to the nuclear matrix and to be anchored to a residual skeleton (Razin 2001). These DNA sequences tightly bound to the nuclear matrix have been named Scaffold/Matrix Attachment Regions (S/MARs). The S/MARs are repetitive AT-rich sequences of few hundreds bp (Razin 2001; Liebich et al. 2002), and correspond to portions of DNA with a propensity for curvature (reviewed in Fiorini et al. 2006) and base unpairing (Kohwi-Shigematsu and Kohwi 1990). The S/MARs are often found in introns of large genes, in the boundaries of transcription units and near regulatory sequences (Cockerill and Garrard 1986; Gasser and Laemmli 1986; Käs and Chasin 1987; Keaton et al. 2011).

The identity of the proteins which are of functional importance for the nuclear scaffold is still debated. Nevertheless, some proteins exhibit keys properties allowing them to tether S/MARs on the nuclear matrix. Here we will focus on SATB1, SAF-A, SMAR1 and Topoisomerase II.

SATB1 displays a strong affinity for AT-rich sequences which are over-represented among S/MARs (Nakagomi et al. 1994) and the sequences bound by SATB1 are tightly anchored on nuclear matrix (de Belle et al. 1998). SATB1 also displays a "cage-like" distribution that embraces chromatin-dense regions and forms a network in the nucleus resistant to the high-salt extraction (Cai et al. 2003). All these properties clearly position SATB1 as a potential structural component of the nuclear matrix.

The knockout of SATB1 not only leads to a release of chromatin anchored loops from the matrix but also affects the expression of genes close to SATB1-binding sites. Indeed, at loci where SATB1 should be bound, *satb1*-null thymocytes display a modification of the high-order chromatin structure as well as an aberrant histone post-translational modifications pattern (Cai et al. 2003). This aberrant histone modification pattern is probably due to the ability of SATB1 to recruit in normal cells the chromatin-modification complexes (Yasui et al. 2002). This role SATB1 for anchoring chromatin loops has been clearly established on IL4/5/13 locus (Cai et al. 2006), but also on beta-globin locus (Wang et al. 2009) and MHC class I locus (Kumar et al. 2007). In the last three examples, SATB1 knock-down alters both the high-order folding of chromatin and gene expression.

Taken together, the anchoring property and the ability to recruit complexes that modify chromatin accessibility allow to consider SATB1 (and the nuclear matrix in general) as a structural framework that can simultaneously maintain the chromatin loops and control the expression of looped genes by epigenetic modifications. These findings reinforce the model (yet unproved) in which SATB1 functions as a scaffold where anchored sequences can access to a bound and common regulatory machinery.

The creation on nuclear matrix of a local microenvironment mediating gene transcription has been frequently reported. SAF-A (scaffold attachment factor A), which has also been identified as a component of the nuclear matrix (Romig et al. 1992), has been involved in the recruitment of p300/CBP on the S/MARs of the topoisomerase I locus prior to its transcription (Martens et al. 2002). The authors propose that the acetylation of nucleosomes by p300/CBP on the S/MAR sequences of topoisomerase I locus might create a localized chromatin state favouring the ensuing transcription of topoisomerase I gene.

Tethering to nuclear matrix has also been implicated in the establishment of repressive chromatin states. SMAR1, a MAR-binding protein, has been shown to repress BAX and PUMA expression upon mild DNA damage. This repression is mediated through the recruitment of HDAC1 by SMAR1 on the S/MAR sequences of PUMA and BAX genes. Interestingly, upon severe DNA damage, SMAR1 is sequestered into PML-bodies and no longer binds to the S/MARs of both genes. This alleviates the repression of BAX and PUMA genes and promotes apoptosis (Sinha et al. 2010).

Besides modulating gene transcription, anchoring sequences might also predispose to DNA recombination. First, anchored sequences on nuclear matrix often coincide with nuclear recombination hot-spots (Iarovaia et al. 2004a, b). This might be due to the presence of topoisomerase II, another MAR-binding protein which is a structural and ubiquitous component of nuclear matrix (Adachi et al. 1989). The interferon locus contains two S/MARs and is frequently recombined in various cancers. Interestingly, topoisomerase II is able to tether the two S/MARs of the interferon locus (Eivazova et al. 2009). This physical proximity of the S/MARs combined to the endonuclease activity of the topoisomerase II might favour the illegitimate recombinations frequently observed on this locus. Second, topoisomerase II and the nuclear matrix have also been involved in the caspase-independent excision of DNA during apoptosis (Solovyan et al. 2002). The last two examples illustrate that the roles of the nuclear matrix, although not fully understood, are more diverse than a passive anchoring of chromatin loops.

16.2 Chromatin Folding – From the Surrounding Environment to the Chromosome Territory

It is well established that a selected gene does not show the same expression pattern on its natural context and in an ectopic chromosomal location. Indeed, a gene in ectopic position can be expressed in only a fraction of cells (position effect variegation) or can show lower expression (stable position effect) (Yokoyama et al. 1990; Dobie et al. 1996). These findings are particularly critical for transgenic approaches and one of the strategies developed to overcome these position effects on transgenes generally consists in the addition of insulator sequences (Mlynarova et al. 1994; Alami et al. 2000; Potts et al. 2000). Besides its impact on transgenes expression, the position effects attest of the importance of genomic neighbourhood (or chromatin domain) on gene regulation.

16.2.1 Genomic Neighbourhood

Increasing evidence strongly suggests that the genomic DNA can been segmented in domains exhibiting similar gene expression levels or epigenetic marks (for a recent review, see De and Babu 2010). Indeed, transcriptome data in various species including human showed that genes are linearly clustered in adjacent groups showing similar levels of expression (Caron et al. 2001; Spellman and Rubin 2002; Versteeg et al. 2003; Mijalski et al. 2005). In human, clusters of highly expressed genes were called "Ridge" and those of weakly expressed genes "Anti-ridge" (Versteeg et al. 2003).

In addition, the profiling of different histone marks also shows that chromatin can be divided in adjacent domains exhibiting either abundance or scarcity in specific histone marks. This has been observed for the content in H3K9me2 (Wen et al. 2009), H2K27me3 (Pauler et al. 2009) and H2AK5ac (Cuddapah et al. 2009). Enrichment in CTCF has been specifically observed at the junctions from high to low content domains. These domains which are cell type-specific and thus might allow a cell-specific gene expression are conserved between mouse and human on orthologs regions (Wen et al. 2009).

The question that we should now address is the functional meaning of this segmentation in domains, and whether it really affects gene expression. Bringing clues to answer this question, Gierman et al. (2007) correlated the expression level of the same GFP transgene with its integration site in 90 different clones. Remarkably, they observed that the expression level of GFP was higher when integrated in Ridge domains (Gierman et al. 2007). Moreover, they analyzed GFP expression according to the expression level of the two closest neighbouring genes (in 5' and 3'). Strikingly, when integrated in Ridge domains, they found that highly expressed transgenes can be surrounded by either a low- or a high-expressed neighbour. This finding ruled out the possibility that GFP expression only reflects local chromatin configuration directed by the two neighbouring genes and rather emphasized the importance of a domain-wide influence.

In addition to the functional segmentation of the genome by gene expression or histone modifications, a structural segmentation has also been observed. For instance, using FISH, Ridge and Anti-ridge domains are found in different nuclear sub-compartments with very little intermingling (Goetze et al. 2007). This suggests that chromatin fiber is packed in such a way that different functional domains occupy different nuclear locations. Moreover, by coupling 3C-like assays with high-throughput sequencing, the 3D-organization of the human chromatin has been uncovered (Lieberman-Aiden et al. 2009). Using this method, called Hi-C, each interacting partners of any portion of chromatin have been identified with a resolution of 100 kb to 1 Mb. It resulted from this analysis that the bulk genome can be divided in two structural compartments showing mutual exclusions with only few cross-interactions: one correlating with open-chromatin and the other with closed-chromatin. Interestingly, at the mega-base scale, the interaction probability is compatible with an organization in fractal globules segmenting the chromatin in adjacent structural compartments. From both studies, a stimulating concept emerged which states that genomic neighbourhood might be one of the parameters driving the 3D-organization of chromatin fiber.

16.2.2 Gene-Expression in the Context of Chromosome Territory (CT)

In the nuclear space, each chromosome occupies a defined volume which has been called chromosome territory (CT) (Cremer and Cremer 2001). It has been observed that chromosomes are positioned according to their gene-density, with small, gene-rich chromosomes (chr 16, 17, 19, 20, 21, 22) exhibiting a central position in the nucleus whereas gene-poor chromosomes show a more peripheral location. This property tends to be evolutionary conserved (Tanabe et al. 2002) even if some variations have been observed from one cell type to another (Parada et al. 2004). The borders of CTs are not well defined and some intermingling between chromosomes occurs at these junctions (Branco and Pombo 2006). Combined to the non-random radial organization of CTs, the chromosomes intermingling allows specific *in-trans* contacts with neighbouring chromosomes.

Evictions of genes from the core of the spatially constrained CTs have been described. Most of the time, evictions have been correlated with high gene expression or induction (reviewed in Heard and Bickmore 2007; Fraser and Bickmore 2007). For instance the Major Histocompatibility (MHC) class II locus loops out of the chromosome 6 territory while genes become expressed (Volpi et al. 2000; Branco and Pombo 2006); a cluster of more than 40 genes required for keratinocytes differentiation is evicted from the chromosome 1 territory in keratinocytes (Williams et al. 2002); *Hoxb* gene-rich locus from mouse chromosome 11 relocates away from its original CT upon locus activation (Chambeyron and Bickmore 2004; Chambeyron et al. 2005); and a gene-rich cluster of constitutively expressed genes on 11p15.5 frequently shows extraterritorial location (Mahy et al. 2002). In each example, eviction from CT is correlated with high gene expression, either through gene activation (MHC, *Hoxb* and mouse keratinocytes differentiation cluster) or through constitutive high-expression level (mouse 11p15.5 locus). Besides, inhibition of RNA polymerase II decreases the frequency of extraterritorial location (Mahy et al. 2002), which again suggests that gene transcription is the major process driving CT eviction. Nevertheless, gene activation cannot be the only eviction force since for the *Hoxd* locus, a significant looping out has been observed in tail bud but not in limb bud even if both tissues highly express *Hoxd* genes (Morey et al. 2007). Moreover, a *Hoxb1* transgene inserted in *Hoxd* locus can promote an eviction of the CT even if the transgene is not transcribed (Morey et al. 2008). Both findings thus imply that transcription activation and eviction can be at least partially uncoupled.

The intermingling of chromosome territories and the motion of giant chromatin loops outside their original CT allow the establishment of interchromosomal (trans) contacts. A *trans*-interaction has been described in naïve T-cells between Ifng gene (a TH1-cytokine gene, mouse chromosome 10) and the TH2-cytokine genes locus (mouse chromosome 11). Upon differentiation in either TH1-lymphocytes or TH2-lymphocytes, this trans-interaction is lost concomitantly with the expression of the appropriate cytokine genes (Spilianakis et al. 2005). Furthermore, the deletion of a hypersensitive site on TH2 locus both decreases the *trans*-contact and alters *in trans* the expression program of Ifng, providing functional relevance to this physical interaction. Similar *trans*-interactions have been extensively depicted including between alpha- and beta-globin locus (Osborne et al. 2004; Brown et al. 2006, 2008), between X-inactivation centers (Bacher et al. 2006; Xu et al. 2006, 2007; Augui et al. 2007), between Igf2/H19 and Wsb1/Nf1 loci (Ling et al. 2006), between cMyc and Igh loci after the stimulation of resting B-cells (Osborne et al. 2007), and between H enhancer and olfactory receptor genes (Lomvardas et al. 2006). These chromatin *trans*-associations have been detected by both 3C-like methods and FISH experiments. The major question that remains to be addressed is about the functional relevance of these chromatin *trans*-interactions and whether they do not only reflect topological constraints inherent to chromosome folding. If the long-range *trans*-contacts do actually represent functional interactions, much care should be taken before pointing out the nuclear processes mediating these interactions (as commented in Williams et al. 2010).

The long-range contacts between loci tend to colocalize at specific subnuclear compartments (reviewed in Ferrai et al. 2010). The main nuclear compartments involved are the focal concentration of active RNA polymerase II called "transcription factories" (Cook 1999, 2002; Sutherland and Bickmore 2009). The transcription factories are believed to contain the cellular machinery required for the proper expression of active genes. The observation of active alpha- and beta-erythroid genes sharing the same transcription factory (Osborne et al. 2004; Schoenfelder et al. 2010b) has provided an attractive framework to explain their co-regulation. Nevertheless, proofs still have to be clearly established to demonstrate that an interaction is required for the proper coregulation of alpha- and beta-globin genes. Furthermore, the measured distances between "colocalized" alpha- and beta-gene is significantly above the diameter of a transcription factory, which led other authors to rather propose a localization of erythroid genes at the same nuclear splicing speckle (Brown et al. 2006, 2008).

16.3 Setting Up the High-Order Chromatin Folding During Development

The non-random distribution of CT and of genomic loci described earlier may be further exemplified by the extreme rearrangement of chromatin observed in the rod photoreceptor cells of nocturnal mammals (Solovei et al. 2009). Here, an inverted organization is found with heterochromatin occupying the interior-most part of the nucleus. The chromatin, in a very striking example of sub-cellular, lineage specific evolution, then acts as a lens, increasing photons collection and thus the overall efficiency of the eye. In mice, this peculiar architecture is established during the first month of life after the rod cells has stopped dividing.

This immediately raises the question of the nature of the driving forces able to establish and maintain this organisation during development. The movement of particular loci during development has been studied in *C. elegans* (Meister et al. 2010).

One of the main results of this study is that promoters of tissue specific genes are able to somehow impose the subnuclear position of their loci, bringing them, upon developmental activation, towards an inner position. Different hypothesis may be put forward in this regard. In their study, Meister et al. demonstrate that activation of the gene precedes its movement. It can be proposed that the ensuing modification of chromatin, induced by the recruitment of remodelling machines, helps to tether the activated loci in the interior of the nucleus, for example on speckles or on sites of active transcription. Since activation precedes relocation in this context, subnuclear localisation can only be viewed as one of the mechanisms activating or repressing genes. It should also be remembered that there is not a strict correlation between the position of a locus within the nucleus and its activity. As an evidence, it was indeed shown that only some genes display a significant decrease in transcription when experimentally tethered to the nuclear envelop (reviewed in Towbin et al. 2009).

At a larger scale, and as described earlier in this review, chromosomes with high density of genes tend to occupy a more internal position within the nuclear volume compared to their gene poor counterparts. Koehler et al. carried out elegant series of *in situ* experiments to establish the time during development at which this peculiar organization is set up for the gene-rich chr19 and the gene-poor chr20 (Koehler et al. 2009). Using *in vitro* fertilized bovine embryos, they showed that this gene density correlated positioning is not seen at 4- to 8-cell stages of bovine development, and become observable at the time of the major genome activation, i.e. 8- to 16-cell stage. They showed furthermore that this corresponds to an internalisation of chr19 rather than a displacement of chr20 towards the periphery. This study clearly establishes a correlation between transcription and large scale organisation of the nucleus. It is presently not clear if the movement of a CT is the sum of the elementary movements of each loci linked to the CT, or if a force acts on the chromosome as an entity. These possibilities, though not exclusive, are conceptuality different. To account for the first possibility, we can imagine a scenario in which individual loci will associate to nuclear structures according to, for example, their transcriptional status. This association would be established after the diffusion of the loci within the nucleus and would be stabilised thanks to the affinity of the loci (and, say, its associated transcriptional machinery) towards the nuclear structures. The sum of individual movements will eventually impose the global architecture of the nucleus. In this scenario, no molecular motor are required. Regarding the second possibility, one has to summon ATP, motors, and particular loci where generated forces should be applied.

Indeed, there are different arguments in favour of the existence of force generating, energy-consuming phenomena able to move loci or entire chromosomes. Although not in the context of development, Mehta et al. recently provided data showing that, in primary cultures of fibroblasts, chromosomes were quickly repositioned upon serum starvation (Mehta et al. 2010). This phenomenon is shown to require energy, and seems to be dependant on actin and nuclear myosin 1β. The energy dependence of nuclear organisation may find its source in transcription itself. Consistent with this view are the results of numerical simulations that suggest that entropic effects, due to the interactions of active genes in transcription factories, may be the driving force of non random CT positioning in the nucleus (Dorier and Stasiak 2010, and references therein). In this context, during development, the modification of the genes transcriptional status would thus affect the position of whole chromosome territory. On the other hand, and because actin is required for chromosome movement, it can be hypothesized that molecular motors more or less specifically rearrange the nuclear structure at different scales. In a recent review, clues in favour of this view have been presented (Gieni and Hendzel 2009).

Clearly, if common themes can be unveiled, much more remain to be fully described to get a better understanding of the establishment and maintenance of the high level of nuclear architecture.

16.4 High-Order Folding of Chromatin and Diseases

Several pathologies have been linked to the epigenetic alteration of specific genes (Fig. 16.2a) and, accordingly, a new and important research field deals with the epigenetic origins and development of diseases. The diseases caused by the mutation of genes involved in epigenetic mechanisms have been extensively reviewed (Portela and Esteller 2010; Allis et al. 2007). As this review is focused on the large-scale analysis of chromatin structure, we will only describe in the following sections how large-scale chromatin conformation can explain/be affected by diseases development (Fig. 16.2).

16.4.1 Spatial Proximity and Translocation Frequency

The Chronic Myelogenous Leukaemia (CML) was one of the first cancers linked to a genetic abnormality, namely a t(9;22) translocation that fuses part of the BCR gene to the ABL gene and leads to the formation of an oncogenic protein (Nowell 2007). Interestingly, in the haematopoietic lineage of healthy donors, BCR and ABL loci are more often close to each other than randomly expected (Lukášová et al. 1997)

Fig. 16.2 Chromatin organization and its alteration in diseases. (**a**) at the nucleosome level: loci can undergo abnormal compaction (like p16/INK4a in various cancer (Herman et al. 1995)) or abnormal decompaction (like satellite 2 in leukaemia (Fraga et al. 2005)), leading to altered expression pattern. The black lollipops indicate methylated CpG and the white ones indicate unmethylated CpG. The tick on green hexagon indicates permissive histone marks whereas the cross on red hexagon indicates repressive histone marks. (**b**) at the chromatin loop level: Some loci (as for instance Dlx5/Dlx6 locus in Rett Syndrome (Horike et al. 2005)) can undergo major changes in the long-range conformation of the chromatin. (**c**) at the chromatin domain level: a whole chromatin domain can see its histone content modified in cancerous cells (2q14.2 locus in colorectal cancer (Frigola et al. 2006)). The depicted long-range epigenetic silencing of a whole chromosome domain is likely to undergo a structural compaction. (**d**) at the nucleus level: the genomic rearrangements occurring in cancers preferentially happen between physically close genomic loci (Lukášová et al. 1997; Osborne et al. 2007; Lin et al. 2009). Here we depict a translocation occurring between two neighbouring chromosomes

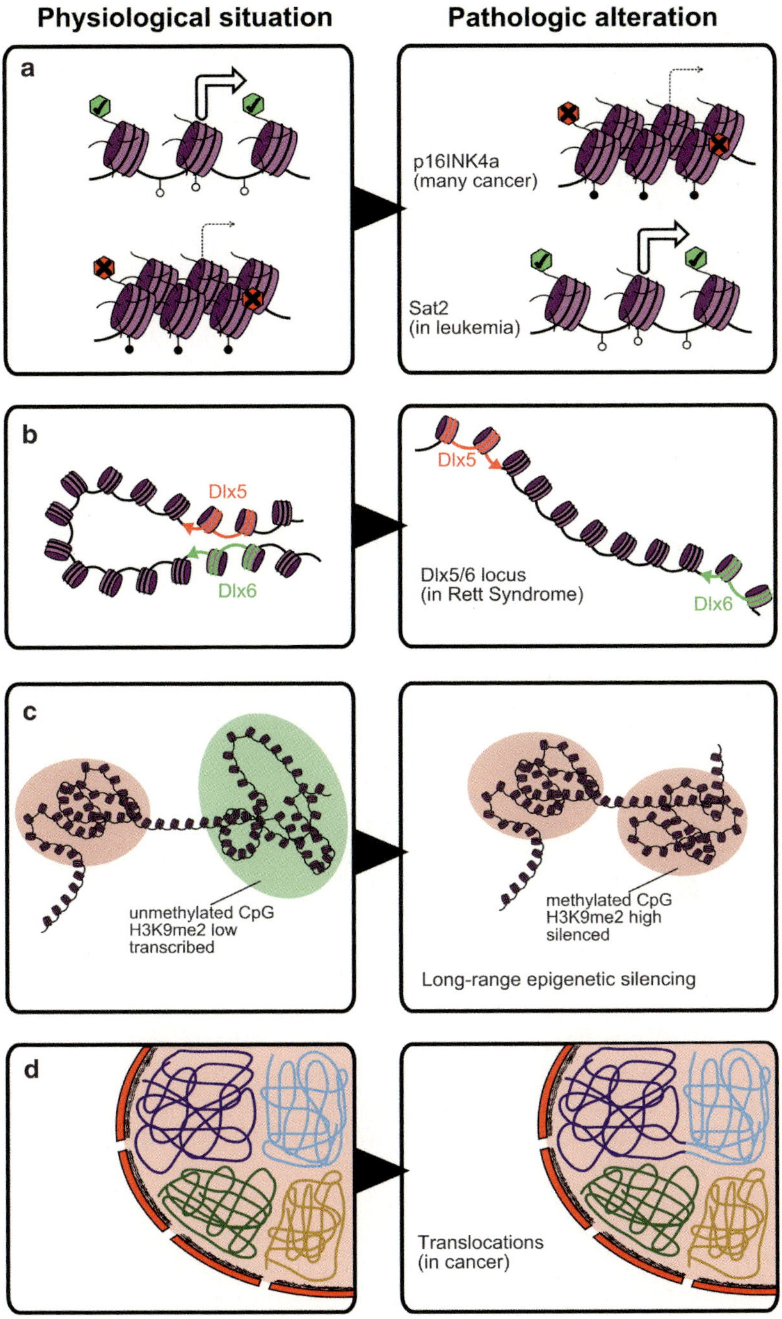

Physiological situation | **Pathologic alteration**

a

p16INK4a
(many cancer)

Sat2
(in leukemia)

b

Dlx5
Dlx6

Dlx5/6 locus
(in Rett Syndrome)

Dlx5
Dlx6

c

unmethylated CpG
H3K9me2 low
transcribed

methylated CpG
H3K9me2 high
silenced

Long-range epigenetic silencing

d

Translocations
(in cancer)

thus increasing the probability of chromatid exchange. Similarly, a close proximity has been observed between MYC locus and IGH locus (Osborne et al. 2007) facilitating the t(8;14) translocation observed in Burkitt's lymphoma.

To get insight into molecular mechanisms, Lin et al. used a prostate cell line and showed that the combination of a testosterone treatment and a genotoxic stress induces chromosomal rearrangements, with some of them often found in prostate cancer (Lin et al. 2009). Using this method, they observed an androgen-dependent interaction between two loci followed by an androgen-receptor mediated recruitment of the Non-Homologous-End-joining machinery to these loci, likely contributing to the observed chromosomal rearrangements. Since the expression of nuclear receptors is tissue-specific, we can speculate that the nuclear receptor-targeted DNA-recombination could explain the tissue-specific chromosome rearrangements observed in different cancers.

To reinforce the idea that the fusion proteins observed in cancer result from chromatid exchange between spatially close loci, ~50% of human endometrial stromal sarcomas express the fusion protein JAZF1-JJAZ1 (resulting from t(7;17) translocation) which confers anti-apoptotic properties (Li et al. 2007). Interestingly, in normal endometrial stromal cells without chromosomal rearrangement, a chimeric mRNA has been observed whose translation results in the JAZF1-JJAZ1 fusion protein (Li et al. 2008). This chimeric mRNA is believed to result from trans-splicing between the jazf1 and the jjaz1 mRNAs. One would speculate that the spatial proximity between both loci might be the key point to explain the trans-splicing in healthy cells and the chromosomal rearrangement in cancerous cells (Fig. 16.2d).

16.4.2 Change in Euchromatin/Heteochromatin Frontiers in Cancer

Epigenetic alterations play a major role in cancer progression. Abnormal DNA methylation pattern was the first epigenetic change described in cancerous cells with a global hypomethylation on repetitive sequences and a local hypermethylation of some usually unmethylated CpG islands (on tumour-suppressor genes). These epigenetic changes were linked to the mutation and/or overexpression of the DNMT proteins and of the methyl-CpG binding proteins and were extensively reviewed (Esteller 2005; Feinberg 2007; Kanwal and Gupta 2010; Berdasco and Esteller 2010; Portela and Esteller 2010).

Regarding the post-translational histone modifications, a common hallmark of cancerous cells seems to be a global loss in the acetylated H4K16 and trimethylated H4K20 (Fraga et al. 2005). These depletions are found on repetitive sequences and might contribute with DNA-hypomethylation to the expression and instability of repeated-sequences that is observed in various cancers (Gaudet et al. 2003; Ehrlich 2005). To our knowledge, the other global maps of histone modification profiles are tissue-specific and can be use as prognosis factors on specific cancer development or clinical outcome (Seligson et al. 2005, 2009). These global changes in histone modification pattern result from the alteration (both genetic and epigenetic) of every classes of histone-modifying enzymes (for recent reviews, see Kanwal and Gupta 2010; Ellis et al. 2009; Berdasco and Esteller 2010; Portela and Esteller 2010; Chi et al. 2010).

Until recently, DNA hypermethylation found in cancer was thought to be restricted to selected CpG promoters thus affecting the expression level of a single gene. However, Frigola et al. (2006) discovered a 4 Mb genomic locus that undergoes a global silencing in colorectal cancer associated to an increase in H3K9me2 histone mark. This large portion of genomic DNA contains numerous CpG islands which many of them display an abnormal hypermethylation pattern in colorectal cancer. Strikingly, most of the genes of this 4 Mb genomic locus exhibit a lowered expression in cancer even if some of them remain unmethylated (Frigola et al. 2006). This finding indicates that the lower expression of those genes is mainly due to a global silencing of the whole chromatin domain rather than to an iterative local silencing (Fig. 16.2c). Accordingly, this long-range epigenetic repression brings clues towards the functional importance of the genomic neighbourhood / chromatin domains in the comprehension of global gene regulation.

16.4.3 Change in Euchromatin/Heterochromatin Frontiers in Laminopathies

The laminopathies are genetic diseases caused by a mutation in one of the genes coding for lamina proteins, namely *LMNA* gene or *LMNB1/LMNB2* gene. Lamina proteins are principally located at the nuclear periphery that is thought to be a repressive compartment. Lamina proteins interact with chromatin (reviewed in Dorner et al. 2007; Dechat et al. 2008, 2009). Several pathologies have been linked to mutation in lamina genes (extensively reviewed in Worman et al. 2010); some of them associated with major changes in chromatin structure. For instance, in Hutchinson-Gilford Progeria Syndrome (HGPS) that is mainly caused by the C1824T mutation in *LMNA* gene, abnormalities of the nuclear periphery have been reported, including a lobulation of the nucleus and a thickening of lamina (Sandre-Giovannoli et al. 2003; Eriksson et al. 2003; Goldman et al. 2004). In addition, peripheral heterochromatin is progressively lost in an age-dependent manner. This loss is accompanied by a global decrease in H3K27me3 (related to a decrease in EZH2 level), a depletion in H3K9me(1/3) (related to a decrease in Suv39h1 and Suv39h2 levels), as well as an increase in H4K20me3 (Columbaro et al. 2005; Shumaker et al. 2006). In cells expressing this mutant version of LMNA, both H3K27me3 foci and H3K9me3 foci are redistributed throughout the nucleoplasm (Shumaker et al. 2006), leading to the conclusion that the bulk of the chromatin is reorganized.

The autosomal dominant form of Emery–Dreifuss muscular dystrophy (EDMD) due to the R453W missense mutation in *LMNA* gene, another laminopathy, shows similar alteration of the whole chromatin organization with a decrease in H3K27me3 and a decorrelation between H3K9me3 and the dense DAPI-stained pericentric heterochromatin (Håkelien et al. 2008). Interestingly, myoblasts carrying EDMD mutation fail to differentiate in myotube, and show improper expression of myogenin gene associated with altered histone marks (Håkelien et al. 2008).

Besides the global reorganization of histone marks, cells with mutations in lamin genes exhibit a relocation of selected chromosome territories. For instance, chr18 territory has been shown to exhibit a more internal location in human cells carrying mutation in *LMNA* gene or in mouse cells deficient in lamin B1 (Meaburn et al. 2007; Malhas et al. 2007). Similarly, human chr13 territory position can change in cells carrying a mutation in *LMNA* gene, preferentially toward a more internal position (E161K, R482L or R527H), even if a more peripheral location has also been described in *LMNA* D596N mutant (Meaburn et al. 2007; Mewborn et al. 2010).

Taken together, these findings point out that laminopathies can be associated with major changes in the large-scale structure of the chromatin; and thus shed light on the important role played by lamina proteins as chromatin/nucleus organizer.

16.4.4 The Case of Aging

Similarly to what happens during tumorigenesis, DNA-methylation pattern changes in an age-related manner: aged cells exhibit lower DNA-methylation level except on some specific genes exhibiting a local hypermethylation (reviewed in Fraga and Esteller 2007; Gonzalo 2010). These changes in DNA-methylation pattern likely reflect the impaired control of expression of the DNMT proteins observed in aged fibroblasts: the global loss in methyl-DNA would be related to the loss in the maintenance DNA methyltransferase DNMT1, whereas the hypermethylation might be explained by the overexpression of the *de-novo* DNA methyltransferase DNMT3b (Casillas et al. 2003). In addition to the modifications in DNA-methylation level, aged cells exhibit changes in histone modifications patterns. For instance, while comparing aged and young samples, it has been reported a global augmentation in H3K27me3, H3K79me(1/2), H4K20me3 and a global decrease in H3K36me3, H3K9me3 (Sarg et al. 2002; Scaffidi and Misteli 2006; Wang et al. 2010). In coordination with the altered pattern of histone modifications, HP1 protein and components of the nucleosome remodelling and deacetylase (NURD) complex including HDAC are lost (Scaffidi and Misteli 2006; Shen et al. 2008; Pegoraro et al. 2009), thus contributing to the global rearrangement of chromatin.

A global chromatin rearrangement is also observed in senescent cells, which leads to the formation of the so-called Senescence-Associated Heterochromatic Foci (SAHF). These foci correspond to a form of facultative heterochromatin enriched in H3K9me3, HP1, and macroH2A but depleted in open-chromatin marks (Narita et al. 2003; Zhang et al. 2005; Narita 2007). Cellular senescence is mainly studied in cultured cells, and its link with organism aging thus remains to be clearly established. Nevertheless, since senescent cells accumulate in aged animals (Herbig et al. 2006), we can hypothesize that the formation of SAHF might be an important phenomenon during aging and might reflect the global epigenetic changes associated with aging.

16.5 DNA Transcription in Coordination with DNA Replication

The duplication of the genetic material takes place while genes keep being transcribed. Although we mostly focused on the link between the high-order folding of chromatin and the gene transcription, we will now underline the interrelation between replication and transcription and their common impact on nuclear architecture.

The analysis of the nucleotide content along the genome highlighted local DNA composition asymmetries consisting in A toward T, and G toward C skews. These compositional asymmetries have been widely observed including in mammals and probably result from asymmetric mutation/repair associated with DNA transcription and replication (Green et al. 2003; Touchon et al. 2003, 2005; Brodie Of Brodie et al. 2005). Thanks to a genome-wide analysis of these asymmetries, hundreds of replication origins (ORI) were predicted in mammals (Brodie Of Brodie et al. 2005; Touchon et al. 2005). Interestingly, around the putative origins, genes are abundant, broadly expressed, mainly divergent, and in a context of open chromatin. All these features progressively weaken while increasing the distance from the origin (Huvet et al. 2007; Audit et al. 2009). This striking observation of a non-random organization of genes/chromatin in the vicinity of these predicted replication origins allows one to consider these detected ORI as privileged loci partitioning the genome in linear successive functional units.

In addition to this linear segmentation of the genome, DNA replication is also involved in some aspects of the 3D-organization of the genome. DNA replication and DNA transcription processes do not take place uniformly in the nucleus: both occur at hundreds of distinct discrete sites, called respectively replication and transcription foci (Cook 1999). Those sites can be easily labelled using analogs of (desoxy)ribonucleotides. Replication foci colocalize with replication factors (reviewed in McNairn and Gilbert 2003) and transcription foci with elongating DNA polymerase II (Iborra et al. 1996; Brown et al. 2008), which consolidates the idea that these foci represent the assemblage of replication/transcription machinery (Cook 1999, 2002). According to the classical model, each of those foci contained several sequences undergoing simultaneous transcription or replication, the sequences forming giant chromatin loops anchored on the factory (Berezney et al. 2000; Berezney 2002; Frouin et al. 2003; Chakalova et al. 2005; Sutherland and Bickmore 2009). Thus in these models, both replication and transcription machineries provide the structural basis and acting forces required to organize the genome in 3D (Chakalova et al. 2005; Fraser and Bickmore 2007; Ottaviani et al. 2008). Interestingly, if both replication and transcription are somehow able to drive the 3D-organization of the chromatin, one would expect that they could share structures. It is not fully the case since it has been observed that replication foci from early S-phase do not colocalize with transcription foci. Nevertheless, both kind of foci exhibit a high degree of proximity since it has been shown that the transcription foci mainly localized at the borders of the replication ones (Malyavantham et al. 2008). This observation suggests that the replication and the transcription could be spatially coordinated in the nucleus.

The interconnection between DNA replication and transcription is also well-established by the fact that most of the genes replicated in early S-phase are actively transcribed whereas the late-replicated ones are mostly silenced (Schübeler et al. 2002; MacAlpine et al. 2004; Woodfine et al. 2004, 2005; White et al. 2004). Moreover changes in replication timing during differentiation have been also correlated to changes in nuclear positioning, with early-to-late changes leading to a more peripheral location and a decreased expression level (Hiratani et al. 2008, 2010). It should also be noticed that chromatin assembly is slightly different in early S-phase and in late S-phase. Indeed, plasmids microinjected in early S-phase (respectively late S-phase) are rapidly assembled into a hyperacetylated (respectively hypoacetylated) chromatin (Zhang et al. 2002), while plasmids which change their replication timing also switch from one chromatin state to the other (Lande-Diner et al. 2009). These findings depict well how replication and transcription can be intimately connected. Furthermore, the self-interacting structural domains described in Hi-C experiment were interpreted in terms of open/close chromatin reminiscent of gene transcription (Lieberman-Aiden et al. 2009). However, those self-interacting structural domains also tightly correlate with the replication timing, better than any other histone marks (Ryba et al. 2010). Taken together, these results put forward the idea that DNA replication can be as important as DNA transcription to organize chromatin at different scales, from the histone modifications landscape to the structure of chromosome domains.

Acknowledgments We apologize to those whose work could not be discussed here due to space limitations.

The author's work is supported by grants from Agence Nationale de la Recherche (ANR-07-BLAN-0062-01), Région Rhône-Alpes MIRA 2007 and 2010, Association pour la Recherche sur le Cancer n° ECL2010R01122, CEFIPRA n° 3803-1, and CNRS.

References

Adachi Y, Käs E, Laemmli UK (1989) Preferential, cooperative binding of DNA topoisomerase II to scaffold-associated regions. EMBO J 8(13):3997–4006

Alami R, Greally JM, Tanimoto K, Hwang S, Feng YQ, Engel JD, Fiering S, Bouhassira EE (2000) Beta-globin YAC transgenes exhibit uniform expression levels but position effect variegation in mice. Hum Mol Genet 9(4):631–636

Allis CD, Jenuwein T, Reinberg D, Caparros ML (2007) Epigenetics. Harbor Laboratory Press, Cold Spring

Audit B, Zaghloul L, Vaillant C, Chevereau G, d'Aubenton Carafa Y, Thermes C, Arneodo A (2009) Open chromatin encoded in DNA sequence is the signature of 'master' replication origins in human cells. Nucleic Acids Res 37(18):6064–6075

Augui S, Filion GJ, Huart S, Nora E, Guggiari M, Maresca M, Stewart AF, Heard E (2007) Sensing X chromosome pairs before X inactivation via a novel X-pairing region of the Xic. Science 318(5856):1632–1636

Bacher CP, Guggiari M, Brors B, Augui S, Clerc P, Avner P, Eils R, Heard E (2006) Transient colocalization of X-inactivation centres accompanies the initiation of X inactivation. Nat Cell Biol 8(3):293–299

Berdasco M, Esteller M (2010) Aberrant epigenetic landscape in cancer: how cellular identity goes awry. Dev Cell 19(5):698–711

Berezney R (2002) Regulating the mammalian genome: the role of nuclear architecture. Adv Enzyme Regul 42:39–52

Berezney R, Dubey DD, Huberman JA (2000) Heterogeneity of eukaryotic replicons, replicon clusters, and replication foci. Chromosoma 108(8):471–484

Branco MR, Pombo A (2006) Intermingling of chromosome territories in interphase suggests role in translocations and transcription-dependent associations. PLoS Biol 4(5):e138

Brodie Of Brodie EB, Nicolay S, Touchon M, Audit B, d'Aubenton Carafa Y, Thermes C, Arneodo A (2005) From DNA sequence analysis to modeling replication in the human genome. Phys Rev Lett 94(24):248103

Brown JM, Leach J, Reittie JE, Atzberger A, Lee-Prudhoe J, Wood WG, Higgs DR, Iborra FJ, Buckle VJ (2006) Coregulated human globin genes are frequently in spatial proximity when active. J Cell Biol 172(2):177–187

Brown JM, Green J, Das Neves RP, Wallace HAC, Smith AJH, Hughes J, Gray N, Taylor S, Wood WG, Higgs DR, Iborra FJ, Buckle VJ (2008) Association between active genes occurs at nuclear speckles and is modulated by chromatin environment. J Cell Biol 182(6):1083–1097

Cai S, Han HJ, Kohwi-Shigematsu T (2003) Tissue-specific nuclear architecture and gene expression regulated by SATB1. Nat Genet 34(1):42–51

Cai S, Lee CC, Kohwi-Shigematsu T (2006) SATB1 packages densely looped, transcriptionally active chromatin for coordinated expression of cytokine genes. Nat Genet 38(11):1278–1288

Caron H, van Schaik B, van der Mee M, Baas F, Riggins G, van Sluis P, Hermus MC, van Asperen R, Boon K, Voûte PA, Heisterkamp S, van Kampen A, Versteeg R (2001) The human transcriptome map: clustering of highly expressed genes in chromosomal domains. Science 291(5507): 1289–1292

Carter D, Chakalova L, Osborne CS, feng Dai Y, Fraser P (2002) Long-range chromatin regulatory interactions in vivo. Nat Genet 32(4):623–626

Casillas MA, Lopatina N, Andrews LG, Tollefsbol TO (2003) Transcriptional control of the DNA methyltransferases is altered in aging and neoplastically-transformed human fibroblasts. Mol Cell Biochem 252(1–2):33–43

Chakalova L, Debrand E, Mitchell JA, Osborne CS, Fraser P (2005) Replication and transcription: shaping the landscape of the genome. Nat Rev Genet 6(9):669–677

Chambeyron S, Bickmore WA (2004) Chromatin decondensation and nuclear reorganization of the HoxB locus upon induction of transcription. Genes Dev 18(10):1119–1130

Chambeyron S, Silva NRD, Lawson KA, Bickmore WA (2005) Nuclear re-organisation of the Hoxb complex during mouse embryonic development. Development 132(9):2215–2223

Chi P, Allis CD, Wang GG (2010) Covalent histone modifications–miswritten, misinterpreted and mis-erased in human cancers. Nat Rev Cancer 10(7):457–469

Cobb RM, Oestreich KJ, Osipovich OA, Oltz EM (2006) Accessibility control of V(D)J recombination. Adv Immunol 91:45–109

Cockerill PN, Garrard WT (1986) Chromosomal loop anchorage of the kappa immunoglobulin gene occurs next to the enhancer in a region containing topoisomerase II sites. Cell 44(2):273–282

Columbaro M, Capanni C, Mattioli E, Novelli G, Parnaik VK, Squarzoni S, Maraldi NM, Lattanzi G (2005) Rescue of heterochromatin organization in Hutchinson-Gilford progeria by drug treatment. Cell Mol Life Sci 62(22):2669–2678

Cook PR (1999) The organization of replication and transcription. Science 284(5421):1790–1795

Cook PR (2002) Predicting three-dimensional genome structure from transcriptional activity. Nat Genet 32(3):347–352

Cremer T, Cremer C (2001) Chromosome territories, nuclear architecture and gene regulation in mammalian cells. Nat Rev Genet 2(4):292–301

Cuddapah S, Jothi R, Schones DE, Roh TY, Cui K, Zhao K (2009) Global analysis of the insulator binding protein CTCF in chromatin barrier regions reveals demarcation of active and repressive domains. Genome Res 19(1):24–32 de Belle I, Cai S, Kohwi-Shigematsu T (1998) The genomic sequences bound to special AT-rich sequence-binding protein 1 (SATB1) in vivo in Jurkat T cells are tightly associated with the nuclear matrix at the bases of the chromatin loops. J Cell Biol 141(2):335–348

de Laat W, Grosveld F (2003) Spatial organization of gene expression: the active chromatin hub. Chromosome Res 11(5):447–459

De S, Babu MM (2010) Genomic neighbourhood and the regulation of gene expression. Curr Opin Cell Biol 22(3):326–333

Dechat T, Pfleghaar K, Sengupta K, Shimi T, Shumaker DK, Solimando L, Goldman RD (2008) Nuclear lamins: major factors in the structural organization and function of the nucleus and chromatin. Genes Dev 22(7):832–853

Dechat T, Adam SA, Goldman RD (2009) Nuclear lamins and chromatin: when structure meets function. Adv Enzyme Regul 49(1):157–166

Dobie KW, Lee M, Fantes JA, Graham E, Clark AJ, Springbett A, Lathe R, McClenaghan M (1996) Variegated transgene expression in mouse mammary gland is determined by the transgene integration locus. Proc Natl Acad Sci U S A 93(13):6659–6664

Dorier J, Stasiak A (2010) The role of transcription factories-mediated interchromosomal contacts in the organization of nuclear architecture. Nucleic Acids Res 38(21):7410–7421

Dorner D, Gotzmann J, Foisner R (2007) Nucleoplasmic lamins and their interaction partners, LAP2alpha, Rb, and BAF, in transcriptional regulation. FEBS J 274(6):1362–1373

Drissen R, Palstra RJ, Gillemans N, Splinter E, Grosveld F, Philipsen S, de Laat W (2004) The active spatial organization of the beta-globin locus requires the transcription factor EKLF. Genes Dev 18(20):2485–2490

Ehrlich M (2005) DNA methylation and cancer-associated genetic instability. Adv Exp Med Biol 570:363–392

Eivazova ER, Gavrilov A, Pirozhkova I, Petrov A, Iarovaia OV, Razin SV, Lipinski M, Vassetzky YS (2009) Interaction in vivo between the two matrix attachment regions flanking a single chromatin loop. J Mol Biol 386(4):929–937

Ellis L, Atadja PW, Johnstone RW (2009) Epigenetics in cancer: targeting chromatin modifications. Mol Cancer Ther 8(6):1409–1420

Eriksson M, Brown WT, Gordon LB, Glynn MW, Singer J, Scott L, Erdos MR, Robbins CM, Moses TY, Berglund P, Dutra A, Pak E, Durkin S, Csoka AB, Boehnke M, Glover TW, Collins FS (2003) Recurrent de novo point mutations in lamin A cause Hutchinson-Gilford progeria syndrome. Nature 423(6937):293–298

Esteller M (2005) Aberrant DNA methylation as a cancer-inducing mechanism. Annu Rev Pharmacol Toxicol 45:629–656

Feinberg AP (2007) Phenotypic plasticity and the epigenetics of human disease. Nature 447(7143):433–440

Ferrai C, de Castro IJ, Lavitas L, Chotalia M, Pombo A (2010) Gene positioning. Cold Spring Harb Perspect Biol 2(6):a000588

Fiorini A, Gouveia F de S, Fernandez MA (2006) Scaffold/Matrix attachment regions and intrinsic DNA curvature. Biochemistry (Mosc) 71(5):481–488

Fraga MF, Esteller M (2007) Epigenetics and aging: the targets and the marks. Trends Genet 23(8):413–418

Fraga MF, Ballestar E, Villar-Garea A, Boix-Chornet M, Espada J, Schotta G, Bonaldi T, Haydon C, Ropero S, Petrie K, Iyer NG, Pérez-Rosado A, Calvo E, Lopez JA, Cano A, Calasanz MJ, Colomer D, Piris MA, Ahn N, Imhof A, Caldas C, Jenuwein T, Esteller M (2005) Loss of acetylation at Lys16 and trimethylation at Lys20 of histone H4 is a common hallmark of human cancer. Nat Genet 37(4):391–400

Fraser P (2006) Transcriptional control thrown for a loop. Curr Opin Genet Dev 16(5):490–495

Fraser P, Bickmore W (2007) Nuclear organization of the genome and the potential for gene regulation. Nature 447(7143):413–417

Frigola J, Song J, Stirzaker C, Hinshelwood RA, Peinado MA, Clark SJ (2006) Epigenetic remodeling in colorectal cancer results in coordinate gene suppression across an entire chromosome band. Nat Genet 38(5):540–549

Frouin I, Montecucco A, Spadari S, Maga G (2003) DNA replication: a complex matter. EMBO Rep 4(7):666–670

Gasser SM, Laemmli UK (1986) Cohabitation of scaffold binding regions with upstream/enhancer elements of three developmentally regulated genes of *D. melanogaster*. Cell 46(4):521–530

Gaudet F, Hodgson JG, Eden A, Jackson-Grusby L, Dausman J, Gray JW, Leonhardt H, Jaenisch R (2003) Induction of tumors in mice by genomic hypomethylation. Science 300(5618):489–492

Gieni RS, Hendzel MJ (2009) Actin dynamics and functions in the interphase nucleus: moving toward an understanding of nuclear polymeric actin. Biochem Cell Biol 87(1):283–306

Gierman HJ, Indemans MHG, Koster J, Goetze S, Seppen J, Geerts D, van Driel R, Versteeg R (2007) Domain-wide regulation of gene expression in the human genome. Genome Res 17(9):1286–1295

Goetze S, Mateos-Langerak J, Gierman HJ, de Leeuw W, Giromus O, Indemans MHG, Koster J, Ondrej V, Versteeg R, van Driel R (2007) The three-dimensional structure of human interphase chromosomes is related to the transcriptome map. Mol Cell Biol 27(12):4475–4487

Goldman RD, Shumaker DK, Erdos MR, Eriksson M, Goldman AE, Gordon LB, Gruenbaum Y, Khuon S, Mendez M, Varga R, Collins FS (2004) Accumulation of mutant lamin A causes progressive changes in nuclear architecture in Hutchinson-Gilford progeria syndrome. Proc Natl Acad Sci U S A 101(24):8963–8968

Gonzalo S (2010) Epigenetic alterations in aging. J Appl Physiol 109(2):586–597

Green P, Ewing B, Miller W, Thomas PJ, Program NISCCS, Green ED (2003) Transcription-associated mutational asymmetry in mammalian evolution. Nat Genet 33(4):514–517

Håkelien AM, Delbarre E, Gaustad KG, Buendia B, Collas P (2008) Expression of the myodystrophic R453W mutation of lamin A in C2C12 myoblasts causes promoter-specific and global epigenetic defects. Exp Cell Res 314(8):1869–1880

Heard E, Bickmore W (2007) The ins and outs of gene regulation and chromosome territory organisation. Curr Opin Cell Biol 19(3):311–316

Herbig U, Ferreira M, Condel L, Carey D, Sedivy JM (2006) Cellular senescence in aging primates. Science 311(5765):1257

Herman JG, Merlo A, Mao L, Lapidus RG, Issa JP, Davidson NE, Sidransky D, Baylin SB (1995) Inactivation of the CDKN2/p16/MTS1 gene is frequently associated with aberrant DNA methylation in all common human cancers. Cancer Res 55(20):4525–4530

Hiratani I, Ryba T, Itoh M, Yokochi T, Schwaiger M, Chang CW, Lyou Y, Townes TM, Schübeler D, Gilbert DM (2008) Global reorganization of replication domains during embryonic stem cell differentiation. PLoS Biol 6(10):e245

Hiratani I, Ryba T, Itoh M, Rathjen J, Kulik M, Papp B, Fussner E, Bazett-Jones DP, Plath K, Dalton S, Rathjen PD, Gilbert DM (2010) Genome-wide dynamics of replication timing revealed by in vitro models of mouse embryogenesis. Genome Res 20(2):155–169

Horike S, Cai S, Miyano M, Cheng JF, Kohwi-Shigematsu T (2005) Loss of silent-chromatin looping and impaired imprinting of DLX5 in Rett syndrome. Nat Genet 37(1):31–40

Huvet M, Nicolay S, Touchon M, Audit B, d'Aubenton Carafa Y, Arneodo A, Thermes C (2007) Human gene organization driven by the coordination of replication and transcription. Genome Res 17(9):1278–1285

Iarovaia OV, Bystritskiy A, Ravcheev D, Hancock R, Razin SV (2004a) Visualization of individual DNA loops and a map of loop domains in the human dystrophin gene. Nucleic Acids Res 32(7):2079–2086

Iarovaia OV, Shkumatov P, Razin SV (2004b) Breakpoint cluster regions of the AML-1 and ETO genes contain MAR elements and are preferentially associated with the nuclear matrix in proliferating HEL cells. J Cell Sci 117(Pt 19):4583–4590

Iborra FJ, Pombo A, Jackson DA, Cook PR (1996) Active RNA polymerases are localized within discrete transcription 'factories' in human nuclei. J Cell Sci 109(Pt 6):1427–1436

Kadauke S, Blobel GA (2009) Chromatin loops in gene regulation. Biochim Biophys Acta 1789(1):17–25

Kanwal R, Gupta S (2010) Epigenetics and cancer. J Appl Physiol 109(2):598–605

Käs E, Chasin LA (1987) Anchorage of the Chinese hamster dihydrofolate reductase gene to the nuclear scaffold occurs in an intragenic region. J Mol Biol 198(4):677–692

Keaton MA, Taylor CM, Layer RM, Dutta A (2011) Nuclear scaffold attachment sites within ENCODE regions associate with actively transcribed genes. PLoS One 6(3):e17912

Keys JR, Tallack MR, Zhan Y, Papathanasiou P, Goodnow CC, Gaensler KM, Crossley M, Dekker J, Perkins AC (2008) A mechanism for Ikaros regulation of human globin gene switching. Br J Haematol 141(3):398–406

Koehler D, Zakhartchenko V, Froenicke L, Stone G, Stanyon R, Wolf E, Cremer T, Brero A (2009) Changes of higher order chromatin arrangements during major genome activation in bovine preimplantation embryos. Exp Cell Res 315(12):2053–2063

Kohwi-Shigematsu T, Kohwi Y (1990) Torsional stress stabilizes extended base unpairing in suppressor sites flanking immunoglobulin heavy chain enhancer. Biochemistry 29(41):9551–9560

Kumar PP, Bischof O, Purbey PK, Notani D, Urlaub H, Dejean A, Galande S (2007) Functional interaction between PML and SATB1 regulates chromatin-loop architecture and transcription of the MHC class I locus. Nat Cell Biol 9(1):45–56

Kurukuti S, Tiwari VK, Tavoosidana G, Pugacheva E, Murrell A, Zhao Z, Lobanenkov V, Reik W, Ohlsson R (2006) CTCF binding at the H19 imprinting control region mediates maternally inherited higher-order chromatin conformation to restrict enhancer access to Igf2. Proc Natl Acad Sci U S A 103(28):10684–10689

Lande-Diner L, Zhang J, Cedar H (2009) Shifts in replication timing actively affect histone acetylation during nucleosome reassembly. Mol Cell 34(6):767–774

Li H, Ma X, Wang J, Koontz J, Nucci M, Sklar J (2007) Effects of rearrangement and allelic exclusion of JJAZ1/SUZ12 on cell proliferation and survival. Proc Natl Acad Sci U S A 104(50):20001–20006

Li H, Wang J, Mor G, Sklar J (2008) A neoplastic gene fusion mimics trans-splicing of RNAs in normal human cells. Science 321(5894):1357–1361

Lieberman-Aiden E, van Berkum NL, Williams L, Imakaev M, Ragoczy T, Telling A, Amit I, Lajoie BR, Sabo PJ, Dorschner MO, Sandstrom R, Bernstein B, Bender MA, Groudine M, Gnirke A, Stamatoyannopoulos J, Mirny LA, Lander ES, Dekker J (2009) Comprehensive mapping of long-range interactions reveals folding principles of the human genome. Science 326(5950):289–293

Liebich I, Bode J, Reuter I, Wingender E (2002) Evaluation of sequence motifs found in scaffold/matrix-attached regions (S/MARs). Nucleic Acids Res 30(15):3433–3442

Lin C, Yang L, Tanasa B, Hutt K, Ju BG, Ohgi K, Zhang J, Rose DW, Fu XD, Glass CK, Rosenfeld MG (2009) Nuclear receptor-induced chromosomal proximity and DNA breaks underlie specific translocations in cancer. Cell 139(6):1069–1083

Ling JQ, Li T, Hu JF, Vu TH, Chen HL, Qiu XW, Cherry AM, Hoffman AR (2006) CTCF mediates interchromosomal colocalization between Igf2/H19 and Wsb1/Nf1. Science 312(5771):269–272

Lomvardas S, Barnea G, Pisapia DJ, Mendelsohn M, Kirkland J, Axel R (2006) Interchromosomal interactions and olfactory receptor choice. Cell 126(2):403–413

Lukášová E, Kozubek S, Kozubek M, Kjeronská J, Rýznar L, Horáková J, Krahulcová E, Horneck G (1997) Localisation and distance between ABL and BCR genes in interphase nuclei of bone marrow cells of control donors and patients with chronic myeloid leukaemia. Hum Genet 100(5–6):525–535

MacAlpine DM, Rodríguez HK, Bell SP (2004) Coordination of replication and transcription along a Drosophila chromosome. Genes Dev 18(24):3094–3105

Mahy NL, Perry PE, Bickmore WA (2002) Gene density and transcription influence the localization of chromatin outside of chromosome territories detectable by FISH. J Cell Biol 159(5):753–763

Malhas A, Lee CF, Sanders R, Saunders NJ, Vaux DJ (2007) Defects in lamin B1 expression or processing affect interphase chromosome position and gene expression. J Cell Biol 176(5):593–603

Malyavantham KS, Bhattacharya S, Alonso WD, Acharya R, Berezney R (2008) Spatio-temporal dynamics of replication and transcription sites in the mammalian cell nucleus. Chromosoma 117(6):553–567

Martens JHA, Verlaan M, Kalkhoven E, Dorsman JC, Zantema A (2002) Scaffold/matrix attachment region elements interact with a p300-scaffold attachment factor A complex and are bound by acetylated nucleosomes. Mol Cell Biol 22(8):2598–2606

McNairn AJ, Gilbert DM (2003) Epigenomic replication: linking epigenetics to DNA replication. Bioessays 25(7):647–656

Meaburn KJ, Cabuy E, Bonne G, Levy N, Morris GE, Novelli G, Kill IR, Bridger JM (2007) Primary laminopathy fibroblasts display altered genome organization and apoptosis. Aging Cell 6(2):139–153

Mehta IS, Amira M, Harvey AJ, Bridger JM (2010) Rapid chromosome territory relocation by nuclear motor activity in response to serum removal in primary human fibroblasts. Genome Biol 11(1):R5

Meister P, Towbin BD, Pike BL, Ponti A, Gasser SM (2010) The spatial dynamics of tissue-specific promoters during C. elegans development. Genes Dev 24(8):766–782

Mewborn SK, Puckelwartz MJ, Abuisneineh F, Fahrenbach JP, Zhang Y, MacLeod H, Dellefave L, Pytel P, Selig S, Labno CM, Reddy K, Singh H, McNally E (2010) Altered chromosomal positioning, compaction, and gene expression with a lamin A/C gene mutation. PLoS One 5(12):e14342

Mijalski T, Harder A, Halder T, Kersten M, Horsch M, Strom TM, Liebscher HV, Lottspeich F, de Angelis MH, Beckers J (2005) Identification of coexpressed gene clusters in a comparative analysis of transcriptome and proteome in mouse tissues. Proc Natl Acad Sci U S A 102(24):8621–8626

Mlynarova L, Loonen A, Heldens J, Jansen RC, Keizer P, Stiekema WJ, Nap JP (1994) Reduced position effect in mature transgenic plants conferred by the chicken lysozyme matrix-associated region. Plant Cell 6(3):417–426

Morey C, Silva NRD, Perry P, Bickmore WA (2007) Nuclear reorganisation and chromatin decondensation are conserved, but distinct, mechanisms linked to Hox gene activation. Development 134(5):909–919

Morey C, Silva NRD, Kmita M, Duboule D, Bickmore WA (2008) Ectopic nuclear reorganisation driven by a Hoxb1 transgene transposed into Hoxd. J Cell Sci 121(Pt 5):571–577

Murrell A, Heeson S, Reik W (2004) Interaction between differentially methylated regions partitions the imprinted genes Igf2 and H19 into parent-specific chromatin loops. Nat Genet 36(8):889–893

Nakagomi K, Kohwi Y, Dickinson LA, Kohwi-Shigematsu T (1994) A novel DNA-binding motif in the nuclear matrix attachment DNA-binding protein SATB1. Mol Cell Biol 14(3):1852–1860

Narita M (2007) Cellular senescence and chromatin organisation. Br J Cancer 96(5):686–691

Narita M, Nunez S, Heard E, Narita M, Lin AW, Hearn SA, Spector DL, Hannon GJ, Lowe SW (2003) Rb-mediated heterochromatin formation and silencing of E2F target genes during cellular senescence. Cell 113(6):703–716

Nativio R, Wendt KS, Ito Y, Huddleston JE, Uribe-Lewis S, Woodfine K, Krueger C, Reik W, Peters JM, Murrell A (2009) Cohesin is required for higher-order chromatin conformation at the imprinted IGF2-H19 locus. PLoS Genet 5(11):e1000739

Nowell PC (2007) Discovery of the Philadelphia chromosome: a personal perspective. J Clin Invest 117(8):2033–2035

Oestreich KJ, Cobb RM, Pierce S, Chen J, Ferrier P, Oltz EM (2006) Regulation of TCRbeta gene assembly by a promoter/enhancer holocomplex. Immunity 24(4):381–391

Osborne CS, Chakalova L, Brown KE, Carter D, Horton A, Debrand E, Goyenechea B, Mitchell JA, Lopes S, Reik W, Fraser P (2004) Active genes dynamically colocalize to shared sites of ongoing transcription. Nat Genet 36(10):1065–1071

Osborne CS, Chakalova L, Mitchell JA, Horton A, Wood AL, Bolland DJ, Corcoran AE, Fraser P (2007) Myc dynamically and preferentially relocates to a transcription factory occupied by Igh. PLoS Biol 5(8):e192

Ottaviani D, Lever E, Takousis P, Sheer D (2008) Anchoring the genome. Genome Biol 9(1):201

Palstra RJ, Tolhuis B, Splinter E, Nijmeijer R, Grosveld F, de Laat W (2003) The beta-globin nuclear compartment in development and erythroid differentiation. Nat Genet 35(2):190–194

Parada LA, McQueen PG, Misteli T (2004) Tissue-specific spatial organization of genomes. Genome Biol 5(7):R44

Patrinos GP, de Krom M, de Boer E, Langeveld A, Imam AMA, Strouboulis J, de Laat W, Grosveld FG (2004) Multiple interactions between regulatory regions are required to stabilize an active chromatin hub. Genes Dev 18(12):1495–1509

Pauler FM, Sloane MA, Huang R, Regha K, Koerner MV, Tamir I, Sommer A, Aszodi A, Jenuwein T, Barlow DP (2009) H3K27me3 forms BLOCs over silent genes and intergenic regions and specifies a histone banding pattern on a mouse autosomal chromosome. Genome Res 19(2):221–233

Pegoraro G, Kubben N, Wickert U, Göhler H, Hoffmann K, Misteli T (2009) Ageing-related chromatin defects through loss of the NURD complex. Nat Cell Biol 11(10):1261–1267

Portela A, Esteller M (2010) Epigenetic modifications and human disease. Nat Biotechnol 28(10):1057–1068

Potts W, Tucker D, Wood H, Martin C (2000) Chicken beta-globin 5'HS4 insulators function to reduce variability in transgenic founder mice. Biochem Biophys Res Commun 273(3):1015–1018

Qiu X, Vu TH, Lu Q, Ling JQ, Li T, Hou A, Wang SK, Chen HL, Hu JF, Hoffman AR (2008) A complex DNA looping configuration associated with the silencing of the maternal Igf2 allele. Mol Endocrinol 22:1476–1488

Razin SV (2001) The nuclear matrix and chromosomal DNA loops: is their any correlation between partitioning of the genome into loops and functional domains? Cell Mol Biol Lett 6(1):59–69

Romig H, Fackelmayer FO, Renz A, Ramsperger U, Richter A (1992) Characterization of SAF-A, a novel nuclear DNA binding protein from HeLa cells with high affinity for nuclear matrix/scaffold attachment DNA elements. EMBO J 11(9):3431–3440

Ryba T, Hiratani I, Lu J, Itoh M, Kulik M, Zhang J, Schulz TC, Robins AJ, Dalton S, Gilbert DM (2010) Evolutionarily conserved replication timing profiles predict long-range chromatin interactions and distinguish closely related cell types. Genome Res 20(6):761–770

Sandre-Giovannoli AD, Bernard R, Cau P, Navarro C, Amiel J, Boccaccio I, Lyonnet S, Stewart CL, Munnich A, Merrer ML, Lévy N (2003) Lamin a truncation in Hutchinson-Gilford progeria. Science 300(5628):2055

Sarg B, Koutzamani E, Helliger W, Rundquist I, Lindner HH (2002) Postsynthetic trimethylation of histone H4 at lysine 20 in mammalian tissues is associated with aging. J Biol Chem 277(42):39195–39201

Sayegh CE, Sayegh C, Jhunjhunwala S, Riblet R, Murre C (2005) Visualization of looping involving the immunoglobulin heavy-chain locus in developing B cells. Genes Dev 19(3):322–327

Scaffidi P, Misteli T (2006) Lamin A-dependent nuclear defects in human aging. Science 312(5776):1059–1063

Schoenfelder S, Clay I, Fraser P (2010a) The transcriptional interactome: gene expression in 3D. Curr Opin Genet Dev 20(2):127–133

Schoenfelder S, Sexton T, Chakalova L, Cope NF, Horton A, Andrews S, Kurukuti S, Mitchell JA, Umlauf D, Dimitrova DS, Eskiw CH, Luo Y, Wei CL, Ruan Y, Bieker JJ, Fraser P (2010b) Preferential associations between co-regulated genes reveal a transcriptional interactome in erythroid cells. Nat Genet 42(1):53–61

Schübeler D, Scalzo D, Kooperberg C, van Steensel B, Delrow J, Groudine M (2002) Genome-wide DNA replication profile for *Drosophila melanogaster*: a link between transcription and replication timing. Nat Genet 32(3):438–442

Seligson DB, Horvath S, Shi T, Yu H, Tze S, Grunstein M, Kurdistani SK (2005) Global histone modification patterns predict risk of prostate cancer recurrence. Nature 435(7046):1262–1266

Seligson DB, Horvath S, McBrian MA, Mah V, Yu H, Tze S, Wang Q, Chia D, Goodglick L, Kurdistani SK (2009) Global levels of histone modifications predict prognosis in different cancers. Am J Pathol 174(5):1619–1628

Sexton T, Bantignies F, Cavalli G (2009) Genomic interactions: chromatin loops and gene meeting points in transcriptional regulation. Semin Cell Dev Biol 20(7):849–855

Shen S, Liu A, Li J, Wolubah C, Casaccia-Bonnefil P (2008) Epigenetic memory loss in aging oligodendrocytes in the corpus callosum. Neurobiol Aging 29(3):452–463

Shumaker DK, Dechat T, Kohlmaier A, Adam SA, Bozovsky MR, Erdos MR, Eriksson M, Goldman AE, Khuon S, Collins FS, Jenuwein T, Goldman RD (2006) Mutant nuclear lamin A leads to progressive alterations of epigenetic control in premature aging. Proc Natl Acad Sci U S A 103(23):8703–8708

Simonis M, Kooren J, de Laat W (2007) An evaluation of 3c-based methods to capture DNA interactions. Nat Methods 4(11):895–901

Sinha S, Malonia SK, Mittal SPK, Singh K, Kadreppa S, Kamat R, Mukhopadhyaya R, Pal JK, Chattopadhyay S (2010) Coordinated regulation of p53 apoptotic targets BAX and PUMA by SMAR1 through an identical MAR element. EMBO J 29(4):830–842

Skok JA, Gisler R, Novatchkova M, Farmer D, de Laat W, Busslinger M (2007) Reversible contraction by looping of the Tcra and Tcrb loci in rearranging thymocytes. Nat Immunol 8(4):378–387

Solovei I, Kreysing M, Lanctôt C, Kösem S, Peichl L, Cremer T, Guck J, Joffe B (2009) Nuclear architecture of rod photoreceptor cells adapts to vision in mammalian evolution. Cell 137(2):356–368

Solovyan VT, Bezvenyuk ZA, Salminen A, Austin CA, Courtney MJ (2002) The role of topoisomerase II in the excision of DNA loop domains during apoptosis. J Biol Chem 277(24):21458–21467

Spellman PT, Rubin GM (2002) Evidence for large domains of similarly expressed genes in the Drosophila genome. J Biol 1(1):5

Spilianakis CG, Flavell RA (2004) Long-range intrachromosomal interactions in the T helper type 2 cytokine locus. Nat Immunol 5(10):1017–1027

Spilianakis CG, Lalioti MD, Town T, Lee GR, Flavell RA (2005) Interchromosomal associations between alternatively expressed loci. Nature 435(7042):637–645

Sutherland H, Bickmore WA (2009) Transcription factories: gene expression in unions? Nat Rev Genet 10(7):457–466

Tanabe H, Müller S, Neusser M, von Hase J, Calcagno E, Cremer M, Solovei I, Cremer C, Cremer T (2002) Evolutionary conservation of chromosome territory arrangements in cell nuclei from higher primates. Proc Natl Acad Sci U S A 99(7):4424–4429

Tolhuis B, Palstra RJ, Splinter E, Grosveld F, de Laat W (2002) Looping and interaction between hypersensitive sites in the active beta-globin locus. Mol Cell 10(6):1453–1465

Touchon M, Nicolay S, Arneodo A, d'Aubenton Carafa Y, Thermes C (2003) Transcription-coupled TA and GC strand asymmetries in the human genome. FEBS Lett 555(3):579–582

Touchon M, Nicolay S, Audit B, Of Brodie EBB, d'Aubenton Carafa Y, Arneodo A, Thermes C (2005) Replication-associated strand asymmetries in mammalian genomes: toward detection of replication origins. Proc Natl Acad Sci U S A 102(28):9836–9841

Towbin BD, Meister P, Gasser SM (2009) The nuclear envelope–a scaffold for silencing? Curr Opin Genet Dev 19(2):180–186

Vakoc CR, Letting DL, Gheldof N, Sawado T, Bender MA, Groudine M, Weiss MJ, Dekker J, Blobel GA (2005) Proximity among distant regulatory elements at the beta-globin locus requires GATA-1 and FOG-1. Mol Cell 17(3):453–462

Versteeg R, van Schaik BDC, van Batenburg MF, Roos M, Monajemi R, Caron H, Bussemaker HJ, van Kampen AHC (2003) The human transcriptome map reveals extremes in gene density, intron length, GC content, and repeat pattern for domains of highly and weakly expressed genes. Genome Res 13(9):1998–2004

Volpi EV, Chevret E, Jones T, Vatcheva R, Williamson I, Beck S, Campbell RD, Goldsworthy M, Powis SH, Ragoussis J, Trowsdale J, Sheer D (2000) Large-scale chromatin organization of the major histocompatibility complex and other regions of human chromosome 6 and its response to interferon in interphase nuclei. J Cell Sci 113(Pt 9):1565–1576

Wang L, Di LJ, Lv X, Zheng W, Xue Z, Guo ZC, Liu DP, Liang CC (2009) Inter-MAR association contributes to transcriptionally active looping events in human beta-globin gene cluster. PLoS One 4(2):e4629

Wang CM, Tsai SN, Yew TW, Kwan YW, Ngai SM (2010) Identification of histone methylation multiplicities patterns in the brain of senescence-accelerated prone mouse 8. Biogerontology 11(1):87–102

Wen B, Wu H, Shinkai Y, Irizarry RA, Feinberg AP (2009) Large histone H3 lysine 9 dimethylated chromatin blocks distinguish differentiated from embryonic stem cells. Nat Genet 41(2):246–250

White EJ, Emanuelsson O, Scalzo D, Royce T, Kosak S, Oakeley EJ, Weissman S, Gerstein M, Groudine M, Snyder M, Schübeler D (2004) DNA replication-timing analysis of human chromosome 22 at high resolution and different developmental states. Proc Natl Acad Sci U S A 101(51):17771–17776

Williams RRE, Broad S, Sheer D, Ragoussis J (2002) Subchromosomal positioning of the epidermal differentiation complex (EDC) in keratinocyte and lymphoblast interphase nuclei. Exp Cell Res 272(2):163–175

Williams A, Spilianakis CG, Flavell RA (2010) Interchromosomal association and gene regulation in trans. Trends Genet 26(4):188–197

Woodfine K, Fiegler H, Beare DM, Collins JE, McCann OT, Young BD, Debernardi S, Mott R, Dunham I, Carter NP (2004) Replication timing of the human genome. Hum Mol Genet 13(2):191–202

Woodfine K, Beare DM, Ichimura K, Debernardi S, Mungall AJ, Fiegler H, Collins VP, Carter NP, Dunham I (2005) Replication timing of human chromosome 6. Cell Cycle 4(1):172–176

Worman HJ, Ostlund C, Wang Y (2010) Diseases of the nuclear envelope. Cold Spring Harb Perspect Biol 2(2):a000760

Xu N, Tsai CL, Lee JT (2006) Transient homologous chromosome pairing marks the onset of X inactivation. Science 311(5764):1149–1152

Xu N, Donohoe ME, Silva SS, Lee JT (2007) Evidence that homologous X-chromosome pairing requires transcription and Ctcf protein. Nat Genet 39(11):1390–1396

Yasui D, Miyano M, Cai S, Varga-Weisz P, Kohwi-Shigematsu T (2002) SATB1 targets chromatin remodelling to regulate genes over long distances. Nature 419(6907):641–645

Yokoyama T, Silversides DW, Waymire KG, Kwon BS, Takeuchi T, Overbeek PA (1990) Conserved cysteine to serine mutation in tyrosinase is responsible for the classical albino mutation in laboratory mice. Nucleic Acids Res 18(24):7293–7298

Zhang J, Xu F, Hashimshony T, Keshet I, Cedar H (2002) Establishment of transcriptional competence in early and late S phase. Nature 420(6912):198–202

Zhang R, Poustovoitov MV, Ye X, Santos HA, Chen W, Daganzo SM, Erzberger JP, Serebriiskii IG, Canutescu AA, Dunbrack RL, Pehrson JR, Berger JM, Kaufman PD, Adams PD (2005) Formation of MacroH2A-containing senescence-associated heterochromatin foci and senescence driven by ASF1a and HIRA. Dev Cell 8(1):19–30

Zlatanova J, Caiafa P (2009) CCCTC-binding factor: to loop or to bridge. Cell Mol Life Sci 66(10):1647–1660

Part IV
Epigenetics and Disease

Chapter 17
Cancer: An Epigenetic Landscape

Karthigeyan Dhanasekaran, Mohammed Arif, and Tapas K. Kundu

17.1 Introduction

Cancer is not a single disease, rather a group of abnormality generally associated with uncontrolled cell growth. The characteristics of cancer is determined by its tissue of origin. In humans during the development of cancer the tumor tissue acquires several physiological abilities, termed as "Hallmarks of cancer", through which cancer cells overcome the check points of a cell cycle, avoid the immune surveillance system, disobey the growth regulatory signals and induce the assembly of new blood vessels in the tumor. Later these cells become metabolically hyperactive to harness the energy required for maintaining the various cancer hallmarks. However, apart from these cellular characteristics, physiologically cancer growth and progression is significantly dependant on the "tumor microenvironment". All cancers are genetic but a very few are hereditary. Somatic mutations are considered to be the point of initiation. Nevertheless, the fine tuning of cancer progression, more precisely the establishment of a complex network among the genes expressed in a cancer cell is mediated by the epigenetic reprogramming, which could be affected by the tumor microenvironment (Fig. 17.1). In this chapter we shall discuss about the present understanding of the possible contribution of chromatin modifications and remodelling in cancer manifestation.

K. Dhanasekaran • M. Arif • T.K. Kundu (✉)
Transcription and Disease Laboratory, Molecular Biology and Genetics Unit,
Jawaharlal Nehru Centre for Advanced Scientific Research,
Jakkur P.O., Bangalore 560064, India
e-mail: tapas@jncasr.ac.in

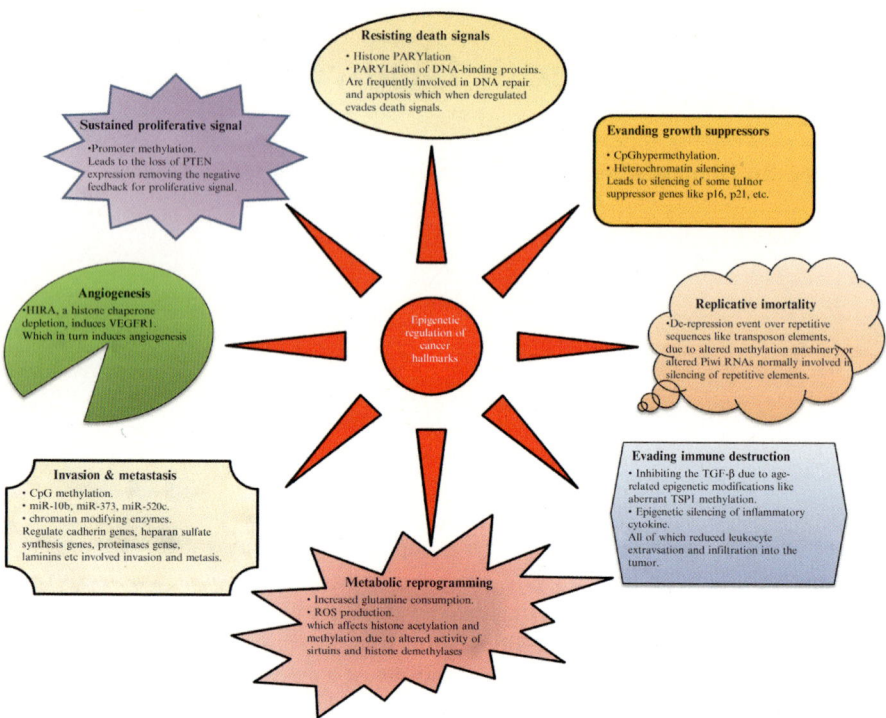

Fig. 17.1 Are alteration of epigenetic network causally related towards the establishment of cancer? Chromatin configuration changes during the acquisition of eight distinguished hallmarks of cancer. These epigenetic alterations are found in both cancer cells as well as tumor associated stromal cells. Here a few representative examples have been depicted

17.2 The Epigenetic Machinery

17.2.1 Histone Post-Translational Modifications

Different posttranslational modifications of histones establish the epigenetic landscape of a functional genome and there by play a key role in the maintenance of cellular homeostasis. In malignancy these histone modifications dramatically alters to support the unusual cancer cell physiology. For example, overproduction the key methyltransferases that catalyze the methylation of either H3-K4 or H3-K27 residues are frequent events in neoplasia. Global reductions in monoacetylated H4-K16 and trimethylated H4-K20 are general features of cancer cells. Recently, it was reported that changes in global levels of individual histone modifications are associated with several cancer. Trimethylation of lysine 27 on histone H3 (H3K27me3) is a target of Polycomb group of proteins, which has been implicated in the formation of repressive chromatin domains. H3K27me3 spreads over large regions harboring many target genes and negatively regulates transcription by promoting a compact chromatin structure (Francis et al. 2004; Ringrose et al. 2004). In human, H3K27me3 is considered as a prognostic marker in prostate, breast, ovarian, pancreatic and esophageal cancers where the expression levels of H3K27me3 are significantly higher (Füllgrabe 2011). Enhancer of zeste homolog 2 (EZH2) is the catalytic subunit of the polycomb repressive complex 2 (PRC2), which mediates H3K27me3 (Hansen et al. 2008). Overexpression of EZH2 is found in diverse cancers, including prostate, breast, renal and ovarian cancers, as well as glioblastoma multiforme (Füllgrabe 2011). Overexpression of EZH2 has been associated with the invasion and progression of cancers, especially with the progression of prostate cancer (Yu et al. 2007). Generally, EZH2 overexpression in cancer cells seems to result in an EZH2-dependent increase in H3K27me3. However, in some cases of breast, ovarian and pancreatic cancers, no association is drawn between EZH2 and H3K27me3, instead such cases indicate correlation between H3K27me3 and altered HDMs (Wei et al. 2008) JMJD3/KDM6B and UTX (Xiang et al. 2007). Loss of H3K18 acetylation (H3K18ac) is a general marker for active transcription. The H3K18 acetylation has been correlated with poor prognosis in patients with prostate, pancreatic, lung, breast and kidney cancers and loss of this modification is an important event in tumor progression (Seligson et al. 2005; Elsheikh et al. 2009).

Studies have shown that hypoacetylation of histone H4 lysine 12 (H4K12Ac) can be used as predictive biomarkers for cancer recurrence in the prostate (Seligson et al. 2005) and in non-small-cell lung cancer (Barlesi et al. 2007; Van Den Broeck et al. 2008). Global hypoacetylation of H4K12 was even considered to be informative of tumor stage for colorectal cancer (Ashktorab et al. 2009). Histone hyperacetylation has been observed in hepatocellular carcinoma and oral cancer (Bai et al. 2008; Arif et al. 2010).

Recently, increased H3K56ac, as well as upregulated expression of its positive regulator ASF1A, has been observed in many cancers. In humans, H3K56 is the target of both CBP/p300 and hGCN5 (Das et al. 2009; Tjeertes et al. 2009; Vempati et al. 2010) and is found to be deacetylated by different HDACs: SIRT1, SIRT2, SIRT3, HDAC1 and HDAC2 (Miller et al. 2010; Yuan et al. 2009). In addition to this, promoter CpG-island hypermethylation in cancer cells is known to be associated with a particular combination of histone marks: deacetylation of histones H3 and H4, loss of histone H3 lysine K4 (H3K4) trimethylation, and gain of H3K9 methylation and H3K27 trimethylation In another study, the loss of monoacetylated Lys 16 acetylation and trimethylated Lys 20 of H4 were found to be associated with primary tumors and tumor cell lines and they appear to follow pattern in which CpG-island hypermethylation precedes Kirsten rat sarcoma viral oncogene homolog (KRAs) mutations in small colorectal adenomas . Furthermore, loss of CpG island hypermethylation mediated inactivation of tumour suppressor, such as p21/WAF1, has been observed when hypoacetylated and hypermethylated histones H3 and H4 are present at the promoters (Robertson and Jones 2000; Chuang et al. 1997).

17.2.2 Histone Variants

In addition to the altered histone marks, aberrant incorporation of specific histone variants is observed in cancer pathogenesis. The histone cell cycle regulation-defective homolog A (HIRA), a histone chaperone, incorporates lysine 56 acetylated H3.3 over the VEGFR1 gene on triggering angiogenesis within the tumor tissue (Dutta et al. 2010). On HIRA depletion, the induction of VEGFR1 and other angiogenic genes are impaired. A direct link between histone variant expression and cancer development has recently been drawn by Khare et al. They showed that during sequential development of hepatocellular carcinoma, H2A and H2A.1 are overexpressed, whereas H2A.2 is decreased. The increased expression of H2A.1 has been linked to hyperproliferation (Khare et al. 2011). In addition, the histone variant macroH2A appears to suppress the progression of malignant melanoma through the regulation of CDK8 (Kapoor et al. 2010). Furthermore the expression of the human histone variant macroH2A shown to predict lung cancer recurrence and could therefore serve as a useful prognostic biomarker. A strong negative correlation was observed between the macroH2A1.1 and macroH2A2 expression and tumor proliferation rate. Tumors with a low proliferating index shows higher expression of macroH2A1.1 and macroH2A.2, whereas tumors with a high proliferative index often have lower or no expression in most of the clinically aggressive tumors (Sporn et al. 2009).

In a very recent report, it was demonstrated that the expression of macroH2A1.1 is reduced in several types of cancers due to the changes in the alternative splicing of macroH2A1 pre-mRNA mediated by splicing factor, QK1. QKI expression was reduced with the concomitant reduction in the macroH2A1.1 splicing in many of the same cancer types. Furthermore, it was found that the reduction of macroH2A1.1 levels enhances the proliferation of lung and cervical cancer cells with the reduction in the protein levels of PARP-1 (Novikov et al. 2011).

A possible role for H2A.Z in cancer development has first been reported by genome wide gene expression profiling studies. It was shown that H2A.Z was overexpressed in sporadic colorectal tumors. Overexpression of H2A.Z occurred in the later stages of breast cancer progression in a metastatic breast carcinoma (MET) but not in invasive ductal carcinoma (INV). H2A.Z has also been proposed for its role in estrogen receptor signaling and breast cancer progression. Furthermore, overexpression of H2A.Z has been implicated in the destabilization of chromosomal boundaries, which in turn leads to the spreading of repressive chromatin domains and the de novo hypermethylation of tumor suppressor gene promoters in cancer cells (Dalvai et al. 2010 and Santisteban et al. 2011).

Additionally, the highly conserved histone variant H2AX is a sensor of the DNA damage. It rapidly gets phosphorylated by ataxia telangiectasia mutated (ATM) and ATM-Rad3-related (ATR) kinases in the PI3K pathway upon double strand DNA breaks (DSB). It was found that phosphorylation of H2AX and recruitment of repair factors were deregulated in MCF-7 cells compared to normal cell lines like MCF10A cells, indicating the roles of this phosphorylated Histone variant in maintaining the genomic stability and its perturbations associated with the causal factors for tumorigenesis (Dalvai et al. 2010 and Santisteban et al. 2011).

CENP-A, the centromere-specific H3 variant, is essential for centromeric function and hence chromosome segregation becomes defective in colorectal cancer and it has been suggested that the overexpression of CENP-A is one of the responsible factors leading to aneuploidy (Tomonaga et al. 2005). The possible role of other variants in cancer progression remains to be explored (Henikoff and Furuyama 2010).

17.2.3 Non-Histone Chromatin Proteins and Cancer

Another distinguishing feature of cancerous tissue is the altered expression of histone modifying enzymes due to gene amplification, chromosomal translocations or due to perturbations in the stability of these enzymes and their associated binding partners. For example in leukemic patients after chromosomal translocations involving HATs and HMTs are reported to alter the global histone acetylation and methylation. Especially translocation of HAT and HAT related genes leads to the formation of deleterious fusion proteins. Gene amplification of HMTs and HDMs are a common scenario in solid tumors. The gain of PcG (polycomb) and loss of TrxG (Trithorax) is a common theme in human cancer, demonstrating the oncogenic and tumour suppressive roles, respectively, of these complexes. The oncogenic potential of PcG and its roles in transcription regulation is well established in cancers. Most of these cancers have compromised cellular memory where the stem cell associated genes like Hox genes which are the known targets of PcG and TrxG are reactivated (Mills et al. 2010).

The expression of PRC2 components like EZH2 is upregulated in various cancers such as melanoma, lymphoma, and breast and prostate cancer. EZH2 is presently regarded as an oncogene (Tong et al. 2011). The role of PRC2 in tumour progression is not well distinguished whether PRC2 is required for the process of de-differentiation of somatic cells or it is involved more in the epithelial to mesenchymal transition. PcG-mediated repression is also reversible in cancer cells, as TrxG proteins can override PcG-mediated repression, leading to the reactivation of tumour suppressors.

17.2.4 Chromatin Remodeling and Cancer

Despite intricate packaging, DNA as chromatin, it remains accessible during specific spatio-temporal windows for critical cellular processes such as transcription, replication, recombination, and repair which are facilitated by two classes of enzymes, ATP-dependent nucleosome remodelers and histone modifying enzymes. Chromatin remodelers are macromolecular machines and possess multisubunit components. Chromatin remodeling complexes (CRCs) utilize the energy of ATP to disrupt nucleosome DNA contacts, move nucleosomes along DNA, and remove or exchange nucleosomes making the DNA/chromatin accessible during cellular processes. In the adult stem cell, deletion or mutation of these proteins often leads to apoptosis or tumorigenesis as a consequence of dysregulated cell cycle control (Hargreaves and Crabtree 2011). ATP-dependent chromatin remodellers can be further divided into families on the basis of subunit composition and biochemical activity, and these families include SWI/SNF, ISWI, INO80, SWR1 and NURD/Mi2/CHD complexes. SWI–SNF chromatin remodelling complexes contain an ATPase subunit (either BRM or BRG1) as well as the non-catalytic subunit (SNF5) that is common to various SWI–SNF-like complexes. Loss of SNF5, BRM or BRG1 has also been associated with human cancer. Malignant rhabdoid tumours (MRTs) a rare but extremely aggressive form of childhood cancer which is caused by biallelic deletion or truncating mutations of SMARCB1 (which encodes SNF5) (Versteege et al. 1998). Among the remodelers, SWI/SNF complexes in particular are emerging as bonafide tumour suppressors (Wilson and Roberts 2011). Several CRCs, most prominently the BAF complex, have been implicated in cancer initiation or progression. Early studies demonstrated that many cell lines have lost both BRG1 and hBRM expression and that introduction of BRG1 or hBRM results in slower or arrested growth.

The vertebrate Mi-2/NuRD complex is a multi-subunit protein complex containing both histone deacetylase and nucleosome-dependent ATPase subunits involved in transcriptional repression. The Mi-2/NuRD complex that is involved in transcriptional repression provides a physical link between histone deacetylation and ATP-dependent chromatin remodeling (Kunert et al. 2009). The MTA (metastasis-associated) proteins represent one class of alternative subunits of the human Mi-2/NuRD complex. The members of this family in human cells are differentially expressed. MeTastasis Associated protein 1 (MTA1), the founder of this family, was discovered in a differential display screen comparing mRNA from rat breast cancer cell line. As the mRNA was expressed at higher levels in cell lines with an increased potential for invasion and metastasis following injection into nude mice, the protein was named as MeTastasis Associated protein 1. Human MTA1 has a discreet correlation between high level expression and invasive growth properties. In addition, a splice variant of MTA1, known as MTA1s, lacks a nuclear localization sequence, is found predominantly in the cytoplasm, and is unlikely to be a component of Mi-2/NuRD. The MTA3 locus in humans also codes for more than one protein based on alternative splicing. In human cancer cell lines, the shorter of the two isoforms (known as MTA3) is more abundant than the longer protein known as MTA3L (Bowen et al. 2004).

Like most chromatin remodelling complexes, INO80 subfamily complexes have been identified as transcriptional regulators. The effect of the INO80 complex on transcription is undoubtedly linked to the function of the transcription factors yin yang 1 (YY1; in mammals) and Pleiohomeotic (in *D. melanogaster*), which are found in INO80 complexes. These transcription factors, which are involved in cell proliferation, differentiation and embryonic development, can serve to specify the genes that are targeted for INO80-mediated chromatin remodelling. In addition the INO80 complex has also been shown to be involved in DNA damage responses.

Recent advances in the fields of DNA repair and chromatin reveal that both histone modifications and chromatin remodeling are important for the repair of DNA lesions, such as DSBs. Recent studies indicate that yeast phosphorylated-H2AX is also required for the recruitment of the chromatin remodeling complex INO80 to DSB sites, thus establishing a link between chromatin remodeling and DNA repair (Morrison and Shen 2009). Furthermore, it was shown recently by the Osley group (Tsukuda et al. 2005) that histone eviction near DSB sites, mediated by INO80 remodeling activity is dependent on MRX (Mre11-Rad50-Xrs2) complex, and a delayed recruitment of the Rad51 repair protein results from defects in histone loss. Taken together, these data suggest that INO80 complex participates in DNA repair pathways by its nucleosome remodeling ability and by regulating the accessibility of repair proteins around the DSB site. Hence, perturbed functioning of INO80 is well correlated to the altered genomic instability that is seen in cancerous cell types.

17.2.5 Non-Coding RNA, Epigenetics and Cancer

17.2.5.1 MicroRNAs (miRNAs) and Cancer

miRNAs are endogenous small non-coding RNAs which regulate gene expression in a sequence specific manner. miRNAs target the 3'untranslated region (3'-UTR) or the 5'-UTR of the target mRNA resulting in mRNA degradation and/or inhibition of translation. miRNAs are involved in most of the biological events, including development, differentiation, cell cycle regulation and metabolism. The full spectrum of miRNAs expressed in a specific cell type (the miRNAome) varies between normal and pathologic tissue, and specific signatures of deregulated miRNAs with diagnostic and prognostic implications. The versatile miRNAs on one hand are regulated by the same epigenetic mechanisms that affect the protein coding genes and on the other hand, a subgroup of miRNAs (epi-miRNAs) targets, directly or indirectly, the effectors of epigenetic machinery such as DNMTs, HDACs and polycomb genes. Thus, miRNAs can also indirectly regulate gene expression by directly regulating epigenetic processes (Pucci and Mazzarelli 2011).

A causal role for miRNAs in cancer was first suggested in 2002 by Croce (Di Croce et al. 2002) and colleagues with the discovery that miR-15 and miR-16, which were located on chromosome 13q14. This region of the chromosome 13 is frequently deleted in chronic lymphocytic leukemia (CLL) patients which are the sole genetic abnormality in such patients. Thus the deletion of miR-15/16 is considered to be a direct cause for CLL. Deregulation of miRNA expression is involved in the initiation and progression of tumorigenesis and has been investigated in almost all kinds of human cancer. Some miRNAs act mainly as tumor suppressors while others have a well-established role as oncogenes, depending upon their target genes. For example, the miR-15/16-1 cluster, that targets BCL2, acts as a tumor suppressor in CLL, whereas let7, acts as a tumor suppressor in lung cancer (Hatziapostolou and Iliopoulos 2011). Oncogenic miRNAs, which target growth inhibitory pathways, are often upregulated in cancer. MiR-155 and miR-21, are one among the well characterized oncomirs that are induced in several neoplasms. Apart from the above mentioned examples there are multiple other mechanisms which underlie the abnormal miRNA expression in cancers which includes transcriptional deregulation, mutations, DNA copy number abnormalities and defects in the miRNA biogenesis machinery. Epigenetic changes like DNA methylation and histone modifications are also responsible for aberrant miRNA expression. Earlier studies indicate that treatment of different cancer cell lines with the DNA demethylating agent (5-aza-20 -deoxycitidine) and/or HDAC inhibitors is able to alter the expression levels of miRNAs. Several miRNAs (miR-1, miR-124a and miR-127) are under epigenetic control in human cancer due to the fact that they are embedded in CpG island regions and are epigenetically silenced by promoter hypermethylation and histone modifications. Furthermore, it has been proposed that transcription factors can recruit epigenetic effectors at miRNA promoter regions and contribute to the regulation of their expression. (Sharma et al. 2010)

On the other hand, certain miRNAs (epi-miRNAs) are regulators of epigenetic effectors. Epi-miRNAs regulate the expression of DNMT3a, DNMT3b and DNMT1 (miR-29a, -29b, -29c) in lung cancer and AML, RBL2 the inhibitor of DNMT3 genes (miR-290), HDACs (miR-1, miR-140, and miR-449a) in prostate cancer and skeletal muscle tissue and polycomb genes like EZH2 (miR-101) in prostate and bladder cancers. Hence, the involvement of epi-miRNAs and epigenetic regulation of miRNAs introduces additional layers of complexity in understanding the contribution of transcriptional aberration in cancer (Valeri et al. 2009).

17.2.5.2 Long Intergenic Non-Coding RNAs (lincRNAs) and Cancer

Long intergenic non-coding RNAs (lincRNAs) regulate dosage compensation, imprinting, and developmental gene expression by establishing chromatin domains in an allele and cell type specific manner. LincRNAs are intimately associated with chromatin-remodeling complexes, but molecular mechanisms of their functions are still lacking. Accumulating reports of misregulated lncRNA expression across numerous cancer types suggest that aberrant lincRNA expression may be a major contributor to tumorigenesis. Posttranslational modifications of histones recruit DNA-binding proteins and chromatin-remodeling machinery and are often coupled for combinatorial control. For instance, in embryonic stem cells many genes, such as the HOX, that encode developmental regulators are transcriptionally silent but possess bivalent histone H3 lysine 4 (H3K4) and lysine 27 (H3K27) methylation, which are resolved into univalent H3K4 or H3K27 methylation domains upon differentiation. Recently, a lincRNA was shown to coordinate histone modifications by binding to multiple histone modification enzymes (Tsai et al. 2010). lincRNA HOTAIR is known to function as a scaffold for two distinct histone modification complexes. High levels of HOTAIR expression were correlated with both metastasis and poor survival rate. Though, the precise mechanism is not known, it is proven that HOTAIR reprograms chromatin state to promote cancer metastasis. The MALAT1 gene, or metastasis-associated lung adenocarcinoma transcript 1, like HOTAIR is also associated with high metastatic potential and grave prognosis. MALAT1 is unregulated in a variety of human cancers of the breast, prostate, colon, liver and uterus. Also the MALAT1 locus 11q13.1 often harbors chromosomal translocation breakpoints associated with cancer. Though, numerous findings are reported in the area of linc RNAs their mechanistic correlation towards cancer is open for investigation (Gibb et al. 2011).

17.3 Diet/Nutrition/Environmental Factors Dictating Cancer Epigenetics

Gene versus environment is often the most debated etiology of almost all the diseases. Cancer is not an exception to this argument. In fact, the alarming rise in the cancer population is solely because of the rapidly changing lifestyle that includes what a person consumes, inhales and comes across in his/her life span. This gene: environment interactions occurs constantly in every individual and it is corresponded between the two by epigenetic changes like DNA methylation, histone modifications, histone variant exchange and via other epigenetic signals that ultimately turn off certain genes which are tumor suppressor in function or switch on the expression of oncogenes or rather a combination of both to drive them towards cancer. In this scenario cancer researchers across the globe hypothesize that the changing lifestyle contributes to a considerable amount towards cancer predisposition by compromising the "Allostatic load" of an individual (Knox 2010).

Among the environmental factors diet of an individual has a major impact on cancer susceptibility. Recent reports claim that diet taken by parents can influence the epigenetics of their offspring especially the DNA methylation imprint (Wu et al. 2004; Dolinoy et al. 2006). Histone PTMs contributes to a great extent in the epigenetic signaling and the metabolite source for these modifications is dependent on the cellular metabolism. This in turn is maintained by the dietary input of essential nutrients that gives rise to the metabolites required for chromatin modifying enzymes. Often because of this inherent connection diet influences epigenetics from the very beginning of life and deregulated metabolism is one of the important causative for altered epigenetics and cancer predisposition (Borrelli et al. 2008). One of the well studied dietary factors is the amino acid deficiency that can affect the overall methylation status by altering the methyl donor availability to the methylation machinery (Waterland et al. 2006). This can often lead to the demethylation of CpG island and subsequent activation of oncogene expression in the post natal life (Doherty et al. 2000). Folate, is another important nutrient that is known to influence the promoter methylation since, it is essential for the synthesis of S-adenosylmethionine that serves as a methyl group donor (Keku et al. 2002; Piestrzeniewicz-Ulanska et al. 2004). Choline, another nutrient like folate influences the methylation of certain gene promoters that prevents cancer in some cases while it predisposes in other instance, depending on the gene affected by DNA methylation (Zeisel et al. 2007). Alcohol consumption is also known to affect the folate metabolism. This can indirectly influence the methylome of a cell and predispose to carcinogenesis (Hamid et al. 2009).

Cancer cell is known to be associated with metabolic defects that increase the reactive oxygen species (ROS) generation which in turn alters DNA methylation status and histone modifications by oxidizing DNMTs and HMTs or they may induce genetic change through direct oxidation of nucleotide bases leading to mutation. Increased glutamine consumption in cancer cells perturbs histone acetylation and methylation due to a decrease in the NAD$^+$/NADH ratio creating an inhibitory environment for the activity of sirtuins and histone demethylases which ultimately liberates genes from their negative regulation (Hitchler and Domann 2009). Fiber content of diet influences the biotic population and the butyrate production in the gut. Butyrate is a well known inhibitor of HDACs. In colorectal carcinoma distinct correlation exists between dietary fiber content, HDAC activity and cancer predisposition (Corfe et al. 2009). Dietry components like diallyl disulfide from garlic and sulforaphane present in cruciferous vegetables also inhibit type I and II HDACs and prevent cancer occurrences. Curcumin and copper on ingestion induces histone hypoacetylation by inhibiting the HATs. Copper does this by triggering oxidative stress and curcumin by directly inhibiting p300 (Balasubramanyam et al. 2004; Morimoto et al. 2008). Dihydrocoumarin found in sweet clover is a well reported inhibitor of SIRT1 and disrupts heterochromatic silencing (Olaharski et al. 2005).

One other nutrient i.e., Biotin when covalently attached to histones induces gene silencing in the cellular response to DNA damage. A sudden malfunctioning of biotinylation due to biotin deficiency may hence pave way to tumorigenesis. Tryptophan and Niacin together are the precursors of nicotinamide adenine dinucleotide (NAD), that serves as a substrate for poly(ADP-ribosylation). PARYlation of histones and other DNA-binding proteins are frequently involved in DNA repair and apoptosis which when deregulated due to the deficiency of the key nutrients may lead to aberrant PARYlation and transformation (Oommen et al. 2005). The other dietary factor strongly correlated to cancer is consumption of high-fat diet. In case of rats fed with high-fat diet the ER promoter is hypomethylated and hence an increased ER expression occurs that often predisposes their offspring's to cancer (Aguilera et al. 2010).

Hazardous toxicants in the environment are also known to modulate the epigenetics of an individual. Nickel is one such toxicant that binds to heterochromatin and alters the DNA methylation in such regions apart from the decondensation of heterochromatin structures which ultimately results in epimutation leading to cancer genesis (Klein et al. 1991; Conway and Costa 1989). Apart from heretrochromatin binding in some cases people have reported that nickel can induce DNA hypermethylation and heterochromatin silencing of some tumor suppressor genes like p16. Also it can affect the lysine acetylation status of certain region based on its steric hindrance with the acetylation machinery when nickel is bound to the

Histidine residue neighboring to these lysines (Govindarajan et al. 2002; Sutherland et al. 2001). Arsenic, like nickel can also alter the DNA methylation by changing the availability of SAM which is an essential substrate for methyltransferase and also a metabolite involved in the detoxification of arsenic from the cell (Zhao et al. 1997, PNAS 94:10907–10912). Cadmium at high level acute exposures can inhibit DNMTs while the same at low level and chronic exposure can increase the activity of DNMTs (Salnikow and Zhitkovich 2008).

Some of the other contributing factors that can tamper the epigenetic homeostasis are the ageing associated hypometylation of certain gene promoters (Issa et al. 1996), pathogen induced alteration of DNA methylation and histone modifications (Maekita et al. 2006; Kalantari et al. 2004) and many such unknown factors like energy restriction (Hughes et al. 2009), alcohol consumption (Arasaradnam et al. 2008), cigarette smoking (Hobo et al. 2010) etc., have direct or indirect effects over the epigenetic deregulation especially the aberrant methylation status that are claimed to contribute towards tumor progression. Though the effects of diet and environment are well correlated to oncogenesis scientific evidences for these are only a tip of an ice berg which needs more statistically valid experiments to confirm the hypothesis.

17.4 Epigenetic Mechanisms in Cancer Manifestation

17.4.1 Epigenetics of Tumor Progression and Metastasis

The contribution of epigenetics in cancer is equally important as compared to the genetic causes of cancer. In fact it is the epigenetic status of a transformed cell that decides the course of cancer genesis by creating the second hit on the gene expression of a given tumor suppressor. Fist hit is created by the genetic mutation whereas the second hit is provided by epimutation over the second allelic wild type copy of the same gene. Since the first report came on DNA hypomethylation among the patients with colorectal cancer (Feinberg and Vogelstein 1983), the role of epigenetic mutation in various diseases pathogenesis is being widely studied. Today it is well established in the field of cancer biology that epigenetic alterations are involved in almost every step of cancer right from transformation to tumor establishment and from tumor progression to metastasis. Very much like the genetic mutation that gives rise to the property of self-sufficiency in growth signals, insensitivity towards growth inhibitory signals, ability to avert apoptosis, increased proliferation, angiogenesis to support such growth, and the ability to metastasize and invade are aided by parallel epigenetic events in such cancerous cells. As mentioned earlier this epigenetic change could be aberrant DNA methylation over the CpG islands or CpG shores which shuts down the expression of certain tumor suppressors like CDKN2A etc. or it could be a global DNA hypomethylation and activation of certain oncogenes like HRAS and KRAS (Ryan et al. 2010). It can be even a de-repression event over the repetitive sequences like transposon elements due to altered methylation machinery or altered Piwi RNAs (Siddiqi and Matushansky 2011) which are normally involved in silencing of repetitive elements that keeps a check over chromosomal instability. Alterations in the HAT activity, like hypoacetylation over p53 are involved in diverting signals from apoptosis to survival (Sykes et al. 2006). Fusion events like MLL-CBP (Ayton and Cleary 2001), MOZ-CBP (Chan et al. 2007), MLL-p300 (Ida et al. 1997) in certain hematological malignancies are known to alter the gene expression in favor of tumor progression and angiogenesis. Inhibition of certain HDACs, like class II HDACs aid in angiogenesis and migration (Witt et al. 2009). Even an aberrant recruitment of certain HDACs may lead to certain tumor suppressor gene inactivation or an aberrant expression may cause an imbalance in the overall acetylation status within the cell (Glaser et al. 2003; Zimmermann et al. 2007).

Though numerous mutations are reported for the metastasis associated genes there are epimutations like aberrant CpG methylation, recruitment of repressive transcription factors and other chromatin modifying enzymes over these genes like cadherin genes, heparan sulfate synthesis pathway genes, tissue inhibitors of proteinases genes, laminins etc. which ultimately plays essential role in tumor cell invasion and metastasis. Moreover miRNAs are discovered to have important roles in proliferation, differentiation, apoptosis and development. Hence it is definitely not surprising to correlate miRNA and cancer progression right from oncogenesis to metastasis (He and Hannon 2004; Lu et al. 2005). The miRNA, miR-10b promotes cell migration and invasion in breast cancer by upregulating the prometastatic gene RHOC. miRNAs, miR-373 and miR-520c by suppressing CD44 aids in tumor invasion and metastasis. The oncomiR, MiR-21, inhibits multiple metastasis suppressor genes, to bring about tumor invasion and metastasis. The tumor supressor miRNAs, miR-126 regulates cell proliferation while miR-335 regulates tumor invasion. The miR-200 family miRNAs inhibits E-cadherin repressors ZEB1, ZEB2, SIP1 and transcription factor that results in epithelial- to-mesenchymal transition required for tumor metastasis (Peter et al. 2009).

17.4.2 Epigenetics and Cancer Metabolism

Metabolic reprogramming in cancer cell was first discovered by Warbug in 1956 (Warburg et al. 1956), where he found that cancer cells prefer to metabolize glucose by anaerobic glycolysis rather than oxidative phosphorylation. Numerous cell signaling events are initiated in the cancer cell when exposed to the stress created within the tumorous tissue microenvironment. The mutations which contribute towards such aberrant signaling is well studied and researchers are now trying to study the biochemical aspects of such cancer cell metabolism. One other emerging area of interest is the contribution of epigenetic alteration in initiating these aberrant metabolic events. Such precise adaptive response aids cancer cells in better survival and even gain immortality and stemness. Cancer cell metabolism can also reprogramme these cells to promote the "cancer stem cell" phenotype, by altered utilization of metabolites. The high-energy metabolites lactate and ketone when utilized by cancer cells might induce stemness by increasing the Acetyl-CoA pool, and hence might lead to increased histone acetylation, and gene expression to create stemness in the cancer population (Martinez-Outschoorn et al. 2011). The oncogenic Myc is known to be involved in the production of Acetyl-CoA from mitochondrial metabolism by acting as a nutritional sensor during cell cycle. Acetyl-CoA produced on Myc activation is later diverted to a great extent in the acetylation of histones. Since the Myc dependant global acetylation is at the point of cell cycle entry, its deregulation is expected to predispose in cancer development (Morrish et al. 2010). Recently, it was also shown that the ATP-citrate Lyase activity is enhanced growth factor signaling that ultimately increases the nutrient metabolism. The global histone acetylation is determined by glucose availability via ACL activity regulation which leads to enhanced utilization of glucose and hence an energetically favorable state is achieved for cellular proliferation to occur. Probably, this could favor in creating an epigenetically conducive environment for the cancer cell proliferation and tumor progression (Wellen et al. 2009). In support to the same hypothesis acetyltransferases that acetylate protein substrates other than histone which posses gene regulatory function are known to have control over the metabolic genes involved in the production of metabolic fuel. At this juncture these observations encourage us to believe that one of the driving forces for tumor progression other than epigenetic modification of histone might be through regulation of certain transcription factors whose activity is dependent on the acetylation status.

Metabolic defects in most of the cancer cell leads to increased ROS production which might attack the chromatin modifying enzymes that might lose its function and cause global changes in the chromatin structures that are more prone to damage and hence, chromosomal instability ensues. Similarly due to the Warburg effect and increased glutamine consumption, the histone acetylation and methylation are affected due to the altered activity of sirtuins and histone demethylases (Hitchler and Domann 2009).

17.4.3 Epigenetics and Cancer Stem Cells

Cancer "stem/initiating cells" are considered as the seeds of a tumor tissue in most aggressive forms of cancer. These cancer stem cells possess a signature sign of increased polycomb protein group (PcG) complex members, to maintaining a poised, low basal activity, over the genes for which PcG targeting is normally key to maintain bivalent epigenetic status similar to that of the progenitor stem cell like status where both H3K4 methylation and H3K27 tri-methylation marks coexist (Ben-Porath et al. 2008; Bernstein et al. 2006). Embryonic stem cell in general maintains repression over certain sets of promoter region using PcG proteins. Surprisingly, the same sets of regions are also more likely to be methylated to permanently lock them in the stem cell state (Widschwendter et al. 2007). Similarly, *BMI*1 is a part of the polycomb group genes (PcG) that acts as a chromatin modifier well known for its role in embryonic and stem cell self-renewal. *BMI*1 upregulation is also associated with malignant transformation and is involved in the maintenance and propogation of cancer stem cells. The DNA methylation aberrations, like global hypomethylation and promoter hypermethylation, were not rare with tumors displaying cancer stem cell properties (Widschwendter et al. 2007). The side population representing such cancer stem cell often posses DNA methylation defects and it is well correlated with increased tumorigenicity within the side population (Marquardt et al. 2010). The tumor suppressor miRNA, miR-34a negatively regulates the expression of CD44 an important marker for stemness in Prostate cancer. Thus epigenetic defects like downregulation of miR-34a leads to cancer development and manifestation of their migratory, invasive and metastatic properties (Liu et al. 2011).

Accumulation of global epimutations arises from very early alterations in the epigenetic machinery, during neoplastic evolution. Since epigenetic mechanisms are central to maintenance of stem cell identity, its disruption may give rise to a high-risk aberrant progenitor cell population. Later this population transforms upon subsequent genetic gatekeeper mutations and become cancerous (Sharma et al. 2010). On the contrary some key gate keeper epigenetic mutations are also reported to happen during stress response and turn on tumor specific marks (Baylin and Ohm 2006). Still the effects of such epimutations either in the early or late events of tumor initiation and progression has equally important roles to play as compared to the genetic mutations.

17.5 Epigenetics and Therapeutics: Promise and Challenges

Presently, there are several epigenetic drugs like inhibitors of DNMTs and HDACs are at various developmental stages while some are in clinical trials. Several laboratories around the world are busy in designing and developing inhibitors and/or modulators for targeting the enzymes that mediate the epigenetic modifications. The selection of a right target (proteins) at the right place (type of cancer) is important. Targeted delivery of these epigenetic drugs hold great promise by, decreasing there side-effect and toxicity. Further a complete understanding of the epigenetics of cancer cell is also a prerequisite for a more efficient anticancer strategy. With the ever increasing list of new epigenetic modifications on histones and its variants, nonhistone protein and there cross talk makes the task even more challenging. The role of epigenetic drugs on nonhistone proteins is also to be addressed. Combination therapies of epigenetic drugs, with other anticancer therapy provide an additional space against cancer treatment which should be exploited to its maximum benefit. Designing smart small molecule inhibitors or ligands which disrupt protein-protein interaction in various protein complexes (e.g. HAT, HDAC complexes) should also be explored while the effect of known molecules on such complexes needs to be further investigated. In the years to come we hope to understand the characteristics of cancer stem cells from different origin in the epigenetic context. The signal dependent differentiation of the cancer stem cells could be understood more systematically by systems biology approach. Though numerous epigenetic correlations have been drawn to cancer manifestation the cause and effect is not well deciphered, which leaves a humungous task for researchers to knit the finer details in mapping the epigenetic causes of cancer.

Acknowledgments Work done in our laboratory is supported by Department of Biotechnology, Govt. of India; Department of Science and Technology (DST), Govt. of India, and Jawaharlal Nehru Centre for Advanced Scientific Research. TKK is a recipient of Sir JC Bose fellowship (DST). DK and MA are senior research fellows of the Council of Scientific and Industrial Research (CSIR), India.

References

Aguilera O, Fernández AF, Muñoz A, Fraga MF (2010) Epigenetics and environment: a complex relationship. J Appl Physiol 109:243–251
Arasaradnam RP, Commane DM, Bradburn D, Mathers JC (2008) A review of dietary factors and its influence on DNA methylation in colorectal carcinogenesis. Epigenetics 3:193–198
Arif M, Vedamurthy BM, Choudhari R, Ostwal YB, Mantelingu K, Kodaganur GS, Kundu TK (2010) Nitric oxide-mediated histone hyperacetylation in oral cancer: target for a water-soluble HAT inhibitor, CTK7A. Chem Biol 17:903–913
Ashktorab H, Belgrave K, Hosseinkhah F, Brim H, Nouraie M, Takkikto M, Hewitt S, Lee EL, Dashwood RH, Smoot D (2009) Global histone H4 acetylation and HDAC2 expression in colon adenoma and carcinoma. Dig Dis Sci 54:2109–2117
Ayton PM, Cleary ML (2001) Molecular mechanisms of leukemogenesis mediated by MLL fusion proteins. Oncogene 20:5695–5707
Bai X, Wu L, Liang T, Liu Z, Li J, Li D, Xie H, Yin S, Yu J, Lin Q, Zheng S (2008) Overexpression of myocyte enhancer factor 2 and histone hyperacetylation in hepatocellular carcinoma. J Cancer Res Clin Oncol 134:83–91
Balasubramanyam K, Varier RA, Altaf M, Swaminathan V, Siddappa NB, Ranga U, Kundu TK (2004) Curcumin, a novel p300/CREB-binding protein-specific inhibitor of acetyltransferase, represses the acetylation of histone/nonhistone proteins and histone acetyltransferase-dependent chromatin transcription. J Biol Chem 279:51163–51171
Barlési F, Giaccone G, Gallegos-Ruiz MI, Loundou A, Span SW, Lefesvre P, Kruyt FA, Rodriguez JA (2007) Global histone modifications predict prognosis of resected non small-cell lung cancer. J Clin Oncol 25:4358–4364
Baylin SB, Ohm JE (2006) Epigenetic gene silencing in cancer – a mechanism for early oncogenic pathway addiction? Nat Rev Cancer 6:107–116
Ben-Porath I, Thomson MW, Carey VJ, Ge R, Bell GW, Regev A, Weinberg RA (2008) An embryonic stem cell-like gene expression signature in poorly differentiated aggressive human tumors. Nat Genet 40:499–507
Bernstein BE, Mikkelsen TS, Xie X, Kamal M, Huebert DJ, Cuff J, Fry B, Meissner A, Wernig M, Plath K, Jaenisch R, Wagschal A, Feil R, Schreiber SL, Lander ES (2006) A bivalent chromatin structure marks key developmental genes in embryonic stem cells. Cell 125:315–326
Borrelli E, Nestler EJ, Allis CD, Sassone-Corsi P (2008) Decoding the epigenetic language of neuronal plasticity. Neuron 60:961–974
Bowen NJ, Fujita N, Kajita M, Wade PA (2004) Mi-2/NuRD: multiple complexes for many purposes. Biochim Biophys Acta 1677:52–57

Chan EM, Chan RJ, Comer EM, Goulet RJ 3rd, Crean CD, Brown ZD, Fruehwald AM, Yang Z, Boswell HS, Nakshatri H, Gabig TG (2007) MOZ and MOZ-CBP cooperate with NF-kappaB to activate transcription from NF-kappaB-dependent promoters. Exp Hematol 35:1782–1792

Chuang LS, Ian HI, Koh TW, Ng HH, Xu G, Li BF (1997) Human DNA-(cytosine-5) methyltransferase-PCNA complex as a target for p21WAF1. Science 277:1996–2000

Conway K, Costa M (1989) Nonrandom chromosomal alterations in nickel-transformed Chinese hamster embryo cells. Cancer Res 49:6032–6038

Corfe BM, Williams EA, Bury JP, Riley SA, Croucher LJ, Lai DY, Evans CA (2009) A study protocol to investigate the relationship between dietary fibre intake and fermentation, colon cell turnover, global protein acetylation and early carcinogenesis: the FACT study. BMC Cancer 9:332

Dalvai M, Bystricky K (2010) The role of histone modifications and variants in regulating gene expression in breast cancer. J Mammary Gland Biol Neoplasia. 15(1):19–33

Das C, Lucia MS, Hansen KC, Tyler JK (2009) CBP/p300-mediated acetylation of histone H3 on lysine 56. Nature 459:113–117

Di Croce L, Raker VA, Corsaro M, Fazi F, Fanelli M, Faretta M, Fuks F, Lo Coco F, Kouzarides T, Nervi C, Minucci S, Pelicci PG (2002) Methyltransferase recruitment and DNA hypermethylation of target promoters by an oncogenic transcription factor. Science 295(5557):1079–1082

Doherty AS, Mann MR, Tremblay KD, Bartolomei MS, Schultz RM (2000) Differential effects of culture on imprinted H19 expression in the preimplantation mouse embryo. Biol Reprod 62:1526–1535

Dolinoy DC, Weidman JR, Waterland RA, Jirtle RL (2006) Maternal genistein alters coat color and protects Avy mouse offspring from obesity by modifying the fetal epigenome. Environ Health Perspect 114:567–572

Dutta D, Ray S, Home P, Saha B, Wang S, Sheibani N, Tawfik O, Cheng N, Paul S (2010) Regulation of angiogenesis by histone chaperone HIRA-mediated incorporation of lysine 56-acetylated histone H3.3 at chromatin domains of endothelial genes. J Biol Chem 285:41567–41577

Elsheikh SE, Green AR, Rakha EA, Powe DG, Ahmed RA, Collins HM, Soria D, Garibaldi JM, Paish CE, Ammar AA, Grainge MJ, Ball GR, Abdelghany MK, Martinez-Pomares L, Heery DM, Ellis IO (2009) Global histone modifications in breast cancer correlate with tumor phenotypes, prognostic factors, and patient outcome. Cancer Res 69:3802–3809

Feinberg AP, Vogelstein B (1983) Hypomethylation distinguishes genes of some human cancers from their normal counterparts. Nature 301:89–92

Francis NJ, Kingston RE, Woodcock CL (2004) Chromatin compaction by a polycomb group protein complex. Science 306:1574–1577

Füllgrabe J, Kavanagh E, Joseph B (2011) Histone onco-modifications. Oncogene 30:3391–3403

Gibb EA, Brown CJ, Lam WL (2011) The functional role of long non-coding RNA in human carcinomas. Mol Cancer 10:38

Glaser KB, Li J, Staver MJ, Wei RQ, Albert DH, Davidsen SK (2003) Role of class I and class II histone deacetylases in carcinoma cells using siRNA. Biochem Biophys Res Commun 310:529–536

Govindarajan B, Klafter R, Miller MS, Mansur C, Mizesko M, Bai X, LaMontagne K Jr, Arbiser JL (2002) Reactive oxygen-induced carcinogenesis causes hypermethylation of p16(Ink4a) and activation of MAP kinase. Mol Med 8:1–8

Hamid A, Kiran M, Rana S, Kaur J (2009) Low folate transport across intestinal basolateral surface is associated with down-regulation of reduced folate carrier in in vivo model of folate malabsorption. IUBMB Life 61:236–243

Hansen KH, Bracken AP, Pasini D, Dietrich N, Gehani SS, Monrad A, Rappsilber J, Lerdrup M, Helin K (2008) A model for transmission of the H3K27me3 epigenetic mark. Nat Cell Biol 10:1291–1300

Hargreaves DC, Crabtree GR (2011) ATP-dependent chromatin remodeling: genetics, genomics and mechanisms. Cell Res 21(3):396–420

Hatziapostolou M, Iliopoulos D (2011) Epigenetic aberrations during oncogenesis. Cell Mol Life Sci 68:1681–1702

He L, Hannon GJ (2004) MicroRNAs: small RNAs with a big role in gene regulation. Nat Rev Genet 5:522–531

Henikoff S, Furuyama T (2010) Epigenetic inheritance of centromeres. Cold Spring Harb Symp Quant Biol 75:51–60

Hitchler MJ, Domann FE (2009) Metabolic defects provide a spark for the epigenetic switch in cancer. Free Radic Biol Med 47:115–127

Hobo W, Maas F, Adisty N, de Witte T, Schaap N, van der Voort R, Dolstra H (2010) siRNA silencing of PD-L1 and PD-L2 on dendritic cells augments expansion and function of minor histocompatibility antigen-specific CD8+ T cells. Blood 116:4501–4511

Hughes LA, van den Brandt PA, de Bruïne AP, Wouters KA, Hulsmans S, Spiertz A, Goldbohm RA, de Goeij AF, Herman JG, Weijenberg MP, van Engeland M (2009) Early life exposure to famine and colorectal cancer risk: a role for epigenetic mechanisms. PLoS One 4:e7951

Ida K, Kitabayashi I, Taki T, Taniwaki M, Noro K, Yamamoto M, Ohki M, Hayashi Y (1997) Adenoviral E1A-associated protein p300 is involved in acute myeloid leukemia with t(11;22)(q23;q13). Blood 90:4699–4704

Issa JP, Vertino PM, Boehm CD, Newsham IF, Baylin SB (1996) Switch from monoallelic to biallelic human IGF2 promoter methylation during aging and carcinogenesis. Proc Natl Acad Sci U S A 93:11757–11762

Kalantari M, Calleja-Macias IE, Tewari D, Hagmar B, Lie K, Barrera-Saldana HA, Wiley DJ, Bernard HU (2004) Conserved methylation patterns of human papillomavirus type 16 DNA in asymptomatic infection and cervical neoplasia. J Virol 78:12762–12772

Kapoor-Vazirani P, Kagey JD, Powell DR, Vertino PM (2008) Role of hMOF-dependent histone H4 lysine 16 acetylation in the maintenance of TMS1/ASC gene activity. Cancer Res 68:6810–6821

Keku T, Millikan R, Worley K, Winkel S, Eaton A, Biscocho L, Martin C, Sandler R (2002) 5,10-Methylenetetrahydrofolate reductase codon 677 and 1298 polymorphisms and colon cancer in African Americans and whites. Cancer Epidemiol Biomarkers Prev 11:1611–1621

Khare SP, Sharma A, Deodhar KK, Gupta S (2011) Overexpression of histone variant H2A.1 and cellular transformation are related in N-nitrosodiethylamine-induced sequential hepatocarcinogenesis. Exp Biol Med (Maywood) 236:30–35

Klein CB, Conway K, Wang XW, Bhamra RK, Lin XH, Cohen MD, Annab L, Barrett JC, Costa M (1991) Senescence of nickel-transformed cells by an X chromosome: possible epigenetic control. Science 251:796–799

Knox SS (2010) From 'omics' to complex disease: a systems biology approach to gene-environment interactions in cancer. Cancer Cell Int 10:11

Kunert N, Wagner E, Murawska M, Klinker H, Kremmer E, Brehm A (2009) dMec: a novel Mi-2 chromatin remodelling complex involved in transcriptional repression. EMBO J 28:533–544

Liu C, Kelnar K, Liu B, Chen X, Calhoun-Davis T, Li H, Patrawala L, Yan H, Jeter C, Honorio S, Wiggins JF, Bader AG, Fagin R, Brown D, Tang DG (2011) The microRNA miR-34a inhibits prostate cancer stem cells and metastasis by directly repressing CD44. Nat Med 17:211–215

Lu J, Getz G, Miska EA, Alvarez-Saavedra E, Lamb J, Peck D, Sweet-Cordero A, Ebert BL, Mak RH, Ferrando AA, Downing JR, Jacks T, Horvitz HR, Golub TR (2005) MicroRNA expression profiles classify human cancers. Nature 435:834–838

Maekita T, Nakazawa K, Mihara M, Nakajima T, Yanaoka K, Iguchi M, Arii K, Kaneda A, Tsukamoto T, Tatematsu M, Tamura G, Saito D, Sugimura T, Ichinose M, Ushijima T (2006) High levels of aberrant DNA methylation in Helicobacter pylori-infected gastric mucosae and its possible association with gastric cancer risk. Clin Cancer Res 12:989–995

Marquardt JU, Factor VM, Thorgeirsson SS (2010) Epigenetic regulation of cancer stem cells in liver cancer: current concepts and clinical implications. J Hepatol 53:568–577

Martinez-Outschoorn UE, Prisco M, Ertel A, Tsirigos A, Lin Z, Pavlides S, Wang C, Flomenberg N, Knudsen ES, Howell A, Pestell RG, Sotgia F, Lisanti MP (2011) Ketones and lactate increase cancer cell "stemness," driving recurrence, metastasis and poor clinical outcome in breast cancer: achieving personalized medicine via Metabolo-Genomics. Cell Cycle 10:1271–1286

Miller KM, Tjeertes JV, Coates J, Legube G, Polo SE, Britton S, Jackson SP (2010) Human HDAC1 and HDAC2 function in the DNA-damage response to promote DNA nonhomologous end-joining. Nat Struct Mol Biol 17:1144–1151

Mills AA (2010) Throwing the cancer switch: reciprocal roles of polycomb and trithorax proteins. Nat Rev Cancer 10:669–682

Morimoto T, Sunagawa Y, Kawamura T, Takaya T, Wada H, Nagasawa A, Komeda M, Fujita M, Shimatsu A, Kita T, Hasegawa K (2008) The dietary compound curcumin inhibits p300 histone acetyltransferase activity and prevents heart failure in rats. J Clin Invest 118:868–878

Morrish F, Noonan J, Perez-Olsen C, Gafken PR, Fitzgibbon M, Kelleher J, VanGilst M, Hockenbery D (2010) Myc-dependent mitochondrial generation of acetyl-CoA contributes to fatty acid biosynthesis and histone acetylation during cell cycle entry. J Biol Chem 285:36267–36274

Morrison AJ, Shen X (2009) Chromatin remodelling beyond transcription: the INO80 and SWR1 complexes. Nat Rev Mol Cell Biol 10:373–384

Novikov L, Park JW, Chen H, Klerman H, Jalloh AS, Gamble MJ (2011) QKI-mediated alternative splicing of the histone variant MacroH2A1 regulates cancer cell proliferation. Mol Cell Biol 31:4244–4255

Olaharski AJ, Rine J, Marshall BL, Babiarz J, Zhang L, Verdin E, Smith MT (2005) The flavoring agent dihydrocoumarin reverses epigenetic silencing and inhibits sirtuin deacetylases. PLoS Genet 1:e77

Oommen AM, Griffin JB, Sarath G, Zempleni J (2005) Roles for nutrients in epigenetic events. J Nutr Biochem 16:74–77

Peter ME (2009) Let-7 and miR-200 microRNAs: guardians against pluripotency and cancer progression. Cell Cycle 8:843–852

Piestrzeniewicz-Ulanska D, Brys M, Semczuk A, Rechberger T, Jakowicki JA, Krajewska WM (2004) TGF-beta signaling is disrupted in endometrioid-type endometrial carcinomas. Gynecol Oncol 95:173–180

Pucci S, Mazzarelli P (2011) MicroRNA dysregulation in colon cancer microenvironment interactions: the importance of small things in metastases. Cancer Microenviron 4:155–162

Ringrose L, Ehret H, Paro R (2004) Distinct contributions of histone H3 lysine 9 and 27 methylation to locus-specific stability of polycomb complexes. Mol Cell 16:641–653

Robertson KD, Jones PA (2000) DNA methylation: past, present and future directions. Carcinogenesis 21:461–467

Ryan JL, Jones RJ, Kenney SC, Rivenbark AG, Tang W, Knight ER, Coleman WB, Gulley ML (2010) Epstein-Barr virus-specific methylation of human genes in gastric cancer cells. Infect Agent Cancer 5:27

Salnikow K, Zhitkovich A (2008) Genetic and epigenetic mechanisms in metal carcinogenesis and cocarcinogenesis: nickel, arsenic, and chromium. Chem Res Toxicol 21:28–44

Santisteban MS, Hang M, Smith MM (2011) Histone variant H2A.Z and RNA polymerase II transcription elongation. Mol Cell Biol 31:1848–1860

Seligson DB, Horvath S, Shi T, Yu H, Tze S, Grunstein M, Kurdistani SK (2005) Global histone modification patterns predict risk of prostate cancer recurrence. Nature 435:1262–1266

Sharma S, Kelly TK, Jones PA (2010) Epigenetics in cancer. Carcinogenesis 31:27–36

Siddiqi S, Matushansky I (2011) Piwis and piwi-interacting RNAs in the epigenetics of cancer. J Cell Biochem. doi:10.1002/jcb.23363

Sporn JC, Kustatscher G, Hothorn T, Collado M, Serrano M, Muley T, Schnabel P, Ladurner AG (2009) Histone macroH2A isoforms predict the risk of lung cancer recurrence. Oncogene 28(38):3423–3428

Sutherland JE, Peng W, Zhang Q, Costa M (2001) The histone deacetylase inhibitor trichostatin A reduces nickel-induced gene silencing in yeast and mammalian cells. Mutat Res 479:225–233

Sykes SM, Mellert HS, Holbert MA, Li K, Marmorstein R, Lane WS, McMahon SB (2006) Acetylation of the p53 DNA-binding domain regulates apoptosis induction. Mol Cell 24:841–851

Tjeertes JV, Miller KM, Jackson SP (2009) Screen for DNA-damage-responsive histone modifications identifies H3K9Ac and H3K56Ac in human cells. EMBO J 28:1878–1889

Tomonaga T, Matsushita K, Ishibashi M, Nezu M, Shimada H, Ochiai T, Yoda K, Nomura F (2005) Centromere protein H is up-regulated in primary human colorectal cancer and its overexpression induces aneuploidy. Cancer Res 65(11):4683–4689

Tong ZT, Cai MY, Wang XG, Kong LL, Mai SJ, Liu YH, Zhang HB, Liao YJ, Zheng F, Zhu W, Liu TH, Bian XW, Guan XY, Lin MC, Zeng MS, Zeng YX, Kung HF, Xie D (2011) EZH2 supports nasopharyngeal carcinoma cell aggressiveness by forming a co-repressor complex with HDAC1/HDAC2 and snail to inhibit E-cadherin. Oncogene. doi:10.1038/onc.2011.254

Tsai MC, Manor O, Wan Y, Mosammaparast N, Wang JK, Lan F, Shi Y, Segal E, Chang HY (2010) Long noncoding RNA as modular scaffold of histone modification complexes. Science 329:689–693

Tsukuda T, Fleming AB, Nickoloff JA, Osley MA (2005) Chromatin remodelling at a DNA double-strand break site in Saccharomyces cerevisiae. Nature 438(7066):379–383

Valeri N, Vannini I, Fanini F, Calore F, Adair B, Fabbri M (2009) Epigenetics, miRNAs, and human cancer: a new chapter in human gene regulation. Mamm Genome 20:573–580

Van Den Broeck A, Brambilla E, Moro-Sibilot D, Lantuejoul S, Brambilla C, Eymin B, Khochbin S, Gazzeri S (2008) Loss of histone H4K20 trimethylation occurs in preneoplasia and influences prognosis of non-small cell lung cancer. Clin Cancer Res 14:7237–7245

Vempati RK, Jayani RS, Notani D, Sengupta A, Galande S, Haldar D (2010) p300-mediated acetylation of histone H3 lysine 56 functions in DNA damage response in mammals. J Biol Chem 285:28553–28564

Versteege I, Sévenet N, Lange J, Rousseau-Merck MF, Ambros P, Handgretinger R, Aurias A, Delattre O (1998) Truncating mutations of hSNF5/INI1 in aggressive paediatric cancer. Nature 394(6689):203–206

Warburg O (1956) On respiratory impairment in cancer cells. Science 124:269–270

Waterland RA (2006) Assessing the effects of high methionine intake on DNA methylation. J Nutr 136:1706S–1710S

Wei Y, Xia W, Zhang Z, Liu J, Wang H, Adsay NV, Albarracin C, Yu D, Abbruzzese JL, Mills GB, Bast RC Jr, Hortobagyi GN, Hung MC (2008) Loss of trimethylation at lysine 27 of histone H3 is a predictor of poor outcome in breast, ovarian, and pancreatic cancers. Mol Carcinog 47:701–706

Wellen KE, Hatzivassiliou G, Sachdeva UM, Bui TV, Cross JR, Thompson CB (2009) ATP-citrate lyase links cellular metabolism to histone acetylation. Science 324:1076–1080

Widschwendter M, Fiegl H, Egle D, Mueller-Holzner E, Spizzo G, Marth C, Weisenberger DJ, Campan M, Young J, Jacobs I, Laird PW (2007) Epigenetic stem cell signature in cancer. Nat Genet 39:157–158

Wilson BG, Roberts CW (2011) SWI/SNF nucleosome remodellers and cancer. Nat Rev Cancer 11:481–492

Witt O, Deubzer HE, Milde T, Oehme I (2009) HDAC family: what are the cancer relevant targets? Cancer Lett 277:8–21

Wu G, Bazer FW, Cudd TA, Meininger CJ, Spencer TE (2004) Maternal nutrition and fetal development. J Nutr 134:2169–2172

Xiang Y, Zhu Z, Han G, Lin H, Xu L, Chen CD (2007) JMJD3 is a histone H3K27 demethylase. Cell Res 17:850–857

Yu J, Yu J, Rhodes DR, Tomlins SA, Cao X, Chen G, Mehra R, Wang X, Ghosh D, Shah RB, Varambally S, Pienta KJ, Chinnaiyan AM (2007) A polycomb repression signature in metastatic prostate cancer predicts cancer outcome. Cancer Res 67:10657–10663

Yuan J, Pu M, Zhang Z, Lou Z (2009) Histone H3-K56 acetylation is important for genomic stability in mammals. Cell Cycle 8:1747–1753

Zeisel SH (2007) Gene response elements, genetic polymorphisms and epigenetics influence the human dietary requirement for choline. IUBMB Life 59:380–387

Zhao CQ, Young MR, Diwan BA, Coogan TP, Waalkes MP (1997) Association of arsenic-induced malignant transformation with DNA hypomethylation and aberrant gene expression. Proc Natl Acad Sci U S A 94:10907–10912

Zimmermann S, Kiefer F, Prudenziati M, Spiller C, Hansen J, Floss T, Wurst W, Minucci S, Göttlicher M (2007) Reduced body size and decreased intestinal tumor rates in HDAC2-mutant mice. Cancer Res 67:9047–9054

Chapter 18
Epigenetic Regulation of Cancer Stem Cell Gene Expression

Sharmila A. Bapat

Abstract The concept of cancer as a stem cell disease has slowly gained ground over the last decade. A 'stem-like' state essentially necessitates that some cells in the developing tumor express the properties of remaining quiescent, self-renewing and regenerating tumors through establishment of aberrant cellular hierarchies. Alternatively, such capacities may also be reacquired through a de-differentiation process. The abnormal cellular differentiation patterns involved during either process during carcinogenesis are likely to be driven through a combination of genetic events and epigenetic regulation. The role(s) of the latter is increasingly being appreciated in acquiring the requisite genomic specificity and flexibility required for phenotypic plasticity, specifically in a context wherein genome sequences are not altered for differentiation to ensue. In this chapter, the recent advances in elucidating epigenetic mechanisms that govern the self-renewal, differentiation and regenerative potentials of cancer stem cells will be presented.

18.1 Introduction

Understanding the mechanisms that control growth and differentiation in normal cells are an essential requirement to elucidate the origin and reversibility of malignancy. In normal tissues, the generation of new cells through stem cell hierarchies is counter-balanced by apoptosis of differentiated cells, thereby maintaining a constancy of cell numbers. Once established in the normal state, temporal and spatial activation or silencing of specific genes occurs in a cell-type-specific pattern in the hierarchy to secure cell fate. Such processes are stable over several cell generations and long

---

S.A. Bapat (✉)
National Centre for Cell Science, NCCS Complex, Pune University Campus,
Ganeshkhind, Pune 411 007, India
e-mail: sabapat@nccs.res.in

after inductive developmental signals have disappeared. Disruption of cellular homeostasis is believed to be involved in several diseases including cancer (Yamada and Watanabe 2010). Over the last few decades, understanding of such mechanisms has led to the identification of stem-like cells in tumors. The salient defining features of such cancer stem cells (CSCs) that contribute to tumor cell survival and drug resistance include the following:

(i) **Potential to self renew** – which involves a capability of CSCs to undergo asymmetric division in response to microenvironmental signals and maintain themselves in a state of reversible quiescence ensuring continuous regenerative potential.
(ii) **Establishment of a cellular hierarchy** – such a phenomenon maximizes cellular resources towards an efficient cell turnover in the normal as well as transformed tissues. The hierarchies in cancer are considered aberrant due to a differentiation block leading to accumulation of proliferating progenitors. Such a maturation arrest subsequently leads to compromised tissue functioning and could induce resistance to apoptosis during disease progression.
(iii) **Tumor Regeneration** – this reflects on a capability of very few CSCs to form new tumors. This presents two main clinical implications *viz.* the few migrating CSCs regenerate secondary tumors at newer sites and, post-therapy residual CSCs can lead to disease recurrence.

The last few decades have witnessed considerable focused efforts on the identification of genetic events associated with the above cellular processes. CSC research has primarily been supported by the understanding of normal stem cell function and tissue homeostasis. The availability of *in vitro* model systems including embryonic stem (ES) cells and more recently, induced pluripotent stem (iPS) cells have revealed several critical aspects of stem cell functioning that are applicable to the CSC state. Concurrently, a careful examination of diverse gene expression profiles has revealed similar patterns in several aggressive tumors and stem cells, especially of genes that contribute to the above capabilities. Stem cell features (stemness) can also be acquired by tumor cells through a phenomenon often termed as '*de*-differentiation'. This may involve aberrant expression of CSC markers *e.g.* expression of *JARID1B*, an H3K4 demethylase in malignant melanoma (Roesch et al. 2010), or through the process of epithelial-mesenchymal transition (EMT) (Kurrey et al. 2009; Mani et al. 2008). These observations suggest that the CSC phenotype is dynamically regulated, possibly through **stochastic** means that complements the **hierarchical** regenerative mechanisms identified in several tumors. Thereby both cellular models might co-exist and tumor progression could be driven by genomic changes that give rise to other CSCs and their subsequent progeny through clonal evolution (Wani et al. 2006).

In its initial phases, the focus in CSC biology was to study stem cell-specific transcription factors, including Oct4, Nanog, and Sox2 that function in regulatory complexes to determine pluripotency. Integration of these approaches with other regulatory mechanisms to define the various components involved in malignant transformation associated with a CSC phenotype are being initiated (Kashyap et al. 2009).

These increasingly suggest that the requisite genomic specificity and flexibility required to establish phenotypic plasticity and maintain "stemness" profiles involve intricate molecular networks that include epigenetic modulation of the chromatin in both – normal and transformed cell groups (Hochedlinger et al. 2005; Li and Zhao 2008; Melcer and Meshorer 2010). A defining feature of such epigenetic events that lead to heritable changes in gene expression is that they are not based on alterations in the DNA sequence (Khavari et al. 2010). In the context of cancer while a debate continues on whether these can be interpreted to be a cause or effect of the transformation process, a strong role of their contribution is definite and reveals a cross-talk with genetic mechanisms and post-translational modifications. Thereby the regulatory role of epigenetic mechanisms including DNA methylation, chromatin remodeling through histone modifications and nucleosome positioning, non-coding RNA in stem cell biology and its dysfunctional states such as cancer and developmental disorders are receiving key importance. Here, we review the recent evidences that advance our knowledge in epigenetic regulations of cancer stem cells as aberrant derivatives of adult mammalian stem cells.

18.2 Altered DNA Methylation and Cancer Stem Cell Functions

Global DNA hypo- and hypermethylation at promoter regions of genes are common features of human tumours (Ehrlich 2009; Duthie 2011). DNA methylation involves the covalent linkage of a methyl group to carbon 5 of a cytosine in cytosine-phosphate-guanine (CpG) dinucleotides. It remains the most extensively studied epigenetic modification and negatively correlates with gene expression especially in genes with promoter regions rich in CpG islands. Promoter methylation mediated gene silencing provides an alternative to mutational inactivation; the consequent loss of function through either mechanism is known to facilitate tumor initiation and progression. Increasingly several genes which would otherwise not be recognized as being contributory to normal development and cancer are being discovered as a consequence of their altered methylation profiles in tumors. This provides an enhanced understanding of genome-wide changes in the disease that correlate with aberrant cellular functions. Altered methylation is demonstrated in the acquisition of pluripotency, evasion of cell cycle check-points (through suppression of potential key tumor suppressor genes), proliferation, immortalization and genomic instability (expression of oncogenes), as well as aberrant expression of tissue-specific housekeeping and imprinted genes (De Smet et al. 1999).

Self-renewal, a defining feature of stem cells and also CSCs, is distinguished from proliferation by the capacity of one of the daughter cells to exit from the cell cycle to enter a quiescent state without losing its regeneration potential (Bapat 2007). Reprogramming of differentiated cells through transgene expression of Oct4, Sox2, Klf4 and cMyc enables them to re-acquire self-renewal and pluripotency. This is an underlying principle in the generation of induced pluripotent stem cells (iPS) and one that has tremendous therapeutic potential (Takahashi and Yamanaka 2006). Successful reprogramming depends largely on micro-environmental factors and establishment of DNA methylation motifs in some specific CpG sites complemented by demethylation and expression of specific genes critical for acquisition of stemness (Shoae-Hassani et al. 2011). Deviations from efficient reprogramming could result in aberrant states such as transformation and an association with a differentiation block and compromised functioning of regenerated tissues (Ohi et al. 2011). Initial evidence for such a concept was the finding that abnormal promoter methylation in cancer involves several genes involved in maintenance of stem/progenitor populations in embryonic development and adult cell renewal. Moreover, the frequent establishment of these aberrant profiles occurred early during hyperplasia much before malignancy is detected (Baylin and Ohm 2006; Jones and Baylin 2007). More or recently, the introduction of two tumor suppressor genes viz. *HIC1* and *RassF1A* in a methylated (promoter) state into bone marrow-derived mesenchymal stem cells showed potential in generation of CSCs. The stem cells were identified using conventional assays for loss of anchorage dependence, increased colony formation capability, drug resistance, pluripotency, self-renewal, tumor formation, and subsequent serial xenotransplantation (Teng et al. 2011). Further, treatment of the targeted MSC with a DNA methyltransferase inhibitor reversed their tumorigenic phenotype, thereby providing a proof of concept.

Several CSCs also express the stem cell markers associated with stem cells in normal tissues. CD133, a marker associated with neural stem cells which is silenced during differentiation, sometimes presents a hemi-methylated DNA state in neuroblastoma cell lines leading to its expression; consequently it has been applied towards isolation and enrichment of putative CSCs from human tumors (Schiapparelli et al. 2010). An elaborate comparison of promoter DNA methylation patterns of cancer-related genes between human ES cells, different types of terminally differentiated tissues and cancer cell lines revealed a subset of genes methylated in ES cells and cancer (Calvanese et al. 2008). Conversely, several other genes expressed in these two cell groups were methylated in differentiated tissues, suggesting that aberrant methylation effects in cancer could also arise through defects in establishing proper methylation marks early in the cellular tissue regenerative hierarchies, rather than as an anomalous process of *de novo* hypermethylation in differentiating tissue.

At a mechanistic level, methylation patterns are established by proteins with methyltransferase activity. *De novo* methylation marks particularly during embryogenesis and cell differentiation are heritably generated by the cooperative activity of DNMT3A and DNMT3B, while maintenance of pre-existing patterns in the postnatal and adult state is mediated by DNMT1 through a preferential affinity to hemi-methylated DNA during replication. Further gene silencing occurs through either preventing transcriptional activation by blocking transcription factors from accessing target-binding sites or providing binding sites for methyl-binding domain (MBD) proteins such as MeCP2, MBD1, MBD2, etc. that in turn recruit co-repressor complexes to form a repressive chromatin structure involved in transcriptional silencing (Wade 2004).

DNMT1 has been reported to be critical in regulation of the undifferentiated phenotype through suppression of differentiation in highly regenerative tissues (Sen et al. 2010). Such constitutive methylation shows dosage-dependent effects and is essential for self-renewal of multipotent and lineage-restricted progenitors as well as long-term stem cells, but may not be involved in homing, cell cycle control and suppression of apoptosis (Broske et al. 2009; Trowbridge et al. 2009). *DNMT1* is clearly aberrantly regulated in both impaired spermatogenesis and development of embryonal carcinoma (Omisanjo et al. 2007). The *de novo* methyltransferases DNMT3A and DNMT3B are also established to play a critical role in HSC self-renewal (Tadokoro et al. 2007), and in cancer (Ding et al. 2008; Van Emburgh and Robertson 2011). In mouse ES cells, DNMT3A can function as either a positive or negative transcriptional regulator (targeting the genes vitronectin and Oct3/4 respectively), that further suggests a role for ancillary molecules in decision-making (Kotini et al. 2011). Surprisingly, DNMT3L, earlier assigned to be an enzymatically-inactive DNA methyltransferase, appears to regulate the activities of DNMT3A and DNMT3B in cervical cancer cell lines (Gokul et al. 2009). The downstream effects of these changes include altered expression pattern of genes important in nuclear reprogramming, development and cell cycle, as further evinced in an altered phenotype involving increased cellular proliferation (tumor progression) and anchorage-independent growth (stemness). Other molecules such as Lsh (a regulator of repressive chromatin at retrotransposons), also play an important role in silencing of stem cell-specific genes such as *OCT4*. Specifically, Lsh is required for establishment of DNA methylation at promoters of stem cell genes during differentiation, which in part is by regulating access of DNMT3B to its genomic targets (Xi et al. 2009).

DNA methylation was earlier thought to be a relatively stable modification with demethylation occurring as a transient and passive process *e.g.* during early embryogenesis. The frequent global hypomethylation reported in cancer is sometimes associated with loss of genomic imprinting and can lead to altered cell functioning including chromosome instability, activation of transposable elements, etc. (Eden et al. 2003; Karpf and Matsui 2005). Hypomethylation of DNA repeats during tumor progression has also been described as carcinogenesis-associated demethylation since the latter may be construed to involve an exchange of C-residues for m5C residues (Ehrlich 2009). Recently, two mechanisms of DNA demethylase activity have been described – one involving the deamination of the methyl group followed by mismatch repair by activation-induced cytidine deaminase (AID) containing protein complexes, while the other mediates a direct removal and replacement of 5-methylcytosine by cytosine (Rai et al. 2008, 2010). Demethylation catalyzed by the TET (ten-eleven-translocation) proteins Tet1, Tet2 and Tet3 have been demonstrated not only to define ES maintenance and inner cell mass cell specification, but are associated with myeloid malignancies (Mohr et al. 2011). This role has been attributed to their maintenance of Nanog expression, an established marker of the undifferentiated state (Ito et al. 2010). Combined with low fidelity DNA methylation inheritance, such mechanisms could play a dynamic role in regulating cellular processes by generating variable methylation patterns and increased chances of transformation (Xie et al. 2011).

CSCs may undergo DNA demethylation during their generation, in a manner quite akin to nuclear reprogramming and generation of iPS cells. Sorted breast cancer stem cells (CD44$^+$/CD24^{low} that correlate with increased mammosphere forming capabilities), showed a constitutive activation of Jak-STAT pathway through hypomethylation of several gene components (Hernandez-Vargas et al. 2011). Using undifferentiated zebrafish intestinal cells as a model system, it has been demonstrated that loss of the APC tumor suppressor gene causes upregulation of a DNA demethylase system and a concomitant hypomethylation of key intestinal cell fate genes. Mechanistically, the demethylase genes were found to be directly activated by Pou5f1 and Cebpβ, and indirectly repressed by retinoic acid that antagonizing these two genes to induce cell differentiation (Rai et al. 2010). Re-expression of certain embryonic genes e.g. Homeobox family proteins in cancer despite a lack of their expression in normal adult somatic tissues has also been attributed to aberrant demethylation patterns (Li et al. 2011; Rai et al. 2010).

Together, current findings re-emphasize the point that altered methylation status in cancer may not simply reflect promoter DNA methylation gone awry within the tissue stem or progenitor cell(s). In this altered scenario, the identification of genes that may either be hypermethylated or hypomethylated are being considered for their biomarker potential in several cancers including prostate, liver, gastric, etc. cancers (Ammerpohl et al. 2011; Ibragimova et al. 2010; Kwon et al. 2011). Recently, a SRAM (significantly repressed in association with methylation) gene signature that includes *EPCAM, APC, CDH1*, etc. has been applied to distinguish tumors of different lineages in breast cancer epithelial and mesenchymal lineages. The SRAM signature further identified the rare claudin-low and metaplastic tumors in association with mesenchymal characteristics, suggesting aberrant DNA methylation as a marker of cell lineage rather than tumor progression (Han et al. 2009; Sproul et al. 2011). Further, increasing reports on involvement of demethylases leading to a hypomethylated state that influence stem cell functions in tissue homeostasis and cancer reflect on more intricate methylation dynamics that realized earlier. The emerging point of view is of considerable interest in epigenetics and cancer research at present.

18.3 Histone Modifications and Cancer Stem Cell Functions

Post-translational chemical modifications of amino acid residues on histone 'tails' of the nucleosome are established and removed by a network of macromolecular complexes composed of molecules with recognition domains and enzymatic activity including histone acetyltransferases (HATs) and histone methyltransferases (HMTs) that add acetyl and methyl groups respectively, while histone deacetylases (HDACs) and histone demethylases (HDMs) remove these groups. Such complexes also interact with each other and establish cross-talk with other DNA regulatory mechanisms to tightly link the chromatin state and transcription. The specific histone residue modified and the type, location and degree of modification(s) enriched at promoter regions of genes can define the state of expression. This is the crux of the 'histone code'. A general derivation based on their spatial effects on chromatin architecture is that acetylation correlates with transcriptional activation, whereas methylation can lead to either activation or repression. For example, histone acetylation is sufficient to mediate the activation of *NESTIN* transcription in the absence of DNA demethylation (Han et al. 2009). More specific associations of trimethylated lysine 4 on histone H3 (H3K4me3) enrichment at transcriptionally active gene promoters, and trimethylated lysine 9 on histone H3 (H3K9me3) and trimethylated lysine 27 on histone H3 (H3K27me3) with silent gene promoters have also been established (Portela and Esteller 2010; Weishaupt et al. 2010). Stem cells are known to exhibit an association of active chromatin states (defined by H3K4me3) around pluripotency genes and silencing of cell fate- and differentiation- specific genes mediated by H3K27me3 (Boyer et al. 2006; Chi and Bernstein 2009; Mikkelsen et al. 2008). Under-representation of repressive histone marks could be indicative of epigenetic plasticity in stem, young and tumor cells, while committed and senescent (old) cells often display increased levels of these more stable repressive modifications (Kubicek et al. 2006).

The co-existence of two histone modifications with opposing effects at the same promoter region describes 'bivalent domains' that have established their own connotation with biological functions (Chi and Bernstein 2009; Mikkelsen et al. 2008). Genome-wide profiling of histone marks in undifferentiated stem cells identified active H3K4me3 and repressive H3K27me3 marks to be enriched at promoters of developmentally important genes (Bernstein et al. 2006). Such bivalency maintains transcriptional quiescence or poise yet permits the flexibility essential for divergence along an alternate cell fate(s). Subsequently such marks are selectively lost during differentiation to define specific lineage commitment and functionality. Establishment of the H3K4me3-H3K27me3 mark involves activation of two opposing regulator complexes *viz*. trithorax group that catalyzes the H3K4me3 mark and the polycomb group that establishes H3K27me3 marks (Lund and van Lohuizen 2004; Orlando 2003; Ringrose and Paro 2007; Valk-Lingbeek et al. 2004).

The selective transition from bivalent to repressive histone modifications in cell cycle regulatory genes, yet retention of bivalent promoters in developmental genes has been identified to be a key feature in transformation (Huang and Esteller 2010). Cultured mammary epithelial cells routinely enter into senescence and growth arrest after a few passages; however, rare progenitor cell(s) may proliferate and develop into neoplastic clone(s) (Hinshelwood and Clark 2008). Such aberrant progenitors while retaining bivalent features to regulate differentiation-associated genes (leading to lack of lineage commitment and/or maturation arrest), enrich the H3K4me3 mark to activate stemness-associated genes, and the H3K27me3 mark to mediate silencing of tumor suppressor genes. Thus, multipotent renal progenitor cells residing within the nephrogenic mesenchyme, which are under tight control during renal development, undergo sequential epigenetic alterations that mediate silencing of the nephric-progenitor genes such as *SIX2*, leading to tumor initiation and progression (Metsuyanim et al. 2008). In normal stem cells, H3K9 acetylation and demethylation regulates self-renewal through canonical signaling, cell cycle and cytokine related pathways (Tan et al. 2008). In cancer, H3K9me2 / H3K9me3 often serve to enhance H3K27me3-mediated silencing to drive effective transformation (Bloushtain-Qimron et al. 2008; Cheng et al. 2008; Hsu et al. 2009).

Aberrant H3K4me3 marks leading to compromised cell fate decisions and cancer correlates with dysfunctional SET1/MLL histone methyltransferase complex components such as Dpy-30 (Jiang et al. 2011) or complex recruiting factors such as Pygo2 (Duan et al. 2005; Gu et al. 2009). The Gfi1 transcriptional regulator oncoprotein requires histone lysine methyltransferase G9a, histone deacetylase 1 (HDAC1) and H3K9me2 in repressing the cell cycle regulator p21Cip/WAF1 (Duan et al. 2005). While histone demethylases have also been shown to play a key role in eukaryotic gene transcription; the specific molecular networks coordinated by them towards regulation of specific target genes and biological processes remains to be elucidated (Di Stefano et al. 2011).

Histone-modifier genes usually have tissue-type specific patterns of expression in cancer, and the landscape varies when solid tumors or hematological malignancies are compared (Ozdag et al. 2006). The contributions of Polycomb group proteins (PRCs) remain the best studied mechanism associated with histone modifications (Margueron and Reinberg 2011) since its functional versatility contributes significantly to the complexity of regulatory pathways in cancer (Kanno et al. 2008). The PRC2 complex (EZH2, SUZ12, EEX) initiates gene silencing by methylating H3K27, while PRC1 (Bmi1 being a prototype component) maintains gene silencing through mono-ubiquitination of histone H2A lysine 119. Expression of PRC2 components is regulated by the retinoblastoma protein (pRB)–E2F transcription factors, and are consequently overexpressed in various cancers such as melanoma, lymphoma, and breast and prostate cancer (Bracken et al. 2003; Karanikolas et al. 2009; Li et al. 2009; Margueron et al. 2008). On the other hand, depletion of EZH2 leads to enhanced activation of the Ink4/Arf locus that regulates expression of the cell cycle checkpoints p16Ink4a, p19Arf and p15Ink4b (Bracken et al. 2007; Ezhkova et al. 2009). PRC2 is also required for the acquisition of pluripotency, as $Eed^{-/-}$ and $Suz12^{-/-}$ ES cells fail to induce the reprogramming of B cells (Pereira et al. 2010). EZH2 is identified to play a crucial role in neural stem cell self-renewal and proper lineage commitment (Sher et al. 2008). The epithelial cell adhesion molecule (EpCAM) expressed in several progenitor cell populations regulates the reprogramming genes c-MYC, OCT4, NANOG, SOX2, and KLF4 and is thereby critical in maintenance of the undifferentiated state. EpCAM is silenced in some tumors through establishment of a H3K27me3 mark at its promoter by SUZ12 and JMJD3, leading to a perturbed cellular hierarchy (Lu et al. 2010). A novel PcG protein PCL2 has been identified in association with enhanced ES self-renewal, delayed differentiation and altered patterns of histone methylation (Walker et al. 2010). Depletion of PCL2 leads to decreased H3K27me3 and increased expression of the pluripotency transcription factors Tbx3, Klf4, Foxd3, that subsequently activate expression of Oct4, Nanog and Sox2 through a feed-forward gene regulatory circuit, altering the core pluripotency network and driving cell fate decisions towards self-renewal (Walker et al. 2011). Another PcG component L3MBTL is located at an imprinted locus at 20q, and is implied in the pathogenesis of myeloid malignancies associated with 20q deletions (Li et al. 2004).

Co-operation and co-ordination between PRC1 and PRC2 components is also essential for normal stem cell functioning. The p16 locus is reported silenced through association with H3K27me3, BMI1, RING2, and SUZ12. This is mediated by the pRB proteins that recruit PRC2 to trimethylate p16, priming the BMI1-containing PRC1L ubiquitin ligase complex to silence p16 (Kotake et al. 2007). While the role of Bmi1 in stem/progenitor cells self-renewal is well established, in hepatic stem cells Ezh2 additionally plays an essential role in maintenance of not only their proliferative and self-renewal functions, but also full execution of a differentiation program (Aoki et al. 2010). Importantly, Ezh2 depletion (but not Bmi1) promoted up-regulation of several transcriptional regulators leading to differentiation and terminal maturation of hepatocytes; thereby emphasizing a functional non-redundancy of the PRC2 and PRC2 mediated chromatin regulatory mechanisms.

Taken together, the above data emphatically supports the hypothesis that deregulation of the 'histone code' in cells with pluripotent potential may alter defining properties of stem cells, self-renewal and differentiation potential, leading to cancer initiation and progression (Biancotto et al. 2010; Shukla et al. 2008), and this could happen through *de novo* "writing", "erasing" or "misinterpretation" of histone marks by activities of the regulatory complexes (Yoshimura 2009).

18.4 Cross-Talk Between DNA and Histone Methylation

Chromatin thus is dynamically active in regulating transcriptional processes through post-synthetic modifications of DNA and histones. At a higher level, DNA and histone associations also cooperate to impart stability to this regulation with aberrant CpG hypermethylation events in cancer being more frequently associated with the promoters of those genes with enriched PRC occupancy. Such cooperative events increase with age and may contribute to carcinogenesis by irreversibly silencing genes that are suppressed in stem cells, as demonstrated through the identification of an age-PRC target methylation signature present in preneoplastic conditions that drive transformation associated gene expression changes (Teschendorff et al. 2010).

The fact that stem cell associated PRC targets appear to have an increased propensity for cancer-specific promoter DNA hypermethylation than non-targets, supports a stem cell origin of cancer in which reversible gene repression is replaced by permanent silencing, locking the cell into a perpetual state of self-renewal and thereby predisposing it to subsequent malignant transformation (Widschwendter et al. 2007). During renal tumorigenesis in stem cell-like Wilms' tumor xenografts, the PcG components EZH2, BMI1, EED and SUZ12 showed cooperative upregulation along with nephric-progenitor genes WT1, PAX2 and SALL1 although SIX2 was downregulated through promoter hypermethylation leading to loss of renal differentiation. This links polycomb activation and promoter methylation in renal progenitor populations during the initiation and progression of renal cancer (Metsuyanim et al. 2008). Such polycomb-premarking has been also identified in colorectal cancer (Rada-Iglesias et al. 2009).

It is also indicated that the key role of promoter DNA hypermethylation in cancer cells is to tightly repress the transcription of specific genes beyond the repression achieved in the bivalent chromatin state. Higher order repression may further be established by looping in the 3' region of such genes that further encompass multiple, abnormally hypermethylated CpG islands surrounding the entire gene resulting in no basal transcription. This is also evinced by very low levels of Pol2 polymerase at the transcription start site(TSS) and enrichment of the methyl-cytosine binding protein, MBD2 at the CpG islands around the TSS which are classically assessed for DNA methylation (Akiyama et al. 2003; Tiwari et al. 2008). Such stable repression is difficult to revert by simple targeting through demethylation. For example, in breast and colon cancer chromatin immunoprecipation (ChIP)-chip tiling arrays (McGarvey et al. 2006, 2008) have revealed that hypermethylated gene promoters are also enriched with a series of repressive chromatin marks including H3K27me3. While promoter DNA demethylation of these genes and loss of MBD2 either by treatment with the demethylating agent, 5-deoxy-azacytidine (DAC) or by genetic knockout of DNA methyltransferases (*DNMT1* and *DNMT3b*) could be induced (Rhee et al. 2002), repressive H3K27me3 remained and with a concurrent repositioning of the active H3K4me2 mark, led to the formation of a bivalent chromatin state with continuing low level gene expression.

Formation of the above "repressive hubs" draws attention to the dissimilarities between the epigenetic regulatory mechanisms higher order chromatin conformation in CSCs and for the same gene(s) in stem/progenitor cells. In the normal state, activation is induced through dissolution of loops and consequent signal transduction induced cell differentiation. In cancer cells however such genes although appear to be 'poised' for differentiation through the presence of bivalent histone marks, are often further and heritably repressed through additional histone marks and higher order chromatin compaction mechanisms including DNA methylation that effectively renders the gene inaccessible for the transcription required to facilitate conversion to a differentiated state. This effectively establishes the maturation arrested state characteristic of a CSC hierarchy.

18.5 Conclusion and Future Perspectives

The deluge of new discoveries relating to epigenetics in the last few years including establishment of comprehensive DNA methylomes of normal and aberrant states, identification of non-CpG methylation, the definition of CpG island shores, description of new histone modifications and histone variants and their roles, reports of mutations in the epigenetic machinery, the flurry of miRNA-ncRNA studies and the cross-talk between all these mechanisms that would reflect on higher order chromatin regulatory mechanisms, together highlight the recognition of epigenetic mechanisms in homeostatic mechanisms in cells. The acquisition and maintenance of cell fate or "identity" by strict coordination between genetic and epigenetic programs are essential for growth and development. Cancer cells possess traits reminiscent of those ascribed to normal stem cells. Histologically poorly differentiated tumors show preferential overexpression of genes normally enriched in ES cells combined with repression of polycomb-regulated genes and overexpression of activation targets of Nanog, OCT4, SOX2 and c-Myc. On similar lines, an epigenetic stem cell signature may be defined with reference to DNA promoter hypermethylation of polycomb group targets and bivalent chromatin that is described in human tumors. This suggests that understanding the mechanisms by which pluripotency transcription factors, complexes imparting/removing acetylation and methylation modifications, polycomb repressive complexes, histone modifications and higher order chromatin compactions maintain the balance between self-renewal, cellular proliferation and differentiation in both normal and aberrant states is important. Such links are quite compelling for continuing research towards validation or refuting various hypotheses that could enhance our understanding of the biology of cancer initiation.

However, several key questions remain:

- *Are specific epigenetic modifications a cause or a consequence of aberrant cellular differentiation?*
- *What mechanisms convey sequence specificity to the complexes involved?*
- *How can causative (driver) epigenetic changes be distinguished from bystander events?*
- *Can the threshold for accumulation and complementation of genetic and epigenetic changes in a pathological condition be defined?*
- *What restricts formation of "repressive hubs" around developmental genes in normal stem cells?*

The establishment of meaningful stem cell models for addressing these questions will be instrumental towards understanding the dynamics of epigenetics in stem cells, development, cancer and aging. Such a research focus will be key in our progress in tackling tumor formation and disease progression.

References

Akiyama Y, Maesawa C, Ogasawara S, Terashima M, Masuda T (2003) Cell-type-specific repression of the maspin gene is disrupted frequently by demethylation at the promoter region in gastric intestinal metaplasia and cancer cells. Am J Pathol 163:1911–1919

Ammerpohl O, Pratschke J, Schafmayer C, Haake A, Faber W, von Kampen O, Brosch M, Sipos B, von Schonfels W, Balschun K, Rocken C, Arlt A, Schniewind B, Grauholm J, Kalthoff H, Neuhaus P, Stickel F, Schreiber S, Becker T, Siebert R (2011) Hampe J. Distinct DNA methylation patterns in cirrhotic liver and hepatocellular carcinoma, Int J Cancer

Aoki R, Chiba T, Miyagi S, Negishi M, Konuma T, Taniguchi H, Ogawa M, Yokosuka O, Iwama A (2010) The polycomb group gene product Ezh2 regulates proliferation and differentiation of murine hepatic stem/progenitor cells. J Hepatol 52:854–863

Bapat SA (2007) Evolution of cancer stem cells. Semin Cancer Biol 17:204–213

Baylin SB, Ohm JE (2006) Epigenetic gene silencing in cancer – a mechanism for early oncogenic pathway addiction? Nat Rev Cancer 6:107–116

Bernstein BE, Mikkelsen TS, Xie X, Kamal M, Huebert DJ, Cuff J, Fry B, Meissner A, Wernig M, Plath K, Jaenisch R, Wagschal A, Feil R, Schreiber SL, Lander ES (2006) A bivalent chromatin structure marks key developmental genes in embryonic stem cells. Cell 125:315–326

Biancotto C, Frige G, Minucci S (2010) Histone modification therapy of cancer. Adv Genet 70:341–386

Bloushtain-Qimron N, Yao J, Snyder EL, Shipitsin M, Campbell LL, Mani SA, Hu M, Chen H, Ustyansky V, Antosiewicz JE, Argani P, Halushka MK, Thomson JA, Pharoah P, Porgador A, Sukumar S, Parsons R, Richardson AL, Stampfer MR, Gelman RS, Nikolskaya T, Nikolsky Y, Polyak K (2008) Cell type-specific DNA methylation patterns in the human breast. Proc Natl Acad Sci U S A 105:14076–14081

Boyer LA, Plath K, Zeitlinger J, Brambrink T, Medeiros LA, Lee TI, Levine SS, Wernig M, Tajonar A, Ray MK, Bell GW, Otte AP, Vidal M, Gifford DK, Young RA, Jaenisch R (2006) Polycomb complexes repress developmental regulators in murine embryonic stem cells. Nature 441:349–353

Bracken AP, Pasini D, Capra M, Prosperini E, Colli E, Helin K (2003) EZH2 is downstream of the pRB-E2F pathway, essential for proliferation and amplified in cancer. EMBO J 22: 5323–5335

Bracken AP, Kleine-Kohlbrecher D, Dietrich N, Pasini D, Gargiulo G, Beekman C, Theilgaard-Monch K, Minucci S, Porse BT, Marine JC, Hansen KH, Helin K (2007) The Polycomb group proteins bind throughout the INK4A-ARF locus and are disassociated in senescent cells. Genes Dev 21:525–530

Broske AM, Vockentanz L, Kharazi S, Huska MR, Mancini E, Scheller M, Kuhl C, Enns A, Prinz M, Jaenisch R, Nerlov C, Leutz A, Andrade-Navarro MA, Jacobsen SE, Rosenbauer F (2009) DNA methylation protects hematopoietic stem cell multipotency from myeloerythroid restriction. Nat Genet 41:1207–1215

Calvanese V, Horrillo A, Hmadcha A, Suarez-Alvarez B, Fernandez AF, Lara E, Casado S, Menendez P, Bueno C, Garcia-Castro J, Rubio R, Lapunzina P, Aliminos M, Borghese L, Terstegge S, Harrison NJ, Moore HD, Brustle O, Lopez-Larrea C, Andrews PW, Soria B, Esteller M, Fraga MF (2008) Cancer genes hypermethylated in human embryonic stem cells. PLoS One 3:e3294

Cheng Z, Ke Y, Ding X, Wang F, Wang H, Wang W, Ahmed K, Liu Z, Xu Y, Aikhionbare F, Yan H, Liu J, Xue Y, Yu J, Powell M, Liang S, Wu Q, Reddy SE, Hu R, Huang H, Jin C, Yao X (2008) Functional characterization of TIP60 sumoylation in UV-irradiated DNA damage response. Oncogene 27:931–941

Chi AS, Bernstein BE (2009) Developmental biology. Pluripotent chromatin state. Science 323:220–221

De Smet C, Lurquin C, Lethe B, Martelange V, Boon T (1999) DNA methylation is the primary silencing mechanism for a set of germ line- and tumor-specific genes with a CpG-rich promoter. Mol Cell Biol 19:7327–7335

Di Stefano L, Walker JA, Burgio G, Corona DF, Mulligan P, Naar AM, Dyson NJ (2011) Functional antagonism between histone H3K4 demethylases in vivo. Genes Dev 25:17–28

Ding WJ, Fang JY, Chen XY, Peng YS (2008) The expression and clinical significance of DNA methyltransferase proteins in human gastric cancer. Dig Dis Sci 53:2083–2089

Duan Z, Zarebski A, Montoya-Durango D, Grimes HL, Horwitz M (2005) Gfi1 coordinates epigenetic repression of p21Cip/WAF1 by recruitment of histone lysine methyltransferase G9a and histone deacetylase 1. Mol Cell Biol 25:10338–10351

Duthie SJ (2011) Epigenetic modifications and human pathologies: cancer and CVD. Proc Nutr Soc 70:47–56

Eden A, Gaudet F, Waghmare A, Jaenisch R (2003) Chromosomal instability and tumors promoted by DNA hypomethylation. Science 300:455

Ehrlich M (2009) DNA hypomethylation in cancer cells. Epigenomics 1:239–259

Ezhkova E, Pasolli HA, Parker JS, Stokes N, Su IH, Hannon G, Tarakhovsky A, Fuchs E (2009) Ezh2 orchestrates gene expression for the stepwise differentiation of tissue-specific stem cells. Cell 136:1122–1135

Gokul G, Ramakrishna G, Khosla S (2009) Reprogramming of HeLa cells upon DNMT3L overexpression mimics carcinogenesis. Epigenetics 4:322–329

Gu B, Sun P, Yuan Y, Moraes RC, Li A, Teng A, Agrawal A, Rheaume C, Bilanchone V, Veltmaat JM, Takemaru K, Millar S, Lee EY, Lewis MT, Li B, Dai X (2009) Pygo2 expands mammary progenitor cells by facilitating histone H3 K4 methylation. J Cell Biol 185:811–826

Han DW, Do JT, Arauzo-Bravo MJ, Lee SH, Meissner A, Lee HT, Jaenisch R, Scholer HR (2009) Epigenetic hierarchy governing Nestin expression. Stem Cells 27:1088–1097

Hernandez-Vargas H, Ouzounova M, Calvez-Kelm F, Lambert MP, McKay-Chopin S, Tavtigian SV, Puisieux A, Matar C, Herceg Z (2011) Methylome analysis reveals Jak-STAT pathway deregulation in putative breast cancer stem cells. Epigenetics 6:428–439

Hinshelwood RA, Clark SJ (2008) Breast cancer epigenetics: normal human mammary epithelial cells as a model system. J Mol Med 86:1315–1328

Hochedlinger K, Yamada Y, Beard C, Jaenisch R (2005) Ectopic expression of Oct-4 blocks progenitor-cell differentiation and causes dysplasia in epithelial tissues. Cell 121:465–477

Hsu M, Richardson CA, Olivier E, Qiu C, Bouhassira EE, Lowrey CH, Fiering S (2009) Complex developmental patterns of histone modifications associated with the human beta-globin switch in primary cells. Exp Hematol 37:799–806

Huang TH, Esteller M (2010) Chromatin remodeling in mammary gland differentiation and breast tumorigenesis. Cold Spring Harb Perspect Biol 2:a004515

Ibragimova I, de Ibanez CI, Hoffman AM, Potapova A, Dulaimi E, Al Saleem T, Hudes GR, Ochs MF, Cairns P (2010) Global reactivation of epigenetically silenced genes in prostate cancer. Cancer Prev Res (Phila) 3(9):1084–1092

Ito S, D'Alessio AC, Taranova OV, Hong K, Sowers LC, Zhang Y (2010) Role of Tet proteins in 5mC to 5hmC conversion, ES-cell self-renewal and inner cell mass specification. Nature 466:1129–1133

Jiang H, Shukla A, Wang X, Chen WY, Bernstein BE, Roeder RG (2011) Role for Dpy-30 in ES cell-fate specification by regulation of H3K4 methylation within bivalent domains. Cell 144:513–525

Jones PA, Baylin SB (2007) The epigenomics of cancer. Cell 128:683–692

Kanno R, Janakiraman H, Kanno M (2008) Epigenetic regulator polycomb group protein complexes control cell fate and cancer. Cancer Sci 99:1077–1084

Karanikolas BD, Figueiredo ML, Wu L (2009) Polycomb group protein enhancer of zeste 2 is an oncogene that promotes the neoplastic transformation of a benign prostatic epithelial cell line. Mol Cancer Res 7:1456–1465

Karpf AR, Matsui S (2005) Genetic disruption of cytosine DNA methyltransferase enzymes induces chromosomal instability in human cancer cells. Cancer Res 65:8635–8639

Kashyap V, Rezende NC, Scotland KB, Shaffer SM, Persson JL, Gudas LJ, Mongan NP (2009) Regulation of stem cell pluripotency and differentiation involves a mutual regulatory circuit of the NANOG, OCT4, and SOX2 pluripotency transcription factors with polycomb repressive complexes and stem cell microRNAs. Stem Cells Dev 18:1093–1108

Khavari DA, Sen GL, Rinn JL (2010) DNA methylation and epigenetic control of cellular differentiation. Cell Cycle 9:3880–3883

Kotake Y, Cao R, Viatour P, Sage J, Zhang Y, Xiong Y (2007) pRB family proteins are required for H3K27 trimethylation and Polycomb repression complexes binding to and silencing p16INK4alpha tumor suppressor gene. Genes Dev 21:49–54

Kotini AG, Mpakali A, Agalioti T (2011) Dnmt3a1 upregulates transcription of distinct genes and targets chromosomal gene clusters for epigenetic silencing in mouse embryonic stem cells. Mol Cell Biol 31:1577–1592

Kubicek S, Schotta G, Lachner M, Sengupta R, Kohlmaier A, Perez-Burgos L, Linderson Y, Martens JH, O'Sullivan RJ, Fodor BD, Yonezawa M, Peters AH, Jenuwein T (2006) The role of histone modifications in epigenetic transitions during normal and perturbed development. Ernst Schering Res Found Workshop , 1–27

Kurrey NK, Jalgaonkar SP, Joglekar AV, Ghanate AD, Chaskar PD, Doiphode RY, Bapat SA (2009) Snail and slug mediate radioresistance and chemoresistance by antagonizing p53-mediated apoptosis and acquiring a stem-like phenotype in ovarian cancer cells. Stem Cells 27:2059–2068

Kwon OH, Park JL, Kim M, Kim JH, Lee HC, Kim HJ, Noh SM, Song KS, Yoo HS, Paik SG, Kim SY, Kim YS (2011) Aberrant up-regulation of LAMB3 and LAMC2 by promoter demethylation in gastric cancer. Biochem Biophys Res Commun 406:539–545

Li X, Zhao X (2008) Epigenetic regulation of mammalian stem cells. Stem Cells Dev 17:1043–1052

Li J, Bench AJ, Vassiliou GS, Fourouclas N, Ferguson-Smith AC, Green AR (2004) Imprinting of the human L3MBTL gene, a polycomb family member located in a region of chromosome 20 deleted in human myeloid malignancies. Proc Natl Acad Sci U S A 101:7341–7346

Li X, Gonzalez ME, Toy K, Filzen T, Merajver SD, Kleer CG (2009) Targeted overexpression of EZH2 in the mammary gland disrupts ductal morphogenesis and causes epithelial hyperplasia. Am J Pathol 175:1246–1254

Li Q, O'Malley ME, Bartlett DL, Guo ZS (2011) Homeobox gene Rhox5 is regulated by epigenetic mechanisms in cancer and stem cells and promotes cancer growth. Mol Cancer 10:63

Lu TY, Lu RM, Liao MY, Yu J, Chung CH, Kao CF, Wu HC (2010) Epithelial cell adhesion molecule regulation is associated with the maintenance of the undifferentiated phenotype of human embryonic stem cells. J Biol Chem 285:8719–8732

Lund AH, van Lohuizen M (2004) Polycomb complexes and silencing mechanisms. Curr Opin Cell Biol 16:239–246

Mani SA, Guo W, Liao MJ, Eaton EN, Ayyanan A, Zhou AY, Brooks M, Reinhard F, Zhang CC, Shipitsin M, Campbell LL, Polyak K, Brisken C, Yang J, Weinberg RA (2008) The epithelial-mesenchymal transition generates cells with properties of stem cells. Cell 133:704–715

Margueron R, Reinberg D (2011) The Polycomb complex PRC2 and its mark in life. Nature 469:343–349

Margueron R, Li G, Sarma K, Blais A, Zavadil J, Woodcock CL, Dynlacht BD, Reinberg D (2008) Ezh1 and Ezh2 maintain repressive chromatin through different mechanisms. Mol Cell 32:503–518

McGarvey KM, Fahrner JA, Greene E, Martens J, Jenuwein T, Baylin SB (2006) Silenced tumor suppressor genes reactivated by DNA demethylation do not return to a fully euchromatic chromatin state. Cancer Res 66:3541–3549

McGarvey KM, Van Neste L, Cope L, Ohm JE, Herman JG, Van Criekinge W, Schuebel KE, Baylin SB (2008) Defining a chromatin pattern that characterizes DNA-hypermethylated genes in colon cancer cells. Cancer Res 68:5753–5759

Melcer S, Meshorer E (2010) Chromatin plasticity in pluripotent cells. Essays Biochem 48:245–262

Metsuyanim S, Pode-Shakked N, Schmidt-Ott KM, Keshet G, Rechavi G, Blumental D, Dekel B (2008) Accumulation of malignant renal stem cells is associated with epigenetic changes in normal renal progenitor genes. Stem Cells 26:1808–1817

Mikkelsen TS, Hanna J, Zhang X, Ku M, Wernig M, Schorderet P, Bernstein BE, Jaenisch R, Lander ES, Meissner A (2008) Dissecting direct reprogramming through integrative genomic analysis. Nature 454:49–55

Mohr F, Dohner K, Buske C, Rawat VP (2011) TET genes: new players in DNA demethylation and important determinants for stemness. Exp Hematol 39:272–281

Ohi Y, Qin H, Hong C, Blouin L, Polo JM, Guo T, Qi Z, Downey SL, Manos PD, Rossi DJ, Yu J, Hebrok M, Hochedlinger K, Costello JF, Song JS, Ramalho-Santos M (2011) Incomplete DNA methylation underlies a transcriptional memory of somatic cells in human iPS cells. Nat Cell Biol 13:541–549

Omisanjo OA, Biermann K, Hartmann S, Heukamp LC, Sonnack V, Hild A, Brehm R, Bergmann M, Weidner W, Steger K (2007) DNMT1 and HDAC1 gene expression in impaired spermatogenesis and testicular cancer. Histochem Cell Biol 127:175–181

Orlando V (2003) Polycomb, epigenomes, and control of cell identity. Cell 112:599–606

Ozdag H, Teschendorff AE, Ahmed AA, Hyland SJ, Blenkiron C, Bobrow L, Veerakumarasivam A, Burtt G, Subkhankulova T, Arends MJ, Collins VP, Bowtell D, Kouzarides T, Brenton JD, Caldas C (2006) Differential expression of selected histone modifier genes in human solid cancers. BMC Genomics 7:90

Pereira CF, Piccolo FM, Tsubouchi T, Sauer S, Ryan NK, Bruno L, Landeira D, Santos J, Banito A, Gil J, Koseki H, Merkenschlager M, Fisher AG (2010) ESCs require PRC2 to direct the successful reprogramming of differentiated cells toward pluripotency. Cell Stem Cell 6:547–556

Portela A, Esteller M (2010) Epigenetic modifications and human disease. Nat Biotechnol 28:1057–1068

Rada-Iglesias A, Enroth S, Andersson R, Wanders A, Pahlman L, Komorowski J, Wadelius C (2009) Histone H3 lysine 27 trimethylation in adult differentiated colon associated to cancer DNA hypermethylation. Epigenetics 4:107–113

Rai K, Huggins IJ, James SR, Karpf AR, Jones DA, Cairns BR (2008) DNA demethylation in zebrafish involves the coupling of a deaminase, a glycosylase, and gadd45. Cell 135:1201–1212

Rai K, Sarkar S, Broadbent TJ, Voas M, Grossmann KF, Nadauld LD, Dehghanizadeh S, Hagos FT, Li Y, Toth RK, Chidester S, Bahr TM, Johnson WE, Sklow B, Burt R, Cairns BR, Jones DA (2010) DNA demethylase activity maintains intestinal cells in an undifferentiated state following loss of APC. Cell 142:930–942

Rhee I, Bachman KE, Park BH, Jair KW, Yen RW, Schuebel KE, Cui H, Feinberg AP, Lengauer C, Kinzler KW, Baylin SB, Vogelstein B (2002) DNMT1 and DNMT3b cooperate to silence genes in human cancer cells. Nature 416:552–556

Ringrose L, Paro R (2007) Polycomb/Trithorax response elements and epigenetic memory of cell identity. Development 134:223–232

Roesch A, Fukunaga-Kalabis M, Schmidt EC, Zabierowski SE, Brafford PA, Vultur A, Basu D, Gimotty P, Vogt T, Herlyn M (2010) A temporarily distinct subpopulation of slow-cycling melanoma cells is required for continuous tumor growth. Cell 141:583–594

Schiapparelli P, Enguita-German M, Balbuena J, Rey JA, Lazcoz P, Castresana JS (2010) Analysis of stemness gene expression and CD133 abnormal methylation in neuroblastoma cell lines. Oncol Rep 24:1355–1362

Sen GL, Reuter JA, Webster DE, Zhu L, Khavari PA (2010) DNMT1 maintains progenitor function in self-renewing somatic tissue. Nature 463:563–567

Sher F, Rossler R, Brouwer N, Balasubramaniyan V, Boddeke E, Copray S (2008) Differentiation of neural stem cells into oligodendrocytes: involvement of the polycomb group protein Ezh2. Stem Cells 26:2875–2883

Shoae-Hassani A, Sharif S, Verdi J (2011) A neurosteroid, DHEA, could improves somatic cell reprogramming. Cell Biol Int 35(10):1037–1041

Shukla V, Vaissiere T, Herceg Z (2008) Histone acetylation and chromatin signature in stem cell identity and cancer. Mutat Res 637:1–15

Sproul D, Nestor C, Culley J, Dickson JH, Dixon JM, Harrison DJ, Meehan RR, Sims AH, Ramsahoye BH (2011) Transcriptionally repressed genes become aberrantly methylated and distinguish tumors of different lineages in breast cancer. Proc Natl Acad Sci U S A 108:4364–4369

Tadokoro Y, Ema H, Okano M, Li E, Nakauchi H (2007) De novo DNA methyltransferase is essential for self-renewal, but not for differentiation, in hematopoietic stem cells. J Exp Med 204:715–722

Takahashi K, Yamanaka S (2006) Induction of pluripotent stem cells from mouse embryonic and adult fibroblast cultures by defined factors. Cell 126:663–676

Tan J, Huang H, Huang W, Li L, Guo J, Huang B, Lu J (2008) The genomic landscapes of histone H3-Lys9 modifications of gene promoter regions and expression profiles in human bone marrow mesenchymal stem cells. J Genet Genomics 35:585–593

Teng IW, Hou PC, Lee KD, Chu PY, Yeh KT, Jin VX, Tseng MJ, Tsai SJ, Chang YD, Wu CS, Sun HS, Tsai KD, Jeng LB, Nephew KP, Huang TH, Hsiao SH, Leu YW (2011) Targeted methylation of two tumor suppressor genes is sufficient to transform mesenchymal stem cells into cancer stem/initiating cells. Cancer Res 71(13):4653–663

Teschendorff AE, Menon U, Gentry-Maharaj A, Ramus SJ, Weisenberger DJ, Shen H, Campan M, Noushmehr H, Bell CG, Maxwell AP, Savage DA, Mueller-Holzner E, Marth C, Kocjan G, Gayther SA, Jones A, Beck S, Wagner W, Laird PW, Jacobs IJ, Widschwendter M (2010) Age-dependent DNA methylation of genes that are suppressed in stem cells is a hallmark of cancer. Genome Res 20:440–446

Tiwari VK, McGarvey KM, Licchesi JD, Ohm JE, Herman JG, Schubeler D, Baylin SB (2008) PcG proteins, DNA methylation, and gene repression by chromatin looping. PLoS Biol 6:2911–2927

Trowbridge JJ, Snow JW, Kim J, Orkin SH (2009) DNA methyltransferase 1 is essential for and uniquely regulates hematopoietic stem and progenitor cells. Cell Stem Cell 5:442–449

Valk-Lingbeek ME, Bruggeman SW, van Lohuizen M (2004) Stem cells and cancer; the polycomb connection. Cell 118:409–418

Van Emburgh BO, Robertson KD (2011) Modulation of Dnmt3b function in vitro by interactions with Dnmt3L, Dnmt3a and Dnmt3b splice variants. Nucleic Acids Res 39(12):4984–5002

Wade PA (2004) Dynamic regulation of DNA methylation coupled transcriptional repression: BDNF regulation by MeCP2. Bioessays 26:217–220

Walker E, Chang WY, Hunkapiller J, Cagney G, Garcha K, Torchia J, Krogan NJ, Reiter JF, Stanford WL (2010) Polycomb-like 2 associates with PRC2 and regulates transcriptional networks during mouse embryonic stem cell self-renewal and differentiation. Cell Stem Cell 6:153–166

Walker E, Manias JL, Chang WY, Stanford WL (2011) PCL2 modulates gene regulatory networks controlling self-renewal and commitment in embryonic stem cells. Cell Cycle 10:45–51

Wani AA, Sharma N, Shouche YS, Bapat SA (2006) Nuclear-mitochondrial genomic profiling reveals a pattern of evolution in epithelial ovarian tumor stem cells. Oncogene 25:6336–6344

Weishaupt H, Sigvardsson M, Attema JL (2010) Epigenetic chromatin states uniquely define the developmental plasticity of murine hematopoietic stem cells. Blood 115:247–256

Widschwendter M, Fiegl H, Egle D, Mueller-Holzner E, Spizzo G, Marth C, Weisenberger DJ, Campan M, Young J, Jacobs I, Laird PW (2007) Epigenetic stem cell signature in cancer. Nat Genet 39:157–158

Xi S, Geiman TM, Briones V, Guang TY, Xu H, Muegge K (2009) Lsh participates in DNA methylation and silencing of stem cell genes. Stem Cells 27:2691–2702

Xie H, Wang M, de Andrade A, Bonaldo MF, Galat V, Arndt K, Rajaram V, Goldman S, Tomita T, Soares MB (2011) Genome-wide quantitative assessment of variation in DNA methylation patterns. Nucleic Acids Res 39:4099–4108

Yamada Y, Watanabe A (2010) Epigenetic codes in stem cells and cancer stem cells. Adv Genet 70:177–199

Yoshimura A (2009) Stat1 phosphorylation is a molecular switch of Ras signaling and oncogenesis. Cell Cycle 8:1981–1982

Chapter 19
Role of Epigenetic Mechanisms in the Vascular Complications of Diabetes

Marpadga A. Reddy and Rama Natarajan

Abstract Diabetes and metabolic disorders are leading causes of micro- and macrovascular complications. Furthermore, efforts to treat these complications are hampered by metabolic memory, a phenomenon in which prior exposure to hyperglycemia predisposes diabetic patients to the continued development of vascular diseases despite subsequent glycemic control. Persistently increased levels of oxidant stress and inflammatory genes are key features of these pathologies. Biochemical and molecular studies showed that hyperglycemia induced activation of NF-κB, signaling and actions of advanced glycation end products and other inflammatory mediators play key roles in the expression of pathological genes. In addition, epigenetic mechanisms such as posttranslational modification of histones and DNA methylation also play central roles in gene regulation by affecting chromatin structure and function. Recent studies have suggested that dysregulation of such epigenetic mechanisms may be involved in metabolic memory leading to persistent changes in the expression of genes associated with diabetic vascular complications. Further exploration of these mechanisms by also taking advantages of recent advances in high throughput epigenomics technologies will greatly increase our understanding of epigenetic variations in diabetes and its complications. This in turn can lead to the development of novel new therapies.

M.A. Reddy • R. Natarajan, Ph.D. (✉)
Department of Diabetes, Beckman Research Institute of City of Hope,
1500 East Duarte Road, Duarte, CA 91010, USA
e-mail: RNatarajan@coh.org

19.1 Introduction

Diabetes is the leading cause of macro- and microvascular complications such as atherosclerosis, hypertension, nephropathy, retinopathy and neuropathy (Beckman et al. 2002; He and King 2004; Sharma and Ziyadeh 1995). These vascular complications affect all major organs leading to heart failure, kidney failure, blindness and limb amputation and can significantly enhance mortality rates in patients with diabetes relative to the normal population. Current projections predict a significant surge in obesity, diabetes and related metabolic disorders worldwide and in particular among the younger population, likely due to changes in lifestyle. This can greatly increase the economic burden associated with diabetes and its complications. Intense efforts are therefore needed to find more effective therapeutic approaches to curb the progression of diabetic complications. Diabetes increases the blood glucose levels(hyperglycemia) as a result of lack of insulin production (type 1 diabetes, T1D) or increased insulin resistance(type 2 diabetes, T2D), with the latter often being associated with hyperlipidemia. Hyperglycemia has been implicated in chronic inflammation and increased oxidant stress, the major risk factors for the development of vascular complications (Giacco and Brownlee 2010; Sheetz and King 2002; Devaraj et al. 2010). Clinical trials have demonstrated that strict glycemic control is critical to reduce the incidence of diabetic complications. They also showed that diabetic patients are at continued risk for increased vascular complications even long after achieving normal blood glucose levels, suggesting a 'metabolic memory' or 'legacy' effect of prior hyperglycemic exposure (Writing Team DCCT/EDIC Research Group 2002; Colagiuri et al. 2002). These clinical studies were further supported by experimental evidence showing that vascular cells exposed to diabetic milieu retain their pro-inflammatory diabetic phenotype for extended periods even after glucose normalization (Ihnat et al. 2007a; Villeneuve and Natarajan 2010; Pirola et al. 2010). Metabolic memory is a major challenge in the prevention of diabetic vascular complications and it is imperative to examine the molecular mechanisms involved and develop novel therapies.

Transcription regulation plays a central role in the expression of inflammatory and other pathologic genes. In general, the recruitment of transcription factors (TFs) to the cis-elements located in the promoters and enhancers plays a key role in gene regulation. However, it is now clear that epigenetic mechanisms in chromatin, i.e., changes that occur without alterations in the DNA sequence, also play important roles in gene transcription. These mechanisms control chromatin access to transcription regulators in mammalian cells and dictate active or repressed states of genes (Li et al. 2007; Murr 2010; Kouzarides 2007; Bannister and Kouzarides 2011). Furthermore, environmental factors and nutrients can affect epigenetic states and regulate the expression of genes associated with various diseases including cancer and diabetes (Sharma et al. 2010; Liu et al. 2008; Ling and Groop 2009). Recent studies have also implicated epigenetic mechanisms in the phenomenon of metabolic memory. An increased understanding of these changes in chromatin events can yield critical new information about the metabolic memory of vascular complications and aberrant expression of inflammatory genes under diabetic conditions that can lead to the identification of novel new therapeutic targets. The current review discusses the role of epigenetics in diabetic vascular complications and some of the recent developments in this area.

19.2 Inflammatory Gene Expression and Vascular Complications

Diabetes is associated with significantly accelerated rates of vascular diseases like atherosclerosis which is multi-cellular in nature involving interactions between endothelial cells (EC), vascular smooth muscle cells (VSMC) and monocytes in the vessel wall (Ross 1999). Endothelial dysfunction induced by oxidant stress, oxidized lipids and other inflammatory mediators increases monocyte adhesion to EC, which then migrate into subendothelial space where interaction with VSMC promotes their differentiation into macrophages. The uptake of oxidized lipids by macrophages leads to foam cell formation. VSMC proliferation and migration to the sites of lesion also contribute to the formation of the atherosclerotic plaque. In advanced stages of the disease plaque rupture releases pro-inflammatory and pro-thrombotic mediators resulting in stroke (Ross 1999). EC dysfunction and macrophage infiltration also play key roles in diabetic nephropathy and retinopathy (King 2008; Giacco and Brownlee 2010; Wang and Harris 2011). All cell types involved in vascular complications produce inflammatory cytokines and chemokines, and regulate the functions of each other through autocrine and paracrine actions. Pro-inflammatory chemokines such as monocyte chemoattractant protein-1(MCP-1), cytokines including interleukin-6 (IL-6) and tumor necrosis factor-alpha (TNF-α), and growth factors such as Angiotensin II(Ang II) and macrophage colony stimulating factor (M-CSF) play key roles in the initiation and progression of vascular complications (Charo and Taubman 2004; Libby et al. 2002; Weiss et al. 2001). Several studies have demonstrated that diabetes exacerbates the production of inflammatory mediators in the vessel wall leading to further acceleration of vascular dysfunction. Enhanced activity of the pro-inflammatory TF NF-κB plays a key role in the increased inflammatory gene expression in VSMC, EC and monocytes under diabetic conditions. These cells exhibit enhanced oxidant stress and inflammatory genes such as TNF-α, MCP-1, Fractalkine (CX3CL1), IL-6, M-CSF and arachidonic acid metabolizing enzymes such as cyclooxygenese-2 (COX-2) and lipoxygenases, and the receptor for advanced glycation end products (RAGE) under diabetic conditions (Devaraj et al. 2006; De Martin et al. 2000; Barnes and Karin 1997; Brownlee 2001; Natarajan and Nadler 2004; Shanmugam et al. 2003a, b; Guha et al. 2000; Li et al. 2006; Reddy et al. 2009; Hatley et al. 2003; Min et al. 2010; Meng et al. 2010; Villeneuve et al. 2008; Yan et al. 2009). While the transcription of inflammatory genes is most often regulated by NF-κB TF (Barnes and Karin 1997; Glass and Witztum 2001), other TFs such as CREB and STATs have also been implicated (Reddy et al. 2006, 2009; Sahar et al. 2005, 2007; Chava et al. 2009). Biochemical studies have also established the role of multiple upstream signal transduction mechanisms involving increased oxidant stress, polyol pathway, AGEs and RAGE, Protein kinase C (PKC), AT1R, oxidized lipids, tyrosine kinases and mitogen activated protein kinases (MAPKs) that can lead to the activation of TFs regulating genes involved in diabetic complications (Brownlee 2005; Natarajan and Nadler 2004; Yan et al. 2008; Marrero et al. 2005; Reddy et al. 2006). However, therapies based on these mechanisms have not been fully adequate to effectively prevent the progression of various diabetic vascular complications suggesting the need to explore additional mediators and drug targets.

19.3 Metabolic Memory of Diabetic Vascular Complications

Landmark clinical trials such as the Diabetes Control and Complications Trial (DCCT) with T1D patients showed that intensive glycemic control is critical for the prevention or reduction of long term vascular complications. In the follow up Epidemiology of Diabetes Interventions and Complications study (EDIC), the DCCT enrollees were subsequently monitored after placing both the conventional and intensive treatment groups on the same intensive glycemic control. Results from EDIC showed that patients previously receiving conventional therapy during DCCT continued to develop micro-and macrovascular complications at a much greater rate relative to those in the intensive treatment group throughout (Writing Team DCCT/EDIC Research Group 2002; Nathan et al. 2005). In other clinical trials involving T2D patients, levels of hyperglycemia at the time of diagnosis correlated with risk for developing complications in T2D patients (Colagiuri et al. 2002). Furthermore, fluctuations in post-prandial glycemic levels have also been implicated in the increased risk for vascular complications (Ceriello et al. 2008). Overall, these clinical studies have suggested that the persisting effects of prior exposure to hyperglycemia termed 'metabolic memory' or "legacy effect" can have long lasting deleterious effects in diabetes patients.

Studies using cell culture and animal models further established the role of metabolic memory in vascular complications. These included demonstration of increased oxidative stress, fibrotic and inflammatory gene expression in high glucose (HG) treated EC even several days after return to normal glucose (Ihnat et al. 2007b; Roy et al. 1990; El-Osta et al. 2008). VSMC derived from leptin receptor deficient *db/db* mice, a model of T2D, exhibited enhanced pro-inflammatory gene expression, monocyte binding and migration even after culturing in vitro for a few passages compared with non-diabetic *db/+* controls (Li et al. 2006; Villeneuve et al. 2008). VSMC from *db/db* mice also showed persistently increased activation of signal transduction pathways such as oxidant stress, tyrosine kinase, MAPK and downstream pro-inflammatory TFs NF-κB and CREB (Li et al. 2006). EC and macrophages from *db/db* mice also exhibited enhanced inflammatory genes and activation of NF-κB, suggesting that diabetes induces a pre-activated phenotype in target cells involved in vascular complications (Hatley et al. 2003; Li et al. 2006; Wen et al. 2006). In retinal EC, HG inhibited the expression of the antioxidant gene superoxide dismutase (*sod2*) and this persisted even 4 days after reversal to normal glucose (Zhong and Kowluru 2011). In streptozotocin injected dogs with T1D, retinal complications persisted even after achieving normoglycemia with islet transplantation (Engerman and Kern 1987). Furthermore, in T1D rats, poor glycemic control for short periods followed by intensive glycemic control prevented the development of retinopathy. However, intensive glycemic control could not prevent retinopathy and related biochemical parameters in animals that had poor glycemic control for much longer durations (Chan et al. 2010; Kowluru 2003). Together these studies demonstrated the role of metabolic memory in diabetic complications and recent studies now suggest that epigenetic mechanisms may be involved in these events.

19.4 Epigenetic Mechanisms of Gene Regulation in Chromatin: DNA Methylation and Histone Post Translational Modifications (PTMs)

In mammalian cells, chromosomal DNA is tightly packaged into chromatin by histone proteins along with other chromatin assembly factors. Chromatin is a highly organized structure consisting of numerous nucleosome particles. Each nucleosome is composed of about 147 bp DNA wrapped around an octamer histone protein complex made up of dimers of core histone proteins H2A, H2B, H3 and H4 (Li et al. 2007; Workman and Kingston 1998). The histone proteins were initially proposed to be passive components supporting chromatin structure. However, it is now clear that they actively participate in transcriptional regulation. Epigenetic mechanisms regulate dynamic switching of chromatin between active (euchromatin) and inactive (heterochromatin) states that determines the transcription status of target genes and the biological outcomes (Li et al. 2007; Kouzarides 2007; Murr 2010). These mechanisms include DNA cytosine methylation, covalent post translational modifications (PTMs) of nucleosomal histones, small non-coding RNAs or microRNAs (miRNAs) and large intergenic noncoding RNAs (lincRNAs) (Murr 2010). While epigenetics usually refers to heritable changes, including those conferred mitotically or meiotically, that occur without changes in DNA sequence, more recently the definition has been modified to include the structural adaptation of chromosomal regions (Bird 2007). Thus, both DNA methylation and histone PTMs can work together to control epigenetic transmission (Bird 2007; Berger et al. 2009; Portela and Esteller 2010). Epigenetic mechanisms genomewide are now termed as the 'Epigenome' and major efforts are underway to understand how alterations in epigenome status can modulate diverse patho-physiolological conditions (Bernstein et al. 2007; Maunakea et al. 2010), especially since epigenetic changes can be induced by environmental factors.

DNA methylation which occurs at cytosine residues in CpG dinucleotides is one of the most stable epigenetic marks (Miranda and Jones 2007). CpGs can occur in groups called 'CpG islands' near promoters and methylation of promoter CpG islands often leads to gene repression. DNA methylation is regulated by DNA methyl transferases that mediate the transfer of methyl groups from S-adenosyl methionine (SAM). Recent studies have also implicated a role for DNA de-methylation in cellular processes but the identity of the enzymes that mediate DNA demethylation is not fully clear (Wu and Zhang 2010). Abnormal DNA methylation or demethylation at the promoters of oncogenes and tumor suppressors is a major mechanism involved in the development of cancer (Miranda and Jones 2007; Sharma et al. 2010). However, only limited information is available on the role of DNA methylation in diabetes, metabolism and vascular complications. Some of these include the demonstration of DNA methylation in *Agouti* gene expression (Morgan et al. 1999), DNA hypomethylation in aortic tissues from atherosclerosis animal models (Turunen et al. 2009) and differential DNA methylation in monocytes of T1D patients with diabetic nephropathy compared with patients who did not develop nephropathy (Bell et al. 2010). Further information on DNA methylation related to vascular complications can be obtained from recent reviews (Dong et al. 2002; Turunen et al. 2009; Reddy and Natarajan 2011).

Histone PTMs can play major roles in the regulation of gene expression by modifying chromatin structure and by providing anchoring sites for co-activators, co-repressors and other chromatin proteins. Histone PTMs in association with other epigenetic mechanisms such as DNA methylation can modulate the euchromatin (accessible) or heterochromatin (inaccessible) status of chromatin (Murr 2010; Kouzarides 2007; Bannister and Kouzarides 2011). Several histone PTMs have been identified including phosphorylation (serine and threonine), acetylation (lysine), methylation (lysine and arginine), ubiquitination (lysine) and sumoylation (lysine) (Kouzarides 2007). Histone PTMs usually act in concert with each other to form a 'histone code' that dictates the transcriptional states of genes leading to biological phenotypes (Jenuwein and Allis 2001). The function of histone acetylation and methylation has been widely studied in the pathogenesis of cancer (Portela and Esteller 2010). Recent studies have examined their role in diabetes and vascular complications.

19.5 Histone Lysine Acetylation and Methylation in Gene Regulation

In general histone lysine acetylation (HKAc) promotes formation of open chromatin and is associated with active promoters. Acetylation of amino-terminal residues in histone tails neutralizes the positive charge weakening DNA-histone and nucleosome-nucleosome interactions (Kouzarides 2007; Roth et al. 2001). Furthermore, acetylated lysines can provide binding sites for chromatin remodeling proteins. These events can relax chromatin structure and make it more accessible to transcription factors and other regulators to promote gene transcription. HKAc is mediated by histone acetyl transferases (HATs). Several HATs including CBP, p300, pCAF, Tip60, SRC-1, SRC-2 and SRC-3 also act as co-activators. HATs can acetylate multiple lysine residues of histone and non-histone proteins (Roth et al. 2001; Kouzarides 2007). Genomewide studies using chromatin immunoprecipitation linked to Next Generation DNA Sequencing (ChIP-Seq) revealed differential regulation of specific promoters by individual HATs (Ramos et al. 2010).

Interestingly, ChIP-Seq and ChIP coupled with microarray analysis (ChIP-on-chip) studies also showed that p300 is recruited to enhancers and its location along with other histone PTMs such as H3K4me1 could be used to predict enhancers genomewide in mammalian cells (Visel et al. 2009; Hon et al. 2009; Jin et al. 2011).

Histone acetylation is removed by histone deacetylases (HDACs), which are sub-divided into four groups (Class I, Class II, Class III and clas IV) depending on structure, function and mode of action. Class III proteins also known as Sirtuins are unique in that they require NAD^+ as co-factor, and respond to changes in metabolic status (Kouzarides 2007; Yang and Seto 2007). Removal of acetyl groups by HDACs restores the positive charge of histone lysines and promotes chromatin condensation, leading to reduced TF accessibility and inhibition of transcription. The functions of HDACs are quite complex mainly due to their low substrate specificity. However, studies in knockout mice showed they could be involved in specific function such as embryonic stem cell differentiation and EC proliferation (Dovey et al. 2010; Margariti et al. 2010). Overall, histone lysine acetylation can play significant roles in gene regulation related to biological and pathophysiological states.

Histone lysine methylation (HKme) on the other hand is associated with both active and inactive promoters depending on the specific lysine residue methylated. Furthermore, these lysines can be mono, di or tri methylated adding another level of complexity (Shilatifard 2006; Kouzarides 2007; Bannister and Kouzarides 2011). Histone H3 lysine 9 methylation (H3K9me2/3), H3K27me2/3, and H4K20me3 are generally associated with gene repression, while histone H3K4me1/2/3 and H3K36me3 are usually associated with active genes and gene bodies. Histone methyl transferases (HMTs) catalyze HKme while histone demethylases (HDMs) mediate removal of methyl groups. HMTs and HDMs are quite specific, e.g., SUV39H1 HMT mediates H3K9me2 and H3K9me3, and the MLL family members mediate H3K4me1-me3. Conversely, lysine specific demethylase 1(LSD1) removes H3K4 -me1 and -me2, and Jhdm2a demethylates H3K9me2 (Kouzarides 2007; Shi and Whetstine 2007; Trojer and Reinberg 2006). Most HMTs contain the SET domain which mediates lysin methyltransferase activity (Kouzarides 2007). The discovery of several HDMs demonstrated that HKme is indeed reversible and established the dynamic regulation of HKme in gene expression. Lysine methylation marks are recognized by chromo, tudor, MBT and nonrelated PHD domain containing proteins, which regulate the function of HKme (Kouzarides 2007).

Increasing evidence supports the critical function of HKme in diverse pathophysiological conditions including cancer (Bhaumik et al. 2007; Portela and Esteller 2010) and recently in diabetes and its vascular complications (Villeneuve and Natarajan 2010; Pirola et al. 2010; Giacco and Brownlee 2010; Reddy and Natarajan 2011). Because HKme is relatively more stable than other modifications, it is likely to be involved in transcriptional memory. Interestingly, studies showed that the polycomb group of proteins that mediate H3K27me3 remained bound to chromatin during DNA replication implicating a potential epigenetic role for such histone marks in the maintenance of transcription memory (Bantignies and Cavalli 2006; Francis et al. 2009). Thus, HKme could be a mediator of metabolic memory resulting from exposure to environment and dietary factors and diabetic stimuli such as HG, AGEs and oxidized lipids.

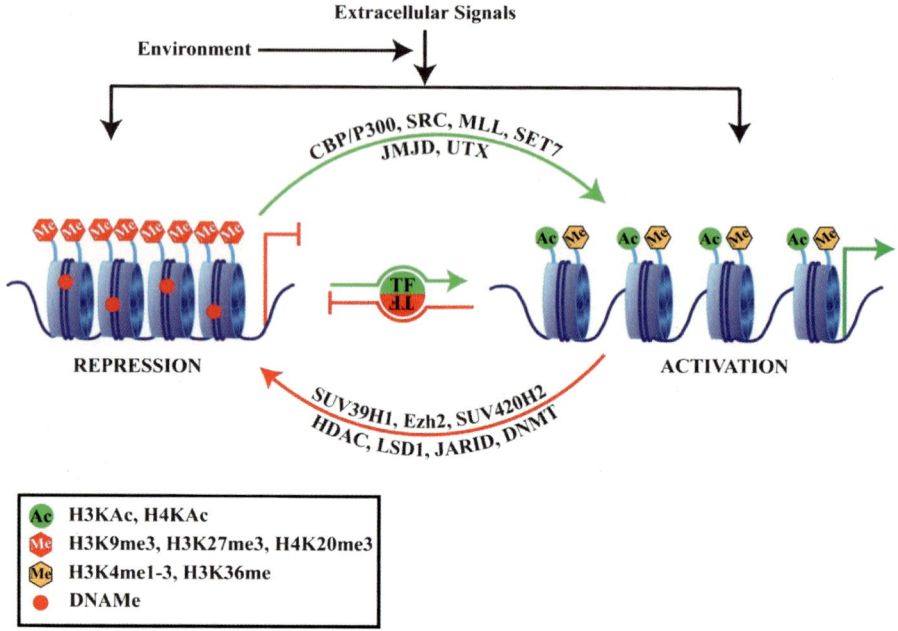

Fig. 19.1 Role of epigenetic mechanisms in gene transcription. Epigenetic mechanisms play central roles in gene regulation mediated by transcription factors (TF) by maintaining active (euchromatin) or repressed (heterochromatin) states of chromatin. These include DNA methylation (DNAMe) and histone post translational modifications (PTMs). In the repressed state, chromatin at repressed genes can be enriched with DNAMe [mediated by DNA methyl transferases (DNMTs)] and repressive histone PTMs such H3K9 methylation (H3K9me), H3K27me and H4K20me mediated by histone methyltransferases (HMTs) SUV39H1, Ezh2 and SUV420H2 respectively. In active states, chromatin is enriched by H3K acetylation (H3KAc) and H4KAc, and H3K4me and H3K36me. Histone acetyltransferases (HAT) such as CBP/p300 and SRC-1 mediate acetylation, while H3K4me is catalyzed by HMTs such as SET7 and MLL family members. Actions of HATs are opposed by histone deacetylases (HDAC). Actions of HMTs are countered by histone demethylases (HDM) such as LSD1 and JARID, which erase H3K4me marks and the JMJD family members, which erase H3K9me and H3K27me marks. This results in gene repression or activation depending on the modification that has been erased. Environmental factors and extracellular signals can affect epigenetic states to modulate the expression of genes associated with various diseases including cancer and diabetes. Ac-lysine acetylation; Me: lysine methylation

Arginine methylation of histones (HRme) can be associated with gene activation or repression. Arginine can be mono or dimethylated, and the dimethylation can be asymmetric or symmetric depending on the methyltransferase involved (Kouzarides 2007). At least 9 protein arginine methyltransferases (PRMTs) have been identified including coactivator-associated arginine methyltransferase 1 (CARM1). HRme is removed by arginine demethylases such as JMJD6. HRme might play key role in inflammation, cell proliferation and differentiation (Wysocka et al. 2006; Miao et al. 2006).

Role of epigenetic mechanisms including histone PTMs and DNA methylation in gene transcription and interaction with environment is summarized in Fig. 19.1.

Recent advances in next generation sequencing (NGS) have significantly enhanced our technical capabilities to analyze genomewide changes in histone PTMs and DNA methylation (Metzker 2010; Hawkins et al. 2010; Maunakea et al. 2010). This area is poised to make tremendous progress towards understanding the function of epigenetic mechanisms under normal and disease states.

19.6 Histone Modifications in Diabetes

Histone modifications were shown to play key role in pancreatic islet specific expression of insulin and related genes in response to changing glucose levels. Under HG conditions, the islet specific transcription factor Pdx1 was shown to recruit a co-activator HAT p300 and a H3K4 methyltransferase SET7/9 at the insulin promoter. This was associated with increased promoter levels of the activation marks H3K9/14Ac and H3K4Me and insulin expression. Conversely, under low glucose conditions, Pdx1could recruit HDAC1 and HDAC2 to the insulin promoter, leading to the inhibition of insulin expression. Interestingly, Pdx1 also induced SET7/9 expression, which in turn increased the expression of genes involved in glucose induced insulin secretion (Deering et al. 2009; Chakrabarti et al. 2003). Recent studies showed that the Polycomb group of proteins including Ezh2 that mediates H3K27me3, JMJD3 that demethylates H3K27me3, and accessory proteins such as Bmi-1 play key roles in pancreatic β-cell proliferation and regeneration through regulation of the tumor suppressor protein p16INK4a (Dhawan et al. 2009; Chen et al. 2009). Histone modifications were also reported in adipocyte differentiation since a regulatory role of H3KAc was identified in the expression of C/EBP-delta, a key transcription factor involved in adipocyte differentiation (Nakade et al. 2007). LSD1(demethylase) and SETDB1(methyltransferase) functions were required for the regulation of H3K4me2 and H3K9me2 respectively during adipogenesis (Musri et al. 2010). Furthermore, knockdown of the H3K9me2 demethylase Jhdm2a (Jmjd1a) resulted in obesity and hyperlipidemia in mice, providing direct evidence for H3Kme in the development of diabetes (Tateishi et al. 2009). The class III HDAC, SIRT1 was found to play an important role in energy metabolism, and SIRT1 activators such as resveratrol can inhibit insulin resistance. Several SIRT1 modulators are being evaluated for the treatment of insulin resistance (Haigis and Sinclair 2010; Blum et al. 2011).

19.7 Histone Modifications in Inflammatory Gene Expression

Evidence shows that NF-κB mediated inflammatory gene expression induced by pro-inflammatory signals was associated with changes in HKAc and HKme in vascular cells and monocytes. Ang II induced IL-6 expression was associated with increased NF-κB activation and promoter H3K9/14Ac in VSMC. Ang II enhanced the recruitment of TFs (NF-κB and CREB) along with co-activator HATs steroid receptor coactivator-1 (SRC-1) and p300/ CBP to the IL-6 promoter. Furthermore, Ang II induced IL-6 expression was inhibited by a p300 mutant lacking HAT activity and a SRC-1 mutant lacking ERK phosphorylation site (Sahar et al. 2007). Oxidized lipids also increased H3KAc at the IL-6 and MCP-1 promoters in VSMC. This increased acetylation and gene expression was dependent on Src kinase activity (Reddy et al. 2009). Role of p300 and p/CAF was also reported in NF-κB mediated chemokine(C-C motif) ligand 11 (CCL11) expression in TNF-α stimulated human airway smooth muscle cells (Clarke et al. 2008).

TNF-α increased H3KAc at the IL-6 promoter, and recruitment of CBP/p300 was required for the optimal induction of NF-κB mediated IL-6 and IL-8 expression in EC (Vanden Berghe et al. 1999). TNF-α induced EC specific E-selectin expression was associated with increased H3K4me2, H3K9/14 Ac and H4K12Ac along with p300/CBP recruitment, which promoted nucleosome remodeling to increase chromatin access to NF-κB (Edelstein et al. 2005). Furthermore, these studies also showed the role of HDAC2 in association with Sin3A in the post-induction repression of these inflammatory genes. EC specific expression of endothelial nitric oxide(eNOS) was associated with H3K4me, H3K9/14Ac and Ser10-phosphorylation at the eNOS core promoter (Fish et al. 2005). Oxidized LDL could induce H3KAc, and recruitment of HATs along with NF-κB at inflammatory gene promoters in EC and these events could be reversed by statin treatment (Dje N'Guessan et al. 2009). Together, these reports demonstrate the role of histone PTMs in regulating the expression of inflammatory genes in vascular cells.

Lipopolysaccharide (LPS) induced the expression of several HDACs along with inflammatory genes in macrophages (Aung et al. 2006). Interestingly, HDAC inhibitors attenuated the expression of some inflammatory genes but also increased the expression of pro-atherogenic genes (Halili et al. 2010). In dendritic cells derived from human monocytes, LPS induced inflammatory gene expression was associated with a rapid decrease in the levels of the repressive mark H3K9me3 followed by an increase to basal levels at later time points. This suggested a regulatory role of H3K9me3 in negative feedback mechanisms associated with post induction repression of inducible inflammatory genes (Saccani and Natoli 2002). Similar decrease in H3K9me3 at early time points and restoration to control levels at later time points was also noted in VSMC stimulated with TNF-α (Villeneuve et al. 2008). LPS mediated inflammatory gene expression was associated with decreases in repressive epigenetic mark H3K27me3 in macrophages. Further studies revealed that LPS also increased the expression of the H3K27me3 demethylase JMJD3 which was shown to regulate inflammatory genes in macrophages (De Santa et al. 2007, 2009). In THP-1 monocytes, TNF-α stimulation increased H3K4me at inflammatory gene promoters and promoted the recruitment of SET7/9, a H3K4 methyl transferase, along with NF-κB at inflammatory gene promoters. SET7/9 gene silencing revealed that a subset of NF-κB regulated inflammatory genes required SET7/9 and its mehyltransferase activity in TNF-α stimulated THP-1 cells (Li et al. 2008). TNF-α induced inflammatory gene expression was also associated with increased H3R17me and recruitment of the arginine methyltransferase CARM1 (Miao et al. 2006). Thus, NF-κB mediated inflammatory gene expression can be fine tuned by histone modifications in both vascular cells and monocytes, key players in the pathogenesis of vascular complications.

19.8 Histone Lysine Modifications in Diabetic Vascular Complications

Several lines of evidence suggest that diabetes and diabetogenic agents promote changes in histone modifications in vascular cells and monocytes to regulate the expression of genes associated with vascular complications. HG induced increases in TNF-α and COX-2 genes in THP-1 monocytes was associated with increased H3KAc at their promoters along with increased promoter recruitment of NF-κB and co-activators CBP, p/CAF and SRC-1, and reduced occupancy of HDAC1 (Miao et al. 2004). Furthermore, increased levels of H3KAc was observed at inflammatory gene promoters in peripheral blood monocytes obtained from T1D and T2D patients, thus establishing in vivo relevance and demonstrating that inflammatory cells from diabetic patients may have a more open chromatin at pathologic genes (Miao et al. 2004). The RAGE ligand S100b also increased H3K9Ac, H3K14Ac and H3R17me as well as recruitment of p300 and CARM1 at the TNF-α promoter in THP-1 monocytes demonstrating that RAGE signaling can alter histone PTMs in chromatin (Miao et al. 2006). Recent studies reported that the anti-inflammatory agent curcumin blocked HG induced cytokine gene expression in THP-1 cells via inhibition of H3KAc and activity of p300 (Yun et al. 2011). These results further confirm the role of HKAc in inflammatory gene expression under diabetic conditions.

TNF-α induced inflammatory gene expression was enhanced in macrophages from T1D mice and this was associated with increased H3K4me and recruitment of SET7/9 at inflammatory gene promoters (Li et al. 2008). Genome-wide location approaches such as ChIP-on-chip showed that HG induced significant changes in H3K9me2 and H3K4me2 patterns at key genes in THP-1 monocytes (Miao et al. 2007). ChIP-on-chip studies using lymphocytes from T1D patients revealed significantly increased H3K9me2 levels at a subset of genes involved in inflammatory and autoimmune regulatory pathways relevant to the pathogenesis T1D and its complications (Miao et al. 2008).

Human VSMC cultured in HG displayed increases in inflammatory genes and this was associated with reduced H3K9me3 at their promoters (Villeneuve et al. 2008). In retinal EC, HG decreased global H3KAc, inhibited HAT activity and increased HDAC expression (Zhong and Kowluru 2010). Recent studies in retinal EC showed that HG mediated inhibition of the antioxidant superoxide dismutase (*sod2*) gene was associated with increased H4K20me3 at its promoter through increased expression of the corresponding HMT SUV420h2 (Zhong and Kowluru 2011). Another study reported decreases in H3K9me and increases in H3K4me along with increased recruitment of SET7/9 at fibrotic gene promoters by HG and the profibrotic growth factor TGF-β in renal mesangial cells which are involved in diabetic nephropathy (Sun et al. 2010). Furthermore, HG induced gene expression and changes in these histone modifications were blocked by a TGF-β antibody (Sun et al. 2010). These studies show that diabetic conditions can alter the levels of histone PTMs in target cells that play key roles in diabetes and its vascular complications.

19.9 Histone Modifications in Metabolic Memory

Metabolic memory has been implicated in the persistence of vascular complications even long after achieving normal glycemic levels in diabetes patients. This was attributed to the sustained increases in the levels of oxidant stress and inflammatory genes in the vasculature (Villeneuve and Natarajan 2010; Giacco and Brownlee 2010; Pirola et al. 2010). Recent studies using cell culture and animal models have suggested that HKme may be involved in metabolic memory of diabetic vascular complications.

EC treated with HG continued to express elevated levels of inflammatory genes even long after restoration to normal glucose medium mimicking metabolic memory. There was a sustained increase in p65 (NF-κB) expression and this was associated with increased levels of the activation mark H3K4me1 and the corresponding HMT Set7/9 at the p65 promoter (El-Osta et al. 2008). In addition, levels of repressive marks H3K9me2 and H3K9me3 were also reduced while the occupancy of the demethylase LSD1 was enhanced in HG treated EC. These changes persisted even after removal of HG suggesting a key role of H3Kme and its regulators in metabolic memory in EC (Brasacchio et al. 2009). Oxidant stress and reactive dicarbonyls such as methylglyoxal were implicated in these changes (El-Osta et al. 2008).

In another model of metabolic memory, aortic VSMC isolated from *db/db* mice were used. These cells displayed enhanced expression of IL-6, MCP-1 and MCSF-1 genes and increased monocyte binding relative to *db/+* control cells, even after culturing for few passages in vitro, mimicking metabolic memory (Li et al. 2006; Villeneuve et al. 2008; Meng et al. 2010). ChIP assays showed that the repressive H3K9me3 mark was significantly reduced at the promoters of these inflammatory genes in VSMC of diabetic *db/db* mice relative to those from genetic control *db/+* mice. Furthermore, TNF-α induced inflammatory gene expression was also significantly enhanced in these *db/db* cells (Villeneuve et al. 2008) and this was associated with persistently reduced levels of repressive H3K9me3 and occupancy of the H3K9me3 methyltransferase Suv39h1 at inflammatory gene promoters suggesting dysregulation of repressive mechanisms in diabetes. Suv39h1 levels were also significantly reduced in *db/db* VSMC and reconstitution by overexpression of Suv39h1 reversed the pro-inflammatory phenotype of *db/db* VSMC (Villeneuve et al. 2008). Thus, H3K9me3 and the corresponding HMT appear to play key roles in this model of metabolic memory related to vascular complications. In addition, recent studies showed that levels of the microRNA-125b were increased in *db/db* VSMC and that Suv39h1 was a direct target of miR-125b in mouse VSMC (Villeneuve et al. 2010).

Furthermore, miR-125b could inhibit Suv39h1 expression, induce inflammatory genes, and reduce H3K9me3 at their promoters and also promote monocyte-VSMC binding in non-diabetic *db/+* cells. In contrast, miR-125b inhibition reversed key pro-inflammatory phenotypes of diabetic *db/db* VSMC (Villeneuve et al. 2010). These results identified a novel role for microRNA mediated mechanisms in inflammatory gene expression in VMSC and possibly metabolic memory through downregulation of key repressive chromatin factors. Interestingly, SIRT1 can activate the methyltransferase activity of Suv39h1 (Vaquero et al. 2007) raising the intriguing question of whether SIRT1 activators could be used for treating diabetic complications or metabolic memory.

Studies with a T1D rat model of diabetic retinopathy have suggested a role for H4K20me3 in metabolic memory (Zhong and Kowluru 2011) . There was persistently reduced expression of the key antioxidant gene *sod2* in retinas from diabetic rats with poor glycemic control that developed retinopathy compared with non-diabetic rats, or diabetic rats with good glucose control that did not develop retinopathy. This was associated with increased levels of H4K20me3, H3K9Ac as well as NF-κB recruitment at the *sod2* promoter and increases in the expression of the H4K20me3 methyl transferase Suv420h2 in diabetic retinas. Interestingly, these changes were sustained in rats displaying memory of retinopathy. Furthermore, in vitro studies with EC cultured in HG demonstrated the role of Suv420h2 in the persistent histone PTMs at the *sod* promoter. Thus, H4K20me3 and Suv420h2 were implicated in this model of metabolic memory associated with diabetic retinopathy.

19.10 Summary

Studies in cultured cells and experimental models have identified the role of HG as well as downstream biochemical, signaling and chromatin mechanisms in vascular complications and metabolic memory (Fig. 19.2). However, diabetes is a multifactorial disease and other factors such as AGEs, oxidized lipids, proinflammatory growth factors, nutrients and lifestyles could also be involved. Recent advances in genomewide approaches such as ChIP-Seq and DNA methylation profiling will greatly accelerate our understanding of histone PTMs and epigenetic mechanisms regulating the expression of genes associated with diabetic complications and metabolic memory (Wang et al. 2009; Metzker 2010; Hawkins et al. 2010; Maunakea et al. 2010). These studies could potentially lead to the identification of several other histone PTMs, HMTs, HDMs and DNA methyl transferases in diabetic complications. Since epigenetic mechanisms and histone PTMs are reversible, there is ample potential for the development of novel therapies. Several candidates including inhibitors of HDACs, HATs and DNA methyltransferases are being tested in epigenetic therapy for cancer (Kelly et al. 2010; Selvi et al. 2010). Some of these are also being considered for treating vascular complications like restenosis (Pons et al. 2009; Natarajan 2011). Recent reports showing that histone modifications and inflammatory genes can be reversed by curcumin in monocytes treated with HG

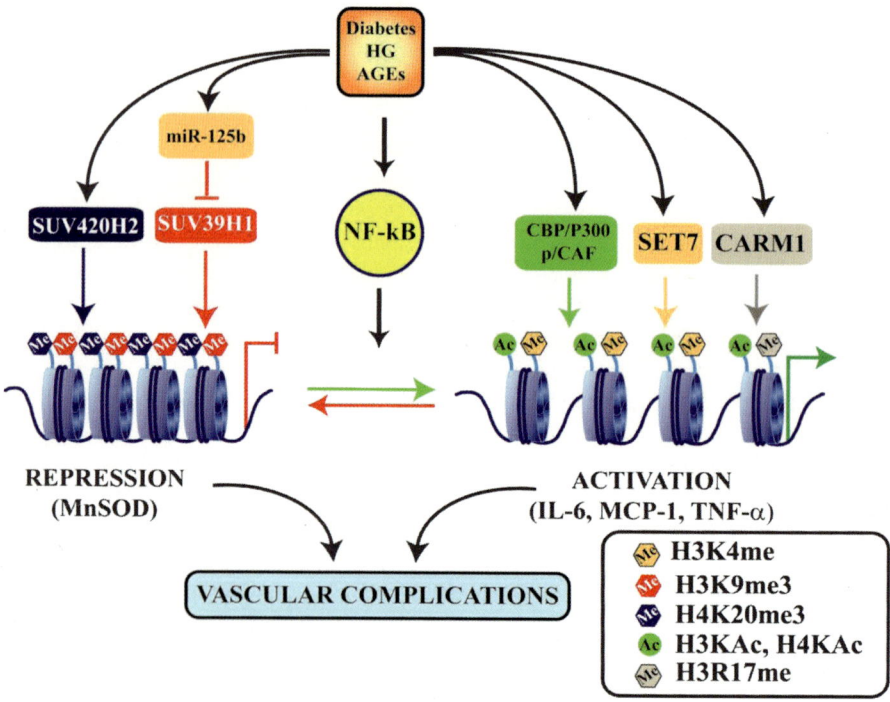

Fig. 19.2 Epigenetic mechanisms of gene regulation and metabolic memory associated with vascular complications of diabetes. Diabetes and diabetogenic agents such as high glucose (HG) and advanced glycation end products (AGEs) activate transcription factors such as NF-κB which regulate gene expression. This is further regulated by epigenetic mechanisms in chromatin activated under these conditions. These include increases in the levels of active marks such as histone H3 or H4KAc and H3K4me or H3R17me through recruitment of co-activator HATs (CBP/p300 and p/CAF) and HMTs (SET7 and CARM1) respectively. Other as yet unidentified HATs, HMTs and HDMs may also be involved. Diabetes can also lead to loss or reduction in repressive marks like H3K9me3 through reduced recruitment of the HMT SUV39H1 as well as inhibition of its expression by miR-125b, which is increased in diabetes. This leads to increased expression of inflammatory genes. In addition, diabetes can also inhibit the expression of protective anti-oxidant genes such as manganese superoxide dismutase (MnSOD) by increasing promoter levels of the repressive mark H4K20me3 and its HMT, SUV420H2. Alterations in gene expression and histone PTMs at their promoters, such as reduced H3K9me3 or increased H3K4me1 and H4K20me3, can persist even after glucose normalization demonstrating the key role of such epigenetic mechanisms in metabolic memory which leads to sustained long term diabetic complications. Ac-lysine acetylation; Me: lysine or arginine methylation

(Yun et al. 2011), and also renal dysfunction in diabetic rats (Chiu et al. 2009) by a TGF-b antibody in renal mesangial cells (Sun et al. 2010) and by statins in EC (Dje N'Guessan et al. 2009) reveal the potential of currently used drugs to directly or indirectly reverse the epigenetic mechanisms involved in diabetic vascular complications. Given the exponential increase in epigenetics research in recent years, epigenetic therapy for the treatment of diabetic complications could become a reality in the near future.

Acknowledgements The authors gratefully acknowledge grant support from the National Institutes of Health (NIDDK and NHLBI), the Juvenile Diabetes Research Foundation, and the American Diabetes Association.

References

Aung HT, Schroder K, Himes SR, Brion K, van Zuylen W, Trieu A, Suzuki H, Hayashizaki Y, Hume DA, Sweet MJ, Ravasi T (2006) LPS regulates proinflammatory gene expression in macrophages by altering histone deacetylase expression. FASEB J 20:1315–1327

Bannister AJ, Kouzarides T (2011) Regulation of chromatin by histone modifications. Cell Res 21:381–395

Bantignies F, Cavalli G (2006) Cellular memory and dynamic regulation of polycomb group proteins. Curr Opin Cell Biol 18:275–283

Barnes PJ, Karin M (1997) Nuclear factor-kappaB: a pivotal transcription factor in chronic inflammatory diseases. N Engl J Med 336:1066–1071

Beckman JA, Creager MA, Libby P (2002) Diabetes and atherosclerosis: epidemiology, pathophysiology, and management. JAMA 287:2570–2581

Bell CG, Teschendorff AE, Rakyan VK, Maxwell AP, Beck S, Savage DA (2010) Genome-wide DNA methylation analysis for diabetic nephropathy in type 1 diabetes mellitus. BMC Med Genomics 3:33–42

Berger SL, Kouzarides T, Shiekhattar R, Shilatifard A (2009) An operational definition of epigenetics. Genes Dev 23:781–783

Bernstein BE, Meissner A, Lander ES (2007) The mammalian epigenome. Cell 128:669–681

Bhaumik SR, Smith E, Shilatifard A (2007) Covalent modifications of histones during development and disease pathogenesis. Nat Struct Mol Biol 14:1008–1016

Bird A (2007) Perceptions of epigenetics. Nature 447:396–398

Blum CA, Ellis JL, Loh C, Ng PY, Perni RB, Stein RL (2011) SIRT1 modulation as a novel approach to the treatment of diseases of aging. J Med Chem 54:417–432

Brasacchio D, Okabe J, Tikellis C, Balcerczyk A, George P, Baker EK, Calkin AC, Brownlee M, Cooper ME, El-Osta A (2009) Hyperglycemia induces a dynamic cooperativity of histone methylase and demethylase enzymes associated with gene-activating epigenetic marks that coexist on the lysine tail. Diabetes 58:1229–1236

Brownlee M (2001) Biochemistry and molecular cell biology of diabetic complications. Nature 414:813–820

Brownlee M (2005) The pathobiology of diabetic complications: a unifying mechanism. Diabetes 54:1615–1625

Ceriello A, Esposito K, Piconi L, Ihnat MA, Thorpe JE, Testa R, Boemi M, Giugliano D (2008) Oscillating glucose is more deleterious to endothelial function and oxidative stress than mean glucose in normal and type 2 diabetic patients. Diabetes 57:1349–1354

Chakrabarti SK, Francis J, Ziesmann SM, Garmey JC, Mirmira RG (2003) Covalent histone modifications underlie the developmental regulation of insulin gene transcription in pancreatic beta cells. J Biol Chem 278:23617–23623

Chan PS, Kanwar M, Kowluru RA (2010) Resistance of retinal inflammatory mediators to suppress after reinstitution of good glycemic control: novel mechanism for metabolic memory. J Diabetes Complications 24:55–63

Charo IF, Taubman MB (2004) Chemokines in the pathogenesis of vascular disease. Circ Res 95:858–866

Chava KR, Karpurapu M, Wang D, Bhanoori M, Kundumani-Sridharan V, Zhang Q, Ichiki T, Glasgow WC, Rao GN (2009) CREB-mediated IL-6 expression is required for 15(S)-hydroxyeicosatetraenoic acid-induced vascular smooth muscle cell migration. Arterioscler Thromb Vasc Biol 29:809–815

Chen H, Gu X, Su IH, Bottino R, Contreras JL, Tarakhovsky A, Kim SK (2009) Polycomb protein Ezh2 regulates pancreatic beta-cell Ink4a/Arf expression and regeneration in diabetes mellitus. Genes Dev 23:975–985

Chiu J, Khan ZA, Farhangkhoee H, Chakrabarti S (2009) Curcumin prevents diabetes-associated abnormalities in the kidneys by inhibiting p300 and nuclear factor-kappaB. Nutrition 25:964–972

Clarke DL, Sutcliffe A, Deacon K, Bradbury D, Corbett L, Knox AJ (2008) PKCbetaII augments NF-kappaB-dependent transcription at the CCL11 promoter via p300/CBP-associated factor recruitment and histone H4 acetylation. J Immunol 181:3503–3514

Colagiuri S, Cull CA, Holman RR (2002) Are lower fasting plasma glucose levels at diagnosis of type 2 diabetes associated with improved outcomes?: U.K. prospective diabetes study 61. Diabetes Care 25:1410–1417

De Martin R, Hoeth M, Hofer-Warbinek R, Schmid JA (2000) The transcription factor NF-kappa B and the regulation of vascular cell function. Arterioscler Thromb Vasc Biol 20:E83–E88

De Santa F, Totaro MG, Prosperini E, Notarbartolo S, Testa G, Natoli G (2007) The histone H3 lysine-27 demethylase Jmjd3 links inflammation to inhibition of polycomb-mediated gene silencing. Cell 130:1083–1094

De Santa F, Narang V, Yap ZH, Tusi BK, Burgold T, Austenaa L, Bucci G, Caganova M, Notarbartolo S, Casola S, Testa G, Sung WK, Wei CL, Natoli G (2009) Jmjd3 contributes to the control of gene expression in LPS-activated macrophages. EMBO J 28:3341–3352

Deering TG, Ogihara T, Trace AP, Maier B, Mirmira RG (2009) Methyltransferase Set7/9 maintains transcription and euchromatin structure at islet-enriched genes. Diabetes 58:185–193

Devaraj S, Glaser N, Griffen S, Wang-Polagruto J, Miguelino E, Jialal I (2006) Increased monocytic activity and biomarkers of inflammation in patients with type 1 diabetes. Diabetes 55:774–779

Devaraj S, Dasu MR, Jialal I (2010) Diabetes is a proinflammatory state: a translational perspective. Expert Rev Endocrinol Metab 5:19–28

Dhawan S, Tschen SI, Bhushan A (2009) Bmi-1 regulates the Ink4a/Arf locus to control pancreatic beta-cell proliferation. Genes Dev 23:906–911

Dje N'Guessan P, Riediger F, Vardarova K, Scharf S, Eitel J, Opitz B, Slevogt H, Weichert W, Hocke AC, Schmeck B, Suttorp N, Hippenstiel S (2009) Statins control oxidized LDL-mediated histone modifications and gene expression in cultured human endothelial cells. Arterioscler Thromb Vasc Biol 29:380–386

Dong C, Yoon W, Goldschmidt-Clermont PJ (2002) DNA methylation and atherosclerosis. J Nutr 132:2406S–2409S

Dovey OM, Foster CT, Cowley SM (2010) Histone deacetylase 1 (HDAC1), but not HDAC2, controls embryonic stem cell differentiation. Proc Natl Acad Sci U S A 107:8242–8247

Edelstein LC, Pan A, Collins T (2005) Chromatin modification and the endothelial-specific activation of the E-selectin gene. J Biol Chem 280:11192–11202

El-Osta A, Brasacchio D, Yao D, Pocai A, Jones PL, Roeder RG, Cooper ME, Brownlee M (2008) Transient high glucose causes persistent epigenetic changes and altered gene expression during subsequent normoglycemia. J Exp Med 205:2409–2417

Engerman RL, Kern TS (1987) Progression of incipient diabetic retinopathy during good glycemic control. Diabetes 36:808–812

Fish JE, Matouk CC, Rachlis A, Lin S, Tai SC, D'Abreo C, Marsden PA (2005) The expression of endothelial nitric-oxide synthase is controlled by a cell-specific histone code. J Biol Chem 280:24824–24838

Francis NJ, Follmer NE, Simon MD, Aghia G, Butler JD (2009) Polycomb proteins remain bound to chromatin and DNA during DNA replication in vitro. Cell 137:110–122

Giacco F, Brownlee M (2010) Oxidative stress and diabetic complications. Circ Res 107:1058–1070

Glass CK, Witztum JL (2001) Atherosclerosis: the road ahead. Cell 104:503–516

Guha M, Bai W, Nadler JL, Natarajan R (2000) Molecular mechanisms of tumor necrosis factor alpha gene expression in monocytic cells via hyperglycemia-induced oxidant stress-dependent and -independent pathways. J Biol Chem 275:17728–17739

Haigis MC, Sinclair DA (2010) Mammalian sirtuins: biological insights and disease relevance. Annu Rev Pathol 5:253–295

Halili MA, Andrews MR, Labzin LI, Schroder K, Matthias G, Cao C, Lovelace E, Reid RC, Le GT, Hume DA, Irvine KM, Matthias P, Fairlie DP, Sweet MJ (2010) Differential effects of selective HDAC inhibitors on macrophage inflammatory responses to the Toll-like receptor 4 agonist LPS. J Leukoc Biol 87:1103–1114

Hatley ME, Srinivasan S, Reilly KB, Bolick DT, Hedrick CC (2003) Increased production of 12/15 lipoxygenase eicosanoids accelerates monocyte/endothelial interactions in diabetic db/db mice. J Biol Chem 278:25369–25375

Hawkins RD, Hon GC, Ren B (2010) Next-generation genomics: an integrative approach. Nat Rev Genet 11:476–486

He Z, King GL (2004) Microvascular complications of diabetes. Endocrinol Metab Clin North Am 33:215–238

Hon GC, Hawkins RD, Ren B (2009) Predictive chromatin signatures in the mammalian genome. Hum Mol Genet 18:R195–201

Ihnat MA, Thorpe JE, Ceriello A (2007a) Hypothesis: the 'metabolic memory', the new challenge of diabetes. Diabet Med 24:582–586

Ihnat MA, Thorpe JE, Kamat CD, Szabo C, Green DE, Warnke LA, Lacza Z, Cselenyak A, Ross K, Shakir S, Piconi L, Kaltreider RC, Ceriello A (2007b) Reactive oxygen species mediate a cellular 'memory' of high glucose stress signalling. Diabetologia 50:1523–1531

Jenuwein T, Allis CD (2001) Translating the histone code. Science 293:1074–1080

Jin F, Li Y, Ren B, Natarajan R (2011) PU.1 and C/EBP(alpha) synergistically program distinct response to NF-kappaB activation through establishing monocyte specific enhancers. Proc Natl Acad Sci U S A 108:5290–5295

Kelly TK, De Carvalho DD, Jones PA (2010) Epigenetic modifications as therapeutic targets. Nat Biotechnol 28:1069–1078

King GL (2008) The role of inflammatory cytokines in diabetes and its complications. J Periodontol 79:1527–1534

Kouzarides T (2007) Chromatin modifications and their function. Cell 128:693–705

Kowluru RA (2003) Effect of reinstitution of good glycemic control on retinal oxidative stress and nitrative stress in diabetic rats. Diabetes 52:818–823

Li SL, Reddy MA, Cai Q, Meng L, Yuan H, Lanting L, Natarajan R (2006) Enhanced proatherogenic responses in macrophages and vascular smooth muscle cells derived from diabetic db/db mice. Diabetes 55:2611–2619

Li B, Carey M, Workman JL (2007) The role of chromatin during transcription. Cell 128:707–719

Li Y, Reddy MA, Miao F, Shanmugam N, Yee JK, Hawkins D, Ren B, Natarajan R (2008) Role of the histone H3 lysine 4 methyltransferase, SET7/9, in the regulation of NF-kappaB-dependent inflammatory genes. Relevance to diabetes and inflammation. J Biol Chem 283:26771–26781

Libby P, Ridker PM, Maseri A (2002) Inflammation and atherosclerosis. Circulation 105:1135–1143

Ling C, Groop L (2009) Epigenetics: a molecular link between environmental factors and type 2 diabetes. Diabetes 58:2718–2725

Liu L, Li Y, Tollefsbol TO (2008) Gene-environment interactions and epigenetic basis of human diseases. Curr Issues Mol Biol 10:25–36

Margariti A, Zampetaki A, Xiao Q, Zhou B, Karamariti E, Martin D, Yin X, Mayr M, Li H, Zhang Z, De Falco E, Hu Y, Cockerill G, Xu Q, Zeng L (2010) Histone deacetylase 7 controls endothelial cell growth through modulation of beta-catenin. Circ Res 106:1202–1211

Marrero MB, Fulton D, Stepp D, Stern DM (2005) Angiotensin II-induced signaling pathways in diabetes. Curr Diabetes Rev 1:197–202

Maunakea AK, Chepelev I, Zhao K (2010) Epigenome mapping in normal and disease States. Circ Res 107:327–339

Meng L, Park J, Cai Q, Lanting L, Reddy MA, Natarajan R (2010) Diabetic conditions promote binding of monocytes to vascular smooth muscle cells and their subsequent differentiation. Am J Physiol Heart Circ Physiol 298:H736–H745

Metzker ML (2010) Sequencing technologies – the next generation. Nat Rev Genet 11:31–46

Miao F, Gonzalo IG, Lanting L, Natarajan R (2004) In vivo chromatin remodeling events leading to inflammatory gene transcription under diabetic conditions. J Biol Chem 279:18091–18097

Miao F, Li S, Chavez V, Lanting L, Natarajan R (2006) Coactivator-associated arginine methyltransferase-1 enhances nuclear factor-kappaB-mediated gene transcription through methylation of histone H3 at arginine 17. Mol Endocrinol 20:1562–1573

Miao F, Wu X, Zhang L, Yuan YC, Riggs AD, Natarajan R (2007) Genome-wide analysis of histone lysine methylation variations caused by diabetic conditions in human monocytes. J Biol Chem 282:13854–13863

Miao F, Smith DD, Zhang L, Min A, Feng W, Natarajan R (2008) Lymphocytes from patients with type 1 diabetes display a distinct profile of chromatin histone H3 lysine 9 dimethylation: an epigenetic study in diabetes. Diabetes 57:3189–3198

Min Q, Bai YT, Jia G, Wu J, Xiang JZ (2010) High glucose enhances angiotensin-II-mediated peroxisome proliferation-activated receptor-gamma inactivation in human coronary artery endothelial cells. Exp Mol Pathol 88:133–137

Miranda TB, Jones PA (2007) DNA methylation: the nuts and bolts of repression. J Cell Physiol 213:384–390

Morgan HD, Sutherland HG, Martin DI, Whitelaw E (1999) Epigenetic inheritance at the agouti locus in the mouse. Nat Genet 23:314–318

Murr R (2010) Interplay between different epigenetic modifications and mechanisms. Adv Genet 70:101–141

Musri MM, Carmona MC, Hanzu FA, Kaliman P, Gomis R, Parrizas M (2010) Histone demethylase LSD1 regulates adipogenesis. J Biol Chem 285:30034–30041

Nakade K, Pan J, Yoshiki A, Ugai H, Kimura M, Liu B, Li H, Obata Y, Iwama M, Itohara S, Murata T, Yokoyama KK (2007) JDP2 suppresses adipocyte differentiation by regulating histone acetylation. Cell Death Differ 14:1398–1405

Natarajan R (2011) Drugs targeting epigenetic histone acetylation in vascular smooth muscle cells for restenosis and atherosclerosis. Arterioscler Thromb Vasc Biol 31:725–727

Natarajan R, Nadler JL (2004) Lipid inflammatory mediators in diabetic vascular disease. Arterioscler Thromb Vasc Biol 24:1542–1548

Nathan DM, Cleary PA, Backlund JY, Genuth SM, Lachin JM, Orchard TJ, Raskin P, Zinman B (2005) Intensive diabetes treatment and cardiovascular disease in patients with type 1 diabetes. N Engl J Med 353:2643–2653

Pirola L, Balcerczyk A, Okabe J, El-Osta A (2010) Epigenetic phenomena linked to diabetic complications. Nat Rev Endocrinol 6:665–675

Pons D, de Vries FR, van den Elsen PJ, Heijmans BT, Quax PH, Jukema JW (2009) Epigenetic histone acetylation modifiers in vascular remodelling: new targets for therapy in cardiovascular disease. Eur Heart J 30:266–277

Portela A, Esteller M (2010) Epigenetic modifications and human disease. Nat Biotechnol 28:1057–1068

Ramos YF, Hestand MS, Verlaan M, Krabbendam E, Ariyurek Y, van Galen M, van Dam H, van Ommen GJ, den Dunnen JT, Zantema A, t Hoen PA (2010) Genome-wide assessment of differential roles for p300 and CBP in transcription regulation. Nucleic Acids Res 38:5396–5408

Reddy MA, Natarajan R (2011) Epigenetic mechanisms in diabetic vascular complications. Cardiovasc Res 90:421–429

Reddy MA, Li SL, Sahar S, Kim YS, Xu ZG, Lanting L, Natarajan R (2006) Key role of Src kinase in S100B-induced activation of the receptor for advanced glycation end products in vascular smooth muscle cells. J Biol Chem 281:13685–13693

Reddy MA, Sahar S, Villeneuve LM, Lanting L, Natarajan R (2009) Role of Src tyrosine kinase in the atherogenic effects of the 12/15-lipoxygenase pathway in vascular smooth muscle cells. Arterioscler Thromb Vasc Biol 29:387–393

Ross R (1999) Atherosclerosis–an inflammatory disease. N Engl J Med 340:115–126

Roth SY, Denu JM, Allis CD (2001) Histone acetyltransferases. Annu Rev Biochem 70:81–120

Roy S, Sala R, Cagliero E, Lorenzi M (1990) Overexpression of fibronectin induced by diabetes or high glucose: phenomenon with a memory. Proc Natl Acad Sci U S A 87:404–408

Saccani S, Natoli G (2002) Dynamic changes in histone H3 Lys 9 methylation occurring at tightly regulated inducible inflammatory genes. Genes Dev 16:2219–2224

Sahar S, Dwarakanath RS, Reddy MA, Lanting L, Todorov I, Natarajan R (2005) Angiotensin II enhances interleukin-18 mediated inflammatory gene expression in vascular smooth muscle cells: a novel cross-talk in the pathogenesis of atherosclerosis. Circ Res 96:1064–1071

Sahar S, Reddy MA, Wong C, Meng L, Wang M, Natarajan R (2007) Cooperation of SRC-1 and p300 with NF-kappaB and CREB in angiotensin II-induced IL-6 expression in vascular smooth muscle cells. Arterioscler Thromb Vasc Biol 27:1528–1534

Selvi BR, Mohankrishna DV, Ostwal YB, Kundu TK (2010) Small molecule modulators of histone acetylation and methylation: a disease perspective. Biochim Biophys Acta 1799: 810–828

Shanmugam N, Kim YS, Lanting L, Natarajan R (2003a) Regulation of cyclooxygenase-2 expression in monocytes by ligation of the receptor for advanced glycation end products. J Biol Chem 278:34834–34844

Shanmugam N, Reddy MA, Guha M, Natarajan R (2003b) High glucose-induced expression of proinflammatory cytokine and chemokine genes in monocytic cells. Diabetes 52:1256–1264

Sharma K, Ziyadeh FN (1995) Hyperglycemia and diabetic kidney disease. The case for transforming growth factor-beta as a key mediator. Diabetes 44:1139–1146

Sharma S, Kelly TK, Jones PA (2010) Epigenetics in cancer. Carcinogenesis 31:27–36

Sheetz MJ, King GL (2002) Molecular understanding of hyperglycemia's adverse effects for diabetic complications. JAMA 288:2579–2588

Shi Y, Whetstine JR (2007) Dynamic regulation of histone lysine methylation by demethylases. Mol Cell 25:1–14

Shilatifard A (2006) Chromatin modifications by methylation and ubiquitination: implications in the regulation of gene expression. Annu Rev Biochem 75:243–269

Sun G, Reddy MA, Yuan H, Lanting L, Kato M, Natarajan R (2010) Epigenetic histone methylation modulates fibrotic gene expression. J Am Soc Nephrol 21:2069–2080

Tateishi K, Okada Y, Kallin EM, Zhang Y (2009) Role of Jhdm2a in regulating metabolic gene expression and obesity resistance. Nature 458:757–761

Trojer P, Reinberg D (2006) Histone lysine demethylases and their impact on epigenetics. Cell 125:213–217

Turunen MP, Aavik E, Yla-Herttuala S (2009) Epigenetics and atherosclerosis. Biochim Biophys Acta 1790:886–891

Vanden Berghe W, De Bosscher K, Boone E, Plaisance S, Haegeman G (1999) The nuclear factor-kappaB engages CBP/p300 and histone acetyltransferase activity for transcriptional activation of the interleukin-6 gene promoter. J Biol Chem 274:32091–32098

Vaquero A, Scher M, Erdjument-Bromage H, Tempst P, Serrano L, Reinberg D (2007) SIRT1 regulates the histone methyl-transferase SUV39H1 during heterochromatin formation. Nature 450:440–444

Villeneuve LM, Natarajan R (2010) The role of epigenetics in the pathology of diabetic complications. Am J Physiol Renal Physiol 299:F14–F25

Villeneuve LM, Reddy MA, Lanting LL, Wang M, Meng L, Natarajan R (2008) Epigenetic histone H3 lysine 9 methylation in metabolic memory and inflammatory phenotype of vascular smooth muscle cells in diabetes. Proc Natl Acad Sci U S A 105:9047–9052

Villeneuve LM, Kato M, Reddy MA, Wang M, Lanting L, Natarajan R (2010) Enhanced levels of microRNA-125b in vascular smooth muscle cells of diabetic db/db mice lead to increased inflammatory gene expression by targeting the histone methyltransferase Suv39h1. Diabetes 59:2904–2915

Visel A, Blow MJ, Li Z, Zhang T, Akiyama JA, Holt A, Plajzer-Frick I, Shoukry M, Wright C, Chen F, Afzal V, Ren B, Rubin EM, Pennacchio LA (2009) ChIP-seq accurately predicts tissue-specific activity of enhancers. Nature 457:854–858

Wang Y, Harris DC (2011) Macrophages in renal disease. J Am Soc Nephrol 22:21–27

Wang Z, Schones DE, Zhao K (2009) Characterization of human epigenomes. Curr Opin Genet Dev 19:127–134

Weiss D, Sorescu D, Taylor WR (2001) Angiotensin II and atherosclerosis. Am J Cardiol 87:25C–32C

Wen Y, Gu J, Li SL, Reddy MA, Natarajan R, Nadler JL (2006) Elevated glucose and diabetes promote interleukin-12 cytokine gene expression in mouse macrophages. Endocrinology 147:2518–2525

Workman JL, Kingston RE (1998) Alteration of nucleosome structure as a mechanism of transcriptional regulation. Annu Rev Biochem 67:545–579

Writing Team DCCT/EDIC Research Group (2002) Effect of intensive therapy on the microvascular complications of type 1 diabetes mellitus. JAMA 287:2563–2569

Wu SC, Zhang Y (2010) Active DNA demethylation: many roads lead to Rome. Nat Rev Mol Cell Biol 11:607–620

Wysocka J, Allis CD, Coonrod S (2006) Histone arginine methylation and its dynamic regulation. Front Biosci 11:344–355

Yan SF, Ramasamy R, Schmidt AM (2008) Mechanisms of disease: advanced glycation end-products and their receptor in inflammation and diabetes complications. Nat Clin Pract Endocrinol Metab 4:285–293

Yan SF, Ramasamy R, Schmidt AM (2009) The receptor for advanced glycation endproducts (RAGE) and cardiovascular disease. Expert Rev Mol Med 11:e9

Yang XJ, Seto E (2007) HATs and HDACs: from structure, function and regulation to novel strategies for therapy and prevention. Oncogene 26:5310–5318

Yun JM, Jialal I, Devaraj S (2011) Epigenetic regulation of high glucose-induced proinflammatory cytokine production in monocytes by curcumin. J Nutr Biochem 22:450–458

Zhong Q, Kowluru RA (2010) Role of histone acetylation in the development of diabetic retinopathy and the metabolic memory phenomenon. J Cell Biochem 110:1306–1313

Zhong Q, Kowluru RA (2011) Epigenetic changes in mitochondrial superoxide dismutase in the retina and the development of diabetic retinopathy. Diabetes 60:1304–1313

Chapter 20
Epigenetic Changes in Inflammatory and Autoimmune Diseases

Helene Myrtue Nielsen and Jörg Tost

Abstract In higher eukaryotic organisms epigenetic modifications are crucial for proper chromatin folding and thereby proper regulation of gene expression. In the last years the involvement of aberrant epigenetic modifications in inflammatory and autoimmune diseases has been recognized and attracted significant interest. However, the epigenetic mechanisms underlying the different disease phenotypes are still poorly understood. As autoimmune and inflammatory diseases are at least partly T cell mediated, we will provide in this chapter an introduction to the epigenetics of T cell differentiation followed by a summary of the current knowledge on aberrant epigenetic modifications that dysfunctional T cells display in various diseases such as type 1 diabetes, rheumatoid arthritis, systemic lupus erythematosus, multiple sclerosis, inflammatory bowel disease, and asthma.

20.1 Introduction

The number of people diagnosed with autoimmune and inflammatory diseases has increased noteworthy in the last 40 years and this fast-growing number has been impossible to explain by Mendelian inheritance only. Epigenetic modifications such

---

H. Myrtue Nielsen
Laboratory for Functional Genomics, Fondation Jean Dausset – Centre d'Etude du Polymorphisme Humain (CEPH), 27 rue Juliette Dodu, 75010 Paris, France

J. Tost, Ph.D. (✉)
Laboratory for Functional Genomics, Fondation Jean Dausset – Centre d'Etude du Polymorphisme Humain (CEPH), 27 rue Juliette Dodu, 75010 Paris, France

Laboratory for Epigenetics, Centre National de Génotypage, CEA-Institut de Génomique, Bâtiment G2, 2 rue Gaston Crémieux, CP 5721, 91000 Evry, France
e-mail: tost@cng.fr

as DNA methylation, histone modifications and chromatin remodeling are closely interwoven and constitute multiple layers of epigenetic modifications to control and modulate gene expression through their joint effects on chromatin structure (Tost 2008). The epigenome of higher eukaryotic organisms changes continuously throughout life (Calvanese et al. 2009; Whitelaw and Whitelaw 2006). In addition to the genetic influences on the epigenome, environmental factors have been found to contribute to the dynamics on the epigenome leading potentially to altered regulation of gene expression (Fraga 2009; Holliday 2006). The influence of non-genetic factors in autoimmune and inflammatory diseases has been demonstrated in studies of genetically-identical (monozygotic/MZ) twins that show a variable degree of discordance with respect to different phenotypic traits including susceptibility to these diseases. MZ twins show a concordance rate for different autoimmune diseases ranging from 12% for rheumatoid arthritis (Aho et al. 1986) up to 70% for psoriasis (Krueger and Duvic 1994). There is substantial epigenetic variation between MZ twin pairs (Ballestar 2010; Meda et al. 2011). This variation increases with age and if twins live in different environments (Fraga et al. 2005). These environment-host interactions might result in such severe epigenetic changes that the epigenetic homeostasis can no longer be maintained leading to aberrant gene expression. Aberrant gene expression in specific cells of the immune system most likely contributes to the loss of immune tolerance, inflammation and autoimmunity, which is characterized by the failure of an organism to recognize self-antigens. Multiple risk factors such as environmental toxins, drugs, hereditary factors, and viruses have been associated with the susceptibility for autoimmune diseases. However, the exact reasons for the breakdown of self-tolerance are poorly understood. Many autoimmune diseases show an increased prevalence in women. This might be possibly due to sex hormones but changes of the epigenetic landscape of T helper cells might also contribute to the higher percentage. In particular alterations in the patterns of epigenetic modifications and/or reactivation of the normally inactive X chromosome are believed to be one of the founders of the imbalance seen between the genders (Brooks 2010). Due to the growing number of incidences of these multifactorial diseases, increased attention and effort has been put in investigating the influence of aberrant epigenetic modifications.

20.2 Th Cell Differentiation

Dysfunction of T helper (Th) cells is the main precursor of autoimmunity as a shift in the balance between effector T cells and the negative regulators of the immune system (Tregs) drives the production of auto-antibodies and leads to the destruction of the respective tissues of affected individuals. T helper cells are important components of the immune system as they are required for optimal host defense against different pathogens. Their major role is to activate and direct other cells of the immune system. However, Th cells are not able to eliminate infected host cells or pathogens by themselves and would be insufficient upon infection if no other immune cells were present.

Interactions between the T cell receptor (TCR) of naïve CD4+ T cells and the major histocompatibility complex (MHC), found on antigen presenting cells (APCs), initiate a network of signaling cascades that result in differentiated effector Th cells. Depending on the mutually exclusive production of specific cytokines, differentiated Th cells can be divided into at least four major subgroups: Th1, Th2, and the more recently found lineages Th17 and regulatory T cells (Treg), reviewed in Zhu et al. (2010). The signature cytokine of Th1 cells is IFN-γ, Th2 cells are characterized by the production of interleukin 4 (IL-4), IL-5 and IL-13 whereas Th17 cells produce IL-17, IL-21, IL-22, and IL-26 (Ouyang et al. 2008). The exogenous presence of signature cytokines for Th1 and Th2 cells, IFN-γ and IL-4 respectively, is necessary for the initiation of differentiation (Le Gros et al. 1990; Lighvani et al. 2001). Th1 cells mediate immunity against intracellular pathogens whereas Th2 cells mediate humoral immunity against extracellular pathogens. Th17 cells are implicated in the defense against extracellular bacteria and fungi at epithelial and mucosal surfaces. Th cells also play a key-role in different diseases as aberrant Th1 and Th17 cytokine expression is associated with autoimmune diseases whereas allergic diseases are associated with aberrant Th2 cytokine expression (Lucey et al. 1996; Murdaca et al. 2011; Robinson et al. 1992). The negative regulators of the immune system are the Treg cells. Their main role is to maintain the homeostatic balance of immune responses through the suppression of activation, proliferation and the specific cellular function of a variety of immune cells including Th1- and Th2 cells, cytotoxic T cells, natural killer (NK) cells, B cells and antigen presenting cells (Sakaguchi et al. 2006, 2008).

For the differentiation into diverse Th cell lineages the molecular changes induced by TCR interaction are passed on to daughter cells through cell expansion. Epigenetic states play an important role when it comes to the transcriptional regulation and maintenance of a given gene. These heritable changes of the cytokine genes themselves and their regulatory elements such as enhancers, promoters, and insulators, that direct the differentiation process, have been found to start at an early stage upon TCR/MHC interaction (Ansel et al. 2003).

Several comprehensive reviews have detailed the multiple steps in Th cell differentiation (Ansel et al. 2006; Murphy and Reiner 2002; Placek et al. 2009; Wilson et al. 2009; Zhu et al. 2010), we will therefore only briefly introduce the pathways and limit this section to key-facts only. Our main focus during Th cell differentiation will be on the involvement of epigenetic modifications.

20.2.1 Th1 Cell Differentiation

IL-12 is the major Th1 differentiation initiator and is released from antigen-presenting cells (APCs) upon TCR/MHC interaction where it induces IFN-γ production through two pathways (Fig. 20.1): (1) The IL-12 induced production and secretion of IFN-γ from natural killer cells activates the transcription factor STAT1 once IFN-γ has bound to its receptor IFNGR. STAT1 activation leads in turn to the expression of the

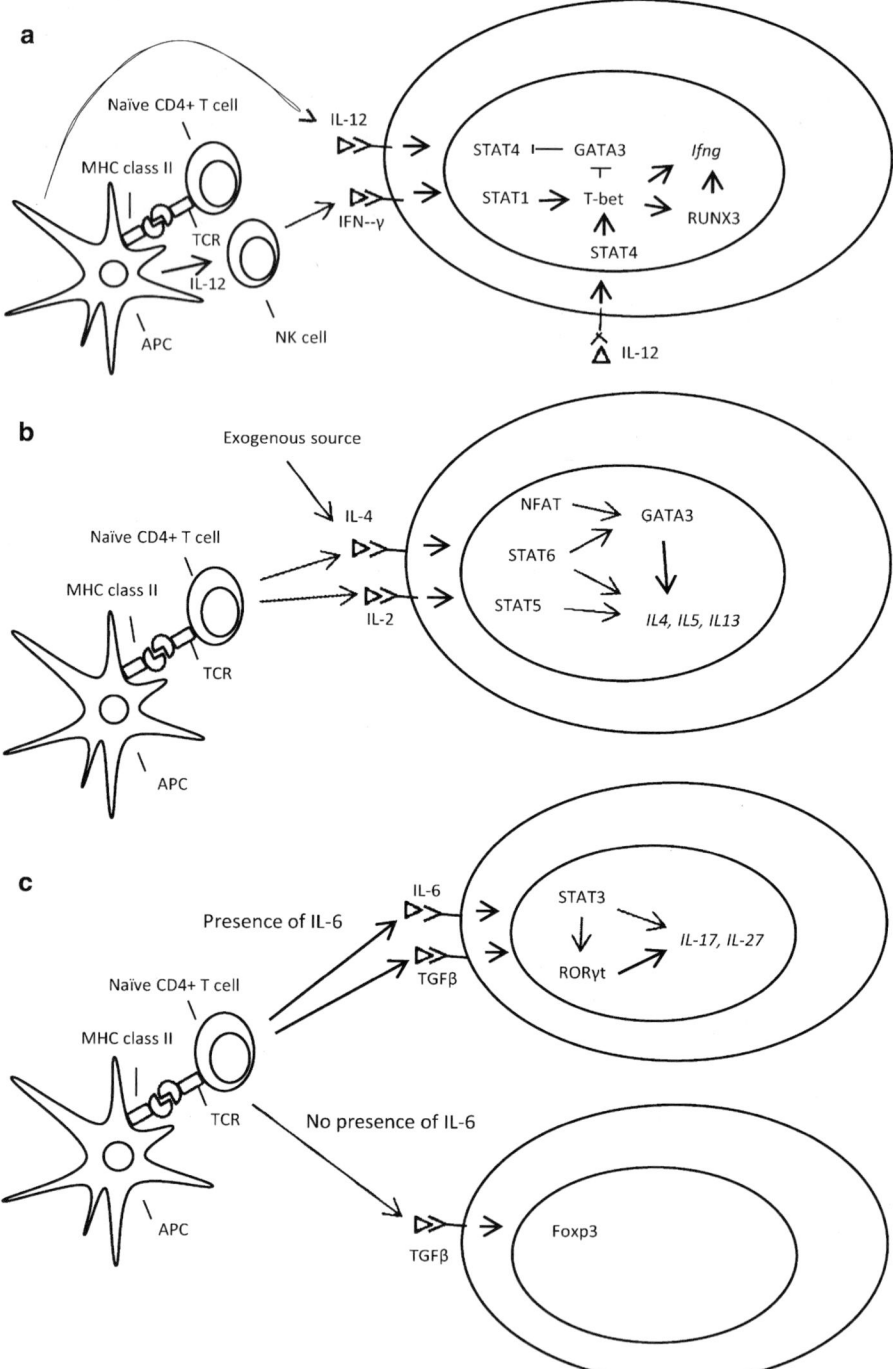

Fig. 20.1 Key cytokines and transcription factors involved in Th cell differentiation. (a) Differentiation of the Th1 cell linage is initiated by TCR/MHC engagement. As IL-12, the major differentiation initiator is released from APCs the transcription of *INF-γ* is achieved through two pathways. The first includes the STAT1 induced transcription of T-bet, the key-regulator of Th1 differentiation.

transcription factor T-bet (Afkarian et al. 2002; Weaver et al. 2007) a key-regulator of the Th1 lineage that regulates multiple genes including *Ifng* (Djuretic et al. 2007), induces the up-regulation of *Runx3* (Djuretic et al. 2007), and negatively regulates GATA3 (Usui et al. 2006). A collaboration between T-bet and Runx3 is further observed as they together have a positive regulatory effect on *Ifng* and a negative regulatory effect on *Il-4* (Djuretic et al. 2007; Naoe et al. 2007). GATA3 is a T cell specific transcription factor with counteracting function on Th1 differentiation as it down-regulates STAT4 (Usui et al. 2003, 2006) the other key-regulator of Th1 differentiation (Fig. 20.1). (2) In the second IL-12 induced Th1 differentiation pathway, IL-12 binds to its cell-surface receptor, IL-12R, which is composed of the two subunits IL-12Rβ1 and IL-12Rβ2 (Presky et al. 1996). The expression of IL-12Rβ2 is dependent on the Brahma-related gene 1 (BRG1) product, which is recruited to the promoter and enhancer of *IL-12Rβ2* rapidly after TCR engagement in resting peripheral T cells (Letimier et al. 2007). BRG1 is the ATPase subunit of the SWI/SNF-like BAF (Brahma-related gene (BRG1)/Brahma (BRM)-associated factor) chromatin remodeling complex (BAF complex) and the recruitment of the BAF complex is associated with histone hyperacetylation and low-level transcription of *IL-12Rβ2* (Letimier et al. 2007). Translocation of the newly synthesized IL-12Rβ2 subunit to the cell surface completes the composition of the IL-12 receptor through which IL-12 induces the activation of STAT4 (Thieu et al. 2008). STAT4 expression has a positive feedback mechanism resulting in high-levels of *IL-12Rβ2* transcription (Letimier et al. 2007) and inducing in turn STAT4 and thereby IFN-γ production. Furthermore, there is *in vitro* evidence that STAT4 positively regulates T-bet at the transcriptional level (Yang et al. 2007) (Fig. 20.1).

20.2.2 Th2 Cell Differentiation

Upon TCR engagement, IL-4 induces the activation of STAT6, which activates GATA3 in cooperation with several TCR-induced signals including the nuclear factor of activated T cells (NFAT) (Ouyang et al. 2000; Zhu et al. 2001) (Fig. 20.1). STAT6 and GATA3 induce synergistically the production of the signature cytokines

---

Fig. 20.1 (continued) In co-operation with Runx3, T-bet up-regulates INF-γ production. The other pathway includes the STAT4 induced transcription of *INF-γ*. GATA3 is a negative regulator of the Th1 cell lineage as it suppresses STAT4. (**b**) Subsequent to TCR/MHC engagement, IL-4 is the major differentiation initiator of Th2 differentiation. IL-4 activates STAT6 and in co-operation with NFAT they induce the activation of GATA3. Together STAT6 and GATA3 result in the production of the Th2 signature cytokines; IL-4, IL-5, and IL-13. The STAT5 induced production of Th2 specific cytokines is initiated as IL-2 binds to its receptor and is GATA3 dependent. (**c**) Differentiation into Th17- and Treg cells is initiated by TGF-β, whereby the presence of IL-6 is decisive in which lineage the cells will differentiate. If present in the external milieu, RORγt will be expressed and the differentiation will go towards Th17 cell linage. In the absence of IL-6 the key transcription factor of Treg differentiation, FOXP3, inhibits the expression of RORγt and encourages differentiation towards Treg cell lineage

IL-4, IL-5, and IL-13. Failure of Th2 differentiation in mice upon *Gata3* knockout in peripheral T cells strongly indicates the importance of Gata3 as a key-regulator of Th2 development (Pai et al. 2004; Zhu et al. 2004). Moreover, the presence of Gata3 is also important for the Stat5 induced production of Il-4 as they collaborate in priming naïve Th cells for the Th2 phenotype (Zhu et al. 2003). Stat5 becomes activated after IL-2 binds to its receptor. GATA3 is thus involved in the Th2 differentiation in both an IL-4 dependent and in an IL-2 dependent manner (Fig. 20.1). GATA3 has an auto-activation effect and stabilizes together with the autocrine and paracrine IL-4 induced activation of STAT6 the Th2 lineage development (Ansel et al. 2006).

20.2.3 Th17 and Treg Cell Differentiation

The differentiation process for these two Th cell lines is initiated by the transforming growth factor β (TGFβ), which simultaneously blocks Th1- and Th2 cell differentiation. The fate of cell differentiation depends on the presence of IL-6 (Kimura and Kishimoto 2010). Il-6 induces in cooperation with TGF-β the expression of the steroid receptor-type nuclear receptor (RORγt), which is required and specific for Th17 development (Ivanov et al. 2007; McGeachy and Cua 2008) (Fig. 20.1). RORγt and STAT3 drive the production of the Th17 signature cytokines; IL-17, IL-21, IL-22, and IL-26. In addition, IL-6 inhibits the expression of FOXP3 through IL-6 induced STAT3 activation (Yang et al. 2008). In the absence of IL-6 the key-transcription factor of Treg differentiation, FOXP3, inhibits RORγt and thereby Th17 differentiation and encourages instead Treg cell line engagement (Ivanov et al. 2007).

20.2.4 Epigenetics and Th Cell Differentiation

Major changes to the patterns of epigenetic modifications within the *IFN-γ* and *IL-4* loci are driven by TCR engagement. Once established these heritable changes in gene expression are passed on to daughter cells where they contribute to Th cell maintenance. Specific histone modifications are associated with heterochromatin and euchromatin (Kouzarides 2007). Heterochromatin, in which the DNA is densely wrapped around the octamer of histone proteins, is associated with modifications such as H3K9-, H3K27-, and H4K20 methylation. On the contrary, euchromatin is associated with epigenetic modifications such as H3K4 methylation and acetylation of lysine residues of H3 and H4.

To permit a prompt and specific response upon infection with a pathogen, naïve T cells have to be poised to enable rapid differentiation in either direction in function of exogenous stimuli. Interestingly, poised CD4[+] T cells have been shown to express low levels of both IFN-γ and IL-4 (Grogan et al. 2001) and the histones located at the genes encoding the cytokines display a bivalent histone modification pattern. They thus display a combination of histone modifications normally associated

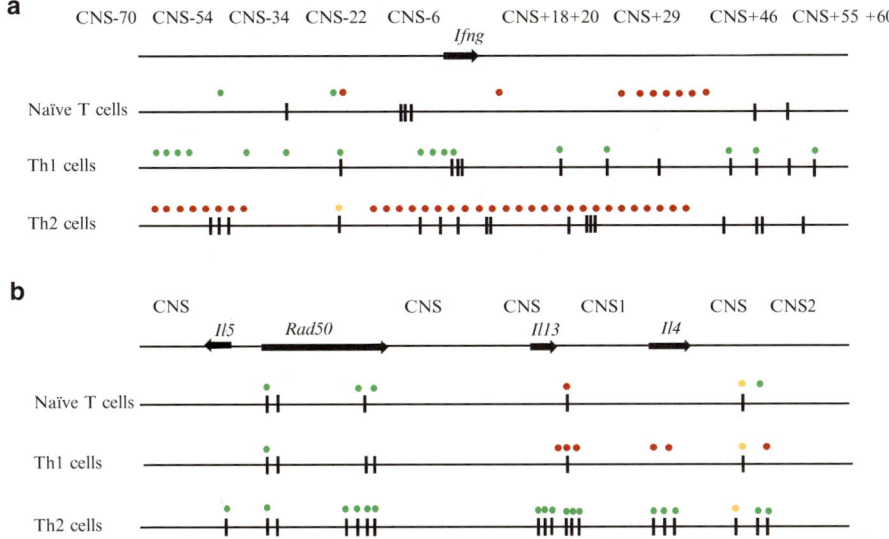

Fig. 20.2 The epigenetic state of the Ifng- and Il-4 locus during Th cell differentiation. Depending on the external cytokine milieu upon TCR/MHC engagement different epigenetic modifications can be found throughout the *Ifng*- and *Il-4* locus. The black bars indicate hypersensitive sites, green dots indicate permissive epigenetic modification, red dots indicate repressive epigenetic modifications, and yellow dots indicate bivalent epigenetic modifications. (**a**) During Th1 cell differentiation the *Ifng* locus is associated with permissive epigenetic modifications contradictory to the *Ifng* locus during Th2 cell differentiation. CNS-22 is believed to be an important regulatory region as consensus binding sites for the key-transcription factors, T-bet, GATA3, STAT4, STAT6, NF-κB, and Ikaros, are found here together with histone hyperacetylation in both naïve-, Th1-, and Th2 cell lineages. (**b**) The *Il-4* locus is associated with permissive epigenetic modifications during Th2 cell differentiation. During Th1 cell differentiation the *Il-4* locus is mostly associated with repressive epigenetic marks

with active transcription (H3K4me2 and H3K9/14 acetylation) combined with modifications normally associated with repressive transcription (H3K27me3) within the two loci (Avni et al. 2002; Baguet and Bix 2004; Fields et al. 2002). High DNA methylation levels (~90%) within the *IL-4* locus contribute to the restrained Th2 cytokine expression in naïve T cells (Lee et al. 2002; Makar et al. 2003). However, the *IL-4* promoter is less methylated in undifferentiated T cells facilitating the observed low-level transcription of the gene (Makar et al. 2003).

Cis-regulatory elements for the genes encoding the cytokines have been identified by *in silico* approaches as conserved non-coding regions (CNSs) (Nardone et al. 2004) or experimentally as DNase I hypersensitivity sites (HSs) (Schoenborn et al. 2007). A region encompassing ~130 kb constitutes the regulatory platform for *IFN-γ* (Barski et al. 2007; Boyle et al. 2008) and within this region several CNSs have been identified (Balasubramani et al. 2010; Hatton et al. 2006; Schoenborn et al. 2007; Shnyreva et al. 2004) (Fig. 20.2). The regulatory region for the *IL-4* locus encompasses a ~85 kb region containing *IL-4*, *IL-5*, *IL-13*, and the ubiquitously expressed

Rad50 which is involved in DNA repair. Six CNSs have been identified within the *IL-4* locus together with a number of HSs (Loots et al. 2000; Nardone et al. 2004; Takemoto et al. 1998; Wilson et al. 2009; Yamashita et al. 2002) (Fig. 20.2).

20.2.5 Epigenetic Changes Within the IFN-γ Locus During Th Cell Differentiation

Several epigenetic marks associated with actively transcribed genes are found throughout the *IFN-γ* locus rapidly after IL-12 and IFN-γ induced Th1 differentiation. One of these permissive epigenetic marks is acetylation of lysines of H3 at individual CNSs (Zhou et al. 2004). The key-transcription factor controlling Th1 differentiation, T-bet, is also involved in the epigenetic regulation through additional events including H4 acetylation at regulatory sites located +20, +30, +46, +50, and +60 kb upstream of the transcription start site of *Ifng*. T-bet further recruits the jumonji-domain-containing protein histone demethylase 3 (JMJD3) which is a part of the H3K27 demethylase complex (Miller et al. 2008). Simultaneously, T-bet associates with the retinoblastoma-binding protein 5 (RbBp5) resulting in H3K4 methyltransferase activity facilitating access to the DNA (Miller et al. 2008). These observations suggest the importance of T-bet in reversing repressive epigenetic marks within the *INF-γ* locus. Permissive marks are also found at the level of histone methylation. H3K4 dimethylation, normally associated with active transcription, is found within the *IFN-γ* locus from CNS +20 to CNS +60 contributing to active transcription of *IFN-γ* (Hamalainen-Laanaya et al. 2007). However, not only permissive marks are found throughout the *IFN-γ* locus during Th1 differentiation. H3K9 dimethylation, normally associated with silenced genes, is also present during the differentiation process (Chang and Aune 2007). These conflicting epigenetic marks of the actively transcribed *IFN-γ* locus may serve as a regulatory mechanism to ensure proper IFN-γ production.

The two key-transcription factors of Th2 differentiation, STAT6 and GATA3, are involved in the establishment of repressive epigenetic marks, notably H3K27 di- and trimethylation, throughout the *IFN-γ* locus (Fig. 20.2). The binding of STAT6 to the *Ifng* promoter is involved in the establishment of the characteristic repressive mark, H3K27 trimethylation (Chang and Aune 2007). GATA3 also binds the *Ifng* promoter to which it recruits EZH2, a component of the Polycomb repressive complex 2, that is associated with H3K27 trimethylation and the formation of heterochromatin (Chang and Aune 2007). Thus during Th2 differentiation GATA3 is associated with both transcriptional activation of genes encoded within the *IL-4* locus and transcriptional repression of *Ifng*. It has been proposed that it is the number of GATA-3 binding sites within the two loci, *IFN-γ* and *IL-4* respectively, which will determine the differences of epigenetic marks and thereby the different outcome for GATA3 binding (Aune et al. 2009).

The multi-functional transcriptional regulator CCCTC-binding factor (CTCF) is implicated in Th2 differentiation as it has been found that transcription of IL-4, IL-5, and IL-13 was strongly reduced in CTCF-deficient Th2 cultures (Ribeiro de Almeida et al. 2009). Interestingly, CTCF-deficiency only had moderate influence on

IFN-γ production and the IL-17 production in Th17 cells was unaffected (Ribeiro de Almeida et al. 2009).

Individual CNSs have important influences on Th cell differentiation. Early remodeling of the *Ifng* locus is suggested to be dependent on CNS −6 (Fig. 20.2) as this region possesses a DNase I hypersensitive site in naïve T cells (Bream et al. 2004; Lee et al. 2004). CNS −22 is believed to be an important regulatory region as it displays a high concentration of consensus binding sites for key-transcription factors involved in T cell development and differentiation such as T-bet, GATA3, STAT4, STAT6, nuclear factor-κB (NF-κB), and Ikaros. Furthermore, histone hyperacetylation in this region has been found in both naïve T cells as well as in Th1 and Th2 cells indicating a key-regulatory effect for *IFN-γ* transcription (Hatton et al. 2006). Insulators of the *Ifng* locus have been mapped to CNS −54 and CNS +46 (Schoenborn et al. 2007).

20.2.6 Epigenetic Changes Within the IL-4 Locus During Th Cell Differentiation

During Th1 differentiation the permissive epigenetic marks of the *IL-4* locus, found in naïve T cells, have to be erased for definite IFN-γ expression. The methyl CpG-binding domain 2 (MBD2) may serve an important function in IL-4 silencing as it has been shown that deficiency of this protein results in heritable aberrant IL-4 expression in Th1 cells lacking GATA3 (Hutchins et al. 2002). MBD2 is known to associate with histone deacetylase (HDACs) complexes resulting in the formation of heterochromatin. H3K27 tri-methylation contributes to silencing of the *IL-4* locus as well (Koyanagi et al. 2005; Wei et al. 2009).

The active *IL-4* locus displays H3K4 tri-methylation (Wei et al. 2009) and the responsible histone methyltransferase MLL has been found to be recruited to *Gata3* and *Il-4* where it serves to maintain the permissive mark throughout cell division (Yamashita et al. 2006). Interestingly, the function of MLL seems to be specific for Th2 cells as Th1 differentiation was not affected in haplo-insufficient mice (Yamashita et al. 2006). GATA3 itself has also been found to be involved in the regulation of several epigenetic modifications as it assists in the formation of accessible DNA by association with H3K4 methyltransferases and recruitment of histone acetyltransferases (HATs) (Ansel et al. 2006; Yamashita et al. 2006). Furthermore GATA3 can displace bound MBD2 and the associated HDAC complexes responsible for the establishment and maintenance of heterochromatin (Hutchins et al. 2002). Additionally GATA3 can displace Dnmt1 thus inhibiting DNA methylation (Makar et al. 2003; Tykocinski et al. 2005). Epigenetic changes are not only seen on the level of histone modifications. DNA methylation is also implicated in Th2 differentiation as the high DNA methylation level in the 5′ region of the *IL-4* gene, in naïve T cells, as well as the hypermethylated intergenic region between *IL-4* and *IL-13* become specifically demethylated in a replication-dependent manner (Lee et al. 2002). The DNA demethylation is strongly associated with high-levels of IL-4 transcription (Lee et al. 2002).

20.3 Epigenetics in Autoimmune and Inflammatory Diseases

Autoimmune- and inflammatory diseases are characterized by a failure of the capability of the immune system to recognize and tolerate the body's own molecular components. In the following paragraphs we will summarize the current knowledge on epigenetic changes in these diseases. Virus associated inflammatory diseases that display profound epigenetic changes such as gastritis or cirrhosis will not be mentioned within this chapter.

20.3.1 Type 1 Diabetes

Type 1 diabetes (T1D) is a complex autoimmune disease affecting more than 30 million people worldwide. It is caused by a combination of genetic and non-genetic factors leading to immune destruction of insulin-secreting islet cells (Eizirik et al. 2009).

A microarray study by Miao et al. established the presence of an aberrant histone methylation profile in lymphocytes of T1D patients (Miao et al. 2008). They identified a cluster of genes with increased H3K9me2, normally associated with transcriptional repression. Many of the genes with increased H3K9me2 were involved in cytokine signaling pathways or encoded transcription factors (Miao et al. 2008). In particular increased H3K9me2 within the promoter region of *CTLA4* revealed a direct link between T1D and aberrant epigenetic modifications in lymphocytes as CTLA4 is involved in repressive T cell response.

In addition dysfunction of Treg cells has been suggested to play a role in T1D pathogenesis (Sgouroudis and Piccirillo 2009). *In vivo* studies support this suggestion as the expression of Foxp3, the key-transcription factor of Treg cell differentiation was induced upon treatment with the DNMT inhibitor decitabine in NOD (non-obese diabetic) mice and the induction was associated with a significant demethylation of the *Foxp3* CpG island (Zheng et al. 2009). DNA methylation also plays a role in the upregulation of LFA-1 as a consequence of hypomethylation of $CD4^+$ T cells which is characteristic for auto-reactive T cells (Richardson 1986). LFA-1 co-stimulates the TCR/MHC interaction (Wulfing et al. 2002), but when overexpressed LFA-1 is thought to stabilize even TCR/MHC interaction with low-affinity and thereby contribute to autoimmune response.

20.3.2 Rheumatoid Arthritis

Rheumatoid arthritis (RA) is an autoimmune disease affecting ~1% of the population caused by a combination of genetic predisposition, deregulated immunomodulation and environmental influences and characterized by chronic inflammation and destruction of synovial joints (Firestein 2003). Global DNA hypomethylation of peripheral blood mononuclear cells (PBMCs), synovial mononuclear cells and synovial tissue isolated from patients diagnosed with RA is likely to cause the aberrant expression of pro-inflammatory genes (Corvetta et al. 1991; Karouzakis et al. 2009; Liu et al. 2011; Neidhart et al. 2000). Activation of the normally inactivated retrotransposable element LINE-1 has been observed in the synovial fluid pellet from RA patients. This is thought to be a consequence of the global hypomethylation that might result in genomic instability. Activated LINE-1 elements can integrate themselves in the genome and affect gene expression and are therefore thought to contribute to the RA phenotype (Neidhart et al. 2000).

Repressed expression of the death receptor 3 (DR3) in synovial cells has been associated with hypermethylation of its promoter region (Takami et al. 2006). This finding provides a direct link between an epigenetically regulated gene and the RA phenotype as *DR3* is a member of the apoptosis-inducing Fas family. DR3 can initiate the intrinsic apoptotic network (Ashkenazi and Dixit 1998). Thus down-regulated DR3 expression contributes to the resistance of synovial cells to undergo apoptosis, which is a hallmark of RA.

Histone deacetylase inhibitors (HDACi) have gained increased interest for the treatment of RA, as they have been associated with decreased levels of pro-inflammatory cytokines (see Table 20.1). At the same time a loss of HDAC1 and HDAC2 activity in synovial tissue from RA patients has been observed compared to normal synovium (Huber et al. 2007). This seems contradictory to the promising strategies using HDACi's for RA treatment in mouse and rat models. However, it is unknown in which specific cell types the HDAC activity is decreased. It might be possible that HDACs are upregulated in other cell types and thereby make these cells possible targets for treatment. In addition, the large number of proteins, which can be modulated by HDACs and the possible overlap of targets between HDACs of the same class, makes it difficult to predict and discern the specific molecular effects of HDAC inhibitors.

20.3.3 *Systemic Lupus Erythematosus*

Systemic lupus erythematosus (SLE) is a chronic autoimmune disease which is characterized by the overproduction of autoantibodies against nuclear self-antigens leading to persistent damage in multiple tissues (Sawalha and Harley 2004).

Lupus is probably the best-studied autoimmune disease at the level of epigenetic modifications and especially DNA methylation. Decreased expression and activity of DNMTs in SLE patients (Richardson et al. 1990) causes a global DNA hypomethylation, which contributes to the aberrant gene expression of a large number of pro-inflammatory genes (Javierre et al. 2010). Very recently it was shown that two miRNAs (miR-21 and miR-148) promote the observed hypomethylation

Table 20.1 The outcome of using HDACi's in different models resembling inflammatory and autoimmune diseases

Disease	Model	Name of HDACi	Effect	Reference
Autoantibody-mediated arthritis	Mouse	FK-228	Decreased expression of TNF-α and IL-1β	Nishida et al. (2004)
Collagen-induced arthritis	Mouse and rat	SAHA and MS-275	Reduced arthritis scores	Lin et al. (2007)
Autoantibody-mediated arthritis	Mouse	Trichostatin A	Reduced arthritis scores and expression of matrix-degrading enzymes	Huber et al. (2007)
Systemic lupus erythematosus	In vitro cultured SLE T cells	Trichostatin A	Down-regulating CD154 and IL-10 while up-regulating IFN-γ	Mishra et al. (2001)
Systemic lupus erythematosus	Mouse	Trichostatin A	Up-regulates Treg cells	Reilly et al. (2008)
Inflammatory bowel disease	Mouse	Valproic acid and vorinostat (SAHA)	Suppression of pro-inflammatory cytokines and increased H3 acetylation Increased apoptosis of lamina propria lymphocytes	Glauben et al. (2006)
Inflammatory bowel disease	Mouse	Trichostatin A	Increased acetylation of Foxp3 and thereby the Treg mediated suppression of inflammation	Tao et al. (2007)

by targeting the DNA methyltransferase 1, interfering thus with maintenance of DNA methylation (Pan et al. 2010). It is also noteworthy, that lupus-like symptoms can also be induced by pharmaceutical regimens such as the vasodilator hydralazine used for the treatment of hypertensive disorders and procainamide administered for the treatment of cardiac arrhythmias. These two non-nucleoside compounds do have demethylating capabilities, although they are much less effective compared to the nucleosides analogues. There are strong analogies in the expression patterns of T-cells treated with demethylating agents such as 5-azacytidine and a subset of T-cells found in patients with active lupus. These methylation-sensitive genes that become hypomethylated upon chemical treatment or during the disease process include notably *CD11a* (*ITGAL*), *CD70*, *PRF1* and *CD40LG*. *CD11a* and *CD70* are both found overexpressed in Th cells (Zhao et al. 2010a, b), which might be due to the global hypomethylation. CD11a is the alpha chain of the integrin LFA-1 which is associated with cell adhesion and co-stimulatory signaling (Wulfing et al. 2002). CD70 is the ligand of the cell-surface receptor CD27, which plays a role in long-term maintenance of T cell immunity and the regulation of B cell activation. Decreased activity of the transcription factor regulatory factor X 1 (RFX1) in Th cells influences the epigenetic state of the promoter region of CD11a and CD70 as RFX1 is involved in the recruitment of DNMT1 and HDAC to the respective promoter region. Decreasing levels of RFX1 are thus associated with hypomethylation and hyperacetylation of the promoters of CD11a and CD70 thus increasing the transcription levels of these autoimmune genes (Zhao et al. 2010a). RFX1 further recruits the histone methyltransferase SUV39H1 as decreased levels of the transcription factor were associated with reduced H3K9 trimethylation within the same region (Zhao et al. 2010b). A strong gender bias exists among SLE patients with a female to male ratio of 9:1 (Soto et al. 2004). CD40LG is encoded on the X-chromosome, one of which is normally silenced in females through multiple (epigenetic) mechanisms. DNA demethylation in Th cells of SLE patients results in expression of the normally inactivated *CD40LG* in women (Lu et al. 2007) and might thereby contribute to the excess production of autoantibodies as CD40LG is a B cell co-stimulatory molecule (Zhou et al. 2009). DNA hypomethylation of the normally inactivated X-chromosome in women is therefore thought to contribute to the female predominance associated with SLE.

Analysis of DNA methylation patterns in MZ twins discordant for the disease phenotype revealed lower methylation levels of rRNA encoding regions in SLE patients (Javierre et al. 2010). As a consequence of the lower DNA methylation levels an overexpression of rRNA was observed which might result in increased assembly of ribosomal complexes that could trigger the autoantibody-production (Javierre et al. 2010).

Although DNA methylation is the best studied epigenetic mechanism in SLE patients an *in vitro* study found that aberrant expression of CD154, IL-10, and IFN-γ could be reversed by treating cells with the HDACi trichostatin A (Mishra et al. 2001). Further, the use of TSA proved to modulate SLE-like disease in mouse model by up-regulating Treg cells (Reilly et al. 2008).

20.3.4 Multiple Sclerosis

Demyelination of the central nervous system (CNS) caused by chronic inflammation is the hallmark of multiple sclerosis (MS).

Hypomethylation of *PAD2*, a gene encoding the peptidyl arginine deiminase type II protein that is responsible for the conversion of the guanidine group of arginines residues into citrulline, results in overexpression of the corresponding gene in MS patients (Mastronardi et al. 2007). The substrate of PAD2 is the myelin basic protein (MBP) but when abnormally citrullinated a conformational change of MBP occurs and it can no longer generate normal myelin (Moscarello et al. 1994).

20.3.5 Inflammatory Bowel Disease

Crohn's disease (CD) and ulcerative colitis (UC) both belong to a heterogeneous group of diseases named idiopathic inflammatory bowel disease (IBD). Despite the early stage of epigenetic studies in IBD, studies have indicated a potentially central role of histone acetylation as improvement of inflammatory conditions has been observed upon HDACi treatment. Inhibition of HDACs suppressed the production of pro-inflammatory cytokines, induced apoptosis, and caused increased acetylation (Glauben et al. 2006). Inhibition of HDACs further resulted in enhanced Treg mediated suppression of IBD *in vivo* indicating the broad implications of histone acetylation in IBD pathogenesis (Tao et al. 2007). The upregulated Treg mediated suppression of inflammation was associated with increased acetylation in the forkhead domain of Foxp3 resulting in its upregulation (Tao et al. 2007). Together these studies indicate that the formation of heterochromatin and thereby induced gene silencing is crucial for IBD pathogenesis. Interestingly, Kryczek *et al.* found that Treg cells can induce the production of IL-1 and IL-6 in colitic cells. The latter finding combined with the presence of IL-17 expressing T cells indicates that a subgroup of T cells combining the Treg and Th17 phenotype might actively contribute to the progression towards early tumorigenesis in ulcerative colitis (Kryczek et al. 2011).

Interestingly it has recently been suggested that treatment of IBD with HAT inhibitors might provide an opportunity to improve inflammatory conditions as well as H4 acetylation was significantly upregulated in the inflammatory tissue of patients diagnosed with CD (Tsaprouni et al. 2011).

20.3.6 Asthma

Asthma is characterized by an inflammatory obstruction of the respiratory airways. DNA methylation is the most extensively studied epigenetic mechanism and has been proposed as a biomarker for environmentally related asthma. Exposure to diesel exhaust particles (DEP) has been shown to change the overall DNA methylation profile in healthy elderly people (Baccarelli et al. 2009) and DNA methylation of the *ACSL3* 5′- CpG island may indicate the extent of transplacental exposure to DEPs (Perera et al. 2009). More specifically, DEP exposure induces hypermethylation of specific sites within the *INF-γ* promoter simultaneously with hypomethylation within the *IL-4* promoter in mice. These epigenetic changes correlate with IgE production (Liu et al. 2008). This most likely contributes to the excessive amounts of Th2 signature cytokines; IL-4, IL-5, Il-9, and Il-13 associated with this chronic disease of the respiratory airways (McGee and Agrawal 2006; Zhu and Paul 2008).

Recently, hypermethylation of *Foxp3*, the key transcription factor required for Treg commitment, was found to be a consequence of ambient air pollution (AAP) exposure in children (Nadeau et al. 2010). The hypermethylation of *Foxp3* was associated with impaired Treg-mediated repression of the overactive immune response. This finding further supports the increasing evidence for a major role of Treg cells in the pathogenesis of asthma (Durrant and Metzger 2010; Larche 2007; Taylor et al. 2005).

The knowledge on the precise influence of aberrant histone modification in asthma is still very limited. However, transcription of the pro-inflammatory genes may be affected by aberrant acetylation of the N-terminal histone tails as increased HAT and reduced HDAC activity have been observed in patients diagnosed with asthma (Cosio et al. 2004; Hew et al. 2006). Natural microbial exposure represents an important environmental component for asthma protection. In a murine model, pregnant mother mice were exposed to the non-pathogenic Gram-negative bacterium *Acinetobacter lwoffii F78* (Brand et al. 2011). Prenatal *A. lwoffii* F78 administration prevented the development of the asthma phenotype in the progeny in an IFN-γ dependent manner. The *IFN-γ* promoter revealed a significant increase in histone H4 acetylation whereas the *IL-4* promoter showed a significant decrease, which closely correlated with IFN-γ and IL-4 RNA/protein levels in CD4$^+$ T-cells in the offspring. Treatment with the histone acetyltransferase inhibitor Garcinol abolished the asthma protective phenotype paralleled by inhibition of H4 acetylation of the *IFN-γ* promoter. These data provide evidence for the importance of epigenetic regulation of TH1/TH2 cytokine genes in transmaternal asthma protection and provides provide a molecular explanation for the protective effects of microbial exposure for the development of the asthma phenotype (Brand et al. 2011).

However, the establishment of direct links between epigenetic states and the disease phenotype is difficult and hampers the use of epigenetic modifications as reliable biomakers: (1) It is so far unknown in which specific cell type epigenetic alterations can be best correlated with the asthma phenotype; are these cells of the immune or of the respiratory system. (2) In the latter case the access to the proper type of tissue may be difficult as it might require unnecessary surgical procedures. (3) Asthma is a heterogeneous group of diseases comprising undoubtedly several clinical phenotypes with similar symptoms but different disease etiology. It is therefore probable that the epigenetic landscape would be different from phenotype to phenotype.

20.4 Epigenetic Therapy of Autoimmune Diseases

The involvement of aberrant regulation of Th cells in autoimmune and inflammatory diseases is today well accepted. Especially the emergence of the two more recently found Th lineages Th17 and Treg has increased our understanding of the process leading towards autoimmunity as a disruption of the homeostatic balance between these two cell types is crucial for the autoimmune conditions (Buckner 2010; Selmi 2009). Very recent findings regarding the plasticity of Treg and Th17 cells might potentially add a new layer of complexity to the common idea that Tregs act solely as autoimmune- and inflammatory repressors as a newly discovered subset of T cells, possessing a mixed phenotype of Treg and Th17 cells, has been suggested to contribute to autoimmune disease instead of counteracting the development (Kryczek et al. 2011). This subset of T cells is believed to develop in inflammatory microenvironments and might constitute a special effector T cell population as they express IL-2, INF-γ, and TNF-α. The phenotypic shift from Treg cells towards Th17 cells has also been shown to take place in skin lesions of patients diagnosed with severe psoriasis (Bovenschen et al. 2011). Interestingly, by treating Treg cells, isolated from severe psoriasis patients, with the HDAC inhibitor TSA they prevented the Th17-like phenotype. This observation has earlier been documented in Tregs isolated from human blood (Koenen et al. 2008) indicating the possibilities for epi-drugs to prevent the expression of pro-inflammatory cytokines generated due to this shift.

Understanding the impact of epigenetics in these multifactorial diseases leads to possibilities for the use of epigenetic drug treatments of affected individuals such as DNA demethylating agents or regimens interfering with some forms of histone modifications. Despite the early stage of epigenetic investigations within the field of autoimmune and inflammatory diseases, interest has focused on the opportunities for reversing patterns of histone acetylation as a number of *in vitro* and *in vivo* studies reported improvements of autoimmune diseases upon HDACi treatment (see Table 20.1). Some of the most widely used HDAC inhibitors for anti-autoimmunity and inflammatory responses are SAHA (Vorinostat), TSA (Trichostatin A), and valproic acid (Table 20.1). SAHA and TSA belong to the group of hydroxamic acids that inhibit zinc-dependent HDACs. With their zinc-binding group (ZBG) hydroxamic acids chelate the zinc ion near the bottom of the cylindrical pocket of HDACs. Opposite the zinc ion they have a capping group that interferes with the amino acid residues lining the entrance of the cylindrical pocket. Thereby the inhibitors block the entrance for other HDAC targets thus inhibiting its function. Along effects on non-histone proteins, this results in histone acetylation and gene expression. Valproic acid is a short-chain fatty acid thus belonging to another group of HDAC inhibitors than the two mentioned above. However like SAHA and TSA, valproic acid also interferes with HDACs catalytical domain and renders substrate recognition impossible, reviewed in (Kristensen et al. 2009). Dysfunction of HATs is thought to be the reason for the increased levels of histone acetylation observed in autoimmune and inflammatory diseases. Especially the two HATs, CBP and PCAF, have been found to be translocated to genes encoding inflammatory factors and thereby influencing their expression. This led to the suggestion that HAT inhibitors could be applied as a new treatment strategy. HATs are diverse and can be divided into three groups: (1) The GNATs (Gcn5 N-acetyltransferases), which include p300/CBP-associated factor (PCAF), Elp3, Hat1, Hpa2, and Nut1. (2) The MYSTs containing a 60 kDa Tat interactive protein including Morf, Ybp2, Sas2, and Tip60. (3) The third group includes the orphan HATs, which do not contain a precise consensus HAT domain. Naturally occurring as well as synthesized HAT inhibitors have been investigated for their capability to decrease the activity of HATs and to what extent they can be used without severe toxicity (Arif et al. 2009; Lau et al. 2000; Stimson et al. 2005). P300 is the most investigated HAT and can be inhibited in a competitive and non-competitive manner (Arif et al. 2009). A hydroxygroup was found to be important for the binding of two competitive inhibitors (garcinol and garcinol based derivatives) to the catalytical acetyl-CoA binding site. However, binding of the acetyl-CoA binding site was not a requirement for the inhibition of p300 as monosubstitution of the C-14 position of isogarcinol created a highly noncompetitive inhibitor of p300 binding to a site outside the acetyl-CoA site but nonetheless decreasing HAT activity (Arif et al. 2009). The importance of HAT inhibitors has been reviewed in (Selvi et al. 2010).

HDAC inhibitors have been shown to suppress the production of pro-inflammatory cytokines (Glauben et al. 2006) even at a significantly lower doses compared to the concentrations used in other diseases such as cancer. This finding might be of great importance as appropriate diet might have a preventive quality for chronic inflammatory disorders (vel Szic et al. 2010). For example, high levels of sulphoraphane, an HDACi naturally found in broccoli sprouts has been associated with H3 and H4 hyperacetylation. Moreover, sulphoraphane has been associated with both global and localized hyperacetylation, G_2/M cell cycle arrest and increased apoptosis in colon cancer (Myzak and Dashwood 2006). However, precaution has to be taken into account as some HDACis have also been found to induce the expression of pro-inflammatory cytokines. The unpredictable outcome of HDAC inhibitors may be associated with the large number of targets for HDACs. Histone proteins as well as non-histone proteins regulated by HDACs are involved in different cellular pathways. Thus, depending on the HDACi of choice, a different outcome might be expected. Therefore it is crucial to further investigate the different HDAC isoforms and their specific targets to ensure the best treatment strategies.

The reactivation of Treg cells in autoimmune diseases has also been thought to be a reasonable therapeutic approach (Buckner 2010) as it has been shown that HDAC inhibitors can promote the generation and function of Treg cells (Tao et al. 2007). Especially inhibition of HDAC9 proved to be important in the regulation of Foxp3-dependent suppression of Il-2 (Tao et al. 2007). This treatment strategy has already demonstrated positive effects in inflammatory bowel disease (IBD) (Mottet et al. 2003). With the identification of specific HDACs important for the suppression of autoimmune- and inflammatory conditions, it will be desirable to have inhibitors specific for the exact HDAC involved at the disposition as previously discussed for applications in cancer treatment (Kristensen et al. 2009). It is not only the use of HDAC inhibitors that has been considered for the treatment of autoimmune diseases, but also the exact opposite treatment strategy has been proposed using HAT inhibitors. HATi's have been proposed for treatment of the autoimmune disease IBD as global H4 acetylation levels normally associated with active gene expression were up-regulated in this disease (Tsaprouni et al. 2011). A comprehensive mapping of the different histone modifications in different relevant cell types is required to fully evaluate the extent to which these promising drugs can be used in the future for the treatment of autoimmune diseases.

20.5 Conclusions

Although the analysis of the genetic component of autoimmune disease has made rapid progress in the last years, the underlying etiology of autoimmune diseases is still poorly understood. We are still far from knowing the exact mechanisms underlying the different diseases phenotypes. It is not yet elucidated which environmental agents contribute to disease development and how they can induce epigenetic and transcriptional deregulation which then translates in inappropriate response of the immune system. Systematic, large scale epigenome projects such as the National Institute of Health Roadmap project in the US or the European Union initiated Human Epigenome Project (HEP) with the latter focusing on different cell types found in human blood will provide new insights into the epigenetic mechanisms. Further the HEP will provide insights in how the response to certain environmental agents will translate into molecular alterations and deregulation of gene expression that will lead to the development of the autoimmune diseases in genetically predisposed individuals. These insights might in the future provide novel therapeutic interventions targeting some of the epigenetic components of the disease process. However, more research is urgently needed to ensure a brighter future for the affected individuals.

References

Afkarian M, Sedy JR, Yang J, Jacobson NG, Cereb N, Yang SY, Murphy TL, Murphy KM (2002) T-bet is a STAT1-induced regulator of IL-12R expression in naive CD4+ T cells. Nat Immunol 3:549–557

Aho K, Koskenvuo M, Tuominen J, Kaprio J (1986) Occurrence of rheumatoid arthritis in a nationwide series of twins. J Rheumatol 13:899–902

Ansel KM, Lee DU, Rao A (2003) An epigenetic view of helper T cell differentiation. Nat Immunol 4:616–623

Ansel KM, Djuretic I, Tanasa B, Rao A (2006) Regulation of Th2 differentiation and Il4 locus accessibility. Annu Rev Immunol 24:607–656

Arif M, Pradhan SK, Thanuja GR, Vedamurthy BM, Agrawal S, Dasgupta D, Kundu TK (2009) Mechanism of p300 specific histone acetyltransferase inhibition by small molecules. J Med Chem 52:267–277

Ashkenazi A, Dixit VM (1998) Death receptors: signaling and modulation. Science 281:1305–1308

Aune TM, Collins PL, Chang S (2009) Epigenetics and T helper 1 differentiation. Immunology 126:299–305

Avni O, Lee D, Macian F, Szabo SJ, Glimcher LH, Rao A (2002) T(H) cell differentiation is accompanied by dynamic changes in histone acetylation of cytokine genes. Nat Immunol 3:643–651

Baccarelli A, Wright RO, Bollati V, Tarantini L, Litonjua AA, Suh HH, Zanobetti A, Sparrow D, Vokonas PS, Schwartz J (2009) Rapid DNA methylation changes after exposure to traffic particles. Am J Respir Crit Care Med 179:572–578

Baguet A, Bix M (2004) Chromatin landscape dynamics of the Il4-Il13 locus during T helper 1 and 2 development. Proc Natl Acad Sci U S A 101:11410–11415

Balasubramani A, Shibata Y, Crawford GE, Baldwin AS, Hatton RD, Weaver CT (2010) Modular utilization of distal cis-regulatory elements controls Ifng gene expression in T cells activated by distinct stimuli. Immunity 33:35–47

Ballestar E (2010) Epigenetics lessons from twins: prospects for autoimmune disease. Clin Rev Allergy Immunol 39:30–41

Barski A, Cuddapah S, Cui K, Roh TY, Schones DE, Wang Z, Wei G, Chepelev I, Zhao K (2007) High-resolution profiling of histone methylations in the human genome. Cell 129:823–837

Bovenschen HJ, van de Kerkhof PC, van Erp PE, Woestenenk R, Joosten I, Koenen HJ (2011) Foxp3+ regulatory T cells of psoriasis patients easily differentiate into IL-17A-producing cells and are found in lesional skin. J Invest Dermatol 131:1853–1860

Boyle AP, Davis S, Shulha HP, Meltzer P, Margulies EH, Weng Z, Furey TS, Crawford GE (2008) High-resolution mapping and characterization of open chromatin across the genome. Cell 132:311–322

Brand S, Teich R, Dicke T, Harb H, Yildirim AO, Tost J, Schneider-Stock R, Waterland RA, Bauer U-M, von Mutius E, Garn H, Pfefferle PI, Renz H (2011) Epigenetic regulation in murine offspring as a novel mechanism for transmaternal asthma protection induced by microbes. J Allergy Clin Immunol 128:618–625

Bream JH, Hodge DL, Gonsky R, Spolski R, Leonard WJ, Krebs S, Targan S, Morinobu A, O'Shea JJ, Young HA (2004) A distal region in the interferon-gamma gene is a site of epigenetic remodeling and transcriptional regulation by interleukin-2. J Biol Chem 279:41249–41257

Brooks WH (2010) X chromosome inactivation and autoimmunity. Clin Rev Allergy Immunol 39:20–29

Buckner JH (2010) Mechanisms of impaired regulation by CD4(+)CD25(+)FOXP3(+) regulatory T cells in human autoimmune diseases. Nat Rev Immunol 10:849–859

Calvanese V, Lara E, Kahn A, Fraga MF (2009) The role of epigenetics in aging and age-related diseases. Ageing Res Rev 8:268–276

Chang S, Aune TM (2007) Dynamic changes in histone-methylation 'marks' across the locus encoding interferon-gamma during the differentiation of T helper type 2 cells. Nat Immunol 8:723–731

Corvetta A, Della Bitta R, Luchetti MM, Pomponio G (1991) 5-Methylcytosine content of DNA in blood, synovial mononuclear cells and synovial tissue from patients affected by autoimmune rheumatic diseases. J Chromatogr 566:481–491

Cosio BG, Mann B, Ito K, Jazrawi E, Barnes PJ, Chung KF, Adcock IM (2004) Histone acetylase and deacetylase activity in alveolar macrophages and blood mononocytes in asthma. Am J Respir Crit Care Med 170:141–147

Djuretic IM, Levanon D, Negreanu V, Groner Y, Rao A, Ansel KM (2007) Transcription factors T-bet and Runx3 cooperate to activate Ifng and silence Il4 in T helper type 1 cells. Nat Immunol 8:145–153

Durrant DM, Metzger DW (2010) Emerging roles of T helper subsets in the pathogenesis of asthma. Immunol Invest 39:526–549

Eizirik DL, Colli ML, Ortis F (2009) The role of inflammation in insulitis and beta-cell loss in type 1 diabetes. Nat Rev Endocrinol 5:219–226

Fields PE, Kim ST, Flavell RA (2002) Cutting edge: changes in histone acetylation at the IL-4 and IFN-gamma loci accompany Th1/Th2 differentiation. J Immunol 169:647–650

Firestein GS (2003) Evolving concepts of rheumatoid arthritis. Nature 423:356–361
Fraga MF (2009) Genetic and epigenetic regulation of aging. Curr Opin Immunol 21:446–453
Fraga MF, Ballestar E, Paz MF, Ropero S, Setien F, Ballestar ML, Heine-Suner D, Cigudosa JC, Urioste M, Benitez J, Boix-Chornet M, Sanchez-Aguilera A, Ling C, Carlsson E, Poulsen P, Vaag A, Stephan Z, Spector TD, Wu YZ, Plass C, Esteller M (2005) Epigenetic differences arise during the lifetime of monozygotic twins. Proc Natl Acad Sci U S A 102:10604–10609
Glauben R, Batra A, Fedke I, Zeitz M, Lehr HA, Leoni F, Mascagni P, Fantuzzi G, Dinarello CA, Siegmund B (2006) Histone hyperacetylation is associated with amelioration of experimental colitis in mice. J Immunol 176:5015–5022
Grogan JL, Mohrs M, Harmon B, Lacy DA, Sedat JW, Locksley RM (2001) Early transcription and silencing of cytokine genes underlie polarization of T helper cell subsets. Immunity 14:205–215
Hamalainen-Laanaya HK, Kobie JJ, Chang C, Zeng WP (2007) Temporal and spatial changes of histone 3 K4 dimethylation at the IFN-gamma gene during Th1 and Th2 cell differentiation. J Immunol 179:6410–6415
Hatton RD, Harrington LE, Luther RJ, Wakefield T, Janowski KM, Oliver JR, Lallone RL, Murphy KM, Weaver CT (2006) A distal conserved sequence element controls Ifng gene expression by T cells and NK cells. Immunity 25:717–729
Hew M, Bhavsar P, Torrego A, Meah S, Khorasani N, Barnes PJ, Adcock I, Chung KF (2006) Relative corticosteroid insensitivity of peripheral blood mononuclear cells in severe asthma. Am J Respir Crit Care Med 174:134–141
Holliday R (2006) Epigenetics: a historical overview. Epigenetics 1:76–80
Huber LC, Brock M, Hemmatazad H, Giger OT, Moritz F, Trenkmann M, Distler JH, Gay RE, Kolling C, Moch H, Michel BA, Gay S, Distler O, Jungel A (2007) Histone deacetylase/acetylase activity in total synovial tissue derived from rheumatoid arthritis and osteoarthritis patients. Arthritis Rheum 56:1087–1093
Hutchins AS, Mullen AC, Lee HW, Sykes KJ, High FA, Hendrich BD, Bird AP, Reiner SL (2002) Gene silencing quantitatively controls the function of a developmental trans-activator. Mol Cell 10:81–91
Ivanov II, Zhou L, Littman DR (2007) Transcriptional regulation of Th17 cell differentiation. Semin Immunol 19:409–417
Javierre BM, Fernandez AF, Richter J, Al-Shahrour F, Martin-Subero JI, Rodriguez-Ubreva J, Berdasco M, Fraga MF, O'Hanlon TP, Rider LG, Jacinto FV, Lopez-Longo FJ, Dopazo J, Forn M, Peinado MA, Carreno L, Sawalha AH, Harley JB, Siebert R, Esteller M, Miller FW, Ballestar E (2010) Changes in the pattern of DNA methylation associate with twin discordance in systemic lupus erythematosus. Genome Res 20:170–179
Karouzakis E, Gay RE, Michel BA, Gay S, Neidhart M (2009) DNA hypomethylation in rheumatoid arthritis synovial fibroblasts. Arthritis Rheum 60:3613–3622
Kimura A, Kishimoto T (2010) IL-6: regulator of Treg/Th17 balance. Eur J Immunol 40:1830–1835
Koenen HJ, Smeets RL, Vink PM, van Rijssen E, Boots AM, Joosten I (2008) Human CD25highFoxp3pos regulatory T cells differentiate into IL-17-producing cells. Blood 112:2340–2352
Kouzarides T (2007) Chromatin modifications and their function. Cell 128:693–705
Koyanagi M, Baguet A, Martens J, Margueron R, Jenuwein T, Bix M (2005) EZH2 and histone 3 trimethyl lysine 27 associated with Il4 and Il13 gene silencing in Th1 cells. J Biol Chem 280:31470–31477
Kristensen LS, Nielsen HM, Hansen LL (2009) Epigenetics and cancer treatment. Eur J Pharmacol 625:131–142
Krueger GG, Duvic M (1994) Epidemiology of psoriasis: clinical issues. J Invest Dermatol 102:14S–18S
Kryczek I, Wu K, Zhao E, Wei S, Vatan L, Szeliga W, Huang E, Greenson J, Chang A, Rolinski J, Radwan P, Fang J, Wang G, Zou W (2011) IL-17+ regulatory T cells in the microenvironments of chronic inflammation and cancer. J Immunol 186:4388–4395
Larche M (2007) Regulatory T cells in allergy and asthma. Chest 132:1007–1014

Lau OD, Kundu TK, Soccio RE, Ait-Si-Ali S, Khalil EM, Vassilev A, Wolffe AP, Nakatani Y, Roeder RG, Cole PA (2000) HATs off: selective synthetic inhibitors of the histone acetyltransferases p300 and PCAF. Mol Cell 5:589–595

Le Gros G, Ben-Sasson SZ, Seder R, Finkelman FD, Paul WE (1990) Generation of interleukin 4 (IL-4)-producing cells in vivo and in vitro: IL-2 and IL-4 are required for in vitro generation of IL-4-producing cells. J Exp Med 172:921–929

Lee DU, Agarwal S, Rao A (2002) Th2 lineage commitment and efficient IL-4 production involves extended demethylation of the IL-4 gene. Immunity 16:649–660

Lee DU, Avni O, Chen L, Rao A (2004) A distal enhancer in the interferon-gamma (IFN-gamma) locus revealed by genome sequence comparison. J Biol Chem 279:4802–4810

Letimier FA, Passini N, Gasparian S, Bianchi E, Rogge L (2007) Chromatin remodeling by the SWI/SNF-like BAF complex and STAT4 activation synergistically induce IL-12Rbeta2 expression during human Th1 cell differentiation. EMBO J 26:1292–1302

Lighvani AA, Frucht DM, Jankovic D, Yamane H, Aliberti J, Hissong BD, Nguyen BV, Gadina M, Sher A, Paul WE, O'Shea JJ (2001) T-bet is rapidly induced by interferon-gamma in lymphoid and myeloid cells. Proc Natl Acad Sci U S A 98:15137–15142

Lin HS, Hu CY, Chan HY, Liew YY, Huang HP, Lepescheux L, Bastianelli E, Baron R, Rawadi G, Clement-Lacroix P (2007) Anti-rheumatic activities of histone deacetylase (HDAC) inhibitors in vivo in collagen-induced arthritis in rodents. Br J Pharmacol 150:862–872

Liu J, Ballaney M, Al-alem U, Quan C, Jin X, Perera F, Chen LC, Miller RL (2008) Combined inhaled diesel exhaust particles and allergen exposure alter methylation of T helper genes and IgE production in vivo. Toxicol Sci 102:76–81

Liu CC, Fang TJ, Ou TT, Wu CC, Li RN, Lin YC, Lin CH, Tsai WC, Liu HW, Yen JH (2011) Global DNA methylation, DNMT1, and MBD2 in patients with rheumatoid arthritis. Immunol Lett 135:96–99

Loots GG, Locksley RM, Blankespoor CM, Wang ZE, Miller W, Rubin EM, Frazer KA (2000) Identification of a coordinate regulator of interleukins 4, 13, and 5 by cross-species sequence comparisons. Science 288:136–140

Lu Q, Wu A, Tesmer L, Ray D, Yousif N, Richardson B (2007) Demethylation of CD40LG on the inactive X in T cells from women with lupus. J Immunol 179:6352–6358

Lucey DR, Clerici M, Shearer GM (1996) Type 1 and type 2 cytokine dysregulation in human infectious, neoplastic, and inflammatory diseases. Clin Microbiol Rev 9:532–562

Makar KW, Perez-Melgosa M, Shnyreva M, Weaver WM, Fitzpatrick DR, Wilson CB (2003) Active recruitment of DNA methyltransferases regulates interleukin 4 in thymocytes and T cells. Nat Immunol 4:1183–1190

Mastronardi FG, Noor A, Wood DD, Paton T, Moscarello MA (2007) Peptidyl argininedeiminase 2 CpG island in multiple sclerosis white matter is hypomethylated. J Neurosci Res 85:2006–2016

McGeachy MJ, Cua DJ (2008) Th17 cell differentiation: the long and winding road. Immunity 28:445–453

McGee HS, Agrawal DK (2006) TH2 cells in the pathogenesis of airway remodeling: regulatory T cells a plausible panacea for asthma. Immunol Res 35:219–232

Meda F, Folci M, Baccarelli A, Selmi C (2011) The epigenetics of autoimmunity. Cell Mol Immunol 8:226–236

Miao F, Smith DD, Zhang L, Min A, Feng W, Natarajan R (2008) Lymphocytes from patients with type 1 diabetes display a distinct profile of chromatin histone H3 lysine 9 dimethylation: an epigenetic study in diabetes. Diabetes 57:3189–3198

Miller SA, Huang AC, Miazgowicz MM, Brassil MM, Weinmann AS (2008) Coordinated but physically separable interaction with H3K27-demethylase and H3K4-methyltransferase activities are required for T-box protein-mediated activation of developmental gene expression. Genes Dev 22:2980–2993

Mishra N, Brown DR, Olorenshaw IM, Kammer GM (2001) Trichostatin A reverses skewed expression of CD154, interleukin-10, and interferon-gamma gene and protein expression in lupus T cells. Proc Natl Acad Sci USA 98:2628–2633

Moscarello MA, Wood DD, Ackerley C, Boulias C (1994) Myelin in multiple sclerosis is developmentally immature. J Clin Invest 94:146–154

Mottet C, Uhlig HH, Powrie F (2003) Cutting edge: cure of colitis by CD4+CD25+ regulatory T cells. J Immunol 170:3939–3943

Murdaca G, Colombo BM, Puppo F (2011) The role of Th17 lymphocytes in the autoimmune and chronic inflammatory diseases. Intern Emerg Med 6:487–495

Murphy KM, Reiner SL (2002) The lineage decisions of helper T cells. Nat Rev Immunol 2:933–944

Myzak MC, Dashwood RH (2006) Histone deacetylases as targets for dietary cancer preventive agents: lessons learned with butyrate, diallyl disulfide, and sulforaphane. Curr Drug Targets 7:443–452

Nadeau K, McDonald-Hyman C, Noth EM, Pratt B, Hammond SK, Balmes J, Tager I (2010) Ambient air pollution impairs regulatory T-cell function in asthma. J Allergy Clin Immunol 126(845–852):e810

Naoe Y, Setoguchi R, Akiyama K, Muroi S, Kuroda M, Hatam F, Littman DR, Taniuchi I (2007) Repression of interleukin-4 in T helper type 1 cells by Runx/Cbf beta binding to the Il4 silencer. J Exp Med 204:1749–1755

Nardone J, Lee DU, Ansel KM, Rao A (2004) Bioinformatics for the 'bench biologist': how to find regulatory regions in genomic DNA. Nat Immunol 5:768–774

Neidhart M, Rethage J, Kuchen S, Kunzler P, Crowl RM, Billingham ME, Gay RE, Gay S (2000) Retrotransposable L1 elements expressed in rheumatoid arthritis synovial tissue: association with genomic DNA hypomethylation and influence on gene expression. Arthritis Rheum 43:2634–2647

Nishida K, Komiyama T, Miyazawa S, Shen ZN, Furumatsu T, Doi H, Yoshida A, Yamana J, Yamamura M, Ninomiya Y, Inoue H, Asahara H (2004) Histone deacetylase inhibitor suppression of autoantibody-mediated arthritis in mice via regulation of p16INK4a and p21(WAF1/Cip1) expression. Arthritis Rheum 50:3365–3376

Ouyang W, Lohning M, Gao Z, Assenmacher M, Ranganath S, Radbruch A, Murphy KM (2000) Stat6-independent GATA-3 autoactivation directs IL-4-independent Th2 development and commitment. Immunity 12:27–37

Ouyang W, Kolls JK, Zheng Y (2008) The biological functions of T helper 17 cell effector cytokines in inflammation. Immunity 28:454–467

Pai SY, Truitt ML, Ho IC (2004) GATA-3 deficiency abrogates the development and maintenance of T helper type 2 cells. Proc Natl Acad Sci U S A 101:1993–1998

Pan W, Zhu S, Yuan M, Cui H, Wang L, Luo X, Li J, Zhou H, Tang Y, Shen N (2010) MicroRNA-21 and microRNA-148a contribute to DNA hypomethylation in lupus CD4+ T cells by directly and indirectly targeting DNA methyltransferase 1. J Immunol 184:6773–6781

Perera F, Tang WY, Herbstman J, Tang D, Levin L, Miller R, Ho SM (2009) Relation of DNA methylation of 5′-CpG island of ACSL3 to transplacental exposure to airborne polycyclic aromatic hydrocarbons and childhood asthma. PLoS One 4:e4488

Placek K, Coffre M, Maiella S, Bianchi E, Rogge L (2009) Genetic and epigenetic networks controlling T helper 1 cell differentiation. Immunology 127:155–162

Presky DH, Yang H, Minetti LJ, Chua AO, Nabavi N, Wu CY, Gately MK, Gubler U (1996) A functional interleukin 12 receptor complex is composed of two beta-type cytokine receptor subunits. Proc Natl Acad Sci U S A 93:14002–14007

Reilly CM, Thomas M, Gogal R Jr, Olgun S, Santo A, Sodhi R, Samy ET, Peng SL, Gilkeson GS, Mishra N (2008) The histone deacetylase inhibitor trichostatin A upregulates regulatory T cells and modulates autoimmunity in NZB/W F1 mice. J Autoimmun 31:123–130

Ribeiro de Almeida C, Heath H, Krpic S, Dingjan GM, van Hamburg JP, Bergen I, van de Nobelen S, Sleutels F, Grosveld F, Galjart N, Hendriks RW (2009) Critical role for the transcription regulator CCCTC-binding factor in the control of Th2 cytokine expression. J Immunol 182:999–1010

Richardson B (1986) Effect of an inhibitor of DNA methylation on T cells. II. 5-Azacytidine induces self-reactivity in antigen-specific T4+ cells. Hum Immunol 17:456–470

Richardson B, Scheinbart L, Strahler J, Gross L, Hanash S, Johnson M (1990) Evidence for impaired T cell DNA methylation in systemic lupus erythematosus and rheumatoid arthritis. Arthritis Rheum 33:1665–1673

Robinson DS, Hamid Q, Ying S, Tsicopoulos A, Barkans J, Bentley AM, Corrigan C, Durham SR, Kay AB (1992) Predominant TH2-like bronchoalveolar T-lymphocyte population in atopic asthma. N Engl J Med 326:298–304

Sakaguchi S, Ono M, Setoguchi R, Yagi H, Hori S, Fehervari Z, Shimizu J, Takahashi T, Nomura T (2006) Foxp3+ CD25+ CD4+ natural regulatory T cells in dominant self-tolerance and autoimmune disease. Immunol Rev 212:8–27

Sakaguchi S, Yamaguchi T, Nomura T, Ono M (2008) Regulatory T cells and immune tolerance. Cell 133:775–787

Sawalha AH, Harley JB (2004) Antinuclear autoantibodies in systemic lupus erythematosus. Curr Opin Rheumatol 16:534–540

Schoenborn JR, Dorschner MO, Sekimata M, Santer DM, Shnyreva M, Fitzpatrick DR, Stamatoyannopoulos JA, Wilson CB (2007) Comprehensive epigenetic profiling identifies multiple distal regulatory elements directing transcription of the gene encoding interferon-gamma. Nat Immunol 8:732–742

Selmi C (2009) What is hot in autoimmunity. Acta Reumatol Port 34:580–588

Selvi BR, Mohankrishna DV, Ostwal YB, Kundu TK (2010) Small molecule modulators of histone acetylation and methylation: a disease perspective. Biochim Biophys Acta 1799:810–828

Sgouroudis E, Piccirillo CA (2009) Control of type 1 diabetes by CD4+Foxp3+ regulatory T cells: lessons from mouse models and implications for human disease. Diabetes Metab Res Rev 25:208–218

Shnyreva M, Weaver WM, Blanchette M, Taylor SL, Tompa M, Fitzpatrick DR, Wilson CB (2004) Evolutionarily conserved sequence elements that positively regulate IFN-gamma expression in T cells. Proc Natl Acad Sci U S A 101:12622–12627

Soto ME, Vallejo M, Guillen F, Simon JA, Arena E, Reyes PA (2004) Gender impact in systemic lupus erythematosus. Clin Exp Rheumatol 22:713–721

Stimson L, Rowlands MG, Newbatt YM, Smith NF, Raynaud FI, Rogers P, Bavetsias V, Gorsuch S, Jarman M, Bannister A, Kouzarides T, McDonald E, Workman P, Aherne GW (2005) Isothiazolones as inhibitors of PCAF and p300 histone acetyltransferase activity. Mol Cancer Ther 4:1521–1532

Takami N, Osawa K, Miura Y, Komai K, Taniguchi M, Shiraishi M, Sato K, Iguchi T, Shiozawa K, Hashiramoto A, Shiozawa S (2006) Hypermethylated promoter region of DR3, the death receptor 3 gene, in rheumatoid arthritis synovial cells. Arthritis Rheum 54:779–787

Takemoto N, Koyano-Nakagawa N, Yokota T, Arai N, Miyatake S, Arai K (1998) Th2-specific DNase I-hypersensitive sites in the murine IL-13 and IL-4 intergenic region. Int Immunol 10:1981–1985

Tao R, de Zoeten EF, Ozkaynak E, Chen C, Wang L, Porrett PM, Li B, Turka LA, Olson EN, Greene MI, Wells AD, Hancock WW (2007) Deacetylase inhibition promotes the generation and function of regulatory T cells. Nat Med 13:1299–1307

Taylor A, Verhagen J, Akdis CA, Akdis M (2005) T regulatory cells and allergy. Microbes Infect 7:1049–1055

Thieu VT, Yu Q, Chang HC, Yeh N, Nguyen ET, Sehra S, Kaplan MH (2008) Signal transducer and activator of transcription 4 is required for the transcription factor T-bet to promote T helper 1 cell-fate determination. Immunity 29:679–690

Tost J (2008) Epigenetics. Caister Academic Press, Norfolk, UK

Tsaprouni LG, Ito K, Powell JJ, Adcock IM, Punchard N (2011) Differential patterns of histone acetylation in inflammatory bowel diseases. J Inflamm (Lond) 8:1

Tykocinski LO, Hajkova P, Chang HD, Stamm T, Sozeri O, Lohning M, Hu-Li J, Niesner U, Kreher S, Friedrich B, Pannetier C, Grutz G, Walter J, Paul WE, Radbruch A (2005) A critical control element for interleukin-4 memory expression in T helper lymphocytes. J Biol Chem 280:28177–28185

Usui T, Nishikomori R, Kitani A, Strober W (2003) GATA-3 suppresses Th1 development by down-regulation of Stat4 and not through effects on IL-12Rbeta2 chain or T-bet. Immunity 18:415–428

Usui T, Preiss JC, Kanno Y, Yao ZJ, Bream JH, O'Shea JJ, Strober W (2006) T-bet regulates Th1 responses through essential effects on GATA-3 function rather than on IFNG gene acetylation and transcription. J Exp Med 203:755–766

Vel Szic KS, Ndlovu MN, Haegeman G, Vanden Berghe W (2010) Nature or nurture: let food be your epigenetic medicine in chronic inflammatory disorders. Biochem Pharmacol 80:1816–1832

Weaver CT, Hatton RD, Mangan PR, Harrington LE (2007) IL-17 family cytokines and the expanding diversity of effector T cell lineages. Annu Rev Immunol 25:821–852

Wei G, Wei L, Zhu J, Zang C, Hu-Li J, Yao Z, Cui K, Kanno Y, Roh TY, Watford WT, Schones DE, Peng W, Sun HW, Paul WE, O'Shea JJ, Zhao K (2009) Global mapping of H3K4me3 and H3K27me3 reveals specificity and plasticity in lineage fate determination of differentiating CD4+ T cells. Immunity 30:155–167

Whitelaw NC, Whitelaw E (2006) How lifetimes shape epigenotype within and across generations. Hum Mol Genet 15(Spec No 2):R131–137

Wilson CB, Rowell E, Sekimata M (2009) Epigenetic control of T-helper-cell differentiation. Nat Rev Immunol 9:91–105

Wulfing C, Sumen C, Sjaastad MD, Wu LC, Dustin ML, Davis MM (2002) Costimulation and endogenous MHC ligands contribute to T cell recognition. Nat Immunol 3:42–47

Yamashita M, Ukai-Tadenuma M, Kimura M, Omori M, Inami M, Taniguchi M, Nakayama T (2002) Identification of a conserved GATA3 response element upstream proximal from the interleukin-13 gene locus. J Biol Chem 277:42399–42408

Yamashita M, Hirahara K, Shinnakasu R, Hosokawa H, Norikane S, Kimura MY, Hasegawa A, Nakayama T (2006) Crucial role of MLL for the maintenance of memory T helper type 2 cell responses. Immunity 24:611–622

Yang Y, Ochando JC, Bromberg JS, Ding Y (2007) Identification of a distant T-bet enhancer responsive to IL-12/Stat4 and IFNgamma/Stat1 signals. Blood 110:2494–2500

Yang XO, Pappu BP, Nurieva R, Akimzhanov A, Kang HS, Chung Y, Ma L, Shah B, Panopoulos AD, Schluns KS, Watowich SS, Tian Q, Jetten AM, Dong C (2008) T helper 17 lineage differentiation is programmed by orphan nuclear receptors ROR alpha and ROR gamma. Immunity 28:29–39

Zhao M, Sun Y, Gao F, Wu X, Tang J, Yin H, Luo Y, Richardson B, Lu Q (2010a) Epigenetics and SLE: RFX1 downregulation causes CD11a and CD70 overexpression by altering epigenetic modifications in lupus CD4+ T cells. J Autoimmun 35:58–69

Zhao M, Wu X, Zhang Q, Luo S, Liang G, Su Y, Tan Y, Lu Q (2010b) RFX1 regulates CD70 and CD11a expression in lupus T cells by recruiting the histone methyltransferase SUV39H1. Arthritis Res Ther 12:R227

Zheng Q, Xu Y, Liu Y, Zhang B, Li X, Guo F, Zhao Y (2009) Induction of Foxp3 demethylation increases regulatory CD4+CD25+ T cells and prevents the occurrence of diabetes in mice. J Mol Med 87:1191–1205

Zhou W, Chang S, Aune TM (2004) Long-range histone acetylation of the Ifng gene is an essential feature of T cell differentiation. Proc Natl Acad Sci U S A 101:2440–2445

Zhou Y, Yuan J, Pan Y, Fei Y, Qiu X, Hu N, Luo Y, Lei W, Li Y, Long H, Sawalha AH, Richardson B, Lu Q (2009) T cell CD40LG gene expression and the production of IgG by autologous B cells in systemic lupus erythematosus. Clin Immunol 132:362–370

Zhu J, Paul WE (2008) CD4 T cells: fates, functions, and faults. Blood 112:1557–1569

Zhu J, Guo L, Watson CJ, Hu-Li J, Paul WE (2001) Stat6 is necessary and sufficient for IL-4's role in Th2 differentiation and cell expansion. J Immunol 166:7276–7281

Zhu J, Cote-Sierra J, Guo L, Paul WE (2003) Stat5 activation plays a critical role in Th2 differentiation. Immunity 19:739–748

Zhu J, Min B, Hu-Li J, Watson CJ, Grinberg A, Wang Q, Killeen N, Urban JF Jr, Guo L, Paul WE (2004) Conditional deletion of Gata3 shows its essential function in T(H)1-T(H)2 responses. Nat Immunol 5:1157–1165

Zhu J, Yamane H, Paul WE (2010) Differentiation of effector CD4 T cell populations (*). Annu Rev Immunol 28:445–489

Chapter 21
Epigenetic Regulation of HIV-1 Persistence and Evolving Strategies for Virus Eradication

Neeru Dhamija, Pratima Rawat, and Debashis Mitra

Abstract Despite the intense effort put by researchers globally to understand Human Immunodeficiency Virus (HIV-1) pathogenesis since its discovery 30 years ago, the acquired knowledge till date is not good enough to eradicate HIV-1 from an infected individual. HIV-1 infects cells of the human immune system and integrates into the host cell genome thereby leading to persistent infection in these cells. Based on the activation status of the cells, the infection could be productive or result in latent infection. The current regimen used to treat HIV-1 infection in an AIDS patient includes combination of antiretroviral drugs called Highly Active Anti-Retroviral Therapy (HAART). A major challenge for the success of HAART has been these latent reservoirs of HIV which remain hidden and pose major hurdle for the eradication of virus. Combination of HAART therapy with simultaneous activation of latent reservoirs of HIV-1 seems to be the future of anti-retroviral therapy; however, this will require a much better understanding of the mechanisms and regulation of HIV-1 latency. In this chapter, we have tried to elaborate on HIV-1 latency, highlighting the strategies employed by the virus to ensure persistence in the host with specific focus on epigenetic regulation of latency. A complete understanding of HIV-1 latency will be extremely essential for ultimate eradication of HIV-1 from the human host.

21.1 Introduction

Viruses are obligate parasites that cannot survive without the host and infect host cells to make more progeny viruses. Some viruses like influenza and rhinovirus acutely infect the target cell to make more viruses resulting in rapid onset of the

N. Dhamija • P. Rawat • D. Mitra (✉)
National Centre for Cell Science, NCCS Complex, Pune University Campus, Ganeshkhind, Pune 411007, India
e-mail: dmitra@nccs.res.in

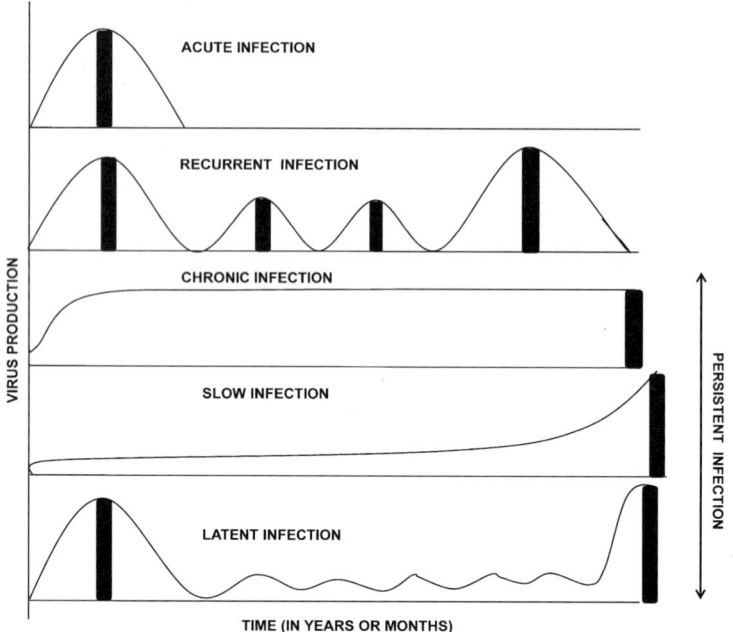

Fig. 21.1 Schematic representation of different types of viral infection and the peak of viral load with reference to time. Two types of viral infection are generally observed: Acute and Persistent infection. Recurrent infection occurs in multiple episodes of acute infection one after the other because of incapability of the immune system to completely clear the pathogen. Persistent infection can be further classified as chronic, slow or latent infection. The black bars represent the peak of the infection in a particular episode of infection

disease and clearance of the pathogen. On the other hand, persistent infections are those in which virus escapes the host's immune surveillance and is not cleared by the immune system. Persistent infections are further classified as chronic, slow and latent infections. Chronic infection is characterized by the continuous presence of the infectious virus following primary infection or recurrence. Prolonged incubation period followed by progressive disease is a hallmark of slow virus infection as seen in measles virus infection. Lack of demonstrable virus between episodes of recurrent disease is indicative of latent infection (Boldogh et al. 1996) (Fig. 21.1). Different viruses employ different strategies to ensure persistence in the host yet commonality exists between them. These strategies include (i) selection of long lived cell populations as reservoirs of virus; (ii) modulation of viral gene expression; (iii) modulation of cellular apoptotic pathway and (iv) evasion of the host immune system. Being a lentivirus, Human Immunodeficiency virus-1 (HIV-1) infects cells of the human immune system and integrates in to the host cell genome thereby leading to persistent infection in these cells. Based on the activation status of the cells, the infection could be productive or could result in latent infection. To combat HIV infection, presently used Highly Active Anti-Retroviral Therapy (HAART) uses a combination of antiviral drugs that usually suppress viremia to levels below the detectable limit by conventional assays, i.e. 50 copies of viral RNA per ml. Cessation of HAART therapy results in viral reappearance in the circulation within a short time due to the persistence of latently infected cellular reservoirs, which has received renewed attention in recent times owing to failure of anti-retroviral therapy to eradicate HIV completely. These latently infected cells are transcriptionally silent or have low level of transcription and are permanent source for virus reactivation after discontinuation of HAART. The present chapter intends to put together our present understanding of the mechanism of HIV persistence in host cells with specific emphasis on epigenetic factors and to examine the evolving strategies to eradicate the virus.

21.2 Strategies for Viral Persistence

Viruses need reservoirs for persistence within the host. Primary requisite of these reservoirs is that they should be long lived. Herpes simplex virus (HSV), for example, uses sensory neurons as sanctuary since they are long lived and terminally differentiated cells and they serve as excellent sites for HSV persistence (Efstathiou and Preston 2005). Another good home for the viruses can be memory cells. Human immunodeficiency virus-1 (HIV-1) for example utilizes resting CD4+ T cells as reservoir (Chun et al. 1995). Stem cells being biologically immortal are sites of infinite storage for murine leukemia virus (Rosenberg and Jolicoeur 1997; Ruddle et al. 1976). Viruses also interplay with the host chromatin for its persistence. The best example to explain this would be Herpes simplex virus. HSV life cycle involves both lytic and latent infection in the host. Lytic infection takes place in the epithelial cells and sensory neurons are sites of latent infection. Latency associated transcript (LAT) is expressed in high amounts to shut off lytic gene expression. Thus viral proteins tend to regulate the latent or lytic infection. Modulation of cellular apoptotic pathways is another strategy used by viruses for persistence. For example, although HIV-1 is known to induce apoptosis in effector cells of immune system like HIV specific CTLs, but it prevents apoptosis in infected cells to allow the production of new pool of viruses from these cells. Inhibition of apoptosis by using viral proteins like Nef, Tat and gp120 or targeting some cellular factors is one of the most common strategies employed by the virus to increase the survival of these infected cells.

HIV-1 Nef has been shown to have dual effect on cellular apoptotic pathways. During the early phase of infection, it is shown to induce apoptosis of bystander cells such as CTLs by increasing the expression of FasL on the surface of infected T cells. Killing of these CTLs helps the virus to escape the host immune response (Geleziunas et al. 2001). However, during the late stage of infection, Nef inhibits apoptosis to increase the survival of HIV-1 infected cells like monocyte derived macrophages (MDMs). It also binds to p53 to protect the infected cells from p53 induced apoptosis (Greenway et al. 2002). Like Nef, Tat and gp120 are also known to modulate cellular apoptosis (Chugh et al. 2007; Swingler et al. 2007). Like viral proteins, some of the host factors also suppress apoptosis in infected cells to promote viral persistence. CTIP2 is one such protein which increases the HIV-1 persistence in microglial cells by functioning as an anti-apoptotic factor (Cherrier et al. 2009; Giri et al. 2009). It also counteracts apoptotic function of Vpr protein. NFκB activity is also induced upon HIV-1 infection and this active NFκB appears to have a role in inhibiting apoptosis in MDMs which helps to maintain latent infection in these cells (Asin et al. 1999; McElhinny et al. 1995). TNFα production is also increased during HIV-1 infection which in turn enhances expression of Bcl-XL and Bcl-2 pro-survival proteins. These proteins further contribute in the apoptosis inhibition (Guillemard et al. 2004). Recently, role of viral RNA have been also identified in modulation of apoptotic events in infected cells (Bennasser et al. 2006; Klase et al. 2009).

Host immune evasion is another strategy used by the viruses to establish persistent infection. Virus infected cells are normally eliminated by the host immune system. Immune system recognise the viral peptides presented by MHC-I molecules on the surface of infected cells and mediate their killing by virus specific CTLs. However, in case of persistent infection, virus tries to hide from the immune system by altering the expression of these surface molecules and escape the recognition of viral antigens by immune system. This job is generally performed by those viral proteins which express early in the virus life cycle and ensure the availability of an environment that is secured from host immune response and thus suitable for establishment of persistent viral infection. In case of HIV-1 infection, Nef decreases the expression of MHC-I from the cell surface and can prevent the recognition by CTL (Schwartz et al. 1996). Tat has been also shown to inhibit the expression of MHCI molecule on the cell surface by its ability to bind TAFII250 (TATA binding protein-associated factor 250), which inhibits cellular HAT activity (Weissman et al. 1998).

21.3 HIV-1 Latency

HIV-1 is capable of infecting a range of cell types, including T lymphocytes, monocytes, macrophages, microglial cells, astrocytes, neurons, microvascular endothelial cells, etc. The virus life cycle involves various steps like virus attachment to the host cell, virus-cell fusion, uncoating, reverse transcription, integration, transcription, translation, assembly, budding and maturation. Following reverse transcription of the HIV-1 genome, linear and circular molecules of proviral DNA containing either one or two copies of the LTR are produced and accumulated as non-integrating by-products within the host cell. Only linear double stranded cDNA integrates into the host cell chromosomes. The HIV provirus can integrate into many different chromosomal locations in the cell and most infected cells harbour more than one provirus. After integration, the LTR-flanked provirus behaves as a cellular gene: the 5′ LTR operates like any eukaryotic promoter (Fig. 21.2) and the 3′ LTR acts as the polyadenylation and termination site. Integration can lead to either latent (transcriptionally inactive) or transcriptionally active forms of infection. The chromosomal milieu encountered by the provirus (integrated virus) likely helps shaping

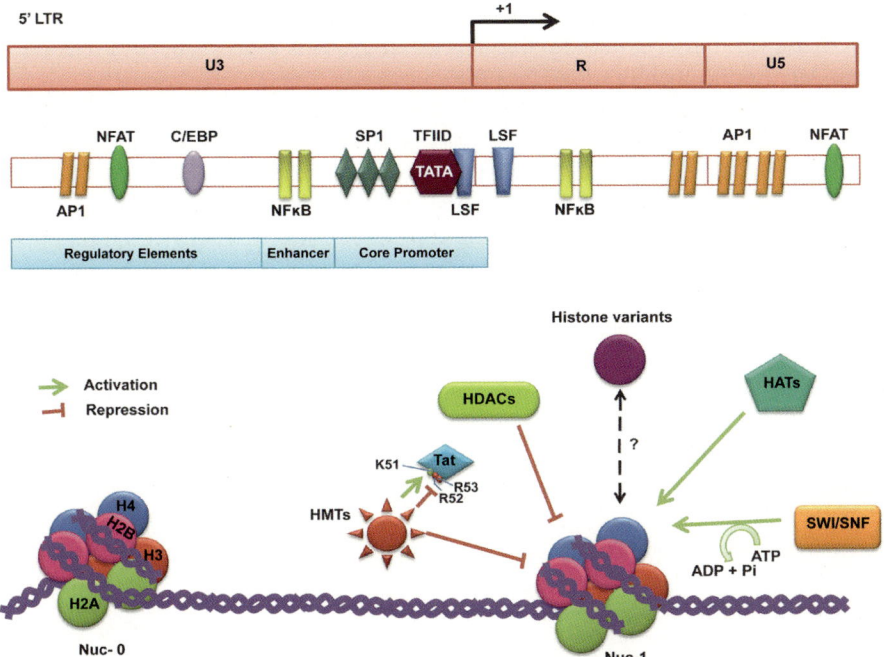

Fig. 21.2 HIV-1 long terminal repeat promoter and epigenetic factors modulating LTR driven transcription. HIV-1 LTR can be subdivided into three regions: U3, R and U5. U3 comprises of the regulatory elements, NFκB enhancer and the core promoter. Irrespective of the site of integration, LTR organizes into two distinct nucleosomes, namely, Nuc-0 and Nuc-1, with fixed relative positions of the regulatory elements. Nuc-1 displacement is necessary for transcription to proceed. Chromatin modifiers like histone acetyl transferases (HAT) and histone demethylases or remodellers like SWI/SNF participate to displace Nuc-1. In their absence, HIV transcription is repressed as is seen in the latent HIV infection. Histone deacetylases (HDACs) and Histone methyl transferases (HMTs) help establish latency in a cell by suppressing the opening of the chromatin at Nuc-1. HMTs and HATs not only function at the Nuc-1 but also on Tat protein to repress or activate Tat function. Lysine 51 methylation, for example, activates Tat's transactivation ability while arginine 52 and 53 methylation suppresses Tat activity resulting in latent situation. The role of histone variants in establishment of HIV latency remains to be elucidated

its transcriptional activity. HIV-1 integration results in proviral genome being packaged into the chromatin. The fundamental unit of chromatin is the nucleosome core/histone octamer comprising of two molecules of each canonical or core histones H2A, H2B, H3, H4 around which 146 bp of DNA wraps around. H1 is a linker histone protein that binds to the nucleosomes and the DNA wrapped around them. Nucleosome is not a rigid entity and its flexibility is needed for the opening of the chromatin and hence transcription. Irrespective of the integration site, HIV-1 LTR incorporates in two distinct nucleosomes, termed Nuc-0 and Nuc-1 with precise location of the regulatory elements as depicted in Fig. 21.2. Remodelling (altering the histone-DNA interaction) at Nuc-1 is needed for successful transcription of the

HIV genome. Chromatin remodelers and modifiers function at Nuc-1 to open the chromatin for transcription to proceed (Fig. 21.2). This aspect will be discussed in detail later in the chapter.

Resting memory CD4+ T-cells are a well-defined latent reservoir of HIV. The reservoir of resting memory CD4$^+$ T cells is established during primary infection. *In vivo* presence of HIV-1 latent reservoir was first time shown in 1995 by Chun and coworkers (Chun et al. 1995), when integrated proviruses were seen in purified resting CD4$^+$ cells from HIV-1 patients. These cells pose various blocks at the transcriptional and post transcriptional level leading to the persistence of transcriptionally silent viral genome in the host cell. During antiretroviral therapy, these latent cells die very slowly with an average half-life of 44 months, indicating that under current therapeutic regimen it will take over 60 years to deplete this reservoir. Despite intense research, the molecular mechanism of HIV-1 latency remains incompletely understood. Thus, it is necessary to unravel the molecular mechanism regulating viral latency, which may lead to development of novel strategies aimed at latent stage of virus and thus complete eradication of HIV.

21.3.1 Types of Latency

21.3.1.1 Pre-integration Latency

This is a stage of virus life cycle represented by HIV infected resting CD4+ T cells which carry partially reverse transcribed unintegrated viral genome in their cytoplasm (Zack et al. 1990). Lower levels of ATP in these resting cells (Korin and Zack 1999; Meyerhans et al. 1994) not only slows down reverse transcription of viral RNA but is also incapable to support ATP dependent nuclear import of large HIV-1 pre-integration complex (Bukrinsky et al. 1992). Hyper-mutation of viral DNA by host restriction factor APOBEC3 restricts viral replication in resting CD4$^+$ T cells (Chiu et al. 2005) and is therefore thought as one of the mechanism for induction of pre-integration latency. It represents a short lived form of viral latency which is commonly seen in untreated individuals. Due to this labile nature, it has gained less clinical concern than the stable form of latency discussed below.

21.3.1.2 Post-integration Latency

It represents a non-productive state of viral infection commonly seen in resting CD4$^+$ memory T lymphocytes carrying a transcriptionally dormant provirus integrated into their genome. It is one of the major regulators of viral persistence in HIV-1 patients receiving HAART therapy (Chun et al. 2003) and is an extremely stable form of latency with a half life of about 44 weeks. This transcriptional dormancy of viral genome can be explained at the molecular level by several different mechanisms:

i. **Nature of integration site:** Chromosomal positioning of a gene has been shown to be an important regulator of gene expression. Initial *in vitro* studies in infected T lymphocytes (Jordan et al. 2003) have shown that viral genome preferentially integrates in the repetitive DNA elements of host genome heterochromatic region. However, some later studies in human T lymphocytic cell lines (Lewinski et al. 2006; Schroder et al. 2002) and in resting CD4+ T cells from patients under HAART therapy (Han et al. 2004), have reported integration of proviral DNA in the intronic regions of actively transcribing gene. All these studies indicate that site of integration in host genome has an important role in regulation of viral latency.
ii. **Transcriptional Interference (TI):** Transcriptional interference is an inhibitory effect of one transcriptional process on a second transcriptional process (Adhya and Gottesman 1982). Integration of viral genome downstream to a highly active host gene promoter can create TI situation leading to latency. In such a situation, the elongating RNA polymerase reads through the downstream promoter preventing its transcription initiation. Thus HIV-1 genome that integrates in the introns of active genes (Han et al. 2004) is transcribed as part of the intron in TI situation and is eventually spliced out and degraded (Lenasi et al. 2008). Recent reports indicate the involvement of chromatin reassembly factors (CRFs) in the transcriptional interference (Gallastegui et al. 2011).
iii. **Differential expression of Transcriptional activators:** HIV-1 gene expression is under the regulation of various cellular factors which has their binding sites on viral LTR like NFκB, NFAT or Sp1 (Nabel and Baltimore 1987). Enhancer region of HIV-1 LTR is also known to respond to T cell activation signals like IL2, TNFα, etc. (Bohnlein et al. 1988; Duh et al. 1989; Tong-Starksen et al. 1987). Differential expression of these transcription stimulators and their limited availability in resting CD4+ T lymphocytes contributes in the maintenance of latency. Murr1, a host restriction factor has been shown to inhibit HIV-1 replication in resting CD4+ T cells by inhibiting NFκB activity and thus contributes in maintenance of HIV-1 latency in quiescent T cells (Ganesh et al. 2003). Resting T cells also show reduced expression of CDK9 and CyclinT1, the two essential components of P-TEFb, and thus pose a barrier to viral transcription (Garriga et al. 1998; Herrmann et al. 1998; Sung and Rice 2006).
iv. **Availability of viral factors like Tat and Rev:** HIV-1 transcription initiates from the LTR promoter and short transcript of approximately 60 bases called TAR (transactivation response element) is made that creates a binding site for Tat. Tat binds to TAR and then recruits pTEFb kinase complex on HIV-1 LTR to promote the phosphorylation of C-terminal domain (CTD) of RNA Pol II (Zhou et al. 2000) and thereby enhancing the production of viral transcripts. Tat is also known to interact with various cellular factors like elongation factor ELL2 (He et al. 2010), acetyltransferase p300/CBP (Benkirane et al. 1998; Bres et al. 2002; Kiernan et al. 1999), ATP dependent chromatin remodelling complex SWI/SNF (Mahmoudi et al. 2006), which either modify Tat or recruit transcriptional activator molecules to induce HIV-1 LTR mediated viral transcription. This predominant role of Tat in HIV-1 transcription suggests that the

absence of Tat or its mutation can result in inefficient transcription leading to HIV-1 latency as is evident by relatively low levels of viral transcription observed in case of latently infected U1 and ACH2 cell lines carrying mutation in Tat and TAR region respectively.

Rev protein is involved in export of viral transcripts from nucleus. A threshold level of Rev expression is required to rescue the unspliced 9.2 kb viral transcript from the nucleus (Pomerantz et al. 1990, 1992). However, the absence of this can lead to latency as is seen in relatively miniscule levels of unspliced 9.2 kb transcript than the singly spliced 4.3 kb or the multiply spliced 2 kb transcripts in latent cell lines (Pomerantz et al. 1992). In addition to Tat and Rev, viral proteins Nef and Vpr have been also proposed to play some role in HIV-1 latency (Levy et al. 1995; Niederman et al. 1989).

21.4 Epigenetic Regulation of HIV Latency

The term "Epigenetic" means a change in the gene expression that does not involve changes in the gene sequence. HIV-1 utilizes a number of epigenetic strategies to regulate latency, the important ones of which are discussed below:

21.4.1 Post-translational Modifications

Various disparate post-translational modifications (acetylation, methylation, ubiquitination, sumoylation, ADP-ribosylation, phosphorylation) http://www.sciencedirect.com/science/journal/0959437X of the proteins called histones that package the DNA to form the "chromatin" play an important role in epigenetic regulation (Berger 2002). In the following sections we will discuss each of these epigenetic modifications in case of HIV-1 infection (Fig. 21.3):

i. **Acetylation and deacetylation**

   Acetylation and deacetylation are known to regulate the accessibility of the transcription complex to the chromatin. Acetylation is associated with opening of the chromatin so that transcription factors can gain access to it. Deacetylation, on the other hand, makes the chromatin inaccessible. Best studied post-translational modification in case of HIV is acetylation-deacetylation of histones. Key participants in this strategy are Histone deacetylases (HDAC) and Histone acetyl transferases (HAT). Four classes of HDACs have been reported in the literature, namely, Class I, Class II, sirtuins and class IV (de Ruijter et al. 2003). HDAC 1, 2, 3 and 8 fall in the category of class I HDACs that are related to the yeast RPD3 HDAC. HDAC 4, 5, 6, 7, 9 and 10 form the class II HDACs which are related to the yeast HDA1 HDAC (Verdin et al. 2003). Sirtuins are Sir2-related NAD$^+$

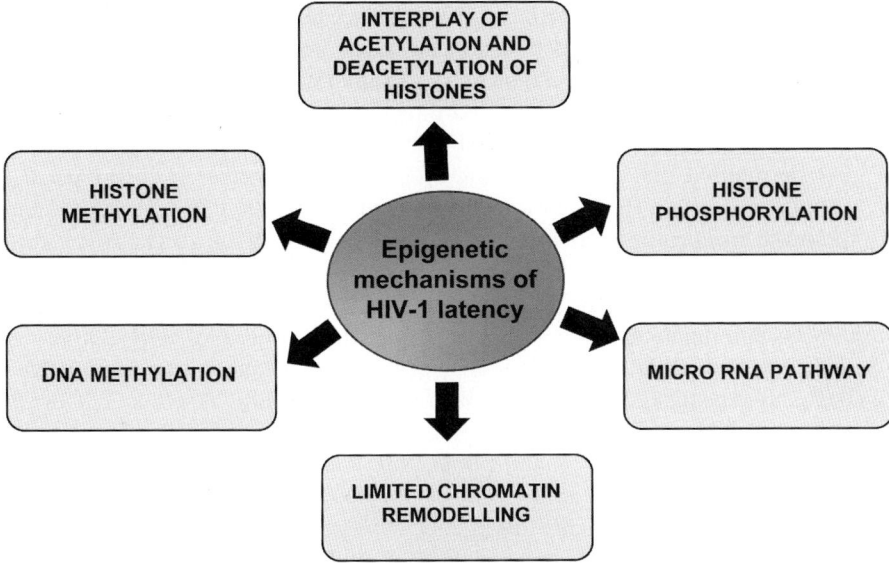

Fig. 21.3 Schematic representation of different epigenetic mechanisms leading to HIV-1 latency. Different mechanisms are responsible for maintenance of HIV-1 latency. Epigenetic mechanisms that function to maintain HIV-1 latency involves post translational modifications of histones like interplay of histone deacetylation and acetylation, histone methylation and phosphorylation. Other epigenetic mechanisms which are contributors of latency are DNA methylation, micro RNA pathway and chromatin remodelling. Absence of the chromatin remodelers limit nucleosome flexibility to remodel, thus restricting active transcription and leading to HIV-1 latency

dependent deacetylases that form the third class of HDACs. Members in this class include SIRT 1–7. HDAC11 shows similarity with class I HDACs but similarity is not very significant so it is a unique member of class IV HDACs (Gao et al. 2002; North and Verdin 2004).

Histone deacetylase HDAC1 has been shown to be recruited by wide array of cellular factors like YY1, LSF, c-Myc, SP1, CTIP2, CBF1, NFκB p50- at the integrated LTR promoter (Coull et al. 2000; Imai and Okamoto 2006; Jiang et al. 2007; Marban et al. 2007; Tyagi and Karn 2007; Williams et al. 2006). A study was performed to look for the specific HDACs which could be plausible contributors to latency. Chromatin immunoprecipitation studies showed that HDAC2 and HDAC3 were recruited at the LTR promoter and have repressive activity (Keedy et al. 2009).

Histone acetyl transferase (HAT) family includes p300/CBP, PCAF (p300/CBP Associating Factor), Tip 60, TAFII250 and GCN5. Although these HATs play an important role in cellular transcription but HIV also exploits these proteins for transcription from the LTR promoter. Tat engages all the five HATs to establish the virus in to the host cell. Tat interacts with and recruits p300/CBP to the LTR promoter (Marzio et al. 1998; Wong et al. 2005) where it acts as coactivator of Tat. Tat actually forms a ternary complex with p300 and PCAF and increases its affinity for PTEFb/CDK9 complex (Benkirane et al. 1998; Deng et al. 2001). Apart from its histone acetylation activity, p300 acetylates viral integrase and Tat. This acetylation at Lys 50 and Lys 51 of Tat is essential for its activity (Ott et al. 1999). Acetylated integrase regulates integration. Vpr also requires help of p300 for its transcriptional activation activity. Thus p300 participates at multiple stages of HIV-1 life cycle. PCAF is another HAT that participates extensively in HIV life cycle. PCAF interacts with Lys 50 acetylated Tat to serve as a coactivator of HIV transcription. Like p300, PCAF also activates transcription from LTR promoter by acting in synergy with Tat protein (Bres et al. 2002; Dorr et al. 2002). GCN5 also acetylates Tat at same residues K50 and K51 as p300 and mediating same function of LTR activation (Col et al. 2001). Tat also interacts with $TAF_{II}250$ (Weissman et al. 2001) to repress transcription from $TAF_{II}250$ dependent MHCI genes (Weissman et al. 1998). Tip60, a HAT originally identified as a Tat interacting cellular protein also assists Tat in transactivation of the LTR promoter (Kamine et al. 1996). Looking at the details of HIV-1 LTR driven gene expression, it becomes obvious that it is a concerted action of many host and viral proteins that brings about successful transcription. A cell can house latent reservoirs and stay dormant lifelong because of the absence of any of these critical factors indispensable for HIV transcription.

ii. **Methylation**

Another post translational modification of histones gaining relevance in HIV biology is methylation. Genomic features decide the integration sites for HIV genome. HIV avoids integration in regions with transcription inhibitory modifications such as H3K27 and DNA CpG methylation (Wang et al. 2007). A recent report suggests that chromatin mediated repression is also mediated by Suv39H1, heterochromatin protein HP1γ and histone H3K9 trimethylation. Histone H3 is tri-methylated at lysine 9 by Suv39H1, a H3K9 methyl transferase (Rea et al. 2000). HP1 (heterochromatic protein 1) is found to be associated with Suv39H1 (Aagaard et al. 1999) and binds to regions methylated by it (Bannister et al. 2001) recruiting more Suv39H1 leading to maintenance of hypermethylation and the chromatin repressive state (Grewal and Moazed 2003; Maison and Almouzni 2004).

Studies performed with *in vitro* cellular models and *in vivo* in PBMCs from HIV-1 infected donors have shown that reactivation can be mediated by silencing of HP1γ (du Chene et al. 2007). Latent proviruses are observed to have heterochromatic markers at the viral LTR. Restrictive chromatin structures at the viral LTR drag HIV into latency (Pearson et al. 2008). Evidences exist in literature about the role of H3 lysine 9 (H3K9) methyltransferase G9a in mediating repressive dimethylation at H3K9 at the LTR promoter. Studies show that G9a inhibitor BIX01294 reactivates expression of HIV-1 from cellular models of latency, ACH2 and OM10.1 (Imai et al. 2010). It is not only methylation of histones that participates in HIV life cycle but methylation of Tat also plays a crucial role for its functional activity. Tat methylated on R52 and 53 by PRMT6 decreases its association with transcription factors and the result is reduced transcription from the LTR (Boulanger et al. 2005; Xie et al. 2007). Tat methylated on K51 by H3-K9 methyl transferase SETDB1 also inhibits its function and hence reduces transcription from the LTR (Van Duyne et al. 2008). Contrary to this, K51 methylation by Set7/9-KMT7 increases Tat activity on the LTR (Pagans et al. 2010). This clearly indicates that specific methylation inhibitors need to be evaluated for their ability to activate the latent reservoirs. Such methylation inhibitors should not be inhibitory to Tat's function leading to latency but should work solely on the LTR activation purging the hidden latent reservoirs.

iii. **Phosphorylation**

Phosphorylation of Histone H1 by PTEFb at the S/TPXK sequence is known to be important in regulating its chromatin binding. This phosphorylation takes place in a Tat specific manner (O'Brien et al. 2010). Mutant analysis of H1 has shown that in absence of this phosphorylation, Tat transactivation is inhibited. Absence of this phosphorylation depletes the H1 mobility and HIV-1 transcription, giving birth to latent reservoir of HIV in these infected cells.

21.4.2 Chromatin Remodelling

Chromatin-modifying complexes can be classified in two groups: first group includes factors that mediate covalent modification of histones. The protruding N-terminal tails of the histone proteins in the nucleosome are substrates of various post translational modifications like methylation, acetylation and phosphorylation. Histone acetyl transferases like GCN5, p300/CBP, PCAF are included in this family. The second group includes proteins that change the location or conformation of nucleosomes utilizing the energy from the hydrolysis of ATP and thus increasing the accessibility to the DNA (Hargreaves and Crabtree 2011). SWI/SNF is an ATP-dependent chromatin remodelling complex belonging to this class that acts to reorganize chromatin structure so as to facilitate binding of transcription factors (Kwon et al. 1994). Integrase interacting protein (Ini1 or SNF5), BRM and BRG1 are the core subunits of SWI/SNF complex, which participate in expression of many eukaryotic genes. Nuc-1 at the LTR requires remodelling via SWI/SNF to initiate transcription. Ini1 and BRG1 interact with Tat to recruit SWI/SNF remodelling complex at Nuc-1 and to activate the transcription from the LTR promoter (Bukrinsky 2006; Mahmoudi et al. 2006; Treand et al. 2006). Inactive or lower levels of SWI/SNF complex in a cell limit this remodelling at Nuc-1 leading to absence of long HIV transcripts and eventually progressing to latency.

21.4.3 Histone Variant Proteins Forming the Nucleosomes

Histone variants (reviewed extensively in Henikoff et al. 2004) for H2A and H3 are being studied since decades for their role in epigenetic regulation. These variants differ in few amino acids. As for example, H3, the core histone and H3.3, the H3 replacement variant vary by four amino acids only. H3.3 replaces H3 soon after the initiation of transcription (Ahmad and Henikoff 2002; Janicki et al. 2004). H2A.Z is a variant associated with active chromatin. Specific protein assembly complexes regulate the exchange of the core histones with the variants in the nucleosome. SWR1 and HIRA complex deposit H2A.Z and H3.3 respectively. SWR1 complex destabilizes the nucleosome, exchanging H2A.Z-H2B for H2A-H2B. CENP-A is yet another H3 histone variant found at the mammalian centromeres (Palmer et al. 1991). Some variants differentiate between the active and the silenced chromatin like the MacroH2A which is enriched on the human inactive X chromosome. Another variant H2A-Bbd, however, is deficient in the inactive X chromosome. Nucleosomes containing H3.3/H2A.Z double variant marks the nucleosome free regions of the active promoters, enhancers and insulators (Jin et al. 2009). Histone variants have elaborate roles in gammaherpesvirus latency and HSV-1 lytic infection, but this area remains less explored in HIV transcription and latency.

21.4.4 Methylation of DNA

DNA methylation also extends repressive effect on the transcription from the LTR (Blazkova et al. 2009; Kauder et al. 2009). HIV-1 encodes for two CpG islands that are seen to be methylated during latency. This methylation recruits transcriptional repressors like MBD2 (methyl CpG binding domain family of proteins) to the HIV-LTR. Latent HIV-1 has been observed to be reactivated by the use of aza-CdR, thereby depicting the role of cytosine methylation in case of HIV latency (Blazkova et al. 2009; Kauder et al. 2009). There are increasing evidences in literature regarding role of histone and CpG methylation in the maintenance of HIV latency.

21.4.5 Micro RNA

Micro RNAs (miRNAs) are 22–25 nucleotide long, endogenous, non coding RNA that have many regulatory functions. miRNAs bind with imperfect complementarity to their targets (3' UTR) and repress the translation of the mRNA which is eventually degraded in the P bodies. Resting T cells that harbour latent HIV-1 provirus are flooded with miRNAs that have predicted binding sites in the HIV-1 genome (Huang et al. 2007). Many host miRNA like miR-28, 125b, 150, 223 and 382 have been reported to be inducers of HIV-1 latency. To circumvent restrictions on their replication by miRNA, viruses use RNA silencing suppressor (RSS). A similar mechanism also exists for HIV where it fights against the killing strategies of the host. Dicer and Drosha, the two key participants of the miRNA processing machinery were found to be inhibitory to the HIV replication both in PBMC from the infected individuals and latently infected U1 cells. HIV, in turn, suppresses the expression of miRNA cluster miR-17/92 to overcome these hindrances for its own survival (Triboulet et al. 2007). Many such see-saw relationships exist in case of HIV. Some of these host proteins and miRNAs are directed to suppress HIV-1 production but HIV-1 is intelligent enough to exploit these factors for its own benefit. Host miR-29a functions to increase the interaction of HIV-1 mRNA with RCK/p54, a P-body protein (Nathans et al. 2009). P-bodies (processing bodies) are cytoplasmic foci that participate in mRNA degradation and translational repression. They contain factors for mRNA turnover like mRNA degradation machinery, decapping enzymes etc. (Parker and Sheth 2007). Thus, host miR-29a-HIV-1 mRNA interaction plunges HIV-1 mRNA to P-bodies possibly targeting it for translational suppression or its decay as is evident by increased virus production upon depletion of the P-bodies or RCK/p54. Alternatively, the virus could be utilizing this interaction to suppress translation of its mRNA and hide itself till favourable conditions reappear (Nathans et al. 2009).

To prepare the cellular milieu conducive for its survival, HIV-1 changes the profile of the miRNAs (Yeung et al. 2005), utilizing key regulatory protein of the RNA silencing pathway, TRBP, for transcription of TAR containing transcripts (Haase et al. 2005). HIV-1 TAR RNA sequesters TRBP, thus suppressing the host RNA silencing machinery (Bennasser et al. 2006). Virus also encodes for miRNA, named "vmiRNA", that functions to fight against the strategies employed by the host (Bennasser et al. 2004; Ouellet et al. 2008). vmiRNA are also suspected to contribute to latency. HIV-1 TAR RNA is processed by Dicer to produce a vmiRNA that recruits HDAC1 to the LTR indicating its role in viral latency (Klase et al. 2007).

Cellular anti HIV-miRNAs restrict HIV-1 to flourish in the monocytes where they are found in abundance (Wang et al. 2009). They are also major contributors of latency (Huang et al. 2007). Anti-miRNA inhibitors are possible therapeutic candidates that can be used to induce these latent reservoirs for clearance by HAART (Zhang 2009). All the above epigenetic contributors to HIV-1 latency have been diagrammatically presented in Fig. 21.3.

21.5 Experimental Models of HIV-1 Latency

Previous studies on HIV-1 latency have depicted the role of various cellular and viral factors in maintaining latency which is suggestive of a multifactorial mechanism behind it. Relative low frequency of these cells in HIV patients and the absence of distinct latency markers which can distinguish between resting CD4$^+$ T cells and the latently infected cells, limits the use of various experimental approaches to study mechanism of latency directly from primary cells. To circumvent this problem, various model systems have been utilized to study HIV-1 latency using chronically infected primary cells or cell lines and animal models. Different *in vitro* and *in vivo* cellular models of HIV-1 latency are listed in Table 21.1.

Table 21.1 Various model system for studying HIV latency

S.No.	Model system	Host	References
1	ACH2	A3.01 CD4+ T cell	Folks et al. (1989)
2	JΔK	Jurkat CD4+ T cell	Antoni et al. (1994)
3	U1	U937 promonocytic cell	Folks et al. (1987)
4	J Lat	Jurkat CD4+ T cell	Jordan et al. (2003)
5	OM10.1	HL60 premyelocytic cell	Butera et al. (1991a, b)
6	SCID-hu (Thy/Liv)	Scid/scid mouse	Brooks et al. (2009)
7	SIV-macaque	macaque	Hazuda et al. (2004)
8	Hu-Rag$^{-/-}$gamma(c)$^{-/-}$	mouse	Choudhary et al. (2009)

21.5.1 Cell Lines as Latency Model System

Various cellular model systems have been developed which resemble with HIV-1 latent state using chronically infected cell lines and clones. These model cell lines normally show very low level of viral gene expression however they can be activated by activator molecules like cytokines, PMA, TNFα, etc., to produce enhanced levels of viral proteins.

U1 (Folks et al. 1987) is a chronically infected clone derived from promonocytic cell line U937 (Pomerantz et al. 1990). It carries two copies of non replicating HIV-1 provirus. Latency in this cell line is a result of reduced activity of Tat which occurs due to mutations in both the alleles of Tat. One of the tat allele lacks initiation codon while the other allele carries an H to L mutation in amino acid 13 (H13L) (Emiliani et al. 1998). ACH2 is also a chronically infected clone derived from A3.01 cell line and carries one copy of HIV-1 provirus (Folks et al. 1989). Latency is established in this clone due to a point mutation in TAR which in turn results in impaired HIV-1 Tat function (Emiliani et al. 1996). JΔK is another chronically infected clone derived from Jurkat cell line (Antoni et al. 1994). Latency is characterized in this clone due to the deletion of NFκB sites from the HIV-1 LTR. J-Lat model was generated from Jurkat cell line infected with a retroviral vector carrying the whole HIV-1 genome lacking the Tat gene. It carries a GFP ORF under the HIV-1 LTR promoter (Jordan et al. 2003). OM 10.1 cells are clonally derived from HL60 premyelocytic cell line that harbors a single latent HIV-1 provirus (Butera et al. 1991a, b).

Latent cellular models have been used to understand the role of histone methylation in HIV-1 latency. BIX01294, an inhibitor of H3K9 methyl transferase, G9a has been shown to activate HIV-1 latent reservoirs from ACH2 and OM10.1 cells (Imai et al. 2010). U1 cells have been used to understand the role of HP1γ in chromatin mediated repression of LTR promoter and HIV latency (Isaure du Chene; EMBO J 424). HIV-1 proviral latency is maintained in J-Lat cells through local histone deacetylation on HIV-1 LTR mediated by NFκB1-p50-HDAC1 complex binding and thus repressing the active transcription (Williams et al. 2006). These cell lines have extensively been used in the HDAC class I and II inhibitor studies (Ylisastigui et al. 2004; Archin et al. 2009; Matalon et al. 2011). J-Lat cells have also been used to understand the role of DNA methylation in maintenance of HIV-1 latency.

LTR promoter is hypermethylated in J Lat cells, MBD2 being the key player binds to the two CpG islands near the transcription start site, maintaining the transcriptionally repressive latent state (Kauder et al. 2009). DNA methylation inhibitor studies in J-Lat cells have experimentally shown to reactivate latent HIV.

Mechanistic studies with these model cell lines revealed the critical role of site of integration in viral persistence. In these latent cell lines, integration was majorly seen in the heterochromatic region of genome (Jordan et al. 2003), however, when similar studies were done with resting CD4+ T cells from HIV-1 infected patients undergoing HAART therapy (Liu et al. 2006) or untreated HIV-1 infected PBMC (MacNeil et al. 2006), the preferred occupancy for proviral integration was found in transcriptionally active genes as reviewed earlier in this chapter. This contrasting picture of proviral integration between these model cell lines and the latently infected primary cells as well as the absence of Tat and TAR mutation during the *in vivo* latent infection have recently raised a concern over the usage of these cellular models to study the *in vivo* HIV-1 latency.

21.5.2 Animal Models of HIV Latency

To study the mechanistic aspects of HIV latency, Zack and colleagues developed a mouse model called SCID (Severe combined immunodeficiency)-hu (Thy/Liv) mouse model (Brooks et al. 2001). It is an immunodeficient CB17 scid/scid mouse which lacks functional T and B cells. It was transplanted with human fetal liver tissue to provide source of hematopoietic progenitor cells and thymus tissues to create a microenvironment for differentiation and proliferation of these cells. These grafted tissues were infected with HIV by injecting the virus into the graft leading to infection of CD4+ thymocytes. When these infected cells undergo several rounds to replication and differentiation, some of the mature resting CD4+ cells carry the latent viral genome as they show minimal HIV-1 LTR mediated transcription.

Due to the close mimicry of these cells with the *in vivo* latently infected quiescent CD4+ cells, this mouse model was used to develop new strategies to eradicate viral latent reservoir. Though this model represents a suitable source of high frequency production of quiescent CD4+ single positive (SP) cells, however, the predominant source of latent infection i.e. infected CD4+ memory T cells are absent in this system. Furthermore, the differences in the distribution, activation threshold and TCR mediated signaling pathways (Dutton et al. 1998; Farber et al. 1997; Jenkins et al. 2001) of naive T cells and memory T cells, suggests that SCID hu Thy/Liv mouse can function as a good model for studying the HIV latency in quiescent T cells but to study the *in vivo* latency, a more efficient model system is required.

To further enhance the mimicry between the latency model system and HIV-1 infected humans, another animal model was developed using Simian immunodeficiency virus infected macaque (SIV-macaque). These SIV-macaques were treated with reverse transcriptase inhibitors to decrease the viral load to undetectable level (<100 copies/ml) (Hazuda et al. 2004). Latently infected CD4+ cells carrying integrated SIV DNA were recovered from lymph node, peripheral blood and spleen of these animals. Later on, various studies have been done in these models using various antiretroviral drugs to mimic the situation of HIV-1 infected latent cells in humans (Hazuda et al. 2004; Hofman et al. 2004). Another humanized mouse, Hu-Rag$^{-/-}$ gamma(c)$^{-/-}$ was also used to study HIV persistence (Choudhary et al. 2009). Although these models are yet to be exploited well in epigenetic studies but they seem to have future potential in unravelling epigenetic mechanisms leading to HIV-1 latency.

21.5.3 Use of Primary Cells as Latency Model

Various recent studies have utilized primary cells as model to study HIV-1 latency. Saleh and colleagues have shown that resting CD4⁺ T cells after incubation with CCL19 and CCL21 can efficiently allow viral integration with restricted viral gene expression representing a model of post integration viral latency (Saleh et al. 2007). Another such model was developed by Yang and colleagues, where resting CD4⁺ T cells were first transduced with a lentiviral vector coding BCL2 cDNA followed by activation with anti-CD3/CD28 and infection with HIV-1. These infected cells moved to a latent state after the removal of cytokines (Yang et al. 2009). This model was used to screen compounds that can reactivate the latent reservoir without activating the T cells globally. Another such model of HIV latency was developed by Planelles group using primary CD4⁺ T cells. This model was used to generate latently infected resting central memory T cells (Bosque and Planelles 2011). Such primary CD4⁺ T cell models have been employed in the study of HDAC inhibitors which have shown good potential in reactivation of latent HIV in cell line models (Sahu and Cloyd 2011; Tyagi et al. 2010). Role of heterochromatic protein HP1γ has been also implicated in maintenance of latency in PBMCs from HAART treated HIV infected individuals (du Chene et al. 2007).

21.6 Eradication of Latent Reservoirs

Latent cells are generally characterized by their resting nature and presence of a dormant virus. These cells also show high resemblance to the uninfected resting T cells and almost appear indistinguishable. This dormant nature of HIV virus in latent cells as well as their high resemblance with uninfected resting T cells poses a great hurdle to understand the molecular nature of HIV latency and to develop highly specific and effective therapeutic strategy. Current strategies aim to activate these latent cells making them more sensitive towards immune response to purge the virus by immune effector cells as well as antiviral agents. However, with the activation of latently infected cells, the threat of new rounds of infection always remains there. So, the new anti-viral therapy should include such agents which can boost up the host immune response and specifically target the latently infected cells. Currently evolving anti-HIV therapeutic strategies involve various ways to activate these latent reservoirs and are being discussed briefly below:

21.6.1 Induction of HIV Transcription from Latent Cells

Intermittent administration of cytokines like IL-2 and IL-7 in HIV patients along with HAART therapy have shown to be effective in reducing the pool of latent cells carrying transcriptionally silent virus and thus seems to be a promising strategy for latent virus eradication (Chun et al. 1999; Scripture-Adams et al. 2002). Modulation of cellular transcription is another common strategy used to induce viral transcription. Cellular transcription is generally modulated by use of HDACs inhibitors like Valproic acid (Blankson et al. 2002; Lehrman et al. 2005), methylation inhibitors (Kauder et al. 2009) and combining these agents with HAART therapy can help to clear the viral reservoirs. It is a promising approach to purge the latent reservoirs which can be then cleared by HAART therapy. Several HDAC inhibitors are being tested for their potency in activating the silent reservoirs. Valproic acid (VPA), a drug in clinical use was tested in resting CD4 cells of aviremic HIV infected patients treated with highly active antiretroviral therapy (HAART). VPA induced acetylation at the integrated HIV promoter and also production of virus from the resting CD4 cells of infected patients on HAART (Ylisastigui et al. 2004). Potency testing was done for various inhibitors of class I and II HDACs, namely, MRK1, MRK4, MRK10, MRK11, MRK12, MRK13, MRK14, apicidin and VPA. Inhibitors of class I HDACs proved to be efficient inducers of virus production from resting CD4 cells of aviremic patients on ART by induction of LTR transcription. Currently the focus is being laid on HDAC inhibitors, already in clinical use for treatment of some other disease. Suberoylanilide hydroxamic acid or Vorinostat (SAHA) (Contreras et al. 2009), Romidepsin (Depsipeptide, FK-228), Belinostat (PXD101) and LAQ824/LBH589 are HDAC inhibitors which can be put to clinical use. Other promising epigenetic targets for anti HIV therapy are extensively reviewed in Varier and Kundu (2006).

Activation of NFκB pathway can also drive the HIV-1 LTR driven gene expression from latently infected cells. In this direction, Prostatin was found to very promising as it can increase the recruitment of active NFκB on HIV-1 LTR (Blankson et al. 2002; Williams et al. 2007) and was also shown to increase viral gene expression and decrease CD4 expression in PBMC from patients on HAART therapy (Kulkosky et al. 2001). NFκB independent activators of HIV-1 transcription were also found to be effective for purging the viral reservoirs. NFκB is a transcription factor which is present in almost all animal cell types and found to be associated with most of the signalling pathways. Therefore, activation of NFκB pathways can result in global activation of T cells. To avoid this, a hunt for transcription factors which can induce HIV-1 transcription in a NFκB independent manner has been started. Recently, Yang and co-workers have shown that resting memory T cells from patients on HAART therapy can be reactivated with Est1 transcription factor to induce virus transcription without activating T cells globally (Yang et al. 2009).

Hexamethylene bisacetamide (HMBA) is a bipolar compound, which can induce Tat independent HIV-1 transcription. It was shown recently that it can activate HIV-1 transcription in latently infected CD4+ cells isolated from aviremic patients under ART therapy by a Tat independent mechanism (Choudhary et al. 2008; Klichko et al. 2006). It seems to activate PI3K/Akt pathway to enhance the recruitment of active P-TEFb complex on HIV-1 LTR to induce viral gene expression (Contreras et al. 2007).

21.6.2 Manipulation of Host miRNA Profile

Many host miRNA which have their binding sites in the viral genome have been shown to be inducers of viral latency (Huang et al. 2007) and have been extensively discussed in an earlier section of this chapter. Manipulation of these miRNAs to prevent their binding on viral mRNA or use of anti-miRNA inhibitors seems to be a possible therapeutic strategy for plunging these latent reservoirs for clearance by HAART (Zhang 2009).

21.6.3 Use of Combinatorial Therapy

Recently, the concept of a combination therapy involving use of various combinations of HIV-1 transcription activators discussed above is gaining interest. Transcriptional activators like Valproic acid or suberoylanilide hydroxamic acid (SAHA) with prostatin have shown synergistic effect on reactivation of latent cell lines like U1 and J-Lat cells, purging the latent HIV more potently from these cells as compared to the individual inhibitor treatment. Similar effects were observed in PBMC isolated from HAART treated aviremic patients (Reuse et al. 2009). It is believed that the future therapeutic regimen development against HIV can be inspired by such combinatorial approach (Fig. 21.4) to eradicate the hidden virus more efficiently from the infected host.

21.7 Concluding Remarks

Latent reservoirs are greatest obstacle in treating HIV patients and ultimate eradication of virus. HIV eradication is a hard nut to crack until these pools are cleared from the infected individual since HAART therapy is incapable of handling and clearing these hidden reservoirs. According to our current understanding, intelligent design of antiretroviral regimen is needed to ensure complete eradication of the virus from the host and not only the active virus. Presently, HAART therapy is a triple therapy

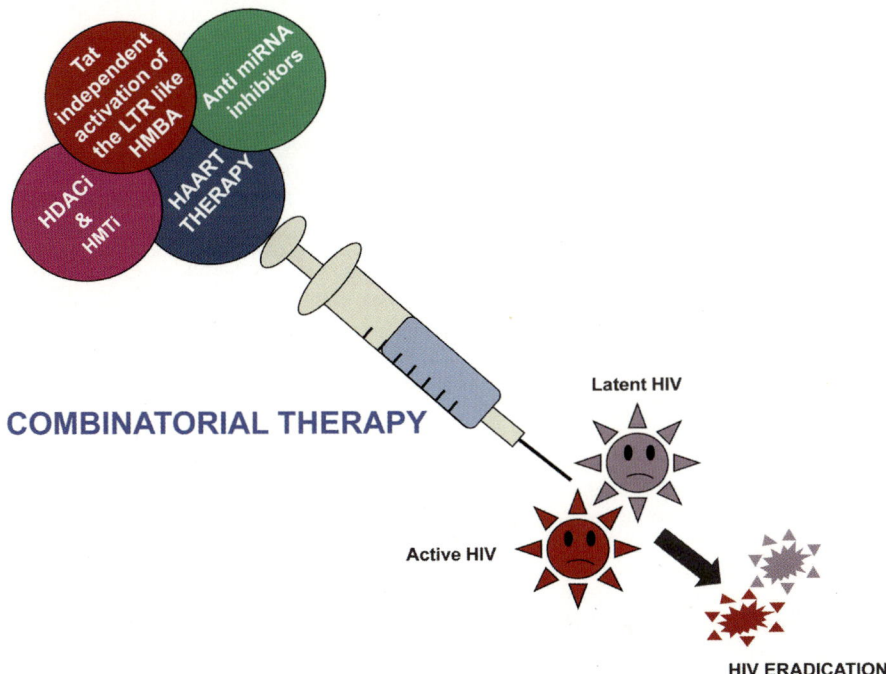

Fig. 21.4 Schematic representation of the design of a future combinatorial therapeutic regimen. HAART therapy is inactive against the latent reservoirs. To make it active for the latent reservoirs, supplementation of HAART with HDAC, HMT, anti-miRNA inhibitors is necessary. Compounds like HMBA can also be included in the combinatorial therapy for the activation of the silent promoter in Tat-independent manner. These supplements would purge the latent reservoirs on which the HAART can act and eradicate the virus

which normally includes one protease inhibitor and two nucleoside or non-nucleoside reverse transcriptase inhibitors or other such combinations. A combinatorial therapy keeping in mind activation of latent virus should be tested for clinical use. The possible design of the therapy can be present triple therapy supplemented with epigenetic silencing inhibitors like HMTi, HDACi, anti-microRNA inhibitors and compounds like HMBA. Lack of complete understanding of the molecular mechanisms operating in maintenance of latency could be another drawback that is a plausible hurdle to purge these latent reservoirs. Histone variants is one such avenue that is unexplored in case of HIV but there are increasing evidences of their roles in latency of other viruses like gammaherpesvirus and HSV-1. Thus, we need to actively work on these aspects of HIV biology to have an AIDS-free future across the globe.

Acknowledgement Research work in DM laboratory was supported by NCCS and Department of Biotechnology (DBT), Government of India. ND is a CSIR senior research fellow (SRF) and PR is a research associate (RA) in DBT funded project.

References

Aagaard L, Laible G, Selenko P, Schmid M, Dorn R, Schotta G, Kuhfittig S, Wolf A, Lebersorger A, Singh PB, Reuter G, Jenuwein T (1999) Functional mammalian homologues of the Drosophila PEV-modifier Su(var)3-9 encode centromere-associated proteins which complex with the heterochromatin component M31. EMBO J 18:1923–1938

Adhya S, Gottesman M (1982) Promoter occlusion: transcription through a promoter may inhibit its activity. Cell 29:939–944

Ahmad K, Henikoff S (2002) The histone variant H33 marks active chromatin by replication-independent nucleosome assembly. Mol Cell 9:191–1200

Antoni BA, Rabson AB, Kinter A, Bodkin M, Poli G (1994) NF-kappa B-dependent and -independent pathways of HIV activation in a chronically infected T cell line. Virology 202:684–694

Archin NM, Keedy KS, Espeseth A, Dang H, Hazuda DJ, Margolis DM (2009) Expression of latent human immunodeficiency type 1 is induced by novel and selective histone deacetylase inhibitors. AIDS 23:1799–1806

Asin S, Taylor JA, Trushin S, Bren G, Paya CV (1999) Ikappakappa mediates NF-kappaB activation in human immunodeficiency virus-infected cells. J Virol 73:3893–3903

Bannister AJ, Zegerman P, Partridge JF, Miska EA, Thomas JO, Allshire RC, Kouzarides T (2001) Selective recognition of methylated lysine 9 on histone H3 by the HP1 chromo domain. Nature 410:120–124

Benkirane M, Chun RF, Xiao H, Ogryzko VV, Howard BH, Nakatani Y, Jeang KT (1998) Activation of integrated provirus requires histone acetyltransferase p300 and P/CAF are coactivators for HIV-1 Tat. J Biol Chem 273:24898–24905

Bennasser Y, Le SY, Yeung ML, Jeang KT (2004) HIV-1 encoded candidate micro-RNAs and their cellular targets. Retrovirology 1:43

Bennasser Y, Yeung ML, Jeang KT (2006) HIV-1 TAR RNA subverts RNA interference in transfected cells through sequestration of TAR RNA-binding protein TRBP. J Biol Chem 281:27674–27678

Berger SL (2002) Histone modifications in transcriptional regulation. Curr Opin Genet Dev 12:142–148

Blankson JN, Persaud D, Siliciano RF (2002) The challenge of viral reservoirs in HIV-1 infection. Annu Rev Med 53:557–593

Blazkova J, Trejbalova K, Gondois-Rey F, Halfon P, Philibert P, Guiguen A, Verdin E, Olive D, Van Lint C, Hejnar J, Hirsch I (2009) CpG methylation controls reactivation of HIV from latency. PLoS Pathog 5:e1000554

Bohnlein E, Lowenthal JW, Siekevitz M, Ballard DW, Franza BR, Greene WC (1988) The same inducible nuclear proteins regulates mitogen activation of both the interleukin-2 receptor-alpha gene and type 1 HIV. Cell 53:827–836

Boldogh I, Albrecht T, Porter DD (1996) Persistent viral infections. In: Baron S (ed) Medical microbiology, 4th edn. University of Texas Medical Branch, Galveston

Bosque A, Planelles V (2011) Studies of HIV-1 latency in an ex vivo model that uses primary central memory T cells. Methods 53:54–61

Boulanger MC, Liang C, Russell RS, Lin R, Bedford MT, Wainberg MA, Richard S (2005) Methylation of Tat by PRMT6 regulates human immunodeficiency virus type 1 gene expression. J Virol 79:124–131

Bres V, Kiernan R, Emiliani S, Benkirane M (2002) Tat acetyl-acceptor lysines are important for human immunodeficiency virus type-1 replication. J Biol Chem 277:22215–22221

Brooks DG, Kitchen SG, Kitchen CM, Scripture-Adams DD, Zack JA (2001) Generation of HIV latency during thymopoiesis. Nat Med 7:459–464

Bukrinsky M (2006) SNFing HIV transcription. Retrovirology 3:49

Bukrinsky MI, Sharova N, Dempsey MP, Stanwick TL, Bukrinskaya AG, Haggerty S, Stevenson M (1992) Active nuclear import of human immunodeficiency virus type 1 preintegration complexes. Proc Natl Acad Sci U S A 89:6580–6584

Butera ST, Perez VL, Wu BY, Nabel GJ, Folks TM (1991a) Oscillation of the human immunodeficiency virus surface receptor is regulated by the state of viral activation in a CD4+ cell model of chronic infection. J Virol 65:4645–4653

Butera ST, Perez VL, Besansky NJ, Chan WC, Wu BY, Nabel GJ, Folks TM (1991b) Extrachromosomal human immunodeficiency virus type-1 DNA can initiate a spreading infection of HL-60 cells. J Cell Biochem 45:366–373

Cherrier T, Suzanne S, Redel L, Calao M, Marban C, Samah B, Mukerjee R, Schwartz C, Gras G, Sawaya BE, Zeichner SL, Aunis D, Van Lint C, Rohr O (2009) p21(WAF1) gene promoter is epigenetically silenced by CTIP2 and SUV39H1. Oncogene 28:3380–3389

Chiu YL, Soros VB, Kreisberg JF, Stopak K, Yonemoto W, Greene WC (2005) Cellular APOBEC3G restricts HIV-1 infection in resting CD4+ T cells. Nature 435:108–114

Choudhary SK, Archin NM, Margolis DM (2008) Hexamethylbisacetamide and disruption of human immunodeficiency virus type 1 latency in CD4(+) T cells. J Infect Dis 197:1162–1170

Choudhary SK, Rezk NL, Ince WL, Cheema M, Zhang L, Su L, Swanstrom R, Kashuba AD, Margolis DM (2009) Suppression of human immunodeficiency virus type 1 (HIV-1) viremia with reverse transcriptase and integrase inhibitors CD4+ T-cell recovery and viral rebound upon interruption of therapy in a new model for HIV treatment in the humanized Rag2−/−{gamma}c−/− mouse. J Virol 83:8254–8258

Chugh P, Fan S, Planelles V, Maggirwar SB, Dewhurst S, Kim B (2007) Infection of human immunodeficiency virus and intracellular viral Tat protein exert a pro-survival effect in a human microglial cell line. J Mol Biol 366:67–81

Chun TW, Finzi D, Margolick J, Chadwick K, Schwartz D, Siliciano RF (1995) In vivo fate of HIV-1-infected T cells: quantitative analysis of the transition to stable latency. Nat Med 1:1284–1290

Chun TW, Engel D, Mizell SB, Hallahan CW, Fischette M, Park S, Davey RT Jr, Dybul M, Kovacs JA, Metcalf JA, Mican JM, Berrey MM, Corey L, Lane HC, Fauci AS (1999) Effect of interleukin-2 on the pool of latently infected resting CD4+ T cells in HIV-1-infected patients receiving highly active anti-retroviral therapy. Nat Med 5:651–655

Chun TW, Justement JS, Lempicki RA, Yang J, Dennis G Jr, Hallahan CW, Sanford C, Pandya P, Liu S, Mc Laughlin M, Ehler LA, Moir S, Fauci AS (2003) Gene expression and viral production in latently infected resting CD4+ T cells in viremic versus aviremic HIV-infected individuals. Proc Natl Acad Sci U S A 100:1908–1913

Col E, Caron C, Seigneurin-Berny D, GraciaJ Favier A, Khochbin S (2001) The histone acetyltransferase hGCN5 interacts with and acetylates the HIV transactivator Tat. J Biol Chem 276:28179–28184

Contreras X, Barboric M, Lenasi T, Peterlin BM (2007) HMBA releases P-TEFb from HEXIM1 and 7SK snRNA via PI3K/Akt and activates HIV transcription. PLoS Pathog 3:1459–1469

Contreras X, Schweneker M, Chen CS, McCune JM, Deeks SG, Martin J, Peterlin BM (2009) Suberoylanilide hydroxamic acid reactivates HIV from latently infected cells. J Biol Chem 284:6782–6789

Coull JJ, Romerio F, Sun JM, Volker JL, Galvin KM, Davie JR, Shi Y, Hansen U, Margolis DM (2000) The human factors YY1 and LSF repress the human immunodeficiency virus type 1 long terminal repeat via recruitment of histone deacetylase 1. J Virol 74:6790–6799

de Ruijter AJ, van Gennip AH, Caron HN, Kemp S, van Kuilenburg AB (2003) Histone deacetylases (HDACs): characterization of the classical HDAC family. Biochem J 370:737–749

Deng L, Wang D, de la Fuente C, Wang L, Li H, Lee CG, Donnelly R, Wade JD, Lambert P, Kashanchi F (2001) Enhancement of the p300 HAT activity by HIV-1 Tat on chromatin DNA. Virology 289:312–326

Dorr A, Kiermer V, Pedal A, Rackwitz HR, Henklein P, Schubert U, Zhou MM, Verdin E, Ott M (2002) Transcriptional synergy between Tat and PCAF is dependent on the binding of acetylated Tat to the PCAF bromodomain. EMBO J 21:2715–2723 du Chene I, Basyuk E, Lin YL, Triboulet R, Knezevich A, Chable-Bessia C, Mettling C, Baillat V, Reynes J, Corbeau P, Bertrand E, Marcello A, Emiliani S, Kiernan R, Benkirane M (2007) Suv39H1 and HP1gamma are responsible for chromatin-mediated HIV-1 transcriptional silencing and post-integration latency. EMBO J 26:424–435

Duh EJ, Maury WJ, Folks TM, Fauci AS, Rabson AB (1989) Tumor necrosis factor alpha activates human immunodeficiency virus type 1 through induction of nuclear factor binding to the NF-kappa B sites in the long terminal repeat. Proc Natl Acad Sci U S A 86:5974–5978

Dutton RW, Bradley LM, Swain SL (1998) T cell memory. Annu Rev Immunol 16:201–223

Efstathiou S, Preston CM (2005) Towards an understanding of the molecular basis of herpes simplex virus latency. Virus Res 111:108–119

Emiliani S, Van Lint C, Fischle W, Paras P Jr, Ott M, Brady J, Verdin E (1996) A point mutation in the HIV-1 Tat responsive element is associated with postintegration latency. Proc Natl Acad Sci U S A 93:6377–6381

Emiliani S, Fischle W, Ott M, Van Lint C, Amella CA, Verdin E (1998) Mutations in the tat gene are responsible for human immunodeficiency virus type 1 postintegration latency in the U1 cell line. J Virol 72:1666–1670

Farber DL, Acuto O, Bottomly K (1997) Differential T cell receptor-mediated signaling in naive and memory CD4 T cells. Eur J Immunol 27:2094–2101

Folks TM, Justement J, Kinter A, Dinarello CA, Fauci AS (1987) Cytokine-induced expression of HIV-1 in a chronically infected promonocyte cell line. Science 238:800–802

Folks TM, Clouse KA, Justement J, Rabson A, Duh E, Kehrl JH, Fauci AS (1989) Tumor necrosis factor alpha induces expression of human immunodeficiency virus in a chronically infected T-cell clone. Proc Natl Acad Sci U S A 86:2365–2368

Gallastegui E, Millan-Zambrano G, Terme JM, Chavez S, Jordan A (2011) Chromatin reassembly factors are involved in transcriptional interference promoting HIV latency. J Virol 85:3187–3202

Ganesh L, Burstein E, Guha-Niyogi A, Louder MK, Mascola JR, Klomp LW, Wijmenga C, Duckett CS, Nabel GJ (2003) The gene product Murr1 restricts HIV-1 replication in resting CD4+ lymphocytes. Nature 426:853–857

Gao L, Cueto MA, Asselbergs F, Atadja P (2002) Cloning and functional characterization of HDAC11 a novel member of the human histone deacetylase family. J Biol Chem 277:25748–25755

Garriga J, Peng J, Parreno M, Price DH, Henderson EE, Grana X (1998) Upregulation of cyclin T1/CDK9 complexes during T cell activation. Oncogene 17:3093–3102

Geleziunas R, Xu W, Takeda K, Ichijo H, Greene WC (2001) HIV-1 Nef inhibits ASK1-dependent death signalling providing a potential mechanism for protecting the infected host cell. Nature 410:834–838

Giri MS, Nebozyhn M, Raymond A, Gekonge B, Hancock A, Creer S, Nicols C, Yousef M, Foulkes AS, Mounzer K, Shull J, Silvestri G, Kostman J, Collman RG, Showe L, Montaner LJ (2009) Circulating monocytes in HIV-1-infected viremic subjects exhibit an antiapoptosis gene signature and virus- and host-mediated apoptosis resistance. J Immunol 182:4459–4470

Greenway AL, McPhee DA, Allen K, Johnstone R, Holloway G, Mills J, Azad A, Sankovich S, Lambert P (2002) Human immunodeficiency virus type 1 Nef binds to tumor suppressor p53 and protects cells against p53-mediated apoptosis. J Virol 76:2692–2702

Grewal SI, Moazed D (2003) Heterochromatin and epigenetic control of gene expression. Science 301:798–802

Guillemard E, Jacquemot C, Aillet F, Schmitt N, Barre-Sinoussi F, Israel N (2004) Human immunodeficiency virus 1 favors the persistence of infection by activating macrophages through TNF. Virology 329:371–380

Haase AD, Jaskiewicz L, Zhang H, Laine S, Sack R, Gatignol A, Filipowicz W (2005) TRBP a regulator of cellular PKR and HIV-1 virus expression interacts with Dicer and functions in RNA silencing. EMBO Rep 6:961–967

Han Y, Lassen K, Monie D, Sedaghat AR, Shimoji S, Liu X, Pierson TC, Margolick JB, Siliciano RF, Siliciano JD (2004) Resting CD4+ T cells from human immunodeficiency virus type 1 (HIV-1)-infected individuals carry integrated HIV-1 genomes within actively transcribed host genes. J Virol 78:6122–6133

Hargreaves DC, Crabtree GR (2011) ATP-dependent chromatin remodeling: genetics genomics and mechanisms. Cell Res 21:396–420

Hazuda DJ, Young SD, Guare JP, Anthony NJ, Gomez RP, Wai JS, Vacca JP, Handt L, Motzel SL, Klein HJ, Dornadula G, Danovich RM, Witmer MV, Wilson KA, Tussey L, Schleif WA, Gabryelski LS, Jin L, Miller MD, Casimiro DR, Emini EA, Shiver JW (2004) Integrase inhibitors and cellular immunity suppress retroviral replication in rhesus macaques. Science 305:528–532

He LM, Hsu J, Xue Y, Chou S, Burlingame A, Krogan NJ, Alber T, Zhou Q (2010) HIV-1 Tat and host AFF4 recruit two transcription elongation factors into a bifunctional complex for coordinated activation of HIV-1 transcription. Mol Cell 38:428–438

Henikoff S, Furuyama T, Ahmad K (2004) Histone variants nucleosome assembly and epigenetic inheritance. Trends Genet 20:320–326

Herrmann CH, Carroll RG, Wei P, Jones KA, Rice AP (1998) Tat-associated kinase TAK activity is regulated by distinct mechanisms in peripheral blood lymphocytes and promonocytic cell lines. J Virol 72:9881–9888

Hofman MJ, Higgins J, Matthews TB, Pedersen NC, Tan C, Schinazi RF, North TW (2004) Efavirenz therapy in rhesus macaques infected with a chimera of simian immunodeficiency virus containing reverse transcriptase from human immunodeficiency virus type 1. Antimicrob Agents Chemother 48:3483–3490

Huang J, Wang F, Argyris E, Chen K, Liang Z, Tian H, Huang W, Squires K, Verlinghieri G, Zhang H (2007) Cellular microRNAs contribute to HIV-1 latency in resting primary CD4+ T lymphocytes. Nat Med 13:1241–1247

Imai K, Okamoto T (2006) Transcriptional repression of human immunodeficiency virus type 1 by AP-4. J Biol Chem 281:12495–12505

Imai K, Togami H, Okamoto T (2010) Involvement of histone H3 lysine 9 (H3K9) methyltransferase G9a in the maintenance of HIV-1 latency and its reactivation by BIX01294. J Biol Chem 285:16538–16545

Janicki SM, Tsukamoto T, Salghetti SE, Tansey WP, Sachidanandam R, Prasanth KV, Ried T, Shav-Tal Y, Bertrand E, Singer RH, Spector DL (2004) From silencing to gene expression: real-time analysis in single cells. Cell 116:683–698

Jenkins MK, Khoruts A, Ingulli E, Mueller DL, McSorley SJ, Reinhardt RL, Itano A, Pape KA (2001) In vivo activation of antigen-specific CD4 T cells. Annu Rev Immunol 19:23–45

Jiang G, Espeseth A, Hazuda DJ, Margolis DM (2007) c-Myc and Sp1 contribute to proviral latency by recruiting histone deacetylase 1 to the human immunodeficiency virus type 1 promoter. J Virol 81:10914–10923

Jin C, Zang C, Wei G, Cui K, Peng W, Zhao K, Felsenfeld G (2009) H3.3/H2AZ double variant-containing nucleosomes mark 'nucleosome-free regions' of active promoters and other regulatory regions. Nat Genet 41:941–945

Jordan A, Bisgrove D, Verdin E (2003) HIV reproducibly establishes a latent infection after acute infection of T cells in vitro. EMBO J 22:1868–1877

Kamine J, Elangovan B, Subramanian T, Coleman D, Chinnadurai G (1996) Identification of a cellular protein that specifically interacts with the essential cysteine region of the HIV-1 Tat transactivator. Virology 216:357–366

Kauder SE, Bosque A, Lindqvist A, Planelles V, Verdin E (2009) Epigenetic regulation of HIV-1 latency by cytosine methylation. PLoS Pathog 5:e1000495

Keedy KS, Archin NM, Gates AT, Espeseth A, Hazuda DJ, Margolis DM (2009) A limited group of class I histone deacetylases acts to repress human immunodeficiency virus type 1 expression. J Virol 83:4749–4756

Kiernan RE, Vanhulle C, Schiltz L, Adam E, Xiao H, Maudoux F, Calomme C, Burny A, Nakatani Y, Jeang KT, Benkirane M, Van Lint C (1999) HIV-1 tat transcriptional activity is regulated by acetylation. EMBO J 18:6106–6118

Klase Z, Kale P, Winograd R, Gupta MV, Heydarian M, Berro R, McCaffrey T, Kashanchi F (2007) HIV-1 TAR element is processed by Dicer to yield a viral micro-RNA involved in chromatin remodeling of the viral LTR. BMC Mol Biol 8:63

Klase Z, Winograd R, Davis J, Carpio L, Hildreth R, Heydarian M, Fu S, McCaffrey T, Meiri E, Ayash-Rashkovsky M, Gilad S, Bentwich Z, Kashanchi F (2009) HIV-1 TAR miRNA protects against apoptosis by altering cellular gene expression. Retrovirology 6:18

Klichko V, Archin N, Kaur R, Lehrman G, Margolis D (2006) Hexamethylbisacetamide remodels the human immunodeficiency virus type 1 (HIV-1) promoter and induces Tat-independent HIV-1 expression but blunts cell activation. J Virol 80:4570–4579

Korin YD, Zack JA (1999) Nonproductive human immunodeficiency virus type 1 infection in nucleoside-treated G0 lymphocytes. J Virol 73:6526–6532

Kulkosky J, Culnan DM, Roman J, Dornadula G, Schnell M, Boyd MR, Pomerantz RJ (2001) Prostratin: activation of latent HIV-1 expression suggests a potential inductive adjuvant therapy for HAART. Blood 98:3006–3015

Kwon H, Imbalzano AN, Khavari PA, Kingston RE, Green MR (1994) Nucleosome disruption and enhancement of activator binding by a human SW1/SNF complex. Nature 370:477–481

Lehrman G, Hogue IB, Palmer S, Jennings C, Spina CA, Wiegand A, Landay AL, Coombs RW, Richman DD, Mellors JW, Coffin JM, Bosch RJ, Margolis DM (2005) Depletion of latent HIV-1 infection in vivo: a proof-of-concept study. Lancet 366:549–555

Lenasi T, Contreras X, Peterlin BM (2008) Transcriptional interference antagonizes proviral gene expression to promote HIV latency. Cell Host Microbe 4:123–133

Levy DN, Refaeli Y, Weiner DB (1995) Extracellular Vpr protein increases cellular permissiveness to human immunodeficiency virus replication and reactivates virus from latency. J Virol 69:1243–1252

Lewinski MK, Yamashita M, Emerman M, Ciuffi A, Marshall H, Crawford G, Collins F, Shinn P, Leipzig J, Hannenhalli S, Berry CC, Ecker JR, Bushman FD (2006) Retroviral DNA integration: viral and cellular determinants of target-site selection. PLoS Pathog 2:e60

Liu H, Dow EC, Arora R, Kimata JT, Bull LM, Arduino RC, Rice AP (2006) Integration of human immunodeficiency virus type 1 in untreated infection occurs preferentially within genes. J Virol 80:7765–7768

MacNeil A, Sankale JL, Meloni ST, Sarr AD, Mboup S, Kanki P (2006) Genomic sites of human immunodeficiency virus type 2 (HIV-2) integration: similarities to HIV-1 in vitro and possible differences in vivo. J Virol 80:7316–7321

Mahmoudi T, Parra M, Vries RG, Kauder SE, Verrijzer CP, Ott M, Verdin E (2006) The SWI/SNF chromatin-remodeling complex is a cofactor for Tat transactivation of the HIV promoter. J Biol Chem 281:19960–19968

Maison C, Almouzni G (2004) HP1 and the dynamics of heterochromatin maintenance. Nat Rev Mol Cell Biol 5:296–304

Marban C, Suzanne S, Dequiedt F, de Walque S, Redel L, Van Lint C, Aunis D, Rohr O (2007) Recruitment of chromatin-modifying enzymes by CTIP2 promotes HIV-1 transcriptional silencing. EMBO J 26:412–423

Marzio G, Tyagi M, Gutierrez MI, Giacca M (1998) HIV-1 tat transactivator recruits p300 and CREB-binding protein histone acetyltransferases to the viral promoter. Proc Natl Acad Sci U S A 95:13519–13524

Matalon S, Rasmussen TA, Dinarello CA (2011) Histone deacetylase inhibitors for purging HIV-1 from the latent reservoir. Mol Med 17:466–472

McElhinny JA, MacMorran WS, Bren GD, Ten RM, Israel A, Paya CV (1995) Regulation of I kappa B alpha and p105 in monocytes and macrophages persistently infected with human immunodeficiency virus. J Virol 69:1500–1509

Meyerhans A, Vartanian JP, Hultgren C, Plikat U, Karlsson A, Wang L, Eriksson S, Wain-Hobson S (1994) Restriction and enhancement of human immunodeficiency virus type 1 replication by modulation of intracellular deoxynucleoside triphosphate pools. J Virol 68:535–540

Nabel G, Baltimore D (1987) An inducible transcription factor activates expression of human immunodeficiency virus in T cells. Nature 326:711–713

Nathans R, Chu CY, Serquina AK, Lu CC, Cao H, Rana TM (2009) Cellular microRNA and P bodies modulate host-HIV-1 interactions. Mol Cell 34:696–709

Niederman TM, Thielan BJ, Ratner L (1989) Human immunodeficiency virus type 1 negative factor is a transcriptional silencer. Proc Natl Acad Sci U S A 86:1128–1132

North BJ, Verdin E (2004) Sirtuins: Sir2-related NAD-dependent protein deacetylases. Genome Biol 5:224

O'Brien SK, Cao H, Nathans R, Ali A, Rana TM (2010) P-TEFb kinase complex phosphorylates histone H1 to regulate expression of cellular and HIV-1 genes. J Biol Chem 285:29713–29720

Ott M, Schnolzer M, Garnica J, Fischle W, Emiliani S, Rackwitz HR, Verdin E (1999) Acetylation of the HIV-1 Tat protein by p300 is important for its transcriptional activity. Curr Biol 9:1489–1492

Ouellet DL, Plante I, Landry P, Barat C, Janelle ME, Flamand L, Tremblay MJ, Provost P (2008) Identification of functional microRNAs released through asymmetrical processing of HIV-1 TAR element. Nucleic Acids Res 36:2353–2365

Pagans S, Kauder SE, Kaehlcke K, Sakane N, Schroeder S, Dormeyer W, Trievel RC, Verdin E, Schnolzer M, Ott M (2010) The Cellular lysine methyltransferase Set7/9-KMT7 binds HIV-1 TAR RNA monomethylates the viral transactivator Tat and enhances HIV transcription. Cell Host Microbe 7:234–244

Palmer DK, O'Day K, Trong HL, Charbonneau H, Margolis RL (1991) Purification of the centromere-specific protein CENP-A and demonstration that it is a distinctive histone. Proc Natl Acad Sci U S A 88:3734–3738

Parker R, Sheth U (2007) P bodies and the control of mRNA translation and degradation. Mol Cell 25:635–646

Pearson R, Kim YK, Hokello J, Lassen K, Friedman J, Tyagi M, Karn J (2008) Epigenetic silencing of human immunodeficiency virus (HIV) transcription by formation of restrictive chromatin structures at the viral long terminal repeat drives the progressive entry of HIV into latency. J Virol 82:12291–12303

Pomerantz RJ, Trono D, Feinberg MB, Baltimore D (1990) Cells nonproductively infected with HIV-1 exhibit an aberrant pattern of viral RNA expression: a molecular model for latency. Cell 61:1271–1276

Pomerantz RJ, Seshamma T, Trono D (1992) Efficient replication of human immunodeficiency virus type 1 requires a threshold level of Rev: potential implications for latency. J Virol 66:1809–1813

Rea S, Eisenhaber F, O'Carroll D, Strahl BD, Sun ZW, Schmid M, Opravil S, Mechtler K, Ponting CP, Allis CD, Jenuwein T (2000) Regulation of chromatin structure by site-specific histone H3 methyltransferases. Nature 406:593–599

Reuse S, Calao M, Kabeya K, Guiguen A, Gatot JS, Quivy V, Vanhulle C, Lamine A, Vaira D, Demonte D, Martinelli V, Veithen E, Cherrier T, Avettand V, Poutrel S, Piette J, de Launoit Y, Moutschen M, Burny A, Rouzioux C, De Wit S, Herbein G, Rohr O, Collette Y, Lambotte O, Clumeck N, Van Lint C (2009) Synergistic activation of HIV-1 expression by deacetylase inhibitors and prostratin: implications for treatment of latent infection. PLoS One 4:e6093

Rosenberg N, Jolicoeur P (1997) Retroviral pathogenesis. In: Coffin JM, Hughes SH, Varmus HE (eds) Retroviruses. Cold Spring Harbor Laboratory Press, Cold Spring Harbor

Ruddle NH, Armstrong MK, Richards FF (1976) Replication of murine leukemia virus in bone marrow-derived lymphocytes. Proc Natl Acad Sci U S A 73:3714–3718

Saleh S, Solomon A, Wightman F, Xhilaga M, Cameron PU, Lewin SR (2007) CCR7 ligands CCL19 and CCL21 increase permissiveness of resting memory CD4+ T cells to HIV-1 infection: a novel model of HIV-1 latency. Blood 110:4161–4164

Sahu GK, Cloyd MW (2011) Latent HIV in primary T lymphocytes is unresponsive to histone deacetylase inhibitors. Virol J 8:400

Schroder AR, Shinn P, Chen H, Berry C, Ecker JR, Bushman F (2002) HIV-1 integration in the human genome favors active genes and local hotspots. Cell 110:521–529

Schwartz O, Marechal V, Le Gall S, Lemonnier F, Heard JM (1996) Endocytosis of major histocompatibility complex class I molecules is induced by the HIV-1 Nef protein. Nat Med 2:338–342

Scripture-Adams DD, Brooks DG, Korin YD, Zack JA (2002) Interleukin-7 induces expression of latent human immunodeficiency virus type 1 with minimal effects on T-cell phenotype. J Virol 76:13077–13082

Sung TL, Rice AP (2006) Effects of prostratin on Cyclin T1/P-TEFb function and the gene expression profile in primary resting CD4+ T cells. Retrovirology 3:66

Swingler S, Mann AM, Zhou J, Swingler C, Stevenson M (2007) Apoptotic killing of HIV-1-infected macrophages is subverted by the viral envelope glycoprotein. PLoS Pathog 3:1281–1290

Tong-Starksen SE, Luciw PA, Peterlin BM (1987) Human immunodeficiency virus long terminal repeat responds to T-cell activation signals. Proc Natl Acad Sci U S A 84:6845–6849

Treand C, du Chéné I, Bres V, Kiernan R, Benarous R, Benkirane M, Emiliani S (2006) Requirement for SWI/SNF chromatin-remodeling complex in Tat-mediated activation of the HIV-1 promoter. EMBO J 25:1690–1699

Triboulet R, Mari B, Lin YL, Chable-Bessia C, Bennasser Y, Lebrigand K, Cardinaud B, Maurin T, Barbry P, Baillat V, Reynes J, Corbeau P, Jeang KT, Benkirane M (2007) Suppression of microRNA-silencing pathway by HIV-1 during virus replication. Science 315:1579–1582

Tyagi M, Karn J (2007) CBF-1 promotes transcriptional silencing during the establishment of HIV-1 latency. EMBO J 26:4985–4995

Tyagi M, Pearson RJ, Karn J (2010) Establishment of HIV latency in primary CD4+ cells is due to epigenetic transcriptional silencing and P-TEFb restriction. J Virol 84:6425–6437

Van Duyne R, Easley R, Wu W, Berro R, Pedati C, Klase Z, Kehn-Hall K, Flynn EK, Symer DE, Kashanchi F (2008) Lysine methylation of HIV-1 Tat regulates transcriptional activity of the viral LTR. Retrovirology 5:40

Varier RA, Kundu TK (2006) Chromatin modifications (acetylation/ deacetylation/ methylation) as new targets for HIV therapy. Curr Pharm Des 12:1975–1993

Verdin E, Dequiedt F, Kasler HG (2003) Class II histone deacetylases: versatile regulators. Trends Genet 19:286–293

Wang GP, Ciuffi A, Leipzig J, Berry CC, Bushman FD (2007) HIV integration site selection: analysis by massively parallel pyrosequencing reveals association with epigenetic modifications. Genome Res 17:1186–1194

Wang X, Ye L, Hou W, Zhou Y, Wang YJ, Metzger DS, Ho WZ (2009) Cellular microRNA expression correlates with susceptibility of monocytes/macrophages to HIV-1 infection. Blood 113:671–674

Weissman JD, Brown JA, Howcroft TK, Hwang J, Chawla A, Roche PA, Schiltz L, Nakatani Y, Singer DS (1998) HIV-1 tat binds TAFII250 and represses TAFII250-dependent transcription of major histocompatibility class I genes. Proc Natl Acad Sci U S A 95:11601–11606

Weissman JD, Hwang JR, Singer DS (2001) Extensive interactions between HIV TAT and TAF(II)250. Biochim Biophys Acta 1546:156–163

Williams SA, Chen LF, Kwon H, Ruiz-Jarabo CM, Verdin E, Greene WC (2006) NF-kappaB p50 promotes HIV latency through HDAC recruitment and repression of transcriptional initiation. EMBO J 25:139–149

Williams SA, Kwon H, Chen LF, Greene WC (2007) Sustained induction of NF-kappa B is required for efficient expression of latent human immunodeficiency virus type 1. J Virol 81:6043–6056

Wong K, Sharma A, Awasthi S, Matlock EF, Rogers L, Van Lint C, Skiest DJ, Burns DK, Harrod R (2005) HIV-1 Tat interactions with p300 and PCAF transcriptional coactivators inhibit histone acetylation and neurotrophin signaling through CREB. J Biol Chem 280:9390–9399

Xie B, Invernizzi CF, Richard S, Wainberg MA (2007) Arginine methylation of the human immunodeficiency virus type 1 Tat protein by PRMT6 negatively affects Tat Interactions with both cyclin T1 and the Tat transactivation region. J Virol 81:4226–4234

Yang HC, Xing S, Shan L, O'Connell K, Dinoso J, Shen A, Zhou Y, Shrum CK, Han Y, Liu JO, Zhang H, Margolick JB, Siliciano RF (2009) Small-molecule screening using a human primary cell model of HIV latency identifies compounds that reverse latency without cellular activation. J Clin Invest 119:3473–3486

Yeung ML, Bennasser Y, Myers TG, Jiang G, Benkirane M, Jeang KT (2005) Changes in microRNA expression profiles in HIV-1-transfected human cells. Retrovirology 2:81

Ylisastigui L, Archin NM, Lehrman G, Bosch RJ, Margolis DM (2004) Coaxing HIV-1 from resting CD4 T cells: histone deacetylase inhibition allows latent viral expression. AIDS 18: 1101–1108

Zack JA, Arrigo SJ, Weitsman SR, Go AS, Haislip A, Chen IS (1990) HIV-1 entry into quiescent primary lymphocytes: molecular analysis reveals a labile latent viral structure. Cell 61: 213–222

Zhang H (2009) Reversal of HIV-1 latency with anti-microRNA inhibitors. Int J Biochem Cell Biol 41:451–454

Zhou M, Halanski MA, Radonovich MF, Kashanchi F, Peng J, Price DH, Brady JN (2000) Tat modifies the activity of CDK9 to phosphorylate serine 5 of the RNA polymerase II carboxyl-terminal domain during human immunodeficiency virus type 1 transcription. Mol Cell Biol 20:5077–5086

Chapter 22
Epigenetics in Parkinson's and Alzheimer's Diseases

Sueli Marques and Tiago Fleming Outeiro

Abstract Neurodegenerative disorders, such as Parkinson's and Alzheimer's disease, are highly complex, due to their multifactorial origin, not only depending on genetic but also on environmental factors. Several genetic risk factors have already been associated with both the diseases, however, the precise way through which the environment contributes to neurodegeneration is still unclear.

Recently, epigenetic mechanisms, such as DNA methylation, chromatin remodeling or miRNAs, which may induce alterations in genes expression, have started to be implicated in both AD and PD. Epigenetic modulation is present since pre-natal stages and throughout lifetime, and depends on lifestyle conditions and environmental exposures, and consequently could represent the missing link between risk factors and the development of sporadic disorders. This chapter will discusses the role of epigenetics in AD and PD.

S. Marques
Cell and Molecular Neuroscience Unit, Instituto de Medicina Molecular,
Av. Prof. Egas Moniz, 1649-028 Lisboa, Portugal
e-mail: touteiro@gmail.com

T.F. Outeiro (✉)
Cell and Molecular Neuroscience Unit, Instituto de Medicina Molecular,
Av. Prof. Egas Moniz, 1649-028 Lisboa, Portugal

Instituto de Fisiologia, Faculdade de Medicina de Lisboa,
Av. Prof. Egas Moniz, 1649-028 Lisboa, Portugal

Department of Neurodegeneration and Restorative Research,
University Medizin Göttingen, Waldweg 33, 37073 Göttingen, Germany
e-mail: touteiro@gmail.com

22.1 Introduction

Neurodegenerative disorders, such as Alzheimer's disease (AD) and Parkinson's disease (PD) affect a growing number of people due to the increase in life expectancy. The world health organization estimates that by 2,025 three-quarters of people over 60 years of age will be living in developing countries (WHO 2002). Although these pathologies have been known for quite some time and some symptomatic therapies exist, it is still not possible to cure or prevent these disorders. Thus, unless novel therapeutic strategies are developed, the economical and social impact will be enormous. The majority of AD and PD cases is sporadic and is thought to arise from the combination of environmental factors and susceptibility genes in ways that are not fully understood. Thus, these disorders are considered highly complex in similarity to cancer, diabetes, cardiovascular and neuropsychiatric disorders. The complexity of these disorders makes the discovery of therapeutics difficult. Thus, it is essential that we understand how the environment can affect the organisms and result in the development of the disease.

The epigenome comprises the heritable but potentially reversible changes in gene expression that occur in the absence of changes to the DNA sequence itself. These changes are brought about by modifications of chromatin, such as acetylation, methylation, phosphorylation, or ubiquitylation of histones, DNA methylation, and microRNA (miRNA) (Dolinoy and Jirtle 2008; Fig. 22.1).

Environmental exposure to nutritional, chemical, physical, as well as intellectual or social factors can alter gene expression, and affect adult phenotype by changes in epigenetic modifications at labile genomic regions (Kovalchuk 2008).

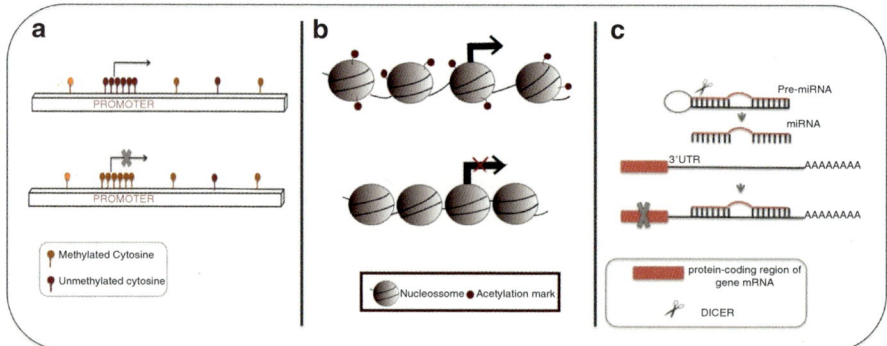

Fig. 22.1 Epigenetic regulation of gene expression. (**a**) DNA methylation. The addition of methyl groups in CpG dinucleotides, that when clustered in CpG islands around the promoters, represses transcription. (**b**) Histone acetylation. One type of histone modifications, which promotes the opening of chromatin around the genes, enables the access of transcriptional machinery. (**c**) miRNA is cut out of a precursor *hairpin-shaped* pre-miRNA to form a mature miRNA, which binds to the 3′ untranslated region (3′ UTR) of a target gene messenger RNA and turns off its activity (*a and b adapted from Marques et al. 2010; c based on an illustration from Victor Ramblos*)

Although it is plausible that epigenetic regulation plays a role in gene expression in AD and PD, the focus on epigenetic effects in these disorders is only now beginning to grow and, to date, very few definitive clues have been discovered.

22.2 Parkinson's Disease

22.2.1 Epidemiology

PD is the second most common neurodegenerative disorder after AD and has an average age of onset of 60 years. According to the Parkinson's Disease Foundation, it currently affects more than four million people worldwide. The prevalence of PD in industrialized countries is generally estimated at about 1–2% of people over 60 years of age. This figure increases to 3–5% in people above 85 years old (Lau and Breteler 2006).

22.2.2 Pathology

PD is characterized by resting tremor, slow and decreased movement (bradykinesia), muscular rigidity, and postural instability. Pathologically, PD patients display loss of dopaminergic neurons in the substantia nigra (SN) pars compacta and frequently have Lewy bodies, which are eosinophilic intracellular inclusions, composed of amyloid-like fibers primarily made up α-synuclein (α-syn) (Weintraub et al. 2008).

The Braak staging hypothesis for PD posits that pathology evolves in 6 stages. In stages 1–2, a pre-motor period, the typical pathological changes, Lewy neurites and Lewy bodies, spread from the olfactory bulb and vagus nerve to lower brainstem regions. The symptomatic period correlates with pathological changes involving the midbrain, including the SN, in stage 3, the mesocortex, in stage 4, and the neocortex in stages 5–6 (Braak et al. 2003).

Other groups have confirmed this pattern of progression of the pathology for the most part. However, some critical evaluation argues that some questions still need to be addressed and additional studies are required (Burke et al. 2008).

22.3 Genetic Linkages

The majority of PD cases are sporadic. However, 5–10% of cases have a defined genetic component, with both recessive and dominant modes of inheritance. Several genes and chromosomal loci, linked to familial forms of parkinsonism and designated as PARK1 to 16, are associated with autosomal dominant, recessive, and

X-linked forms of the disease. There are also loci for which the mode of transmission is still not fully understood due to the limited number of cases identified (Hardy et al. 2009).

The SNCA gene, encoding for α-syn, was the first gene to be associated with familial PD. Three point mutations in the SNCA gene, causing A30P, E46K, or A53T amino acid substitutions in the protein, are known to be linked with autosomal dominant PD. Another mutational mechanism in this gene involves duplication or triplication of the wild-type SNCA gene locus (Klein and Schlossmacher 2007).

Another gene that is responsible for an autosomal dominant form of the disease is the leucine-rich repeat kinase 2 (LRRK2) gene. Some mutations are surprisingly common in certain geographical regions; R1441G for Basque cases, G2019S in Europeans and G2385R and R1628P in eastern Asian people (Farrer 2006). Mutations in both of the dominant genes are associated with the presence of Lewy bodies, the pathognomonic protein inclusions in PD.

Three recessive forms of PD have been identified with mutations in the genes that encode parkin, DJ-1, and PTEN-induced kinase 1 (PINK1). These forms of the disease tend to result in a more selective loss of dopaminergic neurons than that associated with sporadic, late-onset PD. Other genes have been implicated in familial PD, however no pathological identification has been made yet. This is the case for mutations associated with FBXO7, encoding F-box only protein 7 or ATP13A2, encoding probable cation-transporting ATPase 13A2 (Wood-Kaczmar et al. 2006).

22.4 Sporadic Risk Factors

Idiopathic PD, which accounts for about 90–95% of the cases, usually refers to a syndrome characterized by late-onset parkinsonism. The cause of sporadic PD is unknown, but it is considered to result from a combination of environmental and genetic factors. The environmental toxin hypothesis was dominant for much of the twentieth century, and posits that PD-related neurodegeneration results from exposure to a dopaminergic neurotoxin. Living in a rural environment appears to confer an increased risk of PD, perhaps due to increased exposure to pesticide use and wood preservatives. Cigarette smoking and coffee drinking are inversely associated with the risk for development of PD, reinforcing the concept that some environmental factors can modify PD susceptibility (Dauer and Przedborski 2003).

Other factors, such as increased animal fat intake, head trauma, and tobacco smoking have been investigated in relation to PD but no clear correlation has been established yet. Increasing evidence associates the first two factors with an increased risk of PD. In contrast, a negative relationship between tobacco smoking and the development of PD has been observed (Khandhar and Marks 2007).

22.5 The Contribution of Epigenetics

In PD, a direct relation between epigenetics and neurodegeneration has not yet been clearly established, except for miRNA regulation. However, several indirect links between methylation regulators or histone deacetylases inhibitors (HDACi) and the disease mediators have been described.

22.5.1 DNA Methylation

Evidence for DNA methylation role in PD is mainly based on homocysteine (Hcy) cycle dysregulation.

The methyl groups for all biological methylation reactions derive from dietary methyl donors and from cofactors carrying 1-carbon units. In the metabolic cycling of methionine, this compound is converted to the methyl cofactor S-adenosylmethionine (SAM). Subsequent to methyl donation, the product S-adenosylhomocysteine (SAH) becomes Hcy, which is then either catabolized or remethylated to methionine (Finkelstein 2000; Fig. 22.2).

Variability in levels of Hcy among individuals can result from genetic or environmental factors, with dietary folate levels having a major impact such that there is generally an inverse relationship between plasma folate and Hcy levels (Giles et al. 1995).

Increased concentrations of plasma total Hcy have been reported in patients with PD, which will increase SAH and decrease SAM, leading to an overall decrease in methylation potential (SAM/SAH ratio) (Blandini et al. 2001).

Better cognitive function scores and increased SAM/SAH ratio, has been recently described in patients with PD, thus suggesting a role for methylation in the disorder (Obeid et al. 2009).

A mouse model of PD with a dietary folate deficiency exhibited elevated levels of plasma Hcy and was sensitized to 1-methyl-4-phenyl-1,2,3,6-tetrahydropyridine (MPTP) -induced PD-like pathology and motor dysfunction (Duan et al. 2002).

In addition to the Hcy cycle, evidence is now emerging linking DNA methylation levels and genes related to PD, although DNA methylation dysregulation in the disease process was not yet characterized.

A recent study showed that peripheral leukocytes of Japanese PD patients bear fewer short telomeres with constant subtelomeric methylation status in comparison with the healthy controls. Subtelomeric hypomethylation is associated with increased accessibility of DNA-binding proteins for suppression of "telomeric position effect", that is, a mechanism silencing genes neighboring a telomere. The telomeric and subtelomeric regions impaired by oxidative stress progress to become hypomethylated and are open to easy access by oxygen radicals (Maeda et al. 2009).

The PARK2 gene, identified as a mutated target in patients with autosomal recessive juvenile parkinsonism (ARJP), and which has also been accepted as a candidate

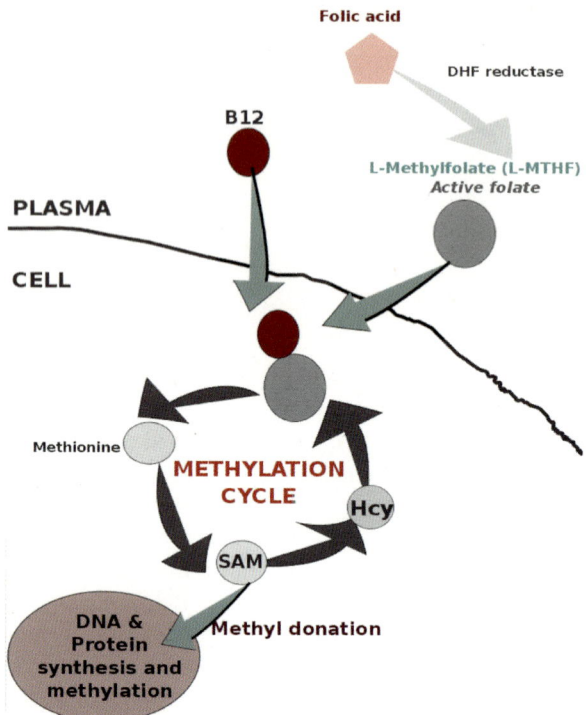

Fig. 22.2 Metabolism of methylation reactions. Folate and vitamin B12 are provided by the diet and are essential for the methylation cycle. Folic acid must be converted into L-methylfolate (L-MTHF) by Dihydrofolate (DHF) reductase in order to be absorbed by the intestinal mucosa. Upon recruitment, L-MTHF associates with Hcy and methionine synthase to produce methionine. This compound is then converted into S-adenosylmethionine (SAM). Through the action of methyltransferases, Hcy is generated and re-enters the cycle or is used to generate methyl groups required for methylation reactions on DNA CpGs or histone lysines/arginines

tumor suppressor gene in several types of cancer, was described to be epigenetically regulated in human leukemia. This could suggest that, similarly to its role on the development of this hematological neoplasm, abnormal methylation and regulation of PARK2 could also play a role in the pathogenesis of PD (Agirre et al. 2006).

Likewise, the SNCA gene is affected by DNA methylation in other pathologies, paving the way to the same dysregulation in PD. An increase in DNA methylation in the promoter of the SNCA gene was described in patients with alcoholism, and was also found to be significantly associated with elevated homocysteine levels and a global DNA hypomethylation and DNA hypermethylation in females with anorexia nervosa (Bönsch et al. 2005; Frieling et al. 2007).

Indeed, the effect of SNCA methylation in PD has recently described. A region of the SNCA CpG island in which the methylation status is altered along with increased SNCA expression was identified in cultured cells. Postmortem brain analysis also revealed regional non-specific methylation differences in this CpG region in the anterior cingulate and putamen among controls and PD. However, in the substantia nigra of PD, methylation was significantly decreased (Matsumoto et al. 2010). Another group reported a decreased methylation of SNCA intron 1 in the SN, putamen, and cortex of sporadic PD patients, pointing again towards epigenetic regulation of SNCA expression in PD (Jowaed et al. 2010). Subsequently, a reduction of nuclear DNA methytransferase 1 (Dnmt1) levels in human postmortem brain samples from PD and DLB patients, as well as in the brains of α-syn, transgenic mice models was demonstrated. The cytoplasmic sequestration of Dnmt1 resulted in global DNA hypomethylation in human and mouse brains, involving CpG islands upstream of SNCA as well as other genes (Desplats et al. 2011).

Epigenetic regulation of SNCA has also been reported in a familial case of PD, SNCA showed monoallelic expression in lymphoblastoid cell line and in the blood cells of a patient heterozygous for p.Ala53Thr and the epigenetic silencing of the mutated allele involved histone modifications but not DNA methylation. The steady-state mRNA levels deriving from the normal SNCA allele in this patient exceeded those of the two normal SNCA alleles combined, in matching, control individuals (Voutsinas et al. 2010).

22.5.2 Histone Modifications

Several evidences have now proven that chromatin remodeling has a central role in some of the long-lasting effects of dopamine on brain function (Brami-Cherrier et al. 2005). Consequently, the interest in the study of histone modifications and their effect on PD-related toxicity has been increasing.

Starting with one of the known risk factors for PD, pesticide exposure, it was observed that dieldrin induced in mesencephalic dopaminergic neuronal cells a time-dependent increase in the acetylation of core histones H3 and H4 indicating that acetylation is an early event in dieldrin neurotoxicity. The hyperacetylation was attributed to dieldrin-induced proteasomal dysfunction, resulting in the accumulation of a key histone acetyltransferase (HAT), the cAMP response element-binding protein -binding protein (CREBBP or CBP). In mouse models, exposure to dieldrin induced histone hyperacetylation in the striatum and SN. An HAT inhibitor, anacardic acid, significantly attenuated dieldrin-induced histone acetylation as well as the apoptotic cascade that includes caspase-3 activation, Protein Kinase Cδ proteolytic activation and DNA-fragmentation, having a neuroprotective effect against dieldrin-induced nigral dopaminergic neuronal degeneration in primary mesencephalic culture models, which was independent of its antioxidant effect (Song et al. 2010).

Parkin, encoded by PARK2, is a ubiquitin E3 ligase which accumulates at the centrosome in a microtubule-dependent manner in response to proteasome inhibition. The centrosome recruitment of parkin is mediated by its direct binding to histone deacetylase 6 (HDAC6), through multiple interaction domains (Jiang et al. 2008).

The most widely used treatment in PD, Levodopa, (L-DOPA) is associated with motor complications, such as motor fluctuations and dyskinesias (Fabbrini et al. 2007). Dopamine depletion and subsequent L-DOPA treatment has also been associated with profound alterations in posttranslational modifications of striatal histones. In animal models of L-DOPA-induced dyskinesia (LID), dopamine depletion was associated with a reduction in histone H3 trimethylation at Lys 4, while chronic L-DOPA therapy leading to hyperkinesia was marked deacetylation of striatal histone H4 at Lys 5/8/12/16. This study highlighted the presence of histone modifications and supported the hypothesis that chromatin remodeling could contribute to the development and maintenance of LID in PD (Nicholas et al. 2008).

Another study regarding L-DOPA-induced dyskinesias, focused on the dopamine D1 receptor, which has been shown to be critical for the development of this side effect in mice. H3 phosphoacetylation was blocked by D1 receptor inactivation, suggesting that inhibitors of H3 acetylation and/or phosphorylation could be useful in preventing or reversing dyskinesia (Darmopil et al. 2009).

Several studies have focused on the protective effect of HDACi in PD pathology.

An important study in laboratory models showed that the neurotoxicity of α-syn in the nucleus could be rescued by the administration of HDACi in both cell culture and transgenic flies. α-syn could bind directly to histones, reducing the level of acetylated histone H3 in cultured cells and inhibiting acetylation in HAT assays (Kontopoulos et al. 2006). Valproic acid (VPA) administered daily in the diet of an animal model of PD, in a dose that was described to lead to a significant inhibition of HDAC activity and to an increase of histone H3 acetylation in brain tissues, resulted in the prevention of the decrease of the dopaminergic marker tyrosine hydroxylase (TH) in SN and striatum. It also significantly counteracted the death of nigral neurons and the 50% drop of striatal dopamine levels caused by rotenone administration. VPA treatment could also counteract alterations observed in α-syn, being the native form decreased in SN and striatum of rotenone-treated rats, while monoubiquitinated one increased in the same regions (Monti et al. 2010). Chen et al. (2006) showed that VPA pretreatment could upregulate the expression of neurotrophic factors and protect midbrain DA neurons from LPS or 1-methyl-4-phenylpyridinium (MPP+)-induced neurotoxicity. The same authors, posteriorly observed that sodium butyrate (SB) and trichostatin A (TSA), mimicked the survival-promoting and protective effects of VPA on DA neurons in neuron-glia cultures. Similarly to VPA, they increased Glial-derived and brain-derived neurotrophic factors (GDNF and BDNF, respectively) transcripts in astrocytes in a time-dependent manner with a simultaneous increment in promoter-associated histone H3 acetylation, being the time-course for acetylation similar to that for gene transcription (Wu et al. 2008).

TSA was also tested in several dopaminergic neuronal cell lines and a single TSA treatment resulted in decreased cell survival and increased apoptosis in dopaminergic neuronal cells. Pre-treatment with TSA resulted in exacerbated neurotoxic damage to dopaminergic neurons induced by MPP+ and rotenone suggesting that HDACi could influence PD pathogenesis by inhibiting survival and increasing vulnerability of dopaminergic neurons to neurotoxins (Wang et al. 2009).

Interestingly, pharmacological inhibition of Sirtuin 2 (SIRT2), a member of the HDAC family, rescues α-syn-induced toxicity and modifies inclusion morphology in cellular and animal models of PD (Outeiro et al. 2007).

22.5.3 miRNAs

Several microRNAs (miRNAs) and their correspondent targets have been identified in PD, providing new insights about the possibility of using RNA interference as a therapeutical approach for the disease.

A study based on familial forms of PD, studying around 800 families, showed a strong association between rs12720208 in the 3′untranslated region (UTR) of FGF20 and the risk of developing PD. This single nucleotide polymorphism (SNP) disrupted a binding site for microRNA-433 (miR-433), increasing translation of FGF20 in vitro and in vivo. In a cell-based system and in PD brains, this increase in translation of FGF20 was correlated with increased α-syn expression. Overexpression of miR-433b in primary midbrain cultures also prevented dopaminergic differentiation whereas its inhibition resulted in increased TH positive cells (Wang et al. 2008). miRNA expression of midbrain from PD patients was determined, also revealing changes in the expression of miR-133b (de Mena et al. 2010).

In the α-syn(A30P) transgenic mice, a decline in miR-132 levels was observed (Gillardon et al. 2008). This miRNA was previously shown to increase following growth factor administration or neuronal depolarization in vitro (Vo et al. 2005). This decrease in PD could thus represent a molecular signature for a concomitant decrease in neurotrophic and/or neuronal activity in the affected brainstem.

microRNA-7 (miR-7), which is expressed mainly in neurons, represses α-synuclein protein levels through the 3′ UTR of α-syn mRNA. miR-7-induced down-regulation of α-syn could protect cells against oxidative stress. Further, in the MPTP-induced model of PD, miR-7 expression decreased, possibly contributing to increased α-syn expression (Junn et al. 2009). miR-7 inhibits cellular susceptibility of neuroblastoma cells to oxidative stress induced by a mutant form of SNCA, providing evidence that miRNAs protect neuronal cells against cellular stress. The presence of miR-7 in the SN is also verified thus supporting a physiological role in dopaminergic neurons (Bak et al. 2008).

As stated above, miRNAs could be explored as therapeutic targets for PD, however for specific targeting of miRNAs, it will be necessary to address the physiological relevance of these molecules.

22.6 Alzheimer's Disease

22.6.1 Epidemiology

Currently, it is estimated that one person in eight (13%) above 65 years old is diagnosed with AD. The prevalence increases to approximately 40% after the age of 85. The number of AD patients is estimated to reach 7.7 million in 2030, constituting a greater than 50% increase from the 5.1 million aged 65 and older who are currently affected. By 2050, this number is projected to reach between 11 million and 16 million if the current state of medical therapeutics is maintained (Alzheimer's Association 2009).

22.6.2 Pathology

The major clinical hallmarks of AD are progressive impairment in memory, judgment, decision-making, orientation, and language. Diagnosis is based on neurologic examination and imaging, along with the exclusion of other causes of dementia. However a definitive diagnosis can be made only at autopsy to detect the histopathological features of the disease, which include senile plaques, composed by insoluble amyloid-β (Aβ) peptide that accumulates extracellularly, intracellular neurofibrillary tangles composed of hyperphosphorylated tau protein and selective neuronal loss. These plaques and tangles are found predominantly in the frontal and temporal lobes, including the hippocampus. In more advanced cases, the pathology extends to other regions of cortex, including the parietal and occipital lobes (Minati et al. 2009).

22.7 Genetic Forms of AD

Early onset familial AD accounts for less than 1% of all AD cases (Lambert and Amouyel 2007). The genes implicated in these forms of the disease are the genes encoding for amyloid precursor protein (APP), located on chromosome 21q21, the gene encoding for presenilin 1 (PSEN1), located on chromosome 14q24.3, and that encoding for presenilin 2 (PSEN2), on chromosome 1q31–q42 (Ertekin-Taner 2007).

APP mutations account for <0.1% of AD patients (13). Mutations in APP are located near the cleavage sites of the protein resulting in increased production of the Aβ peptide. The average age of onset for this mutation is between mid 40s and 50s but can be modified by the apolipoprotein E (APOE) genotype.

Missense mutations within the PSEN1 gene account for 18–50% of the early-onset autosomal dominant forms of AD and lead to a particularly aggressive form of the disease with an age of onset between 30 and 50 years, which is not influenced by the APOE genotype. The majority of PSEN mutations are single-nucleotide substitutions, but small deletions and insertions have been described as well. All mutations within PSEN1 increase production of the Aβ42, the most toxic form of the peptide. Mutations within PSEN2 have a variable age of onset (40–80 years), appear not to be influenced by APOE and result in increased Aβ production as well (Ray et al. 1998; Bettens et al. 2010).

22.8 Sporadic Risk Factors

The majority of AD cases have complex etiology due to both environmental and genetic factors, which alone do not seem sufficient for causing disease.

The APOE gene, on chromosome 19q13, and its variants, is recognized as a major risk factor in AD, modify the risk of Late Onset AD (LOAD).

The APOE gene contains three common polymorphisms – ε2 (cysteines at codon 112 and codon 158), ε3 (cysteines at codon 112), and ε4 (arginine at codon 112). APOE ε3 is the most frequent form (78%), APOE ε4 makes up 15% and APOE ε2 approximately 7% (Tanzi and Bertram 2001). Analysis of these polymorphisms in normal control populations and in patients with AD has shown that ε4 allele frequency in AD is approximately 40%, compared to 15% in normal and the frequency of the ε2 allele decreased from 10% to about 2% in AD. A dose-dependent relationship between the number of copies of ε4, and the age-of-onset of AD such that ε4/ε4 subjects have an earlier onset than do heterozygous ε4 subjects. Subjects with an ε2 allele, on the other hand, have a later onset (St George-Hyslop and Petit 2004).

Nevertheless, only less than 50% of non-familial AD cases are carriers of the APOE ε4 allele. Therefore, APOE ε4 is not a deterministic factor for the development of the disease and other genes must confer susceptibility to AD.

Some candidate susceptibility genes are the ones encoding α2-macroglobulin (α2M), low-density lipoprotein receptor-related protein (LRP), angiotensin converting enzyme (ACE), insulin degrading enzyme (IDE) and others (Rocchi et al. 2003).

In addition to these genetic polymorphisms that increase the risk for developing AD, multiple environmental factors may influence the onset and progression of the disease. The main risk-increasing factors examined in epidemiological studies are associated with life history. In addition to age, which is by far the major risk factor for AD, others have now been extensively correlated, such as hypertension (as a long-term stress of the blood vessel endothelium and walls), diabetes (by vascular changes or insulin deregulation itself), inflammation, obesity, or head injury (Stozicka et al. 2007).

22.9 Epigenetics in AD

In AD, several works are now arising demonstrating direct epigenetic modulation of the disease. However, except for miRNA dysregulation, which has now been extensively described, a lot of research still needs to be done to understand which modifications and in which genes, epigenetic modulation occurs.

22.9.1 DNA Methylation

Some cytosines in the promoter region of the APP gene are frequently methylated in cases ≤70 years old and significantly demethylated in cases >70 years old. These age-related modifications on DNA methylation alter APP expression and consequently can affect the progressive Aβ deposition with aging in the brain (Tohgi et al. 1999).

Several studies are now demonstrating a correlation between dietary factors, the epigenome and AD pathology. The hypothesis was that nutritional deficits could lead to hyperhomocysteinemia (HCY/SAM cycle alteration, which is involved in the transfer of methyl groups) with the consequent decrease of SAM levels. Methyl donor decrease, in turn, could induce demethylation of DNA and this will result in activation and overexpression of genes involved in AD pathology. Indeed, folate, vitamin B12, and SAM are frequently reduced in the elderly. One of the first studies regarding this issue showed that the reduction of folate and vitamin B12 in culture medium could cause a reduction of SAM levels, and an increase in Aβ production (Fuso et al. 2005). The same authors also demonstrated that Vitamin B deprivation could induce hyperhomocysteinemia and an imbalance of SAM and SAH, in association with PSEN1 and BACE up-regulation and Aβ deposition (Fuso et al. 2008).

Vitamin B deficiency also induced hypomethylation of specific CpG moieties in the 5′-flanking region of PSEN1 gene, and this was reverted with SAM supplementation (Fuso et al. 2011a). In line with the changes observed for PSEN1 methylation patterns, DNA methylases (DNMT1, 3a and 3b) and a putative demethylase (MBD2) were differently modulated in the same experimental conditions (Fuso et al. 2011b).

This relation was further explored by developing a triple transgenic/mutant mouse model (APP*/PS1*/CBS*) showing both amyloid deposition and high serum levels of Hcy resultant of deficient CBS (Cystathionine-Beta-Synthase) activity. The study showed that female APP*/PS1*/CBS* mice exhibited significant elevations of Aβ40 and Aβ42 levels in the brain compared with APP*/PS1* double transgenic mice showing that hyperhomocysteinemia can be a risk factor for AD (Pacheco-Quinto et al. 2006).

Another connection with this nutritional regulation of epigenetic alterations described a downregulation of the neuronal PP2A methyltransferase (PPMT), along with a decrease in PP2A methylation in affected brain regions from AD patient (Sontag et al. 2004). Taking this into account, it was recently observed that hyperhomocysteinemia induced in mice by feeding a high-methionine, low folate diet was associated with increased brain SAH levels, with a reduced PP2A methylation levels, and with tau and APP phosphorylation. These results supported the hypothesis that impaired Hcy metabolism and deregulation of critical methylation reactions can trigger the accumulation of phosphorylated tau and APP in the brain, a process that may favor neurofibrillary tangle formation and amyloidogenesis (Sontag et al. 2007).

Dietary and environmental factors, which are present from as early as the prenatal phase, have a profound impact on our epigenome and may affect diseases developed later in life. Expression of AD-related genes (APP, BACE1) was found to be elevated in aged (23-year-old) monkeys exposed to lead (Pb) as infants, along with a decrease in DNA methyltransferase activity and higher levels of oxidative damage to DNA. These data suggest that AD pathogenesis is influenced by early life exposures and argue for both an environmental trigger and a developmental origin of AD, being the intermediate DNA methylation a form of epigenetic imprinting (Wu et al. 2008).

The age-related epigenetic modifications in AD were further explored in two studies. The first one presented a straight evidence of epigenetic involvement in AD pathogenesis by showing an age-specific epigenetic drift in late-onset AD. PSEN1 and APOE, which participate in Aβ processing, methylenetetrahydrofolate reductase (MTHFR) and DNMT1, which are responsible for methylation homeostasis, presented a significant inter-individual epigenetic variability in the brain and lymphocytes of these patients, which could contribute to late-onset AD predisposition (Wang et al. 2008). Longevity related genes were investigated with respect to promoter methylation in peripheral blood in relation to gender, age and AD. Only one of the genes, HTERT, was shown to be hypermethylated in AD compared to aged normal people and is, by opposite to the normal effect of methylation on gene expression, activated by this epigenetic modification. Consequently, these results indicated a higher telomerase activity probably due to telomere and immune dysfunctions involved in AD pathogenesis (Silva et al. 2008).

22.9.2 Histone Modifications

In contrast to DNA methylation studies, very few studies were aimed at characterizing chromatin modifications in AD. A recent study showed that acute treatment of a mouse model of AD with a HDACi had an effect on memory impairment (Francis et al. 2009).

Chronic HDACi (VPA, SB and vorinostat) injections (2–3 weeks) in APPswe/PS1dE9 mice completely restored contextual memory. The newly consolidated memories were stably maintained over a 2-week period. All HDACi affected class I HDACs (HDAC1, 2, 3, 8) with little effect on the class IIa HDAC family members (HDAC4, 5, 7, 9) (Kilgore et al. 2010).

In another study, systemic administration of the 4-phenylbutyrate (PBA) reversed spatial learning and memory deficits in the Tg2576 mouse without altering Aβ burden. However, the phosphorylated form of tau was decreased in the mouse brain after 4-PBA treatment, along with an increase in the inactive form of the glycogen synthase kinase 3beta (GSK3beta) (Ricobaraza et al. 2009). The same authors, then showed that 4-PBA administration reinstated fear learning in the same animal model of AD, independently of the disease stage: both in 6-month-old Tg2576 mice, at the onset of the first symptoms, but also in aged, 12–16-month-old mice, when amyloid plaque deposition and major synaptic loss has occurred. Reversal of learning deficits was associated with a clearance of intraneuronal Aβ accumulation, and alleviation of endoplasmic reticulum (ER) stress. The expression of plasticity-related proteins were also significantly increased by PBA (Ricobaraza et al. 2010).

Another type of chromatin modifiers, HAT such as CBP, have also been implicated in AD pathology. In primary neurons, CBP is specifically targeted by caspases and calpains at the onset of neuronal apoptosis, and CBP was further identified as a new caspase-6 substrate. This ultimately impinged on the CBP/p300 HAT activity that decreased with time during apoptosis entry, whereas total cellular HAT activity remained unchanged. Consequently, histone acetylation levels decreased at the onset of apoptosis. Interestingly, CBP loss and histone deacetylation were observed in two different pathological contexts: APP-dependent signaling and amyotrophic lateral sclerosis model mice, indicating that these modifications are likely to contribute to neurodegenerative diseases (Rouaux et al. 2003).

Aβ accumulation, which plays a primary role in the cognitive deficits of AD, interferes with CREB activity. Restoring CREB function via brain viral delivery of the CBP improved learning and memory deficits in the triple transgenic model of AD. These occur without changes in Aβ and tau pathology, and were linked to an increased level of BDNF (Caccamo et al. 2010).

Despite this link between AD and histone deacetylation, there is no evidence that the expression of AD-related genes is affected by chromatin modulation in AD. To close this gap, it will be important to undertake a more global study looking at the epigenetic control of the expression of AD-related genes by histone modifications in AD samples and in cell and animal models. In this way, we would develop a better understanding of the role of chromatin remodeling in AD pathogenesis.

22.9.3   miRNAs

miRNAs are, by far, the most studied epigenetic modification in AD. The expression of several blood mononuclear cells (BMC) miRNAs was found to increase in AD relative to normal elderly controls levels, and could differ between AD subjects bearing one or two APOE4 alleles. miRNAs significantly upregulated in AD subjects were miR-34a and 181b (Schipper et al. 2007).

miRNAs belonging to the miR-20a family (that is, miR-20a, miR-17-5p and miR-106b) can regulate APP expression in vitro and at the endogenous level in neuronal cell lines. A tight correlation between these miRNAs and APP was found during brain development and in differentiating neurons. This was further corroborated by the observation that a statistically significant decrease in miR-106b expression was found in sporadic AD patients (Hébert and De Strooper 2009).

miR-107 levels decreased significantly even in AD patients with the earliest stages of pathology. Computational analysis predicted that the 3'-untranslated region (UTR) of BACE1 mRNA is targeted multiply by miR-107. BACE1 mRNA levels tended to increase as miR-107 levels decreased in the progression of AD. Cell culture reporter assays performed with a subset of the predicted miR-107 binding sites indicate the presence of at least one physiological miR-107 miRNA recognition sequence in the 3'-UTR of BACE1 mRNA Wang et al. (2008).

miR-29a, -29b-1, and −9 can regulate BACE1 expression in vitro. The miR-29a/b-1 cluster was significantly decreased in AD patients displaying abnormally high BACE1 protein. Similar correlations were found during brain development and in primary neuronal cultures. The same authors provided evidence for a causal relationship between miR-29a/b-1 expression and Aβ generation in a cell culture model (Hébert and De Strooper 2009).

Another study predicted that miR-298 and miR-328 recognize a specific binding site in the 3′-UTR of BACE1 mRNA and exert regulatory effects on BACE1 protein expression in cultured neuronal cells (Boissonneault et al. 2009).

In APPswe/PS_E9 mice and age-matched controls, the expression of miR-34a is inversely correlated with the protein level of bcl2 and its expression directly inhibited Bcl2 translation in SH-SY5Y cells. Higher levels of active caspase-3 were observed in a stable transfectant cell line of miR-34a or in the APPswe/Ps mice compared to controls. Consistently, miR- 34a knockdown increased the level of bcl2 protein in SH-SY5Y cells, which was accompanied by a decrease of active caspase-3. These findings suggested the abnormal expression of miR-34a might contribute to the pathogenesis of AD, at least in part by affecting the expression of bcl2 (Wang et al. 2009).

AD brains presented a specific up-regulation of an NF-kB-sensitive miRNA-146a highly complementary to the 3′UTR of complement factor H (CFH), an important repressor of the inflammatory response of the brain. Up-regulation of miRNA-146a coupled to down-regulation of CFH was observed in AD brain and in interleukin-1β, Aβ42, and/or oxidatively stressed human neural (HN) cells in primary culture (Lukiw et al. 2008).

22.10 Conclusions

In complex disorders, such as AD and PD, which have a multifactorial origin, the understanding of the causes or risk factors underlying the sporadic cases of these disorders remains difficult. Epigenetics may represent the missing link in the interplay between genes and environment, providing possible clues to understand the etiology of these complex diseases. Despite the increasing number of evidences about the involvement of epigenetic mechanisms in the pathology of AD and PD, these are still early days in understanding how epigenetic modulation occurs during the neurodegenerative process and the impact of these changes in the pathological hallmarks and general deregulation occurring in AD and PD. If these changes in the epigenome continue to be unraveled, it may be possible, in the near future, to intervene therapeutically by modulating the epigenetic changes associated with these disorders.

Acknowledgements TFO is supported by a Marie Curie International Reintegration Grant (Neurofold) and an EMBO Installation Grant. SM is supported by a fellowship from Fundação para Ciência e Tecnologia (SFRH/BD/33188/2007).

References

Agirre X, Román-Gómez J, Vázquez I, Jiménez-Velasco A, Garate L, Montiel-Duarte C, Artieda P, Cordeu L, Lahortiga I, Calasanz MJ, Heiniger A, Torres A, Minna JD, Prósper F (2006) Abnormal methylation of the common PARK2 and PACRG promoter is associated with down-regulation of gene expression in acute lymphoblastic leukemia and chronic myeloid leukemia. Int J Cancer 118(8):1945–1953

Association A (2009) Alzheimer's disease facts and figures. Alzheimers Dement 5:234–270

Bak M, Silahtaroglu A, Møller M, Christensen M, Rath MF, Skryabin B, Tommerup N, Kauppinen S (2008) MicroRNA expression in the adult mouse central nervous system. RNA 14(3):432–444

Bettens K, Sleegers K, Broeckhoven CV (2010) Current status on Alzheimer disease molecular genetics: from past, to present, to future. Hum Mol Genet 19(1):R4–R11

Blandini F, Fancellu R, Martignoni E, Mangiagalli A, Pacchetti C, Samuele A, Nappi G (2001) Plasma homocysteine and l-dopa metabolism in patients with Parkinson disease. Clin Chem 47(6):1102–1104

Boissonneault V, Plante I, Rivest S, Provost P (2009) MicroRNA-298 and microRNA-328 regulate expression of mouse beta-amyloid precursor protein-converting enzyme 1. J Biol Chem 284(4):1971–1981

Bönsch D, Lenz B, Kornhuber J, Bleich S (2005) DNA hypermethylation of the alpha synuclein promoter in patients with alcoholism. Neuroreport 16(2):167–170

Braak H, Del Tredici K, Rüb U, de Vos RA, Jansen Steur EN, Braak E (2003) Staging of brain pathology related to sporadic Parkinson's disease. Neurob Aging 24:197–211

Brami-Cherrier K, Valjent E, Hervé D, Darragh J, Corvol JC, Pages C, Arthur SJ, Girault JA, Caboche J (2005) Parsing molecular and behavioral effects of cocaine in mitogen- and stress-activated protein kinase-1-deficient mice. J Neurosci 25(49):11444–11454

Burke RE, Dauer, Vonsattel JPG (2008) A critical evaluation of the Braak staging scheme for Parkinson's disease. Ann Neurol 64:485–491

Caccamo A, Maldonado MA, Bokov AF, Majumder S, Oddo S (2010) CBP gene transfer increases BDNF levels and ameliorates learning and memory deficits in a mouse model of Alzheimer's disease. Proc Natl Acad Sci U S A 107(52):22687–22692

Chen PS, Peng GS, Li G, Yang S, Wu X, Wang CC, Wilson B, Lu RB, Gean PW, Chuang DM, Hong JS (2006) Valproate protects dopaminergic neurons in midbrain neuron/glia cultures by stimulating the release of neurotrophic factors from astrocytes. Mol Psychiatry 11(12):1116–1125

Darmopil S, Martín AB, De Diego IR, Ares S, Moratalla R (2009) Genetic inactivation of dopamine D1 but not D2 receptors inhibits L-DOPA-induced dyskinesia and histone activation. Biol Psychiatry 66(6):603–613

Dauer W, Przedborski S (2003) Parkinson's disease: mechanisms and models. Neuron 39:889–909

de Mena L, Coto E, Cardo LF, Díaz M, Blázquez M, Ribacoba R, Salvador C, Pastor P, Samaranch L, Moris G, Menéndez M, Corao A, Alvarez V (2010) Analysis of the Micro-RNA-133 and PITX3 genes in Parkinson's disease. Am J Med Genet B 153B(6):1234–1239

Desplats P, Spencer B, Coffee E, Patel P, Michael S, Patrick C, Adame A, Rockenstein E, Masliah E (2011) Alpha-synuclein sequesters Dnmt1 from the nucleus: a novel mechanism for epigenetic alterations in Lewy body diseases. J Biol Chem 286(11):9031–9037

Dolinoy DC, Jirtle RL (2008) Environmental epigenomics in human health and disease. Environ Mol Mutagen 49:4–8

Duan W, Ladenheim B, Cutler RG, Kruman II, Cadet JL, Mattson MP (2002) Dietary folate deficiency and elevated homocysteine levels endanger dopaminergic neurons in models of Parkinson's disease. J Neurochem 80(1):101–110

Ertekin-Taner N (2007) Genetics of Alzheimer's disease: a centennial review. Neurol Clin 25:611–667

Fabbrini G, Brotchie JM, Grandas F, Nomoto M, Goetz CG (2007) Levodopa-induced dyskinesias. Mov Disord 22(10):1379–1389

Farrer MJ (2006) Genetics of Parkinson disease: paradigm shifts and future prospects. Nat Rev Genet 7:306–318

Finkelstein JD (2000) Pathways and regulation of homocysteine metabolism in mammals. Semin Thromb Hemost 26(3):219–225

Francis YI, Fà M, Ashrafa H, Zhanga H, Staniszewskia A, Latchmanb DS, Arancioa O (2009) Dysregulation of histone acetylation in the APP/PS1 mouse model of Alzheimer's disease. J Alzheimers Dis 18:131–139

Frieling H, Gozner A, Römer KD, Lenz B, Bönsch D, Wilhelm J, Hillemacher T, de Zwaan M, Kornhuber J, Bleich S (2007) Global DNA hypomethylation and DNA hypermethylation of the alpha synuclein promoter in females with anorexia nervosa. Mol Psychiatry 12(3):229–230

Fuso A, Seminara L, Cavallaro RA, D'Anselmi F, Scarpa S (2005) S-adenosylmethionine/homocysteine cycle alterations modify DNA methylation status with consequent deregulation of PS1 and BACE and beta-amyloid production. Mol Cell Neurosci 28:195–204

Fuso A, Nicolia V, Cavallaro RA, Ricceri L, D'Anselmi F, Coluccia P, Calamandrei G, Scarpa S (2008) B-vitamin deprivation induces hyperhomocysteinemia and brain S-adenosylhomocysteine, depletes brain S-adenosylmethionine, and enhances PS1 and BACE expression and amyloid-beta deposition in mice. Mol Cell Neurosci 37(4):731–746

Fuso A, Nicolia V, Pasqualato A, Fiorenza MT, Cavallaro RA, Scarpa S (2011a) Changes in Presenilin 1 gene methylation pattern in diet-induced B vitamin deficiency. Neurobiol Aging 32(2):187–199

Fuso A, Nicolia V, Cavallaro RA, Scarpa S (2011b) DNA methylase and demethylase activities are modulated by one-carbon metabolism in Alzheimer's disease models. J Nutr Biochem 22(3):242–251

Giles WH, Kittner SJ, Anda RF, Croft JB, Casper ML (1995) Serum folate and risk for ischemic stroke. First National Health and Nutrition Examination Survey epidemiologic follow-up study. Stroke 26(7):1166–1170

Gillardon F, Mack M, Rist W, Schnack C, Lenter M, Hildebrandt T, Hengerer B (2008) MicroRNA and proteome expression profiling in early-symptomatic α-synuclein(A30P)-transgenic mice. Proteomics Clin Appl 2(5):697–705

Hardy J, Lewis P, Revesz T, Lees A, Paisan-Ruiz C (2009) The genetics of Parkinson's syndromes: a critical review. Curr Opin Genet Dev 19:254–265

Hébert SS, De Strooper B (2009) Alterations of the microRNA network cause neurodegenerative disease. Trends Neurosci 32(4):199–206

Jiang Q, Ren Y, Feng J (2008) Direct binding with histone deacetylase 6 mediates the reversible recruitment of parkin to the centrosome. J Neurosci 28(48):12993–13002

Jowaed A, Schmitt I, Kaut O, Wüllner U (2010) Methylation regulates alpha-synuclein expression and is decreased in Parkinson's disease patients' brains. J Neurosci 30(18):6355–6359

Junn E, Lee K-W, Jeong BS, Chan TW, J-Y IM, Mouradian MM (2009) Repression of α-synuclein expression and toxicity by microRNA-7. Proc Natl Acad Sci 106(31):13052–13057

Khandhar SM, Marks WJ (2007) Epidemiology of Parkinson's disease. Dis Mon 53:200–205

Kilgore M, Miller CA, Fass DM, Hennig KM, Haggarty SJ, Sweatt JD, Rumbaugh G (2010) Inhibitors of class 1 histone deacetylases reverse contextual memory deficits in a mouse model of Alzheimer's disease. Neuropsychopharmacology 35(4):870–880

Klein C, Schlossmacher MG (2007) Parkinson disease, 10 years after its genetic revolution: multiple clues to a complex disorder. Neurology 69:2093–2104

Kontopoulos E, Parvin JD, Feany MB (2006) α-Synuclein acts in the nucleus to inhibit histone acetylation and promote neurotoxicity. Hum Mol Genet 15(20):3012–3023

Kovalchuk O (2008) Epigenetic research sheds new light on the nature of interactions between organisms and their environment. Environ Mol Mutagen 49:1–3

Lau LML, Breteler MB (2006) Epidemiology of Parkinson's disease. Lancet Neurol 5:525–535

Lambert JC, Amouyel P (2007) Genetic heterogeneity of Alzheimer's disease: complexity and advances. Psychoneuroendocrinology 32(1):S62–S70

Lukiw WJ, Zhan Y, Guo Cui J (2008) An NF-κB-sensitive Micro RNA-146a-mediated inflammatory circuit in Alzheimer disease and in stressed human brain cells. J Biol Chem 283:31315–31322

Maeda T, Guan JZ, Oyama J, Higuchi Y, Makino N (2009) Aging-associated alteration of subtelomeric methylation in Parkinson's disease. J Gerontol A Biol Sci Med Sci 64(9):949–955

Marques SC, Oliveira CR, Outeiro TF, Pereira CM (2010) Alzheimer's disease: the quest to understand complexity. J Alzheimers Dis 21(2):373–383

Matsumoto L, Takuma H, Tamaoka A, Kurisaki H, Date H, Tsuji S, Iwata A (2010) CpG demethylation enhances alpha-synuclein expression and affects the pathogenesis of Parkinson's disease. PLoS One 5(11):e15522

Minati L, Edginton T, Bruzzone MG, Giaccone G (2009) Current concepts in Alzheimer's disease: a multidisciplinary review. Am J Alzheimers Dis Other Demen 24:95–121

Monti B, Gatta V, Piretti F, Raffaelli SS, Virgili M, Contestabile A (2010) Valproic acid is neuroprotective in the rotenone rat model of Parkinson's disease: involvement of alpha-synuclein. Neurotox Res 17(2):130–141

Nicholas AP, Lubin FD, Hallett PJ, Vattem P, Ravenscroft P, Bezard E, Zhou S, Fox SH, Brotchie JM, Sweatt JD, Standaert DG (2008) Striatal histone modifications in models of levodopa-induced dyskinesia. J Neurochem 106(1):486–494

Obeid R, Schadt A, Dillmann U, Kostopoulos P, Fassbender K, Herrmann W (2009) Methylation status and neurodegenerative markers in Parkinson disease. Clin Chem 55(10):1852–1860

Outeiro TF, Kontopoulos E, Altmann SM, Kufareva I, Strathearn KE, Amore AM, Volk CB, Maxwell MM, Rochet JC, McLean PJ, Young AB, Abagyan R, Feany MB, Hyman BT, Kazantsev AG (2007) Sirtuin 2 inhibitors rescue alpha-synuclein-mediated toxicity in models of Parkinson's disease. Science 317(5837):516–519

Pacheco-Quinto J, de Turco EBR, DeRosa S, Howard A, Cruz-Sanchez F, Sambamurti K, Refolo I, Petancesk S, Pappolla MA (2006) Hyperhomocysteinemic Alzheimer's mouse model of amyloidosis shows increased brain amyloid B peptide levels. Neurobiol Dis 22:651–656

Ray WJ, Ashall F, Goate AM (1998) Molecular pathogenesis of sporadic and familial forms of Alzheimer's disease. Mol Med Today 4:151–157

Ricobaraza A, Cuadrado-Tejedor M, Pérez-Mediavilla A, Frechilla D, Del Río J, García-Osta A (2009) Phenylbutyrate ameliorates cognitive deficit and reduces tau pathology in an Alzheimer's disease mouse model. Neuropsychopharmacology 34(7):1721–1732

Ricobaraza A, Cuadrado-Tejedor M, Marco S, Pérez-Otaño I, García-Osta A (2010) Phenylbutyrate rescues dendritic spine loss associated with memory deficits in a mouse model of Alzheimer disease. Hippocampus 22(5):1040–1050

Rocchi A, Pellegrini S, Siciliano G, Murri L (2003) Review: causative and susceptibility genes for Alzheimer's disease: a review. Brain Res Bull 61:1–24

Rouaux C, Jokic N, Mbebi C, Boutillier S, Loeffler JP, Boutillier AL (2003) Critical loss of CBP/p300 histone acetylase activity by caspase-6 during neurodegeneration. EMBO J 22(24):6537–6549

Schipper HM, Maes OC, Chertkow HM, Wang E (2007) MicroRNA expression in Alzheimer blood mononuclear cells. Gene Regul Syst Biol 20(1):263–274

Silva PNO, Gigek CO, Leal MF, Bertolucci PHF, de Labio RW, Payão SLM, Smith MAC (2008) Promoter methylation analysis of SIRT3, SMARCA5, HTERT and CDH1 genes in aging and Alzheimer's disease. J Alzheimers Dis 13:173–176

Song C, Kanthasamy A, Anantharam V, Sun F, Kanthasamy AG (2010) Environmental neurotoxic pesticide increases histone acetylation to promote apoptosis in dopaminergic neuronal cells: relevance to epigenetic mechanisms of neurodegeneration. Mol Pharmacol 77(4):621–632

Sontag E, Hladik C, Montgomery L, Luangpirom A, Mudrak I, Ogris E, White CL 3rd (2004) Downregulation of protein phosphatase 2A carboxyl methylation and methyltransferase may contribute to Alzheimer disease pathogenesis. J Neuropathol Exp Neurol 63(10):1080–1091

Sontag E, Nunbhakdi-Craig V, Sontag JM, Diaz-Arrastia R, Ogris E, Dayal S, Lentz SR, Arning E, Bottiglieri T (2007) Protein phosphatase 2A methyltransferase links homocysteine metabolism with tau and amyloid precursor protein regulation. J Neurosci 27:2751–2759

St George-Hyslop PH, Petit A (2004) Molecular biology and genetics of Alzheimer's disease. Comptes Rendus Biologies 328:119–130

Stozicka Z, Zilka N, Novak M (2007) Review: risk and protective factors for sporadic Alzheimer's disease. Acta Virol 51:205–222

Tanzi RE, Bertram L (2001) New frontiers in Alzheimer's disease genetics. Neuron 32:181–184

Tohgi H, Utsugisawa K, Nagane Y, Yoshimura M, Genda Y, Ukitsu M (1999) Reduction with age in methylcytosine in the promoter region −224–101 of the amyloid precursor protein gene in autopsy human cortex. Mol Brain Res 70:288–292

Vo N, Klein ME, Varlamova O, Keller DM, Yamamoto T, Goodman RH, Impey S (2005) A cAMP-response element binding protein-induced microRNA regulates neuronal morphogenesis. Proc Natl Acad Sci 102(45):16426–16431

Voutsinas GE, Stavrou EF, Karousos G, Dasoula A, Papachatzopoulou A, Syrrou M, Verkerk AJ, van der Spek P, Patrinos GP, Stöger R, Athanassiadou A (2010) Allelic imbalance of expression and epigenetic regulation within the alpha-synuclein wild-type and p.Ala53Thr alleles in Parkinson disease. Hum Mutat 31(6):685–691

Wang G, van der Walt JM, Mayhew G, Li Y, Zuchner S, Scott WK, Martin ER, Vance JM (2008a) Variation in the miRNA-433 binding site of FGF20 confers risk for Parkinson disease by over-expression of a-synuclein. Am J Hum Genet 82:283–289

Wang S-C, Oelze B, Schumacher A (2008b) Age-specific epigenetic drift in late-onset Alzheimer's disease. PLoS One 3:e2698

Wang WX, Rajeev BW, Stromberg AJ, Ren N, Tang G, Huang Q, Rigoutsos I, Nelson PT (2008c) The expression of microRNA miR-107 decreases early in Alzheimer's disease and may accelerate disease progression through regulation of beta-site amyloid precursor protein-cleaving enzyme 1. J Neurosci 28(5):1213–1223

Wang X, Liu P, Zhu H, Xu Y, Ma C, Dai X, Huang L, Liu Y, Zhang L, Qin C (2009a) miR-34a, a microRNA up-regulated in a double transgenic mouse model of Alzheimer's disease, inhibits bcl2 translation. Brain Res Bull 80(4–5):268–273

Wang Y, Wang X, Liu L, Wang X (2009b) HDAC inhibitor trichostatin A-inhibited survival of dopaminergic neuronal cells. Neurosci Lett 467(3):212–216

Weintraub D, Comella CL, Horn S (2008) Parkinson's disease- part 1: pathophysiology, symptoms, burden, diagnosis, and assessment. Am J Manag Care 14:S40–S48

Wood-Kaczmar A, Gandhi S, Wood NW (2006) Understanding the molecular causes of Parkinson's disease. Trends Mol Med 12:521–528

World Health Organization (2002) Active ageing, a policy framework. Second United Nations World assembly on Aging, Madrid, Spain. www.who.int/hpr/ageing/ActiveAgingPolicyFrame.pdf

Wu J, Basha MR, Brock B, Cox DP, Cardozo-Pelaez F, McPherson CA, Harry J, Rice DC, Maloney B, Chen D, Lahiri DK, Zawia NH (2008a) Alzheimer's disease (AD)-like pathology in aged monkeys after infantile exposure to environmental metal lead (Pb): evidence for a developmental origin and environmental link for AD. J Neurosci 28:3–9

Wu X, Chen PS, Dallas S, Wilson B, Block ML, Wang CC, Kinyamu H, Lu N, Gao X, Leng Y, Chuang DM, Zhang W, Lu RB, Hong JS (2008b) Histone deacetylase inhibitors up-regulate astrocyte GDNF and BDNF gene transcription and protect dopaminergic neurons. Int J Neuropsychopharmacol 11(8):1123–1134

Chapter 23
Cellular Redox, Epigenetics and Diseases

Shyamal K. Goswami

Abstract In the past decade, epigenetic regulation has emerged as the nodal point of modulating gene expression in eukaryotes. It involves structural reorganization of the chromosome, primarily through post translational modifications of the histones and methylation of cytosine residues in the DNA. Other aspects of DNA functions such as DNA replication and repair also come under the larger ambit of epigenetic regulation. Reactive oxygen/nitrogen species ($R^O/_NS$) has long been perceived as deleterious agents affecting cellular functions, causing degenerative diseases. However, during the past decade akin to kinase-phosphatase signaling, oxidative-reductive modifications of proteins by reactive species, especially at lower concentrations, have also emerged as mediators of key regulatory signals. DNA methyltransferases (DNMTs) catalyze methylation of cytosine residues in the DNA while methionine adenosyltransferase (MAT) synthesize the methyl donor, S-adenosyl methionine (SAM). Both the enzymes have redox sensitive cysteine residues in their catalytic sites. Increased generation of ROS, as it occurs in cancer cells; affects activities of these enzymes, altering the status of DNA methylation and the epigenome. Histone methyl transferases, the key determinant of "epigenetic landscape" also require S-adenosyl methionine (SAM) for their activities. Hence, altered pool of SAM in a more oxidized cellular environment also affects the status of histone methylation and the epigenome. Numerous such evidences have accumulated over the past decade suggesting a close interrelation between cellular redox and gene expression wherein various transcription factors, coactivators and chromatin constituents work in tandem mediating the cognate responses. However, a comprehensive understanding of how these events is coordinately regulated by various RO/NS axis is still emerging and the following chapter is an update in this regard.

S.K. Goswami (✉)
School of Life Sciences, Jawaharlal Nehru University, New Delhi 110067, India
e-mail: skgoswami@mail.jnu.ac.in; shyamal.goswami@gmail.com

Since the Central dogma of Molecular Biology was proposed about 40 years ago; our understanding of the intricacies of gene regulation has undergone tectonic shifts almost every decade. It is now widely accepted that the complexity of an organism is not directed by the sheer number of genes it carries but how they are decoded by a myriad of regulatory modules. Over the years, it has emerged that the organizations chromatins and its remodeling; splicing and polyadenylation of pre-mRNAs, stability and localization of mRNAs and modulation of their expression by noncoding and miRNAs play pivotal roles in metazoan gene expression. Nevertheless, in spite of tremendous progress in our understanding of all these mechanisms of gene regulation, the way these events are coordinated leading towards a highly defined proteome of a given cell type remains enigmatic. In that context, the structures of many metazoan genes cannot fully explain their pattern of expression in different tissues, especially during embryonic development and progression of various diseases. Further, numerous studies done during the past quarter of a century suggested that the heritable states of transcriptional activation or repression of a gene can be influenced by the covalent modifications of constituent bases and associated histones; its chromosomal context and long-range interactions between various chromosomal elements (Holliday 1987; Turner 1998; Lyon 1993). However, molecular dissection of these phenomena is largely unknown and is an exciting topic of research under the sub-discipline epigenetics (Gasser et al. 1998).

The term epigenetics was first coined in 1938 by the developmental geneticist Conrad Waddington to explain certain aspects of inheritance in cell differentiation and organogenesis (Waddington 1938). Since then, the operational definition of epigenetics remained a subject of intense debate (Haig 2004). As an example, while molecular biologists try to explain epigenetic regulation of gene function through the modifications of DNA and histones (Riggs et al. 1996), others consider that factors as diverse as hormones and temperature are epigenetic regulators (Herring 1993). According to the most recent definition of epigenetics, as proposed by Berger et al., "an epigenetic trait is a stably heritable phenotype resulting from changes in a chromosome without alterations in the DNA sequence" (Berger et al. 2009). Also, as per their nomenclature, establishment of a stably heritable epigenetic state involves three classes of signals viz., "*Epigenator*, which emanates from the environment and triggers an intracellular pathway; an *Epigenetic Initiator*, which responds to the *Epigenator* and is necessary to define the precise location of the epigenetic chromatin environment; and an *Epigenetic Maintainer* signal, which sustains the chromatin environment in the first and subsequent generations" (Berger et al. 2009).

In that context, various studies over the past decade have shown that reactive oxygen/nitrogen ($R^O/_N S$) species can also influence epigenetic state of a genome. However, since the biology of reactive oxygen/nitrogen species ($R^O/_N S$) is highly complex and diverse (Heo 2011), their influence upon epigenetic events are poorly defined as yet. Therefore, perhaps it is too early to categorize their role in epigenetics in the proposed framework of "*Epigenator-Epigenetic Initiator and Epigenetic Maintainer*". With that background, in the following sections I will summarize some of the recent developments in understanding the role of $R^O/_N S$ in epigenetic regulation. Epigenetic regulation involves structural reorganization of the chromosome that is primarily mediated by the methylation of cytosine residues in the DNA and post translational modifications of the histones. Also, although epigenetic regulation generally implies altered transcriptional activity, other aspects of DNA functions that is DNA replication and repair also come under the purview of epigenetic regulation.

23.1 Intracellular $R^O/_NS$ and Redox Signaling

Intracellular generation of free radicals are attributed to a plethora of pro-oxidant enzymes such as NADPH and xanthine oxidases, nitric oxide synthases, cyclooxygenases, mitochondrial electron transport complexes, and Fenton reaction. In the dynamic cellular milieu, while the pro-oxidant enzymes continuously generate free radicals and $R^O/_NS$, various antioxidant systems attenuate them, maintaining redox homeostasis (Heo 2011; Antelmann and Helmann 2011). Such controlled, limited generation of free radicals and $R^O/_NS$ contribute to normal physiological processes such as cell signaling, cell cycle progression and embryonic development (Heo 2011; Antelmann and Helmann 2011; Boivin et al. 2010). However, under certain pathophysiological conditions, such balance is tilted towards aberrant pro-oxidant activities, resulting in more oxidized intracellular environment, a condition termed as oxidative stress. Oxidative stress has been implicated in a plethora of diseases like cancer, diabetes, inflammatory, neurodegenerative and cardiovascular disorders (Nediani et al. 2011; Pamplona et al. 2008; Weaver 2009). However, despite the existence of ample evidences supporting the oxidative stress theory of diseases, there are many discrepancies in the interpretation of results compiled over the past two decades (Janssen-Heininger et al. 2008). In accordance, antioxidant therapies for curing those diseases have also shown mixed results, further arguing against generalization of oxidative mechanisms of disease progression (Gutteridge and Halliwell 2010). It now appears that rather than causing oxidative damage only, $R^O/_NS$ can also cause transient modifications of specific targets regulating their functions (Weaver 2009; Dansen et al. 2009). However, due to their extremely short half-life and the transient nature of such oxidative/ nitrosative modifications they impart upon their targets, biochemical mechanisms of redox regulation of gene expression are just emerging (Woo et al. 2010). Also, with the availability of newer techniques for the detection and characterization of various $R^O/_NS$, certain earlier observations regarding redox regulation of gene expression have been revisited (Banerjee et al. 2010; Gloire and Piette 2009); while several novel concepts have emerged (Upham and Trosko 2009).

Reversible phosphorylation of serine, threonine and tyrosine residues in signaling proteins play a nodal role in cellular processes like cell proliferation, differentiation, adhesion, survival etc. A plethora of growth factor, cytokines and hormones use such phosphorylation-dephosphorylation mechanisms for transmitting respective signals from the cell surface to the nucleus (Gough and Foley 2010). During the past decade akin to kinase-phosphatase signaling, oxidative-reductive modifications of regulatory proteins have also emerged as mediators of certain signals (Weaver

2009; Dansen et al. 2009; Woo et al. 2010; Upham and Trosko 2009). However, due to inadequate knowledge about the targets and the types of modifications, mechanisms of redox signaling are poorly understood and number of laboratories are developing methodologies for analyzing redoxproteomics under various pathophysiological contexts (Bregere et al. 2008; Dabkowski et al. 2010; Burgoyne and Eaton 2010). Emerging evidences also suggest that amongst various oxidative modifications of regulatory proteins, S-glutathionylation, -sulfenation and -nitrosylation are most prevailing and important in mediating redox-signals under various pathobiological contexts (Dulce et al. 2011; Chen et al. 2010; Maller et al. 2011; Nishida et al. 2011; Kornberg et al. 2010).

23.1.1 Cellular Redox and DNA Methylation

In eukaryotes, post-transcriptional modifications of RNA occur quite extensively, especially in tRNAs. On the contrary, only modification that occurs in DNA bases is the methylation of cytosine residues in the CpG dinucleotides. Overall occurrence of CpG sequences in mammalian genomes is much less (~20%) than that is expected for their random occurrence in the genome. They often occur as clusters (CpG islands) in the regulatory regions of many protein coding genes, especially those which are constitutively expressed with housekeeping functions (Xie et al. 2011). Cells have dedicated enzymatic machinery ensuring methylation of cytosine residues in CpG sequences. Normally, many of the C residues in the CpG islands remain unmethylated or hypomethylated; while under certain conditions their methylation leads to chromatin condensation and transcriptional repression (Easwaran et al. 2010). Methylation of cytosine residues in the CpG islands inhibits the binding of certain transcription factors like CREB but enhances the binding of certain others like C/EBPα and methyl-CpG binding domain proteins like MBP (Pierard et al. 2010; Joulie et al. 2010; Rishi et al. 2010). Upon binding, MBPs act as recruitment points for histone modifying enzymes. Thus, methylation of CpG residues plays a major role in chromosome structure and function. Recent studies suggest that CpG methylation also plays a role in regulating the expression of micro RNAs (Lujambio and Esteller 2009).

DNA methyltransferases (DNMTs) catalyze methylation of cytosine by transferring the methyl groups of S-adenosylmethionine (SAM) to the 5th position of the pyrimidine ring. In mammals, there are three DNA methyltransferases viz., DNMT1, DNMT3a and DMNT3b (Cheng and Blumenthal 2008). Following replication, DNMT1 methylates the newly synthesized strand of DNA, thus carrying out "Maintenance methylation". It is expressed at a high level in the S phase of cell cycle and has high affinity for CpG dinucleotides where only one strand is methylated (hemimethylated DNA). The other two DNA methyl transferases i.e., DNMT3a and DNMT3b can carry out *de novo* methylation of hemimethylated and unmethylated DNA and thus play more active roles in gene silencing. Novel regulatory pathways in plants and mammals are now emerging wherein small interfering RNAs modulate transcriptional silencing via DNA methylation (He et al. 2011).

Since the development and propagation of cancer involves extensive epigenetic dysregulation, cancer cells are often used as a paradigm of understanding epigenetic mechanisms (Brower 2011). In cancer cells, cellular metabolism is substantially altered which in turn perturb the redox homeostasis (Cairns et al. 2011; Tew and Townsend 2011). The ratio of reduced versus oxidized glutathione (GSH/GSSG) is very high in normal cells so that the reduced cellular milieu is maintained. Cancer cells have higher levels of free radicals and reactive oxygen/nitrogen species that substantially diminish the level of reduced glutathione, resulting in a more oxidized environment (Tew and Townsend 2011; Finley et al. 2011). Such oxidized cellular environment affects enzymatic reactions, especially those requiring redox-sensitive cysteine residues (Guttmann 2010). Methionine adenosyltransferase (MAT), the key enzyme involved in the generation of SAM, is inhibited by free radicals (Avila et al. 1998; Lindermayr et al. 2006). Since SAM is the methyl donor to the cytosine residues, decrease in SAM affects DNA (and histone) methylation. DNA methyltransferases also have redox sensitive cysteine residues in their active sites and in oxidized cellular environment, function of DNMT are also directly affected due to the oxidation of those cysteines (Svedružić and Reich 2005).

Taken together, substantial evidences suggest that $R^O/_NS$ might act as a epigenetic regulator via DNA methylation. However, except its potential inhibitory effect on MAT and DNMT, very little mechanistic details are known till date. It is likely that $R^O/_NS$ might regulate DNA methylation by other mechanisms as well. As an example, arsenic, a genotoxic agent; induces oxidative stress and causes hypermethylation of DNA; resulting in cancer, diabetes, cardiovascular and neurological disorders, thereby acting more as an epigenetic regulator than a classical mutagen (Flora 2011). However, whether or not the epigenetic effect of arsenic is a consequence of oxidative injury is not known yet (Coppin et al. 2008). Another lesser known connection between $R^O/_NS$ and DNA methylation is, excessive generation of $R^O/_NS$ leads to DNA damage; extent of which is measured by the formation of 8-oxo-7, 8-dihydroguanine (8-oxoGua). In cancer cells, a positive correlation has been found between level of 8-oxoGua and the extent of hypomethylation of DNA. However, the mechanistic interrelationship is unknown (Guz et al. 2008).

23.1.2 Cellular Redox and Histone Methylation

Epigenetic regulation requires coordination between the cytosine methylation (in CpG islands) and that of lysine[4 & 9] of histone H3. Methylation of these two lysine residues is carried out by histone methyltransferases and their modifications have reciprocal correlations with that of DNA. While methylation of Lys[4] prevents cytosine methylation, that at Lysine[9] facilitates it (Cheng and Blumenthal 2010). Histones can also be methylated at other lysines *viz.*, those at 27, 36 and 79 of histone H3 and at 20 of histone H4. Also, depending upon the methyltransferase, lysine residues are mono-, di-, or tri-methylated, each conferring specific functional properties to that histone molecule they target (Upadhyay and Cheng 2011). Histone and DNA methylation, in conjunction with other histone modifications create an "epigenetic landscape" that facilitates genome wide recruitment of various gene regulatory modules dictating cell proliferation, differentiation, stemness as well as development and progression of various degenerative diseases (Rakyan et al. 2011). Although till date a direct relationship between cellular redox and histone methylation has not been reported, existence of such correlations is very likely. All known histone methyl transferases with the exception of Dot1, contain an evolutionarily conserved SET domain comprising of ~130 amino acids. Like DNA methyl transferases, histone methyl transferases also require S-adenosyl methionine (SAM) for their activities and as discussed above, cellular pool of SAM can be affected by increased levels of reactive species (Svedruzić and Reich 2005). Furthermore, at least a subset of histone lysine methyltransferases i.e., SUV39 subfamily, contains multiple conserved cysteine residues required for maintaining the 3D structure of the catalytic sites (Zhang et al. 2003). So, it is likely that at least some if not all histone lysine methyltransferases might be sensitive to cellular redox. Also, histone lysine methylation is a reversible process wherein demethylation lysine$^{4\ \&\ 9}$ (mono- and di-methyl) of histone 3 is catalyzed by lysine specific demethylase-1, a flavin-dependent amine oxidase. Hene demethylation of lysines is more likely to be affected by changes in cellular redox milleu (Upadhyay and Cheng 2011). Since certain diseases like cancer involves a significant alterations in cellular redox environment as well as extensive epigenetic modifications, understanding correlations between histone methylation and ROS might further enhance our understanding of cancer biology (Varier and Timmers 2011).

23.1.3 Cellular Redox and Chromatin Organization

Structural organization of chromatin is a highly dynamic process involving a myriad of DNA protein and protein-protein interactions. It also involves a large number of post translational modifications of histones, non-histone chromosomal proteins, transcription factors, coactivators etc.; modulating gene expression, gene silencing and DNA replication (Bell et al. 2011). Considering the pervasive role of cellular redox in mediating various physiological responses, it is likely that $R^O/_NS$ also affect chromatin dynamics, especially under various pathophysiological conditions. Exposure of mammalian cells to anticancer drugs, DNA adducts, ionizing and ultraviolet (UV) irradiations cause oxidative injury to chromosomes. Early studies in this regard were directed towards understanding whether or not oxidative stress causes chromosomal organization leading towards various diseases (Mitra et al. 2002; Rahman 2002). While looking the mechanisms of neurodegeneration, Bai and Konat demonstrated that H_2O_2 at lower concentration (5 μM) triggers cytoplasmic signaling followed by higher order chromatin degradation that is reversed upon removal of the oxidant (Bai and Konat 2003). In C6 rat glial cells, depletion of glutathione leads to the generation of giant, high molecular weight and internucleosomal DNA fragments prior to cell death (Higuchi and Yoshimoto 2004). When cell death is induced in renal proximal tubular epithelial cells (LLC-PK1) by 2, 3,

5-tris-(glutathione-S-yl) hydroquinone [TGHQ], it generates ROS followed by premature chromatin condensation that requires activation of ERK, phosphorylation of histone H3 and poly (ADP-) ribosylation (Tikoo et al. 2001). Nitric oxide is a potent reactive nitrogen species and a key mediator of various physiological signals. It has been implicated to many diseases and amongst its targets are protein tyrosine residues which it converts into 3-nitrotyrosine. Neutrophils, a major source of nitric oxide, invade murine mutatect tumors and causes selective nitration of tyrosines in histones (Haqqani et al. 2002). Taken together, there are substantial evidences indicating role of $R^O/_NS$ in altering chromatin structure. However, precise biochemical mechanism by which they impart epigenetic regulations is still obscure and sporadic.

23.1.4 Cellular Redox, Histone Acetylation and Chromatin Remodeling

Various small molecules including metabolites, nutrients and therapeutics affect cellular proteome via epigenetic modulation of gene expression (Selvi et al. 2010). Also, at least some of those modulators exert their effects via alteration of cellular redox, although the mechanisms involved are poorly understood as yet (Cyr and Domann 2011). As an example, arsenic has been in use for centuries as a therapeutic agent against cancer and in recent years there has been resurgence in studying its effects on leukemic cells (Chen et al. 2011). At certain concentrations, it has genotoxic effects due to the generation of reactive oxygen species, resulting in various degenerative diseases like cancer. Arsenic induces hypermethylation of DNA (Coppin et al. 2008) and hyper-acetylation of histones H3 and H4 (Perkins et al. 2000). However, its epigenetic effects are yet to be attributed to its ability to induce oxidative stress (Coppin et al. 2008; Perkins et al. 2000). Nevertheless, since it binds to thiol residues and affects redox metabolism; such association is quite plausible (Flora 2011). Many small molecular inhibitors of HDAC activities also mediate their effects by inducing oxidative stress. Exposure of macrophage like cell line MonoMac6 to cigarette smoke extract increases the $R^O/_NS$ level followed by a decrease in HDAC 1–3 activities due to nitrotyrosine and aldehyde-adduct formation (Yang et al. 2006). Such decreased HDAC activities results in the increased expression of cytokine genes like IL-8 and TNF-α. When A549 epithelial cells are treated with either cigarette smoke-conditioned medium, peroxynitrite or hydrogen peroxide; level of $R^O/_NS$ increases followed by nitration of tyrosine[253] of HDAC2, directing it to proteasomal degradation (Osoata et al. 2009). In neurons, Brain-derived neurotrophic factor triggers NO synthesis which causes S-nitrosylation of histone deacetylase 2 (HDAC2) at Cys^{262} and Cys^{274}. Nitrosylation of HDAC2 does not affect its deacetylase activity but causes its release from chromatin, resulting in increased acetylation of histones which triggers neurotrophin-dependent gene expression (Nott et al. 2008). Nitric oxide-releasing acetylsalicylic acid (NO-ASA) is an anti-inflammatory drug that is cytotoxic to various tumor cell lines. Treatment of human B-lymphoblastoid TK6 cells with NO-ASA leads to the generation of $R^O/_NS$ causing DNA damage and phosphorylation of histone H2AX at Ser^{139} during S-phase. Prolong exposure to NO-ASA also induces atypical apoptosis characterized by highly condensed chromatin but no nuclear fragmentation (Tanaka et al. 2006). Alpha and beta-unsaturated carbonyl compounds like cyclopentenone prostaglandin and 4-hydroxy-2-nonenal which produce reactive carbonyl species (from the peroxidation of arachidonic acid) carbonylate HDAC1, -2, and -3 at two conserved cysteine residues (Cys^{261} and Cys^{273} in HDAC1), attenuating their activities and resulting in changes in histone H3/H4 acetylation followed by the activation of defensive genes like heme oxygenase-1, Gadd45, and HSP70 (Doyle and Fitzpatrick 2010). There are also instances where reactive species increase HDAC activities. When cardiac myocytes are treated with lipopolysaccharides (LPS), HDAC3 activity is increased via mitochondrial ROS and c-Src signaling upregulating TNF-α (Zhu et al. 2010). Treatment L6 myoblasts with insulin under hyperglycemic condition induces methylation (at Lysine$^{4\ \&\ 9}$), phosphorylation (at Ser^{10}) and acetylation of histone H3 by enhanced generation of ROS; highlighting its role in epigenetic regulation of diabetes (Kabra et al. 2009). Taken together, there are ample evidences linking histone acetylation-deacetylation with $R^O/_NS$, but the mechanistic details are yet to be investigated.

23.1.5 Cellular Redox and Transcription Factor Activities

Since cellular metabolism primarily occurs in the cytosol (and mitochondria), earlier it was believed that while cytosol is more susceptible to oxidative insults, nucleus generally remains in a reduced state. However, recent studies demonstrate that the redox environment in the nucleus is also dynamic (Moldovan and Moldovan 2004). At the early stages of cell proliferation, nuclear content of GSH, a key redox buffer; increases and it co-localizes with the DNA (Diaz Vivancos et al. 2010). A reciprocal relationship between level of nuclear GSH and that of certain transcripts encoding stress and defense proteins has also been reported (Markovic et al. 2010). Although our knowledge about the role of $R^O/_NS$ in affecting nuclear events like histone modifications and chromatin remodeling is quite limited; substantial progress has already been made in understanding the modulation of transcription factor activities by cellular redox. The first evidence that a transcription factor activity can be regulated by $R^O/_NS$ came from the observation that oxidation of specific cysteine residues at the DNA binding domain of c-JUN and c-FOS, the two components of AP-1 transcription factor; reduces their DNA binding affinities (Abate et al. 1990). Ref-1, a nuclear redox protein can restore the reduced state of c-JUN and c-FOS, along with the DNA binding functions (Xanthoudakis and Curran 1992). Nitric oxide is a potent reactive nitrogen species and a modulator of various physiological responses. It nitrosylates specific cysteine residues of JUN and FOS and induces S-glutathionylation of c-JUN; affecting their DNA binding functions (Nikitovic et al. 1998; Klatt et al. 1999). Subsequent to these early but seminal observations, numerous studies have shown that $R^O/_NS$ can modulate AP-1 activities by modulating (a) the upstream kinases, (b) expression of its constituent subunits (i.e., *jun* and *fos*) and (c) its interaction with p300/CBP coactivators (Nelson et al. 2006; Jindal and Goswami 2011; Rahman et al. 2004). It now appears that cellular redox might affect the activities of transcription factors in several ways. With increased generation of $R^O/_NS$, certain cysteine residues might switch to more oxidized state, resulting in an alteration in its conformation affecting DNA binding and transactivation functions. Also, increased $R^O/_NS$ might affect their interaction with certain metal ions acting as cofactors; thereby inhibiting their functions (Jindal and Goswami 2011; Tong et al. 2007).

While AP-1 was the first transcription factor shown to be sensitive towards oxidation-reduction *in vitro*; NFκB was the first to be identified eliciting similar responses *in vivo* (Schreck et al. 1991). Since then, NFκB has been a paradigm of understanding role of cellular redox in gene expression (Gloire et al. 2006; Gloire and Piette 2009; Oliveira-Marques et al. 2009). Lung cells are vulnerable to massive oxidative insults as it responds to inhaled pathogens by recruiting ROS-producing macrophages and neutrophils (Fialkow et al. 2007). Inhaled pollutants like cigarette smoke, automobile exhausts etc., also induce oxidative stress in the lung. Excessive generation of $R^O/_NS$ leads to various inflammatory diseases like chronic obstructive pulmonary disease (COPD), asthma, pulmonary fibrosis and cancer; each with characteristic gene expression programs wherein NF-κB plays a pivotal role (Tasaka et al. 2008). In unstimulated cells, members of the NF-κB family are sequestered in the cytoplasm either as their precursors (p100 and p105) or by the three IκB proteins i.e., IκBα, β and ε. Upon stimulation, IκBs are phosphorylated and degraded by 26S proteasome followed by the translocation of NF-κB subunits to the nucleus (Wertz and Dixit 2010). Phosphorylation of IκBs occurs at specific Serine and Tyrosine residues by various canonical and non-canonical pathways (Gloire and Piette 2009; Oliveira-Marques et al. 2009; Wertz and Dixit 2010). Once in the nucleus, subunits of NF-kB undergo multiple post translational modifications regulating its accessibility and binding to target DNA (Gloire and Piette 2009). Post translational modifications of the p65 subunit of NF-κB have been extensively studied as a paradigm of understanding the biological responses mediated by the NF-κB family. Phosphorylation of p65 occurs at multiple sites of which that at Ser[276] induces its interaction with CBP/p300 followed by its acetylation at Lys[310] (Yao et al. 2010; Rajendrasozhan et al. 2008). Increased phophorylation-acetylation of p65 leads to the recruitment of IKKα followed by the phosphorylation of histone H3 at Ser [10] and its acetylation at Lys[9] and Lys[14] by CBP (Yao et al. 2010; Rajendrasozhan et al. 2008). Such phosphorylation-acetylation axis involving p65, IKKα, CBP and HistoneH3 is involved in the activation of pro-inflammatory genes in mouse lung *in vivo* and in human monocyte/macrophage cell line MonoMac6 *in vitro* by pro-oxidant cigarette smoke extract (Rajendrasozhan et al. 2008). Apart from activating numerous genes, NF-κB also represses certain others under specific contexts wherein isoforms of HDAC are recruited by distinct NF-kB complexes (Liu et al. 2010). Pro-oxidant constituents of cigarette smoke extracts decrease the expression level and activity of HDACs, resulting in derepression of those genes (Yang et al. 2006). In agreement with these observations, decrease in HDAC activities due to nitrosylation, nitration and carbonylation (by $R^O/_NS$) has been observed in smokers and asthma patients, (Barnes 2009). Taken together, substantial evidences have accumulated over the past 10 years suggesting a close interrelation between cellular redox and gene expression wherein transcription factors, coactivators and chromatin constituents work in tandem mediating the redox response. However, a comprehensive understanding of how these events are coordinately regulated by various reactive oxygen/nitrogen species are still emerging and coming years are likely to shed more light on this novel aspect of epigenetic control (Shlomai 2010).

23.1.6 Cellular Redox and DNA Replication

As in transcription, DNA repair and replication also involves extensive organization of chromatin, wherein both its accessibility and compactness are tightly controlled (Chagin et al. 2010). Wealth of information accumulated over the past quarter of a century has delineated how in response to various external cues, mammalian cells encompass through highly orchestrated cell cycle pathways ensuring fidelity of genome duplication (Corpet and Almouzni 2009; Chakraborty et al. 2011). In this context, although the nodal role(s) of kinases and phosphatases in cell cycle regulation and DNA replication have been well documented (Yu and Cortez 2011); that of $R^O/_NS$, if any; is largely unexplored. Nevertheless, substantial evidences suggest that upon growth stimuli, low intensity generation of ROS modulate the downstream signals leading towards cell division (Ishimoto et al. 2011). Although an wider role of ROS in cell cycle regulation is yet to be established, evidences accumulated from studies with cancer and other degenerative diseases suggest a close association between generation of ROS and cell division (Matés et al. 2010). Some commentators have even perceived cell cycle as "Redox cycle" (Burhans and Heintz 2009). Apart from limited generation of reactive species modulating physiological signals, excessive generation of $R^O/_NS$ also leads to DNA damage by altering bases, creating single-strand breaks and abasic sites that are repaired via the base excision repair (BER) pathways that involve complex coordination between transcription, histone modifications and replication (Mitra et al. 2002). Taken together, it is thus likely that ROS might also have a broader role in chromatin organization during genome duplication. However, till date, experimental evidences directly correlating ROS and epigenetic regulation of DNA replication is scanty.

23.1.7 Epigenetic Regulation of Embryonic Development and Cellular Redox

Epigenetic regulation plays a nodal role in embryonic development that involves extensive cell differentiation and organogenesis (Banaszynski et al. 2010). During the development of metazoan organisms, based upon the extent of vascularization; different regions of the embryo gets different levels of oxygen which results in the creation of distinct repertoires of R^O/$_N$S like NO, O$_2^-$ and H$_2$O$_2$. It is now believed that such differential distribution of R^O/$_N$S plays a key role in epigenetic regulation of tissue development (Dennery 2010). According to a recent study, during heart development, maturation of mitochondria leads to the closure of the permeability transition pores, resulting in decreased levels of reactive oxygen species which in turn influences differentiation of cardiac myocytes (Drenckhahn 2011). During the early stages of development, ROS generated by Nox2 or Nox4, the two NADPH oxidases involved in signal transduction in non-phagocytic cells are essential for chondrocyte differentiation (Kim et al. 2010). Taken together, although such "Free Radical theory of development" is more of a concept (Hitchler and Domann 2007), emerging evidences support a broader roles of R^O/$_N$S in cell differentiation and tissue development while underlying mechanisms are poorly understood as yet (Ufer et al. 2010).

23.2 Concluding Remarks

Sustenance and propagation of a cell (and an organism) involves a complex choreography of energy generation, metabolic turnover, gene expression, replication and maintenance of the genome, integration of organelle functions etc. Like all other sub disciplines of biology, understanding of cellular redox was initiated almost half a century ago with a compartmentalized approach to understand the energy generation by the mitochondria (Ernster and LEE 1964). However, it evolved with time, and in the process; terms like "Oxidative stress" and "Redox signalling" and Redox homeostasis" emerged with their own connotations. Today, the anticipated role of redox has gone beyond it initial boundary of bioenergetics and has pervaded into territories as diverse as regulation of metabolic enzymes, transcription factors, microRNAs etc. (Shlomai 2010; Simone et al. 2009). In that context, role of epigenetics has also evolved from conceptualization to fine dissection of it mechanisms. As discussed above, the inter connection between cellular redox and epigenetics are still sporadic and coincidental. However, in view of the totality of biological responses wherein both redox homeostasis and epigenetic modulation are key contributors, it is likely that these two are functionally connected. It is thus expected that in coming years more direct and mechanistic details of their interrelationship will be revealed.

Acknowledgement This article is partly supported by the grant awarded to the author by the Council of Scientific and Industrial Research, Govt of India under sanction number 37(1479)/11/EMR II.

References

Abate C, Patel L, Rauscher FJ 3rd, Curran T (1990) Redox regulation of fos and jun DNA-binding activity in vitro. Science 249:1157–1161

Antelmann H, Helmann JD (2011) Thiol-based redox switches and gene regulation. Antioxid Redox Signal 14:1049–1063

Avila MA, Corrales FJ, Ruiz F, Sánchez-Góngora E, Mingorance J, Carretero MV, Mato IM (1998) Specific interaction of methionine adenosyltransferase with free radicals. Biofactors 8:27–32

Bai H, Konat GW (2003) Hydrogen peroxide mediates higher order chromatin degradation. Neurochem Int 42:123–129

Banaszynski LA, Allis CD, Lewis PW (2010) Histone variants in metazoan development. Dev Cell 19:662–674

Banerjee S, Zmijewski JW, Lorne E, Liu G, Sha Y, Abraham E (2010) Modulation of SCF beta-TrCP-dependent I kappaB alpha ubiquitination by hydrogen peroxide. J Biol Chem 285:2665–2675

Barnes PJ (2009) Histone deacetylase-2 and airway disease. Ther Adv Respir Dis 3:235–243

Bell O, Tiwari VK, Thomä NH, Schübeler D (2011) Determinants and dynamics of genome accessibility. Nat Rev Genet 12:554–5564

Berger SL, Kouzarides T, Shiekhattar R, Shilatifard A (2009) An operational definition of epigenetics. Genes Dev 23:781–783

Boivin B, Yang M, Tonks NK (2010) Targeting the reversibly oxidized protein tyrosine phosphatase superfamily. Sci Signal 3(137):l2

Bregere C, Rebrin I, Sohal RS (2008) Detection and characterization of in vivo nitration and oxidation of tryptophan residues in proteins. Methods Enzymol 44:339–349

Brower V (2011) Epigenetics: unravelling the cancer code. Nature 471:S12–S13

Burgoyne JR, Eaton P (2010) A rapid approach for the detection, quantification, and discovery of novel sulfenic acid or S-nitrosothiol modified proteins using a biotin-switch method. Methods Enzymol 473:281–303

Burhans WC, Heintz NH (2009) The cell cycle is a redox cycle: linking phase-specific targets to cell fate. Free Radic Biol Med 47:1282–1293

Cairns RA, Harris IS, Mak TW (2011) Regulation of cancer cell metabolism. Nat Rev Cancer 11:85–95

Chagin VO, Stear JH, Cardoso MC (2010) Organization of DNA replication. Cold Spring Harb Perspect Biol 2:a000737

Chakraborty A, Shen Z, Prasanth SG (2011) "ORCanization" on heterochromatin: linking DNA replication initiation to chromatin organization. Epigenetics 6:665–670

Chen CA, Wang TY, Varadharaj S, Reyes LA, Hemann C, Talukder MA, Chen YR, Druhan LJ, Zweier JL (2010) S-glutathionylation uncouples eNOS and regulates its cellular and vascular function. Nature 468:1115–1118

Chen SJ, Zhou GB, Zhang XW, Mao JH, de Thé H, Chen Z (2011) From an old remedy to a magic bullet: molecular mechanisms underlying the therapeutic effects of arsenic in fighting leukemia. Blood 117:6425–6437

Cheng X, Blumenthal RM (2008) Mammalian DNA methyltransferases: a structural perspective. Structure 16:341–350

Cheng X, Blumenthal RM (2010) Coordinated chromatin control: structural and functional linkage of DNA and histone methylation. Biochemistry 49:2999–3008

Coppin JF, Qu W, Waalkes MP (2008) Interplay between cellular methyl metabolism and adaptive efflux during oncogenic transformation from chronic arsenic exposure in human cells. J Biol Chem 283:19342–19350

Corpet A, Almouzni G (2009) Making copies of chromatin: the challenge of nucleosomal organization and epigenetic information. Trends Cell Biol 19:29–41

Cyr AR, Domann FE (2011) The redox basis of epigenetic modifications: from mechanisms to functional consequences. Antioxid Redox Signal 15:551–589

Dabkowski ER, Baseler WA, Williamson CL, Powell M, Razunguzwa TT, Frisbee JC, Hollander JM (2010) Mitochondrial dysfunction in the type 2 diabetic heart is associated with alterations in spatially distinct mitochondrial proteomes. Am J Physiol Heart Circ Physiol 299(2):H529–H540

Dansen TB, Smits LM, van Triest MH, de Keizer PL, van Leenen D, Koerkamp MG, Szypowska A, Meppelink A, Brenkman AB, Yodoi J, Holstege FC, Burgering BM (2009) Redox-sensitive cysteines bridge p300/CBP-mediated acetylation and FoxO4 activity. Nat Chem Biol 5:664–672

Dennery PA (2010) Oxidative stress in development: nature or nurture? Free Radic Biol Med 49:1147–1151

Diaz Vivancos P, Wolff T, Markovic J, Pallardó FV, Foyer CH (2010) A nuclear glutathione cycle within the cell cycle. Biochem J 431:169–178

Doyle K, Fitzpatrick FA (2010) Redox signaling, alkylation (carbonylation) of conserved cysteines inactivates class I histone deacetylases 1, 2, and 3 and antagonizes their transcriptional repressor function. J Biol Chem 285:17417–17424

Drenckhahn JD (2011) Heart development: mitochondria in command of cardiomyocyte differentiation. Dev Cell 21:392–393

Dulce RA, Schulman IH, Hare JM (2011) S-glutathionylation: a redox-sensitive switch participating in nitroso-redox balance. Circ Res 108:531–533

Easwaran HP, Van Neste L, Cope L, Sen S, Mohammad HP, Pageau GJ, Lawrence JB, Herman JG, Schuebel KE, Baylin SB (2010) Aberrant silencing of cancer-related genes by CpG hypermethylation occurs independently of their spatial organization in the nucleus. Cancer Res 70(20):8015–8024

Ernster L, LEE CP (1964) Biological oxidoreductions. Annu Rev Biochem 33:729–790

Fialkow L, Wang Y, Downey GP (2007) Reactive oxygen and nitrogen species as signaling molecules regulating neutrophil function. Free Radic Biol Med 42:153–164

Finley LW, Carracedo A, Lee J, Souza A, Egia A, Zhang J, Teruya-Feldstein J, Moreira PI, Cardoso SM, Clish CB, Pandolfi PP, Haigis MC (2011) SIRT3 opposes reprogramming of cancer cell metabolism through HIF1α destabilization. Cancer Cell 19:416–428

Flora SJ (2011) Arsenic-induced oxidative stress and its reversibility. Free Radic Biol Med 51:257–281

Gasser SM, Paro R, Stewart F, Aasland R (1998) The genetics of epigenetics. Cell Mol Life Sci 54:1–5

Gloire G, Piette J (2009) Redox regulation of nuclear post-translational modifications during NF-kappa B activation. Antioxid Redox Signal 11:2209–2222

Gloire G, Legrand-Poels S, Piette J (2006) NF-kappaB activation by reactive oxygen species: fifteen years later. Biochem Pharmacol 72:1493–1505

Gough NR, Foley JF (2010) Focus issue: systems analysis of protein phosphorylation. Sci Signal 3(137):eg6

Gutteridge JM, Halliwell B (2010) Antioxidants: molecules, medicines, and myths. Biochem Biophys Res Commun 393:561–564

Guttmann RP (2010) Redox regulation of cysteine-dependent enzymes. J Anim Sci 88:1297–1306

Guz J, Foksinski M, Siomek A, Gackowski D, Rozalski R, Dziaman T, Szpila A, Olinski R (2008) The relationship between 8-oxo-7,8-dihydro-2'-deoxyguanosine level and extent of cytosine methylation in leukocytes DNA of healthy subjects and in patients with colon adenomas and carcinomas. Mutat Res 640:170–173

Haig D (2004) The (dual) origin of epigenetics. Cold Spring Harb Symp Quant Biol 69:67–70

Haqqani AS, Kelly JF, Birnboim HC (2002) Selective nitration of histone tyrosine residues in vivo in mutatect tumors. J Biol Chem 277:3614–3621

He XJ, Chen T, Zhu JK (2011) Regulation and function of DNA methylation in plants and animals. Cell Res 21:442–465

Heo J (2011) Redox control of GTPases: from molecular mechanisms to functional significance in health and disease. Antioxid Redox Signal 14:689–724

Herring SW (1993) Formation of the vertebrate face: epigenetic and functional influences. Am Zool 33:472

Higuchi Y, Yoshimoto T (2004) Promoting effects of polyunsaturated fatty acids on chromosomal giant DNA fragmentation associated with cell death induced by glutathione depletion. Free Radic Res 38:649–658

Hitchler MJ, Domann FE (2007) An epigenetic perspective on the free radical theory of development. Free Radic Biol Med 43:1023–1036

Holliday R (1987) The inheritance of epigenetic defects. Science 238:163–170

Ishimoto T, Nagano O, Yae T, Tamada M, Motohara T, Oshima H, Oshima M, Ikeda T, Asaba R, Yagi H, Masuko T, Shimizu T, Ishikawa T, Kai K, Takahashi E, Imamura Y, Baba Y, Ohmura M, Suematsu M, Baba H, Saya H (2011) CD44 variant regulates redox status in cancer cells by stabilizing the xCT subunit of system xc(−) and thereby promotes tumor growth. Cancer Cell 19:387–400

Janssen-Heininger YM, Mossman BT, Heintz NH, Forman HJ, Kalyanaraman B, Finkel T, Stamler JS, Rhee SG, van der Vliet A (2008) Redox-based regulation of signal transduction: principles, pitfalls, and promises. Free Radic Biol Med 45:1–17

Jindal E, Goswami SK (2011) In cardiac myoblasts, cellular redox regulates FosB and Fra-1 through multiple cis-regulatory modules. Free Radic Biol Med 51:1512–1521

Joulie M, Miotto B, Defossez PA (2010) Mammalian methyl-binding proteins: what might they do? Bioessays 32:1025–1032

Kabra DG, Gupta J, Tikoo K (2009) Insulin induced alteration in post-translational modifications of histone H3 under a hyperglycemic condition in L6 skeletal muscle myoblasts. Biochim Biophys Acta 1792:574–583

Kim KS, Choi HW, Yoon HE, Kim IY (2010) Reactive oxygen species generated by NADPH oxidase 2 and 4 are required for chondrogenic differentiation. J Biol Chem 285:40294–44302

Klatt P, Molina EP, Lamas S (1999) Nitric oxide inhibits c-Jun DNA binding by specifically targeted S-glutathionylation. J Biol Chem 274:15857–15864

Kornberg MD, Sen N, Hara MR, Juluri KR, Nguyen JV, Snowman AM, Law L, Hester LD, Snyder SH (2010) GAPDH mediates nitrosylation of nuclear proteins. Nat Cell Biol 12:1094–1100

Lindermayr C, Saalbach G, Bahnweg G, Durner J (2006) Differential inhibition of Arabidopsis methionine adenosyltransferases by protein S-nitrosylation. J Biol Chem 281:4285–4291

Liu S, Wu LC, Pang J, Santhanam R, Schwind S, Wu YZ, Hickey CJ, Yu J, Becker H, Maharry K, Radmacher MD, Li C, Whitman SP, Mishra A, Stauffer N, Eiring AM, Briesewitz R, Baiocchi RA, Chan KK, Paschka P, Caligiuri MA, Byrd JC, Croce CM, Bloomfield CD, Perrotti D, Garzon R, Marcucci G (2010) Sp1/NFkappaB/HDAC/miR-29b regulatory network in KIT-driven myeloid leukemia. Cancer Cell 17:333–347

Lujambio A, Esteller M (2009) How epigenetics can explain human metastasis: a new role for microRNAs. Cell Cycle 8:377–382

Lyon MF (1993) Epigenetic inheritance in mammals. Trends Genet 9:123–128

Maller C, Schröder E, Eaton P (2011) Glyceraldehyde 3-phosphate dehydrogenase is unlikely to mediate hydrogen peroxide signaling: studies with a novel anti-dimedone sulfenic acid antibody. Antioxid Redox Signal 14:49–60

Markovic J, García-Gimenez JL, Gimeno A, Viña J, Pallardó FV (2010) Role of glutathione in cell nucleus. Free Radic Res 44:721–733

Matés JM, Segura JA, Alonso FJ, Márquez J (2010) Roles of dioxins and heavy metals in cancer and neurological diseases using ROS-mediated mechanisms. Free Radic Biol Med 49:1328–1341

Mitra S, Izumi T, Boldogh I, Bhakat KK, Hill JW, Hazra TK (2002) Choreography of oxidative damage repair in mammalian genomes. Free Radic Biol Med 33:15–28

Moldovan L, Moldovan NI (2004) Oxygen free radicals and redox biology of organelles. Histochem Cell Biol 122:395–412

Nediani C, Raimondi L, Borchi E, Cerbai E (2011) NO/ROS generation and nitroso/redox imbalance in heart failure: from molecular mechanisms to therapeutic implications. Antioxid Redox Signal 14:289–331

Nelson KK, Subbaram S, Connor KM, Dasgupta J, Ha XF, Meng TC, Tonks NK, Melendez JA (2006) Redox-dependent matrix metalloproteinase-1 expression is regulated by JNK through Ets and AP-1 promoter motifs. J Biol Chem 281:14100–14110

Nikitovic D, Holmgren A, Spyrou G (1998) Inhibition of AP-1 DNA binding by nitric oxide involving conserved cysteine residues in Jun and Fos. Biochem Biophys Res Commun 242:109–112

Nishida M, Ogushi M, Suda R, Toyotaka M, Saiki S, Kitajima N, Nakaya M, Kim KM, Ide T, Sato Y, Inoue K, Kurose H (2011) Heterologous down-regulation of angiotensin type 1 receptors by purinergic P2Y2 receptor stimulation through S-nitrosylation of NF-{kappa}B. Proc Natl Acad Sci U S A 108:6662–6667

Nott A, Watson PM, Robinson JD, Crepaldi L, Riccio A (2008) S-Nitrosylation of histone deacetylase 2 induces chromatin remodelling in neurons. Nature 455:411–415

Oliveira-Marques V, Marinho HS, Cyrne L, Antunes F (2009) Role of hydrogen peroxide in NF-kappaB activation: from inducer to modulator. Antioxid Redox Signal 11:2223–2243

Osoata GO, Yamamura S, Ito M, Vuppusetty C, Adcock IM, Barnes PJ, Ito K (2009) Nitration of distinct tyrosine residues causes inactivation of histone deacetylase 2. Biochem Biophys Res Commun 384:366–371

Pamplona R, Naudí A, Gavín R, Pastrana MA, Sajnani G, Ilieva EV, Del Río JA, Portero-Otín M, Ferrer I, Requena JR (2008) Increased oxidation, glycoxidation, and lipoxidation of brain proteins in prion disease. Free Radic Biol Med 45:1159–1166

Perkins C, Kim CN, Fang G, Bhalla KN (2000) Arsenic induces apoptosis of multidrug-resistant human myeloid leukemia cells that express Bcr-Abl or overexpress MDR, MRP, Bcl-2, or Bcl-x(L). Blood 95:1014–1022

Pierard V, Guiguen A, Colin L, Wijmeersch G, Vanhulle C, Van Driessche B, Dekoninck A, Blazkova J, Cardona C, Merimi M, Vierendeel V, Calomme C, Nguyên TL, Nuttinck M, Twizere JC, Kettmann R, Portetelle D, Burny A, Hirsch I, Rohr O, Van Lint C (2010) DNA cytosine methylation in the bovine leukemia virus promoter is associated with latency in a lymphoma-derived B-cell line: potential involvement of direct inhibition of cAMP-responsive element (CRE)-binding protein/CRE modulator/activation transcription factor binding. J Biol Chem 285:19434–19449

Rahman I (2002) Oxidative stress, transcription factors and chromatin remodelling in lung inflammation. Biochem Pharmacol 64:935–942

Rahman I, Marwick J, Kirkham P (2004) Redox modulation of chromatin remodeling: impact on histone acetylation and deacetylation, NF-kappaB and pro-inflammatory gene expression. Biochem Pharmacol 68:1255–1267

Rajendrasozhan S, Yang SR, Edirisinghe I, Yao H, Adenuga D, Rahman I (2008) Deacetylases and NF-kappaB in redox regulation of cigarette smoke-induced lung inflammation: epigenetics in pathogenesis of COPD. Antioxid Redox Signal 10:799–811

Rakyan VK, Down TA, Balding DJ, Beck S (2011) Epigenome-wide association studies for common human diseases. Nat Rev Genet 12:529–541

Riggs AD, Martienssen RA, Russo VEA (1996) Introduction in epigenetic mechanisms of gene regulation. In: Russo VEA et al (eds) Cold Spring Harbor Laboratory Press, Cold Spring Harbor, New York, pp 1–4

Rishi V, Bhattacharya P, Chatterjee R, Rozenberg J, Zhao J, Glass K, Fitzgerald P, Vinson C (2010) CpG methylation of half-CRE sequences creates C/EBPalpha binding sites that activate some tissue-specific genes. Proc Natl Acad Sci U S A 107:20311–20316

Schreck R, Rieber P, Baeuerle PA (1991) Reactive oxygen intermediates as apparently widely used messengers in the activation of the NF-kappa B transcription factor and HIV-1. EMBO J 10:2247–2258

Selvi BR, Mohankrishna DV, Ostwal YB, Kundu TK (2010) Small molecule modulators of histone acetylation and methylation: a disease perspective. Biochim Biophys Acta 1799:810–828

Shlomai J (2010) Redox control of protein-DNA interactions: from molecular mechanisms to significance in signal transduction, gene expression, and DNA replication. Antioxid Redox Signal 13:1429–1476

Simone NL, Soule BP, Ly D, Saleh AD, Savage JE, Degraff W, Cook J, Harris CC, Gius D, Mitchell JB (2009) Ionizing radiation-induced oxidative stress alters miRNA expression. PLoS One 4:e6377

Svedružić ŽM, Reich NO (2005) DNA cytosine C5 methyltransferase Dnmt1: catalysis-dependent release of allosteric inhibition. Biochemistry 44:9472–9485

Tanaka T, Kurose A, Halicka HD, Huang X, Traganos F, Darzynkiewicz Z (2006) Nitrogen oxide-releasing aspirin induces histone H2AX phosphorylation, ATM activation and apoptosis preferentially in S-phase cells: involvement of reactive oxygen species. Cell Cycle 5:1669–1674

Tasaka S, Amaya F, Hashimoto S, Ishizaka A (2008) Roles of oxidants and redox signaling in the pathogenesis of acute respiratory distress syndrome. Antioxid Redox Signal 10:739–753

Tew KD, Townsend DM (2011) Redox platforms in cancer drug discovery and development. Curr Opin Chem Biol 15(1):156–161

Tikoo K, Lau SS, Monks TJ (2001) Histone H3 phosphorylation is coupled to poly-(ADP-ribosylation) during reactive oxygen species-induced cell death in renal proximal tubular epithelial cells. Mol Pharmacol 60:394–402

Tong KI, Padmanabhan B, Kobayashi A, Shang C, Hirotsu Y, Yokoyama S, Yamamoto M (2007) Different electrostatic potentials define ETGE and DLG motifs as hinge and latch in oxidative stress response. Mol Cell Biol 27:7511–7521

Turner BM (1998) Histone acetylation as an epigenetic determinant of long-term transcriptional competence. Cell Mol Life Sci 54:21–31

Ufer C, Wang CC, Borchert A, Heydeck D, Kuhn H (2010) Redox control in mammalian embryo development. Antioxid Redox Signal 13:833–875

Upadhyay AK, Cheng X (2011) Dynamics of histone lysine methylation: structures of methyl writers and erasers. Prog Drug Res 67:107–124

Upham BL, Trosko JE (2009) Oxidative-dependent integration of signal transduction with intercellular gap junctional communication in the control of gene expression. Antioxid Redox Signal 11:297–307

Varier RA, Timmers HT (2011) Histone lysine methylation and demethylation pathways in cancer. Biochim Biophys Acta 1815:75–89

Waddington CH (1938) The epigenetics of birds. University Press, Cambridge, UK, 1952

Weaver AM (2009) Regulation of cancer invasion by reactive oxygen species and Tks family scaffold proteins. Sci Signal 2:e56

Wertz IE, Dixit VM (2010) Signaling to NF-kappaB: regulation by ubiquitination. Cold Spring Harb Perspect Biol 2(3):a003350

Woo HA, Yim SH, Shin DH, Kang D, Yu DY, Rhee SG (2010) Inactivation of peroxiredoxin I by phosphorylation allows localized H(2)O(2) accumulation for cell signaling. Cell 140:517–528

Xanthoudakis S, Curran T (1992) Identification and characterization of Ref-1, a nuclear protein that facilitates AP-1 DNA-binding activity. EMBO J 11:653–665

Xie H, Wang M, de Andrade A, Bonaldo MD, Galat V, Arndt K, Rajaram V, Goldman S, Tomita T, Soares MB (2011) Genome-wide quantitative assessment of variation in DNA methylation patterns. Nucleic Acids Res 39:4099–4108

Yang SR, Chida AS, Bauter MR, Shafiq N, Seweryniak K, Maggirwar SB, Kilty I, Rahman I (2006) Cigarette smoke induces proinflammatory cytokine release by activation of NF-kappaB and posttranslational modifications of histone deacetylase in macrophages. Am J Physiol Lung Cell Mol Physiol 29:L46–L57

Yao H, Hwang JW, Moscat J, Diaz-Meco MT, Leitges M, Kishore N, Li X, Rahman I (2010) Protein kinase C zeta mediates cigarette smoke/aldehyde- and lipopolysaccharide-induced lung inflammation and histone modifications. J Biol Chem 285:5405–5416

Yu DS, Cortez D (2011) A role for cdk9-cyclin k in maintaining genome integrity. Cell Cycle 10:28–32

Zhang X, Yang Z, Khan SI, Horton JR, Tamaru H, Selker EU, Cheng X (2003) Structural basis for the product specificity of histone lysine methyltransferases. Mol Cell 12:177–185

Zhu H, Shan L, Schiller PW, Mai A, Peng T (2010) Histone deacetylase-3 activation promotes tumor necrosis factor-alpha (TNF-alpha) expression in cardiomyocytes during lipopolysaccharide stimulation. J Biol Chem 285:9429–9436

Part V
Understanding of Epigenetics: A Chemical Biology Approach and Epigenetic Therapy

Chapter 24
Stem Cell Plasticity in Development and Cancer: Epigenetic Origin of Cancer Stem Cells

Mansi Shah and Cinzia Allegrucci

Abstract Stem cells are unique cells that can self-renew and differentiate into many cell types. Plasticity is a fundamental characteristic of stem cells and it is regulated by reversible epigenetic modifications. Although gene-restriction programs are established during embryonic development when cell lineages are formed, stem cells retain a degree of flexibility that is essential for tissue regeneration. For instance, quiescent adult stem cells can be induced to proliferate and trans-differentiate in response to injury. The same degree of plasticity is observed in cancer, where cancer cells with stem cell characteristics (or cancer stem cells) are formed by transformation of normal stem cells or de-differentiation of somatic cells. Reprogramming experiments with normal somatic cells and cancer cells show that epigenetic landscapes are more plastic than originally thought and that their manipulation can induce changes in cell fate. Our knowledge of stem cell function is still limited and only by understanding the mechanisms regulating developmental potential together with the definition of epigenetic maps of normal and diseased tissues we can reveal the true extent of their plasticity. In return, the control of plastic epigenetic programs in stem cells will allow us to develop effective treatments for degenerative diseases and cancer.

M. Shah
School of Veterinary Medicine and Science, University of Nottingham,
Sutton Bonington Campus, LE12 5RD Loughborough, Leicestershire, UK

C. Allegrucci (✉)
School of Veterinary Medicine and Science, University of Nottingham,
Sutton Bonington Campus, LE12 5RD Loughborough, Leicestershire, UK

Centre for Genetics and Genomics, University of Nottingham, Queen's Medical Centre,
NG7 2UH Nottingham, UK
e-mail: cinzia.allegrucci@nottingham.ac.uk

24.1 Introduction

How are cells in our body programmed to maintain their identity and function throughout life? The answer to this fundamental question is based on important processes that are initiated during embryo development and maintained in adulthood. This book chapter will describe and discuss the mechanisms that control cell and tissue homeostasis and how these are altered in cancer.

Cell identity is established during embryogenesis when the developmental potential of embryonic cells is restricted by differentiation programs that channel their fate to tissue-specific stem cells and specialised cell types. These dynamic events occur in cells with the same genetic information, thus cell fate depends on the epigenetic regulation of that genetic code. "Epigenetics" can be defined as regulation of gene expression that occurs by modifications imposed on the chromatin without change in the DNA sequence (Bird 2007). It is by changes in chromatin organisation that epigenetic modifications establish heritable transcriptional states responsible for the maintenance of cell function.

Epigenetic regulation includes DNA methylation, modification of histone tails and modulation by non coding RNAs (ncRNAs). Together with chromatin remodelling complexes, these modifications control chromatin organisation and regulate gene transcription (Jaenisch and Bird 2003).

DNA methylation is responsible for gene silencing and occurs at position 5 of cytosine (5mC) within CpG dinucleotides present in repetitive sequences and CpG islands in gene promoters and intragenic regions (Ball et al. 2009; Sharma et al. 2010). DNA methylation is maintained or established *de novo* by the DNA methyltransferases enzymes DNMT1 and DNMT3A/3B/3L, respectively (Bird 2002). Histone modifications comprise a vast range of post-translational modifications, such as acetylation, methylation, phosphorylation, ubiquitylation and ribosylation. These modifications can induce both activation and repression of transcription and their interactions function as a "code" defining cellular states (Turner 2007). Nucleosome remodelling and modulation by ncRNAs are the most important non covalent epigenetic modifications. Non coding RNAs, including microRNAs (miRNAs) and long non coding RNAs (lncRNAs), are single stranded transcripts involved in mRNA degradation and chromatin remodelling (Pauli et al. 2011). While heritable, epigenetic modifications are reversible and their dynamic interplay provides cells with ability to respond to environmental cues. Therefore it is easy to imagine how the epigenetic landscape created by these modifications can regulate phenotype plasticity in different cell types during normal development, but also cause disease if abnormally regulated.

Cancer is a disease characterised by abnormal cell proliferation and it is associated with both genetic lesions and epigenetic abnormalities. Because it can be portrayed as a process of aberrant cell proliferation and differentiation, cancer has been described as "a problem of developmental biology" where a marked resemblance between cancer cells and embryonic cells exists (Pierce and Johnson 1971). Indeed, cancer cells re-initiate epigenetic programs that favour cell growth and survival at the expense of differentiation, thus behaving like undifferentiated embryonic cells and stem cells. As cancer cells depend on those mechanisms that maintain stem cell plasticity (Garraway and Sellers 2006), it is not a coincidence that many tumour suppressor genes that are epigenetically silenced in cancer are developmental genes involved in the regulation of stem cells (Barrero et al. 2010). It is precisely how stem cell plasticity is programmed in development and cancer that will be the focus of our discussion.

24.2 Epigenetics and Development

Epigenetic modifications regulate the acquisition of totipotency and subsequent progressive restriction of totipotent potential during embryonic development. Acquisition of totipotency is associated with two epigenetic reprogramming events: the formation of the zygote and the germ line (Hemberger et al. 2009). Both developmental stages require resetting of a differentiated epigenetic landscape to establish a new state with augmented developmental potency. Differentiation of somatic cells then requires the establishment of specific epigenetic programs that restrict their potential and maintain lineage memory. This section will describe the epigenetic modifications occurring during embryo development and explain how embryonic developmental potential is programmed in embryonic cells, somatic cells and germ cells.

24.2.1 Epigenetic Reprogramming During Embryo Development

Embryo development initiates with the fusion of the male and female pronuclei after fertilisation. The formation of the zygote is followed by epigenetic reprogramming of the specialised gametic genomes to ensure that the embryonic genome acquires totipotency, defined as the ability of a cell to form an entire organism. Immediately after fertilisation, the paternal nucleus undergoes profound chromatin remodelling. This involves exchange of protamines for histones in the nucleosomes and active DNA demethylation (Oswald et al. 2000; Mayer et al. 2000; Santos et al. 2002). Although a specific DNA demethylase enzyme has not been identified, a process involving DNA repair through the intermediate 5-hydroxymethylcytosine (5hmC) has been proposed (Wossidlo et al. 2010, 2011; Hemberger et al. 2009). After fusion, the progressive decline in DNA methylation up to the morula stage is due to passive loss of methylated cytosine marks during DNA replication (Howell et al. 2001). Some genomic sequences escape this demetylation, including some repetitive sequences and most imprinted genes (Meissner 2010). Concurrent with DNA demethylation, reprogramming of histone modifications also takes place. The newly incorporated histones in the paternal pronucleus gradually increase active marks, such as acetylation of histone H3 at lysine 9 (H3K9ac), methylation of histone H3 at lysine 4 (H3K4me3), and repressive marks, e.g. methylated histone H3 at lysine 9 (H3K9me1, H3K9me2) and methylated histone H3 at lysine 27 (H3K27me3) (Meissner 2010). Subsequent to the first cleavage divisions, the embryo undergoes segregation of the first two lineages, the inner cell mass (ICM) and the trophectoderm. The cells of the ICM are pluripotent embryonic cells, able to differentiate to all somatic lineages and the germ line (Wray et al. 2010). Epigenetic programming of ICM cells includes *de novo* DNA methylation, acquisition of H3K9ac, H3K27me3, H3K4me3, H3K9me2 and H3K9me3 (Morgan et al. 2005). Re-establishment of DNA methylation is essential for normal embryonic development, as demonstrated by knockout experiments where deletion of DNMTs and other epigenetic modifiers participating in DNA methylation (LSH and G9a) causes embryonic lethality (Okano et al. 1999; Myant et al. 2011). During gastrulation and differentiation of embryonic cells to somatic lineages, a progressive decrease in plasticity is observed and this is accomplished by a program of epigenetic modifications that restricts cell fate, retains cell memory and confers cellular specialisation.

However, epigenetic restrictions imposed during differentiation are reprogrammed in the germ line. Germ cells derive from embryonic precursors of gametes defined as primordial germ cells (PGC), which are responsible for the development of a new organism in the next generation. Epigenetic reprogramming in the germ line is essential for the generation of a cellular state that will allow totipotency in the newly formed embryo. In addition, reprogramming of PGC ensures an equivalent epigenetic state in both sexes prior to differentiation into mature gametes and erasure of acquired epimutations which could be inherited in the next generation (Allegrucci et al. 2005). PGC are specified in the proximal epiblast and then migrate through the hindgut to the developing gonads. It is during migration and after colonisation of the gonads that extensive epigenetic reprogramming occurs in these cells. This involves loss of H3K9me2 and DNA methylation, and an increase in H3K27me3. It is thought that this epigenetic configuration, enriched in H3K27me3, H3K4me2/me3 and H3K9ac confers PGC with the required plasticity to regain pluripotency (Hemberger et al. 2009). In addition, loss of DNA methylation at imprinted genes ensures erasure of epimutations and correct re-establishment of monoallelic expression for gene dosage in the next generation (Allegrucci et al. 2005; Sasaki and Matsui 2008).

The epigenetic reprogramming and programming of cell plasticity during development is orchestrated by a battery of epigenetic modifiers. Their coordinated action ensures a correct program of cell proliferation and differentiation (Table 24.1).

Table 24.1 Key epigenetic modifiers regulating cell proliferation and differentiation

Epigenetic modifier	Modification/Function	Epigenetic mark	Implicated in cancer
EZH1/2	Histone methylation	H3K27me3	√
SUZ12	PcG-PRC2 complex	–	√
EED	PcG-PRC2 complex	–	√
JARID2	Recruits/stabilise PRC2	–	–
RING1	Histone ubiquitylation	H2AK119u	√
BMI-1	PcG-PRC1 complex	–	√
SET/MLL	Histone methylation	H3K4me3	√
	Histone acetylation	H4K16ac	√
UTX	Histone demethylation	H3K27me3	√
JMJD3	Histone demethylation	H3K27me3	√
JARID1A/B	Histone demethylation	H3K4me3	√
JMJD1A	Histone demethylation	H3K9me2	√
JMJD2C	Histone demethylation	H3K9me3	√
G9a	Histone methylation	H3K9me1/2	√
DNMT1	5mC-DNA methylation	5mC	√
DNMT3A	5mC-DNA methylation	5mC	√
DNMT3B	5mC-DNA methylation	5mC	√
DNMT3L	5mC-DNA methylation	5mC	√
GADD45	DNA demethylation	–	√
AID	DNA demethylation	–	–
LSD1	Histone demethylation	H3K9me2	√
		H3K4me1/2/3	√
SUV39	Histone methylation	H3K9me3	√
BRG1	Chromatin remodelling	–	√
BAF250	Chromatin remodelling	–	√
BAF155	Chromatin remodelling	–	√
CHD1/3/4/7	Chromatin remodelling	–	√
HDAC1/2	Histone deacetylation	–	√
MBD3	Chromatin remodelling	–	–
BPTF	Chromatin remodelling	–	√
TIP60-p400	Chromatin remodelling	–	√
CBP/p300	Histone acetylation	–	√
SETDB1	Histone methylation	H3K9me3	√
DICER	ncRNA processing	–	√
TET1/2/3	DNA hydroxymethylation	5hmC	√

24.3 Epigenetic Regulation of Stem Cells

In the previous section we have reviewed the epigenetic events that regulate development and program cell differentiation in the embryo. Although pluripotent cells exist only for a limited period of time before gastrulation, they can be isolated from the embryo and maintained *in vitro* as embryonic stem cells (ESC). Therefore ESC can be studied as *in vitro* model of naive embryonic cells and differentiated into many different cell types. Differentiation is not limited to embryonic development, but it continues in the adult as continuous supply of specialised cells is needed for tissue turn-over and repair. This is accomplished by lineage restricted multipotent cells, or adult stem cells (ASC). Correct stem cell function is essential during an individual's life starting at the time when tissues are formed and later on, when they need to be regenerated and repaired. By analysing the epigenetic control of ESC and ASC, this section will describe how developmental plasticity of stem cells is programmed for correct function. Knowledge of how stem cell programs are regulated is important not only to advance stem cell-based therapies but also to understand how we can overcome diseases characteristic of stem cell dysfunction.

24.3.1 Control of Embryonic Stem Cells

ESC can be derived from the blastocyst ICM and their pluripotency maintained *in vitro* for many cell generations. ESC can symmetrically self-renew, hence giving rise to two identical stem cells. ESC ability to self-renew and to respond to developmental cues is controlled by a unique gene expression program. The ground state of pluripotent ESC is defined by the expression of a core network of transcription factors that include OCT4, SOX2 and NANOG. These factors act both as transcription activators and repressors, by activating genes involved in cell proliferation and self-renewal while repressing the expression of lineage-specific genes promoting differentiation (Young 2011). This bivalent state of ESC is essential for pluripotency and it is regulated epigenetically by the interplay of core transcription factors and Trithorax (TrxG) and Polycomb (PcG) epigenetic modifiers. TrxG-related proteins (SET/MLL) catalyse H3K4me3 at promoters of active genes, whereas PcG proteins catalyse histone modifications that are associated with gene silencing. PcG proteins include two complexes, PRC1 and PRC2, responsible for H3K27me3 and ubiquitylation of histone H2A at lysine 119 (H2AK119u), respectively (Meissner 2010). While PRC2 is required for initial gene silencing, recruitment of PRC1 stabilises the established transcriptionally repressive state. It is the presence of both active H3K4me3 and repressive H3K27me3 marks (bivalent domain) at developmentally regulated genes that allows ESC to remain in a poised state, ready for activation upon differentiation. Therefore ESC show a global open chromatin structure, with about 75% of gene promoters enriched for H3K4me3. These promoters can be active or inactive, depending on H3K27me3 co-occupancy.

Among silencing mechanisms DNA methylation plays a fundamental role in ESC. ESC present about 60–80% of methylated CpG nucleotides, with a unique distribution (Meissner 2010). Comprehensive maps of DNA methylation in ESC have demonstrated that the majority of high CpG promoters (HCP) are lacking methylation and are enriched in H3K4me3. These represent housekeeping genes, pluripotency genes and key developmental genes. In contrast, tissue specific gene promoters with low CpG density (LCP) are mostly methylated (Mikkelsen et al. 2007; Meissner et al. 2008). Therefore an epigenetic landscape presenting either

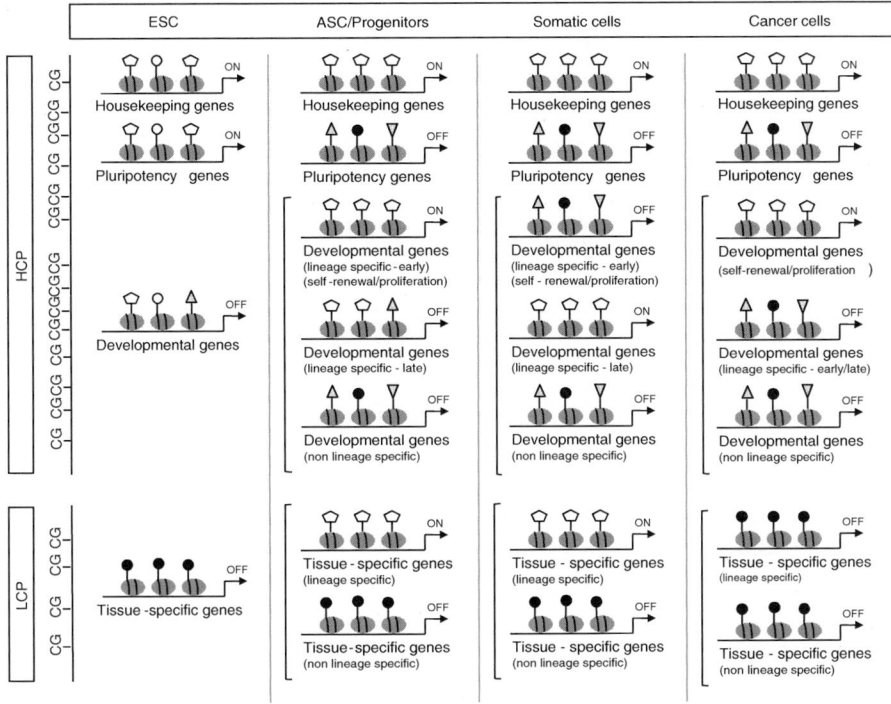

Fig. 24.1 Epigenetic landscapes regulating stem cell plasticity in development and cancer. HCP and LCP gene promoters are enriched for genes with different epigenetic regulation in different cell types. The figure shows how housekeeping genes, pluripotency genes, developmental genes and tissue-specific genes are epigenetically regulated in ESC, ASC, differentiated (somatic) cells and cancer cells. During differentiation there is a decrease in cell plasticity due to loss of bivalent domains and acquisition of repressive chromatin marks that restrict cell fate. Cancer cells reactivate an epigenetic landscape that is more plastic and shifted towards self-renewal and proliferation at the expense of differentiation (△: H3K4me3; ○: 5hmC; ▲: H3K27me3; ●: 5mC; ▽: H3K9me2/3)

unmethylated promoters (HCP with H3K4me3 or bivalent domain with H3K4me3/H3K27me3), or methylated promoters (LCP) define ESC (Fig. 24.1). In addition to methylation of CpG dinucleotides, other modifications of the DNA have been discovered in ESC. These include cytosine methylation in a non CG context (Lister et al. 2009) and cytosine hydromethylation (5hmC). Hydroxylation of 5mC to 5hmC is catalysed by the TET family of enzymes (Koh et al. 2011) and it is believed to be involved in the demethylation of 5mC and prevention of DNMTs activity (Xu et al. 2011). A genome-wide study of 5hmC in ESC revealed that this mark is enriched in gene bodies, transcription start sites of HCP promoters and enhancers. Bivalent or PcG only marked promoters are also particularly enriched for 5hmC (Xu et al. 2011; Pastor et al. 2011). Although a clear function for 5hmC in transcription regulation is still elusive, its distribution suggests a role in preparing genomic loci for activation upon differentiation. Indeed, 5hmC has been shown to be present in ESC, but declines after differentiation (Tahiliani et al. 2009; Ficz et al. 2011). Finally, ncRNAs also participate in the epigenetic regulation of ESC plasticity. MicroRNAs regulate stability and translation of mRNAs involved in stem cell self-renewal and differentiation. ESC express a unique set of miRNA whose transcription is regulated by the core pluripotency factors, and these miRNA are involved both in sustaining self-renewal (e.g mir-290-295/302) and inducing rapid degradation of ESC transcription factors during differentiation (e.g mir-145) (Marson et al. 2008; Tay et al. 2008; Xu et al. 2009). In addition, many lineage-specific miRNA gene promoters are co-occupied by OCT4/NANOG/SOX2 and PcG and these are repressed in ESC, but become active upon differentiation (e.g. let-7, mir-155, mir-124, mir-9) (Young 2011; Pauli et al. 2011). Balance between self-renewal and differentiation is a key characteristic of ESC. Altogether bivalent histone modifications, DNA methylation and miRNAs contribute to the establishment of an open chromatin state that allows undifferentiated cell function and the ability to respond to developmental signals in a timely fashion.

24.3.2 Control of Adult Stem Cells and Somatic Cells

Differentiation of pluripotent cells is associated with a loss of developmental potency that ensures cellular specialisation and committed identity. During lineage specification, ASC are formed and it is from these committed multipotent stem cells that specialised cells are derived. The role of stem cells in the adult is to maintain tissue homeostasis by regenerating aged or damaged cells. ASC reside in tissue-specific niches that control their asymmetrical self-renewal, defined as the ability to form a stem cell and a differentiated progenitor at cell division. ASC are more restricted in their differentiation potential compared to ESC as they can only give rise to multiple cell types within a tissue under physiological conditions. Genome-wide maps of epigenetic modifications in ASC and differentiated cells show that restriction in developmental potential is associated with a resolution of ESC open chromatin into a more restricted configuration (Fig. 24.1). Silencing of pluripotency genes is readily observed during differentiation, by loss of H3K4me3, gain of H3K27me3, H3K9me3, DNA methylation (Barrero et al. 2010) and expression of specific miRNAs (e.g. mir-134, mir-296, mir-470) (Tay et al. 2008). Differentiation into a specific lineage involves expression of genes specific to that cell type and silencing of genes expressed in other tissues. In this way, differentiation into the neural stem/progenitor cells (NSC) is accompanied by a decrease of H3K27me3 at neural genes silenced by bivalent marks, which correspond to increased gene expression. Genes poised or weakly induced retain bivalent marks, while H3K27me3 silencing is increased in non-neural lineage genes, together with H3K9me3 (Hawkins et al. 2010; Bernstein et al. 2006; Mikkelsen et al. 2007; Bracken et al. 2006). The same pattern is also observed in muscle and germ cell differentiation (Caretti et al. 2004; Chen et al. 2005; Asp et al. 2011), suggesting a PcG-mediated regulation of cell fate decisions.

Both H3K4me3 and bivalent HPC remain mostly unmethylated during differentiation. In contrast, resolution to univalent H3K27me3 mark results in an increase in DNA methylation and a complete loss of the bivalent marks results in DNA hypermethylation. A different epigenetic regulation is observed at LCP associated with tissue specific genes, as methylated LCP associated with neural genes gain H3K4me3 and non lineage specific genes retain DNA methylation (Meissner et al. 2008). Although overall DNA methylation levels are similar in pluripotent and differentiated cells, a small subset of genes displays tissue specific methylation. A recent study demonstrated 491 differentially methylated regions (DMR) being more methylated in fibroblasts compared to ESC (Lister et al. 2009), with DMR representing only 6–8% of CpG islands in different tissues (Berdasco and Esteller 2010). Important DNA methylation differences can be observed in ASC compared to differentiated cells of the same lineage. For instance, breast self-renewal and proliferation genes are hypomethylated in $CD44^+/CD24^-$ stem cells compared to differentiated luminal $CD24^+$ cells (Bloushtain-Qimron et al. 2008).

Many studies indicate DNA methylation as a mechanism providing long term gene silencing and epigenetic memory in differentiated cells, however experiments of conditional deletion of DNMT1 suggest that DNA methylation plays also an important role in maintaining ASC self-renewal and suppressing differentiation (Sen et al. 2010). Indeed, loss of methylation causes differentiation alterations in epithelial progenitor cells (EPC) (Sen et al. 2010) and hematopoietic stem cells (HSC) (Trowbridge et al. 2009; Broske et al. 2009). Other epigenetic mechanisms are involved in regulation ASC self-renewal, but their relation to DNA methylation is still unknown. For instance, the PCR1 PcG protein BMI-1 is required for NSC, HSC, mammary and intestinal stem cell proliferation (Molofsky et al. 2003; Lessard and Sauvageau 2003; Pietersen et al. 2008; Sangiorgi and Capecchi 2008), while overexpression of PCR2 PcG protein EZH2 blocks differentiation of myoblasts and EPC (Caretti et al. 2004; Sen 2011) and prevents HSC exhaustion (Kamminga et al. 2006). Because cell memory is set during development and inherent to each tissue-type, it was long assumed that differentiation of ASC is strictly specific to their lineage. However, recent studies demonstrate that under certain conditions, particularly after injury, ASC can trans-differentiate into cells of different tissues (Lotem and Sachs 2006). Therefore ASC, like ESC, show a differentiation plasticity that is conferred by epigenetic programs that can reversibly regulate transcription of genes expressed in different tissue according to physiological and pathological signals. The contribution of ASC plasticity to cancer will be described in the next section.

24.4 Cancer Stem Cells

The idea that cancer is caused by transformed cells with stem cell properties is not novel, but it has received renewed interest among scientists in recent years. The observation that tumours are formed by cells with functional heterogeneity has led to the postulation of two mutually exclusive models for the cellular origin of cancer: the stochastic model and the cancer stem cell hierarchy. The stochastic (or clonal evolution) model predicts that every cell can become tumorigenic under the influence of endogenous (transcription factors) and exogenous (microenvironment) factors that can generate their own heterogeneous sub clones (Nowell 1976). In the stochastic model, cancer cells fluctuate between several states, owing to their plasticity (Wang and Dick 2005). In contrast, the cancer stem cell (CSC) model considers that tumours originate from transformed stem cells and they are organised in a hierarchical manner, whereby CSC lies at the apex and the proliferating progenitors and terminally differentiated cancer cells reside at the bottom of the hierarchy (Bonnet and Dick 1997). Recent studies show that both models can act together depending on microenvironmental signals and that tumour initiating cells can originate both from transformation of normal ASC or epigenetic reprogramming of more differentiated cells (Campbell and Polyak 2007) (Fig. 24.2). Both theories converge on the idea that cancer arises from transformed cells that acquire growth and survival advantage, which is a landmark of stem cells. The theory that cancer could arise from embryo-like cells was proposed about 150 years ago (Virchow 1855) and was later developed by Cohnheim and Durante with the concept of "maturation arrest", according to which cancer could develop from embryonic rudiments remaining in adult organs (Cohnheim 1867; Durante 1974). These theories were proven years later by studies on germ cell tumours demonstrating that teratocarcinomas contain CSC with very similar characteristics to ESC, with self-renewal and differentiation potential (Sell and Pierce 1994; Sperger et al. 2003). For a long time it was known that only a small population of cancer cells is tumorigenic and can propagate the tumour. Single-cell analysis of leukaemia revealed two different populations of cancer cells in terms of proliferative kinetics: the frequent large, fast-cycling cells and the rare, smaller slow-cycling cells with the same properties to that of normal HSC (Clarkson 1974). Through elegant studies, Dick and colleagues proved the existence of CSC by showing that in acute myeloid leukaemia (AML), a rare population of CSC with $CD34^+/CD38^-$ cell surface expression were able to recapitulate the original disease over repeated transplantation into NOD/SCID (non-obese diabetic/severe combined immunodeficiency) mice (Lapidot et al. 1994). Since then, CSC have been identified and isolated in solid tumours including breast (Al-Hajj et al. 2003), brain (Singh et al. 2003), melanoma (Fang et al. 2005), pancreatic (Hermann et al. 2007), prostate (Tang et al. 2007) and ovarian cancers (Bapat et al. 2005) (Table 24.2). CSC are a rare population of cells that resemble normal stem cells. They can self-renew, are long-lasting, remain relatively quiescent, and can generate all heterogeneous cell types comprising the tumour. CSC can lay dormant within their niche and therefore escape chemotherapy, which only targets highly proliferating cells. Their resistance to current cancer therapies (chemotherapy and radiotherapy) is also due to expression of ATP-binding cassette (ABC) transporters (pumping out harmful drugs), increased free radical scavenging and high expression of anti-apoptotic proteins (Visvader 2011). Since CSC retain many features of normal stem cells, their identification often relies on the expression of tissue specific stem cell markers. For example, leukemic stem cells can be

Fig. 24.2 Cancer stem cell-of-origin model. Different cell types from the lineage hierarchy can undergo oncogenic events to transform into tumour cells. CSC can either originate from a transformed normal stem cell, transit-amplifying cell, progenitor cell, or a terminally differentiated cell to give rise to a heterogeneous population of cancer cells

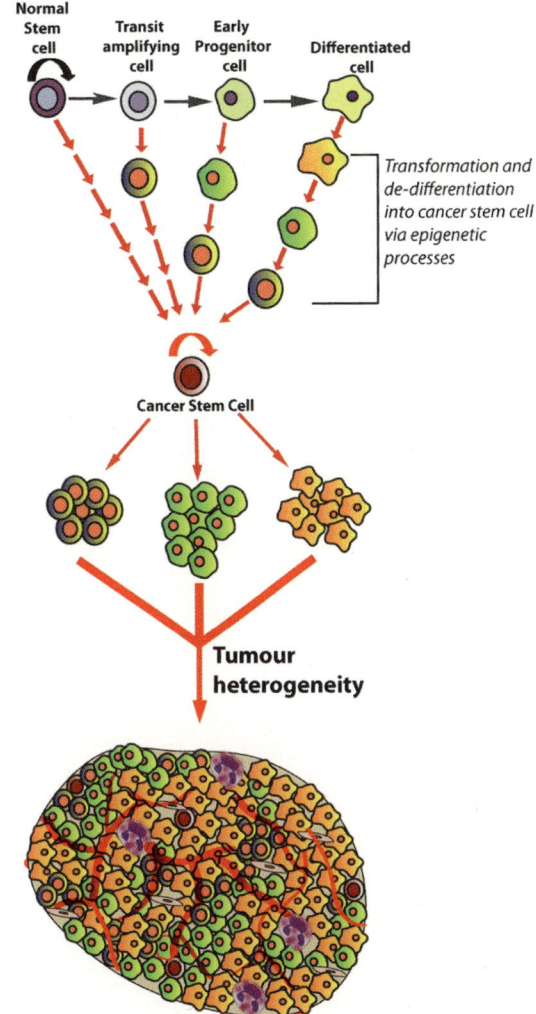

identified by CD34 and CD38 which are expressed on the cells in the HSC hierarchy (Bonnet and Dick 1997). Other universal CSC markers are instead based on their ability to pump out toxicants, which defines them as "side population" cells able to extrude a Hoechst dye, expressing ABC transporters and the detoxifying enzyme ALDH1 (Visvader and Lindeman 2008). Functional assays for CSC identification include xenografts into immunocompromised mice and formation of spheroids in culture. The xenograft assay involves transplanting cancer cells into NOD/SCID mice for tumour formation. Isolated CSC are generally more tumorigenic than differentiated tumour cells and their serial transplantation shows that they can reproduce the original disease through every passage. Sphere forming assays, which involve culturing CSC under stem cell conditions, preserve survival

Table 24.2 CSC markers in human tumours

Cancer	CSC markers	Cell-of-origin	References
Acute myeloid leukemia (AML)	CD34$^+$CD38$^-$ ALDH1	HSC	Bonnet and Dick (1997) Ran et al. (2009)
Acute lymphoblastic leukemia (ALL)	CD34$^+$CD19$^-$CD10$^-$	HSC	Cox et al. (2004)
Breast	CD44$^+$CD24$^{low/-}$ ALDH1	Mammary stem cells	Al-Hajj et al. (2003) Ginestier et al. (2007)
Colon	CD133$^+$ EpCAM^{hi} CD44$^+$	Intestinal stem cells	O'Brien et al. (2007) Dalerba et al. (2007)
Melanoma	JARID1B ABCB5$^+$	Skin stem cell	Roesch et al. (2010) Schatton et al. (2008)
Prostate	CD133$^+$ TRA-160 TRA-160$^+$CD151$^+$CD166$^+$	Basal progenitor cells	Goldstein et al. (2010) Rajasekhar et al. (2011)
Pancreas	CD24$^+$CD44$^+$EpCAM$^+$	–	Li et al. (2007)
Brain	CD133$^+$	NSC	Singh et al. (2003)
Head and neck	CD44$^+$	–	Prince et al. (2007)
Lung	CD133$^+$ ALDH1	Bronchioalveolar stem cells	Eramo et al. (2008) Jiang et al. (2009)
Liver	CD44$^+$ CD90$^+$	–	Yang et al. (2008)
Ovary	CD44$^+$ CD177$^+$	–	Zhang et al. (2008)
Stomach	CD44$^+$	–	Takaishi et al. (2009)
Osteosarcoma	CD133$^+$	Mesenchymal stem cells	Tirino et al. (2008)

of CSC while inducing cell death by apoptosis in non-CSC (Visvader and Lindeman 2008). However, regardless of the assay, a major challenge for studying CSC is their inherent developmental plasticity, which involves the co-existence of different epigenetic states during cancer progression. For instance, CD44$^+$/CD24$^-$ breast CSC exist in a metastable state oscillating between differentiation and de-differentiation, with CSC giving rise to luminal CD24$^+$ cells and luminal cells de-differentiating back into CSC (Meyer et al. 2009). The same has been observed in melanoma, where JARID1B$^+$ CSC generate JARID1B$^-$ cells and vice versa (Roesch et al. 2010). In addition, CSC plasticity can often extend beyond their lineage and they can express genes normally expressed in different tissues. Consistent with the trans-differentiation potential of ASC after injury, CSC show the same plasticity resulting in abnormal tissue regeneration (Lotem and Sachs 2006). CSC plasticity is influenced by embryonic developmental programmes and a similar gene expression signature between highly malignant, poorly differentiated solid tumours and ESC has been reported (Ben-Porath et al. 2008). This is due to the ability of cancer to take control of normal developmental programs for selective advantage, albeit in part related to an upregulated Myc-regulatory network (Kim et al. 2010). For example, the epithelial-to-mesenchymal (EMT) transition, a reversible embryonic programme that allows transition between cellular phenotypes during gastrulation, contributes to CSC plasticity. EMT is recapitulated during tumour progression and metastasis by a transition from an epithelial to a mesenchymal phenotype with acquired cell motility. This is induced by activation of key signalling pathways (TGF-β, Notch, FGF) that drive epigenetic silencing of the adhesion molecule E-cadherin (Thiery et al. 2009). EMT is also important for maintenance of stem cell properties and CSC can hijack this program to regulate their plasticity. In addition to this, CSC establish their own niche by recruiting cells to recreate a similar microenvironment to that of a normal stem cell niche. The niche can induce and expand CSC by enhancing "stemness" features in non tumourigenic cells by overexpressing signals that are important for stem cell renewal and promote EMT through epigenetic alterations (Mani et al. 2008).

24.4.1 Epigenetic Origin of Cancer Stem Cells

Epigenetic alterations are generally observed at early stages of tumorigenesis and are likely candidates for a mechanism of tumour initiation. ASC are long-lived and during their aging process they may undergo epigenetic insult which can induce survival programs and predispose to the onset of cancer after further genetic and epigenetic alterations (Feinberg et al. 2006; Baylin and Ohm 2006). Numerous evidences indicate a role for epigenetic defects in the development of CSC. Normal stem cells are vulnerable to epigenetic alteration when induced to sustained self-renewal. Extensive DNA methylation alterations have been reported in ESC after long term in culture, with changes which are inherited after differentiation and associated with cancer (Allegrucci et al. 2007). Similar alterations have been observed in NSC (Shen et al. 2006), with a recent study demonstrating hypermethylation of HPC after many generations and inherited after differentiation to astrocytes (Meissner et al. 2008). Hypermethylation of bivalent domain genes in stem cells is particularly important for tumorigenesis as tumour suppressor genes have bivalent promoters in ESC and ASC (Barrero et al. 2010) and hypermethylation of tumour suppressor genes is a hallmark of cancer (Jones and Baylin 2007). As bivalent genes are developmental genes and transcription factors that regulate stem cell fate, it seems apparent how their epigenetic silencing in ASC could generate stem cells locked in a self-renewal state with impaired or limited differentiation potential (Fig. 24.1). Several studies have demonstrated that PcG target genes are much more likely to become hypermethylated in cancer (Widschwendter et al. 2007; Ohm et al. 2007; Schlesinger et al. 2007) and a mechanism by which the PcG-H3K27me3 mark could direct DNA methylation has been proposed (Keshet et al. 2006; Vire et al. 2006). In addition, overexpression of PcG proteins BMI-1 and EZH2 are also often found in cancer and they both play a fundamental role in regulating stem cell function (Bracken and Helin 2009). DNA methylation alterations at other genomic regions can also participate in the development of CSC. Hypomethylation of the genome can induce chromosome instability together with aberrant activation of proto-oncogenes associated with stem cell self-renewal and proliferation (Sharma et al. 2010). Chromosome translocations producing MLL fusion proteins are involved in AML, with more than

50 different fusion partners being identified. Importantly, MLL-ENF fusion is able to transform HSC and committed progenitors, thus creating CSC with acquired self-renewal and de-differentiated phenotype (Milne et al. 2005).

Loss of imprinting (LOI) via DNA demethylation can also be associated with growth advantage in stem cells and biallelic expression of IGF2 accounts for half of Wilms tumours and predisposition to colon cancer. Other LOI involved in cancer include PEG1/MEST involved in lung cancer, CDK1C in pancreatic cancer, TP73 in gastric cancer and DIRAS3 in breast cancer (Feinberg et al. 2006). Finally, DNA methylation at promoters of miRNA genes involved in stem cell differentiation can lead to CSC. For instance, silencing of the miRNA *let-7* contributes to breast, colon and lung cancer (Zimmerman and Wu 2011) and silencing of the mir-200 gene family induces EMT and CSC phenotype in breast, lung and ovarian cancer (Brabletz and Brabletz 2010). As epigenetic alterations can result in CSC and plasticity is a fundamental property of stem cells, we should not be surprised that CSC share this characteristic. What becomes apparent is that the effect of the environment is of primary importance for cancer initiation and progression while the behaviour of stem cells is completely dependent on physiological and pathological signals. As controlling environmental conditions is not an easy endeavour, we need to develop treatments that are able to completely eradicate CSC as their plasticity is a likely prospect of tumour recurrence.

24.5 Resetting Cancer by Epigenetic Re-programming

The developmental plasticity demonstrated by CSC and the reversible nature of epigenetic alterations has led to the development of epigenetic therapies as a new treatment option for cancer patients. Several epigenetic drugs that aim at halting tumour progression and restoring normal cell function are being tested in human clinical trials. Because epigenetic drugs can induce differentiation of ESC and ASC, their use for resumption of normal tissue differentiation is also been tested (Berdasco and Esteller 2010). The concept of differentiation therapy as an alternative or adjuvant treatment to chemotherapy has been inspired by many years of research. Landmark experiments have shown that embryonic microenvironments that program cell fate during development are able to reverse malignancy by resetting normal pathways of cell differentiation. For instance, teratocarcinoma cells can be induced to differentiate when transplanted into chimeric embryos, and malignancy can be reverted by injecting cancer cells into zebrafish, chicken and mouse embryos (Telerman and Amson 2009). Nuclear transfer experiments have shown that the epigenotype of cancer cells can be reprogrammed by oocyte molecules (Blelloch et al. 2004; Hochedlinger et al. 2004) and recent experiments with oocyte extracts have shown that this effect is mediated by chromatin remodelling and reactivation of silenced tumour suppressor genes resulting in reduction of tumour growth in mouse xenografts (Allegrucci et al. 2011). New insights have also come from studies of reprogramming to pluripotency. Yamanaka's work demonstrated that the ectopic expression of core pluripotency factors (OCT4, SOX2, MYC, KLF4) in somatic cells can reprogram the epigenetic state of somatic cells to that of pluripotent cells giving rise to induced pluripotent stem cells (iPSC) (Takahashi and Yamanaka 2006). With a similar approach, cancer cells are reprogrammed to induced pluripotent cancer cells (iPC) with re-acquired differentiation potential (Sun and Liu 2011). Reprogramming studies suggest that even if finely regulated, epigenetic landscapes are plastic and reversible and that cellular transformation and de-differentiation share similar epigenetic mechanisms. This view is sustained by the evidence that many factors able to reprogram somatic/cancer cells to induced pluripotent cells act as oncogenes. This is the case for OCT4 and SOX2 (aberrant expression is tumorigenic in epithelial tissues), NANOG (overexpressed in germ cell tumours), KLF4 (associated with colorectal cancer), LIN28 (associated with hepatic cancer) and MYC (potent oncogene involved in many cancer types) (Daley 2008). Therefore, differentiation therapy could be the way to reset CSC to normal function and epigenetic programs that regulate stem cells are of particular interest as novel treatment targets.

24.6 Conclusions

More than 50 years have passed since Waddington proposed the idea that signals from embryonic environments can influence cell fate and behaviour according to an "epigenetic landscape" (Waddington 1940). He proposed that embryonic development can be visualised as a landscape delimited by hills and valley, where cells can take different directions depending on the signals received along the path. However, possible paths in this landscape are restricted by barriers dictated by hills and only defined valleys are available when cells are restricted to a defined initial trajectory. With a modern view, we identify these landscapes as epigenetic modifications regulating expression of developmental genes during differentiation in a coordinated fashion. Although cell fate is developmentally established, a degree of plasticity is retained for tissue turnover or acquired in pathological conditions. Waddington recognised that cancer cells escape the effect of developmental forces (Waddington 1935), introducing the concept of cancer as a defect in the mechanisms that control cell differentiation. We now know that epigenetic alterations are hallmark of cancer and they can transform normal cells to tumour cells with altered proliferation and differentiation. Our knowledge of cell plasticity has greatly increased over the last few years as stem cell technologies have developed and genome-wide mapping of epigenetic modifications of stem cells and somatic cells are being established. Defining the epigenome of cellular states in normal and cancer tissues is therefore a new challenge, but a most beneficial one as it holds the promise to eradicate or control cancer, an expected disease of an aging population.

Acknowledgements The authors would like to acknowledge funding from the University of Nottingham, the Royal Society of London and EvoCell Ltd. We are obliged to Dr Ramiro Alberio for valuable discussion and critical evaluation of the manuscript.

References

Al-Hajj M, Wicha MS, Benito-Hernandez A, Morrison SJ, Clarke MF (2003) Prospective identification of tumorigenic breast cancer cells. Proc Natl Acad Sci USA 100:3983–3988

Allegrucci C, Thurston A, Lucas E, Young L (2005) Epigenetics and the germline. Reproduction 129:137–149

Allegrucci C, Wu YZ, Thurston A, Denning CN, Priddle H, Mummery CL, Ward-van Oostwaard D, Andrews PW, Stojkovic M, Smith N, Parkin T, Jones ME, Warren G, Yu L, Brena RM, Plass C, Young LE (2007) Restriction landmark genome scanning identifies culture-induced DNA methylation instability in the human embryonic stem cell epigenome. Hum Mol Genet 16:1253–1268

Allegrucci C, Rushton MD, Dixon JE, Sottile V, Shah M, Kumari R, Watson S, Alberio R, Johnson AD (2011) Epigenetic reprogramming of breast cancer cells with oocyte extracts. Mol Cancer 10:7

Asp P, Blum R, Vethantham V, Parisi F, Micsinai M, Cheng J, Bowman C, Kluger Y, Dynlacht BD (2011) PNAS Plus: genome-wide remodeling of the epigenetic landscape during myogenic differentiation. Proc Natl Acad Sci USA 108:E149–E158

Ball MP, Li JB, Gao Y, Lee J-H, LeProust EM, Park I-H, Xie B, Daley GQ, Church GM (2009) Targeted and genome-scale strategies reveal gene-body methylation signatures in human cells. Nat Biotechnol 27:361–368

Bapat SA, Mali AM, Koppikar CB, Kurrey NK (2005) Stem and progenitor-like cells contribute to the aggressive behavior of human epithelial ovarian cancer. Cancer Res 65:3025–3029

Barrero MJ, Boue S, Izpisua Belmonte JC (2010) Epigenetic mechanisms that regulate cell identity. Cell Stem Cell 7:565–570

Baylin SB, Ohm JE (2006) Epigenetic gene silencing in cancer – a mechanism for early oncogenic pathway addiction? Nat Rev Cancer 6:107–116

Ben-Porath I, Thomson MW, Carey VJ, Ge R, Bell GW, Regev A, Weinberg RA (2008) An embryonic stem cell-like gene expression signature in poorly differentiated aggressive human tumors. Nat Genet 40:499–507

Berdasco M, Esteller M (2010) Aberrant epigenetic landscape in cancer: how cellular identity goes awry. Dev Cell 19:698–711

Bernstein BE, Mikkelsen TS, Xie X, Kamal M, Huebert DJ, Cuff J, Fry B, Meissner A, Wernig M, Plath K, Jaenisch R, Wagschal A, Feil R, Schreiber SL, Lander ES (2006) A bivalent chromatin structure marks key developmental genes in embryonic stem cells. Cell 125:315–326

Bird A (2002) DNA methylation patterns and epigenetic memory. Genes Dev 16:6–21

Bird A (2007) Perceptions of epigenetics. Nature 447:396–398

Blelloch RH, Hochedlinger K, Yamada Y, Brennan C, Kim M, Mintz B, Chin L, Jaenisch R (2004) Nuclear cloning of embryonal carcinoma cells. Proc Natl Acad Sci USA 101:13985–13990

Bloushtain-Qimron N, Yao J, Snyder EL, Shipitsin M, Campbell LL, Mani SA, Hu M, Chen H, Ustyansky V, Antosiewicz JE, Argani P, Halushka MK, Thomson JA, Pharoah P, Porgador A, Sukumar S, Parsons R, Richardson AL, Stampfer MR, Gelman RS, Nikolskaya T, Nikolsky Y, Polyak K (2008) Cell type-specific DNA methylation patterns in the human breast. Proc Natl Acad Sci USA 105:14076–14081

Bonnet D, Dick JE (1997) Human acute myeloid leukemia is organized as a hierarchy that originates from a primitive hematopoietic cell. Nat Med 3:730–737

Brabletz S, Brabletz T (2010) The ZEB/miR-200 feedback loop–a motor of cellular plasticity in development and cancer? EMBO Rep 11:670–677

Bracken AP, Helin K (2009) Polycomb group proteins: navigators of lineage pathways led astray in cancer. Nat Rev Cancer 9:773–784

Bracken AP, Dietrich N, Pasini D, Hansen KH, Helin K (2006) Genome-wide mapping of Polycomb target genes unravels their roles in cell fate transitions. Genes Dev 20:1123–1136

Broske AM, Vockentanz L, Kharazi S, Huska MR, Mancini E, Scheller M, Kuhl C, Enns A, Prinz M, Jaenisch R, Nerlov C, Leutz A, Andrade-Navarro MA, Jacobsen SE, Rosenbauer F (2009) DNA methylation protects hematopoietic stem cell multipotency from myeloerythroid restriction. Nat Genet 41:1207–1215

Campbell LL, Polyak K (2007) Breast tumor heterogeneity: cancer stem cells or clonal evolution? Cell Cycle 6:2332–2338

Caretti G, Di Padova M, Micales B, Lyons GE, Sartorelli V (2004) The Polycomb Ezh2 methyltransferase regulates muscle gene expression and skeletal muscle differentiation. Genes Dev 18:2627–2638

Chen X, Hiller M, Sancak Y, Fuller MT (2005) Tissue-specific TAFs counteract Polycomb to turn on terminal differentiation. Science 310:869–872

Clarkson BD (1974) The survival value of the dormant state in neoplastic and normal cell populations. In: Clarkson B, Baserga R (eds) Control of proliferation in animal cells, vol 1, Cold spring harbor conferences on cell proliferation. Cold Spring Harbor Laboratory, New York, pp 945–972

Cohnheim J (1867) Ueber entzundung und eiterung. Path Anat Physiol Klin Med 40:1–79

Cox CV, Evely RS, Oakhill A, Pamphilon DH, Goulden NJ, Blair A (2004) Characterization of acute lymphoblastic leukemia progenitor cells. Blood 104:2919–2925

Dalerba P, Dylla SJ, Park IK, Liu R, Wang X, Cho RW, Hoey T, Gurney A, Huang EH, Simeone DM, Shelton AA, Parmiani G, Castelli C, Clarke MF (2007) Phenotypic characterization of human colorectal cancer stem cells. Proc Natl Acad Sci USA 104:10158–10163

Daley GQ (2008) Common themes of dedifferentiation in somatic cell reprogramming and cancer. Cold Spring Harb Symp Quant Biol 73:171–174

Durante F (1874) Nesso fisio-patologico tra la struttura dei nei materni e la genesi di alcuni tumori maligni. Arch Memori ed Osservazioni di Chirurgia Practica 11:217–2226

Eramo A, Lotti F, Sette G, Pilozzi E, Biffoni M, Di Virgilio A, Conticello C, Ruco L, Peschle C, De Maria R (2008) Identification and expansion of the tumorigenic lung cancer stem cell population. Cell Death Differ 15:504–514

Fang D, Nguyen TK, Leishear K, Finko R, Kulp AN, Hotz S, Van Belle PA, Xu X, Elder DE, Herlyn M (2005) A tumorigenic subpopulation with stem cell properties in melanomas. Cancer Res 65:9328–9337

Feinberg AP, Ohlsson R, Henikoff S (2006) The epigenetic progenitor origin of human cancer. Nat Rev Genet 7(1):21–33

Ficz G, Branco MR, Seisenberger S, Santos F, Krueger F, Hore TA, Marques CJ, Andrews S, Reik W (2011) Dynamic regulation of 5-hydroxymethylcytosine in mouse ES cells and during differentiation. Nature 473:398–402

Garraway LA, Sellers WR (2006) Lineage dependency and lineage-survival oncogenes in human cancer. Nat Rev Cancer 6:593–602

Ginestier C, Hur MH, Charafe-Jauffret E, Monville F, Dutcher J, Brown M, Jacquemier J, Viens P, Kleer CG, Liu S, Schott A, Hayes D, Birnbaum D, Wicha MS, Dontu G (2007) ALDH1 is a marker of normal and malignant human mammary stem cells and a predictor of poor clinical outcome. Cell Stem Cell 1:555–567

Goldstein AS, Stoyanova T, Witte ON (2010) Primitive origins of prostate cancer: in vivo evidence for prostate-regenerating cells and prostate cancer-initiating cells. Mol Oncol 4:385–396

Hawkins RD, Hon GC, Lee LK, Ngo Q, Lister R, Pelizzola M, Edsall LE, Kuan S, Luu Y, Klugman S, Antosiewicz-Bourget J, Ye Z, Espinoza C, Agarwahl S, Shen L, Ruotti V, Wang W, Stewart R, Thomson JA, Ecker JR, Ren B (2010) Distinct epigenomic landscapes of pluripotent and lineage-committed human cells. Cell Stem Cell 6:479–491

Hemberger M, Dean W, Reik W (2009) Epigenetic dynamics of stem cells and cell lineage commitment: digging Waddington's canal. Nat Rev Mol Cell Biol 10:526–537

Hermann PC, Huber SL, Herrler T, Aicher A, Ellwart JW, Guba M, Bruns CJ, Heeschen C (2007) Distinct populations of cancer stem cells determine tumor growth and metastatic activity in human pancreatic cancer. Cell Stem Cell 1:313–323

Hochedlinger K, Blelloch R, Brennan C, Yamada Y, Kim M, Chin L, Jaenisch R (2004) Reprogramming of a melanoma genome by nuclear transplantation. Genes Dev 18:1875–1885

Howell CY, Bestor TH, Ding F, Latham KE, Mertineit C, Trasler JM, Chaillet JR (2001) Genomic imprinting disrupted by a maternal effect mutation in the Dnmt1 gene. Cell 104:829–838

Jaenisch R, Bird A (2003) Epigenetic regulation of gene expression: how the genome integrates intrinsic and environmental signals. Nat Genet 33(Suppl):245–254

Jiang F, Qiu Q, Khanna A, Todd NW, Deepak J, Xing L, Wang H, Liu Z, Su Y, Stass SA, Katz RL (2009) Aldehyde dehydrogenase 1 is a tumor stem cell-associated marker in lung cancer. Mol Cancer Res 7:330–338

Jones PA, Baylin SB (2007) The epigenomics of cancer. Cell 128:683–692

Kamminga LM, Bystrykh LV, de Boer A, Houwer S, Douma J, Weersing E, Dontje B, de Haan G (2006) The Polycomb group gene Ezh2 prevents hematopoietic stem cell exhaustion. Blood 107:2170–2179

Keshet I, Schlesinger Y, Farkash S, Rand E, Hecht M, Segal E, Pikarski E, Young RA, Niveleau A, Cedar H, Simon I (2006) Evidence for an instructive mechanism of de novo methylation in cancer cells. Nat Genet 38:149–153

Kim J, Woo AJ, Chu J, Snow JW, Fujiwara Y, Kim CG, Cantor AB, Orkin SH (2010) A Myc network accounts for similarities between embryonic stem and cancer cell transcription programs. Cell 143:313–324

Koh KP, Yabuuchi A, Rao S, Huang Y, Cunniff K, Nardone J, Laiho A, Tahiliani M, Sommer CA, Mostoslavsky G, Lahesmaa R, Orkin SH, Rodig SJ, Daley GQ, Rao A (2011) Tet1 and Tet2 regulate 5-hydroxymethylcytosine production and cell lineage specification in mouse embryonic stem cells. Cell Stem Cell 8:200–213

Lapidot T, Sirard C, Vormoor J, Murdoch B, Hoang T, Caceres-Cortes J, Minden M, Paterson B, Caligiuri MA, Dick JE (1994) A cell initiating human acute myeloid leukaemia after transplantation into SCID mice. Nature 367:645–648

Lessard J, Sauvageau G (2003) Bmi-1 determines the proliferative capacity of normal and leukaemic stem cells. Nature 423:255–260

Li C, Heidt DG, Dalerba P, Burant CF, Zhang L, Adsay V, Wicha M, Clarke MF, Simeone DM (2007) Identification of pancreatic cancer stem cells. Cancer Res 67:1030–1037

Lister R, Pelizzola M, Dowen RH, Hawkins RD, Hon G, Tonti-Filippini J, Nery JR, Lee L, Ye Z, Ngo QM, Edsall L, Antosiewicz-Bourget J, Stewart R, Ruotti V, Millar AH, Thomson JA, Ren B, Ecker JR (2009) Human DNA methylomes at base resolution show widespread epigenomic differences. Nature 462:315–322

Lotem J, Sachs L (2006) Epigenetics and the plasticity of differentiation in normal and cancer stem cells. Oncogene 25:7663–7672

Mani SA, Guo W, Liao MJ, Eaton EN, Ayyanan A, Zhou AY, Brooks M, Reinhard F, Zhang CC, Shipitsin M, Campbell LL, Polyak K, Brisken C, Yang J, Weinberg RA (2008) The epithelial-mesenchymal transition generates cells with properties of stem cells. Cell 133:704–715

Marson A, Levine SS, Cole MF, Frampton GM, Brambrink T, Johnstone S, Guenther MG, Johnston WK, Wernig M, Newman J, Calabrese JM, Dennis LM, Volkert TL, Gupta S, Love J, Hannett N, Sharp PA, Bartel DP, Jaenisch R, Young RA (2008) Connecting microRNA genes to the core transcriptional regulatory circuitry of embryonic stem cells. Cell 134:521–533

Mayer W, Niveleau A, Walter J, Fundele R, Haaf T (2000) Demethylation of the zygotic paternal genome. Nature 403:501–502

Meissner A (2010) Epigenetic modifications in pluripotent and differentiated cells. Nat Biotechnol 28:1079–1088

Meissner A, Mikkelsen TS, Gu H, Wernig M, Hanna J, Sivachenko A, Zhang X, Bernstein BE, Nusbaum C, Jaffe DB, Gnirke A, Jaenisch R, Lander ES (2008) Genome-scale DNA methylation maps of pluripotent and differentiated cells. Nature 454:766–770

Meyer MJ, Fleming JM, Ali MA, Pesesky MW, Ginsburg E, Vonderhaar BK (2009) Dynamic regulation of CD24 and the invasive, CD44posCD24neg phenotype in breast cancer cell lines. Breast Cancer Res 11:R82

Mikkelsen TS, Ku M, Jaffe DB, Issac B, Lieberman E, Giannoukos G, Alvarez P, Brockman W, Kim TK, Koche RP, Lee W, Mendenhall E, O'Donovan A, Presser A, Russ C, Xie X, Meissner A, Wernig M, Jaenisch R, Nusbaum C, Lander ES, Bernstein BE (2007) Genome-wide maps of chromatin state in pluripotent and lineage-committed cells. Nature 448:553–560

Milne TA, Martin ME, Brock HW, Slany RK, Hess JL (2005) Leukemogenic MLL fusion proteins bind across a broad region of the Hox a9 locus, promoting transcription and multiple histone modifications. Cancer Res 65:11367–11374

Molofsky AV, Pardal R, Iwashita T, Park IK, Clarke MF, Morrison SJ (2003) Bmi-1 dependence distinguishes neural stem cell self-renewal from progenitor proliferation. Nature 425:962–967

Morgan HD, Santos F, Green K, Dean W, Reik W (2005) Epigenetic reprogramming in mammals. Hum Mol Genet 14(Spec No 1):R47–R58

Myant K, Termanis A, Sundaram AY, Boe T, Li C, Merusi C, Burrage J, de Las Heras JI, Stancheva I (2011) LSH and G9a/GLP complex are required for developmentally programmed DNA methylation. Genome Res 21:83–94

Nowell PC (1976) The clonal evolution of tumor cell populations. Science 194:23–28

O'Brien CA, Pollett A, Gallinger S, Dick JE (2007) A human colon cancer cell capable of initiating tumour growth in immunodeficient mice. Nature 445:106–110

Ohm JE, McGarvey KM, Yu X, Cheng L, Schuebel KE, Cope L, Mohammad HP, Chen W, Daniel VC, Yu W, Berman DM, Jenuwein T, Pruitt K, Sharkis SJ, Watkins DN, Herman JG, Baylin SB (2007) A stem cell-like chromatin pattern may predispose tumor suppressor genes to DNA hypermethylation and heritable silencing. Nat Genet 39:237–242

Okano M, Bell DW, Haber DA, Li E (1999) DNA methyltransferases Dnmt3a and Dnmt3b are essential for de novo methylation and mammalian development. Cell 99:247–257

Oswald J, Engemann S, Lane N, Mayer W, Olek A, Fundele R, Dean W, Reik W, Walter J (2000) Active demethylation of the paternal genome in the mouse zygote. Curr Biol 10:475–478

Pastor WA, Pape UJ, Huang Y, Henderson HR, Lister R, Ko M, McLoughlin EM, Brudno Y, Mahapatra S, Kapranov P, Tahiliani M, Daley GQ, Liu XS, Ecker JR, Milos PM, Agarwal S, Rao A (2011) Genome-wide mapping of 5-hydroxymethylcytosine in embryonic stem cells. Nature 473:394–397

Pauli A, Rinn JL, Schier AF (2011) Non-coding RNAs as regulators of embryogenesis. Nat Rev Genet 12:136–149

Pierce GB, Johnson LD (1971) Differentiation and cancer. In Vitro 7:140–145

Pietersen AM, Evers B, Prasad AA, Tanger E, Cornelissen-Steijger P, Jonkers J, van Lohuizen M (2008) Bmi1 regulates stem cells and proliferation and differentiation of committed cells in mammary epithelium. Curr Biol 18:1094–1099

Prince ME, Sivanandan R, Kaczorowski A, Wolf GT, Kaplan MJ, Dalerba P, Weissman IL, Clarke MF, Ailles LE (2007) Identification of a subpopulation of cells with cancer stem cell properties in head and neck squamous cell carcinoma. Proc Natl Acad Sci USA 104:973–978

Rajasekhar VK, Studer L, Gerald W, Socci ND, Scher HI (2011) Tumour-initiating stem-like cells in human prostate cancer exhibit increased NF-kappaB signalling. Nat Commun 2:162

Ran D, Schubert M, Pietsch L, Taubert I, Wuchter P, Eckstein V, Bruckner T, Zoeller M, Ho AD (2009) Aldehyde dehydrogenase activity among primary leukemia cells is associated with stem cell features and correlates with adverse clinical outcomes. Exp Hematol 37:1423–1434

Roesch A, Fukunaga-Kalabis M, Schmidt EC, Zabierowski SE, Brafford PA, Vultur A, Basu D, Gimotty P, Vogt T, Herlyn M (2010) A temporarily distinct subpopulation of slow-cycling melanoma cells is required for continuous tumor growth. Cell 141:583–594

Sangiorgi E, Capecchi MR (2008) Bmi1 is expressed in vivo in intestinal stem cells. Nat Genet 40:915–920

Santos F, Hendrich B, Reik W, Dean W (2002) Dynamic reprogramming of DNA methylation in the early mouse embryo. Dev Biol 241:172–182

Sasaki H, Matsui Y (2008) Epigenetic events in mammalian germ-cell development: reprogramming and beyond. Nat Rev Genet 9:129–140

Schatton T, Murphy GF, Frank NY, Yamaura K, Waaga-Gasser AM, Gasser M, Zhan Q, Jordan S, Duncan LM, Weishaupt C, Fuhlbrigge RC, Kupper TS, Sayegh MH, Frank MH (2008) Identification of cells initiating human melanomas. Nature 451:345–349

Schlesinger Y, Straussman R, Keshet I, Farkash S, Hecht M, Zimmerman J, Eden E, Yakhini Z, Ben-Shushan E, Reubinoff BE, Bergman Y, Simon I, Cedar H (2007) Polycomb-mediated methylation on Lys27 of histone H3 pre-marks genes for de novo methylation in cancer. Nat Genet 39:232–236

Sell S, Pierce GB (1994) Maturation arrest of stem cell differentiation is a common pathway for the cellular origin of teratocarcinomas and epithelial cancers. Lab Invest 70:6–22

Sen GL (2011) Remembering one's identity: the epigenetic basis of stem cell fate decisions. FASEB J 25(7):2123–2128

Sen GL, Reuter JA, Webster DE, Zhu L, Khavari PA (2010) DNMT1 maintains progenitor function in self-renewing somatic tissue. Nature 463:563–567

Sharma S, Kelly TK, Jones PA (2010) Epigenetics in cancer. Carcinogenesis 31:27–36

Shen Y, Chow J, Wang Z, Fan G (2006) Abnormal CpG island methylation occurs during in vitro differentiation of human embryonic stem cells. Hum Mol Genet 15:2623–2635

Singh SK, Clarke ID, Terasaki M, Bonn VE, Hawkins C, Squire J, Dirks PB (2003) Identification of a cancer stem cell in human brain tumors. Cancer Res 63:5821–5828

Sperger JM, Chen X, Draper JS, Antosiewicz JE, Chon CH, Jones SB, Brooks JD, Andrews PW, Brown PO, Thomson JA (2003) Gene expression patterns in human embryonic stem cells and human pluripotent germ cell tumors. Proc Natl Acad Sci USA 100:13350–13355

Sun C, Liu YK (2011) Induced pluripotent cancer cells: progress and application. J Cancer Res Clin Oncol 137:1–8

Tahiliani M, Koh KP, Shen Y, Pastor WA, Bandukwala H, Brudno Y, Agarwal S, Iyer LM, Liu DR, Aravind L, Rao A (2009) Conversion of 5-methylcytosine to 5-hydroxymethylcytosine in mammalian DNA by MLL partner TET1. Science 324:930–935

Takahashi K, Yamanaka S (2006) Induction of pluripotent stem cells from mouse embryonic and adult fibroblast cultures by defined factors. Cell 126:663–676

Takaishi S, Okumura T, Tu S, Wang SS, Shibata W, Vigneshwaran R, Gordon SA, Shimada Y, Wang TC (2009) Identification of gastric cancer stem cells using the cell surface marker CD44. Stem Cells 27:1006–1020

Tang DG, Patrawala L, Calhoun T, Bhatia B, Choy G, Schneider-Broussard R, Jeter C (2007) Prostate cancer stem/progenitor cells: identification, characterization, and implications. Mol Carcinog 46:1–14

Tay Y, Zhang J, Thomson AM, Lim B, Rigoutsos I (2008) MicroRNAs to Nanog, Oct4 and Sox2 coding regions modulate embryonic stem cell differentiation. Nature 455:1124–1128

Telerman A, Amson R (2009) The molecular programme of tumour reversion: the steps beyond malignant transformation. Nat Rev Cancer 9:206–216

Thiery JP, Acloque H, Huang RY, Nieto MA (2009) Epithelial-mesenchymal transitions in development and disease. Cell 139(5):871–890

Tirino V, Desiderio V, d'Aquino R, De Francesco F, Pirozzi G, Graziano A, Galderisi U, Cavaliere C, De Rosa A, Papaccio G, Giordano A (2008) Detection and characterization of CD133+ cancer stem cells in human solid tumours. PLoS One 3:e3469

Trowbridge JJ, Snow JW, Kim J, Orkin SH (2009) DNA methyltransferase 1 is essential for and uniquely regulates hematopoietic stem and progenitor cells. Cell Stem Cell 5:442–449

Turner BM (2007) Defining an epigenetic code. Nat Cell Biol 9:2–6

Virchow R (1855) Editorial Archiv fuer pathologische Anatomie und Physiologie und fuer klinisque Madizin 8:23–54

Vire E, Brenner C, Deplus R, Blanchon L, Fraga M, Didelot C, Morey L, Van Eynde A, Bernard D, Vanderwinden JM, Bollen M, Esteller M, Di Croce L, de Launoit Y, Fuks F (2006) The Polycomb group protein EZH2 directly controls DNA methylation. Nature 439:871–874

Visvader JE (2011) Cells of origin in cancer. Nature 469:314–322

Visvader JE, Lindeman GJ (2008) Cancer stem cells in solid tumours: accumulating evidence and unresolved questions. Nat Rev Cancer 8:755–768

Waddington CH (1935) Cancer and the theory of organisers. Nature 135:606–608

Waddington CH (1940) Organisers & genes. Cambridge University Press, Cambridge, UK

Wang JCY, Dick JE (2005) Cancer stem cells: lessons from leukemia. Trends Cell Biol 15:494–501

Widschwendter M, Fiegl H, Egle D, Mueller-Holzner E, Spizzo G, Marth C, Weisenberger DJ, Campan M, Young J, Jacobs I, Laird PW (2007) Epigenetic stem cell signature in cancer. Nat Genet 39:157–158

Wossidlo M, Arand J, Sebastiano V, Lepikhov K, Boiani M, Reinhardt R, Scholer H, Walter J (2010) Dynamic link of DNA demethylation, DNA strand breaks and repair in mouse zygotes. EMBO J 29:1877–1888

Wossidlo M, Nakamura T, Lepikhov K, Marques CJ, Zakhartchenko V, Boiani M, Arand J, Nakano T, Reik W, Walter J (2011) 5-Hydroxymethylcytosine in the mammalian zygote is linked with epigenetic reprogramming. Nat Commun 2:241

Wray J, Kalkan T, Smith AG (2010) The ground state of pluripotency. Biochem Soc Trans 38:1027–1032

Xu N, Papagiannakopoulos T, Pan G, Thomson JA, Kosik KS (2009) MicroRNA-145 regulates OCT4, SOX2, and KLF4 and represses pluripotency in human embryonic stem cells. Cell 137:647–658

Xu Y, Wu F, Tan L, Kong L, Xiong L, Deng J, Barbera AJ, Zheng L, Zhang H, Huang S, Min J, Nicholson T, Chen T, Xu G, Shi Y, Zhang K, Shi YG (2011) Genome-wide regulation of 5hmC, 5mC, and gene expression by Tet1 hydroxylase in mouse embryonic stem cells. Mol Cell 42:451–464

Yang ZF, Ho DW, Ng MN, Lau CK, Yu WC, Ngai P, Chu PW, Lam CT, Poon RT, Fan ST (2008) Significance of CD90+ cancer stem cells in human liver cancer. Cancer Cell 13:153–166

Young RA (2011) Control of the embryonic stem cell state. Cell 144:940–954

Zhang S, Balch C, Chan MW, Lai HC, Matei D, Schilder JM, Yan PS, Huang TH, Nephew KP (2008) Identification and characterization of ovarian cancer-initiating cells from primary human tumors. Cancer Res 68:4311–4320

Zimmerman AL, Wu S (2011) MicroRNAs, cancer and cancer stem cells. Cancer Lett 300:10–19

Chapter 25
Histone Acetylation as a Therapeutic Target

B. Ruthrotha Selvi[*†], Snehajyoti Chatterjee[*†], Rahul Modak,
M. Eswaramoorthy, and Tapas K. Kundu

Abstract The recent developments in the field of epigenetics have changed the way the covalent modifications were perceived from mere chemical tags to important biological recruiting platforms as well as decisive factors in the process of transcriptional regulation and gene expression. Over the years, the parallel investigations in the area of epigenetics and disease have also shown the significance of the epigenetic modifications as important regulatory nodes that exhibit dysfunction in disease states. In the present scenario where epigenetic therapy is also being considered at par with the conventional therapeutic strategies, this article reviews the role of histone acetylation as an epigenetic mark involved in different biological processes associated with normal as well as abnormal gene expression states, modulation of this acetylation by small molecules and warrants the possibility of acetylation as a therapeutic target.

[*†]Equal Contribution

B.R. Selvi
Transcription and Disease Laboratory, Molecular Biology and Genetics Unit,
Jawaharlal Nehru Centre for Advanced Scientific Research,
Jakkur, P.O., Bangalore, 560 064, India

MRC Human Genetics Unit, Institute of Genetics and Molecular Medicine,
University of Edinburgh, Western General Hospital, Crewe Road,
Edinburgh, EH4 2XU, UK

S. Chatterjee • R. Modak • T.K. Kundu (✉)
Transcription and Disease Laboratory, Molecular Biology and Genetics Unit,
Jawaharlal Nehru Centre for Advanced Scientific Research,
Jakkur, P.O., Bangalore, 560 064, India
e-mail: tapas@jncasr.ac.in

M. Eswaramoorthy
Chemistry and Physics of Materials Unit, Jawaharlal Nehru Centre for Advanced Scientific Research, Jakkur, P.O., Bangalore, 560 064, India

25.1 Histone Acetylation in Physiology

25.1.1 Introduction

Epigenetics 'is the structural adaptation of chromosomal regions so as to register, signal or perpetuate altered activity states' (Bird 2007), which results in altered gene expression through the mechanism(s) other than mutation in the underlying DNA sequences. Most of these epigenetic modifications are reversible, therefore not inherited. Some epigenetic modifications like selective histone methylation, methylated DNA regions and altered chromatin structures could be inherited across several cycles of cell division. Both histones and nonhistone proteins undergo epigenetic alterations through post-translational modifications. Several post-translational modifications (PTMs) like acetylation, methylation, phosphorylation, parylation, ubiquitination, sumoylation, acylation etc., ornate the histones in a context dependent manner and in turn regulate various biological processes.

25.1.2 Acetylation and Deacetylation

Protein lysine acetylation is one of the key regulators of biological functions of histones and several nonhistone proteins. Protein acetylation is brought about by the transfer of acetyl group from acetyl coenzyme A (acetyl CoA) to lysine residues. This reaction is catalyzed by a group of proteins called lysine/histone acetyltransferases (KATs/HATs). Based on their cellular localization KATs are classified into nuclear or type A and cytoplasmic or type B KATs. There are only three cytosolic KATs: HAT1 (KAT1), HAT2 and HAT4 (Blackwell et al. 2007; Chang et al. 1997; Takahashi et al. 2006) reported till date and they acetylate nascent histones. Nuclear KATs are further classified into five families based on their structural and functional differences (Fig. 25.1). There are three members of GNAT family- Gcn5 (KAT2A), p300/CBP associated factor (PCAF/KAT2B) and ELP3 (KAT9). Gcn5 and PCAF acetylate both histone and nonhistone substrates and they are part of various complexes involved in diverse physiological functions (Nagy and Tora 2007). There are two homologs in p300/CBP family- p300 (KAT3B) and CREB binding protein (CBP/KAT3A). Both of them are transcription coactivators and they have overlapping as well as distinct biological roles (Kalkhoven 2004). Tip60 (KAT5), MOZ (KAT6A), MOF (KAT8), MORF (KAT6B) and HBO1 (KAT7) are the major members of MYST family of KATs and they play crucial role in DNA damage repair, development and differentiation (Sapountzi and Côté 2011). There are few transcription factors- TFIIIC90 (KAT12), ATF2 and TAF1 (KAT4) that have inherent KAT activity and affect transcription directly. There are few nuclear hormone related KATs like SRC1 (KAT13A) and ACTR (KAT13B), which also act as coactivators. Though they possess histone acetylation activity, they are often part of p300/CBP

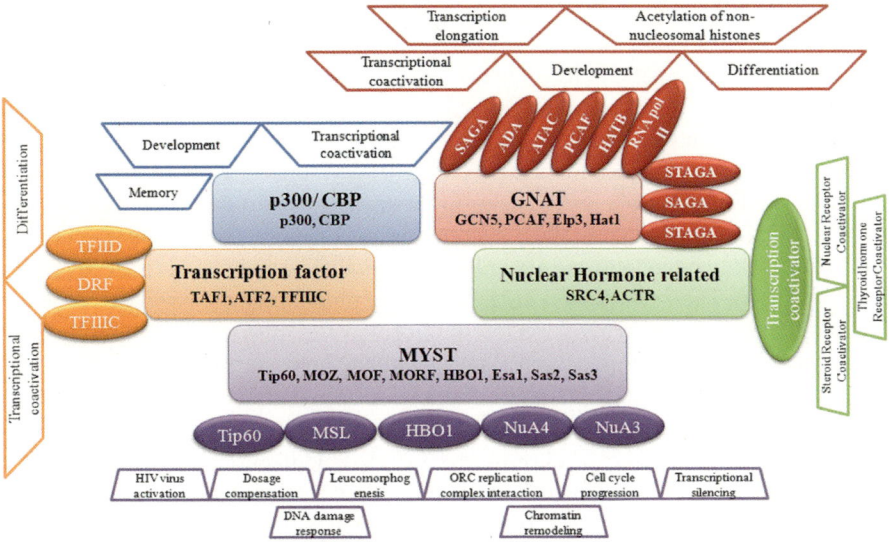

Fig. 25.1 KATs and their physiological roles. Different classes of KATs are involved in specific or overlapping cellular physiology. GNAT family of KATs are present in multiprotein complexes which are designated for different functions ranging from transcription regulation to differentiation. *Rectangular box*: KAT family, *Ovals*: Known complexes, *Trapezoids*: Biological functions

mediated coactivator complexes. Apart from these families there are few KATs: CIITA, CDYL, which does not fall under any family. The reversible acetylation marks on histones are removed by lysine deacetylases (HDACs/KDACs), which is essential for gene silencing. There are three major classes of KDACs (Hildmann et al. 2007). Class I KDACs (KDAC 1, 2, 3 and 8) are nuclear and involved in epigenetic regulations. They are always part of multienzyme complexes. Class-II KDACs (KDAC 4, 5, 6, 7) are characterized by nucleo-cytoplasmic shuttling. They can function both independently as well as part of a complex and are commonly involved in differentiation. NAD dependent KDACs are grouped under class III and they are generally called sirtuins. Sirtuins are involved in both transcription and metabolism.

25.1.3 Role of Histone Acetylation in Various Biological Processes

Specific histone acetylation is prerequisite for many biological processes starting from nuclear shuttling of histones, chromatin organization, replication, to gene expression through transcription activation. Histone acetylation/deacetylation is controlled by various external and internal signalling factors like hormones, growth factors, pathogen interactions, redox/metabolic state of the cells etc.

25.1.3.1 Replication

Replication in eukaryotic cells includes faithful propagation of DNA sequence as well as chromatin state at the different regions of the genome. Replication begins with decompaction of chromatin which allows replication machinery to bind to the replication origin (Falbo and Shen 2006). 'Replication fork' progresses via disruption of nucleosome in the front and transfer of the histones on to the leading or lagging strands. Finally, newly synthesized histones are incorporated during the replication dependent de-novo nucleosome assembly and daughter strands are produced (Groth et al. 2007a, b). The Type B acetyltransferases are predominantly associated with the replication function of acetylation. Newly formed free histone H4 is diacetylated at K5 and K12 by Hat1p/Hat2p complex, which helps in nuclear shuttling and nucleosome incorporation, but these marks are rapidly lost after completion of replication (Parthun et al. 1996; Taddei et al. 1999). Nuclear Hat1p complex is associated with acetylated H4 and H3 (K14 and K23) to help in H3-H4 tetramer assembly (Qin and Parthun 2002). The role of H4K16 and K8 in replication is yet to be ascertained. Further it has been shown that Hat1-RbAp46 complex, which acetylates H4, is essential for H3-H4 predeposition complex formation (Barman et al. 2008, 2006). The exact role of H4K5/K12 diacetylation during replication is yet to be completely understood. Recently, H4K16 acetylation was implicated in the recruitment of mini-chromosome maintenance (MCM) complex during replication initiation (Chiani et al. 2006). HAT1p/HAT2p sub complex is also part of origin recognition complex (ORC) during DNA replication in yeast (Suter et al. 2007). In yeast *Saccharomyces cerevisiae*, upon accumulation of replication stress or S-phase dependent double strand break, there is enrichment of H3K56 acetylation at the break site, indicating a significant role of H3K56 acetylation during replication (Vempati 2012; Vempati et al. 2010). Rtt109 mediated acetylation of H3K56 occurs in the newly synthesized H3-H4 dimer associated with Asf1 complex (Driscoll et al. 2007; English et al. 2006; Han et al. 2007). Recent report suggests that golgi apparatus localized HAT4 specifically acetylates free H4 at K20 (tail) and at K79 and K91 in the globular domain, thus involved in nucleosome assembly (Yang et al. 2011). Other chromatin modifications such as phosphorylation, methylation also cooperates in establishing the network associated with acetylation for the process of replication.

25.1.3.2 Repair

DNA double strand break (DSB) is constantly induced by both endogenous (metabolic) and exogenous (environmental) agents, which invokes an efficient mechanism of repair to maintain genomic stability. Growing body of evidence suggests that both homologous recombination and nonhomologous end-joining pathways are involved in DSB repair, which is further controlled by involvement of several PTMs. DSB induces instantaneous hypoacetylation of H3K9 and H3K56 by SIRT2 and SIRT3 (Tjeertes et al. 2009; Vempati et al. 2010). DSB also induces accumulation of Ku-70 and subsequent repair pathways (Krishnan et al. 2011). Post DSB repair, Nuclear NuB4 (Hat1p/Hat2p- Hif1p/Asf1) complex is involved in acetylated H3-H4 deposition and chromatin reassembly (Ge et al. 2011). Histone chaperone, anti-silencing function-1A (ASF1A), is crucial for post-repair H3K56Ac restoration, which in turn, is needed for the dephosphorylation of γ-H2AX and cellular recovery from checkpoint arrest. Complete restoration of H3K56Ac is monitored by ataxia telangiectasia mutated (ATM) checkpoint kinase (Battu et al. 2011). Tip60 and Mof mediated acetylation of H2AK5 and H4(K5/8/12/16) are also crucial for genomic stability after DSB repair. Glycogen synthase kinase-3 (GSK-3) mediated phosphorylation of Tip60 S86 is crucial for its activation (Charvet et al. 2011), which in turn acetylates p53 and H4 ultimately regulating the expression of PUMA (p53-upregulated mediator of apoptosis). Thus activation of p53 by DNA damage results in cell-cycle arrest, allowing DNA repair and cell survival, or induction of apoptosis. Apart from acetylation, other PTMs like protein methylation (Lake and Bedford 2007), phosphorylation (Charvet et al. 2011), ubiquitination (Wu et al. 2011), ADP-ribosylation (Messner and Hottiger 2011), sumoylation (Goodarzi et al. 2011) play crucial role in DNA DSB repair. The interplay between different PTMs and remodelling factors during repair have been discussed in details in the reviews by van Attikum and Gasser 2009; Lee et al. 2010.

25.1.3.3 Chromatin Organization

Acetylation at the lysine residues on histones leads to charge neutralization, which in turn reduces histone-DNA interaction leading to unwinding of the chromatin. This process is further assisted by ATP dependent remodelling complexes and is a prerequisite for all chromatin dependent phenomena. Histone acetylation in conjunction with other PTMs acts as the mark for site specific recruitment of various complexes and also contributes towards the processivity of the complex. Histone deacetylation precedes repression of active chromatin dependent processes and leads to chromatin compaction and heterochromatinization. Mechanistically, this process is a cascade of several events that include DNA methylation, histone methylation as well as histone deacetylation. These set of modifications are recognized by the repressor modules which further facilitate the compaction of the chromatin. The large scale reorganization is brought about by various chromatin associated proteins as well as the boundary elements that define the regions of transcriptional competence. Thus, acetylation is a very important mark essential for the overall structural organization of chromatin (Fukuda et al. 2006; Wilson and Merkenschlager 2006; Zhu and Wani 2010).

25.1.3.4 Transcription- RNA pol I/II/III

Histone tail acetylation at the gene promoters triggers a set of events that finally lead to the process of transcription. Along with ATP dependent remodelling complexes, H3K9 and K14 acetylation and H3K4Me3 help in recruitment of activator proteins at the promoter sites. This leads to recruitment of general transcription factors followed by the RNA polymerases to start the transcription. The H3K36 methylation,

H3 acetylation and other PTMs also play critical roles in transcription elongation and termination. The roles of epigenetic marks in the process of transcriptional regulation have been reviewed in details in part III of this book.

Histone Acetylation Associated Crosstalk During Transcription and Gene Expression

The histone acetylation associated crosstalk with the phosphorylation and methylation has been well documented in transcriptional regulation. A classical example is the phosphorylation of histone H3S10 which facilitates acetylation of H3K14 and thus activates transcription (Lo et al. 2000), which is further augmented by inhibition of HP1 recruitment at methylated H3K9 (Fischle et al. 2005). Bimodal function of methylated H3K4 is typified by recruitment of NuA3 histone acetyltransferase complex (Martin et al. 2006) and also a histone deacetylase complex (Shi et al. 2006), which shows context dependence of the crosstalks. Since the modification sites for acetylation and sumoylation (another post translational modification) have a considerable overlap, they also happen to reciprocally regulates each other's function (Chupreta et al. 2005). Transient loss of sumoylation leads to increase of acetylation and concurrent switch from a repressed to activated state in mammalian cells in a carbon source dependent manner (Segré and Chiocca 2011). Similar reciprocal role of acetylation and sumoylation has been observed during the regulation of ETS-domain transcription factor PEA3 (Guo et al. 2011) and oncoprotein Krüppel-like factor-8 (KLF8) (Urvalek et al. 2011). Histone H3K18 acetylation by p300/CBP incidentally acts as a mark for the subsequent arginine methylation by CARM1 at histone H3R17. Thus, the acetylation modification also influences the process of arginine methylation and has been identified as a part of the p53 dependent transcriptional activation cassette (An et al. 2004). Recently, it has been identified that acetylation also influences gene expression indirectly by modulating the miRNA expression. This evidence has been obtained by using the deacetylase inhibitors; a more detailed investigation is however needed with respect to the role of acetylation and miRNA expression. The integration of miRNA in the acetylation centred crosstalk, during transcription regulation is yet to be established.

Recent studies have further added another level of complexity by showing the spatio-temporal regulation of the PTMs. Zippo et al. have reported recruitment of oncogenic kinase PIM1 phosphorylates H3S10 at the enhancer region of serum responsive gene *FOSL1*, which then recruits 14-3-3 protein (Zippo et al. 2009). 14-3-3 complex recruits the KAT- MOF and brings about acetylation of H4K16 at the enhancer. Acetylated H4K16 recruits Brd4 and associated kinase P-TEFb, which phosphorylates RNA pol II and facilitates transcription elongation of *FOSL1* gene. This order of recruitment of phosphorylated H3S10 and acetylated H4K16 are in contrast to the previous report (Wang et al. 2001). These observations along with many others have shown that the order of recruitment of PTMs and their readers and writers decide the functional outcome (Lee et al. 2010; Smith and Shilatifard 2010).

25.1.3.5 Development and Differentiation

Histone acetylation plays crucial role in the different developmental stages and the epigenetic landscape during these stages are very different from that in the differentiated cells. Chromatin is in a more 'fluid' state in stem cells and also during the earlier stages of development, which helps in easy accessibility of the genetic information. The process of differentiation has been most well studied in the mouse embryonic stem cell model as well as lineage specific models such as myocyte, adipocyte, neuronal lineages etc. With respect to the developmental stages, the maximum information is from the mouse model. Several studies have also been done with the drosophila model such as the role of polycomb complexes. However, we shall be focusing on the stem cell based data available to review the status of chromatin and its functional modulation during the process of differentiation and development.

Chromatin During the Process of Differentiation and Development

Differentiation generally refers to the formation of functionally mature cell types. In this process there are stages of developmental timelines, wherein the cell attains partial functionality. In general, most of the adult stem cells belong to this category. The totipotent cell in the mammalian life cycle is restricted to the embryo till its cleavage stage. Following this, the cells attain a pluripotent state which signifies their ability to lead to the formation of the three germ layers (ectoderm, endoderm and mesoderm), from wherein, finally the more differentiated cell types gets established. These differentiated cell types also exist either as multipotent (hematopoietic lineage) or the unipotent (neuronal lineage). However, the terminally differentiated cell types are functionally complete. The chromatin, which is the nucleoprotein ensemble of the genetic material, is generally considered to be more fluidic, or in an open state. As the stages of differentiation proceeds, compaction of chromatin proceeds. The electron spectroscopic imaging studies reveals that the structural organization of ES cell chromatin is much more open and mostly more active transcriptionally (Fussner et al. 2010). This has also been validated by the studies looking for heterochromatin proteins whose reorganization has indeed validated this hypothesis that during the process of differentiation the chromatin does undergo a reorganization from an open chromatin state to an inactive, more compact form (Bártová et al. 2008).

Histone Acetylation During the Process of Differentiation and Development

The process of development and differentiation are events which are regulated by distinct transcriptional states and hence the role of acetylation in these processes is crucial. However, there exists stage specificity with respect to the expression profiles of these acetyltransferases and deacetylases. The possible role of acetyltransferases in the process of early development has been obtained from studies on the mouse model knockout systems. For a long time there were speculations that such knockouts could be lethal since the acetyltransferases are widely considered to be redundant and absolutely essential for the transcriptional events. In contrary, till date distinct roles of acetyltransferases have been observed during development. More importantly, knockout of some KATs were not lethal and were shown to be critical for key developmental events. The lysine acetyltransferase family PCAF/ GCN5 have distinct developmental roles. PCAF is expressed at later stages in the mouse development around E12 and hence does not show any phenotype on knockdown. GCN5, which is expressed at very early stages such as E7.5 replaces the PCAF, which is essential for the developmental processes since in the PCAF knockout, increased GCN5 levels have been observed in the liver and lungs (Yamauchi et al. 2000). But, the reverse was not observed when GCN5 was not present in which case the mice died in utero. The double knockout exhibited defects in vascular formation and few showed defects in the neural tube closure. However, it is clear that the expression of GCN5 and PCAF are timed distinctly and hence has potential distinct, and overlapping roles. The other important acetyltransferase, p300 and CBP were so difficult to be knocked out, that for several years their exact roles in the developmental stages could not be identified. Later on advancement in the technology of silencing in a specific manner, led to the specific knockdown at different stages as well as in different tissues. The most important information gained from these studies is that, probably p300 and CBP functions are redundant since one replaces the other (Goodman and Smolik 2000). However, in the neuronal developmental pathway the requirement of CBP has been shown to be absolutely critical. Subsequently, it was also identified that several neurodegenerative disorders were characterised by a loss in CBP acetyltransferase function further strengthening its need in neuronal processes. Recently, it has been very elegantly shown in tissue specific targeted knockouts that CBP acetyltransferase has a role to play in memory formation (Barrett et al. 2011). An independent study also investigated the role of p300 and GCN5 by knocking down both the acetyltransferases (Phan et al. 2005). This study led to the identification of critical roles of these enzymes in the early developmental processes. The MYST family acetyltransferases are also found to be essential for important developmental processes (Voss and Thomas 2009). The MOZ acetyltransferase has been shown to be necessary for the process of hematopoiesis in the mice (Katsumoto et al. 2006). The major hematopoietic organs were affected in this tissue specific knockdown. The Tip60 acetyltransferase has been essentially proven to have a tumor suppressor function. The homozygous knockdown was not viable and was defective in implantation. Therefore, heterozygous knockdown were generated which showed a haploinsufficiency with respect to tumor formation (Gorrini et al. 2007). Most of the acetyltransferases and deacetylases show differential expression levels during the process of early development. Subsequent to this, these enzymes become critical to several processes in gene expression. Hence, their expressions are more or less same during the later stages of development. Their activities do differ, thus the modulation during the process of differentiation is more at the activity level than the expression level itself. For example, the acetyltransferase PCAF which is expressed after E12.5 in mouse is present all through the stages of the myogenic differentiation process. However, the initiation of this process is signified by the PCAF mediated MyoD acetylation (Described in details in Chap. 7 by Bharathy et al. of this collection). The process of differentiation is generally functional either during the later stages of development giving rise to the fully functional cell types or during diseases such as cancer. However, both these phenomena are characterized by almost similar mechanisms. The following section highlights the role of acetylation in different diseases including cancer.

25.2 Histone Acetylation and Diseases

The acetylation of histones and non histone proteins plays an important role in the various aspects of gene expression as discussed in the previous section. These key players also have a significant role to play in the process of disease manifestation and progression which is majorly due to a loss in the acetylation/deacetylation balance. This could be either because of the changes, in their activities or due to a change in their expression itself. The exact mechanism with respect to each of the KAT families, in disease manifestation and progression will be discussed in the next section. Of the different KAT families, we shall majorly focus on the CBP/p300 family, PCAF/GCN5 family and the MYST family.

25.2.1 Involvement of p300/CBP Family in Diseases

The p300/CBP family of KATs is represented by the p300 and CBP acetyltransferases. These are important transcriptional coactivators and are involved in several aspects of gene expression. Their involvement in certain diseases such as cancer, virus associated diseases, diabetes, cardiac hypertrophy and neurodegenerative disorders have been extensively investigated as discussed in the subsequent section.

25.2.1.1 Cancer

Role of p300 and CBP in cancer is highly controversial. There are reports suggesting role of p300 as a tumor suppressor whereas on the contrary p300 is also found to be a tumor promoter. It is also been reported that inhibition of p300 inhibits cancer cell growth, hence strengthening its role as a tumor promoter gene. Down regulation of p300 activity induces senescence and growth inhibition in melanocytes (Bandyopadhyay et al. 2002). In acute monocytic leukemia, MOZ translocates from chromosome 8 to chromosome 22 [t(8;22)(p11;q13)] where it forms a fusion protein with p300 (Kitabayashi et al. 2001). In acute myeloid leukemia, due to chromosomal translocation CBP forms a fusion protein with MOZ (Troke et al. 2006). These fusion proteins are more active in their function and cause aberrant hyperacetylation and transcriptional activation. Expression levels of p300 are found to be very high in resectable esophageal squamous cell carcinoma (ESCC) than in normal esophageal mucosa cells. High p300 expression levels are also associated with higher grades of ESCC (Li et al. 2011b). Recent reports suggest that p300 mRNA and protein levels are very high in patients of hepatocellular carcinoma (HCC) and high expression of p300 is essential for acquisition of the aggressive phenotype (Li et al. 2011a). Studies from our group have also revealed the mechanistic link between the acetyltransferase p300 and the histone chaperone, NPM1 in the manifestation of oral cancer (Shandilya et al. 2009; Arif et al. 2010). p300 is an activator of Glial Fibrillary Acidic Protein *GFAP* gene which is associated with astrocytic differentiation of glioblastoma multiforme (GBM) cells. Silencing of p300 gene by RNAi induces invasion potential of (GBM) cells *in vivo* (Panicker et al. 2010). Histone H3K56 is a common site of acetylation for p300/CBP and is associated with progression of few cancers (Das et al. 2009).

Tumor suppressor function of p300 and CBP is supported by the observation that the monoallelic mutation of CBP locus is one of the major causes of Rubinstein-Taybi syndrome with compromised CBP KAT activity and these patients are more prone to develop malignancy (Troke et al. 2006). Alteration of p300 gene has also been identified in various cancers of epithelial origin (Muraoka et al. 1996; Gayther et al. 2000). Mutation in p300 gene and loss of heterozygosity at p300 locus is reported to be associated with progression of colorectal, gastric, breast and brain cancer.

25.2.1.2 Virus Associated Diseases

p300 regulates integration of HIV in human genome by acetylating viral protein integrase (Cereseto et al. 2005). In HIV infected cells, LTR promoters are transcriptionally inactive due to chromatinization but are regulated by protein acetylation. HIV1 protein Tat relieves this inhibition by recruiting p300 and CBP in the viral LTR (Marzio et al. 1998). p300/CBP interacts with HIV Tat protein and acetylates Tat thereby activates the Tat dependent transcription of HIV genes on the integrated provirus (Deng et al. 2000; Ott et al. 1999). Interaction with Tat enhances the KAT activity of p300. KAT activity of p300 is essential to derepress the HIV-1 chromatin structure and thus activate the transcription of the integrated viral DNA *in vivo*. Apart from several cellular transcription factors, IRF family proteins especially IRF-1 can stimulate HIV1 LTR transcription in absence of Tat. p300/CBP and PCAF can acetylate IRF-1. IRF1 also recruits CBP in the promoter of HIV1 LTR in absence of Tat. Acetylation of the histone chaperone nucleophosmin (NPM1) by p300 is essential for the nuclear localisation of Tat and activates transcription of the integrated HIV provirus (Gadad et al. 2011).

25.2.1.3 Diabetes

Recently KATs and KDACs have been found to be associated with diabetes (Described in detail in Chap. 19 by Reddy and Natarajan of this collection). Diabetes is associated with over expression of various vasoactive factors and extracellular matrix (ECM) proteins leading to alteration of morphology and function of organs which are also affected by complications arising due to chronic diabetes (Brownlee 2001). Glucose induces expression of fibronectin and other vasoactive factors in glucose induced endothelial cells through activation of NF kB and Activating protein 1 (AP1) (Chen et al. 2010). The mechanism of p300 expression by high glucose levels is not yet elucidated but the possible reasons could include activation of PKC and PKB, MAP kinase activation and oxidative stress induced PARP activation (Kaur et al. 2006; Xu et al. 2008; Chen et al. 2010). High glucose levels induce binding of p300 in the promoters of ET-1 and FN gene and cause acetylation of histones. Due to the acetylation of the promoters, various transcription factors are recruited which leads to the expression of vasoactive factors and ECM proteins. siRNA mediated silencing of p300 in human umbilical vein endothelial cells (HUVECs) effectively inhibit glucose mediated upregulation of vasoactive factors and ECM proteins.

25.2.1.4 Cardiac Hypertrophy

Agonist induced hypertrophy of cardiomyocytes is associated with enhanced transcriptional activity of p300 (Gusterson et al. 2002). p300 plays various essential roles in differentiation, growth and apoptosis of cardiac myocytes. During myocardial cell hypertrophy, GATA4 (a zinc finger protein) is upregulated. p300 acetylates GATA4 and enhances its DNA binding ability (Yanazume et al. 2003) which further induces the expression of ANF, ET-1, and α-MHC genes (Dai and Markham 2001). Thus, p300 mediated acetylation of GATA4 plays a crucial role in the development of myocyte hypertropy. Minute increase in the level of p300 is sufficient for the manifestation of myocardial hypertrophy by acetylating MEF2. p300 mediated acetylation and thus activation of MEF2 is also pivotal for driving the expression of hypertrophy associated genes (Wei et al. 2008).

25.2.1.5 Memory Impairment and Neurodegenerative Diseases

Information can be stored in the brain as short term memory which lasts for several minutes to hours. Short term memory can be stabilized to long-term memory which lasts for years. This stabilization of short term memory into long term memory requires new gene expression. CBP KAT activity is crucial for this stabilization of short term memory into long term memory. CBP is also required for transcriptional regulation for long term memory formation. Neurotrophic deprived primary cultured cerebellar neurons show decreased histone H3 and H4 acetylation with subsequent disappearance of CBP. Loss of function of CBP/p300 KAT activity during apoptosis and neurodegenerative diseases suggests its possible role in the disease progression (Rouaux et al. 2003).

25.2.2 Involvement of PCAF/GCN5 Family in Diseases

The PCAF/GCN5 family involvement has been shown indirectly in different diseases; we shall discuss their roles in cancer, HIV, malaria and diabetes.

25.2.2.1 Cancer

Twist1 is a basic helix-loop-helix transcription factor, is associated with tumor growth by the expression of Y-box binding protein 1(YB1) (Shiota et al. 2008). PCAF directly interacts and acetylates Twist1 and controls its intracellular localization and transcriptional activity. Knockdown of PCAF retards cell growth and invasive ability of urothelial cancer cells (Shiota et al. 2010). PCAF also plays important role in the tumor suppressor function of p53. PCAF protein levels are drastically reduced in intestinal type gastric cancer (ITGC). PCAF could efficiently suppress tumorigenicity of gastric cells and inhibit cells entering S phase from G1 phase (Ying et al. 2010).

25.2.2.2 HIV

One of the most essential events of HIV replication cycle is the integration of the viral DNA in the host genome. This function is performed by the viral protein integrase. Acetylation of integrase is one of the deciding factors for this activity. Apart from p300, GCN5 also acetylates HIV integrase protein and hence contributes to HIV infection (Terreni et al. 2010). Tat, a HIV encoded protein gets acetylated by both p300/CBP and PCAF (Kiernan et al. 1999) with functional consequence in the HIV life cycle.

25.2.2.3 Parasite Related Diseases

Plasmodium falciparum GCN5 (PfGCN5) has a conserved activity (Fan et al. 2004) and has preferences for acetylating histone H3K9 and K14 of plasmodium (Miao et al. 2006). Other KAT genes (MYST family member, PF11_0192) and two other KAT proteins (PFL1345 and PFD0795w) have also shown to be encoded by the plasmodium genome. Different acetylation marks are abundant throughout the erythroid cycle. PfGCN5 is actively recruited to various gene promoters to mediate histone acetylation and thus activate transcription of malarial proteins (Cui et al. 2007). Mono allelic expression of *var* gene is regulated by histone acetylation and thereby provides the antigenic switching and virulence of the parasite (Tonkin et al. 2009; Cui and Miao 2010; Petter et al. 2011). Thus acetylation of various malarial histones could be essential for the regulation of genome wide gene expression during erythroid development of the malaria parasite.

25.2.2.4 Diabetes

Type 2 diabetes mellitus (T2DM) is a disease caused by elevated glucose production characterized with insulin resistance, hyperinsulinaemia and hyperglycemia (Mitrakou et al. 1992; Perriello et al. 1997). In normal cells, plasma glucose levels are maintained within a very limited range by tightly maintaining the transcription regulation of phosphoenolpyruvate carboxykinase (PEPCK; gene code *Pck1*), the rate-limiting enzyme of hepatic gluconeogenesis. The peroxisome proliferator-activated receptor-g coactivator-1a (PGC1 α or PPARGC1A) is a crucial effector of pck1 (Yoon et al. 2001; Dentin et al. 2007). Upon activation, PGC1α forms complex with various transcription factors including FOXO1 and hepatic nuclear factor 4a (Hnf4a or Hnf4α), which subsequently induces the transcription of gluconeogenesis genes including pck1 (Yoon et al. 2001; Puigserver et al. 2003; Rhee et al. 2003). GCN5 (or KAT2A) acetylates PGC1α and deactivates which in turn causes suppression of gluconeogenic gene expression (Caton et al. 2010; Lerin et al. 2006).

25.2.3 Involvement of MYST Family KATs in Diseases

MYST is derived from three classical members of KAT family: mammalian MOZ, yeast Ybf2/Sas3 and Sas2, and mammalian Tip60. MORF and HBO1 discovered later are the other members of the MYST family. This family of acetyltransferases have been implicated in the DNA repair pathway and hence have intrinsic association with diseases such as cancer.

25.2.3.1 Cancer

Tip60 is an essential component for the chromatin repair mechanism. Tip60 mutations are associated with defects in DNA double strand break repair and provide resistance for the DNA damaged cells towards apoptosis (Ikura et al. 2000), which is a characteristic feature associated with cancers. Androgen receptor (AR) is a hormone dependent transcription factor which has a strong link with development of prostate cancer. Tip60 acetylates AR and activates its function (Brady et al. 1999; Gaughan et al. 2001, 2002). Tip60 has also been linked with tumor suppressor p14ARF. Upon genotoxic stress, p14ARF directly interacts with Tip60 resulting in activation of the ATM/CHK signaling cascade and cell cycle arrest at the G2 phase (Eymin et al. 2006). Tip60 acetylates retinoblastoma (Rb) tumor suppressor. Reports suggest that in colon and lung carcinoma tissues show significant reduced expression of Tip60 (LLeonart et al. 2006). Tip60 is also reported to acetylate K120 residue of tumor suppressor p53 and is associated with activation of p53 mediated cell cycle arrest and apoptosis. Acetylation of p53K120 by Tip60 is crucial for p53 mediated activation of pro-apoptotic genes BAX and PUMA. K120 acetylation does not confer stability to p53 nor enhances DNA binding, but it occurs upon DNA damage or oncogenic stress (Tang et al. 2006). P53 K120 is also frequently mutated in various human cancers. Upregulation of Tip60 is also associated with onset of epithelial tumorigenesis (Hobbs et al. 2006).

MOZ (monocytic leukaemia zinc-finger protein) forms a fusion protein with CBP by chromosomal translocation in AML (Borrow et al. 1996). MOZ also forms in-frame fusions with CBP and its homologue p300 (Kitabayashi et al. 2001); (Panagopoulos et al. 2002; Rozman et al. 2004) nuclear receptor coactivator TIF2 (Carapeti et al. 1998; Kindle et al. 2005) and NcoA3. The coding region of MOZ is identified as a common retroviral insertion site which could be its reason for translocation in leukemia (Lund et al. 2002). The catalytic domains of these fusion proteins remain functionally active and due to misdirection of the targeting domains the various histone and nonhistone proteins get aberrantly acetylated. Apart from AML, in benign uterine smooth muscle tumours, MORF (MOZ related factor) coding region gets disrupted by another chromosomal translocation (Moore et al. 2004). Loss of MORF and the manifestation of the tumor could be due to decrease in the stability of the ING5-MOZ/MORF-BPRF complex (Doyon et al. 2006) and subsequently affect the ability of ING5 to modulate p53 activity (Shiseki et al. 2003). MOZ is overexpressed in chemical hepatocarcinogenesis but chromosomal translocation has not been identified. Human MOF acetylates histone H4K16 in cells (Smith et al. 2005) and this modification is a critical regulator of various chromatin related processes and is also linked to several diseases (Shia et al. 2006). Most importantly, loss of H4K16 acetylation is reported to be one of the frequent hallmarks of various cancers (Fraga et al. 2005). H4K16 acetylation inhibits chromatin higher order structure (30 nm fiber) formation and also interferes with various chromatin remodelling enzymes (Shogren-Knaak et al. 2006). Loss of H4K16 acetylation by downregulation of HBO1 results in reduced cell cycle arrest (Smith et al. 2005).

25.2.3.2 HIV

HIV1 Tat protein suppresses the transcription of Mn-SOD gene by inhibiting Tip60 KAT activity (Creaven et al. 1999). Downregulation of Mn-SOD gene is one of the factors for the generation of oxidative stress leading to T-cell depletion in AIDS (Westendorp et al. 1995). Tat has also been reported to be responsible for degradation of Tip60 by ubiquitination (Col et al. 2005).

25.2.3.3 Malaria

MYST acetyltransferase has been identified in *Plasmodium falciparum* (malarial parasite) genome. PfMYST preferentially acetylates histone H4 at K5, K8, K12 and K16 in vitro. PfMYST localizes to both nucleus and cytoplasm. These functions are essential for the transcriptional regulation, cell cycle progression and DNA damage repair of the parasite. pfMYST is also reported to be recruited to the *var* gene promoter and acetylates H4 to facilitate transcription initiation (Miao et al. 2010).

The significant contribution of the acetyltransferases towards disease manifestation and progression, has led to increased speculation about considering the modulators of these enzymes as potential therapeutic agents. However, the specific inhibitor or activator for a particular enzyme could prove to be an extremely valuable therapeutic strategy. In this section we will discuss briefly about various KAT inhibitors and activators reported and their mode of action. Lastly we will give a brief overview about the possible role of these modulators in disease therapeutics as well as the potential combinatorial therapy.

25.3 Modulators of Histone Acetylation

25.3.1 Inhibitors of Histone Acetylation

The concept that reversible histone acetylation could be a therapeutic strategy was first established by KDAC inhibitors. Some KDAC inhibitors are already in clinical trials for their potential role in cancer treatment. Suberoylanilide hydroxamic acid (SAHA, vorinostat) have been now approved by U.S. Food and Drug Administration (FDA) for treatment against advanced cutaneous T-cell lymphoma (Mann et al. 2007; Richon et al. 1998). In comparison to KDAC inhibitors, use of KAT modulators against diseases is a relatively new field. Extensive research has been employed to identify various KAT modulators and the strategies followed are: rational design, high-throughput screening of synthetic libraries, and enzymatic screens with natural products (Cole 2008). The inhibitors of acetyltransferases are comparatively more in number with respect to the acetyltransferase activators.

Among different KATi, inhibitors of p300/CBP and PCAF/GCN5 have been more extensively worked on. Among all the KAT inhibitors, the first reported KAT inhibitors are the bisubstrate analogue Lys-CoA and H3-CoA-20 (Lau et al. 2000). These inhibitors were designed on the basis of the knowledge that upon substrate binding the target enzyme employs a ternary complex, involving the H3 peptide and acetyl CoA. Lys-CoA one of the simple analogue of peptide CoA is a potent and specific inhibitor for p300. H3 CoA is a 20 amino acid peptide derived from histone H3 tail and conjugated with Coenzyme A and specifically inhibits PCAF. Bisubstrate analogues were also used to develop inhibitors against Tip60 and Esa1 (yeast orthologue of Tip60). Among others, H3K16 CoA proved to be the most potent inhibitor of Tip60 and Esa1 (Wu et al. 2009). But these bisubstrate inhibitors were not cell permeable and hence their use is restricted.

The first reported natural KAT inhibitor is anacardic acid (nonadecyl salicylic acid) isolated from cashewnut shell liquid. Anacardic acid non specifically inhibits KATs p300, PCAF and Tip60. Due to its non specificity against any particular enzyme and lack of cell permeability, the scaffold has been exploited for the synthesis of better inhibitors. A salicylate derivative of anacardic acid has been developed with specificity towards PCAF, which is reported to have twofold improved potency to inhibit recombinant PCAF enzyme and has also been shown to inhibit PCAF mediated acetylation in HEP G2 cell lines (Ghizzoni et al. 2010). Curcumin, a polyphenolic compound isolated from *Curcuma longa* rhizome is a specific inhibitor of p300. Curcumin possesses anticancer effects by inducing apoptosis in cancerous cells but not in healthy normal cells, whereas one of the major disadvantage of curcumin is that it has a very poor bioavailability. As an improvement, our group has synthesized and characterized CTK7A, a water soluble derivative of curcumin which inhibits p300/CBP and PCAF mediated acetylation and has an anti oral cancer effect (Arif et al. 2010). However, the bioavailability of CTK7A in comparison to curcumin is yet to be investigated.

Another naturally isolated inhibitor of p300 and PCAF is garcinol. Garcinol is polyisoprenylated benzophenone isolated from the edible fruit, *Garcinia indica* or kokum fruit (Balasubramanyam et al. 2004). The IC_{50} values of garcinol for p300 and PCAF is 7 and 5 µM respectively. Intramolecular cyclisation of garcinol yielded isogarcinol. Further substitution of the 14th position yielded 14-isopropoxy IG (LTK-13) and 14-methoxy IG (LTK-14) and disulfoxy IG (LTK-19). LTK13, 14 and 19 could selectively inhibit p300 but not PCAF (Mantelingu et al. 2007b). The mechanistic details of the specificity of LTK14 against isogarcinol and garcinol have also been reported. LTK14 acts as a noncompetitive inhibitor for acetyl CoA and histones unlike non specific inhibitors isogarcinol and garcinol. LTK14 has a single binding site on the p300 KAT domain whereas both garcinol and isogarcinol has two binding sites (Arif et al. 2009). The specificity of LTK14 towards p300 is provided by the formation of the lactone ring upon cyclization of garcinol. The methyl group present on the aromatic phenol of LTK14 reduces its affinity towards PCAF and not p300. Further studies revealed that LTK14 alters the tertiary structure of p300 whereas garcinol and isogarcinol alters the secondary structure of p300. Recently, plumbagin has been reported to be a p300 specific KAT inhibitor (Ravindra et al. 2009). It is a hydroxynaphthoquinone and is isolated from the roots of the medicinal plant, *Plumbago rosea*. Plumbagin follows the non competitive mode of inhibition for p300 and its single hydroxyl group has been shown to be crucial for its KAT inhibitory activity. As most of the KAT inhibitors possess polyhydroxyl groups, it is speculated that this chemical moiety might be a decisive factor for acetyltransferase inhibition.

Derivatization of γ-butyrolactones yielded MB-3 which is an inhibitor of CBP and Gcn5 with an IC_{50} of 100 µM. The length of the aliphatic side chain of MB-3 is critical for its KAT inhibitory activity (Biel et al. 2004). Isothiozalones have also proved to be potent inhibitors of p300 and PCAF. But these compounds are highly reactive and have off targets as they have high chemical reactivity with free thiol groups of other proteins. Recently, epigallocatechin-3-gallate (EGCG), a component present in green tea has been shown to posses KAT inhibitory potential. It showed inhibitory effects towards majority of the KATs but not other chromatin modifying enzymes (Choi et al. 2009). C646, a pyrazolone-containing small molecule obtained after in silico based virtual ligand screening specifically inhibits p300/CBP KAT and also shows anticancer activities (Bowers et al. 2010).

25.3.2 Small Molecule Activators of Acetyltransferases

KAT activators are small molecules that can activate the acetyltransferase activity of a KAT by possibly altering the structural configuration of the enzyme. The first reported and best characterized KAT activator is N-(4-Chloro-3-trifluoromethyl-phenyl)-2-ethoxy-6-pentadecyl-benzamide (CTPB) (Balasubramanyam et al. 2003). CTPB can efficiently induce histone acetylation levels both *in vitro* and *in vivo* (Selvi et al. 2008). Upon binding to p300, CTPB induces structural alteration of p300 and further leads to autoacetylation. Detailed structural and functional studies led to the conclusion that the $-CF_3$ and $-Cl$ at the para position of CTPB is crucial for its KAT activation property (Mantelingu et al. 2007a). Due to its impermeability to cells, CTPB has been conjugated on glucose derived carbon nanosphere (CSP) by simple adsorption to achieve in vivo activation. Another recently reported KAT activator is nemorosone, which belongs to a group of polyhydroxybenzophenones. Like CTPB, Nemorosome also activates p300 KAT activity and the possible mechanism of activation is also by binding to the same region of p300 where CTPB binds (Dal Piaz et al. 2010). Presently, isolation and characterization of more specific activators of KATs are in progress for use as possible future generation therapeutics.

25.4 KAT Modulators as Possible Future Generation Therapeutics

Histone hyperacetylation has been causally associated with several diseases such as inflammatory disorders, viral diseases etc. Hence, the use of histone acetylation inhibitors should in theory be capable of atleast partially reversing these effects. Most of these studies have been done on cell culture based assays. However, the results obtained have provided enough evidence to support the possibility of considering the KAT modulators as a therapeutic strategy.

25.4.1 Histone Acetylation Inhibitors as a New Generation Therapeutic Strategy

Anacardic acid can induce apoptosis in a TNFα or chemotherapeutic agent dependent manner and could efficiently downregulate transcription of genes essential for invasion, proliferation, survival and angiogenesis. Anacardic acid could inhibit constitutive expression of NFκB and also inhibit acetylation and nuclear translocation of p65. Anacardic acid also inhibits PfGCN5 activity and thereby downregulates developmentally regulated genes of *Plasmodium falciparum* (Cui et al. 2008).

Being a specific inhibitor for p300/CBP, curcumin inhibits HIV virus proliferation and acetylation of HIV Tat. Apart from HIV, curcumin has also proved to be a potential anticancer, antiviral, antiarthritic, anti-amyloid, antioxidant, and anti-inflammatory agent (Zhou et al. 2011). CTK7A (Sodium 4-(3,5-bis(4-hydroxy-3-methoxystyryl)-1H-pyrazol-1-yl)benzoate), the water soluble derivative of curcumin induces histone hypoacetylation in oral cancer cell lines and promote senescence and growth arrest. Treatment of CTK7A also reduces tumor size in oral cancer xenografted nude mice (Arif et al. 2010). The polyisoprenylated benzophenone garcinol induces apoptosis in HeLa cells and induces global gene repression (Balasubramanyam et al. 2004). LTK14, a derivative of garcinol inhibits HIV multiplication, inhibits viral syncia formation in HIV infected cells and is nontoxic to T cells. Plumbagin has also proved its therapeutic activity by inhibiting NFκB pathway. Epigallocatechin-3-gallate (EGCG) isolated from green tea has antiproliferative properties against colon, lung, and breast cancers and chronic inflammation especially by its anti-NF-κB transactivation activity and subsequently inhibiting EBV induced B-cell transformation (Choi et al. 2009). C646 has also strong anticancer activities due to its p300 specific inhibition (Bowers et al. 2010). By inhibiting p300 acetyltransferase activity, it decreases AR function and downregulates p65 which further induces apoptosis in prostate cancer cells (Santer et al. 2011). Isothiazolones derivatives have possible implications in colon cancer treatment (Stimson et al. 2005).

25.4.2 *Therapeutic Perspective of KAT Activators*

Histone acetylation is altered in various diseases and the balance between acetylation and deacetylation is hampered. Various diseases including cancers and neurodegenerative disorders are associated with reduced expression of p300 and CBP. KDAC inhibitors have been extensively used to reactivate acetylation and a few have already entered in clinical trials. KDACi treatment has also provided promising results in treatment of glioblastoma. The major limitation of using KDAC inhibitors is the lack of specificity for any particular KDACs. KDACs functions in association with the other members of the family and the loss of one protein generally gets compensated by other. Thus KDAC inhibitors induce global histone acetylation. Various diseases are manifested with alteration of KAT activity of a particular acetyltransferases and thus KAT activators specific for a particular enzyme could induce its activity and as a consequence reactivate its function. KAT activators also induce histone hyperacetylation, but unlike KDACi it specifically activates its target KAT. CTPB specifically activates acetyltransferase activity of p300 and CBP. p300 and CBP activity is reported to be lost in various neurodegenerative disorders and thus CTPB could be a potential therapeutic agent in the treatment of the disease. The lack of cellular permeability limits CTPB to be used in animal model. Also the major hurdle for drugs used for brain targeting for diseases such as glioma is to cross the blood brain barrier (BBB).

25.4.2.1 Nanomaterial as Delivery Vehicle for Activators

Often the failure to access the drugs across the blood brain barrier in therapeutically required quantities is one of the major impediments in looking out for alternative drug molecules to treat many central nervous system related diseases like Alzheimer's, Huntington's, Parkinson's etc. (Jain 2011). The blood brain barrier which segregates the blood compartment of the brain from the extracellular fluid, is composed of a monolayer of endothelical cells connected through tight intercellular junctions. This barrier restricts the diffusion of many hydrophobic drugs (also active peptides and proteins) into the brain (Kreuter et al. 1995); (Brasnjevic et al. 2009). These limitations can be circumvented by the use of nanoparticles whose size range vary from 1 nm to few hundred nm and are perfectly suitable to interact with biological cells, proteins and DNA (Malam et al. 2011). Furthermore, the size dependent optical, magnetic and chemical properties of many of the nanoparticles at the quantum regime facilitate their use in diagnostics, imaging and drug delivery. Some of the critical issues limiting the pharmaceutical design, like the delivery of poorly soluble drugs, targeting the drug at particular site, blood brain barrier, and simultaneous delivery of two or more drugs can be effectively addressed by using nanoparticles. Targeted delivery of drugs using nanoparticles not only alter the pharmacokinetic profile of the drugs but also reduce their systemic toxicity by preventing their accumulation in other parts of the body. Polymeric nanoparticles, solid lipid nanoparticles, cyclodextrin nanoparticles, liposomes, and inorganic nanoparticles such as mesoporous silica, gold, CdSe, Fe_3O_4, fullerene, carbon nanotubes are widely used for diagnostic, imaging and drug delivery applications (Emerich and Thanos 2006). Delivery of drugs in different cell lines using these nanoparticles have been demonstrated, both in *in vitro* and *in vivo* experiments. However, when it comes to brain cells, most of these nanoparticles fail to deliver and demands special modifications as they have to cross the BBB. Receptor mediated endocytosis pathway in which, the strong affinity between a macromolecule (linked to the nanoparticles) and its specific receptor (expressed on the cell surface) is normally exploited for the delivery of the drug into the brain cells through nanoparticles. Large neuropeptides such as insulin, transferrin, and leptin cross the BBB through the receptor mediated endocytosis mediated by the corresponding receptors (Gupta et al. 2005); (Duffy and Pardridge 1987). Decorating such molecules on the surface of the nanoparticles would be expected to help them cross the BBB. For example, PLGA, the poly(lactic-co-glycolic acid) nanoparticles were shown to cross the BBB in rat brain only when it was modified with peptides (Costantino et al. 2005). Similarly, transferrin coated, drug loaded PLGA nanoparticles are shown to deliver the drug across the BBB through receptor mediated endocytosis mechanism as the BBB expresses transferrin receptor excessively (Chang et al. 2009). In an another approach, surfactant (polysorbate 80) coated PLGA nanoparticles(loaded with doxorubicin drug) were shown to enhance the adsorption of apolipoproteins which in turn, facilitated the receptor mediated endocytosis of the PLGA nanoparticles (Gelperina et al. 2010). Solid lipid nanoparticles which are coated with PEG (polyethylene glycol)

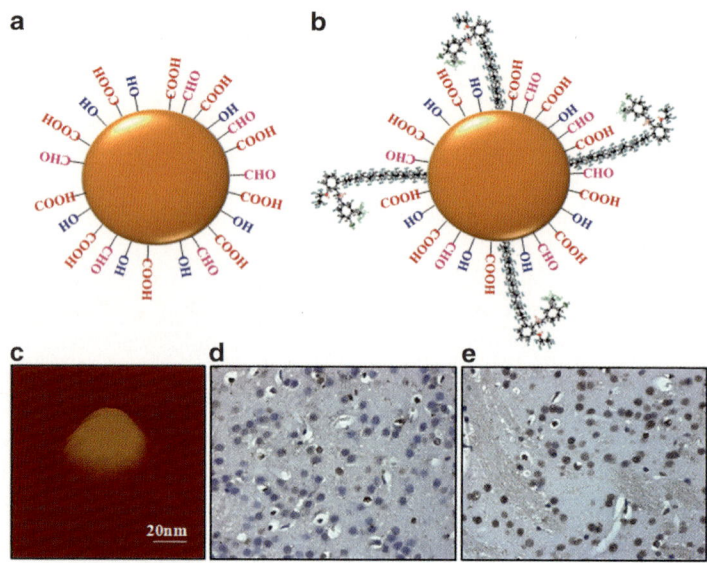

Fig. 25.2 Carbon nanosphere as a vehicle to carry KAT activators to mice brain. (a) Pictorial representation of Carbon nanosphere (CSP) showing the charged residues on the surface, **(b)** CTPB adsorbed on the surface of carbon nanosphere. **(c)** AFM image of chemically conjugated CSP CTPB. CSP CTPB was coated on the surface of freshly cleaved mica plate and was visualized under AFM (Chatterjee et al. 2010) **(d)** Immunohistochemistry images of mice treated with **(d)** CSP and **(e)** CSP CTPB in mice brain using anti-acetylated histone H3 antibody

to avoid the reticuloendothelial system clearance and thiamine ligands which binds with the thiamine receptors are transported to the brain through the receptor mediated endocytosis (Kaur et al. 2008). Recently, carbon nanospheres are shown to cross the blood brain barrier (Selvi et al. 2008). Without any surface modification probably due to the presence of oligosaccharide side chains containing terminal glucose, mannose residues mimicking the receptor mediated endocytosis internalization of such macromolecules.

25.4.2.2 Histone Acetylation Activation Across the BBB and Its Implications

Glucose derived carbon nanosphere CTPB conjugate, easily cross the BBB and induces histone acetylation in the brain (Fig. 25.2). Thus KAT activator conjugated with CSP could specifically activate acetyltransferases activity in brain and be used as a potential therapeutic agent. This delivery system of KAT activators coupled with nanospheres could be extremely useful in the treatment of various neurodegenerative diseases. In Alzheimer's' disease, reduction in levels of histone acetylation (especially H4) in hippocampus is reported (Francis et al. 2009). Thus small molecule activators, in combination with a targeted delivery system (like CSP) could be a potential therapeutic in treatments for Alzheimer's disease or other neurodegenerative diseases.

25.5 Future Perspectives

The modulators of histone acetylation have been tested for their effect on these disease conditions mostly in an in vitro condition. Before embarking on testing the therapeutic potential of these histone acetylation modulators, a detailed study with respect to their pharmacokinetics, bioavailability, the effect on gene expression, miRNA expression etc. needs to be understood. Since, these evidences will help in understanding their specific versus nonspecific effects in the physiology. However, the broader implications of the concept of acetylation being considered as a therapeutic target can also encompass the following aspects.

25.5.1 Combinatorial Therapy

Acetylation forms an important nodal point for several epigenetic crosstalks that regulate the process of transcription and hence, disturbing this centric modification may not necessarily influence the whole pathway. Especially in the case of pathophysiological conditions, several regulatory events are dysfunctional and hence to achieve a therapeutic effect it may be necessary to target not just the acetylation mark but also the other associated modifications such as lysine and arginine methylation marks with their specific modulators.

25.5.2 Targeted Delivery

The acetylation event is a global phenomenon essential for the process of gene expression and hence the modulation may lead to undesirable effects at other parts of the genome. Also, the small molecules have potential pleiotropic effects. Therefore, targeted delivery of these modulators with the help of nanomaterials might prove to be a more efficient strategy.

25.5.3 Novel Therapeutic Concepts

The cancer stem cell (CSC) populations are essentially in a non-replicating state, thus several of the frontline anti cancer drugs, fail to have any effect on these. The targeted acetylation modulators can be used to facilitate the differentiation of this

CSC population and then killed. On the other hand, several regenerative medicinal strategies are exploring the possibility of stem cell based therapies, the major limitation being the availability of these cells. With the advent of the iPS technology, the histone acetylation modulators can also be tested for the dedifferentiation process.

25.5.4 Pathway Alterations

There have been recent evidences implicating the role of histone acetylation in metabolic processes with important enzymes being identified as potential substrates. And it is also becoming evident that there indeed exists an acetylation centric epigenetic language that acts as a signalling event facilitating the process of transcription and gene expression. Hence, the use of modulators as a therapeutic strategy should be done only along with the information of the effect of these modulators on metabolism and other pathways.

Collectively, the modulators of acetylation are important candidates to be considered for their potential therapeutic applications based on the several preclinical investigations highlighted in this chapter. The efficacy and the effectiveness of such an effort is strengthened by the fact that deacetylase inhibitors are presently being used in clinical trials.

Acknowledgements Work in our laboratory is supported by Jawaharlal Nehru Centre for Advanced Scientific Research (JNCASR), Department of Biotechnology, Government of India (Grant Nos. Grant/DBT/CSH/GIA/1752/2011-2012 and Chromatin and Disease: Programme Support Grant No. Grant/DBT/CSH/GIA/1957/2011-2012), Department of Science and Technology (DST) Government of India, Dabur Research Foundation and National Agricultural Innovative Project (NAIP), Indian Council of Agricultural Research, Govt. of India under component 4: Basic and Strategic Research (Grant No. NAIP/Comp-4/C-30017/2008-09). TKK is a recipient of Sir JC Bose national fellowship (DST, Government of India).

References

An W, Kim J, Roeder RG (2004) Ordered cooperative functions of PRMT1, p300 and CARM1 in transcriptional activation by p53. Cell 117:735–748

Arif M, Pradhan SK, Thanuja GR, Vedamurthy BM, Agrawal S, Dasgupta D, Kundu TK (2009) Mechanism of p300 specific histone acetyltransferase inhibition by small molecules. J Med Chem 52:267–277

Arif M, Vedamurthy BM, Choudhari R, Ostwal YB, Mantelingu K, Kodaganur GS, Kundu TK (2010) Nitric oxide-mediated histone hyperacetylation in oral cancer: target for a water-soluble HAT inhibitor, CTK7A. Chem Biol 17:903–913

Balasubramanyam K, Swaminathan V, Ranganathan A, Kundu TK (2003) Small molecule modulators of histone acetyltransferase p300. J Biol Chem 278:19134–19140

Balasubramanyam K, Altaf M, Varier RA, Swaminathan V, Ravindran A, Sadhale PP, Kundu TK (2004) Polyisoprenylated benzophenone, garcinol, a natural histone acetyltransferase inhibitor, represses chromatin transcription and alters global gene expression. J Biol Chem 279:33716–33726

Bandyopadhyay D, Okan NA, Bales E, Nascimento L, Cole PA, Medrano EE (2002) Down-regulation of p300/CBP histone acetyltransferase activates a senescence checkpoint in human melanocytes. Cancer Res 62:6231–6239

Barman HK, Takami Y, Ono T, Nishijima H, Sanematsu F, Shibahara K, Nakayama T (2006) Histone acetyltransferase 1 is dispensable for replication-coupled chromatin assembly but contributes to recover DNA damages created following replication blockage in vertebrate cells. Biochem Biophys Res Commun 345:1547–1557

Barman HK, Takami Y, Nishijima H, Shibahara K, Sanematsu F, Nakayama T (2008) Histone acetyltransferase-1 regulates integrity of cytosolic histone H3-H4 containing complex. Biochem Biophys Res Commun 373:624–630

Barrett RM, Malvaez M, Kramar E, Matheos DP, Arrizon A, Cabrera SM, Lynch G, Greene RW, Wood MA (2011) Hippocampal focal knockout of CBP affects specific histone modifications, long-term potentiation, and long-term memory. Neuropsychopharmacology 36:1545–1556

Bártová E, Galiová G, Krejcí J, Harnicarová A, Strasák L, Kozubek S (2008) Epigenome and chromatin structure in human embryonic stem cells undergoing differentiation. Dev Dyn 237:3690–3702

Battu A, Ray A, Wani AA (2011) ASF1A and ATM regulate H3K56-mediated cell-cycle checkpoint recovery in response to UV irradiation. Nucleic Acids Res 39:7931–7945

Biel M, Kretsovali A, Karatzali E, Papamatheakis J, Giannis A (2004) Design, synthesis, and biological evaluation of a small-molecule inhibitor of the histone acetyltransferase Gcn5. Angew Chem Int Ed Engl 43:3974–3976

Bird A (2007) Perceptions of epigenetics. Nature 447:396–398

Blackwell JSJ, Wilkinson ST, Mosammaparast N, Pemberton LF (2007) Mutational analysis of H3 and H4 N termini reveals distinct roles in nuclear import. J Biol Chem 282:20142–20150

Borrow J, Stanton VP, Andresen JM, Becher R, Behm FG, Chaganti RS, Civin CI, Disteche C, Dubé I, Frischauf AM, Horsman D, Mitelman F, Volinia S, Watmore AE, Housman DE (1996) The translocation t(8;16)(p11;p13) of acute myeloid leukaemia fuses a putative acetyltransferase to the CREB-binding protein. Nat Genet 14:33–41

Bowers EM, Yan G, Mukherjee C, Orry A, Wang L, Holbert MA, Crump NT, Hazzalin CA, Liszczak G, Yuan H, Larocca C, Saldanha SA, Abagyan R, Sun Y, Meyers DJ, Marmorstein R, Mahadevan LC, Alani RM, Cole PA (2010) Virtual ligand screening of the p300/CBP histone acetyltransferase: identification of a selective small molecule inhibitor. Chem Biol 17:471–482

Brady ME, Ozanne DM, Gaughan L, Waite I, Cook S, Neal DE, Robson CN (1999) Tip60 is a nuclear hormone receptor coactivator. J Biol Chem 274:17599–17604

Brasnjevic I, Steinbusch HW, Schmitz C, Martinez-Martinez P, Initiative ENR (2009) Delivery of peptide and protein drugs over the blood-brain barrier. Prog Neurobiol 87:212–251

Brownlee M (2001) Biochemistry and molecular cell biology of diabetic complications. Nature 414:813–820

Carapeti M, Aguiar RC, Goldman JM, Cross NC (1998) A novel fusion between MOZ and the nuclear receptor coactivator TIF2 in acute myeloid leukemia. Blood 91:3127–3133

Caton PW, Nayuni NK, Kieswich J, Khan NQ, Yaqoob MM, Corder R (2010) Metformin suppresses hepatic gluconeogenesis through induction of SIRT1 and GCN5. J Endocrinol 205:97–106

Cereseto A, Manganaro L, Gutierrez MI, Terreni M, Fittipaldi A, Lusic M, Marcello A, Giacca M (2005) Acetylation of HIV-1 integrase by p300 regulates viral integration. EMBO J 24:3070–3081

Chang L, Loranger SS, Mizzen C, Ernst SG, Allis CD, Annunziato AT (1997) Histones in transit: cytosolic histone complexes and diacetylation of H4 during nucleosome assembly in human cells. Biochemistry 36:469–480

Chang J, Jallouli Y, Barras A, Dupont N, Betbeder D (2009) Chapter 1 – Drug delivery to the brain using colloidal carriers. Prog Brain Res 180:2–17

Charvet C, Wissler M, Brauns-Schubert P, Wang SJ, Tang Y, Sigloch FC, Mellert H, Brandenburg M, Lindner SE, Breit B, Green DR, McMahon SB, Borner C, Gu W, Maurer U (2011) Phosphorylation of Tip60 by GSK-3 determines the induction of PUMA and apoptosis by p53. Mol Cell 42:584–596

Chatterjee S, Gadad SS, Kundu TK (2010) Atomic force microscopy: a tool to unveil the mystery of biological systems. Resonance 15:622–642

Chen S, Feng B, George B, Chakrabarti R, Chen M, Chakrabarti S (2010) Transcriptional coactivator p300 regulates glucose-induced gene expression in endothelial cells. Am J Physiol Endocrinol Metab 298:E127–E137

Chiani F, Di Felice F, Camilloni G (2006) SIR2 modifies histone H4-K16 acetylation and affects superhelicity in the ARS region of plasmid chromatin in Saccharomyces cerevisiae. Nucleic Acids Res 34:5426–5437

Choi KC, Jung MG, Lee YH, Yoon JC, Kwon SH, Kang HB, Kim MJ, Cha JH, Kim YJ, Jun WJ, Lee JM, Yoon HG (2009) Epigallocatechin-3-gallate, a histone acetyltransferase inhibitor, inhibits EBV-induced B lymphocyte transformation via suppression of RelA acetylation. Cancer Res 69:583–592

Chupreta S, Holmstrom S, Subramanian L, Iniguez-Lluhi JA (2005) A small conserved surface in SUMO is the critical structural determinant of its transcriptional inhibitory properties. Mol Cell Biol 25:4272–4282

Col E, Caron C, Chable-Bessia C, Legube G, Gazzeri S, Komatsu Y, Yoshida M, Benkirane M, Trouche D, Khochbin S (2005) HIV-1 Tat targets Tip60 to impair the apoptotic cell response to genotoxic stresses. EMBO J 24:2634–2645

Cole PA (2008) Chemical probes for histone-modifying enzymes. Nat Chem Biol 4:590–597

Costantino L, Gandolfi F, Tosi G, Rivasi F, Vandelli MA, Forni F (2005) Peptide-derivatized biodegradable nanoparticles able to cross the blood-brain barrier. J Control Release 108:84–96

Creaven M, Hans F, Mutskov V, Col E, Caron C, Dimitrov S, Khochbin S (1999) Control of the histone-acetyltransferase activity of Tip60 by the HIV-1 transactivator protein, Tat. Biochemistry 38:8826–8830

Cui L, Miao J (2010) Chromatin-mediated epigenetic regulation in the malaria parasite Plasmodium falciparum. Eukaryot Cell 9:1138–1149

Cui L, Miao J, Furuya T, Li X, Su XZ (2007) PfGCN5-mediated histone H3 acetylation plays a key role in gene expression in Plasmodium falciparum. Eukaryot Cell 6:1219–1227

Cui L, Miao J, Furuya T, Fan Q, Li X, Rathod PK, Su XZ (2008) Histone acetyltransferase inhibitor anacardic acid causes changes in global gene expression during in vitro Plasmodium falciparum development. Eukaryot Cell 7:1200–1210

Dai YS, Markham BE (2001) p300 functions as a coactivator of transcription factor GATA-4. J Biol Chem 276:37178–37185

Dal Piaz F, Tosco A, Eletto D, Piccinelli AL, Moltedo O, Franceschelli S, Sbardella G, Remondelli P, Ratrelli L, Vesci L, Pisano C, De Tommasi N (2010) The identification of a novel natural activator of p300 histone acetyltransferase provides new insights into the modulation mechanism of this enzyme. Chembiochem 11:818–827

Das C, Lucia MS, Hansen KC, Tyler JK (2009) CBP/p300-mediated acetylation of histone H3 on lysine 56. Nature 459:113–117

Deng L, de la Fuente C, Fu P, Wang L, Donnelly R, Wade JD, Lambert P, Li H, Lee CG, Kashanchi F (2000) Acetylation of HIV-1 Tat by CBP/P300 increases transcription of integrated HIV-1 genome and enhances binding to core histones. Virology 277:278–295

Dentin R, Liu Y, Koo SH, Hedrick S, Vargas T, Heredia J, Yates J, Montminy M (2007) Insulin modulates gluconeogenesis by inhibition of the coactivator TORC2. Nature 449:366–369

Doyon Y, Cayrou C, Ullah M, Landry AJ, Côté V, Selleck W, Lane WS, Tan S, Yang XJ, Côté J (2006) ING tumor suppressor proteins are critical regulators of chromatin acetylation required for genome expression and perpetuation. Mol Cell 21:51–64

Driscoll R, Hudson A, Jackson SP (2007) Yeast Rtt109 promotes genome stability by acetylating histone H3 on lysine 56. Science 315:649–652

Duffy KR, Pardridge WM (1987) Blood-brain barrier transcytosis of insulin in developing rabbits. Brain Res 420:32–38

Emerich DF, Thanos CG (2006) The pinpoint promise of nanoparticle-based drug delivery and molecular diagnosis. Biomol Eng 23:171–184

English CM, Adkins MW, Carson JJ, Churchill ME, Tyler JK (2006) Structural basis for the histone chaperone activity of Asf1. Cell 127:495–508

Eymin B, Claverie P, Salon C, Leduc C, Col E, Brambilla E, Khochbin S, Gazzeri S (2006) p14ARF activates a Tip60-dependent and p53-independent ATM/ATR/CHK pathway in response to genotoxic stress. Mol Cell Biol 26:4339–4350

Falbo KB, Shen X (2006) Chromatin remodeling in DNA replication. J Cell Biochem 97:684–689

Fan Q, An L, Cui L (2004) Plasmodium falciparum histone acetyltransferase, a yeast GCN5 homologue involved in chromatin remodeling. Eukaryot Cell 3:264–276

Fischle W, Tseng BS, Dormann HL, Ueberheide BM, Garcia BA, Shabanowitz J, Hunt DF, Funabiki H, Allis CD (2005) Regulation of HP1-chromatin binding by histone H3 methylation and phosphorylation. Nature 438:1116–1122

Fraga MF, Ballestar E, Villar-Garea A, Boix-Chornet M, Espada J, Schotta G, Bonaldi T, Haydon C, Ropero S, Petrie K, Iyer NG, Pérez-Rosado A, Calvo E, Lopez JA, Cano A, Calasanz MJ, Colomer D, Piris MA, Ahn N, Imhof A, Caldas C, Jenuwein T, Esteller M (2005) Loss of acetylation at Lys16 and trimethylation at Lys20 of histone H4 is a common hallmark of human cancer. Nat Genet 37:391–400

Francis YI, Fà M, Ashraf H, Zhang H, Staniszewski A, Latchman DS, Arancio O (2009) Dysregulation of histone acetylation in the APP/PS1 mouse model of Alzheimer's disease. J Alzheimers Dis 18:131–139

Fukuda H, Sano N, Muto S, Horikoshi M (2006) Simple histone acetylation plays a complex role in the regulation of gene expression. Biref Funct Genomic Proteomic 5:190–208

Fussner E, Ahmed K, Dehghani H, Strauss M, Bazett-Jones DP (2010) Changes in chromatin fiber density as a marker for pluripotency. Cold Spring Harb Symp Quant Biol 75:245–249

Gadad SS, Rajan RE, Senapati P, Chatterjee S, Shandilya J, Dash PK, Ranga U, Kundu TK (2011) HIV-1 infection induces acetylation of NPM1 that facilitates Tat localization and enhances viral transactivation. J Mol Biol 410:997–1007

Gaughan L, Brady ME, Cook S, Neal DE, Robson CN (2001) Tip60 is a co-activator specific for class I nuclear hormone receptors. J Biol Chem 276:46841–46848

Gaughan L, Logan IR, Cook S, Neal DE, Robson CN (2002) Tip60 and histone deacetylase 1 regulate androgen receptor activity through changes to the acetylation status of the receptor. J Biol Chem 277:25904–25913

Gayther SA, Batley SJ, Linger L, Bannister A, Thorpe K, Chin SF, Daigo Y, Russell P, Wilson A, Sowter HM, Delhanty JD, Ponder BA, Kouzarides T, Caldas C (2000) Mutations truncating the EP300 acetylase in human cancers. Nat Genet 24:300–303

Ge Z, Wang H, Parthun MR (2011) Nuclear Hat1p complex (NuB4) components participate in DNA repair-linked chromatin reassembly. J Biol Chem 286:16790–16799

Gelperina S, Maksimenko O, Khalansky A, Vanchugova L, Shipulo E, Abbasova K, Berdiev R, Wohlfart S, Chepurnova N, Kreuter J (2010) Drug delivery to the brain using surfactant-coated poly(lactide-co-glycolide) nanoparticles: influence of the formulation parameters. Eur J Pharm Biopharm 74:157–163

Ghizzoni M, Boltjes A, Graaf C, Haisma HJ, Dekker FJ (2010) Improved inhibition of the histone acetyltransferase PCAF by an anacardic acid derivative. Bioorg Med Chem 18:5826–5834

Goodarzi AA, Kurka T, Jeggo PA (2011) KAP-1 phosphorylation regulates CHD3 nucleosome remodeling during the DNA double-strand break response. Nat Struct Mol Biol 18:831–839

Goodman RH, Smolik S (2000) CBP/p300 in cell growth, transformation, and development. Genes Dev 14:1553–1577

Gorrini C, Squatrito M, Luise C, Syed N, Perna D, Wark L, Martinato F, Sardella D, Verrecchia A, Bennett S, Confalonieri S, Cesaroni M, Marchesi F, Gasco M, Scanziani E, Capra M, Mai S, Nuciforo P, Crook T, Lough J, Amati B (2007) Tip60 is a haplo-insufficient tumour suppressor required for an oncogene-induced DNA damage response. Nature 448:1063–1067

Groth A, Corpet A, Cook AJ, Roche D, Bartek J, Lukas J, Almouzni G (2007a) Regulation of replication fork progression through histone supply and demand. Science 318:1928–1931

Groth A, Rocha W, Verreault A, Almouzni G (2007b) Chromatin challenges during DNA replication and repair. Cell 128:721–733

Guo B, Panagiotaki N, Warwood S, Sharrocks AD (2011) Dynamic modification of the ETS transcription factor PEA3 by sumoylation and p300-mediated acetylation. Nucleic Acids Res 39:6403–6413

Gupta A, Sharma GG, Young CS, Agarwal M, Smith ER, Paull TT, Lucchesi JC, Khanna KK, Ludwig T, Pandita TK (2005) Involvement of human MOF in ATM function. Mol Cell Biol 25:5292–5305

Gusterson R, Brar B, Faulkes D, Giordano A, Chrivia J, Latchman D (2002) The transcriptional co-activators CBP and p300 are activated via phenylephrine through the p42/p44 MAPK cascade. J Biol Chem 277:2517–2524

Han J, Zhou H, Horazdovsky B, Zhang K, Xu RM, Zhang Z (2007) Rtt109 acetylates histone H3 lysine 56 and functions in DNA replication. Science 315:653–655

Hildmann C, Riester D, Schwienhorst A (2007) Histone deacetylases–an important class of cellular regulators with a variety of functions. Appl Microbiol Biotechnol 75:487–497

Hobbs CA, Wei G, DeFeo K, Paul B, Hayes CS, Gilmour SK (2006) Tip60 protein isoforms and altered function in skin and tumors that overexpress ornithine decarboxylase. Cancer Res 66:8116–8122

Ikura T, Ogryzko VV, Grigoriev M, Groisman R, Wang J, Horikoshi M, Scully R, Qin J, Nakatani Y (2000) Involvement of the TIP60 histone acetylase complex in DNA repair and apoptosis. Cell 102:463–473

Jain KK (2011) Nanobiotechnology and personalized medicine. Prog Mol Biol Transl Sci 104:325–354

Kalkhoven E (2004) CBP and p300: HATs for different occasions. Biochem Pharmacol 68:1145–1155

Katsumoto T, Aikawa Y, Iwama A, Ueda S, Ichikawa H, Ochiya T, Kitabayashi I (2006) MOZ is essential for maintenance of hematopoietic stem cells. Genes Dev 20:1321–1330

Kaur H, Chen S, Xin X, Chiu J, Khan ZA, Chakrabarti S (2006) Diabetes-induced extracellular matrix protein expression is mediated by transcription coactivator p300. Diabetes 55:3104–3111

Kaur A, Jain S, Tiwary AK (2008) Mannan-coated gelatin nanoparticles for sustained and targeted delivery of didanosine: in vitro and in vivo evaluation. Acta Pharm 58:61–74

Kiernan RE, Vanhulle C, Schiltz L, Adam E, Xiao H, Maudoux F, Calomme C, Burny A, Nakatani Y, Jeang KT, Benkirane M, Van Lint C (1999) HIV-1 tat transcriptional activity is regulated by acetylation. EMBO J 18:6106–6118

Kindle KB, Troke PJ, Collins HM, Matsuda S, Bossi D, Bellodi C, Kalkhoven E, Salomoni P, Pelicci PG, Minucci S, Heery DM (2005) MOZ-TIF2 inhibits transcription by nuclear receptors and p53 by impairment of CBP function. Mol Cell Biol 25:988–1002

Kitabayashi I, Aikawa Y, Yokoyama A, Hosoda F, Nagai M, Kakazu N, Abe T, Ohki M (2001) Fusion of MOZ and p300 histone acetyltransferases in acute monocytic leukemia with a t(8;22) (p11;q13) chromosome translocation. Leukemia 15:89–94

Kreuter J, Alyautdin RN, Kharkevich DA, Ivanov AA (1995) Passage of peptides through the blood-brain barrier with colloidal polymer particles (nanoparticles). Brain Res 674:171–174

Krishnan V, Chow MZ, Wang Z, Zhang L, Liu B, Liu X, Zhou Z (2011) Histone H4 lysine 16 hypoacetylation is associated with defective DNA repair and premature senescence in Zmpste24-deficient mice. Proc Natl Acad Sci USA 108:12325–12330

Lake AN, Bedford MT (2007) Protein methylation and DNA repair. Mutat Res 618:91–101

Lau OD, Kundu TK, Soccio RE, Ait-Si-Ali S, Khalil EM, Vassilev A, Wolffe AP, Nakatani Y, Roeder RG, Cole PA (2000) HATs off: selective synthetic inhibitors of the histone acetyltransferases p300 and PCAF. Mol Cell 5:589–595

Lee JS, Smith E, Shilatifard A (2010) The language of histone crosstalk. Cell 142:682–685

Lerin C, Rodgers JT, Kalume DE, Kim SH, Pandey A, Puigserver P (2006) GCN5 acetyltransferase complex controls glucose metabolism through transcriptional repression of PGC-1alpha. Cell Metab 3:429–438

Li M, Luo RZ, Chen JW, Cao Y, Lu JB, He JH, Wu QL, Cai MY (2011a) High expression of transcriptional coactivator p300 correlates with aggressive features and poor prognosis of hepatocellular carcinoma. J Transl Med 9:5

Li Y, Yang HX, Luo RZ, Zhang Y, Li M, Wang X, Jia WH (2011b) High expression of p300 has an unfavorable impact on survival in resectable esophageal squamous cell carcinoma. Ann Thorac Surg 91:1531–1538

LLeonart ME, Vidal F, Gallardo D, Diaz-Fuertes M, Rojo F, Cuatrecasas M, López-Vicente L, Kondoh H, Blanco C, Carnero A, Ramón y Cajal S (2006) New p53 related genes in human tumors: significant downregulation in colon and lung carcinomas. Oncol Rep 16:603–608

Lo WS, Trievel RC, Rojas JR, Duggan L, Hsu JY, Allis CD, Marmorstein R, Berger SL (2000) Phosphorylation of serine 10 in histone H3 is functionally linked in vitro and in vivo to Gcn5-mediated acetylation at lysine 14. Mol Cell 5:917–926

Lund AH, Turner G, Trubetskoy A, Verhoeven E, Wientjens E, Hulsman D, Russell R, DePinho RA, Lenz J, van Lohuizen M (2002) Genome-wide retroviral insertional tagging of genes involved in cancer in Cdkn2a-deficient mice. Nat Genet 32:160–165

Malam Y, Lim EJ, Seifalian AM (2011) Current trends in the application of nanoparticles in drug delivery. Curr Med Chem 18:1067–1078

Mann BS, Johnson JR, Cohen MH, Justice R, Pazdur R (2007) FDA approval summary: vorinostat for treatment of advanced primary cutaneous T-cell lymphoma. Oncologist 12:1247–1252

Mantelingu K, Kishore AH, Balasubramanyam K, Kumar GV, Altaf M, Swamy SN, Selvi R, Das C, Narayana C, Rangappa KS, Kundu TK (2007a) Activation of p300 histone acetyltransferase by small molecules altering enzyme structure probed by surface enhanced Raman spectroscopy. J Phys Chem B 111:4527–4534

Mantelingu K, Reddy BA, Swaminathan V, Kishore AH, Siddappa NB, Kumar GV, Nagashankar G, Natesh N, Roy S, Sadhale PP, Ranga U, Narayana C, Kundu TK (2007b) Specific inhibition of p300-HAT alters global gene expression and represses HIV replication. Chem Biol 14:645–657

Martin DG, Grimes DE, Baetz K, Howe L (2006) Methylation of histone H3 mediates the association of the NuA3 histone acetyltransferase with chromatin. Mol Cell Biol 26:3018–3028

Marzio G, Tyagi M, Gutierrez MI, Giacca M (1998) HIV-1 tat transactivator recruits p300 and CREB-binding protein histone acetyltransferases to the viral promoter. Proc Natl Acad Sci USA 95:13519–13524

Messner S, Hottiger MO (2011) Histone ADP-ribosylation in DNA repair, replication and transcription. Trends Cell Biol 21:534–542

Miao J, Fan Q, Cui L, Li J (2006) The malaria parasite Plasmodium falciparum histones: organization, expression, and acetylation. Gene 369:53–65

Miao J, Fan Q, Cui L, Li X, Wang H, Ning G, Reese JC (2010) The MYST family histone acetyltransferase regulates gene expression and cell cycle in malaria parasite Plasmodium falciparum. Mol Microbiol 78:883–902

Mitrakou A, Mokan M, Bolli G, Veneman T, Jenssen T, Cryer P, Gerich J (1992) Evidence against the hypothesis that hyperinsulinemia increases sympathetic nervous system activity in man. Metabolism 41:198–200

Moore SD, Herrick SR, Ince TA, Kleinman MS, Dal Cin P, Morton CC, Quade BJ (2004) Uterine leiomyomata with t(10;17) disrupt the histone acetyltransferase MORF. Cancer Res 64:5570–5577

Muraoka M, Konishi M, Kikuchi-Yanoshita R, Tanaka K, Shitara N, Chong JM, Iwama T, Miyaki M (1996) p300 gene alterations in colorectal and gastric carcinomas. Oncogene 12:1565–1569

Nagy Z, Tora L (2007) Distinct GCN5/PCAF-containing complexes function as co-activators and are involved in transcription factor and global histone acetylation. Oncogene 26:5341–5357

Ott DE, Chertova EN, Busch LK, Coren LV, Gagliardi TD, Johnson DG (1999) Mutational analysis of the hydrophobic tail of the human immunodeficiency virus type 1 p6(Gag) protein produces a mutant that fails to package its envelope protein. J Virol 73:19–28

Panagopoulos I, Fioretos T, Isaksson M, Mitelman F, Johansson B, Theorin N, Juliusson G (2002) RT-PCR analysis of acute myeloid leukemia with t(8;16)(p11;p13): identification of a novel MOZ/CBP transcript and absence of CBP/MOZ expression. Genes Chromosomes Cancer 35:372–374

Panicker SP, Raychaudhuri B, Sharma P, Tipps R, Mazumdar T, Mal AK, Palomo JM, Vogelbaum MA, Haque SJ (2010) p300- and Myc-mediated regulation of glioblastoma multiforme cell differentiation. Oncotarget 1:289–303

Parthun MR, Widom J, Gottschling DE (1996) The major cytoplasmic histone acetyltransferase in yeast: links to chromatin replication and histone metabolism. Cell 87:85–94

Perriello G, Pampanelli S, Del Sindaco P, Lalli C, Ciofetta M, Volpi E, Santeusanio F, Brunetti P, Bolli GB (1997) Evidence of increased systemic glucose production and gluconeogenesis in an early stage of NIDDM. Diabetes 46:1010–1016

Petter M, Lee CC, Byrne TJ, Boysen KE, Volz J, Ralph SA, Cowman AF, Brown GV, Duffy MF (2011) Expression of P. falciparum var genes involves exchange of the histone variant H2A.Z at the promoter. PLoS Pathog 7:e1001292

Phan HM, Xu AW, Coco C, Srajer G, Wyszomierski S, Evrard YA, Eckner R, Dent SY (2005) GCN5 and p300 share essential functions during early embryogenesis. Dev Dyn 233:1337–1347

Puigserver P, Rhee J, Donovan J, Walkey CJ, Yoon JC, Oriente F, Kitamura Y, Altomonte J, Dong H, Accili D, Spiegelman BM (2003) Insulin-regulated hepatic gluconeogenesis through FOXO1-PGC-1alpha interaction. Nature 423:550–555

Qin S, Parthun MR (2002) Histone H3 and the histone acetyltransferase Hat1p contribute to DNA double-strand break repair. Mol Cell Biol 22:8353–8365

Ravindra KC, Selvi BR, Arif M, Reddy BA, Thanuja GR, Agrawal S, Pradhan SK, Nagashayana N, Dasgupta D, Kundu TK (2009) Inhibition of lysine acetyltransferase KAT3B/p300 activity by a naturally occurring hydroxynaphthoquinone, plumbagin. J Biol Chem 284:24453–24464

Rhee J, Inoue Y, Yoon JC, Puigserver P, Fan M, Gonzalez FJ, Spiegelman BM (2003) Regulation of hepatic fasting response by PPARgamma coactivator-1alpha (PGC-1): requirement for hepatocyte nuclear factor 4alpha in gluconeogenesis. Proc Natl Acad Sci USA 100:4012–4017

Richon VM, Emiliani S, Verdin E, Webb Y, Breslow R, Rifkind RA, Marks PA (1998) A class of hybrid polar inducers of transformed cell differentiation inhibits histone deacetylases. Proc Natl Acad Sci USA 95:3003–3007

Rouaux C, Jokic N, Mbebi C, Boutillier S, Loeffler JP, Boutillier AL (2003) Critical loss of CBP/p300 histone acetylase activity by caspase-6 during neurodegeneration. EMBO J 22:6537–6549

Rozman M, Camós M, Colomer D, Villamor N, Esteve J, Costa D, Carrió A, Aymerich M, Aguilar JL, Domingo A, Solé F, Gomis F, Florensa L, Montserrat E, Campo E (2004) Type I MOZ/CBP (MYST3/CREBBP) is the most common chimeric transcript in acute myeloid leukemia with t(8;16)(p11;p13) translocation. Genes Chromosomes Cancer 40:140–145

Santer FR, Höschele PP, Oh SJ, Erb HH, Bouchal J, Cavarretta IT, Parson W, Meyers DJ, Cole PA, Culig Z (2011) Inhibition of the acetyltransferases p300 and CBP reveals a targetable function for p300 in the survival and invasion pathways of prostate cancer cell lines. Mol Cancer Ther 10:1644–1655

Sapountzi V, Côté J (2011) MYST-family histone acetyltransferases: beyond chromatin. Cell Mol Life Sci 68:1147–1156

Segré CV, Chiocca S (2011) Regulating the regulators: the post-translational code of class I HDAC1 and HDAC2. J Biomed Biotechnol 2011:690848

Selvi BR, Jagadeesan D, Suma BS, Nagashankar G, Arif M, Balasubramanyam K, Eswaramoorthy M, Kundu TK (2008) Intrinsically fluorescent carbon nanospheres as a nuclear targeting vector: delivery of membrane-impermeable molecule to modulate gene expression in vivo. Nano Lett 8:3182–3188

Shandilya J, Swaminathan V, Gadad SS, Choudhari R, Kodaganur GS, Kundu TK (2009) Acetylated NPM1 localizes in the nucleoplasm and regulates transcriptional activation of genes implicated in oral cancer manifestation. Mol Cell Biol 29:5115–5127

Shi X, Hong T, Walter KL, Ewalt M, Michishita E, Hung T, Carney D, Pena P, Lan F, Kaadige MR, Lacoste N, Cayrou C, Davrazou F, Saha A, Cairns BR, Ayer DE, Kutateladze TG, Shi Y, Cote J, Chua KF, Gozani O (2006) ING2 PHD domain links histone H3 lysine 4 methylation to active gene repression. Nature 442:96–99

Shia WJ, Pattenden SG, Workman JL (2006) Histone H4 lysine 16 acetylation breaks the genome's silence. Genome Biol 7:217

Shiota M, Izumi H, Onitsuka T, Miyamoto N, Kashiwagi E, Kidani A, Yokomizo A, Naito S, Kohno K (2008) Twist promotes tumor cell growth through YB-1 expression. Cancer Res 68:98–105

Shiota M, Yokomizo A, Tada Y, Uchiumi T, Inokuchi J, Tatsugami K, Kuroiwa K, Yamamoto K, Seki N, Naito S (2010) P300/CBP-associated factor regulates Y-box binding protein-1 expression and promotes cancer cell growth, cancer invasion and drug resistance. Cancer Sci 101:1797–1806

Shiseki M, Nagashima M, Pedeux RM, Kitahama-Shiseki M, Miura K, Okamura S, Onogi H, Higashimoto Y, Appella E, Yokota J, Harris CC (2003) p29ING4 and p28ING5 bind to p53 and p300, and enhance p53 activity. Cancer Res 63:2373–2378

Shogren-Knaak M, Ishii H, Sun JM, Pazin MJ, Davie JR, Peterson CL (2006) Histone H4-K16 acetylation controls chromatin structure and protein interactions. Science 311:844–847

Smith E, Shilatifard A (2010) The chromatin signaling pathway: diverse mechanisms of recruitment of histone-modifying enzymes and varied biological outcomes. Mol Cell 40:689–701

Smith ER, Cayrou C, Huang R, Lane WS, Côté J, Lucchesi JC (2005) A human protein complex homologous to the Drosophila MSL complex is responsible for the majority of histone H4 acetylation at lysine 16. Mol Cell Biol 25:9175–9188

Stimson L, Rowlands MG, Newbatt YM, Smith NF, Raynaud FI, Rogers P, Bavetsias V, Gorsuch S, Jarman M, Bannister A, Kouzarides T, McDonald E, Workman P, Aherne GW (2005) Isothiazolones as inhibitors of PCAF and p300 histone acetyltransferase activity. Mol Cancer Ther 4:1521–1532

Suter B, Pogoutse O, Guo X, Krogan N, Lewis P, Greenblatt JF, Rine J, Emili A (2007) Association with the origin recognition complex suggests a novel role for histone acetyltransferase Hat1p/Hat2p. BMC Biol 5:38

Taddei A, Roche D, Sibarita JB, Turner BM, Almouzni G (1999) Duplication and maintenance of heterochromatin domains. J Cell Biol 147:1153–1166

Takahashi H, McCaffery JM, Irizarry RA, Boeke JD (2006) Nucleocytosolic acetyl-coenzyme a synthetase is required for histone acetylation and global transcription. Mol Cell 23:207–217

Tang Y, Luo J, Zhang W, Gu W (2006) Tip60-dependent acetylation of p53 modulates the decision between cell-cycle arrest and apoptosis. Mol Cell 24:827–839

Terreni M, Valentini P, Liverani V, Gutierrez MI, Di Primio C, Di Fenza A, Tozzini V, Allouch A, Albanese A, Giacca M, Cereseto A (2010) GCN5-dependent acetylation of HIV-1 integrase enhances viral integration. Retrovirology 7:18

Tjeertes JV, Miller KM, Jackson SP (2009) Screen for DNA-damage-responsive histone modifications identifies H3K9Ac and H3K56Ac in human cells. EMBO J 28:1878–1889

Tonkin CJ, Carret CK, Duraisingh MT, Voss TS, Ralph SA, Hommel M, Duffy MF, Silva LM, Scherf A, Ivens A, Speed TP, Beeson JG, Cowman AF (2009) Sir2 paralogues cooperate to regulate virulence genes and antigenic variation in Plasmodium falciparum. PLoS Biol 7:e84

Troke PJ, Kindle KB, Collins HM, Heery DM (2006) MOZ fusion proteins in acute myeloid leukaemia. Biochem Soc Symp 73:23–39

Urvalek AM, Lu H, Wang X, Li T, Yu L, Zhu J, Lin Q, Zhao J (2011) Regulation of the oncoprotein KLF8 by a switch between acetylation and sumoylation. Am J Transl Res 3:121–132

van Attikum H, Gasser SM (2009) Crosstalk between histone modifications during the DNA damage response. Trends Cell Biol 19:207–217

Vempati RK (2012) DNA damage in the presence of chemical genotoxic agents induce acetylation of H3K56 and H4K16 but not H3K9 in mammalian cells. Mol Biol Rep 39(1):303–308

Vempati RK, Jayani RS, Notani D, Sengupta A, Galande S, Haldar D (2010) p300-mediated acetylation of histone H3 lysine 56 functions in DNA damage response in mammals. J Biol Chem 285:28553–28564

Voss AK, Thomas T (2009) MYST family histone acetyltransferases take center stage in stem cells and development. Bioessays 31:1050–1061

Wang C, Fu M, Angeletti RH, Siconolfi-Baez L, Reutens AT, Albanese C, Lisanti MP, Katzenellenbogen BS, Kato S, Hopp T, Fuqua SA, Lopez GN, Kushner PJ, Pestell RG (2001) Direct acetylation of the estrogen receptor alpha hinge region by p300 regulates transactivation and hormone sensitivity. J Biol Chem 276:18375–18383

Wei JQ, Shehadeh LA, Mitrani JM, Pessanha M, Slepak TI, Webster KA, Bishopric NH (2008) Quantitative control of adaptive cardiac hypertrophy by acetyltransferase p300. Circulation 118:934–946

Westendorp MO, Shatrov VA, Schulze-Osthoff K, Frank R, Kraft M, Los M, Krammer PH, Dröge W, Lehmann V (1995) HIV-1 Tat potentiates TNF-induced NF-kappa B activation and cytotoxicity by altering the cellular redox state. EMBO J 14:546–554

Wilson CB, Merkenschlager M (2006) Chromatin structure and gene regulation in T cell development and function. Curr Opin Immunol 18:143–151

Wu J, Xie N, Wu Z, Zhang Y, Zheng YG (2009) Bisubstrate inhibitors of the MYST HATs Esa1 and Tip60. Bioorg Med Chem 17:1381–1386

Wu J, Chen Y, Lu LY, Wu Y, Paulsen MT, Ljungman M, Ferguson DO, Yu X (2011) Chfr and RNF8 synergistically regulate ATM activation. Nat Struct Mol Biol 18:761–768

Xu B, Chiu J, Feng B, Chen S, Chakrabarti S (2008) PARP activation and the alteration of vasoactive factors and extracellular matrix protein in retina and kidney in diabetes. Diabetes Metab Res Rev 24:404–412

Yamauchi T, Yamauchi J, Kuwata T, Tamura T, Yamashita T, Bae N, Westphal H, Ozato K, Nakatani Y (2000) Distinct but overlapping roles of histone acetylase PCAF and of the closely related PCAF-B/GCN5 in mouse embryogenesis. Proc Natl Acad Sci USA 97:11303–11306

Yanazume T, Morimoto T, Wada H, Kawamura T, Hasegawa K (2003) Biological role of p300 in cardiac myocytes. Mol Cell Biochem 248:115–119

Yang X, Yu W, Shi L, Sun L, Liang J, Yi X, Li Q, Zhang Y, Yang F, Han X, Zhang D, Yang J, Yao Z, Shang Y (2011) HAT4, a golgi apparatus-anchored B-type histone acetyltransferase, acetylates free histone H4 and facilitates chromatin assembly. Mol Cell 44:39–50

Ying MZ, Wang JJ, Li DW, Yu GZ, Wang X, Pan J, Chen Y, He MX (2010) The p300/CBP associated factor: is frequently downregulated in intestinal-type gastric carcinoma and constitutes a biomarker for clinical outcome. Cancer Biol Ther 9:312–320

Yoon JC, Puigserver P, Chen G, Donovan J, Wu Z, Rhee J, Adelmant G, Stafford J, Kahn CR, Granner DK, Newgard CB, Spiegelman BM (2001) Control of hepatic gluconeogenesis through the transcriptional coactivator PGC-1. Nature 413:131–138

Zhou H, Beevers CS, Huang S (2011) The targets of curcumin. Curr Drug Targets 12:332–347

Zhu Q, Wani AA (2010) Histone modifications: crucial elements for damage response and chromatin restoration. J Cell Physiol 223:283–288

Zippo A, Serafini R, Rocchigiani M, Pennacchini S, Krepelova A, Oliviero S (2009) Histone crosstalk between H3S10ph and H4K16ac generates a histone code that mediates transcription elongation. Cell 138:1122–1136

Chapter 26
DNA Methylation and Cancer

Gopinathan Gokul and Sanjeev Khosla

Abstract Cancer has been considered a genetic disease with a wide array of well-characterized gene mutations and chromosomal abnormalities. Of late, aberrant epigenetic modifications have been elucidated in cancer, and together with genetic alterations, they have been helpful in understanding the complex traits observed in neoplasia. "Cancer Epigenetics" therefore has contributed substantially towards understanding the complexity and diversity of various cancers. However, the positioning of epigenetic events during cancer progression is still not clear, though there are some reports implicating aberrant epigenetic modifications in very early stages of cancer. Amongst the most studied aberrant epigenetic modifications are the DNA methylation differences at the promoter regions of genes affecting their expression. Hypomethylation mediated increased expression of oncogenes and hypermethylation mediated silencing of tumor suppressor genes are well known examples. This chapter also explores the correlation of DNA methylation and demethylation enzymes with cancer.

26.1 Introduction

Cancer is a complex and heterogeneous disease characterized by uncontrolled growth and cellular machinery that has gone haywire. In order to account for the diverse molecular changes that occur in different cancer types it has been conceptualized that cancer encompasses many diseases. Pathologists view cancer as acquiring properties of cells belonging to different developmental stages but appearing inappropriately in the tumors (Pitot 1986). The two-hit hypothesis proposed by Knudson (1971) remains the basis of correlating the genetic events with cancer

G. Gokul • S. Khosla (✉)
Laboratory of Mammalian Genetics, CDFD, Hyderabad 500001, India
e-mail: sanjuk@cdfd.org.in

initiation even today. For instance, in retinoblastoma, two genetic events are required in the retinal cell, which would result in the inactivation of both copies of the tumor suppressor – retinoblastoma gene (*RB*). In the familial set up where all cells of a person inherit a mutated allele of *RB*, a single mutation in the normal functional allele can give rise to cancer whereas in non-hereditary cancer, both mutations need to take place in the same cell. This genetic model of cancer has been supported by characterization of a wide array of molecular changes that occur among cancer types including the mutations which in small numbers are seen in benign, non-invasive tumors and the large-scale genetic changes and genetic instability found in invasive and metastatic tumors (Feinberg et al. 2006). However, there are a few shortcomings to the Knudson's two-hit model. Firstly, the model does not take into account the heterogeneity and complexity observed in tumors and secondly, tumors that show aneuploidy, polyploidy, and complex karyotypes are often incompatible with the two-hit hypothesis (Dutrillaux et al. 1991). Moreover, and of late it is becoming amply clear that genetic mechanisms alone cannot account for all the diverse molecular changes seen in tumors. Epigenetic mechanisms which provide an additional layer of control for gene expression have been proposed to be an important factor in cancer development and postulated to provide for one of the hits required for carcinogenesis.

26.2 Cancer as an Epigenetic Disease

The term "epigenetics" defines the study of all meiotically and mitotically heritable changes in gene expression patterns that are not coded in the DNA sequence itself (Egger et al. 2004). Epigenetic modifications like DNA methylation and histone modifications can bring about changes in gene function or dosage similar to those obtained in case of mutations, chromosomal rearrangements and gene duplications. Studies in tumor tissues have revealed that a given gene can exhibit multiple epigenetic changes just as the numerous possible genetic alterations (Feinberg et al. 2006). Thus, integrating the data on epigenetic regulation of gene expression with their genetic alterations is of utmost importance for understanding the mechanisms underlying cancer development. In 1983, through the pioneering efforts of Andrew Feinberg and Bert Vogelstein, loss of methylation at CpG dinucleotides in cancer cells was identified, and can be considered as the first report on the correlation of epigenetics with cancer. They found that a substantial proportion of CpG dinucleotides in the human genome had lost DNA methylation in cancer cells (Feinberg and Vogelstein 1983). Around the same time, Ehrlich's group also found that the global levels of 5-methylcytosine were reduced in cancer cells (Gama-Sosa et al. 1983). Though there was initial skepticism about this correlation, today the role of DNA methylation has become well established in multiple cancers (Jones and Baylin 2002; Feinberg and Tycko 2004). This loss of methylation has been reported in every tumor type studied; both benign and malignant; and interestingly even pre-malignant adenomas exhibit altered DNA methylation (Goelz et al. 1985; Feinberg et al. 1988).

Although hypomethylation of CpG islands was the first epigenetic change identified in cancers, it has not been given as much importance as has been to hypermethylation of CpG islands. Baylin and colleagues in 1986 identified site-specific hypermethylation of *CALCITONIN* gene in small-cell lung cancer (Baylin et al. 1986). The first correlation of CpG island hypermethylation with the inactivation of a tumor-suppressor gene was established for the Retinoblastoma gene (*RB*) when *RB* promoter was found to be methylated in a significant subset of sporadic and hereditary retinoblastomas (Greger et al. 1989). Promoter hypermethylation has now been shown for several other loci to be the cause of gene inactivation in cancer. Hypermethylation is observed within the CpG islands of specific genes and represents a change in 5-methylcytosine distribution at specific loci in the genome rather than an overall increase in the total amount of methylation. In many cases, this *de novo* methylation of CpG islands occurs early in the process of carcinogenesis and surprisingly can even be detected in the apparently normal epithelium of patients-a process that is associated with aging (Issa et al. 1994; Ahuja et al. 1998) and inflammation (Coussens and Werb 2002; Nelson et al. 2004; Lu et al. 2006). In fact, aging is considered the most important risk factor for the development of most malignancies of adulthood as DNA methylation patterns in aging cells have a tendency to be aberrant, similar to what is seen in transformed cells (Neumeister et al. 2002).

Another important link that has broadened the role of epigenetics in cancer is the correlation between chromatin organization and DNA methylation. The earliest of experiments probing this link were done in the Cedar and Graessmann laboratories. They found that pre-methylated, naked DNA templates when transfected or microinjected into cells became transcriptionally silent after getting packaged into a repressive form of chromatin (Keshet et al. 1986; Buschhausen et al. 1987). Foremost amongst the chromatin modifications are the covalent modifications of histones that can change the chromatin conformation and control gene activity. Cytosine methylation can attract methylated DNA-binding proteins and histone deacetylases to the methylated CpG island, leading to chromatin compaction and gene silencing (Jones et al. 1998; Nan et al. 1998). This has further strengthened the link between the two major epigenetic components of gene regulation. In addition, correlation has also been made between covalent histone modifications and nucleosomal remodeling (Esteller 2006). It is now well established that the three processes of DNA cytosine methylation, histone modification and nucleosomal remodeling are intimately linked and that alterations in these processes result in the epigenetic changes which would lead to events like the permanent silencing of cancer-relevant genes and genomic instability. Working on mouse models, Jaenisch and colleagues demonstrated that *Dnmt1* hypomorphic mutation reduces the frequency of intestinal neoplasia when crossed to Apc^{Min} mice (Gaudet et al. 2003) whereas, a high frequency of lymphomas were observed in mice with hypomorphic *Dnmt1* allele (Eads et al. 2002). Together, these data indicated that cancer risk is associated with a disruption in the balance of methylation rather than hypomethylation and hypermethylation *per se*.

Based on the subtle differences observed in the epigenetic profiles of stem cells and the cancer cells, several researchers have proposed the theory of cancer stem cells, which suggests that stem cells are the more likely targets of epigenetic disruption leading to cancer. "Cancer stem cells" (CSCs), which constitute a small minority of neoplastic cells within a tumor, are the abnormal stem cells generated from normal stem cells upon accumulation of series of progressive genetic/epigenetic changes. CSCs are now well characterized in the hematopoietic system and colon cancers (O'Brien et al. 2007; Mani et al. 2008). The incompetence of many of the anti-cancer chemotherapeutic agents has been attributed to the presence of cancer stem cells, against which these drugs are less effective (Ren 2005).

Thus, in the past 20 years, cancer epigenetics has transformed into a full-fledged field with a focus on the different mechanisms involved in epigenetic regulation including DNA methylation, histone modifications and nucleosomal remodeling. In addition, the discovery of epigenetically regulated imprinted genes and their role in cancer has also added another dimension to this field (Jirtle 1999; Jelinic and Shaw 2007; Hirasawa and Feil 2010). While the finer mechanistic detail of the role of epigenetic modifications in cancer progression still needs to be worked out, epigenetic events are recognized as possibly one of the hits required in Knudson's hypothesis (Yu and Shen 2002). Initially, the clonal genetic model of cancer was widely accepted, wherein cancer arises through a series of mutations in oncogenes and tumor suppressor genes. This would give rise to a monoclonal population of tumor cells and epigenetic changes were viewed as surrogate correlates of cancer. But according to the recently proposed epigenetic progenitor model, cancer begins with an epigenetic alteration of stem/progenitor cells within a tissue, which is followed by a gatekeeper mutation involving a tumor suppressor gene or an oncogene, leading to genetic and epigenetic instability (Feinberg et al. 2006).

26.3 DNA Methylation and Cancer

The epigenetic landscape is maintained by interplay between the key modifications: DNA methylation, histone tail modifications, and small RNA molecules. However, in this review we focus on the interplay between DNA methylation machinery and carcinogenesis. In the altered epigenetic setup of a cancerous cell, aberrant DNA methylation is one of the most well studied epigenetic mechanisms affecting gene expression. Collated below is data pertaining to the role of DNA methylation machinery in establishing the normal and altered epigenetic makeup of the human genome.

26.3.1 CpG Islands and CpG Methylation

Methylation of DNA is a stable modification and can be inherited through cell divisions. Methylation of DNA in mammals primarily takes place at the 5' position of cytosine in the context of 5'-CG-3' dinucleotide (Sinsheimer 1955; Bird 2002).

Although non-CpG methylation has been reported in mammals (Ramsahoye et al. 2000), its presence is miniscule as compared to CpG methylation. The haploid human genome contains approximately 50 million CpG dinucleotides, which represents $2^{50,000,000}$ different permutations of CpG methylation. CpG dinucleotides are unevenly distributed across the human genome with vast DNA stretches deficient in CpGs, interspersed by CpG clusters called CpG islands (Laird 2003). CpG islands were identified almost 20 years ago on the basis of the strikingly discordant patterns of digestion of genomic DNA by restriction enzyme isoschizomers that differed only by their sensitivity to cytosine methylation (Singer et al. 1979). CpG islands have been defined by different groups based on various criteria. Initially they were defined as stretches of DNA (200 bp or longer) with a (G+C) content of 0.50 or greater and an observed to expected CpG dinucleotides ratio of 0.60 or greater. Later on, Takai and Jones (2003) found that increasing the size threshold to 500 bp and the (G+C) content threshold to 0.55 biased the definition against repetitive sequences and included only unique CpG sequences. According to computational estimates, there are at least 29,000 CpG islands in the human genome (Lander et al. 2001; Venter et al. 2001). CpGs are vastly under-represented in the genome mainly because deamination of cytosine gives rise to uracil, which is recognized as foreign in the DNA and hence replaced. Moreover, deamination of methylcytosine gives rise to thymine, which is less readily recognized as foreign and therefore prone to mutation (C:G to T:A transition) and depleted in the genome (Duncan and Miller 1980). Interestingly, CpG islands are found at the promoter sites of approximately 50% of the genes in the human genome (Loshikhes and Zhang 2000), most of which are the 'housekeeping' genes, where they are kept free from methylation (Larsen et al. 1992; Ponger et al. 2001). On the other hand, CpG islands within transposable elements are heavily methylated (Yoder et al. 1997). The methyl group in CpG dinucleotides protrudes from the cytosine nucleotide into the major groove of the DNA and has two main effects: it displaces transcription factors that normally bind to the DNA; and it attracts methyl binding proteins, which in turn are associated with gene silencing and chromatin compaction (Fazzari and Greally 2004). Methylation of DNA in the common 'B' form facilitates a conformational change to the 'Z' form, increases the helical pitch of DNA and alters the kinetics of cruciform extrusion (Murchie and Lilley 1989; Zacharias et al. 1990). Promoters with a few exceptions are generally inactive when the CpG island within them are methylated (an exception, for example, is seen in the case of the H-2K gene promoter, Tanaka et al. 1983). The transcriptional machinery can respond to CpG methylation in different ways. Either methylation at specific sites within the CpG island can interfere with the efficiency of expression (as seen in the case of Herpes simplex thymidine kinase and the Epstein-Barr virus latency C promoters) or it can be the density of methylation (human α- and γ-*GLOBIN*, mouse *MyoD1* promoter) that is responsible for interfering with transcription (Zingg and Jones 1997). Moreover, since the structure of 5-methylcytosine is similar to thymine, the methylation of cytosine might lead to generation of new consensus sequences for some transcription factors. This has been seen in the case of methylation at CpG in a low affinity AP-1 binding site that converts it to high affinity site (Tulchinsky et al. 1996) or in the CRE sequence

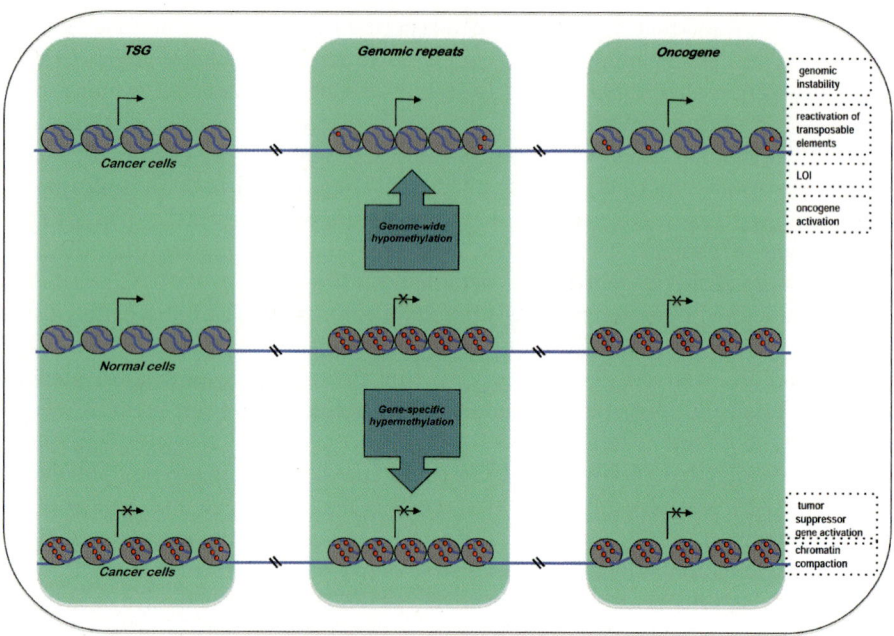

Fig. 26.1 Aberrant methylation during cancer progression. In normal cells, while promoter CpG islands of most genes lack methylation, repeat elements and promoters of oncogenes are kept inactive by DNA methylation. Genome-wide hypomethylation causes activation of oncogenes and activates aberrant transcripts from within repeat elements leading to genomic instability. On the other hand, hypermethylation of tumor suppressor genes (TSG) leads to their silencing. *LOI* loss of imprinting, *TSG* tumor suppressor gene. *Raised arrows* denote transcriptional activation; *crossed arrows* denote transcriptional repression. *Red circles*: Methyl CpG and *circles* with *curved lines* denote nucleosomal organization of chromatin

which enhances the binding of C/EBPα transcription factor (Rishi et al. 2010). Similarly, the binding of several other DNA binding proteins (AP-2, Ah receptor, CREB/ATF, E2F, EBP-80, c-Myc, NF-κB) is inhibited when the CpG within their recognition sequence is methylated (Jones and Gonzalgo 1997).

26.3.2 Aberrant DNA Methylation in Cancer

Global hypomethylation and gene-specific hypermethylation are the hallmarks of most cancers studied (Fig. 26.1 and Feinberg and Tycko 2004). Global alterations in DNA methylation have been observed not only in fully developed cancers, but also in the precancerous stage, including chronic inflammation, persistent viral infection and cigarette smoking. Furthermore, aberrant DNA methylation is significantly associated with aggressiveness of cancers and the poor outcome in cancer patients.

26.3.2.1 Genome-Wide Hypomethylation

The earliest evidence linking DNA methylation to cancer came with the discovery of global hypomethylation in tumors (Riggs and Jones 1983). Many CpG islands are normally methylated in somatic tissues (Strichman-Almashanu et al. 2002). During global hypomethylation, these methylated islands can become hypomethylated in cancers activating the nearby genes. High-throughput genomic DNA methylation studies have identified that the frequency of hypomethylated sites might be quite high in tumors and numerous genes have been identified that lose DNA methylation in different cancers (Adorjan et al. 2002; Lacobuzio-Donahue et al. 2003). The important link between DNA hypomethylation in cancer and chromosomal instability was established by Ehrlich and colleagues (Tuck-Muller et al. 2000). DNA hypomethylation was found to be particularly severe in pericentromeric satellite sequences or other DNA repeat elements, and several cancers including Wilms tumor, ovarian and breast carcinomas frequently contain unbalanced chromosomal translocations with break points in the pericentromeric DNA of chromosome 1 and 16 (Qu et al. 1999; Ehrlich 2009; Yoshida et al. 2011). Hypomethylation of satellite sequences might predispose them to breakage and recombination. Jaenisch and colleagues showed that neurofibromatosis 1 (Nf1)$^{+/-}$ Trp53$^{+/-}$ mice were 2.2 times more prone to loss of heterozygosity (LOH) when a hypomorphic *Dnmt1* allele was introduced (Eden et al. 2003). Similarly, hypomethylation of L1 retrotransposons promote chromosomal rearrangements in colorectal cancer (Suter et al. 2004; Ogino et al. 2008). LINE1 hypomethylation has been implicated in cancers of bladder, ovary, liver and colon (Kim et al. 2009; Dammann et al. 2010; Schernhammer et al. 2010; Wilhelm et al. 2010). The mechanism behind global hypomethylation in cancers largely remains unknown but several experiments point towards the involvement of SWI/SNF chromatin-remodeling complexes. Individuals with the developmental disorder ATRX (α-thalassaemia, myelodysplasia) have mutations in the *ATRX* gene, which encodes a SNF2-family helicase. In mutant ATRX cells, the ribosomal DNA repeats are hypomethylated (Gibbons et al. 2000). *Lsh*, a SNF2-family member is required for maintenance of normal DNA methylation as its knockout leads to global defect in genomic methylation and chromosomal instability (Fan et al. 2003). A common splice variant of the *de novo* DNA methyltransferase-*DNMT3B* (*DNMT3b4*) that was identified in patients with liver cancer, is also associated with hypomethylation of pericentromeric satellite sequences (Saito et al. 2002). Further, it was shown that mice carrying a hypomorphic allele for the maintenance methyltransferase *Dnmt1* (which reduces expression of Dnmt1 to 10% of wild type levels), developed tumors early and showed chromosomal instability (Gaudet et al. 2003). Hypomethylation of CpG islands has been implicated in overexpression of *CYCLIND2* and *MASPIN* in gastric carcinoma, *MN/CA9* in human renal cell carcinoma, *S100A4* metastasis-associated gene in colon cancer and human papillomavirus 16 (*HPV16*) in cervical cancer (Nakamura and Takenaga 1998; Cho et al. 2001; Badal et al. 2003; De Capoa et al. 2003; Lacobuzio-Donahue et al. 2003; Oshimo et al. 2003; Piyathilake et al. 2003; Sato et al. 2003).

Global DNA hypomethylation is associated with progression of multiple types of cancers including cervical, ovarian, multiple myeloma, chronic lymphocytic leukemia and breast cancer (Cho et al. 2010; Missaoui et al. 2010; Fabris et al. 2011; Walker et al. 2011). In case of gastric, tongue and esophageal carcinogenesis, global DNA hypomethylation was negatively correlated with invasiveness (Baba et al. 2009; Tomita et al. 2010).

26.3.2.2 Gene-Specific Hypermethylation

Aberrant transcriptional silencing of genes associated with DNA hypermethylation of their promoter region is probably the most intensely studied epigenetic abnormality in cancers. It is difficult to answer as to why certain genes become methylated during carcinogenesis, but several hypotheses have been proposed to address this issue. The Darwinian theory suggests the selective growth advantage conferred to cells upon inactivation of particular genes as the reason why they become methylated in tumors (Esteller 2005). Also, it has been suggested that genes which are under the control of polycomb proteins are more vulnerable to DNA methylation (Ohm and Baylin 2007). Like DNA hypomethylation, hypermethylation can play a seminal role in neoplastic evolution. Promoter DNA hypermethylation can silence specific genes including tumor suppressor genes in cooperation with histone modifications. For example, hypermethylation was found to be associated with deacetylation of histone H3 and H4, loss of histone H3- lysine4 methylation and gain of H3K9 methylation (Baylin and Ohm 2006).

While Retinoblastoma (*RB*) was the first tumor suppressor gene shown to be silenced by DNA hypermethylation, $p16^{ink4A}$ is one of the most common tumor suppressors exhibiting loss of function following DNA hypermethylation. It exhibits DNA hypermethylation during progression of lung cancers and even in preneoplastic lesions (Belinsky et al. 1998; Nuovo et al. 1999) and its germ line loss leads to increase in hematopoietic stem cell (HSC) life span in terms of their ability for tissue maintenance and repair (Janzen et al. 2006; Krishnamurthy et al. 2006; Molofsky et al. 2006). The silencing of $p16^{ink4A}$ gene has also been reported in preinvasive stages of breast, colon and other cancers. Furthermore, experimental loss of $p16^{ink4A}$ appears to facilitate early tumorigenesis by being permissive for subsequent emergence of genomic instability (Kiyono et al. 1998) and may directly allow for additional epigenetic silencing of other genes (Reynolds et al. 2006). Germ line mutations of many tumor suppressors cause familial forms of cancers. The same tumor suppressor genes have been found to be promoter DNA hypermethylated in subsets of non familial cancers such as *VHL* in renal, *APC* in colon and *BRCA1* in breast cancers (Ting et al. 2006a, b). Till date, a large number of tumor suppressor or candidate tumor suppressor genes have been identified to be DNA hypermethylated in multiple cancer types, including but not limiting to: *p53, RASSF1A, p14ARF, CDKN2A, p16, p21, TIMP3, ECRG4, HIC1* (Cohen et al. 2003; Amatya et al. 2005; Chanda et al. 2006; Gotze et al. 2009; Juhlin et al. 2010; Dadkalos et al. 2011; Radpour et al. 2011).

Tumor suppressors are not the only genes to exhibit DNA hypermethylation; a multitude of other genes like *hMLH1, MGMT, E-CADHERIN, CALCITONIN*, etc. are silenced in cancers (Esteller 2007; Jacinto and Esteller 2007). Apart from individual gene hypermethylation, in some cancers, groups of genes were found to exhibit increased DNA methylation levels. Two such examples identified recently in cancers are, CpG island methylator phenotype (CIMP) and Long Range Epigenetic Silencing (LRES). CIMP defines a group of cancers with a 3–5 fold elevated frequency of aberrant gene methylation especially in case of *INK4A, MLH1* and *THBS1* apart from harboring microsatellite instability (Toyota et al. 1999). The clustering pattern suggestive of CIMP has been confirmed in glioblastoma, gastric cancer, liver cancer, pancreatic cancer, esophageal cancer, ovarian cancer, acute lymphocytic leukemia and acute myelogenous leukemia apart from colorectal cancer (Issa 2004). Though the causes leading to CIMP are largely unknown, its presence is associated with a poor outcome in multiple malignancies (Issa 2003). Age-related methylation, life-style, exposure to epimutagens, chronic inflammations have all been thought of as important contributors of CIMP. Long range epigenetic silencing (LRES) is a phenomenon where large regions of chromosomes can be coordinately suppressed. Typically, LRES can span megabases of DNA and involve broad heterochromatin formation accompanied by the hypermethylation of clusters of contiguous CpG islands within the region. This process is usually associated with DNA and histone hypermethylation and can in turn lead to DNA methylation of flanking, non-methylated genes. Such hypermethylated regions spanning hundreds of kilobases involving gene family clusters have also been found in esophageal, neuroblastoma, breast and colorectal cancers (Frigola et al. 2006; Clark 2007).

26.3.3 The DNA Methyltransferases

DNA methyltransferases (DNMTs) are the enzymes, which catalyze the covalent addition of methyl groups to cytosine in the CpG dinucleotide context. Broadly they have been classified into: maintenance methyltransferases (*DNMT1*) and *de novo* methyltransferases (*DNMT3A, DNMT3B, DNMT3L*). DNMT2, with its weak methyltransferase activity has not been classified into either of the groups. The DNMTs are characterized by a 'C' terminal catalytic domain, which transfers methyl group from S-Adenosyl-L-methionine (SAM) to cytosine and contains five conserved amino acid motifs, namely I, IV, VI, IX and X. Motifs I and X form the AdoMet binding site.

Dnmt1 was the first methyltransferase to be identified. Homologs of *Dnmt1* have been found in nearly all eukaryotes that have DNA bearing 5-methylcytosine, but not in species that lack 5-methylcytosine. It is a large protein of 1,620 amino acids, having an 'N' terminal regulatory region containing an NLS, a region that targets *Dnmt1* to replication foci, PHD like domain, proliferating cell nuclear antigen (PCNA)-binding domain apart from the 'C' terminal catalytic domain. An 'N' terminal truncated but enzymatically active, oocyte-specific isoform of *Dnmt1*

(*Dnmt1o*) has also been identified, which accumulates to high levels in the cytoplasm of embryos (Doherty et al. 2002). Being a maintenance methyltransferase, Dnmt1 is involved in the bulk of DNA methylation and has also been implicated in non-CpG methylation (Grandjean et al. 2007). The DNMT3 family comprises *de novo* methyltransferases: *DNMT3A, DNMT3B* and an enzymatically inactive paralogue *DNMT3L*. Both DNMT3A and DNMT3B have a regulatory 'N' terminal domain containing an ATRX-like Cys-rich domain (PHD domain) and a PWWP domain, which are involved in interaction of these enzymes with other proteins and in targeting them to heterochromatin (Gowher et al. 2005). DNMT3L on the other hand doesn't possess any catalytic activity owing to mutations within all the conserved motifs that contain the catalytic residues of DNA-(cytosine-C5)-methyltransferase. It interacts with and regulates the *de novo* methyltransferases DNMT3A and 3B, stimulating their activity (Chedin et al. 2002; Suetake et al. 2004; Kareta et al. 2006). Recently, it has also been shown that DNMT3L interacts with Histone 3 Lysine 4 (H3K4) when it is unmethylated, providing a link between *de novo* DNA methylation and histone modifications (Ooi et al. 2007).

Homozygous deletion of Dnmts: *Dnmt1, Dnmt3a, Dnmt3b* are lethal in mice (Li et al. 1992; Lei et al. 1996; Okano et al. 1999). The deletion of *Dnmt3a* and *Dnmt3b* abolishes *de novo* methylation, while *Dnmt1* depletion leads to bulk DNA demethylation (Okano et al. 1999). Embryos of $Dnmt1^{-/-}$ mice are stunted, show delayed development, and do not survive past mid-gestation (Li et al. 1992). Though $Dnmt3a^{-/-}$ homozygous mutant mice developed to term and appeared normal at birth, they became runted and died at about 4 weeks of age (Okano et al. 1999). On the other hand, $Dnmt3b^{-/-}$ homozygous mutant mice were not viable, though they appeared to develop normally till E 9.5. These embryos displayed multiple developmental defects including growth impairment and rostral neural tube defects (Okano et al. 1999). Similarly, $Dnmt3a^{-/-}$ and $Dnmt3b^{-/-}$ double homozygous embryos were smaller in size and showed abnormal morphology at E 8.5, E 9.5 and died before E 11.5. A closer analysis of the embryos revealed that their growth and morphogenesis were arrested shortly after gastrulation (Okano et al. 1999).

Conditional deletion of *Dnmt3b* was also shown to result in DNA hypomethylation leading to chromosomal instability and spontaneous immortalization in mouse embryonic fibroblasts (Dodge et al. 2005). The DNMT3 family also plays an important role in genomic imprinting. Pericentromeric satellites are one of the specific targets of *DNMT3B*. Furthermore, inactivation of *DNMT3B* in humans leads to ICF syndrome, which is associated with low methylation in pericentromeric satellite regions. *Dnmt3a* conditional mutant females died *in utero* and lacked methylation and allele-specific expression at many maternally imprinted loci (DMRs of *Snrpn, Igf2r, Peg1*), whereas *Dnmt3a* conditional mutant males showed impaired spermatogenesis and lacked methylation at a few paternally imprinted loci examined in spermatogonia (*H19* DMR, *Dlk1-Gtl2* IG-DMR) (Kaneda et al. 2004). DNMT3A2, a germ-cell-specific isoform of *DNMT3A* is also required for genomic imprinting.

Though enzymatically inactive, DNMT3L is required for DNA methylation in ES cells where it is expressed in high amounts compared to differentiating somatic cells (Ooi et al. 2010). Mice obtained after disruption of *Dnmt3l* by targeted mutation were viable and of normal phenotype in both heterozygous and homozygous conditions (Bourc'his et al. 2001). The male progeny however were found to be sterile and displayed azoospermia as their spermatocytes failed to complete meiosis, apart from exhibiting decrease in methylation at the differentially methylated regions of *H19* and *Rasgrf1* imprinted genes (Webster et al. 2005). On the other hand, the heterozygous progeny of homozygous females died before mid gestation (at 9.5 days post coitum) and these embryos were found to lack methylation imprints at *Snrpn* and *Peg1* imprinted genes apart from other pre-natal growth defects (Bourc'his et al. 2001). The role of Dnmt3l in establishing patterns of DNA methylation for several imprinted genes during gametogenesis has also been established by other groups (Webster et al. 2005; Arima et al. 2006).

26.3.4 The DNA Demethylase Enzymes

In contrast to the well characterized DNA methyltransferases, the enzymes that catalyze the removal of methyl moiety from methylcytosine have remained enigmatic. It has been easy to envisage passive mode of demethylation, wherein cytosine methylation is lost upon replication without being followed by maintenance methylation of hemimethylated DNA. However, active demethylation observed in the zygote (specifically for the paternal genome) and during germ cell formation has been hard to explain (Mayer et al. 2000; Ooi and Bestor 2008; Wu and Zhang 2010). Only very recently, a few proteins have been identified that modify methylcytosine either to thymine or hydroxymethylcytosine, which can be directly or indirectly converted to cytosine by DNA repair enzymes (Morgan et al. 2004; Rai et al. 2008; Kriaucionis and Heintz 2009; Tahiliani et al. 2009; Ito et al. 2010).

Conversion of methylcyotsine to thymine in mammals is catalyzed by Apolipoprotein B RNA-editing catalytic component-1 (Apobec-1) and Activation Induced Deaminase (AID, also known as Apobec2), members of the cytidine deaminase family (Morgan et al. 2004; Rai et al. 2008). The role of these enzymes in active demethylation of methylcytosine has been confirmed by various experiments directly or indirectly. In 2004, it was demonstrated that AID had 5-methylcytosine deaminase activity *in vitro* (Morgan et al. 2004). Moreover, it was found that AID is expressed in primordial germ cells (PGCs), oocytes and cells from early mouse embryos, cells which are known for their active demethylation activity (Morgan et al. 2004; Wu and Zhang 2010). That AID is indeed involved in DNA demethylation during PGC specification was confirmed by Popp et al. (2010) when they showed that erasure of DNA methylation mark during PGC formation is hindered in AID deficient mice. The role of AID in DNA demethylation was further strengthened by the observation that AID deficiency prevents the demethylation of pluripotency specific genes (*OCT4, NANOG*) during the process of converting fibroblasts into pluripotent iPS cells (Bhutani et al. 2010).

It was recently discovered that TET family of proteins including TET1, TET2 and TET3 can convert methylcytosine into hydroxymethylcytosine (Tahiliani et al. 2009; Ito et al. 2010). JBP1 and JBP2 proteins catalyze conversion of methyl-thymine into β-D-glucosyl-hydroxymethyluracil in Trypanosomes. TET proteins were identified to be their homologs in mammals based on iterative sequence profile computational search (Tahiliani et al. 2009). This study also demonstrated that cytosine does exist in the hydroxmethylcytosine form in the mammalian genomes (also shown by Kriaucionis and Heintz 2009) and TET1 does convert methylcytosine to hydroxymethylcytosine *in vitro* (Tahiliani et al. 2009). Like Activation induced deaminase, TET proteins mediated conversion of methylcytosine to hydroxymethylcytosine has also become established as a part of the DNA demethylation process. TET1 has been shown to be involved in maintaining ES cell pluripotency and in the specification of inner cell mass during embryogenesis (Ito et al. 2010). Just after fertilization and before the first cell division, the male pronucleus has been shown to get actively demethylated (Mayer et al. 2000). A very recent study showed that unlike the previous established notion, methylcytosines in male pronucleus gets converted into hydroxymethylcytosine and not cytosine (Iqbal et al. 2011). Moreover, conversion back to cytosine is not an active process as hydroxymethylcytosine stays in the genome even after several cell divisions (Iqbal et al. 2011). The same paper also indicated on the basis of transcriptional profiling that TET3 rather than TET1 was probably responsible for the conversion to hydroxymethylcytosine (Iqbal et al. 2011).

26.3.5 DNMTs and Cancer

Being the effectors of DNA methylation, DNMTs have an important role to play in the aberrant DNA methylation observed in cancer cells (Fig. 26.2). DNMT protein levels and activities were found to be elevated in various cancer types, including gastric, bladder, brain, leukemia, colon and lung (Issa et al. 1993; Belinsky et al. 1996; Melki et al. 1998; De Marzo et al. 1999; Robertson et al. 1999; Ramsahoye et al. 2000; Girault et al. 2003; Li et al. 2003; Xiong et al. 2005; Amara et al. 2010; Qu et al. 2010). Moreover, multiple DNMTs have been found to co-localize to promoters of hypermethylated genes and also have been defined as components of the transcriptional repression complexes (Di Croce et al. 2002; Kim et al. 2002; Datta et al. 2003). However, it has been difficult to correlate the increased expression of DNMTs with CpG hypermethylation in several cancers (Miremadi et al. 2007). While the maintenance methylation activity of DNMT1 is primarily observed in somatic cells, the *de novo* methylation activities are seen in germ cells and at embryonic stages. This paradigm of separation of methylation activity has been challenged in the cancer setting.

Though aberrant DNA methylation patterns observed in cancer is an undisputed fact, it is still not clear as to which of the DNMTs are primarily responsible. The maintenance methyltransferase *DNMT1* has been implicated because its expression

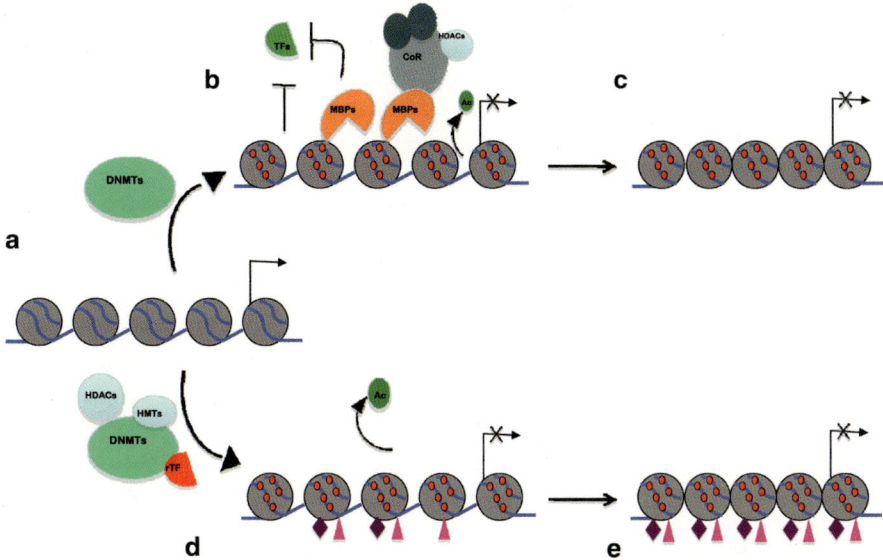

Fig. 26.2 Dual role of DNMTs in silencing of gene expression. *DNMTs* can be targeted directly or with the help of transcriptional repressors (*rTF*) to CpG dinucleotides. (**a**) Nascent chromatin strand (**b** & **c**) DNMTs can get directly targeted to a genomic locus due to a stimulus like hemimethylated DNA and lead to CpG methylation. This could cause inhibition of transcription factor (*TF*) binding either directly or due to the binding of *MBPs* (Methyl binding proteins). The MBPs can recruit co-repressor complexes (*CoR*) and *HDACs* leading to nucleosomal remodeling and chromatin compaction. (**d** & **e**) The DNMTs are also found in complex with histone deacetylase and histone methyltransferases. In this scenario DNA methylation is coupled to setting up of repressive histone marks including H3K9me3 and H3K27me3 eventually leading to condensed and tightly packed chromatin. *Raised arrows* denote transcriptional activation; *crossed arrows* denote transcriptional repression. *Red circles*: Methyl CpG; *Green circles*: H3K9 acetylation; *Circles* with *curved lines* denote nucleosomal organization of chromatin; *Diamonds* represent H3K9me3; *Triangles* represent H3K27me3

levels are higher in cancers and it also exhibits low levels of *de novo* methylation activity against unmethylated substrates (Jair et al. 2006). DNMT1 overexpression resulted in detectable *de novo* methylation of CpG island in human fibroblasts (Vertino et al. 1996) and induced transformation in NIH3T3 cells (Wu et al. 1993). On the other hand, severe depletion of DNMT1 produced only minor decrease in overall methylation, minimal loss of promoter hypermethylation and undetectable re-expression of silenced tumor suppressor genes in colorectal cancer cells (Rhee et al. 2000, 2002; Ting et al. 2004, 2006a, b). However, a double knock out of *DNMT1* and *DNMT3b* (*DNMT1$^{-/-}$ DNMT3b$^{-/-}$*) in HCT116 cells resulted in >95% loss in genomic methylcytosine content and complete promoter demethylation apart from re-expression of aberrantly silenced genes (Rhee et al. 2002; Akiyama et al. 2003; Suzuki et al. 2004). *DNMT1* is thus one candidate that is capable of initiating aberrant CpG island hypermethylation in cancer cells. Indeed, DNMT1 has been reported to repress transcription through its interactions with methyl CpG binding proteins, histone deacetylases and histone methyltransferases (Bogdanovi and Veenstra 2009). On the other hand, global methylation levels were affected indirectly through the regulation of DNMT1 by BRCA1 and miRNA29b (Garzon et al. 2009; Shukla et al. 2010). In the Apc^{Min} mice model, a reduction of *Dnmt1* activity, due to heterozygosity of *Dnmt1* gene, in conjugation with treatment using DNMT inhibitor like 5-Aza-cytidine reduced the average number of intestinal adenomas (Laird et al. 1995). Similar observations in genetically engineered mice involving *Dnmt1* clearly demonstrate a causal relationship between alteration of DNA methylation and cancer. But the requirement of the maintenance methyltransferase *DNMT1* in maintaining promoter hypermethylation and gene silencing in cancer cells is still debatable. Mutational inactivation of *DNMT1* has not been observed in most of the cancers though colorectal cancers infrequently harbored these mutations (Kanai 2008). Among the other DNMTs, germ line single nucleotide polymorphisms (SNPs) in *DNMT3B* have been associated with risk of breast cancer, gastric cancer, hepatocellular carcinoma, lung adenocarcinoma and lung cancer (Cebrian et al. 2006; Ezzikouri et al. 2009; Hu et al. 2010).

Recently data from our own laboratory showed loss of methylation at *DNMT3L* promoter in cervical cancer and few other cancer types (Gokul et al. 2007; Manderwad et al. 2010 and unpublished results). This decrease in methylation was associated with an increased expression of DNMT3L. Moreover, overexpression of DNMT3L in HeLa cells lead to increased cell proliferation and anchorage-independent growth in a time and passage dependent manner (Gokul et al. 2009). Microarray analysis revealed that the expression patterns of genes important in nuclear reprogramming, development and cell cycle were misregulated. Interestingly, among this misregulated gene set, many imprinted genes were found to be downregulated; consistent with the role of DNMT3L in imprinting. DNMT3L was also identified to be essential for the growth of human testicular germ cell tumors (Minami et al. 2010).

26.3.6 DNA Demethylase Enzymes and Cancer

Though the role of AID in DNA demethylation has been identified only recently, its role in creating somatic hypermutations and in class-switch recombination of human immunoglobulin genes has been known for some time now (Revy et al. 2000; Marusawa and Chiba 2010). Several studies have reported unregulated expression of AID in cancer (Matsumoto et al. 2007; Klemm et al. 2009). An interesting correlation that has been investigated is the link between *Helicobacter pylori* infection, aberrant expression of Activation Induced Deaminase and gastric cancer (Matsumoto et al. 2010; Touati 2010; Endo et al. 2011). A few studies have also highlighted the role of aberrant AID expression in causing widespread genomic instability (Klemm et al. 2009; Robbiani et al. 2009).

TET1 is an abbreviation for Ten-Eleven Translocation-1 and was named so because of its involvement in a t(10;11)-associated leukemia (Lorsbach et al. 2003).

It was simultaneously identified by Ono et al. (2002) who found *TET1* to be fused to *MLL* gene in Acute Myeloid Leukemia. Later studies have also reported fusion of *TET1* (also known as *LCX*) with *MLL* suggesting a role for *TET1* in carcinogenesis (Shih et al. 2006; Burmeister et al. 2009, reviewed in Dahl et al. 2011). The fusion of *TET1* and *MLL* is also interesting from an epigenetic perspective as MLL itself is H3K4 histone methyltransferase (Krivtsov and Armstrong 2007). Deletions and mutations have also been reported for *TET2* in several myeloid cancers (Figueroa et al. 2010; Ko et al. 2010; Langemeijer et al. 2011). Along with the mutations in TET proteins, a recent report showed that the levels of hydromethylcytosine were also altered in myeloid cancer (Ko et al. 2010) indicating that the process of DNA demethylation itself might be key to the process of carcinogenesis.

26.3.7 DNA Methylation Binding Proteins and Cancer

The transcriptional repressive activities of DNA hypermethylation are primarily interpreted and mediated by a family of proteins which harbor the methyl-CpG-binding domain (MBD) – the protein motif responsible for binding methylated CpG dinucleotides (Bird and Wolffe 1999; Bird 2002; Burgers et al. 2002). These include MECPs (Methyl CpG binding protein-MECP2), MBDs (Methylcytosine-binding protein MBD1, 2, 3, 4) and the novel protein Kaiso (Prokhortchouk et al. 2001). Except for MBD3, the other methyl binding proteins specifically recognize methyl-CpG (Klose and Bird 2006). These proteins can mediate silencing of gene expression by recruiting other members of the epigenetic machinery, primarily the chromatin remodeling co-repressor complexes (Jones et al. 1998; Ng et al. 1999; Wade et al. 1999; Zhang et al. 1999; Sarraf and Stancheva 2004). The MBDs have a high specificity towards gene promoters and they have been found at hypermethylated and aberrantly silenced cancer gene loci (Ballestar and Esteller 2005; Ting et al. 2006a, b). For example, MBD2 binds to aberrantly methylated promoter of tumor suppressor genes: *p14/ARF* and *p16/ink4A* in colon cancer cell lines (Magdinier and Wolffe 2001; Martin et al. 2008) and suppress their expression. Importantly, MBDs are found to be associated with complexes that contain HDACs; for example the methyl binding proteins, MECP2, MBD1 and MBD2 have been found to associate with transcriptional co-repressors, such as SIN3A, which are known to bind HDACs directly (Jones et al. 1998; Ng et al. 1999). This binding results in compaction of chromatin and stable repression of the target gene. MBD1 interacts with histone H3 methyltransferase (SETDB1) linking epigenetic marks on DNA to histone modifications, whereas MBD4 is thought to act as a thymine DNA glycosylase, repairing G:T or G:U mismatches at CpG sites (Kanai 2008). The newly identified methyl binding protein- Kaiso, associates with the histone deacetylase containing-N-CoR co-repressor complex bringing about repression of methylated genes (Yoon et al. 2003). Interestingly, mice lacking Kaiso have been shown to exhibit resistance to intestinal tumorigenesis (Prokhortchouk et al. 2006). Polymorphisms (SNPs) in many of these MBD proteins (MBD1, MBD2 and MBD4) were found to be associated with an increased risk of cancer (lung and breast). Overexpression of MeCP2 has been observed in breast cancer (Muller et al. 2003) and it is implicated in the silencing of *IL-6* in pancreatic adenocarcinoma cell lines (Dandrea et al. 2009). MeCP2 also plays important roles in gastric and colorectal carcinogenesis (Pancione et al. 2010; Wada et al. 2010). Progeny of a cross between Mbd2-deficient mice and Apc$^{Min/+}$ mice were found to be resistant to the development of intestinal tumors and this resistance was dependent on the dosage of Mbd2 (Sansom et al. 2003). The role of MBD2 has also been positively implicated in the silencing of Nrf2 expression in a mouse prostate adenocarcinoma model (Yu et al. 2010). The hypermethylated region of human telomerase reverse transcriptase (*hTERT*) was also shown to be specifically associated with MBD2 bringing about its transcriptional repression in many cancer cell lines (Chatagnon et al. 2009).

26.3.8 DNA Methylation Inhibitors and Epigenetic Therapy

In spite of being robust, epigenetic modifications like DNA methylation and histone modifications are reversible in nature. This characteristic of epigenetic modifications makes them a useful target for cancer therapy and the enzymes that mediate and maintain DNA methylation (DNMTs) and various histone modifications (HATs, HDACs and HKMTs) are the prime drug targets in the epigenetic therapy proposed for cancer.

Various small molecules have been identified with inhibitory effects on DNA methylation, histone methylation or acetylation and it is possible to reactivate a hypermethylated tumor suppressor gene by the use of DNA demethylating agents or DNA methyltransferase inhibitors (DNMTi). Most of these molecules are nucleoside analogs, which are incorporated into replicating DNA in place of cytosine. Once incorporated, they can sequester DNMTs by the formation of a covalent bond between them, eventually depleting the cell of enzymatic activity resulting in heritable demethylated DNA (Juttermann et al. 1994; Santi et al. 1984). Due to the high replicative potential of cancer cells they are particularly prone to demethylation underscoring the efficacy of DNMTi as antineoplastic drugs. Among the best studied inhibitors are 5-aza-cytidine and 5-aza-2'-deoxycytidine which were found to bring about DNA demethylation (Sorm et al. 1964, reviewed in Momparler and Bovenzi 2000; Gal-Yam et al. 2008; Kwa et al. 2011). Both inhibitors were shown to reactivate genes silenced by aberrant methylation in various human tumor cell lines (Momparler and Bovenzi 2000) apart from inducing in vitro differentiation of human leukemic cells (Pinto et al. 1984; Momparler et al. 1985). They also exhibited potent antineoplastic activity in hematological malignancies and lung cancer (Rivard et al. 1981; Pinto and Zagonel 1993; Momparler et al. 1986; Momparler and Ayoub 2001). Presently, of the four epigenetic-based drugs that have been approved by FDA for cancer therapy, two are inhibitors of DNMTs: 5-azacytidine (Vidaza) and 5-aza-2'-deoxycytidine (Decitabine) (Kwa et al. 2011). Both drugs have been approved by the FDA for the treatment of myelodysplastic syndrome, a preleukemic disease (Pinto and Zagonel 1993). There are few other inhibitors like Epigallocate, Procaine etc. some of which are less toxic compared to 5-aza-cytidine, but their dosage, stability and efficacy have impeded their advancement to clinical trials (Kwa et al. 2011). Zebularine is another promising DNMTi which has the advantage of higher stability and longer half-life together with the convenience of oral administration but is effective only in high doses (Champion et al. 2010; Cheng et al. 2004; Kwa et al. 2011). Recently, derivatives and variants of 5-azacytidine like S110 and 2'-deoxy-5,6-dihydro-5-azacytidine have been characterized and found to be less toxic and more stable demethylating agents (Chuang et al. 2010; Matousova et al. 2011). In spite of concerns relating to non-specific side effects, toxicity, drug resistance and inefficacy against solid tumors, DNMTis remain an attractive choice as therapeutic agent in hematological malignancies. Additionally because of their ability to cross the blood brain barrier they can be used to treat brain malignancies (Diede et al. 2010).

The interplay between histone modifications and DNA methylation in cancer cells has prompted the use of histone deacetylase inhibitors (HDACi) with DNMTi. Such combinatorial therapy was shown to increase the expression of silenced genes to higher levels than with either of the inhibitors alone (Issa 2007; Kwa et al. 2011). Cameron et al. (1999) were the first to show the synergistic transcriptional activation of the *p16CDKN2A* tumor suppressor gene upon treatment of tumor cells with 5-azacytidine and Trichostatin A, a HDACi. We in our lab also have been able to show similar 5-azacytidine transcriptional activation for the *DNMT3L* gene in cervical cancer cells upon treatment with 5-azacytidine and Trichostatin A (Gokul et al. 2009). It was also shown that the treatment of tumor cells with 5-azacytidine decreased levels of repressive histone marks (H3K9me3) while increasing active histone marks (H3K4me3) at promoters of genes that are aberrantly silenced by DNA methylation (Nguyen et al. 2002). These facts not only underscore the importance of such combinatorial therapy but also warrant the need to investigate the nature of synergy between different inhibitors to bring about an optimal response in tumors.

26.4 Concluding Remarks

The abnormalities in cancer change the "epigenetic landscape" of the cell and involve multiple aberrations in virtually every component of the epigenome. Since epigenetic silencing processes are mitotically heritable, they play the same roles and undergo the same selective processes as genetic alterations in the development of cancer. Similar to genetic mutations, epigenetic events follow the Darwin's hypotheses for the evolution of species by which alterations in gene expression induced by epigenetic events which confer cellular growth advantage are selected for in the host, resulting in the progressive, uncontrolled growth of the tumor. Can epigenetic mechanisms initiate tumorigenesis by their own? As described in this chapter there are numerous examples of aberrant epigenetic changes occurring in various cancers as early events. For example, DNA hypomethylation in cancers has been shown to be an early event, but it has not been conclusively shown nor are there examples that show epigenetic processes initiating cancer on their own. However, the evidence accumulated over the last two decades reiterates the complexity of the mechanisms involved in cancer development and emphasizes the fact that there is strong interplay between epigenetic and genetic events in cancer and that understanding the epigenetic basis of carcinogenesis is critical in solving the puzzle that cancer is.

Acknowledgements The basis of several sections in this manuscript was the Introduction chapter from GG's doctoral thesis. Work in SK laboratory is supported by CDFD core support and grants from Department of Biotechnology, Government of India.

References

Adorjan P, Distler J, Lipscher E, Model F, Muller J, Pelet C, Braun A, Florl AR, Gutig D, Grabs G et al (2002) Tumor class prediction and discovery by microarray-based DNA methylation analysis. Nucleic Acids Res 30:e21

Ahuja N, Li Q, Mohan AL, Baylin SB, Issa JP (1998) Aging and DNA methylation in colorectal mucosa and cancer. Cancer Res 58:5489–5494

Akiyama Y, Maesawa C, Ogasawara S, Terashima M, Masuda T (2003) Cell-type-specific repression of the maspin gene is disrupted frequently by demethylation at the promoter region in gastric intestinal metaplasia and cancer cells. Am J Pathol 163:1911–1919

Amara K, Ziadi S, Hachana M, Soltani N, Korbi S, Trimeche M (2010) DNA methyltransferase DNMT3b protein overexpression as a prognostic factor in patients with diffuse large B-cell lymphomas. Cancer Sci 101:1722–1730

Amatya VJ, Naumann U, Weller M, Ohgaki H (2005) TP53 promoter methylation in human gliomas. Acta Neuropathol (Berl) 110:178–184

Arima T, Hata K, Tanaka S, Kusumi M, Li E, Kato K, Shiota K, Sasaki H, Wake N (2006) Loss of the maternal imprint in $DNMT3L^{mat-/-}$ mice leads to a differentiation defect in the extraembryonic tissue. Dev Biol 297:361–373

Baba S, Yamada Y, Hatano Y, Miyazaki Y, Mori H, Shibata T, Hara A (2009) Global DNA hypomethylation suppresses squamous carcinogenesis in the tongue and esophagus. Cancer Sci 100:1186–1191

Badal V, Chuang LS, Tan EH, Badal S, Villa LL, Wheeler CM, Li BF, Bernard HU (2003) CpG methylation of human papillomavirus type 16 DNA in cervical cancer cell lines and in clinical specimen: genomic hypomethylation correlates with carcinogenic progression. J Virol 77:6227–6234

Ballestar E, Esteller M (2005) Methyl-CpG-binding proteins in cancer: blaming the DNA methylation messenger. Biochem Cell Biol 83:374–384

Baylin SB, Ohm JE (2006) Epigenetic gene silencing in cancer: a mechanism for early oncogenic pathway addiction? Nat Rev Cancer 6:107–116

Baylin SB, Hoppener JW, de Bustros A, Steenbergh PH, Lips CJ, Nelkin BD (1986) DNA methylation patterns of the calcitonin gene in human lung cancers and lymphomas. Cancer Res 46:2917–2922

Belinsky SA, Nikula KJ, Baylin SB, Issa JP (1996) Increased cytosine DNA-methyltransferase activity is target-cell-specific and an early event in lung cancer. Proc Natl Acad Sci USA 93:4045–4050

Belinsky SA, Nikulan KJ, Palmisano WA, Michels R, Saccomanno G, Gabrielson E, Baylin SB, Herman JG (1998) Aberrant methylation of p16(INK4a) is an early event in lung cancer and a potential biomarker for early diagnosis. Proc Natl Acad Sci USA 95:11891–11896

Bhutani N, Brady JJ, Damian M, Sacco A, Corbel SY, Blau HM (2010) Reprogramming towards pluripotency requires AID-dependent DNA demethylation. Nature 463:1042–1047

Bird A (2002) DNA methylation patterns and epigenetic memory. Genes Dev 16:6–21
Bird AP, Wolffe AP (1999) Methylation-induced repression-Belts, braces, and chromatin. Cell 99:451–454
Bogdanović O, Veenstra GJ (2009) DNA methylation and methyl-CpG binding proteins: developmental requirements and functions. Chromosoma 118:549–565
Bourc'his C, Xu GL, Lin CS, Bollman B, Bestor TH (2001) Dnmt3L and the establishment of maternal genomic imprints. Science 294:2536–2539
Burgers WA, Fuks F, Kouzarides T (2002) DNA methyltransferases get connected to chromatin. Trends Genet 18:275–277
Burmeister T, Meyer C, Schwartz S, Hofmann J, Molkentin M, Kowarz E, Schneider B, Raff T, Reinhardt R, Gökbuget N et al (2009) The MLL recombinome of adult CD10-negative B-cell precursor acute lymphoblastic leukemia: results from the GMALL study group. Blood 113:4011–4015
Buschhausen G, Wittig B, Graessmann M, Graessmann A (1987) Chromatin structure is required to block transcription of the methylated herpes simplex virus thymidine kinase gene. Proc Natl Acad Sci USA 84:1177–1181
Cameron EE, Bachman KE, Myohanen S, Herman JG, Baylin SB (1999) Synergy of demethylation and histone deacetylase inhibition in the re-expression of genes silenced in cancer. Nat Genet 21:103–107
Cebrian A, Pharoah PD, Ahmed S, Ropero S, Fraga MF, Smith PL, Conroy D, Luben R, Perkins B, Easton DF et al (2006) Genetic variants in epigenetic genes and breast cancer risk. Carcinogenesis 27:1661–1669
Champion C, Guianvarc'h D, Senamaud-Beaufort C, Jurkowska RZ, Jeltsch A, Ponger L, Arimondo PB, Guieysse-Peugeot AL (2010) Mechanistic insights on the inhibition of c5 DNA methyltransferases by zebularine. PLoS One 5:e12388
Chanda S, Dasgupta UB, Guhamazumder D, Gupta M, Chaudhuri U, Lahiri S, Das S, Ghosh N, Chatterjee D (2006) DNA hypermethylation of promoter of gene p53 and p16 in arsenic-exposed people with and without malignancy. Toxicol Sci 89:431–437
Chatagnon A, Bougel S, Perriaud L, Lachuer J, Benhattar J, Dante R (2009) Specific association between the methyl-CpG-binding domain protein 2 and the hypermethylated region of the human telomerase reverse transcriptase promoter in cancer cells. Carcinogenesis 30:28–34
Chedin F, Lieber MR, Hsieh CL (2002) The DNA methyltransferases-like protein *DNMT3L* stimulates de novo methylation by Dnmt3a. Proc Natl Acad Sci USA 99:16916–16921
Cheng JC, Yoo CB, Weisenberger DJ, Chuang J, Wozniak C, Liang G, Marquez VE, Greer S, Orntoft TF, Thykjaer T, Jones PA (2004) Preferential response of cancer cells to zebularine. Cancer Cell 6:151–158
Cho M, Uemura H, Kim SC, Kawada Y, Yoshida K, Hirao Y, Konishi N, Saga S, Yoshikawa K (2001) Hypomethylation of the MN/CA9 promoter and upregulated MN/CA9 expression in human renal cell carcinoma. Br J Cancer 85:563–567
Cho YH, Yazici H, Wu HC, Terry MB, Gonzalez K, Qu M, Dalay N, Santella RM (2010) Aberrant promoter hypermethylation and genomic hypomethylation in tumor, adjacent normal tissues and blood from breast cancer patients. Anticancer Res 30:2489–2496
Chuang JC, Warner SL, Vollmer D, Vankayalapati H, Redkar S, Bearss DJ, Qiu X, Yoo CB, Jones PA (2010) S110, a 5-aza-2'-deoxycytidine-containing dinucleotide, is an effective DNA methylation inhibitor in vivo and can reduce tumor growth. Mol Cancer Ther 9:1443–1450
Clark SJ (2007) Action at a distance: epigenetic silencing of large chromosomal regions in carcinogenesis. Hum Mol Genet 16(Spec No 1):R88–R95
Cohen Y, Singer G, Lavie O, Dong SM, Beller U, Sidransky D (2003) The RASSF1A tumor suppressor gene is commonly inactivated in adenocarcinoma of the uterine cervix. Clin Cancer Res 9:2981–2984
Coussens LM, Werb Z (2002) Inflammation and cancer. Nature 420:860–867
Dadkalos A, Oleksiewicz U, Filia A, Nikolaidis G, Xinarianos G, Gosney JR, Malliri A, Field JK, Liloglou T (2011) UHRF1-mediated tumor suppressor gene inactivation in nonsmall cell lung cancer. Cancer 117:1027–1037

Dahl C, Grønbæk K, Guldberg P (2011) Advances in DNA methylation: 5-hydroxymethylcytosine revisited. Clin Chim Acta. doi:10.1016/j.cca.2011.02.013

Dammann RH, Kirsch S, Schagdarsurengin U, Dansranjavin T, Gradhand E, Schmitt WD, Hauptmann S (2010) Frequent aberrant methylation of the imprinted IGF2/H19 locus and LINE1 hypomethylation in ovarian carcinoma. Int J Oncol 36:171–179

Dandrea M, Donadelli M, Costanzo C, Scarpa A, Palmieri M (2009) MeCP2/H3meK9 are involved in IL-6 gene silencing in pancreatic adenocarcinoma cell lines. Nucleic Acids Res 37:6681–6690

Datta J, Ghoshal K, Sharma SM, Tajima S, Jacob ST (2003) Biochemical fractionation reveals association of DNA methyltransferase (Dnmt)3b with Dnmt1 and that of Dnmt 3a with a histone H3 methyltransferase and Hdac1. J Cell Biochem 88:855–864

De Capoa A, Musolino A, Della Rosa S, Caiafa P, Mariani L, Del Nonno F, Vocaturo A, Donnorso RP, Niveleau A, Grappelli C (2003) DNA demethylation is directly related to tumor progression: evidence in normal, pre-malignant and malignant cells from uterine cervix samples. Oncol Rep 10:545–549

De Marzo AM, Marchi VL, Yang ES, Veeraswamy R, Lin X, Nelson WG (1999) Abnormal regulation of DNA methyltransferase expression during colorectal carcinogenesis. Cancer Res 59:3855–3860

Di Croce L, Raker VA, Corsaro M, Fazi F, Fanelli M, Faretta M, Fuks F, Lo Coco F, Kouzarides T, Nervi C et al (2002) Methyltransferase recruitment and DNA hypermethylation of target promoters by an oncogenic transcription factor. Science 295:1079–1082

Diede SJ, Guenthoer J, Geng LN, Mahoney SE, Marotta M, Olson JM, Tanaka H, Tapscott SJ (2010) DNA methylation of developmental genes in pediatric medulloblastomas identified by denaturation analysis of methylation differences. Proc Natl Acad Sci USA 107:234–239

Dodge JE, Okano M, Dick F, Tsujimoto N, Chen T, Wang S, Ueda Y, Dyson N, Li E (2005) Inactivation of Dnmt3b in mouse embryonic fibroblasts results in DNA hypomethylation, chromosomal instability, and spontaneous immortalization. J Biol Chem 280:17986–17991

Doherty AS, Bartolomei MS, Schultz RM (2002) Regulation of stage-specific nuclear translocation of Dnmt1o during preimplantation mouse development. Dev Biol 242:255–266

Duncan BK, Miller JH (1980) Mutagenic deamination of cytosine residues in DNA. Nature 287:560–561

Dutrillaux B, Gerbault-Seureau M, Remvikos Y, Zafrani B, Prieur M (1991) Breast cancer genetic evolution: data from cytogenetics and DNA content. Breast Cancer Res Treat 193:245–255

Eads CA, Nickel AE, Laird PW (2002) Complete genetic suppression of polyp formation and reduction of CpG-island hypermethylation in Apc(Min/+) Dnmt1-hypomorphic Mice. Cancer Res 62:1296–1299

Eden A, Gaudet F, Waghmare A, Jaenisch R (2003) Chromosomal instability and tumors promoted by DNA hypomethylation. Science 300:455

Egger G, Liang G, Aparicio A, Jones PA (2004) Epigenetics in human disease and prospects for epigenetic therapy. Nature 429:457–463

Ehrlich M (2009) DNA hypomethylation in cancer cells. Epigenomics 1:239–259

Endo Y, Marusawa H, Chiba T (2011) Involvement of activation-induced cytidine deaminase in the development of colitis-associated colorectal cancers. J Gastroenterol 46(Suppl 1):6–10

Esteller M (2005) Aberrant DNA methylation as a cancer-inducing mechanism. Annu Rev Pharmacol Toxicol 45:629–656

Esteller M (2006) The necessity of a human epigenome project. Carcinogenesis 27:1121–1125

Esteller M (2007) Cancer epigenomics: DNA methylomes and histone-modification maps. Nat Rev Genet 8:286–298

Ezzikouri S, El Feydi AE, Benazzouz M, Afifi R, El Kihal L, Hassar M, Akil A, Pineau P, Benjelloun S (2009) Single nucleotide polymorphism in DNMT3B promoter and its association with hepatocellular carcinoma in a Moroccan population. Infect Genet Evol 9:877–881

Fabris S, Bollati V, Agnelli L, Morabito F, Motta V, Cutrona G, Matis S, Recchia AG, Gigliotti V, Gentile M et al (2011) Biological and clinical relevance of quantitative global methylation of repetitive DNA sequences in chronic lymphocytic leukemia. Epigenetics 6:188–194

Fan T, Yan Q, Huang J, Austin S, Cho E, Ferris D, Muegge K (2003) Lsh-deficient murine embryonal fibroblasts show reduced proliferation with signs of abnormal mitosis. Cancer Res 63:4677–4683

Fazzari MJ, Greally JM (2004) Epigenomics: beyond CpG islands. Nat Rev Genet 5:446–455

Feinberg AP, Tycko B (2004) The history of cancer epigenetics. Nat Rev Cancer 4:143–153

Feinberg AP, Vogelstein B (1983) Hypomethylation distinguishes genes of some human cancers from their normal counterparts. Nature 301:89–92

Feinberg AP, Ghrke CW, Kuo KC, Ehrlich M (1988) Reduced genomic 5-methylcytosine genomic content in human colonic neoplasia. Cancer Res 48:1159–1161

Feinberg AP, Ohlsson R, Henikoff S (2006) The epigenetic progenitor origin of human cancer. Nat Rev Genet 7:21–33

Figueroa ME, Abdel-Wahab O, Lu C, Ward PS, Patel J, Shih A, Li Y, Bhagwat N, Vasanthakumar A, Fernandez HF et al (2010) Leukemic IDH1 and IDH2 mutations result in a hypermethylation phenotype, disrupt TET2 function, and impair hematopoietic differentiation. Cancer Cell 18:553–567

Frigola J, Song J, Stirzaker C, Hinshelwood RA, Peinado MA, Clark SJ (2006) Epigenetic remodeling in colorectal cancer results in coordinate gene suppression across an entire chromosome band. Nat Genet 38:540–549

Gal-Yam EN, Saito Y, Egger G, Jones PA (2008) Cancer epigenetics: modifications, screening, and therapy. Annu Rev Med 59:267–280

Gama-Sosa MA, Slagel VA, Trewyn RW, Oxenhandler R, Kuo KC, Gehrke CW, Ehrlich M (1983) The 5-methylcytosine content of DNA from human tumors. Nucleic Acids Res 11:6883–6894

Garzon R, Liu S, Fabbri M, Liu Z, Heaphy CE, Callegari E, Schwind S, Pang J, Yu J, Muthusamy N, Havelange V, Volinia S, Blum W, Rush LJ, Perrotti D, Andreeff M, Bloomfield CD, Byrd JC, Chan K, Wu LC, Croce CM, Marcucci G (2009) MicroRNA-29b induces global DNA hypomethylation and tumor suppressor gene reexpression in acute myeloid leukemia by targeting directly DNMT3A and 3B and indirectly DNMT1. Blood 113:6411–6418

Gaudet F, Hodgson JG, Eden A, Jackson-Grusby L, Dausman J, Gray JW, Leonhardt H, Jaenisch R (2003) Induction of tumors in mice by genomic hypomethylation. Science 300:489–492

Gibbons RJ, McDowell TL, Raman S, O'Rourke DM, Garrick D, Ayyub H, Higgs DR (2000) Mutations in ATRX, encoding a SWI/SNF-like protein, cause diverse changes in the pattern of DNA methylation. Nat Genet 24:368–371

Girault I, Tozlu S, Lidereau R, Bieche I (2003) Expression analysis of DNA methyltransferases 1, 3A, and 3B in sporadic breast carcinomas. Clin Cancer Res 9:4415–4422

Goelz SE, Vogelstein B, Hamilton SR, Feinberg AP (1985) Hypomethylation of DNA from benign and malignant human colon neoplasm's. Science 228:187–190

Gokul G, Gautami B, Malathi S, Sowjanya AP, Poli UR, Jain M, Ramakrishna G, Khosla S (2007) DNA methylation profile at the DNMT3L promoter: a potential biomarker for cervical cancer. Epigenetics 2:80–85

Gokul G, Ramakrishna G, Khosla S (2009) Reprogramming of HeLa cells upon DNMT3L overexpression mimics carcinogenesis. Epigenetics 4:322–329

Gotze S, Feldhaus V, Traska T, Wolter M, Reifenberger G, Tannapfel A, Kuhnen C, Martin D, Muller O, Sievers S (2009) ECRG4 is a candidate tumor suppressor gene frequently hypermethylated in colorectal carcinoma and glioma. BMC Cancer 9:447

Gowher H, Liebert K, Hermann A, Xu G, Jeltsch A (2005) Mechanism of stimulation of catalytic activity of Dnmt3A and Dnmt3B DNA-(cytosine-C5)-methyltransferases by Dnmt3L. J Biol Chem 280:13341–13348

Grandjean V, Yaman R, Cuzin F, Rassoulzadegan M (2007) Inheritance of an epigenetic mark: the CpG DNA methyltransferase1 is required for de novo establishment of a complex pattern of non-CpG methylation. PLoS One 2:e1136

Greger V, Passarge E, Hopping W, Messmer E, Horsthemke B (1989) Epigenetic changes may contribute to the formation and spontaneous regression of retinoblastoma. Hum Genet 83:155–158

Hirasawa R, Feil R (2010) Genomic imprinting and human disease. Essays Biochem 48:187–200

Hu J, Fan H, Liu D, Zhang S, Zhang F, Xu H (2010) DNMT3B promoter Polymorphism and risk of gastric cancer. Dig Dis Sci 55:1011–1016

Iqbal K, Jin SG, Pfeifer GP, Szabó PE (2011) Reprogramming of the paternal genome upon fertilization involves genome-wide oxidation of 5-methylcytosine. Proc Natl Acad Sci USA 108:3642–3647

Issa JP (2003) Methylation and prognosis: of molecular clocks and hypermethylator phenotypes. Clin Cancer Res 9:2879–2881

Issa JP (2004) CpG island methylator phenotype in cancer. Nat Rev Cancer 4:988–993

Issa JP (2007) DNA methylation as a therapeutic target in cancer. Clin Cancer Res 13:1634–1637

Issa JP, Vertino PM, Wu J, Sazawal S, Celano P, Nelkin BD, Hamilton SR, Baylin SB (1993) Increased cytosine DNA-methyltransferase activity during colon cancer progression. J Natl Cancer Inst 85:1235–1240

Issa JP, Ottaviano YL, Celano P, Hamilton SR, Davidson NE, Baylin SB (1994) Methylation of the oestrogen receptor CpG island links ageing and neoplasia in human colon. Nat Genet 7:536–540

Ito S, D'Alessio AC, Taranova OV, Hong K, Sowers LC, Zhang Y (2010) Role of Tet proteins in 5mC to 5hmC conversion, ES-cell self-renewal and inner cell mass specification. Nature 466:1129–1133

Jacinto FV, Esteller M (2007) Mutator pathways unleashed by epigenetic silencing in human cancer. Mutagenesis 22:247–253

Jair KW, Bachman KE, Suzuki H, Ting AH, Rhee I, Yen RW, Baylin SB, Schuebel KE (2006) De novo CpG island methylation in human cancer cells. Cancer Res 66:682–692

Janzen V, Forkert R, Fleming HE, Saito Y, Waring MT, Dombkowski DM, Cheng T, DePinho RA, Sharpless NE, Scadden DT (2006) Stem cell ageing modified by the cyclin-dependent kinase inhibitor p16(INK4a). Nature 443:421–426

Jelinic P, Shaw P (2007) Loss of imprinting and cancer. J Pathol 211:261–268

Jirtle RL (1999) Genomic imprinting and cancer. Exp Cell Res 248:18–24

Jones PA, Baylin SB (2002) The fundamental role of epigenetic events in cancer. Nat Rev Genet 3:415–428

Jones PA, Gonzalgo ML (1997) Altered DNA methylation and genome instability: a new pathway to cancer? Proc Natl Acad Sci USA 94:2103–2105

Jones PL, Veenstra GJ, Wade PA, Vermaak D, Kass SU, Landsberger N, Strouboulis J, Wolffe AP (1998) Methylated DNA and MeCP2 recruit histone deacetylase to repress transcription. Nat Genet 19:187–191

Juhlin CC, Kiss NB, Villablanca A, Haglund F, Nordenstrom J, Hoog A, Larsson C (2010) Frequent promoter hypermethylation of the APC and RASSF1A tumor suppressors in parathyroid tumors. PLoS One 5:e9472

Juttermann R, Li E, Jaenisch R (1994) Toxicity of 5-aza-2′-deoxycytidine to mammalian cells is mediated primarily by covalent trapping of DNA methyltransferase rather than DNA demethylation. Proc Natl Acad Sci USA 91:11797–11801

Kanai Y (2008) Alterations of DNA methylation and clinicopathological diversity of human cancers. Pathol Int 58:544–558

Kaneda M, Okano M, Hata K, Sado T, Tsujimoto N, Li E, Sasaki H (2004) Essential role for de novo DNA methyltransferase Dnmt3a in paternal and maternal imprinting. Nature 429:900–903

Kareta MS, Botello ZM, Ennis JJ, Chou C, Chedin F (2006) Reconstitution and mechanism of the stimulation of de novo methylation by human *DNMT3L*. J Biol Chem 281:25893–25902

Keshet I, Lieman-Hurwitz J, Cedar H (1986) DNA methylation affects the formation of active chromatin. Cell 44:535–543

Kim GD, Ni J, Kelesoglu N, Roberts RJ, Pradhan S (2002) Co-operation and communication between the human maintenance and de novo DNA (cytosine-5) methyltransferases. EMBO J 21:4183–4195

Kim MJ, White-Cross JA, Shen L, Issa JP, Rashid A (2009) Hypomethylation of long interspersed nuclear element-1 in hepatocellular carcinomas. Mod Pathol 22:442–449

Kiyono T, Foster SA, Koop JI, McDougall JK, Galloway DA, Klingelhutz AJ (1998) Both Rb/p16INK4a inactivation and telomerase activity are required to immortalize human epithelial cells. Nature 396:84–88

Klemm L, Duy C, Iacobucci I, Kuchen S, von Levetzow G, Feldhahn N, Henke N, Li Z, Hoffmann TK, Kim YM, Hofmann WK, Jumaa H, Groffen J, Heisterkamp N, Martinelli G, Lieber MR, Casellas R, Müschen M (2009) The B cell mutator AID promotes B lymphoid blast crisis and drug resistance in chronic myeloid leukemia. Cancer Cell 16:232–245

Klose RJ, Bird AP (2006) Genomic DNA methylation: the mark and its mediators. Trends Biochem Sci 31:89–97

Knudson AG (1971) Mutation and cancer: statistical study of retinoblastoma. Proc Natl Acad Sci USA 68:820–823

Ko M, Huang Y, Jankowska AM, Pape UJ, Tahiliani M, Bandukwala HS, An J, Lamperti ED, Koh KP, Ganetzky R, Liu XS et al (2010) Impaired hydroxylation of 5-methylcytosine in myeloid cancers with mutant TET2. Nature 468:839–843

Kriaucionis S, Heintz N (2009) The nuclear DNA base 5-hydroxymethylcytosine is present in Purkinje neurons and the brain. Science 324:929–930

Krishnamurthy J, Ramsey MR, Ligon KL, Torrice C, Koh A, Bonner-Weir S, Sharpless NE (2006) P16(INK4a) induces an age-independent decline in islet regenerative potential. Nature 443:453–457

Krivtsov AV, Armstrong SA (2007) MLL translocations, histone modifications and leukaemia stem-cell development. Nat Rev Cancer 7:823–833

Kwa FA, Balcerczyk A, Licciardi P, El-Osta A, Karagiannis TC (2011) Chromatin modifying agents – the cutting edge of anticancer therapy. Drug Discov Today 16:543–547

Iacobuzio-Donahue CA, Maitra A, Olsen M, Lowe AW, van Heek NT, Rosty C, Walter K, Sato N, Parker A, Ashfaq R et al (2003) Exploration of global gene expression patterns in pancreatic adenocarcinoma using cDNA microarrays. Am J Pathol 162:1151–1162

Laird PW (2003) The power and the promise of DNA methylation markers. Nat Rev Cancer 3:253–266

Laird PW, Jackson-Grusby L, Fazeli A, Dickinson SL, Jung WE, Li E, Weinberg RA, Jaenisch R (1995) Suppression of intestinal neoplasia by DNA hypomethylation. Cell 81:197–205

Lander ES, Linton LM, Birren B, Nusbaum C, Zody MC, Baldwin J, Devon K, Dewar K, Doyle M, FitzHugh W et al (2001) Initial sequencing and analysis of the human genome. Nature 409:860–921

Langemeijer SM, Jansen JH, Hooijer J, van Hoogen P, Stevens-Linders E, Massop M, Waanders E, van Reijmersdal SV, Stevens-Kroef MJ, Zwaan CM et al (2011) TET2 mutations in childhood leukemia. Leukemia 25:189–192

Larsen F, Gundersen G, Prydz H (1992) Choice of enzymes for mapping based on CpG islands in the human genome. Genet Anal Tech Appl 9:80–85

Lei H, Oh SP, Okano M, Jüttermann R, Goss KA, Jaenisch R, Li E (1996) De novo DNA cytosine methyltransferase activities in mouse embryonic stem cells. Development 122:3195–3205

Li E, Bestor TH, Jaenisch R (1992) Targeted mutation of the DNA methyltransferase gene results in embryonic lethality. Cell 69:915–926

Li S, Chiang TC, Richard-Davis G, Barrett JC, Mclachlan JA (2003) DNA hypomethylation and imbalanced expression of DNA methyltransferases (DNMT1, 3A, and 3B) in human uterine leiomyoma. Gynecol Oncol 90:123–130

Lorsbach RB, Moore J, Mathew S, Raimondi SC, Mukatira ST, Downing JR (2003) TET1, a member of a novel protein family, is fused to MLL in acute myeloid leukemia containing the t(10;11)(q22;q23). Leukemia 17:637–641

Loshikhes IP, Zhang MQ (2000) Large-scale human promoter mapping using CpG islands. Nat Genet 26:61–63

Lu H, Ouyang W, Huang C (2006) Inflammation, a key event in cancer development. Mol Cancer Res 4:221–233

Magdinier F, Wolffe AP (2001) Selective association of the methyl-CpG binding protein MBD2 with the silent p14/p16 locus in human neoplasia. Proc Natl Acad Sci USA 98:4990–4995

Manderwad GP, Gokul G, Kannabiran C, Honavar SG, Khosla S, Vemuganti G (2010) Hypomethylation of the DNMT3L promoter in Ocular Surface Squamous Neoplasia (OSSN). Arch Pathol Lab Med 134:1193–1196

Mani SA, Guo W, Liao MJ, Eaton EN, Ayyanan A, Zhou AY, Brooks M, Reinhard F, Zhang CC, Shipitsin M et al (2008) The epithelial-mesenchymal transition generates cells with properties of stem cells. Cell 133:704–715

Martin V, Jorgensen HF, Chaubert AS, Berger J, Barr H, Shaw P, Bird A, Chaubert P (2008) MBD2-mediated transcriptional repression of the p14ARF tumor suppressor gene in human colon cancer cells. Pathobiology 75:281–287

Marusawa H, Chiba T (2010) Helicobacter pylori-induced activation-induced cytidine deaminase expression and carcinogenesis. Curr Opin Immunol 22:442–447

Matousova M, Votruba I, Otmar M, Tloustova E, Gunterova J, Mertlikova-kaiserova H (2011) 2′-deoxy-5,6-dihydro-5-azacytidine – a less toxic alternative of 2′-deoxy-5-azacytidine: a comparative study of hypomethylating potential. Epigenetics 6:769–776

Matsumoto Y, Marusawa H, Kinoshita K, Endo Y, Kou T, Morisawa T, Azuma T, Okazaki IM, Honjo T, Chiba T (2007) Helicobacter pylori infection triggers aberrant expression of activation-induced cytidine deaminase in gastric epithelium. Nat Med 13:470–476

Matsumoto Y, Marusawa H, Kinoshita K, Niwa Y, Sakai Y, Chiba T (2010) Up-regulation of activation-induced cytidine deaminase causes genetic aberrations at the CDKN2b-CDKN2a in gastric cancer. Gastroenterology 139:1984–1994

Mayer W, Niveleau A, Walter J, Fundele R, Haaf T (2000) Demethylation of the zygotic paternal genome. Nature 403:501–502

Melki JR, Warnecke P, Vincent PC, Clark SJ (1998) Increased DNA methyltransferase expression in leukaemia. Leukemia 12:311–316

Minami K, Chano T, Kawakami T, Ushida H, Kushima R, Okabe H, Okada Y, Okamoto K (2010) DNMT3L is a novel marker and is essential for the growth of human embryonal carcinoma. Clin Cancer Res 16:2751–2759

Miremadi A, Oestergaard MZ, Pharoah PD, Caldas C (2007) Cancer genetics of epigenetic genes. Hum Mol Genet 16(Spec No 1):R28–R49

Missaoui N, Hmissa S, Dante R, Frappart L (2010) Global DNA methylation in precancerous and cancerous lesions of the uterine cervix. Asian Pac J Cancer Prev 11:1741–1744

Molofsky AV, Slutsky SG, Joseph NM, He S, Pardal R, Krishnamurthy J, Sharpless NE, Morrison SJ (2006) Increasing p16(INK4a) expression decreases forebrain progenitors and neurogenesis during ageing. Nature 443:448–452

Momparler RL, Ayoub J (2001) Potential of 5-aza-2′-deoxycytidine (Decitabine) a potent inhibitor of DNA methylation for therapy of advanced non-small cell lung cancer. Lung Cancer 34(Suppl 4):S111–S1115

Momparler RL, Bovenzi V (2000) DNA methylation and cancer. J Cell Physiol 183:145–154

Momparler RL, Bouchard J, Samson J (1985) Induction of differentiation and inhibition of DNA methylation in HL-60 myeloid leukemic cells by 5-AZA-2′-deoxycytidine. Leuk Res 9:1361–1366

Momparler RL, Rossi M, Bouchard J, Bartolucci S, Momparler LF, Raia CA, Nucci R, Vaccaro C, Sepe S (1986) 5-Aza-2′-deoxycytidine synergistic action with thymidine on leukemic cells and interaction of 5-AZA-dCMP with dCMP deaminase. Adv Exp Med Biol 195 Pt B:157–163

Morgan HD, Dean W, Coker HA, Reik W, Petersen-Mahrt SK (2004) Activation-induced cytidine deaminase deaminates 5-methylcytosine in DNA and is expressed in pluripotent tissues: implications for epigenetic reprogramming. J Biol Chem 279:52353–52360

Muller HM, Fiegl H, Goebel G, Hubalek MM, Widschwendter A, Muller-Holzner E, Marth C, Widschwendter M (2003) MeCP2 and MBD2 expression in human neoplastic and non-neoplastic breast tissue and its association with oestrogen receptor status. Br J Cancer 89:1934–1939

Murchie AI, Lilley DM (1989) Base methylation and local DNA helix stability. Effect on the kinetics of cruciform extrusion. J Mol Biol 205:593–602

Nakamura N, Takenaga K (1998) Hypomethylation of the metastasis-associated *S100A4* gene correlates with gene activation in human colon adenocarcinoma cell lines. Clin Exp Metastasis 16:471–479

Nan X, Ng HH, Johnson CA, Laherty CD, Turner BM, Eisenman RN, Bird A (1998) Transcriptional repression by the methyl-CG-binding protein MeCP2 involves a histone deacetylase complex. Nature 393:386–389

Nelson WG, De Marzo AM, DeWeese TL, Isaacs WB (2004) The role of inflammation in the pathogenesis of prostate cancer. J Urol 172:S6–S11

Neumeister P, Albanese C, Balent B, Greally J, Pestell RG (2002) Senescence and epigenetic dysregulation in cancer. Int J Biochem Cell Biol 34:1475–1490

Ng HH, Zhang Y, Hendrich B, Johnson CA, Turner BM, Erdjument-Bromage H, Tempst P, Reinberg D, Bird A (1999) MBD2 is a transcriptional repressor belonging to the MeCP1 histone deacetylase complex. Nat Genet 23:58–61

Nguyen CT, Weisenberger DJ, Velicescu M, Gonzales FA, Lin JC, Liang G, Jones PA (2002) Histone H3-lysine 9 methylation is associated with aberrant gene silencing in cancer cells and is rapidly reversed by 5-aza-2'-deoxycytidine. Cancer Res 62:6456–6461

Nuovo GJ, Plaia TW, Belinsky SA, Baylin SB, Herman JG (1999) In situ detection of the hypermethylation-induced inactivation of the p16 gene as an early event in oncogenesis. Proc Natl Acad Sci USA 96:12754–12759

O'Brien CA, Pollett A, Gallinger S, Dick JE (2007) A human colon cancer cell capable of initiating tumor growth in immunodeficient mice. Nature 445:106–110

Ogino S, Nosho K, Kirkner GJ, Kawasaki T, Chan AT, Schernhammer ES, Giovannucci EL, Fuchs CS (2008) A cohort study of tumoral LINE-1 hypomethylation and prognosis in colon cancer. J Natl Cancer Inst 100:1734–1738

Ohm JE, Baylin SB (2007) Stem cell chromatin patterns: an instructive mechanism for DNA hypermethylation? Cell Cycle 6:1040–1043

Okano M, Bell DW, Haber DA, Li E (1999) DNA methyltransferases Dnmt3a and Dnmt3b are essential for de novo methylation and mammalian development. Cell 99:247–257

Ono R, Taki T, Taketani T, Taniwaki M, Kobayashi H, Hayashi Y (2002) LCX, leukemia-associated protein with a CXXC domain, is fused to MLL in acute myeloid leukemia with trilineage dysplasia having t(10;11)(q22;q23). Cancer Res 62:4075–4080

Ooi SK, Bestor TH (2008) The colorful history of active DNA demethylation. Cell 133:1145–1148

Ooi SK, Qui C, Bernstein E, Li K, Jia D, Yang Z, Erdjument-Bromage H, Tempst P, Lin SP, Allis CD, Cheng X, Bestor TH (2007) *DNMT3L* connects unmethylated lysine 4 of histone H3 to de novo methylation of DNA. Nature 448:714–717

Ooi SK, Wolf D, Hartung O, Agarwal S, Daley GQ, Goff SP, Bestor TH (2010) Dynamic instability of genomic methylation patterns in pluripotent stem cells. Epigenet Chromatin 3:17

Oshimo Y, Nakayama H, Ito R, Kitadai Y, Yoshida K, Chayama K, Yasui W (2003) Promoter methylation of cyclin D2 gene in gastric carcinoma. Int J Oncol 6:1663–1670

Pancione M, Sabatino L, Fucci A, Carafa V, Nebbioso A, Forte N, Febbraro A, Parente D, Ambrosino C, Normanno N, Altucci L, Colantuoni V (2010) Epigenetic silencing of peroxisome proliferator-activated receptor γ is a biomarker for colorectal cancer progression and adverse patients' outcome. PLoS One 5:e14229

Pinto A, Zagonel V (1993) 5-Aza-2'-deoxycytidine (Decitabine) and 5-azacytidine in the treatment of acute myeloid leukemias and myelodysplastic syndromes: past, present and future trends. Leukemia 7(Suppl 1):51–60

Pinto A, Attadia V, Fusco A, Ferrara F, Spada OA, Di Fiore PP (1984) 5-Aza-2'-deoxycytidine induces terminal differentiation of leukemic blasts from patients with acute myeloid leukemias. Blood 64:922–929

Pitot HC (1986) Oncogenes and human neoplasia. Clin Lab Med 6:167–179

Piyathilake CJ, Henao O, Frost AR, Macaluso M, Bell WC, Johanning GL, Heimburger DC, Niveleau A, Grizzle WE (2003) Race- and age-dependent alterations in global methylation of DNA in squamous cell carcinoma of the lung (United States). Cancer Causes Control 14:37–42

Ponger L, Duret L, Mouchiroud D (2001) Determinants of CpG islands: expression in early embryo and isochore structure. Genome Res 11:1854–1860

Popp C, Dean W, Feng S, Cokus SJ, Andrews S, Pellegrini M, Jacobsen SE, Reik W (2010) Genome-wide erasure of DNA methylation in mouse primordial germ cells is affected by AID deficiency. Nature 463:1101–1105

Prokhortchouk A, Hendrich B, Jørgensen H, Ruzov A, Wilm M, Georgiev G, Bird A, Prokhortchouk E (2001) The p120 catenin partner Kaiso is a DNA methylation-dependent transcriptional repressor. Genes Dev 15:1613–1618

Prokhortchouk A, Sansom O, Selfridge J, Caballero IM, Salozhin S, Aithozhina D, Cerchietti L, Meng FG, Augenlicht LH, Mariadason JM et al (2006) Kaiso-deficient mice show resistance to intestinal cancer. Mol Cell Biol 26:199–208

Qu GZ, Grundy PE, Narayan A, Ehrlich M (1999) Frequent hypomethylation in Wilms tumors of pericentromeric DNA in chromosomes 1 and 16. Cancer Genet Cytogenet 109:34–39

Qu Y, Mu G, Wu Y, Dai X, Zhou F, Xu X, Wang Y, Wei F (2010) Overexpression of DNA methyltransferases 1, 3a, and 3b significantly correlates with retinoblastoma tumorigenesis. Am J Clin Pathol 134:826–834

Radpour R, Barekati Z, Kohler C, Lv Q, Burki N, Diesch C, Bitzer J, Zheng H, Schmid S, Zhong XY (2011) Hypermethylation of tumor suppressor genes involved in critical regulatory pathways for developing a blood-based test in breast cancer. PLoS One 6:e16080

Rai K, Huggins IJ, James SR, Karpf AR, Jones DA, Cairns BR (2008) DNA demethylation in zebrafish involves the coupling of a deaminase, a glycosylase, and gadd45. Cell 135:1201–1212

Ramsahoye BH, Biniszkiewicz D, Lyko F, Clark V, Bird AP, Jaenisch R (2000) Non-CpG methylation is prevalent in embryonic stem cells and may be mediated by DNA methyltransferase 3a. Proc Natl Acad Sci USA 97:5237–5242

Ren R (2005) Mechanisms of BCR-ABL in the pathogenesis of chronic myelogenous leukaemia. Nat Rev Cancer 5:172–183

Revy P, Muto T, Levy Y, Geissmann F, Plebani A, Sanal O, Catalan N, Forveille M, Dufourcq-Labelouse R, Gennery A et al (2000) Activation-induced cytidine deaminase (AID) deficiency causes the autosomal recessive form of the Hyper-IgM syndrome (HIGM2). Cell 102:565–575

Reynolds PA, Sigaroudinia M, Zardo G, Wilson MB, Benton GM, Miller CJ, Hong C, Fridlyand J, Costello JF, Tlsty TD (2006) Tumor suppressor P16INK4A regulates polycomb-mediated DNA hypermethylation in human mammary epithelial cells. J Biol Chem 281:24790–24802

Rhee I, Jair KW, Yen RW, Lengauer C, Herman JG, Kinzler KW, Vogelstein B, Baylin SB, Schuebel KE (2000) CpG methylation is maintained in human cancer cells lacking DNMT1. Nature 404:1003–1007

Rhee I, Bachman KE, Park BH, Jair KW, Yen RW, Schuebel KE, Cui H, Feinberg AP, Lengauer C, Kinzler KW, Baylin SB, Vogelstein B (2002) DNMT1 and DNMT3b cooperate to silence genes in human cancer cells. Nature 416:552–556

Riggs AD, Jones PA (1983) 5-methylcytosine, gene regulation, and cancer. Adv Cancer Res 40:1–30

Rishi V, Bhattacharya P, Chatterjee R, Rozenberg J, Zhao J, Glass K, Fitzgerald P, Vinson C (2010) CpG methylation of half-CRE sequences creates C/EBP alpha binding sites that activate some tissue-specific genes. Proc Natl Acad Sci USA 107:20311–20316

Rivard GE, Momparler RL, Demers J, Benoit P, Raymond R, Lin K, Momparler LF (1981) Phase I study on 5-aza-2′-deoxycytidine in children with acute leukemia. Leuk Res 5:453–462

Robbiani DF, Bunting S, Feldhahn N, Bothmer A, Camps J, Deroubaix S, McBride KM, Klein IA, Stone G, Eisenreich TR et al (2009) AID produces DNA double-strand breaks in non-Ig genes and mature B cell lymphomas with reciprocal chromosome translocations. Mol Cell 36:631–641

Robertson KD, Uzvolgyi E, Liang G, Talmadge C, Sumegi J, Gonzales FA, Jones PA (1999) The human DNA methyltransferase (DNMTs) 1, 3a and 3b: coordinate mRNA expression in normal tissues and overexpression in tumors. Nucleic Acids Res 27:2291–2298

Saito Y, Kanai Y, Sakamoto M, Saito H, Ishii H, Hirohashi S (2002) Overexpression of a splice variant of DNA methyltransferase 3b, DNMT3b4, associated with DNA hypomethylation on pericentromeric satellite regions during human hepatocarcinogenesis. Proc Natl Acad Sci USA 99:10060–10065

Sansom OJ, Berger J, Bishop SM, Hendrich B, Bird A, Clarke AR (2003) Deficiency of Mbd2 suppresses intestinal tumorigenesis. Nat Genet 34:145–147

Santi DV, Norment A, Garrett CE (1984) Covalent bond formation between a DNA-cytosine methyltransferase and DNA containing 5-azacytosine. Proc Natl Acad Sci USA 81:6993–6997

Sarraf SA, Stancheva I (2004) Methyl-CpG binding protein MBD1 couples histone H3 methylation at lysine 9 by SETDB1 to DNA replication and chromatin assembly. Mol Cell 15:595–605

Sato N, Maitra A, Fukushima N, van Heek NT, Matsubayashi H, Iacobuzio-Donahue CA, Rosty C, Goggins M (2003) Frequent hyomethylation of multiple genes overexpressed in pancreatic ductal adenocarcinoma. Cancer Res 63:4158–4166

Schernhammer ES, Giovannucci E, Kawasaki T, Rosner B, Fuchs CS, Ogino S (2010) Dietary folate, alcohol and B vitamins in relation to LINE-1 hypomethylation in colon cancer. Gut 59:794–799

Shih LY, Liang DC, Huang CF, Wu JH, Lin TL, Wang PN, Dunn P, Kuo MC, Tang TC (2006) AML patients with CEBPalpha mutations mostly retain identical mutant patterns but frequently change in allelic distribution at relapse: a comparative analysis on paired diagnosis and relapse samples. Leukemia 20:604–609

Shukla V, Coumoul X, Lahusen T, Wang RH, Xu X, Vassilopoulos A, Xiao C, Lee MH, Man YG, Ouchi M, Ouchi T, Deng CX (2010) BRCA1 affects global DNA methylation through regulation of DNMT1. Cell Res 20:1201–1215

Singer J, Roberts-Ems J, Riggs AD (1979) Methylation of mouse liver DNA studied by means of the restriction enzymes MspI and HpaII. Science 203:1019–1021

Sinsheimer RL (1955) The action of pancreatic deoxyribonuclease II isomeric dinucleotides. J Biol Chem 215:579–583

Sorm F, Piskala A, Cihak A, Vesely J (1964) 5-Azacytidine, a new, highly effective cancerostatic. Experientia 20:202–203

Strichman-Almashanu LZ, Lee RS, Onyango PO, Perlman E, Flam F, Frieman MB, Feinberg AP (2002) A genome wide screen for normally methylated human CpG islands that can identify normal imprinted genes. Genome Res 12:543–554

Suetake I, Shinozaki F, Miyagawa J, Takeshima H, Tajima S (2004) *DNMT3L* stimulates the DNA methylation activity of *Dnmt3a* and *DNMT3B* through a direct interaction. J Biol Chem 279:27816–27823

Suter CM, Martin DI, Ward RL (2004) Hypomethylation of L1 retrotransposons in colorectal cancer and adjacent normal tissue. Int J Colorectal Dis 19:95–101

Suzuki H, Watkins DN, Jair KW, Schuebel KE, Markowitz SD, Chen WD, Pretlow TP, Yang B, Akiyama Y, Van Engeland M, Toyota M, Tokino T, Hinoda Y, Imai K, Herman JG, Baylin SB (2004) Epigenetic inactivation of SFRP genes allows constitutive WNT signaling in colorectal cancer. Nat Genet 36:417–422

Tahiliani M, Koh KP, Shen Y, Pastor WA, Bandukwala H, Brudno Y, Agarwal S, Iyer LM, Liu DR, Aravind L, Rao A (2009) Conversion of 5-methylcytosine to 5-hydroxymethylcytosine in mammalian DNA by MLL partner TET1. Science 324:930–935

Takai D, Jones PA (2003) The CpG island searcher: a new WWW resource. In Silico Biol 3:235–240

Tanaka K, Appella E, Jay G (1983) Developmental activation of the H2K gene is correlated with an increase in DNA methylation. Cell 35:457–465

Ting AH, Jair KW, Suzuki H, Yen RW, Baylin SB, Schuebel KE (2004) CpG island hypermethylation is maintained in human colorectal cancer cells after RNAi-mediated depletion of DNMT1. Nat Genet 36:582–584

Ting AH, Jair KW, Schuebel KE, Baylin SB (2006a) Differential requirement for DNA methyltransferase 1 in maintaining human cancer cell gene promoter hypermethylation. Cancer Res 66:729–735

Ting AH, McGarvey KM, Baylin SB (2006b) The cancer epigenome-components and functional correlates. Genes Dev 20:3215–3231

Tomita H, Hirata A, Yamada Y, Hata K, Oyama T, Mori H, Yamashita S, Ushijima T, Hara A (2010) Suppressive effect of global DNA hypomethylation on gastric carcinogenesis. Carcinogenesis 31:1627–1633

Touati E (2010) When bacteria become mutagenic and carcinogenic: lessons from H. pylori. Mutat Res 703:66–70

Toyota M, Ahuja N, Ohe-Toyota M, Herman JG, Baylin SB, Issa JP (1999) CpG island methylator phenotype in colorectal cancer. Proc Natl Acad Sci USA 96:8681–8686

Tuck-Muller CM, Narayan A, Tsien F, Smeets DF, Sawyer J, Fiala ES, Sohn OS, Ehrlich M (2000) DNA hypomethylation and unusual chromosome instability in cell lines from ICF syndrome patients. Cytogenet Cell Genet 89:121–128

Tulchinsky EM, Georgiev GP, Lukanidin EM (1996) Novel AP-1 binding site created by DNA-methylation. Oncogene 12:1737–1745

Venter JC, Adams MD, Myers EW, Li PW, Mural RJ, Sutton GG, Smith HO, Yandell M, Evans CA, Holt RA et al (2001) The sequence of the human genome. Science 291:1304–1351

Vertino PM, Yen RW, Gao J, Baylin SB (1996) De novo methylation of CpG island sequences in human fibroblasts overexpressing DNA (cytosine-5-)-methyltransferase. Mol Cell Biol 16:4555–4565

Wada R, Akiyama Y, Hashimoto Y, Fukamachi H, Yuasa Y (2010) miR-212 is downregulated and suppresses methyl-CpG-binding protein MeCP2 in human gastric cancer. Int J Cancer 127:1106–1114

Wade PA, Gegonne A, Jones PL, Ballestar E, Aubry F, Wolffe AP (1999) Mi-2 complex couples DNA methylation to chromatin remodeling and histone deacetylation. Nat Genet 23:62–66

Walker BA, Wardell CP, Chiecchio L, Smith EM, Boyd KD, Neri A, Davies FE, Ross FM, Morgan GJ (2011) Aberrant global methylation patterns affect the molecular pathogenesis and prognosis of multiple myeloma. Blood 117:553–562

Webster KE, O'Bryan MK, Fletcher S, Crewther PE, Aapola U, Craig J, Harrison DK, Anug H, Phutikanit N, Lyle R, Meachem SJ, Antonarakis SE, de Kretser DM, Hedger MP, Peterson P, Carroll BJ, Scott HS (2005) Meiotic and epigenetic defects in *DNMT3L*-knockout mouse spermatogenesis. Proc Natl Acad Sci USA 102:4068–4073

Wilhelm CS, Kelsey KT, Butler R, Plaza S, Gagne L, Zens MS, Andrew AS, Morris S, Nelson HH, Schned AR, Karagas MR, Marsit CJ (2010) Implications of LINE1 methylation for bladder cancer risk in women. Clin Cancer Res 16:1682–1689

Wu SC, Zhang Y (2010) Active DNA demethylation: many roads lead to Rome. Nat Rev Mol Cell Biol 11:607–620

Wu J, Issa JP, Herman J, Bassett DE Jr, Nelkin BD, Baylin SB (1993) Expression of an exogenous eukaryotic DNA methyltransferase gene induces transformation of NIH 3T3 cells. Proc Natl Acad Sci USA 90:8891–8895

Xiong Y, Dowdy SC, Xue A, Shujuan J, Eberhardt NL, Podratz KC, Jiang SW (2005) Opposite alterations of DNA methyltransferase gene expression in endometrioid and serous endometrial cancers. Gynecol Oncol 96:601–609

Yoder JA, Walsh CP, Bestor TH (1997) Cytosine methylation and the ecology of intragenomic parasites. Trends Genet 13:335–340

Yoon HF, Chan DW, Reynolds AB, Qin J, Wong J (2003) N-CoR mediates DNA methylation-dependent repression through a methyl CpG binding protein Kaiso. Mol Cell 12:723–734

Yoshida T, Yamashita S, Takamura-Enya T, Niwa T, Ando T, Enomoto S, Maekita T, Nakazawa K, Tatematsu M, Ichinose M, Ushijima T (2011) Alu and Satα hypomethylation in Helicobacter pylori-infected gastric mucosae. Int J Cancer 128:33–39

Yu JC, Shen CY (2002) Two-hit hypothesis of tumor suppressor gene and Revisions. J Med Sci 22:13–18

Yu S, Khor TO, Cheung KL, Li W, Wu TY, Huang Y, Foster BA, Kan YW, Kong AN (2010) Nrf2 expression is regulated by epigenetic mechanisms in prostate cancer of TRAMP mice. PLoS One 5:e8579

Zacharias W, Jaworski A, Wells RD (1990) Cytosine methylation enhances Z-DNA formation *in vivo*. J Bacteriol 172:3278–3283

Zhang Y, Ng HH, Erdjument-Bromage H, Tempst P, Bird A, Reinberg D (1999) Analysis of the NuRD subunits reveals a histone deacetylase core complex and a connection with DNA methylation. Genes Dev 13:1924–1935

Zingg JM, Jones PA (1997) Genetic and epigenetic aspects of DNA methylation on genome expression, evolution, mutation and carcinogenesis. Carcinogenesis 18:869–882

Chapter 27
Role of Epigenetics in Inflammation-Associated Diseases

Muthu K. Shanmugam and Gautam Sethi

Abstract There is considerable evidence suggesting that epigenetic mechanisms may mediate development of chronic inflammation by modulating the expression of pro-inflammatory cytokine TNF-α, interleukins, tumor suppressor genes, oncogenes and autocrine and paracrine activation of the transcription factor NF-κB. These molecules are constitutively produced by a variety of cells under chronic inflammatory conditions, which in turn leads to the development of major diseases such as autoimmune disorders, chronic obstructive pulmonary diseases, neurodegenerative diseases and cancer. Distinct or global changes in the epigenetic landscape are hallmarks of chronic inflammation driven diseases. Epigenetics include changes to distinct markers on the genome and associated cellular transcriptional machinery that are copied during cell division (mitosis and meiosis). These changes appear for a short span of time and they necessarily do not make permanent changes to the primary DNA sequence itself. However, the most frequently observed epigenetic changes include aberrant DNA methylation, and histone acetylation and deacetylation. In this chapter, we focus on pro-inflammatory molecules that are regulated by enzymes involved in epigenetic modifications such as arginine and lysine methyl transferases, DNA methyltransferase, histone acetyltransferases and histone deacetylases and their role in inflammation driven diseases. Agents that modulate or inhibit these epigenetic modifications, such as HAT or HDAC inhibitors have shown great potential in inhibiting the progression of these diseases. Given the plasticity of these epigenetic changes and their readiness to respond to intervention by small molecule inhibitors, there is a tremendous potential for the development of novel therapeutics that will serve as direct or adjuvant therapeutic compounds in the treatment of these diseases.

M.K. Shanmugam • G. Sethi (✉)
Department of Pharmacology, Yong Loo Lin School of Medicine,
National University of Singapore, Singapore 117597, Singapore
e-mail: phcgs@nus.edu.sg

27.1 Introduction

The term epigenetics was first coined by developmental biologist Conrad H. Waddington in 1942, is defined as heritable changes in gene expression patterns that occur without any changes in the primary DNA sequence but changes that are sufficient to regulate the pattern of gene expression. Recently, this definition has been elaborated to include transient changes in gene expression (Hirst and Marra 2009). Epigenetic regulation plays a pivotal role in development and differentiation though all cells of an organism have the same DNA sequence, they can differentiate into a multitude of diverse cell types (Feinberg 2007). It is now becoming increasingly apparent that the epigenetic changes together with genetic abnormality drives chronic inflammatory diseases that ultimately lead to tumor progression, and that cancer is the manifestation of both genetic and epigenetic modifications (Egger et al. 2004; Esteller 2008). The word inflammation is derived from the Latin word "inflamacio" denoting to 'set a fire'. The classical signs of inflammation are redness (rubor), swelling (tumor), heat (calor) and pain (dalor). The first four classical signs were described by the Roman physician Cornelius Celsus (ca. 30 BC–38 AD), while loss of function was added by a German physician Rudolf Virchow in 1870. Virchow postulated that microinflammation that results from irritation leads to the development of most chronic diseases including cancer. It was Rudolf Virchow who first linked inflammation with cancer, atherosclerosis, diabetes, arthritis, obesity, allergy, and other chronic diseases (Heidland et al. 2006). In naming the pathological conditions, inflammation is usually indicated by adding the suffix '-itis', such as bronchitis, esophagitis, gastritis, colitis, pancreatitis, prostitis, cervicitis, and hepatitis. Some conditions such as asthma and pneumonia, although inflammatory conditions, do not follow this convention (Aggarwal and Gehlot 2009). Since Virchow proposed a connection between inflammation and cancer, it has been estimated that approximately 15% of all cancers are linked to inflammation, including chronic inflammatory bowel disease and colorectal cancer (Balkwill and Mantovani 2001), associations between cervical cancer and the human papillomavirus (HPV); liver cancer and hepatitis B or C virus (Saigo et al. 2008); Barrett's esophagus, and esophageal cancer; and that inflammation alone will not cause cancer; mutations and epigenetic events from environmental exposures or alterations in immunity are also key contributors in the cancer process (Schottenfeld and Beebe-Dimmer 2006). Several proinflammatory mediators, such as cytokines, chemokines, prostaglandins (PGs), nitric oxide (NO) and leukotrienes disrupt normal signaling cascades within cells which contributes to the development of neoplasms (Surh et al. 2005). The primary defense response initiated by the human body is the activation of immune system upon injury via recruitment of circulating cells such as macrophages, monocytes, lymphocytes, neutrophils and leukocytes. The activated cells thus act on the injured site and reduce the inflammation in an acute attack. In cases of severe inflammation, these cells are in turn activated, leading to excessive production of proinflammatory molecules. This phenomenon leads to the process of chronic inflammation which has been implicated in the development of various major human diseases including malignancies (Aggarwal et al. 2006). The importance of the epigenome in the pathogenesis of common human diseases is likely to be almost as significant as that of traditional genetic mutations (Portela and Esteller 2010). The enzymes that are involved in these epigenetic changes are also being currently used as targets for drug development (Eliseeva et al. 2007; Selvi et al. 2010b; Wagner et al. 2010).

Eukaryotic DNA is wrapped around an octamer of the core histones H2A, H2B, H3, and H4, thus building the fundamental unit of chromatin, the nucleosome (Berger 2007). Posttranslational modifications of the histone tails and cytosine methylation of the DNA determine the accessibility of the chromatin and, therefore, the ability of transcription factors to bind and initiate gene expression (Berger 2007). These chromatin marks are unstable and they rapidly change in response to any external stimulus (Berger 2007; Bird 2007) and any permanent changes to the DNA can lead to development of defective organs and in the development of diseases. This primarily happens at specific dinucleotide sites along the genome, i.e., cytosines 5' of guanines, or at CpG sites. In fact, 40% of genes contain CpG-rich islands upstream from their transcriptional start site, and up to 70–80% of all CpG dinucleotides in the genome are methylated (Bird 2002; Wilson 2008). Aberrant DNA methylation has been observed with inflammation, viral infection, in cancer and DNA methylation pattern has been shown to be molecular marker for diagnosis of early cancer (Esteller 2008). It is also related to the effectiveness of therapeutic agents affecting DNA methylation and histone deacetylation. Epigenetic modifications of chromatin and DNA have been recognized as important accommodating and expressive or suppressive factors in controlling the expression of gene transcription (Wilson 2008). Two major epigenetic mechanisms are the posttranslational modification of histone proteins in chromatin and the methylation of DNA itself, which are regulated by distinct pathways (Wilson 2008). The epigenome is influenced by environmental factors throughout life. Nutritional factors can have profound effects on the expression of specific genes by epigenetic modification. Despite global hypomethylation, distinct CpG islands located in regulatory regions of genes can be specifically hypermethylated, leading to a repressed transcription of the associated genes (Das and Singal 2004). Differences in methylation status of CpG sites, monoallelic silencing and other epigenetic regulatory mechanisms have been constantly observed in key inflammatory response genes (Portela and Esteller 2010; Wilson 2008).

Recent literature review strongly supports the hypothesis that a number of critical genes such as those involved in cell cycle control CDKN2 (Otterson et al. 1995), apoptosis DAPK1 (Katzenellenbogen et al. 1999), DNA repair BRCA-1 (Dobrovic and Simpfendorfer 1997), metastasis TIMP-3 (Kang et al. 2000), drug resistance, differentiation, and angiogenesis are silenced by epigenetic modification processes, such as histone hypermethylation. Thus, methylation of CpG dinculeotides is a common biological mechanism for switching off gene expression, and more interestingly, this is not only limited to cancer. While DNA mutations have been closely linked with chronic inflammation driven cancers in certain individuals, evidence also suggests that epigenetic alterations can also contribute to familial cancer risk (Fleming et al. 2008). Autoimmune and neoplastic diseases increase in frequency with increasing age, with epigenetic dysregulation proposed as a potential explanation. Chronic inflammation has long been recognized as instrumental to cancer progression (Brigati et al. 2002; Coussens and Werb 2002), and recent data linking inflammation to initiation of variety of diseases can be explained on the basis of epigenetics dysregulation (Mantovani et al. 2008; Rosin et al. 1994). Another recent study has clearly indicated epigenetic interplay between histone modifications and DNA methylation in gene silencing (Vaissiere et al. 2008) and the selective silencing of the DNA methyltransferase B has been correlated to the acute fatty-acid-induced non-CpG methylation of proliferator-activated receptor gamma (PPAR-gamma) coactivator-1 alpha promoter (Barres et al. 2009). Such modifications could restrain the number of targets that inhibit progression of the process of acute inflammation after an inflammatory insult. Thus, the focus of the current chapter is to highlight the evidences that epigenetic modulation may play a critical role in regulation of various inflammatory mediators and in the development of various chronic inflammation driven diseases.

27.2 Epigenetic Modifications on Tumor Necrosis Factor-α gene

Tumor necrosis factor (TNF-α) was first isolated as an anticancer cytokine more than two decades ago, however, its major role appears to be in early inflammatory responses and primarily mediates acute phase response after an inflammatory insult (Aggarwal 2003; Balkwill 2009; Sethi et al. 2008). TNF-α is secreted by a variety of cells that infiltrate the site of injury such as neutrophils, monocytes, macrophages, lymphocytes and leukocytes upon stimulation of Toll like receptors by lipopolysaccharide (Foster et al. 2007), or activation via cytokines and lipid mediators (Hayes et al. 1995; Lee and Sullivan 2001; Lin et al. 2004; Takeda and Akira 2004). Promoter hypomethylation of the Toll-like receptor 2 (TLR2) gene in bronchial epithelial cells leads to pro-inflammatory response when stimulated with bacterial peptidoglycan (Shuto et al. 2006) and in intestinal epithelial cells epigenetic modifications regulate TLR4 (Takahashi et al. 2009). DNA methylation and histone acetylation play an important role in establishing epigenetic modification across *TNF-α* locus (Sullivan et al. 2007). However, when dysregulated and secreted in the circulation, TNF-α can mediate the development of a wide variety of diseases, including cancer (Aggarwal 2003; Balkwill 2009; Szlosarek et al. 2006). Epigenetic modifications of the *TNF-α* locus occur both developmentally and in response to acute stimulation and, importantly they actively regulate expression by increase in DNA methylation (Kruys et al. 1993). The *TNF-α* locus migrates from heterochromatin to euchromatin in a progressive fashion, reaching euchromatin slightly later in differentiation. Finally, histone modifications characteristic of a transcriptionally competent gene occur with myeloid differentiation and progress with differentiation (Sullivan et al. 2007). Additional histone modifications characteristic of active gene expression are acquired with stimulation. In each case, manipulation of these epigenetic variables altered the ability of the cell to express TNF-α. MLL, a histone methyltransferase has been shown to bind to *TNF-α* promoter and ATF2 and CBP, both histone acetylators have been implicated in the transcriptional regulation of *TNF-α* and H3 lysine 4 methylation appears to be important for transcription (Hess 2004) (Dou et al. 2006; Guenther et al.

2005; Hayakawa et al. 2004; Kang et al. 2007; Pogribny et al. 2007; Tsai et al. 2000). However, the pathways responsible for activation of these molecules differ from cell type to cell type. Majority of the studies are done on myeloid cells as they produce the bulk of TNF-α. Maturation of monocytic cells and high glucose environment have been shown to increase histone acetylation at the *TNF-α* locus and in turn leads to increased TNF-α expression (Lee et al. 2003; Miao et al. 2004). Inter-individual variation in TNF-α production with upto 40-fold differences has been explained by polymorphisms in the *TNF-α* promoter region (Wilson et al. 1997). For example, *TNF-α*-308A allele was associated with more severe outcomes in various infectious diseases (McGuire et al. 1994). In another set of experiments, the methylation status of the *TNF-α* promoter (−310 to +30) was examined in LPS-stimulated human macrophages. This region of the *TNF-α* promoter contains 12 CpG dinucleotides. Lower methylation at two specific CpG sites (−304 and −245) was correlated with high production of TNF-α mRNA, suggesting that the methylation state of the *TNF-α* promoter may be an important factor in driving the level of *TNF-α* gene expression, and it may help to explain the origins in variation in the inflammatory response between individuals (McGuire et al. 1994). In addition, *TNF-α* gene in murine monocytic cell line exhibit significant increase in histone acetylation (Ramirez-Carrozzi et al. 2006) and studies on other cytokines have indicated that tissue specific expression may be controlled epigenetically and that histone modification represents a common regulatory strategy for cytokines (Lee et al. 2006). In contrast, inhibition of histone deacetylases causes increase in global histone acetylation increased the ability of the cells to secrete TNF-α (Ramirez-Carrozzi et al. 2006). These findings may have relevance for inflammatory disorders in which TNF-α is overproduced. The proinflammatory role of TNF-α has been linked to all steps involved in tumorigenesis, including cellular transformation, survival, proliferation, invasion, angiogenesis, and metastasis (Sethi et al. 2008). TNF-α overexpression has been observed in variety of inflammation driven diseases such as a rheumatoid arthritis, Crohn's disease, ulcerative colitis and asthma. Recently, Set7/9 was shown to play an important role in the manifestation of inflammatory disorders where TNF-α is over-expressed (Li et al. 2008). These studies demonstrate the importance of epigenetic regulation in the control of TNF-α expression. Defining the complete role of epigenetic regulation of TNF-α may lead to novel therapeutic strategies for the treatment of various chronic inflammatory diseases.

27.3 Epigenetic Changes in Interleukin genes

Histone acetylation by the lysine acetyltransferases (KATs) activates inflammatory gene transcription while increase in histone deacetylase activity leads to inhibition of inflammatory gene transcription (Duncan et al. 2011). Especially in chronic inflammatory diseases increased acetylation of histones at the promoter region of inflammatory genes is mediated by NF-κB. Inflammatory interleukins overexpression has been linked with the development of diseases, which suggests that inflammation plays a major role in disease progression (Bayarsaihan 2011). Proinflammatory interleukins include IL-1, IL-2, IL-6, IL-8, and IL-12. IL-1α, expressed in both normal tissue and several tumor cells, is a regulatory cytokine that can induce the activation of transcription factors, including NF-κB and AP-1, and promotes the expression of genes involved in cell survival, proliferation, and angiogenesis (Wolf et al. 2001). Histone H4 hyperacetylation is a well known inflammation associated epigenetic mark. IL-1β causes histone acetylation on H4 at K8 and K12 residues (Ito et al. 2000) and has also been reported to rapidly acetylate histone tails mediated by CBP/p300 (Villagra et al. 2010). In contrast, HDAC recruitment leads to histone deacetylation and gene repression. HDACs regulate the transcription of both pro and anti-inflammatory genes by recruiting co-repressor complexes GATAs, ZEB1, and FOXp3 (Villagra et al. 2010). Age dependent changes in immune response increases the risk of infection, promote inflammation associated disease progression. Age associated hypomethylation of the DNA has been proposed as a major cause of chronic inflammation and cancer (Agrawal et al. 2010). Recently, age-dependent upregulation of IL-23p19 gene expression associated with H4K4 methylation was observed in dendritic cells (El Mezayen et al. 2009). In T cells, IL-2 gene demethylates shortly after activation leading to IL-2 production (Bruniquel and Schwartz 2003). In the differentiation of naïve CD4 T cells to T-helper 2 cells is due to rapid H3 acetylation at the IL-4/-13 gene cluster (Baguet and Bix 2004). Thus interleukins regulated inflammatory signaling pathways are either activated or silenced in inflammation driven diseases and are likely candidates that play critical role in the progression of disease to a chronic stage.

27.4 Epigenetic Changes in Oncogenes

Oncogenes also called proto-oncogenes are altered versions of normal cellular genes that play crutial role in the regulation of cell growth (Croce 2008; Weinberg 1994). The discovery of oncogenes, almost 30 years ago, has provided a critical breakthrough in our understanding of the molecular and genetic basis of cancer. Oncogenes have also provided important knowledge concerning the regulation of normal cell proliferation, differentiation, and apoptosis (Croce 2008). Recent reports suggest that Myc oncogene regulates complex inflammatory program (Meyer and Penn 2008) and Myc activation in B cells rapidly induces production and release of IL-1β (Shchors et al. 2006). Pleiotropic effects of oncogenes also include the induction of a pro-tumor microenvironment, through the persistent promotion of an inflammatory milieu (Borrello et al. 2008; Croce 2008; Grivennikov and Karin 2010). Approximately 20 transforming oncogenes (including ras, raf, myc, c-src, EGFR, HER-2) and a large number of tumor suppressor genes (e.g., p53, VHL, PTEN) are now known to directly trigger angiogenesis (Huang et al. 2007; Rak et al. 2000a) by up-regulation of the vascular endothelial growth factor (VEGF) in cancer cells expressing mutant ras oncogene (Grugel et al. 1995; Rak et al. 1995). It is interesting to note that such pro-angiogenic effects can be mimicked and/or amplified by exposing tumor cells to bona fide epigenetic stimuli such as hypoxia (Harris 2002; Laderoute et al. 2000; Mazure et al. 1996; Semenza 2000), hypoglycemia (Semenza 2000), inflammatory cytokines (Cohen et al. 1996) hormonal stimulation (Jain et al. 1998), and altered cell-cell contact (Koura et al. 1996; Rak et al. 2000b; Sheta et al. 2000). Global hypomethylation of genomic DNA has been identified as a participant in the aggressive behavior of cancer cells by upregulating oncogenes such as c-Myc or h-Ras (Das and Singal 2004). Promoter hypomethylation can activate the aberrant expression of oncogenes and induce loss of imprinting (LOI) in some loci. For example, MASPIN (also known as SERPINB5), a tumor suppressor gene that becomes hypermethylated in breast and prostate epithelial cells (Futscher et al. 2004), appears to be hypomethylated in other tumor types. MASPIN hypomethylation, and therefore its expression, increases with the degree of dedifferentiation of some types of cancer cells (Bettstetter et al. 2005; Futscher et al. 2002). Thus, the complex interplay between the histone methylation and DNA methylation and the specificity in epigenetic mechanisms that regulate DNA methylation pattern still remains to be elucidated in detail.

27.5 Epigenetic Changes in the Regulation of Master Transcription Factor NF-κB

The transcription factor nuclear factor-kappa B (NF-κB), discovered by David Baltimore in 1986, is present in the nucleus and binds the promoter of immunoglobulin kappa chain in B cells (Sen and Baltimore 1986, 2006). In the mammalian cells, the NF-κB family of transcription factors is composed of homodimers and heterodimers derived from five distinct subunits, RelA (p65), c-Rel, RelB, p50 (NF-κB1) and p52 (NF-κB2). All family members share a highly conserved Rel homology domain (RHD; ~300 aa) responsible for DNA binding, dimerization domain, and interaction with IκBs, the intracellular inhibitor of NF-κB (Karin 2006; Sethi et al. 2008; Sethi and Tergaonkar 2009). In unstimulated cells, the majority of NF-κB complexes are kept predominantly cytoplasmic and in an inactive form by binding to a family of inhibitory proteins, the IκBs (Karin 2006; Sen and Baltimore 2006). Generally, the inactive NF-κB/IκBα complex is activated by phosphorylation on two conserved serine residues within the N-terminal domain of the IκB proteins. Phosphorylation of these conserved serine residues in response to stimulation, leads to the immediate polyubiquitination of IκB proteins by the SCF-β-TrCP complex (Karin 2006). This modification subsequently targets IκB proteins for rapid degradation by the 26S proteasome (Karin 2006). Activation of the NF-κB signaling cascade can result in phosphorylation and degradation of IκBα, allowing translocation of NF-κB to the nucleus, where it induces gene transcription (Ahn and Aggarwal 2005; Ahn et al. 2007; Shen and Tergaonkar 2009;

Vallabhapurapu and Karin 2009). NF-κB is activated by many divergent stimuli, including proinflammatory cytokines (e.g., TNF-α, IL-1β), T- and B-cell mitogens, bacteria, lipopolysaccharide (LPS), viruses, viral proteins, double-stranded RNA, and physical and chemical stresses (Sethi and Tergaonkar 2009). Activated NF-κB binds to specific DNA sequences in target genes, designated as κB elements, and regulates transcription of over 400 genes involved in inflammation, immunoregulation, tumor cell proliferation, invasion, metastasis, angiogenesis, chemoresistance and radioresistance (Li and Sethi 2010; Mantovani 2010; Wong and Tergaonkar 2009).

In chronic inflammation activation of inflammatory response is mainly mediated through NF-κB which is regulated by acetylation, lysine methylation and arginine methylation (Cheung et al. 2007). The main modification is acetylation which has been shown to be important regulators of inflammatory gene expression (Medzhitov and Horng 2009) in T-cells and monocytes (Lal et al. 2009; Wells 2009; Wierda et al. 2010). Acetylation of distinct lysine residues of RelA regulates NF-κB transcriptional activation, DNA binding affinity, IκB-α assembly and subcellular localization (Chen et al. 2001; Chen and Greene 2003). IKK-α has been shown to bind to NF-κB dependent promoter via polymerase II complex and CREB binding protein (CBP), where it can acetylate histone H3 on Lys9 (Yang et al. 2008) and also can phosphorylate histone H3 at Ser10 (Anest et al. 2003). This IKK-α dependent and cytokine mediated phosphorylation is critical in CBP-mediated acetylation of histone H3 on Lys14 (Yamamoto et al. 2003). Acetylation of histone H3 is often seen in cytokine mediated inflammation and in autocrine and paracrine activation that follows with increased NF-κB driven inflammatory gene expression (Barnes 2009a). Glucocoticoids and HDAC2 can reverse NF-κB driven inflammation (Barnes 2009b). Along with NF-κB poly (ADP-ribose) polymerase-1 (PARP-1) has been show to play prominent role in chronic inflammation driven diseases (Aguilar-Quesada et al. 2007). PARP-1 has been demonstrated as a promoter specific co-activator of NF-κB *in vivo* and independent of its enzymatic activity (Hassa et al. 2003). PARP-1 directly interacts with p300, p50 and p65 and synergistically co-activates NF-κB dependent transcription (Hassa et al. 2005). Under inflammatory conditions, p300 can in turn acetylate PARP-1 at specific lysine residues in variety of cell types. NF-κB mediated activation of histone H3, H4 acetylation by IL-1β, TNF-α, or endotoxins can increase the expression of granulocyte-macrophage colony stimulation factor (GM-CSF)(Aggarwal 2004). Thus reversible acetylation of NF-κB and proteins like PARP-1 and histones play a major role in regulating gene expression during inflammation. Altered methylation pattern such as lysine methylation are also observed in inflammatory diseases. H3K4 methyltransferase SET7/9 can also regulate recruitment of NF-κB-p65 to promoters and thereby modulate inflammatory gene expression (Li et al. 2008). Arginine methylation has not been directly implicated in inflammation; however, PRMT1 and PRMT4 are known transcriptional co-activators of NF-κB (Covic et al. 2005; Hong et al. 2004; Meyer et al. 2007). In hyperglycemic memory, high transcriptional activity of pro-inflammatory transcription factor NF-κB is due to enzymatic deletion of H3K9 methyl marks from its promoter (Brasacchio et al. 2009; El-Osta et al. 2008) and this finding provides novel perspective to the link between NF-κB activity and epigenetic modification in

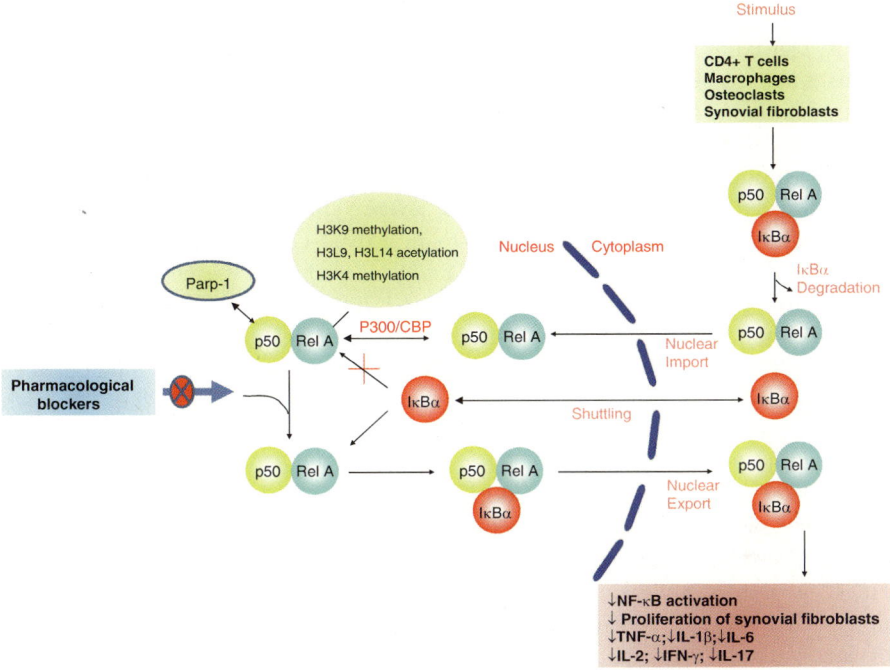

Fig. 27.1 Histone methylation and acetylation mediated aberrant NF-κB transcriptional activation in tumor cells. Epigenetic activation in the gene promoter region and interactions between DNA methylation machinery, chromatin modifiers (HDACs) and PARP-1. Pharmacological inhibition of individual components in the repressive complex with DNMT inhibitors and HDAC inhibitors, either alone or in combination, result in DNA demethylation and deacetylation and complex disintegration leading to reversal of genome-wide epigenetic alterations in cancer cells through reversal of NF-κB mediated gene transcription and cell signaling pathways

the development of chronic inflammation driven diseases (Taylor 2008; Villeneuve et al. 2008). Overall, epigenetic modifications in NF-κB as described briefly in Fig. 27.1 and associated inflammatory molecules leads to the development of major inflammatory diseases which will be now be discussed in detail below.

27.6 Epigenetic Changes During Autoimmune Diseases

Rheumatoid arthritis (RA) is a chronic inflammation driven autoimmune disease of the joints that occurs in 1% of world population (Firestein 2003). Rheumatoid arthritis arises from interplay of inherited genetic predisposition such as HLA-DR allele subtypes and specific gene polymorphism, autoantibody production and dietary factors (Ermann and Fathman 2001; Smith and Haynes 2002). One of the hallmarks of autoimmune disorders such as rheumatoid arthritis is the infiltration of circulating inflammatory cells and synovial fibroblasts. Synovial inflammation is mainly caused by overproduction of cytokines, chemokines and matrix metalloproteinases that eventually lead to the progressive destruction of articular cartilage and bone (Ospelt and Gay 2008). Several studies have shown that treatment of RA with inhibitor of TNF-α or IL-6 confirmed that the disease is driven by differences in cytokine production (Feldmann and Maini 1999). By analyzing DNA methylation status in circulating T-cells of RA patients, severe hypomethylation was observed in the T-cells of RA patients compared to healthy T-cells (Neidhart et al. 2000; Richardson et al. 1990). In addition, Nile et al. evaluated the DNA methylation status in the promoter region of IL-6 gene (−1,200 to +30) in PBMC cells of RA patients. They found that hypomethylation at a single CpG site at −1,181 contributed to elevated expression of IL-6 gene which contributes to sustained inflammatory condition (Nile et al. 2008). Sullivan et al. investigated DNA methylation of CpG-rich sequences within the promoter of the TNF-α gene and found a positive correlation between high expression levels of TNF-α and low methylation status (McInnes and Schett 2007). Epigenetic histone modifications in RA have mostly concentrated on histone acetylation and particularly on the use of HDAC inhibitors as therapeutic agents. There have been a number of reports of beneficial effects from the use of HDAC inhibitors *in vitro* and *in vivo* (Grabiec et al. 2008). Intravenous administration of HDI FK-228 reduces the expression of TNF-α and IL-1β in a mouse model of autoantibody-mediated arthritis (AMA) (Nishida et al. 2004). Prophylactic administration of SAHA and MS-275 has been shown to reduce arthritis score, radiologic score, and bone resorption in the collagen-induced arthritis (CIA) model in mice and rats (Lin et al. 2007). Trichostatin A was shown to reduce the clinical scores of arthritis and synovial inflammation in mouse AMA with concomitant decrease in expression of matrix-degrading enzymes (Nasu et al. 2008). Autoantibodies generated against citrullinated proteins are biomarker proteins and can be found in about 80% of the RA patients (Zendman et al. 2006). Therefore, studying the citrullination of proteins is of major interest in RA. PAD4 is expressed in different cell types of the RA synovium, as well as in blood monocytes (Chang et al. 2005; Vossenaar et al. 2004). Recently, Chang et al. have reported increased expression of PAD4 in RA synovial membranes compared to OA (Chang et al. 2009). However, it is still unclear to what extent PAD4 contributes to epigenetic changes involved in the pathogenesis of RA. In addition by studying the methylation pattern of death receptor 3 (DR3) in synovial cells of RA and OA patients, Takami et al. found significant hypermethylation of CpG dinuclotides in the synovial cells derived from RA. This phenomenon might explain the resistance to apoptosis seen in synovial cells of RA patients (Ashkenazi and Dixit 1998; Takami et al. 2006). Thus, in rheumatoid arthritis, the reduced activity of HDACs plays a key role in regulating NF-κB–mediated gene expression (Huber et al. 2007b). Patients with type 1 diabetes also present a characteristic pattern of histone marks, showing lymphocytes but not monocytes with increased H3K9me2 in a subset of genes associated with autoimmune and inflammatory pathways (e.g., CLTA4, IL6) (Miao et al. 2008).

27.7 Epigenetic Changes in Chronic Inflammation Driven Cancers

Epigenetic changes have been observed in all types of cancers, understanding the regulatory mechanisms that cause severe aberrant induction of DNA methylation is gaining importance. Chronic inflammation, aging and viral infections have been shown to methylate on-core regions in CpG islands of tumor suppressor genes such as in retinoblastoma (Feinberg and Tycko 2004; Ohtani-Fujita et al. 1993), in VHL (Herman et al. 1994), CDKN2A(p16) (Gonzalez-Zulueta et al. 1995; Merlo et al. 1995), CDH1 (E-cadherin) (Graff et al. 1995; Yoshiura et al. 1995) and hMLH1 (Kane et al. 1997) that permanently repress downstream gene expression (Ahuja et al. 1998; Hsieh et al. 1998; Issa et al. 1994, 2001; Jones and Baylin 2002; Kang et al. 2002; Osawa et al. 2002). These aberrant methylation patterns provide an excellent opportunity to be exploited as tumor biomarkers and as targets for chemotherapeutics (Egger et al. 2004; Issa et al. 2004; Laird 2003). There are many comprehensive reviews on the role of epigenetics and cancer (Feinberg and Tycko 2004; Jaenisch and Bird 2003; Jones and Baylin 2002; Laird 2003). Here in this chapter we will discuss the role of epigenome in inflammation driven cancer development. These changes occur in genes involved in cell cycle progression, angiogenesis, matrix adhesion molecules and metastasis thus leading to cancer development. In cancer, the most fundamental characteristic is the heritable changes such as gene silencing due to CpG hypermethylation at the promoter regions of tumor suppressor genes. Loss of tumor suppressor proteins, for eg., RB1, VHL, p16INK4A, MLH1, BRCA1, APC, st7 and nm23 have been shown to disrupt almost all cell signaling pathways that promote cancer development (Hanahan and Weinberg 2000; Pal et al. 2004). Genes such as RB1, VHL, and BRCA1 hypermethylation is restricted to hereditary cancers such as retinoblastoma, renal, breast and ovarian cancer (Dobrovic and Simpfendorfer 1997; Herman et al. 1994; Stirzaker et al. 1997; Yang et al. 2006a). In contrast p16INK4A is mutated in the germline of patients with hereditary malignant melanoma (Pho et al. 2006) and pancreatic cancers (Goldstein 2004), and is often inactivated by promoter hypermethylation in a wide variety of cancers, including breast, gastrointestinal, respiratory tract, gynecological, and hematopoietic cancers (Aniello et al. 2006; Chevillard-Briet et al. 2002; Frietze et al. 2008; Gronbaek et al. 2000; Kondo et al. 2008; Li et al. 2006; Luo et al. 2006; Majumder et al. 2006; Wang et al. 2004; Yang et al. 2006a). Agents that inhibit arginine methylation and histone deacetylation uphold chromatin integrity and prevent chromatin condensation and reestablish demethylation and gene transcription (Cheng et al. 2004; Verbiest et al. 2008). Thus histone modifying enzymes have a critical role in determining the tumor progression and multidrug resistant phenotype of cancer cells. A list of various histone modifying enzymes associated with cancer development is summarized in Table 27.1. In particular DNA methylation and histone modifications are very attractive anti-cancer targets towards development of novel therapeutic approaches (Gronbaek et al. 2007).

Table 27.1 A list of major histone modifying enzymes linked to the development of cancer

Histone methyl transferases/acetylase	Enzymes linked to various cancers	References
Arginine methyl transferases	PRMT1, PRMT2, PRMT4, PRMT5, RPMT7, CARM1	Cheung et al. (2007), Chevillard-Briet et al. (2002), Frietze et al. (2008), Hong et al. (2004), Meyer et al. (2007), Majumder et al. (2006), Pal et al. (2004), Verbiest et al. (2008)
Lysine methyl transferases	MLL1 (KMT2A), MLL4 (KMT2D), EZH2 (KMT6), SUV39H1/2 (KMT1A/B), G9a (KMT1C), Eu-HMTase1 (KMT1D), SETDB1/ESET (KMT1E), SET8/PR-SET7 (KMT5A), SMYD2 (KMT3C), SMYD3, DOT1L (KMT4)	Kang et al. (2007), Pogribny et al. (2007), Aniello et al. (2006), Hess (2004), Kondo et al. (2008), Li et al. (2006), McGarvey et al. (2006), Saigo et al. (2008), Bachmann et al. (2006), Bryant et al. (2007), Collett et al. (2006), Ding et al. (2006), Hamamoto et al. (2004, 2006), Huang et al. (2007), Okada et al. (2005), Shi et al. (2007)
Histone demethylases	LSD1; JHDM1; JMJD2C (GASC1); JMJD6; PADI4	Agger et al. (2008), Chang et al. (2007), Cloos et al. (2006), Metzger et al. (2005), Shi et al. (2004), Tsukada et al. (2006), Wissmann et al. (2007), Cuthbert et al. (2004)
Histone acetyltransferase and histone deacetylases	HATs; HDAC1, HDAC2, SET7 (HMT), SirT1	Portela and Esteller (2010)

27.8 Epigenetic Changes During Chronic Obstructive Pulmonary Disease

Chronic obstructive pulmonary disease (COPD) is a chronic inflammatory disease of the lung which leads to the development of lung cancer (Bowman et al. 2009; Lee et al. 2009; Yao and Rahman 2009) Lymphocytes, macrophages, and neutrophils are found in abundant at the site of inflammation and are the main orchestrators and amplifiers in the progression of COPD. However, these inflammatory cells can be manipulated by immunosculpting and immunoediting pathways in the progression of cancer, and immuno-based therapeutic strategies for cancer to induce immune escape of cancer cells especially in a tumor-promoting microenvironment which is due to chronic inflammation seen in lungs of patients with COPD (Reiman et al. 2007). Steroid resistance observed in COPD patients have been attributed to decreased expression of HDAC in lungs of patients with COPD (Ito et al. 2005). Chromatin remodeling is manifested by post-translational modifications of core histone proteins and DNA methylation which is shown to regulate proinflammatory gene expression during the development of COPD and lung carcinogenesis. High levels of histone acetylation is observed on the promoters of proinflammatory genes in alveolar macrophages and airway epithelial cells in patients with COPD, and the extent of acetylation is significantly correlated with disease severity (Ito et al. 2005). The fundamental mechanism of hyperacetylation of histones/nonhistone proteins in lungs of patients with COPD is associated with reduced HDAC2 level (Adenuga et al. 2009; Ito et al. 2005). This phenomenon is also observed in lungs of rodents exposed to cigarette smoke (Adenuga et al. 2009; Yang et al. 2006b). Increasing HDAC2 activity or levels by phenolic antioxidants and theophylline, are being tested as therapeutic strategies to reduce the lung inflammatory response and attenuates corticosteroid resistance in patients with COPD (Barnes 2009a). Methylation of p16 promoter has been detected in the sputum of patients with COPD and is correlated with heavy cigarette smoking (Adenuga et al. 2009). Genome-wide DNA demethylation with site-specific hypermethylation occurs in lung cancer cells and is associated with silencing of a variety of tumor-suppressor genes by the recruitment of HDACs (Barnes 2009b). The mechanisms underlying these observations may be due to aberrant expression/activity of DNA methyltransferases (DNMTs) and demethylases in cancer cells. Methylation in the promoters of multiple genes is shown in adenocarcinomas and in non -small cell lung cancer (NSCLC), and this methylation is associated with tumor progression and recurrence (Brock et al. 2008). Therefore, determination of DNA methylation on specific gene may provide the useful biomarkers for early detection and/or chemoprotective intervention in lung cancer. Modifications of core histone proteins increase the complexity of epigenetic alterations mediated by aberrant DNA methylation in cancer cells. Increased HDAC1, and decreased HDAC5 and HDAC10 are correlated with advanced stage of disease and adverse outcome in lung cancer patients (Barnes 2009a). DNA demethylating agents and HDAC inhibitors synergistically induce apoptosis in lung cancer cells, and prevent lung cancer development in animals exposed to tobacco carcinogens (Adenuga et al. 2009). However, the specificity on a particular isoform of HDACs, optional therapeutic doses, timing, and mode of administration are still under evaluation for these agents.

27.9 Epigenetic Changes Observed During Neurodegenerative Disorders

Aberrant epigenetic changes have also been observed in neurodegenerative diseases such as Alzheimer's disease, Parkinson's disease and Huntington's disease and in neurological disorders such as epilepsy, multiple sclerosis and in amyolotropic lateral sclerosis which are primarily driven by chronic inflammation (Portela and Esteller 2010). Alzheimer's disease originates from the production of amyloid β (Aβ) via the proteolytic cleavage of amyloid precursor proteins (APP) by presenilin 1 and presenilin 2 play crutial roles in neurodegeneration. Neprilysin (discoidin domain receptor tyrosine kinase 1), is the major Aβ-degrading enzyme in the brain (Iwata et al. 2000) is hypermethylated in cerebral endothelial cells when these cells are treated with high concentrations of Aβ (Chen et al. 2009). However, neprilysin does not seem to be regulated by DNA methylation in NB7 and SH-SY5Y cells, but by histone acetylation (Belyaev et al. 2009). Furthermore overexpression of APP C-terminal peptide in PC12 cells and in rat primary cortical neurons increases acetylation of histones H3K14 and H4K5 (Kim et al. 2004). In addition, neuron-specific overexpression of HDAC2 in mice, but not of HDAC1, can decrease dendritic spine density, synapse number, synaptic plasticity, and memory formation (Guan et al. 2009). S100A2 and SORBS3 genes show differences in DNA methylation in Alzheimer's disease (Hauke et al. 2009; Siegmund et al. 2007). S100A2 has been previously identified as metastatic marker in non small cell lung cancer (Bulk et al. 2009).

Parkinson's disease is characterized by progressive loss substantia nigra dopaminergic neurons and striatal projections which causes tremor, muscle rigidity, bradykinesia and postural instability(Urdinguio et al. 2009). Dopamine depletion observed in this disease is associated with decrease in histone H3K4me3 (Urdinguio et al. 2009). Chronic levodopa therapy leads to deacetylation of histones H4K5, K8, K12, and K16. MPTP (1-methyl-4-phenyl-1, 2, 3, 6-tetrahydropyridine), widely used in Parkinson's disease model, induces H3 acetylation, which is reduced after treatment with levodopa (Nicholas et al. 2008). Thus, epigenetic regulation in this disease might have important role, however, its direct implications remain unknown (Hauke et al. 2009). Multiple sclerosis is a chronic inflammatory disease characterized by a demyelination process followed by degeneration(Urdinguio et al. 2009). For example, hypomethylation at the promoter region of PADl2 (peptidyl arginine deiminase, type II), was also found to be overexpressed in multiple sclerosis. PADl2 catalyzes the citrullination of myelin basic protein and changes the properties of myelin (Moscarello et al. 1986). Trichostatin A is a HDAC inhibitor which has been shown to be a useful alternative treatment drug in experimental autoimmune encephalomyelitis mice. This drug has shown a reduction in spinal cord inflammation, demyelination, neuronal loss and an attenuated disability in the chronic phase (Camelo et al. 2005). It seems likely that epigenetic alteration mechanisms, such as DNA methylation and transcriptional dysregulation, are a marker of disease status in Alzheimer's disease as well as other neurodegenerative disorders.

27.10 Small Molecule Compounds as Inhibitors of Epigenetic Changes Induced by Histone Modifying Enzymes

27.10.1 HDAC Inhibitors

There is growing interest in the various mechanisms that regulate chromatin remodeling, including modulation of histone deacetylase activities (Lee et al. 2008). Evidence suggests that such epigenetic factors may be important in understanding the basic pathological variation in inflammatory response and in recent times the therapeutic potential in epigenetic manipulations in large number of chronic inflammation driven diseases. Several groups of HDAC inhibitors have been well-characterized (Marks and Xu 2009), including short-chain fatty acids, hydroxamic acids, cyclic tetrapeptides, cyclic peptides, benzamides, and electrophilic ketones. Sodium butyrate (NaB) and valproic acid (VPA) are relatively weak short-chain fatty acid HDAC inhibitors and can inhibit histone deacetylation by non-competitive binding to classes I and IIa of HDAC-enzymes (Bolden et al. 2006; Zhang et al. 2008). HDAC inhibitors have been shown to prevent fibrosis associated with systemic sclerosis (Huber et al. 2007a), modulation apoptosis of synovial fibroblasts in RA (Jungel et al. 2006), suppression of proinflammatory cytokine production, and prevention of bone destruction in RA (Chung et al. 2003). Competitive HDAC inhibitors disrupts the cell cycle and/or induce apoptosis via upregulation of genes such as caspase3, p21 and Bax and can sensitize tumor cells to trichostatin A and polyphenol HDAC inhibitory compounds (Dashwood et al. 2006). This apparent selectivity of action in cancer cells makes HDAC inhibitors an attractive target for drug development. Liu et al. reported that the expression levels of HDAC1, HDAC3, and HDAC8 proteins were downregulated following curcumin treatment in Raji cells, whereas Ac-histone H4 protein expression was upregulated after treatment with curcumin (Liu et al. 2005). Other dietary HDAC inhibitors include sulforaphane (SFN), diallyl disulfide (DADS) and polyphenols. SFN has been reported to inhibit HDAC activity in human colorectal cancer (Myzak et al. 2004), prostate cancer (Myzak et al. 2006b) and breast cancer cells (Pledgie-Tracy et al. 2007). *In vivo* SFN retarded the growth of prostate cancer xenografts (Myzak et al. 2007) and suppressed spontaneous intestinal polyps in the Apc^{min} mouse (Myzak et al. 2006a) with evidence for altered histone acetylation status and HDAC inhibition. In general, these dietary agents are weak ligands and inhibit HDAC activity at higher concentrations than trichostatin A or SAHA, which are effective in the nanomolar to low micromolar range (Dashwood et al. 2006). There are several clinical data available for HDAC inhibitors (e.g. vorinostat and N-acetyldinaline) in the treatment of advanced NSCLC (Gridelli et al. 2008), and these agents are being investigated in randomized phase III clinical trials. In addition, synthetic histone mimetic I-BET showed anti-inflammatory activity by interfering with binding of bromodomain-containing BET proteins to acetylated histones, and disrupts the formation of the chromatin complexes essential for expression of inflammatory genes. The suppression of key inflammatory genes by I-BET suggested a potent ability of this compound to treat inflammatory conditions under *in vivo* settings (Nicodeme et al. 2010). Other HDAC inhibitors including depsipeptide and MS-275 are also currently undergoing trials for cancer treatment (Frew et al. 2009; Tan et al. 2010).

27.10.2 HAT Inhibitors

Histone acetyltransferases (HAT) have been implicated in the progression of inflammation driven diseases including various human cancers and thus represent novel, therapeutically relevant molecular targets for drug development. HAT inhibitors obtained from purified from cashew nut shell liquid, anacardic acid, has been shown to inhibit p300 and p300/CBP-associated factor histone acetyltransferase activities (Balasubramanyam et al. 2003). Anacardic acid non-specifically inhibits p300 and PCAF with an IC50 < 10 μM (Balasubramanyam et al. 2003) Anacardic acid has also been shown to have an additive effect on apoptosis induced by TNF-α and chemotherapeutic agents like cisplatin and doxorubicin (Sung et al. 2008). A series of 28 anacardic acid analogues were prepared and tested and all the compounds showed 95% HAT inhibitory activity *in vitro* to a broad variety of cancer cell. Interestingly all the compounds tested were relatively non-toxic to normal cells (Eliseeva et al. 2007). In another study they have shown that inhibition of HAT activity by anacardic acid sensitizes cancer cells to ionizing radiation (Sun et al. 2006). Another polyphenolic compound, curcumin obtained from *Curcuma longa* inhibits p300 while it does not inhibit PCAF (Balasubramanyam et al. 2004b). Recently, a water soluble derivative of curcumin, CTK7A (Sodium 4-(3,5-bis (4-hydroxy-3-methoxystyryl)-1H-pyrazol-1-yl) benzoate) has been shown to have broad spectrum HAT inhibitory activity compared to curcumin (Arif et al. 2010). In this study, using constitutively hyperacetylated oral and liver cancer cells they showed that CTK7A can inhibit cell proliferation, attenuate HAT p300/CBP and PCAF autoacetylation while it did not affect the activity of other histone modifying enzymes like G9a, CARM1, Tip60 and HDAC1 and SIRT2 at 100 μM concentration. *In vivo* using nude mice model they examined the anti-tumor activity of CTK7A. Intra-peritoneal administration of CTK7A at dose of 100 mg/kg b.w. twice a day inhibited oral tumor growth by 50% compared to control group. This reduction in tumor growth was correlated with reduction in H3K9, H3K14 acetylation and p300 expression as determined by immunohistochemistry. Furthermore, the levels of GAPDH, NPM1, iNOS, COX2 and Ki67 levels were downregulated in CDK7A treated mice (Arif et al. 2010).

Garcinol obtained from *Garcinia indica* is a potent inhibitor of p300 acetyltransferase and PCAF (Balasubramanyam et al. 2004a). Isogarcinol, 14-isopropoxy isogarcinol and 14-methoxy isogarcinol and 13, 14 disulfoxy isogarcinol specifically inhibit p300 HAT activity but not PCAF activity (Arif et al. 2009). Most lysine acetyl transferase inhibitors possess poly-hydroxy functional groups (Selvi et al. 2010b). Plumbagin obtained from *Plumbagin ovate* and EGCG, a major phenol in green tea have been shown to inhibit p300 HAT activity (Choi et al. 2009; Ravindra et al. 2009).

27.10.3 Lysine and Arginine Methyltransferases Inhibitors

Lysine methyltransferase inhibitors represent one of the valid targets for discovery of small molecule inhibitors. A fungal derived toxin, chaeotocin was found to inhibit Su(var)3-9, a drosophila melanogaster H3K9 methyltransferase (Greiner et al. 2005). BIX-01294 is another compound that was shown to inhibit histone methyltransferas G9a and reverse H3K9 dimethylation marks in mouse ES cells and

Table 27.2 A list of epigenetic modifying enzyme inhibitors

Histone modifying enzymes	Inhibitors	References
Arginine methyl transferases	AMI-1; AMI-5; Inhibitor 4b; Stilbamidine; Allantodapsone; RM-65	Cheng et al. (2004), Mai and Altucci (2009), Mai et al. (2008), Ragno et al. (2007), Selvi et al. (2010b), Spannhoff et al. (2007)
DNA methyl transferases	Azacitidine; Decitabine; Zebularine; 5-fluoro-2′-deoxycitidine; epigallocatechin-3-gallate; Hydralazine; RG108; Dietary polyphenols	Cheng et al. (2004), Link et al. (2010)
Histone deacetylases	Sinefugin; Sodium butyrate; Sodium phenylbutyrate; Valproic acid; OSU-HDAC42; Trichostatin A; Vorinostat; Panobinostat; Belinostat; Romidepsin; Entinostat; Dacinostat; Givinostat; MGCD-0103; Dietary polyphenols	Huber et al. (2010), Kuendgen and Lubbert (2008), Link et al. (2010), Vedel et al. (1978), Wagner et al. (2010)
Histone acetyltransferases	Anacardic acid; Curcumin; Curcumin analog CTK7A; Garcinol; Plumbagin; EGCG	Arif et al. (2009), Balasubramanyam et al. (2004a, b), Choi et al. (2009), Ravindra et al. (2009), Selvi et al. (2010b)
Lysine methyl transferases	Chaetocin; Bix-01294; Bix-01338	Greiner et al. (2005), Kubicek et al. (2007)

fibroblasts, but did not affect mono and trimethylation levels (Kubicek et al. 2007). In the same study, they have identified another compound that inhibits both lysine and arginine methyltransferase, BIX-01338. DZNep, 3-deazoneplanocin A, was reported to inhibit trimethylation of H3K27 and H4K20 and reactivates silenced genes in cancer (Miranda et al. 2009). Arginine methyl transferase is relatively less studied compared to other targets (Selvi et al. 2010b). The first PRMTs inhibitor was Sinefungin, was shown to inhibit myelin basic protein methylation (Amur et al. 1986). A well designed small molecule screen identified AMI-1 as a potent arginine methylation specific inhibitor with no effect on lysine methyltransferases (Cheng et al. 2004) and a carboxy analogue of AMI-1 was also found to inhibit arginine methyltransferases (Castellano et al. 2010). Ellagic acid has been shown to inhibit specifically CARM1/PRMT4 activity both *in vitro* and *in vivo* (Selvi et al. 2010a). It is know that PRMTs methylate arginine residues in variety of cancers, however, the role of arginine methylation in the initiation and progression of chronic inflammation driven diseases remains to be determined (Spannhoff et al. 2009). A list of various enzyme modifying inhibitors is summarized in Table 27.2 and the structure and plant source of natural agents that can mitigate the epigenetic process is shown in Fig. 27.2.

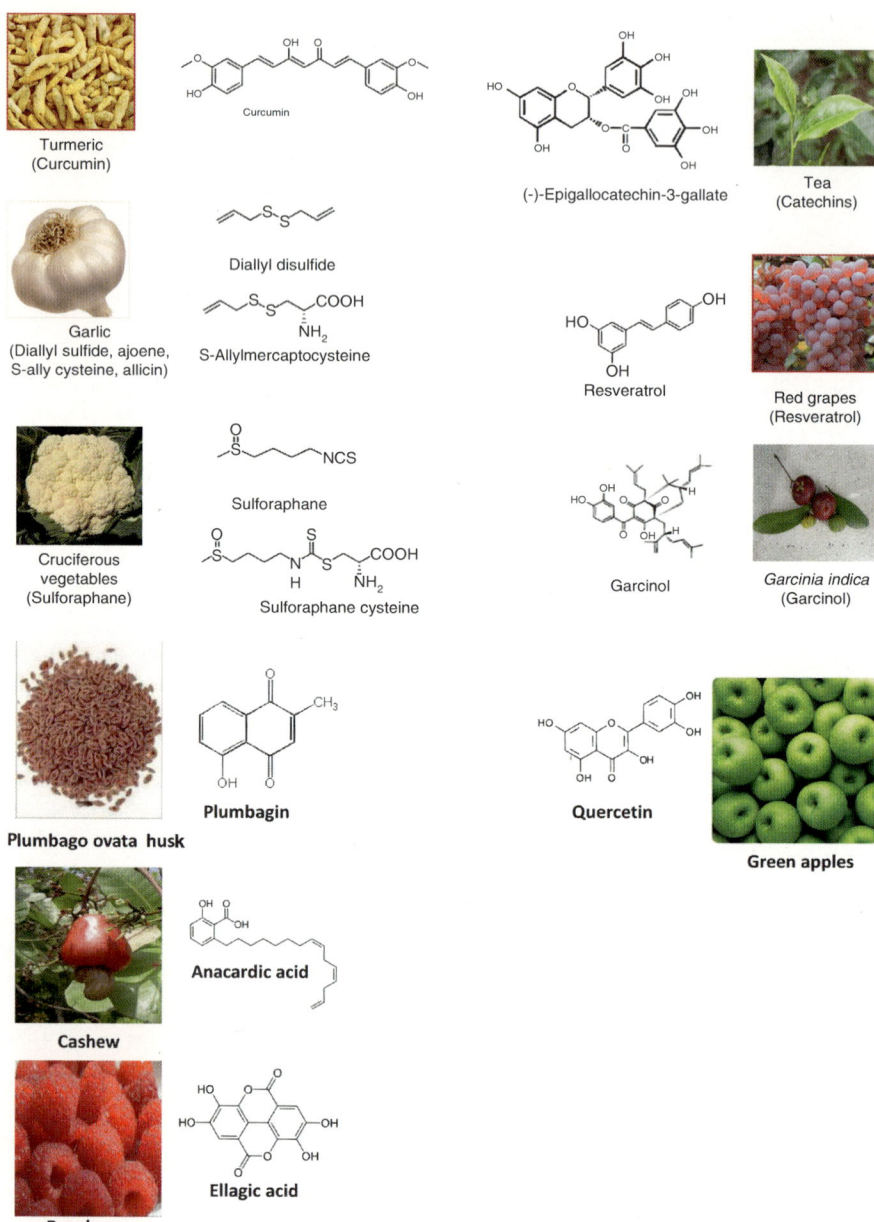

Fig. 27.2 List of small molecule from natural sources that can modulate epigenome in tumor cells

27.11 Perspectives and Conclusion

Epigenetics modulations such as DNA methylation, histone modification and knowledge about the major enzymes that mediate these alterations are central to tissue specific gene expression and development of diseases. A better understanding of these changes would provide a sound epigenetic molecular basis for various chronic inflammatory diseases. Progress in epigenetic mechanisms and alterations during inflammatory responses opens up opportunities for the development of novel HDAC and HAT inhibitors. Notable examples of HDAC inhibitor success include those of vorinostat, trichostatin A, and ITF2357 in the prevention and treatment of inflammatory diseases such as RA and different cancers. HAT inhibitors such as garcinol, anacardic acid, plumbagin, and curcumin have shown significant promise in the pre-clinical studies and well designed clinical trials are needed to validate their activities in humans. Overall, emerging evidence suggests that changes to epigenome may be important in the pathogenesis of chronic inflammation driven diseases. With advanced technology, integration of whole genome microarray technology and ChIP based screening will help in the development of novel and potent drugs that inhibit epigenetic changes with increased specificity.

References

Adenuga D, Yao H, March TH, Seagrave J, Rahman I (2009) Histone deacetylase 2 is phosphorylated, ubiquitinated, and degraded by cigarette smoke. Am J Respir Cell Mol Biol 40:464–473

Aggarwal B (2003) Signalling pathways of the TNF superfamily: a double-edged sword. Nat Rev Immunol 3:745–756

Aggarwal BB (2004) Nuclear factor-kappaB: the enemy within. Cancer Cell 6:203–208

Aggarwal BB, Gehlot P (2009) Inflammation and cancer: how friendly is the relationship for cancer patients? Curr Opin Pharmacol 9:351–369

Aggarwal BB, Shishodia S, Sandur SK, Pandey MK, Sethi G (2006) Inflammation and cancer: how hot is the link? Biochem Pharmacol 72:1605–1621

Agger K, Christensen J, Cloos PA, Helin K (2008) The emerging functions of histone demethylases. Curr Opin Genet Dev 18:159–168

Agrawal A, Tay J, Yang GE, Agrawal S, Gupta S (2010) Age-associated epigenetic modifications in human DNA increase its immunogenicity. Aging (Albany NY) 2:93–100

Aguilar-Quesada R, Munoz-Gamez JA, Martin-Oliva D, Peralta-Leal A, Quiles-Perez R, Rodriguez-Vargas JM, Ruiz de Almodovar M, Conde C, Ruiz-Extremera A, Oliver FJ (2007) Modulation of transcription by PARP-1: consequences in carcinogenesis and inflammation. Curr Med Chem 14:1179–1187

Ahn KS, Aggarwal BB (2005) Transcription factor NF-kappaB: a sensor for smoke and stress signals. Ann N Y Acad Sci 1056:218–233

Ahn KS, Sethi G, Aggarwal BB (2007) Nuclear factor-kappa B: from clone to clinic. Curr Mol Med 7:619–637

Ahuja N, Li Q, Mohan AL, Baylin SB, Issa JP (1998) Aging and DNA methylation in colorectal mucosa and cancer. Cancer Res 58:5489–5494

Amur SG, Shanker G, Cochran JM, Ved HS, Pieringer RA (1986) Correlation between inhibition of myelin basic protein (arginine) methyltransferase by sinefungin and lack of compact myelin formation in cultures of cerebral cells from embryonic mice. J Neurosci Res 16:367–376

Anest V, Hanson JL, Cogswell PC, Steinbrecher KA, Strahl BD, Baldwin AS (2003) A nucleosomal function for IkappaB kinase-alpha in NF-kappaB-dependent gene expression. Nature 423:659–663

Aniello F, Colella G, Muscariello G, Lanza A, Ferrara D, Branno M, Minucci S (2006) Expression of four histone lysine-methyltransferases in parotid gland tumors. Anticancer Res 26:2063–2067

Arif M, Pradhan SK, Thanuja GR, Vedamurthy BM, Agrawal S, Dasgupta D, Kundu TK (2009) Mechanism of p300 specific histone acetyltransferase inhibition by small molecules. J Med Chem 52:267–277

Arif M, Vedamurthy BM, Choudhari R, Ostwal YB, Mantelingu K, Kodaganur GS, Kundu TK (2010) Nitric oxide-mediated histone hyperacetylation in oral cancer: target for a water-soluble HAT inhibitor, CTK7A. Chem Biol 17:903–913

Ashkenazi A, Dixit VM (1998) Death receptors: signaling and modulation. Science 281:1305–1308

Bachmann IM, Halvorsen OJ, Collett K, Stefansson IM, Straume O, Haukaas SA, Salvesen HB, Otte AP, Akslen LA (2006) EZH2 expression is associated with high proliferation rate and aggressive tumor subgroups in cutaneous melanoma and cancers of the endometrium, prostate, and breast. J Clin Oncol 24:268–273

Baguet A, Bix M (2004) Chromatin landscape dynamics of the Il4-Il13 locus during T helper 1 and 2 development. Proc Natl Acad Sci U S A 101:11410–11415

Balasubramanyam K, Swaminathan V, Ranganathan A, Kundu TK (2003) Small molecule modulators of histone acetyltransferase p300. J Biol Chem 278:19134–19140

Balasubramanyam K, Altaf M, Varier RA, Swaminathan V, Ravindran A, Sadhale PP, Kundu TK (2004a) Polyisoprenylated benzophenone, garcinol, a natural histone acetyltransferase inhibitor, represses chromatin transcription and alters global gene expression. J Biol Chem 279:33716–33726

Balasubramanyam K, Varier RA, Altaf M, Swaminathan V, Siddappa NB, Ranga U, Kundu TK (2004b) Curcumin, a novel p300/CREB-binding protein-specific inhibitor of acetyltransferase, represses the acetylation of histone/nonhistone proteins and histone acetyltransferase-dependent chromatin transcription. J Biol Chem 279:51163–51171

Balkwill F (2009) Tumour necrosis factor and cancer. Nat Rev Cancer 9:361–371

Balkwill F, Mantovani A (2001) Inflammation and cancer: back to Virchow? Lancet 357:539–545

Barnes PJ (2009a) Role of HDAC2 in the pathophysiology of COPD. Annu Rev Physiol 71:451–464

Barnes PJ (2009b) Targeting the epigenome in the treatment of asthma and chronic obstructive pulmonary disease. Proc Am Thorac Soc 6:693–696

Barres R, Osler ME, Yan J, Rune A, Fritz T, Caidahl K, Krook A, Zierath JR (2009) Non-CpG methylation of the PGC-1alpha promoter through DNMT3B controls mitochondrial density. Cell Metab 10:189–198

Bayarsaihan D (2011) Epigenetic mechanisms in inflammation. J Dent Res 90:9–17

Belyaev ND, Nalivaeva NN, Makova NZ, Turner AJ (2009) Neprilysin gene expression requires binding of the amyloid precursor protein intracellular domain to its promoter: implications for Alzheimer disease. EMBO Rep 10:94–100

Berger SL (2007) The complex language of chromatin regulation during transcription. Nature 447:407–412

Bettstetter M, Woenckhaus M, Wild PJ, Rummele P, Blaszyk H, Hartmann A, Hofstadter F, Dietmaier W (2005) Elevated nuclear maspin expression is associated with microsatellite instability and high tumour grade in colorectal cancer. J Pathol 205:606–614

Bird A (2002) DNA methylation patterns and epigenetic memory. Genes Dev 16:6–21

Bird A (2007) Perceptions of epigenetics. Nature 447:396–398

Bolden JE, Peart MJ, Johnstone RW (2006) Anticancer activities of histone deacetylase inhibitors. Nat Rev Drug Discov 5:769–784

Borrello MG, Degl'Innocenti D, Pierotti MA (2008) Inflammation and cancer: the oncogene-driven connection. Cancer Lett 267:262–270

Bowman RV, Wright CM, Davidson MR, Francis SM, Yang IA, Fong KM (2009) Epigenomic targets for the treatment of respiratory disease. Expert Opin Ther Targets 13:625–640

Brasacchio D, Okabe J, Tikellis C, Balcerczyk A, George P, Baker EK, Calkin AC, Brownlee M, Cooper ME, El-Osta A (2009) Hyperglycemia induces a dynamic cooperativity of histone methylase and demethylase enzymes associated with gene-activating epigenetic marks that coexist on the lysine tail. Diabetes 58:1229–1236

Brigati C, Noonan DM, Albini A, Benelli R (2002) Tumors and inflammatory infiltrates: friends or foes? Clin Exp Metastasis 19:247–258

Brock MV, Hooker CM, Ota-Machida E, Han Y, Guo M, Ames S, Glockner S, Piantadosi S, Gabrielson E, Pridham G, Pelosky K, Belinsky SA, Yang SC, Baylin SB, Herman JG (2008) DNA methylation markers and early recurrence in stage I lung cancer. N Engl J Med 358:1118–1128

Bruniquel D, Schwartz RH (2003) Selective, stable demethylation of the interleukin-2 gene enhances transcription by an active process. Nat Immunol 4:235–240

Bryant RJ, Cross NA, Eaton CL, Hamdy FC, Cunliffe VT (2007) EZH2 promotes proliferation and invasiveness of prostate cancer cells. Prostate 67:547–556

Bulk E, Sargin B, Krug U, Hascher A, Jun Y, Knop M, Kerkhoff C, Gerke V, Liersch R, Mesters RM, Hotfilder M, Marra A, Koschmieder S, Dugas M, Berdel WE, Serve H, Muller-Tidow C (2009) S100A2 induces metastasis in non-small cell lung cancer. Clin Cancer Res 15:22–29

Camelo S, Iglesias AH, Hwang D, Due B, Ryu H, Smith K, Gray SG, Imitola J, Duran G, Assaf B, Langley B, Khoury SJ, Stephanopoulos G, De Girolami U, Ratan RR, Ferrante RJ, Dangond F (2005) Transcriptional therapy with the histone deacetylase inhibitor trichostatin A ameliorates experimental autoimmune encephalomyelitis. J Neuroimmunol 164:10–21

Castellano S, Milite C, Ragno R, Simeoni S, Mai A, Limongelli V, Novellino E, Bauer I, Brosch G, Spannhoff A, Cheng D, Bedford MT, Sbardella G (2010) Design, synthesis and biological evaluation of carboxy analogues of arginine methyltransferase inhibitor 1 (AMI-1). ChemMedChem 5:398–414

Chang X, Yamada R, Suzuki A, Sawada T, Yoshino S, Tokuhiro S, Yamamoto K (2005) Localization of peptidylarginine deiminase 4 (PADI4) and citrullinated protein in synovial tissue of rheumatoid arthritis. Rheumatology (Oxford) 44:40–50

Chang B, Chen Y, Zhao Y, Bruick RK (2007) JMJD6 is a histone arginine demethylase. Science 318:444–447

Chang X, Zhao Y, Sun S, Zhang Y, Zhu Y (2009) The expression of PADI4 in synovium of rheumatoid arthritis. Rheumatol Int 29:1411–1416

Chen LF, Greene WC (2003) Regulation of distinct biological activities of the NF-kappaB transcription factor complex by acetylation. J Mol Med 81:549–557

Chen L, Fischle W, Verdin E, Greene WC (2001) Duration of nuclear NF-kappaB action regulated by reversible acetylation. Science 293:1653–1657

Chen KL, Wang SS, Yang YY, Yuan RY, Chen RM, Hu CJ (2009) The epigenetic effects of amyloid-beta(1–40) on global DNA and neprilysin genes in murine cerebral endothelial cells. Biochem Biophys Res Commun 378:57–61

Cheng D, Yadav N, King RW, Swanson MS, Weinstein EJ, Bedford MT (2004) Small molecule regulators of protein arginine methyltransferases. J Biol Chem 279:23892–23899

Cheung N, Chan LC, Thompson A, Cleary ML, So CW (2007) Protein arginine-methyltransferase-dependent oncogenesis. Nat Cell Biol 9:1208–1215

Chevillard-Briet M, Trouche D, Vandel L (2002) Control of CBP co-activating activity by arginine methylation. EMBO J 21:5457–5466

Choi KC, Jung MG, Lee YH, Yoon JC, Kwon SH, Kang HB, Kim MJ, Cha JH, Kim YJ, Jun WJ, Lee JM, Yoon HG (2009) Epigallocatechin-3-gallate, a histone acetyltransferase inhibitor, inhibits EBV-induced B lymphocyte transformation via suppression of RelA acetylation. Cancer Res 69:583–592

Chung YL, Lee MY, Wang AJ, Yao LF (2003) A therapeutic strategy uses histone deacetylase inhibitors to modulate the expression of genes involved in the pathogenesis of rheumatoid arthritis. Mol Ther 8:707–717

Cloos PA, Christensen J, Agger K, Maiolica A, Rappsilber J, Antal T, Hansen KH, Helin K (2006) The putative oncogene GASC1 demethylates tri- and dimethylated lysine 9 on histone H3. Nature 442:307–311

Cohen T, Nahari D, Cerem LW, Neufeld G, Levi BZ (1996) Interleukin 6 induces the expression of vascular endothelial growth factor. J Biol Chem 271:736–741

Collett K, Eide GE, Arnes J, Stefansson IM, Eide J, Braaten A, Aas T, Otte AP, Akslen LA (2006) Expression of enhancer of zeste homologue 2 is significantly associated with increased tumor cell proliferation and is a marker of aggressive breast cancer. Clin Cancer Res 12:1168–1174

Coussens LM, Werb Z (2002) Inflammation and cancer. Nature 420:860–867

Covic M, Hassa PO, Saccani S, Buerki C, Meier NI, Lombardi C, Imhof R, Bedford MT, Natoli G, Hottiger MO (2005) Arginine methyltransferase CARM1 is a promoter-specific regulator of NF-kappaB-dependent gene expression. EMBO J 24:85–96

Croce CM (2008) Oncogenes and cancer. N Engl J Med 358:502–511

Cuthbert GL, Daujat S, Snowden AW, Erdjument-Bromage H, Hagiwara T, Yamada M, Schneider R, Gregory PD, Tempst P, Bannister AJ, Kouzarides T (2004) Histone deimination antagonizes arginine methylation. Cell 118:545–553

Das PM, Singal R (2004) DNA methylation and cancer. J Clin Oncol 22:4632–4642

Dashwood RH, Myzak MC, Ho E (2006) Dietary HDAC inhibitors: time to rethink weak ligands in cancer chemoprevention? Carcinogenesis 27:344–349

Ding L, Erdmann C, Chinnaiyan AM, Merajver SD, Kleer CG (2006) Identification of EZH2 as a molecular marker for a precancerous state in morphologically normal breast tissues. Cancer Res 66:4095–4099

Dobrovic A, Simpfendorfer D (1997) Methylation of the BRCA1 gene in sporadic breast cancer. Cancer Res 57:3347–3350

Dou Y, Milne TA, Ruthenburg AJ, Lee S, Lee JW, Verdine GL, Allis CD, Roeder RG (2006) Regulation of MLL1 H3K4 methyltransferase activity by its core components. Nat Struct Mol Biol 13:713–719

Duncan HF, Smith AJ, Fleming GJ, Cooper PR (2011) HDACi: cellular effects, opportunities for restorative dentistry. J Dent Res 90:1377–1388

Egger G, Liang G, Aparicio A, Jones PA (2004) Epigenetics in human disease and prospects for epigenetic therapy. Nature 429:457–463

El Mezayen R, El Gazzar M, Myer R, High KP (2009) Aging-dependent upregulation of IL-23p19 gene expression in dendritic cells is associated with differential transcription factor binding and histone modifications. Aging Cell 8:553–565

Eliseeva ED, Valkov V, Jung M, Jung MO (2007) Characterization of novel inhibitors of histone acetyltransferases. Mol Cancer Ther 6:2391–2398

El-Osta A, Brasacchio D, Yao D, Pocai A, Jones PL, Roeder RG, Cooper ME, Brownlee M (2008) Transient high glucose causes persistent epigenetic changes and altered gene expression during subsequent normoglycemia. J Exp Med 205:2409–2417

Ermann J, Fathman CG (2001) Autoimmune diseases: genes, bugs and failed regulation. Nat Immunol 2:759–761

Esteller M (2008) Epigenetics in cancer. N Engl J Med 358:1148–1159

Feinberg AP (2007) Phenotypic plasticity and the epigenetics of human disease. Nature 447:433–440

Feinberg AP, Tycko B (2004) The history of cancer epigenetics. Nat Rev Cancer 4:143–153

Feldmann M, Maini RN (1999) The role of cytokines in the pathogenesis of rheumatoid arthritis. Rheumatology (Oxford) 38(Suppl 2):3–7

Firestein GS (2003) Evolving concepts of rheumatoid arthritis. Nature 423:356–361

Fleming JL, Huang TH, Toland AE (2008) The role of parental and grandparental epigenetic alterations in familial cancer risk. Cancer Res 68:9116–9121

Foster SL, Hargreaves DC, Medzhitov R (2007) Gene-specific control of inflammation by TLR-induced chromatin modifications. Nature 447:972–978

Frew AJ, Johnstone RW, Bolden JE (2009) Enhancing the apoptotic and therapeutic effects of HDAC inhibitors. Cancer Lett 280:125–133

Frietze S, Lupien M, Silver PA, Brown M (2008) CARM1 regulates estrogen-stimulated breast cancer growth through up-regulation of E2F1. Cancer Res 68:301–306

Futscher BW, Oshiro MM, Wozniak RJ, Holtan N, Hanigan CL, Duan H, Domann FE (2002) Role for DNA methylation in the control of cell type specific maspin expression. Nat Genet 31:175–179

Futscher BW, O'Meara MM, Kim CJ, Rennels MA, Lu D, Gruman LM, Seftor RE, Hendrix MJ, Domann FE (2004) Aberrant methylation of the maspin promoter is an early event in human breast cancer. Neoplasia 6:380–389

Goldstein AM (2004) Familial melanoma, pancreatic cancer and germline CDKN2A mutations. Hum Mutat 23:630

Gonzalez-Zulueta M, Bender CM, Yang AS, Nguyen T, Beart RW, Van Tornout JM, Jones PA (1995) Methylation of the 5' CpG island of the p16/CDKN2 tumor suppressor gene in normal and transformed human tissues correlates with gene silencing. Cancer Res 55:4531–4535

Grabiec AM, Tak PP, Reedquist KA (2008) Targeting histone deacetylase activity in rheumatoid arthritis and asthma as prototypes of inflammatory disease: should we keep our HATs on? Arthritis Res Ther 10:226

Graff JR, Herman JG, Lapidus RG, Chopra H, Xu R, Jarrard DF, Isaacs WB, Pitha PM, Davidson NE, Baylin SB (1995) E-cadherin expression is silenced by DNA hypermethylation in human breast and prostate carcinomas. Cancer Res 55:5195–5199

Greiner D, Bonaldi T, Eskeland R, Roemer E, Imhof A (2005) Identification of a specific inhibitor of the histone methyltransferase SU(VAR)3-9. Nat Chem Biol 1:143–145

Gridelli C, Rossi A, Maione P (2008) The potential role of histone deacetylase inhibitors in the treatment of non-small-cell lung cancer. Crit Rev Oncol Hematol 68:29–36

Grivennikov SI, Karin M (2010) Inflammation and oncogenesis: a vicious connection. Curr Opin Genet Dev 20:65–71

Gronbaek K, de Nully BP, Moller MB, Nedergaard T, Ralfkiaer E, Moller P, Zeuthen J, Guldberg P (2000) Concurrent disruption of p16INK4a and the ARF-p53 pathway predicts poor prognosis in aggressive non-Hodgkin's lymphoma. Leukemia 14:1727–1735

Gronbaek K, Hother C, Jones PA (2007) Epigenetic changes in cancer. APMIS 115:1039–1059

Grugel S, Finkenzeller G, Weindel K, Barleon B, Marme D (1995) Both v-Ha-Ras and v-Raf stimulate expression of the vascular endothelial growth factor in NIH 3T3 cells. J Biol Chem 270:25915–25919

Guan JS, Haggarty SJ, Giacometti E, Dannenberg JH, Joseph N, Gao J, Nieland TJ, Zhou Y, Wang X, Mazitschek R, Bradner JE, DePinho RA, Jaenisch R, Tsai LH (2009) HDAC2 negatively regulates memory formation and synaptic plasticity. Nature 459:55–60

Guenther MG, Jenner RG, Chevalier B, Nakamura T, Croce CM, Canaani E, Young RA (2005) Global and Hox-specific roles for the MLL1 methyltransferase. Proc Natl Acad Sci USA 102:8603–8608

Hamamoto R, Furukawa Y, Morita M, Iimura Y, Silva FP, Li M, Yagyu R, Nakamura Y (2004) SMYD3 encodes a histone methyltransferase involved in the proliferation of cancer cells. Nat Cell Biol 6:731–740

Hamamoto R, Silva FP, Tsuge M, Nishidate T, Katagiri T, Nakamura Y, Furukawa Y (2006) Enhanced SMYD3 expression is essential for the growth of breast cancer cells. Cancer Sci 97:113–118

Hanahan D, Weinberg RA (2000) The hallmarks of cancer. Cell 100:57–70

Harris AL (2002) Hypoxia–a key regulatory factor in tumour growth. Nat Rev Cancer 2:38–47

Hassa PO, Buerki C, Lombardi C, Imhof R, Hottiger MO (2003) Transcriptional coactivation of nuclear factor-kappaB-dependent gene expression by p300 is regulated by poly(ADP)-ribose polymerase-1. J Biol Chem 278:45145–45153

Hassa PO, Haenni SS, Buerki C, Meier NI, Lane WS, Owen H, Gersbach M, Imhof R, Hottiger MO (2005) Acetylation of poly(ADP-ribose) polymerase-1 by p300/CREB-binding protein regulates coactivation of NF-kappaB-dependent transcription. J Biol Chem 280:40450–40464

Hauke J, Riessland M, Lunke S, Eyupoglu IY, Blumcke I, El-Osta A, Wirth B, Hahnen E (2009) Survival motor neuron gene 2 silencing by DNA methylation correlates with spinal muscular atrophy disease severity and can be bypassed by histone deacetylase inhibition. Hum Mol Genet 18:304–317

Hayakawa J, Mittal S, Wang Y, Korkmaz KS, Adamson E, English C, Ohmichi M, McClelland M, Mercola D (2004) Identification of promoters bound by c-Jun/ATF2 during rapid large-scale gene activation following genotoxic stress. Mol Cell 16:521–535

Hayes MP, Freeman SL, Donnelly RP (1995) IFN-gamma priming of monocytes enhances LPS-induced TNF production by augmenting both transcription and MRNA stability. Cytokine 7:427–435

Heidland A, Klassen A, Rutkowski P, Bahner U (2006) The contribution of Rudolf Virchow to the concept of inflammation: what is still of importance? J Nephrol 19(Suppl 10):S102–S109

Herman JG, Latif F, Weng Y, Lerman MI, Zbar B, Liu S, Samid D, Duan DS, Gnarra JR, Linehan WM et al (1994) Silencing of the VHL tumor-suppressor gene by DNA methylation in renal carcinoma. Proc Natl Acad Sci USA 91:9700–9704

Hess JL (2004) Mechanisms of transformation by MLL. Crit Rev Eukaryot Gene Expr 14:235–254

Hirst M, Marra MA (2009) Epigenetics and human disease. Int J Biochem Cell Biol 41:136–146

Hong H, Kao C, Jeng MH, Eble JN, Koch MO, Gardner TA, Zhang S, Li L, Pan CX, Hu Z, MacLennan GT, Cheng L (2004) Aberrant expression of CARM1, a transcriptional coactivator of androgen receptor, in the development of prostate carcinoma and androgen-independent status. Cancer 101:83–89

Hsieh CJ, Klump B, Holzmann K, Borchard F, Gregor M, Porschen R (1998) Hypermethylation of the p16INK4a promoter in colectomy specimens of patients with long-standing and extensive ulcerative colitis. Cancer Res 58:3942–3945

Huang J, Sengupta R, Espejo AB, Lee MG, Dorsey JA, Richter M, Opravil S, Shiekhattar R, Bedford MT, Jenuwein T, Berger SL (2007) p53 is regulated by the lysine demethylase LSD1. Nature 449:105–108

Huber LC, Distler JH, Moritz F, Hemmatazad H, Hauser T, Michel BA, Gay RE, Matucci-Cerinic M, Gay S, Distler O, Jungel A (2007a) Trichostatin A prevents the accumulation of extracellular matrix in a mouse model of bleomycin-induced skin fibrosis. Arthritis Rheum 56:2755–2764

Huber LC, Stanczyk J, Jungel A, Gay S (2007b) Epigenetics in inflammatory rheumatic diseases. Arthritis Rheum 56:3523–3531

Huber K, Schemies J, Uciechowska U, Wagner JM, Rumpf T, Lewrick F, Suss R, Sippl W, Jung M, Bracher F (2010) Novel 3-arylideneindolin-2-ones as inhibitors of NAD+ -dependent histone deacetylases (sirtuins). J Med Chem 53:1383–1386

Issa JP, Ottaviano YL, Celano P, Hamilton SR, Davidson NE, Baylin SB (1994) Methylation of the oestrogen receptor CpG island links ageing and neoplasia in human colon. Nat Genet 7:536–540

Issa JP, Ahuja N, Toyota M, Bronner MP, Brentnall TA (2001) Accelerated age-related CpG island methylation in ulcerative colitis. Cancer Res 61:3573–3577

Issa JP, Garcia-Manero G, Giles FJ, Mannari R, Thomas D, Faderl S, Bayar E, Lyons J, Rosenfeld CS, Cortes J, Kantarjian HM (2004) Phase 1 study of low-dose prolonged exposure schedules of the hypomethylating agent 5-aza-2'-deoxycytidine (decitabine) in hematopoietic malignancies. Blood 103:1635–1640

Ito K, Barnes PJ, Adcock IM (2000) Glucocorticoid receptor recruitment of histone deacetylase 2 inhibits interleukin-1beta-induced histone H4 acetylation on lysines 8 and 12. Mol Cell Biol 20:6891–6903

Ito K, Ito M, Elliott WM, Cosio B, Caramori G, Kon OM, Barczyk A, Hayashi S, Adcock IM, Hogg JC, Barnes PJ (2005) Decreased histone deacetylase activity in chronic obstructive pulmonary disease. N Engl J Med 352:1967–1976

Iwata N, Tsubuki S, Takaki Y, Watanabe K, Sekiguchi M, Hosoki E, Kawashima-Morishima M, Lee HJ, Hama E, Sekine-Aizawa Y, Saido TC (2000) Identification of the major Abeta1-42-degrading catabolic pathway in brain parenchyma: suppression leads to biochemical and pathological deposition. Nat Med 6:143–150

Jaenisch R, Bird A (2003) Epigenetic regulation of gene expression: how the genome integrates intrinsic and environmental signals. Nat Genet 33(Suppl):245–254

Jain RK, Safabakhsh N, Sckell A, Chen Y, Jiang P, Benjamin L, Yuan F, Keshet E (1998) Endothelial cell death, angiogenesis, and microvascular function after castration in an androgen-dependent tumor: role of vascular endothelial growth factor. Proc Natl Acad Sci U S A 95:10820–10825

Jones PA, Baylin SB (2002) The fundamental role of epigenetic events in cancer. Nat Rev Genet 3:415–428

Jungel A, Baresova V, Ospelt C, Simmen BR, Michel BA, Gay RE, Gay S, Seemayer CA, Neidhart M (2006) Trichostatin A sensitises rheumatoid arthritis synovial fibroblasts for TRAIL-induced apoptosis. Ann Rheum Dis 65:910–912

Kane MF, Loda M, Gaida GM, Lipman J, Mishra R, Goldman H, Jessup JM, Kolodner R (1997) Methylation of the hMLH1 promoter correlates with lack of expression of hMLH1 in sporadic colon tumors and mismatch repair-defective human tumor cell lines. Cancer Res 57:808–811

Kang SH, Choi HH, Kim SG, Jong HS, Kim NK, Kim SJ, Bang YJ (2000) Transcriptional inactivation of the tissue inhibitor of metalloproteinase-3 gene by DNA hypermethylation of the 5'-CpG island in human gastric cancer cell lines. Int J Cancer 86:632–635

Kang GH, Lee S, Kim WH, Lee HW, Kim JC, Rhyu MG, Ro JY (2002) Epstein-barr virus-positive gastric carcinoma demonstrates frequent aberrant methylation of multiple genes and constitutes CpG island methylator phenotype-positive gastric carcinoma. Am J Pathol 160:787–794

Kang MY, Lee BB, Kim YH, Chang DK, Kyu Park S, Chun HK, Song SY, Park J, Kim DH (2007) Association of the SUV39H1 histone methyltransferase with the DNA methyltransferase 1 at mRNA expression level in primary colorectal cancer. Int J Cancer 121:2192–2197

Karin M (2006) Nuclear factor-kappaB in cancer development and progression. Nature 441:431–436

Katzenellenbogen RA, Baylin SB, Herman JG (1999) Hypermethylation of the DAP-kinase CpG island is a common alteration in B-cell malignancies. Blood 93:4347–4353

Kim HS, Kim EM, Kim NJ, Chang KA, Choi Y, Ahn KW, Lee JH, Kim S, Park CH, Suh YH (2004) Inhibition of histone deacetylation enhances the neurotoxicity induced by the C-terminal fragments of amyloid precursor protein. J Neurosci Res 75:117–124

Kondo Y, Shen L, Ahmed S, Boumber Y, Sekido Y, Haddad BR, Issa JP (2008) Downregulation of histone H3 lysine 9 methyltransferase G9a induces centrosome disruption and chromosome instability in cancer cells. PLoS One 3:e2037

Koura AN, Liu W, Kitadai Y, Singh RK, Radinsky R, Ellis LM (1996) Regulation of vascular endothelial growth factor expression in human colon carcinoma cells by cell density. Cancer Res 56:3891–3894

Kruys V, Thompson P, Beutler B (1993) Extinction of the tumor necrosis factor locus, and of genes encoding the lipopolysaccharide signaling pathway. J Exp Med 177:1383–1390

Kubicek S, O'Sullivan RJ, August EM, Hickey ER, Zhang Q, Teodoro ML, Rea S, Mechtler K, Kowalski JA, Homon CA, Kelly TA, Jenuwein T (2007) Reversal of H3K9me2 by a small-molecule inhibitor for the G9a histone methyltransferase. Mol Cell 25:473–481

Kuendgen A, Lubbert M (2008) Current status of epigenetic treatment in myelodysplastic syndromes. Ann Hematol 87:601–611

Laderoute KR, Alarcon RM, Brody MD, Calaoagan JM, Chen EY, Knapp AM, Yun Z, Denko NC, Giaccia AJ (2000) Opposing effects of hypoxia on expression of the angiogenic inhibitor thrombospondin 1 and the angiogenic inducer vascular endothelial growth factor. Clin Cancer Res 6:2941–2950

Laird PW (2003) The power and the promise of DNA methylation markers. Nat Rev Cancer 3:253–266

Lal G, Zhang N, van der Touw W, Ding Y, Ju W, Bottinger EP, Reid SP, Levy DE, Bromberg JS (2009) Epigenetic regulation of Foxp3 expression in regulatory T cells by DNA methylation. J Immunol 182:259–273

Lee JY, Sullivan KE (2001) Gamma interferon and lipopolysaccharide interact at the level of transcription to induce tumor necrosis factor alpha expression. Infect Immun 69:2847–2852

Lee JY, Kim NA, Sanford A, Sullivan KE (2003) Histone acetylation and chromatin conformation are regulated separately at the TNF-alpha promoter in monocytes and macrophages. J Leukoc Biol 73:862–871

Lee GR, Kim ST, Spilianakis CG, Fields PE, Flavell RA (2006) T helper cell differentiation: regulation by cis elements and epigenetics. Immunity 24:369–379

Lee MJ, Kim YS, Kummar S, Giaccone G, Trepel JB (2008) Histone deacetylase inhibitors in cancer therapy. Curr Opin Oncol 20:639–649

Lee G, Walser TC, Dubinett SM (2009) Chronic inflammation, chronic obstructive pulmonary disease, and lung cancer. Curr Opin Pulm Med 15:303–307

Li F, Sethi G (2010) Targeting transcription factor NF-kappaB to overcome chemoresistance and radioresistance in cancer therapy. Biochim Biophys Acta 1805:167–180

Li H, Rauch T, Chen ZX, Szabo PE, Riggs AD, Pfeifer GP (2006) The histone methyltransferase SETDB1 and the DNA methyltransferase DNMT3A interact directly and localize to promoters silenced in cancer cells. J Biol Chem 281:19489–19500

Li Y, Reddy MA, Miao F, Shanmugam N, Yee JK, Hawkins D, Ren B, Natarajan R (2008) Role of the histone H3 lysine 4 methyltransferase, SET7/9, in the regulation of NF-kappaB-dependent inflammatory genes. Relevance to diabetes and inflammation. J Biol Chem 283:26771–26781

Lin HI, Chu SJ, Wang D, Feng NH (2004) Pharmacological modulation of TNF production in macrophages. J Microbiol Immunol Infect 37:8–15

Lin HS, Hu CY, Chan HY, Liew YY, Huang HP, Lepescheux L, Bastianelli E, Baron R, Rawadi G, Clement-Lacroix P (2007) Anti-rheumatic activities of histone deacetylase (HDAC) inhibitors in vivo in collagen-induced arthritis in rodents. Br J Pharmacol 150:862–872

Link A, Balaguer F, Goel A (2010) Cancer chemoprevention by dietary polyphenols: promising role for epigenetics. Biochem Pharmacol 80:1771–1792

Liu HL, Chen Y, Cui GH, Zhou JF (2005) Curcumin, a potent anti-tumor reagent, is a novel histone deacetylase inhibitor regulating B-NHL cell line Raji proliferation. Acta Pharmacol Sin 26:603–609

Luo D, Zhang B, Lv L, Xiang S, Liu Y, Ji J, Deng D (2006) Methylation of CpG islands of p16 associated with progression of primary gastric carcinomas. Lab Invest 86:591–598

Mai A, Altucci L (2009) Epi-drugs to fight cancer: from chemistry to cancer treatment, the road ahead. Int J Biochem Cell Biol 41:199–213

Mai A, Cheng D, Bedford MT, Valente S, Nebbioso A, Perrone A, Brosch G, Sbardella G, De Bellis F, Miceli M, Altucci L (2008) epigenetic multiple ligands: mixed histone/protein methyltransferase, acetyltransferase, and class III deacetylase (sirtuin) inhibitors. J Med Chem 51:2279–2290

Majumder S, Liu Y, Ford OH 3rd, Mohler JL, Whang YE (2006) Involvement of arginine methyltransferase CARM1 in androgen receptor function and prostate cancer cell viability. Prostate 66:1292–1301

Mantovani A (2010) Molecular pathways linking inflammation and cancer. Curr Mol Med 10:369–373

Mantovani A, Allavena P, Sica A, Balkwill F (2008) Cancer-related inflammation. Nature 454:436–444

Marks PA, Xu WS (2009) Histone deacetylase inhibitors: potential in cancer therapy. J Cell Biochem 107:600–608

Mazure NM, Chen EY, Yeh P, Laderoute KR, Giaccia AJ (1996) Oncogenic transformation and hypoxia synergistically act to modulate vascular endothelial growth factor expression. Cancer Res 56:3436–3440

McGarvey KM, Fahrner JA, Greene E, Martens J, Jenuwein T, Baylin SB (2006) Silenced tumor suppressor genes reactivated by DNA demethylation do not return to a fully euchromatic chromatin state. Cancer Res 66:3541–3549

McGuire W, Hill AV, Allsopp CE, Greenwood BM, Kwiatkowski D (1994) Variation in the TNF-alpha promoter region associated with susceptibility to cerebral malaria. Nature 371:508–510

McInnes IB, Schett G (2007) Cytokines in the pathogenesis of rheumatoid arthritis. Nat Rev Immunol 7:429–442

Medzhitov R, Horng T (2009) Transcriptional control of the inflammatory response. Nat Rev Immunol 9:692–703

Merlo A, Herman JG, Mao L, Lee DJ, Gabrielson E, Burger PC, Baylin SB, Sidransky D (1995) 5′ CpG island methylation is associated with transcriptional silencing of the tumour suppressor p16/CDKN2/MTS1 in human cancers. Nat Med 1:686–692

Metzger E, Wissmann M, Yin N, Muller JM, Schneider R, Peters AH, Gunther T, Buettner R, Schule R (2005) LSD1 demethylates repressive histone marks to promote androgen-receptor-dependent transcription. Nature 437:436–439

Meyer N, Penn LZ (2008) Reflecting on 25 years with MYC. Nat Rev Cancer 8:976–990

Meyer R, Wolf SS, Obendorf M (2007) PRMT2, a member of the protein arginine methyltransferase family, is a coactivator of the androgen receptor. J Steroid Biochem Mol Biol 107:1–14

Miao F, Gonzalo IG, Lanting L, Natarajan R (2004) In vivo chromatin remodeling events leading to inflammatory gene transcription under diabetic conditions. J Biol Chem 279:18091–18097

Miao F, Smith DD, Zhang L, Min A, Feng W, Natarajan R (2008) Lymphocytes from patients with type 1 diabetes display a distinct profile of chromatin histone H3 lysine 9 dimethylation: an epigenetic study in diabetes. Diabetes 57:3189–3198

Miranda TB, Cortez CC, Yoo CB, Liang G, Abe M, Kelly TK, Marquez VE, Jones PA (2009) DZNep is a global histone methylation inhibitor that reactivates developmental genes not silenced by DNA methylation. Mol Cancer Ther 8:1579–1588

Moscarello MA, Brady GW, Fein DB, Wood DD, Cruz TF (1986) The role of charge microheterogeneity of basic protein in the formation and maintenance of the multilayered structure of myelin: a possible role in multiple sclerosis. J Neurosci Res 15:87–99

Myzak MC, Karplus PA, Chung FL, Dashwood RH (2004) A novel mechanism of chemoprotection by sulforaphane: inhibition of histone deacetylase. Cancer Res 64:5767–5774

Myzak MC, Dashwood WM, Orner GA, Ho E, Dashwood RH (2006a) Sulforaphane inhibits histone deacetylase in vivo and suppresses tumorigenesis in Apc-minus mice. FASEB J 20:506–508

Myzak MC, Hardin K, Wang R, Dashwood RH, Ho E (2006b) Sulforaphane inhibits histone deacetylase activity in BPH-1, LnCaP and PC-3 prostate epithelial cells. Carcinogenesis 27:811–819

Myzak MC, Tong P, Dashwood WM, Dashwood RH, Ho E (2007) Sulforaphane retards the growth of human PC-3 xenografts and inhibits HDAC activity in human subjects. Exp Biol Med (Maywood) 232:227–234

Nasu Y, Nishida K, Miyazawa S, Komiyama T, Kadota Y, Abe N, Yoshida A, Hirohata S, Ohtsuka A, Ozaki T (2008) Trichostatin A, a histone deacetylase inhibitor, suppresses synovial inflammation and subsequent cartilage destruction in a collagen antibody-induced arthritis mouse model. Osteoarthritis Cartilage 16:723–732

Neidhart M, Rethage J, Kuchen S, Kunzler P, Crowl RM, Billingham ME, Gay RE, Gay S (2000) Retrotransposable L1 elements expressed in rheumatoid arthritis synovial tissue: association with genomic DNA hypomethylation and influence on gene expression. Arthritis Rheum 43:2634–2647

Nicholas AP, Lubin FD, Hallett PJ, Vattem P, Ravenscroft P, Bezard E, Zhou S, Fox SH, Brotchie JM, Sweatt JD, Standaert DG (2008) Striatal histone modifications in models of levodopa-induced dyskinesia. J Neurochem 106:486–494

Nicodeme E, Jeffrey KL, Schaefer U, Beinke S, Dewell S, Chung CW, Chandwani R, Marazzi I, Wilson P, Coste H, White J, Kirilovsky J, Rice CM, Lora JM, Prinjha RK, Lee K, Tarakhovsky A (2010) Suppression of inflammation by a synthetic histone mimic. Nature 468:1119–1123

Nile CJ, Read RC, Akil M, Duff GW, Wilson AG (2008) Methylation status of a single CpG site in the IL6 promoter is related to IL6 messenger RNA levels and rheumatoid arthritis. Arthritis Rheum 58:2686–2693

Nishida K, Komiyama T, Miyazawa S, Shen ZN, Furumatsu T, Doi H, Yoshida A, Yamana J, Yamamura M, Ninomiya Y, Inoue H, Asahara H (2004) Histone deacetylase inhibitor suppression of autoantibody-mediated arthritis in mice via regulation of p16INK4a and p21(WAF1/Cip1) expression. Arthritis Rheum 50:3365–3376

Ohtani-Fujita N, Fujita T, Aoike A, Osifchin NE, Robbins PD, Sakai T (1993) CpG methylation inactivates the promoter activity of the human retinoblastoma tumor-suppressor gene. Oncogene 8:1063–1067

Okada Y, Feng Q, Lin Y, Jiang Q, Li Y, Coffield VM, Su L, Xu G, Zhang Y (2005) hDOT1L links histone methylation to leukemogenesis. Cell 121:167–178

Osawa T, Chong JM, Sudo M, Sakuma K, Uozaki H, Shibahara J, Nagai H, Funata N, Fukayama M (2002) Reduced expression and promoter methylation of p16 gene in Epstein-Barr virus-associated gastric carcinoma. Jpn J Cancer Res 93:1195–1200

Ospelt C, Gay S (2008) The role of resident synovial cells in destructive arthritis. Best Pract Res Clin Rheumatol 22:239–252

Otterson GA, Khleif SN, Chen W, Coxon AB, Kaye FJ (1995) CDKN2 gene silencing in lung cancer by DNA hypermethylation and kinetics of p16INK4 protein induction by 5-aza 2'deoxycytidine. Oncogene 11:1211–1216

Pal S, Vishwanath SN, Erdjument-Bromage H, Tempst P, Sif S (2004) Human SWI/SNF-associated PRMT5 methylates histone H3 arginine 8 and negatively regulates expression of ST7 and NM23 tumor suppressor genes. Mol Cell Biol 24:9630–9645

Pho L, Grossman D, Leachman SA (2006) Melanoma genetics: a review of genetic factors and clinical phenotypes in familial melanoma. Curr Opin Oncol 18:173–179

Pledgie-Tracy A, Sobolewski MD, Davidson NE (2007) Sulforaphane induces cell type-specific apoptosis in human breast cancer cell lines. Mol Cancer Ther 6:1013–1021

Pogribny IP, Tryndyak VP, Muskhelishvili L, Rusyn I, Ross SA (2007) Methyl deficiency, alterations in global histone modifications, and carcinogenesis. J Nutr 137:216S–222S

Portela A, Esteller M (2010) Epigenetic modifications and human disease. Nat Biotechnol 28:1057–1068

Ragno R, Simeoni S, Castellano S, Vicidomini C, Mai A, Caroli A, Tramontano A, Bonaccini C, Trojer P, Bauer I, Brosch G, Sbardella G (2007) Small molecule inhibitors of histone arginine methyltransferases: homology modeling, molecular docking, binding mode analysis, and biological evaluations. J Med Chem 50:1241–1253

Rak J, Mitsuhashi Y, Bayko L, Filmus J, Shirasawa S, Sasazuki T, Kerbel RS (1995) Mutant ras oncogenes upregulate VEGF/VPF expression: implications for induction and inhibition of tumor angiogenesis. Cancer Res 55:4575–4580

Rak J, Mitsuhashi Y, Sheehan C, Tamir A, Viloria-Petit A, Filmus J, Mansour SJ, Ahn NG, Kerbel RS (2000a) Oncogenes and tumor angiogenesis: differential modes of vascular endothelial growth factor up-regulation in ras-transformed epithelial cells and fibroblasts. Cancer Res 60:490–498

Rak J, Yu JL, Klement G, Kerbel RS (2000b) Oncogenes and angiogenesis: signaling three-dimensional tumor growth. J Investig Dermatol Symp Proc 5:24–33

Ramirez-Carrozzi VR, Nazarian AA, Li CC, Gore SL, Sridharan R, Imbalzano AN, Smale ST (2006) Selective and antagonistic functions of SWI/SNF and Mi-2beta nucleosome remodeling complexes during an inflammatory response. Genes Dev 20:282–296

Ravindra KC, Selvi BR, Arif M, Reddy BA, Thanuja GR, Agrawal S, Pradhan SK, Nagashayana N, Dasgupta D, Kundu TK (2009) Inhibition of lysine acetyltransferase KAT3B/p300 activity by a naturally occurring hydroxynaphthoquinone, plumbagin. J Biol Chem 284: 24453–24464

Reiman JM, Kmieciak M, Manjili MH, Knutson KL (2007) Tumor immunoediting and immunosculpting pathways to cancer progression. Semin Cancer Biol 17:275–287

Richardson B, Scheinbart L, Strahler J, Gross L, Hanash S, Johnson M (1990) Evidence for impaired T cell DNA methylation in systemic lupus erythematosus and rheumatoid arthritis. Arthritis Rheum 33:1665–1673

Rosin MP, Anwar WA, Ward AJ (1994) Inflammation, chromosomal instability, and cancer: the schistosomiasis model. Cancer Res 54:1929s–1933s

Saigo K, Yoshida K, Ikeda R, Sakamoto Y, Murakami Y, Urashima T, Asano T, Kenmochi T, Inoue I (2008) Integration of hepatitis B virus DNA into the myeloid/lymphoid or mixed-lineage leukemia (MLL4) gene and rearrangements of MLL4 in human hepatocellular carcinoma. Hum Mutat 29:703–708

Schottenfeld D, Beebe-Dimmer J (2006) Chronic inflammation: a common and important factor in the pathogenesis of neoplasia. CA Cancer J Clin 56:69–83

Selvi BR, Batta K, Kishore AH, Mantelingu K, Varier RA, Balasubramanyam K, Pradhan SK, Dasgupta D, Sriram S, Agrawal S, Kundu TK (2010a) Identification of a novel inhibitor of coactivator-associated arginine methyltransferase 1 (CARM1)-mediated methylation of histone H3 Arg-17. J Biol Chem 285:7143–7152

Selvi BR, Mohankrishna DV, Ostwal YB, Kundu TK (2010b) Small molecule modulators of histone acetylation and methylation: a disease perspective. Biochim Biophys Acta 1799:810–828

Semenza GL (2000) Hypoxia, clonal selection, and the role of HIF-1 in tumor progression. Crit Rev Biochem Mol Biol 35:71–103

Sen R, Baltimore D (1986, 2006) Multiple nuclear factors interact with the immunoglobulin enhancer sequences. Cell 46(1986):705–716; Republished in 2006: J Immunol 177:7485–7496

Sethi G, Tergaonkar V (2009) Potential pharmacological control of the NF-kappaB pathway. Trends Pharmacol Sci 30:313–321

Sethi G, Sung B, Aggarwal BB (2008) TNF: a master switch for inflammation to cancer. Front Biosci 13:5094–5107

Shchors K, Shchors E, Rostker F, Lawlor ER, Brown-Swigart L, Evan GI (2006) The Myc-dependent angiogenic switch in tumors is mediated by interleukin 1beta. Genes Dev 20:2527–2538

Shen HM, Tergaonkar V (2009) NFkappaB signaling in carcinogenesis and as a potential molecular target for cancer therapy. Apoptosis 14:348–363

Sheta EA, Harding MA, Conaway MR, Theodorescu D (2000) Focal adhesion kinase, Rap1, and transcriptional induction of vascular endothelial growth factor. J Natl Cancer Inst 92:1065–1073

Shi Y, Lan F, Matson C, Mulligan P, Whetstine JR, Cole PA, Casero RA (2004) Histone demethylation mediated by the nuclear amine oxidase homolog LSD1. Cell 119:941–953

Shi X, Kachirskaia I, Yamaguchi H, West LE, Wen H, Wang EW, Dutta S, Appella E, Gozani O (2007) Modulation of p53 function by SET8-mediated methylation at lysine 382. Mol Cell 27:636–646

Shuto T, Furuta T, Oba M, Xu H, Li JD, Cheung J, Gruenert DC, Uehara A, Suico MA, Okiyoneda T, Kai H (2006) Promoter hypomethylation of Toll-like receptor-2 gene is associated with increased proinflammatory response toward bacterial peptidoglycan in cystic fibrosis bronchial epithelial cells. FASEB J 20:782–784

Siegmund KD, Connor CM, Campan M, Long TI, Weisenberger DJ, Biniszkiewicz D, Jaenisch R, Laird PW, Akbarian S (2007) DNA methylation in the human cerebral cortex is dynamically regulated throughout the life span and involves differentiated neurons. PLoS One 2:e895

Smith JB, Haynes MK (2002) Rheumatoid arthritis–a molecular understanding. Ann Intern Med 136:908–922

Spannhoff A, Machmur R, Heinke R, Trojer P, Bauer I, Brosch G, Schule R, Hanefeld W, Sippl W, Jung M (2007) A novel arginine methyltransferase inhibitor with cellular activity. Bioorg Med Chem Lett 17:4150–4153

Spannhoff A, Sippl W, Jung M (2009) Cancer treatment of the future: inhibitors of histone methyltransferases. Int J Biochem Cell Biol 41:4–11

Stirzaker C, Millar DS, Paul CL, Warnecke PM, Harrison J, Vincent PC, Frommer M, Clark SJ (1997) Extensive DNA methylation spanning the Rb promoter in retinoblastoma tumors. Cancer Res 57:2229–2237

Sullivan KE, Reddy AB, Dietzmann K, Suriano AR, Kocieda VP, Stewart M, Bhatia M (2007) Epigenetic regulation of tumor necrosis factor alpha. Mol Cell Biol 27:5147–5160

Sun Y, Jiang X, Chen S, Price BD (2006) Inhibition of histone acetyltransferase activity by anacardic acid sensitizes tumor cells to ionizing radiation. FEBS Lett 580:4353–4356

Sung B, Pandey MK, Ahn KS, Yi T, Chaturvedi MM, Liu M, Aggarwal BB (2008) Anacardic acid (6-nonadecyl salicylic acid), an inhibitor of histone acetyltransferase, suppresses expression of nuclear factor-kappaB-regulated gene products involved in cell survival, proliferation, invasion, and inflammation through inhibition of the inhibitory subunit of nuclear factor-kappaBalpha kinase, leading to potentiation of apoptosis. Blood 111:4880–4891

Surh YJ, Kundu JK, Na HK, Lee JS (2005) Redox-sensitive transcription factors as prime targets for chemoprevention with anti-inflammatory and antioxidative phytochemicals. J Nutr 135:2993S–3001S

Szlosarek P, Charles KA, Balkwill FR (2006) Tumour necrosis factor-alpha as a tumour promoter. Eur J Cancer 42:745–750

Takahashi K, Sugi Y, Hosono A, Kaminogawa S (2009) Epigenetic regulation of TLR4 gene expression in intestinal epithelial cells for the maintenance of intestinal homeostasis. J Immunol 183:6522–6529

Takami N, Osawa K, Miura Y, Komai K, Taniguchi M, Shiraishi M, Sato K, Iguchi T, Shiozawa K, Hashiramoto A, Shiozawa S (2006) Hypermethylated promoter region of DR3, the death receptor 3 gene, in rheumatoid arthritis synovial cells. Arthritis Rheum 54:779–787

Takeda K, Akira S (2004) TLR signaling pathways. Semin Immunol 16:3–9

Tan J, Cang S, Ma Y, Petrillo RL, Liu D (2010) Novel histone deacetylase inhibitors in clinical trials as anti-cancer agents. J Hematol Oncol 3:5

Taylor CT (2008) Interdependent roles for hypoxia inducible factor and nuclear factor-kappaB in hypoxic inflammation. J Physiol 586:4055–4059

Tsai EY, Falvo JV, Tsytsykova AV, Barczak AK, Reimold AM, Glimcher LH, Fenton MJ, Gordon DC, Dunn IF, Goldfeld AE (2000) A lipopolysaccharide-specific enhancer complex involving Ets, Elk-1, Sp1, and CREB binding protein and p300 is recruited to the tumor necrosis factor alpha promoter in vivo. Mol Cell Biol 20:6084–6094

Tsukada Y, Fang J, Erdjument-Bromage H, Warren ME, Borchers CH, Tempst P, Zhang Y (2006) Histone demethylation by a family of JmjC domain-containing proteins. Nature 439:811–816

Urdinguio RG, Sanchez-Mut JV, Esteller M (2009) Epigenetic mechanisms in neurological diseases: genes, syndromes, and therapies. Lancet Neurol 8:1056–1072

Vaissiere T, Sawan C, Herceg Z (2008) Epigenetic interplay between histone modifications and DNA methylation in gene silencing. Mutat Res 659:40–48

Vallabhapurapu S, Karin M (2009) Regulation and function of NF-kappaB transcription factors in the immune system. Annu Rev Immunol 27:693–733

Vedel M, Lawrence F, Robert-Gero M, Lederer E (1978) The antifungal antibiotic sinefungin as a very active inhibitor of methyltransferases and of the transformation of chick embryo fibroblasts by Rous sarcoma virus. Biochem Biophys Res Commun 85:371–376

Verbiest V, Montaudon D, Tautu MT, Moukarzel J, Portail JP, Markovits J, Robert J, Ichas F, Pourquier P (2008) Protein arginine (N)-methyl transferase 7 (PRMT7) as a potential target for the sensitization of tumor cells to camptothecins. FEBS Lett 582:1483–1489

Villagra A, Sotomayor EM, Seto E (2010) Histone deacetylases and the immunological network: implications in cancer and inflammation. Oncogene 29:157–173

Villeneuve LM, Reddy MA, Lanting LL, Wang M, Meng L, Natarajan R (2008) Epigenetic histone H3 lysine 9 methylation in metabolic memory and inflammatory phenotype of vascular smooth muscle cells in diabetes. Proc Natl Acad Sci USA 105:9047–9052

Vossenaar ER, Radstake TR, van der Heijden A, van Mansum MA, Dieteren C, de Rooij DJ, Barrera P, Zendman AJ, van Venrooij WJ (2004) Expression and activity of citrullinating peptidylarginine deiminase enzymes in monocytes and macrophages. Ann Rheum Dis 63:373–381

Wagner JM, Hackanson B, Lubbert M, Jung M (2010) Histone deacetylase (HDAC) inhibitors in recent clinical trials for cancer therapy. Clin Epigenet 1:117–136

Wang J, Lee JJ, Wang L, Liu DD, Lu C, Fan YH, Hong WK, Mao L (2004) Value of p16INK4a and RASSF1A promoter hypermethylation in prognosis of patients with resectable non-small cell lung cancer. Clin Cancer Res 10:6119–6125

Weinberg RA (1994) Oncogenes and tumor suppressor genes. CA Cancer J Clin 44:160–170

Wells AD (2009) New insights into the molecular basis of T cell anergy: anergy factors, avoidance sensors, and epigenetic imprinting. J Immunol 182:7331–7341

Wierda RJ, Geutskens SB, Jukema JW, Quax PH, van den Elsen PJ (2010) Epigenetics in atherosclerosis and inflammation. J Cell Mol Med 14:1225–1240

Wilson AG (2008) Epigenetic regulation of gene expression in the inflammatory response and relevance to common diseases. J Periodontol 79:1514–1519

Wilson AG, Symons JA, McDowell TL, McDevitt HO, Duff GW (1997) Effects of a polymorphism in the human tumor necrosis factor alpha promoter on transcriptional activation. Proc Natl Acad Sci USA 94:3195–3199

Wissmann M, Yin N, Muller JM, Greschik H, Fodor BD, Jenuwein T, Vogler C, Schneider R, Gunther T, Buettner R, Metzger E, Schule R (2007) Cooperative demethylation by JMJD2C and LSD1 promotes androgen receptor-dependent gene expression. Nat Cell Biol 9:347–353

Wolf JS, Chen Z, Dong G, Sunwoo JB, Bancroft CC, Capo DE, Yeh NT, Mukaida N, Van Waes C (2001) IL (interleukin)-1alpha promotes nuclear factor-kappaB and AP-1-induced IL-8 expression, cell survival, and proliferation in head and neck squamous cell carcinomas. Clin Cancer Res 7:1812–1820

Wong ET, Tergaonkar V (2009) Roles of NF-kappaB in health and disease: mechanisms and therapeutic potential. Clin Sci (Lond) 116:451–465

Yamamoto Y, Verma UN, Prajapati S, Kwak YT, Gaynor RB (2003) Histone H3 phosphorylation by IKK-alpha is critical for cytokine-induced gene expression. Nature 423:655–659

Yang HJ, Liu VW, Wang Y, Tsang PC, Ngan HY (2006a) Differential DNA methylation profiles in gynecological cancers and correlation with clinico-pathological data. BMC Cancer 6:212

Yang SR, Chida AS, Bauter MR, Shafiq N, Seweryniak K, Maggirwar SB, Kilty I, Rahman I (2006b) Cigarette smoke induces proinflammatory cytokine release by activation of NF-kappaB and posttranslational modifications of histone deacetylase in macrophages. Am J Physiol Lung Cell Mol Physiol 291:L46–L57

Yang SR, Valvo S, Yao H, Kode A, Rajendrasozhan S, Edirisinghe I, Caito S, Adenuga D, Henry R, Fromm G, Maggirwar S, Li JD, Bulger M, Rahman I (2008) IKK alpha causes chromatin modification on pro-inflammatory genes by cigarette smoke in mouse lung. Am J Respir Cell Mol Biol 38:689–698

Yao H, Rahman I (2009) Current concepts on the role of inflammation in COPD and lung cancer. Curr Opin Pharmacol 9:375–383

Yoshiura K, Kanai Y, Ochiai A, Shimoyama Y, Sugimura T, Hirohashi S (1995) Silencing of the E-cadherin invasion-suppressor gene by CpG methylation in human carcinomas. Proc Natl Acad Sci USA 92:7416–7419

Zendman AJ, van Venrooij WJ, Pruijn GJ (2006) Use and significance of anti-CCP autoantibodies in rheumatoid arthritis. Rheumatology (Oxford) 45:20–25

Zhang Z, Zhang ZY, Fauser U, Schluesener HJ (2008) Valproic acid attenuates inflammation in experimental autoimmune neuritis. Cell Mol Life Sci 65:4055–4065

Chapter 28
Plasmodium falciparum: Epigenetic Control of *var* Gene Regulation and Disease

Abhijit S. Deshmukh, Sandeep Srivastava, and Suman Kumar Dhar

Abstract *Plasmodium falciparum*, one of the deadliest parasites on earth causes human malaria resulting one million deaths annually. Central to the parasite pathogenicity and morbidity is the switching of parasite virulence (*var*) gene expression causing host immune evasion. The regulation of *Plasmodium var* gene expression is poorly understood. The complex life cycle of *Plasmodium* and mutually exclusive expression pattern of *var* genes make this disease difficult to control. Recent studies have demonstrated the pivotal role of epigenetic mechanism for control of coordinated expression of *var* genes, important for various clinical manifestations of malaria. In this review, we discuss about different *Plasmodium* histones and their various modifications important for gene expression and gene repression. Contribution of epigenetic mechanism to understand the *var* gene expression is also highlighted. We also describe in details *P. falciparum* nuclear architecture including heterochromatin, euchromatin and telomeric regions and their importance in subtelomeric and centrally located *var* gene expression. Finally, we explore the possibility of using Histone Acetyl Transferase (HAT) and Histone Deacetylase (HDAC) inhibitors against multi-drug resistance malaria parasites to provide another line of treatment for malaria.

Plasmodium, one of the most important members of Apicomplexan protozoans imposes a significant economic and health impact on human populations around the world. *Plasmodium* parasite has complex life cycle with morphologically distinct asexual and sexual developmental stages in the human host and Anopheline mosquito vector. The parasites undergo rapid transition between morphological states and antigenic variation in order to sustain chronic infection and immune evasion in human host. Parasites achieve these functions through adopting variety

---

A.S. Deshmukh • S. Srivastava • S.K. Dhar (✉)
Special Centre for Molecular Medicine, Jawaharlal Nehru University,
New Delhi 110067, India
e-mail: skdhar2002@yahoo.co.in

of regulatory pathways like transcriptional and posttranscriptional regulation of gene expression, translational repression and posttranslational modification of proteins. These regulatory modes help to respond to host conditions during an acute stage infection or life cycle transition of the parasite.

The word "Epigenetics" coined by Conard Waddington refers to the change and maintenance of the pattern of gene expression for generations without changing the DNA sequence of the gene (Waddington 1942; Berger et al. 2009). The epigenetic study of eukaryotic organisms helped to understand chromatin mediated regulation of gene expression. Chromatin is physiological substrate for DNA replication, repair and transcription etc. Chromatin based epigenetic mechanism involved DNA methylation, covalent and non-covalent modifications of chromatin and noncoding RNA (Goldberg et al. 2007). In *Plasmodium falciparum*, there is no experimental evidence of DNA methylation so far (Choi et al. 2006; Templeton et al. 2004). Therefore, chromatin alteration is achieved mainly through post translational modifications (PTMs) of histones, chromatin remodelling and replacement of core histones by histone variants. The presence of the several histone modifying enzymes and chromatin remodelling proteins in the parasite genome underlines the significance of epigenetics mechanism in transcriptional regulation. Therefore, epigenetic control may play a fundamental role for the parasite transcriptional regulation to overcome various critical conditions arising due to the presence of the parasites in the different hosts to complete its life cycle.

In *Plasmodium*, the epigenetic mechanism of regulation of gene expression can be divided broadly into three distinct areas based on parasite development (Merrick and Duraisingh 2010).

Firstly, during asexual intra-erythrocytic developmental stages where differential gene expression occurs which are responsible for all clinical symptoms of malaria. *Plasmodium falciparum*, the main *Plasmodium* species responsible for human mortality and morbidity shows unusual mode of gene expression during its 48 h developmental process within the erythrocyte, implying tight and integrated genome-wide regulation of transcription (Bozdech et al. 2003; Le Roch et al. 2003; Llinas et al. 2006). Recently, a battery of proteins like Api-AP2, HP1, histone deacetylases, and histone methylases have been shown to be involved in gene regulation (Balaji et al. 2005; Fan et al. 2004a, b; Iyer et al. 2008). These observations suggest the role of epigenetic mechanism in transcriptional regulation in *Plasmodium*.

Secondly, epigenetics likely play role during sexual and morphological differentiation for the rest of the life cycle. The blood stage parasites differentiate in to gametocytes. These gametocytes mate to form ookinetes followed by formation of sporozoites in the mosquito, leading to the subsequent transmission and development in the human hepatocytes before the release of the merozoites in the asexual erythrocytic cycle. Distinct transcriptional profiling has been reported in gametocytes (Yuda et al. 2009), ookinetes (Silvestrini et al. 2005), oocyst sporozoites (Raibaud et al. 2006), salivary gland sporozoites (Kaiser et al. 2004) hepatocyte stage (Matuschewski et al. 2002; Sacci et al. 2005) and erythrocyte stage (Tarun et al. 2008). All these observations may suggest epigenetic control over life cycle transition and stage differentiation.

Thirdly and most importantly, epigenetic control is involved in the mutually exclusive expression of individual *var* genes involved in the virulence processes such as cytoadherance and variant erythrocyte invasion. The best characterized family of antigen coding gene is the *var* family in *P. falciparum*. This gene family encodes ~60 variants of *P. falciparum* erythrocyte membrane protein 1 (PfEMP1), expressed on infected erythrocytes. PfEMP1 is responsible for the attachment of the infected erythrocytes with the vascular endothelial cells thereby preventing the clearance from the circulatory system. Most of the *var* genes are generally silenced, with only one or a few being expressed at any given time (Le Roch et al. 2003; Chen et al. 1998; Duffy et al. 2002; Fernandez et al. 2002; Merrick et al. 2010; Mok et al. 2007, 2008). The switching of *var* gene expression in an allele specific manner helps parasite to evade from the host immune system leading to the chronic infection. The expressed *var* gene(s) probably does not undergo recombination, as occurs in *Trypanosoma brucei*. The *var* genes are located at the subtelomeric and internal chromosome loci and their expression patterns are marked with differential histone modifications (Scherf et al. 1998; Duraisingh et al. 2005; Freitas-Junior et al. 2005). The repression or silencing of a specific or a set of *var* genes can be reversed when the genes coding for histone deacetylase (PfSir2, HDAC1) enzymes are deleted from *Plasmodium* genome (Lopez-Rubio et al. 2007). More recently, Origin recognition complex (ORC1) protein and Heterochromatin protein1(HP1) have been suggested to have role in *var* gene regulation in *P. falciparum* (Duraisingh et al. 2005; Mancio-Silva et al. 2008; Perez-Toledo et al. 2009; Flueck et al. 2009).

Moreover, invasion proteins present in *P. falciparum*, responsible for new erythrocyte invasion are also regulated epigenetically. All these findings strongly suggest the presence of epigenetic control in *Plasmodium*.

Here we will describe the key players in epigenetic control in general, the status of them in *Plasmodium falciparum* and any significant difference from the conventional systems.

28.1 *Plasmodium* Nucleosome and Histone

The nucleosome is the building block of chromatin, consisting of ~160 bp of DNA wrapped in 1.75 superhelical turns around an octamer of core histones made of one H3/H4 tetramer and two H2A/H2B dimers (Luger et al. 1997). Histones are architectural proteins and play essential roles in DNA replication, repair and transcription. They contain five major classes H1, H2A, H2B, H3 and H4 with some gene variants. Histones H2A, H2B, H3 and H4 are the core histones and form the protein core around which nucleosomal DNA is wrapped. Each core histone contains a C-terminal tail, has a conserved histone fold domain mediating the assembly of heterodimers of specific pairs of histones. The N terminal flexible tail is accessible for several types of posttranslational modifications including phosphorylation, methylation and acetylation on serine and lysine residues, ADP-ribosylation and ubiquitination (Berger 2002). These modifications play essential roles in transcription regulation.

The malaria parasite chromosomes have a typical nucleosomal organization consisting of 155 bp of DNA (Cary et al. 1994; Lanzer et al. 1994). Nucleosomal organization extends into the telomeric repeats (TAREs) but it is totally absent at the

Table 28.1 Histones and their different modifications in *P. falciparum*

Sr. No.	Histones	Gene ID	Chromosome	Length (aa)	PTM(s)
1.	H2A	PFF0860c	11	132	N-term-ac, K3ac, K5ac
2.	H2A.Z	PFC0920w	3	158	N-term-ac, K11ac, K15ac, K19ac, K25ac K28ac, K30ac, K35ac
3.	H2B	PF11_0062	11	117	K112ub
4.	H2Bv	Pf07_0054	7	123	N-term-ac, K3ac, K8ac, K13ac, K14ac, K18ac, T85ph
5.	H3	PFF0510w	6	136	K4me, K4me2, K4me3, K9ac, K9me, K9me3, K14ac, K14me, R17me, R17me2, K18ac, K23ac, K27ac, K36me3, K56ac, K79me3
6.	H3.3	PFF0865w	6	136	K4me, K4me2, K4me3, K9ac, K14ac, R17me, R7me2, K18ac, K23ac, K27ac
7.	H4	PF11_0061	11	103	N-term-ac, R3me, R3me2, K5me, K5ac, K8ac, K12ac, K12me, K16ac, R17me, K20me, K20me2, K20me3
8.	CenH3	Pf13_0185	13	170	

telomere end. In fact, telomere is devoid of histone (Figueiredo et al. 2000). *P. falciparum* genome contains four evolutionary conserved, canonical core histones, H2A, H2B, H3 and H4 and four histone variants H2A.Z, H2Bv, H3.3 and CenH3 respectively (Miao et al. 2006; Trelle et al. 2009). In *P. falciparum*, the linker histone H1 has not been recognized suggesting higher order compaction of nuclear DNA. The details of *P. falciparum* histones are given in Table 28.1. Among the three histone H3 homologues identified, two correspond to the canonical core histone H3 and its variant H3.3, while the other resembles centromeric H3 (CenH3) (Sullivan 2003). The *P. falciparum* H3 and H3.3 protein sequences are 94% identical. PfCenH3 protein shares 61% amino acids identity with H3 and H3.3 (Cui and Miao 2010). Only one histone H4 is present in the *Plasmodium* genome and it is the most conserved among all histones (Beauchamps et al. 1997; Przyborski et al. 2003). The N-terminal tail of H4 is absolutely conserved among the different malaria parasites. Two proteins in *P. falciparum* show homology with H2B (PF07_0054 on chromosome 7 and PF11_0062 on chromosome 11 respectively) (Bennett et al. 1995). Pf11_0062 shows more homology to consensus H2B sequence than Pf07_0054 (Malik and Henikoff 2003). Therefore, Pf11_0062 is considered as Pf H2B and Pf07_0054 is referred to as PfH2Bv.

28.2 Chromosome Boundaries in *P. falciparum*

Eukaryotic genome can be differentiated in dark staining heterochromatin regions and light staining euchromatin. Heterochromatin has a tendency to spread into euchromatin regions. It has been reported that H2A.Z in yeast acts as a boundary element that stops the spread of Sir2 mediated heterochromatin into euchromatin regions of DNA (Meneghini et al. 2003). The natural barriers to spread of heterochromatin into euchromatin are known as boundary elements. H2A.Z maintains an open configuration at promoters regions of different genes and thereby regulates transcriptional regulation of genes. In *P. falciparum*, H2A.Z co-localizes with histone modifications of euchromatin but not in heterochromatin (Petter et al. 2011). Recent reports suggest that histone variant H2A.Z in *Plasmodium* acts as a boundary element. H2A.Z is a variant of H2A and has been conserved throughout the evolution and plays crucial roles in the survival of the organism. PfH2A.Z shows enrichment in the euchromatin compartment of chromatin. In *Plasmodium* two different sirtuin proteins (PfSir2A and PfSir2B respectively) that are responsible for silencing different sets of *var* genes, have been reported. PfH2A.Z is found predominantly at the active *var* gene promoters and competes for binding sites with PfSir2A. Electron microscopy has also revealed the positioning of H2A.Z at the border between dark staining heterochromatin and light coloured euchromatin. H2A.Z deposition takes place at specific *var* gene promoters in ring stage resulting expression of *var* gene. However, in schizont stage, *var* gene promoters are devoid of H2A.Z variant. H2A.Z is the antagonist of PfSir2A but not PfSir2B because this histone variant causes expression of only PfSir2A regulated genes as shown using PfSir2A knock out parasite line where the *var* genes are found in expressed state even in schizont stage.

28.3 Histone Modifications

Canonical and variant histones contain a variety of PTMs located on the N terminal tail. The most common PTMs include acetylation, methylation, phosphorylation, ubiquitination, poly-ADP-ribosylation and sumolyation (Kouzarides 2007). Different histone modifications can alter chromatin structure by modulating the interactions of proteins with DNA and subsequently recruit effector proteins. Specific combinations of these different histone tail modifications create a "histone code" (Jenuwein and Allis 2001).

A lot of efforts have been put to identify various modification of *P. falciparum* histones using mass spectrometry (MS), liquid chromatography-tandem MS (LC-MS/MS), quadruple time-of-fight (Q-TOF) and linear trap quadruple Fourier transform (LTQ-FT). These techniques identified more than 40 post translational modifications on *P. falciparum* histones (Trelle et al. 2009) (Table 28.1).

Majority of *P. falciparum* histone modifications include histone lysine acetylation and methylation marks. Histone acetylation is linked to active genes whereas histone lysine methylation is involved in both activation and silencing. *P. falciparum* H3, H3.3 and H4 contain a number of active gene marks such as acetylated lysine and methylated lysine (H3K4) and arginines (H3R17 and H4R3) and acetylated lysine residues in H3K9. Marks for silent gene include trimethylated H3K9 and H4K20. The PfH2A.Z is acetylated on a repeated GGK motif and involved in transcriptional activation (Miao et al. 2006; Trelle et al. 2009). The PfH2Bv variant of H2B is also acetylated. For other histone modifications, H4 is found to be sumoylated, whereas H2B is ubiquitinated at K112 (Issar et al. 2008a, b; Trelle et al. 2009). There is no evidence of histone phosphorylation yet, as is commonly observed in model eukaryotes.

Histone modifications are controlled by the various enzymes or ATP-dependent remodelers and proteins with PTM-binding modules. Below, we will describe each modification in brief, associated enzyme with that particular type of modification and the status of these modifications and enzymes in *Plasmodium*.

28.3.1 Histone Acetylation and Deacetylation

Histone lysine acetylation is catalysed by histone acetyltransferase (HATs). Five families of HATs have been reported: GNATs (**GCN5 *N*** acetyltransferases), MYSTs (MOZ, Ybf1/Sas3, Sas2 and Tip60), p300/CBP (CREB-binding protein), general transcriptional factor HATs and nuclear hormone-related HATs (Carrozza et al. 2003; Lee and Workman 2007). At least four HATs are found in malaria parasite genomes: PF08_0034, PF11_0192, PFL1345c and PFD0795w (Horrocks et al. 2009). PfGCN5 (PF08_0034) preferentially acetylates H3K9 and K14 *in vitro* and the HAT domain can partially rescue the yeast GCN5 mutant (Fan et al. 2004a, b). The H3K9ac mark enriched in the promoters of active genes helps for gene activation. The recombinant PfMYST (PF11_0192), another HAT protein involved in the regulation of parasite cell cycle, acetylates histone H4 at K5, K8, K12 and K16 (Cui and Miao 2010). There is no evidence of HAT activity of the other two HATs.

Three classes of histone deacetylases (HDACs) have been identified in *P. falciparum*. PfHDAC1 (PFI1260c) is a class I enzyme. PF14_069 (Joshi et al. 1999) and PF10_0078 are Class II HDACs (Horrocks et al. 2009). Two class III enzymes, PfSir2A (Pf13_0152) and PfSir2B (PF14_0489) play role in regulating the mutually exclusive expression of *var* genes. PfSir2A catalyzes NAD-dependent deacetylation of acetylated lysine peptides of H3 and H4. It also shows ADP ribosyltransferase activity on all histones (French et al. 2008; Merrick and Duraisingh 2007). Both PfSir2 paralogues are required for silencing the different *var* gene promoter subsets. PfSir2A is involved in silencing of subtelomeric *var* genes transcribed towards the telomere and the internal *var* genes (promoter types *Ups*A, *Ups*E and *Ups*C), whereas PfSir2B controls *var* genes under the control of *Ups*B promoter (Tonkin et al. 2009). PfSir2A is involved in establishing heterochromatin in the subtelomeric regions and maintenance of telomeric length. Genetic deletion of either PfSir2 leads to a general derepression of subset of the *var* gene family confirming their role in *var* gene regulation (Duraisingh et al. 2005; Merrick et al. 2010; Tonkin et al. 2009).

28.3.2 Histone Methylation

Methylation is catalysed by methyltransferases that attach a methyl group onto its substrate. Bioinformatics analysis of *P. falciparum* genome reveals at least ten members of histone methyltransferases (HMTs) containing a SET [Su(var), E(z),

Trithorax] domain, characteristic of histone lysine methyl transferases. Four well characterized HKMTs (PfSET1, 2, 3 and 8) methylate H3K4, H3K36, H3K9 and H4K20 respectively (Cui et al. 2008a, b). Only recombinant PfSET2 and PfSET8 are enzymatically active and PfSET8 causes H4K20 mono-, di- and tri methylation (Cui et al. 2008a, b; Sautel et al. 2007). Malaria parasites have three arginine methyltransferases (pRMTs): PfRMT1 (PF14_0242), PfRMT5 (Pf13_0323) and PfCARM1 (PF08_0092) (Fan et al. 2009). Only PfCARM1 is well characterized. It catalyzes mono- and di-methylation of H4R3 and some nonhistone substrates (Fan et al. 2009).

The malaria parasite genome also contains two types of histone demethylases (HDMs), the lysine specific demthylases (LSD1) and JmjC (jumionji C) domain containing histone demethylases (JHDMs). There are at least one LDS1 (PFL0575w) and two JHDMs (MAL8P1.111 and PFF0135w) in *Plasmodium* (Shi and Whetstine 2007). The role of histone methylation in gene regulation and maintenance of the subtelomeric heterochromatin needs to be explored further in *P. falciparum*.

28.3.3 Histone Ubiquitination

Malaria parasites encode several proteins involved in ubiquitination pathways. These include paralogs for polyubiquitin (PFL0585w, PY03971), two ubiquitin-ribosomal protein fusions (Ub-S27a and Ub-52) (PF14 0027, PF13 0346), neural precursor cell expressed developmentally and down-regulated 8 (NEDD8) (MAL13p1.64, Pv122475), small ubiquitin-related modifier (SUMO) (PFE0285c, Pv097850), homologs to ubiquitin 1 (HUB1) (PFL1830w, Pv100840), ubiquitin-related modifier 1 (URM1) (Pf11 033, PY06420) and autophagy 8 (ATG8) (PF10 0193, Pv001860) (Ponts et al. 2008). The H2A and H2B are commonly subjected to histone ubiquitination. The H2A ubiquitination is considered as a repressive mark and H2B ubiquitination is involved in both transcriptional activation and silencing (Weake and Workman 2008). The malaria parasite genome contains a number of protein homologues of E1 (ubiquitin/UBL activating enzymes), E2 (ubiquitin/UBL conjugating enzymes) and E3 (ubiquitin/UBL ligases) (MAL8P1.23, PF08_0094, PF08_0020, PFl0470, PFB0440c etc.) enzymes and proteases that might be involved in removing Ubl modifications (Ponder and Bogyo 2007; Ponts et al. 2008). Role of histone ubiquitination in the regulation of parasite gene transcription is not known clearly. Two deubiquitination enzymes PfUCH54 (PF11_0177) and PFUCHL3 (PF14_0576) have been identified in *P. falciparum* (Artavanis-Tsakonas et al. 2006; Frickel et al. 2007).

28.3.4 Histone Sumoylation

In malaria parasites, PfSUMO proteins have been identified which sumoylate H4 (Issar et al. 2008a, b). Histone H4 sumoylation is associated with decreased gene expression (Shiio and Eisenman 2003).

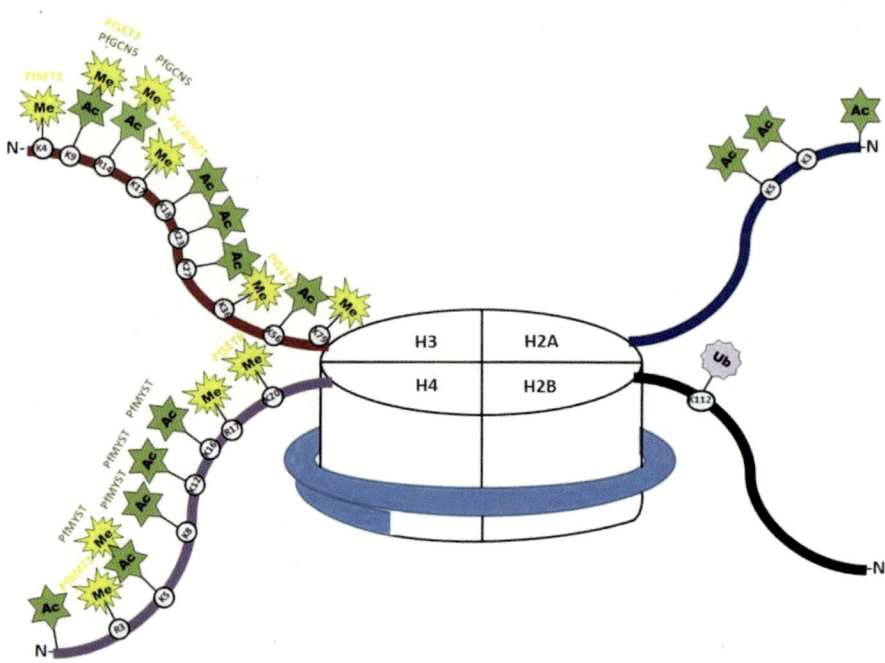

Fig. 28.1 The core histone (*H2A, H2B, H3* and *H4*) and the DNA wrapped with are shown. The lysine residues (*K*) along with their positions are marked. Different types of histone modifications are marked as followed: *Ac* acetylation, *Me* methylation, *Ub* ubiquitination. Several histone code readers like PfSET1, PfSET3, PfMYST etc. are shown (adopted from Cui and Miao 2010 and subsequently modified)

28.3.5 Histone Poly-ADP-Ribosylation

Poly-ADP ribose polymerase homologue has not been identified in *Plasmodium* indicating that histone Poly-ADP-ribosylation may be absent in the malaria parasites.

The nucleosome structure containing the core histones with different types of modifications are shown in Fig. 28.1.

28.4 Histone Code Readers

The arrays of histone tail modification are recognized by a number of conserved protein domains that facilitate downstream events via the recruitment or stabilization of chromatin-related protein complexes. These protein modules are classified into several subgroups like the bromodomain, Royal superfamily, plant homeodomain (PHD) fingers, WD40 repeats and 14-3-3 proteins (Deitsch et al. 2009; Taverna et al. 2007). Bromodomain is an evolutionary conserved acetyl lysine- binding module found in many chromatin-associated proteins (Zeng and Zhou 2002). PHD fingers and the Royal superfamily protein fold, including chromodomain are methyllysine-binding modules (Maurer-Stroh et al. 2003). The WD40 repeat protein WDR5, a component of SET1 complex, binds unmodified H3R2 (Ruthenburg et al. 2006). The 14-3-3 proteins contain phosphoserine-binding modules which bind to H3 phosphoserine with high affinity and are involved in various cellular functions (Macdonald et al. 2005).

The malaria parasites contain a number of proteins with PTM-binding modules. The PfGCN5 HAT and PfSET1 HKMT proteins with a single chromodomain have been identified in *P. falciparum*. *P. falciparum* also contains a number of Royal superfamily and PHD finger domain containing proteins. The PfMYST and heterochromatin protein 1 (PfHP1) contain a single chromodomain. The PfHP1 protein containing a chromodomain and a chromo-shadow domain are involved in H3K9me3 binding and dimerization respectively (Flueck et al. 2009; Perez-Toledo et al. 2009). This protein is associated with the H3K9me3 marks in subtelomeric and intrachromosomal silent *var* genes correlating with silencing status of the *var* gene family. The dimerization is probably responsible for subtelomeric heterochromatin formation by aggregating nucleosomes in subtelomeric region. Ten PHD domain containing proteins, 90 WD40 motif containing proteins and three putative 14-3-3 motif containing proteins have been identified in *P. falciparum* so far. However, these proteins need to be functionally characterized. The details of the histone modifying enzymes and histone code readers are shown in Table 28.2.

28.5 Chromatin Remodelling Complexes

Transcriptional activation and silencing is brought about not only through array of histone post-transcriptional modifications but also through direct activity of chromatin remodelling complexes on nucleosomes. In eukaryotes, four different classes of ATP-dependent chromatin remodelling complexes (Swi/Snf, INO80, ISWI and Mi2/NURD) have been described (Martens and Winston 2003). The functions of Swi2/Snf2 family chromatin remodelers in protozoan parasites are not known clearly. In trypanosomes, a protein related to Swi2/Snf2 is involved in *de novo* synthesis of the modified thymine base 'J' within telomeric DNA, which correlates with the epigenetic silencing of variant surface glycoproteins (VSGs) (DiPaolo et al. 2005). Analysis of the *P. falciparum* genome identifies 11 Swi2/Snf2 type ATPases, characterized by an interrupted P-loop ATPase domain (Horrocks et al. 2009; Templeton et al. 2004). Swi2/Snf2 ATPase are major components of the SNF2 mediated DNA repair mechanism and four of these predicted Swi2/Snf2 ATPases share similarity with components of the Rad54 and Rad5/16 pathways (PF08_0126, MAL13P1.216, PFL2440w and PFF0225w). In *P. falciparum*, *var* gene activation is associated with the alteration of local chromatin structure and reduced nucleosomal occupancy at the promoters (Duraisingh et al. 2005; Voss et al. 2007; Westenberger et al. 2009). Therefore, it is important to identify the role of Swi2/Snf2 ATPase in antigenic variation.

Table 28.2 List of histone modifying enzymes (A) and proteins containing histone PTM-binding modules (B)

A.	
Gene ID	PlasmoDB
Histone acetyl transferase (HAT)	PF08_0034
PfGCN5	PF11_0192
	PFL1345c
	PFD0795
	PFA0465c
	PF14_0350
Histone deacetylase (HDAC)	PF1260c
PfHDAC1	PF14_0690
	PF10_0078
PfSir2	PF13_0152
	PF14_0489
Histone methyl transferase (HKMT)	
PfSET1	PFF1440w
PfSET2	MAL13P1.122
PfSET3	PF08_0012
PfSET4	PFl0485c
PfSET5	PFL0690c
PfSET6	PF13_0293
PfSET7	PF11_0160
PfSET8	PFD0190w
PfSET9	PFE0400w
Histone demethylases	Mal8P1.111
PfmjC1	PFF0135w
PfmjC2	PFL0575w
Protein Arginine transferase (PRMT)	
PfRMT1	PF14_0242
PfRMT4/PfCARM1	PF08_0092
PfRMT5	PF13_0323
Swi2/Snf2 ATPases	PF08_0048
PfSRCAP1	PF11_0053
PfSNFL	PFF1185w
	PFB0730w
	PF10_0232
	MAL8P1.65
	PF13_0308

B.		
Binding module/PTM mark	PlasmoDB	Annotation/PTM mark
Royal superfamily/Kme chromodomain	PF1005c	PfHP1/H3K9me3
	PF11_0192	PfMYST
	PF11_0418	

(continued)

Table 28.2 (continued)

B.		
Binding module/PTM mark	PlasmoDB	Annotation/PTM mark
Bromodomain/Kac	Pf08_0034	PfGCN5
	PFF14440w	PfSET1
	PFA0510w	
	PFL0635c	
	PFL1645w	
	PF10_0328	
	PF14_0724	
Double chromodomain Tudor domain	PF10_0232	PfCHD1
	PF11_0374	PfTSN
	PFC1050w	PfSMN
PHD finger/Kme	PFF1440w	PfSET1
	MAL13P1.122	PfSET2
	MAL13P1.302	
	PFC0425w	
	PF10_0079	
	PF11_0429	
	PFL1011c	
	PF14_0315	
	PFL0575w	
WD40 repeat	PFA0520c	CAF-1
	PFD0455w	CAF-1
14-3-3 protins/S_{ph}	MAL8p1.69	
	MAL13P1.309	
	PF14_0220	

Adapted from Chung et al. (2009), Cui and Miao (2010) and subsequently modified
  K lysine, *me* methylation, *ac* acetylation

28.6 Nuclear Architecture of *Plasmodium falciparum*

Nuclear architecture also plays a crucial role in epigenetic control of transcriptional regulation. Nucleus includes some specific organelles like nucleolus, nuclear speckles, chromatin territories etc. These organelles and the specialized structures may have fundamental role in expressing or silencing specific genes based on their specific location in the nucleus. Proteins present in the nucleus may be either associated with DNA or responsible for the maintenance of nuclear architecture like nuclear envelope proteins or nucleoplasmic proteins.

Nucleolus is the most important sub compartment of nucleus which performs crucial functions like ribosome biogenesis and storage of specific proteins. Like yeast, *Plasmodium falciparum* nucleus shows only one large nucleolus that occupies a large proportion of nucleus at one pole with a hat like or crescent shaped structure (Figueiredo et al. 2005). It harbours proteins like Sir2, ORC1 (origin recognition complex subunit 1), small nucleolar RNP protein NOP1 (fibrillarin), TERT (telomerase reverse transcriptase) and other important proteins (Figueiredo et al. 2005). Although nucleolus contains histone deacetylase Sir2 protein that may be involved in rDNA transcription, its role in epigenetics mediated gene regulation remains to be explored further.

Previous reports suggest chromosome specific cluster formation of telomeric ends in *Plasmodium falciparum* and these clusters are organized at the nuclear periphery (Mancio-Silva et al. 2008, 2010; Freitas-Junior et al. 2005). Telomeric ends accumulate various proteins like Sir2, ORC1, HP1 leading to the heterochromatin formation. However, the protein bound DNA structure slowly reorganizes as the cell cycle progresses (Freitas-Junior et al. 2005). It has been suggested that the region beneath the nuclear envelop acts as a silencing zone and in order to relieve the silencing effect of a specific gene, it has to come out from this silencing zone (Freitas-Junior et al. 2005; Duraisingh et al. 2005; Ralph et al. 2005). The telomere associated virulence gene (*var*) families undergo such kind of modification events. Following transcription, the transcripts are transported into the cytoplasm through the nuclear pores. The silencing zone may be disrupted at the nuclear pore complex region. All the above phenomena are depicted in the schematic diagram as shown in Fig. 28.2.

It has been found that central core area of nucleus beneath the periphery is responsible for active transcription of genes as suggested by co-localization with H3K9Ac that marks the active transcription (Lopez-Rubio et al. 2007; Issar et al. 2008b). However, peripheral circle beneath the nuclear membrane is marked by H3K9me3 suggesting repressive status corresponding to the location of telomeric clusters containing silent *var* genes (Issar et al. 2008b). Any silenced *var* gene may be activated only if its promoter region acquires histone modifications like H3K9Ac and H3K4me3 in the ring stage, the earlier stage of cell cycle (Chaal et al. 2010).

28.7 Pathophysiology of Malaria

Malaria is an intravascular infection. The symptoms of disease appear only due to destruction of erythrocytes and circulation of parasite and erythrocyte specific materials in blood stream of vertebrate host. It used to be the belief that parasite specific toxins are released in the blood stream when schizonts are ruptured. Now, it is established that malaria parasites induce release of some cytokines. Symptomatic malaria is found to be associated with the release of interferon gamma, TNF (tumor necrosis factor alpha), interlukein-1 beta (IL-1 beta) and IL-6 (Grau et al. 1989, 1997). The release of cytokines in response to parasite specific antigens and other debris in the form of lipid or protein is crucial for pathogenesis. Pathogenesis also involves rosette formation and heavy accumulation of large clumps of infected erythrocytes in the capillaries of vital organs like brain, placenta, lungs, and kidneys resulting in multi- organ failure.

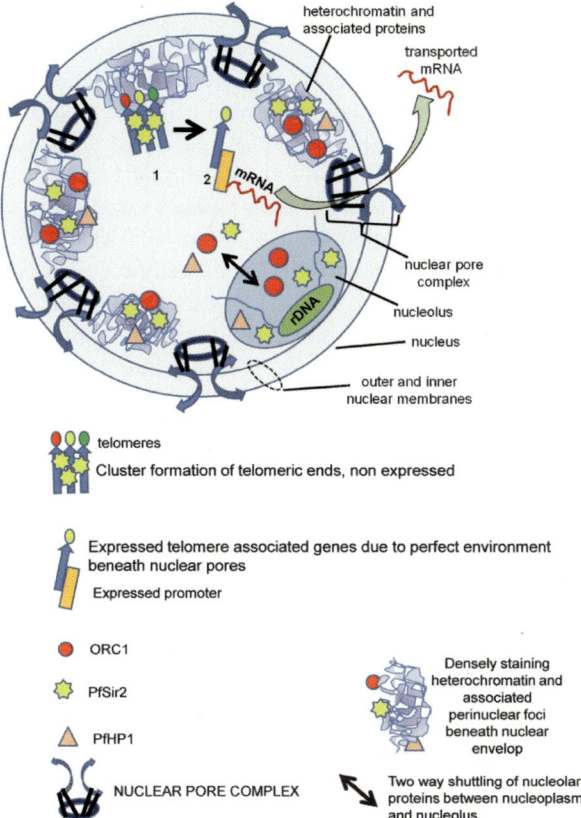

Fig. 28.2 Nuclear architecture of *Plasmodium falciparum*. The architecture of the nucleus with nuclear envelop, heterochromatin region containing clusters of telomeric region, nuclear pores, nucleolus are shown. Chromosomal ends with telomeric regions form clusters at the heterochromatin region containing several proteins like PfSir2, ORC1 involved in the silencing of *var* genes. Activation of *var* promoters require reorganization of the heterochromatin structure that will allow the specific gene to come out of this silencing zone. Nuclear pores may allow the transport of the transcripts following transcription. Nucleolus is involved in the transcription of rDNA that may require the presence of Sir2 and ORC1 in this organelle although their specific roles are not clear yet

28.7.1 Erythrocyte Rosetting

Erythrocyte rosetting is a phenomenon responsible for malarial parasite pathogenesis which involves complex formation of several red blood cells including both parasitized and uninfected cells. It includes arrangement of red blood cells around a central cell in a flower like manner. The surface of parasitized red blood cell shows the presence of *knob* like structures playing a key role in attachment of red blood cells leading to rosette formation (Aikawa et al. 1983; Gruenberg et al. 1983). Electron microscopy reveals the interaction of red blood cells at the *knobs*

(Scholander et al. 1998). Sequestration of parasitized red blood cells in vascular system is a characteristic event of *Plasmodium* pathogenesis. Sequestration is the removal of parasitized red blood cells from blood vascular system followed by specific interactions between vascular endothelial cells and parasite specific surface antigens which decorate the outer surface of red blood cell in a stage specific manner. Sequestration ultimately leads to vital organ failure. Literature shows the involvement of PfEMP1 in sequestration process responsible for escape from immune system. PfEMP1, a polypeptide of ~200 to ~350 kDa is the product of *var* gene family (Howard et al. 1983). PfEMP1 specifically recognizes host cell receptors CD36 and ICAM1 (Smith et al. 1995). PfEMP1 polypeptides are composed of an N-terminal Duffy binding like domain, DBL1, cysteine rich interdomain region (CIDR), a hydrophobic transmembrane region and an internal segment that is acidic in nature. Some of the severe effects of rosetting are described below.

28.7.1.1 Placental Malaria

Pregnant women are found to be highly susceptible to malaria. It has already been established that intense sequestration of parasitized red blood cells takes place in the placenta leading to the thickening of syncytiotrophoblasts (the multinucleated cells of placenta of embryos) and abnormal blood flow exchange between foetus and mother. Syncytiotrophoblast is the major foetal surface in contact with maternal blood. Production of inflammatory cytokines in response to heavy placental accumulation of parasitized red blood cells leads to the growth retardation of foetus and more than 50% women approaches death subsequently. No interaction takes place between uninfected erythrocytes and placental receptors showing the importance of parasite specific cell surface antigens involved in attachment with receptors. As already stated above, PfEMP1 molecules mediate interaction with placental surface receptors. (Salanti et al. 2003) has shown the involvement of a CSA binding parasite ligand var2CSA PfEMP1.

28.7.1.2 Cerebral Malaria

Plasmodium falciparum is the causative agent of cerebral malaria in humans and it is due to the accumulation of infected red blood cell clusters in cerebral blood capillaries followed by massive haemorrhage due to rupture of capillaries (Ho et al. 1991). Cerebral malaria is associated with increased cytokine secretion leading to TH-1 response. Tumor necrosis factor and some interleukins including IL-6, IL-8 etc. show elevated expression in response to parasite infection (Grau et al. 1997). Increased cytokines are responsible for symptoms of cerebral malaria like fever and shivering followed by coma and ultimately death. Rosette accumulation leads to symptoms of cerebral malaria and suggests the role of epigenetics control in *var* gene regulation.

28.7.2 Antigenic Variation

Antigenic variation is defined as a process in which an infectious organism changes its surface protein from one to another form thereby evading host immune response. Antigenic variation also enables the pathogens to reinvade the host immune system because the antigenic peptides are no longer recognized due to the high antigenic variability (Beeson and Brown 2002; Biggs et al. 1991). It is the characteristic event of many pathogenic organisms from viruses to protozoa like influenza virus, HIV, trypanosoma, neisseria, *Plasmodium falciparum* etc. Antigenic variation in *Plasmodium falciparum* involves different *var* gene families that are associated with subtelomeric regions. Regulation of these genes involves three dimensional repositioning of different *var* genes in the nucleus.

In general, antigenic variation can be in the form of (a) antigenic drift involving changes in a few amino acids of surface antigenic peptide or (b) antigenic shift showing acquisition of new surface proteins. *Plasmodium falciparum* shows antigenic drift of surface proteins encoded by *var* gene families. Influenza virus is the classical example of antigenic drift and antigenic shift types of variation involving hemagglutinin and neuraminidase. Antigenic variation in *P. falciparum* involves a multigene family of surface antigens. Literature suggests that every *var* gene consists of a 5' exon followed by a small 3' exon. First exon makes extracellular variable domain of PfEMP1 and small exon is specific for a cytoplasmic domain showing considerable homology among *var* gene repertoire. These multigene families are found to be associated at chromosomal ends in tandem repeats and ectopic recombination is the result of high antigenic diversity of *var* genes.

28.8 Epigenetic Control and *var* Gene Expression

Central to the pathogenesis of *Plasmodium* as described above is the variation in the expression of *var* gene encoded PfEMP1 protein responsible for cytoadherence and antigenic variation. As described above, only one *var* gene is expressed in an allele specific manner while 59 *var* genes remain silenced. The switching of *var* gene expression is the key to evade the host immune system. The link between the *var* gene expression and epigenetic control has been demonstrated nicely. It has been shown that the promoter region of a *var* gene when cloned in a transfection vector leads to constitutive expression rather than stage specific expression in the parasites (Deitsch et al. 1999; Voss et al. 2003).

The epigenetic control of *var* gene regulation was further strengthened when two independent groups showed that the deletion of histone deacetylase Sir2 homolog in *Plasmodium* (Prusty et al. 2008; Merrick and Duraisingh 2007) could result in the derepression of some but not all *var* genes (Duraisingh et al. 2005). Further, the association of PfSir2 with the telomeric region and telomeric associated repeat elements (TARE) was confirmed by chromatin immunoprecipitation (ChIP) experiments and electrophoretic mobility shift assay (EMSA) (Mancio-Silva et al. 2008). Similar results have been reported for another protein involved in heterochromatin formation, *P. falciparum* heterochromatin protein 1 (PfHP1) (Flueck et al. 2009; Perez-Toledo et al. 2009). It is possible that PfSir2 generates a heterochromatin region by spreading from telomeric to TAREs thereby affecting the expression of the *var* genes adjacent to the TARE regions. Another protein that has been implicated in this process is PfORC1, a replication initiation protein also involved in *var* gene silencing possibly by facilitating the heterochromatin formation in coordination with Sir2. Preliminary results indicate that the N-terminal domain of PfORC1 is involved in the process in a Sir2 dependent manner. PfHP1, also implicated in the heterochromatin formation may not be involved directly in this process (Deshmukh et al. 2012).

It has been further demonstrated that the promoter of var2CSA, a specific *var* gene is occupied with Sir2 only when it is inactive. Additionally, the same promoter is occupied with acetylated histone H4 only when it is transcriptionally active and not inactive (Freitas-Junior et al. 2005). Taken together, it can be concluded that var2CSA is regulated by epigenetic control of the promoter by reversible histone acetylation and deacetylation.

A more detailed study was performed later to investigate the histone modification status of the promoter of var2CSA during different stages. It has been shown that the 5′UTR region of active var2CSA is enriched with acetylated H3K9 and di and tri- methylated H3K4 during the ring stage of the parasite development. However, the di-methylated form of H3K4 is predominant during the trophozoite and schizont stage of the parasite life cycle when var2CSA is not active (Lopez-rubio et al. 2007). It has also been shown that H3K9me3 is enriched at the 5′UTR region of var2CSA and throughout the coding region when it is inactive. Thus the trimethylated form of H3K9 and acetylated and trimethylated form of H3K4 are the epigenetic markers that dictates the repressed and active state of var2CSA respectively in *Plasmodium falciparum*. Although the epigenetic control of *var* gene regulation has been established, it is still not clear how a particular type of *var* gene is activated out of many such genes.

Recently epigenome mapping of *P. falciparum* revealed interesting results. The MS analysis of histone modifications revealed the existence of H3/H4 acetylation and H3K4me3. ChIP on chip profiling of different acetylated and methylated form from asynchronous parasites suggest the presence of surface antigen gene families in the heterochromatin region. However, analysis of synchronized parasites revealed unique results. In ring, both H3K4me3 and H3K9me mark both active and inactive genes equally. However, during schizont, only the 5′ upstream regions of the active genes are marked by them. These striking differences during ring and schizont stages clearly indicate that the epigenetic marking in the parasites is a dynamic process that undergoes several changes throughout the erythrocytic stages (Flueck et al. 2009). These results also suggest that epigenetic control in the parasites are far more complex that need to be evaluated thoroughly in order to correlate the transcriptome map and the epigenome map at a very high resolution at the different erythrocytic developmental stages.

28.9 HDAC and HAT Inhibitors for Therapeutics

Histone epigenetic code is a key regulator of eukaryotic gene expression. In eukaryotes, histone deacetylase (HDAC) and histone acetyltransferase (HAT) enzymes control the acetylation of histones (lysine residues within histone tails) and nonhistone proteins and thus are important for various cellular functions, such as transcription, DNA replication and repair, cell signalling, cell cycle regulation and differentiation (Xu et al. 2007; Yang and Seto 2007). Equilibrium between HAT and HDAC activities must be maintained for proper transcriptional activity and above-mentioned cellular functions. Therefore, deregulation of HDAC activity has been recognized as an important therapeutic aspect used in cancer cells. Supberoylanilide hydroxamic acid (SAHA) (Vorinostat) inhibitor of HDAC is approved for the treatment of T cell lymphoma (Grant et al. 2007).

Widespread multi-drug resistance to malaria and ineffective vaccines have made global efforts to eradicate malaria very difficult. Therefore, there is urgent need to develop new classes of antimalarial compounds to combat this disease. Malaria parasite undergoes significant morphological changes during its asexual life cycle in humans, during transmission from the insect vector to the human host. Therefore, appropriate control of histone acetylation is important for parasite survival. The HDACs have been evaluated as promising drug targets, and many HDAC inhibitors possess potent antimalarial activities (Andrews et al. 2009). The HDAC inhibitor elicits an increase in *P. falciparum* histone acetylation with reduced parasite proliferation, which suggests importance of HDAC function (Chaal et al. 2010). The HDACs have been identified as important regulators of transcription in *P. falciparum*. Apicidin, a cyclic tetrapeptide isolated from Fusarium spp., affects both class I and II HDACs in the parasite and it causes profound transcription changes within 1 h of treatment leading to the cell cycle arrest of the parasites (Darkin-Rattray et al. 1996; Chaal et al. 2010). Apicidin also induces expression of stage specific genes that are otherwise suppressed during that particular stage of intraerythrocytic developmental cycle (IDC) in *P. falciparum*. Nicotinamide, an inhibitor of PfSir2A has been reported to affect parasite growth at high concentration (Prusty et al. 2008). Other inhibitors such as SAHA, trichostatin and hydroxamate derivatives have also been shown to exhibit antimalarial effects (Jutta et al. 2011; Andrews et al. 2008; Colletti et al. 2001; Dow et al. 2008).

Histone acetylation in the parasite has been shown to regulate the monoallelic expression of the *var* genes, which mediates the antigenic switching and virulence of the parasite (Duraisingh et al. 2005; Freitas-Junior et al. 2005; Voss et al. 2006, 2007). Variegated expression of genes essential for erythrocyte invasion in different parasite clones are under epigenetic control suggesting conserved epigenetic mechanism for transcriptional regulation in malaria parasites (Cortes et al. 2007). The H3 acetylation by GCN5 plays important role in the parasite gene activation and inactivation of histone acetylation compromises the parasite development suggesting its role for viable drug targets (Cui et al. 2007; Fan et al. 2004a, b). The downregulation and inhibition of PfGCN5 activity by curcumin and anacardic leads to parasite growth inhibition (Cui et al. 2007, 2008a, b). Treatment with Anacardic acid for 12 h induced twofold or greater changes in the expression of ~5% of genes in *P. falciparum* trophozoites, among which 76% were downregulated (Cui et al. 2008a, b). Therefore, the effect of these inhibitors in histone hypoacetylation and downregulation of developmentally regulated genes in the parasite may have great potential for parasite survival and growth.

28.10 Concluding Remarks

Revolutionising advances in the study of epigenetic control of gene expression did not leave apicomplexan *P. falciparum* unaffected especially following characterization of mutually exclusive expression of *var* gene family in 1998. The studies gained momentum in post genomic era when DNA sequences were available for scrutiny. The parasite displays considerable conservation of epigenetic mechanisms in having a typical nucleosome organization, numerous histone posttranslational modifications, an array of conserved chromatin modifications and remodelers with histone binding modules. However, there are some divergences like canonical eukaryotic histone code H3K4me2 that appears to provide a novel heritable mark for expression of *var* gene in the subsequent generations and lack of distinct PTMs like H3K27me, H3K79me in *Plasmodium*. These differences suggest a unique parasite specific epigenetic mechanism with its novel set of histone code writers and their cognate readers.

Progress of epigenetic studies in the parasite is helping to understand the overall pathophysiology of the parasite including the critically important antigenic variation as well as erythrocyte invasion pathways. However, our understanding of the *Plasmodium* epigenetics is marred by several limitations. This includes lifecycle limitations where the studies have been predominantly limited to the erythrocytic stages that too in long term cultured laboratory strains with *in vitro* conditions that are not representative of *in vivo* situations. The significance of erythrocytic stage can never be underestimated considering that it is the stage with disease manifestations and pathogenesis but other stages of the complicated life cycle need attention as it is known to have distinct transcriptional profile. Practical limitations of parasite material in hepatocyte culture and mosquito stages have become a hindrance in studying these stages. Also, the effect of host immune system on the epigenetic regulation of the parasite especially in the context of antigenic variation needs to be probed. Epigenetic studies in the *Plasmodial* parasite are still nascent but will surely further the understanding of the uniquely complex lifecycle of this parasite and in the process may offer some valuable targets for anti-malarial chemotherapeutics.

Acknowledgements Work in Dhar laboratory is funded by Swarnajayanti Fellowship (Department of Science and Technology, Govt. of India), MALSIG project (European Union), Alexander Von Humboldt Fellowship and National Biosciences award for career development awarded by Dept. of Biotechnology, Government of India. AD and SS acknowledge Council of Scientific and Industrial Research (CSIR), India for fellowships. The authors acknowledge Department of Biotechnology, Government of India for providing the Centre of Excellence in Parasitology grant.

References

Aikawa M, Rabbage JR, Udeinya I, Miller LH (1983) Electron microscopy of knobs in *Plasmodium falciparum* infected erythrocytes. J Parasitol 69:435–437

Andrews KT, Tran TN, Lucke AJ, Kahnberg P, Le GT, Boyle GM, Gardiner DL, Skinner-Adams TS, Fairlie DP (2008) Potent antimalarial activity of histone deacetylase inhibitor analogues. Antimicrob Agents Chemother 52:1454–1461

Andrews KT, Tran TN, Wheatley NC, Fairlie DP (2009) Targeting histone deacetylase inhibitors for anti-malarial therapy. Curr Top Med Chem 9:292–308

Artavanis-Tsakonas K, Misaghi S, Comeaux CA, Catic A, Spooner E, Duraisingh MT, Ploegh HL (2006) Identification by functional proteomics of a deubiquitinating/deNeddylating enzyme in *Plasmodium falciparum*. Mol Microbiol 61:1187–1195

Balaji S, Babu MM, Iyer LM, Aravind L (2005) Discovery of the principal specific transcription factors of Apicomplexa and their implication for the evolution of the AP2-integrase DNA binding domains. Nucleic Acids Res 33:3994–4006

Beauchamps P, Tourvieille B, Cesbron-Delauw MF, Capron A (1997) The partial sequence of the *Plasmodium falciparum* histone H4 gene. Res Microbiol 148:201–203

Beeson JG, Brown GV (2002) Pathogenesis of *Plasmodium falciparum* malaria: the roles of parasite adhesion and antigenic variation. Cell Mol Life Sci 59:258–271

Bennett BJ, Thompson J, Coppel RL (1995) Identification of *Plasmodium falciparum* histone 2B and histone 3 genes. Mol Biochem Parasitol 70:231–233

Berger SL (2002) Histone modifications in transcriptional regulation. Curr Opin Genet Dev 12:142–148

Berger SL, Kouzarides T, Shiekhattar R, Shilatifard A (2009) An operational definition of epigenetics. Genes Dev 23:781–783

Biggs BA, Gooze L, Wycherley K, Wollish W, Southwell B et al (1991) Antigenic variation in *Plasmodium falciparum*. Proc Natl Acad Sci 88:9171–9174

Bozdech Z, Llinas M, Pulliam BL, Wong ED, Zhu J, DeRisi JL (2003) The transcriptome of the intraerythrocytic developmental cycle of *Plasmodium falciparum*. PLoS Biol 1:E5

Carrozza MJ, Utley RT, Workman JL, Cote J (2003) The diverse functions of histone acetyltransferase complexes. Trends Genet 19:321–329

Cary C, Lamont D, Dalton JP, Doerig C (1994) *Plasmodium falciparum* chromatin: nucleosomal organisation and histone-like proteins. Parasitol Res 80:255–258

Chaal BK, Gupta AP, Wastuwidyaningtyas BD, Luah YH, Bozdech Z (2010) Histone deacetylases play a major role in the transcriptional regulation of the *Plasmodium falciparum* life cycle. PLoS Pathog 6:e1000737

Chen Q, Fernandez V, Sundstrom A, Schlichtherle M, Datta S, Hagblom P, Wahlgren M (1998) Developmental selection of *var* gene expression in *Plasmodium falciparum*. Nature 394:392–395

Choi SW, Keyes ME, Horrocks P (2006) LC/ESI-MS demonstrates the absence of 5-methyl-2-deoxycytosine in *Plasmodium falciparum* genomic DNA. Mol Biochem Parasitol 150:350–352

Chung DW, Ponts N, Cervantes S, Le Roch KG (2009) Post-translational modifications in *Plasmodium*: more than you think! Mol Biochem Parasitol 168:123–134

Colletti SL, Myers RW, Darkin-Rattray SJ, Gurnett AM, Dulski PM, Galuska S, Allocco JJ, Ayer MB, Li C, Lim J, Crumley TM, Cannova C, Schmatz DM, Wyvratt MJ, Fisher MH, Meinke PT (2001) Broad spectrum antiprotozoal agents that inhibit histone deacetylase: structure-activity relationships of apicidin. Part 2. Bioorg Med Chem Lett 11:113–117

Cortes A, Carret C, Kaneko O, Lim BYSY, Ivens A, Holder A (2007) Epigenetic silencing of *Plasmodium falciparum* genes linked to erythrocyte invasion. PLoS Pathog 3:1023–1035

Cui L, Miao J (2010) Chromatin-mediated epigenetic regulation in the malaria parasite *Plasmodium falciparum*. Eukaryot Cell 9:1138–1149

Cui L, Miao J, Furuya T, Li X, Su X, Cui L (2007) PfGCN5 mediated histone H3 acetylation plays a key role in gene expression in *Plasmodium falciparum*. Eukaryot Cell 6:1219–1227

Cui L, Fan Q, Cui L, Miao J (2008a) Histone lysine methyltransferases and demethylases in *Plasmodium falciparum*. Int J Parasitol 38:1083–1109

Cui L, Miao J, Furuya T, Fan Q, Li X, Rathod PK, Su XZ, Cui L (2008b) Histone acetyltransferase inhibitor anacardic acid causes changes in global gene expression during in vitro *Plasmodium falciparum* development. Eukaryot Cell 7:1200–1210

Darkin-Rattray SJ, Gurnett AM, Myers RW, Dulski PM, Crumley TM, Allocco JJ, Cannova C, Meinke PT, Colletti SL, Bednarek MA, Singh SB, Goetz MA, Dombrowski AW, Polishook JD, Schmatz DM (1996) Apicidin: a novel antiprotozoal agent that inhibits parasite histone deacetylase. Proc Natl Acad Sci USA 93:13143–13147s

Deitsch KW, del Pinal A, Wellems TE (1999) Intracluster recombination and *var* transcription switches in the antigenic variation of *Plasmodium falciparum*. Mol Biochem Parasitol 101:107–116

Deitsch KW, Lukehart SA, Stringer JR (2009) Common strategies for antigenic variation by bacterial, fungal and protozoan pathogens. Nat Rev Microbiol 7:493–503

Deshmukh AS, Srivastava S, Herrmann S, Gupta A, Mitra P, Gilberger TW, Dhar SK (2012) The role of N-terminus of *Plasmodium falciparum* ORC1 in telomeric localization and *var* gene silencing. Nucleic Acids Res. Epub ahead of print

DiPaolo C, Kieft R, Cross M, Sabatini R (2005) Regulation of trypanosome DNA glycosylation by a SWI2/SNF2-like protein. Mol Cell 17:441–451

Dow GS et al (2008) Antimalarial activity of phenylthiazolyl-bearing hydroxamate-based histone deacetylase inhibitors. Antimicrob Agents Chemother 52:3467–3477

Duffy MF, Brown GV, Basuki W, Krejany EO, Noviyanti R, Cowman AF, Reeder JC (2002) Transcription of multiple var genes by individual, trophozoite-stage *Plasmodium falciparum* cells expressing a chondroitin sulphate A binding phenotype. Mol Microbiol 43:1285–1293

Duraisingh MT, Voss TS, Marty AJ, Duffy MF, Good RT, Thompson JK, Freitas-Junior LH, Scherf A, Crabb BS, Cowman AF (2005) Heterochromatin silencing and locus repositioning linked to regulation of virulence genes in *Plasmodium falciparum*. Cell 121:13–24

Fan Q, An L, Cui L (2004a) *Plasmodium falciparum* histone acetyltransferase, a yeast GCN5 homologue involved in chromatin remodeling. Eukaryot Cell 3:264–276

Fan Q, Li J, Kariuki M, Cui L (2004b) Characterization of PfPuf2, member of the Puf family RNA-binding proteins from the malaria parasite *Plasmodium falciparum*. DNA Cell Biol 23:753–760

Fan Q, Miao J, Cui L, Cui L (2009) Characterization of PRMT1 from *Plasmodium falciparum*. Biochem J 421:107–118

Fernandez V, Chen Q, Sundstrom A, Scherf A, Hagblom P, Wahlgren M (2002) Mosaic-like transcription of var genes in single *Plasmodium falciparum* parasites. Mol Biochem Parasitol 121:195–203

Figueiredo LM, Pirrit LA, Scherf A (2000) Genomic organisation and chromatin structure of *Plasmodium falciparum* chromosome ends. Mol Biochem Parasitol 106:169–174

Figueiredo LM, Rocha EPC, Mancio-Silva L, Prevost C, Hernandez-Verdun D, Scherf A (2005) The unusually large *Plasmodium* telomerase reverse-transcriptase localizes in a discrete compartment associated with the nucleolus. Nucleic Acids Res 33:1111–1122

Flueck C, Bartfai R, Volz J, Niederwieser I, Salcedo-Amaya AM et al (2009) *Plasmodium falciparum* heterochromatin protein 1 marks genomic loci linked to phenotypic variation of exported virulence factors. PLoS Pathog 5:e1000569

Freitas-Junior LH, Hernandez-Rivas R, Ralph SA, Montiel-Condado D, Ruvalcaba-Salazar OK, Rojas-Meza AP, Mancio-Silva L, Leal-Silvestre RJ, Gontijo AM, Shorte S, Scherf A (2005) Telomeric heterochromatin propagation and histone acetylation control mutually exclusive expression of antigenic variation genes in malaria parasites. Cell 121:25–36

French JB, Cen Y, Sauve AA (2008) *Plasmodium falciparum* Sir2 is an NAD+-dependent deacetylase and an acetyllysine-dependent and acetyllysine-independent NAD+ glycohydrolase. Biochemistry 47:10227–10239

Frickel EM, Quesada V, Muething L, Gubbels MJ, Spooner E, Ploegh H, Artavanis-Tsakonas K (2007) Apicomplexan UCHL3 retains dual specificity for ubiquitin and Nedd8 throughout evolution. Cell Microbiol 9:1601–1610

Goldberg AD, Allis CD, Bernstein E (2007) Epigenetics: a landscape takes shape. Cell 128:635–638

Grant S, Easley C, Kirkpatrick P (2007) Vorinostat. Nat Rev Drug Discov 6:21–22

Grau GE, Tayler TE, Moleneux ME, Wirima JJ, Vasalli P, Homell M, Lambert PH (1989) Tumor necrosis factor and disease severity in children with *P. falciparum* malaria. N Engl J Med 320:1589–1591

Grau GE, Fajardo LF, Piquet PF, Allet B, Lambert PH, Vasalli P (1997) Tumor necrosis factor (cachectin) as an essential mediator in murine cerebral malaria. Science 237:1210–1212

Gruenberg J, Allred D, Sherman IW (1983) Scanning microscope analysis of the protrusions (knobs) present on the surface of *Plasmodium falciparum* infected erythrocytes. J Cell Biol 97:795–802

Ho M, Davis TME, Silamut K, Bunnang D, White NJ (1991) Rosette formation of *Plasmodium falciparum* infected erythrocytes from patients with acute malaria. Infect Immun 59:2135–2139

Horrocks P, Wong E, Russell K, Emes RD (2009) Control of gene expression in *Plasmodium falciparum* – ten years on. Mol Biochem Parasitol 164:9–25

Howard RJ, Barnwell JW, Kao V (1983) Antigenic variation in *Plasmodium knowlesi* malaria: identification of the variant antigen on infected erythrocytes. Proc Natl Acad Sci 80:4129–4133

Issar N, Ralph SA, Mancio-Silva L, Keeling C, Scherf A (2008a) Differential subnuclear localisation of repressive and activating histone methyl modifications in *P. falciparum*. Microbes Infect 11:403–407

Issar N, Roux E, Mattei D, Scherf A (2008b) Identification of a novel post-translational modification in *Plasmodium falciparum*: protein sumoylation in different cellular compartments. Cell Microbiol 10:1999–2011

Iyer LM, Anantharaman V, Wolf MY, Aravind L (2008) Comparative genomics of transcription factors and chromatin proteins in parasitic protists and other eukaryotes. Int J Parasitol 38:1–31

Jenuwein T, Allis CD (2001) Translating the histone code. Science 293:1074–1080

Joshi MB, Lin DT, Chiang PH, Goldman ND, Fujioka H, Aikawa M, Syin C (1999) Molecular cloning and nuclear localization of a histone deacetylase homologue in *Plasmodium falciparum*. Mol Biochem Parasitol 99:11–19

Jutta M, Ferryanto C, Pak P, Wabiser F, Kenangalem E, Piera KA, Fairlie DP, Tjitra E, Anstey NM, Andrews KT, Price RN (2011) Ex Vivo activity of histone deacetylase inhibitors against multidrug-resistant clinical isolates of *Plasmodium falciparum* and *P. vivax*. Antimicrob Agents Chemother 55:961–966

Kaiser K, Matuschewski K, Camargo N, Ross J, Kappe SH (2004) Differential transcriptome profiling identifies Plasmodium genes encoding pre-erythrocytic stage-specific proteins. Mol Microbiol 51:1221–1232

Kouzarides T (2007) Chromatin modifications and their function. Cell 128:693–705

Lanzer M, de Bruin D, Wertheimer SP, Ravetch JV (1994) Transcriptional and nucleosomal characterization of a subtelomeric gene cluster flanking a site of chromosomal rearrangements in *Plasmodium falciparum*. Nucleic Acids Res 22:4176–4182

Le Roch KG, Zhou Y, Blair PL, Grainger M, Moch CH, Haynes JD, De La Vega P, Holder AA, Batalov S, Carucci DJ, Winzeler EA (2003) Discovery of gene function by expression profiling of the malaria parasite life cycle. Science 301:1503–1508

Lee KK, Workman JL (2007) Histone acetyltransferase complexes: one size doesn't fit all. Nat Rev Mol Cell Biol 8:284–295

Llinas M, Bozdech Z, Wong ED, Adai AT, DeRisi JL (2006) Comparative whole genome transcriptome analysis of three *Plasmodium falciparum* strains. Nucleic Acids Res 34:1166–1173

Lopez-Rubio JJ, Gontijo AM, Nunes MC, Issar N, Hernandez Rivas R, Scherf A (2007) 5′ flanking region of var genes nucleate histone modification patterns linked to phenotypic inheritance of virulence traits in malaria parasites. Mol Microbiol 66:1296–1305

Luger K, Mäder AW, Richmond RK, Sargen DF, Richmond TJ (1997) Crystal structure of the nucleosome core particle at 2.8 Å resolution. Nature 389:251–260

Macdonald N, Welburn JP, Noble ME, Nguyen A, Yaffe MB, Clynes D, Moggs JG, Orphanides G, Thomson S, Edmunds JW, Clayton AL, Endicott JA, Mahadevan LC (2005) Molecular basis for the recognition of phosphorylated and phosphoacetylated histone h3 by 14-3-3. Mol Cell 20:199–211

Malik M, Henikoff S (2003) Phylogenomics of the nucleosome. Nat Struct Biol 10:882–891

Mancio-Silva L, Rojas-Meza AP, Vargas M, Scherf A, Hernandez-Rivas R (2008) Differential association of Orc1 and Sir2 proteins to telomeric domains in *Plasmodium falciparum*. J Cell Sci 121(Pt12):2046–2053

Mancio-Silva L, Zhanga Q, Scheidig-Benatara C, Scherf A (2010) Clustering of dispersed ribosomal DNA and its role in gene regulation and chromosome-end associations in malaria parasites. Proc Natl Acad Sci USA 107:15117–15122

Martens JA, Winston F (2003) Recent advances in understanding chromatin remodeling by Swi/Snf complexes. Curr Opin Genet Dev 13:136–142

Matuschewski K, Ross J, Brown SM, Kaiser K, Nussenzweig V, Kappe SH (2002) Infectivity-associated changes in the transcriptional repertoire of the malaria parasite sporozoite stage. J Biol Chem 277:41948–41953

Maurer-Stroh S, Dickens NJ, Hughes-Davies L, Kouzarides T, Eisenhaber F, Ponting CF (2003) The Tudor domain 'Royal Family': Tudor, plant Agenet, Chromo, PWWP and MBT domains. Trends Biochem Sci 28:69–74

Meneghini MD, Wu M, Madhani HD (2003) Conserved histone variant H2A.Z protects euchromatin from the ectopic spread of silent heterochromatin. Cell 112:725–736

Merrick CJ, Duraisingh MT (2007) *Plasmodium falciparum* Sir2: an unusual sirtuin with dual histone deacetylase and ADP-ribosyltransferase activity. Eukaryot Cell 6:2081–2091

Merrick CJ, Duraisingh MT (2010) Epigenetics in Plasmodium: what do we really know? Eukaryot Cell 9:1150–1158

Merrick CJ, Dzikowski R, Imamura H, Chuang J, Deitsch K, Duraisingh MT (2010) The effect of *Plasmodium falciparum* Sir2a histone deacetylase on clonal and longitudinal variation in expression of the var family of virulence genes. Int J Parasitol 40:35–43

Miao J, Fan Q, Cui L, Li J, Li J, Cui L (2006) The malaria parasite *Plasmodium falciparum* histones: organization, expression, and acetylation. Gene 369:53–65

Mok BW, Ribacke U, Winter G, Yip BH, Tan CS, Fernandez V, Chen Q, Nilsson P, Wahlgren M (2007) Comparative transcriptomal analysis of isogenic *Plasmodium falciparum* clones of distinct antigenic and adhesive phenotypes. Mol Biochem Parasitol 151:184–192

Mok BW, Ribacke U, Rasti N, Kironde F, Chen Q, Nilsson P, Wahlgren M (2008) Default pathway of var2csa switching and translational repression in *Plasmodium falciparum*. PLoS One 3:e1982

Perez-Toledo K, Rojas-Meza AP, Mancio-Silva L, Hernandez-Cuevas NA, Delgadillo DM et al (2009) *Plasmodium falciparum* heterochromatin protein 1 binds to tri-methylated histone 3 lysine 9 and is linked to mutually exclusive expression of var genes. Nucleic Acids Res 37:2596–2606

Petter M, Lee CC, Byrne TJ, Boysen KE, Volz J, Ralph SA, Cowman AF, Brown GV, Duffy MF (2011) Expression of P. falciparum var genes involves exchange of the histone variant H2A.Z at the promoter. PLoS Pathog 7:e1001292

Ponder EL, Bogyo M (2007) Ubiquitin-like modifiers and their deconjugating enzymes in medically important parasitic protozoa. Eukaryot Cell 6:1943–1952

Ponts N, Yang J, Chung DW, Prudhomme J, Girke T, Horrocks P, Le Roch KG (2008) Deciphering the ubiquitin-mediated pathway in apicomplexan parasites: a potential strategy to interfere with parasite virulence. PLoS One 3:e2386

Prusty D, Mehra P, Srivastava S, Shivange AV, Gupta A, Roy N, Dhar SK (2008) Nicotinamide inhibits *Plasmodium falciparum* Sir2 activity in vitro and parasite growth. FEMS Microbiol Lett 282:266–272

Przyborski JM, Bartels K, Lanzer M, Andrews KT (2003) The histone H4 gene of *Plasmodium falciparum* is developmentally transcribed in asexual parasites. Parasitol Res 90:387–389

Raibaud A, Brahimi K, Roth CW, Brey PT, Faust DM (2006) Differential gene expression in the ookinete stage of the malaria parasite *Plasmodium berghei*. Mol Biochem Parasitol 150:107–113

Ralph SA, Scheidig-Benatar C, Scherf A (2005) Antigenic variation in *Plasmodium falciparum* is associated with movement of *var* loci between subnuclear locations. Proc Natl Acad Sci USA 102:5414–5419

Ruthenburg AJ, Wang W, Graybosch DM, Li H, Allis CD, Patel DJ, Verdine GL (2006) Histone H3 recognition and presentation by the WDR5 module of the MLL1 complex. Nat Struct Mol Biol 13:704–712

Sacci JB, Ribeiro JM Jr, Huang F, Alam U, Russell JA, Blair PL, Witney A, Carucci DJ, Azad AF, Aguiar JC (2005) Transcriptional analysis of in vivo *Plasmodium yoelii* liver stage gene expression. Mol Biochem Parasitol 142:177–183

Salanti A, Staalsoe T, Lavstsen T, Atr J, Mpk S et al (2003) Selective upregulation of a single distinctly structured *var* gene in chondroitin sulphate A-adhering *Plasmodium falciparum* involved in pregnancy-associated malaria. Mol Microbiol 49:179–191

Sautel CF, Cannella D, Bastien O, Kieffer S, Aldebert D, Garin J, Tardieux I, Belrhali H, Hakimi MA (2007) SET8-mediated methylations of histone H4 lysine 20 mark silent heterochromatic domains in apicomplexan genomes. Mol Cell Biol 27:5711–5724

Scherf A, Hernandez-Rivas H, Buffet P, Bottius E, Benatar C, Pouvelle B, Gysin J, Lanzer M (1998) Antigenic variation in malaria: in situ switching, relaxed and mutually exclusive transcription of var genes during intra-erythrocytic development in *Plasmodium falciparum*. EMBO J 17:5418–5426

Scholander C, Carlson J, Kremsenr PG, Wahlgren M (1998) Extensive immunoglobulin binding of *Plasmodium falciparum*-infected erythrocytes in a group of children with moderate anemia. Infect Immun 66:361–363

Shi Y, Whetstine JR (2007) Dynamic regulation of histone lysine methylation by demethylases. Mol Cell 25:1–14

Shiio Y, Eisenman RN (2003) Histone sumoylation is associated with transcriptional repression. Proc Natl Acad Sci USA 100:13225–13230

Silvestrini F, Bozdech Z, Lanfrancotti A, Di Giulio E, Bultrini E, Picci L, Derisi JL, Pizzi E, Alano P (2005) Genome-wide identification of genes upregulated at the onset of gametocytogenesis in *Plasmodium falciparum*. Mol Biochem Parasitol 143:100–110

Smith JD, Chitnis CE, Craig AG, Roberts DJ, Hudson-Taylor DE, Peterson DS, Pinches R, Newbold CI, Miller LH (1995) Switches in expression of *Plasmodium falciparum var* genes correlate with changes in antigenic and cytoadherent phenotypes of infected erythrocytes. Cell 82:101–110

Sullivan WJ Jr (2003) Histone H3 and H3.3 variants in the protozoan pathogens of *Plasmodium falciparum* and *Toxoplasma gondii*. DNA Seq 14:227–231

Tarun AS, Peng X, Dumpit RF, Ogata Y, Silva-Rivera H, Camargo N, Daly TM, Bergman LW, Kappe SH (2008) A combined transcriptome and proteome survey of malaria parasite liver stages. Proc Natl Acad Sci USA 105:305–310

Taverna SD, Li H, Ruthenburg AJ, Allis CD, Patel DJ (2007) How chromatin-binding modules interpret histone modifications: lesson from professional pocket pickers. Nat Struct Mol Biol 14:1025–1040

Templeton TJ, Iyer LM, Anantharaman V, Enomoto S, Abrahante JE, Subramanian GM, Hoffman SL, Abrahamsen MS, Aravind L (2004) Comparative analysis of apicomplexa and genomic diversity in eukaryotes. Genome Res 14:1686–1695

Tonkin CJ, Carret CK, Duraisingh MT, Voss TS, Ralph RA, Hommel M, Duffy MF, Silva LM, Scherf A, Ivens A, Speed TP, Beeson JG, Cowman AF (2009) Sir2 paralogues cooperate to regulate virulence genes and antigenic variation in *Plasmodium falciparum*. PLoS Biol 7:e84

Trelle MB, Salcedo-Amaya AM, Cohen AM, Stunnenberg HG, Jensen ON (2009) Global histone analysis by mass spectrometry reveals a high content of acetylated lysine residues in the malaria parasite *Plasmodium falciparum*. J Proteome Res 8:3439–3450

Voss TS, Kaestli M, Vogel D, Bopp S, Beck HP (2003) Identification of nuclear proteins that differentially interact with *Plasmodium falciparum var* gene promoters. Mol Microbiol 48:1593–1607

Voss TS, Healer J, Marty AJ, Duffy MF, Thompson JK, Beeson JG, Reeder JC, Crabb BS, Cowman AF (2006) A *var* gene promoter controls allelic exclusion of virulence genes in *Plasmodium falciparum* malaria. Nature 439:1004–1008

Voss TS, Tonkin CJ, Marty AJ, Thompson JK, Healer J, Crabb BS, Cowman AF (2007) Alterations in local chromatin environment are involved in silencing and activation of subtelomeric *var* genes in *Plasmodium falciparum*. Mol Microbiol 66:139–150

Waddington CH (1942) The epigenotype. Endeavour 1:18–20

Weake VM, Workman JL (2008) Histone ubiquitination: triggering gene activity. Mol Cell 29:653–663

Westenberger SJ, Cui L, Dharia N, Winzeler E, Cui L (2009) Genome-wide nucleosome mapping of *Plasmodium falciparum* reveals histone-rich coding and histone-poor intergenic regions and chromatin remodeling of core and subtelomeric genes. BMC Genomics 10:610

Xu WS, Parmigiani RB, Marks PA (2007) Histone deacetylase inhibitors: molecular mechanisms of action. Oncogene 26:5541–5552

Yang XJ, Seto E (2007) HATs and HDACs: from structure, function and regulation to novel strategies for therapy and prevention. Oncogene 26:5310–5318

Yuda M, Iwanaga S, Shigenobu S, Mair GR, Janse CJ, Waters AP, Kato T, Kaneko I (2009) Identification of a transcription factor in the mosquito-invasive stage of malaria parasites. Mol Microbiol 71:1402–1414

Zeng L, Zhou MM (2002) Bromodomain: an acetyl-lysine binding domain. FEBS Lett 513:124–128

Index

A

Acetylation, 6, 17, 18, 25, 38, 40, 44, 46, 49–51, 106, 112, 115, 124, 125, 127, 143, 144, 146, 147, 152, 155, 156, 160, 162–164, 240, 247–249, 264, 276, 294, 295, 298–300, 321, 324, 329, 331, 333, 334, 360, 376, 401, 403, 408, 410, 411, 425, 429, 440–444, 448, 460–462, 466, 468–470, 472, 486–489, 495, 508, 513, 514, 520, 532–533, 535, 546, 548, 549, 567–588, 609, 612, 630–632, 634–636, 639–642, 661, 663, 664, 666, 669, 674, 675

Acetyl coenzyme A (Ac-CoA), 106, 115–116, 478

Acquired immuno deficiency syndrome (AIDS), 497, 580

ADP-ribosylation, 152, 294, 408, 442, 486, 546, 571, 661

Agger, K., 306

Aging, 151–168, 290, 332, 386, 429, 467, 517, 547, 559, 599

AIDS. *See* Acquired immuno deficiency syndrome (AIDS)

Alberio, R., 559

Albert, B., 218

Al-Hajj, M., 556

Alpha helix, 12

Altaf, M., 269

Alu-SINE, 277–278

Alzheimer's disease, 226, 362, 507–521, 550, 586, 587, 639, 640

Anamika, K., 269, 270

Androgen receptor, 306, 384, 579

Angiogenesis, 349, 402, 409, 410, 583, 629, 631–634, 637

Antoni, B.A., 492

Apoptosis, 25, 111, 151, 156, 166, 295, 351, 352, 377, 404, 408–410, 419, 420, 423, 465, 466, 468, 471, 481, 482, 514, 519, 520, 533, 556, 571, 577, 579, 582–584, 629, 632, 636, 639, 641, 642

Arabi, A., 224

Architectural proteins, 7, 571

Arya, G., 49

Ausio, J., 49

Ayrault, O., 224

B

Bach1, 109, 111, 112

Bai, H., 532

Bannister, A.J., 306

Bates, D.J., 159

Baylin, S.B., 599

Berger, S.L., 528

Bierhoff, H., 219, 222, 224

Black, J.C., 269, 270

Bonnet, D., 556

Boundaries, 184–186, 191, 197, 261, 266, 273–278, 323, 376, 402, 447, 571, 662–663

Bracken, A.P., 155

Bradbury, E.M., 45, 49

Brooks, D.G., 492

Buschbeck, M., 323

Butera, S.T., 492

C

Cameron, E.E., 613
Cancer, 5, 69, 133, 158, 161, 163, 166, 278, 279, 290, 291, 307, 329–330, 333, 352, 363, 382, 384–385, 399–413, 419–429, 440–442, 447, 471, 508, 512, 528, 530–534, 536, 545–559, 575–576, 578–560, 579, 581, 582, 584, 587, 597–614, 628–630, 632, 633, 635, 637–643, 675
Cancer stem cells (CSC), 5, 410–413, 419–429, 545–559, 587, 600
Cardiomyocytes, 226, 577
Cavanaugh, A.H., 224
CCCTC-binding factor (CTCF), 213, 268, 271, 273, 274, 277–279, 293, 296, 360, 361, 374, 375, 378, 462
Cell growth, 212, 224, 226, 307, 399, 547, 575, 578, 632
Cell proliferation, 133, 159, 163, 168, 300, 350, 404, 410, 411, 442, 443, 529, 531, 534, 546, 548–550, 553, 610, 632, 634, 642
Cellular hierarchy, 426
Chen, P.S., 514
Choudhary, S.K., 492
Chromatin
    compaction, 7, 123, 126–127, 157, 248, 324, 330, 428, 429, 571, 600, 602, 609
    loops, 194, 374–377, 379, 382, 387
    modifications, 18, 21, 122–128, 130, 140, 143–146, 190, 267, 308, 376, 399, 519, 570, 600, 676
    structure, 4, 7, 14, 26, 37–52, 57–76, 114, 123, 152, 155, 156, 162, 164, 168, 194, 197, 248, 261–280, 290, 294, 298, 299, 305, 324, 330, 349, 351, 361, 373–388, 411, 422, 439, 440, 456, 488, 489, 532, 550, 568, 576, 663, 667
Chromatin-mediated elements, 72
Chromosome-organizing clamp, 277
Chun, T.W., 484
Chung, D.W., 669
Claypool, J.A., 224
Cloos, P.A., 306
Cohesin, 271–275, 278, 279, 374
Cohnheim, J., 554
Comai, L., 224
Condensin, 271–274, 279
Cox, C.V., 556

CpG island, 3–6, 67, 75, 168, 289–291, 384, 385, 401, 406, 407, 409, 421, 428, 439, 464, 469, 490, 493, 508, 512, 513, 529–531, 546, 553, 599–605, 610, 629, 637
Cross-talk, 48, 116, 164, 168, 421, 424, 427–428
Cui, L., 669
Cuthbert, G.L., 306

D

Dalerba, P., 556
Dang, W., 164
Deoxyribonucleic acid (DNA)
    compositional asymmetry, 59
    methylation, 4, 5, 24–26, 67, 81, 83, 86–88, 90–92, 112, 113, 115, 122, 124, 127, 131–133, 141, 146, 153–155, 159, 161, 162, 168, 194, 225, 226, 277, 289–308, 328, 360, 384, 386, 406–409, 412, 421–424, 428, 439–440, 442, 443, 447, 456, 461, 463–465, 467–469, 487, 490, 492, 493, 508, 511–513, 517–519, 529–531, 546–553, 557, 558, 571, 597–614, 629–631, 633, 635–640, 645, 659
    replication, 18, 58, 61–64, 82, 87, 88, 111, 112, 116, 194, 226, 279, 290, 294, 307, 323, 350, 387–388, 441, 528, 532, 535–536, 548, 570, 659, 661, 675
Development 5, 48
Diabetes, 151, 307, 435–449, 464, 508, 517, 528, 530, 533, 575–579, 628, 636
Di Croce, L., 405
Dick, J.E., 554, 556
Differentiation, 3–26, 64, 72, 111, 116, 119–134, 139–147, 153, 154, 186, 187, 189, 191, 192, 194, 223, 226, 247, 280, 294, 296, 305, 350, 352, 374, 379, 388, 404–406, 410, 413, 419, 420, 422–429, 437, 441–443, 456–464, 493, 515, 527, 529, 531, 546–554, 556–559, 568, 569, 573–577, 587, 613, 628–630, 632, 660, 675
DNA. *See* Deoxyribonucleic acid (DNA)
DNA methyl transferases (DNMT), 25, 153, 160–162, 164, 289–292, 294, 384, 386, 405, 408, 409, 412, 464, 465, 530, 548, 551, 606, 609–613, 635, 639
Doi, A., 290
Duan, Y., 48
Durante, F., 554

Index

E
Earnshaw, W.C., 334
Ectoderm, 120, 177, 178, 302, 573
Ehrlich, M., 603
Embryonic stem cell (ESC), 25, 60, 62, 75, 130, 188, 189, 250, 279, 293, 296, 305, 323, 406, 411, 420, 441, 549–554, 556–558, 573
Enhancer of zeste homolog 2 (EZH2), 401
Eot-Houllier, G., 269
Epigenetic(s)
  control, 25–26, 81, 125–126, 133, 227, 406, 439, 520, 535, 550, 659–676
  regulation, 81–96, 119–133, 139–147, 223, 406, 419–429, 462, 469, 479–497, 508, 509, 513, 527, 528, 531, 533, 536, 546, 549–553, 599, 600, 628, 631, 640, 676
Epigenome, 68, 105–117, 125, 132, 133, 155, 191, 194, 195, 439, 456, 472, 508, 518, 521, 559, 614, 628, 629, 637, 644, 645, 674
Eramo, A., 556
ESC. *See* Embryonic stem cell (ESC)
Evolution, 4, 7, 12–23, 26, 60, 69, 94–96, 107, 182, 195, 325, 326, 345, 380, 412, 420, 554, 604, 614, 662
Extra TFIIIC (ETC) site, 268, 271, 272, 277

F
Faast, R., 323
Feinberg, A., 599
Felsenfeld, G., 334
Fertility, 127
Flavell, R.A., 375
Fodor, B.D., 306
Folks, T.M., 492
Frigola, J., 385

G
Garcia-Bassets, I., 306
Garrick, D., 323
Gatta, R., 270
Gene expression, 4, 40
General transcription factors (GTFs), 239, 241, 242, 247, 248, 262, 299, 353, 357, 571
Genome
  packaging, 3, 7
  replication, 57–76
Genomic imprinting, 290, 292, 307, 360–361, 423, 607

Germ cell, 119–133, 547, 548, 552, 554, 559, 607, 608, 610, 611
  epigenetic reprogramming, 124, 125, 548
Gierman, H.J., 378
Ginestier, C., 556
Glauben, R., 466
Gluconeogenesis, 579
Goldstein, A.S., 556
Grandinetti, K.B., 157
Grandori, C., 224
Grummt, I., 224

H
Halkidou, K., 224
Hannan, K.M., 224
Haploid differentiation, 130
HAT. *See* Histone acetyl transferase (HAT)
Haussler, D., 352
Hayes, J.J., 45
Hazuda, D.J., 492
HDAC. *See* Histone deacetylase (HDAC)
Henikoff, S., 334
Histone
  acetylation, 25, 106, 112, 115, 143, 160, 162–164, 276, 298–300, 403, 408, 411, 425, 440, 441, 468, 470, 487, 488, 508, 513, 520, 532–533, 549, 567–588, 630–632, 636, 639–641, 663, 664, 674, 675
  chaperone, 22, 23, 196, 226, 266, 402, 571, 576
  modifications, 25, 26, 40, 106, 109, 190, 191, 248, 295, 297, 298, 307, 360, 376, 384, 406, 460, 469, 600, 631, 645, 674
  modifying machinery, 24
  tail, 39, 40, 45–51, 601, 666
  tail positioning, 42–44
Histone acetyl transferase (HAT), 147, 162–163, 276, 299, 350, 359, 403, 410, 412, 442–445, 468, 469, 471, 472, 482, 483, 486–488, 513, 514, 519, 520, 641–642, 645, 664, 667–669, 675–676
Histone cell cycle regulation defective homolog A (HIRA), 157, 323, 324, 402, 490
Histone deacetylase (HDAC), 24, 25, 143, 144, 146, 147, 157, 163, 164, 250, 279, 292, 301, 306, 386, 404, 406, 408, 412, 442–445, 463, 465, 467, 469–471, 486, 487, 492, 494, 495, 497, 514, 519, 533, 535, 572, 609, 612, 613, 631, 635, 636, 638–641, 645, 668, 670, 673, 675–676

Hoppe, S., 224
Hox genes, 18, 178–182, 184, 186–188, 190, 191, 193, 195, 197, 198, 403
Huang, J., 157
Huber, L.C., 466
Human genome organisation, 57–76
Human Immunodeficiency Virus (HIV), 354, 479–497, 576, 578, 580, 584, 673
Hypoacetylation, 401, 408, 410, 570, 584, 676

I
Imhof, A., 269, 270
Inflammation, 346, 442, 456, 464, 466, 468, 517, 584, 599, 602, 627–645
Initiation, 58, 60–70, 72, 82, 87, 88, 90, 125, 147, 182, 193, 196, 197, 213–219, 222, 238, 241, 243–245, 247, 249–251, 263, 267, 276, 277, 291, 328, 350, 354–357, 404, 405, 412, 421, 425, 427, 429, 437, 457, 485, 490, 492, 557, 558, 570, 574, 580, 598, 629, 643, 674
Issaeva, I., 306
Iwase, S., 306

J
Jiang, F., 556
Jones, P.A., 601
Jordan, A., 492

K
Khare, S.P., 402
Klose, R.J., 306
Knudson, A.G., 597
Koehler, D., 381
Konat, G.W., 532
Kouzarides, T., 269
Krishnan, V., 162
Kundu, T.K., 495

L
Lamins, 167, 168, 303, 386
Langley, E., 166
LaPenna, G., 48
Large intergenic noncoding RNAs (lincRNAs), 351, 352, 406–407, 439
Latency, 44, 481–494, 496, 497, 602
Lee, K.M., 50
Lee, M.G., 306
Lee, S., 159
Lerner, M.R., 350
Lessard, F., 224

Li, C., 556
Lin, C.Y., 224
Lin, H.S., 466
Lins, R.D., 48
Liu, H., 48
Loewer, S., 351
Long terminal repeat (LTR), 277, 304, 351, 482, 483, 485–493, 495, 496, 576

M
Maf, 109
Maf recognition elements (MARE), 109
Malaria, 578, 580–581, 660, 661, 664, 665, 667, 670–672, 675
Mango, S.E., 323
Mattick, J.S., 363
Mayer, C., 224
Mediator, 67, 162, 163, 240–242, 245–250, 300, 353, 441, 532, 571
Mehta, I.S, 381
Meiosis, 14, 120, 123, 125–127, 129, 131, 607
Meister, P., 381
Metabolite, 106–108, 116, 407, 409, 411, 532
Metastasis, 404, 406, 407, 409–410, 557, 604, 629, 631, 634, 637
Methionine adenosyltransferase (MAT), 107–109, 114, 530
Methyladenine, 82–84, 86, 96
Methylation, 4, 40
Methylcytosine, 67, 82–84, 86, 114, 153, 155, 290, 292, 293, 423, 428, 599, 601, 602, 606–608, 610
Methyltransferase, 5, 40, 81–96, 106–110, 112–115, 122, 124, 125, 141, 144–146, 153, 155, 158, 159, 161, 164, 167, 189, 194, 196, 220, 224, 225, 250, 289, 300, 303, 360–362, 386, 400, 409, 422–424, 426, 428, 441–444, 446, 447, 462, 463, 467, 488, 512, 518, 530, 531, 546, 604, 606–607, 609–613, 630, 634, 639, 642–644, 664
Metzger, E., 306
Miao, F., 464
Miao, J., 669
Michaelson, J.S., 323
Michishita, E., 269
MicroRNA (miRNA), 5, 24, 126, 128–130, 158–160, 307, 348–352, 359, 362, 363, 405–406, 410, 412, 428, 439, 446, 447, 465, 490–491, 496, 497, 508, 511, 515, 517, 520–521, 527, 537, 546, 552, 558, 572, 587, 610
Mirsky, A.E., 38
Mirzabekov, A.D., 46

Mishra, N., 466
Modifications, 17, 18, 25, 26, 38–40, 44, 47, 48, 50, 51, 83, 94, 106, 109, 115, 128, 141, 147, 152, 155, 160, 189–191, 217, 242, 243, 245, 248–251, 267, 269, 270, 278, 289, 294, 295, 297, 298, 304, 307, 344, 347, 360, 376, 381, 382, 384, 385, 401, 406, 411, 421, 423, 424, 442, 460, 461, 469, 486, 488, 489, 519, 520, 529, 546, 549, 572, 580, 586, 587, 600, 601, 629–631, 633–635, 645, 659, 663, 666, 670, 674
Molecular dynamics, 46, 48–49
Monoallelic mutation, 576
Morris, S.A., 269
Multi-scale wavelet analysis, 73–74
Munro, J., 163
Murayama, A., 226
Murray, K., 38
Muth, V., 224

N
Nagy, Z., 269
Nano delivery, 585–586
Narlikar, G.J., 333
Nativio, R., 375
N-domains, 58–70, 72
Nimura, K., 269, 270
Nishida, K., 466
Noncoding RNAs (ncRNAs), 182, 190, 193–194, 225, 290, 343–365, 405–407, 421, 631, 659
Nuclear domains, 347
Nuclear factor-kappaB (NF-κB), 164, 167, 333, 437, 438, 443–448, 461, 463, 521, 534, 535, 577, 584, 602, 632–636
Nuclear organization in disease, 362–363
Nucleoprotein, 132, 573
Nucleosome(s), 4, 7, 8, 12–19, 21–23, 38, 40–46, 49, 51, 65–67, 72, 76, 111, 126, 127, 146, 152, 193, 196, 226, 238, 240, 243, 247–249, 263, 265, 266, 268, 276, 277, 279, 292, 294, 297, 300, 302, 306, 319–329, 331, 333–335, 376, 382, 386, 403–405, 421, 424, 439, 440, 444, 483, 487, 489–490, 546, 547, 570, 661–662, 666, 667, 676

O
O'Brien, C.A., 556
Oncogene, 69, 157, 226, 401, 403, 405, 407, 409, 421, 440, 559–601, 603, 632–633
OncomiR, 364, 410
Oncoprotein, 109, 426, 572
Ono, R., 611
Origin of replication, 58, 59, 61, 62, 72, 82, 88, 387

P
Panova, T.B., 224
Parello, J., 49
Parkinson's disease, 507–521, 585, 639, 640
PARYlation, 408, 568
Patrinos, G.P., 375
Pelletier, G., 224
Phillips, D.M.P., 38
Pistoni, M., 224
Planelles, V., 494
Plasmodium falciparum, 14, 578, 580, 583, 659–676
Plasticity, 296, 421, 425, 470, 519, 545–559, 624, 640
Polycomb, 5, 6, 21, 114, 145, 156, 158, 186–195, 197, 251, 292, 302, 303, 361, 362, 400, 401, 403, 405, 406, 411, 425–427, 429, 441, 443, 462, 550, 573, 604
Polycomb complex, 573
Popp, C., 608
Prieur, A., 163
Prince, M.E., 556
Promoter-proximal pausing, 250

R
Rajasekhar, V.K., 556
Ran, D., 556
Reactive cysteines, 45, 531
Reactive species, 531, 533, 536
Recombination, 83, 91, 92, 94, 120, 122, 125, 153, 162, 164, 278, 375, 377, 384, 403, 570, 603, 611, 660, 673
Redox signaling, 528–529
Reilly, C.M., 466
Reis, E.M., 352
Remodeling, 19, 21, 26, 109, 111, 125–127, 145–147, 188, 196, 239, 243, 246, 266, 291, 292, 294, 299, 306, 361, 403–406, 421, 440, 444, 456, 459, 463, 513, 514, 520, 527, 532–534, 600, 604, 609, 612, 638, 640
Remodelling factor, 3, 571

Replication, 18, 21, 57–76, 82, 86–88, 90, 111, 112, 116, 122, 194, 226, 279, 290, 294, 307, 320–323, 350, 387–388, 422, 441, 463, 484, 485, 490, 493, 528, 530, 532, 535–536, 548, 569, 570, 578, 606, 607, 659, 661, 674, 675
Reprogramming, 122–125, 132, 144, 351, 352, 399, 410, 421–424, 426, 547–549, 554, 558–559, 610
Restriction-modification, 83
Ribosomal RNA synthesis, 212
Ribosome biogenesis, 23, 219, 227, 669
Rinn, J.L., 351
RNA polymerase I, 211–227, 346
RNA polymerase II, 237–252, 266, 296, 349, 353, 379, 380
RNA polymerase III, 261–280
Roberts, C., 323
Roesch, A., 556
Rogina, K., 163
Röthlisberger, U., 48
Rubinstein-Taybi syndrome, 576

S

S-adenosylhomocysteine (SAH), 108, 115, 511, 518
S-adenosylmethionine (SAM), 105–117, 407, 409, 439, 511, 512, 518, 530, 531, 606
SAH. *See* S-adenosylhomocysteine (SAH)
Saleh, S., 494
Santisteban, M.S., 328
Scharf, A.N., 269, 270
Schatton, T., 556
Secondary structure, 37–52, 325, 345, 347, 348, 353, 354, 582
Selectivity factor 1 (SL1), 215, 217, 218, 220, 222, 224
Self-renewal, 133, 412, 419–423, 425–427, 429, 550–554, 557, 558
Senescence, 151–153, 156–163, 165, 166, 168, 332, 386, 425, 575, 584
Sertoli cells, 120, 121, 130
Shi, Y.J., 306
Signalling, 221, 329, 357, 495, 536, 557, 569, 588, 675
Singh, S.K., 556
Sir2, 143, 164–166, 273, 300, 486, 662, 669–671, 673, 674
siRNA. *See* Small interfering RNA (siRNA)
Sirtuin, 160, 163–168, 408, 411, 441, 486, 514, 569, 662
Skeletal muscle, 139–147, 406

Small interfering RNA (siRNA), 113, 128–130, 214, 324, 349, 352, 359, 364, 530, 577
Small nuclear RNAs (snRNAs), 4, 211, 345, 347, 348, 356, 357, 362
Small nucleolar RNAs (snoRNAs), 347, 348, 362, 363
Small RNAs, 122, 128–133, 238, 346, 347, 349, 360, 364, 601
snoRNAs. *See* Small nucleolar RNAs (snoRNAs)
snRNAs. *See* Small nuclear RNAs (snRNAs)
Spermatogenesis, 15, 120–122, 124, 126, 129–130, 423, 607
Spermatozoa, 120–121, 126
Spilianakis, C.G., 375
Stefanovsky, V.Y., 224
Stem cells, 5, 24, 25, 62, 120, 130, 140, 161, 279, 293, 296, 305, 323, 351, 359, 403, 404, 406, 410–413, 419–429, 441, 481, 545–559, 573, 587, 588, 600, 605
Sumoylation, 125, 152, 248, 294, 308, 440, 486, 568, 571, 572, 665

T

Takai, D., 601
Takaishi, S., 556
Tamura, T., 324
Tao, R., 466
Therapeutics, 50, 116, 147, 227, 302, 363–364, 412–413, 422, 436, 437, 471, 472, 484, 491, 494–497, 508, 515, 532, 567–588, 600, 613, 629, 631, 636–642, 675–676
Tie, F., 269
Tirino, V., 556
Transcription
 elongation, 219, 220, 243, 245, 247, 249, 354, 356, 572
 factor, 5, 14, 21, 82, 111, 140, 141, 143, 145, 157, 164, 166, 178, 197, 213, 215, 218, 220, 227, 238, 239, 245, 248, 251, 262–267, 276, 291, 292, 298–300, 303, 306, 333, 351, 353, 354, 357, 359, 380, 381, 404, 406, 410, 411, 420, 422, 426, 429, 436, 440, 442, 443, 448, 457–464, 467, 469, 486, 488, 489, 495, 530, 532–535, 537, 550, 552, 554, 557, 568, 571, 572, 576–579, 602, 609, 629, 632–634
 regulation, 5, 6, 19, 25, 109, 112, 250, 319–335, 343–365, 403, 436, 551, 569, 572, 577, 579, 580, 661
Transcriptome, 121, 344, 377, 674

Index

Transforming growth factor-β (TGF-β), 147, 445, 446, 448, 459, 460, 557
Trithorax, 187, 188, 191, 193, 195, 403, 425, 550, 664
TRNA-boundary element, 266
Tsukada, Y., 306
Tumor necrosis factor-alpha (TNF-alpha), 437, 444–446, 466, 470, 533, 630–631, 634, 636, 642, 670, 672
Tumor regeneration, 420
Tumor suppressor, 5, 111, 154, 156, 166, 168, 290, 299, 402, 405, 407–410, 412, 421, 422, 424, 425, 440, 443, 512, 574–576, 578, 579, 597–601, 603–605, 610, 612, 613, 627, 632, 633, 637, 639

U
Updike, D.L., 323
Upstream binding factor (UBF), 213, 215, 217–220, 222, 224–226

V
var gene, 578, 580, 659–676
Varier, R.A., 495
Vincent, T., 224
Vogelstein, B., 599
Voit, R., 224

W
Waddington, C.H., 306, 527, 559, 659
Wang, Y., 306, 520
Warbug, O., 410, 411
Whestine, J.R., 269, 270
Whetstine, J.R., 306
White, R.J., 266

X
Xhemalce, B., 269

Y
Yamane, K., 306
Yang, D., 49
Yang, H.C., 494, 495
Yang, Z.F., 556

Z
Zack, J.A., 493
Zhai, W., 224
Zhang, C., 224
Zhang, S., 556
Zhang, Y., 293
Zhao, J., 224
Zhu, S., 457

Printed by Printforce, the Netherlands